HANDBOOK
OF
FOOD
ADDITIVES

SECOND EDITION

EDITOR

THOMAS E. FURIA

Technical Development Manager
Food Industry Department
CIBA-GEIGY Corporation
Ardsley, New York

Published by

A DIVISION OF
THE CHEMICAL RUBBER CO.
18901 CRANWOOD PARKWAY
CLEVELAND, OHIO 44128

HANDBOOK OF FOOD ADDITIVES, 2nd ed.

This book presents data obtained from authentic and highly regarded sources. Reprinted material is quoted with permission, and sources are indicated. A wide variety of references are listed. Every reasonable effort has been made to give reliable data and information, but the editor and the publisher cannot assume responsibility for the validity of all materials or for the consequences of their use.

Comments, criticisms, and suggestions regarding the format and selection of subject matter are invited. Any errors or omissions in the data that appear in the handbook should be brought to the attention of the editor and/or chapter authors.

International Standard Book Number 0-87819-542-4

Library of Congress Catalog Card Number 68-21741

AGRICULTURE

Handbook Series

Handbook of Chemistry and Physics, 53rd edition

Standard Mathematical Tables, 20th edition

Handbook of tables for Mathematics, 4th edition

Handbook of tables for Organic Compound Identification, 3rd edition

Handbook of Biochemistry, selected data for Molecular Biology, 2nd edition

Handbook of Clinical Laboratory Data, 2nd edition

Handbook of Food Additives, 2nd edition

Handbook of Laboratory Safety, 2nd edition

Handbook of tables for Applied Engineering Science, 2nd edition

Handbook of tables for Probability and Statistics, 2nd edition

Manual for Clinical Laboratory Procedures, 2nd edition

Fenaroli's Handbook of Flavor Ingredients, 1st edition

Handbook of Analytical Toxicology, 1st edition

Handbook of Chromatography, 1st edition

Handbook of Environmental Control, 1st edition

Handbook of Lasers, 1st edition

Handbook of Radioactive Nuclides, 1st edition

Manual of Laboratory Procedures in Toxicology, 1st edition

*Atlas of Spectral Data, 1st edition

*Handbook of Engineering in Medicine and Biology, 1st edition

*Handbook of Marine Sciences, 1st edition

*Handbook of Microbiology, 1st edition

*Handbook of Materials Science, 1st edition

*Handbook of Spectroscopy, 1st edition

*Currently in preparation.

Division of THE CHEMICAL RUBBER CO.

Editor-in-Chief
Robert C. Weast, Ph.D.
Vice President, Research, Consolidated Gas Service Company, Inc.

Editor-in-Chief, Mathematics
Samuel M. Selby, Ph.D., Sc.D.
Professor of Mathematics
Hiram College

Editor-in-Chief, Medical
Sciences
James W. Long, M.D.
Director, Health Services
National Science Foundation

Editor-in-Chief, Biosciences
Irving Sunshine, Ph.D.
Chief Toxicologist, Cuyahoga
County Coroner's Office, Cleveland, Ohio

HANDBOOK SERIES

BIOCHEMISTRY
Herbert A. Sober, Ph.D.
National Institutes of Health

BIOENGINEERING
David G. Fleming, Ph.D.
Case Western Reserve University
Lester Goodman, Ph.D.
National Institutes of Health

CHEMISTRY
Robert C. Weast, Ph.D.
Consolidated Gas Service Co.

CHROMATOGRAPHY
Joseph Sherma, Ph.D.
Lafayette College
Gunter Zweig, Ph.D.
Syracuse University Research Corp.

CLINICAL SCIENCES
Willard R. Faulkner, Ph.D.
Vanderbilt University Medical Center
John W. King, M.D., Ph.D.
Cleveland Clinic Foundation

ELECTRO-OPTICS
Robert J. Pressley, Ph.D.
Holobeam Corp.

ENGINEERING SCIENCES
Ray E. Bolz, D. Eng.
Case Western Reserve University
George L. Tuve, Sc.D.
Professor Emeritus, Case Institute
of Technology

ENVIRONMENTAL SCIENCES
Richard G. Bond, M.S., M.P.H.
University of Minnesota
Conrad P. Straub, Ph.D.
University of Minnesota

FOOD AND NUTRITION
Nicolo Bellanca, Ph.D.
CIBA-GEIGY Corp.
Giovanni Fenaroli, Ph.D.
University of Milano, Italy
Thomas E. Furia
CIBA-GEIGY Corp.

MARINE SCIENCES
F. G. Walton Smith, Ph.D.
University of Miami

MATERIALS SCIENCE
C. T. Lynch, Ph.D.
Wright-Patterson Air Force Base

MATHEMATICS AND STATISTICS
William H. Beyer, Ph.D.
University of Akron
Brian Girling, M.Sc., F.I.M.A.
The City University, London
Samuel M. Selby, Ph.D., Sc.D.
Hiram College

MICROBIOLOGY
Allen I. Laskin, Ph.D.
Esso Research and Engineering Co.
Hubert Lechevalier, Ph.D.
Rutgers University

ORGANIC CHEMISTRY
Saul Patai, Ph.D.
Hebrew University of Jerusalem
Zvi Rappoport, Ph.D.
Hebrew University of Jerusalem

RADIOLOGICAL SCIENCES
Yen Wang, M.D., D.Sc. (Med.)
University of Pittsburgh

SPECTROSCOPY
Jeanette Grasselli, M.S.
Standard Oil Company (Ohio)
W. M. Ritchey, Ph.D.
Case Western Reserve University
James W. Robinson, Ph.D.
Louisiana State University

TOXICOLOGY
Irving Sunshine, Ph.D.
Cuyahoga County Coroner's Office, Ohio

CRITICAL REVIEW JOURNALS

ANALYTICAL CHEMISTRY
Louis Meites, Ph.D.
Clarkson College of Technology

BIOCHEMISTRY
Gerald Fasman, Ph.D.
Brandeis University

BIOENGINEERING
David G. Fleming, Ph.D.
Case Western Reserve University

CLINICAL SCIENCES
Willard R. Faulkner, Ph.D.
Vanderbilt University Medical Center
John W. King, M.D., Ph.D.
Cleveland Clinic Foundation

ENVIRONMENTAL SCIENCES
Richard G. Bond, M.S., M.P.H.
University of Minnesota
Conrad P. Straub, Ph.D.
University of Minnesota

FOOD AND NUTRITION
Thomas E. Furia
CIBA-GEIGY Corp.

MACROMOLECULAR SCIENCE
Eric Baer, Ph.D.
Case Western Reserve University
Phillip Geil, Ph.D.
Case Western Reserve University
Jack Koenig, Ph.D.
Case Western Reserve University

MICROBIOLOGY
Allen I. Laskin, Ph.D.
Esso Research and Engineering Co.
Hubert Lechevalier, Ph.D.
Rutgers University

RADIOLOGICAL SCIENCES
Yen Wang, M.D., D.Sc. (Med.)
University of Pittsburgh

SOLID STATE SCIENCES
Richard W. Hoffman, Ph.D.
Case Western Reserve University
Donald E. Schuele, Ph.D.
Bell Telephone Laboratories

TOXICOLOGY
Leon Golberg, D.Phil., D.Sc.
Albany Medical College of
Union University

Preface to the Second Edition

Since publication of the first edition of the *Handbook of Food Additives* in late 1968, the food industry has undergone significant changes with respect to additives (ingredients) utilized. Restraints imposed by the FDA based on new safety findings by government and industry, as well as deep consumer concern about the ingredients in foods, has reshaped the entire additives issue. Certainly, the food industry is still within the grips of change, and our current efforts can only highlight some of these. For example, the entire issue of nonnutritive sweeteners has plunged the industry into a state of flux. Cyclamates are now banned . . . saccharin is currently under fire . . . a few new candidates are under development but still quite distant from being sanctioned for use in foods. Under these conditions, we chose to reissue the chapter on nonnutritive sweeteners as originally written by Dr. Salant, since much of the information on saccharin is still useful. Should conditions change to a more stable nature, however, this chapter will be revamped to reflect new data. The reader should note that cyclamates have been deleted from Part II of the current edition.

Additional important changes reflected in the current edition include the following: (1) the lowering of use levels of brominated vegetable oil (BVO) as a clouding agent, (2) possible restrictive action against FD&C Red No. 2, (3) FDA's proposed ban of diethyl pyrocarbonate, and (4) the general revocation by FDA of letters to industry affirming the GRAS status of many functional and supplementary ingredients. But, this seemingly negative side of the ledger has been to some extent offset by the positive nature of additions to the ingredient armament, which the reader will find adequately reflected in the updated chapters and Part II. For example, FEMA has deemed GRAS about 125 new flavor ingredients . . . a new certified food color (FD&C Red No. 40) has been added to the approved list, the first in nearly four decades . . . approval has been obtained for synthetic sources of fatty acids, new emulsifiers, xanthan gum as a new hydrocolloid, and many, many more.

Most of the chapters have been revamped to include new data or uses; reference citations have been checked as carefully as possible. An entirely new chapter, titled Phosphates in Food Processing, has been added. On this point, I wish to thank Dr. Ellinger for contributing what the reader should find to be one of the most comprehensive reviews available on the subject.

Finally, I would like to assure readers that we shall do our utmost to keep abreast of rapid changes and to present these in subsequent editions of the handbook.

Thomas E. Furia

Hartsdale, New York
November 1972

Chapter Authors

Benjamin Borenstein, Ph.D.
Manager, Food Industry
Technical Services Department
Roche Chemical Division
Hoffmann-La Roche Inc.
Nutley, New Jersey

Donald F. Chichester, Ph.D.
Chemical Consultant
1734 NW Fifth Avenue
Gainesville, Florida

R. H. Ellinger, Ph.D.
Manager, Regulatory Compliance
Kraft Foods Company
Division of Kraftco, Inc.
Chicago, Illinois

Thomas E. Furia
Technical Development Manager
Food Industry Department
CIBA-GEIGY Corporation
Ardsley, New York

Wm. Howlett Gardner, Ph.D.
Chemical Consultant
29 Mirriam Avenue
Bronxville, New York

Martin Glicksman
Senior Research Specialist
Corporate Research Department
General Foods Corporation
Tarrytown, New York

William C. Griffin
Associate Director
Product Development
Atlas Chemicals Division
ICI America Inc.
Wilmington, Delaware

Robert E. Klose
Group Leader
Corporate Research Department
General Foods Corporation
Tarrytown, New York

T. D. Luckey, Ph.D.
Professor of Biochemistry
University of Missouri
Medical School
Columbia, Missouri

Matthew J. Lynch
Development Manager
Product Development
Atlas Chemicals Division
ICI America Inc.
Wilmington, Delaware

James E. Noonan
Vice President/Technical Director
Warner-Jenkinson Manufacturing Co.
St. Louis, Missouri

Abner Salant, Ph.D.
Vice President
Monsanto Flavor/Essence, Inc.
New York, New York

Loren B. Sjöström
Vice President
Arthur D. Little, Inc.
Acorn Park
Cambridge, Massachusetts

Ben N. Stuckey, Ph.D.
Manager, Food Chemical Sales
DPI Division
Eastman Chemical Products, Inc.
Kingsport, Tennessee

Robert L. Swaine
Vice President—Technical Operations
Canada Dry Corporation
Greenwich, Connecticut

Fred W. Tanner, Jr., Ph.D.
Chemical Consultant
2714 Norfolk Road
Orlando, Florida

Leland A. Underkofler, Ph.D.
Retired
Miles Laboratories, Inc.
Elkhart, Indiana

O. B. Wurzburg
Vice President—Research
Starch Division
National Starch and Chemical Corporation
Plainfield, New Jersey

Contents

PART I

PART II

Part I

Introduction
to Food Additives

T. D. Luckey, Ph.D.

Professor of Biochemistry
University of Missouri
Medical School
Columbia, Missouri

Introduction

"Where a man can live, there he can also live well." (From *Meditation* of Marcus Aurelius, 108 B.C.). To extend this philosophy to all people of the world is an acceptable goal of our time. Attempts to do this better, more efficiently, or more economically explain the variety of activities of mankind that have resulted in changes of his environment, purposefully or otherwise. Food additives include the most useful, most well tested, most justified, best controlled, most discussed, most legalized, and least maligned of the materials that man adds to his environment in Quixotic maneuvers to keep nutritional supplies ahead of the population explosion. As people are brought closer together in work, play, and social activities, there is a greater emphasis on personal appearance, cleanliness, and health. Food additives, toiletries (4 billion dollars sales annually), and drugs (about 4 billion dollars sales for nonprescription drugs annually) are the purposeful chemicals used to provide increased health and productive life for our megalopolophilic civilization. It

1

should be noted that these purposeful chemicals are in a distinctly different category from the waste products of civilization, which provide the bulk of our pollution problem.

Accepting the fact that man is changing his environment at an ever accelerating rate, we are led to the question of what protection we have against this change. Protective factors that counteract the chemicals in our environment include physical, chemical, biological, and social activities. It is worthwhile to outline these forces that are balanced against the changing environment before presenting a detailed consideration of food additives.

The physical factors of placement and dilution offer limited protection against our changing environment. A small amount of waste pumped into a stream may be harmful to biota of that stream, while the dilution of that material when the stream runs into a river may be great enough to have negligible influence on the river biota. Chemical action may cause material to precipitate and become inactive, or it may degrade the material so that its effective life in the environment may be shortened. Biological processes often speed this action. Biological dilution is also important. It is frightening to learn that there are 20 tons of DDT in the tissues of the people of the United States. This fact seems much less amenable to study than does this same information stated as: the average American has 0.1 grams of DDT in his tissues.

Many other biological factors provide protection against change. The feel, taste, and odor of foods are coupled to rejection by spitting, gagging, or regurgitation. Compounds may be modified by hydrochloric acid in the stomach, alkaline salts of the bile, and enzymes of the alimentary tract. Diarrhea and diuresis decrease body contact for certain compounds. A variety of detoxication mechanisms are well known, and some drugs elicit enzyme induction to provide our tissues with specific metabolic pathways for degrading such compounds. Although evolution is a biological factor to help species adapt to a changing environment, man's manipulation may be changing the environment too fast for slowly reproducing mammals to use this protective factor.

Society is the last protective factor. Research, knowledge, and wisdom should provide a basis for action appropriate to the time and country in order to tranquilize social unrest. This can be capped by laws and judicial action to provide the protective factor that is society itself. Man's ability to change his environment evokes the responsibility on him to regulate this action carefully for the ultimate perpetuation of society and mankind. Evolution goes in one direction; we are committed either to use the fruits of science wisely or to allow our society to disintegrate.

This brief introduction sketches the background and philosophy needed to place the one category of purposeful chemicals—food additives—in proper perspective.

Food

Diverse Views

Some materials, as the pentose sugars, are absorbed into the cells of the human body by diffusion; other compounds, as glucose, require the cell to expend energy for adsorption against a concentration gradient by "active transport." Sometimes the material taken into the cell cannot be utilized. However, in all cases the material taken in must have some effect on the cell and therefore must be considered under the general term *food*. Such a view considers the needs and reactions of the total organism.

Both the philosophic and the cellular views suggest that all material taken into a cell or organism, whether useful, nonutilized, or harmful, should be considered food. Food is all the solids, liquids and gases which are taken into the cell or organisms by any route. This general definition is broader than is desirable for the present purpose. Therefore, the concept will be narrowed by removing three categories of food. Metabolic food is that material synthesized by certain cells and utilized by other cells within the body. One example is the utilization of skin-synthesized vitamin D by other tissues which cannot make it. A less definitive example is the synthesis of material by intestinal microorganisms followed by its adsorption. Although the material is actually produced outside the body, "intestinal synthesis" is not usually separated from tissue synthesis for the first approximation of nutritional consideration. The second category is unintentional food. This includes waste material which inadvertently passes from one cell to another, inactive and noxious gases from the air, and materials other than water and oxygen which are absorbed through the skin. At birth, a baby is transferred from unintentional food, its mother's blood, to intentional food, its mother's milk. The third category is water.

Intentional food is the solid, liquid and gaseous material purposefully taken into the body. In this sense soil is food for the earthworm. Although oxygen is sometimes considered an essential nutrient, oxygen and air may be omitted from the present discussion. Water will also be deleted because it complicates our considerations. Water is often considered in a different institutional department and by different laws than food. However, beverages should be retained in the category. The solids may be simplified by omitting those which are not purposely taken as food. This deletes toothpaste, snuff, mouth washes (including fluoridated ones which actually are good), drugs and lipstick. Chewing gum and chewing tobacco are in a questionable status. Intravenous feeding and antifertility pills are out; vitamin and mineral pills and candy are in. The subcategory remaining includes purposeful food: material ingested for nourishment. This subcategory constitutes a small, manageable portion of the total food concept which corresponds reasonably well with dictionary definitions: Food is "nutritive material absorbed or taken into the body of an organism which serves for the purpose of growth, work or repair and for the maintenance of the vital processes." This would preclude most food additives. Food is "nutriment in solid form, as opposed to *drink*, which may also contain more or less nourishing material." Food is "anything that nourishes, develops, or sustains."[1] In modern concepts, the proof of the pudding is in the subsequent metabolic effects. The FAO/WHO Codex Alimentarius Commission defined food at their November 1966 meeting: "*Food* means any substance, whether processed, semi-processed or raw, which is intended for human consumption, and includes drink, chewing gum and any substance which has been used in the manufacture, preparations, or treatment of 'food,' but does not include cosmetics or tobacco or substances used only as drugs." This definition is broad enough to include food additives.

Food Standards

Food for commercial use is strictly regulated by enforcement of state and federal laws. Exact specifications of each item of concern by commerce and law provide the only equitable means of solving problems. Therefore definitions of foods become a matter of national and international concern and may involve problems of great economic, social, moral, medical, legal and/or political importance. A definitive reference on food standards and definitions in the United States was prepared by Gunderson, Gunderson and Ferguson.[2] Food standards establish an authority for the quantity, weight, value, or quality of specific food items. The standards are concerned with *identity*, *quality* and *fill of container* to promote honesty and fair dealing in the interest of consumers.

Congressional action provides direct standards for butter (1923) and nonfat dry milk (1956). Filled milk is specifically prohibited by a congressional act of 1923. The Agricultural Marketing Act and the Poultry Products Inspection Act provided extension and improvement of standards on both voluntary and mandatory basis. The Food and Drug Administration provides many food definitions and standards of identity. These categories include foods for special dietary uses, pesticide chemicals in or on raw agricultural commodities, and food additives. The Public Health Service has defined a few foods and is concerned with safety to health, primarily by protection from harmful microorganisms. This agency provides drinking water standards, microbiological standards for foods, radiation protection, miscellaneous food standards, and an ordinance and code regulating eating and drinking establishments.

Hundreds of foods are regulated by the Department of Agriculture. The meat inspection activity of this department is one of the oldest food standard and protection activities of the U.S. government. Although other departments have federal standards on fish and shellfish, the Bureau of Commercial Fisheries of the Department of Interior has this prime responsibility under the Fish and Wildlife Act of 1956. The Treasury Department provides standards for wine, beer, tobacco, bottled cocktails and whiskey. The Department of Defense provides food standards which emphasize packaging and storage requirements. Kosher foods and a variety of other foods are regulated by the Federal Trade Commission. The Veterans' Administration takes most of its standards from other federal agencies. The General Services Administration is responsible for publication of food standards in the Federal Register and in the Code of Federal Regulations and the provisions of standards for food which is to be purchased by more than one branch of the government. Coffee and instant coffee standards are provided only by the General Services Administration.

Over 50 countries are working on international food standards as a joint FAO/WHO Codex Alimentarius Commission.[3] Development and acceptance of the Codex, an international food standard, should encourage more free movement of foods from country to country and help developing countries promote adequate legislation, provide consumer protection, and encourage acceptance of their exports. Formal acceptance of a Codex standard by any government would obligate that country to the minimum standards for imported and exported foods. The development of the Codex is expected to be a major factor in future trade which is of great importance to any country with substantial exports or imports. In general the Codex standards may be expected to be somewhat lower than standards already accepted by the United States. In such cases, appropriate labeling would be necessary to clearly indicate that a given food meets Codex standards but not those of the country. The Codex Alimentarius progress was reviewed by Koenig.[4]

Formula Foods

New foods appear on the market with increasing frequency. We are tantalized by ready availability of dehydrated strawberries in cereals, instant mashed potatoes, preseasoned foods in heat resistant plastic, partially-baked rolls and pastry, prepared milk formula in disposable bottles, prepared ingredients for foreign foods, mixed drinks, frozen TV dinners, and a variety of dietetic foods: low calorie, dietetic, salt-restricted, and hypo-allergenic. Supermarkets stock thousands of items never found in the country store of yesteryear. Among those 6–8000 items available are hidden artificial or formulated foods. Synthetic foods, i.e., protein from crude oil and edible carbohydrates from cellulose, are not included here.[5] *Formula food* is the term more pertinently applied to the production of new foods from previously accepted materials. Examples such as artificial whipped cream, frozen "ice cream" and margarine from improved processing suggest the wealth of possibilities. The ingredients of these foods must have been accepted by food standard codes; then a manufacturer may use them to prepare a food which has not previously existed. The Department of Agriculture has formula standards for margarine, chopped ham, corned beef hash, soy protein concentrate, calcium reduced dried skim milk and smoke flavorings. Manufacture and labeling are liable to periodic scrutiny by government officials; the main problem is getting customer acceptance. Archer Daniels Midland has started to use vegetable proteins to produce several new foods.[6] The total of new foods introduced (many of them are new formulations) is 5000 per year; most are short-lived.[7]

Food Supplements

Food supplements are sometimes considered with items such as alcohol, soft drinks, gum and candy. Beverages (alcoholic and non-alcoholic) and candy should be classed as foods since they have nutritive value. Chewing tobacco is a food supplement in some countries. Since many food supplements are specific nutrients, they are sometimes excluded from the strict definition of food additives. A separate category for things such as iron salts, vitamin D, wheat germ meal, or poppy seed seems unwarranted. Therefore materials sometimes listed as food supplements will be considered either as components of formula foods or as food additives where they fit nicely according to their uses.

Food Additives

Definitions

As was true with food, the broadest definition must be reduced considerably before a definition acceptable to our purpose is provided. Thus, anything added to food is not necessarily a food additive. "A *food additive* is a substance or mixture of substances, other than a basic food stuff, which is present in food as a result of any aspect of production, processing, storage or packaging. The term does not include chance contaminants."[8] Inclusion of the second sentence in this definition changes it from a definition with public health as its prime concern to one which is more concerned with regulation. The second sentence makes it a definition of intentional food additives and purposefully omits incidental, unintentional or contaminating material. This ignores pesticides, packaging materials, fertilizers, promotants and material which may inadvertently but consistently become a part of a food. Therefore, the broad meaning without the second sentence is a more acceptable definition. Fortunately, the Food Protection Committee of the National Research Council accepted incidental additives within its domain.[9] These two views of what constitutes a food additive are the scientists' view; the legal-legislative view is more restrictive.

The Food Additives Amendment to the Federal Food, Drug and Cosmetic Act of 1958 contains this definition: "The term food additive means any substance the intended use of which results or may reasonably be expected to result, directly or indirectly, in its becoming a component or otherwise affecting the characteristics of any food (including any substance intended for use in producing, manufacturing, packing, processing, preparing, treating, packaging, transporting, or holding food; and including any source of radiation intended for any such use), if such substance is not generally recognized, among experts qualified by scientific training and experience to evaluate its safety, as having been adequately shown through scientific procedures (or, in the case of a substance used in food prior to January 1, 1958, through either scientific procedures or experience based on common use in food) to be safe under the condition of its intended use; except that such a term does not include—

1. a pesticide chemical in or on a raw agricultural commodity; or
2. a pesticide chemical to the extent that it is intended for use or is used in the production, storage, or transportation of any raw agricultural commodity; or
3. a color additive; or
4. any substance used in accordance with a sanction or approval granted prior to the enactment of this paragraph pursuant to this act, the Poultry Products Inspection Act (21 U.S.C. 451 and the following) or the Meat Inspection Act of March 4, 1907 (34 Stat. 1260), as amended and extended (21 U.S.C. 71 and the following)."

This 1958 amendment recognized three classes of intentional food chemicals: 1. those Generally Recognized As Safe which are known appropriately as GRAS; 2. those with prior sanction; and 3. food additives. This definition is much more restrictive than the first two. The GRAS list of several hundred materials includes components in every functional class of intentional food additives. Pesticides on raw agricultural products and color additives are excluded from the legal-government definition because other laws cover them.

Historical View

Man's first contact with incidental food additives—the materials produced by fire when cooking food—is one of vital concern to us today. Spices and salt were known to the peoples of all civilizations and probably date into prehistorical times. Their accidental use along with heat began the differentiation between those who cook and chefs, between those who eat and gourmets. The ancient English quickly acquired a taste for new spices introduced by the Romans about 50 B.C. The fabulous ransom of Rome paid to Alaric the Visigoth included 1000 kilograms of pepper.[10] Tradesmen searching for spices resulted in the travels of Marco Polo. The search for new trade routes for tea and spices resulted in the discovery of America by Columbus. A prize discovery of Cortez was the vanilla bean. Early in the nineteenth century, Texas settlers made a chili powder when they ground peppers from Mexico.[10] The first U.S. patent on a food additive in 1886 was for a mixture of salt and calcium phosphate as a food condiment.[11] World War II sparked the palates of U.S. soldiers, and subsequent use of imported seasonings multiplied. About the same time synthetic pesticides were introduced: rodenticides, 1946; herbicides, 1939; fungicides, 1940; DDT, 1942; denticides, 1944; and organic phosphate insecticides, 1947. However, inorganic pesticides had been used for a century.[12] The antibiotic stimulation of growth in domestic animals was discovered in 1945, and commercial adaptation of this was begun in 1951.[13]

The Egyptians used food colors 3500 years ago, and a sweet reed called "Khand" (the original candy) was introduced to Europe when Alexander the Great returned from India.[14] Perkin's synthesis of aniline purple in 1856 opened the way to great numbers of synthetic colors, but few synthetic colors have been used in foods. A world review by WHO in 1956 revealed the permitted lists of 43 countries included only 114 synthetic compounds and 50 natural colors.[14]

Saunders' excellent review indicates that 661 million pounds of food additives were used in the United States in 1965, and the wholesale value was $285 million.[7] This amounts to more than 3 pounds of food additives per person per year; the yearly coloring used per person is 4 gm.

Protective Laws

The protective factors previously discussed are applicable to food additives as well as other chemicals in our environment. The last factor, society, provides specific safety laws regulating chemicals. Interestingly enough there are few written laws controlling the use of spices in foods. Governments recognize that a merchant who uses too much spice in his food simply eliminates himself from business. Foods with naturally occurring toxicants generally are not used extensively.[15]

The elaborate designs on weights in ancient Egypt and other civilized countries suggest the existence of laws pertaining to the sale of foods and condiments. An Indian law of 300 B.C. prohibits adulteration of grain, scents and medicine.[16] England had laws (1215 and 1597) protecting the consumer from certain harmful foods.[16] Accum wrote about adulterations in drugs (1798) and food (1820) in London.[17] England, Germany and Sweden passed general food laws in 1860–70; the Sale of Food and Drugs Act of 1875 in England was more effective. The laws allow the public's health to be safeguarded in a manner previously impossible; at the same time they prohibit questionable practices by the less scrupulous which would be injurious to honest merchants.

Until the twentieth century government action in the food problems of the United States was virtually nonexistent. In 1862, the year Pfizer began making cream of tartar, President Lincoln created the Department of Agriculture whose assignment was the interests of the family farmer. The 1905 pest prohibition and the quarantine act of 1912 have served to protect our farms from hoards of invaders. The former act was reinforced by the Federal Plant Pest Act of 1957. In 1965 quarantine procedures intercepted an estimated 32,575 separate plant pests.[18]

The awareness and education of society should be a responsibility of both government and industry to obtain an educated support for acceptable laws. The basic issue is that chemicals are not more harmful than electricity, a pen or money. Each may be used for a good or a bad cause. All chemicals, like the electricity or gas in our homes, are harmful if used in too great quantities. When harnessed and used properly, appropriate chemicals may stimulate growth, combat infection, alleviate pain or facilitate food preparation. Without a responsible society, the works of man, industry or government will be frustrated. Education becomes a necessity for progress. Approximately 30 million people in the United States who are employed in the field of food and agriculture provide about 500 different careers, and there are almost two jobs for each trained person educated each year.[19]

The food industry of this country is one of the most enlightened, progressive industries in the world. It usually perceives and solves potential problems before laws need to be made, for it learned early that sustained profits come only through protection of the customer. One of the main goals in its self-regulation has been providing people with increasingly safe, high quality food at economic cost. Its progress in food preservation is built upon the basic work of Appert, who developed an effective canning procedure between 1796 and 1810, and of Pasteur in the prevention of microbial spoilage.[20] The first shipment of iced produce, butter to Boston in 1851, was followed by the weekly *butter train* across New York State and the first successful refrigerator box car in 1868; in 1889 California shipped its first carload of fresh fruits to New York City.[21] Now shipping cars have controlled temperature, gas and pressure. Transportation by train was supplemented by trucks which exclusively serve 25,000 communities and, lately, by air freight. Inspection, automation, and education are the keys to success. The egg production and broiler plants have mechanized biology into big business. This type of mass production coupled with good transportation and the development of supermarket procedures allow the American housewife 6–8000 items from which to choose in a single store.

Organizations such as The Food and Drug Law Institute, Inc. sponsor annual educational meetings with industry and government, university instruction, seminars, the Food and Drug Cosmetic Law Journal and the Food and Drug Institute Series of research books. The Manufacturing Chemists' Association, Inc., which published numerous education booklets and speeches, and the vigorous activity of industry in the contributions of the Food Protection Committee of the Food and Nutrition Board of the National Academy of Sciences and the National Research Council show the public mindedness and the enlightened education activities of industry. Communication between government and industry will be improved when the official organ of the government, the *Federal Register*, will be supplemented by FDA Papers as the official magazine of the Food and Drug Administration.

State and federal laws and inspection have intermingled. Once a state has solved a problem with a satisfactory law, it is often emulated by other states, and parts of good state laws are often incorporated into federal law. Once a federal law is passed on the regulation of food or a food chemical in interstate trade, this becomes a basic pattern for all states to consider for intrastate commerce. Examples of early action include the 1764 adoption of a sanitary code for slaughter houses by the Massachusetts Bay Colony, the 1850 pure food and drug law of California, and the 1856 Massachusetts law prohibiting milk adulteration. Cooperation between states and federal inspection and control is usually obtained. In 1824, the Flour Inspection Act for Alexandria, D.C., was passed by Congress.

Professor G. W. Wigner won the 1880 Board of Trade prize for an essay recommending legislation against food adulteration. Congress failed to pass an act drafted from this essay, but between 1880 and 1890 several states passed laws based upon Wigner's work and the British 1875 pure food and drug law. These include the 1881 New York food and drug law, the 1884 food sanitation law of Illinois, the District of Columbia food and drug law, and meat inspection in New York in 1891. Another example of state-federal cooperation is the Alabama-U.S. Public Health Service Standard Milk Ordinance of 1924 from which developed the Grade "A" Pasteurized Milk Ordinance.[22] Although they may be more strict, state laws must be presented without conflict with the application or administration of federal laws.

Regulation of manufacturing processes and provision of pure food and drugs is not an area where Congress has constitutional authority. Therefore, the federal laws are aimed primarily at interstate and international commerce. Early Federal legislation of 1890 provided for inspection of salted pork and bacon, a measure aimed to help international shipping since our export meat market was threatened by trichinae, and a series of laws prohibiting import by Italy (1879), other European countries in 1881 and the immediate slaughter measure of Great Britain in 1882.[23] Although removal of the European restrictions was begun in 1891, public health officials and industry were alerted to the problem of providing wholesome food. The animal quarantine law of 1903 helped to control animal and poultry diseases. Upton Sinclair's novel *The Jungle* aroused the public, and the voice of society helped to quickly pass the Meat Inspection Act of 1906; it was supplemented by the Imported Meat Act of 1930, which raised the standards of imported meats to those of domestic meats, and by the Poultry Products Act of 1957. These provided our protection for 112 million animals and 2.2 billion birds slaughtered in one year.[23] Since World War II the emphasis on inspection and assuring food safety has grown from carcass, kidney and liver examination and a search for insect residues to include chemical and pharmacological methods.

Dr. Harvey W. Wiley, Chemist for the Department of Agriculture from 1883–1930, campaigned against misbranded food and adulterated foods. In 1902 the Sherman Act prohibited the false labeling of food and dairy products. The dramatic evidence from Wiley's *poison squad*, who ate at the food additives table 40 days on and 10 days off, was used to help establish the original

Food and Drug Act signed by President Theodore Roosevelt in 1906. Since the series of Paddock's Bills (1886, 1888, 1889, 1890, 1891, 1892) and Brosin's Bills (1897, 1898, 1899, 1900, 1901 and 1902) for a Pure Food and Drug Act failed to pass, Wiley commented that the 1906 act rode on the coat tails of and was passed in Congress on the same day as the Meat Inspection Act.[24] That same year Congress also authorized money to establish pure food standards. The 1906 Act forbids the production of misbranded or adulterated food products in the District of Columbia and in the territories where the authority of Congress is supreme, and it prohibits interstate distribution of fraudulent or unhealthy products. It banished chemical preservatives such as boric acid, salicylic acid and formaldehyde. Food adulteration included the addition of poisons and deleterious materials, the extraction of valuable constituents, the concealment of inferiority, substitution of other articles, and the mixture of substances which would adversely affect health.[16] A supplementary act of 1913 forbids the misbranding of food in any way to mislead the purchaser, particularly as to quantity. The McNary-Mapes Amendment of 1930 authorized standards of quality and fill of containers for canned foods. In 1927 a separate law enforcement agency, the Food, Drug and Insecticide Administration, was formed; this became the Food and Drug Administration in 1931.

The 1939 Federal Food, Drug and Cosmetic Act retained the provisions of the 1906 act and added new provisions; food was defined, food standards of identity and fill were presented, adulteration of food was prohibited and truthful labeling was made mandatory.[16,25] This law restricted the use of most chemicals (even in safe quantities) unless they were required for food manufacturing practice or could not be avoided; tolerance levels were set for the latter. In 1954 the Hale Amendment allowed a simplified method for promulgating food standards, and the Miller Pesticide Amendment set safe amounts of pesticide residues which may remain on agricultural produce in interstate shipment. The manufacturer is required to register his products and their labels with the Department of Agriculture. The Department of Agriculture certifies that the pesticide used is indeed effective if used in accordance to written direction, and the Food and Drug Administration of the Public Health Service sets tolerance for any residues based upon evidence submitted by the manufacturer. Carlson had suggested that government and society insist that the burden of proof of harmlessness falls upon the manufacturer or introducer of a new food material and that such tests include all physiological processes of man.[26] Such a philosophy is becoming accepted. In 1940 FDA transferred from the Department of Agriculture to the Federal Security Agency and was changed to the Department of Health, Education and Welfare in 1953. In 1968 FDA was made a constituent of the Consumer Protection and Environmental Health Service of the Public Health Service branch of the Department of Health, Education and Welfare. Its mission is to protect the public health of the United States from impairment by foods, food additives, drugs, pesticides, and other consumer products.

The Food Additive Amendment of 1958 provides that any substance added to food must have been proven safe by the manufacturer before it is offered for sale. It permits the use of food chemicals which improve the food and advance food technology when they are safe at the level of intended use. It was a clarion announcement that our government had concluded that the use of chemicals is necessary and unavoidable for the provision of an adequate food supply. It was also clear that government subsidy of safety research on compounds for the market was removed. Now the total cost of proving a material safe must be borne by the manufacturer. Much of the $200,000 to $3,000,000 required for testing would have been spent by reliable companies with no legal compulsion.[16] This cost is not unreasonable for successful new products by large companies, and is negligible when considered a means of retaining public confidence in the face of scare books.[7] This amendment excludes those additives recognized as safe by competent experts who consider the conditions of intended usage, those additives generally recognized as safe (GRAS) based on past experience, incidental additives which may or may not make the food illegal according to the basic 1938 law, pesticides (which are covered elsewhere), color additives, and other substances approved previously; i.e., butter is covered by the 1923 act, filled milk by the 1923 act and non-fat dry milk by a 1956 amendment to a 1944 act.[7,16,25] GRAS additives are subject to the conditions that the quantity be an amount reasonably required and that the material be of acceptable grade and be processed as a food ingredient. The 1958 act provides protection from 2400 new products which industry desires for its progress.[7] It includes medications administered to animals which may leave residues after slaughter. It includes not only flavorings, preservers, neutralizers, foaming agents and other material added directly to food, but also substances used in food containers and sources of radiation.[27] Addition of radioactivity to food is forbidden unless a regulation has been issued stating safe conditions for use of that radioactivity.[28] Sterilization radiation is now withdrawn for bacon and is being tested for potatoes, ham and other meats. Low level radiation (2–300,000 rads) is being examined for perishables to provide insect disinfestation and light doses (8000 rads) may be used to interrupt the life cycle of grain insects.[29,30,31]

All colors were brought under one control by the Color Additives Amendment of 1960. All colors will be reexamined; those causing cancer in any animal will be forbidden and limits will be set in the others. The Food Additive and the Color Additive Amendments forbid any regulation permitting the use of any substance in any amount which "is found to induce cancer in man or animals." The interpretation of this as a "zero tolerance" is equivalent to the "no residue" level for pesticides in foods. Both concepts have been seriously criticized on conceptional, operational and scientific grounds. Both are gradually fading into disuse. The "no residue" category has been dropped by the Government and the "zero tolerance" remains in force only for compounds for which no safe level can be set.[32]

Conditions of safe use are set by the Food and Drug Administration or rejection is to be made within 180 (90 plus 90) days of the filing of the petition by the manufacturer. Trade secrets will be protected. The licensing concept considers the additive from the viewpoints of chemistry, safety, quantity to be used, effect produced, condition of use, manufacturing methods, and residue analysis. Criteria of safety will include probable consumption, cumulative effects, safety factors set by animal testing and pretesting to establish with reasonable certainty that no harm will result from the proposed use of the chemical. It is the manufacturer's responsibility to obtain clearance on any additive which is not specifically excluded. The detailed voluminous material will usually take 2–4 years to compile, and much of the 1000–2000 page petition will be new information. Procedures for recourse through a public hearing and a U.S. Court of Appeals are available to protest any action.

Oser, Sanders and Mahoney have presented problems of the law and their impact upon manufacturers.[7,33,34] Since the Food and Drug Administration cannot publish a complete list of GRAS products, a manufacturer must decide whether to petition for a new additive with its implication that the compound is not entirely safe, or to request it be established as a GRAS compound. The decision of both government and industry as to who are the deciding experts presented another problem. And how should the experts be polled? How familiar should they be with the product? What standards a food chemical should have was solved for those materials which appear in the *Food Chemicals Codex.*[35] Interindustrial relations arise such as what guarantee should a paper or a plastic manufacturer give to a merchant who desires to package a variety of food products. Different packaging materials may migrate into the food depending upon the physical-chemical properties of the food. Only interindustrial cooperation can properly handle such problems.

One of the biggest problems is the restriction on potential carcinogens. Since a large proportion of chemicals shows different actions depending upon the quantities used, only specific quantities can be considered to have specific action.[36,37,38] There was no reasonable possibility of a potent carcinogen to be overlooked in the usual testing programs before or after the Food Additives Amendment of 1958; however, ways and means to search for a weak carcinogen are less well developed.[39] Compounds such as cholesterol and phenol are weak carcinogens.[40,41] Cycad plant seeds, a source of starch in tropical areas, have been reported to have a carcinogen by G. Laquer and M. Spats.[42] Dougherty found tumors in nematodes following injection of water.[43] The similarity of reactions of known carcinogens in man to those in rats and mice combined with the overdose of material presented over significant portions of the lifespan of three species with three routes of administration (oral, intravenous and skin painting) give reasonable assurance that negative findings indicate minimal hazards.[39] The question of whether a compound is a primary or secondary (metabolic product) carcinogen has little meaning in the practical sense of public health safety; it does emphasize the uncertainty of species substitution. Will the compound be metabolized in man in the same manner that it was in the test animal? Individual characteristics may be important in determining whether or not a given compound is a carcinogen. Some drugs, as FUDR, are not active unless the body possesses the enzyme to convert this to the carcinogen FUDRP. Lucid reviews of chemical carcinogens have been presented.[42,44,45,46]

Changing a law seems to be easier when the change is against something than when it is for something. The latter suffers from presenting a new or different concept, while the former is based on specific grievances, disappointments, or failures. The FDA showed its fallibility by banning irradiated bacon with no change in essential information about this product that had previously been approved. When information that cyclamates cause cancer and other problems when injected into animals was evaluated by a panel of the National Academy of Sciences, the Department of Health, Education and Welfare ordered immediate stoppage of the use of cyclamates in nonprescription food products. These actions may be contrasted with the swell of public reaction against DDT based on much new information and a growing awareness of the overall ecologic impact of the activity of one segment of our society. The result has been that several states and some countries have banned further use of DDT in spite of its past benefits and potential usefulness.

Enforcement of the laws is another aspect to be considered. The problems encountered in food production, i.e., the cranberry episode of 1959 or the salmonella problems of 1966, became over-emphasized by being news items. The bulk of the work goes without headlines to provide protection in all areas of food. The extensive meat inspection program has been summarized; importation acts keep standards high for food entering this country; federal and state public health agencies keep infection problems under control; and veterinary biologics and chemicals involved in plant and animal production are regulated by state and federal departments of agriculture. Constant monitoring of foods reveals trace residues of agricultural chemicals present in marketed foods; commercial processing removes much of these residues.[47,48] The pesticide residues found are consistently a small fraction of the safe legal tolerances set by laws.[27] Preliminary results suggest no serious pesticide build-up in soil where large amounts of pesticides have been applied for many years.[49] The Food and Drug Administration staff grew from 1400 in 1959 to 3,100 in 1965.[27] This agency oversees 100,000 factories and workhouses and 700,000 other establishments having an annual sales of $117 billion. It collected 19,000 samples from goods imported into this country and took 109,000 samples from domestic goods in 1964. In 1960 about 6000 samples were tested for radioactivity. A relative new activity is the inspection of thousands of mills which mix medicated feeds for livestock. The newness of the laws and the improvements in methods for residue testing have not yet allowed a meaningful pattern to form on the misuse of food additives. It is anticipated that this pattern will be comparable to that in older areas of law enforcement, dealing with fertilizers, pesticides, antibiotics, biologics and vitamin fortification programs; the vast majority of products fall within legal limits. This is quite different from the conditions of a century ago.[50,51,52,53]

The federal cost for reinforcement (FDA budget) for fiscal 1964–65 was $39 million.[54] The total bill for consumer protection in food and related products by state governments was $32 million in fiscal 1961.[55] Every effort is made to assure safe food.[56]

Global View

A global view of food laws pertaining to chemical additives is essential for continued public health protection and the understanding and anticipation of problems of exportation and importation. The early development of many countries followed the British customs and laws. More recently the United States has been emulated by underdeveloped countries. Pride may make an underdeveloped country accept no lower standards for imported food than the exporting country would maintain for itself. International organizations such as WHO, FAO and the Common Market are becoming more important to individual small countries and to the future development of international trade. The FAO/WHO Codex Alimentarius has several sections which will soon be presented for acceptance by any or all nations of the world. The standards set in this will probably become a basis for trade even for nations which did not sign it. The first standards will be minimum standards which may be raised when worldwide acceptance is ready.[4] The food codex of Diepenbrock is helpful.[57]

FAO (Food and Agriculture Organization of the United Nations) was established in 1945 to help its member governments to improve the nutritional level of their people by increasing the food resources and improve production and distribution of food and agricultural products. WHO (World Health Organization of the United Nations) was formed in 1945 to deal with all international matters concerning health, including nutrition. Combined efforts result in committees such as the WHO-FAO Joint Expert Committee on Food Additives.[58] The first meeting of this committee, at Geneva in 1955, proposed to collect and disseminate information. In 1956 it began issuing a periodical entitled *Current Food Additives Legislation*, a series of monographs on *Food Additive Control* for individual nations, and a series of position papers. The FAO Food Additive Control booklets are available for Canada (1959), the United Kingdom (1960), Denmark (1961), The Netherlands (1961), Australia (1961), and Russia (1969). The topics include principles for the use of food additives, methods for establishing safety from the use of food additives, carcinogenic hazards involved in the use of food additives, specifications for antimicrobial preservatives, antioxidants, food colors, emulsifiers, stabilizers, bleaching and maturing agents, and an evaluation of the toxicity and established acceptable daily intake levels for most types of food additives. Their definition of food additives includes animal feed adjuncts which may result in residues in human food and packaging material which may migrate into foods. Pesticide residues and radioactive contaminants are considered in separate programs (Joint FAO/WHO Panel of Experts on the Use of Pesticides in Agriculture and Joint FAO/LAEA/WHO Expert Committee on the Technical Basis for Legislation on Irradiated Food).[59] The committee recognized the usefulness of food additives for maintenance of the nutritional quality of food, the enhancement of keeping qualities, the provision of stability with decreased food wastage, making food attractive and essential aids to food processing. The cases where food additives should not be used include any place where the material would disguise the use of faulty processing and handling techniques, deceive the consumer, or substantially reduce the nutritive value of the food. Nor should they be used when they can be omitted by using good manufacturing processes that are economically feasible, and particularly when there is any doubt about the safety of the additive.

The continuous activity of the joint FAO/WHO Expert Committee on Food Additives provides a cumulative definitive statement of specifications for identity and purity of food additives and their toxicological evaluation. Their pertinent reports are listed below. Summaries of the *Technical Report Series* are given in the monthly publication, the *WHO Chronicle*.

1. "General Principles Governing the Use of Food Additives: First Report," *FAO Nutrition Meetings Report Series*, No. 11, 1956; *World Health Org. Tech. Rep. Ser.*, **129**:1956.
2. "Procedures for the Testing of International Food Additives to Establish Their Safety for Use: Second Report," *FAO Nutrition Meetings Report Series*, No. 17, 1958; *World Health Org. Tech. Rep. Ser.*, **144**:1958.

3. "Specifications for Identity and Purity of Food Additives (Antimicrobial Preservatives and Antioxidants): Third Report." Subsequently revised and published as *Specifications for Identity and Purity of Food Additives*, Vol. I: *Antimicrobial Preservatives and Antioxidants*, Rome, Food and Agriculture Organization of the United Nations, 1962.
4. "Specifications for Identity and Purity of Food Additives (Food Colours): Fourth Report." Subsequently revised and published as *Specifications for Identity and Purity of Food Additives*, Vol. II: *Food Colours*, Rome, Food and Agriculture Organization of the United Nations, 1963.
5. "Evaluation of the Carcinogenic Hazards of Food Additives: Fifth Report," *FAO Nutrition Meetings Report Series*, No. 29, 1961; *World Health Org. Tech. Rep. Ser.*, **220**: 1961.
6. "Evaluation of the Toxicity of a Number of Antimicrobials and Antioxidants: Sixth Report," *FAO Nutrition Meetings Report Series*, No. 31, 1962; *World Health Org. Tech. Rep. Ser.*, **228**:1962.
7. "Specifications for the Identity and Purity of Food Additives and Their Toxicological Evaluation: Emulsifiers, Stabilizers, Bleaching, and Maturing Agents: Seventh Report," *FAO Nutrition Meetings Report Series*, No. 35, 1964; *World Health Org. Tech. Rep. Ser.*, **281**:1964.
8. "Specifications for the Identity and Purity of Food Additives and Their Toxicological Evaluation: Food Colours and Some Antimicrobials and Antioxidants: Eighth Report," *FAO Nutrition Meetings Report Series*, No. 38, 1965; *World Health Org. Tech. Rep. Ser.*, **309**:1965.
9. "Specifications for Identity and Purity and Toxicological Evaluation of Some Antimicrobials and Antioxidants," *FAO Nutrition Meetings Report Series*, No. 38A, 1965; WHO/Food Add/65.24.
10. "Specifications for Identity and Purity and Toxicological Evaluation of Some Food Colours," *FAO Nutrition Meetings Report Series*, No. 38B, 1966; WHO/Food Add/66.25.
11. "Specifications for the Identity and Purity of Food Additives and Their Toxicological Evaluation: Some Antimicrobials, Antioxidants, Emulsifiers, Stabilizers, Flour-Treatment Agents, Acids, and Bases: Ninth Report," *FAO Nutrition Meetings Report Series*, No. 40, 1966; *World Health Org. Tech. Rep. Ser.*, **339**: 1966.
12. "Toxicological Evaluation of Some Antimicrobials, Antioxidants, Emulsifiers, Stabilizers, Flour-Treatment Agents, Acids, and Bases," *FAO Nutrition Meetings Report Series*, Nos. 40A, B, C; WHO/Food Add/67.29.
13. "Specifications for the Identity and Purity of Food Additives and Their Toxicological Evaluation: Some Emulsifiers and Stabilizers and Certain Other Substances: Tenth Report," *FAO Nutrition Meetings Report Series*, No. 43, 1967; *World Health Org. Tech. Rep. Ser.*, **373**:1967.
14. "Specifications for the Identity and Purity of Food Additives and Their Toxicological Evaluation: Some Flavouring Substances and Non-Nutritive Sweetening Agents: Eleventh Report," *FAO Nutrition Meetings Reports Series*, No. 44, 1968; *World Health Org. Tech. Rep. Ser.*, **383**:1968.
15. "Toxicological Evaluation of Some Flavouring Substances and Non-Nutritive Sweetening Agents," *FAO Nutrition Meetings Report Series*, No. 44A, 1968; WHO/Food Add/68.33.
16. "Specifications for the Identity and Purity of Food Additives and Their Toxicological Evaluation: Some Antibiotics: Twelfth Report," *FAO Nutrition Meetings Report Series*, No. 45, 1969; *World Health Org. Tech. Rep. Ser.*, **430**:1969.

Although all countries may never accept the same detailed regulations, there is agreement in principle and there should be international acceptance of experimental data in this area as there is in other areas. An international data pool should be established for all but trade secret compounds. A start in this direction has been made by one FAO/WHO Committee.[60] There should be just compensation to each manufacturer which expended time and money to provide the data.

The general view of chemicals used in foods in Europe was presented in 1958.[61] The internationally sponsored Food Law Research Center of the Institute of European Studies of Brussels University is embarked upon in a comprehensive study of the food laws of Europe.[62] The continuous progress toward the European Economic Community (the Common Market) is reflected in the progress toward common patents as well as food laws.[63] Such progress will surely be reflected in import-export practices of the group as well as in increased trade within the group.

Generally speaking, the philosophy of the United States has been on the liberal end of the countries in the world in admitting new materials to be used. In some countries the acceptance of one or two compounds for any given purpose precludes others from being admitted. The Common Market will probably allow more chemicals to be used than would be individually acceptable to the constituent countries; for example, West Germany now accepts one antioxidant in feeds. This tendency of the Common Market to liberalize practices of individual European countries, when combined with the recent conservative activity of the United States (caused by the costs of the more rigorous requirements imposed in 1958), may provide the basis for these two economic units to come to a common working philosophy regarding the number of compounds allowed for any given purpose.

Types of Food Additives

Of the variety of ways food additives have been classified, most involve functional groupings. Chemical groupings are convenient since they place molecules of similar structure and physical-chemical properties in comparable categories. Toxicologic and metabolic studies also may be correlated with chemical groupings. However, compounds of a single chemical family perform different functions in the food industry. Although single compounds are placed in two or more functional groupings, the functional classification is of more practical use to the feed industry. Moreover, since chemical structure does impart physical-chemical characteristics, many function classes follow reasonable chemical lines; i.e., the enzymes are a major class in either system. The consumer viewpoint is less restricted than is the legal, governmental or scientific; without knowing hows or whys, he wants an attractive product which provides assurance of health. A functional classification is given to place in perspective the specific types discussed in subsequent chapters. The classification presented in detail is comparable to, but provides an overall view less well than, this simplified system of Oser:[64]

A. Intentional Food Additives:
   1. Nutritive
   2. Freshness maintenance
   3. Sensory
   4. Processing aids
B. Incidental Food Additives

Many of the 2–3000 chemicals added to food occur in nature. They may be isolated from rich source material and used at will. Some, i.e., enzymes, will never be commercially available from synthetic sources. However, it is more efficient to make most of them in the laboratory. Thus, although some are isolated in relatively pure form from natural sources, most are man made.

Intentional Food Additives

The first groups of additives are intentional or direct additives. Each compound in this group is purposely added for a specific function in accordance with legal requirements. A reliable listing provides a view of the number of compounds used in functional classes: preservatives, 30; antioxidants, 28; sequestrants, 44; surfactants, 85; stabilizers, 31; bleaching and maturing agents, 24; buffers, acids, alkalies, 60; colors, 35; special sweeteners, 9; nutrient supplements, 116; flavoring compounds, 720; natural flavoring materials, 357; and miscellaneous materials, 158.[65]

Physical state modifiers are used both for consumer appeal and for industrial processing. Smooth, uniform texture with a specific consistency is desirable in many food products. Hydrophilic gums may be used in thickening and jellying processes as presented by R. E. Klose and M. Glicksman in Chapter 7. The use of carbohydrates and starches in consistency changes and as food binders is discussed by O. B. Wurzburg in Chapter 8. Surface active agents and their uses in principle and in practice are reviewed by W. C. Griffin and M. J. Lynch in Chapter 9. The industrial uses of polyhydric alcohols for crystallization modifiers, grain production, humectants, dehydration, and bulking agents are presented by W. C. Griffin and M. J. Lynch in Chapter 10.

Flow properties are important in many food products. Table salt may contain limited amounts of anti-caking agents such as magnesium or calcium silicate or calcium aluminum silicate. Powdered sugars, dry milk powders and baking mixes must be pourable. Emulsifiers and emulsion stabilizers help to maintain consistency and are invaluable ingredients of mayonnaise, candies and salad dressing. Foam stabilizers and antifoaming agents are used industrially in appropriate places. Sequestering and chelating agents used as stabilizers in the food industry are discussed in Chapter 6 by T. E. Furia, while O. B. Wurzburg reviews the role of starches and carbohydrates to stabilize food products in Chapter 8.

Color plays a major role in acceptability of food products. The Color Additives Amendments and certified color additives are presented in Chapter 14 by J. Noonan. Alcohol and other solvents are used in food to dissolve, dilute or carry a variety of additives, including color additives.

Chemical state modifiers are important to the food industry. Moisture is important as a factor in bacterial control, texture and control of flow. Glycerine is used as a humectant in marshmallows to retain the soft fluffy texture. Control of acidity or alkalinity is important in processed foods such as cakes, quick breads, soft drinks, etc. In Chapter 5 W. H. Gardner presents the variety of acidulants available commercially and their use in flavor modification, food preservation, and in the physical and chemical modification of foods. Antioxidants as stabilizers for foods, their uses, regulations, toxicity and analysis are presented by B. N. Stuckey in Chapter 4.

The natural enzymes in food continue to act following harvest and may aid digestion when raw foods are eaten. In Chapter 1 L. A. Underkofler presents the basic properties of enzymes and their uses by the food industry.

Aging acceleration and retardation processes which may be found in both primitive and modern treatment of food include drying, smoking, heating and salting. Maturing and bleaching agents such as chlorine, chlorine dioxide, potassium iodate or potassium bromate are used to improve color and baking qualities in flour. Some are useful in cheeses.

Sensory additives are those materials which are added to enhance, change or maintain the aroma, flavor, taste, texture, or color of foods. Here esthetic and physiological responses are more important than physical, chemical or nutritive properties. Some of these, color and texture, are well defined physical characteristics. Aroma, flavor and taste remain components of one general sensation. Taste includes more than flavor and aroma; sour, sweet, salt and bitter are the basic vectors of taste. In Chapter 13 A. Salant presents the chemistry, pharmacology, toxicology, organoleptic properties and general principles of the application of low calorie sweeteners.

Aroma receives less attention than it deserves in food production. The advent of gas chromatography will help to establish the important parameters of aroma for future development. Aroma and flavor are characteristics of food which separate the gourmet from the gourmand. The quantities of flavor additives are small but their number equals more than two-thirds of all additives. GRAS flavors include the usual spices and flavoring oils which are encountered in spice cakes, sausages and soft drinks. R. L. Swaine reviews both natural and synthetic flavors in Chapter 11. Flavor potentiators, their historical background and their uses are reviewed in Chapter 12 by L. B. Sjöström.

Nutritive additives are used to improve the biological value of certain foods where such supplementation is approved by public health, medical and scientific authorities for the prevention or elimination of nutritional deficiencies. Nutritive additives in staple food products are desirable for populations which may have marginal deficiencies or for replacement of nutrients lost in processing, i.e., thiamine in the milling of rice and wheat. Scrimshaw has reviewed this subject.[66] In Chapter 2 B. Borenstein presents the chemistry of vitamins and amino acids, their losses during food processing and storage and their specific use in food fortification. Vitamins D, A, thiamine, riboflavin, niacin and ascorbic acid are approved additives which have effectively eliminated rickets and pellagra from our population. So many sources of vitamins are presently available that a current problem is to control vitamin D excess in children. The above vitamins plus carotene, tocopherols, pantothenates, pyridoxine, vitamin B-12 and inositol are on the GRAS list.

Mineral additives include iron salts and potassium iodide. These are needed supplements where food is taken primarily from poor mineral soils. Other minerals which may be considered to be useful in animal nutrition are arsenic, chromium, cobalt, copper, manganese, selenium and zinc.[67] Iron supplementation undoubtedly helps incipient anemia among many females in this country. Iodized salt reduces simple goiter; however, many people who need iodine have not been educated to use iodized salt. Many communities fluoridate water as a means to control dental caries.[68] The nature of metals in food and their use as food and feed additives and in food processing is presented by T. E. Furia in Chapter 6.

Standards of identity have been accepted for several materials which may be considered to be food supplements. These include non-fat dry milk solids, cheeses and yeasts.

Preservation of food by drying, heating, high salt and high sugar are some of the oldest methods known. They are still used in a variety of ways today.

Although some pesticides may be used directly on foods, most applications are made for crop growth. Pesticide problems have been presented in detail in both the popular and the scientific presses.[69, 70] A review of the many anti-microbial agents used for food may be found in Chapter 3 by F. W. Tanner and D. F. Chichester. Standard methods of sterilization produce changes in the foods; these include protein denaturation, vitamin and amino acid destruction, and often mineral changes. Each method may be easily misused. Filtration sterilization is little used. Radiation sterilization of dry foods causes fewer undesirable reactions than does heat.[71] Radiation of marketable foods has been explored for two decades, and ultraviolet sterilization of liquid foods (milk) is being explored. An annotated bibliography for 20 years work on the whole-someness of irradiated foods is available; nevertheless the Food and Drug Administration moves slowly on new problems such as this.[72, 73] Ethylene oxide reactions with food have been partially elucidated but when properly applied, this gas has been used to sterilize food for animal colonies through several generations without showing harm.[74] Most of the above methods are used also in sub-sterilization doses to delay spoilage or to disinfest foods.

Unintentional Food Additives

Unintentional additives have no purposeful function in food. This class includes all materials which would not usually become a part of food if man could completely control food production. Indirect additives include those materials purposely added during production, processing or storage of food. In practice indirect additives are found in agricultural produce in quantities well within acceptable and legal tolerances.

Materials from nature which fall into the category of unintentional addi-tives include radionuclides, a variety of material in the water used for prepar-ing foods and the dirty residues, such as insect parts and excreta, which typified food of the "good old days" more than it does today. A number of whole-fish pro-ducts—shrimp, oysters, clams, and sardines—are accepted as GRAS; fish flour was considered dirty until recently. Essential nutrients which are components of fertilizer, such as phosphate, nitrate or calcium, are expected to become in-corporated into the plant and are not considered to be unintentional additives.

Residues and additives in foods of plant origin were reviewed by Goodman and others.[75, 76] The amounts of insecticides found in foods are generally well under tolerances set by different countries, excepting the *no* residue tolerance.[77] The major types of insecticide residues found on plants are heavy metal compounds, such as lead arsenate; chlorinated hydrocarbons, such as DDT and dieldrin; and phosphate derivatives, such as parathion and TEPP (tetraethyl pyrophosphate). Herbicides may be used to kill weeds, suppress growth and increase rooting of cuttings, suppress sprouting of onions and potatoes, deblossom fruit trees, and increase the number of fruits developing per cluster in fruit trees. These plant hormones include indole-3-acetic acid, amino triazole, nitrated and chlorinated phenols, and giberellin. Fungicides include nitrated and chlorinated phenols, such as phygon (2,3-dichloro-1,4-naphtha-quinone); organic heavy metal complexes, such as phenyl mercuric acetate; and heavy metals, such as Bordeaux mixture ($CuSO_4 + Ca(OH)_2$). The use of such compounds has been helpful in eliminating deaths from ergot poisoning by the fungus *Claviceps purpurea*.[76] Antibiotics and other bacteriostatic and bactericidal compounds are also used. Nematocides are all active as gases; thus, they should be present in minimal quantities in any prepared food. Nematocides include simple halogenated hydrocarbons, such as methyl bromide, ethylene dibromide and chloropicrin (tear gas). Rodenticides and other animal pest control chemicals may be used in certain agricultural areas. The ancient Chinese practice of burning kerosene to generate ripening compounds (ethylene and propylene) is one of the oldest practices in the fruit industry. Ripe fruit releases minute quantities of ethylene, which accelerates ripening in green fruit.[12] The pure gas is now used commercially. Other compounds are used to decrease flowering, to increase fruit size, or to induce parthenocarpy.

Residues from foods of animal origin have been reviewed by Luckey and others.[78, 79, 80] We proposed that antibiotic stimulation of animal growth might be mediated either by way of the intestinal microflora or by direct action of the tissues.[81] Since low levels of antibiotics stimulate growth in germfree chicks and poults, both mechanisms are involved. Almost all commerical broilers and swine and many beef cattle processed in this country were fed one of the tetracyclenes or a penicillin. Other antibiotics are not used in feedstuffs in large quantities. The high cost of obtaining permission to use additives has discouraged manufacturers from testing the newer antibiotics for growth stimulating properties. Arsenicals, such as arsenilic acid, promote growth and increased feed efficiency in animals. The action of these compounds may be comparable to that of the direct action of antibiotics. Hormone active compounds, such as iodinated casein or diethylstilbesterol, are used commercially. Estrogens are used in feed-lot cattle, but are omitted from the feed a few days prior to slaughter. Coccidiostats, such as arsenosobenzyne, nicarbizin and sulfaquinoxaline, helped to alleviate one of the worst problems of the poultry industry in the United States.

Vitamin, amino acid, and/or enzyme supplements to animal feeds are a cause of no concern since they are also direct food additives. Other feed additives, except antioxidants, are used sporadically and have less opportunity to accumulate in the meat used by the consumer.

The many intentional additives used in processing are in quite a different category from the unintentional ones. The latter are in the category of "necessary environment pollutants." Automation of the food industry changes processing contaminants from biological filth, detectable with microscope and cultural procedures, to more subtle traces of chemicals which require the tools of the chemist and the physicist for detection. These contaminants may come from vessels, conduits, conveyer belts, trays, valves, lubricants, storage containers, cleaning fluids, and anti-rust compounds. The changes foods undergo during storage depend upon many factors of processing.[82]

Package materials present complex problems associated with the migration of chemicals into foods. They are being approached in a few sophisticated physical-chemical laboratories over the world.[83] The migration of material into food packaging material should also be considered. The following experiment performed by students in my laboratory with adequate controls illustrates the migration problem. Lindane mixed into dried powdered grass was fed to baby crickets housed in plastic containers. The insecticide killed the crickets. The plastic containers were repeatedly cleaned in water, dilute alkali, dilute acid, detergent and alcohol. After each cleaning crickets were placed in the boxes with grass containing no insecticide; they all died in one day. Presumably the insecticide had migrated into the plastic, could not be washed out, and migrated out of the plastic with a high enough vapor concentration to kill the crickets.

Plasticizers, pigments and inks, monomers, and partially reacted polymers may migrate out of plastics. Specific chemical moieties are being investigated from food containers of metal, wood, paper, fabric, rubber, cellophane, polyethylene, polypropylene, nylon, polyurethane, lacquer, waxes, adhesives, and resinous dispersions and coatings.

The number of food products sold in aerosol containers has stabilized into a predictable component of the food industry. The gases used in aerosol bombs that form this category of additives are CO_2, NO, N_2, and their combinations.[84] Although propane and butane are approved for use, they are seldom used because of their taste and odor. The total number of aerosol units used in the United States is slightly over 2 billion; those containing food are almost 5% of the total. The aerosol food units numbered 100 million in 1967, 86 million in 1968, and 88 million in 1969.[85] The total units used in other countries increased threefold between 1960 and 1964.[86]

Microorganisms

Microorganisms contaminate our air, water and food in relatively low numbers. Most of them are innocuous; some are helpful, and a few are harmful. Appropriate application of heat, radiation, or certain chemicals will kill all microorganisms. These may change the food. The antimicrobial agents used in the food industry are presented in Chapter 3. Dawson and others have presented a general review of the microbiological potential and problems in foods production.[87-91] Over one-third of the food poisonings involve salmonellosis, an intestinal infection with one of over a thousand varieties of salmonella. Most of these are preventable if adequate cleanliness and heat precautions have been observed. Poultry and eggs are world-wide sources of salmonellosis. Other harmful bacteria produce specific toxins in food before it is eaten. The most potent is the thermolabile *Clostridium botulinum* toxin. One taste of contaminated food may be fatal. In the United States 10 cases resulted in 10 deaths in 1962.[91] The heat labile toxins of *Clostridium perfringens* are less harmful and rarely reported among food poisoning statistics. *Staphylococcus aureus* produces several heat stabile toxins. These account for about one-third of all ptomaine statistics. In this problem heating can kill the live bacteria but does not destroy the endotoxin. Aflotoxin from *Aspergillus flavus* in certain peanut crops is harmful to poultry and is a hazard to livestock, but it is rarely reported in man.[88] Surprisingly 10% of the pork sausage in large city markets is infected with *Trichinella spiralis* and 25-50 million Americans carry the larvae.[91] Shigellosis (bacillary dysentery) may be transmitted by contaminated food, milk or water; 134 deaths were reported in the 12,443 cases of 1962.[92]

These general characteristics of intentional and unintentional food additives provide a perspective for the detailed considerations of the compounds presented in subsequent chapters.

Summary

The chemicals which are intentionally added to our food constitute a small but important part of the many chemicals which man is adding to his environment. These chemicals are needed to efficiently produce an abundance of high quality food, one mark of the American way of life. All possible aids to food production and distribution will be needed to feed the coming double generation. Early food-inspection laws were strengthened by the 1938 and 1958 acts. Regulation of these laws has slowed the pace of the appearance of new compounds by making industry bear the burden of proof of safety as well as performance. These laws have converged the diverse concepts of food additives to legal, restricted definitions. The global and the scientific view still include unintentional and generally recognized additives as safe materials. Unintentional additives may have been incorporated into the food during its production, processing, storage or marketing. One poultry feed may have 20 chemical additives.[93] These compounds may appear in our food within legal tolerances.

Intentional food additives include chemicals which are used for nutritional purposes, consumer acceptance, preservation of quality and processing. The changing food industry allows the homemaker of today to spend only 90 minutes per day in the kitchen compared with the 5 hours used by her mother.[94] Her convenience foods which save time, effort and sometimes money contain many of the direct additives presented in this book. A single example is the dessert you may have eaten recently. The ready-mix or store-bought cake probably had more than 15 kinds of additives. Some leave no residues in the cake; all have been found to be safe when used properly, and all are useful in making it taste as if it were made, like little girls, of sugar and spice and everything nice.[95] Here is the cake:

Wheat seeds are treated with a mercurial fungicide (1) to avoid rot. Plants may get a dose of an insecticide (2) such as malathion or parathion. To kill insects, harvested grain receives a fumigant (3)—carbon tetrachloride, carbon bisulfide, etc.—and a protectant (4) such as methoxychlor.

Flour requires a chemical for bleaching and aging (5); chlorine or nitrosyl chloride are commonly used. Enrichment chemicals (6)—thiamine, riboflavin, niacin, and iron—are added as weapons against vitamin deficiencies.

Baking powder and salt have an anticaking agent (7), calcium silicate. Its starch has been bleached (8) with potassium permanganate.

Shortening contains an antioxidant (9), such as butylated hydroxytoluene, to keep it from becoming rancid and at least one emulsifier (10) to improve texture, help the cake rise, etc. Among common emulsifiers are mono- and diglycerides of fatty acids and lecithin.

Egg whites in commercial cakes contain other emulsifiers and (11) surface active agents such as cholic acid and triethyl citrate. Then an enzyme is used as a fermenting agent (12) to remove sugar from the white before drying.

Along with natural flavors and spices, synthetic flavors (13) are added as an improvement on nature, and the frosting of the cake may be decorated with a color additive (14).

Finally, to prevent mold, a commercial cake contains one or more preservatives (15)—sodium propionate, calcium propionate or sodium sorbate.

The icing may contain the following ingredients and additives: sugar, dextrose, corn syrup, solids, wheat and cornstarch, nonfat dry milk, water, salt, shortening with freshness preserver, cocoa processed with alkali, natural flavors, artificial flavors, plant lecithins, pectin, or cellulose gum, monoglycerides, diglycerides, polysorbate surfactants, potassium sorbate, sodium phosphate, sodium citrate or sodium stearate, citric acid and possibly artificial color. If a decorator bomb were used, the propellant may not be listed. Equally anonamous are ingredients of some of the "ice creams" which may have complemented the cake. This dessert exemplifies the use of food additives by 50,000 industrial companies with a $100 billion annual product that involves about one-third of the total FDA budget and manpower.[96] These are the chemicals with which we live in this year of 1969.

References

1. *Webster's New International Dictionary of the English Language*, 2nd Ed., 1947, G. and C. Merriam Co., Springfield, Mass.
2. Gunderson, F. L., Gunderson, H. W., and Ferguson, Jr., E. R., 1963, *Food Standards and Definitions in the United States*. Academic Press, New York.
3. Grange, G. R., 1966, International Food Standards: Status Report. Tenth Annual Education Conference of the Food and Drug Legal Institute—Food and Drug Administration, Washington, D.C.
4. Koenig, N., 1966, Food Standards for the World. *Protecting Our Food. The Yearbook of Agriculture 1966*, pp. 312–321. U.S. Government Printing Office, Washington, D.C.
5. Mrak, E. M., 1966, Food Technology. Proceedings, Western Hemisphere Nutrition Congress, American Medical Association, Chicago, pp. 229–234.
6. ADM Produces New Foods, *Chem. and Eng. News, 44:* N14, p. 60, 1966.
7. Saunders, H. J., 1967, Food Additives. *Chem. and Eng. News, 44:* 42, pp. 100–120, and *44:* 43, pp. 108–128.
8. Food Protection Committee, 1959, Principles and Procedures for Evaluating the Safety of Food Additives. NAS-NRC Publ. No. 750, Washington, D.C.
9. Food Protection Committee 1961, The Use of Chemicals in Food Production, Processing, Storage and Distribution. NAS-NRC Publ. No. 887, Washington, D.C.
10. Spices and Herbs Have an Intriguing History. *Todays Health, 43*, p. 31, 1965.
11. Peterson, A. G., 1960, *Salt and Salt Shakers*. Washington College Press, Washington, D.C.
12. Mrak, E. M., 1962, Technical Benefits from Food Additives. *Chemical and Biological Hazards in Food*. Ayres, J. C., Kraft, A. A., Snyder, H. E., and Walker, H. W. Iowa State University Press, Ames.
13. Moore, P. R., Luckey, T. D., Everson, C. A., McCoy, E., Elvehjem, C. A., and Hart, E. B., 1946, Use of Sulfasuxidine, Streptothricin and Streptomycin in Nutritional Studies with the Chick. *J. Biol. Chem., 165*, pp. 437–441.
14. Chapman, R. A., 1961, Nutritional Aspects of the Use of Food Colors. *Fed. Proc., 20:* 7, pp. 253–255.
15. Food Protection Committee, 1966, Toxicants Occurring Naturally in Foods. NAS-NRC Publ. No. 1354, Washington, D.C.
16. *Food Additives*. Manufacturing Chemists' Association, Inc., Washington, D.C., 1963.
17. Accum, F. C., 1798, Discourse on the Genuineness and Purity of Drugs. *Nicholsons' Journal*, London and 1820, *A Treatise on Adulterations of Food*. Ab'm Small, Philadelphia.
18. Rainwater, H. I., and Smith, C. A., 1966, Quarantines—First Line of Defense. *Protecting Our Food, Yearbook of Agriculture*, pp. 216–224. U.S. Government Printing Office, Washington, D.C.
19. Schultz, H. W., 1966, Education for 500 Careers. *Protecting Our Food, The Yearbook of Agriculture*, pp. 236–246. U.S. Government Printing Office, Washington, D.C.
20. Appert, M., 1812, *The Art of Preserving All Kinds of Animal and Vegetable Substances for Several Years*. East-India Company, London.

21. Reynolds, J. E., 1966, Industry Profit and Protection. *Protecting Our Food, The Yearbook of Agriculture*, pp. 247–257. U.S. Government Printing Office, Washington, D.C.
22. Ruppert, E. L., and Mackison, F. W., 1966, Public Health Programs. *Protecting Our Food, The Yearbook of Agriculture*, pp. 306–311. U.S. Government Printing Office, Washington, D.C.
23. Lee, R. J., and Harper, H. W., 1966, Meat and Poultry Inspection. *Protecting Our Food, The Yearbook of Agriculture*, pp. 280–289. U.S. Government Printing Office, Washington, D.C.
24. Wiley, H. W., 1929, *History of a Crime Against the Pure Food Law*. Washington, D.C.
25. Larrick, G. P., 1959, The Pure Food Law. *Food, The Yearbook of Agriculture*, pp. 444–451. U.S. Government Printing Office, Washington, D.C.
26. Carlson, A. J., 1928, The Physiologic Life. *Sci.*, *67*, pp. 355–360.
27. McLaughlin, F. E., 1966, The Food and Drug Administration. *Protecting Our Food, The Yearbook of Agriculture*, pp. 290–296. U.S. Government Printing Office, Washington, D.C.
28. Food Protection Committee, 1962, Radionuclides in Food, Publ. No. 988 of the Food and Nutrition Board NAS-NRC, Washington, D.C.
29. Anellis, A., Grecz, N., Huber, D. A., Berkowitz, D., Schneider, M. D., and Simon, M., 1965, Radiation Sterilization of Bacon for Military Feeding. *Appl. Microbiol.*, *13*, pp. 37–42.
30. Deatherage, J. R., 1966, New Horizons in Research. *Protecting Our Food, The Yearbook of Agriculture*, pp. 367–377. U.S. Government Printing Office, Washington, D.C.
31. Mehrlich, F. P., and Siu, R. G. H., 1966. Military and Space Operations. *Protecting Our Food, The Yearbook of Agriculture*, pp. 191–206. U.S. Government Printing Office, Washington, D.C.
32. No Residue, Zero Tolerance Out. *Chem. and Eng. News*, *44:* A18, p. 27, 1966.
33. Oser, B. L., 1960, Food Additives—New Law Causes Many Problems. *Chem. and Eng. News*, *38:* F15, pp. 108–117.
34. Mahoney, J. F., 1961, Industry Problems in Developing Intentional Food Additives. *Am. J. Clin. Nutr.*, *9*, pp. 277–282.
35. Food Protection Committee, 1966, Food Chemicals Codex, Publ. No. 1406, NAS-NRC, Washington, D.C.
36. Townsend, J. F., and Luckey, T. D., 1960, Hormoligosis in Pharmacology. *J. Am. Med. Assoc.*, *173*, pp. 44–48.
37. Luckey, T. D., 1959, Modes of Action of Antibiotics in Nutrition. *Recent Progress in Microbiology*, pp. 340–349. Almquist and Wiksell, Stockholm.
38. Luckey, T. D. 1968, Insecticide Hormoligosis. *J. Econ. Ent.*, *61*, pp. 7–12.
39. Food Protection Committee, 1959, Problems in the Evaluation of Carcinogenic Hazard from Use of Food Additives. Publ. No. 749, NAS-NRC, Washington, D.C.
40. Hieger, I., 1958, Cholesterol Carcinogenesis. *Brit. Med. Bull. 14*, pp. 159–160.
41. Boutwell, R. K., and Bosch, D. K., 1959, The Tumor-Promoting Action of Phenol and Related Compounds for Mouse Skin *Cancer Res.*, *19*, pp. 413–424, 1966. Research Report. *Lab. Manag.*, *66*, p. 10.
42. Evaluation of the Carcinogenic Hazards of Food Additives. FAO/WHO Meetings Report, Series No. 29, FAO, Rome, 1961.
43. Dougherty, E. C., Ferral, D. J., Brody, B., and Gotthold, M. L., 1963, A Growth Anomaly and Lysis with Production of Virus-like Particles in Axenically Reared Microannelid (Enchytraeus fragmentosus). *Nature, 198*, pp. 973–975.
44. Miller, J. A., 1966, Tumorigenic and Carcinogenic Natural Products. In, Toxicants Occurring Naturally in Foods. NAS-NRC Publ. No. 1354, Washington, D.C.
45. Cromwell, N. H., 1965, Chemical Carcinogens, Carcinogenesis and Carcinostasis. *Am. Sci.*, 53, pp. 213–236.
46. Weisburger, J. H., and Weisburger, E. K., 1966, Chemicals as Causes of Cancer. *Chem. and Eng. News*, *44:* F7, pp. 124–142.
47. Dormal, S., 1961, Nutritional Aspects of Pesticides and the Uses of Agricultural Chemicals. *Fed. Proc.*, *20*, pp. 231–236.
48. McLaughlin F. E., 1967, Pesticides Residues in Total-Diet Samples. *Sci.*, *154*, pp. 101–104.
49. Agricultural Research Service, 1966, Monitoring Agricultural Pesticide Residues. ARS 81-113, USDA, Washington, D.C.

50. Antisell, T., 1869, Report of the Commissioner of Agriculture. U.S. Government Printing Office, Washington, D.C.
51. Wiley, H. W., 1887, Foods and Food Adulterants. Bulletin 13, USDA, U.S. Government Printing Office, Washington, D.C.
52. Wedderburn, A. J., 1890, A Popular Treatise on the Extent and Character of Food Adulterations. Bulletin 25, Division of Chemistry, USDA, U.S. Government Printing Office, Washington, D.C.
53. Hart, F. L., 1952, A History of the Adulteration of Food Before 1906. *Food Drug Cosmetic Law Journal, 7*, pp. 7–18.
54. FDA Publ. No. 1. Rev. 1965. U.S. Government Printing Office, Washington, D.C.
55. Committee on Government Operations, 1963, Consumer Protection Activities of State Governments, Part 2. U.S. Government Printing Office, Washington, D.C.
56. Larrick, G. P., 1960, Applications of Science in Assuring Safety of the Food Supply. *Science and Food*, pp. 57–62, The Food Protection Committee, NAS-NRC, Washington, D.C.
57. Diepenbrock, F., 1964, Gehes Codex. Schwarzek-verlag G.M.B.H., Munich.
58. Second Joint FAO/WHO Conference on Food Additives. FAO Nutrition Meetings Report Series No. 34 WHO, Geneva, 1963.
59. The WHO Program in Nutrition. *WHO Chronicle, 19*, pp. 467–476.
60. Specifications for the Identity and Purity of Food Additives and Their Toxicological Evaluation: Emulsifiers, Stabilizers, Bleaching, and Maturing Agents. WHO Technical Report Series, No. 281, Geneva, 1964.
61. Souci, S. W., and Mergenthaler, E., 1958, Fremd Stoffe in Lebensmitteln besonder er Berucksichtigung der Konserrierung. Verlage Von J. Bergman, Munchen.
62. New Group Will Study Europe's Food Laws. *Chem. and Eng. News, 44:* F7, pp. 78–80, 1966.
63. Common Patents for the Common Market. *Chem. and Eng. News, 42*, pp. 86–106, 1964.
64. Oser, B. L., 1961, Modern Technology as Related to the Safety of Foods. *Fed. Proc., 20: 7*, pp. 224–230.
65. Food Protection Committee, 1965, Chemicals Used in Food Processing. Publ. 1274 NAS-NRC, Washington, D.C.
66. Scrimshaw, N. S., 1962, Specific Nutrients, *Chemical and Biological Hazards in Food*, pp. 27–49. Ed. by Ayres, J. C., Kraft, A. A., Snyder, H. E., and Walker, H. H. Iowa State University Press, Ames.
67. Many Trace Elements Affect Animal Nutrition. *Chem. and Eng. News, 44:* M16, pp. 48–50, 1966.
68. Sognnaes, R. F., 1965, Fluoride Protection of Bones and Teeth. *Sci., 150*, pp. 989–992.
69. Gunther, F. A., 1962, Pesticides, *Chemical and Biological Hazards in Food*, pp. 77–88. Ed. by Ayres, J. C., Kraft, A. A., Snyder, H. E., and Walker, H. W. Iowa State University Press, Ames.
70. Food Protection Committee, 1956, Safe Use of Pesticides in Food Production. Publ. 470 NAS-NRC, Washington, D.C.
71. Luckey, T. D., Wagner, M., Reyniers, J. A., and Foster, F. L., 1955, Nutritional Adequacy of a Semi-Synthetic Diet Sterilized by Steam of Cathode Rays. *Food Res., 20*, pp. 180–185.
72. Reber, E. F., Raheja, K., and Davis, D., 1966, Wholesomeness of Irradiated Foods. *Fed. Proc., 25*, pp. 1529–1577.
73. Apathy Hit in Food Irradiation. *Chem. and Eng. News, 45:* F6, p. 26, 1967.
74. Luckey, T. D., 1963, Germfree Life and Gnotobiology. Academic Press, Inc., New York.
75. Goodman, R. N., 1961, Chemical Residues and Additives in Foods of Plant Origin. *Am. J. Clin. Nutr., 9*, pp. 269–276.
76. Vlitos, A. J., 1962, Plant Growth Regulators, *Chemical and Biological Hazards in Food*, pp. 89–126. Ed. by Ayres, J. C., Kraft, A. A., Snyder, H. E., and Walker, H. W. Iowa State University Press, Ames.
77. 1968 Evaluations of Some Pesticide Residues in Foods. FAO/pl: 1968/M/9/1 and WHO/Food Add/69.35, FAO and WHO, Geneva, 1969.
78. Luckey, T. D., 1959, Antibiotics in Nutrition, *Antibiotics Their Chemistry and Non-Medical Uses*, pp. 174–321. Ed. by H. S. Goldberg, D. Van Nostrand, Inc., Princeton.
79. Bird, H. R., 1961, Additives and Residues in Foods of Animal Origin. *Am. J. Clin. Nutr., 9*, pp. 260–268.

80. Kastelic, J., 1962, Animal Growth Regulators, *Chemical and Biological Hazards in Food*, pp. 127–141. Ed. by Ayres, J. C., Kraft, A. A., Snyder, H. E., and Walker, H. W. Iowa State University Press, Ames.

81. Luckey, T. D., 1956, Mode of Action of Antibiotics—Evidence from Germfree Birds, pp. 135–145. First International Conference on the Use of Antibiotics in Agriculture, Publ. No. 397, NAS-NRC, Washington, D.C.

82. Shipstead, H., and Tarassuk, N. P., 1953, Chemical Changes in Dehydrated Milk During Storage. *J. Agr. and Food Chem.*, *1*, pp. 613–616.

83. Food Protection Committee, 1958, Food-Packaging Materials: Their Composition and Uses. Publ. No. 645, NAS-NRC, Washington, D.C.

84. Anderson, E. V., 1966, Food Aerosols. *Chem. and Eng. News, 44:* M2, pp. 90–99.

85. *Chem. and Eng. News*, May 25, 1970, pp. 13–14.

86. Aerosol Industry Growing Fast in Europe. *Chem. and Eng. News, 43*, pp. 104–106, 1965.

87. Dawson, R. C., 1965, Potential for Increasing Food Production Through Microbiology. *Bact. Rev., 29*, pp. 251–266.

88. Wogan, G. N., 1965, *Mycotoxins in Foodstuffs*. M.I.T. Press, Cambridge.

89. Cockburn, W. C., Tayler, J., Anderson, E. S., and Hobbs, B. C., 1962, Food Poisoning. *Roy. Soc. of Health*, London.

90. Ayres, J. C., Kraft, A. A., Snyder, H. E., and Walker, H. W., 1962, Chemical and Biological Hazards in Food. Iowa State University Press, Ames.

91. Earl, H. G., 1965, Food Poisoning: The Sneaky Attacker, *Todays Health, 43*, pp. 64–88.

92. Van Itallie, P. H., 1964, A Review of Immunology, *Pulse of Pharm., 18*, pp. 3–12.

93. New Bureau to Handle Animal Products of F.D.A. *Chem. and Eng. News, 40:* 011, p. 35, 1965.

94. Campbell, C., 1967, Easy Ways to Delicious Meals. Campbell Soup Co., Camden.

95. Reprinted by permission from *Changing Times*, the Kiplinger Magazine (May 1960 issue). Copyright © 1960 by the Kiplinger Washington Editors, Inc.

96. *Congressional Quarterly*, Aug. 22, 1969.

CHAPTER 1

Enzymes

Leland A. Underkofler, Ph.D.

Retired from Miles Laboratories, Inc.
Elkhart, Indiana

Introduction

Enzymes are important factors in food technology because of the roles they play in the composition, processing and spoilage of foods. Enzymes occur naturally in many food raw materials and can affect the processing of foods in many ways. Sometimes the presence of natural enzymes is advantageous; for example amylases in sweet potatoes assist in curing to give desirable texture and flavor. In other cases natural enzymes may produce undesirable reactions such as rancidity produced by lipases, or browning reactions due to polyphenol oxidases. Sometimes tests for natural enzymes in foods are used as an index of whether heat treatment has been sufficient, such as detection of phosphatase in milk or cheese as evidence of inadequate pasteurization or of catalase or peroxidase in vegetable products as evidence of inadequacy of blanching. Recognition of the function of enzymes and their usefulness in bringing about desirable changes has led to their large-scale use as modifiers of food ingredients. Such commercial enzyme applications have grown from a relatively insignificant role to one of the most important aspects of food processing during the past quarter of a century.

The practical application of enzymes to accomplish certain reactions has been conducted for centuries. The fermentation of foods—beer, wine, bread, cheese—is older than recorded history. All fermentations are conversions produced enzymatically through the metabolism of living organisms, hence these food fermentations are examples of enzymatic modifications. Where a number of enzymes, constituting an enzyme system, are necessary to produce desired changes, it is advantageous to employ intact cells in fermentation processes. Where a single enzyme or a system of only two or three enzymes is involved in a desired reaction, isolated enzyme products are preferable to intact cells. A detailed treatment of fermentations is beyond the scope of this chapter.

Examples of ancient uses of enzymatic action, other than fermentations, are malt in brewing, stomach mucosa in milk clotting for cheese making, papaya juice for meat tenderization, and dung for leather bating. Of course, the ancient peoples using these natural products did not know there were such entities as enzymes. However, crude enzyme preparations extracted from animal tissues such as pancreas and stomach mucosa, or from plant tissues such as malt and papaya fruit, became articles of commerce.

The usefulness of the crude enzyme preparations was established, and when the biocatalytic enzymes responsible for their action became recognized, a search began for better, less expensive, and more readily available sources for such enzymes. Advances in enzymology and microbiology have resulted in more progress in enzyme production and application during the past 75 years than in the preceeding 5000 years. Early pioneers who laid the foundations for the production and use of industrial enzymes were Takamine,[167] Rohm[144] and Wallerstein.[192] The development of methods for large-scale production of enzymes, along with knowledge for controlling enzymatic processes and applications has resulted in a sizeable number of industrial enzyme products. These have been listed by deBecze by trade name, giving for each the producer, enzymes present, primary action on substrate, uses, and recommended pH and temperature ranges.[43, 43a]

Today, commercial enzyme preparations obtained from plant, animal and microbial sources are widely used by industry. Malt amylase and the proteases, papain, ficin and bromelain, from tropical plants are the best known enzymes from plant sources. Proteases, amylases and lipases from pancreas, pepsin and rennet from stomach mucosa, and catalase from liver are useful commercial enzymes from animal tissues. Certain microorganisms have been found to produce abundant amounts of useful enzymes, and these have become a major source for commercial enzymes.

Enzyme Theory

Before considering the specific applications for enzymes in present day food industries, it is desirable to consider briefly simple enzyme theory—what enzymes are, how they work, and factors affecting their action. This is necessary in order to understand and control the practical applications of enzymes.

For an exhaustive treatment of enzyme theory a modern textbook such as that of Dixon and Webb should be consulted.[48]

Chemical Nature of Enzymes

Enzymes belong to the broad class of substances which the chemist calls catalysts. A catalyst is a substance which influences the velocity of a reaction without being used up in the reaction. The catalysts take part in reactions, but reappear in their original form, describing a cycle. Theoretically, a catalyst can convert an unlimited amount of reacting substance. Enzymes are a very special kind of catalysts with very distinctive properties.

The most commonly accepted definition is that an enzyme is a soluble, colloidal, organic catalyst produced by a living cell, even more simply, an enzyme is a biocatalyst produced by a living cell. Enzymes can be produced only by living cells to accomplish specific metabolic needs. But fortunately the enzymes can be separated readily from the cells which produce them and perform their catalytic activities entirely apart from the cells, and hence are available for useful applications.

Since all enzymes are either simple or conjugated proteins, another definition quite widely used is that an enzyme is a protein with catalytic properties due to its power of specific activation.

Enzymes, like all catalysts, affect the rates of chemical reactions, but not the extent of the chemical change concerned. The enzymes accelerate reactions which are, in themselves, thermodynamically possible; that is, are attended by losses of free energy. However, in order for the reaction to occur a certain amount of resistance must be overcome; that is the molecules must be activated by supplying a certain amount of activation energy. We can describe enzyme action by stating that the enzyme lowers the amount of activation energy required by the reaction. Hence, enzymes are able to bring about, under mild conditions near room temperature, reactions which without enzymes would require drastic conditions of high temperature or other high-energy conditions.

All of the highly purified enzymes which have been isolated have been found to be proteins. In some enzymes non-protein prosthetic groups are also present, such as a specific metal like zinc or calcium, or an organic heme or a flavin or other group.

Although the beginnings of enzymology can be traced back to the early nineteenth century, the development of enzymology as a science has come mainly during the last 60 years. Enzyme activities were fairly well understood by 1920, and about this time serious attempts at purification of enzymes began. The first enzyme to be prepared in crystalline form was urease by Sumner in 1926. Even 20 years ago there were very few purified enzymes, whereas now the number of pure and crystalline enzymes is over 100, more than 600 have been somewhat purified, and over a thousand have been studied to some extent. Just how many enzymes may exist is unknown, but there are probably 10,000 or more.

Enzymes differ from other catalysts in several respects. These differences are, of course, due to the protein nature of the enzymes. Being proteins, enzymes are denatured and inactivated when subjected to unphysiological conditions, such as heat or strong chemicals. Hence, one of the distinctive properties of enzymes, in contrast with ordinary chemical catalysts, is their thermal lability, and sensitivity to acids and bases. But the difference between enzymes and other catalysts is most clearly displayed in one respect: enzymes generally have extremely specific actions. Whereas acids, that is hydrogen or hydronium ions, may catalyze hydrolysis of many kinds of substances such as esters, acetals, glycosides (sugars) or peptides (proteins), separate and different enzymes are necessary for the hydrolysis of each of these, and even of specific members of each class. Thus, esterases and lipases can split only esters or fats, without hydrolyzing glycosides or peptides. Carbohydrases act specifically on glycosides and cannot attack ester or peptide bonds. Individual carbohydrases are necessary for each of the individual glycosides. For example, although sucrose, lactose and maltose are all disaccharides, separate and distinct enzymes are necessary for their hydrolysis. The enzyme lactase which hydrolyzes lactose has no effect whatever on sucrose or maltose, and so on. In enzymology two types of enzyme specificity are recognized, substrate specificity and reaction specificity. Examples of substrate specificity are the hydrolysis of lactose by lactase, and the oxidation of glucose by glucose oxidase. Examples of reaction specificity are the actions of proteinases which are capable of splitting particular peptide bonds of proteins. These peptide bonds are amide linkages between carboxyl and amino groups of the constituent amino acids. Usually specific proteinases have narrow reaction specificity as to the exact peptide bonds they can split. For example, the action of pepsin is restricted largely to hydrolysis of peptide linkages to which the carboxyl group has been furnished by phenylalanine or tyrosine, and has little or no effect on peptide bonds formed by other amino acids.

A good scientific definition is that an enzyme is a protein with catalytic properties due to its power of specific activation. Active research is beginning to give information regarding the exact composition of enzymes. They are proteins with molecular weights of individual enzymes ranging from about 13,000 to over a million. The amino acid compositions of a few enzymes have been determined, and even the amino acid sequence in some. But there is nothing in the amino acid analysis or sequence which differentiates enzymes from other proteins. Since enzyme proteins possess specific catalytic functions, special structures must be responsible. All proteins contain reactive groups such as free amino, carboxyl, hydroxyl, sulfhydryl, and imidazole groups, and frequently non-protein prosthetic groups. However, mere presence of reactive groups, or combining sites, in protein molecules does not insure enzyme activity. These must be so arranged that the reactive groups of the substrate may fit the enzyme and be held for further action. The long peptide chains of native protein molecules are known to be folded and arranged in exact positions so in an enzyme the combining sites are suitably arranged to make up active centers necessary for substrate binding and catalytic activity.

Naming and Classification of Enzymes

Enzymes bring about changes in specific compounds. The compound upon which an enzyme acts is known as its substrate. Up until recently there have been no completely systematic methods for naming and classifying enzymes. Names of early enzymes were completely unsystematic—diastase, emulsin, pepsin, ptyalin, trypsin, catalase. Later it became customary, where possible, to name an enzyme after the substrate upon which it acts, with the ending "-ase"; thus, peptidase, esterase, urease, lactase, amylase. In other cases, enzymes were named from the reactions they catalyzed, such as dehydrases, dehydrogenses, transferases, phosphorylases.

In 1961, and modified slightly in 1964, the Commission on Enzymes of the International Union of Biochemistry published an entirely systematic method for classifying and naming enzymes, based on substrate and type of reaction.[10] The enzymes are divided into six main classes, oxidoreductases, transferases, hydrolyases, lyases, isomerases and ligases. Each class is further divided into a number of sub-classes and sub-sub-classes, according to the nature of the chemical reaction catalyzed, and is coded on a four number system intimately connected with this system of classification. Oxidoreductases catalyze oxidations or reductions. Transferases catalyze the shift of a chemical group from one donor substrate to another acceptor substrate. Hydrolases catalyze hydrolytic splitting of substrates. Lyases remove groups or add groups to their substrates (not by hydrolysis). Isomerases catalyze intramolecular rearrangements. Ligases (synthetases) catalyze the joining together of two substrate molecules. Both systematic and trivial names are recommended for the enzymes. The systematic name is formed in accordance with definite rules and will identify the enzyme and indicate its action as precisely as possible. In general the systematic name consists of two parts; the first part names the substrate, and the second, ending in "-ase," indicates the nature of the process. The systematic rules are quite extensive and they are difficult to apply if the substrate composition or the enzyme reaction are not fully understood. The trivial name is sufficiently short for general use, and in the majority of cases is the name already commonly employed. To illustrate, the familiar β-amylase (trivial name) has the systematic name: β-1,4-glucan maltohydrolase. The name indicates the substrate is a glucan (starch), a glucose polymer in which the glucose molecules are joined by β-1,4 linkages, and the reaction is hydrolytic splitting off of maltose.

How Enzymes Act

Enzyme-Substrate Combination   It has been shown that an enzymatic reaction proceeds in two stages. The enzyme (E) unites with substrate (S) to form a labile intermediate complex:

$$E + S \rightleftharpoons ES \tag{1}$$

The result is that the substrate molecules become more chemically reactive by some intramolecular changes. We say that the substrate is activated by the enzyme. The unstable intermediate complex reacts with a reactant (R) and breaks down with formation of the end products (P) of the reaction and regeneration of the enzyme:

$$ES + R \rightleftharpoons E + P \tag{2}$$

The idea that enzymatic action is due to formation of an intermediate complex between enzyme and substrate was expressed by Michaelis and Menton,[113] was actually demonstrated by Chance,[29] and recently an enzyme substrate complex was isolated in crystalline form.[199] The mass action equilibrium constant for the formation and dissociation of enzyme and substrate, as given in Eq. 3, has come to be known as the Michaelis constant.

$$\frac{(E - ES) \times (S - ES)}{(ES)} = K_m \qquad (3)$$

Without going into details of how K_m may be determined, it will suffice here to point out that it is a fundamental constant in enzyme work since its value reflects the affinity of an enzyme for its substrate or substrates. The lower the value of K_m, the higher is the affinity of the enzyme for its substrate.

Enzymes are effective in extremely small amounts because the enzyme substrate complex is very reactive and is present for a very short time. With enzymes, a useful term to describe the amount of substrate converted in unit time by a given quantity of enzyme is the molecular activity or turnover number, which represents the number of moles of substrate converted per mole of enzyme per minute. The molecular activities of different enzymes vary widely (100 to 5,000,000).

Factors Affecting Enzyme Action   There are numerous factors which affect enzyme activity, and these must be taken into account in the use of the enzyme. Among the most important are (1) concentration of enzyme, (2) concentration of substrate, (3) time, (4) temperature, (5) pH, and (6) presence or absence of activators or inhibitors.

For most enzymatic reactions the rate of the reaction is directly proportional to the concentration of enzyme, at least during the early stages of the reaction.

With very low substrate concentration enzymatic reaction velocity is proportional to the substrate concentration. In many practical enzymatic applications, where substrate is present in considerable excess during the early stages, the reaction follows zero-order kinetics, and the amount of product formed is proportional to time:

$$dP/dT = k_0 \qquad (4)$$

For such reactions the amount of end product is doubled if the reaction time is doubled.

However, during the course of an enzymatic reaction there is a continuing decrease in substrate concentration which results in slowing down of the reaction with time as shown in Figure 1. Most enzymatic reactions follow the kinetics of a first-order reaction:

$$dP/dT = k_1 \times (S - P) \qquad (5)$$

where k_1 is the first-order reaction constant and $(S - P)$ is the concentration of substrate remaining at any given time. The rate of the reaction is directly proportional to the remaining substrate concentration. Equal fractions of the remaining substrate are transformed in equal time intervals. For example, if 50% of the substrate is converted in 30 minutes, an additional 25% of the original substrate (50% of the remaining substrate) will be converted in the next 30 minutes, and so on.

Time is a very important factor in practical enzymatic applications. As mentioned above, initial reaction velocities are proportional to enzyme concentration. However, as the reaction proceeds the rate diminishes as shown in Figure 1. The decreased velocity may be due to many reasons. Most important usually are exhaustion of the substrate and inhibition of the reaction by its end products. First-order kinetics postulates slowing up of the reaction as available substrate diminishes in amount. For practical uses, because of this inherent behavior of enzymes, sufficient time must be allowed for enzyme reactions to approach completion. This may involve reaction periods of several hours or even days, for example 48 to 96 hours in the industrial enzymatic process for dextrose production from starch.

Heat may affect enzymes in two ways. One effect is inactivation since high temperatures cause denaturation of the enzyme protein resulting in a loss of catalytic properties. The actual temperatures at which heat inactivation is substantial vary a great deal depending upon the particular enzyme. For many enzymes useful in food processing, inactivation becomes appreciable

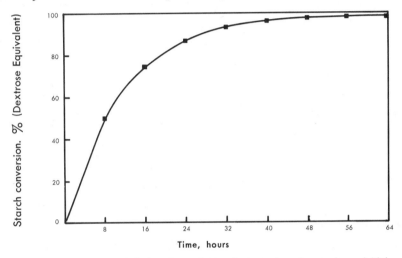

Fig. 1. Enzymatic hydrolysis of starch to dextrose by glucoamylase. Initial starch 30%, glucoamylase 80 units/lb. starch, 60°C, pH 4.0.

and rapid at temperatures above about 50°C. Some enzymes are much more resistant to heat; for example, glucoamylase is effectively employed at 60°C, and bacterial amylases at 80°C or even higher.

The second effect of temperature on enzymatic reactions is in their rate. Like most chemical reactions, enzyme-catalyzed changes are increased in rate by raising the temperature. A rough rule for chemical reactions, including those catalyzed by enzymes, is that every 10°C increase in temperature about doubles the rate; that is, the temperature quotient, Q_{10}, is about 2. Measurements of individual enzyme reactions have given Q_{10} values in the range of 1.2 to 4.

However, with an enzyme reaction as temperature is increased, thermal inactivation of the enzyme may take place so rapidly as to more than offset increased rate of reaction due to the higher temperatures. The so-called optimum temperature is that point of maximum activity above which the rate of reaction decreases because of thermal inactivation. Optimum temperature values must be interpreted with caution, since such factors as substrate concentration and particularly time of an enzymatic reaction have considerable effect on optimum temperature. This is apparent from the temperature-activity curves of Figure 2 which show apparent temperature optima differing by 10°C depending upon whether 30 or 60 minute incubations were used. This is of course due to greater inactivation at the higher temperatures with longer incubation.

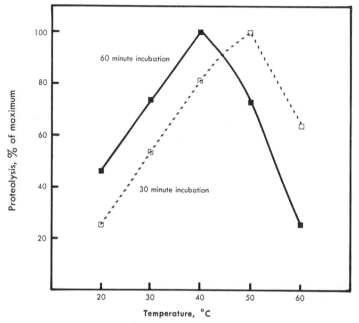

Fig. 2. Temperature—activity curves for bacterial proteinase. Gelatin, pH 7.0.

Source: L. A. Underkofler, 1961, Production and Applications of Plant and Microbial Proteinases. *Production and Application of Enzyme Preparations in Food Manufacture, Soc. Chem. Ind. (London) Monogr.* No. 11, pp. 48–59.

The pH of the system also has a profound effect on enzyme activity. Each enzyme in the presence of its substrate has a characteristic pH at which its activity is highest, known as the optimum pH. For some enzymes the optima are quite sharp, for others there are rather broad optimum pH ranges. Typical pH-activity curves are shown in Figure 3. When the pH is changed the activity decreases rapidly on both sides of the optimum range until the enzyme is completely inactive. Change of the pH toward the optimum will reactivate the enzyme, but change of pH above or below the levels of temporary inactivation may gradually denature the enzyme to permanent inactivity. The pH optimum values for various enzymes vary widely. For example, the optima for the two proteolytic enzymes pepsin and trypsin are about 2.0 and 8.0, respectively. There is also a pH range for best enzyme stability, which is not necessarily the same as for optimum activity.

Some enzymes need cofactors or activators for maximum effectiveness. Frequently metal ions, such as those of calcium, magnesium or manganese, are necessary enzyme activators, either for maximum activity or stability or both.

Enzyme inhibition is a very important area of enzymology. Enzymes are inhibited by a number of conditions, such as by lack of moisture since all enzyme reactions take place in aqueous systems. However, the word enzyme inhibitor usually refers to a substance or chemical which causes inhibition of enzyme reaction. There are reversible and irreversible, specific and non-specific inhibitors.

Since enzymes are proteins, groups such as free carboxyl, amino, and sulfhydryl groups will be common to many enzymes and the blocking of these groups will, of course, give rise to a rather non-specific inhibition. If the enzyme activity is restored by removal of the inhibitor, as by dialysis, the inhibition is reversible. Irreversible inhibition is produced by some poisons which cause irreversible denaturation or destruction of the enzyme.

Specific inhibitors are those substances which block groups conferring specificity on an enzyme; that is to say, which react with combining sites of the active centers. A study of these reversible and specific inhibitions has shown that two types may be distinguished, the competitive and the non-competitive.

Competitive inhibition is shown by substances with structural similarity to the normal enzyme substrate, including frequently products of the enzyme action. The enzyme is capable of combining with such substances, but cannot activate them, and inhibition of the enzyme reaction results from hindered access of the normal substrate to the active centers.

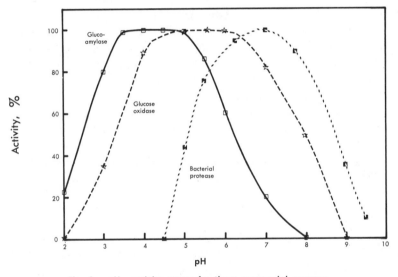

Fig. 3.   pH—activity curves for three commercial enzymes.

In non-competitive inhibition, there is no competition between substrate and inhibitor. Instead, the inhibitor combines with groupings not essential for the formation of the enzyme-substrate complex, but necessary for the substrate activation. There are naturally occurring inhibitors of the kind, such as trypsin inhibitors found in serum, eggs and soybeans. Many poisons involve inhibition of a single enzyme involved in a main metabolic chain, thus rendering the whole chain inoperative and having a profound and even fatal effect upon the organism. The toxicity of cyanide is due to its inhibition of cytochrome oxidase. Recognition of poisons as enzyme inhibitors is of fundamental importance to pharmacologists and toxicologists, and has practical applications. An increasing use of enzyme inhibitor compounds as insecticides is now the basis of a large and expanding industry. The "nerve gases" of military importance are essentially specific enzyme inhibitors.

Methods of Enzyme Assay   An important problem to the enzymologist is to quantitatively measure the activity or concentration of an enzyme.[19a, 43a, 181]

This cannot be done by the ordinary means of analytical chemistry because the only means of detecting an active enzyme is by what it does to its specific substrate. Hence, the only way of measuring the activity or quantity of an enzyme is by determining how fast it changes its substrate under controlled conditions. For example, one can assay the potency of an amylase by adding a solution containing a known weight of the enzyme preparation to a starch solution and allowing it to react for a given time. Careful control of temperature and pH during the digestion period is necessary since these factors so markedly affect the action of the enzyme. Likewise the time chosen must be within the range of linear proportionality of reaction rate with enzyme concentration. The reaction is then stopped by adding a suitable reagent, and the amount of starch hydrolyzed is determined. In the case of the amylase reaction, either the change in viscosity of the starch paste is measured, or the amount of starch not changed is determined (starch-iodine color reaction), or the amount of sugar produced from the starch is determined. In other words, enzymes are assayed by determining the amount of substrate changed under standardized conditions of the procedure, by the change in physical nature of the substrate, or by the amount of a reaction product produced. There are numerous ways of expressing the results of enzyme assays, a common method being in terms of arbitrary units, which are defined for the specific enzyme and substrate used. For amylase, for example, the enzyme potency might be expressed in saccharification units, a unit being defined as the amount of enzyme which produces one gram of maltose when it acts on a 2% starch solution for 30 minutes at 30°C. The Commission on Enzymes of the International Union of Biochemistry has recommended that for all enzymes uniform units be employed, defined as: One unit of any enzyme is the amount which will catalyze the transformation of one micromole (or microequivalent) of substrate per minute under specified conditions of temperature, pH, concentration, etc.[10]

Management of Natural Enzymes and Enzymatic Action

Food materials are of natural plant or animal origin. Since enzymes are biocatalysts produced by living cells, enzymes of many different kinds are present in most fresh food materials. These natural enzymes may contribute to the desirable characteristics of the food or may be an important factor in the deterioration or spoilage of the food. An important aspect of food enzymology therefore has become the management of natural enzymes and enzymatic action.

In a general way spoilage of food from enzyme action may be as varied as that caused by microorganisms; in fact, all microbial food spoilage is caused by the enzymes contained in or elaborated by these organisms. However, microbial food spoilage is a distinct problem in itself, beyond the scope of this chapter.

Many food products contain diastatic enzymes which, when conditions are right, can change starches to sugars. Thus if stored at too high temperatures, potatoes become sweet and soft. Most living cells contain autolytic enzymes that can act on proteins. When animal or plant tissues are stored, these enzymes can bring about extensive changes, which in some cases may be desirable, and in other cases considered as spoilage.

Studies on frozen and dehydrated foods have made it apparent that naturally occurring enzymes can spoil foods. In spite of freezing or removal of moisture the products may undergo marked color or flavor changes. A variety of enzymes may be involved in these undesirable changes. Some act on fats, some on proteins, others are pectolytic, and some are phenolases causing enzymatic browning reactions. These undesirable enzymatic changes can often be avoided by blanching, which consists of preheating the product to the point where the enzyme is inactivated.

The use of heat in pasteurization or sterilization not only destroys microorganisms but also is a convenient way of inactivating enzymes. Enzyme inactivation by heat often takes more time than the attainment of complete sterility from microorganisms. In some cases, however, such as in milk, a test for the absence of phosphatase enzyme is a good indication of adequate pasteurization since the enzyme is completely inactivated at a temperature high enough to destroy undesirable organisms including resistant pathogens such as *Mycobacterium tuberculosis*.

Situations are known in food technology when, although they have been inactivated, certain enzymes may be regenerated and their activity restored after some time. Such enzyme regeneration has been observed with the phosphatase of milk, pectolytic enzymes in citrus juices, and especially peroxidases in fruits and vegetables. In general, the shorter the time of heat treatment the greater are the chances for enzyme regeneration. It is believed this regeneration of enzymes is due to only partial denaturation of the enzyme proteins, which have been somewhat unfolded or uncoiled by the heat treatment, but find themselves after a time partially or completely recombined by the hydrogen or sulfhydryl bonds, and their activity thus restored.

Dairy Products

All dairy products are, of course, based upon milk. Milk itself is one of our most important foods, whereas butter, cheese and ice cream are derived directly from milk.

Milk contains a large number of natural enzymes. Among these are α-amylase, lipase, peroxidase, phosphatase, p-diamine oxidase and xanthine oxidase.[198] There seems no conceivable purpose for the presence of some of the enzymes (α-amylase for example) as far as the digestibility of the milk by the suckling animal is concerned. Some of the enzymes, lipase and phosphatase, are destroyed by pasteurization, others are not even completely destroyed by sterilizing at 115°C for 15 minutes. The peroxidase and xanthine oxidase are inactivated by this sterilization, but 5% of the original α-amylase and 12% of the p-diamine oxidase remain after the sterilization.

The presence of phosphatase in milk is used in quality control to determine whether the milk has been adequately pasteurized. In making the test, a portion of the milk is added to a substrate containing phenyl phosphoric compound and incubated for a short time. An indicator is then added. If the phosphatase has not been inactivated it causes liberation of phenol which forms a colored dye with the indicator.[85] The test is sensitive enough to indicate addition of 0.1% of raw milk to pasteurized milk, a drop of 2°C in the pasteurizing temperature or a slight shortage in holding period.

The natural enzymes in milk seem to have little importance in the quality of dairy products. Milk lipases may lead to undesirable rancidity if freshly drawn milk is cooled too rapidly, or if raw milk is homogenized, agitated, or if foaming or great temperature fluctuations occur. Since lipase is destroyed by pasteurizing, the important lipolytic actions occurring in cheesemaking come from enzymes elaborated by the starter cultures or added pregastric lipases, which will be discussed later.

Meat

In the American way of life meat is a most desired food. The flesh of many types of animals serves as excellent protein food source. Wild game as well as sea food are favored items of our diet. However, meats from cattle, hogs, sheep and poultry have the most importance from the standpoint of quantities consumed. Since flesh from which meat is derived is composed of animal cells, a large number of enzymes, particularly of metabolic types, are present. Some of these enzymes have considerable importance.

A period of time usually elapses between the dispatching of an animal and the consumption of a part of the carcass as food. Normally much of the length of this period is determined by the necessities of commercial practice in making meat available to the consumer. In addition, however, there may be deliberate periods of aging, which have as their purpose the improvement of quality particularly as to tenderness and flavor. Such aging allows favorable enzymatic action and is most commonly employed only for beef in the fashionable quest for the most tender steaks.

Lean meat is essentially voluntary muscle tissue. Muscles are made of a mass of muscle fibers bound together by connective tissue composed of collagen and elastin. The more connective tissue present, the tougher the meat. After an animal carcass is dressed it is chilled. There is solidification of the fat, and simultaneously there is the development of rigor mortis. In rigor mortis the muscles harden and shorten and make the meat quite tough. The phenomenon is accompanied by the formation of lactic acid and other acids from the tissue glycogen through the action of the glycolytic enzyme system present. After reaching a maximum in about 24 hours there is a slow diminution of rigor, and the muscles again become soft and flaccid. This softening occurs mainly inside the muscle fibers, and is catalyzed by native proteolytic enzymes present in the meat known as cathepsins. The cathepsins are present in much greater amount in glandular organs such as spleen, kidney, and liver, but have been demonstrated and their action studied in muscle tissues also. The rate of tendering is a function of the temperature. Usually meat is held just above the freezing point, in order to discourage microbial growth and spoilage, for from 1 to 4 weeks. For choice cuts of beef longer aging periods are used which results in greater disintegration of protein, including connective tissue. Beef aged to this degree has been *tendered*. Such aging is expensive not only because of the expenses involved in holding the meat for such extra-long periods under refrigeration, but also because of losses due to shrinkage, loss of bloom, and microbial growth. Bloom describes the appearance of the surface and is related to the condition of the pigments, connective tissue and fat. Surface microbial growth occurs necessitating trimming. It is believed by some that growth of Thamnidium mold on the surface of the meat contributes materially to the better flavor of properly age-tendered steaks.

The autolytic processes of normal aging can be greatly speeded up if temperatures higher than normal refrigeration at 2–4°C are used. In the "Tenderay" process the meat is hung at 15°C for 3 days, in rooms flooded with ultraviolet light to minimize microbial growth. A comparable increase in tenderness can also be obtained by aging for 24 hours at 40°C in which case antibiotics or irradiation must also be employed to protect the meat from spoilage.[197]

Fish and sea food products of course also contain many enzymes. After death the glycogen in fish muscle is converted into lactic acid and rigor mortis sets in. Autolytic enzyme action may occur in fish and is undesirable, but most fish spoilage is due to microbial action. Fish may be protected from spoilage by freezing, canning, smoking, salting or drying. Quick freezing with a protective ice glaze, and cold storage have made quality sea food products available throughout the country and throughout the year, even though there are alterations during storage, particularly due to oxidation, which affect color and flavor. Enzymatic or microbial factors are of little importance.

Pork and poultry meat products undergo the same enzymatic changes. However, these meats are sufficiently tender after the long, high-temperature cooking times employed that, like fish, tenderness is usually no problem. Microbial spoilage of poultry is now almost always controlled by quick freezing and low-temperature storage.

Cereal Products

Important agricultural crops in which the seeds are utilized for foods belong to the grass or to the legume families. The members of the grass family which are grown for their edible seeds are called cereals or cereal grains, the most important of which are wheat, corn (maize), rice, rye, barley, oats and sorghum grains. Cereal products constitute a major part of the food consumed in the world.

Unlike meat, just discussed, which is dead tissue, the cereal grains as harvested and stored until processed are alive and continue to undergo enzymatic metabolism, such as respiration, although of course at slow rates. Hence, the cereal grains contain the host of enzymes necessary for living organisms. These include the metabolic enzymes necessary for respiration, glycolysis, protein metabolism, etc., as well as the hydrolytic enzymes, of which the amylases, proteases and lipases are of greatest interest in cereal technology.

Ungerminated cereals have considerable β-amylase activity, but relatively little α-amylase. The proteolytic enzymes have been extensively studied, and are of the plant enzyme or papain type as indicated by the inhibitory action of oxidizing agents, and the powerful activating effect of sulfhydryl compounds. At the beginning of germination marked cytological changes occur in the epithelial cells of the scutellum. The scutellar epithelium seems to synthesize enzymes such as cytase, amylases and proteinases.

Most cereal products, such as prepared breakfast foods, are heat processed, and the enzyme content of the cereal ingredients are of little importance. The very important exception is in flour milling. The enzyme content of flours for bread baking is of great importance, and must be properly controlled. The important enzymes are β-amylase, α-amylase and proteases. β-Amylase is present in the unmilled wheat, and hence is present in excess in flour. This excess does no harm, since the degree of saccharification it can bring about is also dependent upon the presence of α-amylase. The amount of α-amylase is controlled by adding malted wheat or malted barley flour to the regular flour to increase its diastatic power to a certain level, as measured by gas-production tests or by so-called maltose number, the number of milligrams of maltose produced in one hour at 30°C by 10 g of flour in suspension.

The control of α-amylase is even more important in baking of breads made largely from rye flours. Where there is insufficient α-amylase the bread is characterized by a dry brittle crumb and the crust becomes cracked and torn upon cooling, whereas an excess results in bread with a wet soggy crumb which frequently falls away from the crust leaving large hollow spaces in the bread. This is due to the thermostability of cereal α-amylase so that considerable dextrin formation may take place during baking before heat inactivation of the enzyme takes place. As baking proceeds the starch swells, undergoes partial gelatinization with a consequent increase in its hydration capacity, whereas the coagulation of the proteins results in a decrease in their water-holding capacity. Control of the quantity of α-amylase in rye flours is therefore very important in Europe, and is generally measured on the Brabender amylograph, which provides a continuous automatic record of the changes in viscosity of a flour-water suspension as the temperature is increased at a uniform rate. Variations in amylase activity are considerably greater in northern Europe and Scandanavia where there is considerable sprout damage as a result of wet harvest weather.

The proteases in flour directly influence the quality or elasticity of the gluten proteins in the dough. If too much proteolytic action is present, too much breakdown of gluten will occur and a "sticky" dough will result. If too little protease activity occurs, the gluten proteins are tough and non-elastic and the baker calls the dough "bucky." When the miller adds malted flour to increase diastatic activity, he also introduces proteases. It is therefore difficult to maintain a protease quantity which is just in balance to give elasticity, and still not be sticky. The amount of protease action is controlled indirectly by means of oxidation, that is the addition of oxidizing agents. The wheat proteases being of the papain type are sensitive to oxidation, and the protease activity can be controlled so that normal elastic doughs can be obtained.

Vegetables and Fruits

The foods we know as vegetables are derived from different parts of plants. Some are botanically bulbs, others are fruits, roots, shoots, tubers, leaves, stems, and flowers. Fresh vegetables and fruits are living and respiring tissues, and have many different enzymes present. Because of this active enzyme content, special precautions are necessary in harvesting, handling and storing vegetables to be marketed in the fresh condition, or in processing for marketing in frozen or dried or canned conditions.

Fresh fruit and vegetables continue to respire after harvest. The respiratory enzymes continue their actions at sometimes greater, and sometimes lesser rates than when attached to the plant before harvest. In general, storage of such fruits and vegetables must be under conditions which minimize respiration and other enzymatic changes. Proper control of temperature, humidity, ventilation, etc., is of paramount importance for successful storage of such commodities as potatoes and apples, for example. Conditions chosen must permit slow respiration, otherwise the tissues die, and spoilage becomes very rapid.

Besides respiratory enzymes, most fruits and vegetables contain a great many other enzymes. These may commonly include carbohydrases (amylases, invertase), proteases, lipases, pectinases, lipoxidases, tyrosinase, peroxidases, catalase, polyphenol oxidases, chlorophyllase, ascorbic acid oxidase, all of which may be responsible for deterioration in the products, as well as a host of other enzymes which may be of little or no significance. The carbohydrases may cause undesirable breakdown of starch or of sucrose. The proteases can produce autolytic action on proteins. The lipases and lipoxidases can cause undesirable rancidity or flavor changes in the lipid components. Pectinases can cause undesirable softening or viscosity changes in fruit products. Tyrosinase, peroxidases, polyphenol oxidases and chlorophyllase can be responsible for undesirable color changes. Ascorbic acid oxidase may destroy valuable vitamin C.

Food technologists have long been aware of the structural considerations in the preservation of fruits and vegetables. When cells remain intact enzymes and substrates are often mechanically separated from each other. When the structure is damaged, as in bruising of fruits, or upon freezing, the various enzymatic reactions proceed at an accelerated pace.

Fresh fruits and vegetables play important roles in our diets. Much attention has been given to the proper time for harvest and to proper storage conditions to ensure as long storage life as possible. Many items are best harvested while still immature or green. The storage conditions chosen permit artificial ripening, which may involve not only many complex physiological respiratory and metabolic enzymatic reactions, but also the disappearance of green chlorophyll, by action of chlorophyllase, to allow the ripe color of the fruit or vegetable to show.

Because of the perishable nature of fresh fruits and vegetables, even under the best devised systems of storage, other methods are commonly used for preserving these food products, including canning, dehydration and freezing. These preservation methods prevent spoilage by quite different means. Canning involves cooking and sterilization which effectively prevents further enzymatic or microbial changes. The texture and flavor of the cooked products are of course usually quite different from those of the fresh products, but still quite desirable in most cases. Dehydration reduces the moisture content to the point where microbial spoilage cannot occur. Sometimes, if permitted to remain, native enzymes can continue their actions in dehydrated food products producing undesirable changes, and certain enzymes are particularly bad if present while the foods are being dried. Frozen fruits and vegetables frequently retain much of the flavor of their fresh counterparts, and are protected from spoilage by the low storage temperatures. Enzyme action can be very important if allowed to be present during the freezing operation, as well as after freezing during storage. Action of the enzymes at freezer temperatures will be slow but will proceed, causing flavor and aroma changes.

In all of these preservation methods blanching has become the established method for destroying the native enzyme systems which might adversely affect the quality of the product, in canning during the relatively slow rise to cooking temperature, in drying during the extended period of reducing the moisture content, and in freezing during the freezing and storage periods. Blanching is carried out in various ways, by heating in boiling water or with steam. The length of time of heating and temperature required depends upon the subsequent process—canning, dehydration or freezing—the medium used, temperature of the medium, temperature of the product, rate of circulation of the medium, and the size of vegetable or pieces of vegetable. Shelled peas, being small, need only about 60 seconds in boiling water or live steam, whereas ears of sweet corn require 8 to 10 minutes in the same medium before freezing.

In general the processes employed in freezing may be summarized as follows. Vegetables or fruit of proper variety and maturity are selected and harvested. The freshly harvested product is carefully washed, cleaned, inspected, and prepared as for cooking. It is then blanched for a period long enough to inactivate all catalase content and immediately cooled to 60°F or lower in cold water. The product is given a final inspection, then packed in moisture-vaporproof cartons and quick-frozen. After freezing the packages are placed in corrugated fiberboard shipping containers and stored at 0°F or lower.

For canning fruits and vegetables preparation is similar to that for freezing through the blanching step, the hot product then being immediately sealed into cans and processed under steam pressure.

For dehydration, the fruit or vegetable must be harvested at the ideal stage for eating, and then handled and processed as carefully and rapidly as possible. The fresh harvested product is thoroughly washed, cleaned, inspected, and may be peeled. Fruits to be dried whole are dipped in weak alkaline solutions. Fruits and many vegetables are sulfured, fruits usually by placing in compartments containing sulfur dioxide from burning sulfur, vegetables usually by immersion in sulfite solutions or by spraying. Cabbage and potatoes are routinely sulfited prior to their dehydration; other vegetables may also be so treated after blanching. Most vegetables and cut fruits are blanched, adequacy being judged by measurements for inactivation of catalase or peroxidase. They are then sulfured and dried in counterflow dehydrators. Whole fruit is frequently sun dried.

It may be noted that adequacy of blanching has usually been determined by ascertaining the absence of peroxidase or catalase, since these enzymes are universally present in fruits and vegetables. In many vegetables peroxidase has been found to undergo self-reactivation. It may be shown to be absent after a heating cycle, but 24 hours later will have in part or wholly recovered its activity. Testing for presence of catalase is easily done with hydrogen peroxide. In the case of vegetables, if blanched sufficiently to inactivate the catalase the frozen pack can be expected to keep reasonably well, but a more severe treatment than that necessary to inactivate catalase is required to insure the best storage properties.

An interesting example of controlling natural enzyme content is in the preparation of tomato juice. Desirable juice of high viscosity with little tendency to separate the suspended solids is dependent upon the pectin content of the tomatoes. If cold pressed, the juice has low viscosity and suspended solids separate because of the action of the natural pectin methylesterase. Hence, the hot break method is employed, in which the fruit is heated to a temperature of about 165°F. This inactivates the pectinase enzyme, and the juice is then pressed.

A great deal of attention has been given to controlling enzymatic browning in cut fruits and vegetables. Darkening of plant tissues when exposed to air is due to oxidation of o-dihydroxyphenol derivatives, such as catechol, protocatechuic acid, caffeic acid and hydroxygallic acid which are abundant in nature, through the action of enzymes originally called oxygenases. The preferred name is now polyphenol oxidase, although the terms phenolase and polyphenolase also are found in use. This entire class of enzymes has copper as the prosthetic group, and comprises a number of different enzymes distinguished by their specific substrates, such as tyrosinase, catecholase, laccase, etc.

The fruits apples, peaches, apricots, bananas, cherries, grapes, pears, strawberries, figs, and vegetables potatoes and red beets contain polyphenol oxidases and substrate compounds which cause enzymatic browning when the cut tissues are exposed to air. For the browning to occur three components must be brought together, enzyme, substrate and oxygen. If any of the three is missing, or prevented from reacting by some means, oxidation and browning will not take place. Natural control exists in many fruits and vegetables, as in cantaloupe and tomato which have neither enzyme or substrate, in the unique Sunbeam peach which has enzyme but no substrate, and most of the berries and citrus fruits which do not have the enzyme.

The polyphenol oxidases are quite sensitive to heat, and the blanching process is a means of control of enzymatic browning. Besides heat these enzymes can be inactivated by heavy metals, halogens, sulfites, cyanides. The heavy metals and cyanides cannot be used in foods because of their toxicity, but halogen salts and sulfites are very frequently used in the food industry. Sulfuring with sulfur dioxide or sulfites is the commonest method employed with most fruits and many vegetables before dehydration, to inhibit the browning reaction. Sodium chloride brines are used for a similar purpose for dipping sliced potatoes before dehydrating. The influence of pH in reducing the rate of browning is widely applied in the food industry. Apricots or peaches, after lye peeling, are immersed immediately into citric acid solution to reduce their pH far below the sharp optimum for polyphenol oxidase activity. A solution of 0.5% citric acid and 0.03% ascorbic acid may be used to prevent browning in cut fruits intended for freezing. In this case the ascorbic acid acts as an antioxidant.

Production of Industrial Enzymes

Industrial enzymes are produced from animal tissues, plant tissues and microorganisms. For example, commercial animal enzymes include pancreatin, trypsin, chymotrypsin and lipase obtained from pancreas, pepsin and rennet from stomach mucosa, and catalase from liver. Commercial plant enzyme products include ficin from fig latex, bromelain from pineapple, papain from papaya, and malt amylase from barley malt. A wide range of enzymes of all classes is obtained from a variety of species and strains of molds, yeasts and bacteria. Microbial production processes afford a degree of control of efficacy, quality and quantity which is difficult to achieve with plant or animal sources. For this reason, together with the greater number of enzymes readily available from microorganisms, microbial enzymes are assuming increasingly predominate roles as industrial enzyme products.

Enzyme preparations are obtained from animal and plant sources by collecting or pressing out the juices, or extracting the tissues with water. Microbial enzyme preparations are produced by cultivating selected organisms. Species and strains of special molds, bacteria and yeasts are selected, developed and cultivated to produce maximum yields of the desired enzymes. Recent publications on production methods are available.[19,179,180]

The crude enzyme solutions obtained from animal or plant tissues or from microbial fermentations are clarified by filtration or centrifugation. Frequently these solutions are concentrated by vacuum evaporation at relatively low temperature. To obtain solid products, salting out or spray drying procedures may be used, but most commonly precipitation by acetone or aliphatic alcohols is employed. The precipitated enzyme concentrates are recovered by filtration or centrifugation, and dried in atmospheric or vacuum driers. The resulting liquid and solid enzyme concentrates are the basic materials for formulating commercial enzyme products. They represent concentrated but not highly purified products. For many commercial applications further purification is not necessary. For some uses it may be undesirable to have contaminating enzymes or other substances present and procedures such as dialysis and chromatographic adsorption methods are employed. Frequently enzymes used for specific analytical purposes, for some pharmaceutical uses, and for research purposes must be highly purified. High-purity enzymes are almost invariably expensive and are used only where high cost is not significant relative to the need they satisfy.

The liquid or solid enzyme concentrates are assayed for potency, and may be sold as produced by the manufacturer on the basis of their potencies. More commonly they are diluted to standard activities. Liquid products may require the addition of stabilizers such as benzoate, glycerol, propylene glycol, sorbitol or sodium chloride to prevent microbial growth or loss of enzyme activity during storage. Solid products are adjusted to standard potencies by addition of such diluents as starch, sucrose, lactose, flour, salts and gelatin. Frequently buffers and other salts are also used in the formulation of either liquid or solid enzyme products to ensure favorable pH conditions, enzyme activity and stability. Frequently manufacturers supply the same basic enzyme standardized to different potency levels or with different diluents depending upon the intended use.

Enzyme Applications

Advantages in Using Enzymes

Enzymes have several distinct advantages for use in industrial processes: (1) They are of natural origin and are non-toxic. (2) They have great specificity of action; hence, they can bring about reactions not otherwise easily carried out, especially without unwanted side reactions. (3) They work best under mild conditions of moderate temperature and near neutral pH, thus not requiring drastic conditions of high temperature, high pressure, high acidity, and the like, which necessitate special expensive equipment, and may cause undesirable side reactions. (4) They act rapidly at relatively low concentrations, and the rate of reaction can be readily controlled by adjusting temperature, pH and amount of enzyme employed. (5) They are easily inactivated when reaction has gone as far as is desired.

Major Classes of Industrial Enzymes and Their Applications

Carbohydrases The carbohydrases are enzymes which hydrolyze poly-saccharides or oligosaccharides. Among the carbohydrases are found the most investigated as well as the most widely used enzymes both for laboratory and industrial applications.

Starch-splitting enzymes Of all the commercial enzymes, amylases, the enzymes which act on starch, have the most numerous applications. Various amylases from plant, animal, fungal and bacterial sources have been in use for many years. There are three types of amylases which hydrolyze starch in different manners:

$$\text{Starch} \xrightarrow{\alpha\text{-amylase}} \text{Dextrins} + \text{Maltose}$$
$$\text{Starch} \xrightarrow{\beta\text{-amylase}} \text{Maltose} + \text{Dextrins}$$
$$\text{Starch} \xrightarrow{\text{glucoamylase}} \text{Glucose}$$

The terms "liquefying" and "saccharifying" amylases are general denoting the two principal types of enzyme action. The α-amylases hydrolyze α-$(1 \rightarrow 4)$ linkages in large starch molecules in random manner, thereby liquefying starch rapidly but also producing extensive saccharification on prolonged action. α-Amylases from different sources—animals, higher plants, fungi and bacteria—vary widely in saccharifying ability, in thermal stability, and in extent of hydrolysis brought about. Since α-amylases cannot hydrolyze α-$(1 \rightarrow 6)$ branching linkages in starch, the ultimate products from high saccharifying α-amylase are maltose, small quantities of other malto-oligosaccharides, a small amount of the trisaccharide panose which contains the original α-$(1 \rightarrow 6)$ linkages of the branched starch fraction, and a little glucose. Considerable amounts of low molecular weight dextrins are generally formed by the low saccharifying or dextrinizing types of α-amylases.

β-Amylase apparently is produced only by higher plants, particularly cereals and sweet potatoes. It is a saccharifying enzyme, producing maltose as the only sugar by splitting maltose units progressively from the non-reducing ends of starch chains. When acting on the branched amylopectin fraction of starch, the action of β-amylase ceases when it reaches an α-$(1 \rightarrow 6)$ linkage, leaving so-called "beta limit dextrins."

Glucoamylase, frequently also called amyloglucosidase, seems to be formed mainly by fungi. It is a saccharifying enzyme producing only glucose by progressive hydrolysis of glucose units from the non-reducing ends of starch chains. Studies by Pazur and coworkers with highly purified glucoamylase have shown that it acts preferentially on longer chains and that it also hydrolyzes α-$(1 \rightarrow 6)$ and α-$(1 \rightarrow 3)$ linkages, although more slowly than the α-$(1 \rightarrow 4)$ linkages.[126,127,128] Hence, this enzyme is capable of converting starch completely to glucose.

Some important commercial uses for amylases have been indicated in Table 1. Amylases have a variety of uses in food processing and also other applications as shown. For example, the first industrial manufacture of the fungal enzyme, Takadiastase, was for a pharmaceutical digestive aid, and this continues to be a major application.

An extremely important use for amylases is in the production of sweet syrups. The hydrolysis of starch to sugars by acids and enzymes has been known and practiced since the early nineteenth century. The first plant for production of syrups by acid conversion was built in France in 1814. When starch slurries are acidified with hydrochloric acid to about pH 1.8 and heated under pressure, random hydrolysis occurs yielding glucose and glucose oligosaccharides of various degrees of polymerization. For any given extent of acid hydrolysis of starch, as measured by copper reducing value and expressed as dextrose equivalent (DE), there is only one composition that will occur.

TABLE 1
Some Commercial Uses of Amylases

	Bacterial	Fungal	Plant
Syrup manufacture	x	x	x
Dextrose manufacture	x	x	
Baking	x	x	x
Saccharification of fermentation mashes			
Distillery	x	x	x
Brewery	x	x	x
Food dextrin and sugar products	x	x	x
Dry breakfast foods	x	x	x
Chocolate and licorice syrups	x	x	
Starch removal from fruit extracts and juices, and from pectin		x	
Scrap candy recovery	x	x	
Starch modification in vegetables	x	x	
Textile desizing	x		
Starch coatings for paper, fabrics	x		
Cold water dispersible laundry starch	x		
Wallpaper removal	x		
Pharmaceutical digestive aid		x	

For example, regular acid conversion syrup of 42 DE has a composition of about 22% glucose, 20% maltose, 20% tri- and tetra-saccharides, and 38% dextrins.[102] Furthermore, acid catalysis at the necessary high temperatures produces recombination or reversion products, and decomposition products. Acid conversion syrups above about 50 DE are found to have objectionable bitter taste, amber color and a tendency to crystallize. Since enzyme hydrolysis is characterized by specificity, it was found that use of commercially available fungal amylases on syrups acid hydrolyzed to within the range of 40 to 50 DE made possible 62 to 65 DE syrups of superior flavor and sweetness. Such high conversion acid-enzyme syrups have been marketed since the early 1940's.[41] A typical high conversion syrup of this kind has the composition: 63 DE, 38% glucose, 34% maltose, 16% tri- and tetra-saccharides and 12% dextrins.[102]

Within recent years manufacturers have learned to take advantage of different commercially available enzyme preparations containing various proportions and combinations of α-amylase, β-amylase and glucoamylase either simultaneously or successively with either acid-liquefied or bacterial amylase-liquefied starch to produce syrups of widely differing compositions, from very low DE-high dextrin syrups, to very high DE-high sugar syrups to meet the demands of different food industries.[183] For example, processes have come into commercial use for producing syrups of about 70 DE and over 80% yeast fermentable with a glucose content below about 44%, the limit for non-crystallizing high-solids syrup. One method employs acid hydrolysis to about 20 DE, converting with malt to about 52 DE and completing the conversion with fungal amylase.[55] Another method uses new fungal enzyme systems which will produce, starting with 15 to 20 DE-acid thinned starch, syrups of 67 to 69 DE of 80 to 82% fermentability containing approximately equal amounts (about 40%) of glucose and maltose.[183]

So-called high-maltose syrups, low in glucose, are also produced commercially for the hard candy industry. By use of malt extract or commercial concentrated malt enzyme preparations either low DE-acid thinned or enzyme-thinned starches are converted to syrups of about 40–42 DE without materially increasing the level of glucose in the final syrup. For example, incubation of 20 DE acid-thinned starch with 0.009% of a 1500° Lintner malt enzyme preparation for 48 hours at pH 5.0 and 55°C gives a DE of about 41 and a carbohydrate composition of about 6% glucose, 47% maltose, 9% triose and 38% higher saccharides.[183] A commercial fungal amylase designed for high-maltose syrup production also is in use. It may be satisfactorily employed with either enzyme liquefied or acid-liquefied starch substrate, the former being preferred because it results in lower glucose and higher maltose contents. With 15–20 DE acid-liquefied corn syrup containing 40% solids at pH 5.3, incubation for 24–28 hours at 54°C using 0.02% of the enzyme gives a DE of about 43 and a composition of 7.5% glucose and 47% maltose. Starting with a 30% starch slurry liquefied with 0.1% bacterial amylase at 88°C with a jet cooker, incubation with 0.06% of the fungal enzyme at pH 5.3 for 40 hours at 53°C gives a syrup of about 45 DE with 3% glucose and 58% maltose.

The most recent extensive application of amylolytic enzymes is the use of fungal glucoamylase for the production of dextrose from starch.[45,63,165,183] All major dextrose manufacturers are now employing glucoamylase instead of the classical acid conversion process. This has permitted more than doubling the concentration of starch in the conversion slurry, has simplified handling, including evaporation, decolorization and crystallization, and has very materially increased yields of recovered dextrose.

Production of dextrose from starch by conversion with glucoamylase was proposed before 1950. However, commercialization was not possible until some ten years later when the glucoamylase became available at economical cost and essentially free from transglucosylases. This was achieved by the discovery of very high glucoamylase-yielding strains of fungi which produce minimal amounts of transglucosylases.[13,26] The presence of transglucosylases is undesirable because of their production of glucose polymers having α-$(1 \to 6)$ linkages which reduce dextrose yields and interfere with dextrose crystallization. Several methods have also been patented for the removal of transglucosylases from glucoamylase preparations.[38,72,73,74,84,96]

Commercial operations employ starch concentrations of 30 to 40%. Glucoamylase gives essentially quantitative conversion to dextrose at low starch concentrations with progressively less complete conversion as starch concentrations are increased.[183] This is due to "back polymerization" catalyzed by the glucoamylase, forming reversion sugars, mainly isomaltose. This polymerizing reaction restricts the maximum concentration of starch which can be efficiently converted. Considering all factors such as dextrose yield and plant throughput, about 30% starch is usually considered the most practical level.

Prior to enzyme conversion the starch must be gelatinized to be susceptible to enzyme attack, and liquefied in order to be handled. Thinning may be accomplished either by heating with dilute acid or by use of thermostable bacterial amylase. Following thinning, the solution is then adjusted to the optimum of about pH 4.0 for the glucoamylase, the enzyme added and the saccharification allowed to proceed to completion in tanks maintained at about 60°C. The conversion times employed vary between 48 and 96 hours, depending upon the enzyme level used. Usually the enzyme concentration employed requires about 72 hours for completion of the conversions. After the conversion, the material is filtered, carbon and ion exchange treated, evaporated and the dextrose crystallized.

Experience has shown that by the acid-enzyme process maximum DE of about 95 and about 92% dextrose based on total solids can be obtained, and the practical yields are about 100 lbs anhydrous dextrose per 100 lbs of starch. The double enzyme process gives about 97 DE and about 95% dextrose, with practical yields of 105 lbs of dextrose per 100 lbs of starch. A practical disadvantage which has hindered general adoption of the double enzyme process is the slow filtration rates for the converted starch solutions. This is due to undigested residual solids. One method which obviates this difficulty is to employ a steam jet heater for the bacterial liquefaction step, with prior addition of optimum amounts of calcium and sodium salts for maximum thermostability of the enzyme.[183] The very rapid heating of the starch slurry in the jet heater causes almost instantaneous gelatinization of the starch and prevents retrogradation to material not susceptible to glucoamylase action. The amount of residual insolubles is greatly decreased and filtration rates are markedly improved by this method.

A recently patented process permits the use of crude starch sources such as corn flour for dextrose production instead of the much more costly separated starch usually employed in the industry.[25] With crude starch sources enzyme thinning must be used since thinning with acids would produce soluble contaminating nitrogenous and other compounds which would make the purification and crystallization of the dextrose liquors difficult or impossible. In this process the starch in the substrate is gelatinized and thinned with bacterial amylase using a steam jet heater, and is then converted with glucoamylase which must be essentially free from protease, lipase and transglucosylase. Following the conversion, the insoluble solids are separated by filtration and marketed after drying as a premium quality high-protein gluten livestock feed. The solution is purified by carbon and ion exchange treatment, and the dextrose can then be recovered in pure form by evaporation and crystallization. One plant operates this process on a very large scale for the production of dextrose in solution which is then used for producing citric acid by fermentation.

Another major food industry using amylases is baking.[5,40,81,177] Two types of amylases are recognized as important in the baking industry, α-amylase and β-amylase. Flour milled from sound wheat contains a relatively high content of β-amylase and very low level of α-amylase.[94] During malting of wheat or barley the β-amylase increases only a little, but the α-amylase increases several thousandfold. Microbial amylases are mainly α-amylases. The differences in thermostability of cereal, fungal and bacterial α-amylases are very important considerations in determining their usefulness in baking processes. The principal function of supplementation with either malt or fungal amylase is to increase the α-amylase content of the flour.

Amylase supplementation affects the fermentation and the bread quality. The panary fermentation by yeast requires sugar both for growth and for the formation of carbon dioxide. The amount of sugar normally present in flour is quite small. Hence, the production of bread depends upon added sugar in the dough, and upon maltose formation from the starch of flour by the amylases present. The enzymes act during fermentation only on the so-called damaged starch granules which constitute a rather small and variable percentage of the total starch content of flours.

During fermentation the β-amylase acts upon the susceptible damaged starch to produce maltose. It is able to act only up to the points of branching. When flour is supplemented with α-amylase, during fermentation this enzyme causes dextrinization of the damaged starch, producing more chain ends for the β-amylase action. The combined action of the excess of β-amylase in flour and the α-amylase contributed by enzyme supplementation results in a rapid and complete saccharification, the major part of the conversion being achieved during the early stages of fermentation. The importance of this increased sugar as a yeast substrate for increased gas production depends upon the level of sugar added to the dough. Where added sugar is limiting, the maltose from starch hydrolysis is particularly important during the proofing period and the first few minutes in the oven when the supply of other sugars may be insufficient for adequate gas production.

The addition of α-amylase not only increases the rate of fermentation but produces enough sugar to increase the sugar content in the baked bread. This is important because of its effect on the flavor of bread, on crust color, and on the toasting characteristics of bread slices.

Another role of amylase during fermentation is its effect on dough consistency. The damaged starch granules have a high water-absorbing and holding capacity, and when this hydrated starch is broken down by amylase the water released causes softening or decrease in consistency of the dough.[82]

The formation of fermentable sugars by the amylases is not their only function in bread baking. α-Amylase has a considerable effect on the viscosity of doughs, on the formation of dextrins, on the grain and texture of the bread, and on the compressibility of bread crumb. Although the action of amylases during fermentation is limited by the content of damaged starch, as the temperature of the dough increases during the baking period, above about 60°C the starch is gelatinized and becomes susceptible to enzyme action. For a few minutes in the oven, until the heat inactivates the amylase, dextrinization and saccharification of the starch can be rapid. This may result in increased gas production, improved crust color, improved moisture and keeping quality of the crumb due to dextrins, and additional sugar which contributes to flavor and caramelized sugar in the crust. The extent of amylase action and starch breakdown depends primarily upon the thermostability of the α-amylase used.[5,115] The most notable and important differences between α-amylases from different sources are their thermostabilities, as shown in Figure 4. Fungal amylase is quite labile, being destroyed rapidly at temperatures above 60°C. Temperatures of 70° to 75°C are required for rapid inactivation of malt α-amylase. Bacterial amylase is the most stable of all and shows little inactivation at temperature up to 85° to 90°C in the presence of starch. The dextrins resulting from the action of cereal α-amylase in the oven have a very important effect on the texture of bread crumb. Small amounts of the dextrins are desirable from the standpoints of bread quality and retarding staling, but larger amounts cause gummy, sticky crumb and loaf fragility.[18,40] Since α-amylase of wheat and barley malt is quite stable over the temperature range for starch gelatinization, it is important that any considerable excess of cereal α-amylase not be present to cause excess dextrinization. On the other hand, baking temperatures inactivate fungal amylase before any appreciable quantities of starch

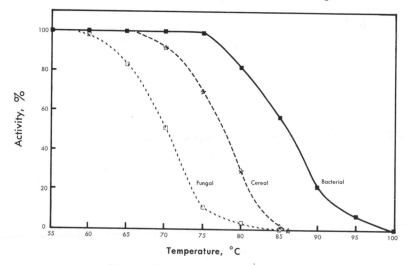

Fig. 4.   Thermal stability of α-amylases.

are gelatinized. Supplementation with fungal amylase therefore offers a considerable margin of safety against over-dextrinization and stickiness in the crumb. More than twenty times the level of fungal amylase necessary for the desirable baking response does not adversely affect the baking results in any way.

Early work led to the conclusion that bacterial amylase had no place in bread making because of its high thermostability, which led to gummy crumb. However, quite recent work has shown that very low levels of bacterial amylase retard staling without harmful effect on the quality of the bread.[148,160] However, lack of tolerance to a slight overdosage of bacterial amylase represents a hazard to its practical use in the baking industry.

The effect of bacterial amylase in dextrinizing flour starch can be used to advantage where a moist, soft, and somewhat tacky crumb is desirable, as in fruit cakes.[164] There is also a place for bacterial amylase in the production of specialty products having soft consistencies similar to pie fillings.[93]

Enzyme supplementation is practiced both by addition to the flour at the mill and by the baker. In Europe both malt flour and fungal amylase may be added to the flour at the mill. In the United States under the Standards of Identity for Bread and Rolls, fungal enzyme preparations from *Aspergillus oryzae* may be used by the baker in bread making, but may not be added at the mill under present Standards of Identity for Flour. The α-amylase content of almost all commercial flour in the United States is increased by supplementation at the mill with wheat or barley malt flours.

Another use for amylases, probably the oldest of all enzyme usages, practiced since antiquity, is the use of malt for the alcoholic fermentation of grains. There are numerous excellent books which should be consulted for the details of malting and of the brewing and distilling industries.[34,136,137,184] In brief, malt is produced by controlled sprouting of grains, usually barley, followed by drying. During malting high levels of α-amylase are formed and some increase in β-amylase occurs. The principal difference in the three general classes of barley malts is in the α-amylase content. Brewers' malt assays 25–30 units, distillers' malt 50–60 units, and gibberellin malt 80–100 units. About 100 million bushels per year of malt are produced in the United States, of which about 6% is used by the distilling industry, 3% by the food industry, and 91% by the brewing industry. The malt is usually ground and used in this form for its enzyme content, but it also provides carbohydrates, proteins and flavoring constituents which are useful as yeast nutrients or as constituents of a product such as beer. The active enzymes are water soluble and can be extracted by water into solution containing about 25% of the malt. By evaporating such extracts, concentrates can be prepared. These are marketed as high-diastatic malt extracts, with considerable maltose content, for many purposes such as brewery and bakery adjuncts.

In brewing, malt alone, or usually along with starchy adjuncts from rice or corn, is mixed with water and heated to boiling to effect gelatinization and liquefaction of the starch. It is then cooled somewhat and additional malt is added to effect saccharification. After a suitable holding period, hops are added and the mash is boiled for a period sufficient to inactivate the enzymes, extract the hops and sterilize the mash. The hot mash is then filtered to remove solids, yeast is added to the wort and fermentation allowed to go to completion. Finishing operations include final filtration, addition of chillproofing protease preparations, carbonation and filling into containers. Bottled and canned beer are pasteurized; draft beer is not and is subject to microbial spoilage. A recent innovation is millipore filtration instead of pasteurization.

In distillery practice ground grain is mixed with water and the mixture heated under pressure to effect gelatinization of the starch and sterilization of the mash. The hot mash is cooled to about 60°C, ground malt equal to approximately 10% of the total grain weight added, the mash cooled and pumped into the fermentors where it is inoculated with pure yeast cultures. A relatively quick partial conversion of the starch to sugars occurs during the brief period at 60°C after the malt addition. Sufficient maltose is produced to promote development of the yeast and a strong fermentation. As the fermentation proceeds the maltose is used up by the yeast and dextrins are continually converted to sugars by the active amylase present, until the conversion and fermentation continues to completion. At the end of the fermentation the mash is distilled to produce industrial alcohol or beverage products such as whiskey or grain neutral spirits. The amount of industrial alcohol produced by alcoholic fermentation of grains is now quite insignificant in the United States; the grain alcohol distilleries in this country exist almost solely for the production of beverage alcohol products. However, in numerous other parts of the world production of industrial alcohol by fermentation continues to be important.

Because of their high thermostability bacterial amylases have found some use in the liquefaction of brewery and distillery mashes. Crude fungal amylase preparations, containing α-amylase and glucoamylase, are employed by a few distilleries instead of malt for the saccharification of their grain fermentation mashes.[64] A process has been patented using fungal enzymes for preparing brewers' worts from barley substrate.[46]

Other uses of amylases in food industries are quite minor with respect to amounts of enzymes involved, but are quite important and frequently almost indispensible to the manufacturers they serve.[180,182] Amylases are utilized in processing cereal products for food dextrin and sugar mixtures, in processing baby foods and some breakfast foods. They are indispensable for the preparation of free-flowing chocolate and licorice syrups in order to liquefy the starch present and so keep the syrups from congealing. They are used for recovering sugars for reuse from scrap candy of high starch content. Fungal amylases are also used for starch removal from fruit juices and extracts, from flavoring extracts, and in preparing clear, starch-free pectin. For the latter use, of course, special amylase preparations completely free from pectinase are requisite. Microbial amylases also have limited use in modifying starch in vegetable purees, and in treating vegetables for canning.

In addition to their important applications in food processing, large amounts of bacterial amylases have other important industrial uses, among which are textile desizing, preparing modified starch coatings for paper and fabrics, and making cold water dispersible laundry starches. Detailed discussion of these applications is beyond the scope of this chapter.

Disaccharide splitting enzymes   Several carbohydrases, invertase (sucrase), lactase and maltase, which hydrolyze disaccharides have considerable importance. These enzymes hydrolyze their corresponding disaccharides with the formation of two molecules of monosaccharide as indicated.

$$Sucrose \xrightarrow{\text{invertase}} Glucose + Fructose$$
$$Lactose \xrightarrow{\text{lactase}} Glucose + Galactose$$
$$Maltose \xrightarrow{\text{maltase}} Glucose + Glucose$$

All of these enzymes may be obtained from various microorganisms, with selected strains of yeast most commonly being used for commercial invertase and lactase. Maltase is not marketed as such.

An interesting example of enzyme specificity is displayed by the invertases. Although yeast and fungal invertases both hydrolyze sucrose, the nature of their actions differs. Yeast invertase is a fructosidase, attacking the fructose end, whereas fungal invertase is a glucosidase, attacking the glucose end of the sucrose molecule. This may be demonstrated by comparing their activities against raffinose. Yeast invertase hydrolyzes raffinose into fructose and melibiose, but fungal invertase causes no reaction since glucose is not terminal in the raffinose molecule:

$$Raffinose \xrightarrow{\text{yeast invertase}} Fructose + Melibiose$$
(fructose-glucose-galactose)   (glucose-galactose)
$$Raffinose \xrightarrow{\text{fungal invertase}} no\ reaction$$

Invertase has several important applications.[30,75,119] It can be employed in manufacturing artificial honey, and particularly for invert sugar which is much more soluble than sucrose. Hence, a large use of crude invertase is to prevent crystallization in high-test molasses. The high solubility of invert sugar is important in the manufacture of confectionaries, liqueurs, and frozen desserts where high sucrose concentrations would lead to crystallization. Invertase is also used in the preparation of chocolate-coated-liquid-center candies. Molding and coating are carried out while the centers are firm, after which the invertase that has been added to the cast centers acts to yield a smooth, stable cream.[75,76,77,78,79]

The solubility of lactose is low as compared with other sugars and it has little sweetness. Hydrolysis of lactose by lactase, a β-D-galactosidase, forming glucose and galactose, increases sweetness, solubility and osmotic pressure of products containing lactose. Lactose is in abundant supply, particularly in by-product cheese whey. Some whey is used for production of crystalline lactose and the heat coaguable lactalbumin. Much whey is concentrated or spray-dried for use in animal feeds and human food products such as baked goods, candies, prepared foods and infant formulas. However, there is still a very large amount of cheese whey which is discarded and presents a disposal problem to the cheese industry.

In spite of the apparent potential for lactase in upgrading whey and various dairy products its actual commercial utilization has been quite limited. A recent review should be consulted by those interested in lactase potentials.[133]

One use for lactase is in preventing lactose crystallization in ice cream, such crystallization causing "grainy" or "sandy" ice cream. Small scale commercial use of lactase has been for hydrolysis of lactose in whey concentrate for animal feeds. The proteins in whey have high biological value, but the presence of lactose limits the amount of whey which can be used in feeds. Hydrolysis of lactose improves the digestibility and palatability, and also prevents crystallization and setting-up in liquid whey concentrates.

Frozen milk concentrates have been produced on a limited scale for reconstitution as a fluid milk beverage, mostly for use on ocean-going vessels and by the armed services. Crystallization of the lactose in the frozen concentrates is a problem from the standpoint of difficulty in reconstitution but more importantly because of its effect on the stability of the frozen concentrate. Presence of crystals causes flocculation and coagulation of casein so that after the frozen concentrates have been stored for several weeks the reconstituted milk is not acceptable. Lactose hydrolysis prevents the formation of lactose crystals and hence the protein instability.[163,174] Lactase would seem to have a real potential in the development of frozen whole milk concentrates of acceptable flavor when reconstituted for the domestic market. Such a product would have many convenience factors for the distributor, retail market, and the housewife from the standpoints of keeping qualities, storage space necessary, and frequency of purchase.[133]

Another potential for lactase is in bread making.[134,135] Skim milk solids are used in bread, contributing about 1.5% of lactose, based on weight of flour used. This lactose is not yeast fermentable and contributes hardly any sweetness to the bread. Hydrolysis by lactase to glucose and galactose would make the glucose available for fermentation and the galactose would still contribute to the formation of crust color.[135]

Pectic enzymes   The pectic enzymes are those which take part in the degradation of pectic substances. These pectic substances are found widely in plant tissues particularly in the fruits. The pectic substances include a multitude of compounds of complex and variable nature.[89] The pectins are polymers made up of chains of galacturonic acid units joined by α-(1 → 4) glycosidic linkages. In native pectins approximately two-thirds more or less depending upon the source, of the carboxylic acid groups are esterified with methanol. These highly esterified pectins yield semi-solid gels with sugar and acid of the kind familiar in jellies and jams. Partial hydrolysis of the methyl esters give low methoxyl pectins which form gels with small amounts of calcium ions. Complete hydrolysis of the methyl esters gives pectic acids. Since crude natural pectins on complete hydrolysis give small amounts of several different sugars and acetic acid along with galacturonic acid, it was once thought that other carbohydrates such as arabans and galactans were present as impurities. It now seems more probable that rhamnose, arabinose, galactose and traces of other sugars are integral parts of the pectin molecules, and some of the free hydroxyl groups of the galacturonic acid are acetylated.[108]

The pectic enzymes may be classified into pectin methylesterases which hydrolyze the methyl ester linkages, the polygalacturonases which split the glycosidic bonds between galacturonic acid molecules, and the pectin transeliminases, or lyases. The latter enzymes, only discovered within the past few years, bring about non-hydrolytic cleavage of α-(1 → 4) linkages forming unsaturated derivatives of galacturonic acid.[3] Some of the enzymes previously considered to be polygalacturonases may turn out to be lyases, since both show the same effect of reducing the viscosity of pectin and producing one molecule of reducing sugar for each α-(1 → 4) linkage broken. Further divisions exist with the pectic enzymes depending upon whether they act on the methylated or on the free polygalacturonic acid substrates, and whether they attack internal linkages. A scheme which shows the nomenclature of the pectic enzymes and their common abbreviations is given in Table 2.[120]

TABLE 2
Nomenclature of Pectic Enzymes

Acting on pectin
Pectin methylesterases (PME)
Polymethylgalacturonases (PMG)
   Endo-PMG   Pectin lyases (PL)
   Exo-PMG   Endo-PL
Acting on pectic acid   Exo-PL
Polygalacturonases (PG)
   Endo-PG   Pectic acid lyases (PAL)
   Exo-PG   Endo-PAL
                            Exo-PAL

Pectic enzymes occur rather widely in many plants, including fruits, and in many microorganisms. The commercial pectinases are derived from fungi, most commonly from strains of the *Aspergillus niger* group. These commercial pectinases, widely employed in processing fruit products, are mixtures of several of the pectic enzymes, particularly pectin methylesterases and polygalacturonases.[44] Differences in the effectiveness of different commercial preparations with different fruits and under different conditions are undoubtedly due to variations in the kinds and amounts of particular pectic enzymes present. Many attempts have been made to correlate the effectiveness of different enzymes in clarifying fruit juices with the contents of the various pectic enzyme fractions, but with little success.[12,83,118] In recent brilliant work by Endo in Japan with a commercial pectic enzyme derived from *Coniothyrium diplodiella*, three endopolygalacturonases, one exopolygalacturonase, and two pectin methylesterases were isolated in highly purified form.[50,51,52,53] None of the individual fractions clarified apple juice satisfactorily. Only combinations of at least one of the endo-PG fractions with at least one of the PME fractions effectively clarified apple juice. When the isolated fractions were all mixed in the proportions in which they were present in the crude mixture the apple juice-clarifying activity was identical with that of the crude preparation itself.[54] This work has demonstrated that other enzymes besides polygalacturonases and pectin methylesterase are not required for clarifying apple juice. Whether this will also be true for pectic enzyme mixtures from other microorganisms applied to other fruit juices and wines, remains to be seen.

Freshly pressed fruit juices contain pectin, the hydrophilic colloidal nature of which makes them viscous and holds dispersed solids in suspension. This is highly desirable in some cases, as in tomato, orange and apricot juices, and the natural pectic enzymes must be destroyed by heat in order to maintain the desired stable cloud and viscosity. In other fruit juices, such as apple and grape juices, a clear product is usually desired. To obtain such clear juices a commercial pectinase is used, and the major application for pectic enzymes is in the production of brilliantly clear apple juice, grape juice, other fruit juices, and wines.

The appropriate commercial pectinase added to crushed apples results in higher yields of the juice. The enzyme added to the juice reduces the viscosity, and degrades the soluble pectin-protective colloid, permitting finely divided insoluble particles to flocculate. Considerable variation exists in practical use of pectic enzymes in clarifying apple juice depending upon season and varieties of apples, and the factors of time, temperature and enzyme concentration employed. The floc formed may be allowed to settle or may be removed by centrifugation or filtration to yield crystal clear brilliant juice. When this juice is pasteurized it is free from the boiled taste characteristic of untreated pasteurized juices, and the pasteurized juice can be marketed the year around.[2,43] Vinegar and jelly produced from depectinized juice are superior in brilliance, color, and aroma.

For Concord grapes, pectic enzymes are indispensible in the hot press process now common in the industry in order to increase the yield of grape juice and improve color extraction. The destemmed and crushed grapes are mixed with the pectinase preparation and heated to 60–65°C. This permits most of the juice to be taken off as free run grape juice with the remaining juice recovered by a continuous screw press.[28] The enzyme action is sufficient to permit high juice yields but not extensive enough to clarify the juice completely or to lower the viscosity unduly.[130]

In making wines, pectic enzymes present a number of processing advantages at different stages of the operation. The addition of pectinases to the crushed grapes will increase the yield of free run juice, reduce pressing time and increase total juice yield. Addition of pectic enzyme increases the extraction of color when the grapes are extracted hot or fermented on the skins. When pectic enzymes are added before or during fermentation the yeast sediment will be more compact at first racking, giving a better yield and clearer wine. Enzyme addition to the fermented wine facilitates filtration and reduces the requirement of bentonite, giving superior clarity and color.[24,39] Pectic enzymes are also widely used in making wines from berries, apples, pears, peaches, and other fruits.

Prune juice is commonly produced by hot extraction of dried prunes and pectic enzymes are not usually employed. However, for production of juice from fresh Italian prunes addition of pectic enzymes to the pulp doubles the yield of juice.[191]

In the United States citrus, apple, grape and prune juices make up the major part of the juice production. However, juices from cranberries, blueberries, strawberries, black currants, cherries, plums and many other fruits are prepared for beverage use or for jelly manufacture, and pectic enzymes are almost universally employed.[57,187] The fruit juices for jelly manufacture are commonly depectinized completely in order to obtain clear juices and make uniform jelly possible by adding back a standard amount of apple or citrus pectin when the jelly is made. The variable quality and quantity of the natural pectins in the original juices do not then interfere.[43]

Pectic enzymes are indispensible for making high-density fruit juice concentrates or purees.[21,190] Enzyme depectinization is required for apple juice concentrate of about 72° Brix, which is made on a large scale, as otherwise a gel would result rather than the desired liquid concentrate. Pectic enzymes are useful in treating orange, grape and prune pulps prior to vacuum puff drying.[121]

A few miscellaneous uses for pectinases exist. One is the production of D-galacturonic acid, di- and trigalacturonic acids, and higher polymers from pectin.[14] Employing pectin methylesterase alone allows production of low methoxyl pectins. Fruit juices may be depectinized with PME and used for making low-sugar jellies for diabetics. Sorbitol as preservative and non-nutritive sweeteners may be added, then calcium ions which with the low-methoxyl pectin causes formation of a stable gel.[43] Pectic enzymes are also used for the recovery and stabilization of citrus oils from lemon and orange peels.[132]

Cellulases and hemicellulases The most abundant carbohydrate in nature is cellulose, the principal structural material of plants. Cellulose is a water-insoluble β-1,4-glucan, a glucose polymer having β-$(1 \rightarrow 4)$ linkages. Cotton is almost pure cellulose, whereas structural parts of plants—wood, straw, stalks, etc.—contain about 50% cellulose. In these plant structural materials the cellulose is bound together by lignin, and there are also present varying amounts of other carbohydrates, hexosans and pentosans, besides cellulose. These other carbohydrate polymers associated with cellulose in nature are of ill-defined composition and may be classified as hemicelluloses. These occur in almost all plant tissues—wood, straw, stalks, hulls, seeds, etc.—in relatively low to relatively high concentrations. Some, the gums, are water-soluble or dispersible, others are insoluble. Mannose, galactose, xylose and arabinose are the most common sugars found in the various hemicelluloses, but other sugars also often occur. Frequently more than one kind of sugar is present in common hemicelluloses. For example, the principal component of guar gum is a complex galactomannan.

Cellulosic materials are degraded in nature by certain microorganisms. These microorganisms obviously produce enzymes which actively degrade cellulose, lignin and hemicelluloses. Active cellulase enzymes have long been sought because of the availability of huge amounts of cellulosic plant wastes as potential sources for glucose for chemical and food industries, the interest by food technologists in upgrading fibrous food products, and the potential as a digestive aid for man and animals and in waste disposal systems. Although commercial cellulase products are on the market, even the best of these are seriously lacking in high activity against native cellulose. A recent qualitative comparison of the efficacy of attack by several fungi on cellulose fibers with the efficacy of cell-free cultures of the same fungi showed almost no correlation between the two activities.[125] One of the most puzzling problems in the field of applied enzymology is why cell-free cultures or isolated cellulases from microorganisms, living cultures of which rapidly digest cellulose, attack cellulose so slowly and so poorly.

A large amount of work has been conducted during the past few years with β-glucanases or cellulases derived from several sources, including *Aspergillus niger*, *Trichoderma viride*, *Trichoderma konongii*, *Myrothecium verrucaria*, various wood rot fungi, and various strains of rumen bacteria. Those especially interested will need to consult the original literature.[61,62,90,91,103,104,141,171] In brief, it seems evident from the extensive work on cellulase fractionation and characterization that the cellulases may be divided into two major types, the C_1-type enzymes which act on native cellulose giving linear, insoluble glucose chains and the C_x enzymes which act upon these or swollen or ground cotton, carboxymethyl cellulose or cellulodextrins to produce soluble, low molecular weight products. Included in the C_x enzymes are various fractions which differ significantly in pH optima and temperature stability. Some of these fractions are endoenzymes and act in a random manner, and some are exoenzymes.

The available commercial cellulases have generally been derived from *Aspergillus niger* and split cellulose to cellulodextrins and glucose. They have rapid action on soluble cellulose derivatives but act much more slowly on insoluble native cellulose. Cellulolytic enzyme preparations from other microorganisms, particularly *Trichoderma viride* and *Trichoderma koningii*, have been developed which have better activity against native cellulose.[61,90,104,171]

Commercial applications for cellulases have been slow in developing, largely because of the poor activity of the available products on the insoluble substrate. There is some use in enzymatic digestive aids. Cellulase has shown some promise in removing vegetable fibers from wool, and in enzymatic drain cleaner formulations. Although there has been some success in increasing the strength of paper by cellulase treatment to dissolve more easily digested constituents in wood pulp, cost of the enzyme and length of treatment have prevented commercialization. Increase of fibrillation in papermaking usually achieved by extensive beating can be obtained by use of cellulase.[27]

In the food industry cellulase treatment has been tried with citrus products prior to concentration, and in the extraction of flavoring materials. The cellulolytic tenderizing of fibrous vegetables and treatment of otherwise indigestible plant materials for foods or feeds may assume increasing importance. For example, fiber digestion may be accomplished by cellulase treatment of cocoanut cake or in common garden vegetables.[111,138,172] Dehydrated vegetable materials which have been pretreated with cellulase enzyme can be rehydrated more readily in water with return to their original shape.[171] It has also been suggested that cellulolytic enzymes be used for predigestion of animal feeds, and for the isolation of starch from potatoes.[172]

Several poorly characterized hemicellulases have been tested for various food applications. Several hemicellulase and "gumase" products are commercially available. The various individual hemicellulases hydrolyze specific types of hexosans and pentosans, including more or less complex mannans, galactans, xylans, arabans, etc. One type of hemicellulase has been found useful commercially in processing products containing galactomannans such as guar and locust bean gums. It has also been found effective in improving the digestibility of guar gum meal for animal feeding.[188] A hemicellulase has been used in preventing gelation in coffee concentrates.[142] Increased yields of high-quality starch from wheat are obtained by use of a pentosanase.[161]

The use of hemicellulases in baking is another area of application which appears to have a good chance of practical development. Wheat endosperm contains about 2.4% of hemicellulose, mainly pentosans containing xylose and arabinose. The insoluble pentosan fraction in wheat flour has a harmful effect on bread, reducing loaf volume and giving coarser texture.[97,98] A suitable pentosanase should be useful in baking technology.

Proteases Next to the carbohydrases, the proteases have the greatest variety of applications. Because of the large number of different amino acids present and the resulting complex nature and high molecular weights of proteins, and the variety of proteases of differing specificity, the action of proteases from a practical standpoint is much more complicated than that of the carbohydrases. A greatly oversimplified picture of proteolytic action may be illustrated as follows:

$$\underset{\substack{|\\R_1\\|\\ \underset{\substack{\text{HO}-\text{H}\\ \text{Exopeptidase}\\ \text{(Aminopeptidase)}}}{}}}{NH_2CHCO}-\underset{\substack{|\\R_2}}{NHCHCO}\cdots\underset{\substack{|\\R_3}}{NHCHCO}\underset{\substack{\text{HO}-\text{H}\\\text{Endopeptidase}}}{-}\underset{\substack{|\\R_4}}{NHCHCO}-\underset{\substack{|\\R_5}}{NHCHCO}-\underset{\substack{|\\R_6\\|\\\underset{\substack{\text{HO}-\text{H}\\\text{Exopeptidase}\\\text{(Carboxypeptidase)}}}{}}}{NHCHCOOH}$$

The proteases break down proteins and their degradation products, polypeptides and peptides, by hydrolyzing the —CO—NH—peptide linkages. The individual proteolytic enzymes are of two types. Endopeptidases hydrolyze internal peptide linkages along the protein chain and do not usually attack terminal units; they produce peptides. Exopeptidases hydrolyze the peptide linkages which join terminal amino acid residues to the main chain, thus producing amino acids. Aminopeptidases act upon the ends of protein chains having terminal amino groups and carboxypeptidases on the ends having terminal carboxyl groups.

A considerable number of commercial proteases of plant, animal and microbial origin are marketed. These commercial preparations usually contain several individual proteolytic enzymes. The individual proteases vary widely in their specific activities, in their pH optima and ranges, and in heat sensitivity. All highly purified proteases which have been studied show specificity for certain peptide bonds and have little or no action on other peptide bonds. For example, trypsin will rapidly hydrolyze only those bonds linking the carboxyl group of the two basic amino acids, lysine and arginine, to the amino group of any other amino acid. Depending upon the individual enzymes present in commercial proteases they may range from quite specific to fairly general in their actions on proteinaceous substrates. Hence, in practical applications of proteases in food processing it is necessary to select the appropriate protease complex or combination of enzymes. Usually this can be determined only by empirical methods. With proper selection of enzymes, and with appropriate conditions of time, temperature and pH, either limited proteolysis or nearly complete hydrolysis of most proteins to amino acids can be brought about. For some uses the proteases of different origin are directly competitive, in other cases the special properties of one may give it advantage over others.

The important uses for proteases in the food industries are shown in Table 3.

TABLE 3
Uses of Proteases in Food Industries

Use	Enzymes Employed
Oriental condiments	Fungal
Dairy—cheese	Animal, bacterial, fungal
Baking—bread, crackers	Fungal, bacterial
Brewing—chillproofing	Plant, fungal, bacterial
Meat—tenderizing	Plant, fungal
Fish—solubles	Bacterial, plant

Fungal protease has been used for centuries in the Orient for production of soy sauce, tamari sauce, and miso, a breakfast food. The application is a rather complex art, although simple in operation. Proteolytic strains of *Aspergillus oryzae* are grown on rice or soybeans to produce a heavily sporulated koji which is added to vats containing roasted soybeans, cereal grains and salt brine. Over a period of many months proteolysis, solubilization and extraction occurs. The soy sauce is then recovered by straining and pressing out the liquid with hydraulic presses.

One of the oldest uses of an enzyme is that of rennet for coagulating milk in cheesemaking. Most proteolytic enzymes will clot milk, but on further contact will dissolve the curd by hydrolysis of the insoluble proteins to peptides and amino acids. Rennin is a unique enzyme which after clotting milk has no further rapid or extensive action on the curd. It seems quite clear, however, that rennin acts as a protease, splitting peptide bonds in one particular casein fraction. The mechanism of curd formation has been postulated to involve limited proteolysis of κ-casein by rennin forming soluble glycopeptide and para-κ-casein. The para-casein precipitates and leads to the precipitation of other casein fractions which have been exposed to calcium ions released as a result of the κ-casein hydrolysis. This is probably an oversimplification of the process and alternate theories have been proposed.

The production of cheese is still very much of an art. The procedures vary depending upon the types of cheese, and somewhat from plant to plant for the same type of cheese. In general the milk is heat treated or pasteurized and then lactic starter culture is added. After a period of about an hour to allow the culture to develop, rennet is added and the curd allowed to form. The curd is then cut into small pieces, and the mass "cooked" at a temperature a little over 100°F. After the whey is drained off, the slabs of curd are piled and turned, and finally salted and pressed. The cheese is then aged or ripened by holding under controlled moderate temperature. During the ripening processes enzymatic reactions occur producing desirable and characteristic flavors in the cheese. Proteolytic activity during the ripening of cheeses is extensive. This proteolysis is in part due to enzymes produced by the lactic starter cultures and in part by continuing slow proteolytic action of the rennet-coagulating enzyme system.[200] The proteolytic breakdown in cheeses results in production of peptides and free amino acids.[110]

A search for substitutes for rennet for milk coagulation in cheesemaking has been underway for many years. In certain countries such as India ritual requirements make a "vegetable" rennet desirable. Also because the source of commercial rennet is the stomach mucosa of young calves, rennet has become increasingly expensive and in short supply. While there is a continuing increase in demand for cheese, and hence rennet, there is at the same time a decrease in the number of young calves being slaughtered and hence a shortage of rennet. Sardinas[149a] has recently reviewed new sources for rennet.

Milk can be clotted by most of the known proteases but with the exception of rennin and pepsin, proteolysis proceeds beyond the slight rise in non-protein nitrogen (NPN) observed with rennet. In an investigation with eight different proteases Tsugo and Yamauchi reported that rennin and pepsin produced the largest amount of curd after 15 minutes and no loss of yield after 60 minutes in contact with the whey, whereas all the other proteases produced less curd and continuing proteolysis led to further loss on holding in the whey.[173] Continued proteolysis also had a profound effect on the strength of the curd. Rennin gave the firmest curd while pepsin and protease from *Pseudomonas myxogenes* gave firmer curds than the other enzymes. Pepsin has in fact come into commercial use in combination with rennet for cheesemaking. There are differences in the behavior of the two enzymes, but pepsin when properly used has proved to be a satisfactory extender for rennet. The effect of rennet extract and commercial pepsin preparation has been followed in great detail throughout the entire production period of cheddar cheeses by Melachouris and Tuckey.[110] The amount of NPN formed during ripening of the cheeses was slightly greater with rennet.

Veringa reported earlier work in India where enzymes from *Ficus carica* (ficin) and from *Streblus asper*, a wild shrub, yielded curds only slightly less in weight than rennet curd. The curd made with ficin was slightly softer than the rennet curd and that with the *S. asper* enzyme was much softer.[186] Of all plant enzymes, ficin has been used more often than any of the other proteases for cheesemaking.

At present, research effort to obtain rennet substitutes has shifted to microbial sources. Milk-clotting enzymes for cheesemaking from the fungi *Mucor pusillus*, *Mucor miehei*, and *Endothia parasitica* have been patented.[11, 29a, 149]

These new microbial enzyme preparations, particularly those from *Mucor miehei*,[137a, 162a] compare favorably with rennet in cheesemaking with respect to curd yield and firmness. There have been reports, however, that some of them lead to more bitterness in young cheese. The bitterness may remain, increase, or decrease during prolonged aging of the cheese, depending upon the enzyme preparation and conditions of use and of aging. Suitable "microbial rennets" are being accepted in the cheese industry.

One of the largest uses for microbial proteases is in baking bread and crackers. The unique and variable properties of bakery doughs are due largely to the insoluble or gluten proteins of the flour. When wheat flour is mixed with water a complex dough system is formed. Because of the hydrated gluten the dough has extremely viscous flow properties and is therefore said to be extensible. However, it is at the same time elastic and resists deformation. With protease enzyme supplementation, gluten hydrolysis can be controlled to achieve better properties in the dough.

Cereals normally contain a very low level of protease. The protease content increases somewhat when wheat or barley is malted. However, the malts are still poor sources of protease, and cereal proteases provided by malt supplementation of flour have relatively little effect on dough properties.

The special viscoelastic properties of doughs are provided by the wheat gluten. The so-called strength of flours is determined by the quantity and quality of the gluten. Strong flours give doughs which have good tolerance to extensive mixing. The use of oxidizing agents—bromates, iodates, peroxides—as dough improvers has a strengthening effect while reducing agents have a weakening effect.[131] The effect of proteolytic enzymes resembles in some respects that of reducing agents. Mixing time and viscosity of doughs are reduced, and the doughs are more pliable and extensible. In practice, the combined use of oxidizing agents and of proteolytic enzymes has shown considerable benefits. In present day baking, flours and doughs which have not been treated with oxidizing dough improvers are very rare. Probably fungal protease derived from *Aspergillus oryzae* is added by the baker for more than half of the bread baked in the United States. Recently, the use of bromelain in bread baking has been permitted but data on its effectiveness have not been published.

A major benefit of adding fungal protease to sponge doughs is the marked reduction in mixing time which results.[32, 140] The proper use of fungal protease reduces the mixing time by about one third, and improves the handling properties of the doughs. The protease increases extensibility of doughs, and is thus valuable in controlling the pliability of doughs, eliminating buckiness, and ensuring proper machinability. Improved loaf characteristics, including better volume, greater symmetry, and improved grain, texture and compressibility are also observed. These probably come as secondary benefits resulting from the greater extensibility and better machinability; the primary benefits are unquestionably the reduction in mixing time and better handling properties of the dough.[140]

The action of protease seems to be limited to the mixing and fermentation periods since the very low inactivation temperatures (50–60°C) preclude any appreciable action during baking.[114] The initial process in dough formation must be hydration of the flour particles to bring out the cohesive properties of the gluten. The amount of mixing required depends upon the gluten quality, more mixing development being required with strong flours than with weak ones. Bucky doughs from very strong flours may require excessively long mixing times for proper development. By judicious use of fungal protease supplementation, the progressive softening induced by the enzyme can effectively substitute for a large part of the mixing development normally required.[32] The baker can thus regulate the mixing time and adapt flours of differing qualities to his standard bakeshop practices. The optimum amount of protease required is quite critical, and depends not only upon flour quality, but also on bakeshop conditions, particularly mixing and fermentation time and temperature. However, excess protease activity is very undesirable since this results in the doughs becoming progressively more sticky and slack, and consequently in poorer rather than better loaf characteristics.[32]

Fungal protease preparations from different strains of *A. oryzae*, or from the same strain cultivated in different manners, vary considerably in their effect on baking performance of flours.[114,140] The modified Ayre-Anderson hemoglobin method is the generally accepted procedure for standardizing the level of fungal protease for breadmaking.[7] However, it has been shown that different fungal preparations of equal hemoglobin assay values may vary widely in their effect on doughs.[140] The only satisfactory way for evaluating *different* protease preparations is by dough handling and baking test methods. The organism strain and conditions of manufacture must be rigidly controlled, as they are of course, by the enzyme manufacturers to insure consistent baking performance. Under these conditions the hemoglobin assay procedure is a useful and reliable quality control method for standardizing an individual product.

The largest application of fungal protease supplementation has been in conventional breadmaking. It is also extensively used in the brew-conventional process. In this process the pre-ferment or brew contains all of the water and water-soluble ingredients, that is, all the ingredients except flour and shortening. After a short 2 to 4-hour fermentation time the flour and shortening are mixed with the brew and the doughs processed through conventional make-up equipment. Since the flour is not exposed to a lengthy fermentation, higher levels of fungal protease are required to achieve the desired mellowing effect on the doughs.

Protease originally found little use in the continuous breadmaking processes employing liquid pre-ferments or brews with which flour and shortening are mixed continuously with the very soft dough being deposited directly into the baking pans for proofing and baking. Reinvestigation of the application of protease in this process is underway, since modifications in the original process such as addition of flour to the brew have been made, and it appears that use of enzymes may be effective in permitting increased output through the continuous mixer with lowered power requirements.[60,189]

The use of microbial protease in cracker, biscuit and cookie doughs has become widespread. Both bacterial and fungal proteases are used in these bakery products. The α-amylase present in the microbial protease preparations contributes partially to the advantages observed. When bacterial protease is employed, the thermal stability of the amylase creates no problem as it does in bread making, since the internal temperature in crackers and cookies rises rapidly in the oven and reaches much higher levels than in bread. The amylase is therefore so quickly inactivated during the baking process that the sticky-crumb problem is eliminated. According to unpublished reports the use of microbial protease in cracker doughs greatly decreases the mixing and holding time necessary to develop the gluten, promotes uniform rolling of the dough without stickiness or "bucking" at the edges, increases spread, gives exceptionally even baking, with uniform and improved browning, produces more open grain with enhancement of tenderness and flavor, and reduces the shortening requirements. Similar favorable results are obtained in baking of sugar wafers, vanilla wafers, and other cookies.

Another major application for proteases is in the brewing industry. A problem which has long been recognized in brewing is the tendency for an undesirable haze to develop in beer and ale when the beverages are cooled. The chill haze consists of a very fine precipitate which forms when beer is cooled below about 10°C. The formation of haze is accelerated by the action of light, heat, oxygen, or the presence of traces of copper or iron. The degree of haziness increases rapidly with extended storage or with higher storage temperatures. The chill haze has a variable composition, but generally contains protein, tannin and carbohydrates.[117] Since protein is a necessary component of chill haze, addition of protease enzymes for its prevention was the logical approach for solving the chill haze problem. Use of proteolytic enzymes has become the dominant method for chillproofing beers.

Almost any protease active at the normal pH of beer (about 4.5)—papain, pepsin, ficin, bromelain, fungal and bacterial proteases—can be used for chillproofing. Papain alone or in combination with other proteases is most commonly employed. The enzymes are added during the cellar operations. The amounts used vary depending upon the raw materials and processing method, and the anticipated length of storage and storage temperature. About 8 ppm of papain seems about an average level of addition to digest enough of the protein to prevent haze formation.[193] Foaming properties of the beer are not adversely affected by nominal levels of chillproofing enzymes.

There is no good correlation between enzyme potency as determined by conventional methods for protease assay, using hemoglobin, casein, gelatin or other proteins as substrates, and the performance of the enzyme in chillproofing. It is therefore necessary to evaluate enzymes for chillproofing by testing them at various levels on untreated, fresh beer as the substrate. Although quite laborious, methods have been developed and published for conducting such evaluations.[170,195] Once an evaluation has been made on beer to determine the proper use level, for a particular enzyme, this enzyme can then be used on the basis of its assay potency determined by a conventional method. For example, if 1 g of papain assaying 1000 units per gram is required per barrel, 2 g of papain assaying 500 units per gram would be necessary. However, with another enzyme such as bacterial protease assaying 1000 units per gram by the same method, no judgment can be made from this assay value as to the required amount to use. This can only be determined by actual evaluation with beer.

Still another use for proteolytic enzymes is in tenderizing meat, particularly beef. There is a great deal of variation in the toughness (or its opposite, tenderness) of beef depending upon the age of the animal and the part of the carcass from which it came. As discussed in a previous section of this chapter, beef is usually aged to improve tenderness by action of the natural enzymes present in the meat. Except for the choice grades of beef, additional enzymatic tenderization is usually necessary or desirable.

A great deal of work has been published on the effect of different proteolytic enzymes on the different beef proteins as determined by chemical and histological methods.[49,109,194] The concentrations of enzymes employed in this work have of necessity been very many times higher than those used for the practical tenderization of beef. The tenderization of meat can be recognized organoleptically long before extensive hydrolysis can be detected by chemical analysis or before any structural changes can be shown histologically under the microscope. However, it is quite clear from this work that almost all proteolytic enzymes have some tenderizing action on meat, and that the several available proteases have differing actions on the various meat proteins. The plant proteases, papain, bromelain and ficin, have considerable effect on connective tissue, mainly collagen and elastin, and show some action on muscle fiber proteins. On the other hand, microbial proteases (bacterial and fungal) have considerable action on muscle fibers, but only slight effect on collagen and none on elastin fibers. Advantage has been taken of these differences in formulating meat tenderizers containing combinations of enzymes, thus taking advantage of the hydrolyzing effect of plant proteases on connective tissue, and the high activity of fungal proteases in hydrolyzing muscle fiber proteins.[176]

Meat tenderizers are applied in the household, in restaurants, by distributors of retail cuts of meat in frozen form, and by meat packers. One report indicated that about one third of the papain sold in the United States (total import about 500,000 lbs per year) was used by consumers for tenderization and 500 million pounds of beef (about 5% of the total) were tenderized by meat packers.[9] The application of proteases in tenderizing meat presents problems. It requires uniform distribution of the enzyme in low concentration to give a limited proteolysis to a particular degree of tenderness and no farther. Too high an enzyme concentration or too long a period of treatment may result in overtenderization, mushiness and even formation of an undesirable hydrolyzate flavor.

Meat tenderizing by the ultimate consumer is usually accomplished by sprinkling a meat cut with a powdered enzyme preparation or by immersion in a solution of the enzyme. Often the meat is pierced with a fork over its entire surface to improve penetration of the enzyme into the meat. In this way tenderization of small steaks is easy to control since the meat is cooked immediately following enzyme treatment. Similarly distributors dip steaks into enzyme solutions, and market them as prepackaged, frozen steaks, which are broiled directly from the frozen state. Most of the tenderizing action occurs during the brief period during cooking before the temperature in the meat rises to the point where the enzyme is inactivated. A typical composition for the surface treatment of beef cuts contains 2% commercial papain or 5% fungal protease, 15% dextrose, 2% monosodium glutamate and salt.[185] It is also possible to use interleaving sheets impregnated with papain in packaging cuts of meat, then freezing, in order to apply a predetermined amount of enzyme to each cut of meat.[8]

The most recent development in enzymatic meat tenderization is injection of proteolytic enzyme solutions into the vascular system of cattle before slaughter (ante-mortem). The vascular system effectively distributes the proteolytic enzyme throughout the tissues. Ante-mortem tenderization is now practiced on a fairly large scale by meat packers using low concentrations of specially purified papain, equivalent to 5 to 30 ppm of commercial papain based on the total weight of the animal.[20] Cows and low-grade steers require the highest concentration of enzyme and prime heifers the lowest.[59] The meat must be refrigerated until used to avoid overtenderization. This ante-mortem tenderization permits production and sale of a much higher percentage of tender cuts from all grades of beef, and particularly from the lower grades. This method also enables tenderization of large roasts which was not possible with surface application. The only real problem encountered seems to be possible overtenderization of highly vascular organs such as the livers, which may develop mushiness and hydrolyzate flavors.

Although the requirement for thorough cooking of chicken or pork at high temperatures usually make enzymatic tenderization unnecessary, some attention has been given to tenderization of poultry and hams. The tenderness of chicken meat can be increased by suitable application of proteases; the results being most dramatic with old roosters and hens. One method is ante-mortem injection into the humoral vein of birds.[20] Another is injection into the peritoneal cavity 6 to 12 hours before slaughter.[70] Immersion in solutions of papain or fungal protease have improved tenderness in freeze-dehydrated chicken meat.[162] A process has also been patented for enzymatic tenderization of hams, employing bacterial protease.[150]

Several minor applications for proteases in food or feed industries, such as in the production of fish solubles may be noted. In order to obtain drying oils from them, Menhaden and similar fish are cooked and pressed, the residue being dried to fish meal and the liquid portion centrifuged to recover the oils. The centrifugate is concentrated to about 50% protein and the viscous solution sold as a feed rich in proteins and vitamins. Undesirable solidification of the fish solubles is prevented by treating the liquid with a crude bacterial protease prior to concentration. Fish can also be ground and treated with proteolytic enzymes before pressing.[106]

Enzyme hydrolysis of renderer's meat scrap with fungal protease, papain or bromelain was superior to acid hydrolysis, yielding recovered protein fractions comparable in amino acid composition with the proteins of meat and bone meal, which could be used as a feed ingredient in, for example, pig starters.[33,37]

Protein hydrolyzates are prepared most commonly by acid hydrolysis of soybean protein, wheat gluten or milk proteins.[22] However, a number of enzyme hydrolyzates are marketed commercially as flavoring agents. Enzyme hydrolyzates can be prepared from almost all available proteins, with the advantage over acid hydrolysis that tryptophan is not destroyed. Enzyme hydrolyzates of vegetable proteins have bland and pleasant flavors. Bitter peptides are often produced by enzymatic hydrolysis of milk proteins.

Industrial non-food uses for proteases are even more numerous and varied than the applications in food industries. These important uses are given in Table 4 as a matter of interest, but will not be further discussed.

Lipases Lipases are widely distributed in nature, in animals, plants and microorganisms. In spite of their great natural importance in digestion and metabolism of fats, lipases have achieved quite limited industrial application. The most important lipases from the standpoint of industrial uses are pancreatic lipase, animal pregastric lipases, and the lipases of certain microorganisms.

True lipases are enzymes which hydrolyze insoluble fats and fatty acid esters occurring in separate, non-aqueous phase. The rate of lipase reaction is dependent upon the surface area of the emulsion on which it acts.[47] Most lipases have considerable specificity with regard to the type of fatty acids, their chain length, and degree of saturation, and with regard to the position of the fatty acids, as 1- or 3- outer chains or the 2-inner chain of glycerides. With pancreatic lipase the enzyme shows preference for the 1- or 3-position, and the relative rates of hydrolysis of triglycerides, diglycerides and monoglycerides are in that order.[31] Milk lipase has similar specificity, as do many, but not all, microbial lipases.[4,80]

TABLE 4
Industrial Non-food Uses for Proteinases

Industry and Use	Enzymes Employed
Animal feeds—ingredient	Bacterial, fungal
Leather—bating hides	Bacterial, fungal
Leather—unhairing hides	Bacterial
Dry cleaning—spot removal	Bacterial
Textiles—degumming, desizing	Bacterial
Photographic—film stripping	Bacterial, plant
Medicine—digestive aid	Fungal, bacterial, plant
Medicine—wound debridement	Bacterial, plant
Medicine—relief of inflammation, bruises, blood clots	Bacterial, fungal, plant

Major application of lipase activity is made in the dairy industry. Controlled lipolysis is necessary in cheesemaking in order to develop characteristic flavors. A high degree of lipolysis is particularly important in Italian type cheeses. Traditionally, Italian cheeses were made using rennet paste as the coagulant. Such paste is made by drying the entire stomach of calves, kids or lambs, including the milk contents. The characteristic piquant flavor of Italian cheeses made with rennet pastes cannot be obtained with rennet extracts, and has been found due to the presence of lipase enzymes which are derived from enzyme secreting glands at the base of the tongue of the animals.[56] These enzymes are extracted from the excised glands of calves, lambs and kids, and are now commercially available. They are variously called oral lipases, oral glandular lipases, or pregastric lipases. It has been conclusively shown that the lipolytic activity of the pregastric lipases is of the same degree and kind as that of rennet pastes from the same species.[65,66,105] Pregastric lipase from calf, kid and lamb are manufactured for and standardized according to their ability to lipolyze butterfat. They produce a specific and reproducible ratio of free fatty acids as a result of the lipolysis of the butterfat. The specific enzyme action on butterfat results in a characteristic flavor for each animal species. The pregastric lipases are now widely used, along with ordinary rennet for coagulation, for production of Italian type cheeses. Efforts to employ pancreatic lipase, plant lipases, or fungal lipases have been unsuccessful since they produce different ratios of free fatty acids giving atypical flavors or even undesirable soapy or rancid flavors.

Pregastric lipase also has wide application in the food industry for treating butterfat-containing products to produce a wide variety of food flavors. Butteroil, cream, half and half, condensed milk, reconstituted whole milk powder or fresh milk are suitable substrates. There are now commercially available dairy-flavor products produced by such pregastric lipase treatment. Lipolyzed butterfat emulsions are used to give enhanced butter flavor improvement to margarines, shortenings, popcorn oils, bakery products, vegetable oils and confections. Lipolyzed cream and lipolyzed cultured cream are employed in candies and confections, and in baked products, margarines, snack coatings and butter sauces. The modified butterfat-containing products are being used to enhance the flavors of cheese dips, cheese sauce, cheese soups, cheese dressings, chocolate, confections, gravies, macaroni and noodle dishes, baked products, etc.

Of particular interest are the significant contributions to flavor of chocolate confections by use of dairy products. Free fatty acids make very appreciable contributions to the flavor of milk chocolate, butter creams, caramels and toffee. Such flavors are obtained by the use of cultured butter or more conveniently by use of pregastric lipase-modified butterfat-containing products. Low levels of free fatty acids enhance the flavor of the confection without creation of new flavor notes. Intermediate levels give buttery flavors, and high levels produce cheesy flavors.

Pancreatic lipase has a minor use in treating egg whites before drying in order to enhance whipping qualities. Minute amounts of lipids from traces of yolk have an adverse effect on whipping properties and are removed by the lipase treatment.[36]

Pancreatic lipase also has use in digestive aids and in therapy of such diseases as cystic fibrosis in which patients have bulky, fatty stools. Lipase of pancreatic or microbial origin is an indispensible ingredient in enzymatic drain cleaner preparations since fats along with other food ingredients are major contributors to stoppages in drain pipes and traps in sanitary plumbing systems.

Lipase activity in flour for baking is undesirable since free fatty acids have a detrimental effect in doughs. On the otherhand lipoxidase, or lipoxygenase, has attained considerable importance in baking. Lipoxidase catalyzes the oxidation of polyunsaturated fatty acids containing *cis*, *cis*-1,4-pentadiene groups by molecular oxygen.[168] Soybean meal appears to be the richest source of lipoxidase.[6] Defatted soybean flours are used commercially as the source of the enzyme. About 0.5 to 1.0% of such soybean flour is used in baking to achieve the desired effects. Lipoxidase is used extensively in the production of bread for bleaching the natural pigments of flour and producing a very white crumb. Lipoxidase also has an effect on the dough mixing properties of flours as well as on bread structure and flavor. These particular effects of lipoxidase are complex and little understood. Addition of lipoxidase gives stronger doughs. It has been suggested that binding occurs between lipids and wheat gluten, and that the effect of lipoxidase during mixing is due to direct action in forming peroxides and indirect effect in catalyzing oxidation of sulfhydryl groups in gluten.[35] The presence of lipoxidase in sponge doughs gives a characteristic desirable nutty flavor. It is now being used commercially by several plants using the brew continuous mix process of breadmaking in order to obtain the desirable nutty flavor.[92]

Glucose Oxidase   A number of fungi produce glucose oxidases which catalyze the reaction of glucose with molecular oxygen forming gluconic acid. Most extensively studied have been the glucose oxidases produced by *Penicillium notatum*,[86,87,88] *Penicillium amagasakiense*,[101] and *Aspergillus niger*.[58,129,166,175,178] The enzymes from each of these organisms have about the same molecular weight (about 150,000), the same isoelectric point (4.2–4.3), the same pH optimum (5.5–5.8), and contain two flavin adenine dinucleotide units per mole of enzyme.

Glucose oxidase shows a high specificity for β-D-glucose.[1,58,129,166] Other hexoses, pentoses, or disaccharides are not oxidized, or are oxidized only at negligible rates. Pazur and Kleppe reported that with their purified *A. niger* preparation mannose and galactose gave 1% and 0.5% of the rate of glucose oxidation.[129] For all practical purposes purified glucose oxidase reacts only with glucose, and this has made the enzyme a valuable tool for analysis, as well as for other purposes.

The important commercial glucose oxidase preparations are derived from *A. niger*[175,178] The usual commercial preparations also contain catalase, which is advantageous in most of the applications, and traces of various carbohydrases. Where the latter are undesirable highly purified preparations are also available. The overall reactions of the commercial glucose oxidase-catalase system are shown in Figure 5. In its industrial applications the glucose oxidase system is used both to remove glucose and to remove oxygen.

The most important application of glucose oxidase for the removal of glucose is from egg albumen and whole eggs prior to drying. Powdered egg products are unstable and deteriorate during storage due to the nonenzymatic browning reaction between glucose and proteins. The first evidence of change is loss in functional properties such as whipping properties and foam stability. On longer storage discoloration, off-flavors and loss of protein solubility occur. The best method for stabilizing dried egg products has been found to be removal of the glucose before drying, and the most satisfactory method is by use of glucose oxidase.[16,155]. In commercial processing of egg whites the pH is adjusted by addition of the required amount of citric acid or hydrochloric acid to about pH 7.4. For egg yolk or whole liquid egg no pH adjustment is necessary. The enzyme is then added and the batch held at about 90°F. Throughout the incubation period gentle agitation is maintained and excess oxygen is supplied by continuously or periodically adding hydrogen peroxide. Completion of the desugaring is determined by negative test with Somogyi reagent, after which the egg product is dried. The time of desugaring can be controlled by the concentration of enzyme employed, and in different operations may vary from about 2 to 7 hours, about 3 hours generally being preferred.

Although the presence of hydrogen peroxide during the desugaring process has a bacteriocidal effect, the requirement for low bacterial count and freedom from Salmonella in egg products has led to introduction of low temperature (7–15°C) desugaring by glucose oxidase followed by pasteurization for yolk and whole egg.[158]

Oxygen is responsible for a wide range of types of deterioration of foods, chief among which are flavor and color changes. In canned acid foods, oxygen also accelerates can corrosion. While somewhat dependent upon the constituents of the foods with which it is used, glucose oxidase has been found effective in protecting certain foods and beverages by removal of oxygen from the foods and containers. The effectiveness of the enzyme in removing oxygen will vary with each product depending upon the pH, level of glucose, and many other factors which may vary widely.

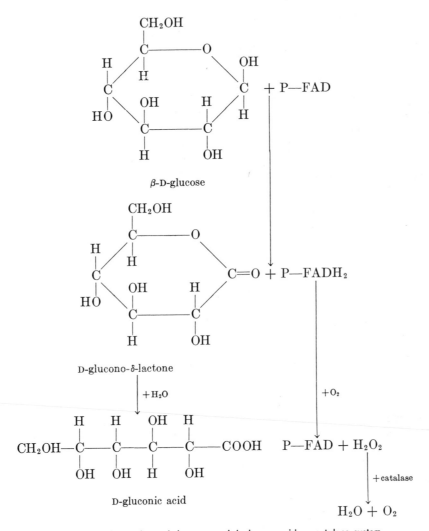

β-D-glucose

D-glucono-δ-lactone

D-gluconic acid

Fig. 5. Overall reactions of the commercial glucose oxidase-catalase system.

Presence of oxygen in canned and bottled beer has very deleterious effects on flavor and stability, and is minimized but not eliminated completely during the filling operations. It has been demonstrated that addition of glucose oxidase is very effective in removing residual oxygen from beer.[123,175] Commercial trial packs employing glucose oxidase in several breweries consistently indicated a beneficial effect of the enzyme on beer stability.[143] Taste panels could detect the flavor differences between treated and untreated beers, and preferred those which had been enzyme treated. Although some breweries employ glucose oxidase it is not used by a majority of brewers. This may be due in part to a reluctance to add glucose to beer which does not always contain enough of the sugar to permit effective oxygen removal by the glucose oxidase system.

Glucose oxidase has been investigated in the treatment of wines. Removal of oxygen from unpasteurized apple wine by glucose oxidase prevented growth of microorganisms and off-flavor development.[201] Addition of 0.1% glucose along with glucose oxidase to white wines to prevent browning and flavor changes was quite effective in depleting the free oxygen content within 24 hours.[124] However, when the enzyme-treated wines were again exposed to air after storage browning occurred in many instances. No reasonable explanation has been advanced for these puzzling observations.

Limited commercial use is made of glucose oxidase to prevent color and flavor changes in bottled or canned soft drinks. Different flavors vary greatly in susceptibility to changes caused by traces of oxygen remaining in the sealed containers. Glucose oxidase may protect in some cases, and have no effect in others. Greatest application for glucose oxidase has been with citrus drinks containing natural juice or oils which are particularly prone to oxidative flavor changes upon exposure to light. The natural cloud stability of the citrus drinks seems to be related to the presence of cellulose. Hence the glucose oxidase employed must be very low in cellulase in order to retain the desirable stable cloud and at the same time protect against flavor changes. Glucose oxidase is effective in protecting cans of carbonated beverages against oxidative corrosion, and is used in canned soda, for example.[178]

Mayonnaise is an outstanding example of a food product which can be protected against deterioration by the addition of glucose oxidase. Mayonnaise is an oil-in-water emulsion containing much air, packed in glass containers and exposed to light and ambient temperatures. Despite the best precautions and use of the highest quality ingredients the product undergoes gradual deterioration noticeable to the expert after a month, and to the general public after 3 or 4 months. The removal of oxygen from mayonnaise by incorporating glucose oxidase has been found to afford significant protection to the product. No detectable change in regard to color, rancidity or increase in peroxide values occurred in 6 months whereas the controls without glucose oxidase had faded perceptibly and were rancid as early as the third month.[23] Glucose oxidase has not yet been added to the list of optional ingredients in the Standards of Identity for mayonnaise, but glucose oxidase is used in some nonstandardized salad dressings. The glucose oxidase preparation must be free of amylase for use in salad dressings which contain starch.

Oxidative deterioration of dry or dehydrated foods such as whole milk powders, roasted coffee, active dry yeast, and white cake mixes as isolated examples, is frequently a serious problem. It is, of course, not feasible to add glucose oxidase directly to such food items since the enzyme is inactive in the dry state. In order to adapt the enzyme for use with dry foods a procedure has been devised in which the enzyme, glucose, buffer, filler and water are placed in a water-impermeable but oxygen-permeable packet. When this is placed in a hermetically sealed container along with a dry food product the contents of the packet rapidly take up the oxygen in the container, leaving an oxygen-free atmosphere.[156,157,178] Studies in the use of the glucose oxidase scavenger packets have been reported for whole milk powder and for whole dry milk or dry ice cream mix but never published.[100,112]

While the enzyme scavenger packet is an intriguing and unique concept, and may find specialized applications, it is not in commercial use. This is largely due to relatively high cost and inconvenience in use. After manufacture the packets must be sealed into their containers immediately since they will rapidly be exhausted if exposed to the atmosphere.

Because of its specificity, glucose oxidase is widely used as an analytical tool for the detection and determination of glucose in the presence of other carbohydrates. It is of particular advantage in biological systems of interest to medical technicians and to food technologists. For example, an important application is in detecting and monitoring diabetic conditions by using it in the form of test strips to detect the presence of glucose in the urine. High purity glucose oxidase, free of undesirable contaminating carbohydrases, is now available commercially for use in quantitative analysis procedures for glucose.[17]

Catalases   Catalases convert hydrogen peroxide to water and oxygen. They are obtained for commercial uses from animal (liver), bacterial (*Micrococcus lysodeikticus*) and fungal (*Aspergillus niger*) sources.

Catalase is employed to remove the last traces of hydrogen peroxide in peroxide bleaching. It finds application in the food industry in conjunction with glucose oxidase as previously discussed. Its principal additional use in the food industry world wide is in connection with cold sterilization or preservation of milk by means of hydrogen peroxide.

In many undeveloped countries of the world cooling facilities are not available on the farms and prompt heat pasteurization is not technically possible. In these circumstances hydrogen peroxide treatment is the only feasible method for preserving milk. For such purposes it has been suggested that 1 ml of 33% hydrogen peroxide per liter of milk be added at the farm, followed by another milliliter at the dairy plant, heating to 50°C for 30 minutes followed by cooling to 35°C and addition of catalase to destroy the residual peroxide.[145]

In the United States hydrogen peroxide treatment instead of heat pasteurization of milk is permitted for cheesemaking. The hydrogen peroxide treatment is quite effective in reducing the counts of pathogenic microorganisms. It inactivates milk catalase and peroxidase, but lipase, proteases and phosphatases are not destroyed. Several publications have described the peroxide-catalase method for cheesemaking and the quality of the cheeses produced.[146,147,169] Both flash and batch methods may be employed. In the flash method about 0.03% of 35% hydrogen peroxide is added to the cool milk which is then passed through a plate heat exchanger where it is heated to 52°C for 25 seconds. It is then cooled to setting temperature for the cheese, and an excess of catalase added to destroy the residual hydrogen peroxide. In the batch method about 0.06% of 35% hydrogen peroxide is added to the vat of milk at normal setting temperature, and the milk held for about 20 minutes prior to addition of the catalase. After the catalase action, lactic starter cultures are added, and the cheese is made in the normal manner.

Glucose Isomerase   A very recent commercial development has been the application of glucose isomerase for converting glucose to fructose for the large-scale production of sweet syrups. In 1957 Marshall and Kooi[107b] reported that *Pseudomonas hydrophila* cultivated in a culture medium containing xylose produced an enzyme that isomerized glucose to fructose, as well as xylose to the corresponding ketose xylulose. Marshall[107a] patented the use of enzymes showing xylose isomerase activity for conversion of glucose to fructose.

Sweet "invert sugar" syrups containing fructose are readily and cheaply obtained by acid or enzymatic hydrolysis of sucrose. Even though glucose is cheaper than sucrose, the high cost of enzyme has prevented the process of converting to fructose by glucose isomerase from being competitive. Although other source organisms giving higher isomerase yields were discovered,[173a, 200a] all of these also required expensive xylose in the growth medium as an inducer of the glucose (xylose) isomerase. Apparently the process became practical only after the discovery by Japanese workers of a strain of *Streptomyces albus* in which the isomerase could be induced by crude xylan sources, such as wheat bran or corn cobs, as well as by separated xylan and by xylose. This organism is now being used industrially both in Japan and in the United States as a source of enzyme for production from glucose solutions of syrups containing fructose.

No details of the exact procedures employed industrially have been published. However, the publication by Takasaki, Kosugi, and Kanbayashi[167a] gives considerable general information regarding the production and utilization of the isomerase, as well as the isolation, crystallization, and properties of the enzyme. From the discussion and flow sheet, it is indicated that the medium contains 3% wheat bran, 2% corn steep liquor, and 0.024% $CoCl_2 \cdot 6H_2O$. After incubation at 30°C with aeration to maximum enzyme production in 25–30 hours, the wheat bran residue is removed by screening and the *Streptomyces albus* cells are recovered by filtration and washed. These cells are then employed with 50% glucose (starch enzyme hydrolyzate) solution containing 0.005 M magnesium sulfate and 0.001 M cobalt chloride with pH maintained at about 7 and temperature between 65–70°C. After the isomerization is completed, the cells are filtered off, the solution decolorized with carbon, subjected to ion exchange purification and evaporated to about 75% solids content. The sugar composition of the finished syrup is 45–50% fructose and 55–50% glucose.

By heat treating the cells filtered from the primary fermentation broth at about 65°C, the intracellular enzyme is rather firmly fixed in the cells. These enzyme-fixed cells can then be used as immobilized enzyme preparations for glucose isomerization either batchwise, being filtered off for reuse, or by continuous column operation. The batch system seems more practical for large-scale use; the loss of perhaps 10% of the enzyme activity in each cycle can be made up by addition of the requisite amount of fresh enzyme-fixed cells.

Flavor Enzymes The use of certain enzymes in the formation of food flavors has previously been mentioned, for example the use of pregastric lipases in dairy products. A recent important development is the availability to the food industry of the specific flavor enhancers, 5′-inosine monophosphate (IMP) and 5′-guanosine monophosphate (GMP). These compounds accentuate meaty flavors, and have application in canned vegetables, sauces, gravies, soup bases, etc. They are produced in large quantities by extraction of yeast ribonucleic acid (RNA) and enzymatic hydrolysis of the RNA to the 5′-nucleotides. The process only became practical when methods for obtaining fungal phosphodiesterases from suitable strains of Penicillium and Streptomyces were developed. This work has been reviewed.[122] Several good review papers provide the background for the use of enzymes in the production of 5'-nucleotide flavor enhancers.[99,159] The details are beyond the scope of the present discussion.

Considerable interest has been shown in flavor improvement of foods by direct enzymatic reactions. It has been shown that many fresh foods contain volatile flavor compounds which have been investigated by gas-liquid chromatographic methods. During processing, for example by dehydration, all or part of the volatile constituents may be lost with resultant change of flavor. However, the foods may also contain non-volatile flavor precursors. By treating the flavorless processed food with an appropriate enzyme the flavor precursors may be converted to flavor compounds, thereby improving or restoring the flavor of the food. These discoveries are of importance in our understanding of the development of flavors in foods. They may lead to practical applications for restoration of flavors in processed foods, but such enzymatic processes are not yet in commercial use. Several reviews have covered our present knowledge of this interesting and promising field.[67,68,95,151]

Most of the reported work has been with vegetables or fruits having strong and readily recognized flavors. For example, members of the cabbage family (Crucifera), cabbage, mustard, horseradish and water cress, have been studied extensively. The flavors of these plants are derived mainly from their mustard oils (isothiocyanates). These are produced from thioglycosides such as sinigrin (black mustard) or sinalbin (white pepper) by the action of thioglycosidases (sinigrinase or myrosinase). The hydrolytic reaction catalyzed by these enzymes produces organic isothiocyanates, sugar and sulfuric acid salt from the thioglycosides. Allylisothiocyanate seems to be a major component, with others such as butenyl-, n-butyl, methyl-, and methylthiopropylisothiocyanates also contributing to the flavor.[15] Dimethyl sulfide and related sulfides from S-methylcysteine sulfoxide also appear important in cabbage flavor.[42] The enzymes which have been studied with the cabbage family have been obtained by extracting the fresh plant materials. The enzymes from cabbage and mustard are somewhat interchangeable. When the mustard enzyme is employed with dehydrated cabbage, the restored flavor is similar to but not identical with the flavor of fresh cabbage. The cabbage enzyme produces a pleasant fresh cabbage flavor initially, but on longer action a strong mustard-like flavor may develop.[107]

Among the flavor precursors of onions are S-substituted cysteine compounds. Sulfoxidase enzyme systems convert these to sulfenic acids, ammonia and pyruvic acid.[152,154] The unstable sulfenic acids decompose further to sulfur compounds responsible for the biting, lacrymatory effect characteristic of onions. The enzymes have been isolated from onion and from an ornamental shrub Albizzia lophanta.[69,153]

Unpublished work from the author's laboratories resulted in enzyme preparations from legume sources and a microbial fermentation product which improved the flavor and taste acceptance of green beans and other canned, frozen and dehydrated legumes. It was necessary to supply oxygen as air or in the form of a peroxide along with the enzyme to generate the flavors from their precursors in the legumes. The flavor changes noted correlated with increasing carbonyl concentration produced.

Similar flavor enzymes have been found with certain fruits. An enzyme produced from the white center cores of raspberries acting upon a raspberry substrate freed from volatile flavors by steam distillation produced characteristic gas chromatographic peaks and the odor of ripe raspberries.[196] Some of the peaks, but not all, were also produced by two commercial enzyme preparations designated as a cellulase and a β-glycosidase.

An enzyme preparation from fresh bananas restored much of the fresh banana odor to a heat processed banana puree. Potential precursors, particularly pyruvic acid, valine, and oleic acid, shortened the time required for development of the fresh fruit flavor. However, such precursors not only accelerated formation of fresh aroma but also the formation of undesirable odors characteristic of overripe bananas.[71]

In most of the considerable amount of work done on enzymatic flavor improvement the enzymes have been isolated from fresh vegetables and fruits by extraction with water and precipitation by an organic liquid such as acetone. In many instances the natural flavor of the processed vegetables or fruits could be restored by such enzyme preparations. In others an off-flavor not characteristic of the particular food product developed. Schwimmer has tabulated rather extensive results with cabbage, horseradish, broccoli, peas, string beans, carrots, tomatoes and onions.[151] It has been claimed that the principle of enzymatic flavor restoration can be applied to most processed foods.[69] It may not be necessary to actually isolate the flavor enzyme in some cases. It has been claimed that the flavors of dehydrated cabbage, potato, tomato, carrots, grapefruit or orange juice can be restored by addition of a small portion of the fresh food after freeze-drying to the dehydrated foods in which the natural flavor-producing enzymes have been destroyed during the blanching or dehydration steps.[116]

The disadvantages which have prevented any significant commercial acceptance of flavor enzymes to date seem to be lack of convenience in use, and the problem of controlling the extent of reaction. For example, to restore the flavor to practically tasteless dehydrated cabbage, the enzyme would need to be added by the ultimate consumer as he reconstitutes and rehydrates the product. Considerable care is necessary as to temperature and time so as to attain the desired activity of the enzyme and thus obtain the desired flavor change without going too far.

Several conditions must be met for flavor restoration to be practical. The essential flavor precursors must be present, stable, and available to enzyme action in the processed food. The necessary enzymes must be found and be economically produced from the fresh foods or other sources. The enzyme actions must be simple hydrolytic or oxidative reactions. The enzyme actions must be controllable and convenient to use. These requirements have been met in part with some vegetables and fruits where flavor depends on hydrolytic or oxidative production of sulfides, isocyanates, alcohols and carbonyl compounds, and strong readily recognized flavors are produced as in cabbage, onion, or some fruits. More subtle changes of flavor by enzymatic action still require attention. Flavor enhancement or flavor production by enzymes is a challenging field, research is very difficult and expensive, but progress can be expected in the future.

New Enzyme Applications and New Enzyme Products

Only a limited number of enzymes, as is apparent from the previous discussion in this chapter, have to date attained commercial significance, even though the number of individual enzymes produced by living cells is in the thousands.

New industrial applications for old and new enzyme products are continually being developed and put to use. In such new developments a first requisite, of course, is that an enzyme exists in nature which can bring about the desired reaction. If a new enzyme product is desired an active search will usually lead to the discovery of a microbial source for an enzyme which can accomplish any reaction within the limits of possible enzymatic catalysis. From the standpoint of industrial enzymology, whether such a search is justified is dictated by the potential importance of the reaction.

In the search for a new enzyme the best source must be intelligently sought from appropriate microorganisms by suitable screening methods. Having found a source organism, extensive laboratory work is required to develop best media and conditions for maximum enzyme yield, and most suitable procedures for isolating, purifying and stabilizing the enzyme. Finally, the fermentation and isolation methods must be scaled up through pilot plant to full manufacturing scale. While production methods are being developed, the enzyme must also be characterized as to its specificity, mode of action, pH and temperature optima, and the like, and best procedures for its use investigated and found. After all these have been accomplished, it then becomes possible for the manufacturer to produce, market and recommend the new enzyme product for its appropriate applications. The book by Reed is an excellent general reference treating both current and prospective applications for enzymes in food processing.[139]

Immobilized Enzymes   One of the most important developments in enzymology within the past few years has been the investigation of methods for producing and using immobilized water-insoluble enzyme systems. The literature has become so extensive that references are given here for only two review articles from which individual references may be obtained.[59a, 159a]

In spite of the inherent advantages of the use of enzyme catalysts, a number of potentially useful industrial applications have not developed because (1) the cost of enzyme isolation and purification is high, making the process uneconomical; (2) the isolated enzymes are too unstable when removed from the living cell; and (3) it is difficult and costly to recover or to remove the soluble enzymes from the reaction mixtures after completion of the enzymatic conversions. Enzymes of enhanced stability immobilized on water-insoluble carriers, which permit easy recovery and repeated reuse, should significantly reduce or eliminate these problems.

Four principal methods have been used for preparing water-insoluble immobilized enzyme preparations. (1) Adsorption on inert carriers or synthetic ion exchange resins. This has inherent limitations because the adsorbed enzyme is weakly bound and is rather easily lost during use. However, it has been successfully applied, for example in resolving DL-amino acid mixtures by acylase. (2) Entrapping enzyme in gel lattices, the pores of which are too small to allow the enzyme to diffuse but large enough to allow the passage of substrate and product. This method has been used mainly for analytical purposes, and in the form of microencapsulated products might have medical significance where attention must be paid to immunological response to enzyme foreign protein or where the enzyme is very fragile and unstable. (3) Covalent binding to a wettable but water-insoluble carrier via functional groups non-essential for their biological activity. (4) Covalent cross-linking of the enzyme protein by an appropriate bifunctional reagent. The covalent linking techniques have been most studied and seem to have greatest potential.

The inert carrier materials used in immobilizing enzymes include glass beads, diazotized cellulose particles, polyaminostyrene beads, polyacrylamide, and other synthetic polymers or copolymers. Various physical properties, such as mechanical stability, hydrophilic or hydrophobic nature, swelling characteristics, as well as electric charge, are considered in the selection of suitable carriers for immobilization of specific enzymes for particular applications.

Goldstein[59a] has briefly considered some of the important factors relating to the properties of insolubilized enzymes, including stability, kinetic behavior, and the physical state such as beads, membranes, sheets, etc. The insolubilized enzymes have advantages over soluble ones: (1) they are chemically and thermally more stable; (2) they are easier to recover and reuse; and (3) they provide a choice of supporting materials that can be tailored to enhance the specific reaction and reactor design. Factors adversely affecting use of immobilized enzymes include: (1) loss of enzyme from the support; (2) slower reaction rate; (3) blocking of enzyme sites and support structure by chemical reaction, by air, or by contaminating solid materials.

Already considerable use has been made on the laboratory scale of immobilized enzymes. Examples are for studying protein structure by obtaining well defined fragments of complex proteins such as immunoglobulins, for isolation of enzyme inhibitors such as trypsin inhibitors from crude animal-organ extracts, for continuous separation of synthetic DL-amino acid mixtures by columns of immobilized aminoacylase, for preparation of polynucleotides by columns of immobilized polynucleotide phosphorylase, and immobilized enzyme columns for analytical applications.

No actual large-scale uses of water-insoluble enzyme derivatives have been reported in the literature. Several have been suggested, including (1) production of dextrose from starch by immobilized α-amylase and glucoamylase, (2) clarification of apple and other fruit juices by immobilized pectinases, (3) chill-proofing of beer using insoluble papain, (4) use of immobilized lactase to hydrolyze lactose in milk and whey, and (5) use of immobilized glucose isomerase to convert glucose to fructose. Enzymatic dextrose production, fruit juice clarification, enzymatic chill-proofing of beer, and glucose isomerization by means of soluble enzymes are currently large-scale commercial operations. Substitution of immobilized enzymes could decrease the cost of enzyme treatment, and simplify removal of the enzymes that is necessary for efficiency or for color and stability in dextrose crystallization, and in fruit juices, beer, and glucose-fructose syrups. The suggested use of *Streptomyces* cells containing glucose isomerase fixed by heat (mentioned previously in this chapter) is actually an example of a possible application of an immobilized enzyme preparation. The use of immobilized lactase would seem to have great large-scale industrial potential for conversion of lactose in milk to reduce human rejection of milk due to lactose intolerance and for conversion of lactose in whey to increase utilization of this nutritious by-product. Only about one-third of the 22 billion pounds of fluid whey produced annually is used in food or animal feeds. Lactose hydrolysis would permit use of higher amounts of whey, with its protein of high biological value, in foods and feeds; the hydrolyzed lactose itself as a cheap by-product could find greater use in foods as a sweetener.

References

1. Adams, E. C., Mast, R. L., and Free, A. H., 1960, Specificity of Glucose Oxidase. *Arch. Biochem. Biophys.*, 91: 230–234.
2. Aitken, H. C., 1961, Apple Juice. *Fruit and Vegetable Juice Processing*, D. K. Tressler and M. A. Joslyn, eds., pp. 619–700, Avi Publ., Westport, Conn.
3. Albersheim, P., and Killias, V., 1962, Studies Relating to the Purification and Properties of Pectin Transeliminase. *Arch. Biochem. Biophys.*, 97: 107–115.
4. Alford, J. A., Pierce, D. A., and Suggs, F. G., 1964, Activity of Microbial Lipases on Natural Fats and Synthetic Triglycerides. *J. Lipid Res.*, 5: 390–394.
5. Amos, J. A., 1955, The Use of Enzymes in the Baking Industry. *J. Sci. Food Agr.*, 6: 489–495.
6. André, E., 1964, Sur la lipoxidase des graines de soja. Etat actuel de nos connaissances. *Oléagineux*, 19: 461–463.
7. Anon., 1962, Proteolytic Activity of Flour. *Cereal Laboratory Methods*, 7th ed., Method 22–60, Am. Assoc. Cereal Chem., St. Paul, Minn.
8. Anon., 1964, Enzyme Interleaving Sheets. *Food Process.*, 25: (Nov.) 84–85.
9. Anon., 1964, Pretenderizers Sizzle. *Chem. Week*, 98: (Jan. 18) 33–34.
10. Anon., 1965, *Enzyme Nomenclature. Recommendations 1964 of International Union of Biochemistry*, Elsevier, Amsterdam.
11. Arima, K., and Iwasaki, S., 1964, *Milk Coagulating Enzyme "Microbial Rennet" and Method of Preparation Thereof*. U.S. Patent 3,151,039.
12. Arima, K., Yamasaki, M., and Ysai, T., 1964, Studies on Pectic Enzymes of Microorganisms. I. Isolation of Microorganisms Which Specifically Produce One of Several Pectic Enzymes. *Agr. Biol. Chem.* (Japan), 28: 248–254.
13. Armbruster, F. C., 1961, *Enzyme Preparation*. U.S. Patent, 3,012,944.
14. Ashby, J., Brooks, J., and Reid, W. W., 1955, Preparation of Pure Mono-, Di,- and Trigalacturonic Acids. *Chem. & Ind.* (London), p. 360.
15. Bailey, S. D., Bazinet, M. L., Driscoll, J. L., and McCarthy, A. I., 1961, The Volatile Sulfur Compounds of Cabbage. *J. Food Sci.*, 26: 163–170.
16. Baldwin, R. R., Campbell, H. A., Thiessen, R., and Lorant, G. J., 1953, The Use of Glucose Oxidase in the Processing of Foods with Special Emphasis on the Desugaring of Egg White. *Food Technol.*, 7: 275–282.
17. Barton, R. R., 1966, A Specific Method for Quantitative Determination of Glucose. *Anal. Biochem.*, 14: 258–260.
18. Beck, H., Johnson, J. A., and Miller, B. S., 1957, Studies on the Soluble Dextrin Fraction and Sugar Content of Bread Baked with Alpha-Amylase from Different Sources. *Cereal Chem.*, 34: 211–217.
19. Beckhorn, E. J., Labbee, M. D., and Underkofler, L. A., 1965, Production and Use of Microbial Enzymes for Food Processing. *J. Agr. Food Chem.*, 13: 30–34.
19a. Bergemeyer, H. U., 1965, *Methods of Enzymatic Analysis*, Academic Press, New York.
20. Beuk, J. F., Savich, A. L., and Goeser, P. A., 1959, *Method of Tenderizing Meat*. U.S. Patent 2,903,362.
21. Blakemore, S. M., 1962, *Puree and Method of Making Same*. U.S. Patent 3,031,307.
22. Blish, M. J., 1953, Protein Hydrolyzates. In Kirk-Othmer *Encyclopedia of Chemical Technology*, Vol. 11, pp. 212–220. Interscience, New York.
23. Bloom, J., Scofield, G., and Scott, D., 1956, Oxygen Removal Prevents Rancidity in Mayonnaise. *Food Packer* 37: (13) 16–17.
24. Blouin, J., and Barthe, J. C., 1963, Utilization pratique d'enzymes pectolytiques en oenologie. *Ind. Aliment. Agr.* (Paris), 80: 1169–1179.

25. Bode, H. E., 1966, *In Situ Dextrose Production in Crude Amylaceous Materials*. U.S. Patent 3,249,512.
26. Bode, H. E., 1966, *Production and Use of Amyloglucosidase*. U.S. Patent 3,249,514.
27. Bolaski, W., and Gallatin, J. C., 1962, *Enzymic Conversion of Cellulosic Fibers*. U.S. Patent 3,041,246.
28. Celmer, R. F., 1961, Continuous Fruit Juice Production. *Fruit and Vegetable Juice Processing*, D. K. Tressler and M. A. Joslyn, eds., pp. 254–277, Avi Publ., Westport, Conn.
29. Chance, B., 1943, The Kinetics of the Enzyme Substrate Compound of Peroxidase. *J. Biol. Chem.*, 151: 553–577.
29a. Charles, R. L., Gertzman, D. P., and Melachouris, N., 1970, *Milk-Clotting Enzyme Product and Process Therefor*. U.S. Patent 3,549,390.
30. Cochrane, A. L., 1961, Invertase: Its Manufacture and Uses. *Soc. Chem. Ind.* (London) Monograph, 11: 25–31.
31. Coleman, M. H., 1963, Rapid Lipase Purification. *Biochim. Biophys. Acta*, 67: 146–147.
32. Coles, D., 1952, Fungal Enzymes in Bread Baking. *Proc. 28th Ann. Meeting, Am. Soc. Bakery Engrs., Chicago, 1952*, pp. 49–53.
33. Connelly, J. J., Vely, V. G., Mink, W. H., Sachsel, G. F., and Litchfield, J. H., 1966, Studies on Improved Recovery of Protein from Rendering Plant Materials and Products. III. Pilot Plant Studies on the Enzyme Hydrolysis Process. *Food Technol.*, 20: 829–834.
34. Cook, A. H., ed., 1962, *Barley and Malt*, Academic Press, New York.
35. Coppock, J. B. M., and Daniels, N. W. R., 1962, Wheat Flour Lipids, Their Role in Breadmaking and Nutrition. *Recent Advances in Processing Cereals, Soc. Chem. Indus.* (London), Monograph 16, pp. 113–127.
36. Cotterill, O. J., 1963, Effect of pH and Lipase Treatment on Yolk-Contaminated Egg White. *Food Technol.*, 17: 103–108.
37. Criswell, L. G., Litchfield, J. H., Vely, V. G., and Sachsel, G. F., 1964, Studies on Improved Recovery of Protein from Rendering Plant Materials and Products. II. Acid and Enzyme Hydrolysis. *Food Technol.*, 18: 1493–1497.
38. Croxall, W. J., 1966, *Process for Removing Transglucosidase from Amyloglucosidase*. U.S. Patent 3,254,003.
39. Cruess, W. V., Quacchia, R., and Ericson, K., 1955, Pectic Enzymes in Winemaking. *Food Technol.*, 9: 601–607.
40. Dalby, G., 1960, The Role and Importance of Enzymes in Commercial Bread Production. *Cereal Sci. Today*, 5: 270–272.
41. Dale, J. K., and Langlois, D. P., 1940, *Syrup and Method of Making Same*. U.S. Patent 2,201,609.
42. Dateo, G. P., Clapp, R. C., Mackay, D. A. M., Hewitt, E. J., and Hasselstrom, T., 1957, Identification of the Volatile Sulfur Components of Cooked Cabbage and the Nature of the Precursors in Fresh Vegetables. *Food Res.*, 22: 440–447.
43. deBecze, G. I., 1965, Enzymes, Industrial. In Kirk-Othmer *Encyclopedia of Chemical Technology*, 2nd ed., Vol. 8, 173–230, John Wiley & Sons, New York.
43a. deBecze, G. I., 1970, Food Enzymes. *CRC Critical Reviews in Food Technology*, 479–518.
44. Demain, A. L., and Phaff, H. J., 1957, Recent Advances in the Enzymatic Hydrolysis of Pectic Substances. *Wallerstein Lab. Commun.*, 20: 119–140.
45. Denault, L. J., and Underkofler, L. A., 1963, Conversion of Starch by Microbial Enzymes for Production of Syrups and Sugars. *Cereal Chem.*, 40: 618–629.
46. Dennis, G. E., and Quittenton, R. C., 1962, *Enzymes in Brewing*. Canada Patent 634,865.
47. Desnuelle, P., and Savary, P., 1963, Specificity of Lipases. *J. Lipid Res.*, 4: 369–384.
48. Dixon, M., and Webb, E. C., 1964, *Enzymes*, 2nd ed., Academic Press, New York.
49. El-Gharbawi, M., and Whitaker, J. R., 1963, Factors Affecting Enzymic Solubilization of Beef Protein. *J. Food Sci.*, 28: 168–172.
50. Endo, A., 1963, Pectic Enzymes of Molds. V. and VI. The Fractionation of Pectolytic Enzymes of *Coniothyrium diplodiella*. *Agr. Biol. Chem.* (Japan), 27: 741–750; 751–757.
51. Endo, A., 1964, Pectic Enzymes of Molds. VII. Turbidity of Apple Juice Clarification and Its Application to Determination of Enzymatic Activity. *Agr. Biol. Chem.* (Japan), 28: 234–238.

52. Endo, A., 1964, Pectic Enzymes of Molds. VIII., IX., X. Purification and Properties of Endo-Polygalacturonase I, II, III. *Agr. Biol. Chem.* (Japan), 28: 535–542; 543–550; 551–558.

53. Endo, A., 1964, Pectic Enzymes of Molds. XI. Purification and Properties of Exo-Polygalacturonase. *Agr. Biol. Chem.* (Japan) 28: 639–645. XII. Purification and Properties of Pectinesterase. *Ibid.*, 757–764.

54. Endo, A., 1965, Pectic Enzymes of Molds. XIII. Clarification of Apple Juice by the Joint Action of Purified Pectolytic Enzymes. *Agr. Biol. Chem.* (Japan), 29: 129–136. XIV. Properties of Pectin in Apple Juice. *Ibid.*, 137–143. XV. Effects of pH and Some Chemical Agents on the Clarification of Apple Juice. *Ibid.*, 222–228. XVI. Mechanism of Enzymatic Clarification of Apple Juice. *Ibid.*, 229–233.

55. Erenthal, I., and Block, G. J., 1962, *High Fermentable Noncrystallizing Syrup and the Process of Making Same.* U.S. Patent 3,067,066.

56. Farnham, M. G., 1950, *Cheese-Modifying Enzyme Product.* U.S. Patent 2,531,329.

57. Fuleki, T., and Hope, G. W., 1964, Effect of Various Treatments on Yield and Composition of Blueberry Juice. *Food Technol.*, 18: 568–570.

58. Gibson, Q. H., Swoboda, B. E. P., and Massey, V., 1964, Kinetics and Mechanism of Action of Glucose Oxidase. *J. Biol. Chem.*, 239: 3927–3934.

59. Goeser, P. A., 1961, Tenderized Meat through Ante-Mortem Vascular Injection of Proteolytic Enzymes. *Proc. Res. Conf. Res. Advisory Council Am. Meat Inst. Found. Univ. Chicago*, 13: 55–58.

59a. Goldstein, L., 1969, Use of Water-Insoluble Enzyme Derivatives in Synthesis and Separation. *Fermentation Advances*, D. Perlman, ed., pp. 391–424, Academic Press, New York.

60. Gross, H., Bell, R. L., and Redfern, S., 1966, Use of Fungal Enzymes with Flour Brews. *Cereal Sci. Today*, 11: 419–423.

61. Halliwell, G., 1965, Hydrolysis of Fibrous Cotton and Reprecipitated Cellulose by Celluloytic Enzymes from Soil Microorganisms. *Biochem. J.*, 95: 270–281.

62. Halliwell, G., and Bryant, M. P., 1963, The Cellulolytic Activity of Pure Strains of Bacteria from the Rumen of Cattle. *J. Gen. Microbiol.*, 32: 441–448.

63. Hansen, V., 1964, Pilot-Versuche zur Dextroseherstellung. *Stärke*, 16: 258–263.

64. Hanson, A. M., Bailey, T. A., Malzahn, R. C., and Corman, J., 1955, Plant Scale Evaluation of Fungal Amylase Process for Grain Alcohol. *J. Agr. Food Chem.*, 3: 866–872.

65. Harper, W. J., 1957, Lipase Systems Used in the Manufacture of Italian Cheese. II. Selective Hydrolysis. *J. Dairy Sci.*, 40: 556–563.

66. Harper, W. J., and Gould, I. A., 1955, Lipase Systems Used in the Manufacture of Italian Cheese. I. General Characteristics. *J. Dairy Sci.*, 38: 87–95.

67. Hasselstrom, T., Bailey, S. D., and Reese, E. T., 1962, Regeneration of Flavors Through Enzymatic Action. *U.S. Dept. Comm., Office Tech. Serv.*, AD-286,640, pp. 285–294.

68. Hewitt, E. J., 1963, Flavor Enhancement Review. Enzymatic Enhancement of Flavor. *J. Agr. Food Chem.*, 11: 14–19.

69. Hewitt, E. J., Hasselstrom, T., Mackay, D. A. M., and Konigsbacher, K. S., 1960, *Natural Flavors of Processed Foods.* U.S. Patent 2,924,521.

70. Huffman, D. L., Palmer, A. Z., Carpenter, J. W., and Shirley, R. L., 1961, The Effect of Ante-Mortem Injection of Papain on Tenderness of Chickens. *Poultry Sci.*, 40: 1627–1630.

71. Hultin, H. O., and Proctor, B. E., 1962, Banana Aroma Precursors. *Food Technol.*, 16: 108–113.

72. Hurst, T. L., and Turner, A. W., 1962, *Method of Refining Amyloglucosidase.* U.S. Patent 3,047,471.

73. Hurst, T. L., and Turner, A. W., 1962, *Method of Refining Amyloglucosidase.* U.S. Patent 3,067,108.

74. Hurst, T. L., and Turner, A. W., 1964, *Method of Refining Amyloglucosidase.* U.S. Patent 3,117,063.

75. Ingleton, J. F., 1963, Use of Invertase in the Confectionary Industry. *Confectionary Prod.*, 29: 773–774 and 776–777.

76. Janssen, F., 1960, Invertase and Cast Cream Centers. *M. Confectioner*, 40: (4) 41–44 and 56–57.

77. Janssen, F., 1961, Factors Affecting Invertase Action in Cast Cream Centers. *M. Confectioner*, 41:(8) 56–58 and 60.

78. Janssen, F., 1962, Composition and Production of Cordial Cherries. *Confectionary Prod.*, 28: 54, 56, 58, 60 and 75.

79. Janssen, F., 1963, Invertase. An Important Ingredient for Cream Centers. *Candy Ind. Confectioners J.*, 121: (6) 30 and 33.

80. Jensen, R. G., Sampugna, J., Parry, R. M., and Shahani, K. M., 1963, Lipolysis of Laurate Glycerides by Pancreatic and Milk Lipase. *J. Dairy Sci.*, 46: 907–910.

81. Johnson, J. A., 1965, Enzymes in Wheat Technology in Retrospect. *Cereal Sci. Today*, 10: 315–319.

82. Johnson, J. A., and Miller, B. S., 1949, Studies on the Role of Alpha-Amylase and Protease in Breadmaking. *Cereal Chem.*, 26: 371–383.

83. Joslyn, M. A., Mist, S., and Lambert, E., 1952, The Clarification of Apple Juice by Fungal Pectic Enzyme Preparations. *Food Technol.*, 6: 133–139.

84. Kathrein, H. R., 1963, *Treatment and Use of Enzymes for the Hydrolysis of Starch.* U.S. Patent 3,108,928.

85. Kay, H. D., and Graham, W. R., 1935, The Phosphatase Test for Pasteurized Milk. *J. Dairy Res.*, 6: 191–203.

86. Keilin, D., and Hartree, E. F., 1948, Properties of Glucose Oxidase (Notatin). *Biochem. J.*, 42: 221–229.

87. Keilin, D., and Hartree, E. F., 1948, The Use of Glucose Oxidase (Notatin) for the Determination of Glucose in Biological Material and for the Study of Glucose-Producing Systems by Manometric Methods. *Biochem. J.*, 42: 230–238.

88. Keilin, D., and Hartree, E. F., 1952, Specificity of Glucose Oxidase (Notatin). *Biochem. J.*, 50: 331–341.

89. Kertesz, Z. I., 1951, *The Pectic Substances*, Interscience, New York.

90. King, K. W., 1965, Enzymatic Attack on Highly Crystalline Hydrocellulose. *J. Ferment Technol.* (Japan), 43: 79–92.

91. King, K. W., and Smibert, R. M., 1963, Distinctive Properties of β-Glucosidases and Related Enzymes Derived from a Commercial *Aspergillus niger* Cellulase. *Applied Microbiol.*, 11: 315–319.

92. Kleinschmidt, A. W., Higashiuchi, K., Anderson, R., and Ferrari, C. G., 1963, Soya Lipoxidase as a Means of Flavor Improvement. *Baker's Dig.*, 37: (5) 44–47.

93. Klis, J. B., 1962, Heat Stable Enzyme. *Food Process.*, 23: (June) 70–71.

94. Kneen, E., and Hads, H. L., 1945, Effects of Variety and Environment on the Amylases of Germinated Wheat and Barley. *Cereal Chem.*, 22: 407–418.

95. Konigsbacher, K. S., and Donworth, M. E., 1965, Beverage Flavors. *Advances in Chemistry* Monograph Series, Am. Chem. Soc., Washington, D.C.

96. Kooi, E. R., Harjes, C. F., and Gilkison, J. S., 1962, *Treatment and Use of Enzymes for the Hydrolysis of Starch.* U.S. Patent 3,042,584.

97. Kulp, K., and Bechtel, W. G., 1963, Effect of Water-Insoluble Pentosan Fraction of Wheat Endosperm on the Quality of White Bread. *Cereal Chem.*, 40: 493–504.

98. Kulp, K., and Bechtel, W. G., 1963, The Effect of Tailings of Wheat Flour and Its Subfractions on the Quality of Bread. *Cereal Chem.*, 40: 665–675.

99. Kuninaka, A., Kibi, M., and Sakaguchi, K., 1964, History and Development of Flavor Nucleotides. *Food Technol.*, 18: 287–293.

100. Kurtz, G. W., and Yonezawa, Y., 1957, The Glucose Oxidase Catalase System as an Oxygen Scavenger for Hermetically Sealed Containers. Institute of Food Technologists, 11th Ann. Mtg., Pittsburgh. *Food Technol.*, 11: (4) abstr. p. 16.

101. Kusai, K., Sekuzu, I., Hagihara, B., Okunuki, K., Yamauchi, S., and Nakai, M., 1960, Crystallization of Glucose Oxidase from *Penicillium amagasakiense*. *Biochim. Biophys. Acta* 40: 555–557.

102. Langlois, D. P., 1953, Application of Enzymes to Corn Syrup Production. *Food Technol.*, 7: 303–307.

103. Li, L., and King, K. W., 1963, Fractionation of β-Glucosidases and Related Extracellular Enzymes from *Aspergillus niger*. *Applied Microbiol.*, 11: 320–325.

104. Li, L., Flora, R. M., and King, K. W., 1963, Purification of β-Glucosidases from *Aspergillus niger* and Initial Observations on the C_1 of *Trichoderma koningii*. *J. Ferment. Technol.* (Japan), 41: 98–104.

105. Long, J. E., and Harper, W. J., 1956, Italian Cheese Ripening. VI. Effect of Different Types of Lipolytic Enzyme Preparations on the Accumulation of Various Free Fatty and Free Amino Acids and the Development of Flavor in Provolone and Romano Cheese. *J. Dairy Sci.*, 39: 245–252.

106. Lumino Feed Co., 1964. *Fish Meal*. German Patent 1,171,248.

107. Mackay, D. A. M., and Hewitt, E. J., 1959, Application of Flavor Enzymes to Processed Foods. II. Comparison of the Effect of Flavor Enzymes from Mustard and Cabbage on Dehydrated Cabbage. *Food Res.*, 24: 253–261.

107a. Marshall, R. O., 1960, *Enzymatic Process*. U.S. Patent 2,950,228.

107b. Marshall, R. O., and Kooi, E. R., 1957, Enzymatic Conversion of D-Glucose to D-Fructose. *Science*, 125: 648–649.

108. McCready, R. M., and Gee, M., 1960, Determination of Pectic Substances by Paper Chromatography. *J. Agr. Food. Chem.*, 8: 510–513.

109. McIntosh, E. N., and Carlin, A. F., 1963, The Effect of Papain Preparations on Beef Skeletal Muscle Proteins. *J. Food Sci.*, 28: 283–285.

110. Melachouris, N. P., and Tuckey, S. L., 1964, Comparison of the Proteolysis Produced by Rennet Extract and Pepsin Preparation Metroclot during Ripening of Cheddar Cheese. *J. Dairy Sci.*, 47: 1–7.

111. Mesnard, P., Devaux, G., Monnier, M., and Fraux, J. L., 1963, Action de la cellulase sur quelques poudres végétales lyophilisées. *Prod. Probl. Pharm.*, 18: 628–629.

112. Meyer, R. I., Jokay, L., and Sudek, R. E., 1960, The Effect of an Oxygen Scavenger Packet, Desiccant in Packet System, on the Stability of Dry Whole Milk and Dry Ice Cream Mix. *J. Dairy Sci.*, 43: 844.

113. Michaelis, L., and Menton, M. L., 1913, Die Kinetik der Invertinwirkung. *Biochem. Z.*, 49: 333–369.

114. Miller, B. S., and Johnson, J. A., 1949, Differential Stability of α-Amylase and Proteinase. *Cereal Chem.*, 26: 359–371.

115. Miller, B. S., Johnson, J. A., and Palmer, D. L., 1953, A Comparison of Cereal, Fungal, and Bacterial α-Amylases as Supplements for Bread Baking. *Food Technol.*, 7: 38–42.

116. Morgan, A. I., and Schwimmer, S., 1965, *Preparations of Dehydrated Food Products*. U.S. Patent 3,170,803.

117. Morton, B. J., Martin, E. G., Dahlstrom, R. V., and Sfat, M. R., 1962, Some Aspects of Beer Colloidal Instability. *Am. Soc. Brewing Chemists Proc.*, pp. 30–39.

118. Neubeck, C. E., 1959, Pectic Enzymes in Fruit Juice Technology. *J. Assoc. Off. Agr. Chemists*, 42: 374–382.

119. Neuberg, C., and Roberts, I. S., 1946, *Invertase Monograph*. Sugar Research Fdn., New York.

120. Neukom, H., 1969, Pectin-Cleaving Enzymes. *Z. Ernährungswiss.*, Suppl. 8: 91–96.

121. Notter, G. K., Brekke, J. E., and Taylor, D. H., 1959, Factors Affecting Behavior of Fruit and Vegetable Juices during Vacuum Puff Drying. *Food Technol.*, 13: 341–345.

122. Ogata, K., Nakao, Y., Igarasi, S., Omura, E., Sugino, Y., Yoneda, M., and Subara, I., 1963, Degradation of Nucleic Acids and Their Related Compounds by Microbial Enzymes. I. On the Distribution of Extracellular Enzymes Capable of Degrading Ribonucleic Acid into 5'-Mononucleotides in Microorganisms. *Agr. Biol. Chem.* (Japan), 27: 110–115

123. Ohlmeyer, D. W., 1957, Use of Glucose Oxidase to Stabilize Beer. *Food Technol.*, 11: 503–507.

124. Ough, C. S., 1960, Die Verwendung von Glucose Oxidase in trockenem Weiszwein. *Rebe Wein*, 10: 14–23.

125. Pal, P. N., and Basu, S. N., 1961, Properties of Some Fungal Cellulases with Particular Reference to Their Inhibition. *J. Sci. Ind. Res.* (India), C20: 336–338.

126. Pazur, J. H., and Ando, T., 1959, The Action of an Amyloglucosidase of *Aspergillus niger* on Starch and Maltooligosaccharides. *J. Biol. Chem.*, 234: 1966–1970.

127. Pazur, J. H., and Ando, T., 1960, The Hydrolysis of Glucosyl Oligosaccharides with α-D-$(1\rightarrow4)$ and α-D-$(1\rightarrow6)$ Bonds by Fungal Amyloglucosidase. *J. Biol. Chem.*, 235: 297–302.

128. Pazur, J. H., and Kleppe, K., 1962, The Hydrolysis of a α-D-Glucosides by Amyloglucosidase from *Aspergillus niger*. *J. Biol. Chem.*, 237: 1002–1006.

129. Pazur, J. H., and Kleppe, K., 1964, Oxidation of Glucose and Related Compounds by Glucose Oxidase from *Aspergillus niger*. *Biochemistry*, 3: 578–583.

130. Pederson, C. S., 1961, Grape Juice. *Fruit and Vegetable Juice Processing*, D. K. Tressler and M. A. Joslyn, eds., pp. 787–815, Avi Publ., Westport, Connecticut.

131. Pence, J. J., Nimmo, C. C., and Hepburn, F. N., 1964, Protein. *Wheat Chemistry and Technology*, I. Hlynka, ed., pp. 227–276, *Am. Assoc. Cereal Chemists*, St. Paul, Minnesota.

132. Platt, W. C., and Posten, A. L., 1962, *Method for Recovering Citrus Oil*. U.S. Patent 3,058,887.

133. Pomeranz, Y., 1964, Lactase (β-D-Galactosidase). I. Occurrence and Properties. *Food Technol.*, 18: 682–687. II. Possibilities in the Food Industries. *Food Technol.*, 18: 690–697.

134. Pomeranz, Y., and Miller, B. S., 1963, Evaluation of Lactase Preparations for Use in Breadmaking. *J. Agr. Food Chem.*, 11: 19–22.

135. Pomeranz, Y., Miller, B. S., Miller, D., and Johnson, J. A., 1962, Use of Lactase in Breadmaking. *Cereal Chem.*, 39: 398–406.

136. Preece, I. A., 1960, *Malting, Brewing and Allied Processes. A Literature Survey*. Heffer, Cambridge.

137. Prescott, S. C., and Dunn, C. G., 1959, *Industrial Microbiology*, 3rd ed., McGraw-Hill, New York.

137a. Prins, J., and Nielsen, T. K., 1970, Microbial Rennet. *Process Biochem.*, 5: (5) 34–35.

138. Ramamurti, K., and Johar, D. S., 1963, Enzymic Digestion of Fiber in Coconut Cake. *Nature*, 198: 481–482.

139. Reed, G., 1966, *Enzymes in Food Processing*. Academic Press, New York.

140. Reed, G., and Thorn, J. A., 1957, Use of Fungal Protease in the Baking Industry. *Cereal Sci. Today*, 2: 280–283.

141. Reese, E. T., ed., 1963, *Advances in Enzymatic Hydrolysis of Cellulose and Related Materials*. Pergamon Press, New York.

142. Reich, I. M., Redfern, S., Lenney, J. F., and Schimmel, W. W., 1957, *Prevention of Gel in Frozen Coffee Extract*. U.S. Patent 2,801,920.

143. Reinke, H. G., Hoag, L. E., and Kincaid, C. M., 1963. Effect of Antioxidants and Oxygen Scavengers on the Shelf Life of Canned Beer. *Am. Soc. Brewing Chemists Proc.*, pp. 175–180.

144. Rohm, O., 1908, *The Preparation of Hides for the Manufacture of Leather*. U.S. Patent 886,411.

145. Rosell, J. M., 1961, Hydrogen Peroxide Catalase Method for Treatment of Milk. *Can. Dairy Ice Cream J.*, 40: (8) 50–52.

146. Roundy, Z. D., 1958, Treatment of Milk for Cheese with Hydrogen Peroxide. *J. Dairy Sci.*, 41: 1460–1465.

147. Roundy, Z. D., 1961, Peroxide Catalase Method of Making Swiss Cheese. *Milk Prod. J.*, 52: (7) 12, 14.

148. Rubenthaler, G., Finney, K. F., and Pomeranz, Y., 1965, Effects of Loaf Volume and Bread Characteristics of α-Amylases from Cereal, Fungal and Bacterial Sources. *Food Technol.*, 19: 239–241.

149. Sardinas, J. L., 1966, *Milk-Curdling Enzyme Elaborated by Endothia parasitica*. U.S. Patent 3,275,453.

149a. Sardinas, J. L., 1969, New Sources of Rennet. *Process Biochem.*, 4: (7) 13–16, 21.

150. Schleich, H., and Arnold, R. S., 1962, *Composition and Methods for Processing Meat Products*. U.S. Patent 3,037,870.

151. Schwimmer, S., 1963, Alteration of the Flavor of Processed Vegetables by Enzyme Preparations. *J. Food Sci.*, 28: 460–466.

152. Schwimmer, S., and Guadagni, D. G., 1962, Relation between Olefactory Threshold Concentration and the Pyruvic Acid Content of Onion Juice. *J. Food Sci.*, 27: 94–97.

153. Schwimmer, S., and Kjaer, A., 1960, Purification and Specificity of the C-S-Lyase of *Albizzia lophanta*. *Biochim. Biophys. Acta*, 42: 316–324.

154. Schwimmer, S., and Weston, W. J., 1961, Enzymatic Development of Pyruvic Acid in Onion as a Measure of Pungency. *J. Agr. Food Chem.*, 9: 301–304.

155. Scott, D., 1953, Glucose Conversion in Preparation of Albumen Solids by Glucose Oxidase Catalase System. *J. Agr. Food Chem.*, 1: 727–730.

156. Scott, D., 1958, Enzymatic Oxygen Removal from Packaged Foods. *Food Technol.*, 12: (7), 7–17.

157. Scott, D., and Hammer, F., 1961, Oxygen Scavenging Packet for In-Package Deoxygenation. *Food Technol.*, 15: 99–104.

158. Scott, D., and Klis, J. B., 1962, Produce Salmonella-Free Yolk and Whole Eggs. *Food Process.*, 23: (Sept.) 76–77.

159. Shimazono, H., 1964, Distribution of 5'-Ribonucleotides in Foods. *Food Technol.*, 18: 303–307.

159a. Silman, I. H., and Katchalski, E., 1966, Water-Insoluble Derivatives of Enzymes, Antigens and Antibodies. *Ann. Rev. Biochem.*, 35: 873–908.

160. Silverstein, O., 1964, Heat Stable Bacterial α-Amylase in Baking. *Baker's Dig.*, 38: (4) 66–72.

161. Simpson, F. J., 1958, *Recovery of Starch.* U.S. Patent 2,821,501.

162. Sosebee, M. E., May, K. N., and Powers, J. J., 1964, The Effects of Enzyme Addition on the Quality of Freeze-Dehydrated Chicken Meat. *Food Technol.*, 18: 551–554.

162a. Sternberg, M. Z., 1971, Crystalline Milk-Clotting Protease from *Mucor miehei* and Some of Its Properties. *J. Dairy Science*, 54: 159–167.

163. Stimpson, E. G., 1954, *Frozen Concentrated Milk Products.* U.S. Patent 2,668,765.

164. Stone, I. M., 1962, *Process of Producing Baked Confections and the Products Resulting Therefrom by α-Amylase.* U.S. Patent 3,026,205.

165. Suzuki, S., 1964, An Overall Look at the Dextrose Industry in Japan. *Stärke*, 16: 285–293.

166. Swoboda, B. E. P., and Massey, V., 1965, Purification and Properties of the Glucose Oxidase from *Aspergillus niger. J. Biol. Chem.*, 240: 2209–2215.

167. Takamine, J., 1894, *Process of Making Diastatic Enzyme.* U.S. Patents 525,823 and 525,820.

167a. Takasaki, Y., Kosugi, Y., and Kanbayashi, A., 1969, *Streptomyces* Glucose Isomerase. *Fermentation Advances*, D. Perlman, ed., pp. 561–589, Academic Press, New York.

168. Tappel, A. L., 1963, Lipoxidase. *The Enzymes*, P. D. Boyer, H. Lardy, and K. Myrbäck, eds., 2nd ed., Vol. 8, pp. 275–283, Academic Press, New York.

169. Tepley, L. J., Derse, P. H., and Price, W. V., 1958, Composition and Nutritive Value of Cheese Produced from Milk Treated with Hydrogen Peroxide and Catalase. *J. Dairy Sci.*, 41: 593–605.

170. Thorne, R. S. W., 1963, The Problem of Beer Haze Assessment. *Wallerstein Lab. Commun.*, 26: 5–19.

171. Toyama, N., 1963, Recent Advances in the Production and Industrial Application of Cellulase in Japan. *Hakko Kyokaishi*, 21: 415–421 and 459–465.

172. Toyama, N., 1965, A Cell-Separating Enzyme as a Complementary Enzyme to Cellulase and Its Application in Processing Vegetables. *Hakko Kogaku Zasshi*, 43: 683–689.

173. Tsugo, T., and Yamauçhi, K., 1960, Comparison of Clotting Action of Various Milk-Coagulating Enzymes. I. Comparison of Factors Affecting Clotting Time of Milk, *15th Intern. Dairy Congr.*, London, 1959, Vol. 2, pp. 636–642. II. Comparison of Curds Coagulated by Various Enzymes. *Ibid.*, 643–647.

173a. Tsumura, N., and Sato, T., 1961, Enzymatic Conversion of D-Glucose to D-Fructose. Part II. Some Properties Concerning Fructose Accumulation Activity of *Aerobacter cloacae*, Strain KN-69. *Agr. Biol. Chem.* (Japan), 25: 616–625.

174. Tumerman, L., Fram, H., and Cornely, K. W., 1954, The Effect of Lactose Crystallization on Protein Stability in Frozen Concentrated Milk. *J. Dairy Sci.*, 37: 830–839.

175. Underkofler, L. A., 1958, Properties and Applications of the Fungal Enzyme Glucose Oxidase. *Proc. Intern. Symp. Enzyme Chem.*, Tokyo, Kyoto, 1957, pp. 486–490, Maruzen, Tokyo.

176. Underkofler, L. A., 1959, *Meat Tenderizer.* U.S. Patent 2,904,442.

177. Underkofler, L. A., 1961, Enzyme Supplementation in Baking. *Baker's Dig.*, 35: (5) 74–78.

178. Underkofler, L. A., 1961, Glucose Oxidase: Production, Properties, and Present and Potential Applications. *Soc. Chem. Ind.* (London), Monograph 11: 72–86.

179. Underkofler, L. A., 1966, Production of Commercial Enzymes. *Enzymes in Food Processing*, G. Reed, ed., pp. 197–200, Academic Press, New York.

180. Underkofler, L. A., 1966, Manufacture and Uses of Industrial Microbial Enzymes. *Chem. Eng. Prog. Symposium Series*, Vol. 62, No. 69, pp. 11–20.

181. Underkofler, L. A., Barton, R. R., and Aldrich, F. L., 1960, Methods of Assay for Microbial Enzymes. *Developments in Industrial Microbiology*, Vol. 2, pp. 171–182, Plenum Press, New York.

182. Underkofler, L. A., Barton, R. R., and Rennert, S. S., 1958, Production of Microbial Enzymes and Their Applications. *Applied Microbiol.*, 6: 212–221.

183. Underkofler, L. A., Denault, L. J., and Hou, E. F., 1965, Enzymes in the Starch Industry. *Stärke*, 17: 179–184.

184. Underkofler, L. A., and Hickey, R. J., eds., 1954, *Industrial Fermentations*, Vol. I, Chemical Publ. Co., New York.

185. Vaupel, E. A., 1958, *Meat Tenderizer*. U.S. Patent 2,825,654.

186. Veringa, H. A., 1961, Rennet Substitutes. *Dairy Sci. Abstr.*, 23: 197–200.

187. Vilenskaya, E. I., 1963, Production of Enzymatically Clarified Juices. *Spirt. Prom.*, 29: (2) 23–26.

188. Vohra, P., and Kratzer, F. H., 1965, Improvement of Guar Meal by Enzymes. *Poultry Sci.*, 44: 1201–1205.

189. Waldt, L. M., 1965, Fungal Enzymes: Their Role in Continuous Process Bread. *Cereal Sci. Today*, 10: 447–450.

190. Walker, L. H., Nimmo, C. C., and Patterson, D. C., 1951, Frozen Apple Juice Concentrate. *Food Technol.*, 5: 148–151.

191. Walker, L. H., and Patterson, D. C., 1954, Preparation of Fresh Italian Prune Juice Concentrates. *Food Technol.*, 8: 208–210.

192. Wallerstein, L., 1911, *Beer and Method of Preparing Same*. U.S. Patent 995,820. *Preparation for Use in Brewing*. U.S. Patent 995,823. *Method of Treating Beer and Ale*. U.S. Patents 995,824, 995,825, 995,826.

193. Wallerstein, L., 1961, Chillproofing and Stabilization of Beer. *Wallerstein Lab. Commun.*, 24: 158–168.

194. Wang, H., Weir, E., Birkner, M., and Ginger, B., 1958, Studies on Enzymatic Tenderization of Meat. III. Histological and Panel Analysis of Enzyme Preparations from Three Distinct Sources. *Food Res.*, 23: 423–438.

195. Weissler, H. E., and Garza, A. C., 1965, Some Physical and Chemical Properties of Commercial Chillproofing Compounds. *Am. Soc. Brewing Chemists Proc.*, pp. 225–238.

196. Weurman, C., 1961, Gas-Liquid Chromatographic Studies on the Enzymatic Formation of Volatile Compounds in Raspberries. *Food Technol.*, 15: 531–536.

197. Wilson, G. D., 1960, Quality Factors in Meat and Meat Foods. *The Science of Meat and Meat Products*, Am. Meat Inst. Fdn., pp. 259–268, Freeman, San Francisco, California.

198. Wüthrich, S., Richterich, R., and Hostettler, H., 1964, Untersuchungen über Milchenzyme. I. Enzyme in Kuhmilch und Frauenmilch. *Z. Lebensm.-Untersuch.-Forsch.*, 124: 336–344.

199. Yagi, K., and Ozawa, T., 1964, Mechanism of Enzyme Action. I. Crystallization of Michaelis Complex of D-Amino Acid Oxidase. *Biochim. Biophys. Acta*, 81: 29–38.

200. Yamamoto, Y., and Yoshitake, M., 1962, Formation of Amino Acids by the Action of Rennet and Lactic Acid Bacteria During the Ripening of Cheese. *Proc. 16th Intern. Dairy Congr.*, Copenhagen, 1961, Vol. B, pp. 395–400, Andelsbogtrykkeriet, Copenhagen.

200a. Yamanaka, K., 1963, Sugar Isomerases. Part II. Purification and Properties of D-Glucose Isomerase from *Lactobacillus brevis*. *Agr. Biol. Chem.* (Japan), 27: 271–278.

201. Yang, H. Y., 1955, Stabilizing Apple Wine with Glucose Oxidase. *Food Res.*, 20: 42–46.

CHAPTER 2

Vitamins and Amino Acids

Benjamin Borenstein, Ph.D.

Manager, Food Industry
Technical Services Department
Roche Chemical Division
Hoffmann-La Roche Inc.
Nutley, New Jersey

Introduction

Vitamins and amino acids are integral components of foods and as such require consideration by the food technologist even when these compounds are not added to foods. Processing and storage of foods can affect these components and the foods containing them nutritionally and organoleptically. A knowledge of the properties of vitamins and amino acids is essential to the understanding of their behavior in foods.

Properties of the Vitamins

To the organic chemist the vitamins are a heterogeneous group of compounds with no common structural attributes (Table 1). To the biochemist

TABLE 1
Primary Compounds with Vitamin Activity

Common Name	Structure	Synonyms	Physiological Forms	Analogs with Vitamin Activity
Thiamine	CH_3 ... $N=C-NH_2$, CH_3C, $C-CH_2-N^+$, $N-CH$ ... $C-CH_2-CH_2OH$, $CH-S$ [CL$^-$]	Vitamin B$_1$ Aneurin Antineuritic factor Antiberiberi factor	Thiamine pyrophosphate (cocarboxylase) Thiamine orthophosphate	Thiamine disulfide Acylated thiamine
Riboflavin	H_2C—C—C—C—C—CH_2OH (OH OH OH), isoalloxazine ring with CH_3, CH_3, $N=C$, $N-H$, $C=O$, $C=O$	Vitamin B$_2$ Vitamin G Lactoflavin	Riboflavin mono- and dinucleotide	7-Methyl-9-(D-1'-ribityl)-isoalloxazine 6-Methyl-9-(D-1'-ribityl) isoalloxazine
Niacin	pyridine ring —COOH	Nicotinic acid P.-P. factor Antipellagra factor	Niacinamide NAD (Co I) NADP (Co II) N'-methylnicotinamide	Esters of niacin
Pantothenic acid	$HOCH_2$—C—CH—CNH—CH_2CH_2COOH (CH_3 OH CH_3 O)	Chick antidermatitis factor	Coenzyme A	Esters Pantothenyl alcohol
Vitamin B$_6$	CH_2OH, CH_2OH, HO, pyridine ring with N, CH_3	Pyridoxine Antiacrodynia factor	Pyridoxal Pyridoxamine Pyridoxal phosphate	

TABLE 1
Primary Compounds with Vitamin Activity (Continued)

Common Name	Structure	Synonyms	Physiological Forms	Analogs with Vitamin Activity
Biotin		Vitamin H Anti-eggwhite injury factor Bios II Coenzyme R	Dethiobiotin	Sulfoxide of biotin Esters
Folic acid		Folacin Pteroylglutamic acid Anti-anemia factor Fermentation L. casei factor	Tri, tetra, hepta conjugates of Pteroyl glutamic acid Tetrahydrofolic acid	Pteroic acid
Vitamin B$_{12}$	C$_{63}$H$_{90}$CoN$_{14}$O$_{14}$P Molecular weight 1357	Cobalamin Cyanocobalamin	Hydroxocobalamin Aquacobalamin	Nitritocobalamin
Ascorbic acid		Vitamin C Antiscorbutic vitamin	Ascorbic acid Dehydroascorbic	
Vitamin A		Retinol Axerophthol	Vitamin A aldehyde Vitamin A acid Vitamin A esters	

TABLE 1
Primary Compounds with Vitamin Activity (Continued)

Common Name	Structure	Synonyms	Physiological Forms	Analogs with Vitamin Activity
Vitamin D	(chemical structure)	Antirachitic vitamin	Vitamin D_2 (Calciferol) (Ergocalciferol) Vitamin D_3 (Cholecalciferol)	Irradiated ergosterol Irradiated 7-dehydro-cholesterol Irradiated 22-dihydro-ergosterol
Vitamin E	(chemical structure)	α-tocopherol Antisterility vitamin	d-α-tocopherol	Esters Racemic tocopherol
Vitamin K	(chemical structure)	Antihemorrhagic vitamin	Vitamin K_1 Vitamin K_2	Menadiol diphosphate Menadione Menadione bisulfite

the vitamins are coenzymes or cofactors necessary to the catalysis of essential biochemical enzymatic reactions. Comprehensive works on the chemistry and biochemistry of the vitamins are *The Vitamins,*[1] *Modern Nutrition in Health and Disease* (reviews the biochemistry and nutrition of vitamins),[2] *Clinical Nutrition,*[3] *Comparative Nutrition of Man and Domestic Animals,*[4] and *Nutrition: A Comprehensive Treatise.*[5] Newer works on the analysis of vitamins are *Vitamin Assay: Tested Methods,*[6] and *Method of Vitamin Assay.*[7]

Water Soluble Vitamins

Vitamin B$_1$ Thiamine is a water-soluble, crystalline compound, stable under normal storage conditions. Its hydrochloride is very soluble in water, forming a strongly acid solution which is fairly stable to oxidation and heat at pH < 5. It is degraded in neutral and alkaline solutions (even at room temperature) and is readily split by sulfite into its pyrimidine and thiazole constituents. Thiamine is absorbed from the small intestine and is phosphorylated in the intestinal mucosa. Although it is widely distributed, there are only a few foods with high vitamin B$_1$ content, i.e., lean pork, peas, beans, whole grains and nuts.

Vitamin B$_2$ Vitamin B$_2$ (riboflavin) is yellow in water with greenish fluorescence. Water solubility is only 10–13 mg/100 ml at 25°C, and it has low solubility in most other solvents. It is stable in acid solution and in the presence of oxidizing agents, but is very sensitive to alkali and light. In neutral solution it is relatively heat stable if protected from light. Vitamin B$_2$ is found in milk, liver, kidney, all meats, leafy green vegetables, tomatoes and yeast. It occurs in tissues as flavin adenine dinucleotide. Vitamin B$_2$ is practically non-toxic; 10 g/kg administered orally to rats resulted in no toxic effects.

Niacin Both niacin (nicotinic acid) and niacinamide are stable to heat and acids and alkali. Niacin is a vasodilator; 50–100 mg produces flushing of ears, face and neck. Niacin is both a carboxylic acid and an amine and reacts accordingly, forming quarternary ammonium salts, esters, and amides. Niacin is non-hygroscopic and is soluble in water and alcohol. Niacinamide is soluble in water, butanol and chloroform and, in contrast to niacin, has a bitter taste.

Nutritionally niacin and tryptophan must be considered simultaneously since tryptophan is a significant precursor of niacin for humans. It is for this reason that high corn diets result in clinical deficiency (pellagra) of niacin, corn being particularly deficient in tryptophan. Sixty mg of tryptophan is approximately equivalent to 1.0 mg of niacin. Horwitt suggested using the term "niacin equivalent" for the total potential niacin value of the diet.[8] The best food sources of niacin are liver, yeast, lean meat, poultry, peanuts and legumes; potatoes and whole wheat are fair. Milk and eggs are not particularly good sources, but both have a high tryptophan content.

Vitamin B$_6$ Vitamin B$_6$ occurs as three active analogs: pyridoxine, pyridoxal and pyridoxamine. Pyridoxine is stable to heat in acidic or alkaline solution. It is light sensitive in neutral and alkaline solution. Vitamin B$_6$ occurs widely in foods—liver, yeast, and egg yolks are particularly good sources. It is synthesized commercially as pyridoxine hydrochloride and it is soluble in water, ethanol and propylene glycol.

The nutritional need for vitamin B_6 is based on its coenzyme function in an unusually large number of metabolic reactions. It is required in the metabolism of amino acids, fats and carbohydrates and for the synthesis of physiologic regulators such as norepinephrine, serotonin and histamine. An excellent overall review of this vitamin is found in the Proceedings of the International Symposium of B_6.[9]

Vitamin B_{12}   With an approximate molecular weight of 1500, the elucidation of this complex structure ranks among the major achievements in chemistry. It occurs in nature, e.g., liver, milk and muscle, as cobalamin, frequently bound to protein. Vitamin B_{12} is produced commercially by fermentation and is readily crystallized as cyanocobalamin, which is the most common form used in human nutrition. Oral absorption of vitamin B_{12} is poor in humans who have pernicious anemia unless "intrinsic factor" is given simultaneously.

The optimum pH for stability of cyanocobalamin in solution is 4.5–5.0.[10] Solutions between pH 4–6 can be autoclaved with minimal losses. Exposure of solutions to light causes degradation. There are many reports in the pharmaceutical literature concerning the adverse effects of ascorbic acid, thiamine and niacinamide on vitamin B_{12} stability in liquid multivitamin preparations.[10,11,12] The degradation induced by vitamin B_1 and niacin occurs only at high concentrations of these vitamins, e.g., 5–10% vitamin B_1 by weight of solution. Riboflavin, pyridoxine and panthenol have no effect on vitamin B_{12} stability. Ferric salts stabilize Vitamin B_{12} in solution, but ferrous salts cause rapid destruction of vitamin B_{12}.[13] Vitamin B_{12} stable, high-concentration multivitamin solutions can be prepared by incorporating iron salts plus EDTA.[14]

Folic Acid   Folic acid or folacin is pteroyl glutamic acid. It is unstable when heated in acid, is light sensitive and is insoluble in most organic solvents. It is slightly soluble in water and methanol. There are several analogs of folic acid found in nature which have folic acid activity for some organisms. It is present in liver, yeast, mushrooms and green leafy vegetables primarily as hepta- and tri-glutamate conjugates and is commercially available as synthetic folic acid. There is some confusion about methods of determining available folate in food for man, partly because the microbiological methods have not been correlated with human availability and conversely, some conjugates are not active for micro-organisms but might be for man.[15]

Pantothenic Acid   Pantothenic acid is an optically active liquid and is stable in cold acid or in cold alkali. Heat in neutral solution is not deleterious. It is synthesized and is available commercially as the calcium salt of the D-isomer or as the D,L racemate. A complex with calcium chloride is available with reputed superior stability in dry products. The alcohol analog of pantothenic acid, i.e., pantothenyl alcohol, is synthesized commercially and has complete biological activity and the advantage of superior stability at pH 3–5 in solutions. Pantothenic acid occurs widely in foods primarily as coenzyme A. The L-isomer has no biological activity. Calcium pantothenate is stable to heat in aqueous solutions at pH 5–7.

Biotin   Biotin is present in liver, kidney, molasses, yeast, milk and egg yolk as well as in nuts, vegetables and grains. The naturally occurring isomer is D-biotin; L-biotin has no biological activity. Biotin is stable to heat and light, but is unstable in strong acids and in alkaline solutions. Optimum stability occurs at pH 5–8.[16] Solubility in water is only 22 mg per 100 ml and is somewhat higher in ethanol. Biotin is insoluble in most organic solvents.

Vitamin C   Ascorbic acid is a white, crystalline substance existing in nature both as the reduced and the oxidized form—dehydroascorbic acid. Biological activity is found in the L-isomer, although D-glucoascorbic and other analogs have slight activity. Ascorbic acid is available commercially as the L-acid and as the sodium salt. It is a moderately strong reducing agent; redox potential +.127v at pH 5. Crystalline ascorbic acid can be stored under normal laboratory conditions for years with little change in activity. Its oxidation is catalyzed by traces of copper and iron. In the absence of oxygen and other oxidizing agents it is heat stable. The copper-catalyzed oxidation of ascorbic acid is a first order reaction in an oxygen saturated solution. Under these conditions stability is superior at pH 3–4.5 compared to pH 6.0–7.0 (Table 2).[17] Degradation of ascorbic acid in concentrated solutions or under anaerobic conditions has lower velocity constants than shown in Table 2 and different pH stability optima.

TABLE 2
Ascorbic Acid Oxidation Kinetics, Cu^{++} 2.8 ppm

Buffer	pH	$k_1 \times 10^4$ (min^{-1}) $25°C$ L-Ascorbic Acid
$H_3PO_4 + Na_2HPO_4$	3.0	120.0
	3.5	250.0
	4.0	530.0
$KH_2PO_4 + Na_2HPO_4$	4.5	955.0
	5.0	1660.0
	5.5	2560.0
	6.0	3340.0
	6.5	2880.0
	7.0	2800.0
	7.5	2860.0
	8.0	2900.0

Source: K. E. Schulte and A. Schillinger, Comparison of the Kinetics of Non-fermentative Oxidation of D-Isoascorbic and L-Ascorbic Acids, *Z. Lebensmittel-Untersuchung*, 94(2):77, 1952. Reprinted by permission.

In propylene glycol, maximum stability is reported to occur at pH 6.0–7.0.[10] Oxidation at alkaline pH is not metal catalyzed and is very rapid. The oxidation to dehydroascorbic acid is reversible; the oxidation of dehydroascorbic acid to 2,3-diketo-L-gluconic acid is not reversible. Plant tissues, milk, and liver are the only foods that have an appreciable ascorbic acid content. Citrus fruits, tomatoes, potatoes, and cabbage are the best dietary sources.

Fat Soluble Vitamins

Vitamin A   In nature vitamin A occurs in several forms—vitamin A alcohol (retinol), vitamin A aldehyde (retinal), vitamin A acid (retinoic acid) and vitamin A palmitate (retinyl palmitate). Vitamin A palmitate predominates since it is the primary liver storage form. Vitamin A_1 in the all *trans* form has a biological activity defined as 0.300 μg = 1 international unit (I.U.).

Cis-trans isomerization of the side chain double bonds of vitamin A occurs readily causing a reduction in potency.

Vitamin A (as distinct from provitamin A carotenoids) is found only in the animal kingdom. Vitamin A_1 occurs in marine fish liver oils and in the liver of most land vertebrates as well as in butter fat and eggs. Vitamin A_2 occurs in fresh water fish oils and contains two double bonds in the ionone ring. The commercial forms of vitamin A are the acetate and palmitate esters. Aqueous emulsions of vitamin A are more efficiently absorbed than are oil solutions. This principle has been used to develop market forms which will be discussed in a later section. Vitamin A oxidizes readily and is light sensitive. A definitive text on vitamin A is Moore's *Vitamin A*.[18]

Provitamin A Carotenoids   Several of the commonly occurring carotenoids exhibit provitamin A activity—α-carotene, β-carotene, cryptoxanthin. The international unit of provitamin A is 0.6 μg β-carotene. Conversion of carotenoids to vitamin A apparently takes place during absorption through the intestinal wall. The widespread occurrence of β-carotene makes it a significant dietary source of vitamin A. In the American diet approximately 60% of the total vitamin A is obtained as provitamin A carotenoids. The carotenoids are sensitive to light, oxidation and heat as are the various forms of vitamin A. Synthetic β-carotene and β-apo-8'-carotenal are available commercially as provitamin A sources. The use of these compounds as colorants is discussed in another chapter of this text and in a recent review.[19]

Vitamin D   The two common compounds of the vitamin D group, D_2 and D_3, are sterols and they are required by most vertebrates having a bony skeleton. Vitamin D may be supplied via the diet or by suitable irradiation of the body. The international unit of vitamin D is represented by 0.025 μg of vitamin D_3. The liver oils of many fishes are abundant sources of vitamin D. Vitamin D_2, calciferol, is the common form used in human nutrition. Both vitamin D_2 and D_3 are available commercially. Vitamin D_2 and D_3 are very soluble in most organic solvents. Crystalline vitamin D_2 shows signs of decomposition when stored for only a few days at room temperature. Storage with inert gas improves stability. Crystalline vitamin D_3 is somewhat more stable than vitamin D_2, possibly because it has one less double bond. Solutions of vitamin D_3 in edible oils or in propylene glycol have good keeping qualities. Because of vitamin D_2's sensitivity to oxygen and light, the U.S.P. allows traces of antioxidants in the crystalline compound.

Vitamin E   Of the eight known tocopherol compounds in nature, several are believed to have some vitamin E activity. Alpha-tocopherol is by far the most potent vitamin E compound and it is now well recognized that "total tocopherol" is not a measure of biological significance. The tocopherols are optically active, liquid alcohols, and only the D-isomers occur in nature. *l*-alpha-Tocopherol has lower vitamin E activity, as shown in Table 3. Alpha- and other tocopherols are antioxidants and are degraded by oxygen and light. The acetate esters prepared by esterifying natural or synthetic tocopherols have excellent stability under almost all conditions, and hence are the preferred commercial form. Vegetable oils are the most important sources of tocopherols, but the total amount of tocopherols and the distribution between the individual tocopherols vary markedly according to species. Wheat germ oil and cottonseed oil are the best sources of vitamin E (alpha-tocopherol). The American dietary supply of vitamin E was recently reviewed.[20]

TABLE 3
Vitamin E Activity of Alpha-Tocopherol Compounds

	International Units/Mg
dl-alpha-tocopheryl acetate	1.0
dl-alpha-tocopherol	1.1
d-alpha-tocopheryl acetate	1.36
d-alpha-tocopherol	1.49
l-alpha-tocopheryl acetate	0.51

Vitamin K   Vitamin K is a group of fat soluble methylnaphthaquinone derivatives necessary for biosynthesis of prothrombin, proconvertin and other factors involved in the blood-clotting system of higher animals. Vitamin K_1, which occurs in green plants, is the predominant natural form associated with foodstuffs. The vitamin K_2 series occurs primarily in bacteria. Vitamin K products are light sensitive.

Several vitamin K derivatives have been synthesized commercially and have vitamin K activity. Menadione and bisulfite derivatives are widely used in animal nutrition. Salts of menadiol diphosphoric acid ester are used in the human field. These compounds are preferred to vitamin K_1 because of their superior stability and, in the case of the diphosphoric acid ester, its water solubility.

Chemistry of Amino Acids

Unlike the vitamins, the amino acids can be readily catalogued chemically. Almost all the amino acids related to foods are alpha amino carboxylic acids. Some are alpha-, epsilon-diamine carboxylic acids; dicarboxylic acids also occur as do sulfur derivatives. The amino acids are white, crystalline compounds and, with the exception of cystine and tyrosine, are water soluble. Only proline is soluble in alcohol and ether. Amino acids are soluble in solutions of strong acid and alkali and are not precipitated by ammonium sulfate or sodium chloride. They form crystalline salts with metallic bases and mineral acids.

To the biochemist the amino acids are structural units for protein synthesis as well as precursors for other amines necessary for metabolism. Amino acids are optically active—except glycine—and only the L-isomers can be incorporated into proteins. Some D-amino acids can be inverted to the L-isomers, e.g., methionine, tryptophan, phenylalanine, histidine and arginine. All proteins synthesized within the body require an amino acid metabolic pool of 20–22 different amino acids. In the adult human, 12–14 of these compounds can be synthesized if sufficient convertible nitrogen is supplied by the diet and if a sufficient concentration of transaminases and other enzymes are present to catalyze synthesis and the interconversion of amino acids. Those amino acids which cannot be synthesized by the animal in question are frequently termed "essential." Human data are still relatively scant compared to animals on the essentiality of amino acids, but lysine, tryptophan, phenylalanine, methionine, threonine, leucine, isoleucine and valine are essential for adults. Infants, in addition, require histidine.

It is difficult to separate the subject of amino acids from that of proteins; the reader is referred to *Processed Plant Protein Foodstuffs*,[21] *Proteins and Their Reactions*[22] and the nutrition texts previously cited for comprehensive treatment of proteins. The major covalent link between amino acids in proteins is the peptide linkage. Lower molecular weight peptide polymers are distinguished from proteins as polypeptides. Free amino acids, the name used for monomer amino acids in this chapter, are of interest because of their formation from peptides and their reactions during food processing as well as their use as food additives. This chapter will briefly discuss the free amino acids found in processed foods and those amino acids of interest as nutritional food additives (Table 4).

TABLE 4
Amino Acids of Particular Importance in Processed Foods

Common Name	Chemical Name
lysine	alpha, epsilon-diaminocaproic acid
tryptophan	alpha amino-3-indolepropionic acid
glutamine	glutamic acid 5-amide
glutamic acid	2-aminopentanedioic acid
methionine	alpha-amino-gamma-methylmercaptobutyric acid
tyrosine	alpha-amino-p-hydroxy-hydrocinnamic acid
threonine	alpha-amino-beta-hydroxybutyric acid
phenylalanine	alpha-amino-beta-phenylpropionic acid
alpha alanine	2-aminopropanoic acid
aspartic acid	aminosuccinic acid
leucine	1-amino-4-methylpentanoic acid
proline	2-pyrrolidinecarboxylic acid
valine	2-amino-3-methylbutanoic acid

Nutrient Losses During Processing and Storage of Foods

Generalizations have been made earlier, in this chapter, concerning the stability of vitamins to heat, light, acid and alkali. This section will discuss the stability patterns of the natural content of vitamins and amino acids in foods as affected by food processing. There are several "caveat lectors" however:

1. The unit operations of food technology have considerable variability in process conditions and the effect of variables is not always predictable with respect to vitamin stability. For example, brine grading of vegetables markedly increases the losses of thiamine and niacin during blanching.[23]

2. The change in varieties of fruits and vegetables grown commercially and the change in growing conditions can be so rapid that unintentional variables are introduced such as pH, metal content, enzyme concentration, etc.

3. A considerable portion of the published data on vitamin stability during canning, blanching, etc., is twenty years old or more and vitamin assay methodology has greatly improved in this interval.

4. Relatively little work has been done to determine velocity constants and kinetics of degradation so that data are difficult to translate to different temperatures.[24]

In general, minerals, carbohydrates, lipids, essential amino acids, vitamins D and K, niacin, riboflavin, pantothenic acid and biotin are stable ($\geq 85\%$ retention) during heat processing and storage of foods. Losses in nutritive value of foods may actually be greater during cooking in institutions and homes, because of leaching of water soluble vitamins, than during commercial processing.

An excellent review of vitamin and amino acid content of wheat and losses during milling is *Wheat Chemistry and Technology*.[25] Little data later than 1940 are available on vitamin stability during storage of fresh foods, fruit maceration and juice filtration; early data are not included in this chapter.

Water Soluble Vitamins

Blanching, Canning and Freezing A comprehensive study of the effects of blanching time, temperature and steam vs. water on ascorbic acid, carotene, riboflavin, niacin, thiamine in peas, spinach, lima beans, green beans is available.[26] High-temperature-short-time water blanching was superior to low-temperature-long-time blanching and steam blanching was superior to water blanching. Ascorbic acid was the most sensitive vitamin retaining 72–93% of the original content under favorable blanching conditions. Thiamine retention was 82–97%, riboflavin 90–100%, carotene 100% and niacin (lima beans only) 75–90%. While losses during blanching are process and food dependent, the above figures can be used as approximate guides. Grading vegetables in brine solution or blanching in brine can markedly affect vitamin losses (see individual vitamins, this section).

A summary of vitamin retention in peas and lima beans after blanching, freezing, processing in tin and in glass, and after cooking is shown in Table 5.[27] These data and the data in Table 6[27] showing storage and cooking losses

TABLE 5
Vitamin Content of Peas and Lima Beans as Affected by Blanching, Processing, and Cooking

	Peas					Lima Beans			
	Percent Retention*					Percent Retention*			
Description of Vegetable	Ascorbic Acid	Thiamine	Riboflavin	Niacin	Carotene	Ascorbic Acid	Thiamine	Riboflavin	Niacin
Fresh, uncooked	100	100	100	100	100	100	100	100	100
Fresh, cooked	81	93	108	93	78	48	89	103	95
Fresh, blanched	67	95	81	90	102	64	68	82	74
Blanched, frozen	55	94	78	76	102	46	62	77	77
Blanched, frozen, cooked	38	63	72	79	103	42	64	88	80
Blanched, processed in tin	47	52	95	70	102	28	42	82	83
Blanched, processed in tin, cooked	27	55	89	70	117	15	33	88	80
Blanched, processed in glass	46	63	85	73	114	33	38	74	77
Blanched, processed in glass, cooked	25	40	84	72	113	21	49	84	71

*Based on the vitamin content of the fresh washed vegetable.
Source: N. B. Guerrant and M. B. O'Hara, *Food Technol.* 7: 473–477, 1953. Reprinted by permission.

TABLE 6

Vitamins Retained by Peas and by Lima Beans, after Blanching, after Freezing, after being Processed in Tin and in Glass, after Storage, and after Cooking. (Retention based on the Vitamin Content of Fresh, Washed Vegetable)

	In the Fresh Vegetable	After Blanching	Frozen				Processed in Tin				Processed in Glass			
			Months in Storage				Months in Storage				Months in Storage			
			0	3	6	12	0	3	6	12	0	3	6	12
Vitamin	Mg/100g	%	%	%	%	%	%	%	%	%	%	%	%	%
Peas—before cooking														
Ascorbic acid	24.40	67	55	48	53	33	47	35	42	36	46	35	34	19
Thiamine	.28	95	94	94	94	79	52	41	38	38	63	38	34	32
Riboflavin	.12	81	78	58	67	72	95	54	61	65	85	65	62	50
Niacin	1.91	90	76	84	87	80	70	73	68	70	73	82	81	70
Carotene	.41	102	102	108	112	107	102	116	109	125	114	126	116	123
Peas—after cooking														
Ascorbic acid	19.90		38	31	32	20	27	21	19	18	25	13	11	6
Thiamine	.26		63	58	58	52	55	41	39	34	40	36	34	30
Riboflavin	.13		72	60	68	72	89	50	65	59	84	64	54	53
Niacin	1.77		79	88	90	81	70	72	69	69	72	82	76	73
Carotene	.32		103	124	115	112	117	108	122	122	113	106	115	123
Lima beans—before cooking														
Ascorbic acid	11.90	64	46	50	45	36	28	28	23	20	33	23	18	6
Thiamine	.27	68	62	59	63	45	42	34	29	19	38	28	26	19
Riboflavin	.07	82	77	80	77	42	82	62	72	42	74	68	65	50
Niacin	1.22	74	77	74	69	57	83	71	64	62	77	75	63	61
Lima beans—after cooking														
Ascorbic acid	5.66		42	23	27	12	15	13	10	8	21	12	8	5
Thiamine	.21		64	63	57	45	33	32	32	16	49	28	24	16
Riboflavin	.08		88	71	71	53	88	63	64	51	84	63	60	50
Niacin	1.16		80	78	70	65	80	78	68	58	71	74	62	57

Source: N. B. Guerrant and M. B. O'Hara, *Food Technol.* 7: 473–477, 1953. Reprinted by permission.

are from a well controlled study in which the vegetables were taken from the same source and prepared in a uniform manner. Heat processing in tin and in glass resulted in losses of ascorbic acid and thiamine. Riboflavin, niacin and carotene were more stable. Freezing caused no changes except for some loss in ascorbic acid. During 12 months storage niacin and carotene were stable, and riboflavin, thiamine and ascorbic acid less stable in that order. Ascorbic acid showed greater losses during storage in glass than in tin. Frozen storage resulted in higher retention of thiamine than did canning. It is obvious that data on fresh vegetables cannot be used as a nutritional guide for stored, processed vegetables.

The stability of ascorbic acid, carotene, niacin, riboflavin, and thiamine in canned fruits and juices stored for 1½ and 2 years was reported.[28] Storage temperature was a significant variable for ascorbic acid and vitamin B_1 but not for carotene, niacin or riboflavin. Two-year storage at 80°F caused losses of 26–47% of ascorbic acid and 14–30% losses of vitamin B_1. At 50°F losses were nominal during the same period. Canning clam meats caused small decreases in niacin and vitamin B_{12} and no changes in vitamins B_1, B_2, B_6 and pantothenic acid. Furthermore, there were no additional changes in 12 months storage at room temperature.[29]

Radiation Stability   Most of the water soluble vitamins are sensitive to ionizing radiation when irradiated in model systems, but are less radiosensitive in foods. Thiamine and ascorbic acid are the most sensitive of the water solubles—niacin the least sensitive—but stability is dependent on the food involved (Tables 7 and 8).[30] Doses of 0.93×10^5 rads destroy 40 to 90% of the ascorbic acid content of strawberries depending on the radiation rate.[31] Doses up to 1.0 megarad have no effect on vitamin B_{12} in milk.

Of the oil soluble vitamins, E and A are very sensitive to radiation; D is not. Vitamin K data are less clear. Added vitamin K_3 is destroyed, but naturally occurring vitamin K in vegetables is reasonably stable. In model systems alpha-tocopherol is much more sensitive than the other oil solubles.

TABLE 7

Vitamin Destruction in Foods Exposed to Varying Doses of Gamma Radiation[30]
(Doses stated in megarads)

Food	Percent Destruction					
	Thiamine		*Riboflavin*		*Niacin*	
	2.79	*5.58*	*2.79*	*5.58*	*2.79*	*5.58*
Bacon	—	93	—	7	—	0
Beef	76	85	5	4	2	1
Haddock	68	76	0	4	14	9
Ham (fresh)	87	96	0	0	2	2
Turkey	76	77	27	50	7	0
Beets	52	75	14	10	0	10
Milk (powdered)	0	0	0	0	35	20
Peaches	94	98	0	0	48	56

Source: M. S. Read, *J. Agric. Food Chem. 8*: 342–349, 1960. Reprinted by permission.

TABLE 8

Effect of Radiation Treatment on Vitamin Content of Foods[30]

Vitamin	Percent Destruction Irradiation (2.79 megarads)
Thiamine	55–65
Riboflavin	6–10
Pyridoxine	24–25
Niacin	0–14
Folic acid	0
Vitamin A	31–70*
Vitamin E	61*

* Dosage 440,000 rads in dairy products.

Source: M. S. Read, *J. Agric. Food Chem. 8*: 342–349, 1960. Reprinted by permission.

In decreasing order of sensitivity are vitamin E, carotene, vitamins A, D and K.[32] Surprisingly vitamins D_3 and K_3 have better stability in oxygen systems than in nitrogen. There are no data comparing tocopherol and tocopheryl acetate sensitivity to radiation.

Radiation of clam meats at varying doses and conditions had little effect on vitamins B_1, B_2, niacin, B_6, pantothenic acid or B_{12}.[29] Frozen whole eggs irradiated at doses of 0.5 and 5.0 Mrad lost no pantothenic acid, biotin or vitamin B_2; vitamin B_1 losses were 24% and 61%, respectively. Doses of 0.02 Mrad for insect disinfestation of wheat have no or little effect on niacin, vitamins B_1, B_2, biotin, B_6 or pantothenic acid. At 0.2 Mrad losses of niacin, pantothenic acid and biotin are about 10%. Vitamin B_1 is surprisingly stable in wheat; 5 Mrad destroys only 24%.[33]

Pasteurization doses, e.g., 0.5 megarad, have only minor effects on vitamins with the possible exception of vitamins A, E and C.

Vitamin B_1   Canned vegetables and fruits stored for one year at 65°F show low losses of thiamine; at 80°F losses are 15–25%. Thiamine and other water solubles distribute between the solids and liquid in canned vegetables. The liquid generally contains approximately 30% of the available thiamine. Brine grading of green baby lima beans prior to blanching by passing the beans through sodium chloride solutions decreased the water-soluble vitamin content, particularly vitamin B_1.[23,34] These workers reported vitamin B_1 overall retention from fresh to frozen beans of 33, 54 and 80%, depending on variety, year, etc.

The sensitivity of thiamine to sulfites is demonstrated by the comparative losses when cabbage is blanched with and without sulfite—45% and 15%, respectively.[35] Dehydration of sulfite-blanched cabbage destroyed the remaining thiamine. Dehydration of steam-blanched cabbage caused an additional 13% loss in thiamine.

The sensitivity of thiamine to sulfite degradation is pH dependent. At low pH bisulfite occurs primarily as the unionized acid and does not degrade thiamine significantly. In canned orange juice the addition of 2 mg/qt bisulfite has no effect on added vitamin B_1 stability (Table 9).[36]

TABLE 9
Effect of Sulfite on the Stability of Thiamine in Canned Orange Juice[36]

	Vitamin B_1—mg/qt	
	Control	$NaHSO_3$ (2 mg/qt)
Initial	4.9	4.9
3 months R.T.	5.2	5.0
6 months R.T.	4.6	4.5
1 month 45°C	2.7	2.8

Thiamine reacts rapidly with nitrite in model systems intended to simulate meat curing conditions, but actual losses in curing experiments with lean pork were only 18–21%.[37] Retention of thiamine in cooked and processed meats ranged from 40% for irradiation products to about 85% for mild cured products.[38] Cooking meat under home preparation conditions produces very variable results; losses from 0–60% have been reported.[39] Roasting temperature of beef and pork is a significant variable. High temperature decreases retention, e.g., 62% retention at low temperature versus 51% at high temperature.[40]

Baking white bread commercially causes losses averaging 20% thiamine. This is fairly constant for natural thiamine and added thiamine since yeast fermentation converts about half of added thiamine to cocarboxylase, which is less stable than thiamine under identical conditions.

Home preparation of rolls (pH 7.) made with baking powder causes losses of 30–50%. Total loss (destruction plus leaching) in cooking fresh peas was 11–40%.[41]

Losses in baking corn bread were approximately 10%. Light has no effect on thiamine stability during bread storage regardless of type of wrapper.[42]

Parboiling rice caused losses of 5–15% thiamine but protected the remaining vitamin from milling losses apparently because the inner bran and scutellum layers became imbedded and were not readily removed during milling.[43]

Vitamin B_1 can catalyze pyruvic acid reaction to acetoin during heat processing of vegetables. This is a potentially important reaction in flavor of canned vegetables.[44] The so-called "Thiamine-Destroying Factor" in soybean does not exist according to Weakley et al.[45] Enzymes capable of destroying vitamin B_1 (thiaminase) occur in some foods—raw fish, clams, shrimp, rice polishings, beans, and mustard seed.

A three-year freezer storage study on vitamins B_1, B_2 and niacin in beef indicated that stability depends on prefreezing aging treatment, but changes were relatively small; i.e., vitamins were stable.

Cow's milk contains approximately 0.4 mg/liter total vitamin B_1 activity, 50–70% thiamine plus thiamine pyrophosphate (cocarboxylase) and protein bound thiamine. Ten percent of the thiamine is destroyed during pasteurization and 30% or more in heat processing evaporated milk. High-temperature-short-time pasteurization decreases thiamine content only 3–4%. Sprayed dried milk powder has minimal losses of 5–15%.

Vitamin B_2 Vitamin B_2 retention during brine grading and blanching of lima beans is superior to and less variable than that of vitamin B_1. Retention in the studies reported above were 76–87%. Retention in blanched cabbage was 80% and losses during cabbage dehydration were negligible. Vitamin B_2 is stable in processing and cooking of meat, ca. 90% retention. Roasting beef and pork results in retention of 70–90% if the drippings are discarded. The drippings contain 15–20% of the original vitamin B_2. Radiation of meat destroyed 25% of the initial vitamin B_2. Since vitamin B_2 is light sensitive, it is important to note that vitamin B_2 is stable in clear cellophane wrapped white bread. Vitamin B_2 also is stable in home cooking procedures. Pasteurization, sterilization, etc., of milk caused less than 10% degradation in milk products.

Light exposure of fluid milk can cause losses of 20–80% in two hours depending on temperature, surface area exposed and light intensity. Fluorescent light is less harmful than sunlight. The greatest destruction is caused by light in the 420–560 mμ range. The light catalyzed degradation of riboflavin in milk also is important because it causes degradation of ascorbic acid and is involved in the degradation of methionine and the formation of "sunlight" flavor (see ascorbic acid and amino acids, this section).

Niacin   Niacin is generally stable in foods. Blanching green peas for three minutes in 99–100°C water caused niacin losses of 15% (probably due to leaching rather than destruction). Similar process in brine caused losses of 27–39%.[23] In the same study, blanching baby green lima beans caused losses of 16–53% niacin. Blanching and dehydration of cabbage caused losses of 5–15%.[35] Niacin is stable during cooking of meat. Drippings, however, contained 16–26% of the initial niacin content.[40] The niacin content of intact pork loin was reduced 30% on storage at 4°C for two weeks, with the largest losses occurring in the first three days.[46] Post-mortem aging (ripening) of beef caused losses of 25–30% niacin in seven days.[47] Radiating meat caused losses of only 10% niacin. Niacin is completely stable in all types of milk processing.

It should be noted that niacin occurs in bound form in some foods, notably corn; biological availability of this form is poor in experimental animals. It is readily made available by treatment with mild alkali. The Central American indians traditionally treat corn with lime, an example of early food technology with a nutritional purpose.

Vitamin B_6   Vitamin B_6 has good stability in most foods. Vitamin B_6 retention in cooking meat ranged from 45 to 80% with an average of 55%.[48] Ionizing radiation of fresh ground beef degraded vitamin B_6 approximately 25%. In cooking vegetables, 20–30% was destroyed and there were similar losses during canning. In milling wheat, 80–90% of the vitamin B_6 is lost. Bread baking losses range from 1 to 17%. Vitamin B_6 is stable during milk pasteurization, and during manufacture and storage of milk powder. Canned evaporated milk retains 50–65% of its initial content during heat processing. Sweetened condensed milk, which has a milder heat treatment, retains about 80%. Vitamin B_6 in milk is light sensitive—21% loss after eight hours in sunlight.

Vitamin B_{12}   Relatively little literature data are available on vitamin B_{12} stability in foods. It appears that vitamin B_{12} is stable in meat processing. Pasteurizing milk and three-day storage had no effect on vitamin B_{12} content.[49] Evaporated milk has losses of 70–90%. Spray dried milk has losses of 20–35%. Peroxide treatment of milk has no effect on vitamin B_{12}. Vitamin B_{12} is stable in haddock fillets stored in ice, frozen at −5°F, and irradiated at doses up to 2.5 Mrads.[50]

Folic Acid   Free folic acid (unconjugated) is stable during milk pasteurization but there is some loss during dehydration.[15] Boiling milk for as little as five seconds decreased free folic acid activity 40–90%; boiling pasteurized milk caused greater losses than boiling unpasteurized milk. The addition of Cu^{++} increased free folic acid degradation and conversely ascorbic acid stabilized folic acid in milk. There are no folic acid losses during blanching of vegetables.

Cooking lentils and other pulses increased folic acid content but vegetables exhibit an irregular pattern—some increase and some decrease. Cooking meat caused some losses.[51] Additional work is necessary on the stability of folic acid conjugates during food processing.

Pantothenic Acid   Post-mortem aging of beef caused slight increases of pantothenic acid after 21 and 42 days.[52] Losses are generally less than 10% during milk processing operations, but cheeses are generally low in pantothenic acid when compared with fresh milk on a nonfat solids basis.

Blanching of baby green lima beans preceded by brine grading causes losses of 27–35%. It is stable to home cooking procedures in the preparation of frozen baby lima beans—90% retention when cooked in a minimum of water and 70% retention when cooked in an excess of water. Roasting meat causes degradation of only 7–10% pantothenic acid, but the meat drippings contain 20–25% of the original content. Cooking rice causes some leaching or destruction of pantothenic acid.

Biotin   While there are little data on biotin stability in foods, all evidence indicates good stability. Evaporated and powdered milk incur losses of 10–15% during processing. Biotin is stable during home cooking of frozen baby lima beans and does not leach in cooking water. In nature, biotin is usually bound. A protein complex "avidin" occurring in egg whites reacts with and inactivates biotin. Avidin is denatured by heat so that cooked egg white is not a biotin antivitamin.

Vitamin C   Probably because of ascorbic acid's sensitivity to oxidation, as discussed previously, its stability in foods has received more attention than that of other vitamins. A comprehensive literature citation is almost impossible. The chemical kinetics of ascorbic acid oxidation have been studied thoroughly.[17] Ascorbic acid oxidase is present in a variety of plant foods and may be responsible for significant losses if not inactivated during tissue maceration. The degradation of ascorbic acid has been suggested as an objective measure of adverse changes in frozen vegetables.[53]

Blanching of cabbage produces losses of 20% ascorbic acid. Dehydration of blanched cabbage degrades about 30% additional ascorbic acid.

Orange juice retains 91% reduced ascorbic acid after 2 weeks at 4–5°C.[54] Orange squash (orange juice, sucrose, citric acid, orange oil), pH 2.5, pasteurized and bottled with sulfur dioxide, retains 100% for 6 months, 80% after 12 months, 50–60% after 18 months and 40% after 24 months. Light has no effect but squash exposed to air in partly filled bottles showed a rapid loss of ascorbic acid.[55]

A valuable ascorbic acid source is acerola juice, the stability of which has been studied intensively. Bottled juice retains 60–70% after 8 months.[56] The mechanism of ascorbic acid degradation and CO_2 evolution in acerola juice has been studied with radioactive tagged L-ascorbic acid-1-^{14}C and L-ascorbic acid-6-^{14}C, and the primary carbon source of CO_2 was found to be C-1.

The stability of ascorbic acid in potato products has been of interest since they are a significant dietary source in many countries. Reduced ascorbic acid decreases 50% in potatoes stored three months—from 25 to 13 mg/100 g. Additional degradation occurs during cooking and washing—from 13 to 6 mg/100 g. After two hours at 150°F on an institutional type steamtable, the reduced ascorbic acid content may be as low as 2 mg/100 g. Commercial dehydration of potatoes to flakes or granules causes losses of 35–45% ascorbic acid. Frozen shoe string potatoes retain 75-85% of raw potato ascorbic acid content; heated, frozen shoe string potatoes retain 43–56% of the initial raw value.[57] Stability of ascorbic acid during storage of frozen shoe string potatoes is excellent. The stability pattern of ascorbic acid in potatoes and sweet potatoes is somewhat confusing during heat processing. The formation of reducing substances which analyze as ascorbic acid is apparently the cause.[58]

Losses in fresh vegetables depend partly on surface area exposed to oxygen. Slicing and other methods of tissue maceration increase vitamin C degradation.

Modification of the storage atmosphere by reducing oxygen content increased ascorbic acid oxidation to dehydroascorbic acid in spinach.[59]

The loss of ascorbic acid during storage of fresh vegetables is temperature dependent as well as rate-of-wilt (dehydration) dependent. Kale, spinach, collards, turnip greens and rape lost 90% of their ascorbic acid when stored at 70°C for seven days vs. 3–30% at 32°F.[60] The same study reported losses in fresh cabbage and snap beans. Storage of winter squash for twenty-five weeks caused losses of 30–60%.[61]

Fresh and frozen broccoli retained about the same percentage of their original ascorbic acid, 65–79%, when cooked. Ascorbic acid was stable in frozen broccoli after thirty-six weeks at 0°F.[62]

Fat Soluble Vitamins

Vitamin A and provitamin A carotenoids have generally good stability in food processing.[19] Vitamin A is stable in milk processing, but prolonged heating of milk or butter at high temperature in the presence of air results in a considerable decrease in vitamin A activity. Frying liver from calves, chickens, etc., causes losses of 10–20%.

Stability of tocopherols in foods was reviewed.[20, 63] Wheat milling losses of alpha-tocopherol were reviewed recently.[64] Alpha-tocopherol is readily oxidized during processing and storage of foods. Vegetable oils are the major dietary source of alpha-tocopherol. Deep fat frying of fresh vegetable oil causes losses of about 10% but storage of fried foods, even at low temperatures, may cause large losses (Table 10). The low content of alpha-tocopherol in frozen foods is surprising and indicates serious degradation even at −12°C.

TABLE 10
Stability of Tocopherol in Fried Foods

	Total Tocopherol Mg %	% Loss
Tocopherol content of oil		
Before frying	82	—
After frying	73	11
Tocopherol content of oil extracted from potato chips		
Initial (after preparation)	75	—
2 weeks at room temperature	39	48
1 month at room temperature	22	71
2 months at room temperature	17	77
1 month at −12°C	28	63
2 months at −12°C	24	68
Tocopherol content of oil extracted from french fried potatoes		
Initial	78	—
1 month at −12°C	25	68
2 months at −12°C	20	74

Source: R. H. Bunnell, J. Keating, A. Quaresimo, and G. K. Parman, *Amer. J. Clin. Nutr. 17*: 1–10, 1965. Reprinted by permission of the publisher, Federation of American Societies for Experimental Biology.

Amino Acids in Foods

The literature emphasizes the effects of food processing on proteins rather than on "free" amino acids. The content of amino acids in a protein may not accurately reflect its nutritive value because some of the amino acids may not be available to the body. Problems involved in determining availability of amino acids in proteins have been reviewed in the literature.[65] The most satisfactory chemical procedure is that for available lysine which is based on the reaction of dinitrofluorobenzene with the epsilon group of lysine.[66] It has been suggested recently that heat treatments less severe than those which cause destruction of amino acids result in lowered protein quality because of changes on the ease with which the protein is hydrolyzed by the digestive enzymes.[67] Blocking of the epsilon amino group of lysine prevents hydrolysis by trypsin, thus lowering availability of other nutritionally limiting amino acids such as methionine.

The total amino acid content of foods obtained by protein hydrolysis has an extensive literature. Comparatively little was known about the free amino acid content of foods until newer analytical methods such as paper and ion exchange chromatography were developed. The separation and determination of individual free amino acids in foods has been accomplished in clam meats,[29] walnuts,[68] potatoes,[69] tomatoes[70] and mushrooms.[71] The distribution within the corn kernel has been reported.[72] Free amino acids are important in food processing primarily because of their organoleptic effects.

Radiation Amino acids are deaminated and decarboxylated by gamma irradiation but are more stable in proteins. Most of the volatiles responsible for off-odors upon irradiation of beef are caused by amino acid degradation. Methyl and ethyl amines are believed produced from alanine and glycine.[73] The major volatile from irradiated beef is methional which may be derived by Strecker degradation of methionine.[74] Methyl mercaptan is produced mainly from methionine.[75] Irradiation of wheat gluten causes a decrease in availability of protein methionine. Radiation of canned clams caused increases in free amino acids.[29]

Browning Reaction The browning reaction (Maillard reaction), which determines color in potato chips and other heat-processed foods, requires both free amino acids and reducing sugars.[76] A summary of model system studies on the browning reaction is available.[77] Sucrose will also react with amino acids to produce the browning reaction at potato chip frying temperatures. The free amino acids in potatoes are phenylalanine, methionine, tyrosine, alanine, lysine, glutamic acid, glutamine aspartic acid and asparagine.[69] The loss of amino acids in potatoes during chip frying has been determined.[78] Generally, all the amino acids decrease except methionine sulfoxides and aspartic acid. The average loss is 50–85% depending on the reducing sugar content. Similarly, free amino acids decrease during storage of freeze-dried apricots due to sugar-amine condensation.[79] The amino acid-deoxyfructoses formed accounted for 90% of the free amino acids lost. The browning reaction is also important in bread baking and in cocoa bean roasting.

Miscellaneous Reactions Free amino acids increase during heat processing of all varieties of tomatoes. The total content was 100–200 mg/100 g juice and fourteen amino acids and two amides were identified.[70]

Free amino acids decreased in bread crust during baking, particularly aspartic and glutamic acids.[80] Leucine produced isovaleraldehyde and threonine produced methyl glyoxal in bread crust.[81] A comprehensive study on changes in free amino acids during fermentation and baking of soda crackers and bread is available.[82] The reaction of proline with dihydroxy acetone is believed important to bread flavor.

Pyrrolidone carboxylic acid (which is not present in fresh fruits and vegetables) is responsible for off-flavor in some processed products, particularly canned beets and tomato juice; it is a reaction product of glutamine during heat processing.[83,84] Glutamic acid is stable during heat processing. Frozen chicken showed free amino acid increase during storage.[85] Free amino acids in clam meats decrease during storage at 33°F, steaming and heat processing. About 11% of the total protein nitrogen of clam meats is free amino acids. The major changes during cocoa bean roasting are the destruction of amino acids and reducing sugars followed by the production of volatile carbonyls. These carbonyls are major aroma constituents and were believed the result of oxidative deamination of free amino acids.[86] Fermentation of cocoa beans caused increases in leucine, phenylalanine, alanine and valine.[87] The free amino acids in beef, pork and lamb have been compared.[88] Free amino acids decreased during storage of freeze-dried raw beef.[89] The effects of fermentation time and heat processing on free amino acids in tempeh were determined.[90]

Polyphenol oxidases (tyrosinase) occur widely in plants and many can oxidize tyrosine forming compounds which polymerize to grey-black color bodies. Foods such as mushrooms and potatoes contain sufficient tyrosine to be a significant source of color problems.

Photolysis of methionine to methional in light exposed milk was believed responsible for "sunlight" flavor.[91]

Addition of Vitamins and Amino Acids to Foods

Vitamins

Purpose With the exception of some uses of ascorbic acid, carotenoids, tocopherol and possibly riboflavin and niacin, the addition of vitamins to foods is practiced for nutritional purposes. The antioxidant-oxygen scavenger functions of ascorbic acid were reviewed recently.[92] The use of ascorbic acid as a dough improver in bread baking was also reviewed.[93] Ascorbic acid is widely used as a dough improver in Europe, and the U.S. Food and Drug Administration has recently been petitioned for this application. A new review of carotenoids as food colorants is available.[19] Tocopherols are used as antioxidants.[94]

Riboflavin is used occasionally as a food colorant. It produces a yellow color somewhat different in character from β-carotene and is used to color milk based products. It has been used in combination with β-carotene to produce a grapefruit juice-like color in beverages. Niacin is occasionally used to improve the color shelf life of both fresh and cured meats but this use is *not approved* in the United States.

Vitamin B₁ Vitamin B_1 is available commercially both as the mononitrate and the hydrochloride. The mononitrate has traditionally been preferred for use in dry mixes with long storage life requirements since it is less hygroscopic and hence is more stable than the hydrochloride. The mononitrate is much less soluble in water than the hydrochloride. Coated vitamin B_1 is available commercially and is used in applications where flavor is a problem.

The stability of vitamin B_1 in dry products is generally excellent; e.g., losses in cocoa powder containing added vitamins B_1, C, D_3 and iron were 13% and 21% after 12 months at 75°F and 98°F, respectively. Degradation during operations requiring heat depends on pH, oxygen content and other factors. Experimental work is generally required to determine the necessary overages for a given heating step. Vitamin B_1 can be added to the dough in preparing toasted cereal products, in some cases. Losses may be as low as 15% in this type of process, but even small losses may cause undesirable flavors and odors. Thiamine degradation odors are objectionable in some cereals and almost unnoticeable in others. The degradation incurred during heating may be avoided by adding thiamine via solution spray after the heating operation. This may cause flavor problems also since the thiamine is concentrated on the surface of the product.

In general, heating in solution at pH 5.5–7.0 is undesirable, but pasteurization of multivitamin milk has no effect on vitamin B_1. Stability is excellent in low pH fruit drinks, e.g., losses in drinks are 6% after 12 months at room temperature. Flavor may be a problem at levels above 4 mg/liter. Drink concentrates containing high levels of vitamins B_1 and C may be less stable than ready-to-serve drinks. The stability problem of vitamin B_1 in solution with vitamin C has already been discussed. Stability in frozen drink concentrates is excellent.

Vitamin B_2 Vitamin B_2 is available as a crystalline product in different crystal habits and particle sizes. The primary difference between these products is the color effects in premixes, which are significant to some users. Because of the poor water solubility of vitamin B_2, riboflavin 5′-phosphate (riboflavin-5′-phosphate ester monosodium salt dihydrate) is also synthesized commercially. It is used primarily in pharmaceutical liquid preparations and can be used to prepare premixes with a deep yellow color. Coated vitamin B_2 is available for uses where flavor is a problem, e.g., chewable vitamin tablets.

Vitamin B_2 is stable in most food applications. For example, grape drink has 100% retention after 12 months at room temperature. It is stable in bread exposed to light and in premixes containing vitamins B_1, B_2, and niacin and iron, designed for baked goods fortification.[42]

Niacin Niacin is available both as nicotinic acid and niacinamide. Coated niacinamide is available for high use level applications in pharmaceuticals. Stability is rarely a problem in foods.

Vitamin B_6 Vitamin B_6 is available as pyridoxine hydrochloride. It is stable when added to foods except when heated. Baking bread losses range from 1 to 17% in a number of literature reports; the mean is about 7%. The retention of natural plus added vitamin B_6 was 90–95% in corn meal and 100% in macaroni stored at 100°F, 50% R.H. for 12 months. Zero loss was encountered on preparing corn bread. When macaroni was cooked, about 50% of the vitamin B_6 was in the macaroni and 50% in the drained cooking water.[95] Bis-4-pyridoxal disulfide was identified in pyridoxal fortified, concentrated, sterilized milk.[96]

Vitamin B_{12} Vitamin B_{12} is available commercially for food use as cyanocobalamin and cobalamin concentrate. Because of its extremely high biological potency, it is normally obtained as standardized dilutions, e.g., 0.1% potency. A variety of diluents are available as triturates as well as a gelatin coated product and an ion exchange resin complex. The latter two products have superior stability under high moisture conditions. Added vitamin B_{12} is stable in frozen drink concentrates—0 to 10% loss in six months at −10°F.

Folic Acid Folic acid is not used frequently in food fortification. It is available in crystalline form.

Pantothenic Acid Pantothenic acid is available in several forms. Panthenol is the most stable in liquid products, but is not recommended for use in foods because of analytical problems. Calcium pantothenate is available as the D-isomer or the D,L racemic mixture. D,L-Calcium pantothenate-calcium chloride complex is less hygroscopic and has superior stability in multivitamin tablets. The L-isomer has no biological activity. Calcium pantothenate has good stability in dry mixes such as grain based complete protein foods. Retention during baking of white bread was 94%.

Biotin Biotin is commercially available as the crystalline D-isomer. It is not widely used in food fortification. Stability in dry products is excellent and work with pharmaceutical liquid products indicates losses of 0–8% at pH 5–8, 6–12% at pH 4.1–4.3 and 15–20% at pH 3.3–3.6 after one year at room temperature.[16]

Vitamin C Vitamin C is available in a variety of mesh sizes both as sodium ascorbate and as ascorbic acid. Stability in solution depends on pH, copper and iron content, exposure to oxygen and temperature. The overages necessary to meet label claim for the required shelf life of products, therefore, vary with the composition of the food. Breakfast cereal fortified with vitamin C had a 40% loss in nine months at room temperature.

Work with model systems resembling blackcurrant juice indicated that anthocyanins greatly accelerated ascorbic acid degradation and 0.85 ppm Cu^{++} was sufficient to accelerate oxidation five fold at pH 2.9.[97,98] The reaction rate was approximately proportional to the square root of the copper concentration. Iron was not as deleterious as copper, but the addition of iron to solutions containing copper accelerated ascorbic acid oxidation. Work with blackcurrant juice substantiated the study on model systems.[97,98] Cysteine was an effective inhibitor of vitamin C oxidation as was EDTA. The optimal level of EDTA to retard oxidation of ascorbic acid in blackcurrant juice was a mole ratio of EDTA/(Cu + Fe) of approximately 2.3.

Stability of vitamin C is temperature dependent, but this is not critical in dry mixes. In cocoa powder fortified with vitamins B_1, C, D_3 and iron, the loss of vitamin C is only 10–15% in 12 months at 75°F or at 98°F. Commercial apple juice fortified with 36–60 mg/100 ml retained 73–81% after 12 months at 70°F vs. 54–60% at 85°F. Light exposure of bottled orange drink caused 35% loss in three months at room temperature.

Fruit drinks stored in the refrigerator after opening lost 10–20% in seven days.[99] Fortified prune juice lost 30–50% of its vitamin C in three months at room temperature. Oxygen permeable containers accelerate degradation of vitamin C in orange drinks. Orange drink concentrate packed in polyethylene lost 100% in 12 months at room temperature. Orange drink packed in paper, milk-type containers lost 50% reduced vitamin C at 45°F in three weeks. In another experiment with paper cartons the loss was 100% in two weeks in orange and grape drink. Ascorbic acid is less stable in prune and grape juice than in other products at the same pH.

In the preparation of gelatin dessert from vitamin C fortified gelatin dessert powder, 25–30% of the reduced vitamin C is oxidized when the powder is dissolved in boiling water, and chilled. However, the oxidation is almost entirely to dehydroascorbic acid, so that the actual loss in biological potency is approximately 3%. Chilled gelatin dessert stored at 40°F lost about 12% per day of reduced ascorbic acid but only 3% per day vitamin C activity.

Vitamin C degradation occurs even in frozen storage of foods. A recent paper has shown that ascorbic acid oxidation is faster in ice than in liquid water.[100] Frozen orange drink concentrate shows losses of about 10% in 12 months at −10°F. Occasional samples show large losses, e.g., 25% of reduced vitamin C between 12 and 18 months, but total biological activity is essentially unchanged. The addition of ferric orthophosphate to supply nutritional levels of iron does not affect ascorbic acid stability in frozen orange drink concentrate.

Fortification of pasteurized, dairy-based, 900 Calorie type foods may require overages as high as 500% to meet label claim, pasteurization causing the major loss.

Exposure of milk in clear bottles to sunlight causes losses of 100% of reduced ascorbic acid and 50% of the total vitamin C activity. Light induced oxidation of vitamin C in milk can be avoided by removing oxygen prior to bottling and by packing in inert gas. Riboflavin is necessary in milk as a light-energy receptor for ascorbic acid oxidation and its removal from milk can also prevent light induced vitamin C oxidation.

Vitamin A Vitamin A is commercially available in a variety of market forms. Vitamin A acetate and palmitate esters are the most important compounds offered. They are synthesized as the all-trans form and the palmitate is also offered as the pre-isomerized equilibrium cis-trans mixture.

Both esters have superior stability to vitamin A alcohol, which is also synthesized commercially. The acetate has similar properties to the palmitate except that it is less stable in the presence of moisture. Both the acetate and palmitate are available in potency standardized oil solutions, emulsions and beadlets. Beadlets are dried emulsions in a water dispersible matrix. Vitamin A is unstable in many applications and antioxidants are necessary to obtain a reasonable shelf life. Dry mix products are generally not a problem. A suitable beadlet is chosen based on flavor, dispersion temperature desired and shelf life requirements. Beadlets containing antioxidants generally lose about 0.5% potency per month. Vitamin A degradation in beadlets follows first order kinetics.

Considerable work has been done on the fortification of nonfat dry milk with vitamin A. Vitamin A oil can be emulsified directly into condensed skim milk prior to spraying, but stability is not satisfactory unless triglycerides are added. Approximately 0.1–0.2% coconut oil or other stable oil by weight of milk solids is required to obtain maximum stability, e.g., 10% loss in 12 months at 75°F.[101,102] In applications of this type it is generally undesirable to dissolve vitamin A beadlets and then dry the final product.

Foods receiving severe heat treatment during processing should preferably be fortified with vitamin A after heating. The vitamin A can be added by spraying an oil/water emulsion onto the surface of the cooked product, e.g., breakfast cereal. The water phase should contain a film forming solute such as sucrose or corn syrup solids capable of forming an oxygen barrier.[103] Better than 90% retention after six months at room temperature can be obtained with this procedure using a vitamin A palmitate emulsion containing suitable antioxidants.

Vitamin A stability studies in frozen drink concentrates have demonstrated somewhat erratic results. Most studies indicate losses of 5–10% in 12 months at −10°F but occasional samples have losses as high as 30%. Work already cited indicates that the stability of natural alpha-tocopherol in frozen foods is rather poor and that ascorbic acid oxidation is faster in ice than in water.[20,100] These findings suggest the possibility that frozen foods may present a general problem with respect to oxidative vitamin stability and additional work is needed on this subject. The problem with respect to vitamin A can be minimized by the use of preisomerized vitamin A and antioxidants.

Stability in oleaginous foods, e.g., margarine, is excellent. Stability is good in vegetable oils with antioxidants but *not* during cooking. Antioxidants have little efficacy when oil is heated at frying temperatures (Table 11).[104]

TABLE 11
Vitamin A Palmitate Stability in Vegetable Oils at 350°F [104]

	Initial	5 Minutes		10 Minutes		15 Minutes		30 Minutes	
Cottonseed Oil									
	USPU/lb	USPU/lb	% Ret	USPU/lb	% Ret	USPU/lb	% Ret	USPU/lb	% Ret
No antioxidant	23,000	21,000	91	20,000	86	16,000	70	8,500	37
5 mg BHA + 5 mg BHT/MU	25,000	22,000	87	20,000	82	16,000	64	7,900	32
25 mg BHA + 25 mg BHT/MU	24,000	25,000	100	21,000	88	21,000	86	9,200	38
Olive Oil									
No antioxidant	31,000	23,000	74	18,000	58	11,000	35	4,200	14
5 mg BHA + 5 mg BHT/MU	29,000	25,000	86	17,000	59	11,000	38	4,000	14
25 mg BHA + 25 mg BHT/MU	30,000	25,000	83	13,000	43	11,000	37	3,300	11

Vitamin A is surprisingly stable in baking bread. Palmitate beadlets can be used with resulting stability of 90–100% after baking white bread by the standard AACC procedure. After five days storage at room temperature retention is 85–95%. Similar results were obtained in baking tortillas and chapaties.[104]

Vitamin D Both vitamin D_2 and D_3 are commercially available and are used in foods. They may be obtained as crystals, standardized oil solutions, emulsions or beadlets. Both compounds are susceptible to oxidation and are usually stabilized with antioxidants. Vitamin D_3 is generally somewhat more stable than vitamin D_2. Pasteurization, boiling, sterilization or treatment with hydrogen peroxide have little effect on vitamin D in fluid milk. Dry milk may be fortified by blending a beadlet form into the dry milk or by homogenizing an oil carrier of vitamin D into the condensed milk before drying. There are few stability problems with vitamin D in foods. Overages are necessary because of the analytical errors involved in assaying low potency vitamin D products.

Vitamin E Both *d*- and *dl*-alpha-tocopheryl acetate are commercially available as oils, standardized beadlets and adsorbates. The esters have excellent stability in all applications. Alpha-tocopherol and mixed tocopherols are also available.

Amino Acids

The most used amino acid in the food industry is monosodium glutamate. The use of this and other amino acids as well as hydrolyzed plant proteins as flavor potentiators and flavors is discussed elsewhere is this text.

The nutritionally limiting amino acids in most human diets are lysine, threonine, tryptophan and methionine. Lysine has received the most attention as a potential nutritional additive. The nutritional efficacy of lysine supplementation has been discussed at length in the literature (Table 12).[105] Lysine and threonine supplementation of rice have been studied.[106] An excellent review of the pros and cons of supplementing the American diet with amino acids is available.[107]

TABLE 12

Protein Efficiency Ratio of Milk Powder and Lysine Supplemented White Bread in Rats[105]

	Weight gain (grams)	PER
White bread control*	19.	0.86
White bread + 4% NFMS	32.	1.33
White bread + 4% NFMS + 0.2% L-lysine	66.	2.00

* 11 % protein diet, 2 groups, 10 rats each diet, 4 weeks.

L-Lysine and D,L-lysine are commercially available as the monohydrochlorides. The stability of added lysine during the baking of white bread depends primarily on the reducing sugar content of the product.[108] In the extreme case of the addition of 25% nonfat milk powder, 40% of the lysine is nutritionally lost. At the usual use rate of skim milk powder in bread, 4–6%, retention of lysine is 80–95%. Processes requiring severe heat treatments of cereal-based foods to gelatinize starch and denature protein may cause severe losses of lysine.

The protein quality of biscuits supplemented with milk powder before baking is considerably lower than that of biscuits plus unbaked milk powder apparently because the heat treated milk protein lysine is less available. Analysis of protein hydrolysates of bread indicated that lysine, phenylalanine, tyrosine and serine were degraded 5–17% during baking.[109] This suggests that if these amino acids were added to bread similar or greater destruction would occur.

Methionine is commercially available as D,L-methionine and as the calcium salt of the hydroxy analog of methionine. These products are used almost exclusively in animal feeds. The relative biological value of these compounds compared to L-methionine has received considerable attention and the evidence suggests equivalency on a mole basis both in rats and man. As discussed, methionine can be degraded producing methyl mercaptan, H_2S, methional and sulfoxide by heat processing and irradiation. Little data are available on the stability of added methionine, threonine or tryptophan in processed foods. The stability of lysine, arginine, histidine, cystine and methionine on autoclaving separately and in amino acid and sugar mixtures has been studied.[110] Cystine was the most heat labile and methionine the least. The loss of added D,L-threonine in baking bread was 20–40% in one experiment.[111]

References

1. Sebrell, W. H., Jr., and Harris, R. S., 1954, *The Vitamins*. Academic Press, New York.
2. Wohl, M. G., and Goodhart, R. S., 1968, *Modern Nutrition in Health and Disease*. 4th Edition, Lea and Febiger, Philadelphia.
3. Jolliffe, N., Editor, 1962, *Clinical Nutrition*. Harper and Brothers, New York.
4. Mitchell, H. H., 1963–1964, *Comparative Nutrition of Man and Domestic Animals*. 2 volumes. Academic Press, New York.
5. Beaton, G. H., and McHenry, E. W. Eds., 1964–1966, *Nutrition: A Comprehensive Treatise*. Academic Press, New York.
6. Strohecker, R., and Henning, H. M., 1965, *Vitamin Assay: Tested Methods*. Verlag Chemie, Weinheim, Germany.
7. The Association of Vitamin Chemists, Inc., 1966, *Method of Vitamin Assay*, 3rd Edition, Interscience Publishers, Division of John Wiley & Sons, Inc., New York.
8. Horwitt, M. K., 1958, Niacin-Tryptophan Requirements of Man. *J. Amer. Dietet. Assn.* 34: 914–919.
9. International Symposium on Vitamin B6, 1964, *Vitamins and Hormones*. 22, pp. 361–885, Academic Press, New York.
10. Bartilucci, A., and Foss, N. E., 1954, Cyanocobalamin (Vitamin B12). I. A Study of the Stability of Cyanocobalamin and Ascorbic Acid in Liquid Formulations. *J. Amer. Pharm. Assn.* (Scient. Ed.) 43: 159–162.
11. Blitz, M., Eigen, E., and Gunsberg, E., 1954, Vitamin B12 Studies—The Instability of Vitamin B12 in the Presence of Thiamine and Niacinamide. *J. Amer. Pharm. Assn.* (Scient. Ed.) 43: 651–653.
12. Blitz, M., Eigen, E., and Gunsberg, E., 1956, Studies Relating to the Stability of Vitamin B12 in B-Complex Injectable Solutions. *J. Amer. Pharm. Assn.* (Scient. Ed.) 45: 803–806.
13. Shenoy, K. G., and Ramasarma, G. B., 1955, Iron as a Stabilizer of Vitamin B12 Activity in Liver Extracts and the Nature of So-called Alkali-Stable Factor. *Arch. Biochem. Biophys.* 55: 293–295.
14. Newmark, H. L., 1958, *Stable Vitamin B12-Containing Solution*, U.S. Patent 2,823,167.
15. Ghitis, J., and Candanosa, C., 1966, The Labile Folate of Milk. *Amer. J. Clin. Nutr.* 18: 452–457.
16. De Ritter, E., and Scheiner, J., 1958, Hoffmann-La Roche Inc., unpublished work.
17. Schulte, K. E., and Schillinger, A., 1952, Comparison of the Kinetics of Non-Fermentative Oxidation of D-isoascorbic and L-ascorbic acids. *Z. Lebensmittel-Untersuchung* 94: (2): 77.
18. Moore, T., 1957, *Vitamin A*. American Elsevier Publishing Company, New York.
19. Borenstein, B., and Bunnell, R. H., 1967, *Carotenoids: Properties, Occurrence and Utilization in Foods in Advances in Food Research*, Vol. 15, pp. 195–264, Academic Press, New York.
20. Bunnell, R. H., Keating, J., Quaresimo, A., and Parman, G. K., 1965, Alpha-Tocopherol Content of Foods. *Amer. J. Clin. Nutr.* 17: 1–10.
21. Altschul, A. M., 1958, *Processed Plant Protein Foodstuffs*. Academic Press, New York.
22. Schultz, H. W., and Anglemier, A. F., 1964, *Proteins and Their Reactions*. Avi Publishing Co., Westport, Conn.
23. Cook, B. B., Gunning, B., and Uchimoto, D., 1961, Variations in Nutritive Value of Frozen Green Baby Lima Beans as a Result of Methods of Processing and Cooking. *J. Agric. Food Chem.* 9: 316–321.

24. Rice, E. E., and Beuk, J. F., 1945, Reaction rates for decomposition of thiamin in pork at various cooking temperatures. *Food Res.* 10: 99–107.

25. Hlynka, I., Editor, 1964, *Wheat Chemistry and Technology.* Am. Assoc. of Cereal Chemists, Inc., St. Paul, Minnesota.

26. Guerrant, N. B., Vavich, M. G., Fardig; O. B., Ellenberger, H. A., Stern, R. M., and Coonen, N. H., 1947, Effect of Duration and Temperature of Blanch on Vitamin Retention by Certain Vegetables. *Ind. Eng. Chem.* 39: 1000–1007.

27. Guerrant, N. B., and O'Hara, M. B., 1953, Vitamin Retention in Peas and Lima Beans After Blanching, Freezing, Processing in Tin and in Glass, After Storage and After Cooking. *Food Technol.* 7: 473–477.

28. Sheft, B. B., Griswold, R. M., Tarlowsky, E., and Halliday, E. G., 1949, Nutritive Value of Canned Foods, Effect of Time and Temperature of Storage on Vitamin Content of Commercially Canned Fruits and Fruit Juices (Stored 18 and 24 Months). *Ind. Eng. Chem.* 41: 144–145.

29. Brooke, R. O., Ravesi, E. M., Gadbois, D. F., and Steinberg, M. A., 1964, Preservation of Fresh Unfrozen Fishery Products by Low-Level Radiation. III. The Effects of Radiation Pasteurization on Amino Acids and Vitamins in Clams. *Food Technol.* 18: 1060–1064.

30. Read, M. S., 1960, Current Aspects of the Wholesomeness of Irradiated Food. *J. Agric. Food Chem.* 8: 342–349.

31. Salunkhe, D. K., Gerber, R. K., and Pollard, L. H., 1959, Physiological and Chemical Effects of Gamma Radiation on Certain Fruits, Vegetables and Their Products. *Proc. Amer. Soc. Horticult. Sci.* 74: 423–429.

32. Knapp, F. W., and Tappel, A. L., 1961, Comparison of the Radiosensitivities of the Fat-Soluble Vitamins by Gamma Irradiation. *J. Agric. Food Chem.* 9: 430–433.

33. Kennedy, T. S., 1965, Studies on the Nutritional Value of Foods Treated with Gamma-Radiation. I. Effects on Some B-Complex Vitamins in Egg and Wheat. *J. Sci. Food Agric.* 16: 81–84.

34. Yamaguchi, M., MacGillivray, J. H., Howard, F. D., Simone, M., and Sterling, C., 1954, Nutrient Composition of Fresh and Frozen Lima Beans in Relation to Variety and Maturity, *Food Research* 19: 617–626.

35. Mallette, M. F., Dawson, C. R., Nelson, W. L., and Gortner, W. A., 1946, Commercially Dehydrated Vegetables, Oxidative Enzymes, Vitamin Content and Other Factors. *Ind. Eng. Chem.* 38: 437–441.

36. Borenstein, B., and Smith, E. G., 1965, Hoffmann-La Roche Inc., unpublished work.

37. Greenwood, D. A., Beadle, B. W., and Kraybill, H. R., 1943, Stability of Thiamine to Heat. II. Effect of Meat-Curing Ingredients in Aqueous Solutions and in Meat. *J. Biol. Chem.* 149: 349–354.

38. American Meat Institute Foundation, 1960, *The Science of Meat and Meat Products.* W. H. Freeman and Co., San Francisco.

39. Mickelsen, O., Waisman, H. A., and Elvehjem, C. A., 1939, The Distribution of Vitamin B_1 (Thiamin) in Meat and Meat Products. *J. Nutrition* 17: 269–280.

40. Cover, S., Dilsaver, E. M., Hays, R. M., and Smith, W. H., 1949, Retention of B Vitamins after Large-Scale Cooking of Meat. II. Roasting by Two Methods. *J. Amer. Dietet. Assn.* 25: 949–951.

41. Neymark, M., and Hellström, V., 1966, *Losses of Thiamine in Foods During Home Preparation.* 7th International Congress on Nutrition, Hamburg, Germany, p. 82.

42. Morgareidge, K., 1956, The Effect of Light on Vitamin Retention in Enriched White Bread. *Cereal Chemistry* 33: 213–220.

43. Rao, P. V. Subba, and Bhattacharya, K. R., 1966, Effect of Parboiling on Thiamine Content of Rice. *J. Agric. Food Chem.* 14: 479–482.

44. Ralls, J. W., 1959, Nonenzymatic Formation of Acetoin in Canned Vegetables. *J. Agric. Food Chem.* 7: 505–507.

45. Weakley, F. B., Eldridge, A. C., and McKinney, L. L., 1961, The Alleged "Thiamine-Destroying Factor" in Soybeans. *J. Agric. Food Chem.* 9: 435–439.

46. Rice, E. E., Squires, E. M., and Fried, J. F., 1948, Effect of Storage and Microbial Action on Vitamin Content of Pork. *Food Research* 13: 195–202.

47. Meyer, B., Thomas, J., and Buckley, R., 1960, The Effect of Ripening on the Thiamine, Riboflavin, and Niacin Content of Beef from Grain-Finished and Grass-Finished Steers. *Food Technol.* 14: 190–192.

48. Lushbough, C. H., Weichman, J. M., and Schweigert, B. S., 1959, The Retention of Vitamin B₆ in Meat During Cooking. *J. Nutrition* 67: 451–459.

49. Hartman, A. M., Dryden, L. P., and Riedel, G. H., 1956, Vitamin B₁₂ Content of Milk and Milk Products as Determined by Rat Assay. *J. Nutrition* 59: 77–88.

50. Brooke, R. O., Ravesi, E. M., Gadbois, D. F., and Steinberg, M. A., 1966, Preservation of Fresh Unfrozen Fishery Products by Low-Level Radiation. 5. The Effects of Radiation Pasteurization on Amino Acids and Vitamins in Haddock Fillets. *Food Technol.* 20: 1479–1482.

51. Banerjee, D. K., and Chatterjea, J. B., 1964, Folic Acid Activity of Indian Dietary Articles and the Effect of Cooking on It. *Food Technol.* 18: 1081–1083.

52. Meyer, B. H., Mysinger, M. A., and Cole, J. W., 1966, Effect of Finish and Ripening on Vitamin B₆ and Pantothenic Acid Content of Beef. *J. Agric. Food Chem.* 14: 485–486.

53. Dietrich, W. C., Lindquist, F. E., Miers, J. C., Bohart, G. S., Neumann, H. J., and Talburt, W. F., 1957, The Time-Temperature Tolerance of Frozen Foods. IV. Objective Tests to Measure Adverse Changes in Frozen Vegetables. *Food Technol.* 5: 109–113.

54. Lamden, M. P., Schweiker, C. E., and Pierce, H. B., 1960, Ascorbic Acid Studies on Chilled, Fresh and Fermented Orange Juice. *Food Research* 25: 197–202.

55. Bender, A. E., 1958, The Stability of Vitamin C in a Commercial Fruit Squash. *J. Sci. Food Agric.* 9: 754–760.

56. Fitting, K. O., and Miller, C. D., 1960, The Stability of Ascorbic Acid in Frozen and Bottled Acerola Juice and Combined with Other Fruit Juices. *Food Research* 25: 203–210.

57. Bring, S. V., 1966, Total Ascorbic Acid of Shoe String Potatoes. *J. Amer. Dietet. Assn.* 48: 112–115.

58. McAfee, J. W., and Watts, J. H., 1958, Effect of Non-Antiscorbutic Reducing Substances Upon the Ascorbic Acid Content of Baked Potatoes. *Food Research* 23: 114–118.

59. McGill, J. N., Nelson, A. E., and Steinberg, M. P., 1966, Effects of Modified Storage Atmospheres on Ascorbic Acid and Other Quality Characteristics of Spinach. *J. Food Sci.* 31: 510–517.

60. Ezell, B. D., and Wilcox, M. S., 1959, Loss of Vitamin C in Fresh Vegetables as Related to Wilting and Temperature, *J. Agric. Food Chem.* 7: 507–509.

61. Hopp, R. J., and Merrow, S. B., 1963, Varietal Differences and Storage Changes in the Ascorbic Acid Content of Six Varieties of Winter Squashes. *J. Agric. Food Chem.* 11: 143–146.

62. Martin, M. E., Sweeney, J. P., Gilpin, G. L., and Chapman, V. J., 1960, Factors Affecting the Ascorbic Acid and Carotene Content of Broccoli. *J. Agric. Food Chem.* 8: 387–390.

63. Harris, R. S., 1962, Stability of E in Storage and Processing of Foods. *Vitamins and Hormones*, Vol. 20, Academic Press, New York, pp. 603–619.

64. Rubin, S. H., 1966, Nutrition Research and Enrichment. *Cereal Sci. Today* 11:(6), 234–239, 280–281.

65. Grau, C. R., and Carroll, R. W., 1958, *Processed Plant Protein Foodstuffs*, A. M. Altschul, Editor. Academic Press, New York, pp. 153–189.

66. Carpenter, K. J., 1960, The Estimation of the Available Lysine in Animal-Protein Foods, *Biochem. J.* 77: 604–610.

67. Lyman, C. M., 1966, Effect of Processing on Nutritional Value of Proteins. Talk at Institute of Food Technologists, National Meeting, May, 1966.

68. Rockland, L. B. and Nobe, B., 1964, Free Amino Acids in English Walnut (*Juglans regia*) Kernels. *J. Agric. Food Chem.* 12: 528–535.

69. Furuholmen, A. M., Winefordner, J. D., Dennison, R. A., and Knapp, R. W., 1964, Isolation, Concentration, Separation and Identification of Amino Acids in Potatoes by Ion Exchange and Paper Chromatography. *J. Agric. Food Chem.* 12: 112–114.

70. Hamdy, M. M., and Gould, W. A., 1962, Varietal Differences in Tomatoes: A Study of Alpha-Keto Acids, Alpha-Amino Compounds, and Citric Acid in Eight Tomato Varieties Before and After Processing. *J. Agric. Food Chem.* 10: 499–503.

71. Kissmeyer-Nielsen, E., McClendon, J. H., and Woodmansee, C. W., 1966, Changes in Amino Acids and Urea in the Cultivated Mushroom, *Agaricus bisporus*, as Influenced

by Nutrient Supplementation of the Compost During the Growth Cycle. *J. Agric. Food Chem.* 14: 633–636.

72. Christianson, D. D., Wall, J. S., and Cavins, J. F., 1965, Location of Nonprotein Nitrogenous Substances in Corn Grain. *J. Agric. Food Chem.* 13: 272–276.

73. Burks, R. E., Jr., Baker, E. B., Clark, P. Esslinger, J., and Lacey, J. C., Jr., 1959, Detection of Amines Produced on Irradiation of Beef. *J. Agric. Food Chem.* 7: 778–782.

74. Wick, E. L., Yamanishi, T., Wertheimer, L. C., Hoff, J. E., Proctor, B. E., and Goldblith, S. A., 1961, An Investigation of Some Volatile Components of Irradiated Beef. *J. Agric. Food Chem.* 9: 289–293.

75. Martin, S., Batzer, O. F., Landmann, W. A., and Schweigert, B. S., 1962, The Role of Glutathione and Methionine in the Production of Hydrogen Sulfide and Methyl Mercaptan during Irradiation of Meat. *J. Agric. Food Chem.* 10: 91–93.

76. Shallenberger, R. S., Smith, O., and Treadway, R. H., 1959, Role of the Sugars in the Browning Reaction in Potato Chips. *J. Agric. Food Chem.* 7: 274–277.

77. Burton, H. S., McWeeny, D. J., and Biltcliffe, D. O., 1963, Non-Enzymic Browning, Development of Chromophores in the Glucose-Glycine and Sucrose-Glycine Systems, *J. Food Sci.* 28: 631–639.

78. Fitzpatrick, T. J., Talley, E. A., and Porter, W. L., 1965, Preliminary Studies on the Fate of Sugars and Amino Acids in Chips Made from Fresh and Stored Potatoes. *J. Agric. Food Chem.* 13: 10–12.

79. Ingles, D. L., and Reynolds, T. M., 1958, Chemistry of Non-Enzymatic Browning. IV. Determination of Amino Acids and Amino Acid-Deoxyfructoses in Browned Freeze-Dried Apricots. *Australian J. of Chem.* 11: 575–580.

80. Linko, Y-Y., and Johnson, J. A., 1963, Changes in Amino Acids and Formation of Carbonyl Compounds During Baking *J. Agric. Food Chem.* 11: 150–152.

81. Linko, Y-Y., Johnson, J. A., and Miller, B. S., 1962, The Origin and Fate of Certain Carbonyl Compounds in White Bread. *Cereal Chem.* 39: 468–476.

82. Morimoto, T., 1966, Studies on Free Amino Acids in Sponges, Doughs, and Baked Soda Crackers and Bread. *J. Food Sci.* 31: 736–741.

83. Mahdi, A. A., Rice, A. C., and Weckel, K. G., 1959, Formation of Pyrrolidonecarboxylic Acid in Processed Fruit and Vegetable Products. *J. Agric. Food Chem.* 7: 712–714.

84. Mahdi, A. A., Rice, A. C., and Weckel, K. G., 1961, Effect of Pyrrolidonecarboxylic Acid on Flavor of Processed Fruit and Vegetable Products. *J. Agric. Food Chem.* 9: 143–146.

85. Khan, A. W., 1964, Changes in Nonprotein Nitrogenous Constituents of Chicken Breast Muscle Stored at Below-Freezing Temperatures. *J. Agric. Food Chem.* 12: 378–380.

86. Pinto, A., and Chichester, C. O., 1966, Changes in the Content of Free Amino Acids During Roasting of Cocoa Beans *J. Food Sci.* 31: 726–732.

87. Rohan, T. A., 1964, The Precursors of Chocolate Aroma: A Comparative Study of Fermented and Unfermented Cocoa Beans. *J. Food Sci.* 29: 456–459.

88. Macy, R. L., Jr., Naumann, H. D., and Bailey, M. E., 1964, Water-Soluble Flavor and Odor Precursors of Meat. I. Qualitative Study of Certain Amino Acids, Carbohydrates, Non-Amino Acid Nitrogen Compounds and Phosphoric Acid Esters of Beef, Pork, and Lamb. *J. Food Sci.* 29: 136–141.

89. El-Gharbawi, M. E., and Dugan, L. R., Jr., 1965, Stability of Nitrogenous Compounds and Lipids During Storage of Freeze-Dried Raw Beef. *J. Food Sci.* 30: 817–822.

90. Stillings, B. R., and Hackler, L. R., 1965, Amino Acid Studies on the Effect of Fermentation Time and Heat-Processing of Tempeh. *J. Food Sci.* 30: 1043–1048.

91. Patton, S., 1954, The Mechanism of Sunlight Flavor Formation in Milk With Special Reference to Methionine and Riboflavin. *J. Dairy Sci.* 37: 446–452.

92. Borenstein, B., 1965, The Comparative Properties of Ascorbic Acid and Erythrobic Acid. *Food Technol.* 19: 1719–1721.

93. Tsen, C. C., 1964, Ascorbic Acid as a Flour Improver. *Baker's Digest* 38 (5), 44–47.

94. Schultz, H. W., Day, E. A., and Sinnhuber, R. O., Editors, 1962, *Symposium on Foods: Lipids and Their Oxidation*. Avi Publishing Co., Inc., Westport, Connecticut.

95. Bunting, W. R., 1965, The Stability of Pyridoxine Added to Cereals. *Cereal Chemistry* 42: 569–572.

96. Wendt, G., and Bernhart, F. W., 1960, The Structure of a Sulfur-Containing Compound with Vitamin B₆ Activity. *Arch. Biochem. Biophys.* 88: 270–272.

97. Timberlake, C. F., 1960, Metallic Components of Fruit Juices. III. Oxidation and Stability of Ascorbic Acid in Model Systems Resembling Blackcurrant Juice. *J. Sci. Food Agric.* 11: 258–268.

98. Timberlake, C. F., 1960, Metallic Components of Fruit Juices. IV. Oxidation and Stability of Ascorbic Acid in Blackcurrant Juice. *J. Sci. Food Agric.* 11: 268–273.

99. Pelletier, O., and Morrison, A. B., 1965, Content and Stability of Ascorbic Acid in Fruit Drinks. *J. Amer. Dietet. Assn.* 47: 401–404.

100. Grant, N. H., and Alburn, H. E., 1965, Fast Reactions of Ascorbic Acid and Hydrogen Peroxide in Ice, a Presumptive Early Environment. *Science* 150: 1589–1590.

101. Bauernfeind, J. C., and Parman, G. K., 1964, Restoration of Nonfat Dry Milk with Vitamins A and D. *Food Technol.* 18: 52–57.

102. Bauernfeind, J. C., and Allen, L. E., 1963, Vitamin A and D Enrichment of Nonfat Dry Milk. *J. Dairy Sci.* 46: 245–254.

103. Bunnell, R. H., and Bauernfeind, J. C., 1961, Hoffmann-La Roche Inc., unpublished work.

104. Bunnell, R. H., and Borenstein, B., 1966, Hoffmann-La Roche Inc., unpublished work.

105. Borenstein, B., Banziger, R., Bunnell, R. H., and Newmark, H. L., 1966, Hoffmann-La Roche Inc., unpublished work.

106. Rosenberg, H. R., Culik, R., and Eckert, R. E., 1959, Lysine and Threonine Supplementation of Rice. *J. Nutrition* 69: 217–228.

107. National Research Council, 1959, *Evaluation of Protein Nutrition.* National Academy of Sciences, Publication 711.

108. Jansen, G. R., and Ehle, S. R., 1965, Studies on Breads Supplemented with Soy, Nonfat Dry Milk, and Lysine. II. Nutritive Value. *Food Technol* 19: 1439–1442.

109. McDermott, E. E., and Pace, J., 1957, The Content of Amino Acids in White Flour and Bread. *Brit. J. Nutrit.* 11: 446–452.

110. Taira, H., 1966, Studies on Amino Acid Contents of Processed Soybean. Part X. The Influence of Added Sugars on the Heat Destruction of the Basic and Sulphur Containing Amino Acids in Soybean Products. *Agric. Biol. Chem.* (Japan) 30: 847–855.

111. Ericson, L.-E., Larsson, S. and Lid, G., 1961, Added Lysine and Threonine During the Baking of Bread. *Acta Physiol. Scand.* 53: 85–98.

CHAPTER 3

Antimicrobial Food Additives

D. F. Chichester, Ph.D.
Consultant
Gainesville, Florida

Fred W. Tanner, Jr., Ph.D.
Retired
Formerly in Administration Division
Pfizer Inc.
New York, New York

Introduction

Satisfactory methods for preserving foods have been a challenge ever since man began to live in groups, and this challenge became greater as he migrated to areas of the globe where crop production is limited to a few months per year. The evolution to an industrial society has not lessened the problem; changes in agricultural practices, dietary shifts toward more perishable foods, increased possibilities of mass contamination through our enlarged distribution systems and the current strong trend toward convenience foods—all are placing greater emphasis upon methods of preservation.

World wide, the growing crisis in the food supply demands that losses be reduced to a minimum. Those countries with the greatest nutritional problems are beset with inadequate production, distribution, transportation and preservation capacity. If and when these problems are solved, and foods other than grain and oil seed proteins become available in quantity, chemical preservatives will have to be applied to the extent of their capabilities, and will forcibly focus attention on the need for more effective antimicrobials.

While, in highly industrialized countries, there have been substantial improvements in the technologies of preservation by heat, freezing and drying, there are many foods to which these processes cannot be applied, or are only partially effective. Hence the use of chemical preservatives, alone or as a supplement to other methods, is essential.

In the case of chemical preservatives, it is unfortunate that the armamentarium has changed but little over many years; some of the oldest compounds remain the most widely used, despite their limitations. However, it has been difficult to find additional simple and economical compounds with good inhibitory power against a wide range of organisms and a low order of toxicity for mammals. This is attested by the experience of the drug industry which has had its difficulties in finding usable anti-infective compounds even though greater latitude for cost and toxicity can be accepted in that area.

This chapter will focus on major compounds that are officially accepted as direct food additives by regulatory agencies of the United States and many other countries. Space limitations require passing over some compounds which appeared to be promising, but for which drawbacks, usually toxicity, were disclosed as investigation proceeded. For instance, β-propionolactone has many desirable attributes as an antimicrobial, but questions of carcinogenicity have been raised.

Many antibiotics have been extensively investigated as antimicrobial additives for foods. There have been technical and practical obstacles as well as the apprehension that those antibiotics used in medicine, if added to foods, might establish populations of resistant organisms. The investigations in this field are of some historical importance and are summarized near the end of the chapter where the current uses of antibiotics as food preservatives are noted.

Some preservatives which have been in use since ancient times, such as acids, salt, sugar and wood smoke, have continued utility in certain products, but with the expansion of other preservation methods, these agents became less subject to study as antimicrobials. Although these will not be included in this chapter, it should be noted that an extensive literature documents their antimicrobial action.[194]

Research on the chemistry of wood smoke has importance from the point of view of flavor, but may warrant further evaluation as a source of information on compounds with preservative action.[62, 72]

Post-harvest spoilage problems and agents for their control also are not part of this presentation. For recent information on this broad field, the reader is referred to the reviews by Smith[222] and by Eckert.[65]

Indirect additives, substances which are used in the manufacture of food packaging materials, are considered here only by being tabulated on p. 177. It will be noted that several of these chemicals also have approval as direct additives.

Table 1 shows in condensed form the officially approved uses in the United States of the chemical antimicrobial agents considered in this chapter, other than antibiotics.* Research and experience have demonstrated the field of applicability of each in relation to the nature of the food products, the types of organisms which it attacks, manufacturing and marketing conditions.

It will be seen that the compounds in the left hand columns of the table have more general applications, whereas those to the right are more specialized. It should be noted also that some of the compounds (sulfites, nitrite and nitrate) have dual roles. Nitrite and nitrate are added primarily as color-fixing agents in processed meats, although the strong anti-clostridial action of nitrite has recently come to the fore. Sulfur dioxide or sulfites are used to prevent discoloration of fruits and vegetables in conjunction with dehydration and some other types of processing.

As will be evident in subsequent parts of this review, these preservatives have some overlapping of their antimicrobial spectra. The benzoates and parabens have the greatest breadth of activity, encompassing many spoilage bacteria, fungi and yeasts. The propionates are active against fungi and essentially inactive against yeast, but against bacteria have a useful ability limited to combating the organisms which cause rope in bread. Sorbates are active against bacteria, yeast and molds but are used mainly against yeasts and molds.

None of these, so far as is now known, has been tried in solving a mold problem of recent origin; that is, the presence of toxin-elaborating fungi on stored grains and oil seed meals.[32] Perhaps the problem can be eliminated by proper crop storage in conjunction with changes in current harvesting methods to reduce moisture. If not, efforts may be needed to extend the range of anti-fungal agents to encompass this new area of contamination.

With the exception of the p-hydroxybenzoate esters (parabens), the most widely used preservatives and several of the more specialized ones are weak acids or salts of weak acids. They exert their greatest activity on the acid side of neutrality. Their activity is attributed to greater relative concentration of undissociated acid at low pH.

In the concentrations used in practice, none of the compounds reviewed here, except diethyl pyrocarbonate and ethylene and propylene oxides, is lethal to microorganisms in foods. Their action is only inhibitory. It should be emphasized that the choosing of a particular preservative must be based on the gross microbial problems, need for selective action, solubility, ease of application and appropriateness of pH range and other conditions. Then empirical factors must be considered; despite theoretical similarities in antimicrobial properties, one preservative may be more effective than another in a specific food. Relative costs, too, cannot be ignored. The benzoates and propionates are relatively inexpensive compounds, whereas sorbates and parabens are relatively more expensive.

---

* See Table 18, page 172, for tabulation of chemical preservatives used in 17 countries outside of the U.S.A. and a brief discussion of variations in permitted used.

TABLE 1

Applications of Principal Antimicrobial Additives to Foods in Interstate Commerce

Types of Food Products		Benzoic Acid and Sodium Benzoate[1]	Ethyl and Propyl Parabens	Sorbates	Propionates	Sulfites	Acetates, Diacetates	Nitrite, Nitrate	Epoxides		Diethyl Pyro-carbonate
									Ethylene Oxide	Propylene Oxide	
Beverages	Carbonated	+	+	+		+					
	Non-Carbonated	+	+	+							
	Beverage Syrups	+[2]	+[2]	+[2]							
	Fruit Drinks	+	+	+							
	Fruit Juices	+	+	+		+					
	Wines and Beer		+[3]	+[3]		+[3]					+[3]
	Purees and Concentrates	+	+	+		+					
Dairy Products	Cheese and Cheese Products			+[4]	+[4]						
Margarine		+[5]		+[5]							
Baked Goods	Yeast-Leavened				+[6]		+[6]				
	Chemically Leavened		+	+	+		+				
	Pie Crust & Pastries		+	+	+		+				
	Pie Fillings	+	+	+	+						
Processed Meat and Fish	Sausage		+[7]	+[7]				+[8]			
	Preserved Fish			+							

TABLE I

Applications of Principal Antimicrobial Additives to Foods in Interstate Commerce (Continued)

Specialties	Salads, Salad Dressings	+	+	+				+
	Dried Fruits & Vegetables	+	+		+			+
	Fresh Fruits & Vegetables		+					+
	Pickles, Relishes, Olives and Sauerkraut	+	+		+	+		+
Ingredients	Spices						+	+
	Starch							+
	Nut Meats					+[9]	+[10]	
	Copra					+		
	Dried Prunes						+	
	Glacé Fruit						+	

[1] A maximum of 0.1 % in all uses is set for sodium benzoate, parabens or combinations, except that 0.2 % of sodium benzoate may be used in orange juice for manufacturing. Orange juice not for manufacturing and other pure fruit juices having Federal Standards of Identity may not contain a preservative.

[2] It is not practical to add sufficient benzoate, paraben or sorbate to highly concentrated beverage syrups to protect the final diluted beverage. Additional preservative must be added to the final product.

[3] Sorbate, sulfite and diethyl pyrocarbonate in wine; *n*-heptyl paraben and diethyl pyrocarbonate in beer.

[4] The maximum limit of propionate and sorbate in regular cheese is 0.3 % by weight; in processed cheese, cheese foods and spreads, the limit for propionate is 0.3 % and for sorbate is 0.2 %.

[5] Sodium benzoate used in salted margarine; potassium sorbate used in unsalted.

[6] Maximum permitted level of propionate in white bread is 0.32 % of flour weight and in whole wheat bread is 0.38 % of flour weight. Maximum level of sodium diacetate in bread is 0.4 part per 100 lb. flour.

[7] Potassium sorbate and propyl-*p*-hydroxy benzoate are permitted on dry sausage. Potassium sorbate is permitted in moist dog-food patties containing meat.

[8] Limits are prescribed for use of nitrite and nitrate in meat and fish. See text.

[9] Black walnut meat only.

[10] All except peanut.

In the text sections dealing with individual preservatives, the pertinent Federal Regulations regarding uses are mentioned, but additional regulations may be applied by states and municipalities. The omission of any statement regarding restrictions does not necessarily mean that there are none, and the user should consult with authorities in the areas where he intends to sell a food product containing additives. He should also inform himself on labeling requirements. In general, under Federal Regulations, preservatives directly added to foods must be declared on the label, but there are certain variations. Nothing within this chapter is intended to imply that any patented inventions can be practiced without a license. Any recommendations or suggestions made on the use of products mentioned herein are without warranty or guarantee.

Not all possible uses of each preservative have been ruled upon by F.D.A. because petitions for all uses have not been presented.

Sodium Benzoate

Benzoic acid, usually in the form of the sodium salt, has long been used as an antimicrobial additive for foods. The sodium salt is preferred because of the low aqueous solubility of the free acid. In use the salt is converted to the acid, the active form.

The pH range for optimum microbial inhibition by benzoic acid is 2.5–4.0 which is lower than that of sorbic acid or propionic acid. Thus, benzoates are well-adapted to the preservation of foods which are acid, or readily acidified, foods such as carbonated beverages, fruit juices, cider, pickles and sauerkraut.

It is of interest that benzoic acid occurs naturally in some foods: cranberries, prunes, greengage plums, cinnamon and ripe cloves.[194]

Physical and Chemical Properties

Sodium benzoate as an article of commerce is in the form of a white powder or flakes. It can be mixed dry into bulk liquids and dissolves promptly. Solubility in water is 50 gm/100 ml at 25°. Solubility in alcohol is 1.3 gm/100 ml. In contrast, the free acid has a solubility in water of only 0.34 gm/100 ml.

COONa

Specifications for sodium benzoate appear in *Food Chemicals Codex*.

Antimicrobial Activity

Sodium benzoate is generally considered to be most active against yeast and bacteria, and less active against molds but it is difficult to obtain substantial evidence on relative activity from available studies. There are few clear-cut comparisons between types of organisms, as the majority of the reported studies have been concerned more with the mode of action and used only on a few organisms.[33,52,54,55,94,137,191,248]

The report by Cruess and Richert contains several curves which depict the relationship of pH to inhibitory concentrations of sodium benzoate for *S. ellipsoideus*, three molds and four species of bacteria.[54]

Cruess stated that at pH 2.3 to 2.4, only 0.02 to 0.03% of sodium benzoate was required to prevent growth of most fermentation organisms studied, and at pH 3.5 to 4.0, the range of most fruit juices, 0.06 to 0.10% was required.[52]

Molin reported that sodium chloride has a considerable synergistic effect with sodium benzoate.[160]

Safety for Use

Subacute toxicity tests have shown that sodium benzoate is relatively more toxic than sodium sorbate. In these tests, rats which received benzoate as 8% of their diet over a period of 90 days showed lessened weight gain and other physiological changes which were not observed in animals which received 8% of sorbate.[60] However, from the extensive human feeding investigations conducted early in this century by the U.S. Department of Agriculture,[46] the following conclusions were drawn:

1. "Sodium Benzoate in small doses (under 0.5 gram per day), mixed with the food, is without deleterious or poisonous action and is not deleterious to health."

2. "Sodium Benzoate in large doses (up to 4 grams per day) mixed with the food has not been found to exert deleterious effects on the general health, nor act as a poison in the general acceptation of the term. In some directions there were slight modifications in certain physiological processes, the exact significance of which is not known."

3. "The admixture of Sodium Benzoate with food in small or large doses has not been found to injuriously affect or impair the nutritive value of food."

There is no danger of accumulation of benzoate in the body.[20] The apparent reason for the high tolerance of the body to benzoic acid is a detoxifying mechanism, whereby the benzoate is conjugated with glycine to produce hippuric acid and is excreted in that form. Griffith found that this mechanism accounted for from 66 to 95% of benzoic acid fed when subjects ingested quantities far in excess of those normally used in foods.[98] He suggested that the remainder of the benzoate not excreted as hippuric acid may have been detoxified by conjugation with glycuronic acid.

Regulatory Status

For a long time benzoates were *generally recognized as safe* (GRAS) in the United States. At this writing the Food and Drug Administration is re-evaluating the GRAS list, and it seems likely that benzoic acid and sodium benzoate, because they have quantitative restrictions, may become subject to a food additive regulation. Standards of identity of some food products exclude preservatives.

In other countries, sodium benzoate is used in some additional types of foods from which it is excluded in the United States, and often much higher levels are permitted. Permitted levels up to 0.2% and 0.3% are common, and for liquid egg yolk the maximum in some countries is 1.25%. On the other hand France limits benzoate to rennet and fish products, and it is excluded from non-alcoholic beverages in Italy and Portugal.

Applications

General As stated previously, sodium benzoate is most suitable for foods and beverages which naturally are in the pH range below 4.0 or 4.5, or can be brought into that range by acid addition. Among the antimicrobial additives, sodium benzoate has the advantage of low cost. On the other hand, when it is incorporated into some foods, for example fruit beverages, there is the possibility of a noticeable taste. If this occurs, benzoate may be used at a lower level in combination with another antimicrobial, such as potassium sorbate or esters of p-hydroxybenzoic acid.

Uses in Various Types of Foods Sodium benzoate has especially wide applicability as an antimicrobial for foods. It is used in carbonated and still beverages, syrups, fruit salads, icings, jams, jellies, preserves, salted margarine, mincemeat, pickles and relishes, pie and pastry fillings, prepared salads and fruit cocktails. Use levels range from 0.05% to 0.10%.

The following information is pertinent with respect to some of these uses:

Carbonated and Non-Carbonated Beverages In general, from 0.05% to 0.1% sodium benzoate is used in carbonated and still beverages. The effectiveness of such levels for carbonated fruit beverages has been demonstrated by Tressler *et al.*[241] and by Fellers.[70]

Fruit Juices and Derivatives This includes the pure juices and juice drinks. Only certain products in this bracket may contain a preservative; the Code of Federal Regulations should be consulted for details.

Jams, Jellies, and Preserves Under Federal Standards of Identity for jams, jellies, and preserves, sodium benzoate is permitted (21 CFR 29.2; 21 CFR 29.3). Benzoate is also permitted in artificially sweetened jams, jellies, and preserves (21 CFR 29.4; 21 CFR 29.5).

Margarine Use of sodium benzoate is permitted under Federal Standards of Identity (21 CFR 45.1; 21 CFR 45.2). Its use is limited to salted margarine, as it is not sufficiently inhibitory in this food in the absence of salt. The benzoate can be added with the salt and any other ingredients that are dissolved in the aqueous mixture before churning.

Pickle Products The effectivness of sodium benzoate as a preservative for these highly acid products has been demonstrated by Fabian and Switzer,[68] by Smith,[221] and by Vaughn and his associates.[269] The last authors demonstrated the possibility of preservation without the use of brine, which creates a disposal problem.

Storage and Handling

Sodium benzoate should be stored in a cool, dry place. Containers should be kept closed as much as possible. This product is not corrosive and presents only slight danger as a toxicant or fire hazard under normal conditions.[207]

Assay

A method of assay for sodium benzoate in foods is found in *Official Methods of Analysis of the A.O.A.C.,* 1965 ed., p. 450. Other methods are noted on pp. 128 and 129 of this chapter under *p*-hydroxybenzoic acid esters. A more recently reported method for benzoate in foods is that of Gantenbein and Karasz.[270] Methods of determining various preservatives considered in this chapter are reviewed by Schuller and Veen[271] of the Institute of Public Health, Utrecht, Netherlands. These include sulfur dioxide, sorbic acid, benzoate, nitrite, and nitrate. The review covers references of the preceding decade and notes the methods which are official in several European countries and in the United States.

Esters of Para-Hydroxybenzoic Acid

The alkyl esters of *p*-hydroxybenzoic acid (parabens) comprise a group of antimicrobial agents which have been used widely in cosmetic and pharmaceutical products. Their utility in foods was later recognized, particularly because they offer a means for extending upward the limiting pH for effective use of benzoic acid.

The antimicrobial action of the parabens was first described in 1924 when Sabalitschka published the first of some 60 reports.[203] European publications appeared in great numbers describing applications in cosmetics, pharmaceuticals and foods. Regulations permitting the use of these preservatives in foods were adopted in many European countries between 1932 and 1938.

In the early 1930's pharmaceutical and cosmetic uses spread to the United States. By 1944 Neidig and Burrell summarized nearly 400 articles pertaining to properties and uses of the parabens.[185] Another definitive review was prepared by Aalto *et al.* on uses, chemical and antimicrobial properties, and analytical methods.[1] Coupled with companion reports on acute and chronic toxicity in dogs, rats, mice and rabbits[148] and the physiology of the parabens in dogs,[132] these three publications comprise an excellent treatise.

Physical and Chemical Properties

The methyl and propyl esters of *p*-hydroxybenzoic acid are most commonly used in the U.S.A., but the ethyl and butyl esters find some applications in other countries.

These compounds, being closely related to benzoic acid, have many properties in common with it, yet differ in ways which enhance their utility. Since it is well recognized that benzoic acid exerts its antimicrobial activity in the undissociated form, it is very likely that the same requirement holds for the hydroxybenzoates. By esterification of its carboxyl group, an undissociated molecule can be retained over a much wider pH range. The weaker phenolic group now provides the acidity rather than the carboxyl group, hence salt formations involve reactions with the phenolic OH.

TABLE 2
Solubilities of the Parabens*

Solubility, gm/100 gm Solvent	Methyl Ester	Ethyl Ester	Propyl Ester	Butyl Ester
Water, 25°C	0.25	0.11	0.04	0.015
Water, 15°C	0.16	0.08	0.023	0.005
Water, 80°C	3.2	0.86	0.45	0.15
Ethanol, 25°C	52	70	95	210
Ethanol, 10%, 25°C	0.5		0.1	
Ethanol, 50%, 25°C	18		18	
Propylene Glycol, 25°C	22	25	26	110
Propylene Glycol 10%, 25°C	0.3		0.06	
Propylene Glycol 50%, 25°C	2.7		0.9	
Glycerin, 25°C	1.7	0.5	0.4	0.3
Peanut Oil, 25°C	0.5	1	1.4	5

* Source: Tenneco Chemicals, Inc., Intermediates Division.

Water solubilities of the esters are inversely related to the number of carbon atoms in the ester group (Table 2), and the solubilities of the methyl and propyl esters are the least for most practical uses.

All of the p-hydroxybenzoic acid esters commonly used in foods are white, free-flowing powders. The methyl ester has a characteristic, though faint odor and a level of taste perception approximately the same as that of sodium benzoate. The propyl ester is virtually odorless. Specifications for the methyl and propyl esters appear in *Food Chemicals Codex*.

The parabens are stable against hydrolysis during autoclaving and they also resist saponification, so that they can be dissolved in 5% sodium hydroxide.

Whereas the esters are usually sold in the form which contains a free phenolic group, some salt forms are provided for convenience to permit getting these preservatives into solution without use of alcohol or heat. The calcium

TABLE 3
Minimum Inhibitory Concentrations of Parabens and Certain Other Preservatives Against Four Types of Molds as Affected by pH*

	At pH 3.0			
	Chaetomonium globosum %	*Alternaria solani* %	*Penicillium citrinum* %	*Aspergillus niger* %
Benzoic acid	0.08	0.10	0.10	0.04
Methyl p-hydroxybenzoate	0.06	0.06	0.05	0.08
Propyl p-hydroxybenzoate	0.008	0.015	0.005	0.02
Propionic acid	0.04	0.04	0.04	0.08
Sorbic acid	0.01	0.005	0.02	0.04
	At pH 5.0			
Benzoic acid	0.10	0.15	0.20	0.20
Methyl p-hydroxybenzoate	0.06	0.08	0.08	0.10
Propyl p-hydroxybenzoate	0.01	0.02	0.01	0.03
Propionic acid	0.04	0.06	0.08	0.08
Sorbic acid	0.06	0.02	0.08	0.08
	At pH 7.0			
Benzoic acid	+	+	+	+
Methyl p-hydroxybenzoate	0.10	0.10	0.15	0.15
Propyl p-hydroxybenzoate	0.04	0.05	0.06	0.05
Propionic acid	+	+	+	+
Sorbic acid	+	+	+	+
	At pH 9.0			
Benzoic acid	+	+	+	+
Methyl p-hydroxybenzoate	0.10	0.10	0.15	0.15
Propyl p-hydroxybenzoate	0.04	0.05	0.06	0.05
Propionic acid	+	+	+	+
Sorbic acid	+	+	+	+

*Source: Technical Data Sheet, Mallinckrodt Chemical Works, Washine Division, Lodi, N.J. Derived from Bandelin (Ref. 19).

salt is stable, but the sodium salt is extremely hygroscopic and unstable. For this reason, some manufacturers offer a physical mixture of the acid and the stoichiometric quantity of sodium hydroxide. The pH of a 0.1% unbuffered solution of the sodium salt is in the range of 10 to 11.

Antimicrobial Activity

The antimicrobial activity of the *p*-hydroxybenzoic esters is directly proportional to the chain length, but solubility also decreases with increasing chain length, hence, in practice, the lower esters are commonly used. However, there are reports of effective inhibitory action of higher esters at concentrations exceeding substrate solubility, a phenomenon not readily explained.

The wider pH range through which the parabens are effective is illustrated in Table 3 where the methyl and propyl parabens are compared with benzoic, propionic and sorbic acids in laboratory media against four types of fungi. In these examples, only the parabens were effective at pH 7 or higher. It should be noted that yeasts and molds are more of a spoilage problem under acid conditions at which they can be controlled by the less expensive sodium benzoate.

The parabens are most active against molds and yeasts. They are less effective against bacteria, especially gram-negative bacteria.[127]

Table 4 shows inhibitory concentrations as reported by Aalto *et al.* of methyl, ethyl, propyl and butyl parabens against molds, yeasts and bacteria.[1]

TABLE 4
Antimicrobial Activities of *p*-Hydroxybenzoates*

Microorganisms	Required for Inhibition, %			
	Methyl	*Ethyl*	*Propyl*	*Butyl*
Aspergillus niger ATCC 10254	0.1	0.04	0.02	0.02
Penicillium digitatum ATCC 10030	0.05	0.025	0.0063	<0.0032
Rhizopus nigricans ATCC 6227A	0.05	0.025	0.0125	0.0063
Trichoderma lignorum ATCC 8678	0.025	0.013	0.0125	0.0063
Chaetomonium globosum ATCC 6205	0.05	0.025	0.0063	<0.0032
Trichophyton mentagrophytes ATCC 9533	0.016	0.008	0.004	0.002
Trichophyton rubrum ATCC 10218	0.016	0.008	0.004	0.002
Candida albicans ATCC 10231	0.1	0.1	0.0125	0.0125
Saccharomyces cerevisiae ATCC 9763	0.1	0.05	0.0125	0.0063
Saccharomyces pastorianus ATCC 2366	0.1	0.05	0.0125	0.0063
Bacillus subtilis ATCC 6633	0.2	0.1	0.025	0.0125
Bacillus cereus var. mycoides ATCC 6462	0.2	0.1	0.0125	0.0063
Staphylococcus aureus (*Micrococcus pyogenes var. aureus*) ATCC 6538P	0.4	0.1	0.05	0.0125
Sarcina lutea	0.4	0.1	0.05	0.0125
Klebsiella pneumoniae ATCC 10031	0.1	0.05	0.025	0.0125
Escherichia coli ATCC 9637	0.2	0.1	0.1	0.4
Salmonella typhosa	0.2	0.1	0.1	0.1
Salmonella schottmuelleri	0.2	0.1	0.05	0.1
Proteus vulgaris ATCC 8427	0.2	0.1	0.05	0.05
Aerobacter aerogenes ATCC 8308	0.2	0.1	0.1	0.04

Precipitation was noted in those instances where the concentration used appreciably exceeded the water solubility of the ester.

* Source: Aalto *et al.*, (Ref. 1). Reproduced by permission of the American Pharmaceutical Association.

Safety for Use

Early work demonstrating the low order of toxicity of the parabens was reviewed by Neidig and Burrell.[165] A definitive study by Matthews et al. determined their acute toxicity in mice, dogs and white rats. Since the methyl and propyl esters are most widely used, emphasis was placed on these products.[148]

All parabens produced the same toxicity symptoms in mice—rapid onset of ataxia, paralysis and deep depression resembling anesthesia. Only rarely was there evidence of increased motor activity. In non-fatal doses, recovery was prompt, usually within thirty minutes. Where death occurred, it was usually within one hour.

Acute toxicities in mice were determined for oral, intravenous and intraperitoneal administrations. Orally the methyl and propyl parabens had an LD_{50} of >8000 mg/kg whether administered with starch, propylene glycol or olive oil. Sodium salts in water had LD_{50}'s of 2000 and 3700 mg/kg, respectively, reflecting their increased rate of absorption. For sodium propyl parabens, the LD_{50} was about 950 mg/kg. Only sodium salts could be tested intravenously. Unesterified p-hydroxybenzoate and the methyl and propyl esters showed intravenous LD_{50}'s of 1200, 170 and 180 mg/kg, respectively.

Upon intraperitoneal administration, the unesterified acid was poorly tolerated, being a strong acid irritating to the peritoneal cavity. Free acids of the esters and all sodium salts were well tolerated.

Intravenous studies in dogs with sodium salts showed that the rate of administration largely controlled the size of the lethal dose. If the rate was relatively slow, i.e., 5 mg/kg/min, the lethal dose for propyl parabens was 600 mg/kg.

Chronic toxicities for white rats were tested by feeding diets containing 2 and 8% each of methyl and propyl parabens for 96 weeks. Animals receiving 2% levels matched the weight gains of the controls. A mild growth retardation was observed at 8% levels.

Mongrel dogs were fed 1 gm/kg and 0.5 gm/kg daily of the methyl or propyl esters for a year or more. All dogs showed satisfactory growth rates and normal healthy conditions at the end of the experiment. Females receiving 0.5 gm/kg/day were mated and delivered healthy litters.

Jones et al. determined the disposition of the parabens in dogs.[231] Urine recoveries ranged from 50–95% except for the butyl ester for which recoveries were 40%. The authors concluded that esters are well absorbed and that hydrolysis of the ester linkage and metabolic conjugation constitute the chief route of elimination. A similar metabolic scheme was observed in man.

Regulatory Status

Under U.S. Food and Drug Administration regulations, methyl and propyl parabens are "generally recognized as safe" (GRAS) when used as chemical preservatives of foods, with total addition limits of 0.1% (21 CFR 121.101). Both of these esters likewise are permitted as indirect additives by prior sanctions as antimycotics in food-packaging materials (21 CFR 121.2001). No limits or restrictions are mentioned. Ethyl paraben is permitted as an indirect additive when used in adhesives for packaging, transport, or holding of food (21 CFR 121.2520). Both methyl and propyl parabens are included among the optional ingredients permitted in artificially sweetened fruit jellies and jams. They may be used alone or in combination with sorbates, propionates and benzoates, such that total preservatives do not exceed 0.1% (21 CFR 29.4; 21 CFR 29.5). The n-heptyl paraben is permitted only in beer (21 CFR 121.1186).

Among the seventeen countries discussed in the section on international uses of antimicrobial food additives (p. 171), about two-thirds of them list the parabens as officially approved for use. In Italy the ethyl, as well as the methyl and propyl, esters may be used as a direct additive. There is a close parallel between uses of parabens and those of sodium benzoate in the countries where the parabens are permitted, with the exception of England and Wales, where the use of parabens is confined to somewhat fewer types of foods. The butyl ester is used in Japan.

Applications

Methods of Incorporation in Foods and Beverages   No unusual procedures are required to incorporate the parabens in foods as long as their aqueous solubilities are kept in mind. The fine powders can be dissolved in 20–30 minutes, with stirring, at room temperature or in shorter times by heating to 70–75°C. The latter method has the added advantage of reducing microbial contamination.

Stock solutions, in ethanol or propylene glycol, may be prepared and carefully added to the product with adequate stirring to avoid precipitation.

The concentrations of aqueous solutions of the parabens can be markedly increased if the free acids are converted to alkali salts—sodium, potassium, calcium or ammonium. The esters are not hydrolyzed by this treatment.

For initial tests, a concentration of about 0.05% of a 2 to 1 ratio of methyl and propyl paraben, or about 0.1% of a 3 to 1 ratio in the case of a fatty food, is suggested. The use of a combination takes advantage of the higher water solubility of the methyl and the greater effectiveness of the propyl and also of the fact that the activities are additive, but the tastes are not. In general, the low concentrations used do not create taste problems, but should these occur, lower levels of the esters combined with sodium benzoate are recommended, particularly in slightly acid foods. There is evidence that the antimicrobial effects of parabens and sodium benzoate are additive.[195]

Representative Uses

Baked Goods   About 0.03 to 0.06% of a combination of 3 to 1 methyl and propyl paraben, sometimes with sodium benzoate, can be used to improve keeping qualities of cakes (particularly fruit cakes), pie crusts, pastries (non-yeast), icings, toppings and fillings, such as fruit jellies and creams. Activity of parabens against yeasts excludes them from bread and rolls.

Beverages   Soft drinks can be preserved with about 0.03 to 0.05% of a 2 to 1 ratio of methyl and propyl parabens. A slightly higher concentration has been suggested for cider.

Beer   The n-heptyl-ester of p-hydroxybenzoic acid can be used in beer to control secondary yeast fermentation, as an alternative to pasteurizing or Millipore filtering. The maximum level is 12 ppm. This ester is added by dissolving in propylene glycol or alkali.

Cheese Parabens are not included in the antimicrobials mentioned in the Standards of Identity for cheese. However, experimentally, control of mold with dips in paraben-containing solutions has been demonstrated.

Creams and Pastes Products of various compositions are reported to be preserved with about 0.1% of a combination of parabens. Foods of high oil content generally require higher ratios of propyl.

Flavor Extracts A combination of parabens at a level of 0.05% is suggested.

Fruit Products Fruit salads, juices, sauces, syrups and fillings are among the foods reported to be preserved by the parabens. Combinations with sodium benzoate are recommended. About 0.05% of a 2 to 1 ratio of methyl and propyl is suggested for trial runs.

Jams, Jellies, Preserves The Federal Standards of Identity for artificially sweetened products of this type include methyl and propyl p-hydroxybenzoates as optional preservatives. About 0.07% of a 2 to 1 ratio of these gives protection against spoilage. They may advantageously be combined with sodium benzoate, with a total limit of 0.1%. The parabens are not authorized in sugar-sweetened jams, jellies and preserves.

Olives and Pickles A level of about 0.1% of parabens is recommended for olives and the same for pickles or a lower level in combination with sodium benzoate.

Syrups A 42.5% sucrose syrup was preserved in laboratory tests by 0.07% methyl or less than 0.02% propyl paraben.[208] In chocolate and other fountain syrups, about 0.02–0.03% propyl is sometimes combined with sodium benzoate.

Experimental use of esters of p-hydroxybenzoic acid in various other foods has been reported: margarine, butter, ices, confections, soy sauce, maple syrup[13] and meat.[77] Propyl p-hydroxybenzoate, 3.5% solution, may be used as a dip or spray for dry sausage casings to control mold (*Fed. Reg. 33*, 5210, 1968).

Storage and Handling

The methyl and propyl parabens and their calcium salts are non-toxic and non-irritating to the skin, so that no special precautions are necessary in their handling. The mixture of sodium hydroxide and an ester of p-hydroxybenzoic acid is hygroscopic. Containers should be stored in a dry place and kept tightly sealed when not in use. Since these mixtures are very alkaline, they should be handled carefully in accordance with standard procedure for such a product.

Assay

A summary of some general principles relating to the determination of p-hydroxybenzoate esters is presented by Aalto *et al.*[1] who also cite several references on analysis in the older literature. A report on the determination of benzoates and hydroxybenzoate esters in foods has been published recently by chemists of the U.S. Food and Drug Administration.[189]

Qualitative determinations of benzoic acid, sorbic acid and the several parabens can be made by thin layer chromatography on kieselguhr-silica gel plates developed with a hexane-acetic acid system. Acidified food samples are steam-distilled, the distillate extracted into solvents and concentrated. Plates are observed under UV light or may be sprayed with one of several identifying reagents. Bromination of the samples provides some refinements; only benzoic acid does not brominate, hence the other compounds exhibit two spots after development. The technique can be made quantitative for benzoic acid by removing the spot, extracting with ethanol and measuring UV absorption.

By combining chromatographic methods with spectrophotometry, Höyem was able to separate and quantitively determine sorbic acid, benzoic acid and the methyl, ethyl, propyl and butyl esters of p-hydroxybenzoic acid.[123]

Sorbic Acid and Its Salts

Knowledge that the straight-chain monocarboxylic acids possess fungistatic action has existed for some time. A greater degree of activity of unsaturated acids, in comparison with saturated ones of the same chain length, was reported in the 1930's by Japanese workers.[136,235] Supporting data were obtained by Wyss *et al.*[264] In 1945, Gooding secured a patent covering the use of unsaturated fatty acids having a double bond in the α-position as fungistatic agents in foods and on food wrappers.[90] He reported that crotonic acid, $CH_3CH{=}CHCOOH$, and its homologs were useful. Sorbic acid, $CH_3CH{=}CHCH{=}CHCOOH$, a diene, was particularly effective, surpassing sodium benzoate in certain applications. Moreover, it was relatively tasteless and odorless. Subsequent evaluation by various investigators confirmed the antifungal properties of sorbic acid and demonstrated further its commercial possibilities. See the reviews by Wyss,[263] Schelhorn,[248] Grubb[103] and York,[266] and report by Wolf.[258] Sorbic acid and its sodium and potassium salts* are established as effective preservatives at low concentration for the control of mold and yeast in cheese products, baked goods, fruit juices, fresh fruits and vegetables, wines, soft drinks, pickles and sauerkraut and certain meat and fish products.

Physical and Chemical Properties

Sorbic Acid, $CH_3CH{=}CHCH{=}CHCOOH$ This white, crystalline powder is only slightly soluble in water—0.16 gm/100 ml at 20°C. In other liquids, solubilities in gm/100 ml at 20°C are alcohol, 14.8; propylene glycol, anhydrous, 5.5; 50%, 0.6; vegetable oil, 0.52. Figure 1 shows solubilities in water over a range of temperatures and in water containing various percentages of acetic acid or alcohol. To incorporate in dry materials, sorbic acid may first be mixed with salt, flour or corn starch. To dissolve in liquids it can be solubilized with sodium or potassium hydroxide, or it may also be dissolved in propylene glycol or ethanol for use in dips or sprays. Sorbic acid sublimes on heating and can be steam-distilled. In food processing, as for example the making of synthetically sweetened jellies, sorbic acid should be added after any stages of prolonged boiling.

Potassium Sorbate, $CH_3CH{=}CHCH{=}CHCOOK$ This is a white, fluffy powder which is very water soluble—139.2 gm/100 ml at 20°C. In alcohol, solubility is 2.0 gm/ml at 20°C. This salt was specifically developed for the preparation of aqueous stock solutions. Solutions of relatively high concentration are necessary for dip, spray and metering applications.

* Called for convenience "sorbates" throughout this chapter.

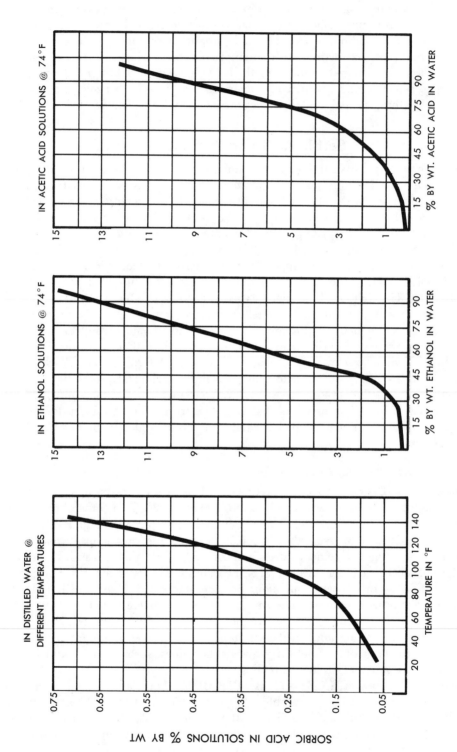

Fig. I.   Increase in solubility of sorbic acid in water with increasing temperature or with increasing concentration of ethanol or acetic acid in water (from Gooding et al., Ref. 91. Reproduced by permission of the Institute of Food Technologists, Copyright © 1955.)

Sodium Sorbate   This is soluble to about 28%. It can be made by reacting sorbic acid with aqueous sodium hydroxide. Stock solutions of 20% sorbic acid equivalence may be prepared in this manner. Specifications for sorbic acid and potassium sorbate appear in *Food Chemicals Codex*.

Antimicrobial Activity

Sorbic acid and its salts have broad-spectrum activity against yeast and molds, but are less active against bacteria. Their range of optimum effectiveness extends up to pH 6.5—considerably above that of the propionates or sodium benzoate, but not so high as that of the parabens. The sorbates have been found generally superior to benzoates for preservation of margarine, cheese, fish, bread and cake.[93]

As with other weak acid microbial inhibitors, activity of sorbate increases as the pH of the medium declines. This is illustrated in Table 3, page 124, derived from the data of Bandelin.[19]

Among studies demonstrating the breadth of activity of sorbates was that of Beneke and Fabian[27] who found that a wide range of fungi isolated from strawberries and tomatoes was inhibited by 0.05% sorbic acid in a tomato juice medium at pH 4.4. In another experiment, 0.075% sorbic acid inhibited the growth of all test fungi when added to strawberry puree at a pH natural to the product.

Another extensive study concerned both with pH and multitude of microbial species was that of Bell, Etchells and Borg.[26] They tested sorbate against 66 species of filamentous fungi, 32 species of yeast and 6 species of lactic acid bacteria and stated: "All of the organisms studied grew in media containing 0.1 per cent sorbic acid at pH 7.0. The yeasts and filamentous fungi were inhibited in media containing 0.1% sorbic acid at pH 4.5. The lactic acid bacteria were inhibited at this concentration of the chemical at pH 3.5."

Extensive tables on antimicrobial spectrum are found also in the thesis by York.[266]

Emard and Vaughn,[67] using a liver infusion-glucose-yeast extract medium and pure cultures, observed a selective inhibition of catalase-positive bacteria in concentrations of sorbic acid from 0.07 to 0.12%.

It has been reported by Emard and Vaughn[67], York and Vaughn[267], and Hansen and Appleman[106] that Clostridia cells and spores resist concentrations of sorbic acid which are inhibitory to the majority of other bacteria tested.

As discussed in the section on Safety for Use, sorbic acid is metabolized like other fatty acids in the mammalian body. This acid also can be metabolized by microorganisms. York and Vaughn further determined in their experiments with Clostridia that sorbic acid disappeared from several cultures of *Clostridium parabotulinum*. They stated that the apparent utilization of sorbic acid after 7 days incubation at 37°C ranged from 86 to 100% for strains of types A and B and was 62% for non-toxic strains.[267]

Melnick *et al.* attributed the consumption of sorbic acid by molds to β-oxidation mechanisms, similar to those in mammals.[154] More recently, Marth and his associates have proposed another mechanism, namely that Penicillia, particularly *P. roqueforti*, degrade sorbic acid by decarboxylation to 1,3 pentadiene, which may account for the hydrocarbon-like odor sometimes associated with sorbates.[147]

These findings on sorbic acid metabolism by *P. roqueforti* are interesting in the light of Lawrence's studies on C_6—C_{12} saturated fatty acids.[139,140] He found that spores of *P. roqueforti* oxidize these acids to methyl ketones with one less carbon atom, resulting from decarboxylation. Thus heptane-2-one was formed from octanoic acid. The diene structure of sorbic acid appears to prevent oxidation to the ketone.

The solubility of sorbic acid in water is diminished by salt, but Gooding and Melnick state that salt has a marked synergistic action with sorbate in fungistasis, and that the same relationship exists with sugar in strong solutions.[91] More recently Rao *et al.*[192] also reported synergism of salt and sorbic acid in a study on the preservation of chapatis. This effect was noticeable with salt contents of 1 to 2.5%

Safety for Use

Sorbic acid has a lower order of toxicity than benzoic acid, because it is probably metabolized like any other fatty acid by man and animals. An interesting dual study of the comparative toxicities of sorbic acid and benzoic acid, conducted in two different laboratories, has been reported by Deuel *et al.*[60]

Acute toxicities were determined in fasted rats in Laboratory A and unfasted rats in Laboratory B. Results are presented in Table 5. Sorbic acid

TABLE 5
LD₅₀ of Sorbic Acid, Sodium Sorbate and Sodium Benzoate in Rats*

	LD_{50} grams/kg	
	Laboratory A	*Laboratory B*
Sorbic acid	—	10.50 ± 1.96 (9.17–12.03)
Sodium sorbate (as sorbic acid)	4.0	5.94 (5.65–6.23)
Sodium benzoate (as benzoic acid)	2.1 (1.7–2.5)	3.45 (3.15–3.74)
Ratio: $\dfrac{LD_{50} \text{ Sodium sorbate}}{LD_{50} \text{ Sodium benzoate}}$	1.90	1.72

* Source: Deuel *et al.*, (Ref. 60). Reproduced by permission of Institute of Food Technologists, Copyright © 1954.

was better tolerated than its sodium salt, a reflection of its rate of absorption. The somewhat greater toxicity for sodium sorbate and sodium benzoate in Laboratory A is explained by the smaller animals employed. However, both laboratories demonstrated that sodium benzoate is nearly twice as toxic as sodium sorbate.

Subacute toxicity studies for rats compared 1, 2, 4 and 8% dietary levels of sorbic acid or sodium benzoate for 90 days. No adverse effects were noted in rats fed sorbic acid at all dietary levels. On the other hand, with sodium benzoate at the 8% level, 40% of the rats did not survive. Survivors' weights were only two-thirds those of the control animals and these exhibited enlarged livers and kidneys. At the 4% level no harmful effects were noted.

Sorbic acid and caproic acid, its saturated fatty acid analog, were fed to puppies at 4% dietary levels for 90 days. An extensive examination of tissues and organs revealed no pathological developments.

The evidence that sorbic acid is metabolized in the body like any normally-occurring fatty acid, such as caproic and butyric, was obtained in rat studies also by Deuel et al.[59]

Regulatory Status

Sorbic acid and potassium sorbate are generally recognized as safe (GRAS) for use in foods under regulations of the U.S. Food and Drug Administration. No upper limits are imposed for foods not under Federal Standards of Identity. Limits are set on use of sorbates in certain foods which are under Standards of Identity. Thus, in regular cheese, the maximum quantity may not exceed 0.3% by weight, calculated as sorbic acid, and the maximum is set at 0.2% in pasteurized blended cheese, pasteurized process cheese, pasteurized process cheese food and spread, pasteurized cheese spread, and cold pack cheese, cheese food and spread. In margarine the maximum is set at 0.1%.

Sorbic acid or its salts may be used in wine or in materials for the production of wine at levels up to 0.1% and need not be declared on the label (Revenue Rulings 58-461 and 59-105, Alcohol and Tobacco Tax Division, Internal Revenue Service, U.S. Treasury Dept.).

Sorbic acid and its salts are permitted in all countries shown in Table 18 in the section on the regulatory status of preservatives in countries other than the United States. Some exceptions from U.S. practice permit sorbates in egg yolk, liquid whole egg, semi-preserved fish, sliced bread, and ice cream.

Applications

General Sorbate can be used as a direct additive, as a spray or dip bath, and as a coating on wrapping material. As a direct additive, sorbic acid can be added dry to cakes, salads and the like; it can be easily mixed with the shortening and dressing and then distributed throughout the product. Potassium sorbate stock solutions of 10 to 20% are used for direct addition to beverages and pickle products. Potassium sorbate solutions may be used for spray and dip bath applications to cheese, dried fruits, smoked fish and similar products. Wrapping materials treated with sorbate have been used principally to protect cheese products.

Cheese and Cheese Products Data showing the activity of sorbic acid against the organisms which cause spoilage in cheese appear in the reports of Wolf[258] and Bonner and Harmon.[31]

Under present Federal Standards of Identity, sorbic acid and/or its sodium and potassium salts may be added to over 40 types of cheese, cheese spreads, and cheese foods. The preservatives are applied to natural cheese by dipping cuts or wedges into solutions of the antimycotic or by spraying them with the solution. For pasteurized process and blended cheese, pasteurized process cheese with fruits, vegetables or meats, cheese food and cheese spreads, required quantities of sorbate may be added to the batch in the melter-mixer. For cold-pack cheese the sorbate in dry form is added to the cheese during grinding or mixing. Also, since mold contamination is a surface phenomenon, sorbic acid may be dusted onto the surface, or protection may be afforded by sorbic-acid-impregnated wrappers. Various·investigations have shown that molds can be effectively controlled by wrappers impregnated with from 2.5 to 5.0 gm of the chemical per 1000 sq in. of thermoplastic-coated cellophane in small packages of American cheese or processed cheese.[109, 153, 218, 220]

Present Federal Standards of Identity do not permit inclusion of sorbates in cottage cheese. However, Geminder found that shelf life was extended by three weeks or more from the use of 0.05% of sorbic acid or its salts.[81] Similar evidence was obtained by Bradley, Harmon and Stine[37] and Perry and Lawrence.[184]

Baked Goods The sorbates affect yeasts as well as molds and therefore cannot be added directly to yeast-raised goods. However, their effectiveness for control of mold in cakes, pies and pie fillings has been demonstrated.[39,92,155,190,211]

Either sorbic acid or the potassium salt can be used in cakes, but the acid is preferred, as addition is direct and solubility is not a factor.

The pH of the cake affects the required sorbate concentration. A more acidic cake generally requires less than does a more alkaline one, such as devil's food cake.[155] Table 6 shows recommended levels of sorbates for bakery products.

Beverages and Beverage Syrups Sorbates may be used alone or in conjunction with sodium benzoate in carbonated and still beverages. The sorbate is as effective as benzoate and is less likely to alter flavor at effective levels. In fountain syrups also benzoate or sorbates may be used alone or in combination. The higher optimum pH range of sorbates makes them preferable over benzoates for chocolate syrups, some of which have a relatively high pH.

Fruit Juices and Derivatives Most of the conditions that apply to beverages apply to fruit juices. Sodium benzoate can be used at 0.05 to 0.10%, potassium sorbate at 0.025 to 0.10%, or they can be used together, each at a lower level. See the *Code of Federal Regulations* for details on the use of sorbates in fruit juices and derivatives.

Studies which have demonstrated the effectiveness of sorbates in fruit juices are those of Ferguson and Powrie[71] and Salunke[206] with apple juice, Pedersen et al.,[182] with grape juice, and Weaver et al., with cider.[251]

Wines Sulfur dioxide is commonly added to the table wines. Although this treatment inhibits bacteria, it may not effectively inhibit certain types of yeast. Moreover, in using sulfur dioxide, successive additions must be made during the storage life of the product. Some producers pasteurize, but this may adversely affect quality.[175] The addition of low levels of sorbate to high sugar wines inhibits yeast effectively during the storage stage whether in bottle or in bulk. The first published report in the United States was that of Auerbach.[15] Reports have appeared from other countries.[58,205,228,246,252]

If a large quantity of sulfur dioxide is added during storage (so that none need be added at bottling), it may be possible to reduce the SO_2 by the use of sorbate. In commercial evaluation 0.04% to 0.05% potassium sorbate permitted a reduction of 25% in the total sulfur dioxide concentration.

More extensive information on the applications of sorbic acid in wine, including its use as an alternate for or in conjunction with sulfur dioxide, is set forth in the papers of Vitagliano, Saller and Kolewa,[205] Ough and Ingraham,[177,178] O'Rourke and Weaver.[175] Technical publications by manufacturers of sorbates also contain directions for use.

TABLE 6
Suggested Addition Levels for Sorbates in Bakery Products*

Product	% Batter Wt	Level of Antimycotic oz/100 lb Batter	Method of Addition
Angel Food Cake	0.03–0.06	0.5–1.0	Dry mix with flour or blend into batter before whipping.
Cheese Cake	0.09–0.125	1.5–2.0	Dry blend with sugar and milk powder.
Chocolate Cake	0.09–0.125	1.5–2.0	Dry mix with flour or blend during creaming.
Devil's Food Cake	0.3	5.0	Dry mix with flour or blend during creaming.
Fruit Cake	0.05–0.10	0.8–1.5	Dry mix with flour or blend in batter. Maximum protection is provided by presoaking fruits in 1% potassium sorbate solutions.
Pound Cake White Cake Yellow Cake	0.075–0.10	1.2–1.5	Dry mix with flour or blend during creaming.
Cake Mixes	0.05–0.10	0.8–1.5	Blend with flour and other dry ingredients.

	% Batch Wt	oz/100 lb Batch	
Bakers Fillings Fudges Icings Toppings	0.05–0.10	0.8–1.5	Add after heating operations when temperature is below 160°F. Allow 5 minutes agitation for distribution or solution.
Pie Crust Dough	0.05–0.10	0.8–1.5	Dry mix with flour or add to dough during mixing.
Pie Filling	0.05–0.10	0.8–1.5	Mix with fruit and syrup. If filling is cooked, add sorbic acid after cooking—on cooling cycle. This is important in custard and pudding-type fillings.
Doughnut Mixes	0.03–0.08	0.8–1.5	Preblend with salt and then blend in mix.

* Source: Chas. Pfizer & Co., 1966, Brochure, "Food Preservatives."

Fruit Butters, Jellies, Jams, Preserves Federal Standards of Identity for artificially sweetened jams, jellies and preserves permit use of sorbic and its salts to maximum levels of 0.1% by weight. These preservatives are not permitted in the sugar-sweetened forms. However, 0.1% sorbate is permitted in sugar-sweetened fruit butters.

Salads and Fruit Cocktails These convenience items, including gelatin salads, are generally kept under refrigeration, which retards bacteria but may permit growth of yeast and molds. The addition of sorbate at 0.05 to 0.1% increases the shelf life of these and similar products.

Potassium sorbate is most conveniently added to the liquid part of salads such as the cover syrup in fruit salads and cocktails, the dressing in cole slaw, potato and macaroni salads, or the hot water in fruit gelatin.

Dried Fruits Consumers desire high moisture in dried fruits such as raisins, prunes and figs, because it means better texture, but this increases their susceptibility to mold• and yeast growth. Potassium sorbate, applied by dip or spray, effectively protects dried fruits, according to industrial evaluations with raisins, prunes and figs.[180,24,169,110]

On the basis of these studies, it is suggested that initial trials be made with solutions containing 2 to 7% potassium sorbate. The deposit of the preservative on the fruit should be from 0.02–0.05% by weight.

Pickles and Pickled Products For some of these products, either sodium benzoate or sorbate can be used. The pH of sweet relishes and olives is more favorable to sorbate than to benzoates. Recommended levels are from 0.025 to 0.05%, the higher level being more appropriate to sweeter products. Tests showing effectiveness at or near these low levels have been reported by Costilow et al., [50,51,214] and Jones and Harper.[131]

The investigation of Phillips and Mundt[188] showed that 0.1% of sorbate in cucumber fermentations controlled scum yeast and did not interfere with lactic acid production by bacteria. In practice even lower levels are effective.

Margarine Under Federal Standards of Identity for margarine, potassium sorbate, but not sorbic acid, may be used as a preservative (21 CFR 45.1). The sorbate can be incorporated by inclusion in the milk. The quantity in or on the margarine must not exceed 0.1% by weight of the finished product. Potassium sorbate is most appropriately used in non-salted margarine as absence of salt makes this type more susceptible to molds.

Meats and Fish Wolf reported in 1948 that sorbic acid is effective against 16 species of molds common to meats.[258] However, under present U.S. Government regulations, preservatives are not permitted in meat products for human consumption with the exception of potassium sorbate and propyl-p-hydroxybenzoate on dry sausage to inhibit surface mold. The sorbate may be applied as a $2\frac{1}{2}$% water solution after stuffing and before drying. Alternatively, casings may be dipped into a $2\frac{1}{2}$% solution of potassium sorbate before stuffing. (Memorandum 326, July 22, 1965, Consumer and Marketing Service, U.S.D.A.). There are certain kinds of salami on which surface mold is considered desirable, so that for these the sorbate treatment would not apply.

Sorbates have been found to increase shelf life of smoked and salted fish by checking mold growth.[35,36,82,237] Depending on processing conditions, potassium sorbate may be applied to whole or eviscerated fish or fillets prior to, or immediately after smoking. One-minute dips in a 5% solution or controlled use of 10% spray solution are satisfactory. In treating salted fish, potassium sorbate is added to the brine solution or mixed in dry salt. There should be no more than 0.1% sorbate on the fish.

Dog food patties, because of their moisture content, are subject to mold growth. For these, potassium sorbate is used at levels up to 0.5%, the usual level being 0.3%.

There are no federal regulations regarding use of sorbates in dry and smoked fish; however, state and local regulations should be consulted.

Miscellaneous Uses The sorbates have been evaluated in various other foods, usually at levels of from 0.025% to 0.1%. Extension of shelf life was noted in such products as tangerine sherbet base, maraschino cherries, strawberry puree and tomato juice, pre-peeled carrots and wax cucumbers.[181,34,27,76,116]

Storage and Handling

Sorbic acid and potassium sorbate should be stored at temperatures below 100°F and should not be exposed to light or heat. Containers should be kept closed. Paper, cloth and other absorbent materials soaked with sorbates may ignite spontaneously so that these substances should be kept away from cellulosic materials. Sorbic acid and potassium sorbate can cause eye irritation. In case of contact with the eyes, flush with water for 15 minutes and get medical attention.

Assay

Assay methods for sorbic acid and its salts in foods are not included in the 1965 edition of *Official Methods of Analysis of the A.O.A.C.* Various methods have been published in the literature.

Alderton and Lewis described a spectrophotometric method which is a modification of one developed earlier by Melnick and Luckmann.[3,152] Ciaccio has applied spectrophotometry to the determination of sorbate in low-calorie salad dressing which also contains propyl-*p*-hydroxybenzoate, and Harrington *et al.* have used this principle in the determination of sorbic acid in apple cider.[47,108] By combining chromatographic methods with spectrophotometry, Höyem was able to separate and quantitatively determine sorbic acid, benzoic acid and the methyl, ethyl, propyl and butyl esters of *p*-hydroxybenzoic acid.[123]

Spanyár and Sánder have reported a titrimetric method for sorbates in foods, and Nury and Bolin have described a colorimetric procedure.[223,168] Gantenbein and Karasz[270] have described a single-extraction ultraviolet spectrophotometric method for sorbates and benzoate.

Other assay methods are mentioned on p. 128 under *p*-hydroxybenzoic acid esters.

Propionic Acid and Its Salts

Propionic acid is a member of the aliphatic monocarboxylic acid series. The antimicrobial action of the members of this series has been known for some time, one of the earliest reports being that of Kiesel in 1913.[134] Another broad-scale study was that by Hoffman, Schweitzer and Dalby.[122] They evaluated the fungistatic properties of the normal saturated fatty acids containing from 1 to 14 carbon atoms, over a pH range from 2 to 8 and found that antifungal effectiveness varied according to chain length, concentration of acid and pH. These same investigators in that year patented the use of propionic acid and certain of its salts as a mold inhibitor in bread.[121] Propionates were selected because the higher homologs, although they have higher antimicrobial activity, also have tastes and odors which would be noticeable, especially in baked goods.

At about this same time others found that propionic acid and propionates are especially effective against *Bacillus mesentericus*, which causes "rope" in bread,[171] and also that propionates are useful to retard the molding of cheese.[159]

Physical and Chemical Properties

Propionic acid, a liquid, has a strong odor and is somewhat corrosive. Therefore, the propionates used in the food industry as antimicrobials are the sodium and calcium salts. These yield the free acid in the pH range of the foods in which they are used. These salts have a slight, cheese-like flavor, which blends with that of most foods.

Both of these salts are white, free-flowing, readily soluble powders. The sodium salt is more soluble, dissolving at the rate of 150 gm per 100 ml of water at 100°C and in alcohol at the rate of 4 gm per 100 ml at 25°C.

Calcium propionate dissolves in water at the rate of 55.8 gm per 100 ml at 100°C and is insoluble in alcohol. Both the sodium and calcium salt blend readily with emulsifying agents used in processed and blended cheese, as well as with basic dough ingredients and in filling for pies.

$$CH_3CH_2COONa \qquad (CH_3CH_2COO)_2Ca$$
Sodium Propionate   Calcium Propionate

Specifications for sodium and calcium propionate appear in *Food Chemicals Codex.*

Antimicrobial Activity

Propionates are more active against molds than sodium benzoate, but have essentially no activity against yeast. They have little action also against bacteria with the notable exception of their ability to inhibit the organism which causes rope. The pH for optimum activity of propionates ranges up to 5.0, although in some foods they are active up to pH 6, or slightly higher. The effect of variations in pH of medium on antimicrobial activity has been studied by Cruess and Richert,[54] Cruess and Irish[53] and Olson and Macy.[172] The relative activity of sodium propionate against 7 species of bacteria, 3 yeasts and 2 molds in broth culture has been reported by Wolford and Anderson.[259]

Safety for Use

Propionic acid occurs naturally in Swiss cheese at levels which may be as high as 1%.[11] It is also a metabolite within the rumen of ruminant animals.[66] Propionate is metabolized like other fatty acids in the mammalian body, as shown by studies in which the acid was labeled with radioactive carbon.[260]

In rat feeding trials, Harshbarger found that sodium or calcium propionate had no effect on growth when fed at levels of from 1 to 3 gm. daily, for 4 to 5 weeks, beginning when the animals were about 4 weeks of age.[111] In later studies by Graham *et al.*, a level of 5% of propionate was placed in bread before baking and the finished product was fed as three-fourths of the diet of rats for one year.[96] There was no discernible toxicity in spite of the simultaneous presence of three other common bread additives at high concentrations in the diet.

Regulatory Status

Under Federal Regulations propionates are generally recognized as safe for use in foods and no upper limits are imposed, except for bread, rolls and cheese which come under Standards of Identity. Sodium and calcium propionate are limited to 0.32% of flour in white bread and rolls and 0.38% in the corresponding whole wheat products (21 CFR 17.1, 17.5). In cheese products they are limited to 0.3%.

Propionates are not mentioned in the food-additive regulations of West Germany, India, Pakistan, the Philippines, and South Africa, but they are included in those of other West European countries for which information is at hand, as well as those of Canada.

Applications

The major uses of propionates are to prevent mold and rope in baked goods and mold in certain types of cheese. Published experimental data in technical journals are meager, but the propionates have received widespread acceptance since they were introduced about 25 years ago. They are suitable for yeast-raised as well as other baked goods, and are reasonable in price. Undoubtedly practical trials in the baking industry have provided the mass of technical information for the successful use of these compounds.

Baked Goods Since most molds are destroyed by heat of baking, the contamination of baked goods occurs afterward and growth can flourish under the wrapper in the humid environment. Higher environmental or storage temperatures assist this process. The serious dimensions of the mold contamination problem in bakeries have been outlined by Ingram et al.[126]

One of the earliest reports of the effectiveness of propionates against molds in baked goods was the patent of Hoffman, Dalby and Schweitzer.[121] Ingram (loc. cit.) has summarized some of his own data on propionic acid and its sodium and calcium salts in mold control. He found that bread which contained 0.2% of propionate (basis total dough weight) showed no visible mold for over eight days in both the wrapped and unwrapped condition at room temperature. Breads without the preservative molded in four days at the lower temperature and in less than four days at the higher.

A recent study by Seiler presented quantitative data on equilibrium relative humidity as a controlling factor in mold growth on bread and cake, as well as information on propionates and sorbates as mold inhibitors.[211] He reported that, for bread, 0.2% of calcium propionate by flour weight extended mold-free life to six days, as compared with four days for untreated loaves. No odor problems were found at this level.

In relation to rope control by propionate there is the early study of O'Leary and Kralovec.[171] Their data showed 0.11% (basis of flour) of calcium propionate to be more effective than 0.25% of calcium acid phosphate in the control of rope in bread inoculated with various levels of B. mesentericus, up to 9,700,000 bacteria per loaf. The propionates appeared to delay the development of color and viscousness more than the phosphate. These workers reported that calcium propionate at a level of 0.188% effectively inhibited rope in bread having a pH as high as 5.8. At pH 5.6, a concentration of 0.15% was effectively inhibitory. Because propionate also inhibits molds and spares yeast, it has superseded the acid phosphate as an anti-rope agent.

More recent comparative data on propionates and other agents for inhibition of rope in baked goods appears in the paper by Ingram and his associates.[126]

Both sodium and calcium propionate can be used in baked goods. The calcium salt is preferred in bread as the calcium contributes to enrichment. The sodium salt is preferred for cakes and unleavened goods where the calcium ion can interfere with chemical leavening. The propionates disperse easily

TABLE 7
Suggested Levels of Propionates in Baked Goods*

Type of Baked Goods	Level of Propionate
White breads, buns, rolls, specialties	2½ to 5 oz per 100 lb flour. Three oz usually should be sufficient, but 5 oz may be needed under extreme conditions, such as warm moist weather.
Dark breads, whole or cracked wheat, rye breads, buns, rolls, etc.	3 to 6 oz per 100 lb flour. Four oz usually should be sufficient, but 6 oz may be needed under extreme conditions.
Angel Food Cake	1½ to 3½ oz per 100 lb batter; generally 2½ oz suffices.
Cheese Cake	2 to 4 oz per 100 lb batter; generally 3 oz suffices.
Chocolate or Devil's Food Cake	5 to 7 oz per 100 lb batter; generally 6 oz suffices.
Fruit Cake	2 to 6 oz per 100 lb batter; the higher the fruit content, the less propionate needed.
Pound Cake, White or Yellow Cake	4 to 6 oz per 100 lb batter; generally 5 oz suffices.
Pie Crust	2 to 5 oz per 100 lb dough.
Pie Fillings	2 to 5 oz per 100 lb filling.

In cakes use only sodium propionate. Add it with the baking powder and take out an equivalent weight of salt from the formula.

In fillings add propionate just before the end of cooking.

In pie crust dissolve propionate in the water to make up the crust.

"Brown 'n serve" rolls. Many bakers today prefer an oil spray on these rolls in order to allow the rolls to expand without tearing and, at the same time, to reduce the formation of a tough skin. Sodium propionate may be incorporated into the spray mixture in the following proportions.

Vegetable oil............................. 15.0 lb
Lard or vegetable shortening............... 5.0 lb
Mono- and diglyceride emulsifiers.......... 0.8 lb
33% solution of sodium propionate......... 1.0 lb

(quantities of vegetable oil and lard or vegetable shortening may be varied according to the spray and flow characteristics desired)

* Source: Chas. Pfizer & Co., Inc., 1966, Brochure, "Food Preservatives."

in basic dough ingredients. They do not alter color, taste, volume or baking time at the levels commonly used.

Table 7 shows quantities of propionate recommended for different types of baked goods. These levels may have to be varied with the season, as microbial growth is accelerated in warm weather.

Dairy Products   The earliest investigations of the mold-controlling action of propionates for a dairy product appear to be those of Miller and Ingle on the dip process for Cheddar cheese.[159,125] Miller reported that immersion of Cheddar cheese cuts in 8% propionic acid solutions increased the mold-free life at 60°F from the usual 3 to 5 days to 12 to 28 days varying with the cheese sampled, contamination, length of immersion and length of drainage. Somewhat more concentrated sodium or calcium salts were required, as might be expected.

For butter, propionate-treated parchment wrappers were found to give considerable protection in experiments by Macy and Olson.[144,172] However, the use of this additive for butter has not been officially accepted.

Currently, sodium or calcium propionate, or combinations of these, are permitted under Federal Standards of Identity in cheese and cheese products shown in Table 8.

Propionates are added to the starting materials in the process cheese cooker. With pasteurized process cheese and cheese products, the propionates may be added before, or along with, the emulsifying salts. When used with cold pack, propionates may be mixed with other ingredients and added in this manner or may be sprinkled into the ground cheese base while it is being agitated and worked.

Miscellaneous Uses   From 0.2% to 0.4% sodium propionate has been said to inhibit the growth of molds on the surface of malt extract. Wolford and Anderson have reported use of propionates to retard development of mold on syrup, blanched apple slices, figs, cherries, blackberries, peas and lima beans.[259] Propionates are optional in artificially sweetened jams, jellies and preserves (21 CFR 29.4; 21 CFR 29.5).

TABLE 8
Types of Cheese and Cheese Products Which May Contain Propionates

Section of Code of Federal Regs., Part 21	Type of Cheese Product
19.750	Pasteurized process cheese
19.751	Pasteurized blended cheese
19.755	Pasteurized process cheese with fruits, vegetables or meats
19.760	Pasteurized process pimento cheese
19.763	Pasteurized blended cheese with fruits, vegetables or meats
19.765	Pasteurized process cheese food
19.770	Pasteurized process cheese food with fruits, vegetables or meats
19.787	Cold-pack cheese food
19.788	Cold pack cheese food with fruits, vegetables or meats

Handling and Storage

Sodium and calcium propionates are neutral salts and present little hazard in storage. They will emit acrid fumes if heated to high temperatures.[207]

Assay

Propionates may be determined in breads and other baked goods by procedures given in *Official Methods of Analysis of the A.O.A.C.*, 1965 ed., p. 204.

Sulfur Dioxide and Sulfites

Sulfur dioxide has been used in foods for many centuries. According to the review by Joslyn and Braverman, fumes of burning sulfur were used as a sanitizing agent in wine-making by the ancient Egyptians and Romans.[133] The practice is continued by the modern wine industry. Sulfiting* has been applied to other foods, both abroad and in the United States, an important use being in dehydrating of fruits and vegetables.[209] During the 19th century sulfiting was extended to meats and fish and the practice became so widespread and irresponsibly applied that it evoked censure by Dr. Harvey W. Wiley so that limitations were imposed in the first of the national food and drug laws in the U.S.A.

In some applications of sulfiting, the antimicrobial action is concurrent with other desirable preservative effects; e.g., the prevention of enzymatic and non-enzymatic discoloration of some foods. The present discussion will focus on the antimicrobial action and incidentally in the other benefits, of sulfiting.

Physical and Chemical Properties

Sulfur Dioxide   This is a colorless, non-flammable gas, with a suffocating odor. It liquefies at $-10°C$ ($+14°F$). SO_2 is used in food preservation as the gas obtained from burning sulfur, or released from the compressed liquid. It can be applied also as a solution of the gas in water. The free gas presumably dissolves in water within plant tissue to yield sulfurous acid and its ions.

Sulfite Salts   There are various sulfite salts which yield free sulfurous acid at low pH. The following salts are legal for use in foods in the United States within the framework of federal regulations.

Sodium Sulfite, Na_2SO_3   A white to tan, or slightly pink, odorless, or nearly odorless powder. One gm dissolves in 4 ml water. Sparingly soluble in alcohol.

Potassium Sulfite, K_2SO_3   White, odorless, granular powder. One gm dissolves in 3.5 ml water.

Sodium Bisulfite, $NaHSO_3$ White, crystalline powder; SO_2 odor. One gm dissolves in 3.5 ml cold water, 2 ml boiling water and about 70 ml alcohol.

Potassium Bisulfite, $KHSO_3$ White crystalline powder having an odor of SO_2. Freely soluble in water; insoluble in alcohol.

Sodium Metabisulfite, $Na_2S_2O_5$ White crystals or powder with an odor of SO_2. Freely soluble in water; slightly soluble in alcohol.

Potassium Metabisulfite, $K_2S_2O_5$ White crystals or powder having odor of SO_2. Freely soluble in water; insoluble in alcohol.

Specifications for all of these compounds, except potassium bisulfite, are in *Food Chemicals Codex.*

The various sulfite salts offer the convenience of handling as dry chemicals. When dissolved in water, these sulfites form sulfurous acid (H_2SO_3), bisulfite ion (HSO_3^-), and sulfite ion ($SO_3^=$). The relative proportion of each form depends on the pH of the solution which is controlled by the starting material or the addition of acid or alkali.

* Sulfiting is used in this chapter to mean the use of either sulfur dioxide or sulfites.

The dry salts tend to decrease in available SO_2 content as a result of oxidation during storage, particularly under humid conditions. The metabisulfites are more stable than the bisulfites, which, in turn, are more stable than the sulfites.

Antimicrobial Activity

Inhibitory Spectrum Sulfurous acid inhibits yeasts, molds and bacteria. But it is selective, in that yeasts are somewhat more resistant than acetic and lactic bacteria, a property applicable in wine-making.[124,126,133,163] The undissociated acid is probably the most effective form; the ratio of ions is pH-dependent. Rehm and Wittman have obtained evidence that the undissociated acid is 1000 times more active than HSO_3^- for *E. coli*, 100–500 times for *S. cerevisiae* and 100 times for *A. niger*.[197] These workers have tabulated the minimum inhibitory concentrations of H_2SO_3 for a large number of bacteria, yeasts and molds on various culture media and varying conditions of pH as reported in 13 studies. According to these data, most bacteria were inhibited by 200 ppm of sulfite, or less. Yeasts were inhibited also by 200 ppm, with a few exceptions. Most strains of mold were susceptible to 200 ppm, but again with notable exceptions. It is the more resistant microbial strains that cause trouble in commercial practice.

Mechanism of Antimicrobial Action As is true of the other acid antimicrobials, pH strongly influences the growth-inhibiting powers of sulfurous acid. This role of pH has been the subject of numerous investigations, among them those reported by Cruess and his associates[52,53,55] and Rahn and Conn.[191]

Cruess *et al.* found that at pH 3.5 two to four times as much sulfur dioxide was needed to inhibit growth as at pH 2.5. At pH 7, sulfur dioxide was without effect on yeast and molds and 1000 ppm was required to inhibit bacteria. Rahn and Conn suggest that at high pH values the HSO_3^- ion functions to inhibit *E. coli*, but this ion is not effective against yeast.

The enhanced effectiveness at low pH may result from more extensive penetration of the cell wall by unionized sulfurous acid. Beyond this, as pointed out by Wyss, the sulfite may react with acetaldehyde in the alcohol-producing reaction, thus making a compound which is not attacked by the fermentative enzyme.[263] More generally, according to this investigator, sulfurous acid may block enzymes of the microorganisms by reducing essential disulfide (—S=S—) linkages. Quantitative studies of the relationship of the formation of carbonyl addition compounds to the antimicrobial activity of sulfurous acid on *S. cerevisiae* and *E. coli* have been made by Rehm, *et al.*[196]

These workers have concluded that steps in the respiratory mechanism which involve nicotinamide dinucleotide are inhibited as a result of formation of certain hydroxysulfonates formed by combinations of SO_2 with ketone groups.

Safety for Use

The use level of sulfites is limited by the fact that, at residual levels above 500 ppm, the taste begins to be noticeable. Ingested quantities are usually less than those initially added to foods because of loss by evaporation in storage and from cooking. Sulfur dioxide and sulfites in the body are oxidized to harmless sulfate and excreted in the urine.[201] It appears that this detoxifying mechanism is adequate to handle the small quantities of sulfites which are likely to be ingested by people on ordinary diets and has considerable excess capacity.

High level feeding tests have indicated great tolerance by dogs, humans and rats. Rost and Franz gave dogs from 0.05 to 1.0 gm of sulfite per day for periods of from one month to more than a year, without evidence of gross or microscopic tissue changes.[202] Larger doses caused vomiting but no other symptoms appeared. In humans, these workers reported that 1 gm of sodium sulfite per day decreased utilization of protein and fat. At this dose there were few, if any, gastrointestinal symptoms, but there were abdominal pains and vomiting at 4 to 5.8 gm per day. Lockett and Natoff gave rats SO_2 at a level of 750 ppm in drinking water in experiments that lasted nearly three years with three generations of animals.[143] They reported no effects on growth, intake of food and fluid, output of feces, fertility, weight of newborn or frequency of tumor development. In a rat study by Fitzhugh et al., feeding levels of sulfite of 0.1% or above yielded some adverse results which are generally regarded as caused mainly by destruction of thiamine in the basal diet by the sulfite.[74]

Regulatory Status

The several sulfite salts and sulfur dioxide are generally recognized as safe for use in foods by the U.S. Food and Drug Administration. There is a proviso, however, that they may not be used in foods which are substantial sources of thiamine (vitamin B_1). Among products presently under Federal Standards of Identity, carbonated beverages may include sulfites or SO_2 (21 CFR 31.1). Under regulations of the U.S. Treasury Department, Alcohol and Tobacco Tax Division, finished wines should not have in excess of 350 ppm of SO_2.

In other countries sulfites are widely permitted in wine, but their use extends also to certain other products where such use is not allowed under regulations in the United States. Examples are liquid and dried egg products, fish fillets, fish paste, and sausage.

Applications

In addition to wine, sulfur dioxide and sulfites are used in a variety of food-products—fruits and vegetables for dehydration, freezing, and brining, fruit juices and purees, syrups and condiments. There are certain limitations that must be borne in mind in these applications. Sulfur dioxide is volatile and tends to disappear from open systems and much may be inactivated by combination with food components.[126] Besides destroying thiamine, it is somewhat corrosive and has a taste which may become apparent at high levels of addition.

Extensive information on the applications of sulfites is found in the bulletins of manufacturers and in several reviews and texts.[6,133,209]

Dehydrated Fruits and Vegetables   Most fruits to be dehydrated, except some cut fruits, are treated by exposure to fumes of burning sulfur or to vaporized liquid sulfur dioxide. Solutions of sulfites are less suitable because they penetrate the fruit poorly and leach sugar. According to Stadtman et al., the storage life of dried apricots is directly proportional to the initial sulfur dioxide level.[226] The gas disappears during storage at a rate approximately proportional to the logarithm of its concentration. Retention depends upon the time of exposure and temperature. The optimum temperature is 110°–120°F when SO_2 is used.

For vegetables a mixture of neutral sulfites and bisulfites is used. The solution is usually applied as a spray during or after blanching and before dehydration. Solutions are used for dipping apple slices and other cut fruits. Application of sulfur dioxide or sulfites to vegetables to be dehydrated increases storage life, preserves color and flavor, and aids retention of ascorbic acid and carotene.

It is not possible to make a general recommendation as to strength of dip solution or the length of time that the product should be immersed, as there are many variables and experimentation is required to obtain satisfactory levels in the fruit. Table 9 presents levels of SO_2 which have been recommended as desirable for dried fruits and vegetables at the start of storage.

Fruit Juices, Syrups, Concentrates and Purees   The preservation of products in these categories by SO_2 is more common in countries outside of the United States, especially in countries with warm climates and those where frozen storage facilities are limited. This chemical method of preservation can be used effectively for bulk juices and purees which are eventually to be processed into consumer products.

Suggested SO_2 levels in these concentrates are from 350 to 600 ppm. Higher levels may be necessary because of the considerable amount of sugar and other SO_2-binding materials. For optimum inhibitory action by the SO_2, acidification of purees, concentrates and similar products may be necessary.

TABLE 9
Recommended Levels of SO_2 in Dried and
Dehydrated Fruits and Vegetables at Start of Storage*

Product	*SO_2 ppm*	*Product*	*SO_2 ppm*
Apricots	2000	Sulfur bleached raisins	1500
Peaches	2000	Apples	800
Nectarines	2000	Cabbage	750–1500
Pears	1000	Potatoes	200–250
Golden, bleached raisins	800	Carrots	200–250

* Source: Monsanto Technical Bulletin 1-250. Used by permission.

If, in their final uses, sulfite-treated bulk fruit materials are to be canned, the sulfite must be reduced to less than 20 ppm. Otherwise, in contact with the metal of the can, sulfite may generate H_2S which will cause a black precipitate.

Various methods have been devised for removing sulfur dioxide from juices, such as heating in vacuum pans, mechanical agitation in conjunction with bubbling of an inert gas, and spraying against baffles under vacuum. It should be noted that sulfite in conjunction with heat processing permits the use of lower temperatures. In the case of jams or other highly sweetened products stored in glass, there is more latitude in the final SO_2 content as the sweetness covers the taste.

Wine-making   Detailed information on the use of sulfite and sulfur dioxide in wine-making is included in the books and reviews of Amerine and Joslyn,[6] Joslyn and Braverman,[133] and Schroeter.[209]

In wine-making, sulfite is important at several stages. First, solutions can be used to sanitize equipment. Before fermentation the fruit juices (musts) are treated to eliminate the natural microbial flora associated with the fruit. Pure cultures of the desired wine-making yeast are then added to carry out the alcoholic fermentation and the other reactions relevant to the delicate flavor and aroma characteristics desired. During the fermentation stage, SO_2 also serves as an antioxidant, clarifier and dissolving agent. During bulk storage of wine after fermentation, a suitable level of sulfur dioxide is maintained to prevent post-fermentation changes caused by microorganisms.

During fermentation, SO_2 levels of 50–100 ppm are usually effective. The quantities to use are influenced by the quality of the grapes, pH, temperature, concentration of sugar and degree of contamination. Sulfite salts, liquid sulfur dioxide or sulfur dioxide solution in water may be used in wine-making. Table 10 shows suggested quantities of liquid SO_2, SO_2 solution, or potassium metabisulfite for use under various conditions.

TABLE 10
Amount of Sulfur Dioxide to Be Added During
Fermentation of Wine Under Various Conditions*

Condition and Temperature of Grapes	Liquid Sulfur Dioxide		6% Sulfurous Acid Solution		Potassium Metabisulfite	
	Per 1,000 gals of Must, oz	Per ton of Grapes, oz	Per 1,000 gals of Must, gals	Per ton of Grapes, pints	Per 1,000 gals of Must, oz	Per ton of Grapes, oz
Clean, sound, cool, and underripe	10	2	1¼	2	20	3½
Sound, cool, optimum maturity	15	2½	2	3	31	5
Moldy, bruised, hot, overripe, low in acid	36	6	3¼	4¼	56	9

* Source: Amerine and Joslyn (Ref. 6). Originally published by the University of California Press; reprinted by permission of The Regents of the University of California.

During bulk storage of wine after fermentation, the level of sulfur dioxide should be maintained at 50–75 ppm to prevent bacterial spoilage. However, the quantity to add at this stage should be related to that already present, as determined by analysis. Renewed applications are necessary during the storage period.

Meats and Fish   Use of sulfites in fresh meats, meat products and fish is generally prohibited in the United States on the ground that they could restore a bright color and appearance of freshness to faded meat.

In England and Wales sulfites are permitted in sausages and sausage meats; and in France, in shrimp and dried fish. Presumably, one purpose of adding sulfite to these products is to check deterioration from oxidative processes. Indeed, in the United States sulfite may be added to shrimp to inhibit an oxidative deterioration known as "black spot."

Storage and Handling

In general, sulfite salts should be stored in a cool, dry place, as in the presence of moisture there is caking and slow release of sulfur dioxide. These salts, if dry, can be handled in iron or steel equipment. Solutions can be handled in type 316 and 347 SS, lead, rubber or wood.

Excessive contact with the skin should be avoided in handling sulfites, especially in the presence of moisture. Breathing of the dust of these salts should be avoided.

Sulfur dioxide vapor is, of course, very irritating to the eyes and respiratory tract, and prolonged exposure to high concentrations should be prevented. A publication of the American Petroleum Institute[14] states that adequate ventilation should exist in any working area to keep the atmospheric concentration below 5 ppm. Workers exposed daily to allowable or somewhat greater concentrations of SO_2 do not suffer chronic ill effects. Workers who must be exposed to concentrations in excess of permissible limits should be provided with approved eye and respiratory protection. The API publication summarizes the acute and chronic effects of SO_2 and recommended treatments and precautionary measures.

Assay

Methods for total sulfurous acid and for free sulfurous acid in food products are published in *Official Methods of Analysis of the A.O.A.C.*, 1965 edition, pp. 353, 446, 467 and 468. A review of methods for the determination of sulfites in foods is presented by Schroeter.[209]

As pointed out by Joslyn and Braverman, sulfur dioxide or sulfites added to fruit and vegetable products exist as undissociated sulfurous acid and as the bisulfite and sulfite ions.[133] The SO_2 exists also in combination with components of the fruit products, such as aldehydes. For purposes of determining efficiency of different sulfuring procedures or products, it would be desirable to have methods for determining all of these chemical entities.

Acetic Acid and Acetates

Empirical knowledge of the preservative action of acetic acid in the form of vinegar extends far back into history. Today, besides vinegar and purified acetic acid, there are the following compounds which yield acetic acid and are available for food use: sodium acetate, calcium acetate, potassium acetate and sodium diacetate. Several of these are interchangeable in use, but some have superior capabilities for certain purposes. Also, matters of flavor and of economics have a bearing on choice of compound. Other later-developed preservatives have intruded upon the sphere of the acetates in some applications. The acetic compounds not only have preservative action, but function also as sequestrants, acidulants and flavoring agents.

Physical and Chemical Properties

Vinegar By definition in the United States, the Food and Drug Administration recognizes six types of vinegar, distinguished by their source. These are: cider vinegar, wine vinegar, malt vinegar, sugar vinegar, glucose vinegar and spirit vinegar. The last type is made from distilled grain alcohol. Vinegar must contain not less than 4 gm of acetic acid per 100 ml.[57]

Acetic Acid, CH₃COOH In pure form this is a colorless liquid which solidifies at 62°F. It is miscible with water, alcohol and glycerine.

Sodium Acetate, Hydrous, CH₃COONa·3H₂O Colorless, transparent crystals or granular crystalline powder. Is odorless or has a faint acetous odor. Effloresces in warm dry air. One gm dissolves in about 0.8 ml of water and about 19 ml of alcohol. The anhydrous form is a white, odorless, granular, hygroscopic powder; 1 gm dissolves in about 2 ml of water.

Calcium Acetate, (CH₃COO)₂Ca A fine, white, bulky, odorless, powder, freely soluble in water and slightly soluble in alcohol.

Sodium Diacetate, CH₃COONa·CH₃COOH·½H₂O A molecular compound of sodium acetate and acetic acid. The acid component is, in a sense, acetic acid of crystallization and is known to be undissociated. Sodium diacetate is a white, hygroscopic crystalline solid having an acetous odor. One gm dissolves in about 1 ml of water.

Specifications for acetic acid, sodium acetate, calcium acetate and sodium diacetate appear in *Food Chemicals Codex.*

Antimicrobial Activity

Among early studies, those of Annheimer and Fabian[10] and Levine and Fellers[141] drew attention to the general effectiveness of acetic acid upon bacteria. The following table is from the paper by the latter.

TABLE 11
Comparative Antimicrobial Effectiveness of Acetic Acid*

Organism	Inhibitory pH[1]	Inhibitory Acidity %	Lethal pH[2]	Lethal Acidity %
Salmonella aertrycke	4.9	0.04	4.5	0.09
Staph. aureus	5.0	0.03	4.9	0.04
Phytomonas phaseoli	5.2	0.02	5.2	0.02
Bacillus cereus	4.9	0.04	4.9	0.04
Bacillus mesentericus	4.9	0.04	4.9	0.04
Sacch. cerevisiae	3.9	0.59	3.9	0.59
Aspergillus niger	4.1	0.27	3.9	0.59

* Source: Levine, A. S. and Fellers, C. R., 1940, J. Bacteriol. *39*; 499–514. Reproduced by permission of Williams & Wilkins Company, American Society of Microbiology.
[1] The pH at which no visible growth occurred, yet the microorganisms remained viable.
[2] The pH at which total destruction took place.

Ingram *et al.*, in their 1956 review, make the general statement that acetic acid is more effective against yeasts and bacteria than against mold.[126] They note that, in their experiments, both acetic acid and its calcium salt were effective against rope, and the salt at high levels (0.4%–1.0%) was effective against mold.

The diacetate is effective against both rope and mold in bread, according to the experiments of Bauer and Glabe.[21] They found that 0.08% of sodium diacetate, based on finished loaf, retarded formation of rope and 0.17% prevented its formation over at least 5 days. Also, mold was retarded by 0.08%, and very effectively retarded by 0.2%. The diacetate at legal use levels is said not to interfere with the fermentation process in yeast-raised products.

Pelshenke presents a comparative table of effective levels of various chemicals against rope which shows 0.15% for sodium diacetate and 0.4–1.0% for calcium acetate.[183]

Acetic acid and its salts have higher antimicrobial activity as pH is lowered, thus increasing the quantity of undissociated acid.[53] Cruess observed similar pH effects with potassium acetate.[52]

Regulatory Status

Under Federal Regulations in the United States, vinegar, acetic acid, sodium acetate, and calcium acetate are generally recognized as safe for use in foods. Sodium diacetate and calcium diacetate also are GRAS.

Some other countries permit the use of the diacetate in baked goods. Canada permits the diacetate also in cheese and allows acetic acid in preserved meat, fish products, and poultry products.

No information has been found on the toxicology of acetic acid and acetates other than that the MLD orally of acetic acid is about 3300 mg/kg for rats and 500 mg/kg for mice.[224]

Applications

Vinegar or acetic acid is used in such foods as catsup, mayonnaise and pickles where acidity and characteristic taste are either desirable or not objectionable. They often may be added for their flavor, but they also exert concurrent antimicrobial action.

Among meat products preserved with vinegar are pickled sausages and pigs feet. At least 3.6% of acetic acid must be present in the water phase of vinegar-pickled sausages. At lower concentrations of acetic acid, and in the presence of a fermentable carbohydrate, lactic acid bacilli and yeast may grow. Pickled pigs feet can be preserved with lower concentrations of acetic acid, as the cooking and washing remove fermentable carbohydrate.[57]

In bread and other baked goods, vinegar, acetic acid and its salts, and sodium diacetate are permitted in the United States. Vinegar may create an undesirable taste at levels required for adequate antimicrobial action. Sodium diacetate has the advantage over the simple acetates, as discussed above, in that it controls both rope and molds. Under Federal Standards of Identity for bread and rolls, sodium diacetate is limited to 0.4 part per 100 parts of flour. In its application in bread, sodium diacetate is being displaced by the propionates.

Recommended use levels of sodium diacetate in white breads and other yeast-raised white-flour products are 3½ to 6 ounces per 100 lb of flour. For dark breads the levels are 4 to 6 ounces per 100 lb. In the sponge and dough method, the diacetate is added in the dough stage with other dry ingredients. In the straight dough method, the preservative is added at the beginning of the mixing stage. For pie crusts, recommended levels are 3 to 5 ounces per 100 lb flour, and for fillings, these same weights per 100 lb filling. Table 12 shows quantities recommended for cakes.

Sodium diacetate has been tested as an antimicrobial for malt syrups and concentrates. Graded levels of sodium diacetate were included in a

TABLE 12
Use Levels of Sodium Diacetate in Cakes,
Ounces per 100 Lb Batter

Chocolate Cake	1	Pound Cake	3
Devil's Food	1	Sponge Cake	3 to 4
Fruit Cake (high fruit)	1	Yeast-raised Coffee Cake	3
Cheese Cake	3	Angel Cake	3.5 to 4.5
Layer Cake	3	Fruit Cake (low fruit)	3.5 to 4.5

series of samples incubated for 15 days at 98°F. At the end of that period no microbial deterioration was found in samples which contained 0.5% or more of the diacetate.[23] A protective effect of sodium diacetate in cheese spread has also been demonstrated. Levels of 0.1% to about 2%, directly incorporated, protected cheese from mold growth for 10 days, while untreated cheese was moldy after 3 days.[22] This same report describes tests of efficacy of diacetate against mold growth on treated parchment wrappers and butter in such wrappers. At 75°F storage temperature, only slight moldiness was found after 28 days under the treatment, whereas there was pronounced mold growth by the 21st day in the untreated samples.

Storage and Handling

Acetic acid is corrosive and should be handled with appropriate care. Breathing of concentrated vapors is harmful. Vinegar also is somewhat irritating.

Acetic acid solidifies at 58–60°F. This, and the fact that its vapors are flammable, should be taken into account when storing this material. Anhydrous sodium acetate is somewhat hygroscopic. Sodium diacetate under ordinary atmospheric conditions slowly loses acetic acid and takes on water. This can be prevented by storage in waterproof containers. In spite of its high acid content there has been no reported ill effect on men working with sodium diacetate.[167]

Assay

Methods for determining acetates in bread and cake are found on pages 204 and 208 of *Official Methods of Analysis of the A.O.A.C.*, 1965 edition.

Nitrites and Nitrates

Nitrites are included in curing mixtures for meats to develop and fix the color. Nitrites decompose to nitric oxide, NO, which then reacts with heme pigments to form nitrosomyoglobin. Nitrates have a somewhat uncertain role; it is generally believed that they may be a reservoir of nitrite formed by reduction in the meat. There seems to be a growing tendency to use only nitrites in curing solutions. Although nitrites (and even nitrates) have been demonstrated to have antimicrobial action *in vitro*, they seldom provide sterilization. These salts always are used in conjunction with sodium chloride, and the combination has been reported to have antimicrobial action despite culturable contamination.[225]

Physical and Chemical Properties

Sodium Nitrite, NaNO$_2$ Pale, to slightly yellow granular powder, or white, or nearly white, opaque, fused masses or sticks. One gm dissolves in about 1.5 ml of water; sparingly soluble in alcohol.

Potassium Nitrite, KNO$_2$ Small white or yellow deliquescent granules or cylindrical sticks. Very soluble in water; sparingly soluble in alcohol.

Sodium Nitrate, NaNO$_3$ Colorless, white granules or powder. One gm dissolves in 1.1 ml water, and in 125 ml alcohol.

Potassium Nitrate, KNO$_3$ White granular or crystalline powder. One gm dissolves in 2.8 ml water, or in 620 ml alcohol.

Nitrite is both an oxidizing and reducing substance, is extremely reactive to organic matter, and is labile to heat. A meat product which contains initially 78 ppm of nitrite may show only 10 to 20 ppm after retorting. Nitrate, on the other hand, is stable, but is subject to reduction to nitrite by action of bacteria.

Specifications for sodium and potassium nitrite, and sodium and potassium nitrate are in *Food Chemicals Codex*.

Antimicrobial Activity

Nitrites Silliker *et al.* expressed the opinion that nitrites contribute to the stability of canned comminuted meats.[216] They emphasized the long history of satisfactory shelf life of these products, even though the heat processing is, of necessity, relatively mild and bacterial spores are found within the finished material. To simulate practical conditions in the experiments reported in their paper, they utilized ground beef and pork trimmings as the medium, studying

TABLE 13
Effect of Individual Curing Ingredients
on the Shelf Stability of Canned Meat*

Brine %	Nitrite ppm	Nitrate ppm	Sound Cans	Non-putrid Swells	Putrid Swells
5	78	156	12	0	0
5	78	0	12	0	0
5	0	156	1	11	0
5	0	0	8	4	0
3.5	78	156	12	0	0
3.5	78	0	12	0	0
3.5	0	156	1	9	2
3.5	0	0	6	6	0
0	0	0	9	3	0

*Source: Silliker *et al.* (Ref. 216) as modified in *Science of Meat and Meat Products* (Ref. 5). Reproduced by permission of Institute of Food Technologists, Copyright © 1958, and by the American Meat Institute, W. H. Freeman and Company, Copyright © 1960.

[1] The brine concentration refers to the percentage of sodium chloride in the water phase of the meat emulsion. Each can received a sub-lethal heat processing (F_0 = 0.08–0.13). The nitrite and nitrate levels refer to the amounts added to the emulsion before heating.

[2] There were twelve cans in each lot, and all were incubated at 38°C.

the spoilage action of natural microbial contaminants. Their results with respect to the antimicrobial effects of sodium chloride, nitrite and nitrate, separately and combined, are shown in Table 13.

However, these same workers and various others have found that the nitrite provides inadequate protection when there is a heavy spore load. Gough and Alford[95] recovered viable *Cl. perfringens* from hams pumped with brine containing 20 spores per ml despite normal curing and smoking. Similarly, Bulman and Ayres found that 3.6% salt with 150 ppm nitrite initially (83 ppm after heat processing) prevented spoilage of ground pork trimmings when there was a natural spore level of less than 1 per gm of meat, but that these agents failed to protect against an inoculum of 50 per gm.[43]

In a recent review on the stability and safety of non-sterile canned cured meats, Spencer mentioned other similar reports on the effect of spore loads. He offers three possible explanations, involving general factors in heat processing and the effect of adjuvants, for the variations encountered in the antimicrobial action of nitrites under heavy spore load.[225]
These are:

(a) Heat processing conditions based on logarithmic death rates are designed for low contamination loads.

(b) The theory of probability of survival may not apply to low spore levels.

(c) The concentration of NaCl and/or nitrite which prevents outgrowth may be related to the number of spores present.

Spencer points out that quantitative supporting data for these possibilities are lacking.

Castellani and Niven reported that tolerance to nitrite varied widely among 16 types of bacteria which they tested.[45] They also described a phenomenon which complicated experiments designed to determine why staphylococci are rarely, if ever, found in large numbers in cured meats. The level of nitrite which was inhibitory to *Staph. aureus, Streptococcus salivarius* and *Streptococcus mitis* was much lower when this salt and glucose were both present in a medium at time of autoclaving, as compared with the levels when the glucose was sterilized separately and added to the nitrite-containing medium. They stated that sulfhydryl compounds restored the tolerance of the organisms to nitrite in media sterilized with glucose present.

Mode of Action of Nitrite As with other acid microbial inhibitors, the effectiveness of nitrite, a weak acid salt, is related to pH as shown in Table 14 from Castellani and Niven.[45] This effect of pH also has been reported by Tarr.[230,231]

Castellani and Niven have considered several older theories in attempting to explain growth inhibitions imposed by nitrites.[45] The well-known Van Slyke reaction of HNO_2 on amino acids could interfere with dehydrogenose structures. Reactions with mono-phenols like tyrosine could alter a cellular component. The reaction with heme pigments might be extended to the cytochrome system of the cells. However, these authors pointed out that nitrite inhibits organisms known to be devoid of heme-containing respiratory catalysts. They preferred to consider a nutritional defect created when nitrites react with sulfhydryl constituents and sulfhydryl-aldehyde condensations not metabolizable under anaerobic conditions. Further information is necessary to support this theory.

Nitrates The role of nitrates is somewhat uncertain. Among the earlier studies of antibacterial action of curing mixtures are those of Tanner and Evans[229] mentioned by Jensen and Hess,[129] and Yesair and Cameron.[265] Tanner and Evans obtained irregular inhibition of seven *Cl. botulinum* cultures with $NaNO_3$ in concentrations from 2.213 to 4.427%, but failed to obtain inhibition of *Cl. botulinum*, *Cl. putrificum* or *Cl. sporogenes* with $NaNO_2$ in concentrations from 0.0588 to 0.392%.

Yesair and Cameron, working with spores of *Cl. botulinum*, reported that in pork infusion agar more than 70% reduction in spore counts was obtained from 0.1% $NaNO_3$, or 0.005% $NaNO_2$, or 2% NaCl. When these salts were combined in amounts representative of curing practices, the death rate approached 100%.

TABLE 14

Effect of pH on the Aerobic Nitrite Tolerance of Staphylococcus Aureus Strain 196, When Glucose Is Autoclaved in a Complex Medium*

pH^1	Nitrite Concentration		Calculated Undissociated[2] HNO_2	
	Growth[3]	No Growth	Growth	No Growth
	ppm	ppm	ppm	ppm
6.90	3,500	4,000	1.12	1.28
6.52	1,800	2,000	1.37	1.52
6.03	600	700	1.38	1.61
5.80	300	400	1.20	1.60
5.68	250	400	1.32	2.12
5.45	140	180	1.25	1.50
5.20	80	150	1.12	2.10
5.05	40	80	0.92	1.84

* Source: Castellani and Niven (Ref. 45). Reproduced by permission of Williams & Wilkins Company, American Society for Microbiology

[1] The pH values were determined on duplicate tubes of medium at the time of inoculation.

[2] The amount of undissociated nitrous acid was calculated from the dissociation curve employing a pKa of 3.4.

[3] Growth was determined after 48 hours at 30°C.

However, Silliker *et al.*[216] concluded that "nitrate played no role in retarding putrid spoilage, but actually stimulated aerobic spoilage." They cite reports of other workers confirming this conclusion. In addition, Spencer, in his 1966 review, says that nitrates were at one time believed to have significant preservative effects, but that this seems to be in doubt in view of more recent studies which he cites.

Safety for Use

Very little information on the effects of nitrates and nitrites on humans is available. The oral MLD in rats is 200 mg/kg for sodium nitrate and 330 mg/kg for sodium nitrite. For dogs, orally, the MLD is 330 mg/kg.[156] Nitrates are said not to be especially toxic until reduced to nitrites in the intestine.[87] The

United States Dispensatory states that the amount of nitrite required to form methemoglobin (in the living organism) is relatively large.[61]

A case is on record of cyanosis from formation of methemoglobin in a 48-year-old man, following consumption of about a pound of Polish sausage. This product is high in nitrite to maintain a red color. Analysis of the residue of the piece from which he ate showed 0.181 mg/gm, close to the legal limit of 0.200 mg/gm. This same man had eaten such sausage many times before in his life without ill effects. It is suggested that there were local high concentrations of nitrite in the portion he ate before the illness. This seems to be an isolated case.

There has been concern over another reaction of nitrites—reaction with secondary or tertiary amines to form nitrosamines, compounds that are both toxic and carcinogenic in a wide variety of species. The possible health hazard from formation of such compounds in nitrate- or nitrite-treated foods was suggested about ten years ago when ruminants and mink in Norway were reported to have malignant liver lesions after eating a ration that contained fish meal preserved by addition of nitrite. Analyses showed nitrosamines in samples of high-nitrate, laboratory-produced toxic meal fed to mink, whereas other non-toxic samples did not contain these compounds according to analytical methods then available. Dimethyl-N-nitrosamine fed to sheep evoked similar lesions.

In attempting to draw inferences from this Norwegian episode regarding nitrite in human foods, one should note that the nitrite level in the fish meal fed to the ruminants and mink was several times that permitted in meat and fish products for human consumption. The extent to which use of nitrites and nitrates (which form nitrites) in foods contributes to possible carcinogenesis in man is under intense investigation at present by the Food and Drug Administration, other agencies of the Federal Government, and the meat industry. A collateral program has arisen from recognition of the powerful effect of nitrite in suppressing outgrowth of *Clostridium botulinum* Type E and toxin formation. Quantitative information on this action is being sought, together with information on the possible occurrence of nitrosamines in nitrite-treated food products. Thus far, there is little to indicate presence of detectable quantities, even with vastly refined analytical methods.

Regulatory Status

Under the Code of Federal Regulations (121.1063–1064), sodium nitrate and nitrite may be used as preservatives and color fixatives in cured meats. The level of nitrate must not exceed 500 ppm and of nitrite 200 ppm. Similar stipulations are made regarding use of these salts in certain types of smoked and cured fish (21 CFR 121.1063–1064). The use of nitrite only to suppress Clostridia in certain types of smoked fish is the subject of another regulation (21 CFR 121.1230).

In some other countries additional uses are legal. In England these salts can be used as preservatives in cheese, other than cheddar or Cheshire types, or soft cheese varieties. Use in cheese is permissible also in the Netherlands, Norway, South Africa, and Sweden. Nitrates and/or nitrites are permitted in fish products in Norway and Sweden and in poultry meat in Canada.

Applications

Since nitrates and nitrites are components of curing salts, the customary procedures in the use of these salts prevail and no special recommendations can be made with reference to their applications as antimicrobials. Information on how to use curing mixtures will be found, of course, in general reference books on food processing and meat technology, such as *The Science of Meat and Meat Products*.[5]

Sodium nitrite alone has been found effective by Tarr and Sunderland as a preservative for fish when incorporated in ice at a level of 0.1 to 0.5%.[232,233,234] Similar findings were reported by Dunn.[64]

Storage and Handling

Both nitrates and nitrites present a fire hazard. When mixed with organic matter they will ignite from friction and at high temperatures they will explode.[207] In practice, of course, these hazards are minimized by the fact that these two salts are marketed in curing mixes, where they form a small percent of the total.

Assay

A method for the determination of nitrates and nitrites in meats appears on p. 347 of *Official Methods of Analysis of the A.O.A.C.*, and on p. 125 of the same publication is a method for nitrites in dry curing mixes or pickle.

A colorimetric method for both nitrate and nitrite in cured meat, which is said to be relatively rapid, has been published by Follet and Ratcliff[75]

Gas Sterilants—Ethylene and Propylene Oxides

Food preservation processes like freezing, drying or the use of chemical preservatives inhibit the action of spoilage organisms and their propagation. Autoclaving eliminates the organisms by death. Chemical preservatives are effective only in foods containing appreciable moisture, and then only if the preservative presence is maintained.

Many low-moisture ingredients, which in themselves are not difficult to preserve, find their way into high-moisture foods, and it is desirable to reduce the microbial load before use. Thermal processes may be unsatisfactory because of flavor volatility or product instability. For these, a "cold sterilization" process entailing simple application, complete sterilization and easy removal of the chemical agent would offer an ideal solution.

The closest approach to this ideal is gaseous sterilization. The sterilizing agent is effective in the vapor state and generally has good penetrating ability, and because of its high vapor pressure, residues are controllable. The ideal agent has yet to be achieved, but two epoxides, ethylene oxide and propylene oxide, have been very useful. Ethylene oxide first came into use because it was recognized as the more effective, and found wide application for non-edible materials such as surgical dressings and bandages, dry chemicals and pharmaceuticals, glass and plastic packaging materials, syringes, sutures and hospital supplies such as blankets, sheets, shoes, leather goods, etc. It is noteworthy that gaseous sterilization can often be accomplished after packaging in paper and some synthetic films.

The concept of gaseous sterilization and the broad applications of epoxide gas sterilization have been reviewed by Phillips[186], Opfell and Miller[174], and Bruch,[41] among others.

Ethylene oxide has been utilized as a fumigant for pest control and by 1935 some 200 references were accumulated in one review.[268] That it also exerted cidal effects on microorganisms was disclosed by early reports from Europe which prompted investigations in the mid-1930's in the U.S. for the purpose of reducing the microbial load of certain food ingredients. The ethylene oxide process was applied during World War II to reduce the bacterial counts in soy flour and cereal products used in processed meat products for the Armed Forces.

Because ethylene glycol and diethylene glycol were encountered as residues, and are suspected of toxicity, ethylene oxide decontamination has been limited to a few low-moisture food ingredients. On the other hand, propylene glycol is generally recognized as safe, and gaseous sterilization by propylene oxide is receiving more attention.

Physical and Chemical Properties

Ethylene oxide is the simplest cyclic ether or epoxide compound, and propylene oxide is the next homolog. Both are widely used raw materials in the manufacture of glycols, polyglycols, polymers and a host of organic chemicals.

Ethylene oxide is a gas at ambient temperatures, a liquid below 10.7°C, and has a freezing point of −111.3°C. Because ethylene oxide is highly reactive, rigid rules for safe handling are mandatory. It is flammable, easily ignited and has vapors which decompose violently when exposed to certain temperatures, pressures and air mixtures. It polymerizes violently with certain catalytic materials.[120] In a mixture with non-toxic inert gases it can be safely and effectively employed as a sterilizing agent. Mixtures of 10–20% ethylene oxide and 90–80% carbon dioxide or other organic diluents are usually recommended for this purpose. Higher concentrations may be used, depending upon the caliber of equipment and personnel.

Propylene oxide is a water-clear liquid with a boiling point of 35°C, and a freezing point of −104.4°C. It is somewhat less reactive than ethylene oxide and has a narrower explosive range (2–22%).

Antimicrobial Action

Ethylene oxide kills all microorganisms, both vegetative cells and spores, as well as viruses. Information on its antimicrobial spectrum is given in the review by Phillips and Kaye[187] and the paper by Whelton et al.[255] Its reduction of thermophiles in spices has been shown by Pappas and Hall.[179]

Propylene oxide is also a broad-range microbicide. Its ability to kill mold spores and bacteria has been reported by Cooper,[49] Phillips,[185] Haines[104] and Whelton et al.[225] The Whelton report states that 0.3 cc of propylene oxide per liter (0.015 lb per cu ft) would kill molds and yeasts on agar slants in 3 hours at 73°F. Bacteria were more resistant.

Because of its lesser activity in comparison with ethylene oxide, propylene oxide requires longer exposure times and is more sensitive to humidity, temperature, and physical structure of the food product. It is less penetrating, and product agitation and air removal demand more attention.

The mechanism by which ethylene oxide and propylene oxide exert their broad antimycotic action remains unknown. Hydration to the respective glycols is not the route. Rather, it seems more reasonable to consider the epoxides' highly reactive nature and the likelihood that an irreversible reaction or addition is made upon one or more vital sites in the cell. Phillips has advanced an alkylation theory whereby labile hydrogens are replaced by a hydroethyl group ($—CH_2CH_2OH$) in essential metabolic systems in the case of ethylene oxide.[185,187] The multitude of labile hydrogens in the intermediary metabolism of cells makes this a plausible theory.

Safety for Use

Both oxides are removed from the treated food by evacuating the reaction chamber and releasing the vacuum with sterile or filtered air. Gentle heat and agitation are helpful to exhaust the oxides as completely as possible.

Residues of the corresponding glycols will be present, but the toxicity of ethylene glycol precludes substantial residues. The LD_{50} of ethylene glycol for mice is 8300 mg/kg and for rats 6100–8500 mg/kg.[224] Most of the toxicologic information available pertains to subcutaneous, intravenous or inhalation routes of administration.[212]

Bruch has expressed the opinion that propylene oxide is about one third as toxic as ethylene oxide by injection or inhalation.[41] The acceptance of propylene oxide as a food additive is based on the premise that propylene glycol is generally recognized as safe.[164,213]

British workers recently have warned that both ethylene and propylene oxides may react with inorganic chlorides in foodstuffs to form chlorohydrins which in themselves are quite toxic.[253] Persistent residues of these were found because of their non-volatility and chemically unreactive nature. However, in a 1950 report by Ambrose of the U.S.D.A., it was stated that rats were fed for 400 days a diet containing 800 ppm of ethylene chlorohydrin and showed no apparent harm.[4] This is, of course, an extremely high intake level of chlorohydrin in comparison with what would be obtained from food.

Phillips has pointed out that the use of ethylene oxide in foodstuffs might entail some nutritional problems.[186] Hawk and Mickelsen noted that rats failed to grow on specially purified diets after gas treatment.[113] This was later shown to be due to large losses of riboflavin, pyridoxine, niacin and folic acid. Vitamin B_{12}, biotin and pantothenic acid were not affected.[18] Others have encountered growth difficulties when treated casein was the sole protein, a situation alleviated by supplementation with histidine and methionine.[257]

In other feeding trials with rats, Oser and Hall found that six samples of yeast treated with ethylene oxide under conditions approximating commercial usage showed slight, insignificant average losses of thiamine, riboflavin, niacin and choline, and possibly significant losses of pyridoxine and folic acid.[176] Similar results were obtained with a natural diet. There were no adverse effects observed in young rats during a 5-week test period when they ate a diet in which was included 10% of yeast treated with ethylene oxide under commercial conditions. Manchon and associates[272] made comparative rat-feeding studies of bread preserved by freezing and by treatment with ethylene oxide, each product being 70% of the diet. In three-generation studies there was no difference in growth or in fertility, nor were differences found in histology of liver and kidney.

Regulatory Status

The U.S. Food and Drug Administration confines the use of ethylene oxide as a fumigant against insects and microbes to black walnut meats, copra and whole spices (21 CFR 120.51). The use of ethylene oxide and other epoxides in foods of high moisture content is prohibited in the United States.

Propylene oxide is permitted in or on food, under certain conditions (21 CFR 121.1076). It can be used as a package fumigant for dried prunes or glacé fruits. It is further permitted as a fumigant for bulk cocoa, gums, processed nutmeats (except peanuts), spices and starch, all of which are to be further processed in final food form. Tolerances of 300 ppm as propylene glycol are permitted (except 700 ppm in prunes and glacé fruits). Dehydrated vegetables and flours are currently excluded.

Applications

Methods of Operation   Ethylene oxide, in diluted form, is applied in closed systems to articles to be sterilized. Pressure vessels similar to laboratory autoclaves are frequently used but other types of pressure vessels may be employed, depending on the material to be treated and the need for subsequent sterile handling.

In general, the material is placed in the pressure vessel and the chamber is evacuated to remove air. An ethylene oxide-carbon dioxide mixture (10:90) is introduced through an expansion chamber or heat exchanger to compensate for the cooling effect of expanding CO_2. Recommended is 2.5 lb of diluted gas per cubic foot for one hour although there are variants of this. This will require a positive pressure of 25–30 psi. At the end of the exposure the gas is vented, followed by evacuation by a suitable pump capable of drawing a vacuum of 26 inches Hg. The vacuum is relieved with sterile or filtered air and the cycle repeated if necessary to remove residues.

Propylene oxide is applied in a similar manner. However to compensate for its higher boiling point some energy input is required. Undiluted propylene oxide can be used safely if precautions are taken to avoid explosive mixtures with air.

The sterilization process with gaseous epoxides is a function mainly of the concentration of the gas, time of exposure and temperature during exposure. There must also be a minimum humidity. The quantity of moisture in "dry" powders or flakes probably suffices to supply enough humidity. These factors and others are evaluated in the literature on the gaseous method and manufacturers' bulletins.

Ethylene Oxide   Ethylene oxide gaseous decontamination of food ingredients originated from a series of patents issued to Griffith and Hall who successfully lowered the microbial counts of dry spices, vegetable gums, dried fruits and spices in cloth and paper bags and the enzyme pancreatin.[98,100,99,101] Baer also applied the method to spices.[16] A review of these practices was published by Hall in 1951.[105]

Kirby et al. obtained control of molds in bread.[135] In 1952 Pappas and Hall described ethylene oxide sterilization of flours of corn, wheat, barley and potato, as well as corn starch.[179] Successful applications have been made also to dried eggs, gelatin and powdered and flaked foods, including cocoa and baby cereals.[151,193,42]

Whelton et al. applied ethylene oxide to packaged dried fruits to minimize microbial spoilage during adverse shipping and storage conditions encountered by the military during World War II.[255] One of their techniques was to add an ethylene oxide solution or "snow" to the outer container of packaged fruit. Gradual permeation throughout the containers essentially sterilized the contents and the gas eventually escaped through the packaging material. Others have used the "snow" application (frozen, ground ethylene oxide in water) in packaged foods.[17,128]

Sair has demonstrated that ethylene oxide treatment of many food components markedly reduces Salmonella counts.[204]

Propylene Oxide   Although applications of propylene oxide are more recent, they can be nearly as effective as ethylene oxide. Sair noted destruction of 90% or more of the bacteria, yeast and molds in cocoa.[204] Total counts as high as 200,000 to 300,000 per gm could be reduced to 10,000. In dried egg yolk original counts of 20,000 per gm were reduced to 200. Similar results were obtained with dried whole egg solids, where putrefactive anaerobes and Salmonellae were reduced below the detection levels. In split green peas, counts of 11,000 to 16,000 were reduced to 40 or less. Final counts were made for yeasts, molds, flat sours and thermophiles. In dried yeast, propylene oxide affected total count, aerobic spore formers and coliforms. Sair reported that propylene oxide treatment is the most practical method of control of Salmonellae in meat scraps.

Bruch and Koesterer examined the cidal action of propylene oxide on powdered and flaked foods inoculated with various organisms.[42] At a relative humidity of 25% and 37°C, 90% of B. subtilis spores were killed in 40 minutes (1250 mg propylene oxide/liter). In cocoa powder, bacterial counts were reduced 5–70% and mold counts 90–99%.

The need for a lower bacterial count in wheat flour to be used in convenience foods prompted Vojnovich and Pfeifer to explore the use of ethylene oxide and propylene oxide as an alternative to heat treatment.[247] Confirming earlier reports, ethylene oxide was more efficient, but satisfactory results could be obtained with propylene oxide. In wheat of 13% moisture an initial bacterial count of 1,000,000/gm was reduced to 500/gm at 118°F by a vapor system containing 1.5 gm/liter of propylene oxide. Treated wheat had an epoxide residue of 300 ppm, but after it was milled the flours contained less.

Several authors have mentioned the use of other gaseous sterilants. Bruch has included methyl bromide and β-propiolactone, as having potentials but with disadvantages.[41] Others rejected for various reasons are epichlorohydrin, epibromohydrin, ethyleneimine, ethyl sulfide, glycidaldehyde, propyleneimine, chloropicrin and ozone.

Storage and Handling

Ethylene oxide is a hazardous chemical because of its flammability and the explosive reactions which it may undergo. Explosive limits in air can range from 3–100%. However, when used as 20% mixtures with carbon dioxide or fluorohydrocarbons, the gas is relatively safe so long as precautions are observed.

Propylene oxide has a narrower explosive range of 2–22%, and is also somewhat less hazardous because it is less reactive. With care it can be used at higher concentrations than ethylene oxide.

Both oxides are vesicants and in the presence of moisture contact with the skin should be avoided. Vapors can cause temporary injury both to eyes and respiratory tract. For ethylene oxide the maximum safe concentration in air for an 8-hour exposure is 50 ppm. For propylene oxide 150 ppm in air can be tolerated.[41]

It is recommended that users of either oxide become thoroughly familiar with and observe rigidly manufacturers' recommendations for handling.

Assay

Ethylene Oxide   A titrimetric method for the determination of ethylene oxide in gas samples which may also contain oxygen, nitrogen, carbon dioxide and water is given in Technology Series Report No. 37-T2 of the Union Carbide Corporation, Chemicals Division, Research and Development Department, Tarrytown, N.Y.

Propylene Oxide   A gas chromatographic method for the determination of residual propylene oxide and propylene glycol in food materials has been developed by Griffith Laboratories, Inc., Chicago (1963). A colorimetric method for both ethylene and propylene glycol residues which utilizes phenylhydrazine has been reported by Brokke et al.[40] The Technical Service Laboratory of Chemical Division, Union Carbide Corporation has described a method for the determination of the 1–2 glycols by sodium hydroxide titration of residues from oxidation of the adjacent hydroxyls with sodium metaperiodate.

Chlorohydrins   A method involving gas chromatography and flame ionization detection for the determination of both ethylene and propylene chlorohydrins is described by Griffith Laboratories, Chicago (1966) and by Wesley et al.[253] The latter also applied the Volhard method for chlorine to a steam distillate of a slurry of the treated food.

Diethyl Pyrocarbonate
(Pyrocarbonic Acid Diethyl Ester)

Diethyl pyrocarbonate (DEPC) disappears in aqueous systems after it has performed its function. It permits a "cold sterilization" or "cold pasteurization" of some food materials as it is cidal to yeasts. This compound was isolated and identified by Boehm and Mehta in 1938.[30] The preservative action was first observed in the laboratories of a German chemical manufacturer, Farbenfabriken Bayer.[28,236] Since then, a substantial amount of research has been done in various countries applying DEPC to preservation problems. These and other historical elements are noted in the 1964 paper by Genth.[83]

Physical and Chemical Properties

Diethyl pyrocarbonate is a colorless liquid, not very soluble in water but soluble in the common organic solvents. It has a fruity, esterlike odor. When added to aqueous solutions, it rapidly hydrolyzes to ethanol and CO_2, leaving only trace residues. It decomposes more rapidly in neutral than in acid solutions. DEPC imparts no taste or odor to wine when used at proper levels.

$$C_2H_5O-\overset{\overset{\displaystyle O}{\|}}{C}-O-\overset{\overset{\displaystyle O}{\|}}{C}-OC_2H_5$$

A specification for diethyl pyrocarbonate appears in *Food Chemical Codex*.

Antimicrobial Activity

Diethyl pyrocarbonate has a broad spectrum of antimicrobial activity with a special effectiveness against yeast. As is often the case with chemical preservatives, the effective levels are somewhat proportional to the microbial load. The static and cidal effects of DEPC are shown in Table 15, where its weaker action against molds also is evident.

TABLE 15
Antimicrobial Action of DEPC*

Organism	Count per ml	DEPC Concentration (ppm or mg/l)	
		Inhibitory	Lethal
Yeasts			
Saccharomyces pastorianus	400	—	100
Saccharomyces pastorianus	500	30	—
Saccharomyces cerevisiae	400	20	100
Saccharomyces cerevisiae	10,000	60	150
Wine Yeast—Champagne Ay	415,000	50	—
	829,000	50	300
Pichia farinosa	400	10	100
	4,000	100	100
Bacteria			
Lactobacillus pastorianus	600	100	300
Mixed lactic culture & yeast	298	170	100–300
Molds			
Mixed Culture[1], normal infection	—	300–500	—
Mixed Culture[1], heavy infection	—	500–800	—

* Source: Adaptation from Hennig (Ref. 119). Reproduced by permission of Sigurd Horn Verlag, KG, Frankfurt am Main, Germany.

[1] *Penicillium luteum, Aspergillus terreus, Trichoderma viride* and *Neurosora sitophila*.

Further information on effectiveness against yeasts is presented in Table 16. The yeasts studied were five typical wine contaminants plus five strains of *Saccharomyces cerevisiae* var. *ellipsoideus*.

The contact time required for DEPC to be effective was also studied by Ough and Ingraham. DEPC was added at levels of 40 and 120 mg/liter to wine inoculated with *Saccharomyces cerevisiae* var. *ellipsoideus* (Montrachet strain). Viable cell count determinations made at intervals up to six hours are summarized in Table 17.

It can be seen that killing times are short, even for the 40 mg/liter level. At a level of 120 mg/liter all of the yeast cells were killed in less than 30 minutes.

TABLE 16
Activity of DEPC Against Yeast*

Yeast	Viable Cell Counts/ml[1]		
	Control		100 ppm DEPC
Pichia fermentans	600	800	0 0
Pichia membranaefaciens	17,200	10,400	0 0
Hansenula anomala	10,800	9,800	0 0
Kloeckera africana	800	1,500	0 0
Rhodotusula rubris	4,300	3,700	0 0
Sacch. cerevis. var. ellip. strain			
Burgundy	4,800	5,000	0 0
Champagne	6,400	5,500	0 0
Cognac	1,900	2,100	0 0
Port	7,300	8,200	0 0
Rhine	2,600	3,000	0 0

* Source: Ough and Ingraham (Ref. 178). Reproduced by permission of American Society of Enologists.
[1] Counts made 2 hours after addition of DEPC; duplicate determinations.

TABLE 17
Killing Times for DEPC Against Yeast*

Time (hr)	Cell Counts/ml at Various DEPC Concentrations (mg/l)					
	0		40		120	
0	480	510	530	330	390	420
½	310	550	400	480	0	0
1	350		240	70	0	0
2	560	430	40	0	0	0
6	550	430	0	0	0	0

* Source: Ough and Ingraham (Ref. 178). Reproduced by permission of American Society of Enologists.

Genth has provided more extensive tabulations of lethal concentrations of DEPC for a variety of organisms as reported by several investigators.[83]

Safety for Use

Because of its rapid hydrolysis no toxicity or residue problems occur in products where DEPC is permitted. Hecht conducted a trial in which drinking water of rats was replaced with grape juice treated with 0.5% DEPC—many times higher than the use level.[115] Over a 59-day period, those rats receiving the treated juice had the same weight gains as the control group.

This same author found that with 2% DEPC there were poor weight gains over a 4-week period, but upon removal of the additive, the normal rate of gain was restored. He also determined with rats that the LD_{50} was 1.1 ml/kg of body weight when the compound was administered in oil by stomach tube.

Inhalation tests were conducted with guinea pigs, rats and mice. When exposed in a chamber to a concentration of 0.1 ml/liter of air, the animals immediately showed irritation of the eyes and nose, followed by head colds, excessive lachrymation and intensified salivation. Several animals died on the third day but survivors slowly recovered. The irritation symptoms recurred when the animals were exposed for 1 hour to 0.0025 ml/liter of chamber space, but most animals survived.

A report of the World Health Organization states that, on the basis of its studies of the biochemical and metabolic aspects, and of the end products of breakdown, as well as products of interaction with constituents of beverages, DEPC is safe in the quantities that are likely to be consumed.[256]

Regulatory Status

In the United States DEPC is permitted in still wines, fermented malt beverages, and non-carbonated soft drinks and fruit-based beverages (21 CFR 121.1117 and amendment). In wines the quantity added before or during bottling to prevent secondary yeast growth and fermentation must not exceed 200 ppm; there shall be no detectable DEPC 5 days or more after bottling. For malt beverages not more than 150 ppm may be added before or during packaging; there shall be no more than 5 ppm 24 hr or more after packaging. In non-carbonated soft drinks and fruit-based beverages, the maximum addition is 300 ppm; there shall be no more than 5 ppm 24 hr after packaging. Products having Standards of Identity that do not permit the use of DEPC are excluded.

Diethyl pyrocarbonate is permitted in non-alcoholic beverages and/or fruit juices in Argentina, Bulgaria, Czechoslovakia, Chile, Greece, Norway, South Africa, and Sweden. In addition, it is permitted in wine by Brazil and West Germany. Its use is limited to wine and beer in Canada, Denmark, and Mexico and to wine only in East Germany and Israel. In Tunisia DEPC can be used in vegetable juices as well as in fruit juices, and in Peru use is limited to acid beverages.[273]

Applications

DEPC is effective only in acid products of low microorganism count. The pH should be less than 4, and the count should be not more than 500/ml. Preliminary treatment, such as flash pasteurization or sterile filtration, may be necessary in many cases. The beverage must not be offered for consumption before the decomposition has reduced the DEPC to the tolerated limit.

For small-scale and experimental work it is best to add DEPC to beverages as a freshly-prepared 10–20% solution in ethyl alcohol. For large-scale operations the DEPC can be added directly, but thorough and efficient mixing is essential.

Uses in Wine   Experiments on applications to wine have been published by Ough and Ingraham,[177] Mayer and Luthi,[150] Thoukis et al.[235] and Henning.[119] DEPC at 50–200 ppm was substituted for pasteurization in these experiments.

Other Uses Other applications of diethyl pyrocarbonate have been suggested. Harrington and Hills observed that 50 ppm DEPC rapidly destroyed most of the yeasts, molds, and bacteria in fresh apple cider; viable counts were reduced by 99%, yet 200 ppm DEPC would not render cider sterile.[107] Molin *et al.* reported that surface mold on fresh strawberries was checked by dipping in solutions of DEPC[162]; the most effective levels were 0.01–0.1%. A minimal level of 0.1% extended the shelf-life of applesauce. Expense may prohibit wide use in low-cost products, even though research can demonstrate effectiveness.

Storage and Handling

Stability DEPC is stable for prolonged periods in the original unopened container. Once the container is opened, however, the DEPC should be used as quickly as possible, since introduction of moisture or foreign material will result in decomposition of the ester. Carbon dioxide produced in these reactions may build up enough pressure to rupture the container. Violent exothermic reactions can result with alumina, iron rust, amines and ammonia. However, when DEPC is diluted to use levels, these reactions do not take place.

Processing Equipment Concentrated DEPC should preferably be handled in equipment made of stainless steel, aluminum or glass. Although certain plastics such as polyethylene and Teflon* do not appear to react with DEPC, others such as vinyl (vinyl acetate, vinyl chloride), cellulose acetate and polystyrene do react. Rubber is also attacked slowly by DEPC, whereas zinc, tin and silver are not.

Safety Measures for Personnel Concentrated DEPC is extremely irritating to the eyes, mucous membranes and skin and should be handled with the utmost care. Areas where it is handled should be well ventilated and inhalation of the vapors avoided as much as possible. Any DEPC dropped on the skin should be immediately washed away with soap and water, since prolonged exposure can cause blistering. If any is spilled on the clothing, the affected garments should be removed and thoroughly cleaned. Spilled material can be wiped up with a rag soaked in diluted ammonia, or can be taken up with baking soda or an inert adsorbent such as cellulose or fuller's earth.

Personnel handling concentrated DEPC should wear goggles. Protective gloves and clothing made of rubber or plastic may be necessary for extended handling of large quantities.

Assay

Diethyl pyrocarbonate in bulk may be determined by the morpholine titration method of Johnson and Funk.[130] This is a general method for acid anhydrides and any anhydride present, in addition to diethyl pyrocarbonate, will interfere. Degradation products do not interfere. However, ethyl chloroformate, the intermediate, does titrate. The intermediate can be determined by chlorine analysis or gas-liquid chromatography.

The anhydride is treated with an excess of morpholine in solution. That portion of the reagent which does not react is back-titrated with a standard solution of hydrochloric acid in methanol. The standard deviation of the method is ±0.5%.

* Trademark, E. I. duPont de Nemours & Co.

Antibiotics

Since the introduction of penicillin in the early 1940's, a large number of other microbially produced antimicrobials have been discovered. As early as 1945, their effectiveness in combating microorganisms in the animal and human body led to exploration of their possible utility as food preservatives. Only five—chlortetracycline, oxytetracycline, nisin, pimaricin and nystatin—currently are in use for this purpose, and most of the applications are limited. However, the extensive research that was conducted on many antibiotics when there was hope of widespread food use is of considerable historical interest.

One property of antibiotics that intrigued the early investigators was their high order of activity on a weight basis—100 to 1000 times greater than that of preservatives already in use. Partially offsetting this, of course, was the higher cost of antibiotics.

As a group, antibiotics, like the conventional preservatives, exhibit selective antimicrobial activity. Some are active against many gram-positive bacteria, others predominantly against gram-negatives, and a few, the "broad-spectrums," are inhibitors of members of both groups. Others are antifungal only.

Within stability ranges, the antimicrobial activity of antibiotics is not influenced by pH as is the activity of the conventional preservatives discussed in the preceding sections. Like the other compounds, however, the antibiotics, at normal use levels, are static and not cidal, and to provide preservative action they must continue to be present, so that stability in food is a factor affecting their utility. In view of their static action, it was early suggested that antibiotics might function largely as adjuncts to other methods of preservation, a concept summarized by Wrenshall[262] as follows:

1. As adjuncts to refrigeration to delay spoilage of perishable foods during storage, transportation and marketing.
2. To prevent microbial buildups during preparation for other processing, such as freezing.
3. As adjuncts to heat processing so that the intensity of the heat treatment might be diminished.

The research which has been conducted on these phases has been summarized in extensive reviews at various dates, among them that in 1955 by Campbell and O'Brien,[44] in 1959 by Farber[69] and by Wrenshall,[262*] and in 1964 by Goldberg.[88] In this section, that which might be termed "historical" is summarized briefly, largely by reference to previous reviews, and this is followed by a more lengthy discussion of phases of research which are currently active.

Applications to Foods

Meats, Poultry and Fish The tetracyclines, especially oxytetracycline (OTC) and chlortetracycline (CTC), were extensively tested and found highly effective for reducing bacterial spoilage. These antibiotics are described as "broad-spectrum"; that is, they attack both gram-positive and gram-negative bacteria of significantly different types. They derive their names from the

* A part of this is an annotated bibliography which includes abstracts of references from 1945 to 1958.

four rings which characterize their structure. Discovery of chlortetracycline was announced in 1948 by Duggar[63] and of oxytetracycline in 1950 by Finlay et al.[63,73]

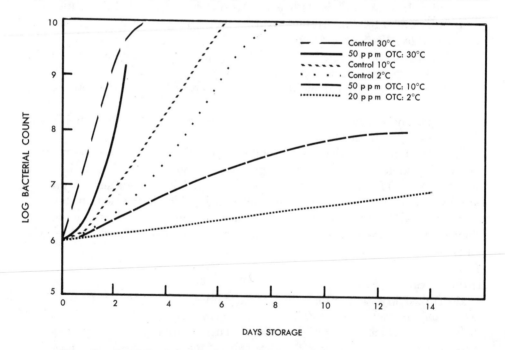

Oxytetracycline   Chlortetracycline

Meats   Chlortetracycline and oxytetracycline have been applied experimentally in several different ways in order to extend edible life of fresh meats and permit tenderizing at higher temperatures. These include dip or spray of whole or divided carcasses or cuts, as well as pre-slaughter injection of antibiotics into the peritoneum or tail or intramuscularly. Whereas, traditionally, carcasses are aged by hanging them at refrigerated temperatures for 2–4 weeks, with antibiotic treatment the same degree of tenderness can be achieved by 48 hours at room temperature followed by a 5-day chilling.

In studies in a tropical country, carcasses of cattle and sheep were treated with OTC by dipping and spraying.[86] These treated carcasses remained edible

Figure 2.   The ability of oxytetracycline to restrict bacterial growth in ground beef stored at various temperatures.

Source: Adapted from the *Science of Meat and Meat Products* by the American Meat Institute Foundation. W. H. Freeman and Company, Copyright © 1960.

for 48 hours in the case of the cattle and 72 hours in the case of the sheep, at temperatures of 90°–98°F and very high humidities. It would seem that this method of preservation might help offset the deficiency of refrigeration in less developed countries, many of which have warm climates, provided assurance could be obtained that food poisoning organisms were controlled as effectively as those causing putrefaction.

Tetracyclines have also been found to extend the shelf life of ground beef and to supplement radiation preservation. (See Figure 2.)

The use of antibiotics such as CTC and OTC in meats can give rise to another problem—the growth of mold, so that simultaneous use of a mold-controlling additive may be necessary.

For a more extensive summary of the substantial number of pieces of research on application of antibiotics in the preservation of fresh meat, see the reviews by Wrenshall and Farber[262,69]. At present these uses are not permitted or are not practiced in any country for which information is available.

Poultry Birds are normally chilled in a slush ice tank after killing, plucking and evisceration. If this bath contains 10 ppm of an antibiotic such as OTC or CTC, the time that carcasses can be held in cold storage is increased by 50 to 100%, according to the research which developed this method of preservation. This treatment has been applied also to poultry parts. As in the case of meat, antibiotic-treated poultry carcasses may permit growth of yeast and molds and means for control of these may be necessary. In general, the use of antibiotics in poultry should be combined with refrigeration.

A few years ago the antibiotic dip treatment of fresh whole poultry carcasses was permitted under regulations of the U.S. Food and Drug Administration, with the restriction that the residue in or on the carcass not exceed 7 ppm for both OTC and CTC. This permission has been revoked by FDA.

Another method investigated for extending edible life of poultry carcasses is that of feeding extremely high levels of antibiotic to the birds for a limited period shortly before slaughter.[8, 217] The antibiotic circulates through the carcass and remains there, continuing to inhibit microorganisms after the bird has been killed. Residues of antibiotic (OTC or CTC) are destroyed by normal methods of cooking.[157]

Fish and Seafood Fresh fish are highly perishable. Although they are commonly kept iced in the trawler and in the market, some of the contaminating organisms are psychrophilic; that is, they are able to flourish at low temperatures. The ability of broad-spectrum antibiotics to retard food spoilage was first observed in tests with fish. The antibiotic at low levels is dissolved in the brine or in the ice used in the storage areas of the fishing vessels. The recommended level for ice is 5 ppm. Whole red fish remained fresh for 20 days under antibiotic ice, while controls, under ordinary ice, spoiled in 14 days. Eviscerated haddock remained fresh for 25 days under antibiotic ice, while the controls spoiled in 13 days.[262]

In a 1966 study it was found that, when the ice used in storing the catch in the trawlers contained 5 ppm of oxytetracycline, the rapid deterioration of fish known to take place after 14 days could be delayed.[261] Four distant-water trawlers used OTC-containing ice for six months during repeated trips. The effectiveness of this treatment in continuing to extend edible life during these

repeated uses would indicate that there was no accumulation of resistant organisms in the storage areas and elsewhere on the boats. At the time of this
report about 70 trawlers operating from Grimsby and Hull used antibiotic
ice.

Freshness of fillets has been extended as much as 300% by dipping in
a brine containing 10–100 ppm of a tetracycline antibiotic.[262] Use of CTC and
OTC for fish continues to be permitted in England and Wales. Use of CTC
only for raw fish, as well as on peeled shrimp and shucked scallops, is permitted
in Canada.

Tylosin, another antibiotic,* was reported to prevent formation of botulism
toxin in smoked fish. In a laboratory experiment, this antibiotic, added at
100 ppm to the brine and present in the fish at 3 to 5 ppm, prevented formation of toxin in fish processed at low temperatures during an incubation period
of 14 days at room temperature.[215]

Antibiotic treatment of crustacea and shellfish, overall, has been somewhat
less successful than treatment of other marine food animals.[262]

Fresh Fruits and Vegetables   The use of preservatives for these falls largely
within the category of "post harvest." Trials have been made with many
antibiotics in efforts to control the destructive bacterial soft rot of leafy vegetables and other diseases caused by bacteria or fungi. Antibiotics tested have
included oxytetracycline, chlortetracycline, streptomycin, neomycin and polymyxin. The antifungal antibiotics, nystatin and pimaracin, also have been
tested. Information on these experiments is found in the reviews by Wrenshall
and by Eckert.[262,65]

Dairy Products†   Generally, the use of preservatives in fresh milk is contrary to the rulings of public health authorities in developed countries. However, antibiotics have been found effective in enhancing keeping qualities of
this product. A few ppm of a tetracycline antibiotic in raw milk permitted
storage at 30°C for up to four days without marked deterioration.[112]

These experiments were conducted some years ago with a view toward
determining whether the use of such preservatives might be feasible and justi-
fiable in less-developed countries with inadequate refrigeration facilities. In
this connection, nisin (see below) is said to be permitted in certain countries
in in-bottle sterilized milks to improve keeping qualities under sub-tropical
conditions.[240]

For control of bacteria in cheese, nisin is effective. Nisin is elaborated by
certain lactic acid organisms and is found naturally in some types of cheese.
This antibiotic was detected as early as 1928 by Rogers[200] and was isolated
and characterized by Mattick and Hirsch.[149] It is a polypeptide with activity
against some gram-positive organisms. Nisin has certain disadvantages which
bar it from medicinal use, which, of course, is favorable to its food use.[117]
It is known to be destroyed in the gastro-intestinal tract and has been judged
practically non-toxic within the usual meaning of the term.[118,79] Nisin-resistant
bacteria are not resistant to antibiotics used in medicine.

A review of the discovery of nisin and its application as a preservative
in cheese has been presented by Hawley as well as by Heinemann et al.[114,117]

* Described under Canned Foods.

† Canned milks are grouped with canned foods in the next section.

A great deal of the spoilage caused by Clostridia which occurs in processed cheese can be prevented by nisin. Eight European countries, as well as India and South Africa, permit use of nisin in cheese. In Sweden nisin may be used also in sterilized condensed milk and cream, and in England and Wales it may be used in clotted cream.

The commercial product in Britain is standardized at 1 million Reading Units per gm and 1 gm is used per 10 kg of the cheese mix.[240] Alternatively, a cheese is produced that contains 15,000 units of nisin per gm, some of which is processed together with ordinary cheese.[89]

Canned Foods   After processing, canned foods may still contain spores of thermophiles, more heat resistant than *Cl. botulinum*. These spores can grow out and cause spoilage. A heat treatment severe enough to provide complete sterility would downgrade flavor or texture or both, and might even impair nutritional value.

A preservative which would attack the spores of thermophiles should permit milder heat treatments, and the possibility that some of the antibiotics might have this capability was brought under investigation in the early postwar period. Obviously, a preservative which would fulfill the desired role must be able to withstand the heat treatment without excessive loss of its antibacterial potency. Besides, it must not lose too much of its remaining potency over the rather extended storage period to which canned goods often are subjected. The tetracycline antibiotics are not suitable for this purpose, as they succumb too readily to heat. Among others tested and not found suitable for one reason or another as adjuvants in canning were: bacitracin, chloramphenicol and gramicidin.

Ultimately, most attention was focused on three antibacterial antibiotics which are not used in human therapy: nisin, already discussed in connection with cheese, subtilin and tylosin. The possible value of nisin and subtilin as adjuvants in heat sterilization was indicated by the screening tests of Anderson and Michener in 1950[7], O'Brien *et al.*,[170] in 1956 and Michener *et al.*, in 1959.[158]

Subtilin, discovered in 1944, is synthesized by a strain of *Bacillus subtilis*. Like nisin, it is a polypeptide, active mainly against gram-positive organisms, including thermophiles and clostridia and their spores. It is of low oral toxicity for it is decomposed in the digestive tract.[7]

In spite of the promising characteristics of subtilin demonstrated in the test tube, it did not satisfy expectations after many years of canning trials. Mainly, it did not suppress Clostridial spores satisfactorily in inoculated packs because these organisms vary in sensitivity. Thus it still left the hazard of botulism if milder heat were used in conjunction with the antibiotic. The National Canners Association, which conducted much of the work, remained unconvinced of its capabilities.

The main characteristics of nisin have been noted under "Dairy Products." The investigation of nisin as an adjunct in heat processing of foods seems to have been initiated by a report of Gillespy following one by Lewis *et al.*, who had found that nisin accelerated the death of spores subject to moist heat.[85,142]

Heinemann and his colleagues reported that nisin added to chocolate milk makes possible a reduction in the processing from a customary F9 to F3, and the flavor is better immediately after processing and maintains this advantage over the storage period.[117] A suitable level of nisin was found to be 80 units per ml. This prevented spoilage for three weeks at 110° or 131°F and for six months at 85°F. Control samples without nisin spoiled at 131°F and 110°F and 10% spoiled at 85°F. About 30% of the added nisin was present after processing.

Nisin has been tested also for control of thermophiles in other canned foods—soups (including mushroom and lentil), carrot puree, peas, and beans in tomato sauce, and tomatoes.[84,117,240]

As Tramer points out, the adequacy of nisin as an adjunct to heat processing depends upon effective quantities remaining in the can after exposure to heat and after the passage of time in storage to check outgrowth of thermophiles.[240] The stability of nisin is affected by pH. It is much less stable near neutral than at pH 3.5. However, it is more stable to short-term, high-temperature processing than to longer, low-temperature procedures. In an appraisal of nisin for use in canning, Goldberg and Barnes[89] state: "Nisin cannot be used indiscriminately in any type of food because of its narrow spectrum of antibacterial activity. It has no activity against the common spoilage organisms of fresh food, e.g. *Pseudomonas*, and may even be destroyed by some groups of bacteria. . . . Although many strains of *Clostridium botulinum* are sensitive to nisin, a few are resistant, so it would be undesirable to use nisin to lower the heat treatment of foods below that required to destroy *C. botulinum* spores in foods where a botulinum hazard exists."

Gibbs and Hurst allude to some of these same limitations and suggest that each individual use of nisin in canning requires careful assessment.[84]

Regulations of England and Wales permit nisin in certain canned goods. Australia permits its use in canned tomato puree.

A third antibiotic which has been investigated more recently as an adjuvant in heat processing is tylosin. This is a macrolide, one of a class of antibiotics characterized by a large lactone ring structure to which rare sugars are attached by glycosidic linkages.

Like subtilin and nisin, tylosin is effective against gram-positive organisms, including Clostridia, its spectrum being somewhat broader than that of the other two. It is said to be more stable to heat than nisin and is essentially non-toxic.[56,9,25]

Tylosin has been tested in various canned foods, including mushrooms, a dog food, meats of various types and foods kept in hot vending machines such as cream style corn, chow mein, cream of chicken soup, tuna and noodle dinner and macaroni creole.[97] It has shown its ability to prolong the sanitary life of these products[254,210].

Tylosin exhibits some cross-resistance with erythromycin and oleandomycin which are somewhat related chemically and are used in medicine. These circumstances have created doubts about its suitability for use in foods.

Other Food Categories   Another antibiotic of more recent origin which is being tested as an additive for several classes of foods is pimaracin, an anti-fungal. This has a macrolide structure, in a class with tylosin, and is elaborated by a mold. First reports on pimaracin appeared in 1958. Its history and properties have been reviewed by Clark, Shirk and Kline.[48] Pimaracin acts apparently by causing microbial cells to lose cellular components such as amino acids and nucleic acids, thus preventing reproduction. It is active over the pH range from 3 to 9, and its toxicity is low.

Pimaracin is only slightly soluble in water, but does dissolve in propylene glycol, glycerol and acetic acid. It can be suspended in water and withstands boiling temperatures at pH greater than 5. However, it is unstable in aqueous suspension if exposed to light.

Pimaracin has been found effective against a wide range of fungal organisms. It has been reported capable of controlling yeasts and molds in fruit juices of several kinds, fruit-flavored carbonated drinks, fresh strawberries and raspberries, dressed poultry, cottage cheese and hard cheese.

This antibiotic can be added by mixing into food, dipping, spraying or dusting, or applying by means of treated wrappers. Effective concentrations are from 10 μg /ml or less in aqueous foods and up to 100 μg/ml in more concentrated foods. These levels are much lower than those of the acid fungistats, but, of course, economics must be taken into account.

Pimaracin is permitted for coating hard cheeses in Norway and Sweden and for surface-coating of cheese and sausage in Belgium.*

Discussion

Aside from public health considerations, one might raise the question whether occurrence of resistant organisms in processing plants would invalidate the preservative action of tetracyclines. Curiously, there are indications that this may not be so.[2,244]

A trial feeding of tylosin to swine showed no increase in the number of resistant organisms in treated animals as compared with controls.[145] Furthermore, in a similar trial with humans over a six month period, the incidence of tylosin-resistant staphylococci was found to be the same in the test group and the controls.[146] Resistant staphylococci were also found prior to tylosin intake, and antibiotic resistance was transitory and erratic.

The prominent role of antibiotics in chemotherapy has stimulated an intensive study of the biochemistry of their growth-suppressive powers. Derived from such studies is a more definitive elucidation of those cellular mechanisms which are affected when microorganisms are inhibited or killed by chemicals. Some of these mechanisms will be briefly treated in the later section dealing with mode of action. Perhaps this knowledge can be applied in the search for new compounds which will be more effective in the control of microbial contamination of foods.

The Use of Antimicrobial Food Additives in Countries Other than the United States

This chapter has discussed chiefly U.S. practices regarding antimicrobial food additives. As each compound was considered, its uses outside the United States were treated briefly. Table 18, covering 17 other countries, groups the major compounds, excluding antibiotics, that may be legally used in these

---

* Another antifungal antibiotic, nystatin, may be used on the skin, but not the flesh, of bananas in England and Wales.

countries to suppress microbial contamination. This information was obtained mainly from official publications.

It is evident that some preservatives permitted in the United States are excluded by the regulations of other countries.* One class, the esters of para-hydroxybenzoic acid, is not legal in Belgium, India, the Netherlands, Pakistan, and South Africa. Propionates are not mentioned for India, Pakistan, and South Africa. Acetates are listed for fewer than half of the countries.

Although they do not appear in Table 18, some compounds that are legal in other countries are barred in the United States. Hexamethylene tetramine can be added to semi-preserved fish and fish products in Austria, Norway and Sweden. Formic acid or sodium formate is legal in Sweden for semi-preserved fish; in Norway, Sweden, and Denmark it is permitted in tonic and cola beverages.

TABLE 18

Major Antimicrobial Additives Legally Permitted in Various Countries
Outside of the United States

Country	Sodium Benzoate	Parabens[1]	Propionates	Sorbates	Acetates[2]	SO$_2$ and Sulfites	Nitrite, Nitrate[3]	Diethyl-pyrocarbonate[4]
Austria	+	+	+	+		+	+	
Belgium	+		+	+		+		+
Canada	+	+	+	+	+	+	+	+
Denmark	+	+	+	+		+	+	+
England and Wales	+	+	+	+		+	+	
Finland	+	+	+	+		+	+	
Germany, West	+	+		+		+	+	+
India	+			+	+	+	+	
Italy	+	+	+	+	+	+	+	
Japan	+	+	+	+	+		+	
Netherlands	+		+	+		+	+	
Norway	+	+	+	+		+	+	+
Pakistan	+			+	+	+	+	
Philippines	+	+	+	+	+	+	+	
South Africa	+			+		+	+	+
Sweden	+	+	+	+	+	+	+	+
Switzerland	+	+		+		+		

[1] Countries vary according to which esters of para-hydroxybenzoic acid are permitted. Salts of parabens usually permitted along with esters.
[2] Diacetates included.
[3] Includes use as meat-curing compounds as well as use as preservatives in certain types of cheese and fish.
[4] See p. 163 of text for details on uses of diethylpyrocarbonate.

* It should be recognized that a given product may have fewer uses in some countries than in others not necessarily because of official rejections but because permission for additional uses has not been sought.

In some overseas jurisdictions preservatives may be used in certain foods that may not be thus treated in the United States. Benzoates, sorbates, or propionates may be used in liquid egg products and in smoked or semi-preserved fish in various countries. Other overseas uses for benzoate are in bread, processed cheese, peeled shrimp, sausages, caviar, and ice used for packing fresh fish.

There is a converse situation—in some countries no preservatives are permitted in some foods for which treatment is accepted in the United States. Some domestic uses of a particular compound are not paralleled. For example, in England and Wales propionates may be used in bread, as in the United States, but they may not be used in cheese.

Mechanism of Antimicrobial Action

A successful food preservative should possess broad antimicrobial inhibitory powers, yet be essentially free of toxicity for man and animals. A differential toxicity of this nature has been difficult to discover. It is the basis for the successful chemotherapy of microbial infections. A student of the history of chemotherapy will be struck by the relatively few compounds which have succeeded as anti-infective agents for mammals, and a majority of these are of relatively recent origin.

From the time Pasteur established that disease and spoilage both were due to microbial invasion, the concept of control of microbes by chemicals has been a goal which until recently seemed just beyond the reach of man. Infectious diseases are complex problems, characterized by the invading microorganisms and the disease reaction of the host. What makes one strain of microorganisms pathogenic, whereas another strain, with the same gross characteristics observable in a laboratory, is not disease-producing? Early treatment of infectious disease amounted to little more than patient management, so that the natural defense mechanisms of man could eventually overcome the situation. As disease became better understood, attempts were made to augment or stimulate such defense mechanisms by the use of vaccines made from dead or attenuated cells or viruses which provoked a greater antibody formation. While only partially successful in treatment, this approach has been highly successful in prevention of infectious disease, including several diseases of virus origin. However, there was little success with fungal infection. The discovery of the sulfa drugs in the 1930's brought about a dramatic change, as these were capable of attacking many widespread and lethal infections in the active state. Chemotherapy was further successful with the development of the antibiotics, starting in the 1940's.

Food spoilage caused by microorganisms is the result of a mass population using the food product as its substrate, and the spoilage is the undesirable change induced by the microbial growth. Another undesirable change is the possible development of a range of toxins excreted by the thriving microorganism. Hence, a preservative may be effective if it prevents multiplication of the low concentration of natural flora during the required shelf life of the product. Of course, this can be attained by complete sterilization (heat, radiation), or total inactivation of the ·flora (freezing). It is seldom achieved by tolerable chemical agents. Thus, one has a gradient of preventive action: static action, where the cells remain viable but do not proliferate, and cidal action, involving death of the cell.

As has been stated by Gale in 1966: "The bacterial cell is a highly organized structure containing finely controlled biosynthetic systems, integrated to produce a self-reproducing living organism. With such a finely constructed and controlled system, it can be expected that damage at specific points can result in such consequent disorganization that the organism becomes non-viable and eventually non-living. It is possible that toxicity could arise from interference with a synthesis or function of a single protein, a single species of nucleic acid, or a particular lipoprotein complex. We can only interpret such possibilities when we know the location and function of our biochemical entities and events within the cell."[80]

Despite the fact that chemical preservatives of various kinds have been used for several decades, their mode of action remains largely unexplained; the mechanism of cellular inhibition and death is only beginning to unravel. The least is known about those antimycotic agents with relatively simple chemical structures, in contrast to the extensive information emerging from studies of highly complex, structurally specific entities like antibiotics.

The general thesis stated by Gale is relatively old, being derived and evolved from a mass of research effort involving circumstantial evidence. Postulated mechanisms could be grouped into: (1) interference with cell membranes; (2) interference with genetic mechanism; and (3) interference with intracellular enzyme activity. The abundant indirect evidence and analogy were well reviewed in 1948 by Wyss.[263]

Interference with the cell membrane and cell wall can have vast effects on permeability for cell nutrients and cellular contents. The various fatty acids which have some degree of antimycotic activity are thought to cause permeability changes by coating the cell surface with a substance that the cell can handle only with difficulty. Such materials would be expected to exert static action rather than cidal, and for the most part static action is observed in practice.

A voluminous literature exists to support theories that preservatives interfere with intracelleular enzyme activity. Isolated enzyme systems were used primarily and many of them can be inhibited *in vitro*. But such observations are not necessarily translatable to *in vivo* situations; the inhibitor may not penetrate the cell wall, or may not attain sufficient concentration at the enzyme site. Furthermore, cellular life comprises a host of interdependent enzyme systems, and inhibition or death is the result of interference with the most sensitive system.

A wealth of finely detailed biochemical data is being generated from studies on antibiotics to explain the process of chemical inhibition of microbial growth; a complete review is beyond the scope of this chapter. However, Newton[166] presented an excellent review in 1965, and a symposium held in England emphasized the rate at which new information is developing.[29] While inhibitory mechanisms of antibiotics are depicted, these are probably general mechanisms of cellular inhibition and death. One might find it interesting to compare the new experimental evidence with the theories deduced a few decades ago.

Each antibiotic has a distinct mode of disrupting cellular life. Several have similar general mechanisms, but differ in their specific site of attack. A few examples can be cited.

Influence on the Cellular Membranes

For an antimicrobial to be effective, it does not appear necessary that it penetrate into the cell. A reaction on the cell wall may be sufficient to alter its permeability. Altered permeability can impair the passage of nutrients into the cell; conversely, there is evidence that the change in permeability causes a leakage of cellular constituents. This latter effect is produced by polypeptide antibiotics such as tyrocidin and polymyxin, and also by surface active agents such as quaternary ammonium compounds. The polyene antifungal antibiotics may create leakage by modifying protoplast membranes.[138] It thus becomes apparent that those organisms not inhibited by such agents must have cell walls lacking the chemical components to provide sites of attachment.

Cell walls are complex polymers of mucopeptides (glycopeptides), lipopolysaccharides, lipoproteins, and proteins. An antimicrobial can affect the synthesis of one of the simple components of these entities, inhibit their polymerization, or bring about the creation of an imperfect cell-wall system which fails to satisfy the cell's requirements. Several antibiotics, such as penicillin, bacitracin, novobiocin and vancomycin, are known to prevent cell-wall synthesis. This in itself is a complex mechanism, and the specific inhibitions exerted by these antibiotics may differ. The antifungal antibiotic, griseofulvin, for example, interferes with chitin synthesis in the cell walls of the hyphae.[199]

Interference with Genetic Mechanisms

Interference with the genetic mechanisms of cells involves chemical activities within the cell. But again, under this general concept, the actual site of involvement may be very difficult to ascertain. Streptomycin exerts its effect by chemically combining with the ribosome at an unidentified site.[39] Chloramphenicol is another antibiotic which attaches to the ribosome, causing RNA accumulation by inhibition of protein synthesis.[245] The specific subunit of the ribosome is known. The important tetracycline group of antibiotics is noteworthy because of their unusual metal-chelation abilities. While metal chelation may have influence on several sensitive sites, inhibition of protein synthesis seems to be the most important, and this occurs at some late stage of synthesis involving the ribosome.[78]

Interference with Cellular Enzymes

While the older literature contains many reports of the inhibitions of various antimicrobial agents on isolated enzyme energy systems, there now is doubt that the inhibition of such systems is an important cause of cell disorganization. At least such a mechanism has not been found *in vivo* among the many antibiotics which have been studied in connection with oxidative phosphorylation and energy-coupling reactions. Cellular concentrations of antibiotics are generally much lower than the concentrations required to inhibit *in vitro* systems. On the other hand, it appears as though other types of enzyme systems are more sensitive to inhibitory action, and these can range from enzymes associated with cell wall construction to those which participate in the multitude of reactions yielding RNA and the double-helical structure of DNA. An interesting start has been summarized by Waring and by Reich, but much remains to be done.[250],[198]

Similarities Between Preservatives and Antibiotics

The known mechanisms which account for the inhibitory actions of antibiotics appear to involve one of two realms of action: (1) effect on the formation, permeability and integrity of the cell wall, or (2) disorganization of the fine structure of the cell's genetic system. There is little comparable data for the commonly-used food preservatives, but it seems likely that they may work through similar mechanisms, if one accepts the data derived from antibiotics as illustrating general biological principles.

That there is cell wall involvement can be deduced from work of Stutzenberger and Bennett[227] and Truby and Bennett[242] in studies with phenols. Inhibitory efficiencies of phenols can be correlated with distribution coefficients between organic and aqueous phases, and distributions favoring the organic phase are the result of hydrogen bonding between the phenolic hydroxyl and the proton-attracting organic phase. Thus, it has been suggested that the bacterial lipid capacity to hydrogen-bond phenolic compounds is a function of the nature and quantity of lipid.

This concept was verified in a study of trichlorophenol and *Staphylococcus aureus* in pure cultures mixed with several species of gram-negative bacteria, such as *Proteus mirabilis, Salmonella schottmuelleri,* and *Aerobacter aerogenes.*[243] In mixed cultures, greater amounts of trichlorophenol were required, slightly lesser amounts if the gram-negative cells were heat killed, and still lesser amounts if the gram-negative cells were defatted. The fat content of the gram-negative cells correlated with trichlorophenol demands in the mixed cultures. These authors further point out that proteins and carbohydrates of the cell wall also determined the cell's sensitivity to phenols, and the relative hydrogen bonding between cell wall constituents and phenols influenced the compound's efficiency. These studies suggest that the paraben compounds do not inhibit by the same biochemical mechanism as does benzoic acid; the former are phenols.

In the case of sorbic acid, Melnick *et al.* have postulated that it exerts its static action by interference with cellular dehydrogenases which normally dehydrogenate fatty acids as the first step in metabolism.[154]

Tonge quotes a theory of Kodicek that unsaturated fatty acids adsorb on the cytoplasmic membrane to interfere with the transport mechanism.[239] But Tonge favors a mechanism in which there is free radical formation by removal of hydrogen from a labile methylene group, α to the double bond. These radicals combine with oxygen to form peroxides which in turn decompose to form new free radicals. The free radicals probably attach to cell membrane sites.

Bosund has critically examined the modes of action of benzoic and salicylic acids. While no definitive theory has emerged, he suggests "their action may be related to their high lipoid solubility, which presumably must cause them to accumulate on the cell membrane or on various structures and surfaces of the bacterial cell."[33]

This then bears on the repeated observation that acidic preservatives such as benzoic, sorbic, propionic acids, etc. are much more active at lower pH's. It is generally thought that the undissociated molecule is more effective than ionized molecules. Bosund notes that undissociated molecules have a great lipoid solubility and low water solubility compared to anions.

Research intended to elucidate a biochemical mode of action for generally-accepted food preservatives is likely to be more difficult than with antibiotics. Preservatives are relatively simple, non-specific compounds in contrast to the complex, highly specific structures encountered in antibiotics. Consequently, the preservatives may be expected to be less reactive with cellular sites of attachment, and even less specific. But is apparent that there is a need for applying the newer biochemical disciplines to studies of these compounds to better understand their functions. Such studies could lead to new preservatives with enhanced properties and values.

Indirect Additives

21 CFR 121.2001   *Antimycotics employed in the manufacture of food-packaging materials*

Calcium propionate
Methyl paraben
Propyl paraben
Sodium benzoate
Sodium propionate
Sorbic acid

21 CFR 121.2505   *Slimicides which may be safely used in the manufacture of paper and paperboard that contact food*

Acrolein
Alkenyl (C_{16}–C_{18}) dimethyl ethyl ammonium bromide
n-Alkyl (C_{12}–C_{18}) dimethyl ethyl ammonium chloride
Bis (1,4-bromoacetoxy)-2-butene
5,5'-Bis(bromoacetoxymethyl) m-dioxane
2,6-Bis(dimethylaminomethyl) cyclohexanone
Bis(trichloromethyl) sulfone
4-Bromoacetoxymethyl-m-dioxolane
2-Bromo-4'-hydroxyacetophenone
Chlorinated levulinic acids
Chloroethylene bisthiocyanate
Chloromethyl butanethiolsulfonate
Cupric nitrate
n-Dialkyl(C_{12}–C_{18}) benzylmethylammonium chloride
3,5-Dimethyl-1,3,5,2H-tetrahydrothiadiazine-2-thione
Dipotassium and disodium ethylenebis(dithiocarbamate)
Disodium cyanodithioimidocarbonate
2-Hydroxypropyl methanethiolsulfonate
2-Mercaptobenzothiazole
Methylenebisbutane thiolsulfonate
Methylenebisthiocyanate
Potassium 2-mercaptobenzothiazole
Potassium N-methyldithiocarbamate

Indirect Additives
(continued)

Potassium pentachlorophenate
Potassium trichlorophenate
Silver fluoride...................... Limit in process water 0.024 lb
                                                  per ton of paper produced
Silver nitrate
Sodium dimethyldithiocarbamate
Sodium 2-mercaptobenzothiazole
Sodium pentachlorophenate
Sodium trichlorophenate
1,3,6,8-Tetraazotricyclo(6,2,1,1,3,6) dodecane
2-(Thiocyanomethylthio) benzothiazole
Vinylene bisthiocyanate

21 CFR 121.2520   *Preservatives for Adhesives in Food Packaging*

n-Alkyl (C_{12}, C_{14}, C_{16}, or C_{18}) dimethyl benzeneammonium chloride
Ammonium benzoate
p-Benzoxyphenol
p-Benzyloxyphenol
Bis(tri-n-butyltin) oxide

1-(3-Chloroallyl)-3,5,7-triaza-1-azoniaadamantane chloride
4-Chloro-3,5-dimethylphenol
4-Chloro-3-methylphenol
Coconut fatty acid amine salt of tetrachlorophenol
Copper 8-quinolinolate
2,6-Di-tert-butyl-4-methylphenol
3,5-Dimethyl-1,3,5,2H-tetrahydrothiadiazine-2-thione
Ethyl-p-hydroxybenzoate
4,4'-Isopropylidenediphenol, polybutylated mixture
2-Mercaptobenzothiazole and dimethyl dithiocarbamic
    acid mixture, sodium salt
2-Mercaptobenzothiazole sodium or zinc salt
Pentachlorophenol
Phenol
o-Phenylphenol
Potassium pentachlorophenate
Quaternary ammonium chloride (hexadecyl, octadecyl derivative)
Salicylic acid
Sodium dehydroacetate
Sodium pentachlorophenate
Sodium o-phenylphenate
Sodium salicylate
Thymol
Tri-$tert$-butyl-p-phenyl phenol
Tri-n-butyltin acetate
Tri-n-butyltin neodecanoate

References

1. Aalto, T. R., Firman, M. C. and Rigler, N. E., 1953, *J. Am. Pharm. Assoc.*, *42*, 449–456.
2. Adams, R., Lerke, P. and Farber, L., 1966, *J. Food Sci.*, *31*, 982–987.
3. Alderton, G. and Lewis, J. C., 1958, *Food Res.*, *23*, 338–344.
4. Ambrose, A. M., 1950, *A.M.A. Arch. Ind. Hyg. & Occup. Med.*, *2*, 591–597.
5. American Meat Institute Foundation, 1960 *The Science of Meat and Meat Products*, W. H. Freeman and Co., San Francisco.
6. Amerine, M. A. and Joslyn, M. A., 1951, *Table Wines: The Technology of Their Production*, University of California Press, Berkeley; 2nd ed., 1970.

7. Anderson, A. A. and Michener, H. D., 1950, *Food Technol.*, *4*, 188–189.

8. Anderson, G. W., Epps, N. A., Snyder, E. S. and Slinger, S. J., 1958, *Poultry Sci.*, *37*, 174–179.

9. Anderson, R. C., Worth, H. M., Small, R. M. and Harris, P. N., 1966, *Food and Cosmetic Toxicol.*, *4*, 1–15.

10. Annheimer, T. D. and Fabian, F. W., 1940, *Am. J. Publ. Health*, *30*, 1040. (cited by Sykes, G., 1965, *Disinfection and Sterilization*, 2nd. Ed., E. & F. N. Spon, Ltd., London).

11. Anon., 1955, Assoc. Food Drug Officials, *U.S.*, *Quart. Bull.*, *19*, 154–156.

12. Anon., 1961, *Oil, Paint and Drug Reptr.*, Dec. 4.

13. Anon., 1960, *National Provisioner*, Apr. 9, 16–17.

14. *A.P.I. Toxicological Review*, 2nd Ed., 1963.

15. Auerbach, R. C., 1959, *Wines and Vines*, *40*, 26.

16. Baer, J. M., 1942, U.S. 2,229,360.

17. Baerwald, F. K., 1945, U.S. 2,370,768.

18. Bakerman, H., Romine, M., Schricker, J. A., Takahashi, S. M. and Mickelsen, O., 1956, *J. Agr. Food Chem.*, *4*, 956–959.

19. Bandelin, F. J., 1958, *J. Am. Pharm. Assoc.*, *Sci. Ed.*, *47*, 691–694.

20. Banfield, E. H., 1952, *Chem. Ind.*, (London), 114–119.

21. Bauer, H. F., and Glabe, E. F., 1942, U.S. 2,271,756.

22. Bauer, H. and Glabe, E. F., 1946, U.S. 2,412,596.

23. Bauer, H. and Glabe, E. F., 1947, U.S., 2,415,070.

24. Baylor, R., 1960, Unpublished data, California Packing Corp.

25. Berkman, R. N., Richards, E. A., Van Duyn, R. L. and Kline, R. M., 1960, *Antimicrobial Agents Annual*, Plenum Press., New York, pp. 595–604.

26. Bell, T. A., Etchells, J. L. and Borg, A. F., 1959, *J. Bacteriol.*, *77*, 573–580.

27. Beneke, E. S. and Fabian, F. W., 1955, *Food Technol.*, *9*, 486–488.

28. Bernhard, H., Thoma, W., and Genth, H., 1956, *Germ.* 1,001,309.

29. *Biochemical Studies of Antimicrobial Drugs*, 16th Symp. of the Soc. for General Microbiol., London, Apr. 1966, Cambridge University Press, Cambridge.

30. Boehm, T. and Mehta, D., 1938, *Ber.*, *71*, 1797–1802.

31. Bonner, M. D. and Harmon, L. G., 1957, *J. Dairy Sci.*, *40*, 1599–1611.

32. Borker, E., Insulata, N. F., Levi, C. P. and Witzeman, J. S., 1966, *Advan. Appl. Microbiol.*, *8*, 315–351.

33. Bosund, I., 1962, *Food Res.*, *11*, 331–352.

34. Bowen, J. F., 1955, *Rept. Can. Comm. On Fruit And Vegetable Pres.*

35. Boyd, J. W. and Tarr, H. L. A., 1954, Fisheries Res. Bd., Can., Rept. Pac. Coast Sta., No. 99.

36. Boyd, J. W. and Tarr, H. L. A., 1955, *Food Technol.*, *9*, 411–412.

37. Bradley, R. L., Harmon, L. G. and Stine, C. M., 1962, *J. Milk and Food Technol.*, *25*, 318.

38. Bradshaw, W., 1958, *Baker's Dig.*, *32*, Oct., p. 58.

39. Brock, T. D., 1966, See No. 29.

40. Brokke, M. E., Kiigermagi, U. and Terriere, L. C., 1958, *J. Agr. Food Chem.*, *6*, 26–27.

41. Bruch, C. W., 1961, *Ann. Rev. Microbiol.*, *15*, 245–262.

42. Bruch, C. W. and Koesterer, M. G., 1961, *J. Food Sci.*, *26*, 428–435.

43. Bulman, C. and Ayres, J. C., 1952, *Food Technol.*, *6*, 255–259.

44. Campbell, Jr., L. L., and O'Brien, R. T., 1955, *Food Technol.*, *9*, 461–465.

45. Castellani, A. G. and Niven, C. F., Jr., 1955, *Appl. Microbiol.*, *3*, 154–159.

46. Chittenden, R. H., Long, J. H. and Herter, C. A., 1909, U.S. Dept. Agric. Bull. 88.

47. Ciaccio, L. L., 1966, *Food Technol.*, *20*, 73–75.

48. Clark, W. L., Shirk, R. J. and Kline, E. F., 1964, See No. 161.

49. Cooper, E. A., 1947, *J. Soc. Chem. Ind.*, *66*, 48–50.

50. Costilow, R. N., 1957, *Food Technol.*, *11*, 591–603.

51. Costilow, R. N., Coughlin, F. M., Robbins, E. K. and Hsu, Wen-Tah, 1957, *Appl. Microbiol.*, *5*, 373–379.

52. Cruess, W. V., 1932, *Ind. Eng. Chem.*, *24*, 648–649.

53. Cruess, W. V. and Irish, J. H., 1932, *J. Bacteriol.*, *23*, 163–166.

54. Cruess, W. V. and Richert, P. H., 1929, *J. Bacteriol.*, *17*, 363–371.

55. Cruess, W. V., Richert, P. H. and Irish, J. H., 1931, *Hilgardia*, *6*, 295–314.

56. Denny, C. F., Sharp, L. E. and Bohrer, C. W., 1961, *Appl. Microbiol.*, *9*, 108–110.
57. Desrosier, N. W., 1959, *The Technology Of Food Preservation*, AVI Publishing Co., Westport, Conn.
58. D'Estivaux, L. B., 1959, *Ann. Fals Fraudes*, *52*, 149.
59. Deuel, Jr., H. J., Calbert, C. F., Anisfeld, L., McKeehan, H. and Blunden, H. D., 1954, *Food Res.*, *19*, 13–19.
60. Deuel, Jr., H. J., Alfin-Slater, R., Weil, C. S., and Smyth, H. F., 1954, *Food Res.*, *19*, 1–12.
61. *Dispensatory of the United States of America*, 1960.
62. Doerr, R. C., Wasserman, A. E. and Fiddler, W., 1966, *J. Agr. Food Chem.*, *14*, 662–665.
63. Duggar, B. M., 1948, *Ann. N.Y. Acad. Sci.*, *51*, 177–181.
64. Dunn, C. F., 1947, *Food Technol.*, *4*, 457–460.
65. Eckert, J. W., 1967, in *Fungicides*, vol. 2, Torgeson, P. C. (ed.), Academic Press, New York.
66. Elsden, S. R., Hitchcock, M. W. E., Marshall, R. A. and Phillipson, A. T., 1946, *J. Exptl. Biol.*, *22*, 191–202.
67. Emard, L. D. and Vaughn, R. H., 1952, *J. Bacteriol.*, *63*, 487–498.
68. Fabian, F. W. and Switzer, R. G., 1941, *Fruit Prod. J.*, *20*, 136.
69. Farber, L., 1959, *Ann. Rev. Microbiol.*, *13*, 125–139.
70. Fellers, C. R., 1929, *Fruit Prod. J.*, *9*, 113–115.
71. Ferguson, W. E. and Powrie, W. D., 1957, *Appl. Microb.*, *5*, 41–43.
72. Fiddler, W., Doerr, R. C., Wasserman, A. E. and Salay, J. M., 1966, *J. Agr. Food Chem.*, *14*, 659–661.
73. Finlay, A. C., Hobby, G. L., P'an, S. Y., Regna, P. P., Routien, J. B., Seeley, D. B., Shull, G. M., Sobin, B. A., Solomons, I. A., Vinson, J. W. and Kane, J. H., 1950, *Science*, *111*, 85.
74. Fitzhugh, D. G., Knudsen, L. and Nelson, A., 1946, *J. Pharmacol. Exptl. Therap.*, *86*, 37–48.
75. Follet, M. J. and Ratcliff, P. W., 1963, *J. Sci. Food Agr.*, *14*, 138–144.
76. Francis, F. J., 1955, *Pur-Pack-Age*, *9*, Apr. 8.
77. Frank, H. A. and Willits, C. O., 1961, *Food Technol.*, *15*, 1–3.
78. Franklin, T. J., 1966, See No. 29.
79. Fraser, A. C., Sharratt, M. and Hickman, J. R., 1962, *J. Sci. Food Agr.*, *13*, 32–42.
80. Gale, E. F., 1966, See No. 29.
81. Geminder, J. J., 1959, *Milk Dealer*, *44*, 44–45; 133–4.
82. Geminder, J. J., 1959, *Food Technol.*, *13*, 459–461.
83. Genth, H., 1964, See No. 161.
84. Gibbs, B. M. and Hurst, A., 1964, See No. 161.
85. Gillespy, T. G., 1957, Nisin Trials, Research Leaflet No. 3, The Fruit and Vegetable Canning and Quick Freezing Res. Assoc., Chipping, Campden, Gloucester, England.
86. Ginsberg, A., Hill, E. C. and Grieve, J. J., 1957, *Vet. Record*, *69*, 983–992.
87. Gleason, M. N., Gosselin, R. and Hodge, H. C., 1963, *Chemical Toxicology of Commercial Products*, 2nd Ed., Williams and Wilkins Co., Baltimore.
88. Goldberg, H. S., 1964, *Advan. Appl. Microbiol.*, *6*, 91–117.
89. Goldberg, H. S. and Barnes, E. M., 1960, in *Antimicrobial. Agents Ann.*, Peter Gray, (ed.), 576–584.
90. Gooding, C. M., 1945, U.S. 2,379,294.
91. Gooding, C. M. and Melnick, D., 1955, *Food Res.*, *20*, 639–648.
92. Gooding, C. M., Melnick, D. and Vahlteich, H. W., 1958, U.S. 2,858,225.
93. Gooding, C. M., Melnick, D., Lawrence, R. L. and Luckmann, E. H., 1955, *Food Res.*, *20*, 639–648.
94. Goshorn, R. H., Degering, E. F. and Tetrault, P. A., 1938, *Ind. Eng. Chem.*, *30*, 646–648.
95. Gough, B. J. and Alford, J. A., 1965, *J. Food Sci.*, *20*, 1025–1028.
96. Graham, D., Teed, H. and Grice, H. C., 1954, *J. Pharm. Pharmacol.*, *6*, 534–545.
97. Greenberg, R. A. and Silliker, J. H., 1964, See No. 161.
98. Griffith, C. L. and Hall, L. A., 1938, U.S. 2,107,699.
99. Griffith, C. L. and Hall, L. A., 1940, U.S. 2,189,948.
100. Griffith, C. L. and Hall, L. A., 1940, U.S. 2,189,949.
101. Griffith, C. L. and Hall, L. A., 1940, U.S. 2,189,947, Reissue 22,284 (1943).
102. Griffith, W. H., 1929, *J. Biol. Chem.*, *82*, 415–427.

103. Grubb, T. C., 1957, *Bacteriol. Rev.*, *21*, 251–254.

104. Haines, E. C., 1944, U.S. 2,354,014.

105. Hall, L. A., 1951, *Food Packer*, Dec., p. 26.

106. Hansen, J. D. and Appelman, M. D., 1955, *Food Res.*, *20*, 92–96.

107. Harrington, W. O. and Hills, C. H., 1966, *Food Technol.*, *20*, 1360–1362.

108. Harrington, W. O., Jahn, A. S. and Hills, C. H., 1962, *J. Food Sci.*, *27*, 15–19.

109. Harris, N. E. and Rosenfeld, D, 1965, *Food Technol*, *19*, 656–658.

110. Harrison, N., 1959, Unpublished data, California Fig Growers and Packers.

111. Harshbarger, K. E., 1942, *J. Dairy Sci.*, *25*, 169–174.

112. Hashida, W., 1953, *Hakko Kogaku Zasshi*, *31*, 15.

113. Hawk, E. A. and Mickelsen, O., 1955, *Science*, *121*, 442–444.

114. Hawley, H. B., 1957, *Food Manuf.*, *32*, 370.

115. Hecht, G., 1961. *Z. Lebensm. Untersuch-Forsch.*, *114*, 292–297.

116. Heiligmann, F. J., 1957, *Am. Vegetable Grower*, *5*, Jan., p. 28.

117. Heinemann, B., Voris, L. and Stumbo, C. R., 1965, *Food Technol.*, *19*, 592–596.

118. Heinemann, B. and Williams, R., 1966, *J. Dairy Sci.*, *49*, 312–313.

119. Hennig, K., 1962, *Weinberg u. Keller*, *9*, 271–278.

120. Hill, Jr., F. S., 1966, *Am. Soc. Safety Engr. J. Tech. Sect.*, *11*, 37.

121. Hoffman, C., Dalby, G. and Schweitzer, T. R., 1939, U.S. 2,154,449.

122. Hoffman, C., Schweitzer, T. R. and Dalby, G., 1939, *Food Research*, *4*, 539–545.

123. Höyem, T., 1962, *J. Assoc. Offic. Agr. Chemists*, *45*, 902–905.

124. Ikeda, R. M. and Crosby, D. C., 1960, *Chemicals and the Food Industry, Manual 26*, Univ. Calif. Coll. of Agr. Sci.

125. Ingle, J. D., 1940, *J. Dairy Sci.*, *23*, 509 (Abstr.).

126. Ingram, M., Ottaway, F. J. H. and Coppock, J. B. M., 1956, *Chem. Ind.*, (London), Oct. 27, 1154–1163.

127. Ingram, M., Buttiaux, R. and Mossel, D. A. A., 1964. See No. 161.

128. Jacobs, M. B., 1947, *Synthetic Food Adjuncts*, D. Van Nostrand Co., New York.

129. Jensen, L. B. and Hess, W. R., 1941, *The Canner*, *92*, 82–87.

130. Johnson, J. B. and Funk, G. L., 1955, *Anal. Chem.*, *27*, 1464–1465.

131. Jones, A. H. and Harper, G. S., 1952, *Food Technol.*, *6*, 304–308.

132. Jones, P. S., Thigpen, D., Morrison, J. L. and Richardson, A. P., 1956, *J. Am. Pharm. Assoc., Sci. Ed*, *45*, 268–273.

133. Joslyn, M. A. and Braverman, J. B. S., 1954, *Advan. Food Res.*, *5*, 97–160.

134. Kiesel, A., 1913, *Ann. Inst. Pasteur* (cited in No. 19).

135. Kirby, G. W., Atkin, L. and Frey, C. N., 1936, *Food Ind.*, *8*, 450, 470, 480.

136. Kitajima, K. and Kawamura, J., 1931, *Bull. Imp. Forestry Sta.*, Tokyo, *31*, 108.

137. Kuroda, T., 1926., *Biochem. Z.*, *169*, 281–291.

138. Lampen, J. O., 1966, See No. 29.

139. Lawrence, R. C., 1965, *Nature*, *208*, 801–803.

140. Lawrence, R. C., 1966, *J. Gen. Microbiol.*, *44*, 393–405.

141. Levine, A. S. and Fellers, C. R., 1940, *J. Bacteriol.*, *39*, 17; *39*, 499–514; *40*, 255–269.

142. Lewis, J. C., Michener, H. D., Stumbo, C. R. and Titus, D. S., 1954, *J. Agr. Food Chem.*, *2*, 298–302.

143. Lockett, M. F., and Natoff, I. L., 1960, *J. Pharm. Pharmacol.*, *12*, 488–496.

144. Macy, H. and Olson, J. C., Jr., 1939, *J. Dairy Sci.*, *22*, 527–534.

145. Malin, B., 1966, Abstr. Papers, 6th Interscience Conf. on Antimicrobial Agents, Philadelphia, p. 55.

146. Malin, B., and Silliker, J. H., 1966, Abstr. papers 6th Interscience Conf. on Antimicrobial Agents, Philadelphia, pp. 55-56.

147. Marth, E. H., Capp, C. M., Hasenzahl, L., Jackson, H. W. and Hussong, R. V., 1966, *J. Dairy Sci.*, *49*, 1197–1205.

148. Matthew, C., Davidson, J., Bauer, E., Morrison, J. L., and Richardson, A. P., 1956, *J. Am. Pharm. Assoc., Sci. Ed*, *45*, 260–267.

149. Mattick, A. T. R. and Hirsch, A., 1944, *Nature*, *154*, 551.

150. Mayer, K. and Luthi, H., 1960, *Mitt. Gebiete Lebensm. Hyg.*, *51*, 132–137.

151. Mayr, G. and Kaemmerer, H., 1959, *Food Manuf.*, *34*, 169–170.

152. Melnick, D. and Luckmann, F. H., 1954, *Food Res.*, *19*, 20–27.

153. Melnick, D. and Luckmann, F. H., 1954, *Food Res.*, *19*, 28–32.

154. Melnick, D., Luckmann, F. H. and Gooding, C. M., 1954, *Food Res., 19*, 44–58.
155. Melnick, D., Vahlteich, H. W. and Hockett, H. W., 1956, *Food Res., 21*, 133–146.
156. *Merck Index*, 1960, 7th Ed.
157. Meredith, W. E., Weiser, H. H. and Winter, A. R., 1965, *Appl. Microbiol., 13*, 86–88.
158. Michener, H. D., Thompson, F. A. and Lewis, J C., 1959, *Appl. Microbiol., 7*, 166–173.
159. Miller, Jr., F. W., 1940, *Proc. Food Conf.*, Inst. Food Technologists.
160. Molin, N., 1961, Presented Ann. Meet., Inst. Food Technologists.
161. Molin, N., 1964 (ed.), *Microbial Inhibitors in food*, Fourth Intl. Symp. on Food Microbiol., Almqvist and Wiksell, Stockholm.
162. Molin, N., Satmark, L. and Thorell, M., 1963, *Food Technol., 17*, 797–801.
163. Monsanto, Tech. Bull. No. 1-250.
164. Morris, H. J., Nelson, A. A. and Calvery, H. O., 1941, *J. Pharm. Exptl. Therap., 74*, 266–273.
165. Neidig, C. P. and Burrell, H., 1944, *Drug Cosmetic Ind., 54*, 408–415.
166. Newton, B. A., 1965, *Ann. Rev. Microbiol., 19*, 209–240.
167. Niacet Bulletin—Sodium Diacetate, 1950, Niacet Division, United States Vanadium Corp.
168. Nury, F. S. and Bolin, R. R., 1962, *J. Food Sci., 27*, 370–373.
169. Nury, F. S., Miller, M. W. and Brakke, J. E., 1960, *Food Technol., 14*, 113–115.
170. O'Brien, R. T., Titus, D. S., Devlin, K. A., Stumbo, C. R. and Lewis, J. C., 1956, *Food Technol., 10*, 352–355.
171. O'Leary, D. K. and Kralovec, R. D., 1941, *Cereal Chem., 18*, 730–741.
172. Olson, J. C., Jr. and Macy, H., 1945, *J. Dairy Sci., 28*, 701–710.
173. Olson, J. C., Jr., and Macy, H., 1946, *J. Dairy Sci., 29*, 173–180.
174. Opfell, J. B. and Miller, C. E., 1965, *Advan. Appl. Microbiol., 7*, 81–102.
175. O'Rourke, C. E., and Weaver, E. A., 1962, *Wines and Vines, 43*, 28.
176. Oser, B. L. and Hall, L. A., 1956, *Food Technol., 10*, 175–178.
177. Ough, C. S. and Ingraham, J. L., 1960, *Am. J. Enol. Viticult,, 11*, 117.
178. Ough, C. S., and Ingraham, J. L., 1961, *Am. J. Enol. Viticult., 12*, 149–151.
179. Pappas, H. J. and Hall, L. A., 1952, *Food Technol., 6*, 456–458.
180. Parrish, G., 1959, Unpublished data, Twining Laboratories.
181. Patrick, R. and Atkins, C. D., 1954, *Proc. Fla. State Hort. Soc., 67*, 194–196.
182. Pedersen, C. S., Albany, M. N. and Christensen, M. D., 1961, *Appl. Microbiol., 9*, 162–167.
183. Pelshenke, P. F., 1954, *Brot und Gebäk, 8*, 27.
184. Perry, G. A. and Lawrence, R. L., 1960, *J. Agr. Food Chem., 8*, 374–376.
185. Phillips, C. R., 1949, *Am. J. Hyg., 50*, 280–289.
186. Phillips, C. R., 1957, See No. 194.
187. Phillips, C. R. and Kaye, S., 1949, *Am. J. Hyg., 50*, 270–279.
188. Phillips, G. F. and Mundt, J. O., 1950, *Food Technol., 4*, 291–293.
189. Pinella, S. J., Falco, A. D. and Schwartz, G., 1966, *J. Assoc. Offic. Anal. Chemists, 49*, 829–834.
190. Pomeranz, Y. and Adler, L., 1958, *Bull. Res. Council Israel Sec., 6*, 220–226.
191. Rahn, O. and Conn, J. E., 1944, *Ind. Eng. Chem., 36*, 185–187.
192. Rao, G. K., Malathi, M. A. and Vijayaghavan, 1966, *Food Technol., 20*, 1070–1073.
193. Rauschen, H., Mayr, G., and Kaemmerer, H., 1959, *Food Manuf., 32*, 169.
194. Reddish, G. F. (ed.), 1957, *Antiseptics, Disinfectants, Fungicides and Sterilization*, 2nd ed., Lea and Febiger, Philadelphia.
195. Rehm, H.-J., 1959, *Z. Lebensm. Untersuch.-Forsch., 110*, 356–363.
196. Rehm, H.-J., 1964, See No. 161.
197. Rehm, H.-J. and Wittmann, H., 1962, *Z. Lebensm. Untersuch.-Forsch., 118*, 413–429.
198. Reich, R., 1966, See No. 29.
199. Reynolds, P. E., 1966, See No. 29.
200. Rogers, L. A. and Whittier, E. O., 1928, *J. Bacteriol., 16*, 211 (Cited in 117.).
201. Rost, E., 1927, Hefftner's Handbuch der Exp. Pharmakologie, *3*, I (Cited by Lehman, A. J., 1957, *Assoc. Food Drug Officials, U.S., Quart. Bull. 21*, 29–30).
202. Rost, E. and Franz, F., 1913, *Arb. kais-Gesundh., 43*, 187 (Cited by Lehman, loc. cit.).
203. Sabalitschka, T., See citations in Ref. 165.
204. Sair, L., 1966, Conf. on the Destruction of Salmonellae, Western Regional Research Laboratory, U.S.D.A., Albany, California.

205. Seller, W. and Kolewa, S. R., 1957, Mitt. Klosternueberg, Sec. A, *Rebe Wein*, *7*, 21.
206. Salunke, D. K., 1955, *Food Technol.*, *9*, 590.
207. Sax, M. I., 1963, *Dangerous Properties of Industrial Materials*, 2nd ed., Reinhold Publ. Co., New York.
208. Schimmel, J. and Husa, W. J., 1956, *J. Am. Pharm. Assoc., Sci. Ed.*, *45*, 204–208.
209. Schroeter, L C., 1966, *Sulfur Dioxide, Applications in Food, Beverages and Pharmaceuticals*, Pergamon Press., Long Island City.
210. Segmiller, J. L., Xezones, H. and Hutchings, I. J., 1965, *J. Food Sci.*, *30*, 166–171.
211. Seiler, D. A. L., 1964, See No. 161.
212. Sexton, R. J. and Henson, E. V., 1950, *Arch. Ind. Hyg. and Occup. Med.*, *1*, 549–564.
213. Shaffer, C. B., Carpenter, C. P., Critchenfield, F. H., Nair, J. H., III and Franke, F. R., 1951, *A.M.A. Arch. Ind. Hyg. Occupat. Med.*, *3*, 448–453.
214. Sheneman, J. M., 1965, *J. Food Sci.*, *30*, 337–343.
215. Sheneman, J. M. and Costilow, R. N., 1955, *Appl. Microbiol.*, *3*, 186.
216. Silliker, J. H., Greenberg, R. A. and Schack, W. R., 1958, *Food Technol.*, *12*, 551–554.
217. Silvestrini, D. A., Anderson, G. W. and Snyder, E. S., 1958, *Poultry Sci.*, *37*, 179–185.
218. Smith, D. P. and Rollin, N. J., 1953, *Mod. Packaging*, *28*, 139.
219. Smith, D. P. and Rollin, N. J., 1954, *Food Res.*, *19*, 54–56.
220. Smith, D. P. and Rollin, N. J., 1954, *Food Technol*, *8*, 133–135.
221. Smith, E. E., 1938, *Western Canner Packer*, *30*, 22.
222. Smith, Jr., W. L., 1962, *Botan. Rev.*, *28*, 411–445.
223. Spanyár, P. and Sánder, A., 1958, *Z. Lebensm. Untersuch.-Forsch.*, *108*, 402–405.
224. Spector W. S., 1956 *Handbook of Toxicology*, Vol. I, W. B. Saunders & Co., New York.
225. Spencer, R., 1966, *Food Manuf.*, *41*, Mar., 39–43.
226. Stadtman, E. R., Barkas, H. A., Mrak, E. M. and MacKinney, L. A., 1946, *Ind. Eng. Chem*, *38*, 99–104.
227. Stutzenberger, J. F. and Bennett, E. O., 1962, *Rev. Latinam. Microbiol.*, *5*, 92-105.
228. Sudari, E., 1957, *Chim. Ind.* (Milan), *39*, 811.
229. Tanner, F. W. and Evans, F. L., 1933, *Zentr. Bakteriol. Parasitenk.*, Abt. II, *89*, 48–54; 1934, *91*, 1–13; 136–147.
230. Tarr, H. L. A., 1941, *J. Fisheries Res. Bd. Can.*, *5*, 265–275.
231. Tarr, H. L. A., 1941, *Nature*, *147*, 417–418.
232. Tarr, H. L. A. and Sunderland, P. A., 1938, *Fisheries Res. Bd. Can., Progr. Rept.*, Pacific Coast Stat, No. 37, 7–11.
233. Tarr, H. L. A. and Sunderland, P. A., 1940, *Modern Refrig. 43*, No. 503, 41.
234. Tarr, H. L. A. and Sunderland, P. A., 1940, *J. Fisheries Res. Bd. Can.*, *5*, 244–248.
235. Tetsumoto, S., 1933, *J. Agr. Chem. Soc. Japan*, *9*, 388–397; 563–567.
236. Thoma, W. and Genth, H., 1956, *Med. Chemie.*, *7*, 973.
237. Thompson, M. H., 1962, *Com. Fisheries Rev.*, *24*, 5–11.
238. Thoukis, G., Bouthilet, R. J., Neda, M. and Caperti, A. Jr., 1962, *Am. J. Enol. Viticult,, 13*, 105–113.
239. Tonge, R. J., 1964, See No. 161.
240. Tramer, J., 1966, *Chem. Ind.* (London), Mar. 12, 446–450.
241. Tressler, D. K. and Joslyn, M. A., 1961, *Fruit and Vegetable Juices*, AVI Publishing Co., Westport, Conn.
242. Truby, C. P. and Bennett, E. O., 1964, *Japan. Jour. Microbiol.*, *8*, 149-155.
243. Truby, C. P. and Bennett, E. O., 1966, *Appl. Microbiol.*, *14*, 769-773.
244. Vaughn, R. H., Nagel, C. W., Sawyer, F. M. and Stewart, G. F., 1957, *Food Technol.*, *11*, 426–429.
245. Vazquez, D., 1966, See No. 29.
246. Vetagliano, M., 1958, *Rev. Viticult. Enol.* (Conegliano), *11*, 15.
247. Vojnovich, C. and Pfeifer, V. F., 1967, *Cereal Sci., Today*, *12*, 54–60.
248. Von Schelhorn, M., 1953, *Food Technol.*, *3*, 97–101.
249. Wallnöfer, P. and Rehm, H.-J., 1965, *Z. Lebensm. Untersuch.-Forsch.*, *127*, 195–206.
250. Waring, M. J., 1966, See No. 29.
251. Weaver, E. A., Robinson, J. F. and Hills, C. H., 1957, *Food Technol.*, *11*, 667–669.
252. Weger, B., 1957, *Weinberg u. Keller*, *4*, 488.
253. Wesley, F., Rourke, F. and Darbishire, O., 1965, *J. Food Sci.*, *30*, 1037–1042.

254. Wheaton, E. and Hays, G. L., 1964, *Food Technol.*, *18*, 549–551.

255. Whelton, R., Phaff, H. J., Mrak, E. M. and Fisher, C. D., 1946, *Food Ind.*, *18*, 23–25.

256. W. H. O. Technical Rept. Series, No. 339.

257. Windmueller, H. G. and Engle, R. W., 1956, *Federation Proc.*, *15*, 386; Windmueller, H. G., Ackerman, C. J. and Engle, R. W., *J. Nutrition*, *60*, 527–537.

258. Wolf, F. A., 1948, Report July 1, 1947 to Dec. 1, 1948 to Quartermaster Food and Container Institute of the Armed Forces, Chicago, Illinois.

259. Wolford, E. R. and Anderson, A. A., 1945, *Food Ind*, *17*, 622–624, 726–728, 730, 732, 734.

260. Wood, H. G., 1948, *Symposium on the Use of Isotopes in Biology and Medicine*, Univ. Wisc. Press.

261. Woods, B. M., 1966, *Chem. Ind.* (London), Apr. 9, 615–618.

262. Wrenshall, C. L., 1959, in *Antibiotics—Their Chemistry and Non-Medical Uses*, H. S. Goldberg (ed.), D. Van Nostrand Co., Princeton, N.J., 449–527.

263. Wyss, O., 1948, *Advan. Food Res.*, *1*, 373–393.

264. Wyss, O., Ludwig, B. J. and Joiner, R. R., 1945, *Arch. Biochem.*, *7*, 415–425.

265. Yesair, J. and Cameron, E. J., 1942, *The Canner*, *94*, 89–90.

266. York, G. K., 1960, Inhibition of Microbes by Sorbic Acid, Ph. D. Thesis, Univ. of California.

267. York, G. K. and Vaughn, R. H., 1954, *J. Bacteriol.*, *68*, 739–744.

268. Young, H. D. and Busbey, R. L., 1935, References to the Use of Ethylene Oxide for Pest Control, U.S.D.A., Bur. of Entomol. and Plant Quarantine, Division of Insecticide Investigation.

269. Vaughn, R. J., Martin, M. H., Stevenson, K. E., Johnson, M. G. and Crampton, W. M., 1969, *Food Technol.*, *23*, 832–834.

270. Gantenbein, W. M. and Karasz, A. B., 1969, *J.A.O.A.C.*, *52*, 738–741.

271. Schuller, P. and Veen, E., 1967, *J.A.O.A.C.*, *50*, 1127–1145.

272. Manchon, Ph., Buquet, A. and Atteba, S., 1970, *Food and Cosmetic Toxicol.*, *8*, 17–25.

273. Hilgeland, P. H., 1968, *Food Manuf.*, *44*, May, p. 38.

CHAPTER 4

Antioxidants as Food Stabilizers

Ben N. Stuckey, Ph.D.

Manager, Food Chemical Sales
DPI Division
Eastman Chemical Products, Inc.
Kingsport, Tennessee

Introduction

Antioxidant technology plays an important role in the utilization of fats and oils as raw materials in food processing and in the marketing of foods containing fats under modern conditions. As such, the proper and effective use of antioxidants is dependent on a basic understanding of: (1) the chemistry of fats and oils, (2) the mechanism of oxidation, and (3) the function of an antioxidant in counteracting this type of deterioration. Antioxidants, to be effective, must be used in conjunction with good raw materials, correct processes, and proper packaging and storage conditions. They will not enhance a product of mediocre quality. A good antioxidant system will impart no flavor, odor, or color to a finished food when used under the proper conditions.

Factors affecting the oxidation and flavor stability of major food fats will be described, but more attention will be devoted to general discussion of antioxidant technology to enable the reader to solve problems encountered in the commercial treatment of fats and fatty foods. References to original and review articles have been included for those interested in a more detailed documentation or for discussions of the theoretical aspects of antioxidant usage.

Oxidation

The major portion of most foods consists of carbohydrates, fats, proteins, and water. While our basic interest is in the oxidative deterioration of the flavor and odor of fats and fatty constituents of foods, a general discussion of other types of oxidation will serve to orient the reader so he may diagnose particular problems. The various types of oxidation will be discussed separately; however, they usually occur simultaneously, particularly in complex foods, and are quite often accompanied by non-oxidative spoilage.

Carbohydrates

Oxidation of the carbohydrate portion of most food products is evidenced by discoloration and off-flavors. This discoloration, usually brown or tan although off-shades of gray or yellow are often present, can be traced to two general types of reactions. The first, known as the Maillard reaction or "chemical browning" reaction, is generally considered to be a reaction between carbohydrates and amino acids and/or various nitrogenous components and organic acids. The discoloration of freshly peeled fruits and vegetables, containing high levels of carbohydrates, is normally due to Maillard-type reactions. A similar discoloration is found in some canned fish products and thought to be caused by the reaction of free ribose with protein amino acids. Careful control of processing temperatures is usually employed to reduce brown discoloration in canned fish products. Radiation-sterilized seafood products have been found to discolor badly from non-enzymatic browning reactions. Antioxidants and sulfur dioxide combinations were found to reduce this discoloration in irradiated cod patties.[1] Generally speaking, Maillard-type browning reactions can be inhibited by the use of ascorbic, citric, and other organic acids. (Note discussion of Browning Reaction, p. 103, Chapter 2, "Vitamins and Amino Acids".)

The second type of carbohydrate oxidation is usually associated with various enzymes, such as peroxidase and catalase. These browning reactions are quite often accompanied by off-flavors and odors described as "hay-like." Additives are usually ineffective in preventing enzyme-induced discoloration or off-odors. Reactions of this type are usually inhibited by heat processing to inactivate the enzymes. (Note discussion of enzymes under Vegetables and Fruits, p. 40, Chapter 1, "Enzymes.")

A third type of discoloration often found in carbohydrate-containing foods is the loss of color due to oxidation of natural pigments. These pigments (carotenoids, etc.) undergo oxidation by free radical-type reactions. Considerable evidence has been gathered indicating that a loss of color in foods with high carotene content can be inhibited by the use of food-grade antioxidants. The chief problem, however, is finding a method for achieving adequate contact between the usually oil-soluble antioxidants and the carotene which is dispersed throughout a food containing an extremely high water content.

Carbohydrate oxidation, as well as pigment oxidation, is catalyzed by high temperatures, metals, and in some cases microbiological by-products. Chelation, or in some cases removal of metal and microbiological catalysts, is employed to reduce these types of oxidation.

Fats

While microbiological spoilage is one of the most important factors to be considered in preserving carbohydrate and protein portions of food products, oxidation (particularly atmospheric oxidation) is the chief factor in quality degradation of fats and fatty portions of foods. Fats and fat-like substances undergo oxidative deterioration which results in off-flavors and off-odors. In extreme cases toxic by-products have resulted from oxidative reactions.[2] Generally, fat oxidation does not progress to a point where toxic by-products are a factor. That portion of fat which is reactive at any given time is quite small. It is estimated that no more than about 10 per cent of the unsaturated molecules may undergo breakdown due to atmospheric oxidation. The resulting products of fat oxidation, however, are organoleptically potent and adversely affect the marketability of the food.

Fat deterioration may be divided into four types:

1. HYDROLYSIS: The formation of free fatty acids and glycerol is often characterized by a "soapy" flavor. This reversible reaction is catalyzed by acids, high temperatures, and lipolytic enzymes. Triglycerides with fatty acids having shorter chain lengths (6–12 carbon atoms) normally produce off-flavors upon hydrolysis.
2. RANCIDITY: A term widely used in the food industry. It normally covers a large number of objectionable off-flavors usually formed by the autoxidation of unsaturated fatty acids resulting in a mixture of volatile components.
3. REVERSION: A type of flavor and odor degradation usually associated with vegetable, fish, and other highly unsaturated oils. This flavor degradation is thought to be brought about by oxidation of linolenic-type acids.
4. POLYMERIZATION: A term usually employed to describe the crosslinking of unsaturated fats between two carbon atoms. Polymers are also formed by an oxygen bonding between two fatty acid chains at an unsaturated site. Both types of polymers may contain cyclic structures.

Antioxidants are effective in reducing rancidity and polymerization, but do not affect hydrolysis or reversion.

Proteins

Oxidation of proteins *per se* is not usually considered a factor in off-odor and off-flavor development in food products. Proteins are broken down by proteolytic enzymes and are denatured (scission of electrostatic bonds) by heat and hydrolytic reactions. They do not have double bond structures which are oxygen labile. Pigments normally associated with proteins, such as the heme pigments, do oxidize and change color rapidly. Fox describes the take-up of oxygen and interconversion by the heme pigments so that oxymyoglobin and metmyoglobin are constantly interconverted.[3] Pigment color degradation in this instance is not affected by additives of any type. Since, as will be shown later, hemoglobin is an active catalyst for oxidation in its own right, various anticatalytic agents have been used to prevent discoloration in meats without success. Normally, packaging materials with selective gaseous transfers are used for regulating color in various hemoglobin-containing products.

Lipid Deterioration by Oxygen

Atmospheric oxidation (autoxidation) of fats and fatty-type foods is usually thought of as a reaction dealing entirely with the unsaturated bonds. Saturated fats oxidize slowly, and in some cases, develop odors and flavors which affect the quality of the finished product. Generally, however, lipid oxidation which produces the flavors described as "rancid" or "reverted" by the food technologist deals almost entirely with the unsaturated portion of the fat chain. The following are the major classes of lipids affected by oxidation:

Glycerides

Swern describes the word "fat" as solid or, more correctly, semisolid, triglycerides at ordinary temperatures, whereas the word "oil" is used to describe triglycerides that are liquid under the same conditions.[4] No clear-cut distinction is made between these words by the fat and oil chemist. The triglycerides make up the most important group of fats used by the food industry.

Structurally, a triglyceride is the condensation product of one molecule of glycerol with three molecules of fatty acid to yield one molecule of triglyceride plus water (Figure 1).

It may be noted from the structure in Figure 1 that the preponderant weight in the glyceride molecule is the fatty acid portion. The fatty acids contribute from 94 to 96 per cent of the total weight of the molecule and comprise the reactive portion, particularly from the standpoint of stability. The chemistry of fats and oils is to a very large extent the chemistry of their fatty acids. Note particularly the unsaturated portions of the fatty acid chains shown in Figure 1; these are the most reactive sites that are affected by oxygen and hydrogen.

Natural fats consist mainly of mixed triglycerides, that is, those whose molecules contain more than one fatty acid, such as oleic and palmitic. The proportions of the component fatty acids and the compositions of the component glycerides greatly affect the physical properties including stability to autoxidation. There are numerous tables which describe the composition of various food fats. Swern lists the compositions of some of the major food fats.[5]

Mono- and diglycerides do not occur in gross quantities in fats, except those which have undergone hydrolysis; minute quantities are found as mixtures with triglycerides in practically all fats. Mixtures of mono- and diglycerides are readily prepared from fatty acids or fats and glycerols, and are used widely in the food manufacturing industry as texture modifying agents. Commerical products containing over 90 per cent monoglycerides are currently available. Mono- and diglycerides contain unsaturated fatty acid moieties and,

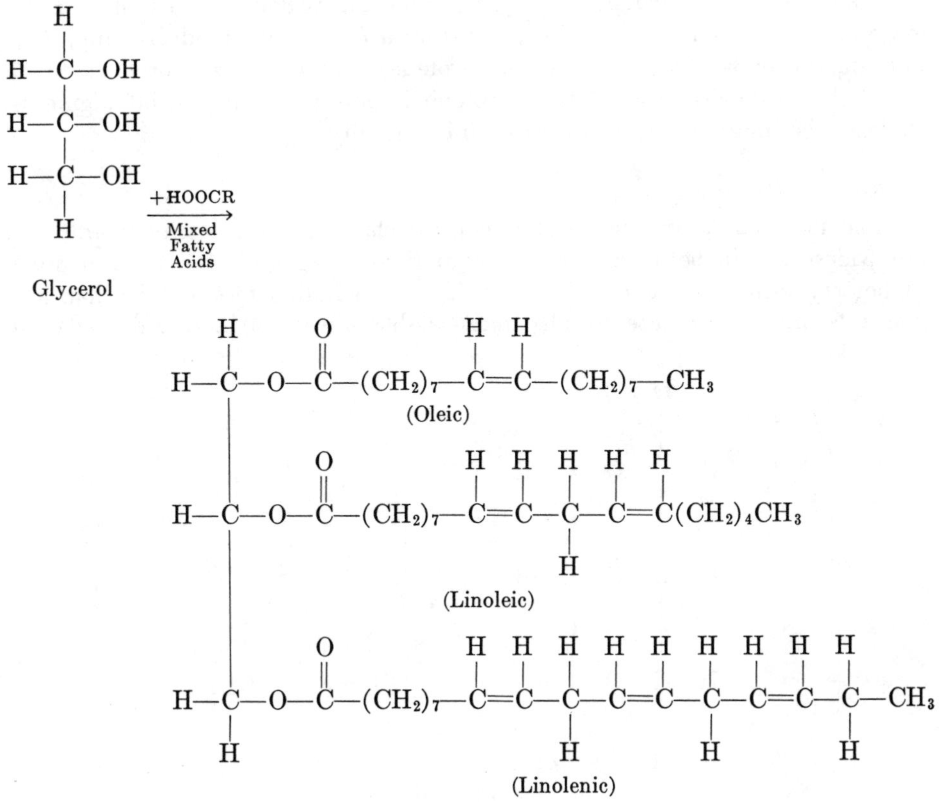

$$
\begin{array}{c}
\text{H} \\
| \\
\text{H—C—OH} \\
| \\
\text{H—C—OH} \\
| \\
\text{H—C—OH} \\
| \\
\text{H}
\end{array}
\quad
\xrightarrow[\substack{\text{Mixed}\\\text{Fatty}\\\text{Acids}}]{\text{+HOOCR}}
$$

Glycerol

$$\text{H—C—O—C—(CH}_2)_7\text{—C=C—(CH}_2)_7\text{—CH}_3$$
(Oleic)

$$\text{H—C—O—C—(CH}_2)_7\text{—C=C—C—C=C(CH}_2)_4\text{CH}_3$$
(Linoleic)

$$\text{H—C—O—C—(CH}_2)_7\text{—C=C—C—C=C—C—C=C—C—CH}_3$$
(Linolenic)

Fig. 1.   Mixed triglyceride.

hence, are subject to oxidation in a manner similar to the triglycerides.

Phosphatides

The phosphatides are associated with fats and oils in both plant and animal tissue. They consist of a polyhydric alcohol which is esterified with a fatty acid and with phosphoric acid. The phosphoric acid is in turn combined with a basic nitrogen-containing compound, such as choline, betaine, or ethanolamine. The two most common phosphatides are lecithin and cephalin (Figure 2).

It may be noted that the structure of the phosphatides is quite similar to the triglycerides (Figure 1). The unsaturated portion of the fatty acid is the focal point for oxygen attack. Phosphatides make up about 1–2 per cent of many crude vegetable oils and higher percentages of animal fats. Egg yolk, for instance, contains approximately 20 per cent phosphatides and is oxidized easily under certain conditions.

Pigments

Considerable information has been developed on the various classes of pigments in foodstuffs. Probably some of the best known are the carotenoids, such as carotene and lycopene. Beta carotene has the structure shown in Figure 3.

Note the unsaturated portion of this molecule which is easily affected by oxygen. Beta carotene changes in color from a fairly deep reddish-orange (the intensity depends upon the form of carotene) to the oxidized product which is a light, yellowish gray. Since carotene is one of the major fat pigments, oxidation becomes important in maintaining quality.

Tocopherols

The tocopherols are the most important class of natural antioxidants and are widespread in both animal and plant tissue. Vegetable oils contain much higher concentrations of natural antioxidants, including tocopherols, than animal fats and, hence, are usually more stable toward oxidative degradation.

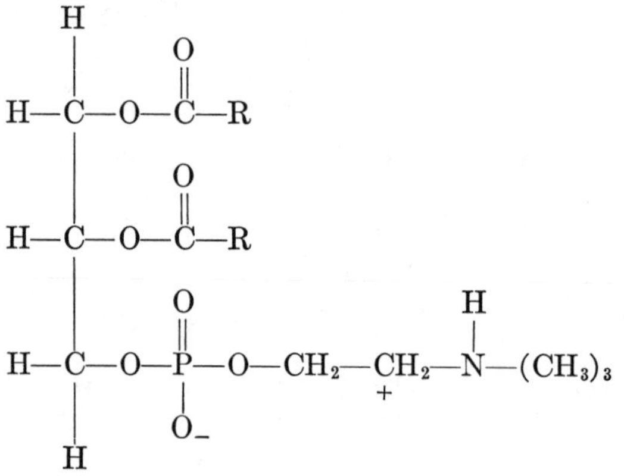

α-Lecithin

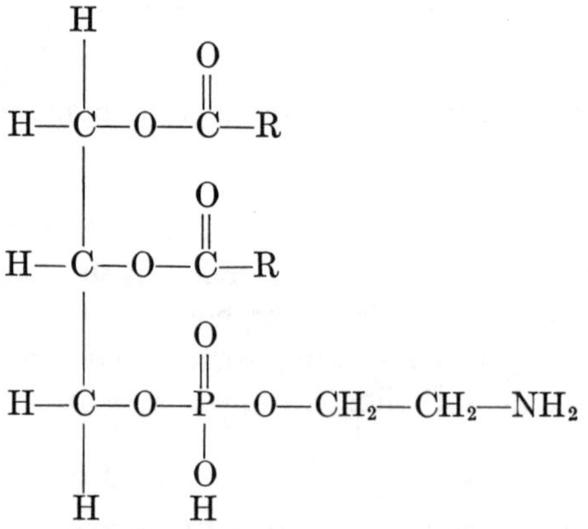

α-Cephalin

Fig. 2.

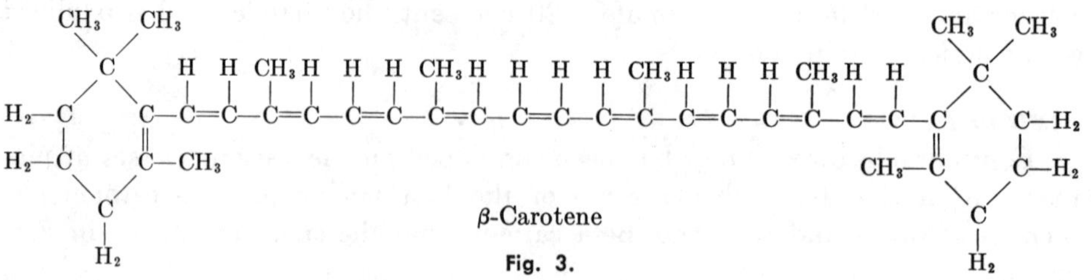

β-Carotene

Fig. 3.

The tocopherols are light yellow oils which are quite fat-soluble due to their long side chains. Their complete nutritional effect is unknown. Considerable information, however, is available on their Vitamin E activity, their sparing action on Vitamin A, and their effect on muscle health in animals. The tocopherols are readily oxidized to tocoquinones, which have no antioxidant properties (Figure 4). Tocopherols are readily degraded by heat, particularly at temperatures employed in refining and processing vegetable fats. (Note Table 10, p. 102, Chapter 2, "Vitamins and Amino Acids.")

Structurally similar to the tocopherols, gossypols and sesamolines have potent antioxidant properties. However, their concentration in oils is usually much lower than the tocopherols.

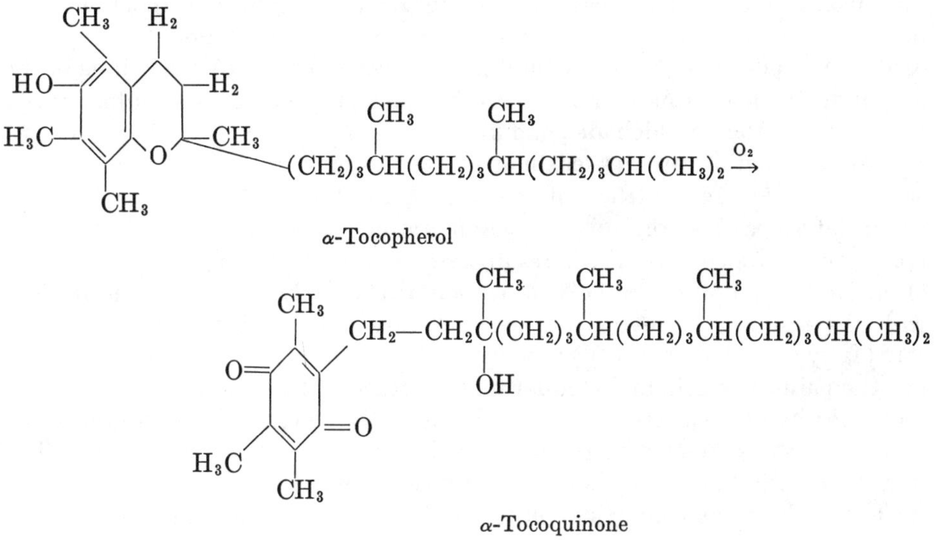

α-Tocopherol

α-Tocoquinone

Fig. 4.

Miscellaneous Components

Animal and vegetable tissue, in addition to the lipids described above, contain many minor constituents which may be quite important to the odor and flavor characteristics of the food, although their total concentration provides less than 1 per cent of the total lipids present. The various essential oils fall into this classification. Citrus oils, for example, are highly unsaturated and produce obnoxious flavors upon oxidation. The sterols (a term normally assigned to high melting, unsaponifiable alcohols) are usually of little concern to the food scientist since they are relatively inert and do not contribute significantly to quality characteristics of food product. Fatty alcohols are found in many marine oils, although they do not occur to any extent in land animals or vegetable oils. They are normally quite stable. On the other hand, most fats contain extremely small quantities of saturated and unsaturated hydrocarbons. The most important of these is the hexaene hydrocarbon squalene, which has a highly unsaturated structure similar to that of the carotenes. Its antioxidant potency is below that of safflower-oil mixed tocopherols in some oleates and linoleates.[6] When reacted with oxygen, squalene produces sharp, rancid odors and flavors. Breakdown of the fat-soluble vitamins (A, D, and E) normally associated with fats and fatty foods is catalyzed by the oxidation products of various fats and oils. Fish oils, particularly, are known for their high Vitamin A potency; and Vitamin A concentrates are sold as oil solutions. Oxidation of oil-soluble vitamins is a classical problem to the food technologist.

Mechanism of Lipid Oxidation

Unsaturated and saturated fatty acids and their esters can be oxidized by the usual chemical oxidizing agents, such as nitric acid, chromic acid, ozone, potassium permanganate, hydrogen peroxide, etc. Such reactions are important industrially and form the basis of certain useful analytical methods, but are not of prime importance to the food technologist. Autoxidation (atmospheric oxidation) under the relatively mild processing and storage conditions of the food industry is of utmost importance because of the resultant malodor- and malflavor-producing aldehydes and ketones.

Autoxidation of unsaturated lipid substances may be divided into two general areas: (1) the oxidation of the highly unsaturated fats, particularly the polyunsaturated fats, resulting in polymeric end products, and (2) the moderately unsaturated fats which result in rancidity, reversion, and other types of off-flavors and off-odors. Polymerization, particularly thermopolymerization, with the resulting formation of carbon-to-carbon bonds between the unsaturated fatty acid chains is of extreme interest to the protective-coating industry, i.e., the paint, varnish, and lacquer manufacturers. The linoleum industry, on the other hand, requires polymerized oils with a high oxygen content, i.e., carbon-to-oxygen-to-carbon bonds as well as carbon-to-carbon linkage. The industrial oxidation of fats is thoroughly reviewed by Swern.[4]

The path of lipid autoxidation, as well as the resulting end products, depends to a large extent upon the conditions of the oxidation; i.e., temperature, catalysts, fatty acids-type, the distribution and geometry of the double bonds, and the amount of oxygen available. Swern notes that the flavor and odor of the more highly saturated animal fats and hydrogenated oils whose unsaturated acids consist largely of mono-unsaturated acids are not significantly altered during the early phases of oxidation.[7] The onset of rancidity in such fats is both sudden and definite. On the other hand, relatively unsaturated oils, such as cottonseed oil or soybean oil, exhibit a more gradual deterioration in flavor and odor and a greater tendency to develop unpleasant flavors and odors different from those of true rancidity. In such oils it is often difficult to determine organoleptically exactly when rancidity actually begins. The mechanisms of oxidation have been observed during breakdown of simple, unsaturated lipid materials. Most of our knowledge on the mechanism of autoxidation has been based on studies of autoxidation of pure ethyl linoleate under closely controlled conditions. Since linoleic acid is the main polyunsaturated acid of commercial fats, its oxidative mechanism is of definite interest to the technologist. The oxidation of fatty substances is autocatalytic and has the characteristics of a "chain reaction," which may be broken into three stages or links:

1. initiation
2. propagation
3. termination

Hydroperoxides generated during propagation are believed to generate additional chain-propagating radicals that further increase the reaction rate. This reaction is particularly important if heavy metals such as iron, copper, nickel, or cobalt are present. It is also believed that these heavy metals are involved in the initiation step. Those who are interested in the basic theories of autoxidation should refer to the works of Farmer, Wexler, Ostendorf, and similar references.[8-10]

Facts which must be considered when solving stability problems are that, in most cases,

1. the primary products from autoxidation reactions are odorless and tasteless;[11]
2. at higher temperatures and/or oxidation levels, secondary direct addition of oxygen may take place at double bonds of oleic and linoleic acids leading to the formation of various non-volatile oxygen-containing compounds;[12]
3. these secondary products are generally odorless but not tasteless;[11]
4. flavor and odor of the ketonic and aldehydic group are extremely intense.

In recent years, gas chromatography has been used to identify and isolate these various off-flavor producing substances. Evans has shown that oxidative off-flavors consist of a number of odoriferous aldehydes, ketones, and other short-chain volatile compounds.[13] Other lipid materials, such as the phosphatides, tocopherols, carotenoids, etc., are believed to oxidize in a similar manner; i.e., a peroxide is formed at the carbon alpha to a double bond producing short-chain volatile monomers with characteristic odors and flavors. Little has been published, however, on the exact mechanisms and resulting end products. Stuckey describes the possible oxidation mechanism of beta-carotene and shows a positive effect of free radical inhibitors which would indicate that the carotenoids oxidize in a manner similar to that described above.[14]

Hoffman provides a good discussion of oxidative off-flavors produced in vegetable oils and a description of various commercial terms.[12] He describes "rancidity" as a widely used term covering many typical (mostly objectionable) off-flavors formed by autoxidation of all unsaturated fatty acids present, but detectable only in an advanced stage of oxidation. "Flavor reversion" is described as the "grassy," "beany," "fishy," "painty" flavors produced by the very early stages of oxidation of oils containing linolenic or higher unsaturated fatty acids, such as linseed oil, soybean oil, marine oils, etc.

Patton, in discussing oxidation of dairy products, states that the classic hydroperoxide mechanism of lipid oxidation seems to be true for oxidation in dairy products.[15] Beyond this, however, he believes the mechanism follows two somewhat different paths. The difference appears to depend on the presence or absence of water as a solvent for the reaction. He describes types of off-flavor which developed in the aqueous system undergoing lipid oxidation as "cardboardy" or "cappy," whereas those observed in an anhydrous milk fat or dry whole milk are described as "oily" or "tallowy." Butter, on the other hand, apparently undergoes oxidation by either or both mechanisms depending upon the amount of water present and can develop a remarkable range of oxidative flavor defects which he describes as "cardboardy," "oily-tallowy," "fishy," "mushroom-like," etc.

Koch, in describing oxidation of dehydrated foods and model systems, states that the problems of lipid oxidation can be conveniently divided into four areas:

1. the relationship of moisture content to oxidative stability;
2. special phenomena associated with the oxidation of lipids in dehydrated model food systems;
3. the relation of enzyme action to oxidative deteriorations;
4. possible interactions between products of fat oxidation and other food constituents, such as proteins, carbohydrates, etc.[16]

Rancidity in pure pork fat (lard) is easily discernible by trained organoleptic panels. In fact, a trained panelist can predict a day or two in advance which samples are at the beginning of rancidity, i.e., developing enough odoriferous aldehydes and ketones to be noticed. Dairy fats, vegetable fats, and complex foods often pose a difficult task for the panelist since they often develop off-flavors and odors not usually associated with "true rancidity."

Mechanism of Antitoxidants

A review of a number of studies on the inhibition of lipid oxidation indicates that more than one antioxidant mechanism may occur, depending upon the conditions of the reaction and the type of system being studied. Shelton described four possible mechanisms by which an inhibitor may function as a chain stopper for the free radical chain mechanism of lipid oxidation:

1. hydrogen donation by the antioxidant;
2. electron donation by the antioxidant;
3. addition of the lipid to the aromatic ring of the antioxidant;
4. formation of a complex between the lipid and the aromatic ring of the antioxidant.[17]

Later studies showed that when the labile hydrogen atom on a typical antioxidant was replaced with deuterium, the antioxidant was not effective. This indicated that the inhibitor donated hydrogen rather than an electron.[18] Stuckey describes similar results with some food products.[14] Some authors believe that the electron or hydrogen donation is a primary reaction, and the formation of a loose complex between the antioxidant and the fat chain is a secondary reaction. In practice, a combination of the various reactions probably occurs, with the antioxidant itself being completely oxidized or inactivated. The inactive antioxidant radical thus produced must be one that cannot initiate further reaction. This is usually depicted as

$$R \cdot + AH \rightarrow RH + A \cdot,$$

where $R \cdot$ is the fat containing a free radical and AH is the antioxidant. Figure 5 illustrates a typical antioxidant reaction.

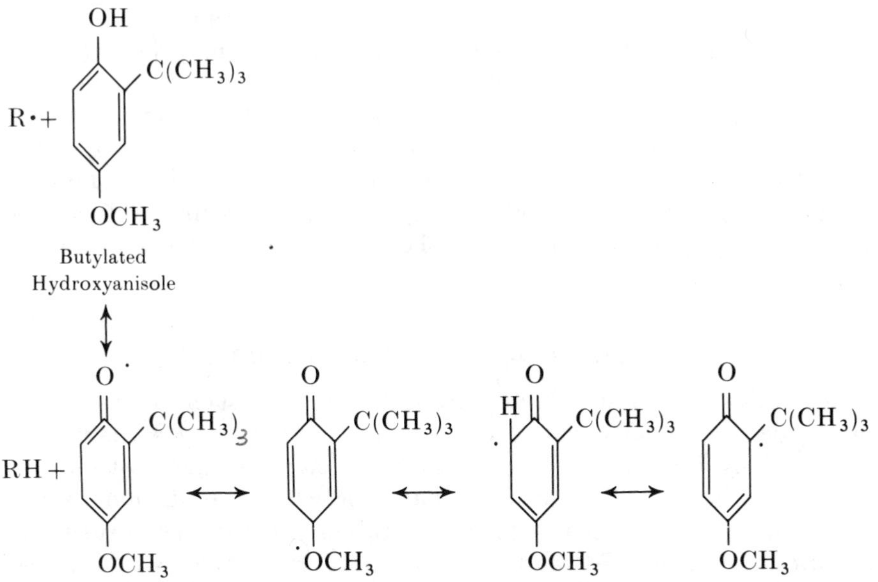

OH

C(CH₃)₃

R·+

OCH₃

Butylated
Hydroxyanisole

Fig. 5. Stable resonance forms of BHA.

Ring substitution with alkyl groups, such as the tertiary butyl group shown above for BHA, reduces the antioxidant activity of the molecule probably because of steric hindrance. These groups stabilize the molecule against outside reactions and make them less volatile and more oil soluble. Antioxidants useful in commercial food application usually are the "hindered" phenols with short side chains, since extremely long side chains apparently reduce the activity of the molecule to a point where it is not too effective in food fats. While various theories have been proposed on the mechanism of free radical acceptors, none have been definitely proved. References of interest include Shelton and Scott and others.[17, 18, 22]

Privett et al. have shown that autoxidation of polyunsaturated lipids is initiated by a discrete reaction occurring prior to the formation of a stable hydroperoxide.[19] The currently available antioxidants are apparently not effective in inhibiting this reaction, and it is believed that this type of reaction is probably responsible for the "flavor reversion" which occurs in polyunsaturated oils. The pro-oxygenic substances formed in this initial reaction apparently catalyze the oxygen up-take at the carbon alpha to the double bond in the fatty acid radical.

Oxidation reactions can be catalyzed by metals, hematin compounds, lipoxidases, etc., to a point where known antioxidants are no longer effective. Ingold and Tappell go into considerable detail on the effect of metals and various biocatalysts.[20, 21] It is believed that these catalysts increase the rate of chain generation to a point where normal concentrations of antioxidants are not effective in stabilizing the free radical. In fact, Ingold shows that in some cases heavy metals tend to reduce the reactivity of the so-called free radicals which would lead to a stabilization of the fat breakdown.[20] Since similar catalysts activate the oxidation of most antioxidants as well as the oxidation of the free fatty acid chain, most catalysts in antioxidant-treated lipids probably catalyze several simultaneous reactions.

Certain compounds, on the other hand, tend to enhance the effectiveness of various antioxidant systems although the opposite may be true in some cases. Water, under certain conditions, apparently acts as an inhibitor of the oxidation reaction. In the case of dehydrated foods, the moisture content has an active bearing on oxidative stability. Koch postulates that a mono-molecular layer of absorbed water on various food products provides the greatest degree of oxidative stability.[16] Results of stability studies at moisture levels both above and below this mono-molecular layer show that water caused decreased flavor stability.

Food-Approved Antioxidants

Antioxidants in most organic chemical reactions (such as oxidation of a triglyceride) are usually classed as those which operate by a chain-breaking mechanism by removing a chain propagation step and those which prevent or retard the introduction of chain initiators or active radicals into the system. Three main influences—heat, light, and metal contamination—are usually considered chain initiators. Metal deactivators and UV inhibitors would, therefore, be classed as inhibitors of these catalysts to the oxidation reaction. The two classes of inhibitors frequently complement each other in their response to environmental conditions. Some compounds may also act in a dual capacity. Scott, in an excellent discussion on the chain-breaking mechanisms of anti-oxidants, shows by both kinetic and chemical evidence that phenolic and amine compounds interfere with the process of autoxidation primarily by transferring either a hydrogen atom or an electron to the most abundant chain-propagating species and are, therefore, classed as chain breakers.[22] Considerable data exist which show the effectiveness of amines, and amino phenols, and phenol-type antioxidants in food fats. The phenylenediamines are extremely potent anti-oxidants for inhibiting autoxidation of most triglycerides as well as other organic compounds subject to autoxidation. Most amine-type compounds, however, are toxic, produce extremely colored bodies, and are not satisfactory as food-grade antioxidants. They are used mainly in gasolines, rubber products, and some packaging materials. Naturally occurring antioxidants as well as synthetic antioxidants used in direct food applications are phenols. Ethoxyquin (6-ethoxy-1,2-dihydro-2,2,4-trimethyl quinoline) is an exception. Figure 6 illustrates the chemical structures of some typical food-approved antioxidants. It may be noted that all of these antioxidants are structurally similar in that they have the unsaturated aromatic ring with either an amine or hydroxyl group to provide an electron or hydrogen atom for free radical satisfaction. All of these antioxidants shown are approved for food or feed applications except monotertiary butylhydroquinone. Food approval for this antioxidant was requested prior to publication of this handbook edition.

To be permissible in foods, an antioxidant must have, in addition to a low order of toxicity, potency in a wide variety of fats, lack of odor and color under use conditions, and approval by a government agency of the country in which the product is used. Antioxidants which have been "Listed as 'Generally Recognized as Safe' by the Food and Drug Administration" in the United States and whose addition is limited to 0.02 per cent (200 ppm)

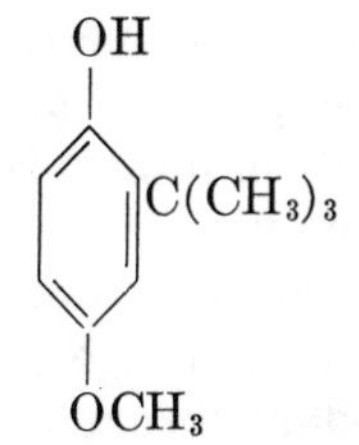

Butylated
Hydroxyanisole
(Mixture of 2-and
3-Isomers)
(Tert. Butyl-4-
Methoxyphenol)

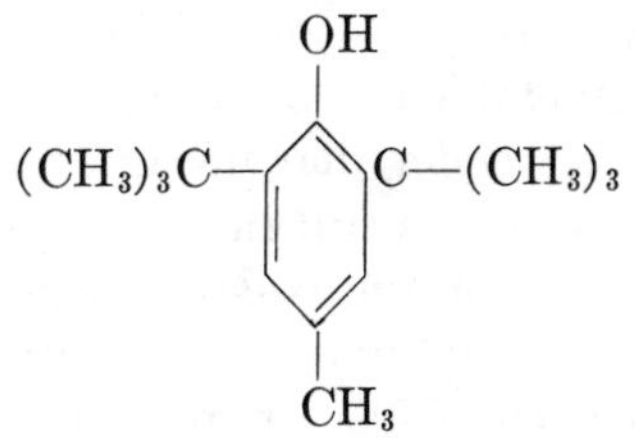

Butylated
Hydroxytoluene
(2,6-Ditert.
Butyl-*p*-Cresol)

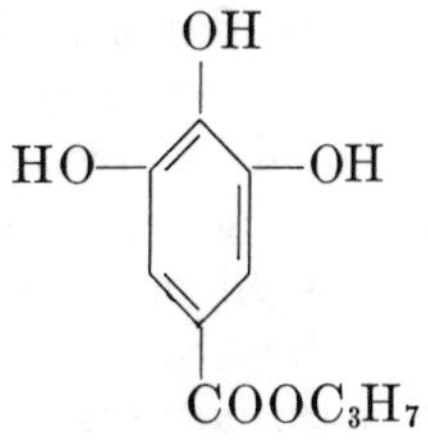

Propyl
Gallate
(Propyl Ester of
3,4,5-trihydroxy-
benzoic Acid)

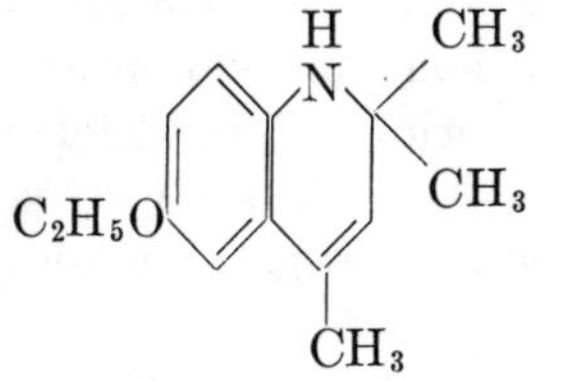

Ethoxyquin
(6-Ethoxy-1,2-dihydro-
2,2,4-Trimethylquinoline)

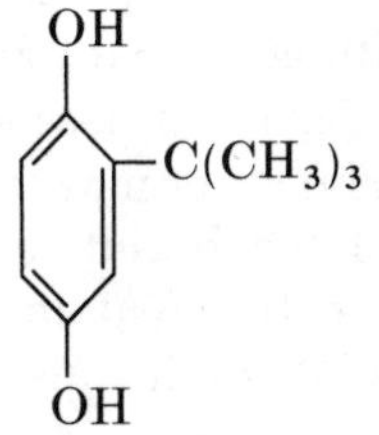

TBHQ
Monotertiary
Butylhydroquinone

Fig. 6.

by weight of antioxidant(s) based on fat or oil content of food are listed in the section "U.S. Government Regulations" of this chapter.

Synergism and Chelation

Considerable evidence has been gathered on the synergistic effect of various compounds in food products. Synergism has been extremely difficult to prove in complicated food products, probably because of the effect of natural inhibitors, catalysts, and other interfering substances, but it has been shown to be evident in simple systems such as methyl linoleate. Uri describes rather pronounced synergistic effects between polyphenolic compounds and certain acidic substances, such as ascorbic, citric, and phosporic acid[23]: these active synergists are also effective metal-chelating agents which would give rise to the theory that their only activity is that of metal chelation. Uri, however, shows that when citric acid is added alone to highly purified methyl linoleate, it exhibits no antioxidant action.[23] Nevertheless, it exhibits a marked synergistic effect when combined with a primary antioxidant, such as hydroquinone. Kraybill *et al.* show that BHA exhibits synergism with certain hydroquinones and phospholipids.[24] Stuckey describes the possibility of hydrogen donation by one of the so-called phenolic or acid-type synergists.[14] Olcott observes that some substances in fish products that have heretofore been considered only synergists are actually powerful antioxidants when used in relatively large amounts.[25]

Cowan, in summarizing work on the flavor of soybean oil, notes the importance various metals play both before and after refining on flavor degradation; i.e., chelators, in combination with good processing techniques, markedly improve the flavor stability of soybean oil.[26] These studies also suggest that citric acid acts solely as a metal chelator and has no synergistic effect.

It should be obvious to the reader that the interplay between antioxidants, synergists, and catalyst suppressors is still rather vague and requires additional study. The effect of metal chelation on oxidation can be shown quite dramatically in fats containing relatively high concentrations of iron, copper, and other metal catalysts.[26] Practically all commercial antioxidant formulations contain citric acid or soluble citrates which are included mainly as insurance against metals in processed fats. Chelators in antioxidant systems are probably not as necessary as they once were, since considerable progress has been made in the elimination of equipment which has contaminated food products with heavy metals and heavy metal salts. Most food products, however, contain sufficient quantities of naturally occurring metals to catalyze oxidation reactions. Also, modern stainless steels provide some metal contamination.[26] Fish fats, particularly, react markedly to phosphoric acid, but their metals are not effectively chelated by citric acid.[14] Some inedible greases, such as yellow grease, require extreme quantities of citric acid to provide adequate stability under commercial conditions.

Hydrogenation

Hydrogenation is the largest single reaction in the edible-fat industry and consists of the direct addition of hydrogen at the double bonds in the fatty acid chain. It is used primarily as a means of converting liquid oils to semisolid plastic fats suitable for shortening or margarine manufacture. However, it also accomplishes other desirable purposes, such as the enhancement of stability and improvement of color. Since the hydrogenation of fat saturates the double bond, hydrogenated fats normally are extremely stable to oxidation depending upon the percentage of double bonds which have been saturated. In recent years, partial hydrogenation of certain triglycerides has become quite popular to produce very satisfactory liquid shortenings. Generally, the higher the percentage of hydrogenation, the lower the unsaturation remaining in the fat, and the lower the possibilities of flavor degradation due to oxidation. Some off-flavors such as "hydrogenation flavor," "catalyst flavor," etc., are due to poor processing and cleanup procedures and are in no way connected with oxidation of the saturated molecule. Hydrogenated oils still contain some unsaturation and are subject to oxidative rancidity (especially when initiated by various catalysts) in the same manner as nonhydrogenated oils. Antioxidants, usually with the addition of chelators, are still necessary to obtain the extreme stability normally required of this type of product.

Measurement of Oxidative Stability

Techniques for measuring storage life of a fat or fatty food are based on subjecting the product to normal or accelerated storage conditions and measuring the progress of oxidation. The exact moment of rancidity is usually arbitrary since it must be based, in the final analysis, on human judgment as to when a product is no longer marketable or acceptable for a given use. Certain chemical tests have become fairly standard in the United States fat and oil industry, but are used only to a limited degree by other disciplines of the food industry. Since they have many inherent advantages, physical tests, such as oxygen absorption, are being used to a greater extent for some products. Some research has been done on vapor phase chromatographic techniques for measuring odoriferous components of fatty foods.[27] No commercial methods using this technique have been published to date, although it does offer considerable promise in some complex foods.

The following summarizes the major commercial methods currently employed. Those interested in more detailed descriptions or in methods other than those listed herein should consult *AOAC Tentative Methods* or Sherwin's *Methods for Stability and Antioxidant Measurement*.[28, 29]

Active Oxygen Method (AOM)

Figure 7 shows typical peroxide content at various stages of fat degradation. The AOM, often called the Swift Stability Test, is a chemical method based

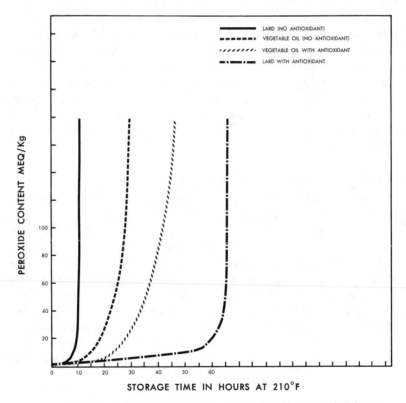

Fig. 7.   Peroxide content versus storage time of some typical fats.

on periodic evaluations of peroxide content of a fat or an oil whose oxidation is accelerated by heat and aeration.

The AOM is the most widely used method for estimating oxidative stability of fats and oils. The method is currently employed for research, quality control, and marketing purposes. It works quite well as a comparative test in relatively simple triglycerides as long as a control sample is included and some judgment is used in evaluating the final results. Some people quote AOM data in terms of factors in which the AOM stability of the treated sample is divided by the AOM stability of the control sample, to arrive at the so-called PF or protection factor.

In the Active Oxygen Method, a liquid sample is placed in a special aeration tube which is inserted in an oil bath.[28] A carefully controlled flow of dry, filtered air is bubbled through the sample while the entire bath and sample are held at a temperature of 97°C. Fat samples are withdrawn periodically and analyzed for peroxide content by reactions using a starch-iodide indicator. In determining the peroxide value, a small portion of fat is dissolved in a chloroform-acetic acid mixture. Potassium iodide is added to the solution, and any peroxide present will cause the liberation of iodine which in turn will complex with a starch indicator to provide a distinctive blue color. This color is titrated to a colorless end point with standard sodium thiosulfate solution. End points or "peroxide values" of 20 milliequivalents per kilogram of fat are normally used for animal fats and 70 milliequivalents per kilogram of fat for vegetable oils. A measure of peroxide value of any given fat will give some indication of its stability. However, since most peroxides formed are quite unstable, particularly at high temperatures, this test should be used only under the proper conditions. Variations of the Active Oxygen Method are used widely as a quality control tool in the fat and oil industry. This method does not correlate well with shelf storage since it is based on a measurement of the peroxides which are formed early in the oxidative reaction. It is also satisfactory only for rendered fats and oils and cannot be used for food products under most conditions.

2-Thiobarbituric Acid Test (TBA)

Various workers have used this chemical test based on the reaction of 2-thiobarbituric acid with the oxidation products of fats and oils to form a red color.[30] It is believed that aldehydes are responsible for the color formation. Since aldehydes are formed later in the oxidative reaction, the TBA test correlates better with organoleptic tests than the AOM. In this test, samples held under various storage conditions are evaluated periodically for the presence of aldehydes by use of the TBA reagent. The formation of a red color which can be measured spectrophotometrically indicates the degree of oxidation in a given sample. This method has been used widely for research studies, particularly of complicated food products, but has not been used extensively as a product control test. When used under very carefully controlled conditions, it will correlate quite well with organoleptic tests.

Free Fatty Acids

The free fatty acid value has been used for years as a test for evaluating fat and oil quality.[28] Normally, a small weighed portion of the fat or oil is dissolved in alcohol, and phenolphthalein indicator is added. The solution is then titrated with standard sodium hydroxide solution to a neutral end point. This method is not satisfactory for determining rancidity or simple free radical-type oxidation, since there is no distinction made between free fatty acids formed through hydrolysis or those from other sources, such as free radicals.

Shelf Storage Test

The technique of storing samples of a given food product under actual storage conditions and periodically evaluating them for development of rancidity is still the most reliable technique available. Its disadvantage is the length of time normally required for making these evaluations. Regardless of the time involved, the only accurate method for determining shelf stability of a fat or an oil or a processed food is to store samples of commercial packages under actual marketing conditions and evaluate them periodically for the development of rancidity. The usual procedure for following oxidative breakdown in shelf storage tests is by organoleptic panels. These panels vary from one or two trained people to large numbers of panelists set up on a statistical basis. While the latter technique gives better reliability, it is not always practical to use a large panel. Adequate results can usually be obtained with three or four trained people who have an interest in organoleptic testing. In some cases, peroxide value is also used to substantiate the organoleptic test in shelf storage samples. This combination of techniques has proved quite valuable, particularly in animal fats where the development of rancidity is rapid and positive.

Oven Storage Test

The oven storage test is merely a technique for reducing the time required for standard shelf storages. Samples are stored at varying temperatures and evaluated periodically using either an organoleptic or various chemical tests to assess rancidity. One technique which has been used rather widely is known as the Schaal oven test.[31] In this method, samples are prepared under standard conditions and stored in ovens at a temperature of approximately 140°–145°F. The samples are evaluated daily until they become rancid. This technique has been used widely for evaluating shortenings in cookies and crackers and in evaluating potato chips and other fried foods.

Oxygen Bomb Test

The oxygen bomb test takes advantage of the fact that during oxidative deterioration of fats and oils there is a measurable uptake or absorption of oxygen from the immediate environment. In the oxygen bomb method the material to be tested is placed in a stainless steel bomb that is connected to a pressure recorder. To accelerate the oxidative process, the bomb is charged with oxygen and heated by immersion in a suitable bath. When oxidation occurs, the pressure within the bomb is reduced through absorption and measured on the chart of a pressure recorder. Stuckey et al. have shown this technique to be quite accurate for various fats and oils and fatty foods.[32] Wintermantle et al. have shown the effectiveness of antioxidants in various cereal products, using this technique.[33] The oxygen bomb test has the advantage of being reproducible, does not require skilled technicians, and can be performed with a minimum of operator time. Petroleum bombs now being used in this method are not too satisfactory, since considerable difficulty is found in maintaining a leak-free system.

Barcroft-Warburg Method

The Barcroft-Warburg apparatus is a closed system in which a sample is held, usually under oxygen, and oxidation is followed manometrically by measurement of oxygen uptake.[34] While the Barcroft-Warburg method, with its various modifications, has proved quite useful for research investigations, it has not found widespread use in routine measurements of oxidative stability. It normally requires considerable attention by trained personnel. It is quite valuable where large numbers of samples are to be compared for a relatively short period of time under closely controlled conditions.

Ekey Method

This method, discussed by Hunter, should probably be described as a modified Barcroft-Warburg.[35] In this test Ekey devised a method, employing rather simple equipment, to measure the time required for a measured quantity of fat to absorb a definite quantity of oxygen under rigidly controlled conditions. The apparatus is attached to a clock or a recorder so that the point at which the predetermined quantity of oxygen has been absorbed can be recorded without the constant attention of a technician. This technique has been used rather widely with satisfactory results for quality control purposes in the fat and oil industry. Some laboratories, however, are replacing this technique with the oxygen bomb method described above.

Modifications of the above methods are used by various laboratories. The American Oil Chemists' Society is investigating various methods for assessing oxidative stability in order to determine which tests should be included in their official methods. To date, only the Active Oxygen Method has been included in the official methods.

Selection of Antioxidants

The selection of antioxidants for food fats is limited in the United States to those compounds which have been approved under applicable regulations. Among other qualifications, these antioxidants must have adequate potency, contribute no off-flavor or odors to the products in which they are used, and be economically attractive when compared to other methods for inhibiting fat oxidation, such as vacuum packaging, low-temperature storage, and others. Ethoxyquin is approved only for use in animal feeds, dehydrated forage crops, and some speciality uses. It has a very low order of potency in rendered animal and vegetable fats compared to other food-approved antioxidants.

The antioxidant formulations most commonly used in edible products contain various combinations of BHA, BHT, and/or propyl gallate together with citric acid in a suitable solvent. The higher gallates, such as dodecyl gallate, while approved for food use in several foreign countries, are not widely used. Several sulfur-containing compounds, such as thiodipropionic acid and dilauryl-thiodipropionate, are approved for use in edible fats, but their relative ineffectiveness has discouraged their use in food. Also, sulfur-containing compounds often produce some odor and flavor problems in most food fats. They are widely used, however, in combination with phenolic-type antioxidants in the stabilization of some food packaging films, particularly the polyolefins.

The tocopherols, gum guaiac, and similar natural antioxidants usually lack potency in most products compared to combinations of BHA, BHT, and propyl gallate.

One of the major problems associated with using antioxidants under commercial conditions has been the failure to achieve complete dispersion. When large quantities of rendered fats and oils are being stabilized, extreme care should be exercised to completely blend the antioxidant into the fat so solution occurs. When adding antioxidants to food products containing relatively small quantities of lipids, the problem is intensified.

The following application methods concerning the commercial usage of antioxidants must necessarily be general since equipment availability, physical nature of the products, and other factors vary widely. All unsaturated fats should be stabilized as soon as possible after being processed, particularly when subjected to heat. The oxidative reaction is autocatalytic, so that even a short delay without antioxidant protection can, under extreme conditions, be injurious to organoleptic quality. Even the best of antioxidant systems employed in food products will not inhibit rancidity in a product which has reached the end of its induction period; i.e., the formation of sufficient quantities of stable hydroperoxides. If possible, antioxidants should be added to a fat prior to heat processing. When heating a fat to increase the solubility of antioxidants, minimum conditions should be used since excessive heat also catalyzes breakdown reactions. Injection systems, where portions of an antioxidant solution are pumped directly into a moving hot fat stream, have proved quite practical in large installations. Smaller operations usually depend on batch processes with subsequent agitation.

Chelators, included in most antioxidant formulations, must be approved from a toxicological standpoint, cause no odor, flavor, or color problems, and be effective at the concentrations used. Citric acid is the most widely used chelator in the food industry. It is used alone and in combination with the phenolic-type antioxidants. This natural acid (which is also used as a flavor modifier) is approved on a "no limit" basis by the U.S. Food and Drug Administration. Since citric acid is only slightly soluble in food fats, propylene glycol is often used as a co-solvent to assist in dissolving the acid in fats. Phosphoric acid is used on a limited basis as a metal chelator, particularly in some vegetable oils and some fish fats. It also has food-additive uses in addition to its metal chelator properties.

Various monoglyceride and lecithin-citric acid reaction products and esters, e.g., isopropyl citrate, have been approved as food additives and are easily incorporated into the fat. Generally, these esters are not as effective chelators as citric acid.

Ethylene diaminetetraacetic acid and its salts are widely used as metal chelators in some food products. They have not been acceptable for addition to fats and oils because of poor solubility characteristics and apparent lack of potency.

Commercial Antioxidants

Commercial use of antioxidants to stabilize fatty foods on a major scale began in 1947 in the United States. Earlier attempts using natural antioxidants had not been too successful. The first satisfactory antioxidant combination was a mixture of BHA, propyl gallate, and citric acid to stabilize lard for shortening purposes. This combination of antioxidants is currently in wide use. Increasing shelf life with antioxidants, however, has spread to practically all major fats and food products which are subject to oxidative deterioration. Choice of antioxidant usage may be broken down in the following categories.

Rendered Animal Fat

These fats, which include both edible and inedible fats and shortenings with relatively high percentages of animal fats, are characterized by a relatively low degree of unsaturation and minimum natural stability. They are commonly used in commercial preparation of baked products and are usually quite responsive to antioxidant treatment. Where AOM-type stability and maximum carry-through protection (particularly under alkaline conditions) are needed, combinations containing high percentages of BHA are usually the most effective. Citric acid is almost always used to chelate trace metals in rendered animal fats. The best all-around combination which has been used for practically all rendered animal fats in commercial practice is a solution containing 20 per cent BHA, 6 per cent propyl gallate, and 4 per cent citric acid. This combination provides both shelf life and carry-through stability and is effective under most conditions. Combinations containing 10 per cent BHA, 10 per cent BHT, 6 per cent propyl gallate, and 6 per cent citric acid are also used, particularly for shortenings containing both animal and vegetable fats. Figure 8 shows initial stability produced by typical antioxidants. Figure 9 shows typical carry-through stability under alkaline conditions.

Several methods have been developed for adding antioxidant solutions to rendered animal fats. The choice naturally depends upon circumstances within the processer's plant, the amount of fat rendered, equipment available, etc. In large operations, proportionate methods are usually used wherein the antioxidant solution is injected by a pump into a pipeline through which the hot fat (about 150°F minimum) is being circulated. The success of the proportioning technique depends upon the length of time that the antioxidant-fat mixture is circulated and the turbulence provided by the circulating pump.

A direct-addition procedure is used to stabilize relatively small quantities of rendered animal fats contained in storage tanks. In this technique, the fat is heated to 145–175°F and vigorously agitated with a suitable stirring device so that the entire body of fat is in motion. Agitation should not be so vigorous, however, as to entrap excessive air in the fat. When the entire body of fat is in motion, the antioxidant is added slowly to the moving fat at the rate of about one gallon per five or ten minutes. The fat is then agitated for about another twenty minutes to insure uniform distribution and true solution. In some cases, antioxidant solutions are added in the form of a fat concentration, i.e., a solution of fat containing about 10 per cent of an antioxidant solution. The hot concentrate is added to the main body of fat following either the proportionate or the direct-addition techniques described above.

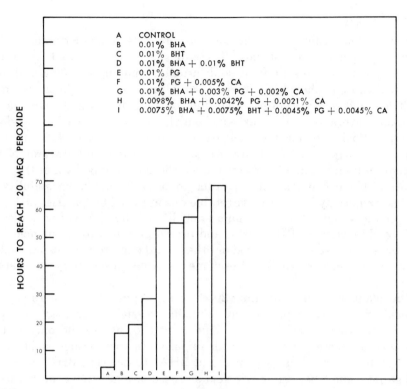

Fig. 8.   Effect of antioxidants on AOM stability of lard.

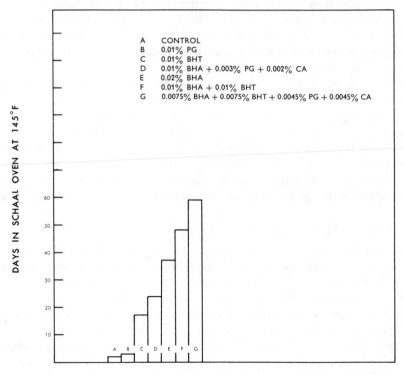

Fig. 9.   Carry-through of antioxidants in pastry prepared with treated lard.

Vegetable Oils

These include edible and inedible oil products and shortenings which contain high percentages of fats. These oils are characterized by a high degree of chemical unsaturation and usually contain some natural antioxidants. Because of their high degree of unsaturation, vegetable oils are quite difficult to stabilize with normal quantities of antioxidants and sometimes do not respond favorably even to extremely high concentrations of antioxidants. The best antioxidants for these highly unsaturated types of fats are those containing multiple hydroxyl groups, such as propyl gallate and NDGA (nordihydroguaiaretic acid). However, even these should be used at the maximum allowable concentrations in order to provide maximum stability. In those cases where hydrogenated vegetable oils are used for pastries and other baked products and where maximum shelf life of the shortening is also required, combinations of BHA, BHT, and propyl gallate with citric acid have proved to be most useful. Citric acid or other metal chelators are usually added as insurance against trace quantities of metals often present in most vegetable fats.

Antioxidant solutions containing 20 per cent propyl gallate and 10 per cent citric acid are widely employed for stabilizing vegetable fats used for frying operations or where shelf life requirements of the fat are the only criteria for stability. In those cases where both initial and carry-through stability are required, solutions containing 10 per cent BHA, 10 per cent BHT, 6 per cent propyl gallate, and 6 per cent citric acid are used at maximum concentrations allowable under government regulations. Figure 10 illustrates typical data produced by addition of commercial antioxidants to vegetable oils. Monotertiary butylhydroquinone (TBHQ), although not food approved at the time of this publication, has been found to be very potent in vegetable oils, particularly highly unsaturated types such as safflower oil. Figure 11 illustrates the effectiveness of TBHQ in some oils.

Antioxidant solutions are added to vegetable oils by one of the two techniques described above for rendered animal fats. In most large operations, the proportion pump technique is used immediately after deodorization. In those cases where vegetable oils must be held prior to deodorization, antioxidants are added to prevent breakdown during storage. Since they are removed by the deodorization process, additional antioxidants must be added after this process to insure maximum stability. Citric acid and other metal chelators are also added during processing to reduce the catalytic effect of various metals.

Food Products With High Fat Content

Food products, such as potato chips, nutmeats, and doughnuts, that fit this category are often prepared by deep-fat frying and sometimes contain as much as 50 per cent fat or oil. Vegetable oils and shortenings are commonly used in these frying operations. Pastries and pie crust mixes with fat content from 8-10 per cent are also placed in this category, but in these instances animal fat shortenings are usually employed in their preparation. The choice of an antioxidant for any of these products must be based upon a consideration of the type of fat used and the processing conditions involved in the preparation of the product. In deep-fat frying, where vegetable oils or shortening are

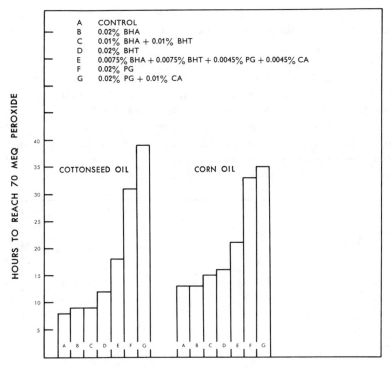

Fig. 10. Effect of antioxidants on AOM stability of vegetable oils.

commonly used, a combination antioxidant, such as a solution of 10 per cent BHA, 10 per cent BHT, 6 per cent propyl gallate, and 6 per cent citric acid

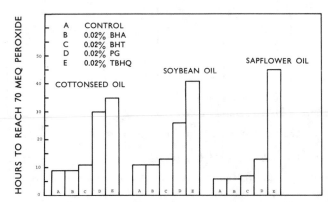

Fig. 11. Comparison of TBHQ with other antioxidants by the active oxygen method. (From Sherwin and Thompson, Ref. 36. Copyright © 1967 by Institute of Food Technologists. Data extracted by permission.)

is usually the most effective. Where animal fats or animal shortenings are employed, and particularly where alkaline conditions are encountered (e.g., various pastries and baked goods), a combination of BHA with propyl gallate and citric acid should be used. (Note Figure 9 for stability tests on pastries prepared with lard.) Figure 12 indicates typical data obtained when various nuts are stabilized with commercial antioxidants.

Wherever possible, the antioxidant should be added to the vegetable oil or animal fat prior to the frying operation or, in the case of the pastry mixes, prior to the mixing operation. For this reason, fats should be purchased with antioxidants added as soon as possible after processing. One of the most important uses of an antioxidant in various frying operations is to protect the fat during the short period of extreme temperatures that occur during frying. Since the phenolic-type antioxidants are steam distillable under conditions of deep-fat frying, they must be continually replenished during this operation. Because the ratio between food product and oil is quite high in modern deep-fat fryers, fresh oil constantly must be added to the frying operation. This technique, of course, also adds fresh antioxidant with the new oil.

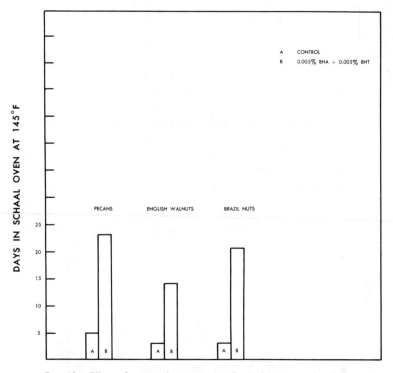

Fig. 12.   Effect of antioxidants in nuts (high fat-content food).

Food Products With Low-Fat Content

Breakfast cereals, dehydrated potatoes, and some cake mixes are typical examples of low-fat food products. The fat content is usually 1–2 per cent or less and of vegetable origin. A major problem encountered in the treatment of low-fat-content products has been the inability to bring the antioxidant in contact with the fat. The fats in these products are usually phospholipids, dispersed throughout the carbohydrates, which in turn are the major components of the product. Even though these fats are normally of vegetable origin, they are satisfactorily stabilized with BHA and BHT. Particularly in the case of breakfast cereals and dehydrated potatoes, it is believed that considerable quantities of natural antioxidants are present, and BHA or BHT act as secondary or synergistic antioxidants. Chelates, such as citric acid, are usually unnecessary in this type of product. Crystalline BHA or BHT is quite often used for this application, as well as solutions of 20 per cent BHA and 20 per cent BHT in combination with a suitable solvent. Propyl gallate is not usually employed for these applications, since the presence of naturally occurring metals together with a high moisture content often causes formation of highly colored metal complexes.

Figure 13 shows typical stability data obtained in low-fat foods.

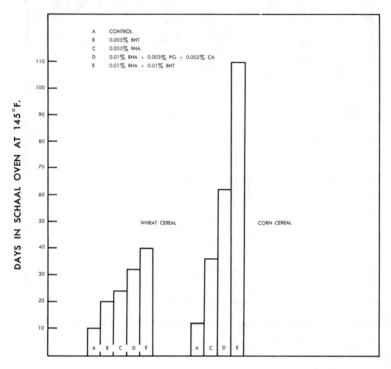

Fig. 13.   Effect of antioxidants in cereal (low-fat content food).

In adding antioxidants to a product of this nature, the volatility of BHA and BHT is brought into play. In some cases, it has been found that antioxidants will be dispersed by volatilization and steam distillation when added to the potato or cereal slurry prior to cooking and dehydration. This technique produced a stable product in processing and subsequent storage. Apparently, the extremely small quantity of antioxidant that finds its way into the fat or oil portion of the product under these conditions is adequate to provide a significant degree of protection. In other cases product protection is achieved by adding relatively high concentrations of antioxidant to the food-packaging material, e.g., inner-waxed liner. In these instances the antioxidant migrates to the surface of the packaging material and vaporizes. In the vapor phase, it disperses throughout the packaged product and provides a sufficient degree of protection to the lipid portion of the packaged food. In some cases, the combination of these techniques has proved successful.

Another approach, which has been quite successful, is to spray an emulsion formulation of an antioxidant solution onto the finished product immediately prior to packaging. This technique, while the most direct method of application, causes some difficulty from a quality control standpoint because of the extremely small quantity of antioxidant which must be dispersed over a rather large area of food product.

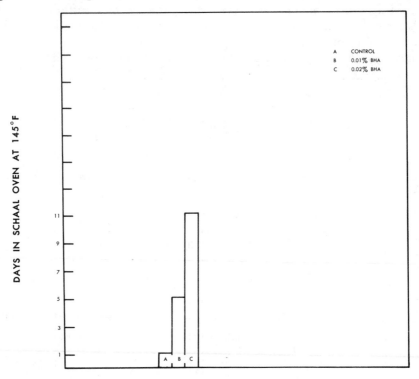

Fig. 14.   Effect of antioxidants added to pork concentration of antioxidant based on weight of fat (28%).

Meat Products

In the United States regulations issued by the Consumer and Marketing Service of the U.S. Department of Agriculture permit the use of lawful antioxidants in dried meats, in dry and fresh pork sausages, and in rendered animal fats or a combination of these with vegetable fats. The meat food products may contain up to 50 per cent of fatty tissue. Although they contain considerable quantities of hemoglobins, meat products have responded quite well to antioxidant treatment. Solutions containing 20 per cent BHA and 10 per cent citric acid have proved quite adequate for this application. Figure 14 shows typical stability data obtained with a meat product.

Application techniques depend upon dispersing the antioxidant throughout the comminuted meat. This is accomplished in commercial processing by using a salt containing BHA and citric acid dispersed on the surface of the crystals. The antioxidant-treated salt is blended into the meat emulsion, and the antioxidant is thereby dissolved in the fatty tissue.

Fishery Products

The use of standard food-approved antioxidants in most marine products has not been too successful from a commercial standpoint. These results are not unexpected, considering (1) the highly unsaturated nature of the triglyceride and phospholipid portion of most fish oils, (2) the presence of natural catalysts (such as the heme pigments) in many fish products, and (3) in some instances inadequate methods of application. Fish products must be divided into fish oils and by-products, from the standpoint of antioxidant application.

Fish Oils   When natural fish liver oils were the main source of Vitamins A and D and it was shown that the vitamin content was susceptible to degradation because of oxidative by-products, accordingly, antioxidants were used to prevent degradation. Combinations of BHA, propyl gallate, and citric acid were found to be effective in stabilizing the degradation of Vitamins A and D. In recent years encapsulating techniques have been employed to protect the vitamin-containing oil in a gel. Encapsulation prevents contact with oxygen and subsequent degradation.

Since the fatty acids in fish oils are highly unsaturated, propyl gallate and some of the higher gallates have been found to be the most effective antioxidants. Fish oils of commerce also contain a high percentage of iron, which cannot be chelated by standard methods. The gallates normally produce discoloration due to the formation of a blue-black iron gallate complex. Accordingly, only small quantities of antioxidants are employed for stabilizing commercial fish oils.

Fish Products   The single, most important deteriorative effect in fresh fish is due to microbiological spoilage. Oxidation is not considered to be of great importance, although it is generally recognized that the first recognizable stages of deterioration are probably due to rancid-type off-flavors produced by oxidation. Frozen fish is subject to two principal oxidative deteriorations: rancidity and rusting. Rusting is a light-yellow to brown discoloration that occurs on exposed surfaces. Olcott postulates that this coloration is due to a Maillard or an aldehyde-amine-type reaction.[25] The data, however, are not conclusive. Considerable work has been done to evaluate the effectiveness of antioxidants on frozen fish fillets. Generally, this work has not been successful. Since most fish oils are contained in several well-defined areas imbedded in the fish flesh below the skin surface, one would expect very little contact between the fatty tissue and the antioxidant by merely dipping a fillet in an antioxidant solution or spraying an emulsion onto the outer portion of the frozen fish fillet. One of the major problems, therefore, in stabilizing frozen fish fillets against oxidative rancidity is an adequate method of application that could be used under commercial conditions. Accordingly, antioxidants are not widely used in frozen fish.

Antioxidants are used in the processing of dehydrated fish meals from oily fish, such as menhaden and anchovetas. BHT and ethoxyquin are used in most commercial fish meals at the rate of approximately one pound and one-quarter pound per ton, respectively. The antioxidant in this case is fed into the meal immediately after the drying operation to prevent oxidation with subsequent heating of the fish meal during storage and transportation. Generally, the antioxidant is fed or blown into the meal as it travels through a screw-type conveyor; the conveyors are depended upon to completely blend the dry antioxidant with the fish meal. The use of antioxidants has changed the fish-meal industries manufacturing techniques from a batch process to a continuous process. In most modern fish-meal production it is no longer necessary to pile fish scrap and turn it at regular intervals to cure it before the grinding and bagging operation. Now heating is prevented by the use of antioxidants, and the dried fish scrap can be ground and bagged in a single line operation.

Miscellaneous Products

Some products which are used in the food industry require stabilization with food-approved antioxidants, although they do not fit the definitions as described above. These antioxidant usages may be divided in the following classifications:

Essential Oils Orange oil, lemon oil, and similar terpene-like flavoring oils undergo free radical oxidation in a manner similar to the phospholipids and triglycerides. Food-approved antioxidants are quite effective in reducing flavor and color degradation in these oils. BHA has proved the most effective of any of the food-approved antioxidants and is widely used for this application. Concentrations of approximately 0.02 per cent, based on the oil, are normally used. The antioxidant is added to the oil at the lowest possible temperature immediately after processing to prevent degradation.

Chewing-gum Base Chewing-gum bases, in addition to the essential oils added as flavoring ingredients, also include petroleum waxes and other polymers with unsaturated sites. Chewing-gum base, when subject to continued oxidation, will become extremely brittle due to cross-linking of the polymers. It will also develop off-flavors and odors. Antioxidants are used quite successfully to prevent this type of breakdown. BHA and BHT are both widely employed for this application. They are blended into the gum base during manufacture.

Paraffin and Mineral Oils These hydrocarbon oils contain unsaturated sites similar to the sites on fatty acids and oxidize in the same manner. BHT is a most effective antioxidant for this usage. It is employed at concentrations of approximately 0.0025 per cent. The solid antioxidant is normally dissolved into the hot paraffin or hot mineral oil without difficulty.

Packaging Materials Antioxidants are often added to packaging materials as a means of stabilizing a low-fatty-food product, such as cereals, nuts, etc. Usually the package contains a waxed inner liner and the antioxidant is added directly to the wax. However, in packages not containing a waxed inner surface, antioxidants can be added to the packaging board in the form of an emulsion. Either BHA or BHT is used for this application. Antioxidants are normally applied as an emulsion after the last calendering operation, since they are readily steam distilled from hot calendering stacks. The concentration of antioxidant in these cases will depend upon the size of package and the type of stability required. Table 1 shows the type of storage life that can be expected from application to packaging paper.

Candy Antioxidants are used in various types of candies to prevent rancidity. The antioxidant application depends upon the type of fat being stabilized. For instance, various types of nuts, such as walnuts, peanuts, etc., should be stabilized prior to addition to the candy. Butter, used as an ingredient in candy, is normally stabilized with combinations of BHA and BHT. Essential oils which are used in candy for flavoring purposes should also be stabilized.

TABLE 1
Typical Laboratory Data Evaluating
Storage Life of Chocolate-Almond Candy in Unwaxed Glassine Paper

Treatment			Life in Days	
Cooking Oil	Nuts	Paper	86°F	76°F
Control	Control	Control	42	46
.02% BHA	Control	Control	46	118
Control	.01% BHA	Control	129+	129+
Control	Control	.05% BHA	129+	129+
.02% BHA	.01% BHA	Control	118	129+
Control	.01% BHA	.05% BHA	129+	129+
.02% BHA	Control	.05% BHA	94	104
.02% BHA	.01% BHA	.05% BHA	129+	129+

Oxidation of Lipids in the Living Cell

There is considerable evidence supporting the view that the oxidation of the lipid constituents of living animals occurs not only in the depot fatty areas but also in non-adipose tissues. *In vivo* oxidation is evidenced by detection of peroxides in the fat as well as by a yellow or brown discoloration of the fat characteristically called yellow-fat disease. Yellow-fat disease is prevalent in mink, cats, rats, and chickens fed a diet of polyunsaturated free fatty acids.[36] Various Vitamin E deficiency symptoms, ·such as encephalomalacia in the chicken, kidney degeneration in the rat, membrane degradation during hemolysis of the red blood cells, and similar symptoms, can be traced to oxidation of lipid portions in the living cell.[37, 38]

Factors which influence the susceptibility of animal fats to oxidation are similar to those catalyzing oxidative breakdown of rendered fats and fatty foods. Machlin points out that trace minerals, sodium bisulfite, dietary peroxides, hemoglobin, and similar pro-oxidants catalyze the *in vivo* oxidation of lipid materials.[37] The structure of the fat molecule also affects its stability. Machlin also notes that ingestion of polyunsaturated fatty acids will result in decreased stability of tissue lipids. It is clear that feeding polyunsaturated fatty acids increases the unsaturation of depot fats of animals, and unless high levels of antioxidants are also ingested, there will occur either *in vivo* oxidation of the depot fat or decreased stability of the fat upon storage. Factors affecting the fatty acid composition of non-adipose tissue are usually considered to be the amount and type of fatty acid in the diet, the age of the animal, duration of feeding, species in question, etc. Saturated fatty acids are usually not considered capable of altering the structure of the tissue lipids. While the long-chain fatty acids (with 16–18 carbons or more) are not readily absorbed, the short chains are transported to the liver where they are readily metabolized. In view of the fact that *in vivo* oxidation (particularly of the depot fat) affects the health of the animal, the quality of the finished carcass, as well as the autoxidation of finished animal products in storage, this subject should be of interest to the quality-minded food technologist.

Excessive oxidation of the cell lipid is usually controlled by natural antioxidants. While other antioxidants affect the stability of the carcass fat, most of our detailed knowledge is concerned with the tocopherols. The amount of tocopherols deposited in a tissue will depend on the particular isomer, the dietary level, duration of the feeding period, the species and tissues concerned, and other factors. Generally, however, the tissue level of the tocopherols will vary directly with the dietary level. A classic example of species difference is reported by Mecchi *et al.* in an experiment in which the same diet was fed to both chickens and turkeys.[39] It was found that the turkey fat was much more susceptible to oxidation than the chicken fat and that the level of tocopherols in the chicken tissue was much higher than that in the turkey tissue.

In addition to natural antioxidants, synthetic antioxidants are widely used in some animal feeds and poultry feeds. BHT and ethoxyquin are used at levels of about 0.025 per cent in poultry feed for this application. Both are absorbed by the animal tissue at relatively low levels. BHA and propyl gallate apparently are not absorbed by the tissue and generally are not considered effective in this application. Alpha tocopherol is widely used to prevent the various Vitamin E deficiency problems in animals.

Tappell has demonstrated that the chief function of Vitamin E is through its lipid antioxidant properties, and that antioxidants merely act as a secondary synergist for this antioxidant factor.[21] Dietary selenium was shown to inhibit *in vitro* lipid oxidation of various animal organ tissues. While Tappell did not report on the mode of action of selenium, he did speculate that selenium probably acts as an inhibitor of some of the pro-oxidant constituents of the cell. Only small amounts of dietary selenium are required, however, which would indicate that it may act as a secondary antioxidant for Vitamin E normally present in the tissue. It should be added here that the selenium compounds evaluated are not food-approved at this date.

U.S. Government Regulations

The use of antioxidants in edible fats, oils, food products, drugs, and cosmetics in the United States is subject to regulation under the Federal Food, Drug and Cosmetic Act. Antioxidants for food products are also subject to regulations under the Meat Inspection Act, the Poultry Inspection Act, and various state and local laws. In general, approved antioxidant compounds may be added to products subject to regulation under the Food, Drug and Cosmetic Act at no more than 0.02 per cent based on the fat content of the food. Principal exceptions to this tolerance exist in the case of standardized foods and products covered by special regulations under the Food and Drug Law. Products covered under the Meat Inspection Act and Poultry Inspection Act can generally be treated with approved types of antioxidants at concentrations up to 0.01 per cent of an individual antioxidant and a combined total of not more than 0.02 per cent of all approved antioxidants based on the weight of the fat. Regulations and tolerances applying to the use of antioxidants are quite complex and are being modified continually. Specific problems relating to the use or labeling of antioxidants should be discussed with an antioxidant supplier and/or the agency responsible for enforcing the regulations applicable to the particular situation. The following antioxidant listing and allowable concentrations are given mainly as a guide and by no means should be considered as a complete up-to-date list.

Antioxidants for Direct Addition to Foods

Most of the food antioxidants are considered GRAS (*Generally Recognized As Safe*) for addition to edible fats and oils and to foods containing them. Following are the food antioxidants listed as GRAS in food-addition regulations, with limitations on their usage.

Antioxidant	*Limitation*
BHA (butylated hyroxyanisole)	
BHT (butylated hydroxytoluene)	0.02% of fat or oil content
Propyl gallate	including essential (volatile)
Dilauryl thiodipropionate	oil content of the food
Thiodipropionic acid	
Gum guaiac	0.1% in edible fats and oils
Tocopherols	No listed limits

Antioxidants that are cleared for use in foods under regulations, i.e., not GRAS, are shown below, with notations on the limit of their usage.

Ethoxyquin (1,2-Dihydro-6-ethoxy-2,2,4-trimethylquinoline)	0.01% (100 ppm) to preserve color of paprika, chili powder, and ground chili
4-Hydroxymethyl-2,6-di-*tert*-butylphenol	0.02% of fat or oil content, including essential (volatile) oil content of the food

BHA, BHT, and propyl gallate are cleared for use in certain prepared foods. The level of addition is based on the total weight of the food. Table 2 summarizes these uses.

TABLE 2
Addition Limits of Antioxidant(s) to Various Foods in ppm Based on Total Weight of Food

Food	*BHA*	*BHT*	*Propyl Gallate*	*Total Permissible*[1]
Beverages and desserts prepared from dry mixes	2	—	—	2
Cereals, dry breakfast	50	50		50
Chewing-gum base	1000	1000	1000	1000
Dry mixes for beverages and desserts	90	—	—	90
Emulsion stabilizers for shortenings	200	200	—	200
Fruit, dry, glacéed	32	—	—	32
Meats, dried	100	100	100	100
Potato flakes	50	50	—	50
Potato granules	10	10	—	10
Potato shreds, dehydrated	50	50		50
Rice, enriched	—	33	—	33
Sausage, dry	30	30	30	60
Sausage, pork, fresh	100	100	100	200[2]
Sweet potato flakes	50	50	—	50
Yeast, active dry	1000	—	—	1000

[1] Combination of lawful antioxidant.
[2] Based on fat content of sausage.

Antioxidants for Food Packaging Materials

Several antioxidants are cleared for use in food packaging by prior sanction.

1. BHA
2. BHT
3. Dilauryl thiodipropionate (DLTDP)
4. Distearyl thiodipropionate (DSTDP)
5. Gum guaiac
6. Nordihydroguaiaretic acid (NDGA)
7. Propyl gallate (PG)
8. Thiodipropionic acid
9. 2,4,5-Trihydroxybutyrophenone (THBP)

Such approval is based on the stipulation that no more than 50 ppm of the antioxidant shall become a part of the packaged food.

Other regulatory sections control the use of other antioxidants in miscellaneous packaging applications. Some of those approved follow.

1. *p-tert*-amylphenol-formaldehyde resins
2. butylated, styrenated cresols
3. 4,4'-butylidenebis(6-*tert*-butyl-*m*-cresol)
4. 4,4'-cyclohexylidenebis(2-cyclohexylphenol)
5. 4,4'-methylenebis(2,5-di-*tert*-butylphenol)
6. 2,2'-methylenebis-6-(1-methylcyclohexyl)-*p*-cresol
7. 4,4'-thiobis(6-*tert*-butyl-*m*-cresol)
8. 1,3,5-trimethyl-2,4,6-tris(3,5-di-*tert*-butyl-4-hydroxybenzyl)benzene

Foreign Regulations

Practically all countries have laws regulating ingredients and packaging materials used in the manufacture of food products. The United States has been the leader in these regulations, and many countries will automatically accept materials approved in this country.

European countries are now attempting to standardize their various laws as an aid to commerce. A standardization of the general food regulations is being studied by a committee of the European Economic Community (common market) with only an advisory capacity. A joint conference of representatives from the Food and Agricultural Organization of the United Nations (FAO) and the World Health Organization (WHO) has resulted in the formation of an expert committee on food additives and the Codex Alimentarius Commission whose task it is to compile a World Food Codex. Portions of the Codex recommending various antioxidants have been published. A similar organization, the Latin American Food Code Council, has published a "Latin American Food Code" which lists limits for various antioxidants.

It is beyond the scope of this chapter to publish details of the various foreign antioxidant regulations. Those interested should contact the Food and Drug Administration or the U.S. Department of Agriculture for additional and up-to-date information.

Toxicity of Food-Approved Antioxidants

The toxicity of the major food-approved antioxidants (butylated hydroxy-anisole, butylated hydroxytoluene, and propyl gallate) has been thoroughly studied in several animal species as well as humans. The most extensive tests have been made on BHA. Hodge *et al.* maintained dogs for one year on diets containing as much as 100 mg per kilo of BHA without adverse effects.[40] There was no storage in the body tissue. Wilder and Kraybill fed rats BHA at levels up to 2 per cent of the rations for six months and 0.12 per cent of the rations for 21 months without deleterious effects.[41] They reported an LD_{50} at 4130 mg per kilo when administered in corn oil and approximately 5000 mg per kilo when fed in a water emulsion. Astill *et al.* shows that 27 per cent to 77 per cent of BHA in man was excreted in the urine mostly within 24 hours.[42] No dealkylation or hydroxylation was found.

Orten *et al.,* reporting on the toxicity of propyl gallate in rats, guinea pigs, and dogs, show no toxicity for extended periods up to two years.[43] They report the LD_{50} of propyl gallate administered to rats orally as 3800 mg per kilo and when administered intraperitoneally as 380 mg per kilo. Heating of the gallate did not affect the LD_{50}.

Deichmann *et al.,* reporting the immediate, subacute, and chronic toxicity of BHT, concluded that it could be used as a food additive at normal concentrations without any public health hazards.[44] Later work cast doubt on the acceptability of BHT because of increased liver weights and abnormal cellular behaviour.[45, 46] However, more extensive studies were unable to verify these findings, although they do show some absorption of BHT by the tissues of the test animals.[47-53] While some toxic effects are shown at extreme levels, no deleterious symptoms developed at normal levels.

Analytical Methods

Antioxidants generally used for direct application to foods are all phenols, with the exception of ethoxyquin, which is an amine-type compound. Analytical methods for determining quantities of antioxidants present in foods are based on (1) *separation* of the phenol usually by extraction, distillation, precipitation, or chromatography, and (2) *quantitative measurement* by either colorimetric, ultraviolet, infrared, or various other techniques. The method used depends on the food product to be analyzed, the degree of accuracy required, equipment available, and other factors. Methods are constantly being improved and refined as better techniques become available. The following methods of separation are the major techniques currently employed.

Extraction Techniques

Steam Extraction This separation technique is the most widely used method for separation of BHA and BHT from fats and oils. It was developed by Anglin, Mahon and Chapman and adapted for various uses by Filipic and Ogg and Sloman *et al.*[54-56] In this method BHA and BHT are vaporized by steam from a heated fat-water mixture and carried along with the steam to a cold water condenser. Salts are added to the distilling flask to increase the temperature of the steam and retain some of the water in the flask to prevent formation of acrolein, which interferes with the test. BHA and BHT are completely separated from propyl gallate, NDGA, and the tocopherols, since these latter antioxidants are not steam distilled.

Liquid-Liquid Extraction   This technique involves a transfer of an antioxidant from one solvent system to a second. The two solvent systems must be immiscible and have different solvent powers for the antioxidants to be separated. Typical laboratory solvents, such as combinations of chloroform and ethanol, hexane and ethanol, and cyclohexane and ethanol, are widely used in this technique. Sahasrabudhe describes solvent solutions for separations of BHA, BHT, propyl gallate, and NDGA.[57]

Hot Solvent Extraction   In complex substrates where the proportion of antioxidant to substrate is quite small and distributed through a large mass, single solvent washing and extraction are not satisfactory. Cyclic washing of a crushed or shredded portion of the substrate with hot solvent, as provided in a Soxhlet unit, will usually extract the fat and antioxidant from the substrate. Austin describes this technique using ethanol as a solvent for analyzing paperboard and paper products.[58]

Cold Solvent Extraction   The washing of some food products with solvents at room temperature, particularly cereals and some dehydrated foods with a fat content of 2 per cent or less, is a simple and effective antioxidant-extraction technique. To assure good contact between the extracting solvent and the substrate, the mixture is usually swirled, shaken on a mechanical shaker, or agitated by some other device. Anderson and Nelson describe this technique for separating BHA and BHT from various breakfast cereals.[59]

Simple Solutions   Solutions of fats and oils in a spectrograde solvent, such as heptane, isopropyl alcohol, and isooctane, have been used for ultraviolet absorption measurements where only one or two phenolic antioxidants are present, or for chromatographic separations. The substrates themselves do not absorb in the ultraviolet region of the spectrum but may contain impurities which do. Buttery and Stuckey describe this technique for analyzing BHA and BHT in potato granules.[60] Whetsel *et al.* used isopropyl alcohol for analyzing BHA and propyl gallate from a simple fat by this technique.[61]

Column Chromatography   Silicic acid and alumina columns have been used for separating various phenolic antioxidants from food products. Caldwell describes this technique for separating BHT from various cereal products and paperboard.[62] Isooctane was used as a solvent to wash the column. The general principles of chromatography and the use of gas-liquid partition, thin-layer, and paper chromatography are also used for separation of antioxidants. They will be described, however, under the following section for *quantitative* methods of identifying and measuring antioxidants, since separation, identification, and measurement are very closely allied in these latter chromatographic techniques.

Quantitative Techniques

A variety of procedures for quantitatively measuring antioxidant content in extracts has been described in many publications. The two techniques which have proved most useful are spectrophotometry and chromatography. Spectrophotometry, particularly colorimetry, has been used widely in the past. Vapor-phase chromatography, a relatively new technique, is becoming an increasingly important method and probably will be used to a much greater degree in the future.

Phenolic antioxidant molecules absorb strongly in the ultraviolet and infrared region of the spectrum, and can be converted to colored complexes which absorb strongly in the visible region. Ultraviolet and infrared techniques have been widely used for identification purposes and antioxidant measurement in concentrated solutions. They are not satisfactory, however, for the extremely minute concentrations of antioxidants normally found in extracts from various food products. Colored complexes are used for this purpose. None of them are particularly new but are adaptations of reactions with well-known analytical reagents. A known volume of extract containing the antioxidant is treated with a color-developing reagent producing maximum absorption in the visible region of the spectrum. The hue and intensity of color are indications of the type and amount of antioxidant present. Naturally occurring phenolic substances also react with the same color reagents, which makes preciseness extremely difficult in the lower antioxidant ranges. Various separation techniques, as described above, are used to reduce error due to interference. The following quantitative techniques have been used for the analysis of food grade antioxidants.

Emmerie-Engels Method   In 1951, Mahon and Chapman made the first practical application of this method of analysis of phenolic antioxidants added to fats.[63] The basis for the color production in this method is a reduction of ferric chloride reagent to the ferrous state by the antioxidant and the formation of an iron-bipyridyl complex. The color of this complex is measured spectrophotometrically.

Gibbs Method   This method, which is specific for BHA among the phenolic antioxidants, makes use of 2,6-dichloroquinonechloroimide which forms a blue indophenol in the presence of the various hindered phenols. Mahon and Chapman found this method very effective for the 3-tertiary butyl hydroxyanisole which is the major isomer of commercial BHA.[64, 65]

Mitchell Method   Anglin, Mahon, and Chapman refined this method (originally used for evaluating pyrogallol tannins) for analysis of both propyl gallate and NDGA.[63] Practically all polyphenols give colored reaction products with the ferrous tartrate reagents, but many of the colored products are insoluble and do not affect the colorimetric determination of the violet color which is probably a chelate of the iron and the gallic acid radical. This method is probably the most widely used for determining propyl gallate concentrations in fats.

Szalkowski-Garber Method   This method (the first colorimetric method of analysis specific for BHT) depends upon the orange-red chromogen formed by a reaction involving dianisidine, sodium nitrate, and BHT.[66] The chromogen is normally extracted into chloroform; the color changes to purple with a maximum color absorbency for accurate determination. Propyl gallate and Santoquin do not affect this technique. It has been used quite satisfactorily for several types of food products, mainly rendered vegetable oils.

Other colorimetric methods are available and have been published in the literature. They are not generally used, however, in commercial quality control operations.

Chromatographic Methods

Chromatographic adsorption analysis is one of the newest and potentially the most important of all methods available for separation, concentration, and identification of phenolic antioxidants. The terms and techniques of column, paper, thin-layer, and gas or vapor-phase chromatography are quite familiar to most food technologists and chemists today. Qualitative methods for phenolic antioxidants are being developed using paper and thin-layer chromatographic techniques. Using these methods, individual antioxidants can be separated and identified with a minimum of technical skill and equipment. Vapor-phase or gas chromatography is rapidly becoming the choice for sensitive and accurate quantitative analysis of these antioxidants, particularly in the lower concentration ranges. This technique requires competent personnel and a considerable investment in laboratory equipment. Sahasrabudhe separated isomers of BHA, BHT, propyl gallate, and NDGA from fats using the thin-layer chromatographic technique and analyzed them by colorimetric methods discussed previously.[57]

Buttery and Stuckey used vapor-phase chromatography for analyzing dehydrated potato products.[60] Jennings *et al.* analyzed paperboard.[67] Anderson and Nelson and Schwecke and Nelson successfully analyzed various cereal products, as well as waxed glassine liners, using this technique.[59, 68] When foods containing higher quantities of fats and oils are anlyzed by vapor-phase chromatography, care must be taken to prevent build-up of the fat and oil on the column. Steam distillation of the antioxidant followed by solvent extraction and concentration was found to provide an antioxidant concentrate which could be detected without difficulty.

Needs of The Industry

Technology in the manufacture, packaging, and storage of most fats and fatty foods has advanced to the stage where rancidity is usually not the limiting factor in shelf life. Other quality factors, such as staleness, enzymatic breakdown, bacterial spoilage, and discoloration often accompanied by rancidity, tend to limit the shelf life more than oxidative rancidity alone. There are, however, still several areas where better methods of application are needed to get more intimate contact of the antioxidant with the unsaturated fat, such as fish fillets, some cereal products, prepared foods, frozen meats, etc. These application problems will have to be solved on an individual basis and probably do not warrant considerable basic research.

One of the major areas which should interest the research-minded food technologist is the stabilization of the lipid portion of the living cell. Sufficient work has been reported to indicate that synthetic antioxidants added to the diet will increase the stability of depot fat in the slaughtered animal. Conversely, feeding of highly unstable fats will reduce the stability of depot fat. Little has been done on *in vivo* stabilization of fats and oils derived from vegetable origin. Considerable knowledge of types of antioxidants which can be ingested by either plants or animals and transported to the fatty portion must be gained. Antioxidant systems which will transfer through the cellular walls and also have sufficient fat solubility to remain in the fatty portion of the cell are needed for meat products.

A basic research need of the industry lies in more information on loss of initial flavor and odor characteristic of freshly processed potato chips, fresh fish, freshly baked bread, etc. Most authorities believe that this initial loss of flavor is probably due to some form of oxidation. Studies should be made to determine the exact mechanism of these reactions. This knowledge would greatly simplify the search for an adequate stabilizing system.

References

1. Yu, T. C., Landers, M. K., and Sinnhuber, R. O., Browning Reaction in Radiation-sterilized Seafood Products. *Food Tech.*, 23: (2), 90–92.
2. Schultz, H. W., Day, E. A., and Sinnhuber, R. O., 1962, *Symposium on Foods: Lipids and Their Oxidation.* The Avi Publishing Company, Inc., Westport, Conn., Ch. 16, pp. 294–320.
3. Fox, Jr., Jay B., 1966, The Chemistry of Meat Pigments. *J. of Fd. and Agr. Chem.*, 14: (3), 207–210.
4. Swern, Daniel, 1964, *Bailey's Industrial Oil and Fat Products*, 3rd Edition. Interscience Publishers, A Division of John Wiley & Sons, New York, Ch. 1, pp. 3–53.
5. *Ibid.* Ch. 6, pp. 165–247.
6. Govind Rao, M. K., and Achaya, K. T., 1968, *J. Am. Oil Chem. Soc.*, 48: (4), 296.
7. Swern, D., 1964, *Bailey's Industrial Oil and Fat Products*, 3rd ed. Interscience Publishers, Division of John Wiley & Sons, New York, Ch. 6, pp. 165–247.
8. Farmer, E. H., Bloomfield, G. F., Sundralingam, A., and Sutton, D. A., 1942, The Course and Mechanism of Autoxidation Reactions in Olefinic and Polyolefinic Substances, Including Rubber. *Trans. Faraday Soc.*, 38: 348–356.
9. Wexler, Herman, 1964, Polymerization of Drying Oils. *Chem. Rev.*, 64: (6), 591–611.
10. Ostendorf, J. P., 1965, Measurement and Prevention of Oxidative Deterioration in Cosmetics and Pharmaceuticals. *J. Soc. Cosmetic Chem.*, 16: 203–220.
11. Henick, A. S., Benca, M. F., and Mitchell, Jr., J. H., 1954, Estimating Carbonyl Compounds in Rancid Fats and Foods. *J. Am. Oil Chem. Soc.*, 31: 88–91.
12. Schultz, H. W., Day, E. A., and Sinnhuber, R. O., 1962, *Symposium on Foods: Lipids and Their Oxidation.* The Avi Publishing Company, Inc., Westport, Conn., Ch. 12, pp. 215–229.
13. Evans, C. D., Frankel, E. N., Cooney, P. M., and Moser, H. A., 1960, Effects of Autoxidation Prior to Deodorization on Oxidative and Flavor Stability of Soybean Oil. *J. Am. Oil Chem. Soc.*, 37: 452–456.
14. Schultz, H. W., Day, E. A., and Sinnhuber, R. O., 1962, Symposium on Foods: Lipids and Their Oxidation. The Avi Publishing Company, Inc., Westport, Conn., Ch. 7, pp. 139–150.
15. *Ibid.* Ch. 10, pp. 190–201.
16. *Ibid.* Ch. 13, pp. 230–251.
17. Shelton, J. R., 1959, Mechanisms of Antioxidant Action in the Stability of Hydrocarbon Systems. *J. Appl. Pol. Sci.*, 2: 345–350.
18. Shelton, J. Reid, and Vincent, David N., 1963, Retarded Autoxidation and The Chain-Stopping Action of Inhibitors. *J. Appl. Pol. Sci.*, 85: 2433–2439.
19. Privett, O. S., Nickell, C., Tolberg, W. E., Paschke, R. F., Wheeler, D. H., and Lundberg, W. O., 1954. Evidence for Hydroperoxide Formation in the Autoxidation of Methyl Linolenate. *J. Am. Oil Chem. Soc.*, 31: 23–27.
20. Schultz, H. W., Day, E. A., and Sinnhuber, R. O., 1962, *Symposium on Foods: Lipids and Their Oxidation.* The Avi Publishing Company, Inc., Westport, Conn., Ch. 5, pp. 93–122.
21. *Ibid.* Ch. 6, pp. 122–138.
22. Scott, G., 1965, *Atmospheric Oxidation and Antioxidants.* American Elsevier Publishing Company, New York, Ch. 4, pp. 115–167.

23. Lundberg, W. O., 1961, *Autoxidation and Antioxidants*, Vol. I, Interscience Publishers, A Division of John Wiley & Sons, New York, Ch. 4, pp. 133–168.

24. Kraybill, H. R., Dugan, L. R., Beadle, B. W., Vibrans, F. C., Swartz, V., and Razabek, H., 1949, Butylated Hyroxyanisole as an Antioxidant for Animal Fats. *J. Am. Oil Chem. Soc.*, 26: 449–453.

25. Schultz, H. W., Day, E. A., and Sinnhuber, R. O., 1962, *Symposium on Foods: Lipids and Their Oxidation*. The Avi Publishing Company, Inc., Westport, Conn., Ch. 9, pp. 173–189.

26. Cowan, J. C., 1966, Key Factors and Recent Advances in the Flavor Stability of Soybean Oil, *J. Am. Oil Chem. Soc.*, 43: 300A–302A, 318A–320A.

27. Nawar, W. W., 1966, Some Considerations in the Interpretation of Direct Headspace Gas Chromatographic Analysis of Food Volatiles. *Food Tech.*, 20: (2), 213–215.

28. American Oil Chemists' Society, Official and Tentative Methods:
    CA 5A-50 Free Fatty Acids
    CD 8-53 Peroxide Value
    CD 12-57 Fat-Stability-Active Oxygen Method.
    The American Oil Chemists' Society, 35 East Wacker Drive, Chicago, Illinois.

29. Sherwin, E. R., 1966, Methods for Stability and Antioxidant Measurement, Presented at AOCS Short Course: *Processing and Quality Control of Fats and Oils*. Michigan State University.

30. Tarladgis, B. G., Pearson, A. M., and Dugan, L. R., 1964, Chemistry of the 2-Thiobarbituric Acid Test for the Determination of Oxidative Rancidity in Foods. II. Formation of the TBA-Malonaldehyde Complex without Acid-Heat Treatment. *J. Sci. Fd. Agr.*, 15: 602–607.

31. Joyner, N. T., and McIntyre, J. E., 1938, The Oven Test as an Index of Keeping Quality. *Oil and Soap*, 15: 184–186.

32. Stuckey, B. N., Sherwin, E. R., and Hannah, Jr., F. D., 1958, Improved Techniques for Testing Fats and Oils by the Oxygen Bomb Method. *J. Am. Oil Chem. Soc.*, 35: 581-584.

33. Wintermantel, J. F., New, D. J., and Ramstad, P. E., 1961, Stability Tests on Cereal Products by Oxygen Bomb Method. *Cereal Sci. Today*, 6: 186–188.

34. Nagy, J. J., Vibrans, F. C., and Kraybill, H. R., 1944, An Application of the Barcroft-Warburg Apparatus to the Study of Antioxidants in Fats. *Oil and Soap*, Dec., 349–352.

35. Hunter, I. R., 1951, Oxygen-Absorption Apparatus for Measuring Induction Periods of Fats. *J. Am. Oil Chem. Soc.*, 28: 160–161.

36. Sherwin, E. R., and Thompson, J. W., 1967, Tertiary-Butylhydroquinone—An Antioxidant for Fats and Oils and Fat-Containing Foods. *Food Tech.*, 21: (6) 106–110.

37. Schultz, H. W., Day, E. A., and Sinnhuber, R. O., 1962, *Symposium on Foods: Lipids and Their Oxidation*. The Avi Publishing Company, Inc., Westport, Conn., Ch. 14, pp. 255–268.

38. *Ibid*. Ch. 15, pp. 269–293.

39. Mecchi, E. P., Pool, M. F., Behaman, G. A., Hamachi, M., and Klose, A. A., 1956, The Role of Tocopherol Content in the Comparative Stability of Chicken and Turkey Fat. *Poultry Sci.*, 35: 1238–1246.

40. Hodge, Harold C., Fassett, David W., Maynard, Elliot A., Downs, William T., and Coye, Jr., Robert D., 1964, Chronic Feeding Studies of Butylated Hydroxyanisole in Dogs. *Toxicol. and Appl. Pharmacol.*, 6: (5), 512–519.

41. Wilder, O. H. M., and Kraybill, H. R., 1948, Summary of Toxicity Studies on Butylated Hydroxyanisole. American Meat Institute Foundation, University of Chicago.

42. Astill, B. D., Mills, John, Fassett, D. W., Roudabush, R. L., and Terhaar, C. J., 1962, Food Additives Metabolism: Fate of Butylated Hydroxyanisole in Man and Dog. *J. of Agr. and Fd. Chem.*, 10: (4), 315–319.

43. Orten, James M., Kuyper, Adrian C., and Smith, Arthur H., 1948, Studies on the Toxicity of Propyl Gallate and of Antioxidant Mixtures Containing Propyl Gallate. *Food Tech.*, 2: 308–316.

44. Deichmann, William B., Clemmer, J. J., Rakoczy, R., and Bianchine, J., 1955, Toxicity of Ditertiarybutylmethylphenol. *A.M.A. Arch. Ind. Health*, 11: (2), 93–101.

45. Brown, W. D., Johnson, A. R., and O'Halloran, M. W., 1959, The Effect of the Level of Dietary Fat on the Toxicity of Phenolic Antioxidants. *Austral. J. Exp. Biol. Med. Sci*, 37: 533–548.

46. Day, A. J., Johnson, A. A., O'Halloran, M. W., and Swartz, C. J., 1959, The Effect of the Antioxidant Butylated Hydroxytoluene and Serum Lipid and Glycoprotein Levels in the Rat. *Austral. J. Exp. Biol. Med. Sci*, 37: 295–305.

47. Tye, R., Engel, J. D., and Rapien, I., 1965, Disposition of Butylated Hydroxytoluene in the Rat. *Fd. Cosmet. Toxicol.*, 3: 547–551.

48. VanStratum, P. G. C., and Vos, H. J., 1965, The Transfer of Dietary Butylated Hydroxytoluene (BHT) into the Body and Egg Fat of Laying Hens. *Fd. Cosmet. Toxicol.*, 3: 475–477.

49. Frawley, J. P., Kay, J. H., and Calandra, J. C., 1965, The Residue of Butylated Hydroxytoluene (BHT) and Metabolites in Tissue and Eggs of Chickens Fed Diets Containing Radioactive BHT. *Fd. Cosmet. Toxicol.*, 3: 471–474.

50. Gilbert, D., and Goldberg, L., 1965, Liver Response Tests. III. Liver Enlargement and Stimulation of Microsomal Processing Enzyme Activity. *Fd. Cosmet. Toxicol.*, 3: 417–432.

51. Gaunt, I. F., Feuer, G., Fairweather, F. A., and Gilbert, D., 1965, Liver Response Tests. IV. Application to Short-Term Feeding Studies with Butylated Hydroxytoluene (BHT) and Butylated Hydroxyanisole (BHA). *Fd. Cosmet. Toxicol.*, 3: 433–443.

52. Gaunt, I. F., Gilbert, D., and Martin, D., 1965, Liver Response Tests. V. Effect of Dietary Restriction on a Short-Term Feeding Study with Butylated Hydroxytoluene (BHT). *Fd. Cosmet. Toxicol.*, 3: 445–456.

53. Daniel, J. W., and Gage, J. C., 1965, The Absorption and Excretion of Butylated Hydroxytoluene (BHT) in the Rat. *Fd. Cosmet. Toxicol.*, 3: 405–415.

54. Anglin, C., Mahon, J. H., and Chapman, R. A., 1956, Determination of Antioxidants in Edible Fats. *J. Agr. Fd. Chem.*, 4: 1018–1023.

55. Filipic, V. J., and Ogg, C. L., 1960, Determination of Butylated Hydroxyanisole and Butylated Hydroxytoluene in Potato Flakes. *J. Assoc. Offic. Agr. Chem.*, 43: 795–799.

56. Sloman, K. G., Romagnoli, R. J., and Cavagnol, J. C., 1962, Trace Analysis of BHA and BHT in Food Products. *J. Assoc. Offic. Agr. Chem.*, 45: 76–80.

57. Sahasrabudhe, M. R., 1964, Food Additives, Application of Thin-Layer Chromatography to the Quantitative Estimation of Antioxidants: BHA, BHT, PG and NDGA. *J. Assoc. Offic. Agr. Chem.*, 47: (5) 888–893.

58. Austin, J. J., 1954, Analysis of Butylated Hydroxyanisole in Paper and Paperboard. *J. Assoc. Offic. Agr. Chem.*, 31: 424–427.

59. Anderson, R. H., and Nelson, J. P., 1963, A Method for the Determination of BHA and BHT in Cereal Products, *Food Tech.*, 17: (7), 95–96.

60. Buttery, R. G., and Stuckey, B. N., 1961, Food Antioxidants: Determination of Butylated Hydroxyanisole and Butylated Hydroxytoluene in Potato Granules by Gas-Liquid Chromatography. *J. Agr. Fd. Chem.*, 9: (4), 283–285.

61. Whetsel, K. B., Krell, M., Johnson, F. E., 1957, Infrared Analysis of Commercial Butylated Hydroxyanisole. *J. Agr. Fd. Chem.*, 5: (8), 602–604.

62. Caldwell, E. F., Nehring, E. W., Postweiler, J. E., Smith, Jr., G. M., and Wilbur, C. J., 1964, Package Treatment versus Direct Application as a Means of Incorporating BHT in Shredded Breakfast Cereals. *Food Tech.*, 18: (3), 125–128.

63. Mahon, J. H., and Chapman, R. A., 1951, Estimation of Antioxidants in Lard and Shortening. *Anal. Chem.*, 23: (8), 1116–1120.

64. Mahon, J. H., and Chapman, R. A., 1951, Butylated Hydroxyanisole in Lard and Shortening. *Anal. Chem.*, 23: (8), 1120–1123.

65. Mahon, J. H., and Chapman, R. A., 1952, Estimation of 2- and 3-Tert.-butyl-4-hydroxyanisole Isomers. *Anal. Chem.*, 24: 534–536.

66. Szalkowski, C. R., and Garber, J. B., 1962, Determination of 2,6-Di-tert.-butyl-4-hydroxytoluene in Edible Fats and Oils. *J. Agr. Fd. Chem.*, 10: 490–495.

67. Jennings, E. C., Curran, J. T., and Edwards, D. G., 1958, Gas-Liquid Partition Chromatography: Determination of 2,6-Di-tert.-butyl-p-cresol on Antioxidant-Treated Paperboard. *Anal. Chem.*, 30: 1946–1948.

68. Schwecke, W. M., and Nelson, J. H., 1964, Determination of Antioxidants (BHA and BHT) in Certain Food Products and Packaging Materials by Gas Chromatography. *J. Agr. Fd. Chem.*, 12: 86–89.

CHAPTER 5

Acidulants in Food Processing

Wm. Howlett Gardner, Ph.D.

Chemical Consultant
Bronxville, New York

Introduction

Acidulants serve numerous purposes in modern food processing, in addition to their major role of rendering foods more palatable and stimulating to the eater.[1] A partial list of their various functions includes their use as:

1. Flavoring agents, where they may intensify certain tastes, blend unrelated taste characteristics, and mask undesirable aftertastes.
2. Buffers, in controlling the pH of food during various stages of processing, as well as of the finished product.
3. Preservatives, in preventing growth of microorganisms and the germina-

tion of spores which lead to the spoilage of food or cause food poisoning or disease.

4. Synergists to antioxidants, in preventing rancidity and browning.
5. Viscosity modifiers, in changing the rheological properties of dough and, consequently, the shape and texture of baked goods.
6. Melting modifiers, for such food products as cheese spreads and mixtures used in manufacturing hard candy.
7. Meat curing agents, together with other curing components in enhancing color, flavor, and preservative action.

Only the most commonly used acidulants will be covered in this chapter, and discussion will be confined largely to their properties and use.

Commercially Available Acidulants

Acetic Acid   Very little pure acetic acid, as such, is used in foods, although it is classified by the Food and Drug Administration (FDA) as a GRAS material (generally recognized as safe).[2] Consequently, it may be employed in products which are not covered by Definitions and Standards of Identity.

Acetic acid, however, is the principal component of vinegars and pyroligneous acid.[3] It is found in unprocessed figs along with citric acid and small amounts of malic acid.[4] In the form of vinegar, more than 100,000 tons are produced annually in the United States for use in food.[5] As the major component of vinegars, acetic acid has been known for centuries—it is one of the earliest flavoring agents.[3]

Vinegars are extensively employed in preparing salad dressings and mayonnaise, sour and sweet pickles, and numerous sauces and catsups.[6-12] They are also used in the curing of meat and in the canning of certain vegetables.[13-16] In the manufacture of mayonnaise, the addition of a portion of the acetic acid (vinegar) to the salt- or sugar-yolk reduces the heat resistance of *Salmonella*, if present.[396, 397] Such a procedure, however, may affect the stability of the product by destroying the emulsifying properties of the yolk during pasteurization.[398] Uniform consistency and freeze-thaw stabilities may be obtained through the addition of solid triglycerides and vinegar to the various emulsified salad dressings.[399] Powdered seasonings may be prepared from sodium glutamate and acetic acid.[400]

Water-binding compositions of sausage often include acetic acid or its sodium salt.[401] Films of acetylated monoglycerides may be used to coat selected foods in order to protect them from oxidation, dehydration, or freeze burn.[402] Calcium acetate, in turn, is used to preserve the texture of sliced, canned vegetables.[403]

Pyroligneous acid extract is permitted by the FDA for the processing of certain smoked products.[17,18]

Pure acetic acid is produced commercially by a number of different processes. As dilute solutions, it is obtained from alcohol by the "Quick-Vinegar Process."[5] Smaller quantities are obtained from the pyroligneous acid liquors acquired in the destructive distillation of hard wood. It is manufactured synthetically in high yields by the oxidation of acetaldehyde and of butane, and as the reaction product of methanol and carbon monoxide.[5]

Vinegars are produced from cider, grapes (or wine), sucrose, glucose or malt by successive alcoholic and acetous fermentations.[19] In the United States, the use of the term "vinegar" without qualifying adjectives implies only cider vinegar. This stems from definitions drawn up by the FDA while it was part of the U.S. Department of Agriculture.[19,22] Although not in the same category as the Definitions and Standards of Identity promulgated for different foods, these definitions are used by the FDA for enforcing various regulations of the Food and Drug Act.[19,20] Although a 4 to 8 per cent solution of pure acetic acid would have the same taste characteristics as cider vinegar, it could not qualify as a vinegar, since it would lack the other components which can readily be detected analytically, as shown in Table 1.[19] In Great Britain, malt vinegar is supplied, by trade agreement, unless the type of vinegar is specified.[23] On the Continent, wine vinegar is the most common variety.

TABLE 1
Composition of Cider Vinegar[8,19,22]

	g per 100 ml
Acidity (as acetic acid)	3.24–9.96
Nonvolatile acids (as lactic acid)	0.05–0.30
Alcohol	0.03–2.00
Glycerol	0.23–0.46
Total solids	1.20–4.45
Reducing sugars (as invert sugar)	0.11–1.12
Volatile reducing substances (as invert sugar)	0.14–0.34
Pentosans	0.08–0.22
Volatile esters (as ethyl acetate)	0.30–0.91
Total ash	0.20–0.57
Soluble ash	0.17–0.51
Phosphoric acid in soluble ash	0.007–0.040
Alkalinity of soluble ash, ml 0.1N acid	21.5–56.0
Polarization in 200-mm tube, V°	−0.2 to −3.6
Sugars in total solids, %	5.3–43.3
Total ash in nonsugar solids, %	12.3–30.0

Propionic Acid   Propionic acid and its salts are primarily used in baked goods to suppress bacteria causing "rope" in the center of bread, and the growth of mold on both bread and cakes.[24-27, 404] The acid is employed to prevent blowing of canned frankfurters without affecting their flavor.[405]

Both sodium and calcium propionates are easily incorporated in dry mixes for baking.[24,28] They also afford protection to cheese against surface mold growth.[34] Dipping containers, caps and wrappers in solutions of these salts is also effective. In a like manner, mold growth on fruits and vegetables can be inhibited by spraying or dipping them in such solutions.[35]

The salts of propionic acid are white, free-flowing powders. Their addition to baked goods at recommended levels does not alter either the color or the taste of the finished product; baking time and other procedures are not affected.[28] The FDA lists the acid and its sodium, potassium and calcium salts as preservatives in their summary of permitted GRAS additives.

Calcium propionate is usually preferred in bread products, where yeast is used, in order to enrich the calcium content.[28] Sodium and potassium propionates are employed in cakes and chemically-leavened products, since calcium ions can alter the action of the leavening agents.[28]

Harshbarger reports that propionates are nontoxic.[29] *In vivo*, they probably undergo metabolism leading to the formation of succinic acid, which is then completely oxidized by the Krebs and associated pathways to carbon dioxide and water.[30-32]

The level of propionate required for preservation is usually a function of the acidity of the food.[28,33] For example, baked products such as cheesecake and angel-food cake, which have a higher acidity (lower pH), require less propionates than do pound cake, chocolate cake or devil's-food cake.

The storage history of baked products has a marked effect upon the rate of mold growth. Temperatures ranging from 32° to 37.5°C favor the multiplication of "bakery molds."[24] Since most molds do not survive baking temperatures, propionates prevent reinfection. With "rope" propionates prevent bacterial infection before the product is baked.[24] There is some evidence in both cases that it is the undissociated propionic acid which serves as the primary inhibitor.

The physical characteristics of propionic acid are summarized in Table 2.

TABLE 2
Physical Properties of Propionic Acid

Chemical Formula	CH_3CH_2COOH
Molecular Weight	74.08
Physical Aspect	Pure acid, colorless liquid
Boiling Point	141.1°C
Specific Gravity, d_4^{20}	0.9934
Refractive Index, n_D^{25}	1.3848
Solubility	Completely miscible with water
Odor	Pungent
Taste	Sour and mildly cheese-like

Source: *Kirk-Othmer Encyclopedia of Chemical Technology*, Vol. II, Interscience Encyclopedia, Inc., 1953. Reprinted by permission of the publisher, John Wiley & Sons, Inc.

Several processes exist for the production of propionic acid and its salts. These include the Reppe process from ethylene, carbon monoxide and steam, and the Larson process from ethanol and carbon monoxide using boron trifluoride as a catalyst.[27] It is also obtained by oxidation of propionaldehyde, as a by-product in the Fischer-Tropsch process for the synthesis of fuel, and in wood distillation. Besides these, propionic acid is produced by the action of microoganisms on a variety of materials.[27]

Sorbic Acid   Generally, most fatty acids containing 1 to 14 carbon atoms display mold inhibition.[35] Several of them, however, have objectionable odors or less desirable properties than propionic acid. Those with branched chains are lower in fungistatic effectiveness. Double bonds in the fatty-acid molecule, on the contrary, increase the fungistatic activity. Hence, it is not surprising to find that sorbic acid, which contains two conjugated unsaturated bonds, and its potassium salt are extensively used as antimycotics in foods.[406-408]

Sorbic acid, as indicated in Table 3, is an almost odorless, white, crystalline

TABLE 3
Physical Properties of Sorbic Acid[34,36,37]

Chemical Formula	$CH_3CH{=}CH{-}CH{=}CH{-}COOH$
Molecular Weight	112.12
Physical Aspect	White crystalline powder
Melting Point	135–137°C
Solubility	
Water, 20°C	0.16 g/100 ml
Ethanol, 25°C	14.8 g/100 ml
Propylene glycol, 20°C	5.5%
Fats and oils, 20°C	0.5–1.0%
Heat of Combustion, ΔH°_{c}	−11,927 Btu/lb
Ionization Constant, 25°C	1.73×10^{-5}
Flash Point (Open Cup)	260°F
Odor	Odorless
Taste	Slightly acidic

powder with a slightly acidic taste.[34,36] When used correctly, it does not affect either the smell or taste or leave any aftertaste in foods. The potassium salt is also a practically odorless, white powder, decomposing above 270°C (518°F); powdered calcium sorbate resembles talc in texture.[37]

Sorbic acid has only very limited solubility in water at ordinary temperatures, but its potassium salt is very soluble (0.16 vs. 58.23 g/100 ml at 20°C). The latter, when added to acidic foods, hydrolyzes to the antimycotically active acid.[36] At a pH of 4.8, about 50 per cent of sorbic acid is undissociated, which may explain its ability to penetrate mold cells and inhibit their growth.[34] The highest pH at which it is effective is 6.5. Sorbic acid is soluble in 95 per cent alcohol to the extent of 12.6 per cent by weight, while only 6.5 per cent of the potassium salt dissolves.[34] In anhydrous propylene glycol, the solubilities are 5.5 and 20 per cent and in fats and oils, 0.6–0.8 and 0.01 grams per 100 grams, respectively.[36] Calcium sorbate is sparingly soluble in water, in organic solvents, and in fats and oils.[37] For this reason, calcium sorbate is particularly suitable for protecting the surfaces of foods and in producing fungistatic wrappers.

There is evidence that sorbic acid acts as a fungistat by arresting the metabolic processes in molds through inhibiting the function of the dehydrogenase enzymes.[34,38-40] Sorbic acid and its salts do not reduce the amount of live mold or yeast, but do retard further growth of the microorganisms, provided the existing degree of contamination is not too high.[34-41] With dense growth, sorbic acid can be metabolized by the microorganisms. Sorbic acid does not inhibit the growth of bacterial organisms causing food poisoning, such as Clostridia, causing botulism.[42]

Experiments indicate that sorbic acid is harmless to rats and dogs when incorporated in their diets to the extent of 5 per cent on a moisture-free basis. It has been shown that in rats the metabolism is identical with that of normally occurring fatty acids. Under normal conditions of alimentation, it is completely oxidized to carbon dioxide and water.

The FDA lists not only the acid as a GRAS chemical preservative but also the calcium, potassium and sodium salts.[2] Their use in certain foods, however, is subject to Federal and State Definitions and Standards of Identity as well as rulings of the U.S. Department of Agriculture under the Meat, Seafood and Poultry Acts. Specific uses may also be subject to various federal, state and local limitations.[36] It is also allowed for many food uses in Canada, most South American countries, most of Europe, South Africa, Japan and other countries of the world.[37]

Sorbic acid and its salts are widely used in margarine in conjunction with antioxidants and other acidulants to prevent rancidity and mold and bacterial contamination.[43] Addition of milk, egg white or coconut oil does not decrease their effectiveness.[37] Sorbic acid is the only preservative currently permitted in any natural cheese and one of the few preservatives permitted in processed cheese.[44-48] In cakes, it is reported to be four times more effective than sodium propionate.[36,49,50] Sorbic acid and its salts are also used as fungistats in pies and pie fillings, bread, and brown-and-serve products.[36,50-53] They are highly effective preservatives for pickles and pickled products, mayonnaise, delicatessen salads, condiments, aspics, sherbet bases, fruit pulps and juices, jams and jellies, dried fruits, refrigerated salads, soft drinks, fruit syrups, beer and wine, and confections.[36,37,54-64,409] Fresh vegetables may be preserved by coating them with a wax containing sorbic acid.[65] Peeled vegetables are protected by a sorbic acid treatment.[47,66] Post-harvest rot of strawberries and tomatoes is also reported to be inhibited by sorbic acid.[36,37,67] Sorbic acid also stops reinfection by yeast, mold and bacteria following gamma radiation treatment to protect meat.[68] Its effectiveness for skinless comminuted meats, poultry, smoked and dried fish, and fish products has also been demonstrated, although still subject to various regulations in this country.[36,37,69,410]

These preservatives should be protected from light and permanent heat effect. Potassium sorbate, in particular, must also be kept dry during storage, by closing the container as soon as possible after use.[38] Solutions of sorbic acid, potassium sorbate, or both, in propylene glycol are sold for the preparation of soft-moist pet foods.[70-71] Mixed anhydrides with saturated fatty acids are claimed to be outstanding preservatives for bread and pastry.[411]

There are several methods of synthesis for sorbic acid and its salts,[72] but only one is used for large-scale manufacture.[73,74] This consists of the oxidation of 2,4-hexadienal which in turn is produced through the trimerization of acetaldehyde.[73,75,76] Synthetically, sorbic acid may also be prepared from ketene and crotonaldehyde, or crotonaldehyde and malonic acid.[77,78]

Other Fatty Acids None of the other fatty acids are employed to any appreciable extent in foods, as either the acid or one of its salts. This can be readily understood when careful consideration is given to their various properties. Several, like butyric acid, have objectionable odors, which greatly limit both their use as food acidulants and antimycotics.[79] Those of higher molecular weight (which do not have the disadvantage of butyric acid) are practically insoluble in water. A wide variety of these acids, however, are found in natural products as mixed glycerol esters.[80] Several of them including butyric acid are important chemical reactants in the manufacture of synthetic flavoring, shortening, emulsifying, and other edible additives employed in a variety of foods.[17,81-83] Some of the acids themselves are ingredients of imitation flavors.[17,82-83]

Succinic Acid Succinic acid is a relatively new nonhygroscopic acidulant in the food field. It is one of the natural acids found in broccoli, rhubarb, beets, asparagus, fresh meat extracts, sauerkraut and cheese.[84,85,412] All of these foods have distinct and marked flavors which may be due, in part, to flavor enhancement by this acid. Its apparent taste characteristics in foods appear to be very similar to the other acidulants of this type, although pure aqueous solutions have been described as slightly bitter.[86] There is some evidence that the taste build-up of succinic acid is rather slow. This should be an advantage in flavoring certain types of food. Certainly, succinic acid is worthy of much greater evaluation in foods than it has received to date, in spite of its present higher cost compared to similar acidulants.

The physical characteristics of succinic acid are summarized in Table 4. Since it is more soluble in water at room temperature than other nonhygroscopic food acids (Figure 1), some greater latitude is had in formulating powdered foods and beverages.

Succinic acid is listed by the FDA as a GRAS additive for miscellaneous and/or general-purpose uses.[2] It is not mentioned specifically in any of the Definitions and Standards of Identity but is covered by those which specify "other edible organic acids" as optional ingredients. Wider use of this terminol-

TABLE 4
Physical Properties of Succinic Acid [84,87,467]

Chemical Formula	CH_2COOH | CH_2COOH
Molecular Weight	118.09
Physical Aspect	White minute monoclinic prisms
Melting Point	188°C
Specific Gravity, d_4^{15}	1.564
Solubility	
Water	See Figure 1
Ethanol, 25°C	9.0 g/100 ml
Ether, 25°C	0.66 g/100 ml
Chloroform, 25°C	0.02 g/100 ml
Bulk Density	55 lbs/cu ft
Ionization Constants	
K_1	6.5×10^{-5}
K_2	2.3×10^{-6}
Heat of Combustion, ΔH°_c	-356.32 kcal/mole
Std. Free Energy of Anion Formation (ΔF°_f), 25°C	-164.97 kcal for aqueous solutions
Buffering Index	2.90
Odor	Odorless
Taste	Tart; slightly bitter in aqueous solutions

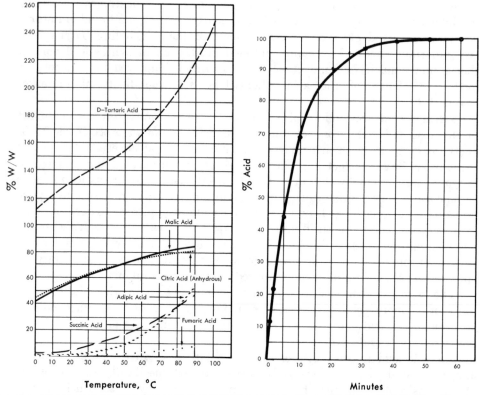

Fig. 1. Solubilities in water at different temperatures. (Courtesy Allied Chemical Corporation.[84])

Fig. 2. Rate of hydrolysis of succinic anhydride at 70°F. (Courtesy of Allied Chemical Corporation.[84])

ogy will undoubtedly be employed as some of the newer acidulants find wider uses in food products.

Succinic acid readily combines with proteins in modifying the plasticity of bread doughs.[88] It is also an excellent dibasic acid for producing edible synthetic fats with desirable thermal properties.[89,90] Succinoyl monoglycerides are optional ingredients of the amended Bakery Products Standards of Identity.[91] Succinic acid is one of the food acids suggested for dry gelatin desserts, for preparing cake flavorings, and for additives such as stearoyl propylene glycol succinates.[413-415] In the growing of food, it is a biogenic stimulant leading to faster plant growth and increased yields.[416-420]

Succinic acid is manufactured by the catalytic hydrogenation of maleic or fumaric acid.[92,93] It has also been produced commercially by aqueous alkali or acid hydrolysis of succinonitrile. Succinonitrile is derived from ethylene bromide and potassium cyanide.[94]

Succinic Anhydride   Succinic anhydride is the only commercially available anhydride for food use.[84] Even as an anhydride, it has some unique properties. As indicated in Figure 2, it hydrolyzes very slowly, requiring more than 30 minutes to convert an agitated aqueous suspension to succinic acid at 21°C. Other features are thermal stability and low melting point of 118.3°–

119°C (244.9°–246.2°F), which permits easy incorporation in products at comparatively low temperatures (Table 5).

TABLE 5
Physical Properties of Succinic Anhydride [84,87,467]

Chemical Formula	$\begin{array}{c} O \\ \| \\ CH_2C \\ \diagup \\ \| \qquad O \\ CH_2C \diagup \\ \diagdown \\ O \end{array}$
Molecular Weight	100.07
Physical Aspect	White orthorhombic prisms
Melting Point	118.3–119°C
Specific Gravity, d_4^{20}	1.503
Solubility	
Water	See Figure 2 for rate of hydrolysis
Ethanol, 25°C	2.56 g/100 ml
Ether, 25°C	0.64 g/100 ml
Chloroform, 25°C	0.87 g/100 ml
Bulk Density	47.2 lbs/cu ft
Heat of Combustion, ΔH°_c	−369.0 kcal/mole
Odor	Odorless
Taste	Burning tart

Succinic anhydride should be an ideal leavening acidulant for baking powders.[84,95] The slow rate at which it reacts with water at room temperatures would lend stability to these products in their dry form. Any moisture which comes in contact with the powder would react with the anhydride to produce succinic acid, which in turn is nonhygroscopic. The low rate of hydrolysis of succinic anhydride to the acid is also an advantage during mixing of the dough, since one of the requirements of a leavening acidulant is not to react with the soda in the mixture until the product is placed in the oven. Steady evolution of carbon dioxide should take place rapidly at elevated temperatures, since the rate of hydrolysis increases rapidly under these conditions, leaving a salt residue with nutrient properties.

Succinic anhydride is also a very useful dehydrating agent for the removal of small amounts of moisture in all types of dry food mixtures which are later to be admixed with water. As such, it is capable of taking up 20 per cent of its weight of water, and the acid formed is a desirable flavoring agent in many foods.

Succinic anhydride is manufactured by dehydrating the acid.[92]

Adipic Acid Like succinic and fumaric acids, adipic acid is practically non-hygroscopic.[84] Its addition to foods imparts a smooth, tart taste. In grape-flavored products, it adds a lingering supplementary flavor and gives an excellent set to food powders containing gelatin.[96] As a result, adipic acid has found a wide number of uses during the past 15 years as an acidulant in dry powdered food mixtures, especially in those products having delicate flavors and where addition of a tang to the flavor is undesirable.[84,97–108]

The physical characteristics of adipic acid are summarized in Table 6.

TABLE 6
Physical Properties of Adipic Acid[84]

Chemical Formula	$\begin{array}{c} CH_2CH_2COOH \\ | \\ CH_2CH_2COOH \end{array}$
Molecular Weight	146.14
Physical Aspect	White monoclinic crystals
Melting Point	153°C
Specific Gravity, d_4^{25}	1.380
Solubility	
Water	See Figure 1
Ethanol, 25°C	16.1 g/100 ml
Ether, 25°C	0.92 g/100 ml
Chloroform, 25°C	<0.01 g/100 ml
Bulk Density	40.5 lbs/cu ft
Heat of Combustion, ΔH°_c	−669.0 kcal/mole
Ionization Constants	
K_1	3.7×10^{-5}
K_2	2.4×10^{-6}
Odor	Odorless
Taste	Tart

Its aqueous solutions have the lowest acidity of any of the common food acids (Figure 3). For concentrations from 0.5 to 2.4 g per 100 ml, the pH of its solutions varies less than half a unit. Hence, it can be used as a buffering agent to maintain acidities within the range of 2.5 to 3.0. This is highly desirable in certain foods, yet the pH is low enough to inhibit the browning of most fruits and other foodstuffs.

Adipic acid is classified by the FDA as a miscellaneous and/or general-purpose food additive for use as a buffer and neutralizing agent.[2] When ingested, it is apparently only partially metabolized in the human system, the balance being eliminated unchanged in the urine.[109] Several years of animal-feeding and other studies show that it is as safe to use as the two most common acidulants, citric and tartaric acids.[110]

Adipic acid is employed as a leavening acidulant in baking powders and as the acidulant in powdered concentrates for fruit-flavored drinks and bottled beverages.[95,111] It may also be used for improving the melting characteristics and texture of processed cheese and cheese spreads, as an agent for increasing the whipping quality of products containing egg white and as a gel-inducing agent in imitation jams and jellies.[111,112] It can be used in the canning of vegetables, as an acidulant in candies and flavoring extracts, as a sequestrant in edible oils, and as an acidulant in throat lozenges. Its propyl ester adds a soysauce flavor to foods.[113] Combined with sodium metaphosphate, it has been used in the preservation of sausages and other meats.[114] Adipic acid is also used as the acidulant in algin gel desserts. When combined with dextrose as a melt, it yields a suitable additive for beverage powders.[421,422]

Adipic acid is also employed as a reactant in the synthesis of edible fats having special thermal properties.[89]

It is produced commercially by nitric acid oxidation of cyclohexanol or a mixture of cyclohexanol and cyclohexanone.[115,116] The former is obtained by reduction of phenol, while the latter is produced by oxidizing cyclohexane.

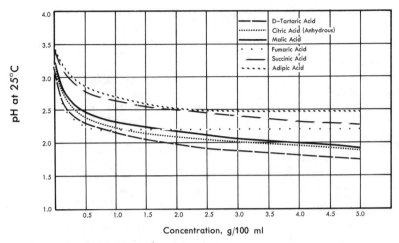

Fig. 3. pH of aqueous solutions at various concentrations. (Courtesy of Allied Chemical Corporation.[84])

Fumaric Acid   Fumaric acid is one of the most economical of the solid food acids, both from the standpoint of cost and the quantities required for imparting certain acid tastes.[84] It is limited, however, in some uses by its slow rate of solution and its relatively low solubility in water (Figure 1). Like adipic and succinic acids, it has an extremely low rate of moisture absorption. This makes it a valuable ingredient for extending the shelf life of many kinds of powdered food products, where solubility is not as important a factor as in powdered beverages.[117-119] Pudding powders containing this acid have been stored for more than a year under adverse conditions without appreciable change.[107,120]

Fumaric acid blends readily with other food acidulants without giving a "burst" to the taste. It also shows an affinity for certain flavoring agents like those in grapes, serving to supplement the overall flavor effect.[84]

The physical properties of fumaric acid are summarized in Table 7.

Appreciable quantities of fumaric acid are used in fruit-juice drinks, gelatin desserts, pie fillings, refrigerated biscuit doughs, maraschino cherries and wines.[95,97,121] Fumaric acid increases the strength of gelatin gels and acts as a calcium liberator when incorporated in alginate preparations.[104,105,120] It is also a valuable chemical reactant for preparing various edible coatings for candy, water-in-oil emulsifying agents, reconstituted fats, and dough conditioners.[122,123]

Fumaric acid displays good antioxidant properties when used to prevent rancidity from developing in lard, butter, cheese, powdered milk, sausage, bacon, frankfurters, nuts, and potato chips.[124-130,423] It can be used in the preservation of green foods and fish to supply the acidity required when sodium benzoate is used as a preservative.[125] It has been suggested also as an ingredient in combination with magnesium carbonate in baking powder. It may also be used for improving the whipping properties of gelatin and egg white and products containing them.[131,132]

The FDA now classifies fumaric acid and its salts as additives permitted in food for human consumption.[133]

TABLE 7
Physical Properties of Fumaric Acid[31,84,87]

Chemical Formula	HOOCCH ‖ HCCOOH
Molecular Weight	116.07
Physical Aspect	White monoclinic crystals
Melting Point	286–287°C
Specific Gravity, d_4^{20}	1.635
Specific Gravity, Sat. Soln., d_4^{25}	1.000
Solubility	
Water	See Figure 1
Ethanol, 25°C	4.3 g/100 ml
Ether, 25°C	0.56 g/100 ml
Chloroform, 25°C	0.02 g/100 ml
Bulk Density	32.6 lbs/cu ft
Heat of Combustion, ΔH°_{c}	−320.0 kcal/mole
Ionization Constants	
K_1	1×10^{-3}
K_2	3×10^{-5}
Std. Free Energy of Anion Formation (ΔF°_{f}), 25°C	−144.41
Buffering Index	3.46
Odor	Odorless
Taste	Tart

Sodium stearyl fumarate may be used as a conditioning agent in starch- and flour-thickened foods.[424] Long-chain alkyl lactyl fumarates have advantages as emulsifying agents in bakery products.[425]

To overcome the slow solubility rate of fumaric acid in water at low temperatures (Figure 1) the acid is admixed with 0.3 per cent dioctyl sodium sulfosuccinate (DOSS) and 0.5 per cent calcium carbonate.[134,135] This is sold as a modified acid for use in dry beverage powders, frozen fruit concentrates, frozen desserts and similar applications where government regulations permit.[136] The FDA has ruled that dioctyl sodium sulfosuccinate may be safely used as a wetting agent in fumaric acid-acidulated dry beverage bases and in fumaric acid-acidulated fruit-juice drinks, when standards of identity do not preclude such use, in accordance with the following conditions:

1. It meets the specifications of the National Formulary, Eleventh Edition, 1960.
2. The labeling of the dry beverage base bears adequate directions for use.
3. The additive is used in such an amount that the finished beverage or fruit-juice drink will contain not in excess of 10 parts per million of the additive.[136]

Dioctyl sodium sulfosuccinate is also permitted for use as a processing aid in the manufacture of unrefined sugar and as a solubilizing agent for gums and hydrophilic colloids.[136] Calcium carbonate is listed by the FDA as a GRAS additive for general-purpose uses.

Numerous evaluations of a variety of food products show that there is no significant difference between the flavor of mixtures of fumaric acid with DOSS and calcium carbonate and that of pure fumaric acid.[135] Rapid rates of solution are claimed for the mixture, even in ice water, and the fumaric acid remains as a homogeneous suspension in frozen concentrates for long periods of time.[135] Furthermore, fumaric acid and its salts have a tendency to stabilize the suspended matter in both flash-sterilized and frozen fruit concentrates and to inhibit the development of off-odors and darkening.[137]

Similar rapid-dissolving acidulants are prepared by mixing pulverized fumaric acid with surfactants and sugar or with citric acid and other hydroxy carboxylic acids.[426, 427]

Fumaric acid is obtained commercially as a by-product in the manufacture of phthalic and maleic anhydrides and by isomerization of maleic acid with heat or a catalyst.[138,139] It has also been produced by the fermentation of glucose or molasses with certain strains of *Rhizopus nigricans* and *Rhizopus japonicus*.[138]

Lactic Acid Lactic acid is one of the most widely distributed acids in nature and one of the earliest used in foods. Unlike the other acids of this class, it is a viscous, nonvolatile liquid. It is included in the FDA list of GRAS additives for miscellaneous and/or general purposes.[2]

Food-grade DL-lactic acid is available as 88- and 50-per cent aqueous solutions, both of which are colorless and practically odorless (Table 8). The variation of density and viscosity of aqueous solutions is shown in Table 9.

TABLE 8
Physical Properties of Lactic Acid[87,140,141]

Chemical Formula	$CH_3CH(OH)COOH$
Molecular Weight	90.08
Physical Aspect	Viscous, colorless, nonvolatile liquid
Melting Point	Pure acid, 16.8°C
Available Forms	88% and 50% aqueous solutions
Solubility	Very soluble in water
Ionization Constant	1.374×10^{-4}
Heat of Combustion, $\Delta H°_c$	−326.0 kcal/mole
Specific Heat, 20°C	0.505 cal/g °C
Density	
88% solution	10.0 lbs/gal
50% solution	9.4 lbs/gal
Odor	Available forms very slightly acrid
Taste	Acrid

Lactic acid is very soluble in water. One of the unusual chemical properties of lactic acid is that it readily undergoes self-esterification even in aqueous solution. When a solution is heated, dehydration takes place between the α-hydroxyl group of one molecule and the carboxyl of another to form a series of polylactic acids which include:

(1) lactyllactic acid, $CH_3CH(OH)COOCH(CH_3)COOH$;

(2) the linear trimer, $CH_3CH(OH)COOCH(CH_3)COOCH(CH_3)COOH$

(3) higher polymers.[142,143]

These condensation products occur in all solutions containing more than 18 per cent lactic acid. Hydrolysis to monomeric lactic acid, however, occurs

TABLE 9
Densities and Viscosities of Aqueous Lactic Acid at 25°C (77.0°F)

Lactic Acid % Concentration	Density g/ml	Viscosity Centipoises
9.16	1.0181	1.15
24.35	1.0545	1.67
45.48	1.1054	3.09
64.89	1.1518	6.69
75.33	1.1748	13.03
85.32	1.1948	28.50

Source: R. A. Troupe, W. L. Aspy, and P. R. Schrodt, *Ind. Eng. Chem.*, 43: 1143, 1951. Reprinted by permission.

upon dilution with water. The relative amounts of the different molecular species is a function of the temperature. Condensation also proceeds intramolecularly to form the cyclic dilactone, lactide: $\lceil OCH(CH_3)COOCH(CH_3)CO$.[140]

Lactic acid is used in packing Spanish-type olives, where it insures clarity of the brine by inhibiting spoilage and further fermentation.[140] It also provides a most suitable acid flavor.[140] Lactylated mono- and diglycerides of fatty acids are widely used in prepared cake mixes, other bakery products and liquid shortenings.[144-154] Both lactic acid and calcium stearyl-2-lactylate are approved optional ingredients in the FDA Definition and Standard of Identity for Bread.[155] Stearyl-2-lactylate is permitted as an emulsifier in leavened bakery products.[155]

Lactic acid, when added to pan-dried egg whites to adjust the pH range from 4.8 to 5.1, permits rapid settling of any shell fragments, improves protein dispersion, and aids in producing a more stable dried-egg powder with superior whipping properties.[156] A milder, more subtle taste is obtained when it is added to the vinegar in preparing certain pickles and relishes.[157-160] It is also used to adjust the acidity and as a flavoring agent in the manufacture of cheese and dried food casein.[161-166] Lactic acid is reported to be excellent, when used in small amounts, for acidifying fruit juice in the production of wine, and to give improved flavor to carbonated fruit juices when used in combination with other acidulants.[167-171] Lactic acid is also used in certain frozen desserts to provide a mild, tart flavor without masking that of the natural fruit.[140] It is employed in some jams and jellies, in the manufacture of beer, in mincemeat, mayonnaise, and for solubilizing pepper oleoresin.[140,172,173] Together with sorbic acid, it is employed in stabilizing sangria.[428]

Calcium lactate is used to preserve the firmness of apple slices during processing, to inhibit the discoloration of fruits and vegetables, as a gelling agent for demethylated pectins, and to improve the properties of dry-milk powders, condensed milk and baked food products.[174-178] An edible, solid lactate product is prepared by adding a 75% solution of lactic acid to calcium acetate and permitting the solution to gel.[429]

Edible lactic acid is produced by the controlled fermentation of highly refined sucrose. It is purified by converting it to crystalline calcium lactate, which then is decomposed with sulfuric acid to give solutions of the pure acid.[179,180] Other commercial processes employ less expensive carbohydrates, such as potato starch, molasses, corn sugar or milk whey, using a similar type of purification or solvent extraction methods. *Lactobacillis delbrückii* cultures are employed in the production of lactic acid from starch or molasses. With whey, *L. bulgaricus* cultures are grown on pasteurized skimmed milk for the fermentation.

Edible lactic acid is also manufactured synthetically by the hydrolysis of lactonitrile (acetaldehyde cyanohydrin or 2-hydroxypropionitrile). In the United States, this compound is obtained as a by-product of acrylonitrile manufacture.[181] In Japan, it is produced for this purpose, from acetaldehyde and hydrocyanic acid.[181]

Malic Acid Malic acid is a general-purpose acidulant. It has a smooth, tart taste which builds up and disappears gradually without giving any sensation of a "burst" in flavor.[182-184] With many flavoring materials, it gives a taste effect which closely resembles that of citric acid. With others, it tends to accentuate the flavor, leaving an impression of a very natural overall savor.[185] It has unusual taste-blending characteristics and in some instances appears to have flavor-fixing qualities, as well as serving to overcome undesirable aftertastes.[182,184] Although its degree of ionization in water, for most practical purposes, is the same as citric acid (Figure 3), it has a much stronger apparent acidic taste. Consequently, smaller amounts are usually required to obtain the same taste effect.[182-191] It does not have as strong an apparent taste, however, as fumaric acid.

Malic acid is a white triclinic crystalline powder. The commercial synthetic product is a racemic mixture of the D- and L-isomers, while the acid found in natural products is the levorotatory L-malic acid (Table 10). Convenient stock solutions containing 50 per cent of this acid can readily be prepared for use at room temperatures, as in the manufacture of carbonated beverages.

L-Malic acid is widely distributed in small amounts in many natural food products, especially in fruits and vegetables. It is the predominant acid in apples, apricots, bananas, cherries, grapes, orange peels, peaches, pears, plums, quinces, broccoli, carrots, peas, potatoes, and rhubarb.[192] It is the second-largest acid in citrus fruits, many berries, figs, beans and tomatoes. This might be expected, since it has one of the most important roles in the respiration of both plants and animals.[193]

Malic acid is one of the miscellaneous and/or general-purpose food additives in the FDA list of GRAS substances.[2,194] It is an optional ingredient in the FDA standards for sherbets and water ices, fruit butters, preserves, jams and jellies, nonalcoholic carbonated beverages, and proposed fruit-flavored noncarbonated beverages.[195-198] It is included under "other edible acids" in the proposed standard for orange nectar[199] and is one of the "any edible organic acid" in the standard for canned tomatoes.[200] Malic acid is also one of the ingredients specified by the Internal Revenue Service for rendering volatile fruit-flavored concentrates nonpotable when they contain 6 to 15 per cent alcohol and have to be transferred from place of manufacture to a winery.[201]

DL-Malic acid has been used for several decades for its flavor-contributing characteristics in nonalcoholic beverages, in imitation jams and jellies and in candy. Its relatively low melting point compared to other solid acidulants

TABLE 10
Physical Properties of Malic Acid[31,84,87]

Chemical Formula	HOCHCOOH $\vert$ CH₂COOH
Molecular Weight	134.09
Physical Aspect	White triclinic crystals
Melting Point	Racemic DL-form, 128.5°–129°C
Forms	Synthetic—DL-racemic mixture Natural—L-levorotatory form
Specific Gravity, d_4^{20}	1.601
Specific Gravity, Sat. Soln., d_4^{25}	1.250
Solubility	
Water	See Figure 1
Ethanol, 25°C	39.16 g/100 ml
Ether, 25°C	1.41 g/100 ml
Chloroform, 25°C	0.04 g/100 ml
Ionization Constants	
K_1	4×10^{-4}
K_2	9×10^{-6}
Heat of Combustion, ΔH°_c	−320.1 kcal/mole
Heat of Solution	−4.0 kcal/mole solute
Std. Free Energy of Anion Formation (ΔF°_f), 25°C	−201.98 kcal for aqueous solutions
Viscosity, 50% Aqueous Solution, 25°C	6.5 cps
Buffering Index	3.26
Odor	Odorless
Taste	Smoothly tart

and high solubility in water cause it to dissolve very readily when it is used in the manufacture of hard candies.[202]

Its ideal blending properties are employed to advantage when it is used in fruit-flavored foods and drinks, syrups, essences and ginger.[95] It does not affect the natural or artificial coloring or the permanence of color of food products or cause a sediment with ordinary "acid-proof" caramel.[95] Malic acid has been recommended as the ideal additive for stabilizing the color of apple, grape and other fruit-juice drinks, and for bringing out the overall taste of fruit-flavored carbonated drinks, including cream sodas.[203-205]

It has been employed in extracting pectin, in the production of sour dough, and in the synthesis of emulsifying agents which inhibit the development of rancidity in oils.[206-208] It is also used as an acidifying agent in canning tomatoes.[200,209] Its ethyl and isopropyl esters are recommended for improving the whipping properties of egg white and of gelatin in foamed types of desserts.[131,210] Fatty alcohol monoesters are useful antispattering agents in cooking fats.[211] DL-Malic acid may also be used to advantage both as a flavoring agent and a discoloration inhibitor in preparing pie and other fruit fillings, in bakers' jellies and preserves, frozen pies, and frosting and meringue mixes and powders.[212] It can serve as a synergist with antioxidants in a wide variety of foods to prevent fats and oils from becoming rancid, and as an agent for releasing oil from gum contaminants in the refining of edible fatty oils.[213-216]

DL-Malic acid is manufactured by hydrating maleic and fumaric acids in the presence of suitable catalysts and separating the malic acid from the equilibrium product mixture.[217] In the United States, the acid is now produced by a continuous process which is much more economical than the older batch type.

Tartaric Acid   Tartaric acid has a strong, tart taste and an ability to augment the flavors of fruits in which it is a natural constituent.[184] This is especially true with both natural and synthetic grape flavors, and to a prominent extent with currant, gooseberry, raspberry, lemon and orange flavoring materials.[218-219]

Tartaric acid of commerce has the L-configuration, but is dextrorotatory (Table 11). It is the most water-soluble of the solid acidulants (Figure 1).

TABLE 11
Physical Properties of Tartaric Acid[84]

Chemical Formula	HOCHCOOH
	HOCHCOOH
Molecular Weight	150.09
Physical Aspect	Translucent monoclinic sphenoidal prisms
Melting Point	168°–170°C
Forms	Commercial acid has the L-structure, but is dextrorotatory
Specific Rotation, $[\alpha]_D^{20°}$	+12.0
Solubility	
Water	See Figure 1
Ethanol, 25°C	19.6 g/100 ml
Ether, 25°C	0.59 g/100 ml
Ionization Constants	
K_1	1.04×10^{-3}
K_2	4.55×10^{-5}
Heat of Combustion, $\Delta H°_c$	−275.1 kcal/mole
Buffering Index	3.53
Odor	Odorless
Taste	Extremely tart

Tartaric acid is approved by the FDA as a GRAS compound for miscellaneous and/or general-purpose food and beverage use.[2] It is listed as an optional ingredient in standards for fruit butters, fruit jellies, preserves and jams, in the artificially sweetened jellies and preserves, and fruit sherbets.[195,196] It is permitted under the "any edible organic acid" clause of the nonalcoholic carbonated beverage standard.[220] Published results are highly conflicting with respect to the amount of tartaric acid, if any, that is assimilated by the human body.[221]

Tartaric acid is widely used in grape- and lime-flavored beverages of all types because of its effect on the flavor.[222] It also adds an optical note to the deep purple color of the grape-flavored products.[223] For similar reasons, it is selected as the acidulant for grape-flavored and for tart-tasting jams, jellies and candies. Blends of tartaric acid with citric acid are used in hard candies to obtain the popular sour-apple, wild-cherry and other especially tart-tasting varieties.[187]

Both tartaric acid and its acidic monopotassium salt (cream of tartar) are common ingredients of baking powders and leavening systems.[224] The limited solubility of cream of tartar in cold water prevents premature leavening, such as is required during the dough mixing. Various amounts of the acid are incorporated to accelerate the leavening reactions during baking.

Tartaric acid is also one of the food acids which may be used as a synergist with antioxidants in preventing rancidity.[125,133,225,226] Added to ground, dry spices, it serves as a stabilizing agent.[227] It also acts to prevent discoloration in cheese and as a chelating agent in food containing animal or vegetable fats and oils.[228,229] The loss of ascorbic acid is the least when tartaric acid is used in the bleaching of cauliflower.[430] Addition of diacetyl glyceryl tartrate to wheat dough improves both the baking properties and the volume of the bread.[431]

Tartaric acid and cream of tartar are manufactured entirely from the waste products of the wine industry.[230] These waste products consist of the following:

(1) Press cakes from unfermented or partially fermented grape juice
(2) Lees (dried slimy sediments from wine vats)
(3) Argols (crystalline crusts from wine vats used in second fermentations).

Several procedures are employed for recovering the tartrate content either as the acid, cream of tartar or Rochelle salt ($KNaC_4H_4O_6 \cdot 4H_2O$). In the manufacture of the acid, the by-products are first extracted with hot water, treated with hydrochloric acid and then lime. The calcium tartrate is then crystallized and decomposed with 70 per cent sulfuric acid to obtain the acid.

Citric Acid Unlike the other hydroxy acids, citric acid is tribasic. It has been used in foods in the United States for more than 100 years.[223] As a result, it is often employed as the standard for comparison in evaluating the effects of other acidulants in various food products.[231] Like malic acid, it is found in numerous natural products and it is one of the important acids involved in both plant and animal respiration.[184,232] Several fresh fruits such as lemons and limes owe their tangy taste to the presence of citrate ions. Its major advantages as an acidulant are its high solubility in water (Figure 1), the appealing effects on flavor and its potent metal-chelating action (Table 12).

Citric acid and its potassium and sodium salts are classified by the FDA as GRAS for miscellaneous, general purpose or both types of uses.[2] Sodium citrate is listed as an optional ingredient in ice cream and the acid is listed as an ingredient for fruit sherbets and water ices.[195,234] Both the acid and its salts are allowed in various fruit-juice drinks, diluted-juice beverages, and nonalcoholic carbonated beverages.[16,220] Sodium citrate is permitted in evaporated milk.[235] The acid may be employed in preparing mayonnaise, salad dressing and French dressing.[236] Citric acid, isopropyl citrate or stearyl citrate may be incorporated in the fat or oil ingredients of margarine.[237] Citric acid and sodium citrate may be used in fruit butters, fruit jellies, and preserves.[196] Artificially sweetened jellies, preserves and jams can also contain calcium citrate.[196] Recently citric acid and sodium citrate have also been permitted for use, both in curing and freeze-drying of certain meats.[238,432,433] With amino acids, its chelating action serves to increase the water retention of products treated with these two substances.[434] Aqueous solutions of citric acid with gelatin, salt, ascorbic acid, glucose and carrageenates have been injected into meat as a preservative.[435] An excellent acidulant for hams, sausage, and soy bean curds is obtained by coating the acid with suitable animal or plant oils.[436] Calcium citrate is listed by the FDA as a GRAS sequestrant.[2]

Canned vegetables (other than those specifically regulated) may contain citric acid as an acidulant.[239] Citric acid or vinegar are prescribed to give a pH of 4.5 or less in the canning of artichokes, while calcium citrate is permitted for firming peppers, potatoes, tomatoes and lima beans during processing.[240] Added acidulants, however, are specifically *prohibited* in the canning of mushrooms.[240] Citric acid is also an optional ingredient in canning of prune juice and figs and may be used to prevent off-flavors in preparing fried potatoes.[241,242]

Citric acid is a valuable acidulant for dairy products. It is permitted in creamed cottage cheese, pasteurized process cheese, pasteurized-cheese spreads and foods and cold-pack cheese.[44] Most cheeses may contain citrates as preservatives, where either citric acid or sodium citrate are employed alone or with other acidulants.[228,243-254] Citric acid is used both as a flavoring agent and to lower the pH of dry-cured cottage cheese. Sodium citrate is added to processed cheese as an emulsifying and aging agent.

Citric acid is also added to improve and protect both the flavor and the aroma of a variety of dairy products besides cheese.[229,244,255-260] A more flavorful buttermilk, for instance, is obtained when citric acid is added to the culture.[244,255] In a like manner, the amount of volatile acids is greatly increased when 0.2 per cent citric acid is incorporated in butter cultures.[244,255] Treating milk for infant feeding with citric acid renders it more digestible by its forming a softer curd.[256] Citric acid can be used for removing radioactive strontium from milk during a fallout emergency.[261]

Citric acid is extensively employed in the preparation of carbonated beverages to bring out the flavor and impart a "tang" to many of the beverages.[262-265] In addition, like other acidulants, it acts as a preservative, both in the syrup and finished product stages. It aids in imparting a desired bouquet by modifying some of the sweet flavor components and inactivates trace metals which may cause haze or deterioration of color and flavor.[264,266] Sodium citrate is used in club soda to impart a cool, saline taste and to enhance retention of the carbonation.

Citric acid continues to be one of the most popular acidulants for artificially flavored noncarbonated, or "still" drinks, and beverage powders.[264,437] It is employed in freeze-drying of fruit juices, in preserving bananas, and in drying raisins.[438-440] In wine, it is added to adjust acidity, prevent cloudy precipitates, and inhibit oxidation.[262,264,267-274]

In jams, jellies, fruit butters and preserves, citric acid is used to control the pH for optimum gel formation, and at the same time serve as flavoring agents.[233] It is also employed to adjust the acidity of relishes, sauces, and other food products requiring particular flavor enhancement.

Citric acid serves several interrelated purposes in the processing of frozen fruits. By inactivating trace metals, it preserves the ascorbic acid as a natural antioxidant for inhibiting color and flavor changes, and at the same time neutralizes residual lye left from peeling operations which would tend to destroy ascorbates.[264] Mixtures of citric acid and erythorbic acid (D-erythroascorbic acid) are used to retard browning of bananas packed 4 to 1 with dry

TABLE 12
Chemical and Physical Properties of Citric Acid[24,84,233,262]

Chemical Formula	CH₂COOH $\mid$ HOCCOOH $\mid$ CH₂COOH
Molecular Weight	192.12
Physical Aspects	From cold water—colorless, translucent ortho-rhombic From hot water—anhydrous, colorless, translucent holohedral class of monoclinic crystals
Melting Point	Anhydrous, 153°C; monohydrate softens at 70–75° when heated slowly and melts completely at 135–152°C. With rapid heating the monohydrate liquifies at 100°C
Solubility	
Water	See Figure 1
Ethanol, 25°C	58.9 g/100 ml
Ether, 25°C	1.84 g/100 ml
Density	Monohydrate, 1.542; anhydrous, 1.665
Molecular Refraction	Monohydrate, 67.11
Refractive Indexes, n_D^{20}	1.493, 1.498, 1.509 (hyd)
Heat of Combustion, ΔH°_c	Monohydrate, -471.4 kcal/mole Anhydrous, -474.5 kcal/mole
Bulk Density	Anhydrous, 56.2 lbs/cu ft
Ionization Constants	
K_1	8.2×10^{-4}
K_2	1.8×10^{-5}
K_3	3.9×10^{-6}
Heat of Solution, 25°C	-3.9 kcal/mole
Viscosity, 50% Aqueous Solution, 25°C	6.5 cp
Std. Free Energy of Anion Formation (ΔF°_f), 25°C	-278.8 kcal for aqueous solutions
Buffering Index	2.46
Odor	Odorless
Taste	Tart

sugar.[264,275] It is also used for similar purposes in fruit-pie fillings and in the dehydration and canning of apples.[264,276-278]

It is incorporated in various types of candy, including fondants, to enhance the flavor of fruit, berries and other ingredients, to prevent crystallization of the sugar, to invert the sucrose and to prevent oxidation of ingredients like nuts.[279,280]

Citric acid is a synergist for antioxidants employed in inhibiting randicity in foods containing fats and oils and in preventing loss of color and flavor of canned fruits and fish.[125-128,208,225,226,264,281-317,441,442] It is also used in cold-storage and saline preservation of herring fillets.[443,444] A mixture of citric and ascorbic acids is used as a dip for oily fish to prevent surface tissue from becoming brownish and gummy (a condition known as "rusting"). A similar dip is used for shrimp. Less ascorbic acid is required for this purpose in the presence of citric acid, since ascorbic acid is more stable in the acid environment. In addition, citric acid inactivates certain naturally occurring enzymes and sequesters trace metals which can act as prooxidant catalysts.[308-317] Citric acid solutions (1 per cent) are also used in the canning of crab meat, which has a low fat content in contrast to fish. Here it is used to inhibit discoloration and the development of odors and off-flavors.[308] Similar treatment is suitable for the canning or freezing of lobster meat, scallops and oysters.[313] Dilute citric acid solutions have been suggested for preventing "black spot" formation in shrimp; this may be caused by copper-catalyzed polyphenol oxidase reactions.[314]

Inedible fats and greases are treated with antioxidants and acidulants, in a manner similar to edible fats, to prevent deteriorative changes.[308] Where restaurant and catch-basin greases are the source for inedible fats, high concentrations of citric acid are usually employed (3 lbs citrate/200 lbs water/10,000 lbs tallow).[308]

Other uses of this versatile acid and its salts include their incorporation in enzyme preparations for clarifying fruit juices, in dry dough, baked farinaceous and crusty bakery products, and in antioxidants for chocolate and cocoa.[125,318-322] Citric acid is also employed to prevent crystallization in honey, as a stabilizing agent for spices and onion powder, in the synthesis of rearranged fats for shortening, in oleomargarine and similar spreads, and in various salad dressings.[145,307,323-331] Miscellaneous applications include its addition to porous caramel products, the production of foods from seaweed, prevention of the discoloration of onions, and its use with antioxidants in wrapping materials.[445-448]

Citric acid is produced commercially by the fermentation of pretreated molasses solutions which are inoculated with various pure strains of *Aspergilli niger*.[332] The acid is also obtained from lemons and from canned pineapple waste.[233]

Two processes are employed for the production of citric acid by mycological fermentation. The earlier involves the use of sterile, shallow aluminum or stainless-steel pans, while the more recently developed procedure makes use of a suspension of fungal mycelia throughout large vats. Since direct crystallization of citric acid from filtered fermentation liquors has never been proven successful on a large scale, except for crude acid, the citric acid formed is usually precipitated with lime for purification and recovery, and sold either as the calcium salt or, more frequently, converted to the acid with 0.2 per cent sulfuric acid. Either the monohydrate or anhydrous crystals are obtained depending upon whether the evaporation is carried out at 60°–70°C (140–158°F) or at 25°–30°C (77–86°F). Careful control of the anions and cations present during fermentation, the amount of enzyme formed, the pH of the sugar solution, and the rate at which nutrients are added are important factors in this type of process. If the pH falls below 3.5, oxalic and gluconic acids will be formed at the expense of citric acid. Some manufacturers operate their process to obtain deliberately small amounts of these acids in order to supply market demands for them. The calcium sulfate by-product recovered from liquor treatment is of a purity suitable for pharmaceutical use, while ergosterol is obtained by extracting the filtered mycelia. The citric acid yield from fermentation processes ranges from 50 to 70 per cent based on sugar conversion.

Citric acid is recovered from pineapple waste by first washing and peeling the fruit. It is then shredded and pressed to obtain the juice. The juice is then subjected to a spontaneous fermentation from natural yeast to destroy pectin, sugar, and proteinaceous materials, clarified, and the citric acid precipitated with lime to give "citrate of lime." The acid is recovered as previously described. With lemons, the peel is used for producing oil of lemon and the inner white portion for extraction of pectin. With pineapple canning wastes, some sucrose and bromelin are obtained as the by-products.[332]

Phosphoric Acid Phosphoric acid is the only inorganic acid extensively employed as a food acidulant. It accounts for about 25% of the weight of all the acids used in food, compared to 60% for citric acid, and 15% for the rest of the food acids.[231] The bulk of phosphoric acid is used in cola, root beer, sarsaparilla and similar flavored carbonated beverages.[333] Its acid salts, such as monocalcium phosphate, dicalcium phosphate, sodium aluminum phosphate, and sodium acid pyrophosphate are ingredients of baking powders and other leavening mixtures.[231,333,334] Phosphoric acid is the least costly of all the food-grade acidulants; it is also the strongest, giving the lowest attainable pH.[231]

Food-grade phosphoric acid is supplied as 75%, 80% and 85% aqueous solutions (Table 13) to meet Food Chemicals Codex specifications.[335]

TABLE 13
Physical Characteristics of Food Grade Phosphoric Acid[335,336,337]

Characteristics	75% Phosphoric Acid	80% Phosphoric Acid	85% Phosphoric Acid
Form	colorless liquid	colorless liquid	colorless liquid
Freezing Point	−21°C	−10°C	21°C
Specific Gravity, d_4^{20}	1.581	1.635	1.691
Weight/Gallon	13.2 lbs		14.1 lbs
Viscosity, 25°C (relative to H_2O)	21.5		43.5

The FDA lists phosphoric acid and a number of its salts as GRAS compounds.[2] The miscellaneous and/or general purpose group of this list includes:

1. The acid itself
2. Mono- and dibasic ammonium phosphates
3. Mono-, di- and tribasic calcium phosphates
4. Sodium acid pyrophosphate
5. Sodium aluminum phosphate
6. Mono-, di- and tribasic sodium phosphate
7. Sodium tripolyphosphate

The following are grouped as nutrients and/or dietary supplements:

1. Calcium glycerophosphate
2. Mono-, di- and tribasic calcium phosphate
3. Ferric phosphate
4. Ferric sodium pyrophosphate
5. Di- and tribasic magnesium phosphate

6. Manganese glycerophosphate
7. Mono-, di- and tribasic sodium phosphates

The following are listed as sequestrants:

1. Calcium hexametaphosphate
2. Monobasic calcium phosphate
3. Dipotassium phosphate
4. Disodium phosphate
5. Sodium acid phosphate
6. Sodium hexametaphosphate
7. Sodium metaphosphate
8. Mono-, di- and tribasic sodium phosphates
9. Sodium pyrophosphate
10. Tetrasodium pyrophosphate
11. Sodium tripolyphosphate

Monosodium phosphate derivatives of mono- and diglycerides of edible fats or oils, or edible fat-forming fatty acids are included under emulsifying agents.

Disodium phosphate is cited in the standard for ice cream as an optional ingredient where chocolate or cocoa is added for flavor.[234] The same applies with frozen custards.[338] Phosphoric acid has been suggested as an optional ingredient in both the proposed fruit-flavored noncarbonated beverage standard and in the nonalcoholic carbonated beverage standard.[198,220] Calcium, magnesium, potassium or sodium phosphates are included as buffering agents. Specifications for artificially sweetened fruit jellies permit the use of sodium phosphate, disodium phosphate and trisodium phosphate.[196] Phosphates are optional emulsifying agents in the pasteurized process cheese standard while phosphoric acid is listed as one of the permitted acidulants.[44] Monocalcium phosphate is a permitted acidulant in cereal flours, and calcium phosphate, sodium acid pyrophosphate or sodium aluminum phosphate are optional ingredients for self-rising flours.[339]

Phosphoric acid is used in cheeses and in brewing to adjust pH.[231] Besides its major use in soft drinks, it is employed to enrich and preserve fodder, as an ingredient of bread dough, as a yeast stimulant, to neutralize the caustic in peeling of fruit, to clarify and acidify collagen in the production of gelatin, in the purification of vegetable oils, and to a small extent in the manufacture of jams and jellies.[336,337] Trisodium phosphate is employed in the cleaning of food and dairy equipment. Phosphoric acid and sodium acid pyrophosphate are also used for this purpose.[337] The latter is also added to cake premixes and cake doughnuts in addition to baking powders.[334] Phosphates have been added to foods to inhibit caries, when they are used either alone or in conjunction with fluorides and calcium salts.[449] Both phosphates and the acid reduce molasses losses in refining sugar and are supplements to animal feeds.[450-453] Various sodium phosphates enhance the water-holding capacity of meat. Alkyl dipolyphosphates show bactericidal properties as food preservatives, while mixed esters, such as hexadecyl bis(monolactoyl glycerol) phosphate, exhibit excellent emulsifying properties in making mayonnaise, meringue, and frosting mixes.[454-456]

Phosphoric acid is produced by two types of processes.[340,341] The older and more economical involves treatment of phosphate rock with sulfuric acid. Except where the product is to be used for fertilizer, it is necessary to further purify the acid slurry to remove various impurities such as fluorides, calcium, iron, vanadium, aluminum, arsenic, and sulfates. During concentration with superheated steam, much of the fluorine, most of the calcium, and some of the iron, aluminum and sulfate impurities are removed as precipitates. The balance of the iron and vanadium are removed by potassium ferrocyanide treatment and the arsenic with hydrogen sulfide.[340]

The other type of process consists of reducing the phosphate rock to elementary phosphorus in an electric or blast furnace, burning phosphorus to phosphorus pentoxide with air, hydrating the oxide by taking it up in 75 to 85 per cent phosphoric acid, and purifying the product with hydrogen sulfide to remove arsenic. Oxidation and hydration are carried out as either one or two-step processes.[340,341]

Mono- and disodium phosphates are produced by treating orthophosphoric acid with soda ash. Tetrasodium pyrophosphate is produced by calcining anhydrous disodium orthophosphate.[340]

Flavor Modification

An important function of most foods (besides supplying nutritive value) is to impart pleasure during eating and to leave a feeling of satisfaction and well-being.[342,343] Diets lacking these attributes are totally inadequate, since they rapidly lead to a loss of appetite. This, in time, can even result in undernourishment. Flavor contributes significantly to the attractiveness of a food. One of the contributions of acidulants in this case is that they add a subtle life to the flavor and avoid a sense of any blandness.

Flavor is the net effect of several physiological reactions involving various combinations of taste, odor, mouth-feel, pain, sensibility to temperatures, and kinesthetic sensations attributed to the muscular effort of chewing.[344,345] The properties are so complex that investigators are usually forced to define flavor in terms of stimulation of taste and odor receptors, often totally ignoring the factors of feel, except when they have a dominant effect on the overall taste reaction.[346,347]

Anything which favorably affects any of the sensory organs can add greatly to the value and marketability of a food. This applies not only to taste but also to such properties of the food as its appearance, including color, its aroma and its mouth-feel. Their total psychological effect may be such as to influence the measurements of a single property, such as a particular taste. For example, consider the psychological effects on taste of being served purple mashed potatoes, even though they have been colored with a tasteless dye.[346] As already noted, acidulants and their salts are often used to preserve both the color and structures of foods besides modifying their flavor.

Odor, in addition to being part of flavor, is closely associated with the stimulation of appetite. Sitting down to a meal is usually preceded by an increased acuity of olfaction. During the meal, there is a decrease in the acuteness of smell and the generation of a feeling of satiety.[348,349] There is evidence that various food acids have a tendency to maintain the olfactory acuteness along with a desire for more food.[348] This would account for the commonly ascribed appetite-stimulating effects of bitters and of dry wine. It is the food acids in these beverages which probably impart this property, since alcohol is known to decrease the olfactory threshold and to give rise to a sensation of satisfaction.[350,351]

An important part in the use of acidulants is their proper and skillful addition along with other adjuncts to food to produce certain desirable physiological and psychological effects. The amounts and combinations used are very important.. The results of these additions cannot be determined by common physicochemical methods, but must be ascertained through the services of taste panels and in some instances by clinical tests. Since humans are employed in place of test equipment, various human reactions and conditions must be taken into account in order to obtain meaningful results. These include such factors as fatigue, prejudices, emotional state, amount of sleep, and degree of hunger of each of the individuals used in the testing.[352] Subjects employed for this purpose must also be free from excessive smoking habits, chronic alcoholism, tooth decay, head colds, hay fever and other allergies and diseases. Conditions under which measurements are made and the manner in which samples are presented can materially affect results.[353] Since individuals sometimes differ in their reactions to the same stimulus, the panel should be sufficiently large to permit statistical analysis of results.[354] Early studies with flavor panels did not recognize the importance of many of these factors.

The fate of any commercial food product continuously rests with the consuming public. Formal study of their various reactions is only a comparatively recent development, but one which is increasing in importance with the growth of competition in the food field.[255-358] Although consumer-acceptance tests have become almost essential to the marketing of new and improved products, they do not supply all the information needed for sensory testing.[359-361] This has led to the development of different types of test panels for specific purposes, where attempts are made to study certain properties or groups of properties in determining what effect they may have on the food product as a whole.

The number of distinct tastes is very large.[352] Most individuals with a sensitive sense of taste can detect four basic sensations. This had led many to believe that all tastes are a combination of sour, bitter, sweet, and salty stimuli.[352] On the basis of recent electrophysiological data, however, there seems to be very little evidence that most individual taste receptors can be placed in only one of these categories. Discrimination of the four basic tastes appears to depend upon different patterns of stimulation of the sense organs within the oral cavity.[352,362-370] The four-modality classification, however, still appears useful from a behavioral standpoint.[352]

Depending on concentration, pleasant or unpleasant sensations are obtained for most of the four fundamental tastes.[371] The effect of concentration differs with various stimuli. Contrast effects may also be obtained by placing different substances on each side of the tongue. Placing salt on one side of the tongue, for example, causes distilled water on the other side to taste sweet. Various stimuli also differ with respect to the relative taste intensity and time in which they produce a taste sensation.[352] Taste sensations increase rapidly after application of the stimulant to the tongue, then more slowly and finally show no further increase.[372] With salts and acids, response as shown by electrophysiological studies rises rapidly to a peak and then falls off to some steady-state level. This peak is absent with sugar and quinine. Food acids differ as to the shape of their apparent intensity curves. The taste of malic acid, for example, builds up more slowly than that of citric acid but persists much longer.[182,183,346] Smaller amounts of malic acid are required to produce the same total sensation of sour taste.[182,183] It has been suggested that the area under the intensity curve is in part a measure of the total acidic sensation.[373] Unfortunately, these observations are based largely on qualitative data, but have an important influence on the balance of the overall taste sensation. Citric acid, for example, is often used to impart a desirable flavor "burst" or rapid build-up in taste, while malic acid produces a more natural and rounded taste in several products. Interesting new effects can be expected from the greater use of blends of the different acidulants, since the rate with which the various taste sensations are detected, once a food is taken into the mouth, has a marked effect on a subject's total reaction to the food.[374,375] Unmasked retention of a single taste often leads to an undesirable aftertaste and can be overcome by a better balance of the individual taste sensations.[373]

Most acids stimulate a sour taste, which investigators have come to associate with the hydrogen ions produced in aqueous solutions.[352,376] Some acids, however, give rise to other tastes.[352] Amino acids, for example, are often sweet and picric acid is very bitter.[352] The apparent sour taste of carbonic acid may be entirely nongustatory in character.[352] Sour sensations apparently depend on a number of factors which include not only pH but the total titratable acidity, buffering effects of salts, and the presence of other stimulants such as salt and sugar. Some sour-tasting compounds have been observed to elicit a sense of pain.[377] The last is a factor which might be more thoroughly investigated.[352]

Taste thresholds for strong acids occur at a pH range of 3.4 to 3.5, while weak organic acids can be detected at dilutions giving a pH of 3.7 to 4.1.[352] Buffer mixtures can extend the tart threshold over a pH range of 5.6 to 7.1, indicating that not only the hydrogen ions, but also the anions and undissociated acids may have a role in stimulating the acidic taste receptors. Marked interactions occur when substances are present which stimulate different basic taste sensations, as is usual with most foods and beverages. Sodium chloride at subthreshold levels, for example, has been shown to moderately reduce the sourness of acetic, hydrochloric, and citric acids and markedly that of lactic, malic, and tartaric acids.[378] Hydrochloric acid does not affect the saltiness of sodium chloride while all other acids increase it. The sweetness of glucose is reduced by acetic and hydrochloric acids, but is unaffected by other acids. Sweetness of sucrose is increased by lactic, malic, citric, and tartaric acids but is unaffected by hydrochloric and acetic acids. Inversion was not a factor in these experiments. Sugars reduce both salty and sour tastes, sucrose having a greater effect than the other sugars upon the sourness of malic and tartaric acids. Opinions vary as to the cause of these effects.[352] Individuals differ radically in their physical and psychological ability to detect difference in acidic taste and in identifying acids.[379] Some tasters are nearly twice as efficient as others in their ability to rank sourness.[379] There seems to be little question about the pure sour taste of most commercial acids at sensitivity threshold concentrations. This is not true for the upper supersaturation taste levels, the so-called suprathreshold ranges.[379] Several different series of comparative acidic tastes have been reported.[86,362,380,381] In pure aqueous solution, the orders given are (1) tartaric > acetic > lactic > citric acid and (2) tartaric > lactic > acetic > citric acid.[379,380]

In another study, the order fumaric > malic > lactic > citric > adipic acid was found to hold for both aqueous solutions and carbonated water.[86] Comparative taste intensity ratings, however, for the different acids differ with concentration. Adding a series of the various acids to dry wine to yield a pH of 3.3 gave the order of malic > lactic > citric > tartaric acid and for the same titratable acidity of malic > tartaric > citric > lactic acid.[379] In a similar study using dry wine as a solvent, the results were tartaric > citric > fumaric > adipic acid.[381] Reported differences in acidic taste of individual acids have been described as follows: tartaric (hard), malic (green), citric (fresh), succinic (less intense but with salty and bitter notes) and lactic (sourish but tart).[382] It has not been thoroughly established, however, that even persons with a highly developed acidic taste can, with any certainty, identify the various acids by taste alone at equal taste concentrations. This is not the case where the acids are volatile or have other effects than sourness in modifying the flavor. Marked individual preferences for certain degrees of sourness and undetected variations in pain effects can influence test results, unless detected and eliminated by statistical methods.[379]

It is quite obvious that there is still much to be learned fundamentally regarding the role of acidulants in flavor modification. At the very least, flavor formulation is still so complex that questions can only be answered by empirical means. In many instances, several of the acidulants appear to give the same sourness at equivalent acidic taste levels. Slight changes in the flavoring agents present, however, can lead to marked differences in flavor effects. Formulators are very fortunate in having several available food acids of different combinations of properties from which to choose, in developing new and improved flavor effects and in developing new processing economies for both solid foods and beverages.

Food Preservation

Food acidulants serve several purposes in the preservation of foods, besides contributing to their flavor.[383-388,453-456] In canning of fruits and vegetables, for example, they act both as sterilizing aids and as antibrowning agents in helping to maintain the normal flavor, color and texture of several of these natural products.[383,386-388]

Bacteria and many types of other microorganisms are more easily killed in acid media.[385] Adding an acidulant to adjust the pH to a standard value often permits a shortening of sterilization time. In some cases, lower sterilization temperatures are possible. This is important since prolonged heating of fruits, vegetables, and their juices not only destroys their natural structure but may also reduce nutritive value.[317]

Contamination from vegetative types of microorganisms is usually accompanied by their spores.[387] Spores are known to be highly resistant to heat. During sterilization only a portion of the spore population may be killed by the heat, thus permitting growth of the organisms during storage. Most types of spores, however, are prevented from germinating by the presence of acid. It is in this role that acidulants contribute significantly to sterilization. Organic acids prevent spore germination at a much higher pH than do inorganic acids. This would indicate that the undissociated acid as well as the anions may partake in the killing of microorganisms at elevated temperatures.

Various fruits, including tomatoes, normally contain large amounts of organic acids but even here, additional amounts of acid may be required for proper sterilization during canning since drought, other weather conditions, or the variety leads to a lowered acid content.[387] Addition of food acids, however, is only an aid to proper food processing. It is *not* a means for overcoming poor sanitary or processing conditions.

A pH of 5 is suitable for the canning of most fruit and vegetable products.[386] *Clostridium botulinum*, however, has been known to grow at a pH of 4.9 on canned pimentos and there is a possibility it may even multiply at a pH of 4.7–4.8. Hence, a pH of 4.5 has been recommended as a safe level in canning this vegetable.[386] A powerful toxin is produced when *Clostridium botulinum* is permitted to grow in food, causing the feared poisoning known as botulism. This organism is a common spore-forming, anaerobic, soil saprophyte.[384] It is noninfectious and does not produce symptoms of poisoning in the absence of the toxin. The toxin is destroyed by heat. Botulism, however, is readily encountered when improperly processed canned goods or poorly cured meats are eaten.[384]

Federal regulations require that canned figs have a pH of 4.9.[241] Increasing the acidity to a pH of 4.3 with citric acid or lemon juice, in this case, does not have to be noted on the label as an additive unless greater amounts are incorporated. Acidulation is also mandatory in the canning of artichokes and is permitted for a number of fruits and vegetables as previously described. It is prohibited in the canning of mushrooms.[240]

Browning is one of the major problems in the dehydration of both fruits and vegetables. Remedies include the soaking of the cut or peeled products in either a solution of ascorbic acid and an acidulant or one containing common salt in place of the ascorbic acid.[386] Sliced apples are particularly susceptible in this respect when dried or frozen. Acidulants in these cases tend to inhibit the oxidation reactions which lead to the brown discoloration by displacing any redox equilibrium in the direction of reduction; acting as a synergist to ascorbic acid, which is often employed as the antioxidant; and by removing through chelation, any traces of metals that might serve as catalysts in the oxidative browning reactions.[317]

Several of the preservatives used in foods are incorporated as their salts. Examples of these are the sodium salts of benzoic, propionic and sorbic acids. Acidulants are also added, unless the product contains sufficient natural acids to produce the acids of these preservatives. In most cases, it is the acids themselves that are the active antimycotics.[383] Vinegar or citric acid is used in the curing of meat in a similar manner to yield nitric oxide which preserves the natural color.[317]

The nutritive value of certain foods is also enhanced by the addition of acidulants when they contain ascorbic acid as one of their components. Unless present in appreciable amounts, it can be preferentially oxidized to dehydro-ascorbic acid with rapid loss of vitamin activity. High acidity tends to reverse this reaction by maintaining a low redox potential, especially during the initial stages of oxidation. Thus, the addition of food acids tends to preserve, in part, some of the nutritive value of these products.[317]

Acidulants are also widely employed as synergists for antioxidants employed in the preservation of fats and oils and in food products containing fatty components. The synergistic effect may, in part, be due to a lowering of the redox potential but more so to sequestering traces of metals.[317,457-459]

Other Functions

Acidulants are also added to foods for several specific purposes besides imparting flavor or aiding in their preservation. They are used in products having gelled structures to produce either proper gelation or to modify the rate of set.[389-392] In formulating food products of this type, the amount of acid required to give the desired physical properties is the determining factor. Flavor effects have to be controlled by the proper selection of the acid. Where sucrose is used as the sweetening agent consideration must be given to the rates of inversion catalyzed by the various acidulants at different temperatures.

Acidulants can also markedly affect the rheological properties of dough when added specifically for this purpose or contained in the leavening agents as an acidulant for generating carbon dioxide.[212,460,461] The use of mixtures of various food acids in leavening is a relatively unexplored field which could yield some valuable results from the standpoint of improved baking products. Acidulants affect the softening point, melt properties and texture when added to foods such as cheese and margarines.[111]

Reaction rate of acidulants with carbonates is an important property in the formulation of beverage powders and other effervescent preparations. The rate at which the carbon dioxide is generated can influence the size of the bubbles, which in turn affects mouthfeel and the sour taste. Solid acidulants showing little or no tendency to take up moisture add to the shelf-life of a wide variety of powdered products, especially mixtures for baking and desserts.

Food acidulants are also used for specific purposes in acidic dairy cleaners for their bactericidal properties and their ability to attack milkstone through chelation.[393,394] Further, they have several special advantages as ingredients of preservatives incorporated in food wrappings of various types. Acidulants are often used in the control of certain properties peculiar to specific foods. Typical examples are the whipping properties of aerated foods, the peeling performance of frankfurters, the consistency of tomato products, and the settling properties of impurities in refining edible oils for salads and cooking.[462-465]

One outstanding characteristic of food acidulants as a whole is their great versatility. This may well account for their extensive use as food additives and their increased growth in the future.[395,466]

References

1. Gardner, W. H., 1966, Role of Acidulants in Food. *Food Acidulants,* Allied Chemical Corporation, New York, N.Y., Ch. 1.
2. Food and Drug Administration, Food Additives. *Code of Federal Regulations,* U.S. Government Printing Office, Washington, D.C., Title 21, §121.101.
3. Jacobs, M. B., 1947, Flavoring Materials: Acids, Alcohols and Hydrocarbons. *Synthetic Food Adjuncts,* D. Van Nostrand Company, Inc., New York, N.Y., p. 73.
4. Lee, F. A., 1951, Fruits and Nuts. *The Chemistry and Technology of Food and Food Products,* M. B. Jacobs, Ed., Vol. II, 2nd. ed., Interscience Publishers, Inc., New York, N.Y., p. 1552.
5. Faith, W. L., Keyes, D. B., and Clark, R. L., 1957, Acetic Acid. *Industrial Chemicals,* 2nd. ed., John Wiley & Sons, Inc., New York, N.Y., pp. 11–26.
6. Worrell, L., 1951, Flavors, Spices, Condiments and Essential Oils. *The Chemistry and Technology of Food and Food Products,* M. B. Jacobs, Ed., Vol. II, 2nd. ed., Interscience Publishers, Inc., New York, N.Y., pp. 1728–1738.
7. Sackrin, S. M., 1945, *Salad Dressings, Mayonnaise and Related Products.* U.S. Department of Commerce, Industrial Series 60, Washington, D.C.
8. Merory, J., 1960, Meat, Fish and Salads. *Food Flavorings: Composition, Manufacture and Use,* The Avi Publishing Company, Inc., Westport, Conn., pp. 232–233.
9. Desrosier, N. W., 1963, Principles of Food Preservation by Fermentation and Pickling. *The Technology of Food Preservation,* Rev. ed., The Avi Publishing Company, Inc., Westport, Conn., pp. 250–259.
10. U.S. Department of Agriculture, Consumer and Marketing Service, 1966, U.S. Standards of Grades of Pickles. *Code of Federal Regulations,* U.S. Government Printing Office, Washington, D.C., Title 7, §§52.1681–52.1684. *Federal Register* 31: 10235–10236 (July 29, 1966).
11. U.S. Department of Agriculture, Consumer and Marketing Service, 1965, U.S. Standards for Grades of Tomato Sauce. *The Almanac of the Canning, Freezing, Preserving Industries,* Edward E. Judge, Westminster, Md., p. 281.
12. Food and Drug Administration, Tomato Products. *Code of Federal Regulations,* U.S. Government Printing Office, Washington, D.C., Title 21, §53.10.
13. Urbain, W. M., 1951, Meat and Meat Products. *The Technology of Food and Food Products,* M. B. Jacobs, Ed., Vol. III, 2nd ed., Interscience Publishers, Inc., New York, N.Y., p. 2275.
14. Meat Inspection Division, 1966, Approval of Substances for Use in the Preparation of Meat Food Products. *Code of Federal Regulations,* U.S. Government Printing Office, Washington, D.C., Title 9, §318.7.
15. Gardner, W. H., 1966, Meat, Fish, Fats and Oils. *Food Acidulants,* Allied Chemical Corporation, New York, N.Y., p. 170.
16. Food and Drug Administration, Canned Fruits and Fruit Juices. *Code of Federal Regulations,* U.S. Government Printing Office, Washington, D.C., Title 21, §§27.3, 27.20, 27.25, 27.30, 27.45, 27.50.
17. Food and Drug Administration, Food Additives. *Ibid.* Title 21, §121.1164.
18. Agricultural Research Service, Addition of Smoke Flavorings in Meat Food Products. *Ibid.* Title 9, Parts 16 and 18, *Federal Register* 27: 10536–10537 (Oct. 30, 1962).
19. Jacobs, M. B., 1958, Vinegar. *The Chemical Analysis of Foods and Food Products,* 3rd ed., D. Van Nostrand Company, Inc., Princeton, N.J., pp. 614–616.
20. U.S. Department of Agriculture, 1928, Definitions and Standards for Foods. Food, Drug and Insecticide Administration, *Service and Regulatory Announcements 2 rev. 1,* p. 20.
21. U.S. Department of Agriculture, 1939, Definitions and Standards for Food. Food and Drug Administration. *Service and Regulatory Announcements FD&C 2.*
22. Balcom, M. S., 1909, Proceeding of the Association of Official Agricultural Chemists— Report on Vinegar. *U.S. Dept. Agr., Bu. Chem., Bull. 132* (1910).
23. Cox, H. E., and Pearson, D., 1962, Fermentation Products. *The Chemical Analysis of Foods,* Chemical Publishing Company, Inc., New York, N.Y., pp. 293–296.

24. Cathcart, W. H., 1951, Baking and Bakery Products. *The Chemistry and Technology of Food and Food Products*, M. B. Jacobs, Ed., Vol. II, 2nd ed., Interscience Publications, Inc., New York, N.Y., pp. 1195–1203.

25. Matz, S. A., 1960, Minor Ingredients. *Bakery Technology and Engineering*, The Avi Publishing Company, Inc., Westport, Conn., pp. 231–232.

26. Food and Drug Administration, Bakery Products—Definitions and Standards of Identity. *Code of Federal Regulations*, U.S. Government Printing Office, Washington, D.C., Title 21, §§ 17.1(a)(13)(iii), 17.3, 17.4.

27. Goos, A. W., 1953, Propionic Acid, *Encyclopedia of Chemical Technology*, R. E. Kirk and D. F. Othmer, Eds., Vol. II, 1st. ed., Interscience Encyclopedia, Inc., New York, N.Y., p. 174.

28. Chas. Pfizer & Co., Calcium and Sodium Propionate. *Pfizer Products for the Food Industry—Products for the Baking and Cereal Industries*, Chemical Sales Division, New York, N.Y., p. H-1.

29. Harshbarger, K. E., 1942, Report of a Study on the Toxicity of Several Food Preserving Agents, *J. Dairy Sci.*, 25:169.

30. Gardner, W. H., 1966, Food Acids in Living Organisms, *Food Acidulants*, Allied Chemical Corporation, New York, N.Y., p. 65.

31. Krebs, H. A., and Lowenstein, J. M., 1960, The Tricarboxylic Acid Cycle. *Metabolic Pathways*, Vol. 1, Academic Press, New York, N.Y., pp. 184–185.

32. Flavin, M., and Ochoa, S., 1957, Propionic Acid Metabolism in Animal Tissues, *J. Biol. Chem.*, 229:965.

33. O'Leary, D. K., and Kralovec, R. D., 1941, Development of B. Mesenterious in Bread and Control with Calcium Acid Phosphate and Calcium Propionate. *Cereal Chem.*, 18:730.

34. Chas. Pfizer & Co., Preservatives. *Pfizer Products for the Food Industry—Products for the Dairy Industry*, Chemical Sales Division, New York, N.Y., pp. H-1—H-4.

35. Jacobs, M. B., 1947, Chemical Preservatives and Stabilizers. *Synthetic Food Adjuncts*, D. Van Nostrand Company, Inc., New York, N.Y., pp. 229–232.

36. Union Carbide Chemicals Company, a Division of Union Carbide, 1961, Sorbic Acid, Potassium Sorbate. *"Sentry" Food Preservatives*, New York, N.Y.

37. Hoechst Farbwerke, A.G., vormals Meister, Lucius & Brüning, 1965, *Sorbic Acid, the Preservative Akin to Food.* Hoechst Chemical Department 1, Frankfurt, W. Germany. In English.

38. Union Carbide Corporation, 1963, *A Summary of the Properties and Uses for "Sentry" Sorbic Acid and Potassium Sorbate*, Chemical Division, New York, N.Y.

39. Smith, D. P., and Rollin, N., 1954, Sorbic Acid as a Fungistatic Agent for Foods, VII. Effectiveness of Sorbic Acid in Protecting Cheese. *Food Res.*, 19:59.

40. Wysso, O., 1948, Microbial Inhibition by Food Preservatives. *Advan. Food Res.*, 1:373.

41. Melnick, D., Luckman, F. H., and Gooding, C. M., 1954, Sorbic Acid as a Fungistatic Agent for Foods. VI. Metabolic Degradation of Sorbic Acid in Cheese by Molds and the Mechanism of Mold Inhibition. *Food Res.*, 19:44.

42. Emard, L. O., and Vaughn, R. H., 1952, Selectivity of Sorbic Acid Media for the Catalase Negative Lactic Acid Bacteria and Clostridia. *J. Bact.*, 63:(4), 487.

43. Becker, E., and Roeder, I., 1957, Sorbic Acid a Preservative for Margarine. *Fette, Seifen, Anstrichmittel*, 59:(5), 321.

44. Food and Drug Administration, Cheese, Processed Cheese, Cheese Foods, Cheese Spreads and Related Foods. *Code of Federal Regulations*, U.S. Government Printing Office, Washington, D.C. Title 21, Part 19.

45. Melnick, D., and Luckmann, F. H., 1954, Sorbic Acid as a Fungistatic Agent for Foods IV. Migration of Sorbic Acid from Wrappers into Cheese. *Food Res.*, 19:28.

46. Perry, G. A., and Lawrence, R. L., 1960, Preservative Effect of Sorbic Acid on Creamed Cottage Cheese. *J. Agr. Food Chem.*, 8:374.

47. Smith, D. P., and Rollin, N. J., 1954, Sorbic Acid as a Fungistatic Agent for Foods. VIII. Need and Efficacy in Protecting Packaged Cheese. *Food Technol.*, 8:(3), 133.

48. Bradley, R. L., Harmon, L. G., and Stine, C. M., 1962, Effect of Potassium Sorbate on Some Organisms Associated with Cottage Cheese Spoilage. *Food Technol.*, 25:318.

49. Melnick, D., Vahlteich, H. W., and Hackett, A., 1956, Sorbic Acid as a Fungistatic Agent for Foods. XI. Effectiveness in Protecting Cake. *Food Res.*, 21:(1), 133.

50. Melnick, D., Gooding, C. M., and Vahlteich, H. W., 1956, Sorbic Acid as a Fungistatic in Bakery Production with Emphasis on a Novel Fungistatic Shortening. *Baker's Dig.*, 30:(5), 46.

51. Bradshaw, W., 1958, Sorbic Acid, a Useful Fungistat for Cakes and Pies. *Baker's Dig.*, 32:(10), 58.

52. Pomeranz, Y., and Adler, L., 1958, Effectiveness of Sorbic Acid in Inhibition of Microbial Spoilage of Bread. *Bull. Res. Council Israel*, Sect. C. 6, p. 220.

53. Feldberg, C., 1957, Instant Bakery Specialties. *Food Process.* 21:(10), 50.

54. Costilow, R. N., Robbins, C. D., and Coughlin, F. M., 1957, Sorbic Acid as a Selective Agent in Cucumber Fermentations. II. Effect of Sorbic Acid on the Yeast and Lactic Acid Fermentations in Brined Cucumbers. *Appl. Microbiol.*, 5:(6), 373.

55. Costilow, R. N., Fersuson, W. E., and Ray, S., 1955, Sorbic Acid as a Selective Agent in Cucumber Fermentations. *Ibid.* 3:(6), 341.

56. Sheneman, J. M., and Costilow, R. N., 1955, Sorbic Acid as a Preservative for Sweet Cucumber Pickles. *Ibid.* 3:(3), 186.

57. Borg, A. F., Etchells, J. L., and Bell, T. A., 1955, The Influence of Sorbic Acid on Microbiological Activity in Commercial Cucumber Fermentations. *Bact. Proc.*, Abstract 19.

58. Costilow, R. N., 1957, Sorbic Acid as a Selective Agent in Cucumber Fermentations. III. Evaluation of Salt Stock from Sorbic Acid Treated Cucumber Fermentations. *Food Technol.*, 11:(1), 591.

59. Patrick, R., and Atkins, C. D., 1954, Effectiveness of Sorbic Acid as a Preservative for Tangerine Sherbet Base. *Proc. Florida State Hort.*, 67:194.

60. Salunkhe, D. K., 1955, Sorbic Acid as a Preservative for Apple Juice. *Food Technol.*, 9: 11, note, 590.

61. Ferguson, W. E., and Powrie, W. D., 1957, Studies on the Preservation of Fresh Apple Juice with Sorbic Acid. *Appl. Microbiol.*, 5:41.

62. Saller, W., and Kolewa, S. R., 1958, The Use of Sorbic Acid for the Preservation of Wine. *Rebe Wein*, Ser. A, 7:(1), 21 Mitt. (Klosterneuberg).

63. Vitagliano, M., 1958, Sorbic Acid in Wine Making. *Riv. viticott. enol.* (Conegliano), 11:15.

64. Auerbach, R. C., 1959, Sorbic Acid as a Preservative Agent in Wine. *Wines and Vines*, 40:26.

65. Heiligman, F., 1957, New Wax Cuts Losses. *Am. Vegetable Grower*, 5:(1), 28.

66. Francis, F. J., 1955, Prepeeled Carrots. *Pre-Pack-Age*, 9:(4), 8.

67. Beneke, E. S., and Fabian, F. W., 1955, Sorbic Acid as a Fungistatic Agent at Different pH Levels for Molds Isolated from Strawberries and Tomatoes. *Food Technol.*, 9:(10), 486.

68. Niven, Jr., F. C., and Chesbro, W. R., 1956, Complementary Action of Antibiotics and Irradiation in the Preservation of Fresh Meats. *Antibiot. Ann.* p. 855.

69. Boyd, J. W., and Tarr, H. L. A., 1955, Inhibition of Mold and Yeast Development in Fish Products. *Food Technol.*, 9:(8), 411.

70. Doherty, C. J., *Soft Moist Pet Foods.* Chemicals Division, Union Carbide Corporation, New York, N.Y.

71. Union Carbide Corporation, 1965, *Storage, Handling and Assay of "Sentry" Preservatives and Humectant in Pet-Food Applications.* Chemicals Division, Research and Development Department, Tarrytown, N.Y.

72. Field, J. A., 1951, Fatty Acids (Unsaturated). *Encyclopedia of Chemical Technology*, R. E. Kirk and D. F. Othmer, Eds., Vol. 6, 1st ed., Interscience Encyclopedia, Inc., New York, N.Y., p. 272.

73. Montagna, A. E., 1957, Sorbic Acid. *Ibid.* First Supplement, p. 848.

74. Anon., 1955, Synthesis Process Brightens Lactic Acid Picture. *Chem. Week*, 77:(7), 73–76.

75. Baumgarten, P., and Glatzel, G., 1925, Action of Alkalies on the Sodium Salt of Enolglutaconic Dialdehyde—Sorbic Aldehyde. *Ber.* 59:2633.

76. Montagna, A. E., and McQuillen, L. V., 1957, *Oxidation of Organic Aldehydes*. Brit. Patent 782,430.

77. Boese, Jr., A. B., 1949, *Unsaturated Organic Compounds*. U.S. Patent 2,484,067.

78. Allen, C. F. H., and Van Allan, J., 1955, Sorbic Acid, *Org. Syn.*, Coll. Vol. 3, John Wiley & Sons, New York, N.Y., p. 783.

79. Desrosier, N. W., 1963, Chemical Additives. *The Technology of Food Preservation*, The Avi Publishing Company, Inc., Westport, Conn., p. 295.

80. McMichael, C. E., and Bailey, A. E., 1951, Edible Fats and Oils. *The Chemistry and Technology of Food and Food Products*, M. B. Jacobs, Ed., Vol. II, 2nd ed., Interscience Publishers, Inc., New York, N.Y., pp. 1151–1152.

81. Jacobs, M. B., 1947, Flavoring Materials. *Synthetic Food Adjuncts*, D. Van Nostrand, New York, N.Y. pp. 73–75.

82. Food and Drug Administration, Synthetic Flavoring Substances and Adjuvants. *Code of Federal Regulations*, U.S. Government Printing Office, Washington, D.C., Title 21, §121.1183.

83. Merory, J., 1960, Imitation Flavors. *Food Flavoring: Composition, Manufacture and Use*, The Avi Publishing Company, Inc., Westport, Conn., pp. 163–192.

84. Gardner, W. H., 1966, *Food Acidulants*, Allied Chemical Corporation, New York, N.Y.

85. Allied Chemical Corporation, Choosing an Acidulant. *Tech. Bull.*, TS-23:9–11, New York, N.Y.

86. Monsanto Chemical Co., Wisconsin Alumni Research Foundation Research Results. *Monsanto Food Processor's Tech. Data FPD*-2, St. Louis, Mo., p. 3.

87. Kharasch, M. S., 1929, Heats of Combustion of Organic Compounds. *Bu. Std. J. Res.*, 2:359.

88. Bennett, R., and Ewart, J. A. D., 1962, Reactions of Acids with Dough Proteins, *J. Sci. Food Agr.*, 13:15.

89. Ward, T. L., Gros, A. T., and Feuge, R. O., 1959, New Fat Products: Glyceride Esters of Adipic Acid. *J. Am. Oil Chemists' Soc.*, 36:667.

90. Feuge, R. O., and Ward, T. L., 1960, Additional Physical Properties of Diglyceride Esters of Succinic and Adipic Acids. *Ibid.*, 37:291.

91. Food and Drug Administration, 1966, Bakery Products. *Federal Register* 31:5432 (Apr. 6.)

92. Gardner, W. H., and Flett, L. H., 1954, Succinic Acid and Succinic Anhydride. *Encyclopedia of Chemical Technology*, R. E. Kirk and D. F. Othmer, Eds., Vol. 13, 1st ed., The Interscience Encyclopedia, Inc., New York, N.Y., p. 187.

93. Flett, L. H., and Gardner, W. H., 1952, Succinic Acid. *Maleic Anhydride Derivatives— Reactions of the Double Bond*, John Wiley & Sons, Inc., New York, N.Y., pp. 56–57.

94. Gergel, M. G., and Revelise, M., 1952, Nitriles and Isocyanides. *Encyclopedia of Chemical Technology*, R. E. Kirk and D. F. Othmer, Eds., Vol. 9, 1st ed., Interscience Encyclopedia, Inc., New York, N.Y., p. 356.

95. Allied Chemical Corporation, 1961, National® Food Acids. National Aniline Division, *Tech. Bull.*, TS-10:8, New York, N.Y.

96. Pintauro, N. D., and Hall, B. J., 1962, *Gelatin Desserts Containing Adipic Acid*. U.S. Patent 3,067,036.

97. Anon., 1964, Citric Challenged as Top Flavor. *Chem. Week*, 94:(11), 82.

98. Gardner, W. H., 1966, Desserts. *Food Acidulants*, Allied Chemical Corporation, New York, N.Y., pp. 136, 141.

99. Gardner, W. H., 1966, Beverage Powders. *Ibid.*, p. 117.

100. Allied Chemical Corporation, 1966, Jellies and Preserves. *Tech. Bull.* TS-26:21–23.

101. Block, H. W., and Touher, P. B., 1951, *Increasing the Solution Rate of Adipic Acid in Cold Water*. U.S. Patent 2,982,653.

102. Raffensperger, S. P., and Takashima, T. T., 1962, *Adipic Acid Composition*. U.S. Patent 3,016,300.

103. Raffensperger, S. P., and Takashima, T. T., 1961, *Adipic Acid Composition for Fruit-Flavored Beverages*. U.S. Patent 3,009,811.

104. Gibsen, F. K., 1959, *Algin-Gel Composition*. U.S. Patent 2,918,375.

105. Rocks, J. K., 1960, *Cold Water Desserts*. U.S. Patent 2,925,343.

106. Campbell, A. D., 1962, *Irish Moss Food Product*. U.S. Patent 3,031,308.

107. Common, J. L., and Campbell, H. A., 1955. *Starch Composition*. U.S. Patent 2,698,803.

108. Handleman, A. R., 1960, Consistency Modification of Dairy Products. *Food Technol.*, 14:(12), 662.

109. Gardner, W. H., 1966, Food Acids in Living Organisms. *Food Acidulants*, Allied Chemical Corporation, New York, N.Y., pp. 78–79.

110. Horn, H. J., Holland, E. G., and Hazelton, L. W., 1957, Safety of Adipic Acid as Compared with Citric and Tartaric Acids. *Agr. Food Chem.*, 5:(10), 759–761.

111. Monsanto Chemical Company, *Monsanto Food-Grade Adipic Acid.* St. Louis, Mo.

112. Gardner, W. H., 1966, Jellies and Preserves. *Food Acidulants*, Allied Chemical Corporation, New York, N.Y., p. 129.

113. Ebihara, K., 1954, *Synthetic Aroma for Soy Sauce.* Japan. Patent 5250'54.

114. Bickel, W., 1956, *Preservation of Meat, Sausage Products and Intestines.* U.S. Patent 2,735,776.

115. Faith, W. L., Keyes, D. B., and Clark, R. L., 1957, Adipic Acid. *Industrial Chemicals*, 2nd ed., John Wiley & Sons, New York, N.Y., pp. 47–52.

116. Standish, W. L., and Abrams, S. V., 1963, Adipic Acid. *Kirk-Othmer Encyclopedia of Chemical Technology*, A. Standen, Ed., 2nd ed., Interscience Publishers, a Division of John Wiley & Sons, Inc., New York, N.Y., pp. 411–412.

117. Raffensperger, S. P., and Takashima, T. T., 1961, *Fumaric Acid Composition for Fruit-Flavored Beverages.* U.S. Patent 3,009,810.

118. Raffensperger, S. P., and Takashima, T. T., 1962, *Fumaric Acid Composition.* U.S. Patent 3,016,299.

119. Block, H. W., and Touher, P. B., 1961, *Increasing the Solubility of Fumaric Acid in Cold Water.* U.S. Patent 3,011,894.

120. Stokes, W. E., and Kennedy, M. H., 1946, *Gelatin Food Product.* U.S. Patent 2,412,305.

121. Erekson, A. B., and Duncan, R. E., 1960, *Refrigerated Biscuit Dough.* U.S. Patent 2,942,988.

122. Feuge, R. O., and Ward, T. L., 1958, 1,3-Diolein and 1,3-Distearin Esters of Fumaric, Succinic and Adipic Acids. *J. Am. Chem. Soc.*, 80:6338.

123. Bertram, S. H., 1951, *Emulsifying Agent*, U.S. Patent 2,552,706.

124. Greenbank, G. R., 1933, *Inhibiting Oxidation of Unsaturated Fats, Oils and Fatty Acids and Substances Containing Fatty Material Having a Tendency to Become Rancid.* U.S. Patent 1,898,363.

125. Hall, L. A., 1950, *Synergistic Antioxidant.* U.S. Patent 2,511,802.

126. Hall, L. A., 1950, *Antioxidant Flakes.* U.S. Patent 2,511,803.

127. Hall, L. A., 1950, *Antioxidant Salt.* U.S. Patent 2,511,804.

128. Hall, L. A., 1952, *Synergistic Antioxidants Containing Amino Acids.* U.S. Patent 2,518,233.

129. Heimann, W., 1948, The Action of Fumaric and Maleic Acids as Fat Antioxidants. *Z. Lebensm. Untersuch.-Forsch.*, 88:586.

130. Frandsen, L., 1952, *Bacteriostatic and Fungicidal Solution for Retarding Food Spoilage.* U.S. Patent 2,622,032.

131. Conrad, L. J., and Stiles, 1954, *Improving the Whipping Properties of Gelatin and Gelatin Products.* U.S. Patent 2,692,201.

132. Abbott, Jr., J. A., Crawford, D. B., and Kelley, K. M., 1957, *Method of Producing Angel Food Batter and Cake.* U.S. Patent 2,781,268.

133. Food and Drug Administration, Food Additives Permitted for Human Consumption. *Code of Federal Regulations*, U.S. Government Printing Office, Washington, D.C., Title 21, §121.1130.

134. Van Ness, J. H., 1964, *Free-Flowing Fumaric and Adipic Acid Compositions.* U.S. Patent 3,151,986.

135. Monsanto Company, CWS® Fumaric Acid. Organic Chemicals Division, *Tech. Bull.*, O/FI-1.

136. Food and Drug Administration, Food Additives Permitted for Human Consumption. *Code of Federal Regulations*, U.S. Government Printing Office, Washington, D.C., Title 21, §121.1137.

137. McColloch, R. J., and Gentile, B., 1958, *Stabilization of Citrus Concentrates.* U.S. Patent 2,845,355.

138. Gardner, W. H., and Flett, L. H., 1952, Maleic Acid, Fumaric Acid and Maleic Anhydride. *Encyclopedia of Chemical Technology*, R. E. Kirk and D. F. Othmer, Eds., Vol. 8, 1st ed., The Interscience Encyclopedia Inc., New York, N.Y., pp. 691–692.

139. Flett, L. H., and Gardner, W. H., 1952, Fumaric Acid. *Maleic Anhydride Derivatives*, John Wiley & Sons, Inc., New York, N.Y., pp. 250–251.

140. Monsanto Company, Water White Food Grade Lactic Acid. Organic Chemical Division, *Tech. Bull.*, O/FI-2.

141. Troupe, R. A., Aspy, W. L., and Schrodt, P. R., 1951, Viscosity and Density of Aqueous Lactic Acid Solutions. *Ind. Eng. Chem.*, 43:1143.

142. Filachione, E. M., and Fisher, C. H., 1944, Lactic Acid Condensation Polymers. *Ibid.*, 36:223.

143. Montgomery, R., 1952, Acidic Constituents of Lactic Acid-Water Systems. *J. Am. Chem. Soc.*, 74:1466.

144. Iveson, H. T., Radlove, S. B., and Julian, P. L., 1954, *Edible Shortening Agent*. U.S. Patent 2,690,971.

145. Julian, P. L., Iveson, H. T., and Radlove, S. B., 1956, *Hydroxylation of Vegetable Oils and Products Therefrom*. U.S. Patent 2,752,376.

146. Schulman, G., 1958, *Liquid Shortening*. U.S. Patent 2,864,703, U.S. Patent 2,864,705.

147. Radlove, S. B., Iveson, H. T., and Julian, P. L., 1960, *Oil-Soluble Emulsifying Agents*. U.S. Patent 2,957,32.

148. Chang, S. S., DeVore, F. L., and Friedman, M. A., 1960, *Shortening Emulsifiers for Use in Icings*. U.S. Patent 2,966,410.

149. Houser, C. J., 1961, *Liquid Shortening*. U.S. Patent 2,968,562, U.S. Patent 2,968,563.

150. Gleason, J. J., 1996, *Triglyceride Shortening Composition*. U.S. Patent 2,970,055.

151. Julian, P. L., Iveson, H. T., Radlove, S. B., Slutkin, R., and Davis, P. F., 1961, *Shortening Composition and Emulsifier System Therefor*. U.S. Patent 3,004,853.

152. Shapiro, S. H., 1961, *Method of Making Mixed Glycerol Esters of Fatty and Lactic Acids*. U.S. Patent 3,012,048.

153. Babayan, V. K., and Comes, G. N., 1962, *Glyceryl Lacto Esters in Coatings*. U.S. Patent 3,051,577.

154. Kuhrt, N. H., and Swicklik, L. J., 1962, *Bread Additive*. U.S. Patent 3,068,103.

155. Food and Drug Administration, Bakery Products. *Code of Federal Regulations*, U.S. Government Printing Office, Washington, D.C., Title 21, §§17.1(a)(11), 17.1(a)(13)(v), 17.3, 17.4.

156. Epstein, A. K., and Shaffer, B. M., 1953, *Treatment of Liquid Egg Albumin*. U.S. Patent 2,630,387.

157. Fabian, F. W., and Wadsworth, C. K., 1939, Experimental Work on Lactic Acid in Preserving Pickles and Pickle Products. I. Rate of Penetration of Acetic Acid and Lactic Acid in Pickles. *Food Res.*, 4:499.

158. Fabian, F. W., and Wadsworth, C. K., 1939, Experimental Work on Lactic Acid in Preserving Pickles and Pickle Products. II. Preserving Value of Acetic Acid and Lactic Acids in the Presence of Sucrose. *Ibid.*, 4:511.

159. Wadsworth, C. K., and Fabian, F. W., 1930, How to Use Lactic Acid in Finishing Pickles and Pickle Products. *Food Ind.*, 11:252, 324.

160. Olthof, E., 1959–1960, Bitter Gherkins. *Conserval (The Hague)*, 8:215; *Chem. Absts.*, 55:841h.

161. Roundy, Z. D., Hills, C., and Ormond, N. R. H., 1960, *Cheese Products and Method for the Manufacture Thereof*. U.S. Patent 2,956,885.

162. Stuart, G. H., Howard, H., Watson, J. T., Clickner, F. H., and Sommer, W. A., 1954, *Food Product and Method of Making*. U.S. Patent 2,682,469.

163. Kennedy, J. G., and Bernhart, F. W., 1953, *Casein-Lactic Acid Composition*. U.S. Patent 2,639,235.

164. Foster, Jr., H. G., Crest, H., and Cornwell, E. H., 1962, *Manufacture of Cheese Curd*. Can. Patent 635,192.

165. Kleismeier, E. W., 1958, *Manufacture of Cheese Curd*. U.S. Patent 2,851,363.

166. Hammond, E. G., and Deane, D. D., 1961, *Cheese Making Process*. U.S. Patent 2,982,654.

167. Nowotny, F., Rzedowski, W., and Kroppowa, M., 1954, Possibility of Substituting Lactic Acid for Citric Acid in the Production of Wine. *Prezemysl Rolnyi Spozywczy*, 8:122–124. English Summary; *Chem. Absts.*, 49:2017f.

168. Sivetz, M., 1949, Acids Play Important Roles in Flavor. *Food Inds.*, 21:1384.

169. Olthof, E., 1961, Lactic Acid in Fruit Drinks. *Soft Drinks Trade J.*, 15:254–255.

170. Olthof, E., 1961, The Use of Lactic Acid in The Preparation of Syrups and Carbonated Fruit Juices. *Conserve Deriv. Agrumari (Palermo)*, 10:223; *Chem. Absts.*, 57:7692e.

171. Olthof, E., 1961, Use of Lactic Acid in The Preparation of Syrups and Carbonated Beverages From Fruits. *Rev. Agroquim. Technol. Alimentos*, 1:31; *Chem. Absts.*, 56:13312a.

172. Iszard, M. S., 1927, The Value of Lactic Acid in the Preservation of Mayonnaise and Other Dressings. *Canning Age*, 8:(4), 434.

173. Schumm, F. R., and Johnstone, C., 1954, *Liquid Pepper Composition and Process for Producing Same*. U.S. Patent 2,680,690.

174. Johnston, C. D., and Thomas, McC. J., 1961, *Composition and Method for Inhibiting Discoloration of Cut Organic Materials*. U.S. Patent 2,987,401.

175. Hansen, F. F., 1951, *Milk Powder and Its Preparation*. U.S. Patent 2,553,578.

176. Hansen, F. F., 1951, *Composition for the Baking of Bread*. U.S. Patent 2,557,283.

177. Hansen, F. F., 1951, *Condensed Milk Composition and Its Preparation*. U.S. Patent 2,570,321.

178. Hansen, F. F., 1952, *Shortening Composition for and Method of Baking Yeast-Raised Baked Goods*. U.S. Patent 2,602,748.

179. Faith, W. L., Keyes, D. B., and Clark, R. L., 1957, Lactic Acid. *Industrial Chemicals*, 2nd ed., John Wiley & Sons, Inc., New York, N.Y., pp. 479–483.

180. Filachione, E. M., 1953, Lactic Acid. *Encyclopedia of Chemical Technology*, R. E. Kirk, and D. F. Othmer, Eds., Vol. 8, 1st ed. Interscience Encyclopedia, Inc., New York, N.Y., pp. 172–173.

181. Anon., 1964, Synthesis Process Brightens Lactic Acid Picture. *Chem. Eng.*, 71:(2), 82, 84.

182. Allied Chemical Corporation, Use of POMALUS™ Food-Grade Malic Acid in Carbonated Beverages, National Aniline Division, *Tech. Bull.*, TS-17:6, New York, N.Y.

183. Allied Chemical Corporations, Use of POMALUS™ Food-Grade Malic Acid in Food Products, National Aniline Division, *Tech. Bull.*, TS-18:7, New York, N.Y.

184. Gardner, W. H., 1966, Sensory Effects. *Food Acidulants*. Allied Chemical Corporation, New York, N.Y. p. 40–45.

185. Sausville, T. J., 1964, The Use of POMALUS™ Acid (Malic Acid-Food Grade) in Jellies and Preserves. Paper presented at Quality Control Session of the National Preservers Association Meeting, Chicago, Ill., p. 9, (Oct. 5).

186. Gardner, W. H., 1966, Jellies and Preserves. *Food Acidulants*. Allied Chemical Corporation, New York, N.Y., p. 128.

187. Gardner, W. H., 1966, Candy. *Ibid.*, p. 150.

188. Sherman, M. B., 1965, Malic Acid in Modern Candy Production. *Mfg. Confectioner* 65:(4), 40.

189. Sausville, T. J., 1965, A New Look at the Acidulant Malic Acid. *Food Technol.* 19:(3), 67–68.

190. Anon., 1965, Malic Acid Now Offers Economy. *Food Process* 26:(1), 101–104.

191. Sausville, T. J. and Carr, D. R., 1965, The Use of Malic Acid in Carbonated Beverages. *Proceedings 12th Ann. Meeting Soc. Soft Drink Technologists*, pp. 89–95.

192. Gardner, W. H., 1966, Sensory Effects. *Food Acidulants*, Allied Chemical Corporation, New York, N.Y., p. 44.

193. Gardner, W. H., 1966, Food Acids in Living Organisms. *Ibid.*, Ch. 5.

194. Gardner, W. H., 1966, Government Regulations. *Ibid.*, pp. 85–86.

195. Food and Drug Administration, Frozen Desserts. *Code of Federal Regulations*, U.S. Government Printing Office, Washington, D.C., Title 21, §§20.4, 20.5.

196. Food and Drug Administration, Fruit Butter, Fruit Jellies, Fruit Preserves and Related Products. *Ibid.*, Part 29.

197. Food and Drug Administration, 1964, Fruit Flavored Beverages. *Federal Register* 29:11627.

198. Food and Drug Administration, 1963, 1966, Nonalcoholic Carbonated Beverages. *Ibid.*, 28:9988; 3:1033.
199. Food and Drug Administration, 1964, Diluted Fruit Juice Beverages. *Ibid.*, 29:11621.
200. Food and Drug Administration, 1966, Canned Tomatoes. *Ibid.*, 31:1154; *Code of Federal Regulations*, U.S. Government Printing Office, Washington, D.C., Title 21, Part 53.
201. Internal Revenue Service, 1964, Production of Volatile Flavor Concentrates—Wine. *Ibid.*, 29:15023; *Code of Federal Regulations*, U.S. Government Printing Office, Washington, D.C., Title 26, Part 240.
202. Sherman, M. B., 1965, Malic Acid in Modern Candy Production. *Mfg. Confectioner*, 65: (4), 39.
203. Ponting, J. D., 1960, *Fruit Juices*. U.S. Patent 2,928,744.
204. Johnston, G. L., 1922, *Grape-Juice Product*. U.S. Patent 1,427,903.
205. Sivetz, M., 1949, Acids Play Important Role in Flavor. *Food Indus.*, 21:1384.
206. Doell, T. W., and Maes, L. A. F., 1921, *Pectin Production*. U.S. Patent 1,385,525.
207. Huber, H., 1960, Developing Sour Dough with Basic Sour, Without Full Sour. *Brot u. Gebäck*, 14:(4), 61.
208. Asnina, F. L., 1959, New Antioxidants and Emulsifiers in Nutritive Fats. *Trudy Ukrain. Nauch.-Issledovatel. Inst. Pishch. Prom.*, 2:113; *Chem. Absts.*, 56:2744e.
209. Pray, L. W., and Powers, J. J., 1966, Acidification of Canned Tomatoes. *Food Technol.*, 20:(1), 87.
210. Harris, R., 1959, *Egg Products with Improved Whipping Properties*. U.S. Patent 2,902,372.
211. N. V. Koninklijke Stearine Kaareenfabrieken "Gouda-Apollo," 1964, *Acid Esters of Malic Acid*. Brit. Patent 952,849.
212. Gardner, W. H., 1956, Baked Goods. *Food Acidulants*, Allied Chemical Corporations. New York, N.Y., Ch. 14.
213. Gardner, W. H., 1966, Fruits, Vegetables and Allied Products. *Ibid.*, pp. 165–166.
214. Gardner, W. H., 1966, Meat, Fish and Oils. *Ibid.*, p. 173.
215. Gardner, W. H., 1966, Noncarbonated Beverages (Chocolate and Cocoa). *Ibid.*, p. 112.
216. Sadler, F. S., 1954, *Refining Fatty Oils*. U.S. Patent 2,666,074.
217. Irwin, W. E., Lockwood, L. B., and Zienty, M. F., 1967, Malic Acid. *Kirk-Othmer Encyclopedia of Chemical Technology*, A. Standen, Ed., Vol. 12, 2nd ed., Interscience Publishers, a Division of John Wiley & Sons, New York, N.Y., pp. 837,849.
218. Chas. Pfizer & Co., Tartaric Acid. *Pfizer Products for the Food Industry—Products for the Beverage Industry*, Chemical Sales Division, New York, N.Y., p. B-1.
219. Jacobs, M. B., 1947, Flavoring Materials; Acids, Alcohols and Hydrocarbons. *Synthetic Food Adjuncts*, D. Van Nostrand Co., Inc., New York, N.Y., p. 78.
220. Food and Drug Administration, Nonalcoholic Beverages—Soda Water. *Code of Federal Regulations*, U.S. Government Printing Office, Washington, D.C., Title 21, §31.1.
221. Gardner, W. H., 1966, Food Acids in Living Organisms. *Food Acidulants*, Allied Chemical Corporation, New York, N.Y., p. 79.
222. Anon., 1962, Tasty Bonus for Fumaric. *Chem. Week*, 90:(24), 125.
223. Chas. Pfizer & Co., 1955, *Pfizer and Food*. Chemical Sales Division, New York, N.Y., p. 5.
224. Chas. Pfizer & Co. Tartaric Acid. *Pfizer Products for the Food Industry—Products for the Baking and Cereal Industries*, Chemical Sales Division, New York, N.Y., pp. C-1—C-2.
225. Henkel, & Cie, G.m.b.H., 1955, *Autoxidation Inhibitors*. Ger. Patent 936,646.
226. Griffith, C. L., and Sair, L., 1956, *Fat-soluble Synergistic Antioxidants for Food*. U.S. Patent 2,768,084.
227. Buchholz, K., 1957, *Stabilization of Ground, Dry Spices*. Ger. Patent 1,000,226.
228. Jon A. Benckiser Chemische Fabrik G.m.b.H., 1956, *Prevention of the Discoloration of Cheese*. Ger. Patent 938,581.
229. Borden Company, 1962, *Stabilizers for Fats and Milk Products*. Brit. Patent 886,519.
230. Pasternack, R., 1951, Tartaric Acid. *Encyclopedia of Chemical Technology*, R. E. Kirk and D. F. Othmer, Eds., Vol. 13, 1st ed., Interscience Encyclopedia, Inc., New York, N.Y., pp. 649–652.
231. Sanders, H. J., 1966, Food Additives, Part II. *Chem. Eng. News*, 44:(43), 115–116.

232. Gardner, W. H., 1966, Food Acids in Living Organisms. *Food Acidulants*, Allied Chemical Corporation, New York, N.Y., pp. 51–61.

233. Stone, G. B., 1949, Citric Acid. *Encyclopedia of Chemical Technology*, R. E. Kirk and D. F. Othmer, Eds., Vol. 4, 1st ed., Interscience Encyclopedia, Inc., New York, N.Y., pp. 8–23.

234. Food and Drug Administration, Ice Cream. *Code of Federal Regulations*, U.S. Government Printing Office, Washington, D.C., Title 21, §20.1.

235. Food and Drug Administration, Evaporated Milk. *Ibid.*, Title 21, §18.1520.

236. Food and Drug Administration, Dressings for Food. *Ibid.*, Title 21, Part 25.

237. Food and Drug Administration, Oleomargarine, Margarine. *Ibid.*, Title 21, Part 45.

238. Agricultural Research Service, 1964, Meat Inspection Regulations—Use of Citric Acid or Sodium Citrate. *Ibid.*, Title 9, Part 18.

239. Food and Drug Administration, Canned Vegetables. *Ibid.*, Title 21, Part 51.

240. Food and Drug Administration, Canned Vegetables Other Than Those Specifically Regulated. *Ibid.*, Title 21, §§51.990(a)(1), 51.990(a)(3)(i).

241. Food and Drug Administration, Canned Fruits and Juices. *Ibid.*, Title 21, §§27.60, 27.70, 27.73.

242. Neal, R. H., Gooding, C. M., and Vahlteich, H. W., 1949, *Prevention of the Development of Off-Flavors in Fried Potatoes.* U.S. Patent 2,485,635.

243. Gooding, C. M., Neal, R. H., and Vahlteich, H. W.. 1949, *Stabilization of Cheese.* U.S. Patent 2,485,637.

244. Chas. Pfizer & Co., Citric Acid. *Pfizer Products for the Food Industry—Products for the Dairy Industry*, Chemical Sales Division, New York, N.Y., pp. E-1—E-2.

245. Elliker, P. R., 1954, Fine Points of Sanitation That Up Cottage Cheese Quality. *Food Eng.*, 26:79.

246. Schafer, J. G., 1958, Cottage Cheese Sanitation Practices. *Am. Milk Rev.*, 20:101.

247. Angevine, N. C., 1959, Recent Development on Cottage Cheese. *J. Dairy Sci.*, 42:2005.

248. Templeton, H. L., and Sommer, H. H., Cheese Spreads. *J. Dairy Sci.*, 15:155.

249. Babel, F. J., 1954, A Method for Improving Creamed Cottage Cheese. *Am. Milk Rev.*, 24:34, 42.

250. Nelson, J. A., and Brence, J. L., 1953, Improving the Flavor of Cottage Cheese by the Addition of Citric Acid to Milk. Mont. Agr. Exp. Sta., *Tech. Bull.* No. 488.

251. Ostra Chemie G.m.b.H., 1956, *Cheese.* Ger. Patent 943,632.

252. Stull, J. W., 1958, Flavor Deterioration in Cottage Cheese. *Am. Milk Rev. Milk Plant Monthly*, 20:(6), 61.

253. Jacobs, M. B., 1951, Cheese. *The Chemistry and Technology of Food and Food Products*, Vol. 2, 2nd ed., Interscience Publishers, Inc., New York, N.Y., pp. 870–890.

254. Mair-Waldburg, H., 1957, Use of Condensed Phosphates in Milk Products. *Kondensierte Phosphate in Lebensmitteln*, Symposium. Mainz, Germany, pp. 104–121.

255. Templeton, H. L., 1937, The Use of Citric Acid and Sodium Citrate in Milk and Milk Products. Wisc. Agr. Exp. Sta., *Res. Bull.* No. 133.

256. Gonce, J. E., and Templeton, H. L., 1930, Citric Acid Milk in Infant Feeding. *Am. J. Dis. Child.*, 39:265.

257. Frandsen, J. H., and Arbuckle, W. S., 1961, *Ice Cream and Related Products.* The Avi Publishing Co., Inc., Westport, Conn., p. 204.

258. Lineweaver, H., Anderson, J. D., and Hanson, H. L., 1952, Effect of Antioxidant on Rancidity Development in Frozen Creamed Turkey. *Food Technol.*, 6:1–4.

259. Vahlteich, H. W., Neal, R. H., and Gooding, C. M., 1949, *Prevention of 'Weeping' of Margarine and Butter.* U.S. Patent 2,485,634.

260. Masek, J., 1956, Changes in Butter Stored in Packs Impregnated with Antioxidants. *Obaly*, 2:75–99; *Chem. Absts.*, 54:377ef.

261. Murthy, G. K., Masurovski, E. B., and Campbell, J. E., 1961, Method for Removing Cationic Radionuclides for Milk. *J. Dairy Sci.*, 44:2158.

262. Chas. Pfizer & Co., 1962, Citric Acid. *Pfizer Products for the Food Industry—Products for the Beverage Industry*, Chemical Sales Division, New York, N.Y., pp. A-1—A-2.

263. Chas. Pfizer & Co., 1962, Sodium Citrate. *Ibid.*, p. A-3.

264. Miles Chemical Co., Food Applications. *Citric Acid U.S.P.*, Elkhart, Ind., pp. 10–17.

265. Jacobs, M. B., 1959, Citric Acid. *Manufacture and Analysis of Carbonated Beverages*, Chemical Publishing Co., New York, N.Y., p. 72.

266. Nebesky, E. A., Esselen, Jr., W. B., McConnell, J. E. W., and Fellers, C. R., 1949, Stability of Color in Fruit Juices. *Food Res.*, 14:261–274.

267. Berg, H. W., Filipello, F., Hinreiner, H., and Webb, A. D., 1955, Evaluation of Thresholds and Minimum Difference Concentrations for Various Constituents of Wines. I. Water Solutions of Pure Substances. *Food Technol.*, 9:23–26.

268. Berg, H. W., Filipello, F., Hinreiner, H., and Webb, A. D., 1955, Evaluation of Thresholds and Minimum Difference Concentrations for Various Constituents of Wines. II. Sweetness: The Effect of Ethyl Alcohol, Organic Acid and Tannin. *Ibid.*, pp. 138–140.

269. Hinreiner, H., Filipello, F., Webb, A. D. and Berg, H. W., 1955, Evaluation of Thresholds and Minimum Difference Concentrations for Various Constituents of Wines. III. Ethyl Alcohol, Glycerol and Acidity in Aqueous Solution. *Ibid.*, pp. 351–353.

270. Berg, H. W., and Akiyoshi, M., 1956, Some Factors Involved in Browning of White Wines, *Am. J. Enology.*, 7:1–7.

271. Luthi, H., 1957, The Malolactic Retrogression in Wines and Ciders. *Rev. Ferment. Ind. Aliment.*, 12:15–21.

272. Joslyn, M. A., and Amerine, M. A., 1964, *Dessert, Appetizer and Related Flavored Wines—The Technology of Their Production*. Division of Agricultural Sciences, University of California, Berkeley, Calif., pp. 136–140, 158–159, 360, 413.

273. Ough, C. S., 1963, Sensory Examination of Four Organic Acids Added to Wine. *J. Food Sci.*, 28:(1), 101–106.

274. Amerine, M. A., Roessler, E. G. and Ough, C. S., 1965, Acids and the Acid Taste. I. The Effect of pH and Titratable Acidity. *J. Enol. Viticult.*, 16:(1), 29–37.

275. Hohl, L. A., 1949, Further Studies on Frozen Fruit and Vegetable Purees. *Food Technol.*, 3:100–110.

276. Powers, M. J., Talburt, W. F., Jackson, R., and Lazar, M. E., 1958, Dehydrocanned Apples. *Food Technol.*, 12:417–419.

277. Archer, R. W., 1962, Firming Processed Apples with Calcium. *Canner Packer*, 131:(9), 28.

278. J. A. Benckiser G.m.b.H. (P. Michels), 1962, *Prevention of Discoloration of Fruits and Vegetables*. Ger. Patent 1,140,804.

279. Stuckey, B. N., 1954, Antioxidants in Candy and Candy Packaging Materials. *Mfg. Confectioners*, 34:(6), 47.

280. Leon, S. I., 1959, *An Encyclopedia of Candy and Ice Cream Making*. Chemical Publishing Co., Inc., New York, N.Y.

281. Hall, L. A., 1961, *Antioxidant Composition*. U.S. Patent 2,981,628.

282. Mattil, K. F., and Sims, R. J., 1956, *Stabilization of Edible Fats and Oils*. U.S. Patent 2,759,829.

283. Hall, L. A., 1956, *Antioxidant Composition*. U.S. Patent 2,772,170.

284. Emanuel, N. M., Knorre, D. G., Lyaskovskaya, Y. N., and Piul'skaya, V. I., 1958, Antioxidants to Improve Stability of Animal Fat. *Myasn. Ind. S.S.S.R*, 29:(2), 52–55.

285. Marino, J. P., 1957, Use of Antioxidants in Hog Lard of National (Argentine) Production. *Rev. Fac. Cienc. Quim., Univ. Nacl. La Plata*, 30:39–45.

286. Cecil, S. R., and Woodroof, J. G., 1951, BHA Ups Shelf Life of Salted Nuts. *Food Ind.* 23:(2), 81–84, 223–224.

287. Dugan, Jr., L. R., Kraybill, H. R., Ireland, L., and Vibrans, F. C., 1950, Butylated Hydroxyanisole as an Antioxidant for Fats and Food Made with Fat. *Food Technol.* 4:457–460.

288. Dutton, H. J., Schwab, A. W., Moser, H. A., and Cowan, J. C., 1948, The Flavor Problem of Soybean Oil IV. Structure of Compounds Counteracting the Effect of Pro-oxidants Metals. *J. Am. Oil Chem. Soc.*, 25:385–388.

289. Dutton, H. J., Schwab, A. W., Moser, H. A., and Cowan, J. C., 1949, The Flavor Problem of Soybean Oil V. Considerations in the Use of Metal Scavengers in Commercial Operations. *Ibid.*, 26:441–444.

290. Hartman, L. and White, M. D. L., 1953, Stability of Edible Tallow from Mutton and Beef. II. Suppression of the Deleterious Effect of Trace Metals. New Zealand. *J. Sci. Technol. Ser. B.*, 35:(3), 254–258.

291. Hanley, J. W., Everson, C. W., Ashworth, L. Q., and Morse, R. E., 1953, Antioxidant Treatment of Bacon. *Food Technol.* 7:(11), 429–431.

292. Kraybill, H. R., and Dugan, Jr., L. R., 1954, New Developments for Food Use—Acid Synergists as Metal Scavengers. *J. Agr. Food Chem.*, 2:(2), 81–84.

293. Morris, S. G., Meyers, J. S., Kip, M. L., and Riemanschneider, R. W., 1950, Metal Deactivation in Lard. *J. Am. Oil Chem. Soc.*, 27:105–107.

294. Neal, R. H., Gooding, C. M., and Vahlteich, H. W., 1949, *Retardation of the Development of Rancidity in Soybean Oil*. U.S. Patent 2,485,631.

295. Gooding, C. M., Vahlteich, H. W., and Neal, R. H., 1949, *Retardation of the Development of Rancidity in Soybean Oil*. U.S. Patent 2,485,633.

296. Neal, R. H., Vahlteich, H. W., and Gooding, C. M., 1949, *Stabilization of Nut Meats*. U.S. Patent 2,485,636.

297. Vahlteich, H. W., Gooding, C. M., and Neal, E., *Retardation of the Development of Rancidity in Fats and Oils*. U.S. Patent 2,485,640.

298. Lindsey, F. A., and Maxwell, W. T., 1949, *Stabilizing Edible Fatty Oil*. U.S. Patent 2,486,424.

299. Black, H. C., 1950, *Stabilization of Fatty Materials*. U.S. Patent 2,494,114.

300. Lindsey, F. A., and Maxwell, W. T., 1950, *Oxidative Stabilization of Edible Fatty Oil with Citric Acid*. U.S. Patent 2,513,948.

301. Gribbins, M. F., and Dittmar, H. R., 1951, *Food Antioxidants*. U.S. Patent 2,564,106.

302. Hall, L. A., 1957, *Fatty Monoglycerides and Antioxidants for Edible Fats Therefrom*. U.S. Patent 2,813,032.

303. Zanzucchi, A., and Delindati, G., 1958, The Stabilization of Lard with Antioxidants. *Ind. Conserve (Parma)*, 33:11–15.

304. Nippon Fats & Oils Co., 1959, *Water-in-oil-type Shortening*. Japan. 8886'59.

305. Duin, H. J., and Schaap, J. A., 1959, *Edible Oil Emulsion*. U.S. Patent 2,919,197.

306. Cochran, W. M., Ott, M. L., and Lantz, C. W., 1958, *Rearranged Basestock Triglyceride Products*. U.S. Patent 2,859,119.

307. Arenson, S. W., 1950, Shortenings for Frying and Baking. *Food Ind.*, 22:(6), 1015–1020.

308. Chas. Pfizer & Co., 1965, Citric Acid. *Pfizer Products for the Food Industry—Products for the Meat Industry*, Chemical Sales Division, New York, N.Y., pp. B-2—B-3.

309. Brown, W. D., Venolia, A. W., Tappel, A. L., Olcott, H. S., and Stanby, M. E., 1957, Oxidative Deterioration in Fish and Fishery Products. II. Progress on Studies Concerning the Mechanism of Oxidation of Oil in Fish Tissues. *Com. Fisheries Rev.* 19:(5a), 27–31.

310. Stoloff, L. F., and Crowther, R. E., 1948, Curb Mackerel Fillet Rancidity. *Food Ind.*, 20:1130.

311. Dyers, W. J., Sigurdson, G. J., and Wood, A. J., 1944, A Rapid Test for the Detection of Spoilage in Sea Fish. *Food Res.*, 9:183.

312. Dugan, L. R., and Wilder, O. H. M., 1955, Stabilization During Rendering. *Am. Meat Inst. Found., Circ.*, No. 18.

313. Ganucheau, J. J., 1948, *Preserving Canned Crustacean Meat*. U.S. Patent 2,448,970.

314. Gardner, W. H., 1966, Meat, Fish, Fats and Oils. *Food Acidulants*, Allied Chemical Corporation, New York, N.Y., Ch. 17.

315. Thompson, M. H., and Water, M. E., 1960, Control of Iron Sulfide Discoloration in Canned Shrimp (Xiphopeneus, sp.) Part I. *Com. Fisheries Rev.*, 22:1.

316. Bailey, M. E., and Fieger, E. A., 1957, Chemical Prevention of Black Spot (Melanogenesis) in Iced Stored Shrimp. *Food Technol.*, 8:317–319.

317. Gardner, W. H., 1966, Inhibiting Undesirable Chemical Reactions. *Food Acidulants*, Allied Chemical Corporation, New York, N.Y., Ch. 3.

318. C. H. Bochringer Sohn (K. Haupmann), 1952, *Enzyme Preparation for Clearing Fruit Juices*. Ger. Patent 857,785.

319. Grindel, F., 1958, *Paste-type Food*. U.S. Patent 2,819,969.

320. Schmidt, C., 1958, *Dry Dough-acidifying Agent*. Ger. Patent 1,046,535.

321. Kirby, G. W., 1952, *Flavoring Material for Baked Farinaceous Goods*. U.S. Patent 2,609,298.

322. Wolf, G., 1961, Preparation of Crusty Bakery Products. *Nahrung* 5:663–679.

323. White, Jr., J. W., 1959, *Inhibiting Crystallization in Honey*. U.S. Patent 2,902,370.

324. Peat, M. R., 1963, *Stabilizing Spices*. U.S. Patent 3,095,306.

325. Yanick, N. S., 1958, *Liquid Pepper Solution*. U.S. Patent 2,860,054.

326. Endo, N., 1963, *Powdered Onion*. Japan. Patent 2760'63.

327. Jones, E. P., and Lancaster, E. B., 1956, *Edible Spreads from Vegetable Oils*. U.S. Patent 2,754,213.

328. Corn Products Co. (D. Melnick & C. M. Gooding) 1963, *Margarine*. Brit. Patent 927,534.

329. Kimball, M. H., Harrel, C. G., and Brown, R. O., 1949, *Dry Salad Dressing*. U.S. Patent 2,471,434.

330. Kimball, M. H., Harrel, C. G., and Brown, R. O., 1949, *Ingredients for Preparing Food Mixes*. U.S. Patent 2,471,435.

331. Kuroda, C., 1961, *Powdered Preparation to be Used for Mayonnaise*. Japan. Patent 2889'61.

332. Lockwood, L. B., and Irwin, W. E., 1964, Citric Acid. *Kirk-Othmer Encyclopedia of Chemical Technology*, A. Standen, Ed., Vol. 5, and ed., Interscience Publishers, a Division of John Wiley & Sons, Inc., New York, N.Y., pp. 528–531.

333. Jacobs, M. B., 1959, Phosphoric Acid. *Manufacture and Analysis of Carbonated Beverages*, Chemical Publishing Co., New York, N.Y., pp. 82–85.

334. FMC Corporation, *Sodium Acid Phosphate*. Phos-file No. 7, New York, N. Y.

335. National Academy of Sciences and National Research Council, 1966, Phosphoric Acid. *Food Chemicals Codex*, 1st ed., Publication 1406, Washington, D.C., pp. 518–519.

336. Virginia-Carolina Chemical Co., A Division of Socony Mobil Oil Co., 1961–1964, Phosphoric Acid—75% Food Grade, —80% Food Grade and 85% N.F. *V-C Inorganic Chemicals* 2-33, 2-31; 2-23; 2-13, 2-11, Richmond Va.

337. FMC Corporation, *Phosphoric Acid*. Phos-file No. 13, New York, N.Y.

338. Food and Drug Administration, Frozen Custard, French Custard Ice Cream; Identity; Label Statement of Optional Ingredients. *Code of Federal Regulations*, U.S. Government Printing Office, Washington, D.C., Title 21, §20.2.

339. Food and Drug Administration, Cereal Flours and Related Products. *Ibid.*, Title 21, Part 15.

340. Van Wazer, J. R., 1953, Phosphoric Acids and Phosphates. *Encyclopedia of Chemical Technology*, R. E. Kirk and D. F. Othmer, Eds., Vol. 10, 1st ed., Interscience Encyclopedia, Inc., New York, N.Y., pp. 424–434.

341. Faith, W. L., Keyes, D. B., and Clark, R. L., Phosphoric Acid (Orthophosphoric Acid). *Industrial Chemicals*, 2nd ed., John Wiley & Sons, New York, N.Y., pp. 592–601.

342. Jacobs, M. B., 1947, Taste, Odor and Flavor. *Synthetic Food Adjuncts*, D. Van Nostrand Co., Inc., New York, N.Y., p. 2.

343. Gardner, W. H., 1966, Sensory Effects. *Food Acidulants*, Allied Chemical Corporation, New York, N.Y., p. 38.

344. Brožek, J., 1957, Nutrition and Behavior. *Nutrition Symposium Ser. 14*, National Vitamin Foundation, Inc., New York, N.Y., pp. 1–124; *Am. J. Clin. Nutr.*, 5:332–343.

345. Jacobs, M. B., 1951, Appearance, Odor, Taste and Flavor. *The Chemistry and Technology of Food and Food Products*, Vol. 1, 2nd ed., Interscience Publishers, Inc., New York, N.Y., pp. 15–16.

346. Beidler, L. M., 1966, Chemical Excitation of Taste and Odor Receptors. *Flavor Chemistry, Advances in Chemistry Ser. 26*, American Chemical Society, Washington, D.C., pp. 1–28.

347. Caul, J. F., Cairncross, S. E., and Sjöström, L. B., 1958, The Flavor Profile in Review. *Perfumery Essent. Oil Record*, 49:130.

348. Goetzl, F. R., and Stone, F., 1947, Diurnal Variations in Acuity of Olfaction and Food Intake. *Gastroenterology*, 9:444.

349. Goetzl, F. R., Abel, M. S., and Ahokas, A. J., 1950, On the Occurrence in Normal Individuals of Diurnal Variations in Olfactory Acuity. *J. Appl. Physiol.*, 2:553.

350. Irwin, D. L., Durra, A., and Goetzl, F. R., 1951, Influence of Tannic, Tartaric, and of Acetic Acid Upon Olfactory Acuity and Sensations Associated With Food Intake. *Am. J. Digest. Diseases*, 20:17.

351. Irwin, D. L., Ahokas, A. J., and Goetzl, F. R., 1950, The Influence of Ethyl Alcohol in Low Concentrations Upon Olfactory Acuity and the Sensation Complex of Appetite and Satiety. *Permanente Found. Med. Bull.*, 8:97.

352. Amerine, M. A., Pangborn, R. M., and Roessler, E. B., 1965, The Sense of Taste. *Principles of Sensory Evaluation of Food*, Academic Press, New York, N.Y., Ch. 2.

353. Jones, F. N., 1958, Prerequisites for Test Environment. *Flavor Research and Food Acceptance*, Reinhold Publishing Corporation, New York, N.Y., pp. 107–111.

354. Wadsworth, G. P., 1958, Application of Statistics in Experimental Testing. *Ibid.*, pp. 112–118.

355. Amerine, M. A., Pangborn, R. M., and Roessler, E. B., 1965, Consumer Studies. *Principles of Sensory Evaluation of Food*, Academic Press, New York, N.Y., Ch. 9.

356. Caul, J. F., 1958, Pilot Consumer Product Testing. *Flavor Research and Food Acceptance*, Reinhold Publishing Corporation, New York, N.Y., pp. 150–161.

357. Carroll, M. B., 1958, Consumer Product Testing Statistics. *Ibid.*, pp. 162–174.

358. Pettersen, E. A., 1958, Market Testing and Analysis. *Ibid.*, pp. 175–187.

359. Peryam, D. R., 1958, Sensory Difference Tests. *Food Technol.*, 12:(5), 231.

360. Amerine, M. A., Pangborn, R. M., and Roessler, E. B., 1965, Laboratory Studies: Types and Principles: Difference and Directional Difference Tests; Quantity-Quality Evaluation. *Principles of Sensory Evaluation of Food*, Academic Press, New York, N.Y., Chs. 6–8.

361. Gardner, W. H., 1966. Flavor Determination. *Food Acidulants*. Allied Chemical Corporation, New York, N.Y., Ch. 7.

362. Pfaffmann, C., 1959, The Sense of Taste. *Handbook of Physiology*, Vol. 1, American Physiological Society, Washington, D.C., pp. 507–534.

363. Pfaffmann, C., 1954, Sensory Mechanisms in Taste Discrimination. *Abstr. Papers Am. Chem. Soc.*, 126th Meeting, Div. Agr. Food Chem., New York, N.Y., p. 17A.

364. Pfaffmann, C., 1955, Gustatory Nerve Impulses in Rat, Cat and Rabbit. *J. Neurophysiol.* 18:429–440.

365. Kimura, K., and Beidler, L. M., 1956, Microelectrode Study of Taste Bud of the Rat. *Am. J. Physiol.*, 187:610–611.

366. Kimura, K. and Beidler, L. M., 1961, Microelectrode Study of Taste Receptors of Rat and Hamster. *J. Cellular Comp. Physiol.*, 57:131–139.

367. Pfaffmann, C., 1964, Taste, Its Sensory and Motivating Properties. *Am. Scientist*, 52: 187–206.

368. Adrian, E. D., 1953, Flavour Assessment. *Chem. Ind.* (London), 48:1274–1276.

369. Wenger, M. A., Jones, F. N., and Jones, M. H., 1956, *Physiological Psychology*. Holt, New York, N.Y., pp. 133–142.

370. Erickson, R. P., 1958, Responsiveness of Single Second Order Neurons in the Rat to Tongue Stimulation. Ph.D. Thesis Brown Univ., Providence, R.I.; *Dissertation Absts.*, 19:1835 (1959).

371. Engel, R., 1928, Study of the Dependence of Pleasure and Repulsion upon the Sense of Taste. *Arch. Ges. Psychol.*, 64:1–36.

372. Bujas, K. and Ostojcic, A., 1939, Gustatory Response as a Function of Time from Excitiation. *Acta. Inst. Psychol. Univ. Zagreb.* 3:(1), 1–24.

373. Gardner, W. H., 1966, Sensory Effects. *Food Acidulants*, Allied Chemical Corporation, New York, N.Y., pp. 41–42.

374. Gardner, W. H., 1966, Candy. *Ibid.*, pp. 150–151.

375. Neilson, A. J., 1957, Time-Intensity Studies. *Drug Cosmetic Ind.*, 80:(4), 452.

376. Jacobs, M. B., 1947, Taste, Odor and Flavor. *Synthetic Food Adjuncts*, D. Van Nostrand Co., New York, N.Y., pp. 62–68.

377. Skramlik, E. von, and Schwarz, G., 1959, Taste Sensations of Various Solutions in the Oral Cavity. *Z. Biol.*, 111:99–127.

378. Fabian, F. W., and Blum, H. B., 1943, Relative Taste Potency of Some Basic Food Constituents and their Competitive and Compensatory Action. *Food Res.*, 8:179–193.

379. Amerine, M. A., Roessler, E. B., and Ough, C. S., 1965, Acids and the Acid Taste. I. The Effect of pH and Titratable Acidity. *Am. J. Enol. Viticult.*, 16:(1), 29–37.

380. Pangborn, R. M., 1963, Relative Taste Intensities of Selected Sugars and Acids. *J. Food Sci.*, 28:726–733.

381. Ough, C. S., 1963, Sensory Examination of Four Organic Acids Added to Wine. *J. Food Sci.*, 18:(11), 555–559.

382. Ribéreau-Gayon, J. and Peynaud, E., 1961, *Treatise on Enology. II. Composition, Transformations and Treatment of Wines.* Paris Librarie Polytechnic Ch. Béranger, p. 83.

383. Desrosier, N. W., 1963, *The Technology of Food Preservation.* The Avi Publishing Co., Inc., Westport, Conn., pp. 174, 243, 293–297, 304.

384. Halvorson, H. O., 1951, Food Spoilage and Food Poisoning. *The Chemistry and Technology of Food and Food Products*, M. B. Jacobs, Ed., Vol. 1, 2nd ed., Interscience Publishers, Inc., New York, N.Y., Ch. 11.

385. Williams, O. B., 1951, Preservation of Food in Hermetically Sealed Containers. *Ibid.* Vol. 3, 2nd ed., p. 1874.

386. Gardner, W. H., 1966, Fruits and Vegetables and Allied Products. *Food Acidulants*, Allied Chemical Corporation, New York, N.Y., Ch. 15.

387. Gardner, W. H., 1966, Role of Acidulants in Foods. *Ibid.*, p. 4.

388. Allied Chemical Corporation, 1965, Pomalus® Food-Grade Malic in the Canning Industry. National Aniline Division *Tech. Data* D-44, New York, N.Y.

389. Sunkist Growers, 1964, *Preservers Handbook.* Product Sales Division, Ontario, California, pp. 71–81, 105–136.

390. Gardner, W. H., 1966, Jellies and Preserves. *Food Acidulants*, Allied Chemical Corporation, New York, N.Y., Ch. 11.

391. Gardner, W. H., 1966, Desserts. *Ibid.*, pp. 135–142.

392. Gardner, W. H., 1966, Baked Goods. *Ibid.*, pp. 157–159.

393. Corash, P. 1951, Milk and Milk Products. *The Chemistry and Technology of Food and Food Products*, M. B. Jacobs, Ed., Vol. 3, 2nd ed., Interscience Publishers, Inc., New York, N.Y., p. 2235.

394. Parker, M. E., Acid Detergents in Food Sanitation. *Ind. Eng. Chem.* 35:(1), 100–105.

395. Sanders, H. J., 1966, Food Additives, Part II. *Chem. Eng. News*, 44:(43), 115, 118, 119.

396. Garibaldi, J. A., 1968, Acetic Acid as a Means of Lowering the Heat Resistance of *Salmonella* in Yolk Products. *Food Technol.*, 22:(8), 1031-1033.

397. Levine, A. S., and Fellers, C. R., 1940, Action of Acetic Acid on Food Spoilage Organisms. *J. Bacteriol.*, 39:499.

398. Palmer, H. H., Ijichi, K., Cimino, S. L., and Roff, H., 1969, Salted Egg Yolks. 2. Viscosity and Performance of Acidified, Pasteurized and Frozen Samples. *Food Technol.*, 23:(11), 1486-1488.

399. Japikse, C. H., 1969, *Emulsified Salad Dressings Containing Oleaginous Gels.* U.S. Patent 3,425,843.

400. Shinshin Food Industrial Co., Ltd., 1967, *Powdered Seasonings.* Japanese Patent 67-21,742; *Chem. Abstr.*, 68: 28637r.

401. Flesch, P., and Bauer, G., 1966, *Novel Additives for Sausages.* French Patent 1,446,678.

402. Luce, G. T., 1967, Acetylated Monoglycerides as Coatings for Selected Foods. *Food Technol.*, 21: (11), 1462-1463, 1466, 1468.

403. Mishima, S., and Hotta, K., 1970, *Canned Vegetables.* Japanese Patent 70-22,502; *Chem. Abstr.*, 74: 63336n.

404. Seiler, D. A. L., 1964, Application of Fungicides in Baking of Bread and Cakes. *Intern. Symp., Food Microbiol.*, 4th, Goteborg, Sweden, 211-220. In English.

405. Rigler, F., and Sojer, B., 1966, Prevention of Blowing of Canned Frankfurters by Means of Propionic Acid. *Technol. Mesa.*, 7: (12), 352-353; *Chem. Abstr.*, 66: 64466b.

406. Lishmund, R. E. J., 1969, Sorbic Acid. *Food Process. Ind.*, 38: (458), 51-53.

407. Lueck, E., 1969, Use of Sorbic Acid in Food Preservation. *Food Process. Ind.*, 38: (459), 53-54.

408. Lueck, E., 1967-69, Sorbic Acid as a Food Preservative. *Ind. Obst Gemueseverwert.*, 52: (7), 199-206; 53: (9), 241-249; 54: (12), 337-344.

409. Katz, M. H., 1968, *Depectized Cocoa and Its Use in Frosting Compositions.* U.S. Patent 3,397,061.

410. Nasedkina, E. A., and Teplitskaya, A. M., 1967, Preservation of Salmon Caviar. *Ryb. Khoz.*, 43: (2), 51-53; *Chem. Abstr.*, 67: 20730z.

411. Farbwerke Hoechst A.-G., 1968, *Anhydrides of Sorbic Acid and Fatty Acids as Preservatives for Bread and Pastry.* French Patent 1,509, 385.

412. Saito, S., and Kano, F., 1964, Biochemical Studies on Fruits and Vegetables After Harvest. I. Variation of Content of Free Amino Acids, Nonvolatile Acids and Ascorbic Acid during Short-Period Storage of Vegetables. *Tokyo Nogyo Daigaku Shuho* 10: (2), 32–37; *Chem. Abstr.,* 66: 75,092e.

413. N. V. Lijm-en Gelatinefabriek "Delft", 1970, *Dry Gelatine Dessert.* Netherlands Patent Application 68,09670; *Chem. Abstr.,* 72: 99389e.

414. Onishi, I., Nishi, A., and Kakizawa, T., 1969, *Food Flavorings.* U.S. Patent 3,478,015.

415. Martin, J. B., 1968, *Acidic Lipid Anhydrides.* U.S. Patent 3,371,102.

416. Zhylina, V. S., 1966, Effect of Pretreatment of Seeds with Growth-Promoting Substances on the Metabolism and Yield of Sugar Beets. *Vestsi Akad. Navuk Belarus, S.S.R., Ser. Sel'ska-gaspad. Navuk,* 3, 24–27; *Chem. Abstr.,* 66: 14812j.

417. Maurina, H., Ezerniece, L., and Gauja, B., 1965, Effect of Succinic Acid on Corn Physiology. *Uch. Zap., Latv. Gos. Univ.,* 71: 99–118; *Chem. Abstr.,* 66: 64581k.

418. Manohar, M. S., and Mathur, M. K., 1966, Effect of Succinic Acid Treatment on the Performance of Pearl Millet. *Advan. Frontiers Plant Sci.,* 17: 143–147 (in English); *Chem. Abstr.,* 66: 27886t.

419. Artemova, E. K., 1968, Presowing Stimulation of Cucumber Seeds by Succinic Acid at Various Temperatures. *Sb. Nauch. Rab. Aspir. Voronezh. Gos. Univ.,* No. 4, 209–213; *Chem. Abstr.,* 74: 30925a.

420. Grinberg, I. P., 1968, Effect of Biogenic Stimulants on Chemical Composition and Other Indexes of Corn Kernel Quality. *Dokl. Nauch. Konf. Molodykh Uch., Kishinev. Sel'skokhoz. Inst., Agron. Fak.,* 1968: 16–18; *Chem. Abstr.,* 72:89087r.

421. Mitchell, W. A., and Seidel, W. C., 1969, *Cold-Water-Soluble Adipic or Fumaric Acid Dry-Beverage Acidulants.* U.S. Patent 3,480,444.

422. Miller, A., and Rocks, J. K., 1969, *Algin Gel Compositions.* U.S. Patent 3,455,701.

423. Meusel, J. A., and Brunn, R. A., 1968, *Fumaric Acid as a Color-Improving Agent for Meat.* U.S. Patent 3,413,126.

424. Anon., 1968, Food Additives. Sodium Stearyl Fumarate. *Federal Register,* 33:816–817 (June 18).

425. Kichline, T. P., and Conn, J. F., 1971, *Emulsifying Additives for Bakery Products.* U.S. Patent 3,582,354.

426. Allied Chemical Corporation, 1966, *Preparation of Compositions Containing Fumaric Acid.* Netherlands Patent Application 6,516,003; *Chem. Abstr.,* 66: 10084e.

427. Reitman, P. D., and Hamilton, E. K., 1968, *Fumaric Acid Composition with Improved Cold-Water Solubility.* U.S. Patent 3,370,956.

428. Ruiz Hernandez, M., 1969, Combined Use of Lactic and Sorbic Acids in the Production of Sangria (A Wine-Based Beverage). *Sem. Vitivinic.,* 24: (1.202), 2.879, 2.881; *Chem. Abstr.,* 72: 88897t.

429. Higuchi, R., 1970, *Solid, Edible Lactic Acid Composition.* Japanese Patent 70–36,160, *Chem. Abstr.,* 74:52310y.

430. Lempka, A., Prominski, W., Strasburger, K., and Sulkowska, J., 1966, Quantitative Changes of L-Ascorbic Acid in Cauliflowers Dehydrated by Lyophilization. *Pr. Zakresu Towarozn. Chem., Wyzsza Szk. Ekon.* Poznanin Zesz. Nauk., Ser. I, No. 26: 7–21; *Chem. Abstr.,* 66: 45578y.

431. Jensen, N., and Vrang, C., 1971, Rheologic Properties of Wheat Doughs with Added Diacetylated Tartaric Acid Ester. *Brot. Gebaeck,* 25: (2), 36–40.

432. Petelinsek, A. and Ponikvar, M., 1970, Effect of Curing and Salting of Meat with Various Additives before Freezing on the Quality of Pork Sausage. *Technol. Mesa,* 12: (2), 46–49, *Chem. Abstr.,* 75: 4122n.

433. Kastornykh, M. S., Fastovskii, I. S., Khomutov, B. I., and Khramov, V. V., 1969, *Additives for Dehydrated Meats.* U.S.S.R. Patent 250,032; *Chem. Abstr.,* 72: 53886m.

434. Vaessen-Schoemaker Holding N. V., 1966, *Additives for Protein Foods.* Netherlands Patent Application 6,501,845; *Chem. Abstr.,* 66: 1744h.

435. Arles, P., 1966, *Meat Preservation.* French Patent 1,461,204.

436. Ueno, R. Miyazaki, T., and Inamine, N., 1970, *Manufacture of Oil-Coated Organic Acids.* Japanese Patent 70–32,217; *Chem. Abstr.,* 75:19005a.

437. Pillsbury Co., 1966, *Instant Beverage Mixtures.* Netherlands Patent Application 6,508,320; *Chem. Abstr.,* 67: 2285d.

438. Koyama, E., and Goto, H., 1970, *Drying Fruit Juice*. Japanese Patent 70-20,933; *Chem. Abstr.*, 74: 75385r.

439. Ballerini, J., 1966, *Preserved Fruits*. French Patent 1,443,016.

440. Winterberg, A., 1970, *Liquid Coating Composition for Preserving Perishable Foods*. French Patent 1,595,925.

441. Nandakumaran, M., Chaudhuri, D. R., and Pillai, U.K., 1970, Blackening of Canned Prawn and Its Prevention. *Fish. Technol. (India)*, 7: (2), 120-128; *Chem. Abstr.*, 74: 75212h.

442. Pirati, D., and Guidi, G., 1970, Effects of Sterilization Temperature and Citric Acid on the Preservation of Natural Color in Canned *Venus gallina* (Clam). *Ind. Conserve*, 55: (3), 215−219.

443. Kuwsi, T., and Kytokangas, R., 1971, Distribution of Citrate in Baltic Herring Fillets Kept in Saline Solution. *Lebensm.-Wiss. Technol.*, 4: (1), 1-6 (in English).

444. Voss, J., and Muenker, W., 1966, Application of Antioxidants and Bactericides in Cold Storage of Herring Fillets. *Fischereiforschung*, 4: (4), 73-84.

445. Viktorova, G. K., Malysheva, G. A., Ivanova, T. P., Vannakh, A. E., Arkhangel'skii, P. M., Kudryavtsev, P. N., Korchinskaya, R. A., Borodulin, D. T., and Yakovets, I. F., 1968, *Caramel Products with Porous Structure*. U.S.S.R. Patent 244,111; *Chem. Abstr.*, 71: 79947m.

446. Fukinbara, I., and Toyama, N., 1968, Production of Food from Seaweed with the Use of Cellulase. *Nippon Shokukin Kogyo Gakkaishi*, 15: (2), 79-82; *Chem. Abstr.*, 69:75731x.

447. Li, K. H., Bundus, R. H., and Noznick, P. P., 1967, *Prevention of Pink Color in White Onions*. U.S. Patent 3,352,691.

448. McKillip, R. H., and Henderson, R., 1968, *Transparent Flexible Packaging Material for Food*. German Patent 1,269,874.

449. Stralfors, A., 1965, Caries-Protective Additives to Foods. *Symp. Swed. Nutr. Found.*, 3:75-82.

450. Oommen, T. T., and Gurumurthy, B. S., 1968, Use of Phosphates and Phosphoric Acid in Reducing Losses in Waste Molasses. *Proc. Ann. Conv. Sugar Technol. Ass. India*, Article 12, 9 pp. in English; *Chem. Abstr.*, 71:92900k.

451. Tomme, M. F., and Sten'kin, D. N., 1969, Use of Monosodium, Disodium, and Monocalcium Phosphates in the Rations of Calves. *Dokl. Vses. Akad. Sel'skokhoz. Nauk*, 9: 10-12; *Chem. Abstr.*, 72: 40279r.

452. Kerscher, U., 1969, Phosphates in Animal Nutrition, Production, and Use. *Wien. Tieraerztl. Monatsschr.*, 56: (7), 280-290.

453. Kuster, W., 1968, *Low-Ash Content Food Supplement Method*. U.S. Patent 3,370,954.

454. Ohashi, T., and Ando, N., 1967, Water-Holding Capacity of Meat. I. Determination of Water-Holding Capacity of Meat and the Effects of Various Food Additives on Water-Holding Capacity of Meat. *Miyazaki Daigaku Nogakubu, Kenkyu Jiho*, 14: (2), 381-389; *Chem. Abstr.*, 70: 95524f.

455. Kohl, W. F., and Ellinger, R. H., 1968, *Bactericidal Polyphosphates for Use in Food Preservation*. S. African Patent 67-07804.

456. Fearing, R., and Souby, J. C., 1971, *Emulsifier for Oil and Water Foods*. U.S. Patent 3,567,466.

457. Rao, M. K. G., and Achaya, K. T., 1968, Unsaturated Fatty Acids as Synergists for Antioxidants. *Fette, Seifen, Anstrichm.*, 70: (4), 231-234.

458. Rajkovic, V. D., Dordevic, J. D., and Banjac, J. V., 1966, Stabilization of L-Ascorbic Acid with Complexion III in Solutions Containing Some Organic Acids. *Glas. Hem. Drust. Beograd*, 31: (7-8), 351-359; *Chem. Abstr.*, 70: 31648s.

459. Loury, M., Bloch, C., and Francois, R., 1966, Use of Tocopherol as Antioxidant in Fats. *Rev. Fr. Corps Gras*, 13: (12), 747-754.

460. Hulse, J. H., and Hannah, R. E., 1969, *Dough-Maturing Agents for Bread Making*. British Patent 1,151,985.

461. Knight, G. S., and Lehman, H., 1968, Application of Fatty Acids in Food. *Fatty Acids and Their Industrial Application*, E. S. Patterson, Ed., Marcel Dekker, Inc., New York, N.Y., pp. 259-281.

462. Hunter, A. R., 1968, *Aerated Food Products*. U.S. Patent 3,365,305.

463. Chipley, J. R., and Saffle, R. L., 1968, Various Acid Effects on the Peeling Performance, Shelf-Life, Color, and Flavor of Frankfurters. *Food Technol.*, 22: (11), 1462-1464.

464. Wagner, J. R., Miers, J. C., Sanshuck, D. W., and Becker, R., 1968, Consistency of Tomato Products. 4. Improvement of the Acidified Hot Break Process. *Food Technol.*, 22: (11), 1484-1488.

465. Velan, M., 1968, *Purification of Edible Oils*. French Patent 1,510,842.

466. Hughes, R. L., 1971, The Future of Chemical Additives in Foods. *Chemical Additives*, Division of Chemical Marketing and Economics, American Chemical Society Symposium, Los Angeles, California (March 29–31), pp. 32–40.

467. Domalski, E. S., 1972, Selected Values of Heats of Combustion and Heats of Formation of Organic Compounds Containing C, H, N, O, P, and S. *J. Phys. Chem. Ref. Data 1* (in press).

CHAPTER 6

Sequestrants in Food

Thomas E. Furia

Technical Development Manager
Food Industry Department
CIBA-GEIGY Corporation
Ardsley, New York

Introduction

Once considered chemical curiosities, sequestrants (chelating agents) have taken a significant place among other stabilizer food additives as vital tools of the food industry. As such, sequestrants help to establish, maintain, and enhance the integrity of many food products. For the most part, sequestrants react with metals to form complexes which, depending on the stability of the metal complex, tend to alter the properties and effects of metal in a substrate. Many of the sequestrants employed by the food industry occur naturally in food including: polycarboxylic acids (oxalic, succinic), hydroxycarboxylic acids (citric, malic, tartaric), polyphosphoric acids (ATP, hexametaphosphate, pyrophosphate), amino acids (glycine, leucine, cysteine) and various macromolecules (porphyrins, peptides and proteins). Metals in food occur mainly as complexes. For example, magnesium in plants occurs mainly as the metal complex chlorophyll, while iron is bound in animals as ferritin and the porphyrin complex hemoglobin. Other examples include cobalt as Vitamin B_{12}, iron as rutin, and copper, zinc and manganese in various enzymes.

From a food manufacturing viewpoint, sequestrants serve to stabilize the numerous properties identified with wholesome food including color, flavor and texture. In this chapter, the chemical and physical properties of various sequestrants will be discussed together with the effect they produce in food.

Chemistry

Although a detailed review of the chemistry of chelation is beyond the scope of this chapter (several excellent monographs on this subject are available), some basic considerations should be emphasized.[1-4,10]

Chelation is an equilibrium process in which the formation of a metal complex may be represented by the general equation:

$$M + L \rightleftharpoons ML$$

where:

$$M = \text{metal ion}$$
$$L = \text{ligand (sequestrant, chelating agent)}$$
$$ML = \text{metal complex}$$

In the above equation the ligand can be an organic compound, such as citric acid, EDTA, etc., or an inorganic, such as one of the polyphosphates.

For chelation to occur two general conditions must be satisfied: (1) the ligand must have the proper steric and electronic configuration in relation to the metal being complexed and (2) the surrounding milieu (pH, ionic strength, solubility, etc.) must likewise be conducive to complex formation. To illustrate the first condition, silver has a bond angle of 180°. In order to be chelated by an organic compound this distance must be capable of being spanned—a difficult requirement for most common sequestrants. On the other hand, most other metals such as calcium, magnesium, copper, iron and zinc have bond angles conforming to square planar (90°), octahedral (90°) or tetrahedral (109°) configurations which are much more receptive to chelation. Usually the ability of a ligand to form a stable 5 or 6 membered ring system with a metal is considered ideal for chelation. Glycylglycine and salicylic acid are examples of chelating agents forming 5 and 6 membered rings with divalent metals. The structural formula for the metal chelate of ethylenediaminetetraacetic acid (EDTA) shows the proper bond relationships between the metal and the donor groups (solid lines).

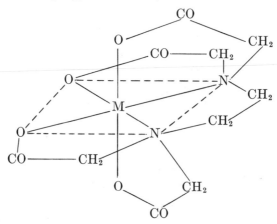

While EDTA has six donor groups (hexadentate), ligands with as few as four donor groups are capable of forming 1:1 metal complexes. With polydentate ligands such as EDTA, all available coordinating atoms do not need to be used simultaneously for chelation to occur.[5]

In some instances, isomeric chelates are obtained with ligands having donor groups capable of forming chelate rings in more than one way. For example, glycylglycine forms a single ring copper (II) chelate (Type A) and a double ring cobalt (II) chelate (Type B); equilibrium mixtures of both types are possible in solution with one species usually predominating.[6]

(Type A)   (Type B)

Assuming steric and electronic conditions are satisfied, other factors, such as pH, significantly influence the formation of a metal complex. For example, while the ionized carboxylate ion is an excellent donor group, the nonionized carboxylic acid group is not. Compounds such as citric acid and EDTA become increasingly dissociated as pH rises and the quantity of metal complexed increases accordingly. However, increasing the pH beyond that needed to completely dissociate the donor group in the presence of metal does little to increase complex formation, although the ionic charge of the metal also determines the net effect of pH on complex formation. The ferric-EDTA complex $[Fe(EDTA)]^-$ is more stable and can be formed in more acidic media than the ferrous species $[Fe(EDTA)]^=$. Some preformed metal chelates (prepared and isolated in conducive media) show good stability in acid solutions; the zinc, iron, copper, bismuth, and cobalt complexes of EDTA are relatively stable in acid.

Frequently, hydroxyl ions will also compete with sequestrants for metal ions. EDTA will prevent the precipitation of ferric hydroxide by forming the

TABLE 1
Acid Ionization Constants for Some Aqua Ions

Ion	pKa @ 25°C
Ba^{2+}	13.2
Ca^{2+}	12.6
Mg^{2+}	11.4
Ni^{2+}	10.6
Ni^{2+}	9.7
Fe^{2+}	8.3
Cu^{2+}	8.0
Al^{3+}	4.9
Fe^{3+}	2.2

soluble ferric-EDTA complex only if the hydroxyl ion concentration is kept low. Metals with low ionic charges form stable complexes in highly alkaline media, e.g., ferrous-EDTA.

The rate at which a metal complex is formed is usually associated with bond rupture of a preceding complex. Although usually neglected in considering metal complex formation, metals in solution exist as the aqua-complex $M(H_2O)^{m+}$. The acidity of metal-aqua complexes is usually significant while the corresponding hydroxo complexes are weak bases:

$$M(H_2O)^{m+} \rightleftharpoons MOH^{(m-1)+} - H^+$$

Thus, the rate at which citric acid complexes iron depends, in part, on the ease with which the iron-aqua bond is first dissociated. Consequently, the higher the acid ionization (hydrolysis) constant of the aqua ion the more rapid the reaction rate for the formation of a metal complex in solution.[7] The acid ionization constants of some aqua ions are shown in Table 1.

In well defined systems, the relative efficiency of sequestrants can be compared by inspecting the stability constants (formation constant, equilibrium constant) for a given metal.[8,9,12] In general terms, the stability constant of a metal complex can be calculated as follows:

$$K = \frac{[ML]}{[M][L]}$$

As the stability constant (K) increases, more of the metal will be complexed, leaving less metal in cation form. Conversely, the higher the stability constant the more stable the metal complex under equivalent conditions. The stability constants of various metal chelates of sequestrants frequently employed in food processing or found naturally in foods are shown in Table 2; those with outstanding chelate performance include EDTA, citrate and pyrophosphate.

The determination of stability constants, even under the most precisely defined conditions, is an arduous task, and meaningful comparisons can be made only when experimental conditions are similar.[11] Stability constants measured in well defined systems are therefore of limited value in predicting their performance in complex food systems containing numerous interfering components.

A notion still prevalent in some technical circles is that sequestrants can achieve "selectivity" in reactions with metals. This notion is incorrect when only one ligand is reacted in a medium containing several metal ions since equilibrium conditions prevail. While it is not usually possible for a single ligand to "selectively" complex a metal ion forming less stable complexes than other metals present, the introduction of a second sequestrant can provide some degree of "selectivity." For example, in a system containing calcium and iron ions, EDTA could begin chelating calcium only after first sequestering iron. If salicylic acid were now added to the system, two complexes would be formed, i.e., an iron salicylate complex and a calcium-EDTA complex. The net result is the "selective" chelation of calcium by EDTA in the presence of iron. However, one cannot expect to find significant quantities of unbound iron ions in this system. In some instances even small differences in the

TABLE 2

Stability Constants (log K_1) of Various Metal Chelates

	Al(III)	Ba	Ca	Co(II)	Cu	Fe(II)	Fe(III)	Mg	Mn	Ni	Sr	Zn
Acetic acid		0.39	0.53	2.24				0.51		0.74	0.43	1.03
Adenine												
Adipic acid		1.92	2.19									
ADP		2.36	2.82	3.68	3.35			3.11	3.54	4.50	2.50	4.28
Alanine		0.80	1.24	4.82	5.90				3.24	5.96	0.73	5.16
β-Alanine					8.18					4.63		4
Albumin			2.20		7.13				2.0			
Arginine						3.20						
Ascorbic acid			0.19								0.35	
Asparagine			0								0.43	
Aspartic acid		1.14	1.16	5.90	8.57			2.43	3.74	7.12	1.48	5.84
ATP		3.29	3.60	4.62	6.13			4.0	3.98	5.02	3.03	4.25
Benzoic acid		0.31			1.6			0.53		0.9	0.36	0.9
n-Butyric acid			0.51		2.14							1.00
Casein			2.23									
Citraconic acid			1.3									
Citric acid		2.3	3.5	4.4	6.1	3.2	11.85	2.8	3.2	4.8	1.3	4.5
Cysteine				9.3	19.2	6.2		<4	4.1	10.4	2.8	9.8
Dehydracetic acid					5.6					4.1		
Desferri-ferrichrysin							29.9					
Desferri-ferrichrome							29.0					
Desferri-ferrioxamin E				11.8	13.7		32.5			12.2		12.0
3,4-Dihydroxybenzoic acid			3.71	7.96	12.79			5.67	7.22	8.27		8.91
Dimethylglyoxime				6.6	11.94					14.6		7.7
O,O-Dimethylpurpurogallin			4.5		9.2			4.9		6.7		6.8
EDTA	16.13	7.78	10.70	16.21	18.8	14.3	25.7	8.69	13.56	18.56	8.63	16.5

TABLE 2 (Continued)

	Al(III)	Ba	Ca	Co(II)	Cu	Fe(II)	Fe(III)	Mg	Mn	Ni	Sr	Zn
Formic acid		0.60	0.80		1.98		3.1				0.66	0.60
Fumaric acid		1.59	2.00		2.51				0.99		0.54	
Globulin			2.32									
Gluconic acid		0.95	1.21		18.29			0.70		5.9	1.00	1.70
Glutamic acid		1.28	1.43	5.06	7.85	4.6		1.9	3.3		1.37	5.45
Glutaric acid		2.04	1.06		2.4			1.08			0.6	1.60
Glycine		0.77	1.43	5.23	8.22	4.3	10.0	3.45	3.2	6.1	0.91	5.16
Glycolic acid		0.66	1.11	1.60	2.81		4.7					
Glycylglycine			1.24	3.00	6.7	2.62	9.1	1.34	2.19	4.18		3.91
Glycylsarcosine				3.91	6.50				2.29	4.44		
Guanosine				3.2	6	4.3		3.0		3.8		4.6
Histamine				5.16	9.55	9.60	3.72			6.88		5.96
Histidine				7.30	10.60	5.89	4.00		3.58	8.69		6.63
3-Hydroxyflavone				9.91	13.20							9.70
Inosine			3.76	2.6	5	3		4.04	4.57	3.3		
Inosine triphosphate				4.74								
Iron-free ferrichrome							24.6					
Isovaleric acid			0.20		2.08							
Itaconic acid			1.20		2.8					1.8	0.96	1.9
Kojic acid	7.7		2.5	7.11	6.6		9.2	3.0		7.4		4.9
Lactic acid		0.55	1.07	1.89	3.02	3.42	6.4	0.93	1.19	2.21	0.70	1.86
Leucine				4.49	7.0		9.9		2.15	5.58		4.92
Lysine							4.5		2.18			
Maleic acid		2.26	2.43		3.90				1.68	2.0	1.1	2.0
Malic acid		1.30	1.80		3.4				2.24		1.45	2.80
Methionine						3.24	9.1	1.55		5.77		4.38
Methylsalicylate					5.90		9.77					
NTA	>10	4.82	6.41	10.6	12.68	8.84	15.87	5.41	7.44	11.26	4.98	10.45

	Al(III)	Ba	Ca	Co(II)	Cu	Fe(II)	Fe(III)	Mg	Mn	Ni	Sr	Zn
Ornithine	7.26			4.02	6.90	3.09	8.7		<2	4.85		4.10
Oxalic acid		2.31	3.0	4.7	6.3	>4.7	9.4	2.55	3.9	5.16	2.54	4.9
β-Phenylalanine					7.74	3.26	8.9					
Pimelic acid												
Pivalic acid			0.55						1.08			
Polyphosphate			3.0		2.19	3.0	10.0	3.2	5.5	3.0		2.5
Proline					3.5	4.07	3.45		3.34			
Propionic acid		0.34	0.50		2.2			0.54			0.43	1.01
Purine					6.90					4.88		
Pyrophosphate			5.0		6.7		22.2	5.7		5.8		8.7
Pyruvic acid			0.8		2.2							
Riboflavin				3.9	<6				3.4	4.1		<4
Salicylaldehyde	14.11			4.67	7.40	4.22	8.70	3.69	3.73	5.22		4.50
Salicylic acid				6.72	10.60	6.55	16.35	4.7	2.7	6.95		6.85
Sarcosine			1.43	4.34	7.83	3.52	9.7			5.41		
Serine			1.20			3.43	9.2			5.44		
Succinic acid		1.57		2.08	3.3		7.49	1.2	2.11	2.36	0.9	1.78
(±)-Tartaric acid		1.95	1.80		3.2		7.49	1.36		3.78	1.94	2.68
Tetrametaphosphate		4.9	5.2		3.18			5.17		4.95	2.8	
Threonine						3.30						
Trimetaphosphate		6.3	2.50		1.55		8.6	1.11	3.57	3.22	1.95	
Triphosphate			6.5		9.8			5.8			3.80	9.7
Tryptophan							9.0					
Uridine diphosphate			3.71	4.55				3.17				
Uridine triphosphate								4.02	4.78			
n-Valeric acid		0.20	0.30		2.12							
Valine					7.92	3.39	9.6		2.84	5.37		5.0
Xanthosine				2.8	3.4	<2				3.0		2.4

stability constants of certain metal chelates provide a convenient method for the separation of metals. The separation of rare earth metal mixtures on ion-exchange resin columns can be effected by elution with chelating agents, e.g., EDTA, DTPA, etc.

The Use of Sequestrants in Food Systems

Basic Considerations

Numerous sequestrants with high metal complexing efficiency are available, though few are applicable for use in foods. Foremost in the minds of food manufacturers, government regulatory agencies, and the consumer is the question of safety. Those sequestrants with a proven record of safety can be considered as direct food additives. The pharmacological aspects of various sequestrants have been studied extensively and critical evaluations of available information indicate that those currently approved for use in food do not constitute a health hazard.[13-26] However, many of the applications described in this chapter are not yet permitted by the U.S. Food and Drug Administration or foreign regulatory agencies. The reader is therefore cautioned to exercise due care by checking local, state and federal laws governing the use of sequestrants in specific food substrates; Part II of this handbook will serve as a quick reference to U.S. regulatory status.

In addition to toxicological considerations, other chemical and physical characteristics often preclude the use in food of some otherwise efficient sequestrant. Color, flavor and substrate compatibility are just a few of the characteristics important in the selection of a suitable sequestrant for use in foods. Where chelate efficiency and compatibility are equivalent, one may be selected over another on the basis of economic factors.

The sequestrants listed in Table 3 will acquaint the reader with some of the most popular sequestrants used for food processing in the United States. Of those listed, the citrates, pyrophosphates and EDTA are by far the most popular and technically useful. Details on the specifications for food grade sequestrants can be found by consulting sequestrant manufacturers of the Codex.[29]

Stabilization of Fats and Oils

While sequestrants are not considered antioxidants in the same sense in which hindered phenols arrest oxidation by chain termination, or like ascorbates by scavenging oxygen (see Chapter 4), they are extremely valuable assistants in antioxidant systems. In almost all oxidation-sensitive food substrates thus far investigated, trace metals serve as pro-oxidant catalysts. Trace levels (0.1–5 ppm) of copper and iron in fats and oils significantly enhance oxidation (rancidity, reversion, off-flavor, etc.). By chelating these metals, pro-oxidant catalytic effects are eliminated or minimized and higher efficiency can be derived from the antioxidant. In most cases the combined effect of sequestrants and antioxidants is synergistic; sequestrants are popularly termed "synergists for antioxidants". The present utilization of sequestrant/antioxidant blends in foods containing fats and oils is reflected by superior initial products with extended shelf-life.

Antioxidants synergized by EDTA, citrate, citrate esters, and phosphates

include BHT & BHA

TABLE 3
Sequestrants Used in Food Processing

Type	Form
Acetate	calcium calcium, di potassium sodium
Citrate	calcium citric acid monoisopropyl monoglyceride potassium sodium stearyl triethyl
Ethylenediaminetetraacetate, (EDTA)	disodium calcium disodium dihydrogen
Gluconate	calcium sodium
Oxystearin	
Phosphate, ortho	monocalcium acid phosphoric acid potassium, dibasic sodium aluminum sodium, dibasic (disodium orthophosphate) sodium, monobasic (monosodium orthophosphate) sodium, tribasic (trisodium orthophosphate)
Phosphate, meta-	calcium hexameta (calcium metaphosphate) sodium hexameta (sodium metaphosphate)
Phosphate, pyro	sodium pyro (sodium tetrapyrophosphate) (tetrasodium pyrophosphate) (sodium acid pyrophosphate)
Phosphate, poly	sodium tripolyphosphate
Phytate	calcium
Sorbitol	
Tartrate	sodium sodium potassium (Rochelle salts) tartaric acid
Thiosulfate	sodium

include: butylated hydroxytoluene (BHT), butylated hydroxyanisole (BHA), nordihydroguaiaretic acid (NDGA), propyl gallate, ascorbic and isoascorbic acids, tocopherols, phospholipids and various thiopropionates. In some rare instances metal-chelates may tend to increase the catalytic effect of metals by making them more available, e.g., solubilization to achieve homogeneous reaction conditions. The inactivation of metal-catalyzed oxidation of ascorbic acid by sequestrants is apparent under acidic conditions, while there is some evidence to suspect that pro-oxidant effects occur in alkaline media.

Sequestrants have been successfully employed together with antioxidants to accomplish the following:

1. stabilize edible fats and oils (tallow, lard, etc.)[28,33,37,38,52–55]
2. stabilize soybean oil[30–32,42,43]
3. extend the shelf-life and retain flavor of salad dressings and spreads containing emulsified fats and oils[56–60,63]
4. inhibit rancidity of oils in roasted nuts and nut meat[44–46]
5. prevent the rancid odor and flavor of fried and baked goods[29,61,62]
6. retain good flavor in oleomargarine[30,48,49]
7. prevent "weeping" in oleomargarine and butter[50]
8. inhibit autoxidation of essential oils[47]
9. inhibit the copper-catalyzed autoxidation of linoleic acid[41]
10. improve oxidative stability of partial glyceride fatty esters prepared by ester interchange[40]
11. remove metallic hydrogenation catalysts from fats and oils[39] and
12. retain flavor and color of meat products.[28,51]

The solubility characteristic of sequestrants in fats and oils is an important practical aspect of stabilization which often governs the choice of sequestrants. While Na_2EDTA and $Na_2CaEDTA$ are readily incorporated into emulsion systems such as salad dressings, sauces, spreads, margarine, etc. to produce finished products with remarkable stability, their rate of solubility in homogeneous fats and oils is poor. Citric acid and citrate esters on the other hand are readily solubilized in fats and oils from propylene glycol solution concentrates; formulations consisting of various mixtures of antioxidant and citrate in propylene glycol/oil solutions are commercially available. However, using laboratory methods for incorporating additives, it was recently shown that Na_2EDTA produced a 43% improvement in lard stabilization over citric acid in the presence of BHT (Figure 1).[64,65]

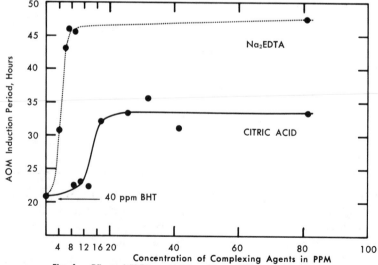

Fig. 1.   Effect of EDTA and citric acid on BHT-treated lard.[65]

Flavor is another important aspect governing the selection of sequestrants slated for fat and oil stabilization. The distinct acidic flavor of citrates and phosphates is without effect in homogeneous fats and oils since their effective use levels (20–200 ppm) are diluted upon incorporation into the finished food. Salad dressings, margarine, sauces, and spreads require as much as 200 ppm sequestrant based on the weight of the *finished product*. At these levels, sequestrants with strong flavor notes can be a source of serious concern; Na_2EDTA and $Na_2CaEDTA$ do not contribute flavor notes to emulsified oil systems.

Vitamin Stabilization

From a food processing viewpoint, vitamins are relatively unstable natural constituents in food. They require stabilization in order not to lose potency or to render the finished food product deficient in nutritional value. Vitamins are prone to oxidative deterioration catalyzed by metal. Most losses in the potency of naturally occurring vitamins in food and vitamin-supplemented food products happen during processing (blanching, retorting, etc.), although significant losses also take place during storage.

While antioxidant/sequestrant mixtures are of considerable value in preventing or inhibiting the decomposition of oil soluble vitamins, sequestrants alone have demonstrated a remarkable utility for stabilizing vitamins in aqueous systems. For example, the copper-catalyzed oxidation of ascorbic acid is inhibited by EDTA.[28,67–69] The effectiveness of sequestrants as stabilizers for ascorbic acid apparently increases with decreasing pH; below pH 3.6 normally sensitive solutions of ascorbic acid can be boiled without serious losses in the presence of sequestrants.[68] Liquid formulations of ascorbic acid suitable for pharmaceutical purposes or as concentrated stock solutions used in many food processing operations are stabilized by EDTA, citrate and to a lesser extent phosphates.[70–73,77] The vitamin C content of tomato, grapefruit, cranberry, and black currant juices is protected by adding sequestrants early in the processing operation.[74–76,78] Food products containing oil-soluble vitamins A, D, E, and K are often stabilized by mixtures of sequestrants and antioxidants; blends of EDTA, citrate, and phosphates with BHT or NDGA have been successfully used for this purpose.[80,81] Carotene is stabilized by sequestrants such as EDTA in the presence of ascorbic acid or 2-amino-3-hydroxybenzoic acid; in this instance ascorbate is utilized as the antioxidant.[28,79] Vitamin B_{12}, folic acid and thiamine are also stabilized by sequestrants.[77,82–85]

Processed Fruits and Vegetables

The deleterious changes associated with fruit and vegetable processing are well known and include loss of color, localized discoloration, off-flavor, and changes in texture. For example, the advent of frozen fruit and vegetable packs has accentuated certain disadvantages of canning, e.g., comparative colors of canned and frozen peas, although freezing processes are not without deleterious effects. In many instances, the elevated temperatures encountered in blanching and retorting permit the naturally occurring metals in food to be liberated from their sites and combine in an unnatural fashion with organic components. The addition of sequestrants prior to blanching and/or retorting inactivates the naturally occurring metals and usually prevents undesirable color changes.

The sensitivity of vegetables to trace metals is probably best illustrated with canned whole-kernel and cream-style corn. Trace quantities of copper, iron and chromium (0.5–5 ppm) produce a typical greenish-gray discoloration. By treating the pack with 50–100 ppm disodium EDTA prior to retorting, discoloration is prevented. Similarly, the surface darkening of sweet potatoes, yams, cauliflower, eggplant, asparagus, brussel sprouts, and turnips is often due to the interaction of iron with rutin. By sequestering the iron after cooking, gray discoloration is inhibited.[86] The rapid surface darkening of canned sliced beets soon after opening is associated with air-oxidation but the addition of citrate or EDTA alone inhibits discoloration.[87] Similarly, the discoloration of canned mushrooms can be minimized by the addition of $Na_2CaEDTA$ to the brine prior to retorting.[176] Levels of up to 200 ppm of $Na_2CaEDTA$ are allowed by the FDA on a temporary permit basis. All attempts to significantly inhibit the discoloration of canned green vegetables resulting from chlorophyll decomposition have thus far been unsuccessful.

The use of ascorbic acid as an antioxidant permits the canning of oxygen-sensitive fruits without "browning". For example, apple sections can be packed without first removing excess oxygen by adding approximately 1000 ppm ascorbate to the pack.[88] Sequestrants synergize the antioxidant activity of ascorbate. Thus, while "browning" of canned apple sections persists when less than 1000 ppm ascorbate is employed, the addition of only 250 ppm Na_2EDTA permits the use of 250 ppm ascorbate. In the absence of ascorbate, levels as high as 1000 ppm EDTA, citrate, or phosphate do not inhibit "browning".

The pink discoloration often encountered in canned pears is associated with high acid content, low pH, and high tannins.[89] Trace quantities of copper, iron and zinc also contribute to "pinking" of canned pears. By treating the packs with 100–500 ppm of Na_2EDTA, citrate or phosphate prior to retorting, "pinking" is significantly inhibited. In a similar fashion, the discoloration of cores in canned peach halves is inhibited by sequestrants added to the packs from hot syrups.

The various species of legumes are a staple canned product sensitive to discoloration and off-flavor development after retorting and during storage. The interaction during retorting of naturally occurring metals (beans are known to accumulate zinc) with organic components is responsible for most of the discoloration. Kidney beans packed without added sequestrants turn dark red while chick peas and lima beans take on a grayish cast. Darkening of canned legumes continues in storage (which in many instances can be as long as 2–3 years) and during this period, off-flavor development also occurs. The addition of sequestrants such as Na_2EDTA and $Na_2CaEDTA$ to the rehydrating soak-tank waters or directly to the packing brine prevents discoloration, off-flavor, and, interestingly, also minimizes the incidence of skin breakage. While it is usually necessary to maintain a level of 0.5% Na_2EDTA in the rehydrating soak water, only 100–500 ppm is required when added from brines.[93] Pre-weighed salt tablets containing the appropriate levels of Na_2EDTA and $Na_2CaEDTA$ for brine preparations are commercially available. This technique has been successfully employed to treat kidney beans, pinto beans, lima beans, black-eyed peas, chick peas, red-eyed beans, and Great Northern beans.

Sequestrants such as citric acid have found considerable use together with antioxidants in processing fresh-frozen fruits.[90] Since the protective action of the naturally occurring ascorbate in fruits is short-lived, due to the action of certain enzymes, additional stabilization is often required to prevent discoloration resulting from oxidation of color-producing substances such as catechols. While blanching inactivates ascorbate destroying enzymes, this can impart a cooked flavor and may promote discoloration in heat sensitive fruits. By dipping fruits in 1–2% citric acid solution prior to freezing, residual lye from the peeling process is neutralized and further destruction of ascorbate is prevented. However, this type of treatment is not usually sufficient to prevent deteriorative effects during freezing, and an additional sequestrant/antioxidant dose must be furnished the pack. Thus, the addition of 0.5% citric acid and 0.02% D-erythroascorbic acid prevents "browning" of bananas during freezing and defrosting.[91] The preparation of freeze-dried banana slices added to ready-to-eat cereals is accomplished without discoloration by employing EDTA while acidified banana purees can be prepared using citric acid.[92] The use of EDTA in freeze-dried vegetable processing also appears to be of value. For example, EDTA inhibits the discoloration (browning) during storage of freeze-dried kohlrabi, while discoloration is enhanced by ascorbic acid.[177]

Potato Products

Potatoes are especially prone to various forms of discoloration.[94] The pre-cooking darkening of certain tuber species ("black spot") is associated with a high phenolic content; this decreases as the level of potassium fertilization increases.[95] On the other hand, the after-cooking gray discoloration of white potatoes ("stem end grayness") is rather non-specific and no means has yet been found to inhibit grayness during cultivation or for accurately predicting its appearance. Heating in the form of blanching, retorting or frying accentuates graying and is responsible for an unacceptable consumer product. Potato products affected by "stem end grayness" include canned whole white, frozen white, cut white, and oil-blanched French-fries.[96] Discoloration is also evident in dehydrated white and sweet potato flakes, purees and candied yams.

Pretreatment of potatoes with citric acid, EDTA, sodium acid pyrophosphate, bisulfite, ascorbate, and alum are of some value in preventing discoloration, but pyrophosphate and Na_2EDTA are the sequestrants of choice.[97–99] Pre-blanching potatoes in 0.1% solution of Na_2EDTA or 1% sodium acid pyrophosphate prevents the after-cooking darkening of oil-blanched frozen French-fries.[98] Treatment with EDTA or pyrophosphate does not alter the flavor of potatoes. In a similar way, after-cooking darkening, poor texture, and blistering of frozen French-fries can be prevented by blanching with solutions of EDTA and bisulfite or alum.[86]

Pre-peeled raw potatoes have poor shelf-life and are sensitive to attack by molds. Washing the peeled potato in a solution containing bisulfite and EDTA shows a 20% increase in shelf-life and good mold resistance.[101] After cooking, discoloration of pre-peeled potatoes is also inhibited by treatment with sequestrants.[102,103] The after-cooking darkening of potatoes can be inhibited by foliar application of sequestrants.[104] Greening of potato tubers exposed to fluorescent light is reduced by spraying with EDTA or by the application of coatings containing 0.1–0.5% $Na_2CaEDTA$.[105,106]

Fish and Shellfish Products

Fish, particularly shellfish, contain unusually high concentrations of metals, e.g., 100 ppm iron, 400 ppm copper, 600 ppm zinc.[107] Metals, together with organic components, react under processing conditions and during storage to produce product discoloration, off-flavor, and unacceptable odors. For example, a discoloration called "black spot" (melanogenesis) occurs in ice-stored shrimp.[108] Canned clams, clam chowder, crabmeat, lobster, and salmon also discolor as the result of retorting to an unacceptable blue-green or grayish cast. Discoloration of the meat results from the interaction of metal with amines (blue-green) or sulfhydryl compounds (brown-black).[109] Rancidity of oily fish fillets occurs rapidly (even during storage in ice) as the result of oxidation catalyzed by the great variety of metals present at the tissue surface.[110]

Another chemical problem associated with processed fish and shellfish is struvite formation. Struvite, or crystalline magnesium ammonium phosphate, forms during the cooling stage following retorting and continues during storage. Although harmless, struvite crystals can grow to lengths of 1–2 inches and give the product the appearance of being glass-contaminated to the consumer. Struvite formation is canned lobster, shrimp, crabmeat, salmon and haddock can be effectively controlled with sequestrants.[115–116] While citrate is without notable effect, the strong binding of magnesium by EDTA prevents the crystal structure of struvite from forming. EDTA can be added to the packs from brine solutions, salt tablets, or introduced in the blanching bath.[117–119]

Sequestrants have been extremely successful in preventing the various deteriorative changes associated with fish and shellfish processing; EDTA and citric acid have been most frequently and successfully employed for this purpose. The blue-green and grayish discoloration of canned, frozen, iced, and fresh shellfish and crustacea is prevented by treatment with solutions of EDTA and alum.[111,112] Alum/EDTA treatment also maintains firmness of shrimp packs for several years while untreated shrimp softens in a few months. The packing procedure usually consists of dipping the meat in a brine solution (2–5% NaCl) containing 0.5% Na_2EDTA. An alternate procedure consists of a brine dip containing 0.02% aluminum (as aluminum sulfate) and 0.1–1.0% Na_2EDTA. In canned shellfish products, addition of 0.7% citric acid to a 1% salt/1–2% sugar brine prevents discoloration.[113] In addition to generally preserving the color of frozen and canned shellfish, the use of EDTA quickly restores the original (native) color to processed oysters.[114] In processing canned clams, the addition of $Na_2CaEDTA$ to the cooking water prevents graying; the total allowable level of $Na_2CaEDTA$ in the finished pack is 370 ppm.

Recent developments concern the use of EDTA as a preservative for fresh fish fillets.[120] The enzymatic conversion of odorless trimethylamine oxide in the fish tissue to trimethylamine is partially responsible for rendering iced fish inedible after only 4–5 days storage. To date, the shipping of fresh fish fillets in ice is accomplished by dipping the fillets in tetracycline antibiotics, but this practice is of serious concern to various drug agencies. Studies with haddock fillets show that dipping in EDTA solutions can extend storage in ice (0°F) for 11 days as compared to untreated fillets.[179] The texture of treated fillets remains acceptable during the entire storage period, and an increase in trimethylamine during storage is reduced by treatment with EDTA. (Taste panels rejected as spoiled fish that showed 3-9.5 mg of TME/100 g.) Similar studies show that dipping fillets for 30-60 seconds in 0.8-1.5% solutions of Na₂CaEDTA extends storage life in ice from 12-14 days; the pick-up of Na₂CaEDTA in the fillets is approximately 200-300 ppm.[121] In salmon muscle 100-500 ppm of Na₂EDTA was effective in retarding odor development due to oxidative rancidity and in reducing discoloration; in the treatment of lemon sole with Na₂EDTA, no noticeable putrid odor was observed after 10 days storage.[178] Preservation can also be achieved by spraying, dusting, or even injecting EDTA solutions into the vascular system of the fish as well as introducing the EDTA in the packing ice.

EDTA has been found to retard the pro-oxidant activity of sodium chloride in blended cod muscle stored at 0°C.[180] The off-flavor of freeze-dried oysters was recently found in part related to a loss in water-absorbing capacity during storage; treatment with EDTA and antioxidants reduced losses in the water-absorbing capacity, thereby maintaining good flavor.[181] Additional reports show that EDTA inhibits IMP dephosphorylation in a wide variety of refrigerated fish products; sensory evaluations of EDTA-treated fish muscle show significantly more accepted flavor as compared with untreated muscles.[182]

Beverages

While certain sequestrants such as citrate and phosphate are extensively employed in beverages as acidulants and flavor modifiers, they are also employed to a significant extent for preventing or inhibiting the deleterious effects of processing and storage. The "gushing" or "wildness" in beer, when caused by trace metals, is controlled by the addition of sequestrants such as EDTA.[122,123] In this instance EDTA does not contribute flavor to the finished product. While the sequestrant can be introduced at almost any stage of the brewing process, gushing is stabilized for longer periods if the sequestrant is used to treat the malt.[124,125] Na₂CaEDTA is reported to eliminate the astringent taste of beer treated with zinc salts, without adversely affecting the foam properties.[186] Another effect of sequestrants in beer is to prevent "chill haze". This is achieved by the sequestrant counteracting the inhibitory effect of copper on proteolytic enzymes.[126] Deleterious oxidative effects in beer such as haze, poor clarity on storage at room temperature, and oxygen-enhanced gushing are inhibited by the use of sequestrants.[125,127,128] Also, Na₂CaEDTA protects beer stabilizers such as isoascorbic acid against oxidation in the presence of trace quantities of metal catalysts.

In wine, cider, and nondistilled vinegar a chronic problem of turbidity (*casse*) is often encountered. While there are a variety of reasons for precipitate formation, this problem is generally due to the insoluble metal-tannin and metal phosphate complexes; iron and, to a lesser extent, copper are involved.[129-131] In most instances *casse* can be prevented by adding small quantities of sequestering agents to the product; citric acid and EDTA are excellent for this purpose. *Preformed casse* in wines is eliminated and haze inhibited by the addition of eight parts EDTA per part of dissolved iron or copper.[132-135]

Citric acid and sodium hexametaphosphate have also been found useful for preventing *casse* in wines and cider.[136,139] *Casse* formation in white wines is associated with an increase in redox potential but this is effectively lowered by sequestrants.[131] Oxidative changes causing browning and darkening in wines (especially white wines) are often catalyzed by metals.[140] EDTA, citrate and sodium hexametaphosphate are of value for this application.[134] $Na_2CaEDTA$ added to distilled spirits at 10–25 ppm maintains color, flavor, and clarity on storage. Natural color can be restored to an aged, discolored liquor, such as bourbon, by adding 2–20 ppm of Na_2EDTA/1 ppm of dissolved metal.[187] $Na_2CaEDTA$ is also favored for use in cordials, liqueurs, and sours that are prone to off-flavor and color development, as the introduction of trace metals cannot generally be avoided.

In the manufacture of carbonated and still beverages, acidic sequestrants such as citrate and phosphate serve the dual role of acidulant and sequestrant (see Chapter 5). Citric acid is added to beverage syrups primarily as an acidic flavor component, but it will also permit pasteurization at a lower temperature (190°F vs. 212°F). In addition, the citrate speeds inversion. In other instances, the lowered pH resulting from the addition of citrate stabilizes sensitive colors employed in fountain syrups such as strawberry. Dye fading occurs on exposure to sunlight in bottled carbonated beverages containing certain certified food colors supplemented with Vitamin C. The addition of 36 ppm EDTA to carbonated beverages containing FD&C Yellow No. 6 permits approximately 50% retention of the color strength after 90 minutes exposure to sunlight.[65] Other FD&C dyes protected from sunlight-fading by EDTA include: Yellow No. 6, Yellow No. 5, Red No. 2, Red No. 4, and Blue No. 1.[141] The elevated temperatures normally encountered during transportation and storage will also cause fading of food dyes. Canned carbonated and still beverages prepared in 1% citric acid and containing 12.3 mg ascorbic acid/120 ml show color fading after two weeks storage at 100°F; colors affected include Orange 1, Sunset Yellow, Brilliant Blue, Ponceau, Amaranth, and Tartrazine.[142] While the citrate concentration (1%) is apparently insufficient to exert a protective effect on the dyes, addition of as little as 33 ppm $Na_2CaEDTA$ inhibits fading. Citrate, phosphate, and EDTA also aid in the flavor retention of stored beverages and EDTA has been shown effective in controlling corrosion in canned carbonated beverages. In canned club soda combinations of stannous chloride and $Na_2CaEDTA$ result in significantly less iron pick-up and longer shelf-life than in untreated club soda.[183]

The antimicrobial activity of certain sequestrants has been used to good advantage in beverage processing. In addition to lowering pasteurization temperatures, the incorporation of citric acid in tomato juice inhibits the growth of "flat-sour" bacteria responsible for poor flavor. Similar "flat-sour" bacteria destroys the storage quality (even under refrigeration) of orange and other citrus juices.

Dairy Products

Sequestrants are frequently employed in the manufacture of dairy products as stabilizers of flavor and odor. Homogenization has reduced most off-flavor problems in whole milk but the poor flavor caused by copper-catalyzed oxidation remains a problem.[144] Experimentally, off-flavor has been shown to be suppressed by adding to milk at least five parts of EDTA per part copper. Since the copper concentration in off-flavored milk is seldom higher than 1.0 ppm, about 5 ppm EDTA, Na$_2$EDTA, or Na$_2$CaEDTA is usually effective.[145,146] The thickening of stored condensed milk and the control of heat coagulation of milk used for manufacturing confections is controlled by the use of sequestrants; EDTA appears excellent for these applications.[147,148] Sequestrants such as citrate, EDTA and polyphosphates are of little value in preventing the oxidized flavor of milk exposed to sunlight or in preventing coagulation by pepsin.

Some sequestrants such as sodium citrate are frequently employed in cheese making as an emulsifier, while EDTA enhances the foaming properties of reconstituted skim milk.[150] Polyphosphates aid in producing stable foams from milk proteins.[151]

In the enzymatic desugaring of egg whites, the pH is usually adjusted with citrate to 7.0–7.5 allowing glucose oxidase to react efficiently (see Chapter 1). The prefreeze heat sterilization of whole eggs tends to produce undesirable coagulation; this is reduced by adding sequestrants. Calcium "salts" (chelates) added to treated eggs after sterilization will inhibit oxidation and other deteriorative changes occurring during freezing.[78] Also, the use of EDTA prevents after-cooking greening of eggs prepared from acidified and glucose-free whole egg powders.[152]

An easily manufactured egg custard with fine mouth feel, smooth appearance, and excellent mold stability can be prepared by using as little as 0.5% Na$_2$EDTA; without EDTA the dessert is grainy, "weeps," and loses mold height on standing.[185]

Meat Products

The anticoagulation properties of certain sequestrants such as EDTA are based on the deactivation of calcium-activated coagulation enzymes. This property can be used advantageously in processing blood used for sausage manufacture.[153] Processed meat products, such as ham, bacon, frankfurters, etc., can be cured with the aid of magnesium salts (usually magnesium phosphate). Na$_2$EDTA will solubilize the magnesium for a more efficient process.[154] In addition, sequestrants such as citrate, pyrophosphate and EDTA will retard nitric oxide—hemoglobin formation in cured meats by controlling the effects of metal ions. Borenstein has shown that EDTA, in conjunction with ascorbic acid and nitrous oxide or sodium nitrite, will significantly accelerate the curing of cooked sausage products.[184]

Gray surface discoloration of meat (e.g., pet foods, in slack-filled containers) is inhibited by sequestrants. A surface spray of 0.05–0.1% Na$_2$EDTA, preferably with some nitrite added, controls this problem. Mixtures of ascorbate and Na$_2$EDTA stabilize the color and flavor of frozen ground beef while combinations of citrate and phenolic antioxidants control rancidity in frozen creamed turkey.[156,157]

Future Trends

In relation to other food ingredients, sequestrants occupy a small but significant portion of the additive spectrum. In addition to the major application areas already discussed, sequestrants serve a vital role in numerous support efforts to the food industry. Food sanitization formulations containing technical grade sequestrants are extensively employed for beerstone removal, dairy cleaners and scale removal from sugar-refining equipment.[158-169] As processes are pushed to increasing loads, sequestrants can be expected to make rapid gains in maintenance programs. Improved processing is also expected; small levels of sequestrants such as EDTA improve sugar crystallization by decomposing carbohydrate-metal complexes.[170-172] Although spice extraction technology is quite old, new products with improved color and flavor are achieved by using mixtures of antioxidants and sequestrants; levels as low as 60 ppm Na$_2$CaEDTA in the presence of BHA have been shown to be extremely effective.[173] Flavor potentiators, currently enjoying the concentrated attentions of the food industry, are other areas in which sequestrants are expected to make significant contributions; 100 ppm of Na$_2$EDTA added to sodium inosinate produces an unusually stable flavoring agent.[174]

Another area of high interest where sequestrants are expected to prove valuable is in instant desserts. The clarity and whipping quality of gelatin are improved by sequestrants, while the rancidity of products containing pregelatinized starches is inhibited by antioxidants/sequestrant blends.[175]

The action of sequestrants as additives in food systems is now being seriously exploited. No doubt their use will be expanded, as greater demands are made on the food industry to produce more stable food for the growing world population and to provide imaginative new products for the consumer.

References

1. Smith, R. L., 1959, *The Sequestration of Metals.* Macmillan Co., New York.
2. Martell, A. E., and Calvin, M., 1952, *Chemistry of the Metal Chelate Compounds.* Prentice-Hall, New York.
3. Bailar, J., 1956, *The Chemistry of the Coordination Compounds.* Reinhold Publishing Co., New York.
4. Chaberek, S., and Martell, A. E., 1959, *Organic Sequestering Agents.* John Wiley and Sons, New York.
5. Bush, D. H. and Bailar, J. C., 1953, *J. Am. Chem. Soc.* 75: 4574.
6. Rabin, B. R., 1956. *Trans. Faraday Soc.* 52: 1130.
7. Basolo, F. and Pearson, R. G., 1958, *Mechanisms of Inorganic Reactions.* John Wiley and Sons, Inc. New York.
8. Bjerrum, J., Schwarzenbach, G., and Sillen, L. G., 1957, *Stability Constants of Metal-ion Complexes, with Solubility Products of Biorganic Substances.* I. *Organic Ligands.* Chem. Soc. Burlington House, London.
9. Bjerrum, J., Schwarzenbach, G., and Sillen, L. G., 1958, *Stability Constants of Metal-ion Complexes, with Solubility Products of Biorganic Substances.* II. *Inorganic Ligands.* Chem. Soc., Burlington House, London.
10. Ringbom, A., 1963, *Complexation in Analytical Chemistry.* Interscience Publishers, New York.
11. Rossotti, F. J. C. and Rossotti, H., 1961, *The Determination of Stability Constants.* McGraw-Hill Book Co., Inc. New York.
12. Sillen, L. G. and Martell, A. E., 1964, *Stability Constants of Metal-Ion Complexes.* Special Publication No. 17, The Chemical Society, Burlington House, W. I. London.
13. Seven, M. J., and Johnson, L. A., 1960, *Metal-Binding in Medicine.* J. B. Lippincott Co., Philadelphia, Pa.
14. Johnson, L. A. and Seven, M. J., 1961, Proc. Conference on Biological Aspects of Metal-Binding. *Federation Proc.* 20: 3, Part II. Supplement 10.
15. Frederick, J. F., 1960, Chelation Phenomena. *Ann. N.Y. Acad. Sci.* 88: (2), 281–532.
16. Schubert, J., 1966, Chelation in Medicine. *Sci. Am.* 214: (5), 40–50.

17. Oser, B. L., Oser, M., and Spencer, H. C., 1963, Safety Evaluation Studies of Calcium EDTA. *Toxicity and Applied Pharmacology.* 5: (2), 142–162.
18. Catsch, A., 1964, zur Toxikologie der Diaethylentetraminpentaessigsäure. *Arch. exptl. Pathol. u. Pharmakol. Naunyn-Schmiedeberg's* 264: 4.
19. Foreman, H. and Trujillo, T. T., 1954, The Metabolism of C¹⁴-Labeled Ethylene-diamine Tetraacetic Acid in Human Beings. *J. Lab. Clin. Med.* 43: 566.
20. Gruber, Jr., C. M., and Halbeisen, W. A., 1948, A Study on the Comparative Toxic Effects of Citric Acid and Its Sodium Salts. *J. Pharmacol. Exptl. Therap.* 94: 65–67.
21. Levey, S., Tasichack, A. G., Brimi, R., Orten, J. M., Smyth, C. J., and Smith, A. H., 1946, A Study to Determine the Toxicity of Furmaric Acid. *J. Am. Pharm. Assoc. Sci. Ed.* 35: 295–304.
22. Locke, A., Locke, R. B., Schlesinger, H., and Carr, H., 1942, The Comparative Toxicity and Cathartic Efficiency of Disodium Tartrate and Fumarate, and Magnesium Fumarate, for the Mouse and Rabbit. *J. Am. Pharm. Assoc.* 31: 12–14.
23. Salant, W. and Swanson, A. M, 1918, Observations on the Action of Tartrates, Citrates, and Oxalates. A Study in Tolerance, Cumulation, and the Effect of Diet. *J. Pharmacol. Exptl. Therap.* 11: 133–145.
24. Packman, E. W., Abbott, D. D., and Harrison, J. W. E., 1963, Comparative Subacute Toxicity for Rabbits of Citric, Fumaric, and Tartaric Acids. *Toxicol. Appl. Pharmacol.* 5: 163–167.
25. Marshall, G. L., 1950, Citrate Tolerance. *California Med.* 73: 494–496, Dec.
26. Dwyer, F. P., and Mellor, D. P., 1964, *Chelating Agents and Metal Chelates.* Academic Press, Inc. New York.
27. Food Chemical Codex, 1966, 1st Edition. National Academy of Science and National Research Council, Washington, D.C.
28. Watts, B., and Wong, R., 1951, Factors Affecting Behavior of Ascorbic Acid with Unsaturated Fats. *Arch. Biochem.* 30: 110.
29. Mahon, J. H., and Chapman, R. A., 1954, Behavior of Antioxidants During the Baking and Storage of Pie Crusts. *J. Am. Oil Chemists' Soc.* 31: 108.
30. Schwab, A. W., Cooney, R. M., Evans, C. D., and Cowan, J. C., 1953, The Flavour Problem of Soyabean Oil. XII: Nitrogen Coordination Compounds Effective in Edible-Oil Stabilization. *J. Am. Oil Chem. Soc.* 30: 177.
31. Dutton, H. J., Schwab, A. W., Moser, H. A., and Cowan, J. C., 1948, The Flavour Problem of Soybean Oil. IV. Structure of Compounds Counteracting the Effect of Pro-Oxidant Metals. *J. Am. Chem. Soc.* 25: 385.
32. Dutton, H. J., et al., 1949, The Flavor Problem of Soybean Oil. V. Considerations in the Use of Metal Scavengers in Commercial Operations. *J. Am. Oil Chem. Soc.* 26: 441–444.
33. Hartman, L. and White, M. D. L., 1953, Stability of Edible Tallow from Mutton and Beef. II. Suppression of the Deleterious Effects of Trace Metals. *New Zealand J. Sci. and Technol. Ser. B.* 35: (3), 254–258.
34. Kraybill, H. R. and Dugan, Jr., L. R., 1954, Antioxidants—New Developments for Food Use—Acid Synergists as Metal Scavengers. *J. Agri. & Food Chem.,* 2: (2), 81–84.
35. Vahlteich, Hans W., et al., 1954, Esters of Citric Acid in Stabilizing Edible Oils. *Food Technol.,* 8: 1, 6–9.
36. Hall, Lloyd A., 1950, *Synergistic Antioxidant Containing Amino Acids.* U.S. Patent 2,518,233.
37. Lindsey, F. A., and Maxwell, W. T., 1950, *Oxidative Stabilization of Edible Fatty Oil with Citric Acid.* U.S. Patent 2,515,948.
38. Morris, S. G., et al., 1950, Metal Deactivation in Lard. *J. Am. Oil Chem. Soc.* 27: 105–107.
39. Bersworth, F. C., 1949, *Method of Treating Animal and Vegetable Oils* U.S. Patent 2,463,015.
40. Kuhrt, N. H., 1956, *Stabilized Unsaturated Composition and Stabilizer Thereof.* U.S. Patent 2,732,386.
41. Schuler, W., and Meiser, R., 1956, *Arch. Expt. Pathol. Pharmakol.* Naunyn Schiedeberg's 228: 474.

42. Neal, R. H., et al., 1949, *Retardation of the Development of Rancidity in Soybean Oil*. U.S. Patent 2,485,631.

43. Gooding, C. M., et al., 1949, *Retardation of the Development of Rancidity in Soybean Oil*. U.S. Patent 2,485,633.

44. Miers, J. C. and Owens, H. S., 1953, *Process for Coating Nuts and the Resulting Products*. U.S. Patent 2,631,938.

45. Cecil, S. R. and Woodroof, J. G., 1951, BHA Ups Shelf Life of Salted Nuts. *Food Indust.* 23: (2), 81–84, 223–224.

46. Neal, R. H., et al., 1949, *Stabilization of Nut Meats*. U.S. Patent 2,485,636.

47. Fryklof, L. E., 1954, Autoxidation of Etheric Oils Used in Pharmacy. *Farm. Revy* 53: 317–335, 361–74 (*Chem. Abstr.* 48: 8486).

48. Melnick, D., 1961, *Flavor Stabilized Salted Margarine*. U.S. Patent 2,983,615.

49. Gooding, C. M., et al., *Stabilization of Margarine*. U.S. Patent 2,485,638.

50. Valteich, H. W., et al., 1949, *Prevention of Weeping of Margarine and Butter*. U.S. Patent 2,485,634.

51. Hanley, J. W., et al., 1953, Antioxidant Treatment for Bacon. *Food Technol.*, 7: (11), 429–431.

52. Lindsey, F. A. and Maxwell, W. T., 1949, *Stabilizing Edible Fatty Oil*. U.S. Patent 2,486,424.

53. Valteich, H. W., et al., 1949, *Retardation of the Development of Rancidity in Fats and Oils*. U.S. Patent 2,485,640.

54. Black, H. C., 1950, *Stabilization of Fatty Materials*. U.S. Patent 2,494,114.

55. Dugan, Jr., L. R., et al., 1950, Butylated Hydroxyanisole as an Antioxidant for Fats and Oils. *Food Technol.* 4: 457–460.

56. Melnick, D., 1959, *Salad Stabilization*. U.S. Patent 2,910,367.

57. Melnick, D. G., Perry, A., and Akerboom, J., 1959, *Vegetable Salad and Method for Preparing Same*. U.S. Patent 2,910,368.

58. Stapf, R. J., 1959, *Emulsified Salad Dressings*. U.S. Patent 2,885,292.

59. Gribbins, M. F. and Dittmar, H. R., 1951, *Food Antioxidants*. U.S. Patent 2,564,106.

60. Akerboom, J., Melnick, D., and Perry, G. A., 1962, *Food Composition Containing an Auxiliary Additive and a Fungistat*. U.S. Patent 3,038,810.

61. Arenson, S. W., 1950, Shortenings for Frying and Baking. *Food Indust.* 2: (6), 1015–1020.

62. Neal, R. H., et al., 1949, *Prevention of the Development of Off-Flavors in Fried Potato Products*. U.S. Patent 2,485,635.

63. Anderson, B. B., Bailey, D. W., and Ash, J. C., 1963, *Method of Fat Extraction*. U.S. Patent 3,074,982.

64. Pines, R., and DiBattista, A., 1963, Private communication. Geigy Chemical Corporation, Ardsley, N.Y.

65. Furia, Thomas E., 1964, EDTA in Foods. *Food Technol.* 18: (12), 50–58.

66. Gardner, Wm. Howlett, 1966, *Food Acidulants*. Allied Chemical Corp. New York, N.Y.

67. Niadas, E. and Robert L., 1959, Kinetics of Inhibition of Cupric Oxidation of Ascorbic Acid by Complexing Agents. *Experimentia* 14: 399.

68. Rao, M. V. L., Sastry, L. V. L., Srinivasan, M., and Subrahmanyan, V., 1959, Inhibition of Oxidation of Ascorbic Acid by EDTA. *J. Sci. Food Agr.* 10: 436.

69. Wolf, P. A., 1960, Some Factors Affecting Inhibition of Copper Oxidation of Ascorbic Acid by EDTA. *Food Technol.* 14: (4), 23. *Proc. Inst. Food Technol.*, San Francisco, May, 1960. (Unpublished)

70. Erdey, L., and Bodor, E., 1952, Stabilization of Ascorbic Acid. *Z. Anal. Chem.* 136: 109.

71. Coleman, R. D., 1951, Evaluation of Inhibitors of the Copper Catalyzed Oxidation of Ascorbic Acid. *Abstr. 119th Meeting Am. Chem. Soc. 17A*. (Unpublished)

72. Jager, H., 1948, Stabilization of Ascorbic Acid. *Pharmazie* 3: 536.

73. Sohne, A. G., 1949, *Solutions with Stable Vitamin C. Content*. Swiss Patent 262,670.

74. Licciardello, J. J., Esselen, W. B., and Fellers, C. R., 1952, Stability of Ascorbic Acid During Preparation of Cranberry Products. *Food Res.* 17: 338.

75. Timerlake, C. F., 1960, Oxidation and Stability of Ascorbic Acid in Black Currant Juice. *J. Food Sci. Agr.* 11: 268.

76. Aamoth, H. L. and Butt, F. G., 1960, Maintaining Food Quality with Chelating Agents. *Ann. New York Acad. Sci.* 88: 526.

77. Bartilucci, A., and Foss, N. E., 1953, Stability of Cyanocobalamin and Ascorbic Acid in Liquid Formulations. *J. Amer. Pharm. Assoc.* 43: 159.

78. Berworth, F. C. and Rubin, M., 1958, *Preservation of Foods.* U.S. Patent 2,862,852.

79. Olcott, H. S., and Campbell, T. W., 1953, *Stabilization of Lipoidal Substances.* U.S. Patent 2,652,332.

80. Dunn, H. J. 1959, *Stable Powdered Oil Soluble Vitamins and Method of Preparing Stable Powdered Oil Soluble Vitamins.* U.S. Patent 2,897,119.

81. Cannalonga, M. F., 1958, *Stabilized Vitamin D Composition.* U.S. Patent 2,862,852.

82. Freedman, L., Blitz, M., Sabine, D., and Eigen E., 1960, *Stable Solutions Containing Vitamin B_{12}.* U.S. Patent 2,939,821.

83. Krulizova, Z., Kucharsky, J., Pohorsky, J., and Spaleny, J., 1959, *Stabile Solutions of Thiamine Salts.* Czech. Patent 90,756.

84. Sahyun, M. 1957, *Iron-Vitamin Composition.* U.S. Patent 2,804,423.

85. Weidenheimer, J. F. and Carstensen, G. T., 1954, *Stabilized Pteroyl Solutions.* U.S. Patent 2,695,860.

86. Fellers, C. R. and Morin, E. L., 1962, *Process for the Prevention of After Cooking Gray Discoloration of Potatoes and Other Vegetables.* U.S. Patent 3,049,427.

87. Clark, W. L. and Moyer, J. C., 1955, Darkening of Sliced Beets. *Food Technol.* 9: 308.

88. Hope, G. W., 1961, The Use of Antioxidants in Canned Apple Halves. *Food Technol.* 15: 548.

89. Luh, B. S., Leonard, S. J., and Patel, D. D., 1960, Pink Discoloration in Canned Bartlett Pears. *Food Technol.* 14: 53.

90. Strachan, C. C. and Moyls, A. W., 1949, Ascorbic, Citric, and Dihydroxy Maleic Acids as Antioxidants in Frozen Pack Fruits. *Food Technol.* 3: 327–332.

91. Hohl, L. A., et al., 1949, Further Studies on Frozen Fruit and Vegetable Purees. *Food Technol.* 3: 100–110.

92. Guyer, R. B. and Erickson, F. B., 1954, Canning of Acidified Banana Purees. *Food Technol.* 8: 165–167.

93. Lennon, W. J., 1962, Private Communication. Geigy Chemical Corp., Ardsley, N.Y.

94. Talburt, W. G., and Smith, O., 1959, *Potato Processing.* Avi Publ. Co., Westport, Conn.

95. Mondy, N. I., Mobley, E. O., and Gedde-Dahl, S. B., 1967, *J. of Food Science* 32: 378–381.

96. Hawkins, W. W., Chipmand, M. E. G., and Leonard, V. G., 1959, After Cooking Darkening in Oil-Blanched French-Fried Potatoes. *Am. Potato J.* 36: 255.

97. Antle, L. L. and Bohn, R. M., 1953, *Method of Preventing Discoloration of Sliced Organic Materials.* U.S. Patent 2,628,905.

98. Makower, R. U., and Schwimmer, S., 1954, Inhibition of Enzymic Color Formation in Potato by ATP. *Biochim. et. Biophys. Acta* 14: 156–7.

99. Hawkins, W. W., Leonard, V. G. and Armstrong, J. E., 1960, Effectiveness of Ascorbic Acid in Preventing the Darkening of Oil-Blanched French-Fried Potatoes. *Food Technol.* 15: 410.

100. Olson, R. L. and Treadway, R. H., 1949, Pre-Peeled Potatoes for Commercial Use. *U.S. Dept. Agr. Bur. Agr. and Ind. Chem. Publ. AIC* 246.

101. Greig, W. S. and Smith, O., 1960, *Factors Affecting Quality and Shortage Life of Pre-Peeled Potatoes and Quality of French-Fries.* N.Y. State Coll. Agr. Cornell Univ., Agr. Expt. Sta. Mem. No. 370, Ithaca, N.Y.

102. Hunsader, M. L. and Hanning, F., 1958, Effect of Complexing Agent on After-Cooking Discoloration of Potatoes. *Food Res.* 23: 269.

103. Greig, W. S. and Smith, O., 1955, Use of Sequestering Agents in Preventing After-Cooking Darkening in Pre-Peeled Potatoes. *Am. Potato J.* 32: 1.

104. Smith, O., and Muneta, P., 1954, Effect of Foliar Application of Sequestering Agents on After-Cooking Darkening. *Am. Potato J.* 31: 404.

105. Gull, D. D., and Isenberg, F. M., 1958, *Proc. Am. Soc. Hort. Sci.* 71: 446.

106. Kitzke, E. D., 1962, *Composition and Method for Treating Potatoes.* U.S. Patent 3,051,578.

107. Monier-Williams, G. W., 1949, *Trace Elements in Food*. John Wiley and Sons, New York.

108. Bailey, M. E. and Fieger, E. A., 1957, Chemical Prevention of Black Spot (Melanogenesis) in Ice Stored Shrimp. *Food Technol.* 8: 317–319.

109. Nielson, K., 1956, Canned Shellfish Discoloration. *World Fisheries Abstr.* 7: 41.

110. Brown, W. D., et al., 1957, Oxidative Deterioration in Fish and Fishery Products II. Progress on Studies Concerning Mechanism of Oxidation Oil in Fish Tissues. *Com. Fisheries Review* 19: (5a), 27–31.

111. Ladenburg, K., 1959, *Salt Tables Containing EDTA and Process for Making Same*. U.S. Patent 2,868,655.

112. Feiger, E. A., et al., 1956, Preserving Shrimp Quality and Beauty. *Southern Fisherman* 16: (12), 21–23, 81–82.

113. Ganucheau, J. J., 1948, *Preserving Canned Crustacean Meat*. U.S. Patent 2,448,970.

114. Fellers, C. R., 1954, *Color Preservation of Frozen and Canned Shellfish*. U.S. Patent 2,669,520.

115. Yamada, H., 1957, Struvite in Canned Marine Products. VI. Effectiveness of Sequestering Agents for Preventing Struvite Formation in Canned Salmon. *Chem. Abstr.* 51: 15032.

116. McFee, E. P., and Swaine, R. L., 1953, Stops Struvite with New Technique. *Food Eng.* 25:(1), 67.

117. McFee, E. P. and Peters, J. A., 1951, *Process of Canning Seafood and Resultant Product*. U.S. Patent 2,554,625.

118. Swaine, R. L., 1954, *Process of Canning Fish and Products Thereof*. U.S. Patent 2,680,076.

119. Fellers, C. R., 1961, *Salt Tablets*. U.S. Patent 3,013,884.

120. Levin, R., 1967, The Effectiveness of EDTA as a Fish Preservative. *J. Milk Food Tech.* 30: 277–283.

121. Furia, Thomas E., 1967, Unpublished Work., Geigy Chemical Corp., Ardsley, N.Y.

122. Gray, P. P. and Stone, I., 1956, *Wallerstein Labs. Communs.* 19: 345.

123. Gray, P. P. and Stone, I., 1960. *Wallerstein Labs. Communs.* 23: 181.

124. Gray, P. P., 1955, *Malt Beverages*. U.S. Patent 2,711,963.

125. Kneen, E., 1956, *Manufacture of Stabilized Beer*. U.S. Patent 2,748,002.

126. Harvey, and Toope, 1958, *Proc. Fifth Convention, Australian Sect., Inst. Brewing*, p. 230.

127. Hudson, J. R., 1958, Metal in Beer Haze. *J. Inst. Brewing* 64: 157.

128. Beckley, 1958, Gushing Beer. *Proc. Fifth Convention, Australian Sect. Inst. Brewing* 51: 63

129. Amerine, M. A., Berg, H. W., and Cruess, W. V., *Technology of Wine Making*, 2nd Ed. Avi Publishing Co., Westport, Conn.

130. Marino, E. F., 1951, Algunos Enturbiamientos de los Vinso (Some Turbidity in Wines) (I) *Siembra* 7:(6) 5–8.

131. Joslyn, M. A. and Lukton, A., 1956, Mechanism of Copper Casse Formation in White Table Wine. *Food Res.* 21: 1.

132. Cantarelli, C., 1955, Sodium EDTA for Stabilizing Wines with Metal-Induced Diseases. *Riv. viticolt. e enol.* 8: 200. (*Chem. Abstr.* 50: 1259).

133. Debiner, L. and Bouzigues, H., 1954, Action of Iron-Removing Agents Employed in Oenology. *Inds. Agr. et Aliment* (Paris) 71: 833–837 (*Chem. Abstr.* 49: 15167).

134. Joslyn, M. A., Lukton, A., and Cane, A., 1953, Removal of Excess Copper and Iron from Wine. *Food Technol.* 7: 20.

135. Krum, J. K. and Fellers, C. R., 1952, Clarification of Wine by a Sequestering Agent. *Food Technol.* 6: 103.

136. Luthi, H., The Treatment of Wines with Citric Acid. *Schweiz. Z. Obst Weinbau*.

137. Creff, R. and Jacquin, P., 1953, The Ferric Casse of Cider: A Critical Study of Several Curative and Preventative Methods of Removal of Iron. *Ann. Inst. Nat'l Recherche Agron. Ser. E., Ann Technol. Agr.*, 2: 153–176.

138. Berg, H. W., et al., 1955, Evaluation of Thresholds and Minimum Difference Concentrations for Various Constituents of Wine. I. Water Solutions of Pure Substances. *Food Technol.* 9: 23–26.

139. Joslyn, M. A., and Lukton, A., 1953, Prevention of Copper and Iron Turbidities in Wine. *Hilgardia* 22: 451.

140. Berg, H. W. and Akiyoski, M., 1956, Some Factors Involved in Browning of White Wines. *Am. J. Enology* 7: 1–7.

141. Noonan, J., 1963, Private Communication. Warner-Jenkinson Mfg. Co., St. Louis. Mo.

142. Johnson, H. T., and Daly, Jr., J. J., 1957, Use of Antioxidants in Canned Soft Drinks. *Ann. Meeting Soc. Soft Drink Technologists.*, April, 1957.

143. Murdock, D. I., 1950, Inhibitory Action of Citric Acid on Tomato Juice Flat-Sour Organisms. *Food Res.* 14: 261–274.

144. Pierpont, P. F., Trout, G. M., and Stine, C. M., 1963, Effectiveness of Nitrogen and Sulfur-Chelating Compounds in Inhibiting Development of Oxidized Flavor in Milk. *J. Dairy Sci.* 46: 1044.

145. Arrington, L. R. and Krienke, W. A., 1954, Inhibition of Oxidized Flavor of Milk with Chelating Compounds. *J. Dairy Sci.* 37: 819.

146. King, R. L. and Dunkley, W. L., 1959, Role of a Chelating Compound in the Inhibition of Oxidized Flavor. *J. Dairy Sci.* 42: 897.

147. Malowan, L. S., 1953, Effect of EDTA on Coagulation of Milk. *Ciencia* (Mex.) 13: (1), 24.

148. Maeno, M., Saito, A., Aradate, F., and Tanaka, S., 1956, Anticoagulation of Condensed Milk. I. Application of EDTA. *Dairy Sci. Abstr.* 18: 93.

149. Sargent, J. S. E., Briggs, D. A., and Irvine, D. M., 1959, Effect of Hard Water on Heat Stability of Skim Milk Powder. *J. Dairy Sci.* 42: 1800.

150. Gooding, C. M., et al., 1949, *Stabilization of Cheese.* U.S. Patent 2,485,637.

151. Lewis, M. A., Marcelli, V., and Watts, B. M., 1953, Stable Foams from Food Proteins with Polyphosphates. *Food Technol.* 7: 261.

152. Kline, L. T., Sonoda, T., Hanson, H. L., and Mitchell, J. K., 1953, Relative Chemical, Functional and Organoleptic Stabilizers of Acidified and Glucose-Free Whole Egg Powders. *Food Technol.* 7: 456.

153. Faust, W. and Ender, W., 1940, *Process for Preventing the Congealing of Blood.* U.S. Patent 2,193,717.

154. Levy, I. F., 1961, *Method of Curing Fresh Meat Products.* U.S. Patent 3,003,883.

155. Weiss, T. J., Green, F., and Watts, B., 1953 Effect of Metal Ions on Formation of Nitric Oxide-Hemoglobin. *Food Res.* 18: 11.

156. Caldwell, H. M. and Glidden, M. A., Kelley, G. G., and Mangel, M., 1960, Effect of Addition of Antioxidants on Frozen Ground Beef. *Food Res.* 25: 131.

157. Lineweaver, H., et al., 1952, Effect of Antioxidant on Rancidity Development in Frozen Creamed Turkey. *Food Technol.* 6: 1–4.

158. Benin, G. S., 1955, Accelerated Method for Cleaning Evaporators. *Sakharnaya Prom.* 29:(4), 22–3 (*Chem. Abstr.* 50: 2998).

159. Bennet, M. C., Connorley, F. H., and Schmidt, N. O., 1955, Further Developments in the Use of Versene in Evaporator Cleaning. *Proc. Brit. West Indies Sugar Technologists Assoc.*, p. 83.

160. Buckley, G. D. and Thurston, E. F., 1956, Descaling of Evaporators in Cane Sugar Factories by EDTA. A Method of Recovery. *Chem. & Ind.* (*London*) p. 493.

161. Hamilton, R., 1954, Evaporator Scale Inhibition. Reprints Hawaiian Sugar Technol. Meeting, pp. 13: 82–85. (*Chem. Abstr.* 49: 5870).

162. Holland, I. V., Massiah, B. V., and Mayers, J. C., 1954, The Use of Versene in Evaporator Cleaning: A Factory Trial. *Proc. Brit. West Indies Sugar Technologists Assoc.*, p. 155.

163. Jensen, I. M., and Claybaugh, G. A., 1951, Ethylenediamine Tetrasodium Acetate Used in Detergency. *J. Dairy Sci.* 34: 865.

164. Relf, E. T., and Foster, D. H., 1959, Chemical Removal of Evaporator Scale. *Proc. Queensland Soc. Sugar Cane Technologists*, p. 51.

165. Roche, M., 1955, The Incrustation of Heating Surfaces of Evaporator Equipment in Sugar Refineries by Calcium Oxalate. *Inds. Agr. et Aliment.* (Paris) 72: 267–72 (*Chem. Abstr.* 49: 11305).

166. Roderig, H. L., Clegg, F. L., Chapman, H. R., Rook, J. A. F., and Hoy, W. A., 1956, Experiments in Descaling Farm Utensils. *J. Soc. Dairy Technol.* 9: 75.
167. Schmidt, N. O. and Wiggins, L. F., 1954, Descaling of Evaporator Heating Surfaces in Cane Sugar Factories. *Ind. Eng. Chem.* 36: 867.
168. Schmidt, M. O. and Wise, W. S., 1958, Chemical Solution of Evaporator Scale. *Ind. Eng. Chem.* 50: 811.
169. Wiggins, L. F., 1956, Process for the Descaling of Sugar Factory Evaporators and Other Heat Transfer Equipment. U.S. Patent 2,774,694.
170. Rao, S. N. G. and Ramaiah, N. A., 1957, Possible Use of Versene in Increasing Sugar Recovery. *Sugar* 52:(12), 46.
171. Gundarao, S., Ramiah, N. A., and Agarwal, S. K. D., 1958, The Use of Versene—Possibility of Increasing Exhaustibility of Molasses. I. Details of Experimentation and Theory. *Indian Sugar* 8: 191. (*Chem. Abstr.* 53: 15610).
172. Tomic, U., 1959, Complex Compounds in First Saturation. Listy Cukrovar 75: 1–14. (*Chem. Abstr.* 53: 10813).
173. Peat, M. R., 1963, *Stabilization of Spice Material.* U.S. Patent 3,095,306.
174. Toi, B., Ikeda, S., and Furukawa, K., 1963, *Flavoring Agent.* U.S. Patent 3,109,741.
175. Korth, J. A., 1959, Pregelatinized Starch. U.S. Patent 2,884,346.
176. Suk, H. Y., *et al.*, 1965, Preventing Discoloration of Mushrooms. *Kungnip Kongop Yonguso Pogo* 15: 85–90; *Chem. Abstr.* 67: 42709b.
177. Engel, R., 1967, Influence of Blanching on the Quality and Storage Stability of Dry Vegetables. III. Storage Browning with Dried Kohlrabis. *Deut. Lebensm. Rundsch.* 63: 35–40.
178. Boyd, J. W., and Southcott, B. A., 1968, Comparative Effectiveness of Ethylenediaminetetraacetic Acid and Chlortetracycline for Fish Preservation. *J. Fish. Res. Board. Can.* 25: 1753–1758.
179. Power, H. E., Sinclair, R. and Savagaon, K., 1968, Use of EDTA Compounds for the Preservation of Haddock Fillets. *J. Fish. Res. Board Can.* 25: 2071–2082.
180. Castell, C. H., *et al.*, 1965, Rancidity in Lean Fish Muscle. *J. Fish. Res. Board Can.* 22: 229–244.
181. Yamasaki, H., Sumagawa, M. and Imai, H., 1967, Storage of Freeze-Dried Foods. VI. Water-Absorbing Capacity and Hygroscopicity of Freeze-Dried Oysters. *Nippon Shokuhin Kogyo Gakkaishi* 14: 1–6; *Chem. Abstr.* 66: 1147a.
182. Groninger, H. S. and Spinelli, J., 1968, EDTA Inhibiting of IMP Dephosphorylation in Refrigerated Fishery Products. *J. Agr. Food Chem.* 16: 97–99.
183. Martin, L., 1965, Use of Antioxidants and Sequestering Agents to Increase Product Container Life of Carbonated Beverages. *Amer. Soft Drink J.* March.
184. Borenstein, B., 1968, *Process for the Production of Cured Meat Products.* U.S. Patent 3,386,836 (to Hoffmann-LaRoche).
185. Polya, E. and Green, J., 1968, *Egg Custard Composition.* U.S. Patent 3,409,443 (to General Foods).
186. Stone, I., 1968, *Foam Stabilization in Fermented Malt Beverages.* U.S. Patent 3,202,515.
187. Nickol, G. B., 1968, *Color Restoration of Distilled Liquors.* U.S. Patent 3,411,918 (to National Distillers & Chem. Corp.).

CHAPTER 7

Gums

Robert E. Klose
Group Leader
and
Martin Glicksman
Senior Research Specialist

General Foods Corporation
Corporate Research Department
Tarrytown, New York

Introduction

The importance of gums in food products is based on the hydrophilic properties of the gums which affect the food structure, texture, and related functional properties. Gum constituents, which are present in most natural foods, are indispensable to the modern processor of convenience foods as additives that provide thickening, gelling, suspending, emulsifying, stabilizing, and film-forming properties.[290] The consumption of hydrocolloids in foods was estimated at approximately 300 million pounds for 1967, with a total dollar value of $75–80 million; starch and gelatin accounted for 250 of the 300 million pounds. Among the leading materials, in decreasing order of use in food, were gum arabic, guar gum, carboxymethylcellulose, carrageenan, locust bean gum, agar, and methylcellulose.[7] A list of properties of gums and their typical applications in the food industry is given in Table 1. Recent estimates of consumption of hydrocolloids in foods and of market prices are given in Table 2.

TABLE 1
Use of Gums in Food Products

Property	Use	Examples
Adhesive	Bakery glaze	Agar
Binding agent	Sausages	Locust bean gum
Bulking agent	Dietetic foods	Gum arabic
Crystallization inhibitor	Ice cream, sugar syrups	Sodium alginate, CMC
Clarifying agent	Beer, wine	Agar
Cloud agent	Fruit juice	Gum arabic
Coating agent	Confectionery	Gum arabic
Emulsifier	Salad dressings	Propylene glycol alginate
Encapsulating agent	Powdered fixed flavors	Gum arabic
Film former	Sausage casings, protective coatings	Sodium alginate
Flocculating agent	Wine	Sodium alginate
Foam stabilizer	Whipped toppings, beer	Propylene glycol alginate
Gelling agent	Puddings, desserts, aspics, mousses	Seaweed extracts
Mold release agent	Gum drops, jelly candies	Gum arabic
Protective colloid	Flavor emulsions	Gum arabic
Stabilizer	Beer, mayonnaise	CMC
Suspending agent	Chocolate milk	Carrageenan
Swelling agent	Processed meats	Guar
Syneresis inhibitor	Cheese, frozen foods	Guar
Thickening agent	Jams, pie fillings, sauces, gravies	Furcellaran
Whipping agent	Toppings, icings	Methylcellulose

TABLE 2
Recent Estimates for Consumption and Market Price of Hydrocolloids

Gum	Estimated Consumption,[a] Millions of Pounds	Market Price,[b] dollars/lb
Natural		
Agar	0.5	2.40–2.80
Algin	7.5	1.13–2.05
Arabic	12.5	0.42–0.60
Carrageenan	6.0	1.50–2.00[a]
Furcellaran	0.1	1.75[a]
Ghatti	0.2	0.38[a]
Guar	6.0	0.38–0.40
Karaya	0.5	0.80–0.90
Locust bean	3.0	0.52–0.58
Tragacanth	1.0	2.50–7.00
Cellulose derivatives		
Carboxymethylcellulose	8.0	0.64
Hydroxypropylcellulose	⎫	1.25–1.35[a]
Hydroxypropylmethylcellulose	⎬ 1.0	0.89
Methylcellulose	⎭	0.89

[a] M. Glicksman, in *Gum Technology in the Food Industry*, Academic Press, 1969.

[b] Unless otherwise indicated, values were taken from the *Oil, Paint, and Drug Reporter*, September 13, 1971. Reprinted by permission of the publisher, Schnell Publishing Co.

Terminology and Classification

Definition

The term *gum* originally applied to sticky, gummy, natural plant exudates; it was later extended to water-insoluble materials, such as chicle and latex, as well as to other water-soluble materials. At present the term *gum* applies to water-soluble thickening and gelling agents; water-insoluble materials are referred to as *resins*. The use of *mucilage* to describe certain types of gums has been essentially discontinued.[8] The current, accepted, technical definition of a gum is a polymeric material that can be dissolved or dispersed in water to give a thickening and/or a gelling effect.[290] Since these materials are colloidal in nature, they are also referred to as *hydrophilic colloids* or *hydrocolloids* in more scientific circles.

Classification

The above definition of a gum, in addition to the natural plant polysaccharides such as tree exudates, seaweed extracts, pectin, and starch, includes proteins such as gelatin and casein, chemical derivatives of cellulose and starch

TABLE 3
Classification of Gums

Natural Gums	Modified Gums (Semi-Synthetic)	Completely Synthetic Gums
Tree Exudates and Extracts:	Cellulose Derivatives:	Vinyl Polymers:
Arabic	Carboxymethylcellulose	Polyvinylpyrrolidone
Tragacanth	Methylcellulose	Polyvinylalcohol
Karaya	Hydroxypropylmethylcellulose	Carboxyvinyl Polymer
Larch	Hydroxypropylcellulose	Acrylic Polymers:
Ghatti	Hydroxyethylcellulose	Polyacrylic Acid
Seed or Root:	Ethylhydroxyethylcellulose	Polyacrylamide
Locust Bean	Microcrystalline Cellulose	Ethylene Oxide Polymers
Guar	Starch Derivatives:	
Psyllium Seed	Carboxymethylstarch	
Quince Seed	Hyroxyethylstarch	
Seaweed Extracts:	Hydroxypropylstarch	
Agar	Microbial Fermentation Gums:	
Algin	Dextran	
Carrageenan	Xanthan Gum	
Furcellaran	Others:	
Others:	Low Methoxyl Pectin	
Pectin	Propylene Glycol Alginate	
Gelatin	Triethanolamine Alginate	
Starch	Carboxymethyl Locust Bean Gum	
	Carboxymethyl Guar Gum	

Source: M. Glicksman, in *Gum Technology in the Food Industry*, Academic Press, 1969.

such as methylcellulose and dextrans, and completely synthetic chemical products such as polyvinylpyrrolidone (PVP) or ethylene oxide polymers. For the sake of convenience, these materials may be classified into three groups, as follows:

1. *Natural Gums*—Those found in nature, such as gum arabic or alginates.
2. *Modified Gums or Semi-Synthetic Gums*—Chemical derivatives of natural material, such as cellulose or starch, and gums derived by microbial fermentation of natural materials.
3. *Synthetic Gums*—Completely synthesized chemical products, such as PVP.

Table 3 lists many of the commonly used gums in each of these classifications. Of the gums listed in Table 3, only a dozen or so are extensively used in the food industry. Most of the synthetic and many of the semi-synthetic gums are either not approved for use in foods or are approved for very limited use; the major exceptions are modified gums, such as carboxymethycellulose, methylcellulose, low methoxyl pectin and propylene glycol alginate. Certain natural gums, such as psyllium seed or quince seed gums, are GRAS but find little use because of limited availability, price, etc. Other materials, such as starch, pectin and gelatin, are not included in this chapter because their use is so widespread and their technology so extensive that they are treated independently (see Chapter 8, "Starch in the Food Industry").

Table 4 lists the gums considered appropriate for inclusion in this chapter, along with their botanical origin, major geographical sources, and major structural components. Included are eleven natural plant gums, five cellulose derivatives, and one microbial gum. The natural plant gums are usually further classified into (a) seaweed extracts comprising agar, algin, carrageenan and furcellaran, (b) tree exudates and extracts comprising arabic, ghatti, karaya, tragacanth, and larch, and (c) seed gums comprising locust bean and guar.

Structure and Chemical Composition

Most of the gums discussed in this chapter are complex polysaccharides, anionic or neutral, often associated in nature with metallic cations such as calcium, potassium, or magnesium. There is a structural relationship between many of the gums. In cellulose and its derivatives, glucose units in the beta configuration are joined by a 1 → 4 linkage. In starch glucose units are in the alpha position with primarily 1 → 4, but some 1 → 6 linkages. Pectic acid is a polymer of galactose with beta 1 → 4 linkages. Alginic acid is also a 1 → 4 linked polymer in the beta position, composed of mannuronic and guluronic acids. The red seaweed extracts, agar, carrageenan and furcellaran, have a common basic structure consisting of a chain of galactose units alternately linked alpha 1 → 3 and beta 1 → 4. The natural plant exudates have complex structures that have not been fully defined but generally consist of various sugar components in a multitude of structural shapes and sizes. The seed extracts, locust bean and guar, are neutral polysaccharides of galactose and mannose, differing only in the proportions of galactose and mannose. Smith and Montgomery have published an extensive review of the chemistry of plant gums.[5] Structure and chemical composition are further discussed in the sections on the individual gums. Major structural components of the various gums are shown in Table 4.

TABLE 4

Botanical Origin, Sources, and Structural Components of Imported Food Gums

Gum	Botanical Origin[a,b,d]	Major Geographical Source[a,b,d]	Sugar Units[c,d]
Seaweed Extracts			
Agar	*Gelidium* species	Japan, U.S.	D-Galactose β-(1→4), 3,6-Anhydro-L-galactose-(1→3), + Sulfate Acid Ester Group
Algin	*Macrocystis pyrifera* *Laminaria* species	U.S., U.K.	D-Mannuronic Acid β-(1→4), L-Guluronic β-(1→4)
Carrageenan	*Chondrus crispus* *Gigartina* species *Euchema* species	U.S., Europe Europe Asia	D-Galactose, 3,6-Anhydro-D-galactose, + Sulfate Acid Ester Groups
Furcellaran	*Furcellaria fastigiata*	Denmark	D-Galactose, 3,6-Anhydro-D-galactose, + D-Galactose-4-sulfate
Tree Exudates and Extracts			
Arabic	*Acacia senegal*	Sudan	L-Arabinose, D-Galactose, L-Rhamnose, D-Glucuronic Acid
Ghatti	*Anogeissus latifolia*	India	L-Arabinose, D-Xylose, D-Mannose, D-Glucuronic Acid
Karaya	*Sterculia urens*	India	D-Galactose, L-Rhamnose, D-Galacturonic Acid
Arabinogalactan (Larch)	*Larix occidentalis*	U.S.	D-Galactose, L-Arabinose
Tragacanth	*Astragalus gummifer*	Iran	D-Galactose, D-Xylose, D-Glucuronic Acid
Seed Gums			
Guar	*Cyamopsis tetragonolobus*	India, Pakistan	D-Mannose β-(1→4), D-Galactose-(1→6) Branches
Locust Bean	*Ceratonia siliqua*	Mediterranean	D-Mannose β-(1→4), D-Galactose-(1→6) Branches
Cellulose Derivatives			
Carboxymethylcellulose	⎫	U.S.	D-Glucose β-(1→4)
Methylcellulose and Hydroxypropylmethylcellulose	} Wood Pulp and Cotton Linters	U.S.	D-Glucose β-(1→4)
Hydroxypropylcellulose		U.S.	D-Glucose β-(1→4)
Microcrystalline Cellulose	⎭	U.S.	D-Glucose β-(1→4)
Microbial Gums			
Xanthan		U.S.	D-Glucose, D-Mannose, D-Glucuronic Acid

[a] Data from "Natural Water-Soluble Gums," *Oil, Paint, and Drug Reporter*, January 27, 1964.

[b] Data from *Industrial Gums*, R. L. Whistler, Ed., Academic Press, 1959.

[c] Data from "Gums," *Encyclopedia of Science and Technology*, Vol. 6, McGraw-Hill Book Company, 1960, p. 297.

[d] Data from M. Glicksman, *Gum Technology in the Food Industry*, Academic Press, 1969.

Physical Properties

A brief, introductory discussion of the most outstanding physical properties of gums is given below. These properties are more fully discussed in the sections on the individual gums.

Dispersibility

One of the common problems encountered with gums, particularly those that form very viscous solutions, is that of dispersibility. It is often difficult to disperse fine particle size gum in water so that hydration takes place quickly. If care is not taken, the gum will take up water to form lumps or gel-like masses which are wet on the outside, but dry or gel-like in the center, and are very difficult to break up and dissolve. Several well known techniques are commonly used to facilitate dispersion and avoid lumping as follows:

1. Add the gum slowly, with sifting if possible, while vigorously agitating the water.
2. When possible, mix the gum thoroughly with other dry ingredients in the formula before adding to the water.
3. Dispersion of gums soluble only in hot water may be facilitated by first wetting with cold water. The converse is true for methylcellulose, which is soluble in cold water but not in hot water.
4. If practical, disperse the gum in a retardant before adding to water. Alcohols, acetone, liquid sugar, and glycerine may be used for this purpose.

These methods are adequate for most laboratory or production situations. Several other methods have been patented which also may be useful. These include: coating the gum with wetting agents,[9] pretreating the gum by lyophilization,[10] adjusting of bulk density, particle size, and incorporation of salts,[11] or the agglomeration of fine particle size gum into larger agglomerate particles.[12] Pittet, in discussing the dissolution of polysaccharides which form viscous solutions, mentions agitation, particle size, preblending with an inert solid, use of an inert water-miscible liquid, and pH adjustment as techniques that may be utilized to form homogeneous, lump-free solutions.[13]

Solubility

The gums commonly used as food additives have very limited solubility in alcohol or other organic solvents. While all are soluble in water, by definition, the degree of solubility and the solution temperature vary. Most of the gums are commonly used at 1–2% concentration, and form solutions above 5% concentration with difficulty. The exceptions to this are gum arabic and larch gum, which are soluble up to approximately 50% concentration. Relatively high concentrations of low viscosity cellulose derivatives can also be prepared. Some gums such as arabic, larch, guar, carboxymethylcellulose, and methylcellulose are completely, or almost completely soluble in cold water. Others, such as locust bean gum and tragacanth, will swell somewhat in cold water, but require heating to attain maximum hydration. Still others, such as agar, are insoluble in cold water and are best dissolved in boiling water. Finally, methylcellulose, which is soluble in cold water, is insoluble in hot water.

Viscosity

Besides water solubility, the other necessary attribute of a gum is that it produces viscous solutions or dispersions in water. Here again there is a great deal of variability between gums in regard to solution viscosities. Appreciable viscosity is only attained at concentrations from 10–20% for arabic and larch, but 1% solutions of tragacanth, guar, and locust bean gum are very viscous. The time it takes individual gums to attain maximum viscosity also varies widely. CMC and guar hydrate comparatively rapidly in cold water. Other gums, such as tragacanth, require long standing or heating to attain maximum viscosity. Table 5 shows some representative viscosities for the various gums. In general, the viscosity of a gum solution depends, apart from the type of gum itself, on temperature, concentration, degree of polymerization of the gum, and other substances in solution. These factors are discussed in the sections on the individual gums. In addition, the method of measuring viscosity must be standard for comparison of data to be meaningful.

Gelation

Only a comparatively few gums have the ability to form gels, i.e., rigid textures that do not flow. Some gums, such as tragacanth, form thick, heavy pastes at high concentrations that are sometimes called gels, but are not true gels. Of the gums discussed in this chapter, only the seaweed extracts form gels that are important in the food industry. Guar gum, for example, forms a gel when treated with borax, but this gel is not edible. Solutions of CMC can form gels when treated with aluminum ions, but as yet these have no important food applications. Three important hydrocolloids not covered in this chapter, gelatin, pectin, and starch, are also gel formers; along with the seaweed extracts, agar, algin, carrageenan and furcellaran, they make up the important gelling materials used in foods. However, the conditions under which these materials form gels and the type of gels formed are sufficiently different so that they cannot be indiscriminately substituted one for another. The major attributes of each of these materials are discussed below.

1. Gelatin—Hot water soluble. Gives thermally reversible, clear, elastic gels, but does not set up at room temperature under normal conditions.
2. Pectin—Hot water soluble. Forms smooth, high solid gels in the presence of sugar and acid which can be spread smoothly, as in jams and jellies. Low methoxyl pectin is cold water soluble and forms irreversible, brittle chemical gels, in a manner similar to alginates.
3. Starch—Raw starch is hot water soluble, and forms long, smooth, cloudy gels, usually cooked with milk. Precooked, or pregelatinized starches form gels in cold milk in the presence of phosphates.
4. Alginates—Form irreversible chemical gels in hot or cold water that set up at room temperature, but are not elastic.
5. Carrageenan and Furcellaran—Form thermally reversible gels when dissolved in hot water and cooled to room temperature.
6. Agar—Similar to carrageenan, but must be dissolved in near boiling water; gels have extreme hysteresis lag. Gels form at 40–50°C and do not melt unless reheated to 80–85°C. Gels are extremely brittle in texture.

TABLE 5
Viscosities of Common Gums (centipoise)

Per cent	Bacteriological Agar[a]	Carrageenan[b]	Sodium Alginate[b]	Furcellaran[d]	Locust Bean Gum[b]	Gum Guar[b]	Gum Arabic[b]	Gum Tragacanth[b]	Gum Karaya[c]	Sodium Carboxymethylcellulose[e]	Methylcellulose MC (25 cps)[f]	Hydroxypropylcellulose (Klucel MF)[g]	Xanthan Gum[h]	Arabinogalactan (Larch)[i]
0.5		24	86		20	1,389				28				
1.0	4	57	214		59	3,025		54	3,000	69	8	40	1,000	
1.5			1,102	850										
2.0	25	397	3,760		1,114	25,060		906	8,500	1,160	25	5,000	4,000	
2.5			8,300							2,840				
3.0		4,411	29,400		8,260	111,150		10,605	20,000	5,330	65	30,000		
4.0	400	25,356	39,660		39,660	302,500		44,275	30,000	34,400	150			
5.0		51,425			121,000	510,000	7	111,000	45,000	115,500	400			
6.0			121,000					183,500						
10.0							17							3
20.0							41							9
30.0							200							14
35.0							424							
40.0							936							41
50.0							4,163							

[a] Selby, H. H. and Selby, T. A., 1959. Agar. *Industrial Gums* (R. L. Whistler, Ed.), Academic Press, N.Y., pp. 15–50.
[b] Glicksman, M., 1962. Utilization of Natural Polysaccharide Gums in the Food Industry. *Advances in Food Research*, Vol. II, Academic Press, p. 109.
[c] Goldstein, A. M. and Alter, E. N., 1959. Gum Karaya. *Industrial Gums* (R. L. Whistler, Ed.), Academic Press, N.Y., pp. 343–360.
[d] Schachat, R. E. and Glicksman, M., 1959. Furcellaran, A Versatile Seaweed Extract. *Econ. Botany 13*, 365.
[e] Mantell, C. L., 1947. *The Water Soluble Gums*, Reinhold Publ. Co., N.Y., p. 153.
[f] Dow Chemical Company, 1962. Methylcellulose.
[g] Hercules Inc., 1968. Klucel Bulletin.
[h] Kelco Co., 1969. Keltrol Bulletin.
[i] Glicksman, M., 1969. *Gum Technology in the Food Industry*, Academic Press, N.Y., p. 193.

Emulsification, Stabilization, and Suspension

Gums have been used as emulsifiers, emulsion stabilizers, and suspending agents since ancient times. All of these properties are related to viscosity. Gums are effective primarily in oil-in-water emulsions, and the concentration must be high enough to have a marked thickening effect on the aqueous phase. Though they have been and are used as primary emulsifying agents, they are considered to operate principally as auxiliary emulsifiers, or emulsion stabilizers by increasing the viscosity, since a good primary emulsifier must have both strong lipophilic and hydrophilic properties.[14,15]

The ability of gums to act as suspending agents, or protective colloids, is also important in foods. The Zsigmondy gold number, which is often used as a measure of the suspending power of a colloid, is defined as the weight of the protecting colloid (in mg); it is barely insufficient to prevent a color change of 10 ml of a red-gold solution to violet when 1 ml of a 10% sodium chloride solution is added. It has been proposed that in place of the Zsigmondy number the protective value (PV) be used, which is defined as the number of grams of red-gold solution that can barely be protected by one gram of the protective agent against flocculation by a 1% sodium chloride solution.[16]

Stability of Solutions

The stability of gum solutions is greatly dependent on pH and the presence of electrolytes and other materials. In this respect, however, the stability of the different gums varies widely, and this phase of stability is discussed under the properties of the individual gums. On the other hand, all gum solutions are susceptible to degradation by bacteria, and the use of preservatives is necessary if long term stability is desired. In general, benzoic acid, sodium benzoate, sorbic acid, potassium or sodium sorbate at 0.1%, or a combination of methyl and propyl parabens are used, depending on the type of gum and the pH of the solution. The benzoates are generally used below pH 4.0, and the parabens for higher pH solutions.[17]

Seaweed Extracts

Agar

Background   Agar is the extract of *Gelidium* and other species of red algae. Chemically, agar is believed to be composed of 3,6-anhydro-L-galactose and D-galactopyranose residues in varying proportions.[5,18] The term agar, which had its origin in the Malayan word agar-agar, has been used to designate both the dried extract and the seaweed source, but it has been proposed that the term "agarophyte" be used for the weed.[4] Agarophytes have been used in the orient for centuries as a basis for dessert gels and other food preparations. Agar was also the first seaweed known to be extracted, purified, and dried when the process was accidentally discovered by a Japanese innkeeper about 1658. It was introduced to Europe and the United States from China in the nineteenth century, where it was initially used as a gelatin substitute in the making of desserts. It soon became widely used as a solid bacteriological culture medium after its use by Robert Koch in his famous experiments on tuberculosis bacteria.[18] Its major uses in the food industry of today are in bakery products, confectionery, dairy products, and canned meat and fish. It is also used in microbiology, dentistry and medicine.[19]

Sources and Harvesting   The *Gelidium* species *amansii* and *cartilagineum* are the major sources of agar, although many species of *Rhodophyceae* are used.[20] Selby and Selby give an extensive listing of algae-yielding agar and agaroids.[21] The weeds used in the commercial production of agar grow from the tide line out to depths of 120 feet, and are harvested by waders along the shore at low tide or raked from small boats, or picked by skin divers or divers in suits. Japan is the largest producer, and up to World War II had a virtual monopoly on the agar trade. When the war shut off the supply of Japanese agar, many countries produced agar in small amounts, including Mexico, Australia, New Zealand, France, Spain, the United States, Russia, South Africa, and some South American countries. Several of these countries still produce agar mainly for domestic consumption. Most of the manufacturers that attempted agar production in the United States during World War II discontinued operations when Japanese agar again became available. The American Agar and Chemical Company of San Diego is currently the only U.S. producer.[21,22]

About two-thirds of the approximately one million pounds of agar used in the U.S. is still imported from Japan.[6] The remainder is domestically produced by American Agar, primarily from *Gelidium cartilagineum* harvested off Southern and lower California. The weed grows on rocks in fast-moving water, from the tide level to depths of 40 or 50 feet. It is usually harvested from May to October by skin divers or divers using suits. Other species such as *Gelidium arborescens* and *Gelidium nudifrons* are sometimes used. After gathering, the seaweed is cleaned, washed with fresh water, and spread in thin layers to dry for 4 to 20 days, then pressed into bales and shipped to the processor.[23,24]

Processing   The discovery that agar could be purified and dried is said to have occurred in 1658. A Japanese innkeeper threw out some agar jellies that had been prepared by boiling seaweed, and later noted that they had evidently frozen, then thawed and dried in the sun to translucent flakes. When boiled in water, these flakes reproduced the original agar gel. Present commercial methods of producing agar are still based on the fundamental principles of hot water extraction, cooling to form a gel, freezing, thawing, and drying.[25] Since agar is soluble in hot water but relatively insoluble in cold water, it is extracted by boiling the agarophyte in water, filtered, cooled to form a gel, cut into pieces and frozen, then thawed to free the agar from salts and other impurities which are soluble in cold water. The wet agar is repeatedly washed with cold water, and finally dried. The original process developed depended entirely on natural means for freezing, thawing, and drying, and this traditional method is still used by many small Japanese producers. However, many Japanese producers, and the American producer use more mechanical, scientific processing. Chemical treatment of the weed, pressure extraction, artificial freezing and drying, chemical bleaching, and many more advanced methods are used.[1,18,21,25] American and Japanese agar are graded according to published specifications. The high quality American agar is divided into bacteriological, medicinal and dental grades, and the Japanese agar into three grades and two subgrades. Selby and Selby give the specifications for these grades.[21]

Structure   The structure of agar is not completely known. It is composed of D-galactose, and 3,6-anhydro-L-galactose, with a small amount of ester sulfate. Smith and Montgomery suggest that the term agar represents a spectrum of polysaccharides with different proportions of 3,6-anhydro-L-galactose and D-galactopyranose residues linked through $C_1 \rightarrow C_4$ and $C_1 \rightarrow C_3$, respectively.[5] Selby and Selby also feel that the term agar should be used only to indicate a galactoglycan possessing certain properties common to a group of substances, each of which may differ structurally from the others.[21] Araki, in reviewing work on the Japanese agar weed *Gelidium amansii*, concluded that agar was a mixture of at least two polysaccharides.[26] These include agarose (the gelling component) which is a neutral non-sulfated linear molecule composed of alternating residues of 1,3-β-D-galactopyranose and 3,6-anhydro-α-L-galactopyranose, and agaropectin (a non-gelling or very weak gelling component) composed of a complicated acidic polymer containing ester sulfate groups and organic acid groups in addition to the two sugar residues.

Properties and Uses   The U.S. Pharmacopoeia defines agar as a hydrophilic colloid, extracted from certain marine algae of the class *Rhodophyceae*, which is insoluble in cold water and soluble in boiling water.[27] A 1.5% aqueous solution should be clear, and forms a firm, resilient gel when cooled to 32–39°C which does not melt below 85°C. It is this extreme hysteresis lag, (the ability of agar to gel at temperatures much lower than the gel-melting temperature) that makes agar unique, and many of its uses depend on this characteristic. Solutions of up to 5% can be readily made by heating to 95°–100°C with agitation. Higher concentrations can be dispersed by autoclaving. Agar solutions have relatively low viscosities for seaweed extracts. Viscosity is dependent on temperature and pH, but is fairly constant from pH 4.5 to 9.0. Agar solutions of very low concentration can form gels. The usual concentration range for gel forming is 1–2%. Agar gels in this range are strong, resilent, somewhat elastic, relatively transparent, thermally reversible, and exhibit syneresis. Solution viscosity, gelling temperature, gel strength, degree of syneresis and gel clarity may vary with the weed source of the agar.[20,21,23] Major food uses of agar are as gelling agents in the bakery industry, in confectionery, in meat and fish products, and in dairy products.[19,21,28]

Algin

Background   Alginic acid was first extracted and isolated from brown algae by the British chemist E. C. C. Stanford in 1881. It is a major constituent of the cell wall of all species of *Phaeophyceae*, or brown algae. Chemically, alginic acid is a high molecular weight, linear polyuronide made up of D-mannuronic and L-guluronic acid. All derivatives of alginic acid are designated by the generic term algin. The food industry is one of the major users of alginates, along with the pharmaceutical, cosmetic, rubber, textile, and paper industries. The most commonly used algin compound is sodium alginate, and algin and sodium alginate are often used interchangeably. However, ammonium, potassium, and propylene glycol alginates are also used.[29]

Sources, Harvesting and Producing Areas   Although all species of brown algae contain algin, only a few species are abundant or accessible enough to be used commercially. The major commercial species are *Macrocystis pyrifera*, *Laminaria digitata*, and *Laminaria cloustoni*. *Ascophyllum nodosum* and some species of *Fucus* are also used. The algin content of the weeds varies from species to species, and is also affected by the season of the year and the locality in which they grow. For example, *Macrocystis pyrifera* has been reported to have an average algin content of 16–18%, and a range of 14–19%, while for *Laminaria digitata* an average of 32–33% and a range of 15–40% have been given.[24,29-31]

The United States is the largest producer of algin, most of which is derived from the giant kelp *Macrocystis pyrifera*. The kelp is harvested from extensive beds in the Pacific Ocean off the coast of southern California by means of large ocean-going vessels with mechanical cutting and loading equipment. No other commercial seaweed lends itself to mechanical harvesting so readily. Systematic harvesting of the beds, which may be as large as a mile in width and several miles in length, is carried out three to four times a year. The giant kelp is a perennial which lives 8–10 years. The mature stipes range in length from 50–200 feet. It grows in water 25–80 feet deep in areas having a rocky bottom and strong currents, and is attached to the ocean floor by a hold-fast, or root-like structure. Other sources of supply for the U.S. algin industry are the coasts of Maine and Nova Scotia where the smaller sublittoral kelps, *Laminaria digitata*, *Laminaria cloustoni* and *Laminaria saccharina* are used. These species range from 5–15 feet in length. Seaweed cast on the shore is gathered, as well as the growing plant which is manually harvested close to the shore from small boats.[23,24,29-31]

Britain, which accounts for over a quarter of the world production, is the second largest algin producer. *Laminaria cloustoni* and *Ascophyllum nodosum* are the primary raw materials. The west coast of Scotland, the Hebridian and Orkney Islands and Ireland are the major suppliers. Both cast and cut weed are used. Other algin producers are Norway, France and Japan. *Laminaria digitata*, *Laminaria cloustoni and Ascophyllum nodosum* are used in Norway, mainly *Laminaria digitata* in France, and *Ecklonia cava* and various species of *Laminaria* in Japan.[29,32,33]

Processing All present methods of producing algin are based on the original process developed by Stanford. This consisted essentially of macerating *Laminaria* for 24 hours with 10% sodium carbonate, which completely distintegrated the plants to a viscous, semigelatinous mass; then filtering, and treating with sulfuric or hydrochloric acid to precipitate the alginic acid. After filtering and washing, it was sold as alginic acid, or converted to sodium alginate.[30,32,34] Algin was made by the Hercules Powder Company during World War I as a by-product of the fermentation of kelp to produce calcium acetate and acetone, using a method similar to Stanfords.[25] Current methods of production in the U.S. are based on the patented processes of Green,[35] (Greens Cold Process) used by the Kelco Company of San Diego, California; and of LeGloahec and Herter,[36] used by Marine Colloids, Inc. of Rockland, Maine. These processes are given in detail by Tseng and others.[25,34] Methods used in the algin industries of Britain, Norway, France and Japan are all very likely based on variations of the Stanford process.[29]

Structure Stanford first thought that alginic acid was a nitrogenous material, and proposed the formula $C_{76} H_{76} O_{22} (NH_2)_2$, but improved methods of isolation showed no nitrogen to be present. Until recently, alginic acid was thought to be primarily a polymer of anhydro-1,4,β-D-mannuronic acid, based on the work of Hirst, Jones, and Jones.[37] The presence of L-guluronic acid in the products of hydrolysis of alginic acid was established by Fischer and Dörfel[38] in 1955. However, alginic acid is currently regarded as a polyuronide comprising 1,4-linked D-mannuronic and L-guluronic acid units, with the possibility of other linkage in addition. Commercial alginic acid has an equivalent weight range from 194 to 215, and commercial sodium alginates have molecular weights between 32,000 and 200,000, with a degree of polymerization from 180 to 930.[20,39-43]

Properties Alginic acid, and also its calcium salt, have very limited solubility in water. The sodium, potassium, and ammonium salts, and the propylene glycol ester are, however, readily soluble in hot or cold water. The viscosity of these solutions is dependent on temperature, concentration, molecular weight, and the presence of polyvalent metal cations. Commercial alginates may vary in viscosity from 10 cps for a 1% solution of a low viscosity material to 2,000 cps for a high viscosity material at the same concentration. Alginate solutions decrease in viscosity on heating, and may be degraded if held at high temperatures for extended periods, but will regain their original viscosity on cooling if degradation is avoided. The viscosity of alginate solutions is generally unaffected over the pH range of 4-10. Below pH 4, the viscosity tends to increase because of the lower solubility of the free acid, and eventually alginic acid is precipitated. Propylene glycol alginate solutions, however, are more acid stable. Viscosity is increased by the presence of polyvalent metal ions. Gels may also form if calcium or other polyvalent metal cations are present in sufficient concentration, and also in acidic solution. Gels are usually formed by the gradual release of calcium or hydrogen ions, or both, and can be controlled by the presence of a sequestrant, such as a phosphate or polyphosphate. Gels formed in this manner are clear, transparent, and do not melt at room temperature. Algin also has excellent film forming properties. Films can be formed by drying a thin layer of a soluble alginate solution; by treating a soluble film with a di- or tri-valent metal, or acidic solution; or by extruding into a precipitating bath which produces an insoluble alginate.[19,29,44]

Food Uses Principle uses for alginates in the food industry are as a stabilizer in ice cream, ice milk, water ices and sherbets, and cheese; as a gelling agent for water dessert gels and milk puddings; as a suspending and thickening agent in fruit drinks and other beverages; as a foam stabilizer in beer; as an emulsifier in salad dressings; and as a film-forming agent in coatings for meat, fish and other products.[19,29,44-47]

Carrageenan

Background The use of dried, bleached Irish moss in food and medicinal preparations was known for hundreds of years in Ireland and elsewhere in Europe and in the United States from 1835. However, it was not until World War II that a dried, purified extract was produced in an appreciable quantity as a replacement for Japanese agar. This extract was named carrageenan after the Irish town of Carragheen where it was first exploited. Legally carrageenan is the extract of *Chondrus crispus*, *Gigartina stellata*, and other related species of red algae. The use of carrageenan has expanded tremendously; it is used widely in the food industry because of its viscosity, gelling, and protein reactivity properties.[5,19,23,48,290]

Sources and Harvesting   The U.S. Federal Register describes the food additive carrageenan as the refined hydrocolloid prepared by an aqueous extraction from members of the families *Gigartinaceae* and *Solieraceae* of the class *Rhodophyceae*, including *Chondrus crispus, Chondrus ocellatus, Eucheuma cottonii, Eucheuma spinosum, Gigartina pistillata, Gigartina radula,* and *Gigartina stellata.*[49] Other sources have been given, including species of *Iridaea*, also of the *Gigartinaceae*.[3] However, *Chondrus crispus* is the most important source in the United States, and *Chondrus* and *Gigartina* in Europe. *Chondrus crispus* is a bushy plant, about 3–6 inches tall, growing in clumps and attached by holdfasts to rocks in the intertidal regions and down to 15 feet or more in depth. It is usually harvested from April to September by hand raking from small boats at ebb tide. Cast weed is also collected on the shore. The weed is dried and baled before shipping to processors. In the United States the major sources of seaweed for carrageenan production are the coasts of Maine, Massachusetts, and the Maritime Provinces of Canada. Ireland, France and Norway supply most of the weed used by European producers. Also, raw materials from Spain, Portugal, and North Africa are used by both American and European processors.[1,23,48,50]

Processing   The dried weed is first washed in cold water to remove soluble impurities. It may then be ion exchanged if monovalent salts are desired. Extraction of carrageenan from the weed is carried out with hot water. Usually, one part of weed is extracted with 50 parts of dilute alkaline solution for one to four hours at approximately 180°F. The crude extract, containing about 1% solids, is treated with adsorbents to remove soluble impurities, then filtered and vacuum concentrated to 2–3% solids. Drum drying or alcohol precipitation is used to recover the carrageenan from the concentrated extract. The drum dried material contains whatever impurities remained in the concentrated extract. The alcohol precipitation method leaves most of these impurities behind in solution. A stringy precipitate is recovered and must be further dried to remove residual water and solvent.[25,48,51]

Structure and Composition   Carrageenan is a strongly charged anionic polyelectrolyte of large size. It is generally regarded as containing two major fractions, which may be separated from dilute aqueous solution by the precipitation of one fraction, by the addition of potassium ions. The precipitated fraction, or gelling fraction, is designated kappa-carrageenan, and is composed of residues of D-galactose and 3,6-anhydro-D-galactose in a molar ratio of 1.1–1.5 to 1 and contains one sulfate half-ester group for every 2-2.5 monosaccharide units, depending on the source. Lambda-carrageenan, the non-gelling fraction, is composed primarily of 1,3-linked D-galactose-2-sulfate and 1,4-linked D-galactose-2,6-disulfate.[52-55] The sulfate content of the two fractions differ substantially with values of approximately 24% and 33% having been given for the κ and λ fractions, respectively.[5]

As an exception, carrageenan from *Eucheuma spinosum* cannot be fractionated with potassium ions. The structure of this type of carrageenan has been clarified by Mueller and Rees[313] and has been designated as ι-carrageenan. It is composed primarily of 1,3-linked galactose-4-sulfate with 1,4-linked 3,6-anhydro-D-galactose-2-sulfate. The sulfate content is similar to λ-carrageenan.

Properties   Carrageenan is readily soluble in water, but it has few other solvents. Generally, it is necessary to heat most carrageenan solutions to attain complete solubility. Normally, temperatures of 50–80°C are necessary, depending on the presence of gelling cations such as potassium or other factors.[51] Carrageenan dispersions are quite viscous, with concentration and temperature having a logarithmic effect on viscosity. Plant source, molecular weight, and the presence of metallic ions also affect viscosity. Two per cent solutions of commercial carrageenan may vary in viscosity from 50–3000 cps at 40°C when no metallic ions are present, with usual viscosities in the 500–1000 cps range.

In the presence of specific metallic cations, such as potassium, calcium, ammonium, and others, carrageenan solutions will form short, inelastic, thermally reversible gels on heating and cooling. The ability of carrageenan to form gels with potassium ions is the basis for many food applications. A level of 0.2% potassium chloride is frequently used. The temperature of gel formation is determined by the concentration and type of ions in solution, but is usually 45–55°C. The gels can be melted at approximately 10°C above setting temperature. The inelastic character of carrageenan can be modified by the presence of other hydrocolloids, particularly locust bean gum, which will also reduce the syneresis usually occurring in carrageenan gels. Carrageenan solutions and gels are fairly stable over a wide pH range at room temperature or lower, but are rapidly degraded under conditions of low pH and high temperature.

The most unique property of carrageenan as a hydrocolloid is its high degree of reactivity with certain proteins; its reactivity with milk protein in particular is the basis for a number of the applications of carrageenan in foods. This reaction between casein and carrageenan called "milk reactivity" makes it possible to suspend cocoa or other particles in milk with the use of a very small amount of carrageenan (0.025%) by creating a thixotropic system and forming a weak gel, but only slightly increasing the viscosity of the milk. If a higher level of carrageenan is used (0.15%), strong gels with the consistency of a custard or flan are formed.[3,20,23,48,51,57]

Until recently the commercial carrageenan was primarily an extract of *Chondrus crispus*, comprising approximately 60% κ-carrageenan (gelling) and 40% λ-carrageenan (non-gelling) fractions. Lately the industry has been utilizing other seaweed sources of carrageenan with different proportions of κ and λ components, with concomitant differences in gelling and solubility properties. In addition, *Eucheuma spinosum*, whose extract has been designated *i*-carrageenan, has been commercialized. This extract is characterized by particular sensitivity to calcium ions and its ability to form thermally reversible, elastic, gelatin-like gels.[290]

Major Food Uses[290]   Major uses of carrageenan in the food industry include the previously mentioned use in chocolate milk to suspend cocoa particles and the use in milk puddings, pie fillings, and water dessert gels as a gelling agent. It is also used in ice cream to prevent whey-off; in dietetic foods, as a bodying agent; in ice pops and variegated ice cream, to prevent color migration; in salad dressings, to give body and act as a suspending agent; and in beer as a clarifying and stabilizing agent. It is also used to improve body and mouthfeel in soups, sauces, soft drinks, fruit drinks, syrups, and toppings.[19,48,51] A major use for *ι*-carrageenan is in the production of shelf-stable, non-refrigerated, canned dessert gel products.[314,315]

Furcellaran

Background   Furcellaran, formerly known as Danish agar, is an extract of the red algae *Furcellaria fastigiata*. It is found mostly in Danish waters, and was not commercially produced until World War II, when the shortage of Japanese agar made it necessary to find substitutes. A composition of 3,6-anhydro-D-galactose and D-galactose sulfate has been suggested for furcellaran. Gelling properties are intermediate to agar and carrageenan. An increasingly large market has developed for furcellaran since World War II, primarily as a gelling agent in the food industry where its ability to form smooth, relatively elastic gels make it valuable.[58]

Sources, Harvesting and Producing Areas   Furcellaran is produced in Denmark from the red algae *Furcellaria fastigiata*. The weed is harvested in Danish, Norwegian, and Baltic waters, from small vessels, using large trawl nets. The harvest is made up of loose plants which have been deposited on the ocean floor by the current, particularly along the east coast of North Jutland. About 100 tons of wet seaweed may be gathered per day by a small trawler. After unloading from the ship, the weed may be bleached in the sun or shipped directly to the factory.[58-60]

Processing   Upon arrival at the factory the weed is treated with an alkaline solution for two or three weeks, then neutralized, washed, and boiled to extract the furcellaran. The extract is filtered or centrifuged, and the clear extract squirted into a 1% potassium chloride solution to form long, gelled threads. The gel is frozen at about −16°C, thawed, and compressed to remove water and impurities. The compressed residue may then be bleached with dilute hypochlorite or hydrogen peroxide solution, and finally tunnel dried at 70°C and ground. Approximately 2.5–3.5% furcellaran is obtained from *Furcellaria fastigiata* by such methods.[59,60]

Structure   The molecular weight of furcellaran is believed to be in the range of 20,000–80,000. Its structure has not been clearly defined, and contradictory findings have been reported, but it appears to be similar to carrageenan in that it has both a kappa gelling and a lambda non-gelling component. The gelling fraction is potassium sensitive, as for carrageenan. However, furcellaran has a lower sulfate content than carrageenan. A composition of 3,6-anhydro-D-galactose and D-galactose sulfate has been suggested with the sulfate attached at the C_6 or C_4 position.[19,61,62]

Properties   Furcellaran is soluble in water at about 75–77°C, as well as in boiling milk, and forms firm, smooth, relatively elastic gels in both media. Concentrations of up to 3% are dispersible in water. Solutions are quite viscous at certain temperatures. The viscosity of a 1.5% dispersion will increase on heating at about 37°C, reach a maximum at 43°C, and decrease with further heating. Upon cooling, viscosity gradually increases until the gelling point is reached. Solutions can be autoclaved for several hours without degrading, as can agar, but degradation occurs on heating under acidic conditions. Furcellaran forms gels on heating and cooling to about 40–45°C. A 1% gel exhibits very little hysteresis, forming and melting between 35–45°C but higher concentrations melt at higher temperatures than the gelling point. Gels are quite firm, being similar to agar in strength, and stronger than carrageenan, but like carrageenan, gel strength is increased by the presence of potassium ions, and gel texture modified by the addition of locust bean gum. The gels formed by furcellaran are more elastic than either agar or carrageenan. Gel strength is pH dependent, and reaches a maximum at pH 8. Gelling ability may be destroyed or decreased by heating at low pH.[19,58,60,62]

Major Food Uses   The principle use of furcellaran is in the food industry, where it is used as a gelling agent in milk and water systems. The largest use is in milk puddings, flan, or blanc mange mixes. Other uses are in jams, jellies, marmalade, dietetic products, bakers jellies, and in meat and fish preservation.[6,19,58,60,63]

Tree Exudates and Extracts

Gum Arabic

Background   Gum arabic, or gum acacia, is the dried exudate of various species of the genus *Acacia*, subfamily *Mimosoideae* and family *Leguminosae* The use of gum arabic by man dates back at least 4000 years to the ancient Egyptians, who used it in making paint colors. Gum arabic from the Kordofan Province of Sudan has been an item of commerce since the first century A.D. The Republic of the Sudan is still the leading producer of gum arabic, in terms of both quantity and quality. Chemically, gum arabic is a neutral or slightly acidic salt of a complex polysaccharide containing calcium, magnesium, and potassium cations. Its most distinguishing property among the natural gums is its extreme solubility in water. Solutions of over 50% concentrations may be prepared. Major uses of gum arabic in foods are as a fixative for flavors, as a foam stabilizer in beverages, as an adhesive for icings and toppings, and as an emulsifier and stabilizer in confectionery and ice cream. It is also widely used in the pharmaceutical, cosmetic, paper, textile, paint, ink, and lithography industries.[1,64]

Sources, Harvesting and Producing Areas   Gum arabic is obtained from trees of the genus *Acacia*, subfamily *Mimosoideae*, family *Leguminosae*. Some 400–500 species of *Acacia* grow in semi-arid and arid regions throughout the world, including Africa, the Near East, India, Australia, and even in the southern United States, Mexico, and Central America. However, relatively few species are used as commercial sources, and the bulk of the world supply comes from the Republic of Sudan. Senegal, Mauretania, Nigeria, and Tanzania (formerly Tanganyika) are the next ranking producers, but all are much smaller producers. Almost all of the gum arabic used in the United States is Kordofan or Sudan gum, from the Republic of Sudan, where the major source is *Acacia senegal* (*Acacia verek*). This tree has a life span of 25–30 years, and attains a height of 15–20 feet. Gum arabic is the result of an infection, either bacterial or fungoidal. It is exuded only by unhealthy trees; heat, poor nutrition, and drought stimulate its production. The infection takes place through wounds in the tree which may be accidental or purposely made to stimulate gum production. The gum is exuded through these wounds in the bark in the form of tears, or drops which rapidly harden due to evaporation. Most of the gum arabic production is from wild trees, but some is from privately owned, cultivated gardens which are tapped and collected on a systematic basis. This gum, called Hashab geneina (garden gum), is the cleanest and lightest grade, and is most preferred for the U.S. market. The wild gum (called Hashab wady) is collected on a part time basis in the dry season, from October to May or June, by natives whose main occupation is usually farming. After gathering, it is taken to central collecting stations where it is auctioned under government supervision, graded by hand and dried, before exporting to gum suppliers in all parts of the world. There it is resorted, ground, processed, and graded to various specifications.

In West Africa the gum is also collected from *Acacia senegal* and is usually called Senegal gum. It is more yellow, or reddish, than Kordofan or Sudan gum and is regarded as being less adhesive and more viscous. It is used extensively in Europe. In Nigeria, gum arabic is usually collected from *Acacia senegal;* in Tanzania the bulk of the gum is from *Acacia drepanolobium.*

Besides being called Kordofan, Sudan, or Senegal gum, gum arabic has a great many other names, based on geographical or botanical origin, or physical characteristics of the gum itself. The name gum arabic is due to the fact that for hundreds of years the gum collected in the Sudan was bought by Arabic traders who transported it to various Arabian ports for reshipment.[64–67] Mantell[1] and Glicksman and Schachat[65] give extensive listings of names for gum arabic along with geographical and botanical origin.

Structure Gum arabic is best described as "heteropolymolecular", *i.e.*, a polymer system having either a variation in monomer (galactose, arabinose, rhamnose, glucuronic acid, and 4-*O*-methylglucuronic acid) composition and/ or a variation in the mode of linking and branching of the monomer units, in addition to a distribution in molecular weight.[299,300] Specifically, the backbone chain of gum arabic is composed of D-galactopyranose units joined by β-D-$(1 \rightarrow 4)$ and β-3-$(1 \rightarrow 6)$ linkages. Side chains composed of D-galactopyranose units are attached usually by β-D-$(1 \rightarrow 3)$ linkages. To these side chains L-arabinofuranose or L-rhamnopyranose residues are attached as end units. Frequently D-glucuronic acid units are attached by β-D-$(1 \rightarrow 6)$ linkages to D-galactose units, and often L-arabinofuranose units are attached to the D-glucuronic acid units by $(1 \rightarrow 4)$ bonds. Most of the L-rhamnose is attached to D-glucuronopyranosyl units as 4-*O*-α-L-rhamnopyranosyl nonreducing terminal units.[299–300] The qualitative breakdown of sugars present has been reported as 30.3% L-arabinose, 11.4% L-rhamnose, 36.8% D-galactose, and 13.8% D-glucuronic acid.[5]

Gum arabic is not very viscous, implying that the gum molecules are essentially globular and close-packed in shape rather than linear. Although molecular weights reported for gum arabic have been in the 240,000–300,000[302] range, the most recent, careful study of the exudate from *Acacia senegal* has shown an average molecular weight of about 600,000.[299]

Properties The most distinguishing property of gum arabic is its extreme solubility in water. Solutions of up to 50% can be made as compared to maximum concentrations of 5% or less for most gums. Only larch gum, a recently developed material, can rival gum arabic in solubility. Solutions of good grades of gum arabic are practically odorless, colorless, and tasteless. Poorer quality (dark) grades of gum arabic have an unpleasant flavor and odor, which is probably due to tannic acid, and should not be used in food products. Gum arabic is insoluble in alcohol and most organic solvents. Aqueous solutions are slightly acidic with a pH range of 4.5–5.5. Viscosities of gum arabic solutions are relatively low with a viscosity of 200 cps having been reported for a 30% solution. Maximum viscosity is attained at pH 6–7, with only a gradual change over the pH range 5–10. Partial hydrolysis may occur at low pH. Viscosity increases gradually as the concentration is increased up to 20–25%, at which point a more marked increase takes place. The viscosity of gum arabic solutions decreases with temperature, the viscosity being inversely proportional to the temperature. The viscosity of gum arabic solutions is also lowered by electrolytes.[17,19,65,69,290]

Major Food Uses Gum arabic is widely used in foods and beverages to make citrus oil emulsions, flavor emulsions, and to prepare spray dried flavors. It is used in the confectionery industry, principally due to its ability to retard sugar crystallization, and as a thickener in candies, jellies, glazes, and chewing gums. It is used as a foam stabilizer in beer, and as an adhesive in icings, glazes, and toppings.[19,65,290]

Gum Ghatti

Background Gum ghatti is an exudate from the tree *Anogeissus latifolia*, family *Combretaceae*, a large tree widely distributed in India and Ceylon. It appears to be a calcium salt of an acidic polysaccharide, ghattic acid, with hydrolysis products of L-arabinose, D-galactose, D-mannose, D-xylose, and D-glucuronic acid. It is harvested and graded in a manner similar to that for gum karaya. Its properties are usually considered to be similar to gum arabic, but its solutions are more viscous and less adhesive. Food uses are based primarily on oil-water emulsifying properties.[1,67,68,70]

Sources, Harvesting and Processing The tree *Anogeissus latifolia*, from which gum ghatti is exuded, is also used as a source of timber, and its tannin-rich leaves are used for tanning. It grows extensively in India and Ceylon. Production, collection, and processing methods are very similar to that for gum karaya. The gum is exuded when the bark is damaged, probably to act as a sealant. It is gathered by hand in the dry season by natives whose usual occupation is farming. The largest crop is normally picked in April. After drying for several days, the gum is transported to Bombay, usually having to be transported through mountain passes, or ghats, hence the name gum ghatti. In Bombay the crude gum is sold by auction to exporters, who first have it classified by hand according to color and purity. In the United States, the processor grinds the tears to a fine particle size. Impurities are reduced by sifting and aeration.[1,67,68,70]

Structure Gum ghatti is essentially a calcium salt of ghattic acid, a polysaccharide with a molecular weight of approximately 12,000. Hydrolysis products are L-arabinose (5 moles), D-galactose (3 moles), D-mannose (1 mole), D-xylose (0.5 mole), D-glucuronic acid (1 mole), and traces of 6-deoxy-hexose. Two aldobiouronic acids, 6-0-(β-D-glucopyranosyluronic acid)-D-galactose and 2-0-(β-D-glucopyranosyluronic acid)-D-mannose, are obtained on graded hydrolysis.[5,71] The gum contains chains of 1,6-linked β-D-galactopyranose residues, to which the aldobiouronic acid units are attached, either directly or through one or more 1,6 linked galactose residues.[72] The neutral and acidic oligosaccharides formed on partial acid hydrolysis have also been further characterized by Aspinall and others.[73,74]

Properties and Uses Gum ghatti can be dispersed in water to form a colloidal dispersion. Only about 80–90% of the gum is actually soluble. It is quite stable over the pH range 3.5–10.0. It is a natural buffer and small amounts of acid or alkali will not effect viscosity.

Use of gum ghatti in foods is based primarily on its effectiveness as an emulsifier for oil and water emulsions. Its use as a substitute for gum arabic has been reported, based on similar properties. However, solutions of ghatti are more viscous and less adhesive than those of arabic.[70]

Gum Karaya

Background Gum karaya is the dried exudate of *Sterculia urens*, a tree native to India. Chemically, karaya is a complex, partially acetylated polysaccharide with an extremely high molecular weight, consisting of L-rhamnose, D-galactose, and D-galacturonic acid. Karaya came into general use in the United States during World War I, as a cheaper substitute for tragacanth. Other names for karaya are Sterculia gum, Indian tragacanth, and India gum. Food uses for karaya include use as a stabilizer in whipped cream products, salad dressings, meringue toppings, ice cream, ice pops, and sherbets, and as a binder in sausages and other meat products. It is also used in the pharmaceutical, cosmetic, paper, and textile industries. Large uses are as a bulk laxative, as a denture adhesive, and in hair wave lotions.[75]

Sources, Harvesting and Producing Areas Gum karaya is the exudate of *Sterculia urens*, family *Bixaceae*, a large, bushy tree native to India, where it grows in forests in the mountainous eastern and central region. The trees, which grow up to 30 feet in height, are cultivated, and may be either privately or government owned. The gum is collected by natives, usually on a part time basis. Collection may be made throughout the year, except in the rainy season, but the best quality gum is collected in the March–June hot spell preceding the monsoon. The trees are tapped, or drilled, and exudation begins immediately and continues for several days in the form of large, irregular tears which may weigh up to several pounds. The average tree can be tapped about five times during its lifetime, and the yield per tree is estimated at 2–10 pounds per season. The native collectors take the gum to their villages, where it is sold to dealers in Bombay, who clean the tears, break them into fragments, then sort and grade the gum according to color and purity before selling to importers and processors in the United States and Europe.[1,68,75,76]

Processing Crude gum karaya is received by the processors in the form of tears, or fragments, in grades which are based on purity, color and viscosity. Bark is the main impurity, but wood fiber and sand may also be present. Purification is the physical removal of adhering bark, wood, soil, or fiber by blowing air through the coarse lumps, then further cleaning, grinding, sizing and blending to obtain uniform grades of gum. The processor sells the gum in food grades based, again, on color and impurities. Colors range from white to tan, with the lightest being the best grade. Impurities range from 0.1 to 3.0% of bark and foreign organic matter. Grades with higher amounts of impurities are sold as technical grades.[75,76]

Structure Gum karaya is a complex polysaccharide with a very high molecular weight, in the range of 9,500,000. It has a high acetyl content, with acid numbers from 13.4 to 22.7 having been reported, depending on the source and age of the sample. The gum has the tendency of splitting off free acetic acid on aging, the rate of splitting depending on temperature, humidity, and particle size. The gum karaya molecule contains L-rhamnose, D-galactose, and D-galacturonic acid units. Ratios of 43% D-galacturonic acid, 13% D-galactose, and 15% L-rhamnose have been obtained by hydrolysis. Also, molecular ratios of 5:6:4, respectively, have been obtained for the above components.[75-78]

Properties Gum karaya is one of the least soluble of the gums. It does not dissolve in water to give a clear solution, but absorbs water rapidly to form viscous solutions at low concentrations. Up to 4% may be hydrated in cold water to form heavy, viscous pastes which have the appearance of gels, but are not true gels. The utility of karaya is based on its viscosity. A 1% solution may reach a viscosity of 3,300 cps. Viscosity decreases with heating, with a concomitant increase in solubility. The normal pH of a 1% solution is approximately 4.5–4.7. Above pH 7 the short-bodied karaya solution becomes irreversibly transformed into a ropy, stringy mucilage, probably due to chemical deacetylation of the molecule. Solutions lose viscosity on aging, and may develop an acetic, or vinegary taste. Electrolytes and excess acid can also cause a drop in viscosity. Karaya is compatible with most other gums as well as proteins and carbohydrates.[5,75-78]

Major Food Uses Food uses of karaya are based on its cold water swelling, viscosity, and suspending powers. It is used in salad dressings as an emulsifier; in the manufacture of ice pops and sherbets to prevent migration of free water and the formation of large ice crystals; in cheese spreads to prevent water separation and to facilitate spreading; and as a binder in meat products. It is also used as a stabilizer in meringues, toppings, and whipped cream products.[19,75,290]

Gum Tragacanth

Background Gum tragacanth is the exudate from several species of *Astragalus*, principally *Astragalus gummifer*, which is found in the dry, mountainous regions of Iran, Syria, Turkey, and Asia Minor. Tragacanth is regarded as being composed of a soluble portion called tragacanthin consisting of glucuronic acid and arabinose, and an insoluble, but swellable, portion called bassorin, which appears to be a complex of polymethoxylated acids. Tragacanth is one of the oldest natural emulsifiers known. Its history predates the Christian era by several centuries, and it has been recognized in the U.S. Pharmaceopia since 1820. It is particularly effective as a suspending agent, due to its long shelf life and acid resistance and is used in the food industry as a stabilizer and thickening agent for salad dressings, citrus oil emulsions, ice cream, candies, and sauces. Other industrial uses are in pharmaceuticals, cosmetics, tobacco, paper, textiles, and dyes.[1,7,79,80]

Sources, Harvesting and Processing Tragacanth is derived from several species of *Astragalus*, principally *Astragalus gummifer*, of the *Leguminosae* family. The plants are small, thorny bushes with relatively large tap roots, that grow wild in dry, mountainous areas of Iran, Syria, Turkey, and Asia

Minor. Iran is the largest producer, and supplies the best quality gum. The plants are systematically tapped by making careful incisions in the tap root, or to a lesser extent, the bark of branches of the plant. The best gum is obtained from the tap roots of smaller bushes, usually 3–12 inches in height. The branches of larger shrubs, 3–6 feet in height, are also tapped, but are regarded as yielding inferior grades of gum. The gum exudes in the form of curled ribbons (Maftuli) or flake (Kharmani) which become horny on drying. The name tragacanth is derived from the Greek words tragos (goat) and akantha (horn), probably referring to the shape of the ribbon form of the gum. Collection is made by hand from May to October. Ribbons are long, flat, flexible, curled, and almost opaque, and are 2–4 inches in length. Approximately 3 grams of ribbon are yielded per tapping. Flakes are oval, thick, brittle, 0.5–2 inches in diameter, and yield about 20 grams per tapping. Collections are brought to trading centers, then to wholesale markets, usually in Teheran, Hamadan, and Isfahan, where they are sorted, graded, packed and shipped. The processor then further grades cleans, mills and blends the gum, and sells it in the form of ribbons, flakes, granules, or powder.[1,79,80]

Structure Tragacanth is a complex polysaccharide containing D-galacturonic acid, D-galactose, D-xylose, and L-arabinose, with associated calcium, magnesium, and potassium cations, and a molecular weight of 310,000. It is regarded as being composed of two major components. Bassorin, the larger component which comprises 60–70%, swells in water but does not dissolve. Tragacanthin, the other component, is water soluble and consists of a ring containing three molecules of glucuronic acid and one molecule of arabinose, with a side chain of two molecules of arabinose. Bassorin, which appears to be a complex structure of polymethoxylated acids, probably yields tragacanthin on demethoxylation. Tragacanth also contains small amounts of cellulose, protein, and starch.[5,17,79,80]

Properties Gum tragacanth swells in cold water to give extremely viscous colloidal solutions, probably the most viscous of all the plant gums. The viscosity is the most important factor in evaluating tragacanth, and is regarded as a measure of the quality and uniformity of the gum as well as a guide to its behavior as a suspending agent, stabilizer, or emulsifier. A viscosity of a 1% solution of a high grade tragacanth is approximately 3400 cps. Maximum viscosity is attained after 24 hours at room temperature, or by heating at 50°C for 2 hours. Sols are not thixotropic. A 1% solution exhibits a pH of approximately 5.1–5.9; maximum viscosity is attained at pH 8.0, but maximum stability of viscosity with aging has been reported at pH 5. Gum tragacanth is quite acid resistant compared to other gums, being reasonably stable down to pH 2.[19,68,78,79,81,82]

Major Food Uses Tragacanth, because of its high viscosity at low concentration, its ability to produce highly stable emulsions, and its acid resistance, finds use in foods as a stabilizer in low pH salad dressings, as a thickener and binder in confectionery, as a stabilizer in ice cream, and as a thickener in assorted bakery products.[19,79,290]

Arabinogalactan (Larch Gum)

Background Larch gum is obtained by the water extraction of the Western Larch tree. The extract is a complex, highly branched polymer of arabinose and galactose in a 1:6 ratio. Although this gum was first reported in 1898, successful commercial production was only begun in the last few years. Larch gum is similar to gum arabic in that it is extremely soluble and its solutions have very low viscosities, making it possible to prepare solutions of approximately 50% concentration. Since larch gum has been commercially available for such a short time, its uses are not as extensive as for many other gums. However it has been approved for use in foods as an emulsifier, stabilizer, binding or viscosity agent for essential oils, flavors, non-nutritive sweeteners, salad dressings, and pudding mixes. At present, it is used most extensively in the lithography industry.[83]

Sources and Production   Several species of *Larix* are known to contain arabinogalactan, but *Larix occidentalis* contains the largest amounts and is presently the sole commercial source. The Western Larch is indigenous to the Northwestern United States, and is particularly plentiful in Montana, Idaho, Washington and Oregon. It is a deciduous tree of the pine family, and is a valuable commercial source of lumber. The gum exudes as a result of injuries and collects under the bark of the tree. It may be collected by hand but this is a difficult procedure and the supply of this exudate is limited. The commercial method is based on the extraction of wood chips which are a by-product of the lumber industry. The gum may be obtained by water extracting the bark-free wood, then concentrating the extract and precipitating the arabinogalactan with cold alcohol. The commercial method of extraction was developed by Washington State University under sponsorship of St. Regis Paper Company. Production is at the site of the J. Neils Lumber Division of St. Regis, at Libby, Montana. The gum is sold by Stein, Hall and Company under the trade name of "Stractan."[83]

Structure   Arabinogalactan is a complex, highly branched polymer of arabinose and galactose in a ratio of 1:6. It has been reported to be composed of a major, highly branched, β-galactan fraction with a molecular weight of approximately 100,000 and a minor arabogalactan fraction with a molecular weight of approximately 16,000 in a 4:1 ratio. Overall molecular weights based on this data would be in the 70,000–90,000 range. However, when methods that do not employ fractionation were used, overall molecular weight values of 30,000–35,000 were obtained.[83-85]

Properties and Uses   Larch gum is readily soluble in water to give light amber colored solutions of over 40% concentration. Solutions are stable over a wide pH range, and are relatively unaffected by electrolytes. The pH of solutions from 10–40% are in the 4–5 range. The viscosity of solutions decreases on heating. Since larch gum is a recent commercial item, its use in food is not extensive at present. Although not a "GRAS" material, to date larch gum has been approved for use as an emulsifier, thickening agent, stabilizer, and binder for essential oils, flavor bases, dressings, puddings, and non-nutritive sweeteners. Larch gum could very likely be used as a substitute for gum arabic in many applications if it were economically feasible.[83-85,290]

Seed Gums

Locust Bean Gum

Background   Locust bean gum is the refined endosperm of the seed of the carob tree, *Ceratonia siliqua*, a large leguminous evergreen which is widely cultivated in the Mediterranean area. Structurally, locust bean is a neutral galactomannan polymer consisting of a main chain of D-mannose units with a side chain of D-galactose on every fourth or fifth unit. The history of the cultivation of the carob tree dates back more than two thousand years. The ancient Egyptians used locust bean gum as an adhesive in mummy binding. It is also believed that the "locusts and wild honey" which sustained John the Baptist in the wilderness were wild carobs. The carob pod is sometimes called St. Johns Bread, and carob seed gum is another common name for locust bean gum. The carob pods are used today as a stock feed because of their high protein content, and sometimes serve as human food. Locust bean gum is extensively used in the food industry, particularly as an ice cream stabilizer, a stabilizer in sauces, salad dressings, pie fillings, bakery products, and in soft cheese manufacture. It is also used in the cosmetic, pharmaceutical, paper, and textile industries.[2,19]

Sources, Harvesting and Producing Areas   Locust bean gum is derived from the fruit of the carob tree, or *Ceratonia siliqua*, of the family *Leguminosae* and subfamily *Caesalpiniaceae*. The carob is a large evergreen tree which may grow 40–50 feet high. The fruit is a dark, chocolate colored pod, rich in sugar and protein, about 4–12 inches long and containing about 10 seeds the size and shape of watermelon seeds. The source of the gum is the seed endosperm which makes up about one third of the seed. The fruit is harvested in the fall by shaking it from the tree with poles, and raking it from the ground. Giant trees have been known to produce up to a ton of fruit. The carob is cultivated throughout the Mediterranean area, and to a small extent in California. Cyprus, Spain, Italy, Greece and Syria are the most important producing areas, but Algeria, Portugal, Turkey and Morocco are also sources.[2,86,87]

Processing   As described above, the carob fruit is a pod containing 10–12 seeds covered by a tough, dark brown husk, under which is a layer of white, semi-transparent endosperm, with the hard, yellow germ in the center. Production of locust bean gum involves the removal of the seed from the pod, the dehusking of the seed coat, and the separation of the endosperm from the germ. The seed comprises 8–10% of the weight of the pod, and the endosperm about one-third of the seed. The pods are cleaned, then mechanically broken into pieces (kibbled) and the seed removed. The seeds are dehusked by passing through mechanical rollers. A pretreatment with dilute alkali is sometimes used. The dehusked seeds are split lengthwise, the germ separated from the endosperm by differential grinding, and the endosperm ground into a fine flour. The flour is graded according to accepted standards of color, impurities, viscosity and germ content. Highest grades are a near-white, speck free powder.[2,19,86,87] An alternate procedure involves the further purification of the endosperm by dispersing it in boiling water, filtering it to remove impurities, then recovering the gum by evaporation of the solution and final tray or roll drying.[1,88]

Structure   Carob gum contains D-mannose and D-galactose as structural units with mannose contents of 73–82% having been reported, depending on the methods used and the geographical origin of the gum.[5] The structure is regarded as that of a neutral galactomannan polymer consisting of a main chain of 1,4-linked D-mannose units with a side-chain of D-galactose on every fourth or fifth unit, attached through 1,6-glycosidic linkages to the polymannose chain. No uronic acid or pentose is present. A molecular weight of 310,000 has been reported.[89]

Properties Locust bean swells in cold water, but heating is necessary for maximum solubility. Optimum viscosity is attained by heating and cooling. Solutions of locust bean gum are cloudy with a white opacity, due to the presence of insoluble impurities such as protein and cellulose. Locust bean dispersions are very viscous. A 1% solution can attain a viscosity of 3,000–3,500 cps. A slight increase in viscosity on aging is exhibited. The normal pH of a 1% solution is 5.3, but viscosity is little affected over a range of pH 3–11. Locust bean gum solutions have no gelling properties, but have the unusual ability to impart a desirable elastic character to carrageenan and agar gels, and to retard syneresis in these gels. Locust bean gum is compatible with carbohydrates, protein, and other plant gums. Its solutions are relatively unaffected by neutral salts, but can be precipitated by electrolytes such as lead acetate and tannic acid.[2,17,19,68,69,89]

Major Food Uses One of the principal food uses of locust bean gum is as an ice cream stabilizer, where it imparts a smooth melt-down and desirable freeze-thaw shock resistance to the ice cream. It is also used in the baking industry to produce softer doughs with greater resiliency. Other uses are as a stabilizer in sauces, salad dressings, and pie fillings, or as a texture modifier in soft cheese, and a binding and lubricating agent in sausages.[2,19,86,]

Guar Gum

Background Guar gum, like locust bean gum, is a galactomannan derived from the seed of a leguminous plant. The source of guar, *Cyamopsis tetragonolobus*, is widely grown in Pakistan and India as a cattle feed, and was introduced to the United States as a cover crop in 1903. It was not until 1953, however, that guar gum was produced on a commercial scale, primarily as a replacement for locust bean gum in the paper, textile and food industries. The most important property of guar is the ability to hydrate rapidly in cold water to attain a very high viscosity. In addition to the food industry, guar is used in the mining, paper, textile, ceramic, paint, cosmetic, pharmaceutical, explosive, and other industries.[90,91]

Sources, Harvesting, and Producing Areas Guar gum is derived from the seed of the guar plant, *Cyamopsis tetragonolobus*, a leguminous plant which resembles the soybean plant. India and Pakistan are the major sources of supply, and the United States is also a producer. The guar plant has been grown for thousands of years in India and Pakistan as animal and human food. It was introduced to the southwestern United States as a possible cover crop in 1903, but little interest was shown until World War II when a concerted effort was made to find a domestic replacement for locust bean gum. Commercial scale production began in the United States in 1953, and now foreign producers of locust bean gum also process guar.

The guar is a hardy and drought-resistant plant which grows three to six feet high with vertical stalks. The guar pods, which grow in clusters along the vertical stems, are about six inches long and contain 6 to 9 seeds, which are considerably smaller than locust bean seeds. As in the case of locust bean gum, the endosperm, which comprises 35–42% of the seed, is the source of the gum. The seeds are harvested in the United States with mechanical grain harvesting equipment adapted for this use. In India and Pakistan harvesting is done by hand. It is harvested before the first rain following the first frost to obtain maximum yield and purity.[1,88,91,92]

Processing In processing guar, as in locust bean gum, the seed must be removed from the pod, the hull of the seed removed, and the endosperm separated from the gum. The hull is loosened and cleaned by treatment with 55% sulfuric acid and washing, or by water soaking and flame-charring. Another method used is differential grinding and sifting, which is also used to separate the endosperm from the germ. The endosperm obtained, containing approximately 80% galactomannan, is ground to a fine particle size. In the United States basic flour milling equipment is used for this process.

Structure Guar is composed of a straight chain of D-mannose with a D-galactose side chain at approximately every other mannose unit. The mannose units are β-$(1 \to 4)$ linked, and the single D-galactose units are joined to this chain by α-$(1 \to 6)$ linkages. The mannose-galactose ratio is about 2:1, and the molecular weight approximately 220,000–250,000.[91,93]

Properties The ability of guar gum to hydrate rapidly in cold water to form very viscous colloidal dispersions is its most important characteristic. The viscosity attained is dependent on time, temperature, concentration, pH and particle size of the gum used. Maximum viscosity is reached in approximately two hours in cold water. The rate of hydration and viscosity are increased at higher temperatures. Solutions are slightly cloudy due to the presence of a small amount of insoluble fiber and cellulose. A 1% solution of a typical commercial guar gum may reach a viscosity of 2700 cps, have a pH of 5.5–6.1, and tend to become more acidic while standing. It will also be relatively stable over the range of pH 4–10.5, and exhibit a slight buffering action. Solutions are thixotropic, and the viscosity is relatively unaffected by the presence of electrolytes. Guar is compatible with starch, gelatin, and other water soluble gums. Gels can be formed by the addition of borax to cross-link guar gum solutions, but these, of course, are not edible.[19,69,77,91,94,95]

Major Food Uses. Guar gum has important use in the food industry as an ice cream stabilizer, where its rate of hydration and water binding properties make it very effective, particularly in high-temperature short-time processes. It is also used to improve the yield of curds in soft cheese and to modify texture; to increase the yield and give greater resiliency to doughs and baked products; and as a binder and a lubricant in sausage. It is often used as a thickening or viscosity control agent in beverages, salad dressings, and relishes.[19,91]

Cellulose Derivatives

Sodium Carboxymethylcellulose

Background Sodium carboxymethylcellulose, commonly referred to as CMC, or cellulose gum, is a synthetic, water-soluble ether of cellulose. Commercial preparation of CMC is carried out by reacting alkali cellulose with sodium monochloroacetate. CMC was initially developed in Germany during World War I as a gelatin substitute, but due to technical difficulties and high production costs commercial development was not realized until World War II, when the shortage of water soluble gums in Germany provided the incentive for large scale manufacture. The development of CMC in the United States was inspired by the discovery that it improved the performance of detergents. Common food uses of CMC are in ice cream, ice pop, and sherbert stabilizers, milk beverages, bakery products, toppings, and salad dressings. It is also widely used in pharmaceuticals, cosmetics, textiles, paper, adhesives, insecticides, paints, ceramics, lithography and detergents.[96,97,290,303]

Manufacture and Structure   CMC is manufactured by a relatively simple chemical reaction. Cellulose, either wood pulp or cotton linters, is treated with sodium hydroxide solution, and the alkali cellulose is reacted with sodium monochloroacetate or monochloroacetic acid, as follows:

$$R—OH + NaOH \rightarrow R—ONa + H_2O$$
$$R—ONa + Cl—CH_2COONa \rightarrow R—O—CH_2COONa + NaCl$$

One such process has been described wherein the aqueous sodium hydroxide and monochloracetic acid are concurrently sprayed onto the powdered cellulose as it passes slowly through a revolving drum. The treated cellulose is aged to allow the reaction to be completed, then it is neutralized and oven dried. Food grade CMC is washed with an alcohol-water mixture to remove salt.

The reaction is controlled to give the desired degree of substitution (D.S.), degree of polymerization (D.P.) and uniformity of substitution, which in turn determine the properties of the finished product. In theory, a D.S. of 3 would be possible, since each anhydroglucose unit contains three available hydroxyl groups with which the monochloroacetate can react. However, materials with a much lower D.S. have the best solubility and general physical properties, and commercial materials usually range from 0.4 to 1.2 D.S., with food grade materials restricted to a maximum of 0.9 D.S. By varying these properties of D.S., D.P., and uniformity of substitution, materials with a wide variety of solution properties are produced. In addition, materials with various particle sizes, shapes, and densities are produced.[96,98,303]

Properties   As previously mentioned, the water solubility and solution properties of CMC are dependent on the D.P., D.S., and uniformity of substitution of the carboxymethyl groups on the polymer. In general, material with a D.S. of 0.3 or less is soluble in alkali, but not in water. For water solubility a D.S. of 0.45 or higher is usually required. CMC with a D.S. of 0.65–0.85 is most commonly used. The D.P. of the cellulose molecule greatly affects the viscosity of solutions. The higher the D.P. the greater the viscosity. The uniformity of substitution determines the thixotropic properties of solutions; the more uniform the distribution the smoother and less thixotropic the solution. Food grade CMC is soluble in hot or cold water. Viscosities of a 2% solution, depending on the CMC type, may range from 10 cps to 50,000 cps, or higher. A reversible loss of viscosity on heating is exhibited. Also, solution viscosity is reduced below pH 5, and below pH 2–3 insoluble free carboxymethylcellulose acid may be precipitated. However, CMC is fairly stable over a wide pH range of 5 to 11, with best viscosity stability at pH 7–9. CMC is compatible with most other water soluble gums, and is generally unaffected by cations that yield soluble salts of CMC. Usually monovalent cations form soluble salts, divalent cations are borderline, and trivalent cations, such as aluminum or ferric, produce precipitates or gels. CMC has good film forming properties, forming clear films which are resistant to oils and most organic solvents.[78,96,98–102,303]

Major Food Uses   Food uses of CMC include use as a stabilizer for ice cream, sherbets, ice pops, and other frozen confections to prevent ice crystal growth; in icings, meringues, jellies, pie fillings, and in puddings to prevent syneresis; in cakes and other baked goods to increase volume and retain moisture; and in salad dressings and flavor emulsions as a protective colloid.[96,98-102,290]

Methylcellulose and Hydroxypropylmethylcellulose

Background. Methylcellulose is the synthetic, water soluble methyl ether of cellulose prepared by reacting alkali cellulose with methylchloride to produce the dimethyl ether of cellulose. For hydroxypropylmethylcellulose, ether groups are substituted along with methyl groups by reacting propylene oxide as well as methyl chloride with the alkali cellulose. Methylcellulose was first manufactured in the United States by the Dow Chemical Company in 1938. The most unusual property of methylcellulose is that it is soluble in cold water and insoluble in hot water, and solutions increase in viscosity and eventually gel on heating, rather than decreasing in viscosity. Common food uses are in bakery products, flavor emulsions, dietetic foods, ice cream and other dairy products, salad dressing, whipped toppings, and frozen foods. It is also used in the pharmaceutical industry as a bulk laxative, in ointments and lotions, emulsions, suspensions, and tablets. Other uses are in the paper, textile, cosmetic, adhesive and ceramic industries.[102,103,290,305]

Manufacture and Structure Methylcellulose is prepared by reacting cellulose fibers (either cotton linters or wood pulp) with caustic soda to produce alkali cellulose, which is then reacted with methyl chloride to give the methyl ether of cellulose.

$$R\text{—}OH + NaOH \rightarrow R\text{—}ONa + H_2O$$
$$R\text{—}ONa + CH_3Cl \rightarrow R\text{—}OCH_3 + NaCl$$

Sufficient aqueous solution of sodium hydroxide (35–60%) is normally used to provide three moles of sodium hydroxide per anhydro-D-glucopyranose unit.

Since there are three available hydroxyl groups per anhydroglucose unit, the maximum degree of substitution (D.S.) is three. The D.S. of the methylcellulose product is very important in determining its properties. Maximum water solubility is exhibited in the D.S. range of 1.6–2.03. Lower D.S. products are soluble only in alkali, and higher D.S. products are soluble only in organic solvents.

To prepare hydroxypropylmethylcellulose, in addition to the methoxyl substitution described above, a further reaction with propylene oxide is carried out, yielding a mixed ether of cellulose:

$$R\text{—}ONa + H_2C\underset{\diagdown O \diagup}{\text{———}}CH\text{—}CH_3 \rightarrow R\text{—}OCH_2\text{—}\underset{|\atop OH}{CH}\text{—}CH_3$$

The amounts and ratios of methoxyl and hydroxypropoxyl groups have an effect on the properties of the product, principally the thermo-gel point temperature, and solubility in organic solvents. Uniformity of substitution and D.P. also effect the properties of the product. By controlling the intrinsic viscosity of the alkali cellulose, materials with a wide range of viscosities are produced.[102,105]

Properties and Uses Methylcellulose and hydroxypropylmethylcellulose are soluble in cold water. Solution viscosity, which is expressed as cps per 2% solution at 20°C, is dependent on the D.P. Methylcellulose is produced in viscosity types ranging from 10 to 8,000 cps for a 2% solution. When heating, methylcellulose solutions increase in viscosity, and eventually gel at 50–55°C. This gel point can be raised by the introduction of hydroxypropyl groups. Depending on the amounts and ratios of methyl and hydroxypropyl groups, the gel point is raised as high as 85°C in commercial products. The gel point is also affected by additives. Most electrolytes, as well as sucrose, glycerol, and sorbitol, depress the gel point, while ethanol and propylene glycol have an elevating effect. Methylcellulose solutions are neutral, and are relatively stable over a pH range of 3–11. Since it is nonionic, methylcellulose is not affected by ordinary concentrations of electrolytes or other solutes, but can be salted out if certain limits are exceeded. Water soluble films can be prepared by casting methylcellulose from water solutions or mixed solvents such as methanol-water.

Methylcellulose has been approved by the F.D.A., and methylcellulose U.S.P. with a methoxyl content of 27.5–31.5 appears on the "GRAS" list. Hydroxypropylmethylcellulose is permitted as a food additive for use as an emulsifier, film former, protective colloid, stabilizer, suspending agent, or thickener. Both methylcellulose and hydroxypropymethylcellulose are used in many food products, including bakery products, salad dressings, ice cream and other dairy products, and flavor emulsions.[78, 103–106, 290, 305]

Hydroxypropylcellulose

Background   Hydroxypropylcellulose is a surface-active, nonionic hydroxypropyl ether of cellulose prepared by reacting propylene oxide and alkali cellulose. Food-grade materials are manufactured by Hercules under the trade name of Klucel.® [304] Hydroxypropylcellulose received clearance by the Food and Drug Administration as an intentional additive for food for human consumption in 1964 for use as an emulsifier, film former, protective colloid, stabilizer, suspending agent, or thickener.[290]

Manufacture and Structure   Hydroxypropylcellulose is prepared by reacting cellulose with alkali to produce alkali cellulose, which is then reacted with propylene oxide at elevated temperature and pressure and finally purified. Propylene oxide appears to be substituted through an ether linkage at all three of the primary hydroxyls on each anhydroglucose unit of the cellulose. Etherification probably takes place so that the hydroxypropyl substituent groups contain almost entirely secondary hydroxyls that are available for further reaction with the propylene oxide, resulting in the formation of side chains containing more than one mole of combined propylene oxide. It is therefore necessary to consider the molar substitution (M.S.) as well as the D.S. The D.S. is the average number of hydroxyl groups substituted per anhydroglucose unit; the M.S. is the average number of molecules of reactant combined with the cellulose per anhydroglucose unit. The M.S.—D.S. ratio represents the average length of the side chains formed.[5, 6]

Properties and Uses   Hydroxypropycellulose is soluble in water below 40°C, but it is insoluble above this temperature. On heating, solutions of hydroxypropylcellulose gradually decrease in viscosity, until a temperature of 40–45°C is reached, where a rapid drop in viscosity occurs as the polymer begins to precipitate as a highly swollen floc. Hydroxypropylcellulose is also soluble in polar organic solvents, either hot or cold. Propylene glycol, ethyl alcohol, and methyl alcohol are good solvents for hydroxypropylcellulose.

The water solutions of hydroxypropylcellulose are virtually Newtonian at low shear rates but become more thixotropic at high shear rates. Viscosity increases rapidly with concentration. Hydroxypropylcellulose is available in several viscosity types and is compatible with most common inorganic salts at low .salt concentration and with most natural gums and synthetic water-soluble polymers. Solutions are generally stable in the pH range of 3–10. Hydroxypropylcellulose has good film-forming properties and forms films with excellent flexibility and heat-sealing properties. Hydroxypropylcellulose is highly surface-active and facilitates the formation of O/W emulsions. Suggested food uses are in whipped toppings, salad dressings, and edible films.[290,304] It also is being utilized in various dairy and imitation dairy products, such as whipped aerosol toppings, frozen toppings and desserts, and related products.[292,293]

Microcrystalline Cellulose

Background Microcrystalline cellulose[306,307,317] is a partially depolymerized form of cellulose prepared by the acid hydrolysis of wood pulp under a patented process;[316] it differs from normal α-cellulose in that it is non-fibrous and has water-absorbing properties.[290] It has been marketed by the American Viscose Division of FMC Corporation under the trade name of Avicel® since 1961. The food-grade material, called Avicel RC, contains 8–12% CMC to improve its organoleptic properties. It is a white, fine powder low in ash, metals, and soluble organic materials. Suggested food uses are in frozen desserts, toppings, spreads, confectionery, and soft drinks.

Manufacture Microcrystalline cellulose is prepared by the controlled hydrolysis of α-cellulose with hydrochloric acid. The α-cellulose is converted into an acid-soluble fraction and an acid-insoluble crystalline material, or microcrystalline cellulose. The amorphous regions of the cellulose are completely hydrolyzed, and the crystalline regions remain as isolated microcrystallates. These are defined as the level-off degree of polymerization cellulose, a D.P. cellulose; that is, if the hydrolysis were continued, the degree of polymerization would not change.

Mechanical shearing in a water slurry is used to free the microcrystals from their fibrous, packed structure in the form of short, rod-like particles. These are dried and ground to a fine, white, free-flowing crystalline powder. Microcrystalline cellulose manufactured for use in wet or colloidal applications in food and pharmaceuticals contains 8–12% CMC, added before drying. This material has improved dispersibility, hydration, and mouthfeel and is sold under the Avicel RC trade name.[290]

Properties and Uses Microcrystalline cellulose occurs as a fine, white, odorless, crystalline powder of free-flowing, non-fibrous particles. It is insoluble in water, in dilute alkali, and in most organic solvents. It is slightly soluble in sodium hydroxide solution.[107] Microcrystalline products containing CMC disperse to form colloidal solutions below 1% concentration or thixotropic gels above 1% concentration. Applications or suggested applications for microcrystalline cellulose in food include use as a non-caloric filler and/or a stabilizer in canned, shelf-stable spreads and salads, frozen desserts, aerosol toppings, and meat, dairy, and bakery products.[290,291]

Microbial Gums

Xanthan Gum

Background[290]   Xanthan gum was developed by the Northern Regional Laboratories of the United States Department of Agriculture at Peoria, Illinois, as part of a program to find new uses for corn products. Originally designated as B-1459, this polymer is produced by the fermentation of dextrose by the bacterium *Xanthomonas campestris*. Structurally it is a complex polysaccharide containing D-glucose, D-mannose, and D-glucuronic acid. Xanthan gum was recently approved for food use.[310] Food-grade material is currently sold by the Kelco Company under the trade name of Keltrol®[308] and by General Mills as Biopolymer XB-23.[309] Suggested uses are as a stabilizer for salad dressings,[296] beer foam, flavor oil emulsions, pickle sauce, and relishes.[290] It is useful in the formulation of milk puddings[311] and similar gelled products,[312] as well as in the formulation of various low-calorie products.[297]

Manufacture and Structure   Xanthan gum is a polysaccharide derived from *Xanthomonas campestris* by a pure-culture fermentation process on a medium containing glucose, distiller's solubles, dipotassium hydrogen phosphate, and trace elements. Only a 3–5% concentration of glucose can be used effectively, because of the high viscosity attained by the culture fluids.

The polysaccharide is isolated by precipitation with alcohol, purified by reprecipitation with isopropyl alcohol, and dried.[294] The resulting product is a complex polysaccharide with a molecular weight of more than one million. The structure is linear with a β-linkage backbone containing D-glucose, D-mannose, and D-glucuronic acid in a 2.8:3.0:2.0 molar ratio, with one D-mannose side chain for every 8 sugar residues and one D-glucose side chain for every 16 sugar residues. The polymer also contains about 4.7% O-acetyl groups and 3.0–3.5% pyruvic acid present as a ketal on the glucose unit.[294]

Properties and Uses   Xanthan gum is a cream-colored powder that is readily soluble in hot or cold water to form viscous, non-thixotropic solutions. It is cleared for use as a stabilizer, emulsifier, thickener, suspending agent, bodying agent, or foam enhancer in foods.[310] It has very unusual and useful properties in that it gives high viscosity solutions at low concentration, exhibits little change in viscosity with variation in temperature, and has excellent stability over a wide pH range. It also has good freeze-thaw stability and suspension properties.[290, 294]

Food Applications

Water Dessert Gels

In the United States gelatin is by far the most commonly used hydrocolloid for the preparation of water dessert gels. Solutions containing 1–2% gelatin form thermally reversible, elastic, clear, sparkling gels when prepared with boiling water and cooled below room temperature. In spite of the popularity of gelatin in this application, dessert gels based on seaweed extracts are also used, primarily because of their ability to set up to gels without refrigeration at room temperature or above. In countries where refrigeration is not common, particularly in tropical countries, this is an important advantage. The use of seaweed extracts in dessert gels is discussed below.

Alginates The soluble salts of algin have the ability to enter into controlled chemical reactions with calcium or other polyvalent ions to form edible gels. This important and unique property has been utilized in formulating dry powdered mixes which are cold water soluble and set up to gels at room temperature. The gels formed are chemically set, non-thermally reversible, and therefore will not melt at room temperature. Algin concentrations of 0.4–1.0% of the finished dessert are used. Advantages are quick setting time and the ability to prepare and use in warm climates without refrigeration. Disadvantages, which have limited the use of calcium alginate gels in dry mix formulations, are due to the fact that they are chemically set gels, and do not melt down in the mouth to give a smooth feel as does gelatin. Also, the set rate and final texture of the gels are dependent on the total calcium ion concentration, and it is therefore difficult to formulate a mix which will give optimum gel texture and setting time in both hard and soft water.

The rate of reaction is very important in the formulation of these gels. A soluble sodium or potassium alginate is usually reacted with a sparingly soluble calcium salt under controlled pH and concentration to form an insoluble calcium alginate gel. Too rapid a reaction brings about the formation of a grainy, discontinuous gel, with inclusion of entrapped air, resulting in poor appearance and eating qualities. If the rate of reaction is too slow, a weak, slow setting gel results. The following factors are important in regulating the release of calcium ions and forming a smooth gel.

1. Selection of the proper, sparingly soluble calcium salt. For food use, choice is usually limited to citrate, tartrate, phosphate, gluconate, sulfate, or carbonate.

2. Control of pH to regulate the release of calcium ions by choice of the proper concentration of an acid, such as citric or adipic, or the use of materials, such as glucono-delta-lactone, will hydrolize in water to liberate acid.

3. Use of a sequestering agent, such as a citrate or phosphate, to allow a controlled release of calcium ions.[19, 29, 39, 45, 107, 320]

Numerous patents have been issued relating to the preparation of alginate based dessert gels. Steiner described a four component gelling system that covered a wide variety of alginate gels for use in hot or cold water, to prepare dessert gels as well as many other products.[108] This system included (a) a water soluble alginate such as sodium alginate, (b) a slightly ionizable calcium salt such as tricalcium phosphate, (c) a weak acid or a substance that liberated free acid on solution in water, such as citric acid or glucono-delta-lactone, and (d) a gel-retarding agent such as sodium hexametaphosphate, to repress ionization of the gel forming salt until the acid was added, and then control the availability of the calcium ions so that a smooth gel formed. A subsequent patent claimed an edible alginate gel preparation from a dry mix containing calcium alginate and a water soluble orthophosphate such as trisodium phosphate, which converted most of the calcium into tricalcium phosphate and solubilized the resulting sodium alginate.[109] An acid ingredient such as glucono-delta-lactone was also included, or a free acid subsequently added. Another patented dry mix for preparing an edible alginate gel consisted of calcium alginate and an alkali metal phosphate as the gelling system.[110] Improved dispersibility by co-drying sodium alginate with sodium hexametaphosphate or sodium citrate has been claimed[111] as well as improved gels by using slightly soluble dibasic acids such as fumaric acid, with a water-soluble alginate, a calcium salt, and an alkali metal salt of a weak acid.[112]

A four component dessert gel system containing a water-soluble alginate, a sparingly soluble calcium salt, a water-soluble alkali metal salt of a weak acid or a substance that liberates acid in water has been patented.[113] An improved gel using a hot aqueous solution of an alkali metal alginate with citric acid which does not gel until cooled, resulting in a smoother, air-free gel has also been claimed.[114] Other patents have covered heat reversible alginate gels made with partial amides of alginic acid[115] and a less calcium sensitive gel based on a high-viscosity carboxymethyl alginate,[116] but to date neither of these products are commercially available.

A dry mix for preparing food gels which do not deteriorate upon freezing and thawing has been patented.[117] This is composed essentially of a water soluble alginate, a sodium salt of CMC, a salt whose cations form a water insoluble salt with alginic acid and a saccharide.

Carrageenan   0.5–1.0% water solutions of carrageenan form gels on heating and cooling, which set up at room temperature. Gel strength is enhanced by the presence of metallic cations, particularly potassium. The application of these gels in food products was long limited because of the undesirable brittle character of the typical carrageenan water gel. However, Baker[118] found that the character of carrageenan gels could be modified to strong, elastic type gels by the use of a neutral polymer, preferably locust bean gum. He proposed a three component blend of 50% carrageenan, $33\frac{1}{3}$% locust bean gum, and $16\frac{2}{3}$% potassium chloride or other potassium salt. Another patented gelling composition based on carrageenan and locust bean gum consists of 86 parts sucrose, 1.36 parts carrageenan, 1.4 parts fumaric acid, 1.6 parts sodium fumarate, and 0.64 parts potassium chloride.[119]

While locust bean gum imparts elasticity to the texture of carrageenan gels, these gels lack the clarity and sparkle desirable in this type of product. This lack of clarity is due to insoluble impurities present in the locust bean gum. The use of a clarified locust bean gum has been proposed to overcome this difficulty.[120]

Another method claimed for improving the texture and elasticity of carrageenan water gels is the use of a potassium salt of a sequestrant, preferably potassium metaphosphate, to bind the extraneous sodium, magnesium, and calcium ions present in tap water, which cause shortness and brittleness in the gel.[121]

In the past several years another carrageenan gel system has been developed based on combinations of κ- and ι-carrageenan.[318] The ι-carrageenan alone produces a clear, resilient gel with no syneresis. The κ-carrageenan is usually added to impart rigidity and unmolding properties to the gel. This system has been effectively used in the preparation of ready-to-eat gelled desserts for infants and individual-serving desserts for adults. It has the advantage of a relatively high gel temperature, so that gels do not melt at room temperature. The gels also do not toughen on standing, can be sterilized in jars or cans, and can suspend fruit readily. This system has been suggested for use in dry-mix dessert gel products in tropical countries where refrigeration is not common.[290,319] Another application of ι-carrageenan, as a dessert gel with superior freeze-thaw properties, has been patented.[123]

Milk Puddings

In the United States, the most popular milk puddings are those containing starch as a gelling agent. These are generally of two types: a cooked milk pudding in which a raw starch is the sole gelling agent, and instant puddings in which the gel is formed with cold milk by the synergistic effect of a pre-gelatinized starch and a milk protein-phosphate salt reaction. However, as in the case of water gels, milk gels or milk puddings can also be formed by the seaweed extracts to form blanc mange, flan, or custard type milk desserts which have a great deal of popularity throughout the world. Milk gels of this type are discussed below.

Algin Until recently the use of algin in dry milk pudding powder mixes was limited because of such defects as poor solubility in milk, lack of firmness, and a granular gel texture due to the difficulty of dissolving the alginate in the presence of calcium ions. These products have recently been improved, and rapid setting milk pudding products based on the reaction of a soluble alginate (such as sodium alginate) with the calcium of the milk, plus a calcium salt component of the mix, have recently been sold in several countries.[124] A preparation for a good quality milk pudding based on a blend of a water-soluble alkali metal alginate (sodium alginate), a mild alkali (sodium carbonate) and a calcium salt (calcium lactate) has been patented by Gibsen.[125] This blend was prepared by interacting alginic acid with 1-3 equivalents of sodium carbonate in alcohol, adding 1.5–3 equivalents of calcium salt, then drying and grinding the residue to a powder. A good quality gel was prepared by dissolving a mixture of this alginate blend with 66.75 grams of sugar, 13 grams cocoa, 1 gram salt and 0.2 grams vanilla in 2 cups of boiled milk and allowing to cool. A similar instant pudding was prepared in cold milk by incorporating a buffering agent (tetrasodium pyrophosphate) and additional calcium salt (calcium gluconate) in the mix.

A composition for cold milk puddings based on 1.0–1.5% sodium alginate (D.P., 80-300), 0.1-3% calcium ions (dicalcium phosphate), and 0.6-1.2% of a complexing agent (sodium hexametaphosphate) has also been patented which could be made up with cold milk, or with cold water if milk solids were included in the mix.[126]

McDowell and Boyle claimed a dry powdered mix of alkali alginate (sodium alginate), alkali carbonate (sodium carbonate), and a substance (calcium sulfate) which would react slowly with the carbonate.[127,128] This dry mix, when dissolved in cold milk, set up to form a pudding.

A composition made up of a sodium alginate-trisodium phosphate mixture, an acidic ingredient, and either whole or skim milk solids was claimed by Hunter and Rocks.[129] This composition produced a firm pudding gel in one-half hour, when dissolved in cold tap water.

Carrageenan Carrageenan has been used as a gelling agent in milk puddings for hundreds of years, initially by boiling Irish moss with milk and sugar to form gels. Currently, dry prepared mixes containing carrageenan as a gelling agent can be used to prepare light, eggless "custards" or "flans", or carrageenan can be combined with starch to give a wide variety of textures, from custard-like to heavy starch-like. Stoloff states that carrageenan is most effective in milk puddings and pie fillings when combined with starch in these various combinations.[48] He gives a typical combination for one pint of milk for a custard as 0.7 grams carrageenan and 4 grams tapioca starch; 0.2 grams carrageenan, 12 grams corn starch, and 6 grams tapioca starch for heavy pudding. Stoloff has also patented a pudding mix based on carrageenan alone consisting of 18.8 grams cocoa, 63 grams sugar, 1.25 grams carrageenan, 1.09 grams salt, and 0.18 gram vanillin, which is prepared by dissolving in one pint of boiling milk.[130] Stoloff also patented a dry composition based on carrageenan, a soluble calcium or magnesium salt, and a soluble phosphate, carbonate, or sulfate salt that could be dissolved in cold milk to form a quick-setting pudding.[131] The use of carrageenan to prepare cold-set milk desserts[321] and flan-type milk puddings[322] has recently been described by Moirano.

The following advantages have been attributed to carrageenan-based puddings and pie fillings[132]:

1. Puddings and pie fillings have a lower viscosity and body during cooking, which aids heat transfer and prevents scorching.
2. Preparations do not have to be brought to a full boil, as for starch puddings, and texture is not as affected by under or over cooking.
3. Gels are fast setting.
4. Puddings are non-running and are easily unmolded.
5. Surface skin is greatly reduced, particularly with chocolate flavor.
6. Puddings have no pastiness, or starchy flavor.

Furcellaran   The largest single use of furcellaran is in "blanc mange" puddings. A dry blend of furcellaran, sugar, flavor, and color is added to cold milk or cream and heated. Boiling is not required. Products of this type are very popular in Europe. Like carrageenan, furcellaran can be used either alone or combined with starch, which reduces syneresis. Levels of approximately 0.4% furcellaran and 1.0% starch (basis finished dessert) give a smooth, firm gel with a glossy surface and a rigid texture.[19,60,133]

Xanthan Gum   A recently issued patent discloses the use of xanthan gum with locust bean gum in the preparation of milk gels.[311]

Dairy Products

Gums are extensively used in a wide variety of dairy products, including ice cream, sherbets, cheese, milk products, etc. Typical applications of the various gums in dairy products are shown in Table 6 and discussed in the following section.

Ice Cream and Ice Milk   The function and application of stabilizers in ice cream and ice milk have frequently been reviewed.[134-138] It is generally agreed that the basic functions of an ice cream stabilizer are as follows:

1. to regulate the formation and growth of ice crystals during freezing and storage, preventing the growth of large, grainy crystals,
2. to prevent the separation or uneven distribution of fats and other solids,
3. to provide body, smoothness, and uniformity,
4. to prevent melt-down.

TABLE 6

Application of Gums in Dairy Products

Product	Agar	Algin	Carrageenan	Furcellaran	Arabic	Xanthan	Tragacanth	Hydroxypropylcellulose	Locust Bean	Guar	CMC	Methylcellulose	Karaya
Ice Cream Stabilizer	X	X	X	X	X		X		X	X	X	X	X
Ice Milk	X	X	X	X	X		X		X	X	X	X	X
Milk Shake	X	X	X				X		X		X		
Sherbets	X	X	X	X	X		X		X	X	X		X
Ice Pops and Water Ices	X	X	X		X		X		X	X	X	X	X
Chocolate Milk Drink		X	X	X	X	X	X		X	X	X		
Pudding	X	X	X	X	X	X			X	X			X
Variegating Syrups			X								X		
Cottage Cheese		X	X		X				X	X			
Cream Cheese	X	X	X		X				X	X			
Cheese Spread		X	X		X	X	X	X	X	X	X		X
Neufchâtel Process Cheese	X	X	X		X		X	X	X	X			X
Whipped Cream		X				X		X	X	X			X
Yogurt	X		X		X				X				

The primary role of a stabilizer is to bind free water in the mix as water of hydration, or entrapment in a gel structure. It is the ability of small percentages of gums to absorb and hold large amounts of water which make them effective in producing good body, smooth texture, slow melt-down, and heat shock resistance in the resultant products. Potter and Williams considered the following factors to be important in choosing a stabilizer[134]:

1. ease of incorporation in mix,
2. effect on viscosity and whipping properties of the mix,
3. type of body produced in the ice cream,
4. effect on melt-down characteristics,
5. ability to retard ice-crystal growth,
6. amount required to produce the stabilization,
7. cost.

Almost every gum has been used in ice cream stabilization at one time or another, and many different gums and combinations of gums are used in present day ice cream stabilizers. In general, better results are obtained from mixtures of gums than from any single gum. The use of the various gums in ice cream is discussed below.

Sodium alginates are widely used in ice cream stabilizers because of their good water-binding properties, ready dispersibility, and low cost. They are usually used at 0.1–0.5% and are particularly suitable in modern continuous horizontal freezers.[124] The effectiveness of sodium alginate has been described as being due to reaction with calcium in the ice cream mix, preventing the tendency of soluble calcium salts to cause clumping of fatty globules.[137] The desirable properties of alginates as ice cream stabilizers have been described as follows[139]:

1. aids in whipping and prevents ice crystal formation during processing and storage,
2. prevents structural breakdown due to temperature variations in transport and retailing,
3. gives a smooth, even meltdown during eating.

Sodium alginate is usually used in conjunction with phosphates to prevent precipitation due to the high calcium content of ice cream. A sodium alginate-phosphate blend at a level of 0.3–0.4%, in combination with a glyceryl monosterate emulsifier functions as an ideal stabilizer.

Many stabilizers containing alginates have been patented. A stabilizer prepared by drying a wet paste of alginic acid, sodium carbonate, sodium phosphate, and a dispersing agent such as sugar or dextrin was claimed by Steiner.[140] A similarly prepared vacuum dried composition of sodium carbonate, alginic acid, sodium phosphate and casein has also been patented.[141] Miller discovered that the addition of glyceryl monostearate and glyceryl monomyristate to ice cream stabilizers such as alginates improved the dispersibility of the alginates as well as the dryness and shipping stability of ice cream.[142]

Propylene glycol alginates are also used in ice cream stabilizers. Propylene glycol alginate as a stabilizer produces a thin mix that does not clump or gel and is more resistant to calcium and other salts, as well as providing good heat shock protection to the finished ice cream.[29]

The use of CMC as an ice cream stabilizer was first evaluated during World War II, when traditionally used stabilizers were in short supply.[143] CMC was found to be an efficient stabilizer at 0.15–0.27% concentration of the mix, imparting good body, chewy texture, and enhancing the whipping properties of the mix. CMC is more effective when used with one or more other stabilizers, such as locust bean gum or carrageenan, since CMC causes whey separation in ice cream mixes when used alone. This separation can be reduced or eliminated by the incorporation of balancing colloids such as carrageenan at lower levels.[144,145] A combination of 1–12 parts CMC and 1 part carrageenan has been recommended.[146] Commonly used CMC stabilizer combinations are CMC-carrageenan, CMC-gelatin, CMC-gelatin-carrageenan and CMC-carrageenan-locust bean gum.[15] A CMC-carrageenan blend at 0.35% has been recommended for a typical ice cream mix containing 10.2% milk fat, 12% serum solids, 13% sucrose, and 0.15% monodiglyceride type emulsifier.[147] CMC, because of its cold water solubility, is particularly suitable for use in modern high-temperature-short-time continuous processing methods.[135,136]

Carrageenan alone is not a satisfactory stabilizer for ice cream, since it greatly increases mix viscosity, making it difficult to incorporate a sufficient quantity for proper stabilization.[18,148,149] However, it is extremely useful as a secondary stabilizer at approximately 0.03% in preventing the "wheying off" caused by primary stabilizers such as CMC, locust bean gum, or guar.[48] Mixtures of 1 part carrageenan and 5.5 parts guar[145,150] and 1 part carrageenan to 1–12 parts CMC have been recommended.[146]

Carrageenan is also used in variegated ice cream products, which are made by incorporating high-solids chocolate syrup, or fruit purees or syrups into an ice cream base to make contrasting streaks of color and flavor. The variegated syrup or puree is stabilized with carrageenan so that ice crystal growth is controlled and melt down characteristics are the same in both phases.[19,48]

Locust bean gum is widely used as a primary stabilizer in ice cream mixes because of its excellent water-binding and swelling qualities, and the smooth melt-down and excellent heat shock resistance it imparts to the final product.[134,151] Wheying off tendencies of locust bean gum are prevented by using it in conjunction with carrageenan. Improved performance has been claimed for locust bean gum when used in conjunction with a protein-stabilizing salt such as sodium hexametaphosphate.[151]

Guar has water binding and hydration properties similar to locust bean gum, and imparts similar texture, body, chewiness, and heat shock resistance. In addition, guar hydrates rapidly in cold water, making it more suitable than locust bean gum for high-temperature-short-time processes. Use of guar has been suggested in combination with calcium sulfate as an emulsifier;[152] with carrageenan to prevent whey-off;[150] and with sodium hexametaphosphate and sodium citrate.[153] A level of approximtely 0.3% of the ice cream mix has been recommended.[25,75,91]

Most of the other gums have at one time or another been used as ice cream stabilizers, but for reasons of price or functionality have generally been replaced by alginates, CMC, carrageenan, guar, and locust bean gum. The use of gum arabic as an ice cream stabilizer was reported to produce a fine texture by inhibiting the formation of ice crystals by its water binding properties.[154] However, ice cream stabilized with gum arabic has inferior melt down properties.[155] A coacervate of gum arabic and carrageenan has also been patented for use as an ice cream stabilizer.[156]

Tragacanth is mentioned in earlier reviews as an effective ice cream stabilizer, imparting good body and texture.[155,157] Potter and Williams found tragacanth to be an effective stabilizer at 0.20–0.35%, particularly when used in combination with other gums, in producing and maintaining a smooth body and texture.[134] The ability of gum tragacanth to maintain and increase viscosity during heat processing has also been reported.[158]

Water Ices and Sherbets   Water ices and ice pops are relatively simple systems consisting of sugar, acid, flavor, and color. Alternately, natural fruit juices may be frozen in the form of ice pops. Sherbets are similar products which also contain milk solids. In these systems, the functions of a stabilizer, and the stabilizers used, are similar to ice cream. However, since these products have lower total solids, the selection of a proper stabilizer is perhaps even more critical. Consideration is usually given to overrun, syrup drainage, body, availability, and convenience of use.[138] In addition, one problem is encountered in water ices and sherbets that is not a consideration in ice cream. During the freezing of ices and sherbets, water tends to crystallize out first, leaving a more and more concentrated solution of other ingredients among the ice crystals. On the other hand, when eaten, the color and flavor tend to migrate to the outer surfaces that are first melted. The stabilizer therefore has the function of keeping color and flavor evenly distributed during freezing and eating, as well as improving texture by preventing the growth of ice crystals.[137] Carrageenan has been found to be an effective stabilizer for artifically colored and flavored ice pops, most of which contain aniline dyes, since the negatively charged carrageenan reacts with the positively charged food grade aniline dyes to form a chemically bound, evenly distributed compound, preventing migration of color. However carrageenan is not particularly effective in natural fruit products.[19]

Alginates have been used in ice pops to impart smoothness of texture by insuring that the fruit flavorings are uniformly distributed throughout the ice crystals during freezing, stabilizing the structure by retaining flavor and color, and preventing drip.[124] Propylene glycol alginate, at levels of 0.15–0.25% is generally used in this application, since sodium alginate is more affected by acids and salts.

Tragacanth has also been used as a stabilizer for ice pops, sherbets, and water ices, at a level of about 0.5% to prevent the syrup from settling out of the ice matrix during storage.[159] Karaya at 0.4% has been used for similar purposes, and has also been used at a 0.15% level in combination with 0.15% locust bean gum.[160]

CMC also has been reported as an effective stabilizer for flavored ice products when used at about 0.1–0.2%.[161,162] Methylcellulose has been used to minimize fast meltdown and flavor drainage in ice confections. High viscosity hydroxypropylmethylcellulose was used at 0.5–2.0%.[163]

Cheese   Locust bean and guar are used in the manufacture of soft cheese. They function to increase the yield of curd solids up to 10%, facilitate separation and removal of the curd, and impart a desirable soft, compact texture to the curd, while producing a limpid whey.[2] Guar is also used in cold pack cheese foods to prevent weeping.[164]

Agar has previously been used in Neufchâtel process cheese and cream cheese, as well as in fermented milk products such as yogurt, but has been replaced in these products by the less expensive locust bean and guar.[165,166] Soft cheese spreads and processed cheese are blends of cheese and other substances such as sodium phosphate and sodium citrates, or other softening substances, to give a softer, spreadable cheese texture. Stabilizers such as sodium alginate are used in these spreads to prevent oil-water separation, and to improve texture and minimize surface hardening. Carrageenan is also used in cheese spreads, along with locust bean gum, to prevent whey separation and improve spreading properties. Also, in cottage cheese, carrageenan is used to induce small curd formation at low pH.

Milk Products   Carrageenan has the ability to prevent or alter the agglomeration of casein particles in milk. This unique ability, which is not yet clearly understood, is one of the most important properties of carrageenan, and gives rise to most of its commercial applications. One of these applications is the suspension of cocoa or other insoluble flavoring material in milk at very low use levels. In hot processed chocolate milk, about 0.03% carrageenan by weight alters the state of aggregation of casein to suspend the cocoa, without substantially changing viscosity. In cold processed chocolate milk, about 0.05% carrageenan dissolved in the flavoring syrup serves to suspend the cocoa. Many variations of this use have been patented.[167-169] Also, carrageenan is incorporated in dry powdered mixes containing cocoa, which are added to milk. In milk shake powders of this type carrageenan has been used in combination with CMC to give a final product with a stable foam, and rich body and mouthfeel.[102]

A pilot plant method for determining the effectiveness of various carrageenan stabilizers in milk has been reported:[170] Smith[171] demonstrated that κ-carrageenan was a more effective chocolate milk stabilizer than λ-carrageenan or unfractionated carrageenan.

Other gums have also been used as milk stabilizers. Furcellaran, being very similar to carrageenan, also has good stabilizing properties in chocolate milk at 0.05% or less. A phosphate-alginate blend, at 0.15%, was used in the preparation of chocolate milk.[139] Methylcellulose has been used in spray dried chocolate milk drink powders as a bodying agent and suspending agent for cocoa solids.[172]

Whipped Cream and Whipped Toppings   Sodium alginate is used in cream at a 0.15% level to act as a viscosity and stabilizing agent.[139] Sodium alginate is also used in synthetic cream at the same level in conjunction with 0.5% methylcellulose, or another foam-forming colloid, to give a quicker whip, greater tolerance to overwhipping, stabilize overrun, and prevent syneresis.[124] Locust bean, carrageenan, and karaya have also been effective stabilizers for whipped cream products, both natural and imitation.[173]

Beverages

Fruit Drinks, Juices, and Nectars   Alginates, particularly propylene glycol alginates, are used at 0.1–0.2% to suspend pulp in fruit drinks.[174] Carrageenan is used in tomato juice to maintain an even suspension of fine pulp, and improve the mouthfeel of juices and nectars.[48] Furcellaran is also used, at about 0.05%, to stabilize fruit pulp in fruit juices and soft drinks.[175] Guar is used as a thickening or viscosity control agent in fruit nectars.[19] Gum arabic is used in the foam stabilization of soft drinks.[176] It is also used as an emulsifier in many flavors used in soft drinks, such as fruit flavors, cola, and root beer.[177,178] Methylcellulose[179] and alginates[29] are used to give body to artifically sweetened soft drinks.

Beer   Gum arabic is used as a foam stabilizer in beer, where 2.5–3 lbs can stabilize 50 barrels of beer.[180] Propylene glycol alginate is another widely used beer foam stabilizer. Its low pH stability, hydrophilic, lipophilic, emulsifying and film forming properties make it useful in this application by controlling bubble size and protecting the foam against fats, soap, and other washing compounds. A level of 40–80 ppm is generally used.[181] Sodium alginate is used in the brewing industry as an auxiliary fining agent, and CMC has been reported to form highly stable foams in beer.[182]

Wines   Agar has been used at 0.05–0.15% in the fining of wines and vinegar, where it was reported to be more effective than gelatin and remove less tannins.[183] The use of sodium alginates for wine clarification has been investigated in Russia, where it was reported to remove nitrogenous substances as well as tannin and coloring materials.[184,185] Best results were obtained when 0.085–0.01 g/l were used with a supplementary clarifier such as gelatin.[186] The mechanism of reaction has been described as the hydrolysis of sodium alginate in the acidic wine to give sodium ions and an insoluble precipitate of alginic acid with salts of iron and copper.[187]

Gum arabic is also used in combination with other gums, for wine clarification.[188] Table 7 lists applications of gums in beverages.

Confectionery

The natural gums are not as extensively used in the confectionery industry as they were in the past. At one time, candies such as jellies and marshmallows were commonly made with agar, and gumdrops were prepared exclusively from gum arabic.[3,25,189] The functions of gum in confectionery, aside from forming gels, or "jellies", is to retard or prevent sugar crystallization, and to emulsify fat and keep it evenly distributed throughout the product. Many of the uses of agar, arabic, and other natural gums to a large extent have been replaced by pectin, gelatin, and thin boiling starches. In reviewing the use of gums in confectionery in 1955, Martin concluded that natural gums were used to a negligible extent in candy production.[190] However, frequent mention of the use of natural gums is still found in the literature. Gum arabic is still used as a component of chewing gum, cough drops, and candy lozenges. The Pine Brothers Division of Beech-Nut Life Savers, Inc. has used gum arabic in cough drop manufacture for years, and has recently adapted the material to automatic continuous processing.[191] The use of gum arabic as a binding agent for the paste base in making lozenges, and a coating for centers of panned goods has been claimed.[192] Low calorie fruit drops and dietetic gum drops are also based on gum arabic.[19] Krigsman describes Turkish Delight jellies prepared by blending sodium alginate, calgon, and a sparingly soluble calcium salt into a sugar syrup.[139] Ferri reports the use of gum tragacanth in a number of confectionery applications, based on its resistance to hydrolysis by food acids, including use as a thickener in cream centers containing natural fruit, and as a binder in lozenges prepared by the cold-press process.[193] In

TABLE 7
Application of Gums in Beverages

Product	Agar	Algin	Carra-geenan	Furcel-laran	Arabic	Locust Bean	Guar	CMC	Methyl-cellulose
Fruit Juices and Nectars		X	X	X			X	X	
Soft Drinks		X	X		X	X		X	X
Soft Drinks with Fruit Pulp		X		X		X		X	
Beer Foam Stabilizer		X	X		X			X	X
Beer Clarification		X	X						
Fining Wines, Juices and Vinegar	X	X	X		X				

TABLE 8
Application of Gums in Confectionery

Product	Agar	Algin	Carra-geenan	Arabic	Karaya	Traga-canth	CMC
Candy Gels and Jellies	X	X		X			
Caramels, Nougats, Taffy			X	X		X	X
Candy Glaze				X			
Chewing Gum				X			
Cough Drops and Lozenges				X		X	
Gum Drops, Pastilles, Jujubes				X	X	X	
Candy Mints and Wafers					X		
Marshmallows	X			X			X

this process, a small amount of gum binds the powdered sugar together, due to the effect of pressing and the slight amount of heat generated. In extrusion processing, a thick gum solution is mixed with the powdered sugar and then extruded into sheets. The use of CMC in controlling sucrose crystal growth in confectionery has been investigated, and claimed to be more effective than corn syrup or invert sugar in model systems.[89,194,195] Also, low viscosity CMC lengthened the life of all-sucrose fondants, and fondants containing 3% corn syrup.[195] In high corn syrup (20%) or invert syrup (10%) fondants, a much less noticeable effect was obtained.

The use of gums in confectionery has recently been discussed by Smith[196] and Alikonis.[197] Table 8 lists typical applications of gums in confectionery.

Bakery Products

Almost all of the natural gums have uses in the baking industry, and many of the gums have diverse uses, ranging from modifying dough properties to stabilizing fillings, icings, meringues, and glazes. Often the function of the gum is to prevent sugar crystallization, retard syneresis, contribute gelling or thickening properties, improve body, or enhance surface glaze. Applications of the various gums in bakery products are listed in Table 9, and discussed below.

Icings and Glazes   The appearance of icings and glazes on baked goods contribute greatly to the acceptability and appeal of the product. Icings and glazes on cakes and other sweet bakery products are basically sugar and water products with other ingredients such as whipping agents, flavor, shortening, milk solids, and stabilizers, which vary according to the type and use of the icing. Icings have been divided into the following general categories[160]: (a) flat icings, for sweet goods, (b) fudge icings for cakes, with low fat and partial aeration, (c) cream icings, with higher fat and aeration, and (d) coating icings, similar to candy coatings, with very high levels of hard fat.

Flat icings, which are usually applied to sweet rolls and packed in cellophane, are the most difficult to stabilize. These icings tend to melt in hot, humid weather, and either stick to the wrapper or are converted to syrup. Under other conditions the icings crack, or lose their gloss. These problems are due to moisture migration in the sealed package, and the solubilization of the sugar. Moisture from the baked product vaporizes and is condensed on the cellophane wrapping. This excess moisture dissolves the sugar, turns the icing into a syrup, and causes the icing to stick to the wrapper. These problems are usually minimized by using hot flat icings that contain a minimum amount of water and a hydrocolloid to bind free moisture. An icing base containing 0.1–0.5% of a gelling or highly viscous gum such as agar, locust bean gum, sodium alginate, carrageenan, pectin, or karaya, and 20 parts sugar is combined with 20 parts of water and brought to a boil. The hot syrup is mixed with 100 parts icing sugar and applied hot (120–140°F) to the rolls. The product can be wrapped within a minute, since water evaporates rapidly from its surface.

In the preparation of a good doughnut glaze, a different moisture problem exists. Here, loss of water causes sugar crystallization and cracking, chipping, or sweating of the glaze. This problem is also minimized by using a low moisture hot glaze containing a hydrocolloid stabilizer, usually at 0.5–1.0% of

TABLE 9
Application of Gums in Bakery Products

Product	Agar	Algin	Carra-geenan	Furcel-laran	Arabic	Xanthan	Karaya	Traga-canth	Hydroxy-propyl-cellulose	Locust Bean	Guar	CMC	Methyl-cellulose
Bread Doughs and Mixes	X		X	X	X					X	X	X	X
Cake Batter and Mixes	X		X	X	X					X	X	X	X
Fruit Cakes			X										X
Yeast-Raised Doughnuts			X								X	X	X
Pie Fillings	X	X	X	X			X			X			X
Fruit Fillings		X	X	X		X				X			
Bakery Jellies	X	X	X										
Boiled Cream Fillings		X	X						X			X	
Doughnut Glaze	X	X	X		X							X	
Flat Icings	X	X	X		X		X			X	X	X	
Meringues	X		X		X		X	X	X	X		X	
Cookies	X						X	X		X			
Citrus Oil Emulsions	X	X	X		X	X	X					X	
Cake Fillings and Toppings	X	X	X	X	X				X			X	X

TABLE 10
Application of Gums in Meat, Fish and Poultry Products

Product	Agar	Algin	Carra-geenan	Furcel-laran	Karaya	Xanthan	Locust Bean	Guar	CMC
Sausage Casings	X	X							
Fish Preservation		X					X		
Canned Meat, Fish and Poultry	X	X	X	X		X	X		X
Coated Jellied Meat	X	X	X						
Antibiotic Ice			X						
Sausage Binder			X		X	X	X	X	
Coating For Meat and Poultry	X		X					X	X

the sugar level. The gum increases the glaze viscosity, facilitates adherence to the doughnut, imparts quick setting properties, provides flexibility to prevent chipping, and reduces melting of the glaze.[19]

Agar, because of its good gelling properties, especially the ability of its gels to withstand high temperature, has had a large use in icings and glazes. However, as for many other uses, agar has tended to be replaced by other hydrocolloids such as carrageenan. An improved icing stabilizer using agar at 1.2–2.4% of water content, in combination with surface active agents such as sorbitan esters of the higher fatty acids, polyoxyethylene derivatives of the esters of stearic acid, and the sodium sulfoacetates of mono- and diglycerides of the fatty acids was patented by Steiner and Rothe.[198]

The stabilization of cream and whipped type icings is improved by the addition of about 2% carrageenan based on water used. A typical boiled-creme icing formula of 3% cornstarch, 0.5% carrageenan, 0.15% potassium chloride, and water to make up 100% has been recommended.[48] Also, glazes can be made by mixing 43° Baumé corn syrup with an equal weight of a 2% carrageenan solution. This can be brushed on warm to give a highly glossy glaze which resists sticking. Alginates, which have also been used as bakery icing stabilizers, have the advantage of being soluble without heating or boiling. Alginates have been used in icings for sweet buns to prevent sticking to the wrapper and to improve appearance, as well as for stabilization of layer cake icings.[199] In addition, the use of 3–8 ozs of alginate per 100 lbs of sugar has been recommended to prepare more adherent icings that do not crack or dry out during normal shelf life.[46]

The stabilizing, free flowing, and adhesive properties of gum arabic solutions have been used in glazes for baked goods, which are applied hot, and adhere firmly when cooled.[148,149] An icing base containing one part gum arabic, one part calcium sulfate, five parts sodium aluminum sulfate, 17 parts powdered sugar, 35 parts egg albumin, and 39 parts starch has been patented.[200] CMC is used to retard sugar crystallization in icings, glazes, and similar toppings. Icings containing CMC are smooth, glossy, and prevent moisture loss from the baked goods.[19] The thickening properties of CMC have been used to prepare a stable boiled icing base dry mix containing egg albumin, calcium sulfate, sodium aluminum sulfate, starch, sugar, and CMC.[200] Other uses are as a stabilizer in a no-cook, instant icing containing CMC and sugar,[201] or CMC and protein.[202]

Toppings and Meringues   Meringues, which consist basically of sugar, water, and egg albumin, with an overrun of about 300%, are very popular toppings for baked goods. A consistent texture from time of preparation to consumption requires the use of a stabilizer in most formulations.

Carrageenan has been reported as an effective stabilizer in hot and cold process meringues. A typical formulation for a cold-process meringue powder, to be prepared with 100 parts tap water and 85 parts powdered sugar is 4 parts egg albumin, 20 parts powdered sugar, 0.33 parts carrageenan, and 0.5 parts guar gum. Hot process or boiled syrup meringues are stabilized by adding 0.5% of a carrageenan-locust bean gum mixture.[48]

A whipped sugar topping based on a sodium alginate-soluble calcium salt system containing sugar and methylcellulose has been patented.[203] Alginates also have been found to be effective as stabilizers in prepared frozen whipped cream toppings for bakery products, to prevent syneresis and to maintain foam height and aerated structure.[204] Propylene glycol alginate (0.2%) is effective in reducing syneresis in meringues, speeds whipping time, give a whiter product, and replaces commonly used additives such as alum and acid salts which may impart an off taste.[19]

Locust bean gum has also been suggested for use as a stabilizer in meringues.[205] Gum tragacanth has been suggested to improve stability in cold process meringues and increase shelf life, alone or in combination with tapioca starch. Karaya has been reported to improve stability and overrun of meringues.[76] Starch and agar meringues stabilized with CMC have excellent volume, a tender, smooth-textured surface and less syneresis. A typical hot-process meringue will contain 15% high viscosity CMC, 63% dextrose, 15% cornstarch, and 7% agar. A typical cold-process formula contains 69% dextrose, 18% cornstarch, and 13% high viscosity CMC. A suggested dry mix formulation contains 30% dried egg albumin, 32% powdered sugar, 13.1% dextrose, 20% cornstarch, 2% high viscosity CMC, 0.75% gum karaya, 1.25% salt, and 0.9% citric acid.[206]

Pie Fillings and Jellies   Starch is used as a primary thickener in most pie fillings. However, all-starch fillings have a firm, starchy appearance that is not always desirable, and may be subject to syneresis. By the selection of the proper starch, usually a modified one, and the proper gum a variety of textures ranging from a semi-fluid to a firm consistency can be prepared. In canned or frozen fruit pie fillings, gums are used as a partial or complete replacement for starch, since starches tend to retrograde in liquid media.[19]

CMC is used at approximately 0.5% to thicken fruit juices and to prevent floating or settling of fruit during preparation as well as impart a clearer, brighter appearance, produce a desirable gel texture, and reduce syneresis in the finished product. In lemon pie fillings, containing about 4% starch, the addition of 0.5% high viscosity CMC gives firm body, and prevents cracking, shrinking, or syneresis.[19] The use of CMC has been recommended in canned berry, berry-apple, and frozen peach or cherry pie filling at 0.2–0.5% alone or in combination with modified waxy maize starch to give a less opaque milky appearance on storage.[207,208] Methylcellulose is used in pie fillings to reduce absorption of water by the crust during baking, to protect flavor during baking, and to stabilize the gel after baking.[104] Greminger and Savage suggest the use of 0.5% hydroxypropylmethyl cellulose (1500 to 400 cps) to improve clarity and sheen of cherry, blueberry, and raspberry pie fillings.[103]

The addition of 0.3–0.5% algin to cooked starch pie fillings prevents separation and cracking, and imparts a soft, smooth gel body with improved clarity and gloss.[46,209]

The incorporation of 0.1% algin in bakery jellies and piping jellies reduces syneresis and improves spreading qualities and storage life.[29] A novel instant chiffon pie filling, prepared with cold water and composed of a water-soluble alginate, a whipping agent, and a calcium salt was prepared by Hunter.[210]

Algin at approximately 0.3% also reduces boil-off of juices during the preparation of fruit pies[211] and controls weeping and spreading after preparation.[212] Furcellaran is used in the preparation of high sugar flan jelly, used for the decoration and covering of fruit cakes. Usually about 70% sugar and 0.8% furcellaran is used. A "tortenguss" jelly sold as a dry powder mix contains 1.5–2.0 grams furcellaran, 10 grams sugar, 0–10 grams starch, and is prepared by mixing with 40 grams sugar and 250 ml water, then heated to boiling. It is used as a jelly covering, or layer for cakes. Furcellaran is used in piping jelly; a cake decoration jelly having a viscous consistency which is extruded from flexible tubes. It is usually a mixture of furcellaran, sugar, citric acid, water, color, and flavor, and is prepared by boiling in water with additional sugar for 10–15 minutes to allow the citric acid to cause a hydrolytic breakdown of the furcellaran, which gives the right consistency to the final product.[213]

Locust bean gum has been used as a stabilizer for canned berry pie fillings[207] and in frozen pie fillings in combination with starch.[205] Guar-starch combinations, also, have been used in frozen pie fillings to prevent dehydration, shrinking, and cracking,[214] as have tragacanth-starch combinations to provide thickening, clarity, and brilliance.[205]

Bread Doughs The use of gums in bread as anti-staling agents, and as stabilizers and texture modifiers has been suggested. Schuurink reported that the water retaining capacity and volume of flours and doughs were increased by the addition of 1–2% CMC.[215] CMC also influenced the mixing behavior of dough, and improved water retention and softness on storage.[216] Methylcellulose has been claimed to increase water absorption and improve crumb texture in low-gluten nonstandardized breads.[217] Methylcellulose strengthens the dough during gas evolution without toughening, due to its reversible thermal gelling characteristics, and appears to retard moisture migration to the surfaces during storage.

Carrageenan, at 0.1%, has been used as a flour conditioner during baking to improve finished texture.[218] Also, carrageenan and hydroxylated lecithin combinations improved dough strength, loaf volume and shape, and finished texture by exerting a synergistic action on flour protein in continuously mixed doughs with average or higher quantities of non-fat dry milk.[219,220]

Locust bean gum is used to complement flour in bread and other leavened bakery products to produce dough with good water-holding and functional properties, giving higher yields.[19] It has also been claimed that the addition of guar solution during kneading increases yield and gives the dough better resiliency, a drier appearance, and the result is a better textured baked product which stays fresh longer.[221]

Gum arabic has been reported to improve the baking properties of rye and wheat flour when 0.08-0.20 parts are added to 1000 parts of flour.[222] Hydrocolloids, including gum arabic, were also claimed to extend the shelf life of white bread.[216]

Sweet Dough Products Methylcellulose is claimed to have several functional properties that make it useful in broadening the tolerance of doughs and batters as follows[217]:

1. Gelation and Binding Characteristics—Methylcellulose has several properties similar to gluten, but is not affected by acids, mineral salts, or proteases, and is more resistant to common oxidizing agents. Also, the thermally reversible gelling properties of methylcellulose aids in gas retention during baking, but does not increase the roughness of the finished product.

2. Emulsifying Properties—Methylcellulose is an effective emulsifying agent when used in batters, and appears to retard starch retrogradation and staling.
3. Moisture Retention—Methylcellulose can bind up to 40 times its weight of water, making it effective in improving mixing tolerances and producing a finished product with better water retention and resistance to staling.
4. Oil Resistance—Methylcellulose is insoluble in fats and oils. Doughnuts and fried cakes containing methylcellulose therefore absorb less fat.

In cake and doughnut batters hydroxypropylmethylcellulose has been used to partially replace the normal egg content. The use of 0.2–0.4% Methocel 65HG, 4000 cps, in a low-egg doughnut formula was found to increase water absorption of the dough, make batter viscosity more uniform, and give gelling character similar to the starch and protein in the batter.[223] The standing tolerance of the dough was also improved, and the final product had a smoother surface and a more tender texture. The use of similar levels of the same type of Methocel increased the specific volume of yeast-raised doughnuts by 10%, and decreased fat absorption by 17% due to the improved gel retention quality of the dough, and the oil insolubility of the methylcellulose product.[217] In cake batters gas retention characteristics depend on the formulation of a stable, uniform, emulsified mixture. Methylcellulose has suitable emulsifying and water retention properties to improve these qualities in batters and yield cakes with longer shelf life. Also, the whipping properties of methylcellulose can partially replace from 10–50% of the egg white in cake mixes, according to a patent by Weaver and Greminger.[224]

Carboxymethylcellulose, at 0.25%, is used to improve doughnut mixes, to increase yield and product tenderness, glaze adherence, shelf life and to reduce fat absorption.[19]

Carrageenan has been recommended for use at 0.1% in cake batters, fruit cakes, and yeast-raised doughnuts to improve texture and appearance in cake batters, reduce fat absorption and improve texture in doughnuts, and give a more moist texture and even distribution of fruit in fruit cakes.[48] Locust bean and guar gum have been suggested as a partial egg replacement in cake and biscuit doughs imparting better texture and longer shelf life to these products, and better unmolding and slicing properties.[19]

Flavor Emulsions Ferri reported preparing good flavor emulsions for bakery products, containing citric acid, lemon oil, glycerol, and water, with gum arabic as the emulsifying agent.[178] Karaya, used in bakers citrus oil emulsions in place of tragacanth gave fairly stable products, but stability was even better when a mixture of karaya and arabic were used. Werbin reported superior flavor emulsion properties for baking with a gum arabic-gum tragacanth mixture.[148,149]

Meat, Fish, and Poultry

Canned Meat and Fish Gelling hydrocolloids are used in canned products such as tongue, fish, poultry or other soft meats to prevent textural breakdown due to moisture pickup and damage during transportation, and to act as a gel filler in solid meat packs or as a gel binder in ground meat packs.

Agar has had use in this application because of its high gel strength and high gel melting temperature. Agar, and agar-gelatin combinations have been used in England in jellied eel packs[225] and in Japanese canned tuna.[226] Carrageenan gels are used as a filler and binder in such items as pet foods and caviar. In this application carrageenan gels are modified by potassium salts and locust bean gum to control melt temperature and gel firmness.[48] Furcellaran gels, which are uniform, and somewhat more elastic than agar gels, are extensively used for fish and meat canning in Europe.[19] Alginates have been used in canned herring at a 2% level to absorb cell water during cooking.[227]

Preservative Coating and Films   A process which utilizes alginates in the preservation of fatty fish, called the Protan method, has recently been developed in Norway. It is used for herring, mackerel, and other fatty fish which could not be processed by conventional quick freezing, since oxidative rancidity takes place even at low temperatures. To prevent this, fish are block-frozen in alginate jelly which forms an air-tight coating, preventing oxidative rancidity. It is also claimed that this process is advantageous in maintaining original appearance and moisture content, reducing freezing time and inhibiting salt migration during storage and damage on thawing. Herring fillets prepared by this process have a substantial market in Norway, the United States, and West Germany.[228,233] The process has been extended to sardines, shrimp, and other seafood, with further extentions to other products envisioned.[139,234,235]

A similar process has been used to coat meat and poultry before freezing by dipping into successive solutions of 10–15% sodium alginate, 3–5% calcium chloride, and 10–20% glycerol, which is used as a plasticizer to obtain a better retention of juices on freezing and thawing.[236,237] Sodium alginate coating for sausages, alone and in combination with ethyl cellulose, has been used to prevent salt rust and increase storage stability.[238,239]

Carrageenan has been suggested as another means of preventing oxidative rancidity after freezing. Solutions of 0.5% carrageenan, sometimes containing antioxidants such as ascorbic acid and lecithin, are used to coat the meat.[240,241,20] Similarly, coatings containing carrageenan, or agar, and tetracycline, have been used as an edible coating for poultry preservation.[242]

Carrageenan, and also CMC, are used in fish preservation in another manner. It was discovered that fish could be preserved for longer periods when stored in ice containing 1 ppm of chlortetracycline (CTC). However, during normal preparation of ice blocks the CTC tended to migrate during freezing and give an unevenly distributed antibiotic concentration. Carrageenan or CMC was found to greatly improve the uniformity of distribution of the CTC.[243] More conventional methods of fish preservation utilize CMC solutions containing ascorbic acid,[244] or salt solutions thickened with CMC.[245] The excellent film forming properties of methylcellulose have similarly been utilized in coating meat products.[246]

Processed Meats   Various gums are used in the manufacture of processed meats such as bologna, salami, and other sausages, acting as binding and stabilizing agents. A good binder must hold moisture during such processing steps as chopping, curing, smoking, cooking, and chilling, as well as during storage, and also must serve to emulsify protein, fat, and water in the product.[247]

The addition of locust bean gum to such products gives a more homogeneous product of improved texture, lubricates the mix and facilitates extrusion and stuffing. Guar, which has the same properties, is normally premixed at 0.2–0.5% with the salts going into the mix, then added rapidly into the meat mix at the beginning of the cycle. The guar gum acts to bind free moisture and retard shrinkage.[19] In addition, karaya and carrageenan have been reported to act as emulsifiers, adhesives, and water binding agents in processed meats, and to improve product appearance.[247,76]

Synthetic Sausage Casings Synthetic sausage casings have been prepared from alginates which are more hygienic and easier to handle and fill than natural products. In a typical process, a 6% sodium alginate solution is extruded into a 10% calcium chloride bath, then plasticized with glycerol and calcium lactate.[248] These casings have two important shortcomings. The casings, which are primarily insoluble calcium alginate, are susceptible to solubilization by reaction with sodium salts normally present in the sausage. Also, they do not shrink along with the rest of the sausage during cooking, as natural casings do. Proposed methods to overcome these shortcomings include preventing the undesirable calcium-sodium ion exchange by incorporating excess calcium ions in the meat casing, or replacing part of the salt in the meat with calcium salts.[19] Langmaack patented a procedure to tie up calcium by forming complexes with micromolecular bases such as ethyleneimine and related condensation products such as melamine and formaldehyde, or urea and formaldehyde, or with proteins.[249] Hartwig claimed a 40% reduction in swelling of calcium alginate casings by transforming the calcium alginate skin into free alginic acid, then reconverting to calcium alginate with calcium chloride.[250] Some typical uses of gums in meat, fish and poultry products are shown in Table 10.

Jams, Jellies and Preserves

Most jellies, jams, and preserves produced in the United States utilize pectin as the gelling agent. However, at one time or another all the phycocolloids have been used or proposed for use in these products. Humm reports the use of agar in fruit butter, jam and preserves.[3] Marmalade, jams, and jellies are one of the more important uses of furcellaran, where it has the advantage of not requiring the prolonged boiling necessary with pectin, thus retaining many of the flavor volatiles in the fruit, giving a fresh flavored product. In a typical preparation of marmalade using furcellaran, 100 lbs of prepared fruit or pulp is mixed with 100 lbs of sugar and boiled for 10–15 minutes. Furcellaran, at 0.2–0.5% of total weight is heated to solution in approximately 10–15 gallons of water, then added to the marmalade, mixed, and packed. Furcellaran is also the gelling base for prepared jam powders sold in Europe, such as "Jam in Ten Minutes."[60]

Flavor Fixation

Gum arabic is almost exclusively used as the fixative in spray dried flavors, which are extensively used in dry packaged mixes such as beverage powders, puddings, gelatin desserts, cake mixes, etc., where flavor stability and long shelf life are important.[251] These spray dried powdered flavors are simply produced by preparing an oil in water emulsion of citrus oils or other fruit or imitation flavors, with gum arabic as the emulsifier. The emulsion is then dried by spraying a fine mist into a current of hot air. In this manner the water is evaporated from the minute droplets very rapidly exposing the flavor to heat for only a brief period. Also, because of the rapid evaporation the temperature of the dried particle is immediately lowered far below the temperature of the surrounding air. The flavor material is encapsulated in the gum arabic, resulting in a dry, free flowing powder, and preventing evaporation of volatile constituents or oxidative deterioration of the flavor. A list of variables which must be considered in the production of good spray dried flavors is as follows:

1. percentage of solids in the emulsion
2. ratio of gum to oil
3. use of an emulsifier with the gum
4. viscosity of the emulsion
5. quality variation in the gum
6. drying temperature
7. maximum ΔT
8. variation in solvents
9. effect of homogenization
10. volatility of the various flavors
11. solubility of the flavors

The preparation of the emulsion is very important. The emulsion can contain a range of particle sizes from 1–50μ, and as a general rule, the better the emulsion, the more stable the dried particles.[252] A typical emulsion may contain one part flavor component, four parts gum arabic, and four parts water before drying. It has been suggested, however, that materials with a lower flavor content are more stable.[252,253]

A process has been described which illustrates the practicality and economy of spray drying as a method of producing large quantities and varieties of dry flavors. The liquid flavor is emulsified with gum arabic (20% by weight) to make a solution of approximately 35% total solids, then fed to the atomizer at a rate of 1,265 lbs per hour. The dryer evaporates 800 lbs per hour recovering 375 lbs of dried flavor, along with 90 lbs of fines which are reprocessed. Total evaporation loss of flavor is less than 5%.[254]

The effectiveness of gum arabic fixed flavors was evaluated by Stoll[255] in a series of tests in which dispersions of a volatile flavoring material (diacetyl, b.p. 88°C) and various gums were heated 15 minutes at 140°C, and analyzed for diacetyl retention. All the diacetyl evaporated from the control, while gum containing samples retained 12.4–20.5%, with gum arabic retaining 18.3%. Wenneis reported that spray dried emulsions of hydroxycitronellal, benzaldehyde, and other easily oxidizable aldehydes were stable for years, as opposed to seconds for unprotected materials.[252]

It has been suggested that eventually derivatives of sugar, starch, or cellulose may prove suitable for fixation of spray dried flavors, and are more economical, but to date gum arabic has no equal.[256] Larch gum, which has similar high-solubility low-viscosity characteristics has been patented as a flavor fixative in a gelatin dessert product, where it affords a clearer, more sparkling gel, since larch gum is more compatible with gelatin than is gum arabic.[257]

Salad Dressing, Sauces, and Relishes

French-type salad dressings are typical oil-in-water emulsions which contain vegetable oil, vinegar, salt, sugar, spices, and flavoring material. The emulsions are prepared by mechanical means and stabilized with a vegetable gum.

The requirement for a gum to act as a stabilizer and thickener in this type of product is that it have relatively good stability to heat and acidity. This is true for other similar products, such as relish sauces, condiment bases, and mayonnaise. In the past gum tragacanth had widespread use in these applications since it is the most stable of the natural gums at low pH. The following steps in the use of gum tragacanth in products of this type have been described as follows[258]:

1. put the gum into the condiment at the end of the boiling period,
2. cool rapidly, by heat exchanger, if possible,
3. make up the condiment in two fractions so that the gum, part of the water, and part of the sugar are in one fraction, and the remaining ingredients in the other fraction.

While gum tragacanth is still used in these products, it is being replaced in many instances with alginate esters and cellulose derivatives.[259,260] The use of propylene glycol alginate in French-type salad dressings has been described by Werbin.[261] About 0.8% gum, based on the aqueous phase, is wetted with a small amount of oil to inhibit lumping, then thoroughly dissolved in water either by heating with agitation at 160°F for twenty minutes or by standing overnight. The aqueous solution of the gum is then added to the oil and then mechanically emulsified in a colloid mill or homogenizer.

CMC has been used as a thickener and emulsifier in salad dressings, mayonnaise, and similar products.[262,263] It is permitted as an optional emulsifying ingredient for salad dressings under the standard of identity for food products of this type. CMC is also recommended in combination with carrageenan for preventing syrup separation in sweet-pickle relishes.[264] About 0.25–0.75% CMC and 0.15–0.45% carrageenan was found to be effective.

Carrageenan is also used to replace part of the raw starch in some salad dressings. Replacement of one-half the normal amount of starch with one-tenth its weight of carrageenan eliminated much of the undesirable taste and retrogradation of the full starch product.[265]

Another gum which has been used in the stabilization of French dressings is karaya, which functions by increasing the viscosity of oil-in-water emulsions, and preventing or retarding separation.[19,76] Xanthan gum has also been used to improve French dressing formulations because of its superior emulsion stability.[323]

Some typical applications of gums in dressings, sauces and relishes are given in Table 11.

Low-Calorie Foods

The market for low-calorie foods has had a rapid growth in recent years, with sales of 365 million dollars estimated for 1963. Low-calorie versions of a wide variety of food products are sold, including soft drinks, salad dressings, confectionery, cookies, puddings, and gelatin desserts. In addition, metered calorie products are also sold.[266] In most of these products sucrose or other sugars are replaced by artificial sweeteners such as saccharin or cyclamate. In other products, such as salad dressings, much lower amounts of oil may be used to substantially lower caloric content. The role of gums in these products is primarily to replace the viscosity and mouthfeel lost when sugar or oil is removed. In still other products, such as puddings and pie fillings, gums may be used to replace all or part of the starch normally used.[267] The gum

TABLE 11

Application of Gums in Salad Dressings, Sauces and Relishes

Product	Algin	Carrageenan	Arabic	Karaya	Tragacanth	Xanthan	Locust Bean	Guar	CMC	Methyl-cellulose
French Dressing	X	X		X	X	X	X	X	X	X
Salad Dressing	X	X		X	X	X	X	X	X	X
Syrups and Toppings	X	X			X		X			
Relish	X	X			X	X	X	X	X	
White Sauces and Gravies	X	X			X	X	X	X	X	
Mustard	X	X		X	X		X	X		
Catsup, Spaghetti Sauce	X	X	X		X		X	X		

TABLE 12

Application of Gums in Low-Calorie Foods

Product	Algin	Carrageenan	Furcellaran	Arabic	Xanthan	Karaya	Tragacanth	Larch	Locust Bean	Guar	CMC	Methyl-cellulose
Carbonated Beverages, Soft Drinks	X	X		X	X						X	X
Puddings	X	X	X	X	X			X				
Water Gel Desserts	X	X	X		X				X			
Salad Dressing	X				X		X			X	X	X
French Dressing	X				X							X
Fruit Juice Drinks				X								
Syrups and Toppings		X	X			X					X	X
Jelly and Jams		X			X		X		X	X	X	X
900 Calorie Foods		X									X	X
Canned Fruit	X										X	

may also function as a bulking agent, a flavor blending media, a masking agent for the aftertaste caused by artificial sweeteners, or to provide a feeling of fullness or satiety. The choice of a gum, or combination of gums, to provide the best possible duplication of the naturally sweetened product is important. One technique for the proper selection of gum is based on "viscosity versus rpm" curves, discussed by Glicksman and Farkas.[268] By this method the rheological properties of gums are compared to those of sugar. For the purpose of simplification, all liquid and semisolid foods may be classified into three rheological groups. In the Newtonian, or ideal system, there is a simple linear relationship between force and flow. The viscosity is independent of the rate of shear and is constant through any range of rpm. Sugar is the classic example of this group.

Non-Newtonian systems, which are dependent on rate of shear, are divided into sub-categories of shear-thinning and shear thickening. The shear thinning, or thixotropic materials decrease in viscosity with an increasing rate of shear, and in general have less desirable organoleptic properties which vary as they deviate from Newtonian behavior. The shear-thickening or dilatant materials increase in viscosity as rate-of-shear increases. These materials, in general, have more desirable, less slimy organoleptic properties.[269,271] However, the rheological and organoleptic properties of gum solutions vary with concentration, temperature, pH, other reactive constituents, etc., so that the selection of the gum or gum combination for a particular application may be quite complicated and difficult. The use of gums in various low calorie food products is discussed below. Typical applications are shown in Table 12.

Bakery Products   CMC has been used in low-calorie biscuits to provide a feeling of fullness and satiety. The biscuits each contain 175 calories, and a CMC level of 2.6%.[272,273] Methylcellulose has been used in simlar crackers or wafers designed for reducing diets, where its functions are to produce a feeling of fullness and satiety due to its water binding ability, and to act as a bulking agent that stabilizes moisture content during passage of the food through the digestive tract.[274]

Milk Puddings and Gelatin Desserts   The use of carrageenan as a gelling agent in low-calorie milk puddings and water dessert gels has been suggested.[275] These formulations consist of carrageenan, a potassium salt, sodium saccharin, calcium cyclamate, color, flavor, and gum arabic as a bulking agent. The low-calorie puddings contain a small amount of starch, and the dessert gel an acidulant. The low-caloric value, gelling and thickening properties of alginates suggest its use in low-calorie puddings. Algin is claimed to have only 1.4 calories/gram, and only a small amount is needed to form a gel, making possible a dessert gel with less than 5 calories/serving.[45]

Jams and Jellies   The use of CMC has been reported to improve stability and inhibit synersis in dietetic jams and jellies. A combination of 0.37% high viscosity CMC, 0.8% low-methoxyl pectin, and 0.25% calcium chloride has been recommended.[276] Non-ionic methylcellulose can be used in similar products with calcium salts of cyclamate sweetener and low-methoxyl pectin.[104] Similarly, Meer and Gerard suggest a low-calorie fruit jelly based on 0.4% locust bean gum, and 1.0% low-methoxyl pectin.[275]

Salad Dressings   Tragacanth has been reported as an effective stabilizer in low-calorie salad dressings, imparting a desirable mouthfeel to these products. Also, the smoothness and texture of high oil content cream salad dressings of the mayonnaise and French-type can be obtained in low calorie versions by replacement of oil with hydroxypropylmethylcellulose, thereby reducing the caloric values of salad dressings as much as 90%. Formulations can be made with caloric values as low as one calorie/gram.[179] Hydroxypropylmethyl-cellulose with a 90°C gel point and 4000 cps viscosity, at 1.0–1.5% use level is suggested.[103] A French dressing containing 1.2–1.5% methylcellulose, with 0.2–0.35% and 0.2–0.5% pectin has been patented,[277] as well as an Italian dressing containing 0.5–0.8% methylcellulose and 0.25–0.40% agar or other gelling gum.[278] A formulation for a low-calorie mayonnaise type salad dressing has also been recommended. This formula is based on a combination of 0.8% low methoxyl pectin, 0.37% high viscosity CMC, 0.25% calcium chloride, and added starch to produce the heavy, smooth, paste-like consistency that is desired.[276]

900 Calorie Food Preparations   Many 900 calorie food products are sold today in various forms and flavors, including dry mixes, beverages, wafers, and crackers. These products are meant to provide complete meal nutrition while only providing a relatively low amount of calories for people interested in weight control. The first successfully marketed product of this type was Metrecal, but now many such products are on the market, usually in liquid form, with a large variety of flavors. In these products, gums serve to provide good organoleptic qualities such as body and mouthfeel, and to keep insoluble flavor particles and other materials in suspension, in addition to providing a feeling of fullness and satiety after consumption. Several gums, including CMC, carrageenan, and methylcellulose have been used in this application.[19]

Beverages   Soft drinks normally contain large amounts of sugar, which, in addition to sweetness and flavor, provide viscosity, body, texture, and mouth-feel. These qualities must be replaced by gums in sugarless low-calorie or no calorie versions of these products. CMC and methylcellulose are often used in these applications. Methylcellulose with a 65°C gel point, either 400 or 4000 cps, has been recommended at levels of 0.15–0.20%. Methylcellulose is compatible with all sweeteners and does not cause cloudiness during chilling.[179]

Dehydrated Foods

The rehydration properties of dehydrated foods such as fruit juices, vegetable powders, and dry soup powders are often improved by the addition of a gum prior to drying. Windover reports that CMC is used in several processes for the preparation of dehydrated fruit and vegetable juices to improve the resuspension and reconstitution properties.[104] Greminger and Savage describe the use of low viscosity methylcellulose (15–25 cps) at 0.5–1.0%, solids basis, in dehydrated fruits and vegetable products to aid in redispersion and provide better consistency in the reconsititued product.[103] The reversible thermal gel point and the viscosity and suspending properties of methylcellulose combine to improve product performance.

The use of methylcellulose as a drying aid in the preparation of spray-dried grapefruit or orange juice at 0.1–4.0%, solids basis, was reported to facilitate the production of free-flowing powders.[279]

Methylcellulose is also used in foam-mat drying. In this drying method stable foams are made by incorporating air or other gases by whipping a liquid concentrate of the food product containing a foam stabilizer. The foam is air-dried or oven-dried in belt-type dryers to give porous mats that are readily reduced to free-flowing powders. This procedure was originally used for orange juice and tomato juice, but has been extended to milk, coffee, lemonade, and other fruit juices. The requirement of a good foam stabilizer is that it permits the formation of a stable foam that can be handled without breakdown, and will reconstitute to give a satisfactory product. Originally, egg albumin, fatty acid monoglycerides, mixtures of mono- and diglycerides, and fatty esters of sucrose at levels of about 1% solids basis were used.[280] Subsequently, it was found that 1.5% Methocel 60HG, 50 cps, was effective in developing and stabilizing foam in tomato paste and apple sauce.[102] More recently combinations of soya protein and methylcellulose have been used to form and stabilize foam in orange juice for drying.[281] Bissett *et al.* identified two promising stabilizers for foam mat drying; a glyceryl monostearate and a modified soybean protein in combination with one-quarter its weight of methylcellulose.[282] The soybean-methylcellulose combination had the advantage of giving a denser foam, permitting a faster belt loading rate, reconstituting more rapidly, producing a more natural color in the reconstituted product, and not contributing any appreciable flavor.

Similarly, for foam mat drying of orange juice, a 0.8% soya protein and 0.2% methylcellulose combination has been recommended as a foam stabilizer.[283] Bissett *et al.* reported a preferred stabilizer for foam-mat drying of Valencia orange juice at 60 grams of a 16.7% solution of modified soya protein and 60 grams of a 4.8% suspension of 10 cps methylcellulose added to 2,386 grams of frozen Valencia orange juice concentrate of 59° Brix.[284] CMC has also been recommended for addition to foods before dehydration. CMC is said to improve the retention of natural flavor as well as aid in the reconstitution and texture of the finished product.[285–287]

Aqueous fat and oil systems were converted to dry, stable, non-greasy solid food products when dried with CMC.[288,289]

References

1. Mantell, C. L., 1947. *The Water Soluble Gums*, Reinhold Publ. Co., N.Y.
2. Rol, F., 1959. Locust Bean Gum, *Industrial Gums*, (R. L. Whistler, ed.) Academic Press, N.Y., 361–376.
3. Humm, H. J., 1951. The Red Algae of Economical Importance: Agar and Related Phycolloids. *Marine Products of Commerce*, (D. K. Tressler and J. M. Lemon, eds.) 2nd ed., Reinhold Publ. Corp., N.Y., 47–93.
4. Chapman, W. J., 1950. *Seaweeds and Their Uses*. Methuen and Company, London.
5. Smith, F. and Montgomery, R., 1959. *The Chemistry of Plant Gums and Mucilages*, Reinhold Publ. Corp., N.Y.
6. Anon., 1964. Natural Water Soluble Gums. *Oil, Paint, and Drug Reporter*, Jan. 27.
7. Sanders, H. J., 1966. Food Additives. *Chem. Eng. News 44:*(43), 108.

8. Whistler, R. L., 1959. *Industrial Gums* (R. L. Whistler, ed.), Academic Press, N.Y., 2–6.

9. Whelan, K., 1966. Faster Production of Stabilized Foods, *Food Process. 27:*80.

10. Lachman, L. and Chavkin, L., 1957. A Study of the Lyophilization of Several Pharmaceutical Gums and Suspending Agents. *J. Amer. Phar. Assoc. Sc. Ed., 46:*412.

11. Henry, J. E., 1956. *Water Soluble Gum Having Improved Properties*, U.S. Patent 2,768,143.

12. Jordan, W. A. and Skagerberg, W. E., 1961. *Process for Making Dispersible Vegetable Gums*, U.S. Patent 2,970,063.

13. Pittet, A. O., 1965. Dissolution of Polysaccharides. *Methods in Carbohydrate Chemistry* (R. L. Whistler, Ed.), Vol. V, Academic Press, N.Y., 3–5.

14. Becher, P., 1965. *Emulsions: Theory and Practice*, 2nd Ed., Reinhold, N.Y., 227–230.

15. Schwarz, T. W., 1962. Emulsion with Hydrocolloids. *Amer. Perf. Cosm. 77:*85.

16. Thiele, H. and Von Levern, H. S., 1965. Synthetic Protective Colloids. *J. Colloid Sci., 20:*(7). 679.

17. Meer, Jr., G., and Meer, W. A., 1962. Natural Plant Hydrocolloids, Part I. *Amer. Perf. 77:*(2), 34.

18. Tseng, C. K., 1947. Agar. *Kirk-Othmer Encyl. Chem. Tech. 1:*232, 1st ed.

19. Glicksman, M., 1962. Utilization of Natural Polysaccharides Gums in the Food Industry, *Advances in Food Res. 11:*109.

20. Stoloff, L., 1947. Seaweed Colloids. *Kirk-Othmer Encyl. Chem. Tech. 1:*118, 1st ed.

21. Selby, H. H. and Selby, T. A., 1959. Agar. *Industrial Gums* (R. L. Whistler, ed.) Academic Press, N.Y., 15–50.

22. Humm, H. J., 1947. Agar—A Pre War Japanese Monopoly. *Econ. Bot. 1:*317.

23. Tseng, C. K., 1947. Seaweed Resources of North America and Their Utilization. *Econ. Bot. 1:*79.

24. Sanford, F. B., 1958. Seaweeds and Their Uses. *Fishery Leaflet 469*, U.S. Dept. of the Interior, Washington, D.C.

25. Tseng, C. K., 1946. Phycocolloids: Useful Seaweed Polysaccharides *Colloid Chemistry* (J. Alexander, ed.) Vol. 6, Reinhold Publishing Corp., N.Y., 629–734.

26. Araki, C., 1965. Some Recent Studies on the Polysaccharides of Agarophytes, Abstracts, *Fifth International Seaweed Symposium*, 3.

27. Anon., 1965. Agar. *U.S. Pharmacopeia XVII*. Mack Publ. Co., Easton, Pa., 17.

28. Tseng, C. K., 1945. America's Agar Industry. *Food Ind. 17:*(10) 141, 258.

29. McNeely, W. H., 1959. Algin. *Industrial Gums*, (R. L. Whistler, ed.) Academic Press, N.Y., 55–82.

30. Tressler, D. K. and Lemon, J., 1951. The Brown Algae-Algin from Kelp and Fucoids. *Marine Products of Commerce*. 2nd ed., Reinhold Publ. Corp., 94–106.

31. Guberlet, M. L., 1956. *Seaweeds at Ebb Tide*. Univ. of Washington Press, Seattle 62.

32. Newton, L., 1951. *Seaweed Utilization*. Sampson Low, London.

33. Webber, E. R., 1962. Alginates from Seaweed. *Food Manuf. 37:*328.

34. Tseng, C. K., 1947. Algin. *Kirk-Othmer Encycl. Chem. Tech.* 343, 1st ed.

35. Green, H. C., 1936. U.S. Patent 2,036,934.

36. LeGloahec, V. C. E. and Herter, J. R., 1938. U.S. Patent 2,138,551.

37. Hirst, E. L., Jones, J. K. N. and Jones, W. O., 1939. *J. Chem. Soc.,*

38. Fischer, G. F. and Dörfel, H. Z., 1955. *Z. Physiol. Chem., Hoppe-Seylers 302:*186.

39. Steiner, A. B. and McNeely, W. H., 1954. Algin in Review. *Natural Plant Hydrocolloids*. Advances in Chemistry No. 11, American Chemical Society, Washington, D.C., 68–82.

40. McDowell, R. H., 1955. *Properties of Alginates*. Alginate Industries Ltd., London.

41. Hirst, E. L., Percival, E. and Wold, J. K., 1964. The Structure of Alginic Acid, Part IV. Partial Hydrolysis of the Reduced Polysaccharide. *J. Chem. Soc.,* 1493.

42. Hirst, E. L. and Rees, D. A., 1965. The Structure of Alginic Acid. Part V., Isolation and Unambigious Characterization of Some Hydrolysis Products of the Methylated Polysaccharide. *J. Chem. Soc.,* 1182.

43. Haug, A., 1959. Fractionation of Alginic Acid. *Acta Chem. Scand. 13:*601.

44. McDowell, R. H., 1960. Applications of Alginates, *Rev. of Pure and Appl. Chem. 10:*1.

45. McDermott, F. X., 1962. Algin, Multi-Use Colloid, *Food Eng. 34:*66.

46. Gibsen, K. F. and Rothe, L. B., 1955. Algin, Versatile Food Improver. *Food Eng. 27:*87.

47. Maass, H., 1959. *Alginsäure und Alginate*, Strassenbau, Chemie, Technik, Verlagsgesellschaft m.b.H.

48. Stoloff, L., 1959. Carrageenan. *Industrial Gums* (R. L. Whistler, ed.), Academic Press, N.Y., 83–116.
49. Anon., 1961. U.S. Federal Register, Oct. 6.
50. Stoloff, L., 1949. Irish Moss. *Econ. Bot. 3:*429.
51. Christensen, O., 1964. Carrageenan, A Useful Food Additive. *Food Mfg. 39:*(3), 44.
52. Painter, T. J., 1965. *Methods in Carbohydrate Chemistry*, Vol. V. (R. L. Whistler, ed.), Academic Press, N.Y., 98.
53. Rees, D. A., 1963. The Carrageenans, Part I. The Relation Between the κ- and λ-Components. *J. Chem. Soc.* 1821.
54. Dolan, T. C. S. and Rees, D. A., 1965. The Carrageenans, Part II, The Positions of the Glycosidic Linkages and Sulfate Esters in λ-Carrageenans. *J. Chem. Soc.* 3534.
55. Conway, E. and Young, E. G., 1965. Seaweed Research. *Nature 208:*840.
56. Black, W. A. P., Blakemore, W. R., Colquhoun, V. A. and Dewar, E. T., 1965. The Evaluation of Some Red Marine Algea as a Source of Carrageenan and of Its κ- and λ-Components. *J. Sci. Fd. Agric. 16:*573.
57. Stoloff, L., 1954. Irish Moss Extractives. *Natural Plant Hydrocolloids*, Advances in Chemistry No. 11, American Chemical Society, Washington, D.C. pp. 92–100.
58. Schachat, R. E. and Glicksman, M., 1959. Furcellaran, A Versatile Seaweed Extract. *Econ. Bot. 13:*365.
59. Lund, S. and Pjerre-Petersen, E., 1953. Industrial Utilization of Danish Seaweed. *Proc. First International Seaweed Symp.*, 85–87.
60. Christiansen, E., 1959. Danish Agar. *Industrial Gums* (R. L. Whistler, ed.), Academic Press, N.Y., 51–54.
61. Dillon, T., 1952. Some Seaweed Mucilages. *Proc. First International Seaweed Symp.*, 44.
62. Painter, T. J., 1960. The Polysaccharides of *Furcellaria fastigiata* I. Isolation and Partial Mercaptolysis of a Gel Fraction. *Can. J. Chem. 38:*112.
63. Schachat, R. E. and Glicksman, M., 1959. Some Lesser-Known Seaweed Extracts. *Industrial Gums*, (R. L. Whistler, ed.), Academic Press, N.Y., 155–160.
64. Mantell, C. L., 1949. The Water Soluble Gums. *Econ. Bot. 3:*3.
65. Glicksman, M. and Schachat, E., 1959. Gum Arabic. *Industrial Gums* (R. L. Whistler, ed.), Academic Press, N.Y., 213–298.
66. Mantell, C. L., 1954. Technology of Gum Arabic. *Natural Plant Hydrocolloids*. Advances In Chemistry, No. 11, American Chemical Society, Washington, D.C., 20–32.
67. Beach, D. C., 1951. Gums and Mucilages. *Kirk-Othmer Encycl. Chem. Tech. 7:*328, 1st ed.
68. Glicksman, M., 1966. Gums, Natural. *Kirk-Othmer Encycl. Chem. Tech. 10:*741, 2nd ed.
69. Patel, B. N., 1964. Hydrocolloids in Cosmetic and Pharmaceutical Dispersion. *Drug Cosm. Ind. 95:*(3), 337.
70. Fleisher, J., 1959. Gum Ghatti, *Industrial Gums* (R. L. Whistler, ed.), Academic Press N.Y., 311–320.
71. Aspinall, G. O., Hirst, E. L. and Wickstrom, A., 1955. Gum Ghatti. The Composition of the Gum and the Structure of Two Aldobiouronic Acids Derived from It. *J. Chem. Soc.*, 1160.
72. Aspinall, G. O., Auret, B. J. and Hirst, E. L., 1958. Gum Ghatti, Part II. The Hydrolysis Products Obtained from the Methylated, Degraded Gum, and the Methylated Gum. *J. Chem. Soc.*, 221.
73. Aspinall, G. O., Auret, B. J. and Hirst, E. L., 1958. Gum Ghatti, Part III. Neutral Oligosaccharides Formed in Partial Acid Hydrolysis of the Gum. *J. Chem. Soc.*, 4408.
74. Aspinall, G. O. and Christensen, T. B., 1965. Gum Ghatti, Part IV. Acid Oligosaccharides From the Gum. *J. Chem. Soc.*, 2673.
75. Goldstein, A. M. and Alter, E. N., 1959. Gum Karaya. *Industrial Gums* (R. L. Whistler, ed.), Academic Press, N.Y., 343–360.
76. Goldstein, A. M., 1954. Chemistry, Properties and Applications of Gum Karaya. Natural Plant Hydrocolloids, *Advances in Chemistry*, No. 11, American Chemical Society, Washington, D.C., 33–37.
77. Meer, Jr., G., and Meer, W. A., 1962. Natural Plant Hydrocolloids Part II. *Amer. Perf. 77:*(4), 49.

78. Patel, B. N., 1964. Hydrocolloids in Cosmetic and Pharmaceutical Dispersions. *Drug Cosm. Ind. 95:*(4), 509.

79. Ferri, C. M., 1959. Gum Tragacanth, *Industrial Gums* (R. L. Whistler, ed.), Academic Press, N.Y., 511–516.

80. Gentry, H. S., 1957. Gum Tragacanth in Iran. *Econ. Bot. 11:*40.

81. Beach, D. C., 1954. History, Production, and Uses of Tragacanth. Natural Plant Hydrocolloids, *Advances in Chemistry*, No. 11, American Chemical Society, Washington, D.C., 38–44.

82. Schwarz, T. W., Levy, G. and Kawagoe, H. H., 1958. Tragacanth Solutions III, The Effect of pH on Stability. *J. Amer. Pharm. Assoc. 47:*(10), 695.

83. Stout, A. W., 1959. Larch Araboglactan. *Industrial Gums* (R. L. Whistler, ed.), Academic Press, N.Y., 307–319.

84. Stein Hall & Co., 1966. Stractan Brochure.

85. Bouveng, H. O. and Lindberg, B., 1958. Studies on Arabogalactans II. Fractionation of the Arabogalactan from *Larix Occidentalis* Nutt. A Methylation Study of One of the Components. *Acta Chem. Scand. 12:*(10), 1977.

86. Coit, J. E., 1951. Carob or St. John's Bread. *Econ. Bot. 5:*82.

87. Block, F. and Binder, R. L., 1960. Report on Processing Trials with Fruits and Nuts Not Common in U.S. *Chemurgic Dig. 19:*10 (October).

88. Whistler, R. L., 1954. Guar Gum, Locust Bean Gum, and Others. Natural Plant Hydrocolloids, *Advances In Chemistry* No. 11, American Chemical Society, Washington, D.C., 45–50.

89. Deuel, H. and Neukon, H., 1954. Some Properties of Locust Bean Gum. Natural Plant Hydrocolloids. *Advances In Chemistry* No. 11, American Chemical Society, Washington, D.C.

90. Poats, F. J., 1960. Guar, A Summer Row Crop for the Southwest. *Econ. Bot. 14:*241.

91. Goldstein, A. M. and Alter, E. N., 1959. Guar Gum. *Industrial Gums* (R. L. Whistler, ed.), Academic Press, N.Y., 321–342.

92. Esser, J. A., 1958. Guar: Its Development and Uses. *Chemurgic Dig. 17:*9 (April).

93. Hui, P. A. and Neukon, H., 1964. *Tappi 47:*(1), 39 C.A.60:94651.

94. Goldfrank, H., 1960. Guar: A Plant Gum With Many Uses. *Chemurgic Dig. 19:*10 (July).

95. Carlson, W. A., Ziegenfuss, E. M. and Overton, J. D., 1962. Compatability and Manipulation of Guar Gum. *Food Technol. 16:*(10), 50.

96. Batdorf, J. B., 1959. Sodium Carboxymethylcellulose. *Industrial Gums* (R. L. Whistler, ed.), Academic Press, N.Y., 643–674.

97. Hader, R. N., Waldeck, W. F. and Smith, F. W., 1952. *Ind. Eng. Chem. 44:*2803.

98. Baird, G. S. and Speicher, J. K., 1962. Carboxymethylcellulose, *Water Soluble Resins* (R. L. Davidson and M. Sittig, eds.), Reinhold Publ. Corp., N.Y., 69–87.

99. Batdorf, J. B., 1964. How Cellulose Gum Can Work for You, *Food Eng. 36:*(8), 66.

100. Ganz, A. J., 1966. Cellulose Gum—A Texture Modifier, *Mfg. Conf. 46:*(10), 23.

101. Hercules Powder Company, 1963. *Cellulose Gum—Properties and Uses.*

102. Glicksman, M., 1963. Utilization of Synthetic Gums in the Food Industry. *Advances in Food Res. 12:*283.

103. Greminger, G. K. and Savage, A. B., 1959. Methylcellulose and Its Derivatives. *Industrial Gums* (R. L. Whistler, ed.), Academic Press, N.Y., 565–596.

104. Windover, F. E., 1962. Alkyl and Hydroxyalkylcellulose Derivations. *Water Soluble Resins* (R. L. Davidson and M. Sittig, eds.), Reinhold Publ. Corp., N.Y., 52–68.

105. Dow Chemical Co. 1964. *Methocel Brochure.*

106. Young, A. E. and Kin, M., 1946. Methylcellulose. *Colloid Chemistry*, Vol. VI, (J. Alexander, ed.), Reinhold Publ. Corp., N.Y., 926–933.

107. Kelco Co. 1961. *Kelco Algin.*

108. Steiner, A. B., 1948. *Algin Gel-Forming Compositions.* U.S. Patent 2,441,729.

109. McDowell, R. H., 1954. *Manufacture of Alginic Materials.* U.S. Patent 2,686,127.

110. Angermeier, H. F., 1951. *Edible Alginate Jelly.* U.S. Patent 2,536,708.

111. Poarch, A. E. and Tweig, G. W., 1957. *Gel Forming Compositions.* U.S. Patent 2,809,893.

112. Gibsen, K. F., 1959. *Algin Gel Composition and Method.* U.S. Patent 2,918,375.

113. Merton, R. R. and McDowell, R. H., 1960. *Powdered Alginate Jellies.* U.S. Patent 2,930,701; British Patent 828,350.

114. Alginate Industries, 1960. *Jelly Production Using Alginates.* British Patent 845,024.
115. Kohler, R. and Diericks, W., 1959. *Product and Process for the Production of Aqueous Gels.* U.S. Patent 2,919,198.
116. Rocks, J. K., 1960. *Method and Composition for Preparing Cold Water Desserts.* U.S. Patent 2,925,343.
117. Glicksman, M., 1962. *Freezable Gels.* U.S. Patent 3,060,032.
118. Baker, G. L., 1949. *Edible Gelling Composition Containing Irish Moss Extract, Locust Bean Gum, and an Edible Salt.* U.S. Patent 2,466,146.
119. Standard Brands Inc., 1960. *Irish Moss Food Product.* British Patent 841,973.
120. A/S Kobenhavns Pectinfabrik, 1966. Specifications for Genugel Type LC 1.
121. Stoloff, L., 1958. *Method and Compositions for the Preparation of Carrageenan Water Gels.* U.S. Patent 2,864,706.
122. Marine Colloids, Inc., 1964. *How To Make Sparkling Clear Water Gel Desserts With Gelcarin DG and Gelcarin HWG.*
123. Glicksman, M., 1966. *Frozen Gels of Eucheuma.* U.S. Patent 3,250,621.
124. Anon., 1958. Emulsifying and Stabilizing Agents. *Food Manuf. 33:*159.
125. Gibsen, K. F., 1957. *Alginate Composition for Making Milk Puddings.* U.S. Patent 2,808,337.
126. Ohling, R. A. G., 1959. *Food Preparation.* Canadian Patent 574,261.
127. McDowell, R. H. and Boyle, J. L., 1960. *Powdered Alginate Jelly Composition and Method of Preparing Same.* U.S. Patent 2,935,409.
128. McDowell, R. H. and Boyle, J. L., 1960. *Gelling of Milk with Alginates.* British Patent 839,767.
129. Hunter, A. R. and Rocks, J. K., 1960. *Cold Milk Puddings and Method of Producing Same.* U.S. Patent 2,949,366.
130. Stoloff, L., 1952. Seaweed Extractives Find Versatile Uses in Foods. *Food Eng. 24:*(5) 114.
131. Stoloff, L., 1957. *Composition for Preparing Puddings.* U.S. Patent 2,801,923.
132. Anon., 1958. Puddings and Pie Fillings. *Seakem Extracts 5:*(1), 2. Published by Marine Colloids Inc., N.Y.
133. Litex Co., Denmark, 1960. *Danish Agar.*
134. Potter, F. E. and Williams, D. H., 1950. Stabilizers and Emulsifiers in Ice Cream. *Milk Plant Monthly 39:*(4), 76.
135. Moss, J. R., 1955. Stabilizers and Ice Cream Quality. *Ice Cream Trade J., 51:*22.
136. Sperry, G. D., 1955. Stabilizers and HTST. *Ice Cream Field 65:*(5), 10.
137. Boyle, J. L., 1959. The Stabilization of Ice Cream and Ice Lollies. *Food Technol. in Australia 11:*543.
138. Frandsen, J. H. and Arbuckle, W. S., 1961. *Ice Cream and Related Products.* Avi Publishing Co., Westport, Conn.
139. Krigsman, J. G., 1957. Alginic Acid and the Alginates Applied to the Food Industry. *Food Technol. in Australia 9:*183, 253.
140. Steiner, A. B., 1949. *Alginate Ice Cream Stabilizing Composition.* U.S. Patent 2,485,934.
141. Shiotani, H. and Hara, M., 1955. *Ice Cream Stabilizer.* Japan Patent 3031 (C.A. 51: 132631).
142. Miller, A., 1960. *Composition and Method for Improving Frozen Confections.* U.S. Patent 2,935,406.
143. Josephson, D. V. and Dahle, C. D., 1945. A New Cellulose Gum Stabilizer for Ice Cream. *Ice Cream Revs. 28:*(11), 32.
144. Gould, A. A., 1949. Recent Development in Ice Cream. *Ice Cream Revs. 32:*6.
145. Julien, J. P., 1953. A Study of Some Stabilizers in Relation to Pasteurization of the Ice Cream Mix. *Ice Cream Trade J., 49:*(9), 92.
146. Blihovde, N., 1952. *Process for Stabilizing Foodstuff and Stabilizing Composition.* U.S. Patent 2,604,406.
147. Keeney, P. G., 1962. Effect of Some Citrate and Phosphate Salts on Stability of Fat Emulsions in Ice Cream. *J. Dairy Sci., 45:*(3), 430.
148. Werbin, S. J., 1953. Recent Advances in the Field of Stabilizers and Emulsifiers. *Southern Dairy Prods. J. 53:*38.
149. Werbin, S. J., 1953. Vegetable Gums. *Bakers Dig., 24:*(4), 21.
150. Werbin, S. J., 1950. *Ice Cream and Stabilizer Therefor.* U.S. Patent 2,502,397.

151. McKiernan, R. J., 1957. The Role of Gums in Stabilizers. Speech at the Michigan Dairy Manufacturers Ann. Conv., Michigan State Univ., East Lansing, Michigan.

152. Julien, J. P., 1953. A Study of Some Stabilizers in Relation to HTST Pasteurization Of Ice Cream. *Ice Cream Trade J., 49:*(9), 44.

153. Weinstein, B., 1958. *Stabilizers for Ice Cream Desserts.* U.S. Patent 2,856,289.

154. Walder, W. O., 1949. The Polysaccharides III. Chemical and Physical Tests. *Food, 18:* 86.

155. Mack, M. J., 1936. Sodium Alginate as a Stabilizer in Manufacturing Ice Cream. *Ice Cream Trade J., 32:*(11), 33.

156. LeGloahec, V. C. E., 1951. *Carrageenate-arabate Coacervate.* U.S. Patent 2,556,282.

157. Josephson, D. V., Dahle, C D., and Patton, S., 1943. A Comparison of Some Ice Cream Stabilizers. *Southern Dairy Prods. J., 33:*(4), 34.

158. Wegener, H., 1954. Dependence of Viscosity of Ice Cream Mixes Containing Various Stabilizers on Heating Time, *Milchwissenschaft 9:*123.

159. Sommer, H. H., 1946. *The Theory and Practice of Ice Cream Making,* 5th ed. Publ. by H. H. Sommer, Madison, Wisconsin.

160. Nash, N. H., 1960. Functional Aspects of Hydrocolloids in Controlling Crystal Structure in Foods. *Physical Functions of Hydrocolloids.* Edited by Staff of ACS Applied Publications, American Chemical Society, Washington, D.C., 45–58.

161. Burt, L. H., 1951. *Frozen Confections.* U.S. Patent 2,548,865.

162. Burt, L. H., 1956. *Frozen Confections.* Canadian Patent 527,490.

163. Windover, F. E., 1960. *Frozen Confection and Method of Making the Same.* U.S. Patent 2,950,199.

164. Klis, J. B., 1966. *Food Proc. 27:*(11), 58.

165. Dahlberg, A. C., 1927. *J. Dairy Sci. 10:*106.

166. Marquardt, J. C., 1930. Making Cream Cheese to a Formula. *Food Inds., 2:*76.

167. Berndt, L. H. and Klein, R. A., 1958. *Irish Moss.* U.S. Patent 2,830,903.

168. Stoloff, L., 1958. *Preparation of Flavored Milk Drinks.* U.S. Patent 2,834,679.

169. Wilcox, D. F., 1958. *Process for Preparing Stabilized Concentrated Milk and Product Produced Thereby.* U.S. Patent 2,845,350.

170. Willard, H. S. and Thomas, W. R., 1956. Studies on Stabilizers in Chocolate Drink as Related to Pilot-Plant Operations. *Proc. Western Div., Am. Dairy Sci. Assoc.,* 37th Ann. Meeting.

171. Smith, D. B., 1958. *Fractionation of Geloses.* Canadian Patent 561,448.

172. Love, D. H., 1959. Reconstitutes Without Solids Separating. *Food Proc. 20:*(1).

173. Alikonis, J., 1960. Practical Aspects of Foam Stabilization. *Physical Functions of the Hydrocolloids.* Edited by the Staff of ACS Applied Publications. American Chemical Society, Washington, D.C., 73–78.

174. Moncrieff, R. W., 1953. Stabilizing Fruit Drinks. *Food 22:*498.

175. Anon., 1961. Food Additives Federal Register, October 6.

176. Anon., 1957. Gum Type Stabilizers Perform New Vital Functions. *Food Eng. 25:*105.

177. Thurston and Braidich Co., 1950. *Water Soluble and Bulking Gums,* New York.

178. Ferrit, C. M., 1947. Stabilization of Emulsions with Vegetable Gums. *Food Inds. 19,* 784.

179. Dow Chemical Co., 1960. *Methocel in Dietetic Foods.*

180. Morningstar-Paisley Inc., 1953. Gum Arabic. *Tech. Service Bull. No. G-101,* N.Y.

181. Nordman, H. E. and Mohr, W. H., 1963. Algin, Beer Foams Best Friend. *Am. Brewer 96,* 22.

182. Wallerstein, J. S., Schade, A. L. and Levy, H. B., 1949. Process for Improving the Foam of Fermented Malt Beverages. U.S. Patent 2,478,988.

183. Veiguinke, A. S., 1951. The Portugese Agar-Agar Industry and the Application of the Product in the Agricultural Industries. *Ind. Agr. It. Aliment* (Paris) *68,* 261 (C.A. *46,* 4701C).

184. Kulzhinskaya, N. A., 1951. Clarification of Wine by Sodium Alginate. *Vinodelie i Vinogradarstvo S.S.S.R. 11,* 30 (C.A. *49:* 1275c).

185. Kulzhinskaya, N. A., 1951. Removing Nitrogeneous Substances from Wine. *Vinodelie i Vinogradarstvo S.S.R. 11,* 11 (C.A. *49:* 1277b).

186. Kulzhinskaya, N. A., 1951. Clarification of Wine by Sodium Aliginate. *Vinodelie i Vinogradarstvo S.S.R. 11,* 23 (C.A. *49:*1275d).

187. Gvelesiani, U. P., 1953. Mechanism of Clarification of Wine with Sodium Alginate. *Vinodelie i Vinogradarstvo S.S.R. 13,* 18 (C.A. *48:* 417e).

188. Bruno, P. J., 1954. *Substitute for Animal Albumins.* Spanish Patent 212,553 (C.A. 49: 13550i).

189. Langwill, K. E., 1939. Colloids: Their Application in Confectionery Manufacture. *Mfg. Confectioner 19:37.*

190. Martin, L. F., 1955. Applications of Research to Problems of Candy Manufacture, *Advances in Food Res. 6:169.*

191. Allured, S. E., 1959. Continuous Production of Gum Arabic Candies. *Mfg. Confectioner 39:23.*

192. Williams, C. T., 1961. The Function of Gum Acacia in Sugar Confectionery. *Confectionary Mfg., 6:299.*

193. Ferri, C. M., 1959. Factors in Selecting Water-Soluble Gums. *Mfg. Confectioner, 39:37.*

194. Desmarais, A. J. and Ganz, A. J., 1962. The Effect of Cellulose Gum on Sugar Crystallization and Its Utility in Confections. *Mfg. Confectioner, 42:33.*

195. Desmarais, A. J., 1962. Several Ways to Limit Sugar Crystallization. *Candy Ind. and Confectioner J. 119:* (4), 51.

196. Smith, P. S., 1965. Natural Gums and Their Uses in Confectionery Products. *Conf. Prod. 31:*(12), 967.

197. Alikonis, J., 1966. Foam Stabilization with Carbohydrate Colloids. *Mfg. Confectioner 46:*(9), 33.

198. Steiner, A. B. and Rothe, L., 1949. *Stabilizer for Icing.* U.S. Patent 2,471,019.

199. Glabau, C. A., 1943. How to Overcome Sticky Icings. *Bakers Weekly*, Dec. 20, 49.

200. Wagner, W. W., 1954. *Icing Base for Package Distribution.* U.S. Patent 2,682,472.

201. Butler, R. W., 1959. *Stabilized Icings and Preparations.* U.S. Patent 2,914,410.

202. Ganz, A. J., 1961. *Stabilized Icings and Process.* U.S. Patent 23,009,812.

203. General Foods Corp., 1958. *Confectionery Product.* British Patent 760,765.

204. Glabau, C. A., 1953. How Bakery Products React to Low Temperature Freezing. *Bakers Weekly*, Jan. 19, 47.

205. Carlin, G. T., Allsen, L. A., Becker, J. A., Logan, P. P. and Ruffley, Jr., J., 1954. Pies—How to Make, Bake, Fill, Freeze, and Serve. *Tech. Bull #121*, National Restaurant Assoc., Food and Equipment Research Dept., Chicago, Ill.

206. Hercules Powder Co., 1962, *Tech. Data Bull. No. VC-406.*

207. Moyls, A. W., Atkinson, F. E., Strachan, C. C. and Britton, D. 1955. Preparation and Storage of Canned Berry and Berry-Apple Pie Fillings. *Food Technol. 9:629.*

208. Kunz, C. E. and Robinson, W. B., 1962. Hydrophilic Colloids in Fruit Pie Fillings. *Food Technol. 16:*(7), 100.

209. Steiner, A. B. and McNeely, W. H., 1951. Organic Derivatives of Alginic Acid. *Ind. Eng. Chem., 43:2073.*

210. Hunter, A. R., 1961. *Neutral-type Instant Chiffon Pie Filling and Method of Producing Same.* U.S. Patent 2,987,400.

211. Anon., 1959. *Food Process 20:*(11), 69.

212. Strachan, C. C., Moyls, A. W., Atkinson, F. E. and Britton, D., 1960. Commercial Canning of Fruit Pie Fillings. *Can. Dept. Agr., Publ. 1062.*

213. Litex Co., Denmark., 1960. *Danish Agar.*

214. Werbin, S. J., 1950. Natural Gums in the Baking Industry. Presented to the Am. Assoc. of Cereal Chemists, Dec. 5, 1950.

215. Schuurink, F. A., 1947. *Improving Flour or Dough.* Dutch Patent 59, 870 (C.A. 41: 7014g).

216. Bayfield, E. G., 1958. Gums and Some Hydrophilic Colloids as Bread Additives. *Bakers Dig., 32:*(6), 42.

217. Dow Chemical Co., 1962. *Methocel in Bakery Goods.*

218. Glabe, E. F., Goldman, P. F. and Anderson, P. W., 1957. Effects of Irish Moss Extractive on Wheat Flour Products. *Cereal Sci. Today, 2:159.*

219. Glabe, E. F., 1964. Carrageenan and Hydroxylated Lecithin: A Stabilizer for Continuous Process Bread. *Bakers Dig., 38:*(3), 42.

220. Glabe, E. F., Anderson, P. W. and Jertson, E. C., 1964. Carrageenan and Hydroxylated Lecithin Applied to Continuous Mix Bread. *Cereal Science, 9:*(7).

221. Burtonite Co., 1960. *Technical Bulletin on Burtonite V-7-E.*

222. Stemmer, M., 1959. *Improving the Baking Properties of Rye and Wheat Flour.* German Patent 1,051,755 (C.A. 55:8694i).

223. Weaver, M. A., Bacon, K. D. and Greminger, G. K., 1957. *Dried Cake Mix Containing Water-Soluble Cellulose Ethers.* U.S. Patent 2,802,740.

224. Weaver, M. A. and Greminger, G. K., 1957. *Cake Mixes Containing Water-Soluble Cellulose Ethers.* U.S. Patent 2,802,741.

225. Anon., 1957. Jellied Eels for Bottling. *Food Mfg., 32:*447.

226. Anon., 1958. Canned Tuna in Agar Jelly. *Com. Fisheries Rev., 20:*(10), 56.

227. Henning, W., 1957. *Canned Herring:* German Patent 1,004,470.

228. Helgerud, O. and Olsen, A., 1955. *A Method for the Preservation of Food.* British Patent 728,168.

229. Helgerud, O. and Olsen, A., 1958. *Block Freezing of Foods (Fish).* U.S. Patent 2,763,557.

230. Olsen, A., 1959. Freezing Fish in Alginate Jelly. *Food Mfg., 30:*(7), 267.

231. Anon., 1955. Alginate-Jelly Mixture Protects Fatty Fish. *Refrig. Eng.,* 57:38.

232. Anon., 1955. Alginate-Jelly Mixture Protects Fatty Fish. *Food Eng.,* 27:(8), 109.

233. Anon., 1957. *Norway Exports 3:*22. Published by the Norwegian Export Council, Oslo, Norway.

234. Anon., 1957. Quick Frozen Scampi. *Food Trade Rev.* 27:(8), 15.

235. Anon., 1957. Protan Jelly Use Extended to Other Seafood Products and Strawberries. *Quick Frozen Foods, 20:*(9), 238.

236. Berlin, A., 1957. Calcium Alginate Films and Their Application for Meats used for Freezing. *Myasnaya Ind. S.S.S.R.* 28:44 (C.A. 51:1700b).

237. Slepchencnko, I. R., Knizhnik, E. B. and Pireva, L. A., 1956. Production of Calcium Alginate Films and their Utilization in the Freezing of Meat. *Sbornik Stud. Rabot. Moskov. Teknol. Inst. Myasnoi i Moloch. Prom.* 39:(4) (C.A. 53:13440f).

238. Childs, W. H., 1957. *Coated Sausage.* U.S. Patent 2,811,453.

239. Weingand, R., 1959. *Process for Producing Synthetic Sausage Casing from Alginates.* U.S. Patent 2,897,546.

240. Stoloff, L., Puncochar, J. F. and Crowther, H. E., 1948. Curb Mackerel Fillet Rancidity with New Dip-and-Coat Technique. *Food Inds.* 20:1130.

241. Stoloff, L., 1951. *Preservation of Foodstuffs.* U.S. Patent 2,567, 085.

242. Meyer, R. C., Winter, A. R. and Weiser, H. H., 1959. Edible Protective Coating for Extending the Shelf-Life of Poultry *Food Technol* 13:146.

243. Boyd, J. W., Bissett, H. M. and Tarr, H. L. A., 1955. Further Observation on the Distribution of CTC Throughout Ice Blocks. *Fisheries Research Board Can. Progs. Repts., Pacific Coast Stas. 102:*14.

244. Inagaki, C. and Iechika, K., 1953. Application of Vitamin C, I. Preservation of Fresh Fish with Ascorbic Acid. *Eiyô to Shokuryô 6:*82 (C.A. 51:18375a).

245. Aktieselskabet Protan and Helgerud, O., 1955. *Preserving Food Products by Freezing in Thickened Salt Solutions.* Norwegian Patent 86,861 (C.A. 50:7348a).

246. Lyaskovskaya, Y., Ivanova, A. and Poletaev, T. 1955. Polyvinyl Alcohol and its Possible Uses in the Meat Industry. *Myasnaya Ind. S.S.S.R. 26:*(1), 52 (C.A. 49:7770e).

247. Kraybill, H. R., 1955. Sugar and other Carbohydrates in Meat Processing. *Use of Sugars and Other Carbohydrates in the Food Industry* Edited by Staff of Ind. and Eng. Chem., American Chemical Society, Washington, D.C.

248. Weingand, R., 1957. *Method of Improving the Adhesion of Synthetic Sausage Casing to the Filling.* U.S. Patent 2,785,974.

249. Langmaach, L., 1961. *Method of Producing Synthetic Sausage Casings.* U.S. Patent 2,973,274.

250. Hartwig, M., 1960. *Method of Reducing the Swelling Capacity of Synthetic Alginate Skins.* U.S. Patent 2,965,498.

251. Wellner, G., 1953. New Coating Process Traps Full Flavor. *Food Eng.* 25:(8), 94.

252. Wenneis, J. M., 1956. Spray Dried Flavor. *Proc. Flavoring Extract Manufacturer's Assoc. of the U.S. 47,* 91.

253. Broderick, J. J., 1954. Superior Spray-Dried Flavors. *Food Eng., 26:*(11), 83.

254. Anon., 1956. Advances in Spray Drying Improve Flavor Quality. *Food Eng.* 28:(2), 76.

255. Stoll, A. C., 1952. A Study of the Effect of Various Substances on the Volatility of Diacetyl at an Elevated Temperature. *Food Res.* 17:278.

256. Anon., 1961. Spray-Dried Flavors. *The Givaudan Flavorist,* No. 3, Givaudan Delawanna Inc., New York.

257. Glicksman, M. and Schachat, R. E., 1966. *Gelatin-Type Jelly Dessert Mix.* U.S. Patent 3,264,114.

258. Burrell, J. R., 1958. Pickles and Sauces *Food Manuf. 33:*10.

259. Burrell, J. R., 1957. Pickles and Sauces. *Food Manuf. 32:*17.

260. Burrell, J. R., 1960. Pickles and Sauces. *Food Manuf. 35:*14.

261. Werbin, S. J., 1960. Practical Aspects of Viscosities of Natural Gums. *Physical Functions of Hydrocolloids,* Edited by the Staff of the ACS Applied Publications, American Chemical Society, Washington, D.C. 5–10.

262. Anon., 1952. Other New Ingredients, CMC as a Stabilizer. *Food Eng. 24:*(5), 116.

263. Fouassin, A., 1957. Alginates, CMC, and Other Thickening Agents in Derivatives of Milk and Mayonnaises. *Rev. fermentations et inds. Aliment 12:*169 (C.A. 52:2294e).

264. Anderson, E. E., Blank, A. P. and Esselen, W. B., 1954. Quell Separation In Pickle Relish. *Food Eng. 26:*(4), 131.

265. Anon., 1956. Modified Starch Substitutes. *Seakem Extracts 3:*(1), 3. Published by Marine Colloids, Inc., N.Y.

266. Salant, A., 1966. Marketing Aspects of Low Calorie Foods. *Food Technol. 20:*(1), 37.

267. Moirano, A., 1966. Hydrocolloids in Dietetic Foods. *Food Technol. 20:*(2), 55.

268. Glicksman, M. and Farkas, E., 1966. Gums in Artificially Sweetened Foods. *Food Technol. 20:*(2), 58.

269. Eirich, F. R., 1956. *Rheology, Theory and Applications,* Vol. I, Academic Press, N.Y., 1–8.

270. Frey-Wyssling, A., 1952. *Deformation and Flow in Biological Systems,* Interscience Publ., N.Y. 59–68.

271. Szczesniak, A. S. and Farkas, E., 1962. Objective Characterization of the Mouthfeel of Gum Solutions. *J. Food Sci., 27:*(4), 381.

272. Pfizer, Charles & Co., Inc., 1962. *Limits. A Satisfying Food for Losing Weight.* Family Products Division, N.Y.

273. Pfizer, Charles & Co., Inc., 1962. Report of 1962, Ann. Meeting, Groton, Conn., April, 30.

274. Dow Chemical Co., 1956. *The Uses of Methocel in Foods.*

275. Meer, Jr., G. and Gerard, T., 1963. Stabilizers, Bodying Agents for Improved Low-Calorie Foods. *Food Proc.,* May.

276. Hercules Powder Company. Use of Cellulose Gums in Dietetic Foods.

277. Nasarevich, L. S. and Spitzer, J. J., 1959. *Low Caloric French Dressing.* U.S. Patent 2,916,383.

278. Bondi, H. S. and Spitzer, J. G., 1959. *Low Caloric Italian Dressing.* U.S. Patent 2,916,384.

279. Eddy, C. W., 1950. *Process of Drying Fruit or Vegetable Materials Containing Added Methylcellulose.* U.S. Patent 2,496,278.

280. Morgan, Jr., A. I., Ginnette, L. F., Randall, J. M. and Graham, R. P., 1959. Technique for Improving Instants. *Food Eng. 31:*(9), 86.

281. Anon., 1962. Foam-mat Drying Report From Florida. *Food Proc., 23:*(12), 57.

282. Bissett, O. W., Tatum, J. H. and Wagner, Jr., C. J., 1961. Foam-mat Drying of Orange Juice: Progress in Foam Preparation, Drying, and Storage Studies., *Proc. Eleventh Citrus Proc. Conf.,* Winter Haven, Florida, Sept. 20, 9.

283. Wagner, Jr., C. J., 1961. Equilibrium Moisture Content of Orange Juice Powders at Low Relative Humidities., *Proc. Florida State Hort. Soc. 74:*287.

284. Bisset, O. W., Tatum, J. H., Wagner, Jr., C. J., Veldhuis, M. K., Graham, R. P. and Morgan, Jr., A. I., 1963. Foam-mat Dried Orange Juice I. Time-temperature Drying Studies. *Food Technol. 17:* (2), 210.

285. Karabinos, J. V. and Hindert, M., 1954. Carboxymethylcelulose. *Advances in Carbohydrate Chem. 9:*285.

286. Perech, R., 1946. *Vegetable or Fruit Juice Concentrate.* U.S. Patent 2,393,561.

287. Perech, R., 1946. *Beverage Concentrates.* U.S. Patent 2,393,562.

288. Bogin, H. H. and Feick, R. D., 1951. *Solid Flavoring Compositions.* U.S. Patent 2,555,464.

289. Bogin, H. H. and Feick, R. D., 1951. *Stable, Fatty-Food Compositions.* U.S. Patent 2,555,467.

290. Glicksman, M., 1969. *Gum Technology in the Food Industry*, Academic Press, New York.

291. McCormick, R. D., 1970. *Control of Viscosity and Emulsion Stability in Foods Using Modified Microcrystalline Cellulose.*

292. Ganz, A. J., 1969. CMC and Hydroxypropylcellulose—Versatile Gums for Food Use. *Food Prod. Dev. 3*(6): 65-68, 72-76.

293. Knightly, W. H., 1969. Dreaming About Dairy Products. Part I. *Food Prod. Dev. 3*(7): 24, 26, 28.

294. Rocks, J. K., 1971. Xanthan Gum. *Food Tech. 25*(5): 22-23, 26, 28, 31.

295. Anon., 1971. Unique Gum Improves Heat, Acid Stability. *Food Proc. 32*(7): 31.

296. Anon., 1971. Xanthan Gum in Salad Dressing Improves Emulsion Stability, Flavor. *Food Proc. 32*(9): 28.

297. Anon., 1971. Produces Low-Calorie Foods with Good Texture, Flavor. *Food Proc. 32*(7): 33.

298. Rusch, D. T., 1971. Developing Dairy Substitutes. *Food Tech. 25*(5): 32, 34, 36.

299. Anderson, D. M. W. and Dea, I. C. M., 1971. Recent Advances in the Chemistry of *Acacia* Gums. *J. Soc. Cosmet. Chem. 22*: 61-76.

300. Anderson, D. M. W. and Stoddart, J. F., 1966. Studies on Uronic Acid Materials. Part XV. *Acacia senegal* Gum. *Carb. Res. 2*: 104-113.

301. Anderson, D. M. W., Hirst, E. L., and Stoddart, J. F., 1966. Studies on Uronic Acid Materials. Part XVII. Some Structural Features of *Acacia senegal* Gum (Gum Arabic). *J. Chem. Soc.* (C), 1959.

302. Schultz, H. W., Ed., 1969. *Symposium on Foods: Carbohydrates and Their Roles.* Avi Publishing Co., Westport, Conn., p. 89.

303. Hercules, Inc., 1971. *Cellulose Gum.*

304. Hercules, Inc., 1971. *Klucel Hydroxypropylcellulose.*

305. Dow Chemical Co., 1968. *Methocel Product Information.*

306. Battista, O. A., 1971. Colloidal Polymer Microcrystals—New Compositions for Cosmetics. *J. Soc. Cosmet. Chem. 22*(9): 561-569.

307. Battista, O. A. and Smith, P. A., 1962. Microcrystalline Cellulose. *Ind. Eng. Chem. 54*(9) : 20-29.

308. Kelco Co., 1969. *Keltrol-Xanthan Gum.* DB # 18.

309. General Mills, 1969. *Biopolymer XB-23 Data Sheet.*

310. Xanthan Gums. *Federal Register 34*(53): March 19, 1969.

311. Schuppner, H. R., Jr., 1970. *Milk Gel Composition.* U.S. Patent 3,507,664.

312. Schuppner, H. R., Jr., 1971. *Heat Reversible Gel and Method for Preparing Same.* U.S. Patent 3,557,016.

313. Mueller, G. P. and Rees, R. A., 1967. Current Structural Views of Red Seaweed Polysaccharides. Proceedings, "Drugs from the Sea" Conference, University of Rhode Island, Kingston, Rhode Island, August 28-29.

314. Anon., 1969. R & D's Exciting Dessert Gel. *Food Eng. 41*(2): 57-59.

315. Moirano, A., 1969. *Dessert Gel and Composition Therefore.* U.S. Patent 3,445,243.

316. Battista, O. A. and Smith, P. A., 1961. *Level-off D.P. Cellulose Products.* U.S. Patent 2,978,446.

317. Battista, O. A., 1965. Colloidal Macromolecular Phenomena. *Amer. Sci. 53*: 151-173.

318. Foster, S. E. and Moirano, A., 1967. *Dessert Gel.* U.S. Patent 3,342,612.

319. Guiseley, K. B., 1968. Seaweed Colloids. *Kirk-Othmer Encyclopedia of Chemical Technology*, 2nd ed., Vol. 17, Interscience Publ., New York, N.Y., pp. 763-784.

320. Andrew, T. R. and Macleod, W. C., 1970. Application and Control of the Algin-Calcium Reaction. *Food Prod. Dev. 4*(5): 99, 102, 104.

321. Moirano, A. L., 1970. *Cold-Set Milk Desserts and Composition Therefore and Methods of Making Same.* U.S. Patent 3,499,768.

322. Moirano, A. L., 1970. *Flan-Type Milk Puddings and Method and Gelling Agent for Making Same.* Canadian Patent 847,287.

323. Stanislav, L. R. and Sheets, J. K., 1971. Improved French Dressing Formulation Using Xanthan Gum Stabilization. *Food Prod. Dev. 5*(6): 52.

CHAPTER **8**

Starch in the Food Industry

O. B. Wurzburg

Vice President—Research
Starch Division
National Starch and Chemical Corporation
Plainfield, New Jersey

Introduction

Starch, in its native or modified form, is used extensively throughout the food industry as a carbohydrate source, extender, processing aid, thickener, stabilizer, and texture modifier. In this chapter we will not be concerned with it as a primary nutrient but rather as an agent for modifying foods.

The availability and relatively low cost are major factors governing the use of starch and its modifications in foods. In addition, starches possess many unique properties which, in combination with their nutritive character, make them of particular value for modifying foods.

Native starches are produced commercially by extraction from the seeds of plants such as corn, wheat, sorghum or rice; the tubers, or roots, of plants like cassava, potato or arrowroot; and the pith of the sago palm. The character of the starch varies with the plant source from which it is derived.

The major source for commercial starch is corn. It represents an economically-sound raw material. This is the result of a combination of factors including abundance of crop, high yield per acre, high starch content, stability of grain in storage, value of by-products, and ease of processing. Because of these characteristics of corn the starch made from it can be marketed at a relatively low price.

Starch is extracted from corn by a wet milling process.[1] The kernels are steeped in warm water containing trace amounts of sulfur dioxide until they are softened. They are then cracked open to permit the removal of the germ by a liquid cyclone. The germ is then pressed to remove the corn oil and the residue is combined with other by-products to form feed. The balance of the kernel is ground, the fiber removed by screening, and the remaining starch-gluten mixture in water is passed through Merco centrifuges and Dorr Clones to separate the heavier starch from the gluten. The latter goes into feed and the starch water slurry is dewatered. The starch is dried for sale as native starch. The starch slurry may also be treated prior to drying so as to produce modified starches or it may be converted by hydrolysis to syrups, sugars, or dextrose.

While the bulk of starch produced in this country is from corn including waxy hybrids and high amylose corn, starches are also milled from sorghum and waxy sorghum as well as potato and wheat. The properties of those produced from sorghums are very similar to those from corn, and the subsequent discussions on corn starches will apply to sorghum starches. Potato starch and wheat starch represent a minor part of starch production. In addition to the above starches, tapicoa, which is milled in the East Indies and Brazil, is also imported into this country for use as is, or after modification to specialty products.

Physical Nature-Granules[2,3]

Starch is deposited in the plant in the form of minute, cold-water insoluble, granules. Depending upon the plant source, the granules range from about 3–100 microns in diameter. The shape of the granules will also be influenced by the plant source.

The hilum is a characteristic feature in many starch granules. It appears as a spot on the granule and is believed to be the nucleus about which the granule has grown. Starch granules exhibit birefringenee as evidenced by polarization crosses which may be observed when the granules are examined under polarized light. The intersection of the polarization cross is at the hilum.

Striation marks are another characteristic feature in some granules. These are not as readily discerned as the hilum but they can be detected in some cases as lines arranged concentrically about the hilum. Potato starch granules show these striations clearly, but in most other starches they are not apparent (in some cases they can be developed by etching with acid).

Microscopic examination of starches in water represents an effective technique for identifying the plant source of a starch. Magnification of about 400 times is usually adequate.

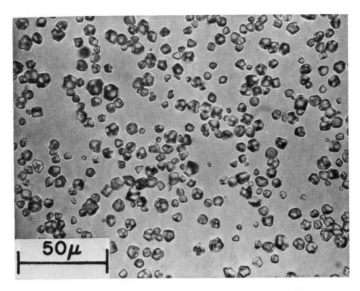

Fig. 1. Photomicrograph of rice starch granules.

The smallest granules found in commercially available starches are rice. They are about 3–8 microns in diameter and polygonal in shape. Rice granules tend to aggregate in clusters as shown in Figure 1.

Corn and sorghum starches have polygonal and rounded granules. The average diameter is about 15 microns. Corn granules range from about 5–25 microns in diameter. The hilum in corn granules is located in the center. Corn and sorghum granules are shown in Figures 2 and 3, respectively.

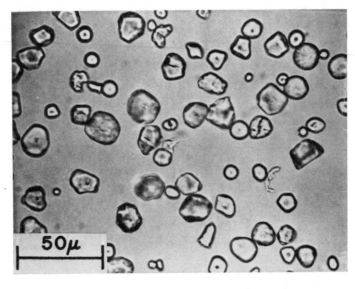

Fig. 2. Photomicrograph of corn starch granules.

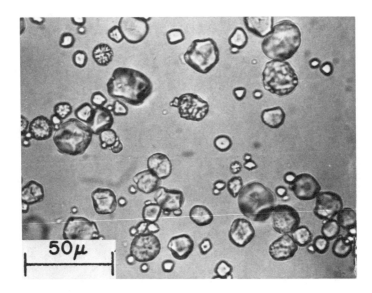

Fig. 3.   Photomicrograph of sorghum starch granules.

Granules of the waxy variety are similar to those of regular corn or sorghum. They may be distinguished from the regular types by staining with a dilute iodine solution. Granules of regular starches stain a deep blue, while those of waxy varieties do not.

Tapioca starch granules are rounded and truncated at one end to form kettle drum shapes. The average size is about 20 microns with a range of 5–35 microns. Many of the granules show a distinct hilum located at the center. Figure 4 shows typical tapioca granules.

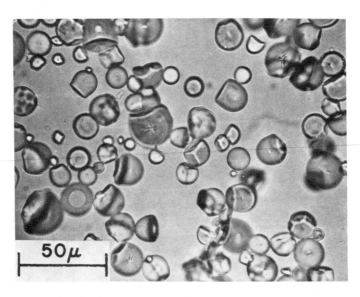

Fig. 4.   Photomicrograph of tapioca starch granules.

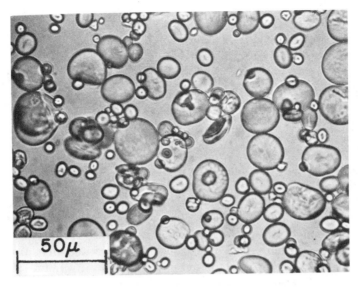

Fig. 5. Photomicrograph of wheat starch granules.

Wheat starch appears to have flat, round or elliptical granules which tend to cluster in two size ranges. The small ones range from about 2–10 microns and the large ones from 20–35 microns. They are shown in Figure 5.

Potato starch has the largest granules of any of the common commercial starches. They range from about 15–100 microns in diameter and are oval and egg-shaped. Unlike corn or tapioca granules, the potato granule has the hilum located toward one end. Its granules also contain detectable striations. Figure 6 shows typical potato starch granules.

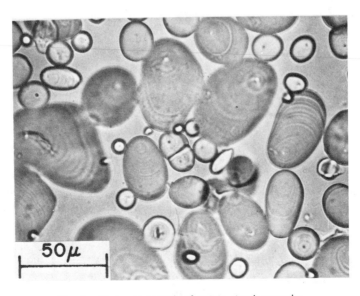

Fig. 6. Photomicrograph of potato starch granules.

Chemical Characteristics[4,5]

Chemically, starch is a carbohydrate. It is a polymer comprising a range of molecular sizes. The basic monomeric unit is D-anhydroglucose. If starch is hydrolyzed with dilute acid, it will give D-glucose. The predominant linkage is the 1,4-alpha glucosidic bond.

Two basic types of polymers are present in most starches—amylose and amylopectin. Both are polymers made up of anhydroglucose units. They differ, however, in size, the way the basic monomeric units are linked together, and in shape.

Amylose is a linear polymer in which essentially all of the anhydroglucose units are linked through 1,4-alpha glucosidic bonds. It may contain anywhere from about 200–2,000 anhydroglucose units. Each of the monomeric units contain one primary and two secondary hydroxyl groups, except the terminal units. At one end of the molecule the anhydroglucose unit contains three secondary and one primary hydroxyl groups. This is commonly called the non-reducing end. At the other end, the anhydroglucose unit contains one primary and two secondary hydroxyls, as well as an aldehydic reducing group in the form of an inner hemiacetal. This is usually called the reducing end.

Figure 7 shows a schematic representation of an amylose molecule. The

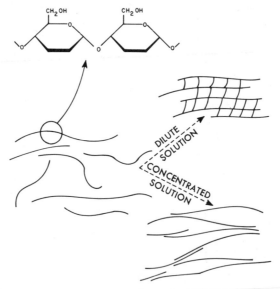

Fig. 7. Amylose molecules.

hydroxyl groups impart hydrophilic properties to the polymer which lead to an affinity for moisture and dispersibility in water. However, because the molecules are linear and contain hydroxyls, they have a tendency to become oriented parallel to one another and approach each other closely enough to permit association of one polymer with another through hydrogen bonding. When this occurs, the affinity of the molecule for water is reduced and the aggregate size increased to the point where precipitation may occur at low concentrations and gelling at higher concentrations, when only partial orientation may occur because of steric factors. The gels are three dimensional networks held together by hydrogen bonding at those sections where the molecules can become closely aligned. This phenomenon of association between amylose molecules is commonly referred to as retrogradation. The tendency of corn starch dispersions to become opaque and to gel on cooling is caused by the retrogradation of amylose molecules present in corn starch. The linearity of amylose molecules contributes toward the formation of strong, flexible films.

Amylose also has an affinity for iodine and large molecules containing hydrophobic and hydrophilic portions, such as fatty acids and various surfactants. It is believed that amylose complexes with these molecules by forming a helix around them. The complex with iodine gives a characteristic blue color which is used to establish the presence of amylose-containing starch and can be used to give a quantitative estimate of the amylose content of starch.

Amylopectin (the other polymer in starch) has a highly branched structure. Each branch contains about 15–25 anhydroglucose units interconnected by the alpha 1,4-linkage. At the branch point, the branches are connected by linkages (containing the hemiacetal aldehyde) attaching carbon 1 of the anhydroglucose unit at the start of the branch to carbon 6 (Figure 8). Amylo-

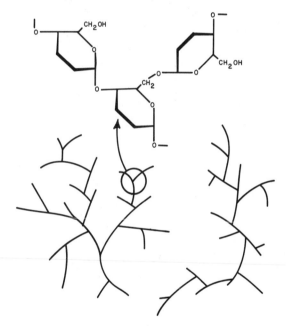

Fig. 8.   Amylopectin molecules.

pectin is usually a larger polymer than amylose. The exact size in most starches is not known, but light scattering measurements indicate it has a molecular weight in the millions. The size and branched nature of amylopectin interferes with the mobility of its molecules and their tendency to become oriented closely enough to permit the extensive hydrogen bonding necessary for retrogradation to occur. As a result, aqueous sols of amylopectin and amylopectin starches are characterized by good clarity and stability, or resistance to gelling or changing with age. Because it is a highly branched molecule, amylopectin does not form films as strong and flexible as those from amylose. It also does not give a blue color complex with iodine. The stability of amylopectin sols is a major factor in the use of modified amylopectin starches; e.g., waxy corn, as thickeners and in those uses where solution stability and freedom from loss in water-holding properties are desired.

Gelatinization Phenomena[6,7]

Starch granules are insoluble in cold water. When wetted or exposed to high humidities they will absorb water and swell slightly. The swelling is reversible, however, and on drying the granules shrink. The average increase in the diameter of granules of various starches going from anhydrous conditions to a water-saturated atmosphere is indicated in Table 1, together with the observed increase in water sorption.[8] The abundance of hydroxyl groups in the starch granules is a primary factor in their tendency to absorb moisture. Where starches are equilibrated under normal atmospheric conditions they usually contain about 10–18% moisture. Commercial corn starch has about 11%, tapioca about 14%, and potato about 18% moisture.

When a slurry of starch in water is heated beyond a critical temperature (which ranges from 56°C up, depending on the starch), the hydrogen bonds holding the granules together begin to weaken, permitting them to swell tangentially to as many times their original size. This process is referred to as gelatinization. As this swelling occurs, the granules first lose their birefringence; then, as they begin to imbibe water, the clarity of the slurry and viscosity increases. Some of the molecules leach out of the swollen granules, which eventually reach their maximum hydration and then begin to rupture and collapse, yielding a dispersion of granule fragments, starch aggregates, and molecules. As this occurs, the viscosity decreases and then tends to stabilize.

TABLE 1
Effect of Water-Saturated Atmosphere on Granule Diameter and Water Sorption of Various Starches[8]

Starch Source	Average Increase in Granule Diameter	Water Sorption/100 Grams Dry Starch
Corn	9.1%	39.9 grams
Potato	12.7%	50.9 grams
Tapioca	28.4%	42.9 grams
Waxy Corn	22.7%	51.4 grams

On cooling, the clarity of the starch cook will sometimes decrease, and the viscosity rises to varying degrees. In extreme cases, the starch sol will form an opaque, rigid gel. The exact way in which a starch behaves during the process of gelatinization depends upon numerous factors; plant origin of the starch is one of the most important. In general, root or tuber starches swell more rapidly in a narrower temperature range than do the common cereal starches. The temperature at which gelatinization takes place depends upon the criteria by which the measurement is made. The most sensitive and reproducible means for measuring the temperature of initial gelatinization is with a polarizing microscope equipped with a Kofler hot stage. The temperature at which the granules lose birefringence is noted.[9,10] Measurements can also be based on the increase in light transmission or viscosity of a starch slurry. These methods, however, are dependent upon the starch concentration and usually lack sensitivity. However, measurements based on viscosity are particularly valuable in terms of developing information to guide in setting conditions of use.

Not all the granules in a given starch will gelatinize or swell at the same time and temperature. Some are substantially more resistant than others and will differ by as much as 10°C or so in gelatinization temperature when viewed under a Kofler microscope hot stage.

Table 2 shows the Kofler gelatinization temperature of most common starches.[9] Note that root and tuber starches, such as tapioca and potato, tend to swell at somewhat lower temperatures than cereal starches.

TABLE 2
Kofler Gelatinization Temperature of Common Native Starches[9]

Corn	62°C –70°C
Waxy Maize	62.5 –72°C
Sorghum	68 –75°C
Waxy Sorghum	67.5 –74°C
Idaho Potato	56 –67°C
Maine Potato	59 –67.5°C
Dominican Tapioca	58.5 –70°C

The root and tuber starches also show a sharper rise in viscosity during cooking than do cereal starches. They also rise to a much higher peak viscosity and show greater relative breakdown with continued cooking (Figure 9). For this comparison the starches were cooked at 6% anhydrous solids at pH 5.5 in a continuous recording Brabender Viscosimeter in which the starch is

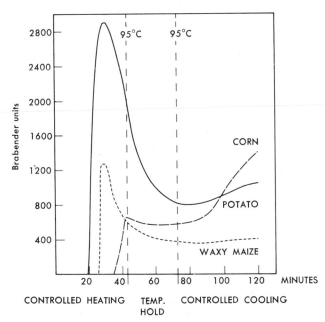

Fig. 9.   Brabender viscosities of corn, waxy maize, and potato starch at pH 5.0 (6% solids), Brabender Cartridge used —350.

agitated continuously. Initially, the starch is heated from 30°C to 95°C and held at this temperature. Note particularly (Figure 9) that waxy corn starch does not behave like a typical cereal starch but more like a root or tuber starch. High amylose starches are quite resistant to swelling, requiring super atmospheric temperatures for complete gelatinization.

The source of the starch is also a governing factor in the behavior of the starch sol on cooling. The amylose content and the molecular size of the amylose fraction are critical. In general, the tendency of most starch sols to thicken or increase in viscosity or gel on cooling and to become opaque is caused by the presence of amylose molecules which can associate through hydrogen bonding. Corn starch (which contains 28% amylose) forms rigid opaque gels when its sols are cooled. Waxy corn starch (which is essentially free of amylose) gives sols which show relatively little thickening on cooling. Tapioca and potato with amylose contents of about 18% and 23%, respectively, give relatively stable, clear sols on cooling in spite of their amylose content. This is because the amylose molecules in these starches have a substantially higher molecular weight than corn amylose and possibly because they may contain a few branch points that prevent them from becoming oriented and approaching each other closely enough to permit hydrogen bonding.

The concentration at which a starch is cooked influences its behavior particularly, as it relates to the swelling power of the granules. Leach, McCowen, and Schoch have determined the swelling power of various starches by heating low concentrations of starch at 95°C for 30 minutes and centrifuging to separate the swollen sediment from the supernatant solution.[11,12] The weight of the swollen sediment is determined as well as its dry weight. This latter is corrected for the solubles by determining the solubles in the supernatant liquid and calculating the amount remaining in the swollen sediment. The swelling power is defined as the swollen sediment weight per unit of dry sediment weight corrected for solubles. When the concentration is above this level, the swollen granules will be capable of forming a continuous phase of swollen granules in which essentially all available water is entrapped. Below this concentration there will be free water, and phase separation of the hydrated granules from the aqueous solution will occur. In Table 3 the critical concentrations of common starches are listed.

TABLE 3
Critical Concentrations of Common Starches

Species	Critical Concentration Value*
Potato	0.1
Sago	1.0
Tapioca	1.4
Corn	4.4
Waxy Maize	1.6
Wheat	5.0
High Amylose Corn	20.0

* These values represent the grams of dry-basis starch required per 100 ml of water in order to produce a paste at 95° in which the swollen granules occupy virtually the entire volume; thus, there is almost no "free" water between the swollen granules.

Source: H. W. Leach, in *Starch Chemistry and Technology*, Vol. 1, Academic Press, 1965, Chapter XII.

In most food applications the concentration of starch in water will be substantially higher than the critical concentration so that the swollen granules will form a continuous phase in which all the water is tied up. Water will only be released if, through intermolecular association, the hydroxyl groups become bound losing their ability to remain hydrated and thus releasing water. This phenomena, which may be observed in starch gels, is called syneresis.

When the starch concentration is high enough to seriously inhibit hydration of the granules, the rheological properties of the paste will be significantly different than at lower concentration. Also, when starches are cooked above the critical concentration the granules will become increasingly susceptible to rupture by shear.

The behavior of starch on cooling is also influenced by concentration. In general, the higher the concentration the greater the viscosity and the tendency to thicken and gel on cooling.

The pH at which a starch is cooked can have a marked effect on its behavior. Most starches will range in pH between 5 and 7 and variations in cooking properties are relatively minor. Cooking below pH 5 or above 7 tends to lower the gelatinization temperature, and accelerate the whole cooking process. In addition, under highly acidic conditions, hydrolysis of the glucosidic bonds may occur. Figure 10 illustrates the effect of pH on the Brabender viscosity curves of waxy corn starch.

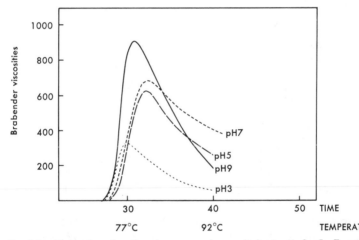

Fig. 10. Brabender viscosity of waxy maize cooked at pH 3, 5, 7 and 9. Description of Test: pH adjusted on starch suspension prior to putting into Brabender Visco/Amylo/Graph; heated 1.5°C per minute to 95°C. Continuous agitation throughout. Brabender Cartridge used —350.

Agitation is another factor governing the starch cooking process. Up to a certain point, the faster and more effective the agitation, the more rapid the heat transfer and the more rapid the cooking process. In addition to accelerating the rate of cooking, increasing the shear during agitation will promote granule rupture causing faster breakdowns in viscosity and yielding lower final viscosities. Agitation during the cooling process will also interfere with association or retrogradation of amylose giving softer gels. The strength or firmness of a starch gel is also influenced by the rate of cooling. Slow cooling favors stronger gels than very rapid cooling because it allows more time for the amylose molecules to become oriented and associated.

The behavior of starches during cooking and on cooling is also influenced by a number of auxiliary additives. Substantial levels of sucrose, dextrose or other water-soluble hydroxyl-containing substances will exert an inhibiting action on the swelling of starch granules. Such materials being water-soluble and containing hydroxyl groups have a strong affinity for water and interfere with the hydration and swelling of the granules. The rate of gelatinization and hydration is retarded and the tendency of the granules to rupture and collapse is decreased. Figure 11 shows the effect of cooking a modified waxy corn starch at 6.5% concentration in the presence of 20% and 40% sucrose on the weight of the starch.

This inhibitory action, in some cases, prevents starch from being cooked sufficiently to become properly hydrated. For this reason sugar is often added to a formulation after the starch has been cooked.

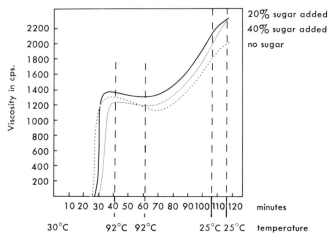

Fig. 11.   Effect of added sugar on viscosity, etc. . . . Description of Test: pH adjusted on starch suspension prior to putting into Brabender Visco/Amylo/Graph; heated 1.5°C per minute to 95°C. Continuous agitation throughout. Brabender Cartridge used —350.

Besides their effect on the gelatinization and viscosity breakdown of starches, sugars and similar hydroxyl-containing compounds tend to stabilize starch pastes and decrease their tendency to set back or retrograde. Presumably, the sugar prevents the hydroxyls on the starch molecules from associating with each other.

Natural and hydrogenated fats show minor effects on the gelatinization of starch. They tend to accelerate the cooking and lower the temperature at which the starch develops its maximum viscosity.[13]

Surface active agents often have a marked effect on the cooking behavior of starch. The exact pattern, however, varies widely depending upon the chemical composition of the surfactant and the type of starch.[14] Many compounds possessing a hydrophobic tail and a hydrophilic group such as long chain fatty acids, glycerol mono fatty acid esters, etc. will complex with amylose, the latter forming a helix about the hydrophobic moiety of the surfactant.

This stabilizes the amylose and reduces its tendency to associate and retrograde. If the starch has not been cooked and the granules are still intact, such surfactants inhibit the gelatinization process raising the gelatinization temperature and retarding the rate of swelling. In some cases, they also will cause marked increases in hot viscosity.

As might be anticipated, this effect is most pronounced with starches which contain significant amounts of amylose. However, it also occurs, though to a much lesser degree, with amylopectin starches such as waxy corn starch. Figure 12 illustrates the effect of glycerol monostearate on the cooking behavior of corn starch. Since these surfactants interfere with the tendency of amylose to retrograde, they will stabilize amylose-containing starches such as corn starch against setting to a gel.

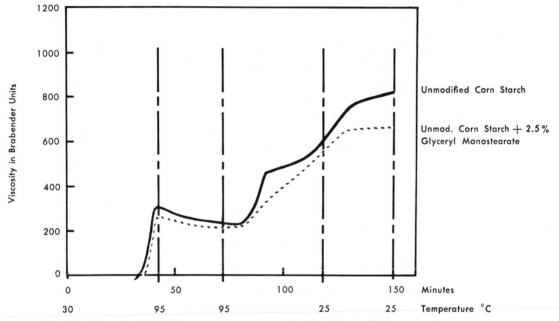

Fig. 12.   Effect of added glyceryl monostearate on viscosity of unmodified corn starch at pH 6.3 to 6.5 (6% starch solids). Brabender cartridge used —700.

The use of monoglycerides as additives to processed foods which contain naturally occurring starches is based on this interaction of monoglyceride with amylose. By complexing with amylose, it reduces the tendency of the amylose to be leached out of the granule. It also stabilizes the starch. In this way, monoglyceride addition improves the consistency of dehydrated potatoes, retorted spaghetti products, etc.

Modification Techniques

To expand the usefulness of starch, the wet milling industry makes wide use of various techniques for modifying one or more of the characteristics which govern the properties of starch.

Genetic Modification of Starch[15]

Genetic control of corn represents one of the most fundamental methods for modifying starch. In addition to most of the plant characteristics, the type of starch deposited in a corn plant is governed genetically. The genic systems which control the ratio of amylose to amylopectin have been under extensive study over the past quarter of a century. This has led to the development of a waxy corn starch, 95–100% amylopectin, which was commercialized in the mid 1940's and high amylose types, 55% and 70% apparent amylose, in the late 1950's and early 1960's. A waxy sorghum which, like waxy corn, gave an amylopectin starch was also developed.

These differences in amylopectin and amylose ratios are reflected in variations in starch properties. Waxy corn starch which contains essentially pure amylopectin, cooks readily to give sols which are characterized by excellent stability or resistance to gelling on cooling or aging. Corn starch, which contains 28% amylose, on the other hand, gives sols which have poor stability and form rigid, opaque gels on cooling. High amylose types which contain 55% apparent amylose or higher are much more resistant to cooking than corn or waxy corn starch. Apparently the higher amylose content reinforces the granules so they are much more resistant to the swelling action of hot water. Temperatures in excess of 100°C ranging up to 160°C are required to disrupt the granules. These starches must be cooked under pressure for proper dispersion. The resulting sols are very unstable and will form opaque gels which are much firmer than those from corn starch and will form far more rapidly. Because of their high amylose content such starches are capable of giving stronger and tougher films than those from regular or waxy corn starch.

The manner in which the ratio of amylose to amylopectin can influence the cooking properties of starch may be seen in Figure 9 by comparing the curves for waxy corn (100% amylopectin) and regular corn (28% amylose-72% amylopectin). As can be seen, waxy corn starch cooks more rapidly than corn starch. High amylose types are much more resistant, failing to develop significant viscosity under these cooking conditions.

Modification of Starch by Conversion

Starch conversions are essentially processes for reducing the viscosity of raw starch by degradative attack on the starch molecules. This involves scission of the molecules at the glucosidic linkages. Depending upon the method, they may also involve chemical alterations such as oxidation of some hydroxyl groups to aldehyde, ketone, or carboxyls. Molecular rearrangements may also occur.

Conversions are widely used to reduce the viscosity of raw starch so it may be cooked and used at higher concentrations than possible with native starch. They are also of value in modifying other properties such as cooking behavior, cold water solubility and tendency to form gels or pastes.

They may be run on cooked starch or on starch in the granule form. The former approach is used in enzyme conversions, thermal conversions and in the production of syrups and glucose. Most of the conversions used for making the so-called "converted starches" sold by the starch industry are made on the intact granule. Such conversions permit the use of efficient low cost techniques which are suited for manufacture of products which can be recovered with the granule still intact and sold in the dry form. These include thin-boiling or fluidity starches made by acid conversion, oxidized starches, white dextrins, British gums and yellow dextrins. Figure 13 shows the viscosity ranges of the various types of converted starches compared to those of raw starches.*

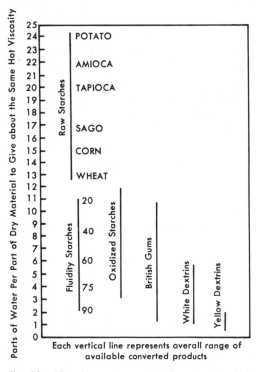

Fig. 13.   Viscosity ranges of native & converted starches.

Thin-boiling starches are made by controlled acid hydrolysis of starch in the granule state in an aqueous suspension. The conversions are run on concentrated suspensions of starch at temperatures in the neighborhood of 125°F, which is below the gelatinization temperature of starch. Sulfuric or hydrochloric acid is most commonly used as the catalyst. The basic reaction involves scission of the 1,4 alpha-D-and 1,6 alpha-D-glucosidic bonds with subsequent reduction in molecular size. The granule is weakened but not to the extent that it is solubilized to any degree. When the desired viscosity is reached, as measured by cooking the starch, the acid is neutralized and the converted starch recovered by filtration and drying.

When thin-boiling starches are cooked the granules do not swell as much as those of raw starch. Instead they disintegrate after only partial swelling giving dispersions which, depending upon the degree of conversion, will have fairly high to relatively low viscosities. The hot cooks are clear and fluid

*The fluidity scale referred to in Figure 13 is based on an empirical scale reflecting the fluidity of the hot starch. The higher the fluidity the lower the viscosity.

but, in the case of corn, thin-boiling starches will become opaque and form rigid gels on cooling. Such products are of particular value in imparting a rigid texture to a product as in gum candies. Thin-boiling starches from waxy corn starch do not show the gelling tendency of corn thin boilings but give fairly stable fluid sols of particular value for adhesive applications and as coatings.

Oxidized starches are also made by reactions run on aqueous suspensions of starch using sodium hypochlorite as the converting agent. These are sometimes called chlorinations which is a misnomer since oxidation not chlorination takes place. The oxidation process is a random reaction involving oxidation of hydroxyls to aldehydes, ketones and carboxyl groups as well as molecular scission. Since carboxyls are substantially more bulky than hydroxyls, their introduction in the amylose fraction reduces its tendency to associate or retrograde. They also impart a charge which may result in repulsion between molecules. As a result, oxidized starches show greater stability or resistance to pasting or gelling than thin-boiling starches of comparable viscosity. Since acid conversions and oxidations are both run on aqueous suspensions of starch, the extent of conversion which can take place is limited. If the starch is degraded too far it will become solubilized and cannot be recovered easily. As a result, acid fluidities and oxidized starches have minimal solubility in cold water.

When converted products which are partially or completely soluble in cold water are desired, dry conversions or dextrinizations are used. Since these involve the use of heat to promote conversion they are sometimes called "pyroconversions." The starch is usually converted in the dry pulverized state. Initially, the moisture content ranges from equilibrium down to about 2%–5%. The conversion is initiated and maintained by heat. In addition, the direction and extent of conversion is controlled through auxiliary factors such as acids, buffers and the moisture content of the starch. Typical conversion involves spraying powdered starch containing anywhere from about 5%–20% moisture with a dilute solution of a volatile acid such as hydrochloric acid, or treating it with a buffer such as ammonium carbonate or trisodium phosphate and heating according to a set schedule to the desired end point. Conversion is generally halted by cooling the starch. Usually five steps are involved:

1. acidification or buffering
2. drying the treated starch
3. pyroconversion
4. cooling
5. packing off

The order of the first two steps may be varied. Each may involve a distinct operation in different pieces of equipment or they may be performed in the same unit. The drying and converting step may also be combined or may be done separately.

The changes which occur during pyroconversions are complex.[16] It appears that three major reactions may be involved. These include:

1. Hydrolysis—Scission of alpha 1,4 and alpha 1,6 glucosidic linkages between glucose units. This reaction is favored by heat, acid, and moisture.

It results in a reduction in the size of the molecules and a decrease in the viscosity as well as an increase in reducing value.

2. Transglucosidation—This involves breaking glucosidic linkages and reforming them at other points to yield more highly branched molecules. This reaction is favored by heat and apparently does not predominate until the moisture has been removed. It does not alter the molecular weight or reducing value, but improves the stability of the dextrin solution by reducing the amount of linear component.

3. Repolymerization—Glucose is capable of polymerizing at high temperatures in the presence of catalytic amounts of acid. During this process the reducing value decreases.

In addition levoglucosan formation occurs at high temperatures.

Depending upon which of these reactions or combinations of reactions are involved, three major types of products may be obtained—white dextrins, yellow dextrins or British gums. In white dextrin conversions, hydrolysis is the dominant reaction. Moisture, high acidity, and relatively low converting temperatures are used in making them. They cover a wide range of viscosities and cold water solubility depending upon the degree of conversion. The most highly converted types have an average degree of polymerization of about 20 and are quite soluble. For a given level of conversion (as measured by viscosity), they tend to have light color, relatively low cold water solubility, and their dispersions tend to have limited stability.

Yellow dextrins are made with relatively higher amounts of acid, low moisture, and fairly high converting temperatures. During the initial stage of the conversion, hydrolysis predominates, but in the later stages transglucosidation and repolymerization reactions are dominant. All yellow dextrins are low in viscosity and are essentially completely soluble in cold water. Their degree of polymerization ranges from 20–50. They are more soluble, darker in color, and more stable than a white dextrin of comparable viscosity.

Little or no acid, in some cases buffers, and high temperatures characterize British gum conversions. While hydrolysis initially occurs, transglucosidation is the predominant reaction. Since hydrolysis is minimized, British gums are not as low in viscosity as yellow dextrins or the thinnest white dextrins. For the same viscosity they are substantially darker in color, more soluble in cold water, and lower in reducing value than white dextrins. In addition, their solutions are much more stable than those of white dextrins.

Modification of Starch by Cross-Linking

The characteristics of a starch dispersion change markedly during the cooking process. The unswollen starch behaves as a suspension in water. However, when the granules begin to swell the dispersion becomes a short, salve-like paste. This quickly changes to an elastic, and in some cases, rubbery texture when the swollen granules rupture forming dispersions of molecules and molecular aggregates.

The short, salve-like consistency of the highly swollen granules is usually associated wth the stage at which the starch reaches its maximum viscosity and thickening power. It also provides textural qualities which are palatable, and particularly suited for use in pie fillings, sauces, puddings, etc. However, in raw starches this condition is very tenuous being readily destroyed by overcooking, variations in pH and shear. Moreover, because of variations between granules in the rate at which they swell, the period of maximum viscosity or hydration is very brief, particularly with root type starches.

Cross-linking is widely used to overcome this problem and provide textural characteristics necessary for use in food systems where starch acts as a thickening agent and stabilizer. The basic idea behind cross-linking is the toughening of starch granules by treatment with di or poly functional reagents capable of reacting with the hydroxyl groups in the starch molecules under conditions in which the granules are not swollen.

Compounds such as mixed anhydrides of acetic and citric acid or adipic acid, metaphosphates, phosphorus oxychloride, epichlorohydrin, etc. are examples of cross-linking reagents. The reactions are normally run on the intact granules in aqueous suspensions under conditions conducive for reaction. A cross-linking molecule is believed to react with hydroxyls on two different starch molecules forming a chemical bridge which reinforces the hydrogen bonds responsible for granule integrity. Only a few cross-links are necessary to toughen the granule so the amount of cross-linking chemical which is normally used is very small. Often only one cross-bond per 2000 anhydroglucose units is sufficient. After reaction, unreacted chemicals are removed from the starch by washing, and the starch is filtered and dried.

When a cross-linked starch is heated in water, the granules may swell, but even though the hydrogen bonds are ruptured, the swollen granules may be kept intact by the chemical cross-link. The exact manner in which a cross-linked starch behaves on cooking will depend upon the extent of cross-linking. A very slight treatment will toughen the granules sufficiently to prevent them from rupturing prematurely. It will extend the period during which the starch has its maximum hydration and may actually increase the maximum viscosity

Fig. 14.   Texture of waxy corn starch and lightly cross-linked waxy corn starch.

and decrease breakdown. Light cross-linking of a starch, such as waxy corn starch or tapioca, whose granules have a strong tendency to rupture and yield elastic, rubbery sols, will transform their sols to short, salve-like pastes. This marked change in texture is illustrated in Figure 14, which shows cooks of waxy corn starch and lightly cross-linked waxy corn starch. The raw starch has a runny, elastic texture, whereas the same starch with less than one cross-link per 2000 anhydroglucose units has a heavy, salve-like texture.

As the level of cross-linking rises, a point will be reached where the swelling of the granules becomes so inhibited that the granules will no longer be able to hydrate fully and their viscosity will begin to drop. Eventually, the granules will be resistant to boiling water. The effect of increasing levels of cross-linking treatment on the viscosity curves for waxy corn starch is illustrated in Figure 15.

The tendency of starch to break down in viscosity in acid media and the overall sensitivity of starch to pH variations can be minimized by cross-linking.

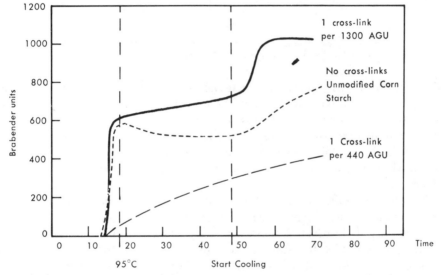

Fig. 15. Effects of cross-linking on the viscosity of corn starch at 6% solids. Description of Test: pH adjusted on starch suspension prior to putting into Brabender Visco/Amylo/Graph; heated 1.5°C per minute to 95°C held 30 minutes, cooled. 1.5°C per minute to 25°C. Continuous stirring throughout. Brabender Cartridge 350. The number of cross-links above are only estimates.

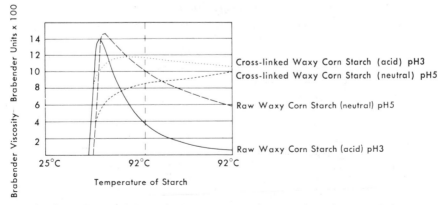

Fig. 16.   Effect of mild cross-linking on viscosity of waxy corn starch at pH 3 & 5.

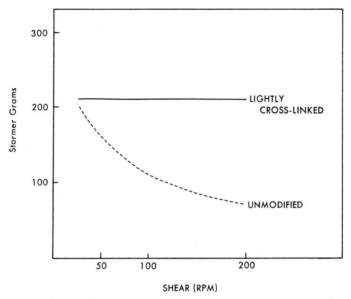

Fig. 17.   Effect of shear on swollen granules of waxy maize (25°C.
standard varnish paddles, 5% solids).

This is illustrated in Figure 16 which shows the effect of cooking unmodified
and lightly cross-linked waxy starches at pH 6 and pH 3. Note first that
the cross-linked starch is much less sensitive to breakdown at pH 3 and that
it is less influenced by variations in the acidity of the cook. By adjusting
the level and type of cross-linking it is possible to tailor starches to show
their maximum viscosity or hydration at any of the pH levels normally found
in foods.

Cross-linking will also make the granules more resistant to high temperatures. Starches which break down extensively on retorting may be made to
withstand any desired retort temperature by cross-linking to the proper level.

Granules of unmodified starch, when swollen, are sensitive to shear, rupturing readily and losing their viscosity. Cross-linked starches are more resistant
to shear and by proper adjustment of the level of treatment, it is possible
to modify starch to cope with the mechanical breakdown encountered in specific food processing systems. The effect of cross-linking in reducing the sensi-

TABLE 4
Influence of Cross-Linking on Granule Swelling Power (GSP) of Corn and Waxy Corn

	AGU per Cross-link	GSP
Corn Starch:		33.8
	1300	17.0
	440	12.1
	220	8.4
Waxy Corn Starch:		100.0
	940	21.4
	550	16.4

* 80% reaction efficiency assumed.

tivity of swollen granules to shear is shown in Figure 17. Unmodified and lightly cross-linked waxy corn starch were both cooked to the same viscosity and subjected to three different rates of shear. Note that the cross-linked starch is significantly more resistant to shear effects.

Granule swelling power is a reflection of the extent of cross-linking. As the level of cross-linking increases, the granule swelling power of starch decreases. This is shown in Table 4, which describes the effect of cross-linking at three levels on the swelling power of corn and waxy corn starch.

Modification by Derivatization

Strictly speaking, cross-linking is a form of derivatization. However, we choose to treat cross-linking separately since the number of hydroxyl groups involved in the reaction is extremely small, and the nature of the chemical moiety introduced on the granule has little or no effect on the properties of the starch beyond their sensitivity to chemicals which might remove them. Here we are concerned with generally much higher levels of substitution than those found in cross-linking.

Most anhydroglucose units in starch contain two secondary and one primary hydroxyl. These are capable of participating in a wide variety of chemical reactions which permit the introduction of innumerable chemical substituents, often under nondegradative reaction conditions. Included among the reactions are esterifications, etherifications and formation of acetals and carbamates as well as grafts.

Most of the derivatives which are produced commercially are made by reactions run on starch in the granule form. This imposes restrictions on the extent of derivatization which is possible, but it permits modification at much lower costs than derivatizations on ruptured granules. Often the reactions are run on starch in aqueous suspension, which greatly simplifies purification by removal of unreacted chemicals and by-products through washing.

The properties of derivatives depend upon the nature of the derivative and the level of substitution. At low levels of derivatization the major effect of the substituent group is to disrupt the linearity of the molecular segment in which substitution occurs, thus interfering with the tendency of the molecules to associate. Low levels of substitution exert a marked stabilizing effect on amylose in particular, reducing or preventing retrogradation. The effect of esterifying corn starch with acetyl groups is shown in Figure 18. Corn starch and acetylated corn having levels of substitution of 0, .030, and .065 acetyl groups per anhydroglucose units were cooked at 7% concentration and poured into molds. After aging overnight to permit the amylose in the starch to retrograde, the gels were removed from the molds. Note that unmodified corn starch formed a rigid gel. As the acetyl content increases, the amylose is progressively stabilized and gelling reduced.

Low levels of substitution, by preventing association through hydrogen bonding, actually increase the ability of starch to hold water. However, since acetyl groups are less hydrophilic than hydroxyls, they will reduce the affinity of starch for water at higher levels of substitution and will eventually make the starch insoluble in water. When the degree of substitution is higher than two, the acetates begin to develop solvent solubility.

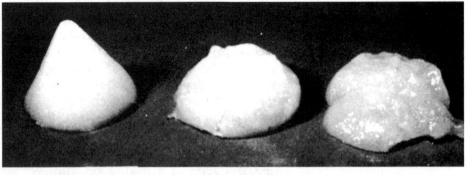

| Unmodified corn | 0.03 Acetyl/A.G.U. | 0.065 Acetyl/A.G.U. |

Fig. 18.   Influence of acetyl groups on gelling tendency of corn starch.

A second effect of derivatization on granular starch is to loosen the forces holding the granule intact. This is reflected in a lowering of the gelatinization temperature and an increase in the rate at which the starch cooks. If the substitution is high enough the granules will begin to swell in cold water.

Depending upon the nature of the substituent groups, derivatization can impart a variety of chemical and physical properties to the starch molecules. By introducing substituents containing hydrophilic groups such as carboxyls, it is possible to reduce the gelatinization temperature quite markedly and increase the viscosity. Anionic groups such as carboxylates or phosphates will also make the starch more reactive with polyvalent metallic groups. By introducing hydrophobic groups, emulsion-stabilizing properties can be imparted to starch.

Modification of Starch by Physical Means

The properties of starch may also be modified by purely physical means. This would include redrying starch to increase the moisture holding ability of the granules. It also includes incorporating additives in starch to modify the flow characteristics of the dry powder. Thus, the mobility of powdered starch may be increased by blending with tricalcium phosphate and other inorganics which will coat the granules and promote lubricity. The tendency of powdered starch to receive an impression and form a mold is improved through the addition of mineral oil or triacetin which promotes adhesion between granules reducing the powder mobility.

The granules may also be precooked to produce a starch which will swell in cold water. Such starches are known as pregelatinized starches. They include drum dried starches which may be made by feeding aqueous suspensions of ungelatinized native or modified starches onto hot drums which cook the starch and evaporate the water yielding a dry flake, which may be ground to give a dry powder which will disperse in cold water. They also include spray dried starches which may be made by cooking and spray drying the cooked starch.

Modification to Syrups and Sugars

Over half of all the corn starch milled in this country is converted to syrups or sugar. The corn sugar and syrup industry is based on the hydrolysis of the glucosidic bonds of uncooked starch to yield glucose, maltose, and higher oligosaccharides which may be marketed in the form of syrup, spray-dried syrup or sugars, lump sugar, or crystalline dextrose.

The basic measure of the degree or extent of conversion is the D.E. or dextrose equivalent. It is an indication of the reducing sugar content calculated as per cent anhydrous dextrose on the total dry substance. Pure dextrose has a D.E. of 100.

Two types of reactions are utilized in the conversion of starch to syrups or sugars. In one, the hydrolytic reaction is catalyzed by acid and in the other, by amylolytic enzymes, or in the case of glucose manufacture, a glucosidic enzyme.

Acid conversions are usually run using hydrochloric acid as the catalyst. The conversions are run on starch slurries containing about 35–40% solids which are simultaneously cooked and hydrolyzed with acid and steam under pressure. When the desired degree of conversion is reached, the acid is neutralized. The fatty acids from the starch are removed by flotation and centrifugation. After concentration the syrup is clarified and decolorized with activated carbon. Ion exchange resins are sometimes used to remove salts. The syrup is then evaporated under vacuum to the desired final concentration.

In acid conversions the starch is attacked in a relatively random fashion, producing a range of molecular species running from dextrose to oligosaccharides. If the conversion is carried to completion, the bulk of the material is converted to dextrose. At the same time an undesirable side reaction occurs; some of the dextrose polymerizes, forming higher molecular-weight materials that are known as reversion products. These do not become a significant factor until the conversion reaches 56–58 D.E., when they begin to impart a bitter taste to the syrup.

Amylolytic enzymes contain alpha and beta amylase. The latter is specific in its hydrolytic attack. It initiates hydrolysis at the nonreducing end, progressively clipping off maltose units. Since enzymatic conversions tend to be more costly than acid conversions for syrup production, they are usually used as supplemental conversions after an initial acid hydrolysis. Because of the specificity of the attack their hydrolytic product pattern is different than that of acid conversions, being much richer in the maltose fraction.

Glucosidases may be used to supplement acid or amylolytic enzyme conversions and give higher portions of dextrose or glucose in the product mix. Their main use is in the manufacture of glucose. Processes based on acid or amylolytic enzyme conversion followed by glucosidase treatment give higher yields and have lower costs than those based on acid conversion alone.

Corn or glucose syrups may be subdivided into four classes, depending on the D.E., as follows:

Type I	20–37 D.E.
Type II	38–57 D.E.
Type III	58–72 D.E.
Type IV	73 D.E. and higher

Some of these, such as Types II and III, may be further subdivided, depending upon whether the conversions are made by acid alone or acid-enzyme. These will differ in the saccharide composition, the acid enzyme having a higher proportion of maltose or disaccharide. The exact composition for an acid-enzyme syrup will depend upon the degree of initial acid conversion and the nature of the enzyme. An idea of how the composition varies between acid and acid-enzyme converted syrups, as well as syrups of varying D.E., may be obtained from Table 5.

TABLE 5
Carbohydrate Composition of Commercially Available Corn Syrups*

Type of Conversion	Dextrose Equiva-lent	Per cent Saccharides							
		Mono-	Di-	Tri-	Tetra-	Penta-	Hexa-	Hepta-	Higher
Acid	30	10.4	9.3	8.6	8.2	7.2	6.0	5.2	45.1
Acid	42	18.5	13.9	11.6	9.9	8.4	6.6	5.7	25.4
Acid-Enzyme[1]	43	5.5	46.2	12.3	3.2	1.8	1.5	—	29.5[2]
Acid	54	29.7	17.8	13.2	9.6	7.3	5.3	4.3	12.8
Acid	60	36.2	19.5	13.2	8.7	6.3	4.4	3.2	8.5
Acid-Enzyme[1]	63	38.8	28.1	13.7	4.1	4.5	2.6	—	8.2[2]
Acid-Enzyme[1]	71	43.7	36.7	3.7	3.2	0.8	4.3	—	7.6[2]

[1] The carbohydrate composition of acid-enzyme syrups will vary as a result of different processes used. The values given in the table are to be considered only as examples of ranges of values that are available commercially.
[2] Includes heptasaccharides.
*Source: Corn Industries Research Foundation, Inc., *Corn Syrups and Sugars*, (1965), p. 15.

In addition to basic differences in D.E. and composition, syrups may also vary in solids content, ranging from about 77–88% dry substance. Regular and low-conversion types are also available in the dehydrated form as corn syrup solids. Recently Clinton Corn Processing has made available a new corn syrup known as Isomerose 100. It is manufactured by a unique enzyme process that causes the isomerization of part of the dextrose to fructose. The approximate carbohydrate composition on a dry basis is 50% dextrose, 42% fructose, and 8% other saccharides. This level of fructose increases the sweetness of the syrup so that it is equivalent to that of a sucrose syrup.

Crude corn sugars are more highly hydrolyzed products than corn syrups. After conversion they are recovered by crystallization from the concentrated conversion liquor. Two classes are available—crude corn sugar 70 and 80, which have D.E.'s (based on dry substance) of about 83 and 91 respectively. The numerical designations of 70 and 80 refer to the approximate percentage of reducing sugars on an "as-is," not a dry substance, basis.

Dextrose is produced from starch by complete hydrolysis, followed by purification and crystallization. It is available as the monohydrate or in the anhydrous form.

Use of Starch and Modified Starches as Food Additives†

Fundamental in any consideration of starches as food additives are their edibility and nutritious nature. One gram of starch is capable of supplying about four calories, which is comparable to that furnished by sugar. Starches, along with sugars, supply a major part of the caloric value of our diet.

†The term *food additives* is used in the sense of the definition of the Food Protection Committee of the Food and Nutrition Board of the National Research Council, which defines a food additive as "a substance or mixture of substances, other than a basic foodstuff, which is present in a food as a result of any aspect of production, processing, storage or packaging."

In considering starches as food additives we are not concerned with those uses where their primary function is to supply caloric values or where they are a natural constituent of foods. Our interest is in applications where starch is used as an additive to facilitate processing or impart specific properties to food systems.

The food industry consumes over 300,000,000 pounds of starch or modified starches annually in a wide variety of applications. The bulk of this is corn starch and waxy corn. Sorghum, waxy sorghum and wheat starch are also used in the food industry. In addition, limited amounts of potato, tapioca and arrowroot are used in the United States.

Uses of Dry Granular Starch

As indicated earlier, native starch granules are insoluble in cold water. Those of corn and waxy corn starch, which are most commonly used in the food industry, range in diameter from 3-25 microns. Hellman *et al.* have measured the surface area of starch granules by nitrogen adsorption and found that one gram has a surface area of about 0.7 square meters.[17] The value correlates with that which is observed microscopically. In terms of commercial usage measurements, one pound of corn starch granules would have an area of about 3,500 square feet.

Besides having a substantial surface area, starch granules have a strong affinity for moisture. As mentioned earlier, this is attributed to the abundance of hydroxyls present in the granule which may act as sites for adsorbing moisture. Figure 19 shows the equilibrium moisture content for corn starch

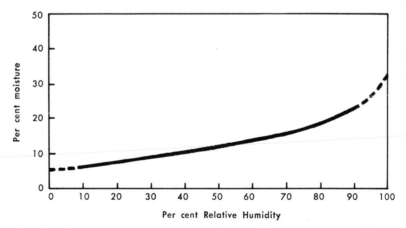

Fig. 19. Moisture content of corn starch at different relative humidities*.

*Source: The Corn Industries Research Foundation, Inc., *Critical Data Tables*, p. 232 (1957).

at 23°C and varying relative humidities. By drying starch to a level which is below its equilibrium moisture content, it will preferentially withdraw moisture from other systems having a lower affinity for moisture.

This combination of properties along with edibility and cost are major factors in the use of starch powders as diluents, bulking agents, fluidifying agents, moulds and moisture adsorbing agents throughout the food industry. Depending upon the application and auxiliary properties which are needed, native powdered starch, redried starch or one of the modified starches may be used.

Starch redried to 7.5%, 5% or even to 1.5% moisture is used as a diluent and moisture-sorbing medium to protect active powders from the deleterious effects of moisture. One example of such an application is the use of powdered starch as a diluent in baking powder. Two types of dry active ingredients are present: an acidic powder such as sodium aluminum sulfate, sodium acid tartrate or calcium acid phosphate, and a carbonate such as sodium bicarbonate. If these two types of ingredients come in contact with each other in the presence of moisture prior to their addition to a batter, they are apt to interact destroying or reducing the efficiency of the powder. Powdered corn starch and redried powdered starch are blended with the active ingredients at levels of 15–40% of the total powder to serve as a diluent to separate the active components from each other, and to preferentially adsorb moisture, keeping the ingredients from losing their activity.

In confectioners sugar, redried powdered corn starch is added at a 3% level to preferentially adsorb moisture and keep the sugar from lumping during storage. Since whiteness is an important factor in confectioners sugar, extra white powdered starch is frequently used. This may be made by special processing to lower the protein or by the use of one of the bleaching treatments cleared by the FDA under Food Starch-Modified.[18]

In some applications where starch powder serves as a diluent or bulking agent for active ingredients, one of the main functions is to permit the use of larger unit measurements and facilitate standardization of dosage or potency. Enzyme preparations and flavorings are examples of such uses. If the additive is used in cold or warm water, and must give a clear solution, soluble canary dextrins or pregelatinized starches may be used.

A starch which has been chemically modified by introducing hydrophobic ester groups at low levels of substitution is of particular interest where it is desirable to impart flow properties to powders. This modified food starch gives very mobile, free-flowing powders which are remarkably water repellent. While powdered corn starch loses its flow properties rapidly and clumps on picking up moisture, this product remains free flowing and water repellent even at 16% moisture content.

This material is also of interest in applications where starch is used as a dusting powder. The mobility of the powder makes it particularly suited for dusting at low levels. Its water repellency prevents it from being wetted in various applications. Where water repellency is not needed powdered starches blended with tricalcium phosphate or other inorganic powders, which coat the granules increasing their lubricity, are used. The use of starches to dust on bread dough to prevent its sticking to adjacent surfaces is an example of the use as a dusting powder.

Powdered corn starch and modified types are used by confectioners in moulding beds as a processing aid in the manufacture of gum candies, centers and other cast pieces. The prime purpose of the starch is to form moulds containing the desired impressions which reflect the shape of the die. The moulds must be resistant to collapse when subject to the vibrations encountered during and after the forming operation. While powdered corn starch may work in some cases, it usually is too mobile, particularly if dried, to give a stable impression. As a result, special moulding starches are frequently used. In these the adhesion between granules is increased by spraying on trace amounts of mineral oil or other oily additives such as acetylated monoglycerides. Since the candy is cast into the starch moulds as a high solids fluid, it is important that the starch is insoluble in the liquid confection (usually sugar, corn syrup, and water) and free of materials which might impart an off taste to the finished candy.

Another function of the moulding starch is to help reduce the moisture content of the confection. The hygroscopicity of starch makes it particularly suitable for this purpose. Prior to formation of the mould and casting of the candy, the starch is usually redried to about 5 to 8% moisture. It is then loaded onto trays or starch boards, the moulds formed by pressing a die into the starch bed, and the liquid confection poured into the impressions. The filled boards are then transferred to the drying room which is maintained at 70°–140°F depending upon the manufacturer and type of candy. The temperature, relative humidity and moisture content of the starch govern the rate at which the confection loses moisture. In normal operation the starch may pick up 2% or so of its weight in moisture from the confection. When the starch is too dry it will harden the surface of the candy when it contacts the starch. If it has not been dried enough the candy will take too long to set.

After the candy has developed enough set and is dry enough to process, the boards are dumped, the pieces of candy removed on screens, and the starch redried and used again. The starch loses some of its capacity to hold moisture on repeated use, and it picks up sugars and other ingredients from the confections. When this occurs, fresh starch may be added to restore its performance.

Uses Based on the Properties of the Starch Pastes

Most of the uses of starch as an additive in the food field are based on utilization of the properties of its sols to thicken or modify the texture of foods. Within the term sol or paste properties, we are including all those associated with colloidal dispersions of starch—those properties related to the swollen hydrated granules as well as the fragmented granules and molecular dispersions which may be in the form of fluid or thixotropic sols or rigid gels. The selection of the starch for a given application will depend upon the texture and rheology which is desired as well as the conditions of use, aging properties needed and taste requirements.

Native starches are used primarily as thickening or gelling agents and processing aids. However, they have a number of serious limitations for most food applications. Corn starch, sorghum or wheat starch may be used to thicken a limited number of foods where the opacity of their sols and their retrogradation and gelling characteristics will not be a serious drawback. However, they are not ideal because of the tendency of their gels to exhibit syneresis or release water on standing.

Waxy corn and tapioca may be used where thickening is desired. They have a greater influence on hot viscosity than corn starch and are used to thicken foods to provide temporary viscosity during filling operations. In such applications the starch serves as a thickening agent to keep ingredients suspended. The sensitivity of the starch to thinning on continued cooking and the elasticity of its sols are not a problem since the starch is not added to contribute to the final viscosity. It is simply an aid to processing to provide enough viscosity to keep ingredients uniformly suspended until retorting.

Cross-linked starches are used in the majority of food applications where the starch serves as a thickening agent to modify texture. They will consistently provide superior thickening power and offer a range of textures which are palatable and free of the elasticity found in many native starches.

The level of cross-linking in a starch used in a given application will depend upon the desired texture and thickening under the specific conditions of use; i.e., acidity, presence and level of sugar, etc., and temperatures of preparation and processing. Usually in neutral or mildly acidic foods a lightly cross-linked starch is used. Acidic foods often require somewhat more highly cross-linked types, and foods subjected to vigorous processing conditions may use even more highly cross-linked types.

Cross-linked corn or sorghum starches are used in applications where clarity and resistance to setback or gelling are not required. Tapioca or potato types generally provide more thickening ability and improved clarity. In the case of tapioca they may also give a blander taste permitting better development of food flavors. Cross-linked waxy starches may be used where excellent thickening ability, clarity and tolerance to conditions of use are desired.

Cross-linked types stabilized by substituent groups such as acetyl, hydroxy-propyl or phosphate groups are used where the foods are subjected to storage at low temperatures. Most starches, including waxy types, tend to lose their hydrating ability and clarity under these conditions. However, these stabilized types show much less change under such conditions and find wide use in retorted and frozen foods.

Canned foods represent a substantial outlet for starch as a thickening agent. The canning of foods requires that they be prepared for filling into a container, sealed in the container, and heat processed at temperatures and times to insure sterilization. Most of the canned foods which use starches are specialty types which often involve preparative steps, such as cooking the starch and formulating prior to the filling step.

Two basic processes are used in the sterilization step. The still retort involves the use of a pressurized chamber which heats the sealed cans in stationary positions by steam to sterilizing temperature. In the other process, which is the continuous retort, the sealed cans are fed into a continuous cooker in which the cans are in motion throughout the process. The movement of the cans facilitates heat transfer so that processing times can be shortened.

The heat treatment destroys bacteria and molds which might still be present in the prepared raw foods. The temperature and time of the heat treatment is also determined by the nature of the food. Those having a pH below 4.5, after being prepared by cooking and sealed into cans, often simply require heating to about 200°F. In fact, some acid products may be processed by simply filling at sufficiently high temperatures, sealing and holding for a period and then cooling without further heat treatment. Those foods whose pH is above 4.5 require higher temperatures for sterilization. These can range up to 260°F depending upon the food.

The time required to reach sterilizing temperatures will depend upon the rate of heat transfer which is governed by the nature of the food, the viscosity during the heat treatment, and the sterilizing process. Still or stationary retorts take longer times than continuous retorts because they do not subject the cans to agitation and hence have slow heat transfer. Systems in which the heat transfer is slow often have localized overheating which may have a deleterious effect on flavor.

The canning industry is continually trying to shorten processing times and improve quality by accelerating the rate of heat penetration. One example of a move in this direction is asceptic canning. In this process the food itself, prior to filling, is sterilized by exposure to high temperatures for short periods in a continuous heat exchanger. It is then poured under aseptic conditions into previously sterilized cans with covers. The containers are sealed in an atmosphere of supersaturated or saturated steam.

In the canning industry starches are used primarily in specialty type foods such as baby foods, soups, sauces, pie fillings, gravies, vegetables in sauces, specialty dinner products like chow mein, chili, and spaghetti, stews, cream style corn, etc.

Native and modified starches are used as processing aids for thickening foods and suspending solids during the filling process. Cross-linked corns are used to thicken opaque foods such as cream style soups, gravies, sauces, cream style corn, stews, etc. Cross-linked waxy corn starch, waxy sorghum and tapioca are used in many of the same applications as cross-linked corns. In addition, they are used in baby foods and canned pie fillings. Where excellent stability to storage at low temperature or freezing and thawing conditions is desired, stabilized, cross-linked starches are used.

By selecting the proper level of cross-linking it is possible to delay gelatinization so the starch contributes little viscosity in the early stages of heat treatment, thus facilitating heat penetration. At later stages and high temperature the starch will thicken and resist breakdown to give much higher viscosities than native starch, which swells more rapidly and then breaks down in viscosity.

In some canning applications low thermophile starches are required. For these uses, native (as well as cross-linked and modified starches) may be given an auxiliary treatment with heat, peroxide or peracetic acid to destroy any thermophiles which may be in the starch. The National Canners Association has set standards for thermophile counts. Starches falling within these specifications are sold as low thermophile types.

The baking industry, aside from applications involving starch as a component of flour, makes wide use of the paste properties of starch. Here again, the starch acts as a thickener and represents a means for modifying texture. Cross-linked waxy starches, cross-linked stabilized starches, as well as pregelatinized versions are widely used as thickeners where stability, sheen, clarity and smooth textures are desired as in fruit pie fillings, fruit fillings, imitation jellies, whipped topping stabilizers, icing stabilizers, cream fillings, etc. Ordinary corn starch is used where a gel texture is desired in custard and cream fillings. In addition, pregelatinized types are sometimes used as processing aids. In custard fillings it is customary to use a formula based on milk, eggs, and starch which is agitated, filled into shells, and baked in the oven. The heat penetration is slow, so there is a tendency for the uncooked starch to settle before gelatinization causing an apparent double bottom or soggy crust.

Pregelatinized starch added at levels of about $1\frac{1}{2}\%$ of the total custard weight is used to supply filling viscosity and maintain a homogeneous mix until the bulk of the starch gelatinizes. After baking, the pregelatinized starch has little effect on the final viscosity or set.

Starches are also used as additives to modify the textural properties of wheat flour. Sweet goods and crackers require a flour having a protein content close to 7% to develop the proper texture and eating qualities. Often such flours are not available and it becomes necessary to use higher protein flour. To prevent the development of doughy textures it is necessary to lower the protein content by raising the shortening and sugar, or else by adding starch. Since starch is less costly than shortening and sugar, it is frequently used to adjust the flour. The amount will vary with the type of bakery goods and protein content of flour. It can range from about 5 to 15% of the flour weight. Corn or wheat starch is most commonly used, although sometimes tapioca is used.

In one case arrowroot, an imported starch somewhat resembling tapioca, is used in making Arrowroot Biscuits. The choice of arrowroot appears to be dictated largely by the desire to specify arrowroot on the label. It constitutes about 10% of the flour used in the biscuit. Over the years it has been regarded as contributing easy digestibility to the biscuits making them particularly suitable for children and invalids. However, there is little scientific data to indicate that arrowroot possesses any unusual properties compared to tapioca or waxy corn starch.

Salad dressings of the mayonnaise type are another important outlet for starch where its paste properties are utilized. In these, starch pastes serve as stabilizers for oil emulsions. Mayonnaise is a semi-solid emulsion of vegetable oil and egg yolk or whole egg with vinegar and/or lemon juice with one or more of the following: salt, spice and sugar. By definition, it must contain a minimum of 65% edible vegetable oil.[19] Since egg and egg yolk are the only emulsifiers permitted in mayonnaise, they must be present in sufficient quantity to give a stable emulsion.

Salad dressing is a semi-solid emulsion of vegetable oil and egg yolk or whole egg and starch paste with acidifying ingredients. It has a lower oil content which is set at 30% minimum. The higher the oil level the higher the quality of the salad dressing. Low quality contains about 30% and a high quality dressing 50% oil.

In general, a salad dressing is a mayonnaise which has been diluted with a cooked starch paste. Roughly half of the finished salad dressing is a starch paste, so starch represents an important ingredient in salad dressing. The starch functions as an auxiliary stabilizer for the emulsion. As a result, since egg yolk is no longer the exclusive stabilizer as in mayonnaise, the level of egg yolk in salad dressing is lower than that in mayonnaise. Standards, however, specify a minimum of 4% egg yolk in salad dressing.[19] Other stabilizing ingredients totaling not more than 0.75% by weight may also be used.

Salad dressing manufacturers may take different routes. The mayonnaise and starch pastes can be made separately and then mixed by a batch process. Another way is to make the starch paste, blend in the eggs, oil and spices, and then emulsify the entire mix in a colloid mill. In large plants continuous systems may be used in which water, sugar, starch, salt and vinegar are preblended and cooked automatically and continuously in a scraped surface heat-exchanger unit and then cooled. The egg and oil are added and the mixture is then passed through a colloid mill for emulsification.

The level of starch in the finished dressing runs in the order of 4–6%. The type of starch is an important factor in determining the properties of the salad dressing. Ideally, its paste properties should simulate those of mayonnaise in the finished product. The pastes must also be resistant to thinning by the vinegar which is present in salad dressings. They must also be reasonably stable. Setback or thickening may be acceptable, but the starch should not gel or exhibit syneresis on aging. The pastes must also be resistant to the shear to which they might be exposed during the emulsification stage. The necessary tolerance to shear will vary widely depending upon whether a batch or continuous process is involved and the particular equipment used in the emulsification step.

A variety of special modified starches are offered for salad dressing manufacture. They are tailored for the different processes and offer a range of properties suited for the various conditions of use.

Large volumes of starches are used by the candy industry which utilizes the paste properties of starch to impart texture to confections, particularly gum types. These include soft gums such as Turkish delights, jelly beans, gumdrops, orange slices and hard types. Basically, all contain some type of starch or modified starch or gum arabic, sugar, syrup, flavorings, color and water. Acid hydrolyzed corn starch converted to about 60 fluidity is the most commonly used starch. In this application the starch is cooked with the sugar and syrup to a fluid sol which is suitable for casting into starch moulds. The key property of the starch which contributes to the texture is its ability to form a rigid gel on cooling. As noted earlier, this is the result of the tendency of the amylose in the starch to retrograde or associate. A fluidity starch is used because of its strong gel forming tendency in relation to its viscosity. At the same viscosity it forms a stronger gel than corn starch and can be used at a higher solids. Soft gums normally use about 10–14% starch. They are cast at solids ranging from about 72–80%, and require aging for 1–4, days in starch moulds to develop set and dry to about 18% moisture.

Carefully controlled tapioca dextrins are also used to impart texture to hard gums replacing gum arabic in this application. They do not gel like corn fluidity types, but because of their lower viscosity they can be used at much higher solids. Depending upon the type of candy, they may be used at levels ranging up to 40% of the piece.

High amylose corn starches are also showing promise in the manufacture of gum confections. Because of their high amylose content they have a strong tendency to gel and are capable of forming gels much more rapidly than corn fluidities. The resulting gels are much firmer than those from corn. By replacing a portion of the fluidity starch with high amylose corn starch, the gel formation can be accelerated to the point where the time in the starch mould can be cut to a fraction of that normally used. Moreover, it is possible to handle the candy at a higher moisture content thus improving its storage life. Since high amylose starch is very resistant to swelling, temperatures of 320°F are necessary in cooking confections in which it is an ingredient.

Uses Based on the Film Forming Properties of Starch

Depending upon the nature of the starch or its modification, a starch sol may contain highly swollen or ruptured granules, fragments of granules, molecules and molecular aggregates in water. As the water evaporates these associate with each other forming a network which, if resting on a surface which will permit the escape of water by evaporation and/or absorption, may produce a film. If the starch has an affinity for the surface, it may bond itself to it forming a coating.

The strength or toughness of the film is governed by the size of the molecules and their shape and the conditions under which the film forms. In general, under the proper conditions of film formation, sols containing amylose will form the toughest films. High molecular weights, in general, also tend to favor film strength. However, the molecular weight correlates directly with viscosity and inversely with concentration. Thus, starches which are potentially excellent film formers must be cooked at relatively low concentrations which means that substantial amounts of water must be removed before the films are formed. It also makes it difficult to deposit sufficient solids to form thick films or coatings. As a result, it is sometimes necessary to sacrifice film strength for operational needs leading to the use of converted starches, such as acid fluidities or dextrins for coating or binding applications.

The ability of starch to form films is utilized by the food industry in covering foods with protective and decorative coatings, binding foods and providing a matrix to carry food substances.

Pan coatings on chewing gum and other confections having a hard sugar coating often are improved by the addition of cooked solutions of fluidity starches or specialty dextrins. In this application the confection is placed in rotating kettles and tumbled while a saturated sugar solution or syrup is slowly added. The sugar crystallizes on the surface of the confection. By repeated application the coating is built up to the desired thickness. The addition of fluidity starch or dextrins to the pan-coating syrup mix protects the coating against chipping. It also inhibits crystallization giving smooth even surfaces. Typical base formulations for pan coating use about $2\frac{1}{2}$–4% dextrin or fluidity starch in the sugar syrup.

Besides providing physical protection, starch coatings also are oil or grease resistant. This is an important factor in the use of starch or modified starches as coatings for nuts and chocolate confections where it is desirable to prevent the migration of oils. This property is also used in the manufacture of French fries in which the potatoes are dipped into a specialty starch sol which coats their surface and reduces oil penetration during the frying step thus improving crispness.

Starches and starch containing products are used as binders in processed meat products such as scrapple, sausages, etc. to maintain body and keep the meat, fat and juices together during processing and storage. The level of starch varies widely depending upon the type of product and government regulations. They are also used as binders in extruded pet foods and snack items where their level of use and the specific starch product involved will vary widely depending upon the nature of the food and the process of manufacture.

Dextrins and specialty starches are also used as a matrix for carrying water-insoluble flavor oils and clouding agents. In this application the dextrin or starch is dispersed in water at a concentration of about 30%. The oil, which is to be encapsulated in the starchy matrix, is then added and emulsified by mechanical agitation in the dextrin or starch solution. The amount of oil usually runs from 20–40% on the weight of the dextrin. The resulting emulsion is then spray dried yielding a dry dextrin or starchy matrix in which the flavor oil is imbedded. In this way liquid flavors may be transformed into dry powders for formulation in dry mixes, powders, etc. By selecting the proper dextrin it is possible to minimize flavor losses through volatization or oxidative action. When the resulting encapsulated oil is added to water, it will form an emulsion of the oil in water, producing, in the case of a clouding agent, an opaque solution.

The use of starches and modified starches in the food industry has been almost limitless. As new foods, new starch modifications and new techniques for utilizing the properties of starch are developed, new applications are being continually discovered. It is an extremely dynamic ever-changing field of interaction between the food and starch technologists.

Uses of Corn Syrups and Sugars in the Food Industry

Corn syrups and sugars find extensive use throughout the food industry. The low cost of the syrups and sugars is a factor behind their use in many areas as a sweetener in place of sucrose. However, they possess a number of other properties which represent a unique contribution to many foods, making them more of a supplement to the use of sucrose rather than a replacement.

At 2% concentration dextrose has about two-thirds the sweetening power of sucrose. However, as the concentration increases the differences in sweetness decrease until at 40% concentration it is hard to differentiate between the sweetness of dextrose and sucrose. The sweetness of corn syrups in general correlates with the D.E.; the higher the D.E., the sweeter the syrup.

If corn syrup or dextrose is used in combination with sucrose, the sweetness of the mix is greater than would be expected. Thus, at 45% solids a mix of 25% 42 D.E. syrup and 75% sucrose is about equal in sweetness to sucrose at 45% solids. This property permits partial replacement of sucrose in products such as the syrup of canned fruits, which has a fairly high sweetener concentration, without loss in sweetness. The previously mentioned speciality syrup, Isomerose 100 corn syrup that contains fructose, should find a unique role in applications demanding a high degree of sweetness. With this level of sweetness, it should be suitable in uses where conventional syrups have been limited. The level of use of corn sweeteners as optional sweetening ingredients is limited in some canned foods by Federal standards of identity.

Another valuable property of corn syrups in particular is their ability to inhibit crystallization of sucrose and other sugars. Acid-enzyme syrups are of special value here because of their low dextrose and high maltose content which increases their inhibitory effect on crystallization. This is an important factor in the use of syrups, in confections such as hard candies, in jams, jellies and preserves, and in ice cream.

Both corn syrups and dextrose lower the freezing point of solutions. The extent of lowering increases with the D.E. Dextrose has the most pronounced effect. Intermediate conversion syrups lower the freezing point to about the same extent as sucrose. This property is utilized in the ice cream industry where both corn syrups and dextrose are used as partial replacements for sucrose to control sweetness and improve the texture and quality of the ice cream by reducing the formation of ice crystals and preventing sugar crystallization. In this connection acid-enzyme 42 D.E. syrups are of interest because of their high maltose and low dextrose content, which give the ice cream good texture and body characteristics.

Bodying or thickening properties are also found in syrups. This is dependent upon the amount of higher molecular weight material in the syrup and increases with decreasing D.E.

Fermentability is an important factor in the use of dextrose and syrups in the baking industry for bread making and in the brewing industry. Dextrose and maltose are readily fermented by baker's yeast. The amount of material in syrup which is fermented by yeast is called fermentable solids. In general, as the D.E. increases the fermentable solids rises.

Dextrose is of particular advantage to the baking industry for its quick and complete fermentability and also its browning action. Dextrose is capable of reacting with nitrogenous compounds, such as proteins, under heat to produce compounds which have a brown color. This effect, known as the Maillard reaction, contributes to the attractive crust color of bakery products as well as caramel color and flavoring.

Since dextrose and corn syrups have reducing groups, they have the ability to inhibit undesirable oxidative reactions in foods. This property helps maintain the bright color when corn sweeteners are used in strawberry preserves and cured meat products, as well as catsup.

Hygroscopicity, or the ability to pick up and hold moisture, is another important factor in using corn sweeteners. In general, for the same type of conversion as the D.E. increases, the hygroscopicity rises. Dextrose is more hygroscopic than maltose. These are important factors in the confection industry which uses low conversion syrups in hard candies to keep them from sticking. For the same D.E., acid-enzyme syrups which are low in dextrose and high in maltose are preferred, since they are less prone to pick up moisture.

In addition to the above, there are a number of other properties which are of value in the use of syrups and sugars, such as the adhesiveness of lower D.E. products, the ability of all corn sweeteners to impart sheen to food surfaces, and the cohesiveness and viscosity of lower D.E. products which contribute to their ability to stabilize foams. These, along with the previously mentioned properties, contribute to the broad outlet for corn sweeteners in the canning, dairy, beverage, baking and confectionery fields, as well as in the domestic syrup market.

Acknowledgment

I would like to express my thanks to Mr. W. Herbst, Dr. L. Kruger, Mr. N. Marotta, and Mr. N. Soine for their assistance in assembling data for use in the tables and figures, and to Dr. J. M. Newton of Clinton Corn Processing Company for suggestions regarding corn syrups and sugars.

References

1. Kerr, R. W., 1950. *Chemistry and Industry of Starch*, 2nd edition, Chapter II, Academic Press, New York.
2. Kerr, R. W., 1950. *Chemistry and Industry of Starch*, 2nd edition, Chapter I, Academic Press, New York.
3. Badenhuizen, N. P., 1965. *Starch Chemistry and Technology*, Vol. I, 1st edition, Chapter V, Academic Press, New York.
4. Wolfrom, M. L. and El Khadem, H., 1965. *Starch Chemistry and Technology*, Vol. I, 1st edition, Chapter X, Academic Press, New York.
5. Foster, J. F., 1965. *Starch Chemistry and Technology*, Vol. I, 1st edition, Chapter XV, Academic Press, New York.
6. Leach, H. W., 1965. *Starch Chemistry and Technology*, Vol. I, 1st edition, Chapter XII, Academic Press, New York.
7. Myers, R. R., and Krause, C. J., 1965. *Starch Chemistry and Technology*, Vol. I, 1st edition, Chapter XVI, Academic Press, New York.
8. Hellman, N. N., Boesch, T. F., and Melvin, E. H., 1952. Starch Granule Swelling in Water Vapor Sorption *J. Am. Chem. Soc.,* 74: 348–350.
9. Schoch, T. J. and Maywald, E. C., 1956. Microscopic Examination of Modified Starches *Anal. Chem.,* 28:(3), 382–387.
10. Watson, S. A., Determination of Starch Gelatinization Temperature, *Methods in Carbohydrate Chemistry IV Starch*, 1st edition pp. 240–242, Academic Press, New York.
11. Leach, H. W., McCowen, L. D., and Schoch, T. J., 1959. Structure of the Starch Granule I. Swelling and Solubility Patterns of Various Starches *Cereal Chem.,* 36: 534–544.
12. Schoch, T. J., Swelling Power and Solubility of Granular Starches. *Methods of Carbohydrate Chemistry IV Starch*, 1st edition, pp. 106–108, Academic Press, New York.
13. Osman, E. M., and Dex, M. R., 1960. Effects of Fats and Nonionic Surface Active Agents in Starch Pastes *Cereal Chem.,* 37: 464–475.
14. Schoch, T. J. and Gray, V. M., 1962. Effects of Surfactants and Fatty Adjuncts on the Swelling and Solubilization of Granular Starches *Die Stärke,* 14:(7), 239–245.
15. Zuber, M. S., 1965. *Starch Chemistry and Technology*, Vol. I, 1st edition, Chapter IV, Academic Press, New York.
16. Horton, D., 1965. *Starch Chemistry and Technology*, Vol. I, 1st edition, Chapter XVIII, Academic Press, New York.
17. Hellman, N. N. and Melvin, E. H., 1950. Surface Area of Starch and Its Role in Water Sorption *J. Am. Chem. Soc.,* 72: 5186–5191.
18. Food Additives Regulations, §121.103, *Food-starch Modified*.
19. Federal Food Standards, Definitions and Standards for Foods, Title 21, Part 25, §25.1.

CHAPTER 9

Surface Active Agents

William C. Griffin
Associate Director
Product Development

and

Matthew J. Lynch
Development Manager
Product Development

ICI America Inc.
Atlas Chemicals Division
Wilmington, Delaware

Introduction

Surface active agents present in most, if not all, natural foods play a major role in the growth processes of foods, and assist in the metabolism of converting them into sources of energy. Modern technology utilizing surfactants has brought about a revolution in food preparation and supply.

At the start of the twentieth century most foods were prepared at, or close to, home. Since prepared foods were perishable, most were eaten shortly after they were prepared. As a result, variety was limited to the foods available within a few miles of the home.

Changes in food handling have occurred with increasing rapidity in the past few decades. Preparation is being carried out on larger and larger scales, and more recently on a continuous basis with several items. Mass marketing techniques, initially of staple items, are now being employed for partially prepared foods. Storage conditions a few decades ago were principally ambient superceded by controlled refrigeration and freezing, and now further assisted by processing and packaging innovations, including freeze-drying and unit packaging. One result of these advances in technology has been the introduction of convenience foods.

Today, foods in rich variety are transported long distances and are stored for months before being eaten. Rapid transportation has contributed to these advances; even products as perishable as ice cream and frozen pies are shipped hundreds of miles. The transportation of prepared foods does not, by itself, impair product quality; improper treatment of food in transit can be damaging. The use of surface active agents helps maintain proper quality.[29,30,38,42,43,44,53]

The seasonal foods of yesterday are now available on a year-round basis. This progress represents a major breakthrough for the food scientists. As a part of this advance, great responsibility has been placed on the food manufacturer to provide products that will exhibit consistent quality despite ambient conditions and even raw material variations.

Food scientists have achieved significant results in upgrading quality and reducing handling problems of perishable foods. Additives are routinely used to control or retard microbiological spoilage. The physical properties of "as made" viscosity, texture and mouthfeel are preserved by surface active agents. By forming chemical complexes with the starch molecule, surfactants can extend the shelf life of bread. Emulsifiers and stabilizers achieve and maintain the desired texture of frozen desserts. These are a few examples of the role of surface active agents in providing perishable foods that retain desired properties during extended storage and shipping periods.

Naturally Occurring Surface Active Agents

A wide variety of natural surface active agents are present in foods. The best recognized of these is lecithin, present in eggs and soybeans in large quantities and in many other foods in smaller quantities. Bile salts are largely surface active agents. Licorice contains highly surface active ingredients such as saponin. Some ingredients in mustard and milk exhibit a high degree of surface activity. Wool wax contains lanolin, a good W/O emulsifier. These are just a few illustrations of the way in which nature surrounds us with products exhibiting surface activity.

Definition—Description of Surface Active Agents

In food applications surface active agents exert a variety of effects. They act as emulsifiers, wetting agents, solubilizers, detergents, suspending agents, crystallization modifiers (both aqueous and non-aqueous systems), complexing agents and in other ways. Of course, many of the uses may overlap in any given application.

Surface active agents are organic or organic-metal molecules that exhibit polar and solubility behavior that result in the phenomenon known as surface activity. The most commonly recognized phenomenon in this respect is the reduction of the boundary tension between two immiscible fluids. Under special conditions, certain finely divided solids can react in this fashion, though they are not commonly known as surface active agents.[1,3]

Surface active agents may be divided according to their structure and behavior into ionic and nonionic compounds. The ionic type of emulsifier is composed of an organic lipophilic group and a hydrophilic group. Ionic types may be further divided into anionic and cationic, depending upon the nature of the ion-active group. In soap, for example, the surface active fatty acid portion of the molecule represents the anion in the molecule, therefore, soap is classed as an anionic surface active agent. As would be expected, anionic and cationic surface active agents are not compatible. Owing to opposing ionic charges, they tend to react with each other and their surface active effect is nullified.

Nonionic surface active agents are completely covalent and show no tendency to ionize. They may, therefore, be combined with other nonionic surface active agents and with either anionic or cationic agents. The nonionic emulsifiers are likewise more immune to the action of electrolytes than the anionic surface active agents.

In addition to the types of surface active agents discussed above, some synthetic and natural gums and clays are employed as protective colloids. Emulsion and suspension stabilizers that act as protective colloids and, in some instances, as thickeners include acacia, tragacanth, alginates, starch and starch derivatives, casein, glue, eggs, mustard, methyl cellulose, sodium carboxy methyl cellulose, hydrated magnesium aluminum silicate, sodium salts of condensed alkylated aryl sulfonic acids, and silica bentonite, activated carbon and alumina gel. These products show varying degrees of sensitivity to acids, alkalies and monovalent and multivalent salts. Without exception, emulsions formed with protective colloids require vigorous mixing for their production.

The principal types of food surface active agents are tabulated in Table 1.

Wetting Wetting in food handling involves a variety of applications. For example, the wetting of a powder, such as a cocoa mix or non-fat milk solids, is different from the rehydration of coconut or dry vegetables. Three distinct systems can be postulated:

1. *The Wetting of Waxy Surfaces*. Many natural foods such as apples and cabbage leaves have a wax coating. A drop of water placed on these surfaces will form small droplets instead of a film. Since it does not wet, the water cannot penetrate into the body of the food. A surface active agent, predominantly lipophilic in character but sufficiently hydrophilic to disperse in water is usually a most efficient product to blend with water when attempting to wet a waxy surface.[8]

2. *Capillary Wetting*. This is typified by the reconstitution or rehydration of foods with longitudinal pore structure. In this case, the addition of a surfactant is frequently detrimental. In capillary action, the higher the surface tension the greater the capillary rise or penetration; therefore, it is desirable to have as high a surface tension as possible to gain the greatest penetration in such a pore structure.

3. *Powder Wetting Clumps*. The wetting or dispersing of a powder is only partially controlled by the addition of surfactants. If wetting is too rapid, portions of the powder mixture often become coated with the liquid

on the outside, leaving a pocket of trapped air on the inside. The choice of a surfactant with proper surface tension reduction properties is helpful, but far more important and generally used are the adjustment of the powder particle size and the inclusion of a portion of a soluble material such as sugar.

TABLE 1
Outline of Surface Active Agents

I. Naturally Occurring
  A. Ionic
    Bile salts
    Phospho lipids—lecithin
    Inositol phosphate
  B. Nonionic
    Cholesterol
    Saponin
II. Synthetic
  A. Ionic
    Soaps
    Dioctyl sodium sulfo succinate
  B. Nonionic
    Propylene glycol mono esters
    Glycerol mono esters
    Sorbitan esters
    Sucrose esters
    Polyglycerol esters
    Polyoxyethylene esters
    Polyoxyethylene sorbitan esters
    Complex esters (lactate, tartarate, etc.)

The phenomenon of wetting is complex and almost always is concerned with more than one application. In many instances a combination of a waxy surface and capillary structure will be present. As noted, these two situations pose opposite requirements. Sometimes the best compromise is to use a surfactant that will provide only the amount of surface tension reduction needed to wet the waxy surface involved, and will result in the highest possible residual surface tension after the preliminary wetting step. Other combination properties often desired include wetting combined with emulsification, as in cake mixes, or with suspending action, as in chocolate drink powders.

Dispersion Surface active agents are usually used for their dispersing action. Dispersions of solids, liquids and gases depend on the reduction of interfacial energy by a surfactant. Disperse systems can involve all three principal phases—liquids, solids and gases (see Table 2).

An emulsion is usually used to provide stability or to achieve desired texture or other properties. "Aerosol" foams are generally used for convenience in handling. True aerosols (fogs) are seldom used in food applications while suspensions and solid dispersions are often employed.

Emulsions An emulsion is a two-phase system consisting of two immiscible liquids, the one being dispersed as finite globules in the other. The dispersed, discontinuous or internal phase is the liquid that is divided into globules. The surrounding liquid is known as the continuous or external phase.[3]

Emulsions may be classified according to particle size.[19] Table 3 compares the major classes of emulsions according to particle size.

TABLE 2
Disperse Systems*

Type	Internal Phase	External Phase
Emulsion	liquid	liquid
Foam	gas	liquid
Aerosol (fog, smoke)	liquid or solid	gas
Suspension (sol)	solid	liquid

*Reprinted from the *Kirk-Othmer Encyclopedia of Chemical Technology*, Vol. 8, 2nd. edition, by permission of the publisher, John Wiley & Sons, Inc., New York.

TABLE 3
Effect of Particle Size on Appearance*

Particle Size	Appearance
macroglobules	two phases may be distinguished
greater than 1 micron	milky-white emulsion
1 to approx. 0.1 micron	blue-white emulsion
0.1–0.15 micron	gray semitransparent (dries bright)
0.05 micron and smaller	transparent emulsion

*Reprinted from the *Kirk-Othmer Encyclopedia of Chemical Technology*, Vol. 8, 2nd. edition, by permission of the publisher, John Wiley & Sons, Inc., New York.

SOLUBILIZATIONS—Solubilized oils (or solubilized emulsions) are characterised by their sparkling clarity, similar to that of a true solution. This effect is achieved by using a large amount of a carefully selected hydrophilic emulsifier that will result in a particle size of colloidal dimension, less than the wavelength of visible light. However, if a beam of light is passed through a solubilization, a Tyndall cone of reflected light will be seen from the fine particles, showing that it is not a true solution. The micelles of the emulsifier expand sufficiently to absorb molecules of the oil. Usually several times as much emulsifier as oil is required to effect solubilization.[10,11,39,42,47,49] However, the end product when properly formulated is stable to various temperature and dilution conditions. One type of commercially available solubilized product is the water soluble flavor oil.

MACRO EMULSIONS—The term "macro" refers to the particle size of the dispersed phase in comparison with that of solubilized oils. Macro emulsions have a particle size ranging from 0.5 to 10.0 microns, though most are from about 2–5 microns.

Macro emulsions are more widely used than solubilized emulsions. When properly formulated they are quite stable. They generally are milky or opaque because of the difference in refractive index of the two phases and because the emulsion particles are larger than the wavelength of light. Milk is one of the most common macro emulsions. At the extremely large particle size end of macro emulsions are the very coarse separating salad dressing emulsions that are shaken before each use.

DEMULSIFICATION—Many surface active agents behave as emulsifiers but no one agent can meet the special requirements of all emulsions. A surfactant that will act as an emulsifier for one formula may cause instability in others.

Such instability may be observed as oil separation or churning, in the case of fat separation.

Foams and Aerosols   Since both involve the mixture of a liquid and a gas, foams and aerosols can be considered together. Surface active agents are of vital importance. With respect to both the character and stability of foams, foams are of importance in a wide variety of food products. Aeration is vital in baked goods, frozen desserts, as well as in low density products such as whipped cream.

On the contrary, aerosols (fogs) are not appreciably influenced by the presence of surfactants. The formation of the aerosol particle occurs so rapidly that the surfactant does not have adequate time to migrate to the surface of the particle and change its surface tension.

Suspensions   A very common food product class is the suspension, which may be defined as a dispersion of finely divided, insoluble material in a liquid medium, usually water. The size of the dispersed phase may vary from 0.1 micron to flocculates or aggregates of particles as large as 100 microns. An example is chocolate milk. In formulating a suspension, the main theoretical aspects that must be considered are flotation, crystal growth, redispersibility and ionic charge.

In suspensions, surface active agents are used in combination with hydrophilic macromolecular stabilizers or thickeners such as sodium carboxymethylcellulose, gum acacia, or aluminum magnesium silicate. Surface active agents solve wetting problems, promote uniformity in these non-Newtonian systems, and aid redispersion. In carrying out these functions, such properties as ionic charge and flocculating characteristics must be considered to achieve the desired effects.

Thickening or Viscosity Modifiers   Liquid foods can usually be thickened most economically through the use of gums and synthetic polymeric materials. A high viscosity can also be obtained by using appropriate emulsion techniques. In mayonnaise, for example, viscosity is largely attained by emulsification. The principle involved is one of structural viscosity because of particle crowding, which occurs when the dispersed phase volume is considerably greater than the continuous phase volume.[13] Mayonnaise is an oil-in-water emulsion. Because there is so much oil, the mayonnaise exhibits a higher degree of thickening than that of the viscosity of the outer phase (water and vinegar). This is illustrated by the series of emulsion photographs in Figure 1 (a through h).

Some dilute O/W emulsions, usually based on stearate emulsifiers, have quite a high viscosity. However, the emulsifier system must be carefully selected and even then this kind of formula frequently lacks long term viscosity stability. It should be pointed out that some nonionic surfactants show a thickening effect on dilution with water. However, this usually requires such a high concentration of surface active agents that it is impractical.

Lubrication   In the manufacture of food products, it is sometime desirable to impart "slip" or lubrication properties. Various materials and oils are frequently used to lubricate processing equipment. Mineral oil is the principal product used to which, in some instances, surfactants are added to gain a higher degree of lubricity.

Surfactants may also be added to a caramel or to peanut butter to reduce stickiness while eating.

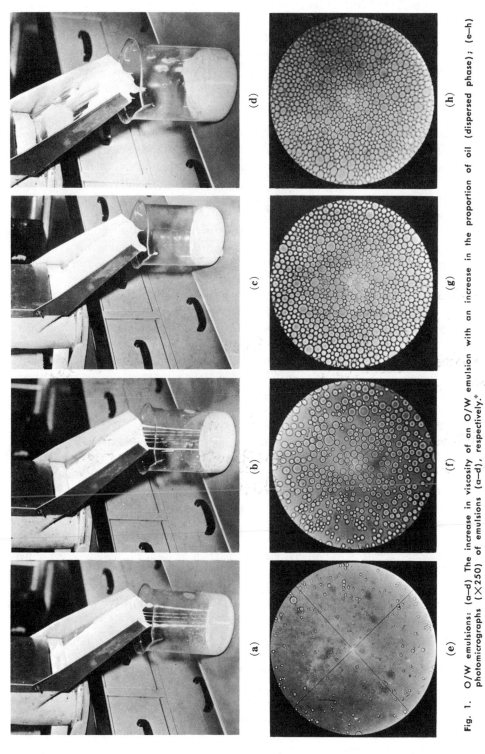

Fig. 1. O/W emulsions: (a–d) The increase in viscosity of an O/W emulsion with an increase in the proportion of oil (dispersed phase); (e–h) photomicrographs (×250) of emulsions (a–d), respectively.*

*From *Kirk-Othmer Encyclopedia of Chemical Technology*, 2nd ed., Vol. 8, John Wiley & Sons, Inc., 1965, p. 122. Reprinted by permission of the publisher.

Complexing The literature cites many examples that indicate certain surfactants have the ability to complex starch.[5,16,17,37,40,41,95] This appears as a change in viscosity of a starch paste. Most widespread use of this phenomenon is in bread baking where the effect may range from a softer crumb to a loaf that has greater resistance to staling. Complexing with surfactants has been reported in systems other than starch.

Selection Methods

Surfactants differ in their functions and efficiencies in improving food products. The functionality of a surfactant depends to a large extent on the affinity of a portion of its molecules for oil or water. All surfactants consist of molecules that combine lipophilic or oil-loving groups (such as fatty acid groupings) and hydrophilic or water-loving groups (such as the OH or hydroxyl groupings.

Various methods have been developed for selecting the proper emulsifier or emulsifier blend to perform a particular task. In selecting an emulsifier system the desired effects, flexibility of plant equipment and economic factors must be considered.

The purpose of a selection method is to meet the application requirements with existing plant equipment with the least costly surfactant system and the lowest selection cost. While there are many surfactants, only a few have met FDA regulations for food applications. For this reason the first selection requirement must be that the emulsifier to be examined has or is reasonably sure to get FDA approval. In foods one of the most common applications of surfactant is as an emulsifier. We will utilize emulsification as an example system to illustrate surfactant selection.

In most emulsion systems the best answer will be a blend of surfactants. One selection system, based largely on empirical findings, is the HLB (Hydrophile-Lipophile Balance) system. It is based on the fact that O/W emulsions are best prepared with water-soluble emulsifiers and W/O emulsions are best prepared with oil-soluble emulsifiers, and that there is little chance of success if a trial is made with the wrong kind of emulsifier. A rational means of selecting emulsifiers with the right solubility can eliminate many trials. This is the purpose of the HLB surfactant selection system.[19]

HLB Surfactant Selection System The HLB system is most easily applied with emulsifiers of known HLB values. The selection system consists of three steps: (1) determining the required HLB of the desired combination of ingredients in the product; (2) trying a variety of different emulsifier chemical types at the "required HLB" determined in step 1; and (3) making a final adjustment in HLB value.

When two or more surfactants are to be blended (and blending is usually to be preferred), the HLB of the combination is easily calculated. If X is the proportion of one surfactant having an HLB of A, and the other surfactant has an HLB of B, the HLB of the combination can be expressed for all practical purposes as $XA + (1 - X)B$. Since this is a straightline relationship, it can easily be computed graphically. However, it should be noted that the straight-line relationship is not precisely true.

In this procedure, the many emulsifiers and blends that have HLB numbers different from the required values may be automatically dismissed. Trial-and-error is not eliminated, but it is reduced to a narrow band and is directed toward selection of surfactants of the best chemical type. This is much more direct than the original trial-and-error method in which the search was for both the right chemical type and the right HLB.

The HLB of an emulsifier is related to its tendency to dissolve in oil or water since emulsifiers are totally soluble in neither. Low HLB (2-8) emulsifiers tend to be oil soluble; high HLB (14-18) emulsifiers tend to be water soluble.

The detailed steps in the procedure are:

a. Determination of Required HLB: Select any matched pair of emulsifiers, one lipophilic and one hydrophilic, of known HLB values (for example, SPAN® 60 sorbitan monostearate HLB = 4.7, and TWEEN® 60 polyoxyethylene sorbitan monostearate HLB = 14.9). A trial run is made first so the selection of emulsifiers at this point need not be perfect for the particular formula.

The first series might consist of seven test emulsions with the emulsifiers blended to give seven different HLB values ranging from 4.7–14.9, or any narrower range to be adequate from previous experience. Use an excess of emulsifier (approximately 10–12% of the weight of the oil phase) and dissolve or intimately disperse the emulsifier into the oil phase, melting ingredients together if necessary.

While simple mixing of ingredients and emulsifiers might be sufficient at this point, it is best to use preparation methods as close as possible to plant methods.

Using appropriate methods of comparison and evaluation based on the product requirements, including emulsion type (O/W or W/O), one or more of the seven emulsifier combinations will usually immediately provide a better emulsion than the others, even though it probably will not exhibit all the desired properties. If all the emulsions seem fairly good, with not much noticeable difference, then repeat the seven tests using less emulsifier. Conversely, if all the emulsions are poor and show no great difference, repeat the tests using more emulsifier.

The emulsions should now be compared for qualities desired: stability, freeze-thaw cycle, clarity, viscosity, ease of preparation, ease of application, etc. It should be possible to select, within about two units, the HLB range that will function best for this application.

Suppose it is found that an HLB of approximately 12 is optimum. Further tests should be made around this figure to establish the most accurate HLB value; i.e., these same two emulsifiers might be blended to try making emulsions at HLB values ranging between 11 and 13.

In this preliminary test, it is possible that a fairly good emulsion might be formed at about HLB 5 as well as the one at HLB 12. If this occurs, the low HLB emulsion is probably a W/O emulsion (does not dilute readily with water and does not conduct electricity), and the high HLB emulsion is an O/W emulsion (easily water dispersible, conducts electricity) However, since the type of emulsion O/W or W/O has already been selected, the reverse type can be rejected. Therefore by this one set of trials, the field for further emulsifier trials has been narrowed. Valuable time will be lost if emulsifiers are tried that have HLB values far from the value selected above.

®   Registered trademark, Atlas Chemical Industries, Inc.

Table 4a
Properties and Functional Uses of Surfactants in Specific Foods

See page 410 for explanation. To identify coded surfactants see Table 4b.

	Whipping Agent	Volume Improver	Tenderness Improver	Solubilizer	Dryness Promoter	Oiling-off Preventer	Palatability Improver	Grain Improver	Flavor and/or Spice Oil Dispersant	Reduce Fat Absorbance	Dough Conditioner and Improver	Defoamer	Bloom Inhibitor	Batter Aerating Agent	Anti-sticking Agent	Anti-staling Agent
BAKERY PRODUCTS																
Bakers' Cakes		3,4,6,7,11,12,13	3,4,6,7,11,12,13				11,12,13	3,4,6,7,11,12,13	2,11,12,14					3,4,6,7,11,13		
Cake Doughnuts		3,4,6,7,11,12,13	2,3,4,6,7,11,12,13						2,11,12,14	2						2,8,11,12,13
Cake Mixes		3,4,5,6,7,8,9,10,11,12,13	3,4,5,6,7,8,9,10,11,12,13				11,12,13	3,4,5,6,7,8,9,10,11,12,13	2,11,12,14					3,4,6,7,12,13		
Bread and Rolls		4,9,10	4,9,10				4,5,8,9,10	9,10	2,11,12,14		4,9,10					4,5,8,9,10
Yeast-raised Sweet Goods		9,10	9,10				5,8,9,10	9,10	2,11,12,14		9,10					5,8,9,10
Yeast-raised Doughnuts		9,10	9,10						2,11,12,14		9,10					5,8,9,10
Cookies									2,11,12,14							
Icings							12		2,11,12,14							
Cream Fillings									2,11,12,14							
CARBONATED BEVERAGES									2,11,12,14							
CHEWING GUM							5								5	
CONFECTIONS																
Caramel and Nougats															1,5,8	1,5,8
Licorice																1,5,8
Starch Jellies																
Mints		1,8					1,8									

TABLE 4a

Properties and Functional Uses of Surfactants Specific Foods (Continued)

	Anti-staling Agent	Anti-sticking Agent	Batter Aerating Agent	Bloom Inhibitor	Defoamer	Dough Conditioner and Improver	Reduce Fat Absorbance	Flavor and/or Spice Oil Dispersant	Grain Improver	Palatability Improver	Oiling-off Preventer	Dryness Promoter	Solubilizer	Tenderness Improver	Volume Improver	Whipping Agent
CONFECTIONERS COATINGS				11,12						11,12						
DAIRY PRODUCTS																
Ice Cream												1,13,14 15,16,17				1,13,15 16,17
Ice Milk												1,13,14, 15,16,17				1,13,15 16,17
Soft Serve												1,13,14, 15,16,17				1,13,15 16,17
Flavored Drinks								2,14								
Instant Beverage Mixes																
Margarine											5,8		1,2			
FLAVOR AND/OR SPICE OILS													2,11, 12,14			
JAMS, JELLIES, FRUIT JUICES					2											
COFFEE WHITENERS											1,11,12, 13		1,2			
WHIPPED TOPPINGS												1,11,12, 13				1,11,12 13
PEANUT BUTTER											1,5,8					
PICKLES								14								
SUGAR REFINING					2,14											
VITAMIN OILS													14			

Source: *Current FDA Status of Atlas® Surfactants and Polyols*, LG-115, Copyright 1969, Atlas Chemical Industries Inc.—now ICI America Inc. Revised January 1, 1971.

TABLE 4b

Surfactant	Code Number	Color & Form	Melting Point Vis. @ 25°C	HLB Rating	FDA Status
Mono- and diglycerides from the glycerolysis of edible fats or oils (52–56% alpha mono, 61–66% total mono)	1	Ivory-white colored solid (bead form)	135°–142°F	3.2	GRAS*
Mono- and diglycerides of fat forming fatty acids (47–50% alpha mono, 54–59% total mono)	2	Clear, light amber liquid	App. 150 cp	2.8	GRAS*
Mono- and diglycerides from the glycerolysis of edible fats or oils (40–44% alpha mono, 48–52% total mono)	3	Ivory-white colored votated plastic solid	115°–122°F	2.8	GRAS*
Mono- and diglycerides from the glycerolysis of edible fats or oils (40–44% alpha mono, 48–52% total mono)	4	Ivory colored votated plastic solid	122°–124°F	2.8	GRAS*
Mono- and diglycerides from the glycerolysis of edible fats or oils (40–44% alpha mono, 48–52% total mono)	5	Ivory-white colored solid (bead or flake)	135°–142°F	2.8	GRAS*
Mono- and diglycerides from the glycerolysis of edible fats or oils (40–44% alpha mono, 48–52% total mono)	6	Ivory colored, votated plastic solid	91°–95°F	2.8	GRAS*
Mono- and diglycerides from the glycerolysis of edible fats or oils (54–58% alpha mono, 65–69% total mono)	7	Ivory colored votated plastic solid	125°–127°F	3.5	GRAS*
Mono- and diglycerides from the glycerolysis of edible fats or oils (52–56% alpha mono, 61–66% total mono)	8	Ivory-white colored solid (bead form)	139°–143°F	3.5	GRAS*
Mono- and diglycerides from the glycerolysis of edible fats or oils (54–58% alpha mono, 65–69% total mono)	9	Ivory colored, votated plastic solid	129°–135°F	3.5	GRAS*
Mono- and diglycerides from the glycerolysis of edible fats or oils (54–58% alpha mono, 65–69% total mono)	10	Ivory colored votated plastic solid	128°–132°F	3.5	GRAS*
Sorbitan monostearate	11	Ivory colored solid (bead form)	127°F	4.7	Approved†
Polyoxyethylene (20) sorbitan monostearate	12	Yellow oily liquid (may gel on standing)	600 cp	14.9	Approved†
Polyoxyethylene (20) sorbitan tristearate	13	Tan waxy solid	App. 92°F	10.5	Approved†
Polyoxyethylene (20) sorbitan monooleate (Polysorbate 80)	14	Yellow oily liquid	400 cp.	15.0	Approved†
Mono- and diglycerides from the glycerolysis of edible fats or oils (80%) and polysorbate 80 (20%)	15	Ivory-white colored solid	130°–135°F	5.2	Approved†
Mono- and diglycerides from the glycerolysis of edible fats or oils (80%) and polyoxyethylene (20) sorbitan tristearate (20%)	16	Ivory-white colored solid (bead form)	130°–135°F	4.3	Approved†
Mono- and diglycerides from the glycerolysis of edible fats or oils (40%) and polyoxyethylene (20) sorbitan tristearate (60%)	17	Ivory-white colored solid (bead form)	130°–135°F	5.9	Approved†

* GRAS—*Generally Recognized As Safe* in food applications by FDA.
† Cleared by FDA for uses in many food products.

b. Determination of Best Chemical Type: The right chemical type is as important as proper HLB; the two go hand in hand. Suppose that it was found that a blend of SPAN 60 and TWEEN 60 stearate emulsifiers at an HLB of 12 gave a better emulsion than any other HLB of these two emulsifiers in the first test series. That particular HLB of about 12 will be best for any chemical type. Now it must be determined whether some other blend at HLB 12 (laurates, palmitates or an oleate, for example) would be better or more efficient than the stearates. By blending two emulsifiers the exact HLB needed can be attained. The emulsifier chemical type should be selected to suit the oil or other active ingredients.

It has been found that the most stable emulsion systems usually consist of blends of two or more emulsifiers, one portion having lipophilic tendencies, the other hydrophilic. (For example, glycerol monostearate, self-emulsifying grade is actually a blend of lipophilic non-self-emulsifying glycerol monostearate with a hydrophilic soap or other substance to make it more water soluble.) A single emulsifier is rarely as effective as a blend of the same HLB.

Of course, the chemical type is at times dictated by application requirements. For example, in an instance where the surface active agent must be destroyed, one means of doing this would be to utilize a soap that is reactive with and destroyed by a polyvalent salt or an acid. If acid stability is required, a nonionic emulsifier is suitable. Food emulsions require FDA approved emulsifiers. Each product will have its own requirements.

After making the preliminary screening, choose a variety of low-HLB emulsifiers of different chemical types for which corresponding high-HLB emulsifiers are available. For example, low- and high-analog blends might include: sodium stearate ranging from a medium to high pH; a mono glyceride and TWEEN 60; a liquid monoglyceride and TWEEN 80; propylene glycol monostearte and polyoxyethylene stearate (Myrj 52); a monoglyceride and a polyglycerol stearate, etc. The objective is to select several pairs of related emulsifiers, and to select pairs covering a suitably wide area of chemical nature.

A test blend of each pair is then prepared in a weight ratio so that its HLB will be 12, the required HLB for the emulsion. An evaluation of these will usually show wide differences for different chemical types; selection based on the criteria for the emulsion system should be relatively easy.

c. Final Adjustment of HLB: After selecting the best chemical type, it is usually desirable to try blends of the selected emulsifier having incremental HLB values close to the indicated required HLB. In this example where the value was 12.0, tests at 11.0, 11.2, 11.4, 11.6, 11.7, 11.8, 11.9, 12.0, 12.1, 12.2, 12.3, 12.4, 12.6, 12.8, and 13.0 will confirm the exact blend with the chosen pair. This is necessary because HLB values are not precise, since they are based on approximate calculations and on empirical tests.

The HLB method is an incomplete system for selecting surface active agents. Its use is often further complicated in food systems by their complexity. In a formula composed of oil, water and emulsifier, correlations are possible in most instances. However, the addition of flour, starch, sugar, milk, salt, eggs and similar ingredients, some of which contain natural emulsifier, raises many complications. Selection of emulsifiers for cake mixes has been subjected to a statistical approach[35,50] that can certainly be used as a model for other food product problems.

A tabulation of seventeen food emulsifiers performing seventeen different functions in thirty-two different food products is presented in Table 4a. The emulsifiers are listed by number, the key for which is in Table 4b. These suggestions can provide a basis for primary choice aided by HLB and statistical means. Table 4b shows the surfactant form, color, melting point, HLB and FDA status.

Manufacture of Emulsions

Formulation should, if possible, take into consideration the intended manufacturing procedure. The emulsifier usually promotes ease of formation of the emulsion as well as stability. As a general rule, the greater the mechanical energy input the less demand there is on the emulsifier. Frequently, the amount of emulsifier may be reduced with increased energy input. Two extremes may illustrate this; an emulsifiable concentrate which requires a minimum of agitation and a sizable amount of emulsifier, and the "mayonnaise" dormant spray mineral oil emulsion concentrate made with repeated homogenization where a few tenths of a per cent emulsifier is sufficient.

Laboratory Preparation   The laboratory scale of operation deserves special consideration because it is often difficult, yet necessary, to duplicate plant manufacturing techniques. For example, in preparing an emulsion of moderate viscosity in a typical laboratory beaker-motor-driven propeller apparatus the actual work input may be surprisingly high. In equipment scale-up, surface-to-volume ratios differ markedly; peripheral speeds of agitators are different; tendencies to maelstrom, suck in air and produce foam differ; rates of heating and especially rates of cooling are significantly changed; and chances for local overheating are greater. This combination of altered factors increases the complexity of the problem.

Laboratory scale preparations should consider and attempt to duplicate plant conditions and err, if possible, on the side of too little energy input. A batch kettle either heated or unheated, is approximated by a beaker-motor-driven propellor of appropriate size (slow motor speed is preferred and a simple baffle is usually best). A plant homogenizer is approximated by a hand homogenizer, or by a gear pump with a pressure relief valve. Emulsification of a concentrate is more uniformly handled by motor-timed controlled shaking.[4,23]

A blender, when used to prepare emulsions, imparts considerable amounts of energy and also incorporates large quantities of air. For this reason, it often gives results that are not achieved in subsequent plant scale-up.

An additional major problem in the laboratory occurs with emulsions that are prepared hot and then cooled. A laboratory beaker of emulsion will cool from 60–30°C in a few minutes. A 1,000-gallon tank requires much longer, even with forced cooling. For emulsions that contain waxy components the rate of cooling throughout the melting range can be critical. Hence, the best procedure is: to determine in the laboratory the best practical cooling schedule; then determine the deviation leeway allowable; and finally, set plant conditions to satisfy those criteria.

Plant Preparation   If laboratory and pilot plant formulation is done with careful attention to the duplication of plant conditions, scale-up will present only the usual problems that in many cases can be anticipated. Based on the laboratory formulation work, it is wise to establish a specific procedure for preparing an emulsion on a plant scale. Many times small deviations in procedure can result in totally different end products.

Addition of Ingredients   Order and rate of addition of ingredients is often of no concern, yet in some cases a given order and rate of addition is necessary to get an acceptable product. An illustration is the preparation of an O/W emulsion by the inversion technique. One usually good procedure is to blend the oil phase and emulsifier in the tank, then add the water to the oil phase slowly while stirring. The initially hazy oil and emulsifier blend usually clarifies at first and then again becomes cloudy. As more water is added, the emulsion assumes a milky cast and the viscosity increases. At some point, called the inversion point, the viscosity suddenly decreases. This signifies that the emulsion has changed from W/O to O/W. Further additions of water may be made more rapidly.

Specific procedures must be worked out for each formulation. Generally, all the oils and oil-soluble ingredients are best combined as the oil phase. Polysols are usually added with the water, and salts are best added with the last half to quarter of the water after a good primary emulsion has been developed.

Temperature of Ingredients   In most liquid-liquid emulsifications, ambient temperature is preferred. With some equipment, it may be necessary to be equipped to remove the heat of friction generated by the mechanical agitation.

Heating of ingredients may be necessary to effect emulsification; it may also be necessary to pasteurize or sterilize the product. In making wax emulsions, the wax or the oil phase should be heated to at least 5°C over the melting point of the highest-melting wax, and the aqueous phase should be heated to at least 2–3°C over the temperature of the oil phase. This procedure will avoid shocking the wax by cooling during mixing.

Rate of cooling of a wax emulsion especially at the melting point of the wax is critical. Each emulsion must be studied to determine whether the best rate of cooling is fast or slow. Fast cooling may be done with a cooling board or means of a heat-exchange unit.

When an emulsion system is heated, emulsifier co-solubility is altered and this may change its emulsifying properties. Care must also be taken to avoid undesirable chemical reactions.

Processing Equipment

Since surface active agents are used in a wide variety of foods, hence equipment discussion must be of a general nature. In most instances, such as adding "emulsifiers" to peanut butter, caramels, chocolate and chocolate coatings, coffee whiteners and similar products using equipment that is normally employed, the "emulsifier" is added at a suitable step in the processing.

The following general comments concerning equipment are offered for illustrative purposes. Emulsification is emphasized since this is the major use of surface active agents. Equipment for emulsification, in general, has been discussed elsewhere.[3,19]

Wetting   In food processing, the problem is usually to produce a powdered food that has the desired rate of wetting in a cold or hot liquid. The problem is one of distributing liquid or melted surfactant evenly onto a large volume of finely divided powder. Ribbon mixers or V-cone blenders with provision for spray injection of the surfactant may be used. It is seldom desirable to use a solvent such as water or alcohol to extend the surfactant, even though this would ease the distribution problem appreciably. At times it suffices to dry-mix the surfactant onto just a portion of the ingredients again using ribbon mixes or V-cone mixes with suitable sprays.

Dispersion *Emulsification Solubilizations* In solubilized systems, the particle size is sufficiently small that mechanical equipment is used only to assure thoroughness and uniformity of mixing, not actual particle size reduction. Hence, propeller type agitation is most frequently employed.

Macro Emulsions Emulsification equipment covers a wide range and selection is related to the characteristics desired in the final products. It is probably best to consider a few typical products as shown in Table 5.

TABLE 5
Typical Food Emulsion Systems

Product	Type Emul.	O/W Ratio	Air/O&W Ratio	E/O Ratio	Energy Input	Mixing Stages— Order
1. Milk, Homogenized reconstituted	O/W	0.02/1	0/1	*0/1&0.05/1	high	Two-mixing, homogenization
2. Mayonnaise	O/W	2/1	0/1	*0/1	high	Two-mixing, homogenization
3. Margarine	W/O	10.20/1	0/1	app. 0.01/1	high	
4. Cake Batter	O/W					
5. Ice Cream	O/W	0.1/1	1/1	0.005/1–0.02/1	high	Two-mixing, homogenization

* An E/O ratio of 0/1 is never solved since there are natural emulsifiers present. This relates to *added* emulsifier.

Milk, Homogenized—Milk is passed through a homogenizer to reduce the emulsion particle size from approximately 10 microns to 1–2 microns. Milk and milk products also may be reconstituted from milk solids, water, and butterfat. In reconstitution the nonfat dry milk solids must be wetted, dispersed, and dissolved in the proper amount of water; the melted fat is added in a premix, and this mixture is pasteurized and passed through a homogenizer. Imitation milk and imitation milk products likewise may be prepared with similar equipment.

Mayonnaise—This emulsion is commercially processed through homogenization equipment. Because of the high oil content, mayonnaise is frequently pre-mixed quite carefully and/or passed through a series of homogenization steps.

Margarine—Margarine is the only large volume "inverse" (W/O) emulsion. Its manufacture is highly specialized.[22,48]

Demulsification It is generally not recognized that demulsification is frequently as important as emulsification. One of the best examples of this is in the preparation of ice cream.

Ice Cream—The body and type of frozen ice cream is largely achieved by a carefully controlled partial demulsification of the fat globules in the mix.[13,14,23,24,25,26,28,32,51] Of course, aeration is occurring at the same time, controlled either by heating or by injection.

Aeration—Aeration is required in many foods. Air or nitrogen may be incorporated by heating through high shear agitation or, in more modern and totally enclosed equipment, it may be meter-injected.

1. Cake batters represent a formulation that combines aeration of the batter and emulsification (of the shortening). The batter usually exhibits a moderately high viscosity and generally no special mixing equipment is required to incorporate air and emulsified shortening. Standard equipment ranges from the batch type planetary beater to in-line continuous mixers where ingredients including air are added through a closed injection arrangement.

2. Pressure packaged "aerosols." Whipped toppings lean heavily on proper surface active agent choice for good foam production and stability without "weeping." No special equipment is required to handle the surface active agents except for possible handling of solid, waxy materials.

Lubrication Food machinery lubricators are prepared by blending an oil and desired surfactants with propeller agitation. Heat is usually desirable to assure complete melting of the ingredients. Lubricants are usually added to a final food-stuff with standard equipment.

Thickening or Viscosity Inducers Thickeners are usually added either to dry powder food preparations or to thin liquids. Dry powders are best handled in blending equipment such as the twin-cone blender. Addition to liquids may incur problems because many of the thickeners have an annoying tendency to clump. Pretreatment of the thickeners with a dry (liquid or solid) diluent is highly recommended; regular processing equipment can be used. When this is not feasible, high-speed mixing of a concentrated slurry in alcohol followed by dilution with water is recommended.

Complexing This is generally accomplished in dilute solutions. Mixing is seldom a problem and regular equipment is employed.

Food Application of Surface Active Agents

Cereal Products

Yeast Raised Baked Goods Bread represents one of the major uses of surface active agents in food. Modern production methods, packaging techniques and distribution channels create demands on bread formulations that did not exist a few decades ago.

Surfactants have been employed in bread for many years. Their value in retarding crumb firming, conventionally described as "staling," has long been noted, as well as their beneficial effect on volume, grain and machineability.[5,16,17,37,40,41,45]

A consideration of the chemistry of starch helps explain the function of surfactants. Starch is composed of two primary fractions: amylose and amylopectin. Amylose, sometimes referred to as the "A" fraction is found at levels of 15–30% depending on the type of starch. The amylopectin or "B" fraction constitutes the remainder or about 70–85% of the total.

When the starch is gelatinized during the baking process, the amylose, or linear polymer of glucose (averaging about 300 molecules of the monosaccharide per mole of polymer) is irreversibly retrograded or crystallized, and provides a major portion of the structural framework of the bread. The retrogradation involves the parallelization of the polymers and consequent association through hydrogen bonding. The amylopectin fraction, which is a branched molecule (averaging only 30 moles of glucose per branch with hundreds of branches per mole) begins to crystallize slowly after the cooling of the bread, and this crystallization or coacervation continues for several days after the bread is baked. Coacervation of the amylopectin causes a gradual increase in the firmness of the bread with the passage of time and is the most important of the three factors involved in staling: 1) a firming of the crumb; 2) loss of flavor; and 3) loss of crispness in the crust with resultant leatheriness. The addition of a small quantity of specific surfactants, 0.25–0.5%, based on the flour weight will retard the firming of the crumb and loss of crispness in the crust.

The exact chemical or physical action of the surfactant in inhibiting starch crystallization is not fully understood. Some cereal chemists are attributing the action to clathration, the action whereby the surfactant enters into the helix of the starch molecule and prevents or retards association by hydrogen bonding. A graphic illustration of this phenomenon is presented in Figure 2 (A and B), with and without emulsifiers respectively.[5,16,17,37,40,41,45]

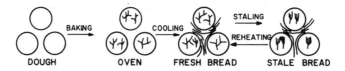

(A) CONTAINING NO SURFACTANT

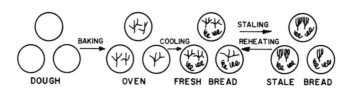

(B) CONTAINING A MONOGLYCERIDE SURFACTANT

Fig. 2. Graphic illustration of the role of starch fractions during the baking and aging of bread.

In the Figure 2(A) with bread containing no surfactant, the linear molecules (amylose) dissolve and diffuse from the swollen starch granule and set up a permanent gel network between granules. Staling is attributed to association of branched molecules within the swollen granules.

In Figure 2(B) the monoglyceride forms an insoluble helical complex with linear molecules and immobilizes them within the swollen granules. Hence, no gel structure develops between the granules, and the fresh bread is deformable and plastic.

During gelatinization, moisture is trapped at the interstices of the amylopectin branches. As coacervation proceeds, the bound water is released as free water that can migrate to the crust causing leatheriness. Because the surfactant retards coacervation and the subsequent release of free moisture, it also aids in retaining the crispness of the crust.

The surfactant type most used in bread is "mono- and diglycerides." Commercial "mono-" contains alpha-mono, beta-mono di- and tri- esters of mixtures of fatty acids. These parameters are controlled to gain performance characteristics desired for different applications.

For example, in bread, performance of the alpha and beta fractions are similar, and far superior to either the di- or tri- esters. Hence, the total mono-ester content is a controlling factor in the choice of monoglyceride emulsifiers to be used in bread.

A second and more subtle factor that can affect the performance of a monoglyceride is the composition of the fatty acids on which it is based. In attempting to understand why the composition of the fatty acids is an important consideration, the development of gluten should be considered as shown in Figure 3(A to D).

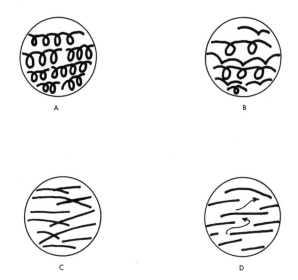

Fig. 3.   Development of gluten during the kneading of bread dough.

During kneading, the gluten in the dough becomes plastic and pliable. As mixing continues, the gluten is stretched into a thin, uniform film. As shown here, the gluten looks like coiled springs. The first drawing, Figure 3(A), represents these gluten springs. Figure 3(B) shows what happens during mixing: the gluten springs are stretched out and some are uncoiled. As mixing proceeds to optimum development the strands, in Figure 3(C), become arranged in parallels and overlap to form a smooth, extensible film. Many times, overmixing, or an excessively high level of unsaturated fatty acids in the surfactant or shortening, weakens the film permitting fermentation gases to escape, as shown in Figure 3(D). This causes open grain, poor texture and reduced volume. Highly unsaturated esters such as glyceryl mono oleate, for instance, have been shown to open the grain of bread and result in poor volume and shelf life. It is important, therefore, that in a mono- and diglyceride for bread, the degree of unsaturation should be controlled to the degree necessary to provide uniformly good results. An iodine value of 48–50 is close to the optimum for this sort of product. At this value, good uniform performance is obtained with a product of sufficient plasticity to be readily incorporated into the dough. An emulsifier of this type not only retards firming of the crumb, but also insures against loss of volume, weakening of side walls, and loss of "oven spring" often associated with the use of a poorly designed mono- and diglyceride.[31]

A modified mono and diglyceride, the diacetyl tartaric acid esters of mono and diglycerides, and calcium stearyl lactylate are the only other emulsifiers employed to any significant extent. Calcium stearyl lactylate is primarily employed as a dough-conditioning agent to increase dough strength and to insure good volume, particularly in continuous process bread. Since this product does little to retard crumb firming, it is used in conjunction with mono and diglycerides.

While this discussion has concentrated on bread, the concepts developed apply to yeast-raised baked goods in general. However, the concentration of surfactant required will vary; for example, in yeast-raised sweet goods the concentration of the preferred mono- and diglyceride is 2.5–3.0% based on flour weight, while as little as 0.25–0.5% is effective in bread. It has been postulated that the requirement is higher in sweet goods because more shortening is used.

Cakes, Cake Mixes and Doughnuts   The primary function of surfactants in cakes is to increase volume and improve texture, eating quality and keeping quality. Photomicrographs of cake batter show that air bubbles which contribute to volume are enclosed in films of protein in which the fat is dispersed. The role of surfactants is to improve the generation and uniformity of the initial air bubbles (foam), and the ability of the protein film to entrap air in the batter during creaming or mixing. The air bubbles have a leavening effect and help control grain size by serving as foci for gas evolution.

Mono and diglycerides have been used in bakers' cakes since the early 1930's when they first made their appearance in so-called high ratio shortenings*—and most of the surface active agents used in bakers' cakes today are supplied in the form of formulated bakers' shortenings.

The effect of increasing emulsifier level in both a 130-ratio white and a 130-ratio chocolate cake is shown in Figure 4. A blended emulsifier is used consisting of mono and diglycerides, sorbitan monostearate and polysorbate 60. As emulsifier usage increases, cake volume increases to a maximum point beyond which volume abruptly decreases. Over-emulsification is usually apparent by a dip in the center of the cake caused by the collapse of weakened air cells.

Much effort has been expended in investigating the value of various surfactants in cakes, and several have been approved by the Food & Drug Administration and are being used today. These include the modified mono- and diglycerides such as the lactylated products, stearyl lactate, propylene glycol monostearate, sorbitan monostearate and polysorbate 60. These surfactants are frequently used in conjunction with mono and diglycerides and with each other. Through selection of the proper emulsifier system, significantly improved batter aeration can be obtained, giving cakes greater volume, improved grain and texture. In cake formulation of all types, a blend of emulsifiers generally produces a better cake than does a single surfactant. Statistically designed methods have been developed for finding the most functional emulsifier blend for a specific cake formula.[35,50]

* Emulsified shortening suitable for baking cakes with high (130) ratios of sugar to flour.

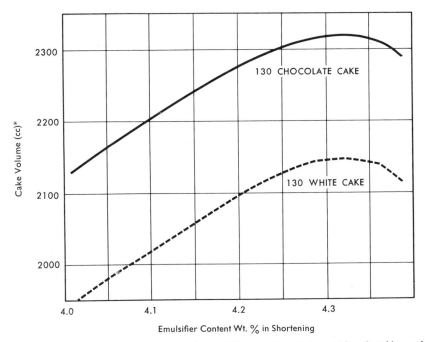

Fig. 4. The effect of increasing the emulsifier level in both a 130 ratio white and a 130 ratio chocolate cake.

* As measured by volume with low-density rapeseeds.

Of the several emulsifiers available for bakers' cakes, sorbitan monostearate and polyoxyethylene sorbitan monostearate have been found especially effective. These emulsifiers are almost always employed in combination to gain significant improvements in cake quality.

Mono and diglycerides usually find their way into a bakers' cake via the shortening route, while the sorbitan esters are most commonly added apart from the shortening—usually in a water concentrate form (hydrate) for convenience and ease of handling.

The series of photographs in Figure 5 show a white cake in which the

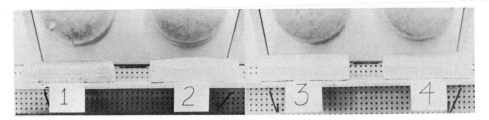

Fig. 5. Pictorial representation of data in Figure 4 (white cake).

amount of emulsifier is increased. Note the increase in volume to a maximum point with improved symmetry and texture. After the maximum level, note the center depression and poorer texture.

Cake mixes usually contain a higher level of emulsifier than bakers' cakes. In addition, the emulsifiers are selected on different parameters. While volume and texture are important, staling rate is less important, and tolerance is of great importance. In contrast to the baker, who follows a recipe carefully with specially designed mixing equipment, the housewife mistreats a cake mix fantastically and expects, and gets, a good cake. This is accomplished through statistical testing to design the emulsifier blend to give the highest tolerance to amount of liquid, level of mixing, and baking temperature.[35,50]

A comparison of a bakers' cake and a cake mix formula, Table 6, reveals that the chief difference is in the selection of emulsifier.

TABLE 6

Ingredient	Bakers' Cake	Cake Mix
Cake Flour	191 gm	210 gm
Sugar	249 gm	243 gm
Non-fat Dry Milk Solids	19.2 gm	—
Milk Solids	—	18 gm
Salt	4.7 gm	6.0 gm
Baking Powder (Double acting)	10.4 gm	—
Dicalcium Phosphate	—	3.3 gm
Mono Calcium Phosphate	—	2.2 gm
Soda	—	2.85 gm
Cream of Tartar	2.6 gm	—
Plastic Shortening	82.4 gm	51.00 gm
Glycerol Monostearate (42% alpha mono)	7.6 gm	4.8 gm
Glycerol Lacto Palmitate	—	4.2 gm
Water	140 gm	237.0 gm
Whole Eggs	14 gm	66.0 gm
Egg Whites	129 gm	—
Batter Weight (approximate)	850.0 gm	850.0 gm

Doughnuts are almost universally marketed as commercial or institutional mixes. Emulsifier level in these mixes is usually moderately high. The most important emulsifier selection parameter for cake type doughnut mixes is fat absorption. Of course, texture and volume are also important. In yeast raised doughnuts, bread type surfactants are often employed.

Noodles, Cereals, and Puddings In recent years emulsifiers, primarily monoglycerides have been found very effective in modifying starch gels. Monoglycerides will produce changes in viscosity by complexing free starch as discussed under cakes in this chapter. The changes induced by the surfactant improve the texture and creaminess, inhibit the formation of clumps in noodles and cereals, reduce sticking (canned) and enhance total appeal of the finished product. A level of approximately 1% monoglycerides in noodles and 2–4% in cereals is generally sufficient.

The problem of dispersing the surfactant in the products has been resolved by adding the monoglyceride in the form of a water hydrate.

Dairy and Related Products

Ice Cream and Soft Serve Surfactants function in two different ways in frozen desserts. In the mix before freezing, they help stabilize the fat emulsion keeping the fat dispersed and in suspension; in the freezer, surfactants produce the effect of dryness by aiding the controlled destabilization of the fat emulsion, thus promoting agglomeration of fat globules. The drying action of surfactants on ice cream in the freezer is entirely different from other types of surfactant applications in which they cause oil and water to mix.

To make ice cream dry, various types of surfactants have traditionally been added to formulas to emulsify the fats and water in the mix. It was believed that by reducing interfacial tension, they aided the dispersion of fats and thereby assisted in the production of fine air cell structure and improved whipping properties. Dryness and stiffness in ice cream was also attributed to the finer air cell structure and improved whipping induced by the surfactant.

However, newer data indicates [13,14,23,24,25,26,28,32,51] that surfactants play a somewhat different role in making ice cream dry. This research has shown that dryness is achieved by getting the finely dispersed fat globules to agglomerate properly.

Under a microscope an ice cream mix appears as a mass of emulsified fat globules, each of which is about one to two microns in diameter. As the mix is agitated and frozen, the surfactant induces the dispersed globules of fat to clump together like grapes. This "bunches of grapes" effect, or fat agglomeration, determines the degree of dryness. Making a dry ice cream with good body and texture, then, requires controlled agglomeration of the fat globules, achieved through proper choice of amount and type of emulsifier.

The dryness obtained with different emulsifiers in a vanilla ice cream is illustrated in Figure 6. The rate of attaining maximum dryness and the degree of dryness varies considerably with choice of emulsifier.

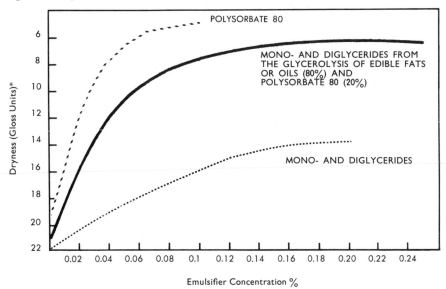

Fig. 6. The effect of emulsifier level and chemical type on the dryness obtained in ice cream.

* Determined with a Gardner P4 glossmeter, Gardner Instruments, Bethesda, Maryland.

Coffee Whiteners, Liquid and Dry Whipped Toppings, Imitation Cream   The use of coffee whiteners based on vegetable or animal fat is increasing in homes, restaurants, institutions and vending machines. The market for these products is growing at a rapid rate primarily because they offer lower cost and better shelf life than conventional dairy products.

At the present time, coffee whiteners are marketed in three physical forms: (a) Liquid Coffee Whiteners—manufactured, transported and marketed in the liquid state. These are used in homes and restaurants; (b) Fozen Liquid Coffee Whiteners—liquid, shipped to the ultimate consumer in a frozen state, and (c) Spray-Dried Coffee Whiteners—manufactured as free-flowing dry powders.

Of the three physical forms available today, spray-dried powder offers the manufacturer a product with the best stability and widest consumer appeal. Both the liquid and the liquid-frozen forms are widely used but are significantly less stable and must be kept under constant refrigeration.

Each form is unique in its formulation and production techniques:

A. Liquid Coffee Whitener—In preparing a liquid coffee whitener that is to be transported and sold in a liquid state, the following requirements should be considered:

1. Stability—Because of the physical form of the product, an unusually high degree of stability must be achieved in the liquid coffee whitener. The product must:

a. remain in a uniform physical state on standing after preparation and prior to sale.

b. be sufficiently well formulated to prevent oiling-off or feathering when added to hot coffee.

. . . withstand freeze-thaw cycles without separating.

2. Viscosity—The viscosity must be accurately controlled to simulate, as closely as possible, the natural dairy products currently being used, such as milk and cream. A heavy bodied product will not disperse in the coffee and is highly unacceptable. A viscosity standard, once achieved, must be faithfully duplicated from batch to batch.

3. Whitening Ability—The coffee whitener must provide uniform whitening ability which is controlled by the total amount of solids present and the fineness of the dispersed phase.

4. Flavor—Because of the multitude of applications, the coffee whitener must have a bland flavor and be odor-free.

Typical formulation ranges for a liquid coffee whitener are shown below:[54]

Ingredient	Range % (on total wt)
Fat	3.0–18.0
Protein (i.e. sodium caseinate)	1.0–3.0
Corn Syrup Solids	1.5–3.0
Sucrose	1.0–3.0
Glycerol Monostearate*	0.3–0.5
Carrageenan	0.1–0.2
Stabilizer Salts (i.e. sodium citrate)	0.1–0.3
Flavor	q.s.
Color	q.s.
Water	q.s. to 100%

* Mono- and diglycerides from the glycerolysis of edible fats or oils (52–56% alpha mono, 61–66 total mono).

The dry ingredients are blended with the fat and liquid ingredients and heated to pasteurizing temperature. (This temperature varies depending on the particular method used.) Once pasteurized, the mix is pumped directly to the homogenizer and homogenized at 2,000 to 2,500 total psi (1,500 to 2,000 psi on first stage and 500 psi on second stage) on a two stage homgenizer, and 1,500 to 2,000 psi on a single stage homogenizer. The homogenized product is cooled rapidly to 38°F and stored in a refrigerator or freezer until ready for use.

B. Frozen Liquid Coffee Whitener—The characteristics and manufacturing techniques for this type of product are similar to those for the liquid type coffee whitener. Because this product is maintained in the frozen state prior to use, the demands of initial stability are not as great as in the liquid product; but sufficient freeze-thaw stability must be present in the formulation so that separation will not occur when the whitener is thawed.

The recommended formulation for liquid coffee whiteners may be used for this type of whitener. Freezing the prepared whitener should be as rapid as possible.

C. Spray-Dried Coffee Whiteners—The spray-dried, free-flowing powders are generally preferred by the ultimate consumer because they need no refrigeration. While the principal use of these products is as replacement for cream in coffee, they are also reconstituted and used with cereals, as whipped toppings and in other foods. The dry products are added directly to the hot coffee and need not be reconstituted.

In addition to the requirements listed for liquid products, the following additional factors must be considered:

1. The dry product must exhibit good flow properties. Clumping and caking must be avoided.
2. It must disperse easily in hot liquids.
3. It must be non-hygroscopic.
4. It must be packaged properly for convenient use and, if necessary, to protect against moisture build-up.

A typical formulation for a spray-dried coffee whitener is as follows:[54]

Ingredient	Levels (%, dry basis)
Vegetable Fat	35–40%
Corn Syrup Solids (42 D.E.)	55–60
Sodium caseinate	4.5–5.5
Dipotassium phosphate	1.2–1.8
Glyceryl monostearate	0.15–0.3
Polyoxyethylene sorbitan tristearate	0.05–0.1
Color	q.s.
Flavor	q.s.
Anticaking Agent	q.s.

To prepare a powdered whitener, an emulsion concentrate is formed prior to spray drying by dissolving and/or dispersing the various dry ingredients in just enough water to: (1) maintain the solids in solution and (2) impart sufficient fluidity to the concentrate so that it may be pumped. The dissolved solids of the concentrate are usually in the range of 50–60%, the higher percentage being strongly recommended.

Once the concentrate has been prepared, it is homogenized in such a manner that the fat particles in the dried emulsion will average about 1 micron in diameter. Under normal circumstances this will require about 2,000–2,500 pounds total pressure on a two-stage homogenizer. However, it should be remembered that spraying will affect the emulsion and, therefore, each manufacturer will have to adjust the pressures to suit his spraying requirements, the viscosity of the emulsion, and the solids content of the system.

At the spray tower, operating conditions should be such that the final product will:

1. have a moisture content not in excess of 1%.
2. have a particle size of 125–150 microns in diameter.
3. have entrapped fat globules no larger than 1–3 microns in diameter.

Care should be taken to make certain that the heat of crystallization of the fat is removed from the powder prior to packaging, or clumping of the product in the package will result.

Shortening, Edible Oils

The raw materials for the production of shortenings may be derived from either vegetable or animal sources, with the finished products consisting either of all vegetable, all-animal or a combination of both types of fat. The primary vegetable oils are derived from cottonseed, soybeans, peanuts and corn. The main animal fats are lard and tallow. The edible oil and shortening suppliers have developed manufacturing processes including selective hydrogenation, rearrangement, winterization and the addition of surfactants to a high degree to supply specific food needs ranging from those of the large manufacturer to those of the ultimate consumer. Shortenings are divided into at least four types according to the major use requirements: Bakers' Cake Shortening, Bakers' Specialty Shortening, Fluid Shortening, and Household Shortening. Each is formulated to serve a special purpose:

Bakers' Shortening   These products are formulated to be functional in icings and cakes. This is achieved by balancing the level of surfactants so that while good quality cakes are produced, the emulsifier level is not sufficiently high to impair icing volume and promote weeping. Figure 7 depicts the approach that is used to achieve the desired results. A cursory study of these data reveals the role of surfactants in shortening. Surfactants help trap air in a system. However, if a high level is added (3.0%) icing volumes are critically affected. A typical Bakers' Shortening sometimes called a "high ratio shortening" can be formulated by adding 2.5% mono- and diglyceride with an iodine value of 60. The 2.5% refers to alpha mono content. A more complete discussion of cake volume appears in the section on Cakes, Cake Mixes and Doughnuts.

Bakers' Specialty Shortenings   In addition to the Bakers' Cake Shortening above, there is considerable interest in formulating specialty shortenings for specific large uses. While the regular Bakers' Shortening is a good general product for cakes, icings and fillings, it is a compromise and greatly improved performance in any one application can be obtained by special formulation.

For example, sorbitan monostearate-polysorbate 60 -monoglyceride blends produce excellent cakes. Volume and texture are aided by the inclusion of

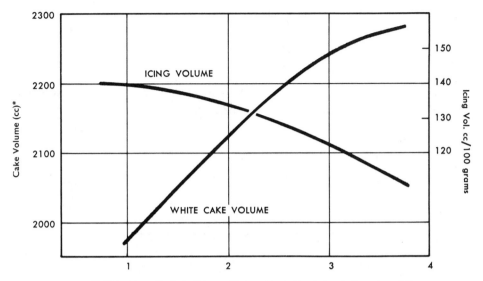

% Alpha Mono Content of Mono-, Diglycerides in Shortening (iodine value of 60)

Fig. 7.   The effect of emulsifier level in bakers' shortening on icing and cake volumes.

* As measured by volume with low-density rapeseeds.

sorbitan monostearate and polysorbate 60 and a fine grain is maintained with the proper level of a soft plastic mono. Both the mono- and the polysorbate 60 contribute to tenderness. These various parameters are best studied by means of statistically designed experiments.[35]

Another specialty product gaining in popularity is a shortening for icings and fillings. A single shortening can be formulated to give excellent icings and fillings using a blend of polysorbate 60 and mono- and diglyceride. Figure 8 shows how various surfactants can be used in shortening for icings and fillings.

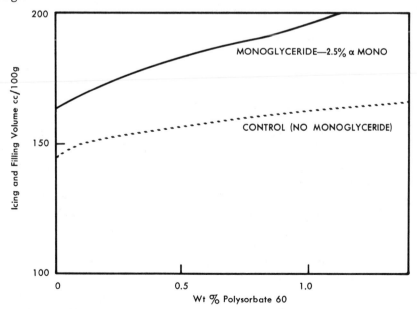

Fig. 8.   The effect of surfactant blends in specialty shortenings on icing and filling volumes.

Fluid Shortenings Fluid Shortenings are desired by both industry and the home because of handling and measuring convenience. A wide range of products are available in this special category.[36,46] They may exist as fluid suspensions,[2] fluid emulsions on clear liquids. The products have developed in this order and represent a progressive improvement.

The suspension type contains a dispersion of hard stock and emulsifiers in triglyceride oils, commonly soybean or cottonseed. The hard stock provides nuclei for aeration in batter preparation. Although much work has been done to perfect these products, the solid and liquid phases tend to separate during storage causing the shortening to be non-uniform when used.

A typical liquid suspension type shortening is[36,50]:

Ingredient	%
Fat and oil	89.75
Monoglyceride	5.0
Polysorbate 60	0.25
Glycerol lacto palmitate	2.0
Stearyl lactylic acid	1.5
Propylene glycol monostearate	1.5

The next development, the stable O/W emulsion, exhibited the creaming ability of a plastic fat and overcame the separation problem of the suspensions. But the high water content of the fluid shortening emulsion makes it expensive to ship. However, the emulsion is highly functional. A typical fluid emulsion formulation is.[15,36]

Ingredient	%
Vegetable oil	45.60
Mono- and diglycerides	2.29
Sorbitan monostearate	1.45
Polysorbate 60	0.62
Water	49.79
Sodium benzoate and sodium propionate, 50/50	0.25

For a clear liquid shortening product to remain transparent under varied conditions the emulsifiers must be completely soluble in the oils. Although the shortening may solidify at refrigeration temperatures, when it is brought back to ambient conditions for baking it must become clear and homogeneous. Propylene glycol monostearate, properly designed, has the required solubility in vegetable oil to be used at the levels necessary to provide adequate batter aeration. Figure 9 demonstrates the functionality of clear liquid shortenings containing propylene glycol monostearate.

Household Shortening The household or consumer shortening is formulated to be most suitable for frying, the principal consumer use. The shortening manufacturer aims for a product that will have a high smoke point and that will not "spatter." The elevated smoke point is achieved by using low

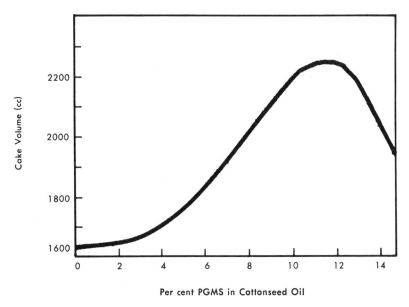

Fig. 9. The relationship of cake volume to emulsifier level using a specialty purpose propylene glycol monostearate in a fluid vegetable oil.

levels, 1.5-1.75% of specially selected surfactants that are low in volatiles, such as free fatty acid, free glycerine, and water.

Meat Products

1. Surfactants play a major role in helping the formulator build quality and improved appearance into the increasingly popular processed meat products.

Surfactants are used to emulsify the meat fats and assist the binders in sausage products.

A good meat binder should have: (1) good water absorption, (2) suitable color, (3) economy, (4) neutral flavor, (5) good physical and chemical stability, (6) ability to emulsify fat.

The common binders used today are cereal flours, dry milk powder, casein, potato flour, soy flour and cracker and bread crumbs. These binders do not all meet these required properties; most fail to emulsify fat adequately or hold enough water. It helps to supplement the binders with low levels of mono and diglycerides (0.5-1.0%). In investigating other surfactants it is important to establish that the product being considered has been or is reasonably sure to obtain Food & Drug Administiration approval. As most establishments producing meat products are subject to inspection by the Meat Inspection Division of the U.S. Department of Agriculture, it is also necessary to determine that the product under consideration has been or could be approved by the Meat Inspection Division. Surface active agents are also regularly used in emulsifying and solubilizing meat flavor.

Miscellaneous

The principal uses of surfactants have been discussed in detail; however, they are extremely important to the preparation, aging or eating properties of many other food products.

Confectioners' Coating   In confectioners' coatings emulsifiers perform two functions. Certain surfactants increase the palatability of the product, while others inhibit "bloom." A blend of two surfactants seem to act synergistically to perform better than each one would if used separately.

Confectioners' coatings resemble chocolate, but most of the cocoa fat has been replaced with another edible fat, usually one with a higher melting point than that of the cocoa fat. Palatability is a problem in these products, because at body temperature the fat used in place of cocoa fat may melt incompletely leaving a greasy or waxy after-taste. This can be minimized by adding low levels of polysorbate 60 and other hydrophilic surfactants to form an emulsion between the fat and the mouth saliva.

"Bloom" may be defined as a mottled discoloration of confectioners' coating preceded by loss of gloss. This defect occurs when the coating starts to melt and a part of the fat migrates to the surface. The cocoa fibers, which impart the color, are left behind inside the coating. On cooling the melted fractions resolidify on the surface to form lighter colored blotches. Accompanying the discoloration is a loss of gloss, brought about by destruction of the fine fat crystal structure previously obtained by tempering.

Sorbitan monostearate inhibits the migration of fat retarding "bloom" and extending the shelf life of coatings. It has recently been found that the surfactant's ability to retard "bloom" is derived from its power to entrap the fat by adsorption. The surfactant forms a monolayer on the cocoa fibers, impeding fat migration. Sorbitan monostearate alone or blended with polysorbate 60, when added to the coatings up to a level of one per cent, provides a greater resistance of the coating to loss of gloss when the coating is "heat shocked."[6,7,9]

Caramels   It has been found that the addition of 0.5 to 1.0% of a hard mono and diglyceride to caramels will improve the chewing quality and reduce sticking to equipment, wrappers and teeth, giving excellent lubrication.

Panned Confections   Panned or coated confectionery products can be improved by adding as little as 0.2% polysorbate 60 to the coating syrup. The emulsifier modifies the sugar crystals to reduce panning time and give coatings with increased opacity.[27]

Protective Coatings   Acetostearin products consisting of di and triglycerides containing one and two acetyl groups respectively, solidify to unique waxlike solids and are used in some protective food coatings.[12] It has been shown that this unusual physical property is associated with the alpha polymorphic form of the fats.[21,52] Spontaneous transformation to the nonwaxy form occurs readily only if the product is a single compound of high purity. At room temperature and below, technical grade products will remain in the waxy form for several years. Their physical properties make the acetylated monoglyceride products potentially valuable as protective coatings for processed meats and for dressed meats that are to be stored at low temperatures. They also show potential as coatings for cheese, fruits, nuts, candy and other foods.

Laboratory tests have shown that these coatings, when properly applied, are effective against moisture transfer and the action of atmospheric gases such as oxygen, nitrogen and carbon dioxide.[33,34] The coatings are normally applied by dipping. Nuts may be coated by spraying or mixing, as long as a thorough continuous coating is obtained and nutmeats are not broken.

Peanut Butter   Peanut butter contains approximately 50% peanut oil suspended in peanut fibers. On standing, the peanut oil separates from the peanut fibers. The separated oil impairs the product's appearance and palatability.

Mono and diglycerides can be added to crystallize part of the free oil during processing and thus prevent the oil and peanut fibers from separating during storage. Other benefits obtained by adding mono- and diglycerides to peanut butter are:

1. improved oil stability
2. gloss appearance
3. excellent spreadability over a wide temperature range
4. versatility in production
5. improved palatability
6. the suggested surfactant concentration is from 1–2.5% depending on the type of monoglyceride selected.

Method of Application—If the monoglyceride is a hard bead it can be dry mixed with sugar and salt and added as a combination blend to the unground peanuts. It is also possible to melt the monoglyceride with an immersion heater, then to pump the melted emulsifier into the mill. Certain types of peanut butter incorporating higher levels of surfactant (2–2.25%) sacrifice spreadability and gloss to obtain extreme stability. An example is peanut butter used in confectionery manufacture.

Margarine   It has been found that the addition of low levels of mono-glycerides improves the stability and palatability of margarine. Normally, bakers' margarine requires a mono-diglyceride-type surfactant to maintain a stable emulsion of the vegetable fat and water during storage, and to improve its performance in baked goods.[22,48]

Salad Dressing, Non-standardized   When a salad dressing of the non-standardized type without egg yolk is desired, a low level of a hydrophilic emulsifier such as polysorbate 60 (0.3%) can be utilized to stabilize the emulsion.

Pet Foods   Soft moist pet foods are rapidly gaining in popularity. It has been found that 1–2.5% monoglyceride will contribute firming and prevent fat separation. The surfactant also acts as a lubricant and aids in extruding the product during processing.

Calf Milk Replacer   Products of this type are being formulated for weaning animals. It has been found that surfactant blends of polysorbate 60 or polysorbate 80 with mono- and diglycerides are functional at 0.2–0.5% based on the weight of fat. Both emulsion stability and palatability are improved. Because of the similarity of the end products, reference may be made to liquid coffee whiteners previously discussed in this chapter.

Flavoring Oils   Flavoring oils are often used in emulsified or solubilized form to reduce the need for alcohol or other solvents and improve dispersion of the flavor. The solubilized flavors, which give clear dilutions, may even show enhanced flavoring action.[10,11,39,42,47,49]

Antioxidants   Surfactants are used to solubilize oil and fat soluble antioxidants to allow the treatment of cereals.[20] This is basically a solubilization problem and as such is related to flavor oil dispersion and solubilization. Usually between one to two times as much surfactant an antioxidant is employed, depending upon the desired degree of dispersion, stability and other additives.

Yeast   Plastic yeast, during formation, must be properly granulated to release the correct amount of water and be properly lubricated to be in a form that can be cut. A low level, 0.1–0.3%, of a moderately oil soluble surfactant controls these properties.

Antifoaming Surface active agents create foam in certain instances and decrease foam in others. General recommendation for aqueous foam systems include liquid monoglycerides.

References

1. Adam, N. K., 1941. *The Physics and Chemistry of Surfaces*, 3rd Ed., London, Oxford University Press p. 214.
2. Andre, J. R. and Going, L. R., 1957. (Proctor & Gamble), U.S. Patent 2,815,286.
3. Becher, P., 1965. *Emulsions—Theory and Practice*, 2nd Ed., New York, Reinhold.
4. Behrens, R. W., 1958. *J. Ag. & Food Chem.*, 6:(1), 20.
5. Brokaw, G. Y., 1962. *Distilled Monoglycerides for Food Foaming and for Starch Complexing*, Canadian Food Industries.
6. Cross, S. T., 1952. U.S. Patent 2,586, 615.
7. Cross, S. T., 1954. U.S. Patent 2,671,027.
8. Cupples, H. L., 1935. *Ind. and Eng. Chem.*, 27:1219.
9. DuRoss, J. W. and Knightly, W. H., July, 1965. Relationship of Sorbitan Monostearate and Polysorbate 60 to Bloom Resistance in Properly Tempered Chocolate, *Manufacturing Confectioner*, 45:50–56.
10. Fabian, F. W., 1949. *Fruit Products Journal*, 28:234.
11. Fabian, F. W. and Kragt, M. N., April 11, 1953. *The Canner*, p. 9.
12. Feuge, R. O., Vicknair, E. J. and Lovegren, N. V., 1952. *J. Am. Oil Chemists' Soc.*, 29:11–14.
13. Frazeur, D. R., 1959. *Ice Cream Field* 73:(5)32.
14. Frazeur, D. R., 1959. *Ice Cream Field* 73:(3)18.
15. Geisler, A. S., 1961. *J. Am. Oil Chemists' Soc.*, 38:306.
16. Gortner, R. A. and Gortner, W. A., 1949. *Outlines of Biochemistry*, John Wiley and Sons, New York, pp. 631–640.
17. Gray, Virginia and Schock, T. J., February, 1963. Effects of Surfactants and Fatty Adjuncts on the Swelling and Solubilization of Granular Starches, *Baker's Digest*, p. 78.
18. Griffin, W. C. and Behrens, R. W., 1952. *Anal. Chem.*, 24:(6)1076.
19. Griffin, W. C., 1965. *Kirk-Othmer Encyclopedia of Chemical Technology*, Vol. 8, 2nd ed., pp. 117–154.
20. Henthorn, L. J., 1967. A Stable Salt-Water Suspension of BHA, *Cereal Science Today*, 12:(2), 49–50.
21. Jackson, F. L. and Lutton, E. S., 1952. *J. Am. Chem. Soc.*, 74:4827–9.
22. Jacobs, M B., 1951. *Food and Food Products*, Vol. II, 2nd Ed., Interscience Publishers, New York.
23. Keeney, P. G., 1960. *J. Dairy Sci.*, 45:(3), 430.
24. Keeney, P. G., 1958. *Ice Cream Review*, 42:(1), 26.
25. Keeney, P. G., 1958. *Ice Cream Field*, 72:(1), 20.
26. Keeney, P. G. and Josephson, D. J., 1958. *Ice Cream Trade J.*, 54:(5), 32.
27. King, G. J., 1952. U.S. Patent 2,591,704.
28. Kloser, J. J. and Keeney, P. G., 1959. *Ice Cream Review*, 42:(10), 36.
29. Knightly, W. H., Lynch, M. J., 1966. The Role of Surfactants in Baked Goods, *Baker's Digest*, 40:(1), 28–31.
30. Knightly, W. H. and J. B. Klis, 1965. 17 Ways to Improve Foods, *Food Processing*.
31. Knightly, W. H., 1963. Surfactants in Food Manufacturing, *Food Manufacturing*, pp. 661–665.
32. Knightly, W. H., 1959. *Ice Cream Trade J.*, 55:(6), 24.
33. Lovegren, N. J. and Feuge, R. O., 1956. *J. Agr. Food Chem.*, 4:634–38.
34. Lovegren, N. J. and Feuge, R. O., 1954. *J. Agr. Food Chem.* 2:558–63.
35. MacDonald, I. A. and Bly, D. A., The Determination of the Optimal Levels of Several Emulsifiers in Cake Mix Shortenings, Presented at the AACC Annual Meeting, Minneapolis, Minnesota, May 2, 1963.

36. MacDonald, I. A., and Lensack, G. P., 1967. Fluid-Liquid Shortening: Formulation and Evaluation in Bakers Cakes, *Cereal Science Today*, *12:*(1).

37. Matz, S. A., 1962. *Food Texture*, *20*, The Avi Publishing Co., Inc.

38. Moncrieff, James, 1966. Emulsifiers in Food Products, *Bakers Digest*, 54.

39. Monteboui, A. J., Halpern, A. and Koretsky, H., 1951. Solubilization of Peppermint Oil, *J. Amer. Ph. Assoc.*, (Pract.) *12:*(1), 1725-6.

40. Osman, *et al.*, 1960. Effects of Fats and Non-ionic Surface-Active Agents on Starch Pastes, *Cereal Chemistry*, *37:*(4), 464.

41. Osman, *et al.*, 1961. Complexes of Amylose with Surfactants, *Cereal Chemistry*, *38:*(5), 449.

42. Pratt, C. D., 1952. *Food Technology*, *6:*425.

43. Sanders, H. J., October 10, 1966. Food Additives, *C & EN News.*

44. Schick, M. J., 1967. *Nonionic Surfactants 1*, Chapter 9, Marcel Dekker, Inc., New York.

45. Schoch, T. J., April, 1965. Starch in Bakery Products, *Bakers Digest*, pp. 48–57.

46. Schwain, F. R., 1965. *Cereal Sci. Today*, *10:*279.

47. Strianse, S. J. and Lanzet, M., December, 1960. Application of the HLB System to the Solubilization of Essential Oils. Proceedings of the Scientific Section of the Toilet Goods Assoc., No. 34, 8–18.

48. Swern, D., 1964. *Bailey's Industrial Oil and Fat Products*, 3rd Ed., Wiley-Interscience, New York.

49. Taylor, W. A., 1947. U.S. Patent 2,422,145.

50. Truax, H. M. and MacDonald, I. A., 1960. A Determination of the Effects of Several Variables on the Performance Characteristics of Shortening Using Statistical Experimental Designs, *J. Amer. Oil Chemists' Soc.*, *37:*(12), 651–657.

51. Valaer, E. P. and Arbuckle, W. S., 1961. *Ice Cream Field*, *77:*(1), 10.

52. Vicknair, E. J., Singleton, W. S. and Feuge, R. O., 1954. *Phys. Chem.*, *58:*94–6.

53. *Atlas Products for the Food and Beverage Industries*, ILG-91, Atlas Chemical Industries, Inc.—now ICI America Inc.

54. *Atlas Emulsifiers for Coffee Whiteners*, Atlas Chemical Industries, Inc.—now ICI America Inc., July 1965.

55. *The Revolutionary Baking Ingredient That Gives Cell Wall Control: ATMUL® 500*, LG-66, Copyright 1959, Atlas Chemical Industries, Inc.—now ICI America Inc.

56. *Dryness in Ice Cream*, LG-68, Copyright 1959, Atlas Chemical Industries, Inc.—now ICI America Inc.

57. *Formulation of Improved Icing and Filling Shortenings with Atlas Food Emulsifiers*, LG-82, Copyright 1964, Atlas Chemical Industries, Inc.—now ICI America Inc.

58. Improved Liquid Shortening, 1969. *Food Eng.*, *41:*(12), 9.

CHAPTER 10

Polyhydric Alcohols

William C. Griffin
Associate Director
Product Development

and
Matthew J. Lynch
Development Manager
Product Development

ICI America Inc.
Atlas Chemicals Division
Wilmington, Delaware

Introduction

Polyhydric alcohols, or polyols, are valuable aids in formulating a wide range of food products. When present naturally, or when added during processing, they can impart one or more of several beneficial characteristics. These effects include crystallization retardation, improvement of stability on aging, control of viscosity or bodying, preservation, solvency, moisture retention, and others.[39,99–102,120,170]

Natural Occurrence

Many of the commonly used polyhydric alcohols occur in nature. The frequency of occurrence appears to be directly related to the carbon-chain length of the polyol. Two- and three-carbon polyols rarely, if ever, occur in nature; four- and five-carbon polyols occur occasionally; six-carbon polyols have been found in many instances; and seven-carbon polyols have been observed in only a few cases. This frequency distribution is probably related to the corresponding sugars and their presence in nature, with some specific exceptions. Only one of the tetritols and two of the pentitols have not been observed in natural foods.[106]

Butylene glycol and glycerine are both reported as occurring in fermented products. The presence of 2,3-butylene glycol in tomato conserves has been reported.[38] Glycerine has been reported to occur in fermented products such as wines and beers.[116] The isomer erythritol occurs in algae,[12,152] grasses,[81] and lichens.[62,79] In the group of polyols known as pentitols, d-arabitol and ribitol are found in nature. Arabitol is reported in lichens[104] and in a species of mushroom,[58] and ribitol is reported in plants.[124,165] Xylose (wood sugar) is one of the most abundant natural sugars in the plant world. However, xylitol, its corresponding polyhydric alcohol, has not been found.[107]

The hexitols, or six-carbon polyols, are the most widespread.[108] Sorbitol was reported in 1877, when it was isolated from the juice of the berries of the mountain ash tree.[27,123] It has since been reported in other fruit berries[80] and in many fruits, including pears, apples, cherries, prunes, peaches, and apricots,[156] in red seaweed,[75] and in *Sorbus commixta*.[8] Strain has presented a listing of a large number of plants in which the sorbitol content was determined.[146] Mannitol was the first crystalline polyol discovered;[126] and, since it is crystalline, it frequently occurs in plant exudates.[83,109] The sap exudate from a tree was the Italian commercial source of mannitol for many years.[21,150,151] Mannitol is also present in seaweed[21] and grasses.[74]

The seven-carbon heptitols are considerably less prevalent in nature. Volemitol is found in algae,[105] lichen,[104] mushrooms,[25] and plants of the primrose family.[24] Perseitol is found in avocados.[112,113]

Definition

The term *polyols* in this discussion will be restricted to those molecules that have two or more hydroxyl groups and have only hydroxyl groups. This will exclude sugars, although they have many hydroxyl groups and, in some instances, exhibit similar properties.[17] The chief difference noted between polyols and sugars is related to the aldehyde linkage present in the sugars.

As a class polyols are more stable chemically and thermally than sugars. They are usually more expensive than sugars; hence, the food industry demands that a polyol contribute a notable desired property to the final product. For these reasons we shall not treat sugars as polyhydric alcohols.

Polyhydric alcohols for our use then are defined as derivatives of aliphatic hydrocarbons formed by the replacement of two or more hydrogen atoms with two or more monovalent hydroxyl groups, each being attached to a different carbon atom.

Both glycols that we shall consider—propylene glycol and butylene glycol—have a longer carbon-chain length than the number of hydroxyl groups. The balance of the polyhydric alcohols to be discussed—triols to heptitols—have an equal number of hydroxyl groups and carbon atoms.

While the polyhydric alcohol family continues to challenge the curiosity of the scientist, only a few polyols are of actual commercial importance in the food industry—propylene glycol, glycerol, sorbitol, and mannitol.

Properties of Polyhydric Alcohols

The properties of polyhydric alcohols may be summed up quite briefly by stating that they are generally water-soluble, hygroscopic materials that exhibit a moderate viscosity at high concentrations in water. For the most part polyols exhibit a sweet taste ranging in sweetness from less than half that of sugar to slightly higher. These properties are shown schematically in Figure 1.

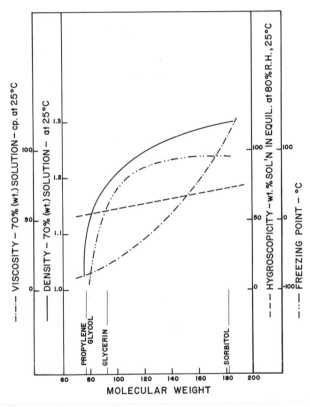

Fig. 1. Polyhydric alcohol behavior vs. molecular weight.

Table 1 shows a comparison of the properties of the four commercial polyhydric alcohols approved for use in foods.* They are all permitted in unstandardized foods, provided the amount used does not exceed that reasonably required to accomplish the intended physical or technical effect. Of course, they are not permitted in standardized foods, unless the standard of identity permits their use. Each case for a standardized food must be investigated separately.

Data in Table 1 show that as the molecular weight increases, the melting points, boiling points and viscosities generally increase. It has also been observed that with increasing molecular weight solvent properties for non-polar

* In addition (A) §121.1057: Polyethylene glycol 6000 is permitted (1) as a binder in plasticizing agents, (2) as an adjuvant in tablet coatings, and (3) as an adjuvant to improve flavor and body work with non-nutritive sweeteners; (B) §121.1114: Xylitol is permitted in some special dietary uses.

TABLE I
Properties of Polyhydric Alcohols and Sugars

	Propylene Glycol	Glycerine	Sorbitol	Mannitol
Molecular Weight	76	92	182	182
Melting Point °C	Supercools	18.6	Metastable 93	166
Boiling Point °C 760 mm	187	290°C Decomposes	Decomposes	Decomposes
Density—25°C	1.036	1.2613	1.49	1.49
Viscosity cp. 25°C	44.0	954	Solid	Solid
Viscosity 70% solution 25°C	10	17	110	Insoluble (Crystallizes)
Hygroscopicity	High	Med-High	Med-Low	Low
Solvency (for oils)	Good	Fair	Poor	Poor
Solubility* in water @25°C	Infinite	Infinite	71%	22%
High Temperature Resistance	Stable, Volatile	Stable, Sl. Volatile	Stable	Stable
Taste	Bitter	Sl. Sweet	Cool, Sweet	Sweet

*grams/100 g. water.

materials decrease. While generalizations of this type are useful, detailed comparative studies have been made and it is now possible to recommend, with some assurance, a specific polyol to contribute a certain function or property to a food. These studies have revealed that polyhydric alcohols act in a special way, each having its own functionality profile.

Most sugars have properties quite similar to that of high molecular weight polyols. However, though they are polyhydroxy compounds they also contain aldehyde linkages that adversely affect their high temperature stability.

Of the various properties of polyols described in Table 1, those most important in food processing are water solubility, hygroscopicity, viscosity, and taste.

Water Solubility Water solubility of the typical polyhydric alcohols appears to be correlated with molecular weight, as well as with crystalline structure and melting point. For example, the lower-molecular-weight polyols, glycerine and, although it is not generally considered in this discussion because of toxicity, ethylene glycol, are infinitely water soluble. As the molecular weight increases, the tendency toward crystallization increases and inherent solubility decreases; common hexitols, for example, range in water solubility from sorbitol at a maximum of 70-75% weight at room temperature to dulcitol at less than 10% at room temperature. The known solubilities of polyols in water are presented in Table 2. The chance to combine polyols of different water solubilities and different viscosities, as well as those with different crystallization characteristics, is a prime tool in modifying crystallization in foods.

Hygroscopicity The hygroscopicity, or the ability to absorb and retain water under conditions of medium and high relative humidity, also varies partially according to molecular weight. Generally, the higher the molecular weight, the less hygroscopic the polyol, although xylitol appears to be an

TABLE 2
Solubility of Polyols in Water

Polyol	Water Solubility, g/100 ml water, 25°C
Propylene Glycol	∞
Glycerol	∞
Erythritol	75
Sorbitol	251
Mannitol	22

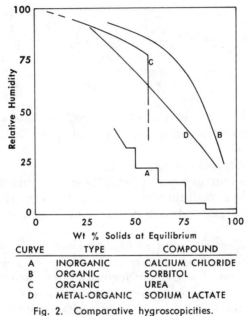

CURVE	TYPE	COMPOUND
A	INORGANIC	CALCIUM CHLORIDE
B	ORGANIC	SORBITOL
C	ORGANIC	UREA
D	METAL-ORGANIC	SODIUM LACTATE

Fig. 2. Comparative hygroscopicities.

Source: W. C. Griffin, R. W. Behrens, and S. T. Cross, 1952, *J. Soc. Cosmet. Chem.*, 3:5.

exception. A non-hygroscopic material does not imbibe water even at high humidities. Many inorganic substances exhibit a "stepwise" humectant action, depending on water of crystallization (Figure 2). Hygroscopic materials such as polyols show a smooth transition curve holding more and more water as the humidity increases.[64,66,171,176] Many organic materials, such as flour, cellulose, and protein, show a modified stepwise curve, holding about 13% (± 5%) moisture over a large range of humidity.[64]

Figure 3 shows the hygroscopicity for the polyols listed in Table 1. The dotted line, in both instances, is invert sugar, which is only slightly hygroscopic. Cane sugar (sucrose) is even less hygroscopic, because it is relatively more crystalline. The effect of crystallization is shown in Figure 3, in the comparison of sorbitol and mannitol. Mannitol, when crystallized (which applies to most of the humidity range), exhibits a very low hygroscopicity, making it useful as a nutritive powder that is stable at high humidity.[171]

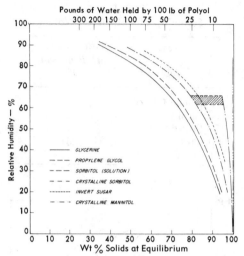

Fig. 3.   Comparative equilibrium hygroscopicities of polyols.

For comparison, consider the usual polyols versus a similar food raw
material, sugar. Equilibrium data for glycerine, invert sugar, and sorbitol are
presented in Figure 3.

Proteins, cellulose, and similar products that exhibit a "shelf" in their
hygroscopicity at about 13% water actually show a diminution in moisture-
holding capacity when small amounts of polyol are added; this is recovered only
when the amount of polyol becomes appreciable with relation to the other
ingredients.[66]

A further consideration with respect to hygroscopicity is whether static
(equilibrium) or dynamic moisture control is the point at issue. Most reported
data, as in Figures 2 and 3, are equilibrium values. Seldom is it practical to
add sufficient polyol to achieve adequate moisture levels or protection from
evaporation at low humidities, or even moderate humidities. This is true primarily because foods inherently have high moisture levels to provide the desired texture, taste, and mouthfeel.

On the other hand, polyols in many instances can help control the rate
of moisture gain and loss.[66, 171] Generally, the higher the molecular weight,
the lower the rate of change (Figure 5). Note, though, that these rate change
times are short term hours, not days or weeks; and moisture levels over longer
periods of time are not controlled under these circumstances.

Dynamic hygroscopicity, the rate with which moisture is gained or lost, is
far more difficult to measure accurately. Hygroscopicity is often studied by
placing a small amount of liquid, unstirred, in a crystallizing dish or beaker
and exposing it to a higher or lower humidity (Figure 4A). If the sample is carefully examined after this exposure, striation lines will be observed showing
that the surface concentration has changed because of either gain or loss
(Figure 4B). At this point change in weight of the sample is related not only to
the rate of gain or loss of moisture at the surface of the sample but also to the
rate of moisture transfer within the bulk of the sample, which is, in many
instances, the controlling function. The ideal situation would be to present

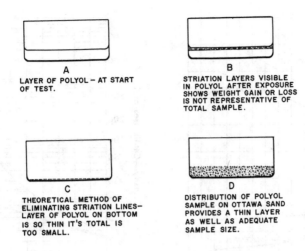

A
LAYER OF POLYOL – AT START OF TEST.

B
STRIATION LAYERS VISIBLE IN POLYOL AFTER EXPOSURE SHOWS WEIGHT GAIN OR LOSS IS NOT REPRESENTATIVE OF TOTAL SAMPLE.

C
THEORETICAL METHOD OF ELIMINATING STRIATION LINES– LAYER OF POLYOL ON BOTTOM IS SO THIN IT'S TOTAL IS TOO SMALL.

D
DISTRIBUTION OF POLYOL SAMPLE ON OTTAWA SAND PROVIDES A THIN LAYER AS WELL AS ADEQUATE SAMPLE SIZE.

Fig. 4. Hygroscopicity determinations.

essentially a layer of polyol only a few molecules thick, as in Figure 4C. Since the sample would be too small to note meaningful weight changes, it appears most expedient to disperse a small sample on Ottawa sand and observe the rate of moisture change (Figure 4D).[173] Even with this method, it is necessary to take precautions that the samples are exposed to a uniform draft, maintained, of course, at a constant temperature and constant relative humidity. Following these techniques, valid dynamic hygroscopicity information can be obtained. Figure 5 shows data obtained on sorbitol and glycerine by this technique.

Viscosity   The viscosity of polyhydric alcohols in aqueous solution is of importance in food applications because of the bodying effect that is conveyed. Viscosity generally increases with increasing molecular weight. Viscosity is, of course, a function of concentration and temperature. Room-temperature viscosity data for a number of the major polyols are presented in Figure 6.

Taste   Although the polyhydric alcohols are a relatively homogeneous family of compounds, their taste characteristics can vary from the decided bitterness of propylene glycol to the sweetness of sugar. Isomers in any one of the higher classifications can show tremendous differences in sweetness, although their overall molecular weight and many other characteristics are similar. However, none of these polyhydric alcohols should be classed as synthetic sweeteners. They range in sweetness from half as sweet to essentially three-quarters as sweet as sugars. The results of a recent study of relative sweetness of polyols and sugars are presented in Table 3[161] with a comparison of some earlier published data. This comparative property has been the subject of many studies.[32, 35, 52, 57, 69, 135, 153, 160]

Reasons for Using Polyhydric Alcohols in Food

Polyhydric alcohols are added to manufactured foods either to promote the retention of the original quality of the food on aging and shipment to

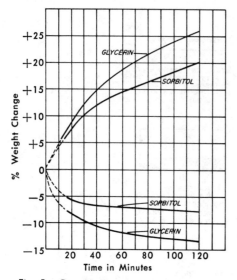

Fig. 5. Comparative dynamic hygroscopicities of polyols.*

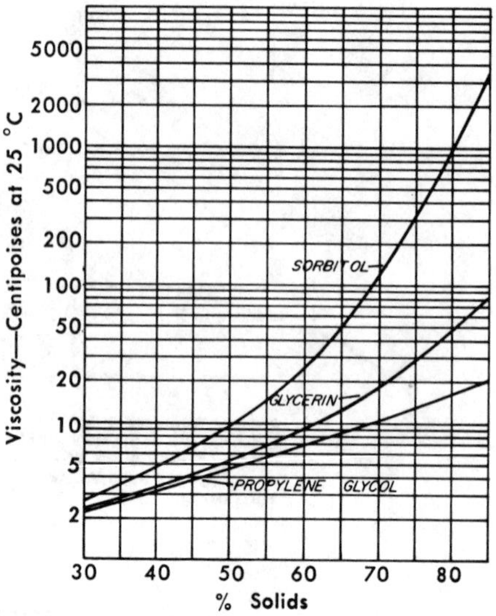

Fig. 6. Comparative viscosities of polyol
solutions.*

*Source: *Atlas Sorbitol and Related Polyols*, CD-60. Copyright 1951, Atlas Chemical Industries, Inc.—now
ICI America, Inc. Revised June 1953.

TABLE 3
Relative Sweetness of Polyols and Sugars (Sucrose = 1)

Polyol or Sugar	Relative Sweetness
Fructose or Levulose	1.4–1.7
Invert Sugar	1.0–1.3
Sucrose	1
Dextrose, anhydrous	0.7–0.8
Xylitol Iditol Maltitol	0.7
Dextrose, monohydrate	0.6–0.7
Galactose Corn Syrup—enzyme converted	0.6
Glucose Hydrate Sorbitol	0.5–0.6
Mannitol Dulcitol Inositol	0.5
Erythritol Lactitol	0.4–0.5
Xylose	0.4
Corn Syrup, unmixed Maltose	0.3
Raffinose	0.2
Lactose	0.1–0.2
Erythritan	0

the consumer, or to gain a texture or product quality that was not present in the original formula. These qualities are achieved through physical or chemical effects in which the polyols function variously, as follows:

1. Viscosity or Bodying Agents
2. Crystallization Modification
3. Taste or Sweetness
4. Hygroscopicity or Humectancy
5. Solvency
6. Rehydration Aids
7. Sequestering
8. Antioxidant
9. Microbiological Preservation
10. Softening
11. Bulking Agents
12. Dietary Foods

Viscosity or Bodying Agents   While the viscosity effect of a dilute polyol is minimal when compared with viscous liquids, it is apparently an effect that the tongue and other sensory receptors of the mouth are able to discern. Relatively small proportions of polyols added to beverages convey an improvement in mouthfeel that is described as bodying action. Similar effects are obtained with thickening agents, although frequently their overall flavor characteristics are not as desirable as those of polyols.

Polyols may be employed in some foods to increase their viscosity and in others to effect a reduction. Sorbitol is effective in increasing viscosity because of the inherent high viscosity of its aqueous solution. Propylene glycol represents the other end of the viscosity scale. Often the best control of viscosity can be achieved by a blend of glycerine, sorbitol, and sugars.[162]

Crystallization Modification   (related to water solubility) Many foodstuffs are dependent on a semi-equilibrium mixture of sugar crystals and sugar syrup for their texture characteristics, especially in the field of confectionery. These products include the typical creams, fondants, and fudge. Because the crystallization continues in storage, this type of product exhibits a limited shelf-life with reference to texture. It has long been standard practice to add invert sugar as a "doctor" to maintain the desired consistency. It has been found that the addition of glycerine[67,132] and sorbitol,[3,5,13,14,37,47-49,70,88,140] when properly employed, can increase the shelf-life by further complexing the crystalline nature of the confection, thus reducing its tendency to harden. DuRoss has studied the development of sugar crystals with and without the addition of sorbitol and has shown a beneficial difference.[47-49] This reduction in crystallizing tendencies is also of value in the production of marshmallow and nougat where the crystallization inhibitory action provides advantages in processing.[46,47,61,167] Polyhydric alcohols are included in military specifications for shelf-life improvement.[174]

Taste or Sweetness   Taste is an unusually complex property, comprising flavor, texture, temperature, mouthfeel, and many other factors. The actual taste of polyols is generally of little consequence when they constitute a minor additive. When a polyol is a major component, such as in "sugar-free" candies (see p. 448, Special Dietary Foods), it usually is the major source of sweetness. Sorbitol and mannitol are especially good in this application.

On the other hand, even in minor amounts polyols may exert a decided effect on taste. Polyols have also been used to modify the sweetness of a product rather than to create sweetness. Sorbitol has been reported to cause a taste improvement when used with saccharin[157,159] by inhibiting the strong bitter characteristic that is correlated with saccharin.[76,157] In wine a small amount of sorbitol exhibits a distinct smoothing and bodying action, probably due to a combination of viscosity and sequestrant (or complexing) action.[18]

Hygroscopicity or Humectancy Humectancy, hygroscopicity, or moisture-holding power of a polyol added to a confection has been reported to be of importance in maintaining freshness.[67] It is believed that the effect of crystal modification discussed in the previous section is of far greater importance. This belief is based on the premise that the polyols that are employed cannot sufficiently influence the moisture-holding power of the confection to create the indicated effect. The loss of moisture has been shown not to be a primary factor in bread or cake staling.[78] Usually, polyols are added at considerably less than 10% of the weight of the confection; if they were twice as effective at holding moisture, they would tend to raise the moisture-holding power by less than 5%. Humectancy is important in the processing of marshmallows,[46,61,167] where the rate of moisture loss to the casting starch over a specific time interval is reduced.

In a few instances the reverse of hygroscopicity is desired, as in the dusting of chewing gum. In this instance the low hygroscopicity of crystalline mannitol is a distinct advantage and is extensively used, blended with starch. A further advantage is the cool, sweet taste also exhibited by mannitol.[171]

Solvency Solvent action of polyols increases rapidly with decreasing molecular weight; thus, propylene glycol is the most potent solvent of the food-grade polyols. Glycerine is the next best solvent, but it is already high enough in molecular weight to be used only rarely as a solvent.[55,82,97,99-102,143] Sorbitol and mannitol have been used in various ways: (1) as flavor carriers or flavor-encapsulating agents;[42,65,133,147] (2) as flavor enhancers in a wide variety of products, such as coffee concentrates,[53,114] meat-curing compositions,[72,73,86] flavor additives for nuts,[9-11] and pure juice concentrates;[45,149] and (3) in a number of flavor-enhancing compositions.[42,141,166,169]

Rehydration Aids The dehydration of foods is of value in preservation and reduction in weight for shipping. Unfortunately, in many instances food dehydration causes difficulty in rehydration, and the reconstituted food is significantly different from the original foodstuff. Many years ago the use of a hexitol to improve the rehydration and terminal characteristics of vegetables was described.[28] More recently the armed-services laboratories have determined that the inclusion of a small amount of a polyol blend during dehydration will allow a marked improvement in quality of rehydration characteristics.[29] It is probable that the polyol avoids the total collapse of the cellular structure during dehydration and keeps it in a better form for acceptance of water at the time of rehydration.

Sequestering The hexitols have been shown to have a mild sequestering action, although it is not comparable to EDTA.* This sequestering behavior shows up as a reduction in wine precipitate.[18] Sorbitol has also been used to advantage in fruit beverages.[51]

---

* EDTA—Ethylene diaminetetraacetic acid.

Antioxidant The objectionable taste called rancidity may occur as a result of one or both of two chemical reactions. The first is oxidation of double bonds, catalyzed by heavy metals; the second is hydrolytic rancidity.

Hexitols exert a mild sequestering action, as noted in the previous discussion of sequestering; in a few instances, particularly with natural oils present such as butter, a mild resistance to rancidity may be observed.

Polyols may also aid in hydrolytic rancidity, glycerine having been reported to retard free fatty-acid formation and thus reduce the rancidity tendency.[130]

Microbiological Preservation Polyols as well as sugars act as preservatives at high concentrations based generally on osmotic-pressure effects. These concentration levels are usually at greater than 75 weight per cent to be effective. An exception to this is the effectiveness of propylene glycol as a preservative. Often propylene glycol is effective at a level as low as 10 per cent. In many instances combinations of propylene glycol and higher molecular weight polyols are employed.[15]

Softening The softening effect of polyols, also referred to as plasticizing, is primarily related to their moisture-holding power, or humectancy. True softening, or plasticizing, is required to a lesser extent in foods than in other products. Softening is closely allied with hygroscopicity and the ability of the polyol to hold moisture. However, at low concentrations of polyols in aqueous systems, the moisture-holding power of the combination is usually less than the calculated combined holding power of the polyols and the substrate. This is probably best explained by the hypothesis that polyols satisfy some of the hydrogen-bonding capacity of the vehicle, just as water does at higher moisture levels. At low moisture content a good softener or plasticizer will still exhibit a softening effect. In considering a polyol for this use, a general rule to follow is that the lower the molecular weight, the better the polyol will plasticize or soften.

Bulking Agents If artificial sweeteners are used instead of sugar, one of the immediate problems that is encountered is the reduction in solids content, or the change in ratio of solids of sweetener to other ingredients. For example, only a few milligrams of artificial sweetener are equivalent to one ounce of sugar. It is frequently possible to formulate powdered beverage concentrates without a bulking agent, because the beverage concentrates contain acidulants that can act as carriers for the flavor and artificial sweetener. In foods such as ice cream, cakes, cookies, and confections, the problem is pronounced. Elimination of the sugar gives a totally unbalanced formula that does not behave properly and results in an unpalatable end product. Polyols such as sorbitol and mannitol are the most commonly used bulking agents.

Dietary Foods Polyols are used in dietary foods as replacements for sugars. The metabolism of hexitols in comparison with sugars has been studied, although not extensively.[54,122,145] Studies have shown that sorbitol is less readily attacked by *Lactobacillus* than sucrose—thus, possibly reducing potential tooth decay.[33,59,60] At times processors have labeled their products "sugar-free" when using the hexitols in place of sugar. This may have some valid basis on the above, but often the inference leads the consumer to believe the product contains fewer or no calories. This is not true, since basically the hexitols have the same caloric value as sugar—and this is true of the polyols in general. Certain polyols (mannitol and dulcitol) actually afford fewer absorbed calories because of the lack of solubility. The major uses in dietetic foods have been in confectionery products.[4,93,96,98,115,125]

Selection Methods

There is no easily followed set of instructions or method in the selection of polyols for food applications. Property improvements that are desired should be considered and compared with the reasons for use presented above. Based on the desired functionality indicated, polyols should be evaluated at probably what initially would appear to be higher than desirable levels. This type of evaluation will give an indication as to whether or not the inclusion of a polyol will have any effect on the formula. Should a desired effect be observed, retrial at lower levels will allow a choice of concentration that should be suitable.

A brief summarization of behavior characteristics and application data has been assembled in Table 4 for use as a starting reference to simplify the selection of a polyol to perform a specific task. The formulas presented on pages 443-451 will serve as examples to illustrate these points.

TABLE 4
Guide for Choosing Polyhydric Alcohols

	Propylene Glycol	Glycerine	Sorbitol	Mannitol
Crystallization Modifier		x	x	
Humectant	x	x	x	
(Moisture Resistant Dust)				x
Plasticizer		x		
Bodying Agent			x	
Solvent	x			
Bulking Agent			x	x
Rehydration		x	x	

Equipment

Most of the polyols above the three carbon-chain length are available in either crystalline form or a solution or syrup. The choice of liquid or solid will depend on the economics of handling and the desired moisture content of the final products. Where a low final moisture content is desired, for example, as in some forms of tableted mints, it may be desirable or even necessary to maintain a relatively uniform low humidity in the manufacturing area throughout the year. Seldom is it necessary to go below approximately 40 per cent relative humidity to gain excellent processing characteristics.

Food Uses of Polyhydric Alcohols

Polyols are used in conventional foods for varied reasons, based on the many functionally different effects they exhibit. Applications in conventional foods will be presented according to the type of food or food ingredient in which they are used. Polyols employed in dietary foods are sufficiently different to warrant separate discussion.

Flavor Concentrate

Propylene glycol is the edible polyol of choice to use as a solvent for flavor compounds.[82,97,102,143] A typical formula follows:

SINGLE FOLD VANILLA FLAVOR CONCENTRATE

Ingredients:	
Vanillin	13.5 mg
Propylene Glycol	q.s. to one gallon

Glycerine is employed to some extent as a flavor vehicle, although it exhibits less solvency than propylene glycol. The polyols replace the formerly used alcohol and provide a more uniform flavor level because of lack of volatility of the carrier. Propylene glycol is also used as a carrier for emulsifiers.[94]

Confections

As mentioned previously, polyols contribute significantly to improving both "as made" quality and shelf life of confections based on sugar.[47,85,92] When properly blended with sugar, polyols act as "doctors," or crystallization modifiers, and exert beneficial texturizing and stabilizing influence on the mass. Example formulas follow:

FONDANT

Ingredients:	
Sucrose	75%
Corn Syrup (80% solids) 43 D.E.	15
Sorbitol Syrup (70% Sorbitol)	10

Preparation:
1. Mix all ingredients.
2. Boil to 244°F.
3. Pour onto a water-jacketed fondant mill.
4. Allow to cool quiescently to 100°F, then agitate.
5. Homogenous nucleation begins approximately six to nine minutes after agitation is started.
6. Agitate until a proper fondant consistency is obtained (15–18 minutes).

BAR FUDGE

Ingredients:	
Corn Syrup (42 D.E.)	23.42%
Sweetened Whole Milk	16.60
Sucrose	23.60
Hard Fat (92°F)	3.57
H_2O	approx. 3.00
Salt	1.00
Fondant (80–20)	23.50
Flavor	q.s.
Sorbitol Syrup (70% Sorbitol)	5.00

Preparation:
1. Cook corn syrup, water, sweetened condensed milk and sorbitol syrup in a steam jacketed kettle to 244°F.
2. Cool to 201°F. Add fondant, fat, and flavor; mix well.
3. Cast into foil pans.

HIGH COOK CARAMEL

Ingredients:

A)   Caramel

Sucrose	19.3%
Corn Syrup	36.9
Condensed Milk (Eagle)	29.6
Butter	3.5
Salt	0.8
Caramel Paste	3.3
Lecithin	0.04
Invert Sugar	3.6
Sorbitol Syrup (70% Sorbitol)	3.0

Ingredients:

B)   Caramel Paste

Corn Syrup 63 D.E.	28.8%
Sucrose	24.7
Condensed Milk (Eagle)	11.8
Starch, 60 Gluidity	9.04
Vegetable Oil (70° MP)	6.02
Butter	1.36
Vanilla	.09
Water	18.27

Preparation:

A)   Caramel

Ingredients are weighed into a Groen kettle and heat is gradually applied to slowly warm the batch to 150°F. Care is exercised to prevent any sticking or scorching of ingredients. When the batch is homogeneous and at a temperature of 150°F, heat is applied and the batch cooked to a temperature of 255°F. The batch is cooled to 245°F and maintained at this temperature while being cast into starch (90°F). After 24 hours the caramels are removed from starch, dusted and enrobed with a dark sweet chocolate. The samples are then boxed.

In a high cook caramel, the primary function of sorbitol is to improve the tenderness of the caramel without reducing its machinability, i.e., flattening out when removed from starch, and to increase the shelf life.

B)   Caramel Paste

Weigh up all ingredients in stainless bowl. Warm, using steam, to 150°F (with steam), using agitation to keep starch and condensed milk from sticking and burning.

When paste has been warmed to 150°F and checked to insure it is completely homogeneous, heat is again applied and the paste cooked to 220°F, cooled and held for caramel.

PEPPERMINT PATTIES (Creams)

Ingredients:

1. Bob Syrup

Sucrose	19.94%
Corn Syrup (65 D.E.)	10.97
Sorbitol (70% sorbitol)	1.00
Water	3.96

Cook to 246°F (Keep container covered to wash sides down)

Remelt

Fondant (80/20), 242°F cook)	47.86

Add fondant at 175°F

Mazetta	15.96
Invertase	0.10
Flavor	0.21

2. Mazetta

Corn Syrup	58.19
Sucrose	29.09
Water	7.27
Egg Albumin	1.82
Water	3.63

Cook to 238°F, whip to a density of 4 lb/gal.

Preparation:

Boil bob syrup to required temperature and cool with agitation to 150°F. Add fondant and disperse well. Re-adjust temperature to 150°F. Add egg cream and disperse well. Cast cream in starch mold at 153°F. Let stand in starch trays for 3 hours. Enrobe with chocolate; package and store. MAKE SURE REMELT IS AT PROPER TEMPERATURE (153°F).

CAST MARSHMALLOWS

Ingredients:	%
Sucrose	42.47
Corn Syrup	25.49
Gelatin	0.89/4.30
Fondant	10.93
Sorbitol Syrup (70% sorbitol)	2.97
H_2O	12.94
Color	q.s.
Flavor	q.s.

Preparation:

Heat sucrose, corn syrup, sorbitol syrup, and water to 180°F (just enough heat to dissolve ingredients). Agitate the solution on a Hobart mixer on low speed. Cool the batch to 155°F, so that the fondant is thoroughly incorporated. Agitate the solution again at high speed; slowly add the gelatin-water mixture during agitation. Whip the total batch to a density of 4 lb/gal. Add flavoring and color. Cast the marshmallow into starch at a temperature of 115–120°F.

Chewing Gum

Glycerine has been reported as a lubricant additive to chewing gum as an aid in processing.[87,111,134] Sorbitol and mannitol are used in the preparation of noncariogenic chewing gums[20,30] (discussed in greater detail under Dietary Foods, p. 448).

Dried Roasted Nuts

A process has been developed for the production of a low fat-containing dry-roasted nutmeat confection. The nuts are roasted in a molten mixture of sorbitol and mannitol.[9,10]

Meat Products

Sorbitol has been described as an additive to meat emulsions (sausages) and bacon to improve color and flavor retention.[86] Sorbitol has also been described as an additive for meat-curing compounds[72,86] and as an additive to meat casings.[6,163]

Pet Foods

Polyols are added to the now popular moist pet foods to plasticize the mass, enabling it to retain the desired texture over a long period of time. A typical formulation follows:[31]

SOFT MOIST PET FOOD

Ingredients:	Parts by Weight
Tripe	18.0
Fish (whole cod and smelt)	6.0
Beef cheek trimmings	6.0
Soy flakes	31.5
Dry corn syrup solids (42 D.E.)	21.4
Soy hulls	3.0
Dry non-fat milk solids	2.5
Bone meal	2.1
Dicalcium phosphate	1.4
Propylene glycol	2.0
Sorbitol	2.0
Tallow	2.0
Mono- and diglycerides	1.0
Sodium chloride	0.6
Potassium sorbate	0.3
Minerals, vitamins, color, etc.	0.3

Icings and Toppings

Polyols are of value in the production of icings and toppings as crystallization retarders.[44,168] A suitable formula illustrating this use follows:

CAKE ICINGS (Low-fat)

Ingredients:	Per cent
Sugar	63.1
Shortening (all-purpose)	17.4
Water	12.5
Corn syrup solids (42 D.E.)	5.0
Crystalline sorbitol solution 70%	2.0

NOTE: The corn syrup/crystalline sorbitol solution 70% ratio may be changed as follows:

Corn syrup solids	4–8%
Crystalline sorbitol solution 70%	2–4%

In cake icings and similar systems where sugar crystals exist in a saturated aqueous media, sorbitol at 1–3 per cent, based on sucrose weight, aids in retarding crystal growth and the development of brittleness and coarse texture associated with large crystals.

Coconut

Propylene glycol, glycerine, and sorbitol are used as plasticizing agents for moist, shredded coconut. While this softening action can be gained from a suitable blend of sugars, the polyols have a decided advantage of contributing color stability. Sugars on coconut undergo a gradual darkening or browning that does not occur with the polyols. The inherent mild sweetness of sorbitol provides a most acceptable end product.[89,164]

Miscellaneous Food Products

The use of polyols in cereals and edible containers, such as ice-cream cones,[118] is claimed to improve firmness and friability.[19] Glycerine has been reported as exerting an anti-staling action on bread.[26]

A number of fruit and vegetable coatings have been developed containing polyhydric alcohols[129,155,158] and glycerine has even been reported as an ingredient in an apple spray that favorably influences the color of the apple.[144] A closely related subject is the edible films for wrapping and packaging foods in which glycerine is described as a plasticizer.[172]

The objectionable caking of brown sugar on storage is reported to be relieved by the addition of glycerine[63] and sorbitol.[40]

Dairy products are improved by the addition of propylene glycol, sorbitol, or mannitol to milk powder.[91,127] The keeping qualities of cheese, butter, and dried and condensed milk are reported to be improved by the addition of glycerine.[84]

A pressurized salad dressing containing sorbitol has been described,[2] as has also a low-moisture pastry dough.[95] The addition of glycerine and sorbitol to egg whites for preservation and improved whipping action has been reported.[90,137,138]

Glycerine alone and in combination with sorbitol has been found to be of value in the freeze-drying of foods.[131] Hexitols have also been described as being of value in the preparation of dehydrated foods.[28,34,71,117] A possible similar action of glycerine inhibiting the protein denaturation in frozen fish has been described,[110] and glycerine has been mentioned as an aid in producing frozen fruits and vegetables and edible emulsions.[16,77] At the other extreme glycerine has been proposed as a means of elevating the boiling point of a heat-exchange liquid used for the sterilization of canned goods.[23] Glycerine and sorbitol are claimed to aid in the production of dry and powdered yeast.[1]

Beverages

Fruit Juices   Glycerine, sorbitol, xylitol, and inositol are reported as exhibiting improved flavor and keeping qualities to fruit and vegetable juices.[36,121,128,154] Sorbitol also has been used in the production of a powdered orange juice.[119,148]

Low-Calorie Beverages   In recent years the general population has become more and more concerned about daily caloric consumption. The food industry has attempted to satisfy this need by marketing low-calorie beverages in both carbonated and fruit-drink forms. In these systems the sugar is reduced or omitted, and synthetic sweeteners are substituted. However, the sugar omission from the beverage reduces both sweetness and "body", presenting a different taste sensation. Synthetically sweetened beverages lack this body. Also, at the levels required in the beverage, the synthetic sweeteners have a slightly different aftertaste, sometimes termed "metallic".[157,159] It has been demonstrated that a relatively small amount of sorbitol (on the order of 1.0–3.0 per cent) will provide a sense of body as well as an improvement in residual taste.

A few general principles should be considered in formulating a dietetic beverage:

1. Select an acid or acid blend such as citric-phosphoric to obtain proper pH or taste.
2. Add bodying agents to acquire proper viscosity and mouthfeel.
3. Select proper carbonation level.
4. Formulate the flavor to mask the "sugar-free" shortcomings.

TYPICAL LOW-CALORIE CARBONATED BEVERAGE[175]

Ingredients:

Lemon Concentrate	
Terpeneless Lemon Oil	1 fl. oz.
Potable Alcohol	42 fl. oz.
Citric Acid Crystals	1.5 oz.
Artificial Sweetener	q.s.
Water, sufficient to make	1 gallon

Procedure:
Step I
1. Mix oil and alcohol and 42 fluid ounces of water.
2. Add sweeteners and citric acid.
3. Add enough water to make one gallon and agitate to dissolve all solids.

Step II
   Dilute the above concentrated flavor to use as a bottling concentrate. Take 375cc of flavor concentrates and add water to make one gallon. The "throw" of this solution is one fluid ounce per twelve-ounce bottle.

Step III

Carbonation is achieved by filling each twelve-ounce bottle with carbonated water 3-3½ volumes of gas. Refrigerate at a constant temperature to decrease frothing during bottling and minimize the loss of gas.

NOTE: Bodying agents such as sodium carboxymethylcellulose (0.02–0.05%), sorbitol (1–2%), pectin, and propylene glycol alginate are optional ingredients that can be added to improve viscosity and mouthfeel. They may be incorporated at any step of the formulation.

Wines   It has been reported that a low concentration (1–2 per cent) of sorbitol added to a table wine improves taste and also reduces sedimentation tendencies.[18] This is presumably a combination of bodying action as well as sequestration.

Special Dietary Foods

Special dietary foods have been designed to help reduce the consumption level of protein, fat, carbohydrate, and calories.[56] When conventional foods are modified for dietary purposes, such properties as taste, texture, and body are sometimes adversely affected.

Polyhydric alcohols have been and continue to be investigated as suitable materials that can be added to dietary foods to improve consumer appeal. Over the last several years, such properties as caloric value, sweetness equivalent, noncariogenic potential and the fermentability of polyhydric alcohols have been investigated. Sorbitol, glycerine, and propylene glycol have essentially the same nutritive value as sugars, namely four calories per gram. At least one worker has reported the nutritive value of mannitol at two calories per gram or half that of glucose.[7] The sweetness values have been determined, and, while propylene glycol and glycerine are described as bitter or bittersweet, sorbitol has been reported to be about 60 per cent as sweet as sugar.[135] The comparative sweetness of a number of polyhydric alcohols and sugars has been determined in comparison; these data are presented in Table 3.

SUGARLESS CHOCOLATE

Sugar is replaced with sorbitol and/or mannitol in the manufacture of a sugarless chocolate.[103]

Ingredients:	Range, %
Nonfat Dry-Milk Solids	2–6
Vegetable Fat (102°F)	30–35
Sorbitol	30–40
Cocoa Powder (12% fat)	12–20
Mannitol	3–10
Lecithin	0.25–0.30
Flavor	q.s.

Preparation:
1. Mix sorbitol and mannitol with a portion of the vegetable fat. Heat to 100–110°F. *Caution:* Do not exceed 115°F.
2. Add cocoa powder and nonfat dry-milk solids to the fat mixture.
3. Refine to a maximum particle size of 50 microns.
4. Add remainder of the fat.
5. Add the flavor and sufficient lecithin to bring to desired viscosity.

The noncariogenic properties of sorbitol and glycerine have been investigated by *in vitro* methods. There is some indication that, based on the fact that sorbitol is fermented more slowly than sugars, it may be less cariogenic when ingested orally because it is washed off the teeth or diluted with saliva before the critical pH is reached.[41, 68, 139, 142]

SUGARLESS CHEWING GUM

Ingredients:	Range, %
Gum Base	25–26.0
SORBO (70% Sorbitol Solution)	16–17.0
Crystalline Sorbitol	36–40.0
Mannitol	14–17.0
Glycerine	0.5
Flavor	1.5
Artificial Sweetener	q.s.

Preparation:
1. Feed gum base into horizontal sigma blade mixer and heat with steam for five minutes at 60°C. Avoid overheating gum base.
2. When base has softened, blend in SORBO solution and glycerine.
3. Turn on mixer and steam.
4. Begin adding crystalline sorbitol and mannitol. Add slowly (⅛ at a time), incorporating well after each addition.
5. Mix all ingredients until uniform.
6. Add flavors.
7. Remove gum from mixer after flavors have been added.
8. Roll on proper gum roller.
9. Roll gum to 0.075 inch thickness.
10. Dust with mannitol and store at 75°F, 45% R. H. for 24 hours.
11. Wrap as desired.

SUGARLESS HARD CANDY[50]

Ingredients:	
Crystalline Sorbitol	95.0%
Crystalline Mannitol	5.0
Flavor	q.s.
Color	q.s.
Water	q.s.

Preparation:
   Dissolve in water using one part water for every two parts dry material, i.e., 300 grams dry powder in 150 grams H_2O. Warm on steam bath until dry powders are in solution; then place on heat and heat to 385°F. Cool to 250°F; stir in color, flavor, and cool to 200°F. Set time will be 10-14 hours.

SUGARLESS MINTS

Ingredients:	
Crystalline Sorbitol	96–98.0%
Magnesium Stearate	q.s.
Flavor	q.s.

Preparation:
1. Granulate sorbitol and screen to particle size that will have good flow properties.
2. Dry blend in lubricant and compress on any suitable tablet machine to desired size and hardness.

DIETETIC ICE CREAM

Ingredients:	
Protein	6.0%
Butterfat	10.0
Stabilizer (pure)	0.4
Carbohydrate (milk sugar)	4.7
Sorbitol Crystalline	10.5
Flavors	q.s.
Artificial Sweetener	q.s.

Procedure:
   Normal procedures and equipment for preparing ice cream are satisfactory for the preparation of this product.

SUGARLESS SPONGE CAKE

Ingredient	Grams	Per cent
Flour	150	18.73
Sorbitol	255	31.84
Egg Whites	90	11.23
Egg Yolks	60	7.49
Baking Powder	8	1.0
Vanilla	10 cc	1.25
Margarine	45	5.62
Skim Milk, boiling hot	183	22.84
	801	100.0

Procedure:

1. Sift flour, baking powder, and one-half of the sorbitol.
2. Whip egg whites stiff with one-half of remaining sorbitol.
3. Beat egg yolks with remainder of sorbitol.
4. Melt margarine in hot skim milk and add to No. 3.
5. Add No. 1 to Nos. 3 and 4 and mix well.
6. Fold in egg whites (No. 2).
7. Bake at 375°F for about 22 minutes.

SUGARLESS COOKIE FORMULA

Ingredient	Grams	Per cent
Eggs	100	20.2
Vegetable Oil	55	11.1
Vanilla	10 cc	2.0
Sorbitol	72	14.6
Avicel (Crystalline Cellulose)	19	3.9
Flour (all-purpose)	228	46.2
Baking Powder	8	1.6
Salt	2	0.4
	494	100.0

Procedure:

1. Beat eggs. Stir in oil and vanilla.
2. Sift together sorbitol and Avicel; add to above until mixture thickens.
3. Sift together flour, baking powder, and salt; stir into above mixture.
4. Drop onto ungreased baking sheet.
5. Bake 8–10 minutes or until a light-brown color in 400°F oven.

SUGAR-FREE GELATIN DESSERT

Ingredients (dry blend):	Per cent
Gelatin (225 bloom)	56.2
Sorbitol	19.0 (approx.)
Fumaric Acid	10.6
Artificial Sweetener	q.s.
Sodium Citrate	3.5
Salt	3.0
Flavor and Color, as desired	

Directions for use:

Bring 2 cups (16 oz.) of water to a boil and add package contents (3 oz.). Stir briskly to dissolve completely. Continue stirring for one minute. Gel will be firm in 1 to 1½ hours. Chill in refrigerator.

SUGAR-FREE INSTANT PUDDING[22,43]

Ingredients (dry blend):	Per cent
Nonfat Dry Milk	60.0
Pregelatinized Starch	18.0
Sorbitol	13.0
Disodium Orthophosphate	3.6
Tetrasodium Pyrophosphate	3.6
Artificial Sweetener	q.s.
Carrageenan	0.7
Flavor and Color, as desired	

Directions for use:
Add the contents of one package (4 oz.) to 2 cups of cold water or milk. Mix well and chill until set.

A variety of other dietetic foods, such as gum drops, fruit packs, and others, are also commercially prepared. In general synthetic sweeteners are substituted for sugar, and sufficient polyol is added to provide the necessary bulking or bodying action.

References

1. Aizawa, M., *et al.*, 1968, Process for Preparing an Active, Dry Powdery Yeast, U.S. Patent 3,407,072.
2. Akerboom, J., *et al.*, 1958, Pressurized Food Dressing, U.S. Patent 2,844,470.
3. Alikonis, J., 1957, Aeration in Candy Technology, *Manuf. Confect.*, *37*(5): 35-37.
4. Alikonis, J., 1956, *Food Eng.*, *28*(3): 92-93.
5. Alikonis, J., June 1951, How to Extend Shelf-Life in Confections. *Manuf. Confect.*
6. Allen, W. M., 1966, Preparation of Meat Products, U.S. Patent 3,275,452.
7. Ariyama, T., and Takahashi, K., 1929, *J. Agr. Chem. Soc., Jap.*, *5:* 674.
8. Asahino, Y., and Shimoda, H., 1930, *J. Pharm. Soc. Jap.*, *50*: 1.
9. Avera, F. L., 1958, Process of Manufacturing Coated Nuts, U.S. Patent 2,859,121.
10. Avera, F. L., 1958, Nut Products, U.S. Patent 2,860,053.
11. Avera, F. L., 1960, Nut-flavoring Additives, U.S. Patent 2,955,040.
12. Bamberger, M., and Landsiedl, A., 1900, *Monatsh.*, *21*: 571.
13. Barnett, C. D., 1961, How to Extend Shelf-life of Candies, *Manuf. Confect.*, *41*: 37-40.
14. Barnett, C. D., 1965, Shelf-life Can be Adjusted by Two Means, *Candy Ind.*, *124*: 31-33.
15. Barr, M., and Tice, L. F., 1957, The Inhibitory Concentrations of Various Sugars and Polyols on the Growth of Microorganisms, *J. Amer. Pharm. Ass.*, *46*: 219-21.
16. Benon, M., Roussel, L., and Rebillard, A., 1966, Freezing Fruits and Vegetables, French Patent 1,448,010.
17. Benson, F. R., Polyhydric Alcohols, *Kirk-Othmer Encyclopedia of Chemical Technology*, 2nd ed., Vol. I, John Wiley & Sons, 1963, pp. 569-588.
18. Berg, H. W., and Ough, C. S., 1962, *Wines and Vines*, Hiaring Co., San Francisco, Calif., pp. 27-28.
19. Berg, I. A., 1948, Compressed Cereal, U.S. Patent 2,437,150.
20. Bilotti, A. G., 1967, Sugarless Gum, U.S. Patent 3,352,689.
21. Black, W. A. P., 1948, *J. Soc. Chem. Ind.*, *67*: 165.
22. Block, H. W., and Common, J. L., 1957, Pudding Composition, U.S. Patent 2,784,099.
23. Bonnet, P. L., 1962, Liquid of Elevated Boiling Point Useful for the Sterilization of Canned Foods, French Patent 1,294,803.
24. Bougault, J., and Allard, G., 1902, *Compt. Rend.*, *135*: 796.
25. Bourquelot, E., 1895, *J. Pharm. Chim.*, *2*(6): 285.
26. Bourne, E. J., Tiffin, A. I., and Weigel, H., 1959, Interaction of Anti-staling Agents with Starch, *Nature*, *184*(8): 547.

27. Boussingault, 1872, *Compt. Rend., 74*: 939.

28. Brandner, J. D., 1947, Dehydration of Foods by Means of Hydrophilic Liquids, U.S. Patent 2,420, 517.

29. Brockmann, M. C., 1969, *Compaction of Military Foods*, Research and Development Associates Activities Report, *21*(2): 83–87.

30. Bucher, R. C., 1966, Chewing Gum Product and Method of Making Same, U.S. Patent 3,262,784.

31. Burgess, H. M., and Mellentin, R. W., 1965, *Humectant in Cat and Dog Feed*, U.S. Patent 3,202,514.

32. Cameron, A. T., 1947, The Taste Sense and the Relative Sweetness of Sugars and Other Sweet Substances, *Sugar Research Foundation Scientific Report Series, No. 9*, p. 72.

33. Camien, M. N., Dunn, M. S., and Salle, A. J., 1947, The Effect of Carbohydrates on Acid Production by Twenty-four Lactic Acid Bacteria, *J. Biol. Chem., 168*: 36–42.

34. Camirand, W. M., *et al.*, 1969, *Osmotic Dehydration of Coated Foods*, U.S. Patent 3,425,848.

35. Carr, C. M., Bech, F. F., and Krantz, J. C., 1936, *J. Amer. Chem. Soc.*, 1394–95.

36. Chernishev, V. M., 1966, Investigation of the Mechanism of the Protective Action of Glycerol in the Refrigerated Treatment of Products of Plant Origin, *Bull. Inst. Int. Froid, Annexe 1966*, (1): 443–452, England.

37. Childs, W., 1961, Superior Fudge and Caramel Processing Time Cut to Less Than Four Minutes, *Candy Ind., 117*: 51–52.

38. Ciusa, W., and Pasquinelli, L., 1943, The Content of 2,3-Butylene Glycol in Tomato Conserves as an Index of the State of Conservation, *Ind. Ital. Conserv., 18*: 49–52.

39. Cosler, H. B., 1957, *Manuf. Confect., 37*(6): 37.

40. Cross, S. T., and Knightly, W. H., 1959, Unpublished Report, *SORBO in Soft Sugars*, Atlas Chemical Industries, Inc.

41. Crowley, M. C., Harner, V. Bennet, A. S., and Jay, P., 1956, *J. Amer. Dent. Ass., 52*: 148–154.

42. Dame, C., Jr., 1967, Mannitol Fixed Flavor and Method of Making Same, U.S. Patent 3,314,803.

43. Diamond, H. W., 1952, Dessert Mix and Method for Making the Same, U.S. Patent 2,619,422.

44. Diamond, H. W., 1952, Powdered Topping and Method of Making the Same, U.S. Patent 2,619,423.

45. Dimick, K. P., *et al.*, 1960, Purification of Fruit Essences and Production of Solid Compositions Therefrom, U.S. Patent 2,949,368.

46. Downey, H. A., 1962, Marshmallow Dry Mix, U.S. Patent 3,018,183.

47. DuRoss, J. W., and Knightly, W. H., 1963, *Manuf. Confect. J., 43*: 26–31.

48. DuRoss, J. W., 1968, Functionality and Application of Sorbitol 70% Solution in Confectionery Products, *Pennsylvania Manufacturing Confectioners' Association, 22nd Annual Production Conference*, pp. 18–21.

49. DuRoss, J. W., 1967, How Sorbitol Keeps Soft Candies in Desired State, *Candy Ind. Conf. J., 128*(1): 34.

50. DuRoss, J. W., 1969, Sugarless Hard Candy Formula, U.S. Patent 3,438,787.

51. DuRoss, J. W., 1962, Unpublished Data, PDD-234, Atlas Chemical Industries, Inc.

52. Eisenberg, S., 1955, Use of Sugars and Other Carbohydrates in the Food Industry, *Advan. Chem. Ser., 12*: 78.

53. Epstein, A. A., 1948, Coffee Concentrate and the Process of Producing It, U.S. Patent 2,457,036.

54. Felber, J. P., Renold, A. E., and Zahnd, G. R., 1959, The Comparative Metabolism of Glucose, Fructose, Galactose and Sorbitol in Normal Subjects and in Disease States, *Mod. Probl. Paediat., 4*: 467–489.

55. Fiedler, W. C., 1959, Flavor-masking a New Calcium Syrup, *Amer. J. Pharm., 131*: 217–223.

56. Finberg, A., December 1955, Big Strides of Dietetic Foods, *Food Eng.*, pp. 70–75.

57. Freed, M., February–March 1970, Fructose—The Extraordinary Natural Sweetener, *Food Prod. Devel.*, pp. 38–39.

58. Frerejacque, M., 1939, *Compt. Rend., 208*: 1123.

59. Frostell, G., 1963, A New Type of Presumably Low Cariogenic Sweets, *Sverges Tandlakar-forbunds Tidning, 55*: 529–531.

60. Frostell, G., 1965, The Shape of the Stephan-curve After Ingestion of Different Kinds of Sweets, *Sverges Tandlakarforbunds Tidning, 57*: 696–704.

61. Glabau, C. A., 1956, Formula and Handling of the Ever Popular Marshmallow, *Bakers Weekly, 172*(10): 45–47.

62. Goris, A., and Ronceray, P., 1907, *Chem. Zentr., 78*(I): 111.

63. Graham, C. P., *et al.*, 1966, Free-flowing, Noncaking Brown Sugar, U.S. Patent 3,264,117.

64. Griffin, W. C., 1945, *Ind. Eng. Chem.*, *37*: 1126–30.
65. Griffin, W. C., 1951, *Solid Essential Oil Concentrate and Process of Preparing the Same*, U.S. Patent 2,566,410.
66. Griffin, W. C., Behrens, R. W., and Cross, S. T., 1952, *J. Soc. Cosm. Chem.*, *3*: 5–29.
67. Grover, D. W., 1947, The Keeping Properties of Confectionery as Influenced by Its Water Vapour Pressure, *J. Soc. Chem. Ind.*, *66*: 201–205.
68. Grubb, T. C., 1945, *J. Dent. Res.*, *24*: 31–44.
69. Gutschmidt, J., and Ordynsky, G., 1961, Determination of the Sweetening Strength of Xylitol, *Deut. Lebensm—Rundsch.*, *57*: 321–324.
70. Hachtman, S. J., 1968, Confection Manufacture, U.S. Patent 3,371,626.
71. Hale, J. F., *et al.*, 1961, Process for Preparing Instant Potato, U.S. Patent 2,980,543.
72. Hall, L. A., 1954, Manufacture of Meat-Curing Composition, U.S. Patent 2,668,770.
73. Hall, L. A., and Kalchbrenner, W. S., Meat-Curing Salt Compositions, U.S. Patent 2,770,548.
74. Harwood, V. D., 1954, *J. Sci. Food Agr.*, *5*: 453.
75. Hass, P., and Hill, T. G., 1932, *Biochem. J.*, *26*: 987.
76. Helgren, F. J., Lynch, M. J., and Kirchmeyer, F. J., 1955, A Taste Panel Study of the Saccharin "Off-Taste", *J. Amer. Pharm. Ass., Sci. Ed.*, *44*(6): 353–355.
77. Herlow, A., 1958, Frozen, Edible Emulsions or Dispersions, German Patent 1,037,244.
78. Herz, K. O., December 1965, Staling of Bread—A Review, *Food Tech.*, p. 94.
79. Hesse, O., 1915, *J. Prakt. Chem.*, *92*(2): 425.
80. Hitzemann, B., and Tollens, B., 1889, *Berichte*, *22*: 1048.
81. Hofmann, A. W., 1874, *Ber.*, *7*: 508.
82. Jacobs, M. B., 1945, Solvents for the Flavor Industry, *Amer. Perfum. Essent. Oil Rev.*, *47*(10): 53–54.
83. Jandrier, E., 1893, *Compt. Rend.*, *117*: 498.
84. Jensen, J. P., 1956, Milk Products, U.S. Patent 2,758,925.
85. Jongen, F. K. G. T., 1956, *Confect. Manuf.*, *2*(1): 16; 1957, *Zucker-u. Susswarenwirtsch*, *10*: 592.
86. Kahn, L. E., *et al.*, 1960, Process of Curing Meat and Composition Therefore, U.S. Patent 2,946,692.
87. Kalafatas, N. J., Rosenthal, H., Kramer, F., Steigmann, E. A., and Tole, A. F., 1962, Chewing Gum, German Patent 1,136,058.
88. Kapeller, A. R., 1952 March, Sorbitol and Its Application in the Confectionery and Chocolate Industry, *Int. Choc. Rev.*, pp. 74–76.
89. Kaufman, C. W., and deMayer, C. B., 1952, Use in Shredded Coconut, U.S. Patent 2,615,812.
90. Kidger, D. P., *et al.*, 1961, Whipping Agent and Method of Preparing the Same, U.S. Patent 2,978,335.
91. King, N., and Shimmin, D., 1961, Microscopic Observations on the Dispersibility of Milk Powder Particles in Some Organic Liquids, *J. Dairy Res.* *28*: 277–283.
92. Knechtel, H., 1966, *Candy Ind. Conf. J.*, p. 12.
93. Knechtel, H., 1964, What the Candy Manufacturer Should Know about Dietetic Confectionery. *Candy Ind. Conf. J.*, *123*(11): 37–38.
94. Knightly, W. H., 1959, Fluid Emulsifier for Ice Cream, U.S. Patent 3,017,276.
95. Kooistra, J. A., Jr., 1966, Low Moisture Content Pastry Dough Compositions, U.S. Patent 3,294,547.
96. Kuzio, W., 1964, Estee Finds Dietetic Candy a Growth Field. *Candy Ind. Conf. J.*, *123*(11): 8,26,35.
97. Lakritz, D. E., 1946, Propylene Glycol Can Become the Most Important Flavor Solvent, *Flavors*, *9*(6): 11–12.
98. Langwill, K. E., June 1953, Chocolate for the Diabetic, *Conf. J.*, pp. 16–21.
99. Leffingwell, G., *et al.*, 1941, Glycerine, Added to Nuts, Will Improve Freshness and Looks, *Food Field Rep.*, *9*: 14.
100. Leffingwell, G., *et al.*, 1941, Glycerine in Food Color Preparations, *Amer. Perf.*, *42*: 47–49.
101. Leffingwell, G., *et al.*, 1941, Glycerine in Ice Cream, *Ice Cream Rev.*, *24*: 34.
102. Leffingwell, G., *et al.*, 1942, Glycerine Used in Many Flavorings, *Food Field Reporter*, *10*: 34.
103. Lensack, G. P., talk presented to the Milwaukee Chapter, American Association of Candy Technologists, April 21, 1965.
104. Lindberg, B., Misiorny, A., and Wachtmeister, C. A., 1953, *Acta Chem. Scand.*, *7*: 591.
105. Lindberg, B., and Paja, J., 1954, *Acta Chem. Scand.*, *8*: 81.
106. Lohmar, R. L., 1947, *The Carbohydrates*, p. 242.

107. *Ibid.*, p. 245.

108. *Ibid.*, p. 247.

109. *Ibid.*, p. 249.

110. Love, R. M., *et al.*, 1965, Protein Denaturation in Frozen Fish, *J. Sci. Food Agr.*, *16*: 65-70.

111. Lutz, W. J., 1952, Chewing Gum, U.S. Patent 2,586,657.

112. Maguenne, L., 1888, *Compt. Rend.*, 107: 583.

113. Maguenne, L., 1890, *Ann. Chim. Phys.*, *19*(6): 5.

114. Meyer, A. E., 1955, Aromatic Vegetable Extracts, *C.A.*, *45*: 876.

115. Meyer, E. G., 1967, Process for Producing a Confectionery Cream Candy-Center and the Resulting Product, U.S. Patent 3,325,293.

116. Miner, C. S., and Dalton, N. N., 1953, *Glycerol*, p. 2.

117. Motzel, W. J., and Baur, F. J., 1965, *Preparing Flavor Improved Preserved Fruit*, U.S. Patent 3,224,886.

118. Mundelein, E. L. W., III, 1954, Edible Container for Frozen Confections and Process for Preparing the Same, U.S. Patent 2,694,012.

119. Mylne, A. M., and Seamans, V. S., 1954, Stabilized Orange Juice Powder, *Food Tech.*, *8*(1): 45-50.

120. Newman, A. A., 1964, Glycerol in Food Manufacturing, *Food Mfr.*, pp. 38-42, London.

121. Nobile, L., 1961, Fruit or Vegetable Juice in Powder Form, U.S. Patent 2,970,058.

122. Olmsted, W. H., 1953, *The Metabolism of Mannitol and Sorbitol, Their Use as Sugar Substitutes in Diabetic Therapy*, 2: 132-137.

123. Pelouze, J., 1852, *Ann. Chim.*, *35*(3): 222.

124. Podwyzssozki, W. V., 1889, *Arch. Pharm.*, 227: 141.

125. Pratt, C. D., 1953, Special Uses for Sorbitol and Emulsifiers in Confectionery Products, *Candy Ind.*, *18*: 18,23.

126. Proust, 1806, *Ann. Chim. Phys.*, *57*(1): 144.

127. Raithby, J. W., and Gibson, D. L., 1954, Studies on Improving the Ease of Reconstitution of Skim Milk Powder, *Can. J. Tech.*, *32*: 60-67.

128. Rakieten, M. L., Newman, B., Falk, K. G., and Miller, I., 1962, Comparison of Some Constituents in Fresh-frozen and Freshly Squeezed Orange Juice, *J. Amer. Diet. Ass.*, *28*: 1050-1053.

129. Ramont, R. E., 1949, Antisticking Coating for Fruit, U.S. Patent 2,474,915.

130. Robinson, H. M., 1956, Stabilization of Fats in Candy, *Mfg. Confect. Int. Confect.*, *36*: 21.

131. Rowe, T. W. G., 1962, Water Vapor Removal in Food Freeze Drying, *Vide*, *17*: 516-530.

132. Saussele, H., Jr., 1962, *Mfg. Confect.*, *43*(6): 47-51.

133. Schapiro, A., 1959, Effervescent Drink Concentrate and Method of Making Same, U.S. Patent 2,868,646.

134. Schobess, M., Klaus, H., and Schaum, G., 1967, Synthetic Chewing Gum Base, German Patent 55,407.

135. Schutz, H. G., and Pilgram, F. J., 1957, *Food Res.*, *22*: 206-213.

136. *Ibid.*, p. 206.

137. Schaffer, B. M., 1956, Treatment of Liquid Egg Whites, U.S. Patent 2,744,829.

138. Schaffer, B. M., 1956, Dried Egg Whites, U.S. Patent 2,758,933.

139. Shaw, J. H., and Griffiths, D., 1960, *J. Dent. Res.*, *39*: 377-384.

140. Shearon, W. H., Jr., 1952, Chemistry in Candy Manufacturing, *Chem. Eng. News*, *30*(44): 4606-4610.

141. Shimazono, H., Yasumatsu, K., Matsumura, C., and Jono, K., 1967, Liquid Flavor Enhancers, U.S. Patent 3,326,697.

142. Shockley, T. E., Randles, C. I., and Dodd, M. C., 1956, *J. Dent. Res.*, *35*: 233-264.

143. Smith, B. H., 1941, New Developments in Flavor, *Proc. Inst. Food Tech.*, pp. 192-194.

144. Smock, R. M., 1963, *Red Color Study on Apples*, reprint from Proceedings New York Horticultural Society Annual Meeting.

145. Steinke, J., Wood, F. C., Jr., Domenge, L., Marble, A., and Renold, A. E., 1961, *Diabetes*, *10*: 218-227.

146. Strain, H. H., 1937, *J. Amer. Chem. Soc.,* *59*: 2264.

147. Strashun, S. I., *et al.*, 1958, Full-flavored Dehydrated Food Products, U.S. Patent 2,854,343.

148. Strashun, S. I., and Talburt, W. F., 1954, Stabilized Orange Juice Powder, *Food Tech.*, *8*(1): 40-50.

149. Strashun, S. I., and Talburt, W. F., 1953, *Food Eng.*, *25*(3): 59-60.

150. Tanret, C., 1902, *Bull. Soc. Chim. Fr.*, *27*(3): 963.

151. Thorner, W., 1879, *Berichte*, *12*: 1635.

152. Tischer, J., 1936, *Z. Physiol. Chem.*, *243*: 103.

153. Ulrich, K., 1952, The Sweetness of Synthetic and Natural Products, *Zucker*, *5*: 236-239.

154. Uprety, M. C., and Revis, B., 1964, Elevated Temperature Studies on Stability of Ascorbic Acid in Certain Fruit Juice and Aqueous Vehicles, *J. Pharm. Sci.*, *53*(10): 1248-1251.

155. Vale, W. H., 1953, Coating Composition for Fruit or Vegetables, Australian Patent 153,174.

156. Vincent, C., and Delachanal, 1889, *Compt. Rend.*, *109*: 676.

157. Vincent, H. C., Lynch, M. J., Pohley, F. M., Helgren, F. J., and Kirchmeyer, F. J., 1955, A Taste Panel Study of Cyclamate-Saccharin Mixture and of Its Components, *J. Amer. Pharm. Ass., Sci. Ed.*, *44*(7): 442-446.

158. von Leesen, A., 1967, Treatment of the Surface of Dried Grapes, German Patent 1,232,448.

159. Walker, H. W., 1951, Synergistically Sweetened Canned Fruits and Methods of Making the Same, U.S. Patent 2,608,989.

160. Walton, C. F., 1927, *International Critical Tables*, *1*: 357.

161. Ward, D. R., 1969, unpublished data, Atlas Chemical Industries, Inc.

162. Ward, D. R., Lathrop, L. B., Lynch, M. J., 1966, *Drug Cosmet. Ind.*, *98*(1): 48-157.

163. Weinmann, R. C., and Cotton, R. A., 1958, Coatings for Meat Products, U.S. Patent 2,849,319.

164. Welker, P. L., *et al.*, 1953, Use of Sorbitol in Shredded Coconut, U.S. Patent 2,631,104.

165. Wessely, E., and Wang, S., 1938, *Monatsh.*, *72*: 168.

166. Anon., 1956, Flavor in Citrus Concentrates, *Givaudan Flavorist*, No. 1, p. 3.

167. Anon, 1957, Marshmallow and Nougat-Nonstop, *Food Eng.*, *29*(4): 104-107.

168. Anon., 1955, Producing Marshmallow Topping, *Candy Ind.*, *23*(12): 12.

169. Anon., 1953, Solid Odors and Flavors Have Greatly Improved Stability, *Chem. Eng. News*, *31*(20): 2094-2096.

170. Atlas Chemical Industries, Inc., *Atlas Products for the Food and Beverage Industries*, ILG-91.

171. Atlas Chemical Industries, Inc., *General Characteristics of Atlas Polyols*, CD-156.

172. British Patent 1,042,436, 1966, Improvements in and Relating to Transparent Edible Films for Wrapping and Packaging Foods, Miles Italiana Derivati Amidi.

173. *J. Ass. Off. Agr. Chem.*, 1925, *8*: 255.

174. Military Specification MIL-C-10928B, January 1956.

175. Monsanto Chemical Company, *Non-nutritive Sweeteners for Low-calorie Food and Beverages*, 2-2134-0466-36.

176. Union Carbide Chemical Company, *Glycols Handbook*, F-4763-F.

CHAPTER **11**

Natural and Synthetic Flavorings

Robert L. Swaine

Vice President—Technical Operations
Canada Dry Corporation
Greenwich, Connecticut

Introduction

Flavorings are those materials added to a substance to (1) give it the flavor of the flavorings or (2) to supplement or modify its own flavor or (3) to cover up or mask the original flavor of the material. For example:

1. orange to give orange identity to carbonated water,
2. malt flavoring to modify the flavor of cereal grains,
3. anethole or anise to cover or mask the bitterness of medicinals.

457

Anything can be a flavor by this definition; sugar can be added to coffee either to add its own flavor of sweetness, to modify the coffee flavor, or to mask the sour and bitterness of the coffee.

Although almost anything can be a flavoring (flavor), the public, as well as industry, tends to believe that flavors are substances that, because of their relative flavor strength, are sold and used in small quantities and are relatively expensive. In this chapter, most of the information and remarks are confined to material that is covered by this general belief.

The flavorist for whom this chapter is intended must not forget that good flavorings (flavors) cannot always correct a product containing poor quality raw materials. It may cost an extra $4 of flavor per 100 pounds of finished product because $2 per 100 pounds was saved by using an off-grade raw material.

The listing of any chemical, oil or flavoring in this chapter does not suggest or infer its safety. Questions concerning the legal status of a particular flavoring should be referred to the United States Department of Health, Education and Welfare, Food and Drug Administration.

History

Today the industry that concentrates on supplying flavorings to manufacturers of confections, processed meats, baked goods, soft drinks, liqueurs, etc. has come a long way from the local druggist who sold to the individual housewife one vanilla bean or ½ dram of peppermint extract. Today the flavor industry's sales are over $100,000,000; over 50 per cent of this is probably in the hands of 5 per cent of the suppliers.

When the flavor industry began and when the use of flavors started can only be conjectured. One could guess that the early cave man did not realize that through the mixture of wild honey with wild berries one flavoring had a bearing on the other. He may have realized that when he cooked meat over different varieties of wood, he produced a differently flavored product. Even today, applewood is considered best for producing a barbecue flavor.

Most likely, the industry as we know it today started with the trade route to Africa and the Near East. The reader must recall that there were few staples available to those living in the early middle ages that could be carried from season to season. Local fruits were available for only a short season. Jams could be made, but this meant the use of sugar, which was not readily available. Vegetables, even potatoes, could be kept for only a short time. Meats, except in winter, were either consumed shortly after slaughtering or were heavily salted. However, salt was expensive and still the meat spoiled. Flour from grain could be kept for longer periods, but usually only enough was produced to last from one harvest to the next. If the grain harvest was poor, famine was the result. In short, a person living at this time existed on a monotonous diet. Most of the time, much of what he consumed was close to being spoiled. It is no wonder that when spices were first encountered in Europe (as a result of the Crusades) all Europe became spice conscious. Spices would help keep meats longer and disguise the taste of partially spoiled products. Although we no longer have spoiled food, we still enjoy many of the same spices available then. The early spice traders must be thanked for beginning the flavoring industry.

In the middle ages we also had the development of herbs as flavorings. Although most were home grown, herbs eventually were sold by local healers. As time progressed, the pharmacist, the doctor and the alchemist contributed to the industry's growth. It was not unusual that these three were involved, as many of the flavors used at the time were considered to possess both magical and medicinal properties. We still remember catnip tea and essence of peppermint. The alchemist, through distillation, gave the industry a concentrated product which enabled the manufacturer to use a more desirable flavoring. The chemist made his greatest contribution in producing synthetic substances, some of which have never been found in nature. However, it is the flavorist, with his creative ability, who has been able to produce the finished and complex products that today we know as flavorings.

Basic Relationship

It has been established that flavorings are materials added either to produce or modify flavor. Flavor is a complex sensation. It is the integration of the sensations of odor, basic tastes, feeling factors, texture and possibly even sound.

There are four basic tastes: sweet (sugar), sour (acid), salty (sodium chloride) and bitter (quinine). If these were the only contributors to flavor, then there would be little difference among the flavors of apples, oranges and onions. For example, when the eater has a cold, he recognizes the fact that odor plays an important part in flavor. When a substance (food) is placed in his mouth, vapor from this food travels from the oral cavity to the nasal area where it registers in the brain as the sensation of odor or smell. Certain materials also contribute feeling factors that must be considered a part of the overall complex of flavor. For example, menthol produces a cooling in the mouth, pepper a bite and cassia a warmth. The onion and the apple can be distinguished by their difference in texture as well as their difference in odor. Therefore, texture must be considered as contributing to flavor. Whether or not to include sound as a part of flavor is still debatable. The "crunch" when celery is bitten, or the "crackle" of some cereals when milk is added is an attribute, but whether it should be included as part of the complex sensation of flavor is still questionable. It is the author's opinion that if sound is included, then appearance must also be considered. However, this is stretching a definition too far. The point to remember is that flavor is a complex sensation and flavorings (can) contribute to produce all or part of this sensation.

Definition of Terms

The flavorist and the flavoring industry have developed a series of terms which should be defined as they appear in the literature as well as throughout this chapter, particularly in the tables.

Essential Oil   An oily substance obtained from plant material through various methods. The essential oil normally has the characteristic taste and odor of the plant from which it was derived. An essential oil is still called a volatile oil as differentiated from a fixed oil. The hydraulically pressed sesame seed yields a fixed oil (sesame oil) that has low odor and is not volatile. The anise seed, upon distillation, yields an odorous and volatile oil—oil of anise. Essential oils may have received their name because at one time they were thought to be essential to the life processes of the plant or that they were the essence of the plant.

Bulking   Mixing of one or more lots of the same flavorful material to produce a uniform product. The entire crop of an essential oil may be bulked to assure uniformity.

Essence   Concentrated fragrance or flavorant.

Absolute   A material extracted from a plant that represents a concentrated form of that material and is extremely similar to the starting material in taste and odor.

Extract   A solution obtained by passing alcohol, or an alcohol-water mixture, through a substance. An example would be vanilla extract. Extracts found on the grocer's shelf, such as orange, almond, lemon, etc., are essential oils dissolved in an alcohol-water mixture.

Oleoresin   A resinous-viscous product obtained when a substance is extracted with a non-aqueous solvent such as a hydrocarbon. The solvent is later removed. Spices as a class form most of the oleoresins that the flavorist encounters; an example would be oleoresin pepper.

Note   A distinct flavor or odor characteristic. For example, many raspberry flavors have a seedy note.

Top Note   The first note normally perceived when a flavor is smelled or tasted. Usually a top note is relatively volatile and suggests identity.

Middle Note or Main Note   The substance of flavor; the main characteristic.

Blender   A material that when added to a substance appears to bring various flavor characteristics together. A blender may or may not introduce a flavor of its own. For example, vanilla can act as a blender.

Bottom Note   The characteristic left when top and middle notes disappear; the residue when a flavoring evaporates.

Fixative   Usually applied to perfume, but in flavorings acts to reduce the overall volatility of the flavoring.

Compounds   Not to be confused with the chemist's definition. A compound is a flavoring (flavor) composed of two or more substances. These substances can be natural or synthetic, a chemical or an essential oil, an extract or an oleoresin or combinations. Unlike chemical compounds, which are of known fixed compositions, flavor compounds are usually secret mixtures. Normally they are finished flavorings that can be added directly to a product so that no additional flavorings are needed.

Specialty   Usually similar to a compound, only not finished. A specialty, although not complete, carries the major part of the flavor load, so that only a few other substances are needed to complete the flavoring.

Synthetic   Not naturally produced. Menthol synthetically produced from pinene is chemically identical to menthol obtained from peppermint oil, but it cannot be called natural.

Isolate   A chemical or fraction obtained from a natural substance. For example, citral can be isolated from lemon oil or lemon-grass.

Imitation   Flavor containing all or some portion of non-natural materials. For example, unless an orange flavoring is made entirely from orange, it is imitation.

Artificial   Similar to imitation. It is possible to have a flavor that contains all naturals, but it must be called artificial because it has no counterpart in nature.

Type or Class   The flavorist tends to group similar flavors together. For example: *red flavors* (flavorings): strawberry, cherry, raspberry; *citrus flavors:* lemon, orange, lime, grapefruit, bergamot; *brown flavors:* coffee, malt, caramel; *spice flavors:* cassia, clove, nutmeg. Classes can also be made by volatility, chemical function or end use.

Aromatic (Chemical)   Any chemical that has aroma or flavor properties. Not to be confused with the chemist's definition of a compound containing a benzene ring structure.

Classification of Flavorings (Natural and Synthetic)

The flavorist in industry is concerned with how he can use the raw materials available to him for his flavorings. Each industry tends to work with relatively few raw materials. Thus, the object of this section is to classify the materials available to the trade as a whole. The classification system as used in the laboratories of Arthur D. Little, Inc. (ADL) and the Canada Dry Corporation is not claimed to be original, but it has proved effective; by knowing the class in which a specific material falls, the advantages and limitations of that material are suggested.

Crude Spices and Herbs

Dried Ground Products   Materials in this group are, from the flavorist's point of view, simple and crude. They have poor flavor strength; normally in a finished product they will be used at 0.5–1 per cent. Usually they are principally cellulosic in nature and often colored. An advantage is that frequently they contain natural antioxidants and have antibacterial activity; materials in this class include cloves, cinnamon, thyme and sage. These spices and herbs are picked, dried and usually ground to a particular particle size; they are the common household seasonings. In industry they find heavy usage in the areas of processed meat, pickling and bakery. Spice particles are accepted in pickles, but are a disadvantage in most products. Specks of black pepper are acceptable on mashed potatoes, but undesirable in a canned soup.

There are two more disadvantages. First, there is much batch-to-batch variation in the coloring as well as the flavoring strength, and secondly, it takes time for the ground spice to reach flavor equilibrium with its surroundings. When the finished product is new, it will have high concentrations of flavoring adjacent to the spice particles, leaving other areas in the product with either little or no flavoring. The customer will note a variation in the flavor of the product proportional to its age.

Raw spices and herbs do maintain flavor integrity; they are perfect representations of the flavor they profess to be. As a class, however, they represent our first flavoring and are becoming a smaller percentage of the overall market as time progresses.

Oleoresins Oleoresins offer the flavorist a superior product over the corresponding spices from which they are derived. This means a more uniform and more potent product is available. The normal use range for an oleoresin is from one-fifth to one-twentieth of the corresponding crude spice. Because of the potency of oleoresins, they are frequently bulked for specific customers, thus providing the customer uniformity. Although oleoresins are normally thick, viscous, highly colored substances, they impart less color to the product than the corresponding spices because they are used in small quantities.

An oleoresin is made by percolating a volatile solvent, such as a chlorinated hydrocarbon, through a ground spice. The solvent is vacuum distilled off and recovered for reuse; the residue is the oleoresin. Although the oleoresin is very similar to the spice from which it is derived, it is not identical; not all flavor notes are extracted.

Certain spices are extracted to yield their oleoresins, not for the flavoring effect, but for the intense color that is produced. Examples of this are paprika and turmeric, which are used heavily in the manufacture of French-type salad dressings.

Oleoresins are not equivalent to their corresponding essential oils. Oils of ginger and black pepper have little of the bite of their corresponding crude spices or oleoresins, but they do have a high concentration of their aromatic properties. Conversely, the oleoresins may contain more of the bite effect than the oils. To overcome this differential, some suppliers add essential oils to their oleoresins. In general, it can be said that oleoresins offer the flavorist an opportunity to use the advantages of spices without their disadvantage of bulk (quantity of unwanted cellulosic material).

Plated Flavorings Almost any liquid flavorings can be mixed or plated on to a neutral carrier. The limit of the amount of flavoring is the amount that can be adsorbed before the carrier starts to dissolve. Carriers normally used are salt, dextrose, sucrose or a combination of the three. The value of plated flavorings is that a uniform quantity of a liquid flavor can be made available in a solid form which is easier to measure. Smaller companies, particularly in the baking and processed meat industries, prefer flavors of this type, as it is easier for them to accurately weigh the more dilute flavoring than to measure volumetrically the more concentrated oleoresins or essential oils.

Oleoresins are particularly adaptable to plating because they are concentrated. When plated, they still give a relatively powerful flavoring with the uniformity that is difficult to get with the crude spice or herb. Essential oils are sometimes plated.

One of the most important items in the plated category is the so-called vanilla sugars. Originally, these were made by placing vanilla beans directly in sugar; then the actual beans were ground together with sugar or a suitable carrier. Today many vanilla sugars are made from oleoresin vanilla. (A word of caution to the reader—do not confuse vanilla sugars with mixtures of vanillin and neutral carriers.)

Essential Oils

The largest single category of flavorings available to the flavorist today is the essential oils. Historically, the first essential oil encountered was oil of rose. It was discovered by the Chinese prior to the Christian era. A layer of this oil was found on a pool that was filled with rose water, a favorite fragrance of the emperor. Oils were squeezed from the rinds of the citrus fruit of the Mediterranean area in early A.D.; it was the alchemist of the middle ages who perfected the art of distillation, and hence provided the tool that accounts for the major measures of processing our essential oils of today.

An essential oil is an odorous substance obtained from a plant material having the major odor characteristic of that material. It is normally concentrated, having about 100 times the flavoring strength of the parent plant. Most essential oils are used at about a level of 0.01–0.1 per cent in the finished product. They are often slightly colored and have a specific gravity of about 1. The advantages of essential oils are their flavor concentrations and their similarity to their corresponding sources. The majority of them are fairly stable (notable exception is the citrus oils) and contain a few natural antioxidants. Although most are soluble in high strength alcohol (greater than 90 per cent), they have poor water solubility, and most contain terpenes that contribute to their poor water solubility.

Many essential oils such as peppermint and lemon are produced in modern highly automated plants; others, like wintergreen, are still produced in small quantities under primitive conditions.

Expressed Oils Expressed oils are obtained by pressing the oil from the source. The citrus fruits are our most common source; originally, they came from the Mediterranean area where they were hand pressed.

The rind of the fruit contains the oil bearing cells. These were pierced by a series of needle-like prongs, and the released oil, together with some juice, was adsorbed into a sponge held in the hands of the operator. The oil and the juice mixture were squeezed from the sponge into a container where the oil separated and was decanted. This oil, the so-called hand pressed oil, was and is of the finest quality. Today, very little hand pressed oil of orange and lemon is imported into this country, but we do receive some hand pressed lime oil from the Dominican Republic.

It is apparent that only those fruits having a high (greater than 3 per cent) concentration of essential oil in the outer covering lend themselves to expression. These would include orange, lemon, lime, bergamot, tangerine, mandarin and grapefruit.

The era of hand pressed oils is nearly finished. Today our citrus oils are machine pressed. The growth of the orange juice industry has given us a by-product—orange oil, one of the most popular essential oils. When juice is squeezed from oranges commercially, a quantity of orange oil is present in the juice; it is removed by centrifuge and becomes the orange oil of commerce. If the present market for orange juice continues, the author can see no possible need for an artificial (synthetic) orange oil for many years.

One obvious advantage of expressed oils is that they have not been exposed to high temperatures. Necessarily, they have had no chance for degradation of flavor, but a disadvantage is, that because of their intimacy with water, certain water soluble components have been lost. Weighing the pros and cons, it must be decided that expressed oils give the flavorist some of his finest flavoring with which to work. One point must be remembered. An expressed citrus oil, regardless of its quality, is not a complete replacement for its corresponding juice. But because it does much in common with the juice flavor-wise, it is an ideal starting point in the compounding of citrus flavors. Due to variations from one part of the season to the next, most expressed citrus oils are bulked.

Distilled Oils   The greater portions of essential oils used by the flavorist are distilled, usually steam distilled, although some are water distilled. The distillation may be primitive on the growing site. The product to be distilled may simply be placed in a covered pot with water, and a condenser made of a piece of copper tubing is attached to the pot. The heat source to cause boiling might be an open fire; the condenser is cooled by running it through a near-by stream. The product of this distillation is separated by gravity.

Distillation can also be highly sophisticated and involve the most modern technology. Normally the charge into the still is partially dried (to increase the relative oil content) and ground (to give better contact of the boiling water and to rupture the individual oil cells). Fractionating columns with or without vacuums may also be employed. Whether it is a crude, peasant distillation in the hills of China or the efficient modern distillation of peppermint in Washington, the principle is the same. The essential oil must be quite water insoluble, but volatile. To be practical, the yield of oil must be economical. A definite yield of oil per ton of product cannot be set as a standard, because, although some material gives a small yield, the value of the oil is so great as to make it economically feasible. Ambrette seeds contain a small percentage of oil, but at $1000 a pound it is worth distilling; clove oil is relatively inexpensive at $10 a pound, but the dried clove bud contains 16 per cent oil.

Distilled oils are classed as a concentrated source of flavorings. They are similar to the products from which they are derived, but because the oils are in contact with water during the distillation, some of the water soluble components are lost. As heat is employed, there is also a flavor change due to heat degradation. Many of the natural antioxidants present in the parent source are not carried over in the distillation.

Rose oil (attar of rose) is used as a flavoring in quantities in tobacco products (snuff, chewing tobacco) in the United States, and in food products in some parts of the world (Near and Far East); it is still finding use in the carbonated beverage industry as a modifier for ginger ale flavoring. The layman, who smells attar of rose for the first time, is genuinely disappointed. The concentrated fragrance resembles the smell of the fresh rose but is certainly not identical. In fact, there are times when the layman cannot even recognize the odor as rose. During the heat of distillation some of the delicate top notes are destroyed by heat. One of the major chemicals in the fresh rose, phenyl ethyl alcohol, is lacking; because of its water solubility, it is lost in the distilling waters.

A last point to be remembered when dealing with not only essential oils, but all concentrated flavorings: do not judge the quality of a flavoring by its odor or taste in its concentrated form. All flavorings should be judged in the concentration that they are to be used. A diluted rose oil smells more rosy than a concentrated one.

It has been strongly suggested that essential oils in general are not stable and do not age well. Most of the time this is true; oils of clove and cinnamon darken, and most have a limited shelf life.

When they are first distilled, certain distilled essential oils contain an odor and flavor note known in the trade as a "still" note. Sometimes the oils are referred to as too fresh for use. This still note either quickly ages out of the oil or is blended out by bulking with other oils. Distilled oil of lime normally contains this still note and should not be downgraded because of it.

Oils of patchouly and vetiver are considered by the expert flavorist as almost unusable for about a year after they are distilled because of a harsh green note. Necessarily, certain companies have developed a process for making freshly distilled oil of patchouly appear to be the same as a fully matured two-year-old sample. Patchouly is used as a blender and modifier in cola-type beverages, as well as the main component in Oriental-type fragrances.

Enfleurage   Few, if any, flavoring oils used today are obtained by enfleurage, because solvent extractions can yield a product of similarly high quality. From a historic viewpoint, the process is worth recalling. Enfleurage oils are costly because of the time involved to produce them. However, because no heat is involved in their manufacture, they are delicate and retain the natural aroma and flavor of their parent materials. Plant materials originally used with the enfleurage process contained a small quantity of essential oil at any given time. The blossoms of jasmine, rose and tuberose, after they were harvested, continued to metabolize, as opposed to dried buds, such as clove. Enfleurage takes advantage of this by the continual adsorption of the essential oils from the plant while more is being manufactured by the plant.

Today, the plants (usually flower petals) are picked early in the morning, brought to the factory where they are placed, blossom by blossom, on a thin fatty layer spread on glass plates. The fragrant oils are adsorbed by the fats (similar to the way butter adsorbs food odors in the refrigerator). The flowers are replaced daily until the desired concentration of oil is reached (usually thirty-six days). The oil-fat mixture is separated by dissolving the oil out in cold alcohol, leaving the fatty residue behind. Then the alcohol is removed by vacuum distillation leaving a so-called absolute. This absolute is an almost perfect reproduction of the living flower. With the exception of tobacco flavoring, absolutes are never used in quantity; they do, when used in small amounts, allow the expert flavorist a chance to produce some highly original and acceptable flavorings. The current cost of absolutes produced by enfleurage has almost prohibited their use. This is a result of an increased standard of living in the producing areas. No longer is one content to rise before dawn and pick flowers for a few cents an hour, when a better paying, shorter-houred job is available in one of the new local factories.

Solvent Extractions   Solvent extraction, as a process, is not new. In the 1800's carbon disulfide was used as a solvent for extracting flavor oils. However, it was not until the 1900's that the process was producing acceptable oils in commercial quantities. Because a solvent-extracted essential oil is produced at a low temperature and without water, it is quite similar to its starting material. Basically, a solvent-extracted oil is similar to an oleoresin, except that it will be taken a step further. Continuous extractions using solvents, such as methyl chloride or liquefied isobutane, dissolve out the essential oils and waxes of the parent substance. The solvent is recovered and reused. The oil and wax mixture is called a concrete. It is further processed to remove the waxes by dissolving the mixture in cold alcohol. The waxes have limited solubility in cold alcohol, but the essential oils are soluble. The alcohol is removed from the oil-alcohol mixture by the same vacuum distillation method used in the preparation of an absolute in the enfleurage process.

Because this process is automated, the essential oils produced are less expensive than those produced by enfleurage and are just as good. Oils of varying nuances can be obtained through the choice of the solvent system. A quality oil must, however, have all traces of the extracting solvent removed or they will interfere with the final product's end flavor.

Concentrated Flavoring Oils   The end products of solvent extraction, expression, enfleurage or distillation are essential oils of a given strength. These are a mixture of chemicals; some contribute positively to the flavor of the oils, others contribute negatively. There are chemicals present in these essential oils that appear to neither add to nor subtract from the flavor; these are neutrals acting as diluents. As the flavoring trade has progressed, the need for more highly flavored (higher concentration) oils has appeared. These concentrated forms of essential oils have various trade names, but in the trade are generally known as folded oils. These should not be confused with rectified oils, which are not really concentrated, but simply cleaned up by redistillation (usually only 5–10 per cent of the material is removed). For example, water may be removed from a crude distillate, or a certain top or bottom fraction may be removed at the request of a specific customer. A folded oil is concentrated by removing neutral or unwanted components from the basic oil. This may be accomplished by specific chemical treatment or by fractional vacuum distillation. A two-fold oil is supposed to have twice the flavoring strength as a single-folded oil; likewise, a five-fold oil should have five times the flavoring strength of a single-fold oil. It is theoretically possible to have up to a 60-folded oil with certain essential oils. But unfortunately, during the folding (concentration) process, some of the desirable flavoring chemicals are lost; as with concentrated vanillas, a five-fold oil diluted five times is not equivalent to a single-folded oil. Normally some of both the top and bodying notes are lost.

It has been the author's experience when working with flavorings, that if a true flavor, particularly a citrus type, is desired, then at least 50 per cent of no greater than a two-fold oil must be used. High-folded oils are excellent for original effect as top notes in non-citrus products, as well as modifiers for citrus products of a different type. For example, a five-fold lime produces an interesting effect on a cola product, and a ten-fold orange modifier will blend in nicely with a lemon flavoring. There are times when only highly-folded oils can be used. This is when stability and water solubility are a problem, but for general flavoring use, folded oils are not intended to be the base of a formulation. Because folded oils are concentrated, smaller amounts are needed; hence one has a lesser solubility problem.

Terpeneless oils are an extension of folded oils. They are normally more highly concentrated than folded oils and are theoretically even more stable. Chemically, terpenes are a specific class of chemicals having ten carbons. Many of our highly flavorful products contain terpenes; the principal terpene in all citrus oils (and many other oils) is d-limonene. Not all terpenes contribute to flavor; many, particularly in their oxidized form, contribute to off-flavor. The flavor chemist has been able to remove unwanted terpenes and produce flavors of unusually high stability and concentrations. Similar to the highly folded oils, the terpeneless oils do not carry the full flavor load. They have lost it during their processing from the original source. Terpeneless oils serve a definite purpose. They can be used in low concentration, because of their high flavor potency. When water solubility is a problem, the terpeneless oils are useful because small quantities are used. In tablets, because their surface area presents excessive exposure to oxygen, terpeneless oils, with their increased stability, should be used.

The terpeneless oils can also be used as top notes and modifiers with single folded oils. The concentrated terpeneless oils should also be used in a product that, due to its physical nature, cannot tolerate a large amount of a flavoring. Terpeneless oils of peppermint, spearmint, juniper and petitgrain are in regular use, but terpeneless forms of the citrus oils are by far the most important. Some of the more common natural oils are given in Table 1 (p. 475).

Extracts

The layman is familiar with certain extracts as products common to the home kitchen: lemon, orange, vanilla and almond. These are of a minor overall importance to the flavoring trade. Extracts of raspberry, cherry, peach and strawberry are important to the trade, as essential oils, oleoresins, etc. of these flavors cannot be made. Extracts are weak compared to essential oils. Up to 10 per cent of the extract is needed in some finished products and the resulting flavor fidelity is only fair, although new techniques (essence recovery) have greatly improved the flavor qualities.

Fruit Extracts Fruit juices, a form of extracts, have been available to the flavorist for years, but are unsatisfactory in most products. Notable exceptions are the use of orange, grapefruit, lemon and lime juices in the beverage industry and most juices in the preserves industry. Juices *per se* are weak flavor-wise compared to essential oils and have a short storage and shelf life. The fruit source of the juice presents an even poorer shelf life and, unless frozen or dried, must be consumed within days of picking. Fruit extracts are made by extracting the dried fruit with a water-alcohol mixture. This can be slightly concentrated, but because of the pectin content of the juice and the deleterious effect of the heat during processing, the concentration is slight. Extracts also carry appreciable amounts of color with them. This is unwanted in some products. Fruit extracts cannot be used for hard candy flavoring (a high temperature processing) as they caramelize, causing off-flavor. In general, fruit extracts can be used where low temperatures are involved and large quantities of the extracts can be tolerated; the only real substitutes for them are artificial flavorings.

Concentrated Extracts   To overcome the weak flavor strength of extracts, the flavor trade has developed the reinforced extract, commonly referred to as W.O.N.F. (With Other Natural Flavors) extract. A WONF flavoring is an extract that obtains 51 per cent of its flavor value from the basic extract; the other 49 per cent comes from any other natural combination of flavorings. For example, a raspberry WONF would contain 51 per cent raspberry extract plus 49 per cent flavoring from such things as orris root. The orris root contains some ionone, which is also present in raspberry. The total effect is that a WONF raspberry has several times the flavoring strength as the corresponding straight non-WONF extract. Although the flavor strength is improved, the basic defect of heat susceptibility is still present. Another problem also has occurred. The added flavorings normally detract from the naturalness of the product. By and large, fruit extracts are still a weak point in the flavorists' repertoire.

Vanilla Extracts   In the U.S., vanilla extract from Mexican beans is considered the best by most flavorists. The author cannot accept this statement in its entirety, as vanilla and extracts from which it is made, can vary as much within a given source country, as it can from country to country.

In the trade, four main sources of vanilla beans are important. These include the Mexican bean from Mexico, the Bourbon bean from principally the Reunion Island and the Island of Madagascar, the Java bean from Java and surrounding islands, and the Tahiti bean from the island of Tahiti. There is some experimental planting of vanilla in the Caribbean area as well as in South America, but these, as yet, have not reached commercial importance.

Each type of vanilla bean is subdivided into several different grades, however, and the grade of bean used to make a finished extract does not appear on the label of the extract. It is, therefore, possible for a Java extract made with an excellent high grade bean to be better than a Mexican extract made from a poor grade bean. However, by reputation, Mexican extracts have always been considered superior.

In general, the flavoring characteristics of four extracts made from the same grade of bean are as follows:

Mexican: high top notes and lighter aromatics, moderate covering and blending power.

Bourbon: moderate top notes, heavier aromatic, excellent covering and blending power.

Java: moderate top notes, relatively poor aromatics, moderate blending and covering powers.

Tahiti: top note heliotropin-like, moderate blending and covering powers. Best used in ice cream products.

The flavorist should not confuse vanilla with vanillin. Both will supply aromatic top notes, but only vanilla will have great blending ability and covering power. In finished products vanilla is most useful. Not only does it seem to give identity through its aromatic top notes, but it also imparts a modifying effect. A skilled flavorist will make use of one of vanilla's most important attributes: the ability to act as a bridge. Vanilla can mask or cover up undesirable flavor notes, such as the green graininess of a cereal, so that other flavors can then be built upon them. Malt also acts as a bridge.

By definition, standard vanilla extract contains 13.35 ounces of beans per gallon of finished extract. If a double strength solution is desired, then twice the quantity of beans is required. The standard extract is called a single-fold extract; the double strength, a two-fold extract. If higher than a two-fold extract product is desired, then it is necessary to concentrate a single- or two-fold extract, usually by vacuum distillation. By this method, a 5-, 10- or higher-fold vanilla extract can be made. As previously stated, a direct comparison of 5-fold vanilla diluted back to a single-fold is noticeably different from a corresponding single-fold extract. The comparison will indicate that some of the top identifying odor and flavor notes are missing from the 5-fold vanilla suggesting that these notes are lost during the folding process. This is readily seen when folded extracts are used in flavor compounding.

Concentrated vanilla products called oleoresins are also available to the flavorist. These are made by solvent-extracting vanilla beans. When an extract is obtained from a water-alcohol extraction and compared to an oleoresin vanilla solution, one will note that the flavor properties are somewhat different.

A word of caution to the flavorist—vanilla extract may be reinforced with vanillin or ethyl vanillin. The resulting product cannot be considered natural and must be labeled as containing artificial or imitation flavor. This reinforcing may be intentional, or it may be an attempt to defraud the user. Unless the flavorist is an expert in vanilla, the way to avoid the use of adulterated vanilla extracts is to buy only from reliable suppliers.

Entrapped Flavors

Most flavorings are susceptible to attack by oxidation. The terpeneless oils are less affected by oxidation than other flavorings, but even they are not completely stable. The entrapped flavors were developed to increase product shelf life. A water slurry of flavoring and a gum, such as gum acacia, is spray dried. During the drying each individual flavor droplet is coated with a layer of gum. This gum keeps the air (O_2) from attacking the flavor until the flavor is released by breaking the acacia barrier with moisture. The value in consumer products, such as cake mixes, beverage powders and gelatin desserts, is obvious. Shelf life is increased three-fold if the deterioration of flavor is due only to oxidation. If flavor goes off because of polymerization or chemical interaction, then an entrapped flavor will not help. It is obvious that entrapped flavor only should be used in low-moisture systems.

The concentration of flavoring in an entrapped flavor is 20–25 per cent by weight; therefore in use, four to five times the amount of entrapped flavor must be used to achieve the corresponding normal flavor. Entrapped flavors should not be confused with plated flavors, which offer no advantage in shelf life.

Synthetic vs. Natural Flavorings

The flavorist is faced with the question of which to use in his compounding—natural or synthetic materials. There is no one correct answer. Price, availability, and the type of end product will determine the selection of flavoring agents.

The consuming public no longer automatically rejects a product labeled synthetic, imitation or artificial. Admittedly, the early artificial flavorings were crude and evoked displeasure from the consumer. Today, we can produce many excellent artificial flavors, superior in certain situations to naturals. Synthetic hard candy flavor of the friut types will take heat processing, but we will need good synthetic coffee, banana, cherry, malt and pineapple flavorings.

Regardless of quality, we are being forced to use more synthetics; there are not enough naturals to meet demands. The cost of producing many naturals has reached the point where they are so highly priced that they cannot be used. This price rise is a result of competition for both land and labor in the producing area. Food crops are more necessary and yield the producers a greater profit.

Natural products (essential oils, extracts, oleoresins) vary depending on climate, fertilizer and growing area. One year's crop may be entirely different from another. Political problems may severely reduce a crop. For example, patchouly imports were almost non-existent during the Indonesian crisis. For years, cinnamon and menthol could not be imported from China. These and other problems cause the flavorist definitely to consider using synthetics over naturals.

It must also be remembered that synthetics (imitation flavorings) have been vastly improved in recent years so that quality products can be obtained. It is the author's opinion that synthetics will play an even greater part in producing good flavoring in the future.

Single Chemicals

Natural flavors are all composed of a varying complexity of chemicals. Some, such as oil of wintergreen, contain principally one chemical, 98 per cent methyl salicylate; oil of clove is over 85 per cent eugenol. Others are more complex and contain small quantities of many chemicals, for example raspberry and strawberry. Unless a natural flavor is composed of 100 per cent of a single chemical, a single chemical will not completely substitute for a natural. Even though oil of wintergreen contains 98 per cent methyl salicylate, methyl salicylate cannot be used to substitute exactly for oil of wintergreen. This is not to state that methyl salicylate, when used by itself, will not make an adequate flavoring.

Some naturals obtain much of their flavor from a relatively minor chemical in their total composition. Citral is present at 4–5 per cent in oil of lemon, but it is the main flavor contributor. In most naturals it is not one, but many chemicals that blend to produce the final flavor effect. When the flavor chemist tries to reproduce a natural with a chemical mixture, he usually finds that he must use a relatively large number of single chemicals, or his final compound will be crude.

Table 2 on pages 482 to 511 lists some of the more common chemicals that are used in flavoring. The flavorist must realize that the majority of these chemicals, when used alone, will not produce a finished flavor. For example, cinnamic aldehyde, when used alone, will yield a product that is recognized as cassia. However, a more acceptable cassia-type flavor is produced when cinnamic aldehyde is blended with other chemicals and essential oils. The flavorist will vary his combination to produce the flavor that pleases his taste and the need of his product, but it is not the object of this chapter to list finished formulations. Excellent works are available for this purpose.

Terpenes and Camphors

The terpenes are hydrocarbons of formula $C_{10}H_{16}$, and their oxygenated derivatives ($C_{10}H_{16}O$ or $C_{10}H_{18}O$) are known as camphors. They may be straight-chain or cyclic, with or without double bonds. The flavorist is concerned with these compounds, for they contribute both positively and negatively to the flavoring with which he works. A positive contribution is that they are responsible for many of the flavor characteristics in certain essential oils, for example, citral in lemon. Unfortunately, many of the compounds, particularly the camphors, contribute undesirable "off" notes.

At one time terpenes were considered dimers of isoprene, $CH_2{:}CH{\cdot}C(CH_3){:}CH_2$. The monomer was believed to be formed in the living plant and, by chemical reaction, formed the various terpenes that were characteristic of the essential oil of that plant. Although the concept is no longer proposed to explain the formation of these compounds, it is still useful as an aid in understanding their structure.

It is not the purpose of this chapter to discuss the elements of the terpenes and camphors, but if the flavorist has a basic understanding of the similarity of chemical structures of these compounds, he will appreciate that only a minor change in composition will entirely change the flavor of a product. For example, citral, with the addition of two hydrogen atoms, will be changed to geraniol and thus the odor from lemon to rose. A rosy note may be accepted as a modifier in some products, but will not be appropriate in a lemon flavoring as a main note. Some of the interrelationships of this class of compounds of interest to the flavorist may be seen in the grouping below.

In Chart I some of the compounds similar to α-pinene may be seen. These cyclic compounds all contain two carbon rings.

CHART I

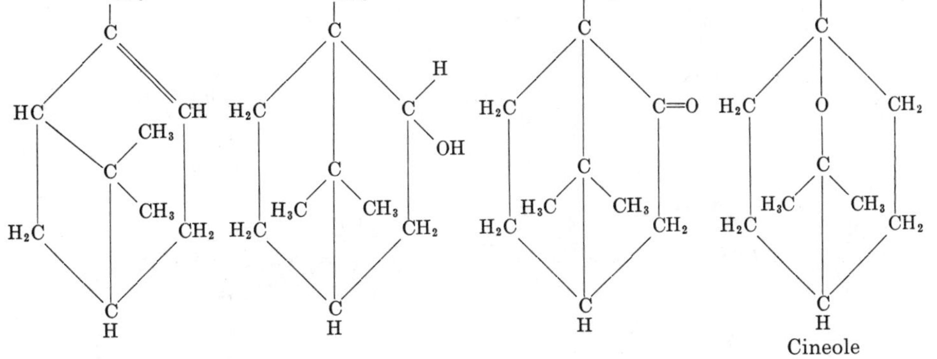

α-Pinene   Borneol   Camphor   Cineole ("Eucalyptol")

In Chart II there are a number of camphors structurally related to limonene. These compounds all contain a single six-carbon ring with methyl and isopropyl configurations in parapositions. The same arrangement of carbon atoms is displayed by thymol and carvacrol but in these cases, the basic structure is a benzene ring; variations on the terpene formula occur through rearrangement or elimination of the double carbon-to-carbon bonds and by addition of hydroxyl or carbonyl groups.

CHART II

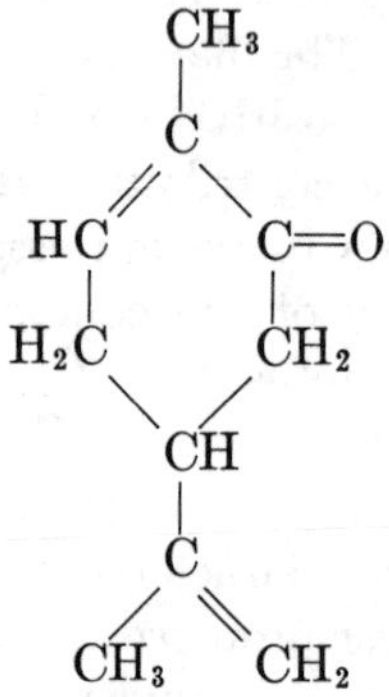

Carvone
(Carvol)

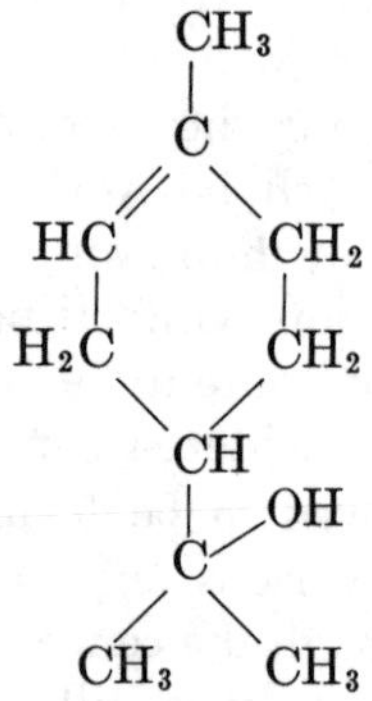

α-Terpineol

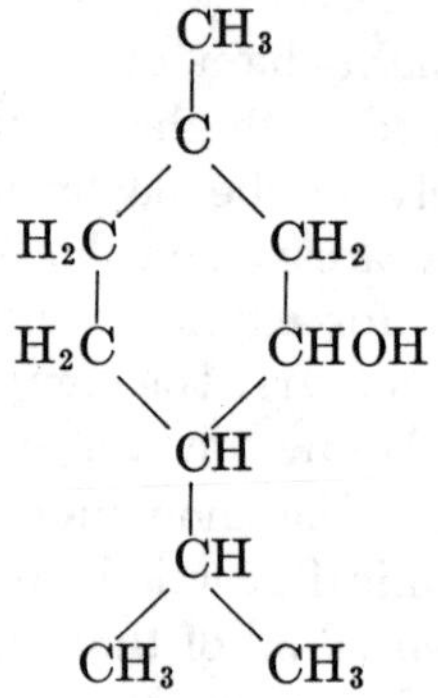

Menthol

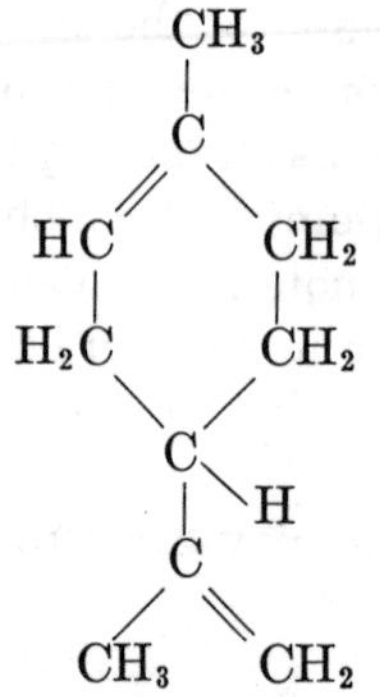

Limonene

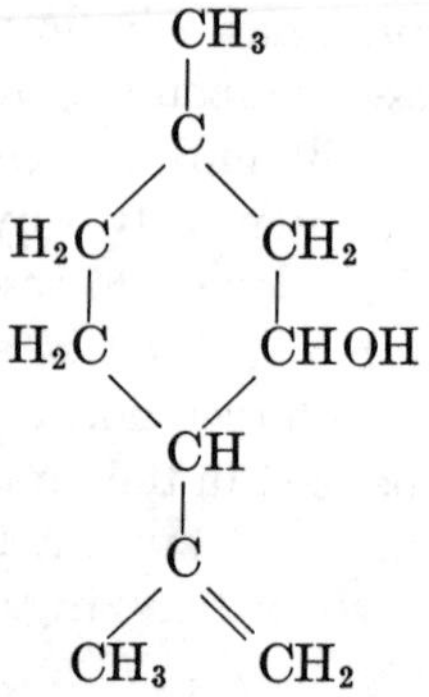

iso-Pulegol

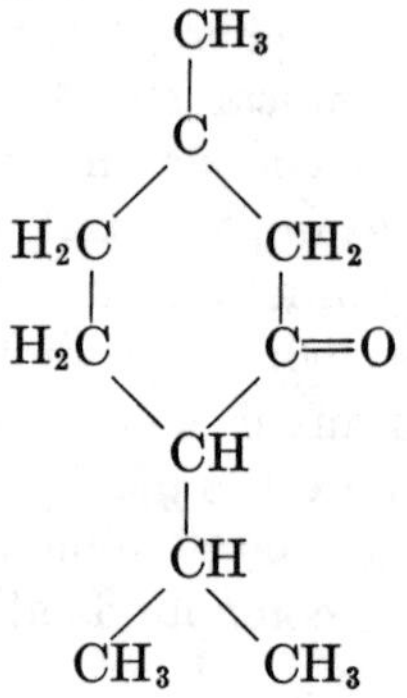

Menthone

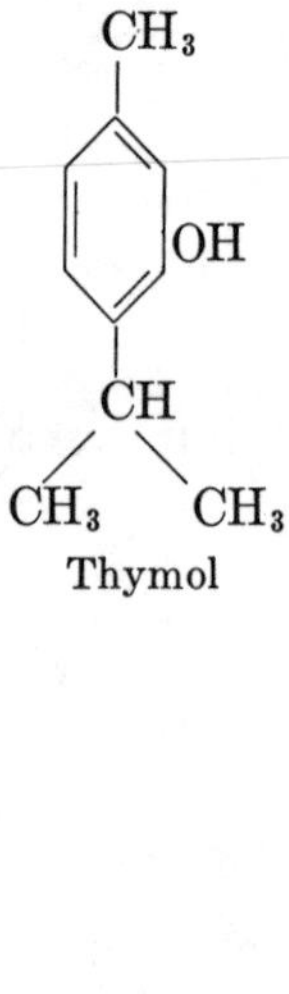

Thymol

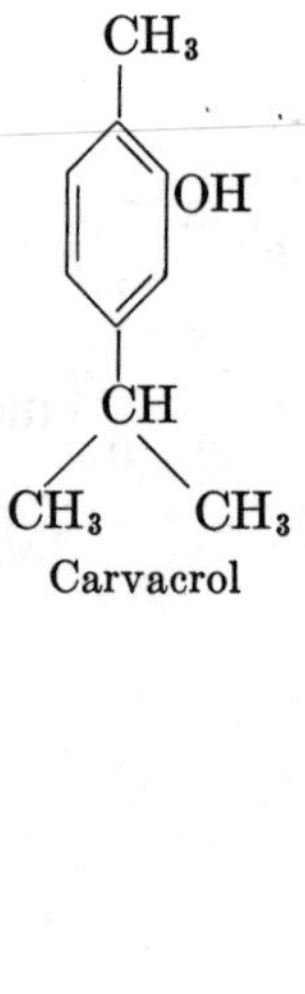

Carvacrol

Related 10-carbon, straight-chain compounds are also of interest. The similarity between the groups is shown if geraniol is depicted in a ring-like configuration (Chart III). The same compound is also shown in the more conventional manner. Citronellal and rhodinal are closely associated in natural occurrence, as well as in chemical structure, and may be identical. Two structures are associated with these chemicals (as shown in Chart III), but the identification of the specific chemical forms with the names citronellal and rhodinal is not generally accepted.

The chemist will recognize that most of the compounds of a given chemical structure illustrated in these charts can occur as two or more isomers. These variations can also influence the properties of the materials available to the flavorist.

CHART III

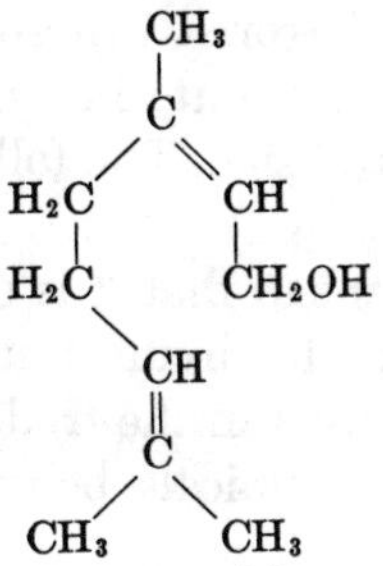

β-geraniol

(or) $CH_3-C=CH-CH_2-CH_2-C=CH-CH_2OH$ with CH_3

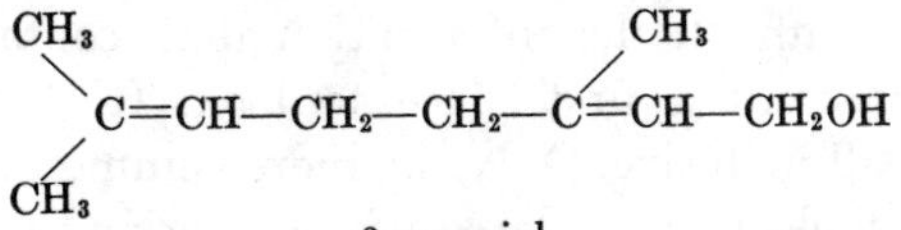

β-geraniol

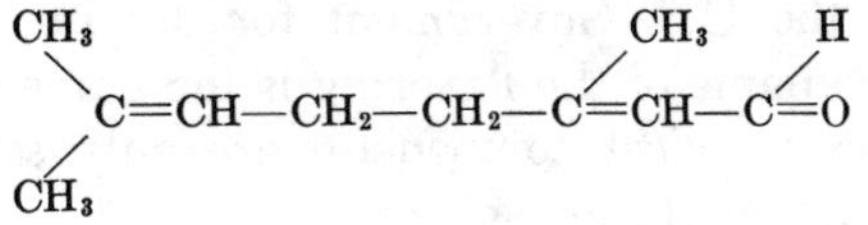

$CH_3-C=CH-CH_2-CH_2-C=CH-C=O$ with CH_3 and H

citral

$CH_3-C-CH_2-CH_2-CH_2-CH-CH_2-CH_2OH$ with CH_2 and CH_3

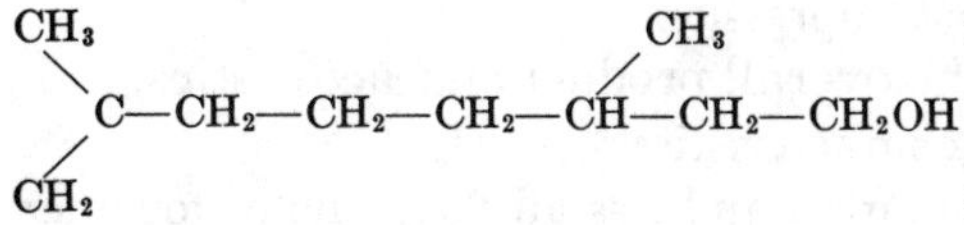

hemonene, or α form
citronellol (rhodinol)

$CH_3-C=CH-CH_2-CH_2-CH-CH_2-CH_2OH$ with CH_3 and CH_3

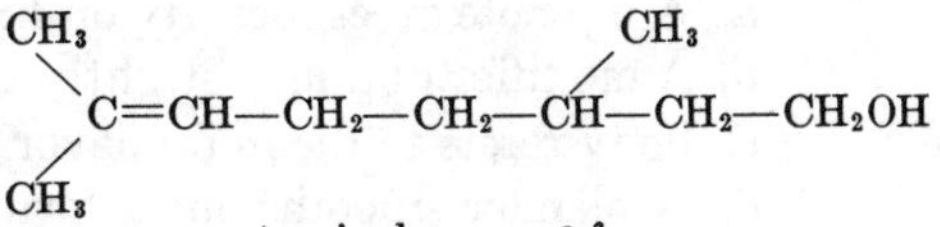

terpinolene, or β form
citronellol (rhodinol)

$CH_3-C-CH_2-CH_2-CH_2-CH-CH_2-C=O$ with CH_2, CH_3, H and O

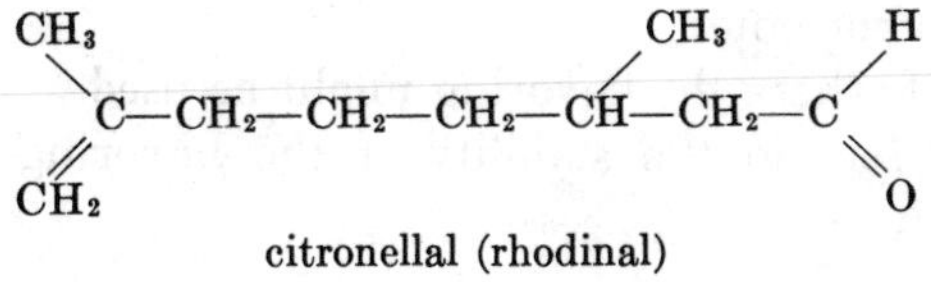

citronellal (rhodinal)

$CH_3-C=CH-CH_2-CH_2-CH-CH_2-C=O$ with CH_3, CH_3 and H

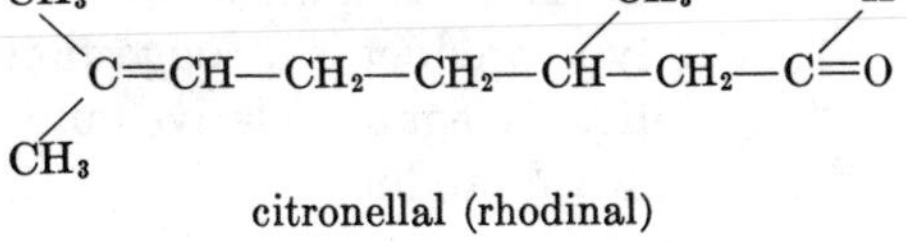

citronellal (rhodinal)

The tables in this chapter are meant to be used by the laboratory-oriented flavorist. The expert with years of experience will not need these as he is familiar with all the information they contain. To the flavorist just starting, they represent a first place to look. It should be pointed out that neither chart is all inclusive. Only the more common materials are listed. The following points should be considered when referring to the tables:

1. A listing of any item does not assure its safety or mean that it is cleared by the U.S. Government for use in foods under all circumstances.
2. The name of the flavoring is the name commonly used in the trade.
3. The chemical components are only some of the chemicals believed to be in the flavoring.
4. The flavor characteristics are those of the flavoring in dilute solution.
5. The solubilities are relative and in high strength ethyl alcohol.
6. The use is the part the material plays in a finished flavoring.
   a. A top note gives identity on first impression.
   b. A modifier changes slightly the overall product characteristics.
   c. Body refers to the main flavor characteristics.
   d. A blender smooths out a formulation and ties all flavor notes together to produce a finished flavor effect.
   e. A fixative reduces the loss of light volatiles.
   f. A mask helps cover unwanted flavor notes.
7. The type product is a suggestion of where the flavoring might be used.
8. Stability is again relative and refers to the stability of the flavoring in finished products.

TABLE 1
Natural Flavoring Materials

Name	Common Chemical Components	Flavor Characteristics	Solubility in Alcohol	Use	Type Product	Stability
Abies sibirica	α- and β-Pinene, bornyl acetate	Pine, camphor, woody	Good	Top note, modifier	Medicinal flavorings	Good
Almond, FFPA	Benzaldehyde	Nutty, almond	Good	Body	Nutty flavors (almond), cherry	Poor
Aloe	Aloin, barbaloin	Bitter	Good	Modifier	Beverages, alcoholic beverages	Good
Althea root	Althein (asparagine)	Sweet, marshmallow	Good	Modifier	Strawberry, cherry, beverages, root beer	Good
Ambergris	Ambra	Benzoin-like	Good	Fixative	Berry, fruit, beverages, candy, ice cream	Good
Ambrette seed	Farnesol	Musk-like	Fair–poor	Fixative	Fruity, beverages, candy, ice cream	Fair–poor
Amyris (W. Indian sandalwood)	Amyrol, amyrolin, cadinene, caryophyllene	Woody	Fair–poor	Modifier	Candy, chewing gum	Good
Angelica	Phellandrene	Bitter, herb-like	Fair	Modifier, body, top note	Alcoholic beverages, gin	Good–fair
Angostura	Galipol, $C_{15}H_{26}O$	Bitter spice	Poor	Body, modifier	Liquors, beverages, bitter spice	Good
Anise	Anethole	Anise	Fair	Body, blender, modifier masking, top note	Beverages, candy	Good
Anise, star	Anethole	Anise-like, licorice	Fair	Body, blender, modifier masking, top note	Pharmaceuticals, liqueurs	Good
Apricot kernel (persic oil)	Benzaldehyde	Bitter, benzaldehyde	Insoluble	Body, modifier	Cherry, beverages, ice cream, candy	Poor
Arnica flowers		Strong aromatic, borneol	Poor	Blender, modifier	Alcoholic beverages, medicinals	Fair
Balsam Peru, oil	Benzyl cinnamate	Woody, spicy	Fair	Fixative, modifier	Pharmaceuticals (cough medicines)	Good

TABLE 1
Natural Flavoring Materials (Continued)

Name	Common Chemical Components	Flavor Characteristics	Solubility in Alcohol	Use	Type Product	Stability
Basil, sweet (oil)	Methyl chavicol, cineole, linalool	Medicinal, herb-like, slight licorice	Fair	Modifier, top note	Spicy flavors, meat products	Poor
Bay leaves (W. Indian oil)	Eugenol	Spicy, slightly bitter	Insoluble	Modifier, top note, body, pickling	Condiments, meats	Fair
Benzoin resin (Siam benzoin, oil)	Coniferyl benzoate	Bitter, chocolate	Good	Fixative, body, modifier	Confections	Good
Bergamot	Limonene	Bitter orange	Fair	Blender, modifier, top note	Citrus flavors, orange, cola	Poor
Birch, sweet betula oil	Methyl salicylate	Wintergreen	Good	Body, modifier	Candies, chewing gum, beverages	Fair–good
Bitter almond, FFPA, oil	Benzaldehyde	Nutty, almond	Good	Body	Confections, nutty, cherry	Poor
Bois de rose, oil	Linalool	Sweet woody, citrus, floral	Good	Top note, modifier	Beverages	Good
Cacao (cocoa)		Chocolate	Poor	Body, modifier	General food flavoring	Good
Cade	Cadinene	Smoky, tarry	Poor	Body	Smoke flavors	Good
Cajeput, oil	Cineole	Camphoric	Good	Modifier	Spice flavors, medicinals, beverages	Good
Calamus	Asarone	Spicy, herb-like	Good	Modifier	Herb and bitter flavors	Fair
Camphor, oil	Camphor, cineole	Woody, camphor, medicinal	Fair	Modifier	Medicinal flavors	Good
Cananga, oil	Cadinene	Woody, spicy	Insoluble	Top note	Cola, spice, fruit, beverages	Good
Capsicum, oleoresin	Capsaicine	Pungent, burning, slight ginger	Fair	Pickling, modifier	Meats, sauces	Good
Caraway	d-Carvone	Caraway	Fair	Body, blender, modifier	Spice flavor, bakery products	Fair
Cardamom	Terpineol, cineole	Spicy, slightly lemon-citrus	Good	Top note, modifier	Processed meats	Good
Carrot, oil	Carotol	Spicy	Fair–good	Blender	Condiments, alcoholic beverages	Good
Cassia	Cinnamic aldehyde	Cinnamon, bite	Good	Top note, body, blender, modifier	Hot, spicy, candy, bakery products	Fair

TABLE 1
Natural Flavoring Materials (Continued)

Name	Common Chemical Components	Flavor Characteristics	Solubility in Alcohol	Use	Type Product	Stability
Cassie flowers, oil	Farnesol, cresols	Bitter, floral, violet	Good	Body	Raspberry	Fair
Catechu black, extract	Catechu tannic acid	Bitter, medicinal	Poor	Modifier	Medicinals, carbonated beverages	Good
Catechu, pale (gambir, extract)	Catechu tannic acid	Bitter, medicinal	Poor	Modifier	Medicinals, carbonated beverages	Good
Cedar leaf	Thujone	Medicinal, camphoric	Good	Blender	Medical products, mouthwashes	Good
Cedar wood	Cedrol	Woody	Fair–poor	Fixative	Medical products	Fair
Celery seed	Limonene, sedenene	Celery, spicy	Poor	Top note, body	Spice blends, carbonated beverages, meat products	Good
Chamomile	—	Pungent aromatic	Fair	Blender	Liqueur flavor	Good
Cherry bark, wild, extract	Benzaldehyde	Bitter almond	Fair	Body	Fruit flavors, medicinals	Good
Cherry laurel, FFPA	Benzaldehyde	Bitter almond	Good	Modifier	Fruit flavors	Good
Chervil, oil	—	Spicy, medicinal, anise	Fair	Modifier	Spice flavors, condiments, processed meats	Fair
Cinnamon	Cinnamic aldehyde	Spicy, hot	Good	Blender, body, top note, modifier	Spice flavors, cola beverages	Good–fair
Cinnamon leaf, oil	Eugenol	Bitter, spicy, pungent	Fair	Modifier, blender	Alcoholic beverages, candies, fruit flavors	Good–fair
Citronella	Geraniol, citronellol	Lemon, citrus	Good	Modifier	Sparingly in citrus flavors	Good
Civet, tincture	Civetone (9-cycloheptadecenone)	Fecal, musky	Good	Fixative	Beverages, candy	Good
Clove	Eugenol	Warm, pungent, spicy clove	Good	Blender, body, top note, modifier	Spicy and medicinal flavors, meat products	Good–fair
Coffee, extract	Caffeine	Bitter, coffee	Fair	Body	General flavoring material	Fair
Cognac	Mixed caproates	Wine-like	Fair	Top note, modifier	Alcoholic beverage flavor	Fair
Cola nut, extract	Caffeine	Bitter, herb-like	Good	Modifier	Pharmaceuticals, cola beverages	Good

TABLE 1
Natural Flavoring Materials (Continued)

Name	Common Chemical Components	Flavor Characteristics	Solubility in Alcohol	Use	Type Product	Stability
Copaiba	Caryophyllene	Bitter, balsamic	Insoluble	Blender, fixative	Medicinal flavors	Fair
Coriander	d-Linalool	Spicy	Good	Body, modifier	General spice flavors, meats	Good
Costus root	Splotnyene	Heavy, slightly violet	Fair	Blender, modifier	Trace amounts in medicinals and liqueurs	Fair, discolors
Cubeb	d-Sabinene	Spice, pepper	Fair	Modifier	Tobacco flavor	Good
Cumin	Cuminic aldehyde	Pungent, sweaty	Fair	Body, modifier, blender	Curries	Good
Curaçao, peel	Limonene	Citrus-orange	Fair	Blender, modifier	Orange liqueurs	Fair
Dill, seed	l-Carvone	Slightly spicy, herb-like	Fair	Body, modifier	Pickling flavor, imitation caraway	Fair
Dill, weed	Phellandrene, carvone	Herb-like, bitter	Fair	Body, blender, modifier	Pickle and spice flavors	Fair
Elemi, oil	Phellandrene	Citrus-lemon, spicy	Poor	Modifier	Pharmaceuticals, cola beverages	Good
Estragon (tarragon)	Methyl chavicol	Tarragon, anise	Fair	Blender, modifier	Liqueur flavoring, spice modifier	Good
Eucalyptus	Eucalyptol	Medicinal	Good	Blender, top note, body	Medicinal flavor (mouthwash), cough drops	Good
Fennel	Anethole	Anise	Good	Top note, blender	Liqueur, salad dressing	Good
Fenugreek, extract	—	Bitter, burnt sugar, maple	Poor	Body	Imitation maple flavor	Excellent
Galbanum, oleoresin	d-α-Pinene, β-pinene	Bitter, acrid	Fair	Modifier	Pharmaceuticals, fruit, nut, spice flavors	Good
Garlic, oil	Allicin (diallyl disulfide?)	Garlic	Poor	Body	Meats, soups, condiments	Good
Geranium	Geraniol	Rose geranium	Good	Blender, modifier	Dentifrices, ginger ale	Good
Ginger	Zingiberone	Spice, hot, slightly soapy	Poor	Body, modifier, fixative, masking agent	Beverages, spice sauces, baked goods	Good
Grapefruit	Limonene	Grapefruit	Insoluble	Body, blender, modifier	Citric products (beverages)	Fair
Guaiac wood, oil	Guaiol	Woody	Fair	Fixative, top note	Chewing gum, alcoholic beverages, medicinals	Good
Hops	Humulone	Fatty, green, oily	Fair	Modifier, top note	Beverage flavors	Fair
Horehound, extract	Tannin	Bitter, slightly woody, pine-like	Good	Body	Candy, pharmaceuticals	Good

TABLE 1
Natural Flavoring Materials (Continued)

Name	Common Chemical Components	Flavor Characteristics	Solubility in Alcohol	Use	Type Product	Stability
Horseradish	Allyl isothiocyanate	Hot, biting, penetrating	Fair	Modifier, body	Hot sauces	Good
Hyssop	l-Pinocamphone	Camphoraceous	Fair	Modifier	Cordials—Benedictine and Chartreuse	Fair
Jasmine, oil	Benzyl acetate, linalool	Floral	Good	Modifier, top note	Raspberry, strawberry, cherry flavors	Poor
Juniper	α-Pinene, camphene	Gin-like, fragrant	Insoluble	Body, top note	Gin base	Fair
Labdanum	Mixed terpenes	Balsamic, bitter, tingly	Fair	Fixative, modifier	Medicinals	Good
Laurel leaf	Cineole, eugenol	Sweet spicy, biting, slightly clove-like	Good	Modifier	Spice flavoring	Fair
Lavandin	Linalyl acetate	Pungent, lavender	Good	Modifier	Dentifrices, chewing gum	Good
Lavender	Linalyl acetate	Lavender	Good	Modifier, blender	Dentifrices, chewing gum, breath products	Good
Lavender, spike, oil	Linalyl acetate	Lavender, floral	Good	Modifier, top note	Mouth fresheners	Good
Lemon	Citral	Lemon	Good	Modifier, blender, body	Citrus products, candy	Fair
Lemongrass, oil	Citral	Lemon	Good	Body, top note	Lemon, fruit flavors	Fair
Licorice, extract	Glycyrrhizin	Sweet, spicy-caramel, licorice	Poor	Body	Pharmaceuticals, candy, chewing gum	Good
Lime	Citral, limonene	Lime	Good	Blender, body, top note, modifier	Citrus products, cola beverages, candy	Poor
Lovage	d-Pinene, myristicin, d-camphene	Celery-like, pungent	Good	Fixative, modifier	Liqueurs, tobacco	Good
Mace		Nutmeg, aromatic, pine	Fair	Body, blender, modifier	Spice flavors	Good
Malt		Malt	Poor	Body, modifier	Cereals	Good
Mandarin	d-Limonene	Citrus-orange	Poor	Modifier, top note	Citrus products	Fair
Marjoram	Terpinene	Spicy, pungent	Good	Modifier, top note	Spice flavors	Good
Mastic	Pinene	Woody, camphoraceous	Insoluble	Modifier, blender	Cordials, gum	Good
Maté	Caffeine, tannic acid	Dry, herb-like, woody	Good	Modifier	Pharmaceuticals	Fair
Mimosa, oil		Floral	Good	Blender	Raspberry, fruit flavors	Fair
Musk, tonquin, tincture	Muskone	Powdery, musky, sweet	Good	Blender	Tobacco flavors, caramel, nut flavors	Good

TABLE 1
Natural Flavoring Materials (Continued)

Name	Common Chemical Components	Flavor Characteristics	Solubility in Alcohol	Use	Type Product	Stability
Mustard	Allyl isothiocyanate	Pungent, sharp	Soluble	Modifier	Relish flavors, salad dressings	Fair
Myrrh		Herb-like, balsamic, bitter	Good	Fixative	Mouthwash flavors	Good
Neroli	Linaloöl, linalyl acetate, methyl anthranilate	Orange blossom	Good	Blender, top note, modifier	Citrus flavors, cola	Fair
Nutmeg	Pinene, myristicin	Spicy, nutmeg	Fair	Blender, top note, modifier, body	General spice flavors, baked goods	Good
Oak wood, extract	Tannin	Bitter, woody, astringent	Fair	Modifier	Alcoholic beverages	Fair
Olibanum, oil	Pinenes, borneol, verbenol	Bitter resinous, soapy	Fair	Blending	Cola, fruit, spice flavors, beverages	Poor
Onion, oil	Diallyl disulfide	Onion	Very poor	Body, modifier	Meats, condiments, soups	Good
Opopanax, oil	Bisabolene	Balsamic, resinous	Fair	Modifier	Medicinals, alcoholic beverages	Poor
Orange, bitter, oil	d-Limonene	Bitter orange	Poor–fair	Body, modifier, top note	Citrus flavors, soft drinks, orange liqueurs	Poor
Orange, sweet	d-Limonene	Sweet orange	Poor–fair	Body, modifier, top note	Soft drinks, candy, citrus flavors, general flavoring	Poor
Origanum	Carvacrol	Sharp, camphor-like, medicinal	Good	Modifier, body, top note	Italian flavors	Good
Orris root	Methyl ionone	Violet	Poor	Top note, blender	Raspberry flavors	Good
Palmarosa	Geraniol	Rose	Good	Blender	Dental preparations	Good
Paprika, oleoresin		Spicy	Fair	Color	Sauces, condiments	Good
Parsley, oil	Apiole	Herb-like, slightly bitter, green	Fair	Modifier	Seasonings, sauces, pickles, meats	Good
Patchouly		Earthy, slightly woody	Fair	Fixative, modifier	Cola beverages	Good
Pepper, black	Piperidine	Warm, spicy	Poor	Modifier, blender, body	General spice flavors, prepared meats	Good
Pepper, white, oil	Phellandrene	Aromatic, biting	Poor	Body, blender, top note, modifier	Baked goods, spice flavors, condiments	Good
Peppermint	Menthol, menthone	Mint	Good	Body, blender, top note, modifier	Minty flavors, gum, dental products, candy, pharmaceuticals	Fair

TABLE 1
Natural Flavoring Materials (Continued)

Name	Common Chemical Components	Flavor Characteristics	Solubility in Alcohol	Use	Type Product	Stability
Petitgrain	Linalyl acetate	Citrus, bitter	Good	Blender	Citrus flavors	Good
Pimenta	Eugenol, cineole	Allspice	Good	Blender, top note	General spice flavors, bakery products	Good
Pine, needle, oil	2-Pinene	Terpentine-like, woody, resinous	Fair	Modifier	Beverages, liquors	Fair
Pipsissewa, extract	Chimaphilin	Minty, green, astringent, bitter	Fair	Modifier	Beverages, root beer, candy	Good
Rose	Geraniol, nerol	Rose	Poor	Blender, modifier	Tobacco, candy, carbonated beverages	Good
Rosemary	Pinene, borneol, cineole	Slightly medicinal, woody	Good	Masking agent, modifier	Herb blends, mouthwashes	Good
Rue, oil	Methyl nonyl ketone	Sharp herbaceous, burning	Good	Blender	Coconut flavors	Good–fair
Saffron, extract	Picrocrocin	Spicy, strong, bitter, medicinal	Good	Blender	Meats, condiments	Good
Sage, Clary, oil	Linaloöl, linalyl acetate	Tea-like, balsamic	Poor	Modifier	Liqueurs, spice compounds	Good
Sage, Dalmatian, oil	Thujone	Spicy, warm, tea-like	Poor	Blender, body	Meat flavors, poultry stuffing	Fair
Sandalwood, East Indian, oil	α- and β-Santalol	Woody, bitter, resinous	Good	Modifier	Breath freshener, carbonated beverages	Poor
Sassafras, extract (safrole-free)	—	Green, herb-like	Good	Body, top note	Soups, pharmaceuticals	Good
Savory, oil	Carvacrol, p-cymene	Herbaceous, medicinal, camphoraceous	Good	Body, modifier	Spice flavors, culinary preparations	Fair
Spearmint, oil	l-Carvone	Mint	Good	Body, modifier, top note, blender	Mint flavors, pharmaceuticals	Fair
Storax, oil	Cinnamic acid	Balsamic	Good	Modifier	Pharmaceuticals	Good
Tamarind, extract	—	Sour	Fair	Modifier	Beverages	Fair
Tangerine, oil	d-Limonene	Citrus-orange	Good	Modifier	Candy, soft drinks, ice cream flavors, orange, lemon-lime flavors	Poor
Tea, extract	Caffeine, tannin	Tea	Fair	Body, modifier	Beverages	Fair

TABLE 1
Natural Flavoring Materials (Continued)

Name	Common Chemical Components	Flavor Characteristics	Solubility in Alcohol	Use	Type Product	Stability
Thyme	Thymol	Medicinal, burnt	Good	Blender, modifier, top note	Medicinal flavors	Good
Tolu balsam, oleoresin	l-Cadinol	Slightly bitter, spicy, cinnamic	Good	Fixative	Pharmaceuticals	Good
Tuberose, oil	Geraniol, nerol	Floral	Good	Modifier	Peach flavor	Poor
Turmeric, oleoresin		Bitter, spicy, woody-sweet	Fair	Coloring agent	Condiments, meats	Good
Vanilla, extract	Vanillin	Vanilla-sweet	Good	Body, modifier, top note	Ice cream, bakery products	Good
Vetiver, oil	$C_{15}H_{24}O$	Slight asparagus	Fair–good	Fixative	Vegetables	Fair
Violet leaves, oil		Floral, violet	Good	Top note	Beverages	Poor
Wintergreen	Methyl salicylate	Minty	Good	Modifier, top note, blender, body	Minty flavors, candy, pharmaceuticals	Good
Yerba santa, extract		Medicinal	Good	Body	Fruit flavors	Good
Ylang-ylang	Benzyl alcohol, linaloöl, cresol, methyl ether	Fragrant, slight orange	Good	Top note	Beverage flavors	Fair

TABLE 2
Synthetic Flavoring Materials

Name	Structure	Flavor Characteristics	Solubility in Alcohol	Use	Type Product	Stability
Acetaldehyde	CH_3CHO	Sharp, pungent	Good	Top note	Fruity and nutty flavors	Fair
Acetal	$CH_3CH(OC_2H_5)_2$	Sharp, slightly hay-like	Good	Top note, modifier	Fruit flavors	Poor
Acetanisole	CH_3O—⬡—$COCH_3$		Moderate	Modifier	Brown and nutty flavors	Good

TABLE 2
Synthetic Flavoring Materials (Continued)

Name	Structure	Flavor Characteristics	Solubility in Alcohol	Use	Type Product	Stability
Acetic acid	CH_3COOH	Sharp, pungent, vinegar	Good	Body, modifier	Relishes, pickles	Good
Acetoin	$CH_3CH(OH)COCH_3$	Buttery	Good	Body, top note	Dairy products	Good
Acetophenone	⟨benzene⟩—$COCH_3$	Bitter, burning	Good	Modifier	Fruit, nutty flavors	Good
ALCOHOLS						
C_6, hexyl alcohol	$CH_3(CH_2)_4CH_2OH$	Sharp, green	Good	Top note	Fruit, coconut flavors	Good
C_7, heptyl alcohol	$CH_3(CH_2)_5CH_2OH$	Fruity	Good	Top note	Fruit flavors	Good
C_8, n-octyl alcohol	$CH_3(CH_2)_6CH_2OH$	Sharp, powerful, slightly orange	Good	Top note	Citrus flavors	Good
C_9, n-nonyl alcohol	$CH_3(CH_2)_7CH_2OH$	Soft citrus	Good	Top note	Floral effects	Good
C_{10}, n-decyl alcohol	$CH_3(CH_2)_8CH_2OH$	Orange, sharp	Good	Top note, body, modifier	Citrus flavors	Good
C_{11}, undecyl alcohol	$CH_3(CH_2)_9CH_2OH$	Slightly citrus, sharp	Good	Top note	Citrus flavors	Good
C_{11}, undecen-1-ol alcohol	$CH_3(CH_2)_8CHOHCH_3$	Slightly fatty, citrus	Good	Blender	Citrus flavors	Good
C_{12}, lauryl alcohol	$CH_3(CH_2)_{10}CH_2OH$	Fatty, slightly orange	Good	Top note, blender	Citrus flavors	Fair
ALDEHYDES						
C_6, hexyl aldehyde	$CH_3(CH_2)_4CHO$	Fatty, fruits	Good	Modifier	Butter, honey flavors	Fair
C_7, heptyl aldehyde	$CH_3(CH_2)_5CHO$	Sweet, burning, sharp, almond	Good	Modifier, blender	Almond flavors	Fair
C_8, octyl aldehyde	$CH_3(CH_2)_6CHO$	Sharp, citrus, orange	Good	Top note, body, modifier	Citrus—orange	Poor
C_9, nonyl aldehyde	$CH_3(CH_2)_7CHO$	Sharp, orange, floral, waxy	Good	Top note, modifier	Citrus—orange, lemon	Fair
C_{10}, decyl aldehyde	$CH_3(CH_2)_8CHO$	Sharp, orange	Fair	Top note, modifier	Citrus, fruity	Fair

TABLE 2
Synthetic Flavoring Materials (Continued)

Name	Structure	Flavor Characteristics	Solubility in Alcohol	Use	Type Product	Stability
ALDEHYDES (Continued)						
C_{10}, methyl nonanal	$CH_3(CH_2)_5CH(CH_3)CH_2CHO$	Sweet, spicy-citrus, fatty	Fair	Top note	Orange flavors	Fair
C_{10}, dimethyl octanal	$(CH_3)_2CH(CH_2)_3CH(CH_3)CH_2CHO$	Bittersweet, lemon, slightly floral	Fair	Top note, modifier	Lemon, orange flavors	Fair
C_{11}, undecylenic aldehyde	$CH_2{=}CH(CH_2)_8CHO$	Bittersweet	Fair	Modifier	Honey flavors	Poor
C_{11}, undecylic aldehyde	$CH_3(CH_2)_9CHO$	Sharp citrus orange, bittersweet	Fair	Top note, modifier, fixative for citrus	Orange citrus, honey flavors	Fair
C_{12}, lauric	$CH_3(CH_2)_{10}CHO$	Bitter, waxy	Fair	Modifier	Honey flavors	Good
C_{12}, MNA	$CH_3(CH_2)_8CH(CH_3)CHO$	Bitter, waxy, slightly citrus	Fair	Modifier	Honey, orange flavors	Good
C_{14}, "peach" (γ-undecalactone)	$CH_3(CH_2)_6CHCH_2CH_2{-}O{-}CO$	Peach, rancid	Fair	Body, top note, blender	Fruity flavors, peach	Good
C_{14}, myristic	$CH_3(CH_2)_{12}CHO$	Sweet amber, slightly citrus	Good–fair	Blender, modifier	Honey flavors	Good
C_{16}, "strawberry" (3-methyl-3-phenyl glycidic acid ethyl ester)	$C_6H_5C(CH_3){-}CHCOOCH_2CH_3$ (epoxide)	Sharp, heavy, strawberry	Fair–good	Body, top note, modifier	Red fruit flavors (strawberry, cherry, raspberry)	Good
C_{18}, aldehyde (γ-nonalactone)	$CH_3(CH_2)_4CHCH_2CH_2{-}O{-}CO$	Coconut	Good	Body, top note, modifier	Nutty flavors	Good
p-Allyl anisole (estragole)	$CH_2{=}CHCH_2{-}C_6H_4{-}OCH_3$	Sweet, anise	Good–fair	Body, modifier	Fruit, licorice, anise, spice flavors	Good–fair
Allyl anthranilate	$C_6H_4(COOCH_2CH{=}CH_2)(NH_2)$	Spicy, slightly grape-like	Good	Top note, modifier	Citrus fruit, grape flavors	Good

TABLE 2
Synthetic Flavoring Materials (Continued)

Name	Structure	Flavor Characteristics	Solubility in Alcohol	Use	Type Product	Stability
Allyl butyrate	$CH_2=CHCH_2OOCCH_2CH_2CH_3$	Sharp, ethereal	Good	Modifier	Butter, fruit, pineapple flavors	Good
Allyl caproate (allyl hexanoate)	$CH_3(CH_2)_4COOCH_2CH=CH_2$	Pineapple, fruity	Fair	Body, modifier, top note	Fruity and tutti-frutti	Fair–good
Allyl caprylate (allyl octanoate)	$CH_3(CH_2)_6COOCH_2CH=CH_2$	Pineapple	Good	Body, modifier	Pineapple flavor	Good
Allyl cinnamate	$\bigcirc\!\!-CH=CHCOOCH_2CH=CH_2$	Fruity	Good	Modifier	Apricot, peach, pineapple flavors	Good
Allyl disulfide	$(CH_2=CHCH_2)S_2$	Sharp, garlic	Good	Body	Onion, garlic	Fair
Allyl heptylate	$CH_3(CH_2)_5COOCH_2CH=CH_2$	Pear, pineapple	Good	Body, modifier	Berry, fruit, brandy	Good
Allyl α-ionone	$CH=CHCOCH_2CH_2CH=CH_2$	Raspberry, violet	Fair–good	Modifier	Fruit flavors	Fair
Allyl isothiocyanate	$CH_2=CHCH_2NCS$	Sharp, mustard oil	Good	Body	Synthetic mustard	Poor
Allyl isovalerate	$(CH_3)_2CHCH_2CO_2CH_2=CHCH_2$	Sharp, ethereal	Good	Modifier	Fruit flavors	Good
Allyl mercaptan	$CH_2=CHCH_2SH$	Coffee, garlic	Good	Modifier	Spice flavors	Good
Allyl pelargonate (allyl nonanoate)	$CH_3(CH_2)_7CO_2CH_2CH=CH_2$	Sharp, ethereal	Good	Modifier	Fruit flavors	Good
Allyl phenoxyacetate	$\bigcirc\!\!-OCH_2CO_2CH_2CH=CH_2$	Sharp, ester-like	Good	Modifier, blender	Fruit flavors	Good
Allyl phenylacetate (allyl α-toluate)	$\bigcirc\!\!-CH_2CO_2CH_2CH=CH_2$	Rosy	Good	Modifier	Pineapple, honey flavors	Good
Allyl propionate	$CH_3CH_2CO_2CH_2CH=CH_2$	Fruity	Good	Top note, modifier	Pineapple flavor	Good
Allyl tiglate	$CH_3CH=C(CH_3)CO_2CH_2CH=CH_2$		Good	Modifier	Fruit flavors	Good

TABLE 2
Synthetic Flavoring Materials (Continued)

Name	Structure	Flavor Characteristics	Solubility in Alcohol	Use	Type Product	Stability
Ambrettolide (ω-6-hexadecen-lactone)	$CH(CH_2)_4C{=}O$ / $CH(CH_2)_8CH_2$ —O—	Powdery, musky	Good–fair	Fixative	Fruit flavors	Good
Amyl alcohol	$CH_3(CH_2)_3CH_2OH$	Sharp, fruity, ester-like	Good	Top note	Fruit and liquor flavors	Good
Amyl butyrate	$CH_3CH_2CH_2CO_2C_5H_{11}$	Ester-like, sharp, pear-like	Good	Body, modifier	Fruit, berry, and butter flavors	Fair
Amyl caproate (amyl hexanoate)	$CH_3(CH_2)_4CO_2C_5H_{11}$	Sharp, ester-like, pear	Good	Modifier	Fruit flavors	Fair
Amyl caprylate	$CH_3(CH_2)_6CO_2CH_2(CH_2)_3CH_3$	Soapy, plum	Good	Modifier	Chocolate, fruit, liquor flavors	Good
Amylcinnamaldehyde	$CH_2(CH_2)_3CH_3$ / C$_6$H$_5$—CH=CCHO	Floral, jasmine	Good	Modifier	Fruit, berry, nut flavors	Fair
Amylcinnamaldehyde dimethyl acetal	$CH_2(CH_2)_3CH_3$ / C$_6$H$_5$—CH=CCH(OCH$_3$)$_2$	Floral	Good	Modifier	Fruit flavors	Fair
Amyl cinnamic alcohol	$CH_2(CH_2)_3CH_3$ / C$_6$H$_5$—CH=CCH$_2$OH	Floral	Good	Modifier	Chocolate, fruit, honey flavors	Good
Amyl cinnamyl acetate	$CH_2(CH_2)_3CH_3$ / C$_6$H$_5$—CH=CCH$_2$O$_2$CCH$_3$	Spicy (cinnamon)	Good	Body, modifier	Fruit, nut, honey flavors	Good
Amyl formate	$HCOOC_5H_{11}$	Sharp, fruity	Good	Body, top note, modifier	Fruity esters	Fair

TABLE 2
Synthetic Flavoring Materials (Continued)

Name	Structure	Flavor Characteristics	Solubility in Alcohol	Use	Type Product	Stability
Amyl heptanoate	$CH_3(CH_2)_5COOC_5H_{11}$	Apple, pear	Good	Modifier	Fruit, coconut flavors	Good
Amyl methyl ketone	$CH_3CO(CH_2)_4CH_3$	Pear	Good	Body, modifier	Fruit, berry flavors	Good
Anethole	CH_3O—⬡—$CH{=}CHCH_3$	Anise, sweet	Good	Body, blender, modifier	Licorice and anise flavors	Good
Anisaldehyde (anisic aldehyde)	CH_3O—⬡—CHO	Soft, flowery	Good	Top note, modifier	Anise flavor	Good
Anisyl alcohol (anisic alcohol)	CH_3O—⬡—CH_2OH	Sweet, peach	Good	Top note, modifier	Apricot, peach flavors	Good
Anisole (methyl phenyl ether)	⬡—OCH_3	Anise, sharp	Good	Body, top note	Licorice, maple	Good
Anisyl acetate	$CH_3CO_2CH_2$—⬡—OCH_3	Bitter, vermouth	Good	Modifier, top note	Vermouth	Good
Anisyl butyrate	CH_3O—⬡—$CH_2OOC(CH_2)_2CH_3$	Fruity, ester-like	Good	Modifier	Fruit, licorice flavors	Fair
Anisyl formate	CH_3O—⬡—CH_2OOCH	Strawberry	Good	Modifier	Fruit flavors	Fair
Anisyl phenylacetate	CH_3O—⬡—CH_2OOCCH_2—⬡	Fruity, floral	Good	Body, top note	Honey flavors	Good
Anisyl propionate	CH_3O—⬡—$CH_2OOCCH_2CH_3$	Fruity, ester-like	Good	Modifier	Raspberry, cherry, licorice flavors	Fair
Aubepine (p-methoxy-benzaldehyde)	(See Anisaldehyde)					
Benzaldehyde	⬡—CHO	Bitter almond	Good	Top note, body	Cherry, fruity, nutty	Poor

TABLE 2
Synthetic Flavoring Materials (Continued)

Name	Structure	Flavor Characteristics	Solubility in Alcohol	Use	Type Product	Stability
Benzaldehyde dimethyl acetal	⬡—CH(OCH$_3$)$_2$	Nutty	Good	Top note, body	Cherry, fruity, nutty	Fair
Benzophenone	⬡—C(=O)—⬡	Apricot, peach	Good	Fixative	Berry, butter, fruit, nut, vanilla flavors	Good
Benzyl acetate	CH$_3$COOCH$_2$—⬡	Synthetic jasmine	Good	Top note	Fruity flavor	Good
Benzyl acetoacetate	CH$_2$OOCCH$_2$COCH$_3$—⬡	Cherry-like, pungent	Good	Modifier	Berry, fruit flavor	Fair–good
Benzyl alcohol	CH$_2$OH—⬡	Low	Good	Top note, fixative	Berry flavor	Fair
Benzyl benzoate	COOCH$_2$—⬡	Chemical	Good	Fixative	Banana, cherry, berry, coffee	Good
Benzyl butyrate	CH$_3$(CH$_2$)$_2$COOCH$_2$—⬡	Pear, sweet	Good	Top note	Fruit flavors	Good
Benzyl cinnamate	CH=CHCOOCH$_2$—⬡	Honey	Good	Fixative, modifier	Spicy, liquor	Good
Benzyl 2,3-dimethyl crotonate	CH$_2$OOCC(CH$_3$)=C(CH$_3$)$_2$—⬡	—	Good	Modifier	Fruit, spice	Good
Benzyl disulfide	CH$_2$—S—S—CH$_2$—⬡	Sharp	Fair	Modifier	Coffee, caramel	Good
Benzyl formate	CH$_2$OOCH—⬡	Apricot, pineapple	Good	Top note	Fruit, berry, orange	Fair
Benzyl isobutyrate	(CH$_3$)$_2$CHCOOCH$_2$—⬡	Sour, strawberry	Good	Modifier	Berry flavors	Good

TABLE 2
Synthetic Flavoring Materials (Continued)

Name	Structure	Flavor Characteristics	Solubility in Alcohol	Use	Type Product	Stability
Benzyl isovalerate	$(CH_3)_2CHCH_2COOCH_2$—	Apple	Good	Modifier	Apple, fruit flavors	Good
Benzyl phenylacetate (benzyl α-toluate)	—CH_2COOCH_2—	Honey	Good	Body, top note	Butter, caramel, fruit, honey	Good
Benzyl salicylate	OH / CH_2OOC—	Raspberry	Good	Fixative, modifier, body	Fruit, berry	Good
Borneol ($C_{10}H_{17}OH$)	OH	Sweet, peppery, woody	Good	Modifier	Medicinal flavors	Good
Bornyl acetate	$OOCCH_3$	Pine-like, camphor, woody	Good	Modifier	Medicinal, spice, modifier	Good
Bornyl formate	$OOCH$	Pine-like, woody	Good	Modifier	Medicinals	Fair
Bornyl isovalerate	$OOCCH_2CH(CH_3)_2$	Woody	Good	Top note	Fruit flavors	Fair–good
Butyl acetate	$CH_3COO(CH_2)_3CH_3$	Pineapple, sweet-sour	Good	Body, modifier	Fruit, berry, butter flavors	Fair
Butyl alcohol	$CH_3(CH_2)_2CH_2OH$	Slightly banana, slightly fusel	Good	Body, modifier	Rum, liquor, butter flavors	Good
Butyl anthranilate	$COOCH_2(CH_2)_2CH_3$ / NH_2	Grape	Good	Modifier	Grape, mandarin, pineapple flavors	Good

TABLE 2
Synthetic Flavoring Materials (Continued)

Name	Structure	Flavor Characteristics	Solubility in Alcohol	Use	Type Product	Stability
Butyl butyrate	$CH_3(CH_2)_2COOCH_2(CH_2)_2CH_3$	Fruity, pineapple	Good	Body	Pineapple, butterscotch, butter, berry flavors	Good
Butyl formate	$HCOOC_4H_9$	Sweet, plums	Good	Body, top note	Fruit, plum, liquor flavors	Fair
Butyl phenylacetate	$CH_2COOCH_2(CH_2)_2CH_3$ (with benzene ring)	Slightly floral	Good	Body, modifier	Butter, caramel, chocolate, fruit flavors	Good
Butyl valerate	$CH_3(CH_2)_3COOCH_2(CH_2)_2CH_3$	Apple	Good	Body, modifier	Butter, fruit, chocolate flavors	Fair–good
Butyric acid	$CH_3CH_2CH_2COOH$	Rancid butter	Good	Body	Butter, cheese, butterscotch, caramel, fruit, nut flavors	Good
Caffeine		Bitter	Good–fair	Modifier	Cola, root beer flavors	Good
Camphene		Woody	Good	Body, modifier	Spice, nutmeg flavors	Fair
d-Camphor		Camphor	Good	Body, top note	Mint flavors	Good
Capric acid	$CH_3(CH_2)_8COOH$	Buttery, oily, sweet	Good	Body, modifier	Coconut, butter, whiskey flavors	Good

TABLE 2
Synthetic Flavoring Materials (Continued)

Name	Structure	Flavor Characteristics	Solubility in Alcohol	Use	Type Product	Stability
Caproic acid	$CH_3(CH_2)_4COOH$	Oily, sweet, soapy, slightly brandy-like	Good	Modifier	Coconut, butter, brandy, whiskey flavors	Good
Caramel	Polymer	Sweet, burnt	Good	Body	General food use	Good
Carvacrol	(structure)	Spicy, bitter, burnt	Good	Modifier	Citrus, fruit, mint, spice, medicinals	Good
4-Carvomenthenol	(structure)	Fruity	Good	Modifier	Citrus, spice	Good
d-Carvone	(structure)	Bitter, caraway	Good	Body	Liqueur flavors, spice	Fair
l-Carvone	(structure)	Spearmint	Good	Body	Mint flavors	Fair
Caryophyllene	(structure)	Clove	Good	Modifier	Spice flavors	Good
Cedrene	—	Woody (cedar), powdery	Good	Modifier	Carbonated beverages	Good

TABLE 2
Synthetic Flavoring Materials (Continued)

Name	Structure	Flavor Characteristics	Solubility in Alcohol	Use	Type Product	Stability
Cedrol		Woody (cedar)	Good	Fixative	Spice flavors	Good
Cineole (eucalyptol)		Eucalyptus, bitter-sweet	Good	Body, blender	Medicinal flavors	Good
Cinnamic acid	$CH=CHCOOH$	Sweet apricot	Good	Body, modifier	Apricot, peach, pineapple flavors	Good
Cinnamic aldehyde	$CH=CHCHO$	Cinnamon	Good	Body, top note	Imitation cassia	Fair–poor
Cinnamyl acetate	$CH=CHCH_2OOCCH_3$	Sweet pineapple	Good	Fixative	Fruit flavors, cinnamon, vanilla	Good
Cinnamyl alcohol	$CH=CHCH_2OH$	Sweet, slightly cassia, warm	Good	Modifier, fixative	Berry and spice flavors	Fair
Cinnamyl anthranilate	$CH=CHCH_2OOC$ (NH_2)	Powdery, fruity (grape)	Good	Body	Grape, honey, cherry flavors	Good–fair
Cinnamyl butyrate	$CH=CHCH_2OOC(CH_2)_2CH_3$	Sweet, honey	Good–fair	Body, top note	Citrus, orange, fruit flavors	Fair
Cinnamyl cinnamate	$CH=CHCOOCH_2CH=CH$	Spicy, cinnamon	Good	Body	Fruit flavors	Good
Cinnamyl phenylacetate (cinnamyl α-toluate)	$CH=CHCH_2OOCCH_2$	Rosy, cinnamon	Good	Modifier	Chocolate, honey, spice flavors	Good

TABLE 2
Synthetic Flavoring Materials (Continued)

Name	Structure	Flavor Characteristics	Solubility in Alcohol	Use	Type Product	Stability
Cinnamyl propionate	$\langle\text{ring}\rangle-CH=CHCH_2OOCCH_2CH_3$	Slightly spicy, sharp	Good	Top note, modifier	Fruit flavors	Fair
Citral	$CH=O$	Citrus, lemon	Good	Body, modifier	Lemon, citrus	Fair–poor
Citral dimethyl acetal	$CH(OCH_3)_2$	Citrus, lemon	Good	Body	Citrus, lemon, fruit	Good–fair
Citronellal (rhodinal)	$CH_2=C(CH_3)(CH_2)_3CH(CH_3)CH_2CHO$	Lemon, rose	Good	Modifier	Beverages	Fair
Citronellol	$CH_3C=CH(CH_2)_2CH(CH_2)_2OH$ $\quad\mid\quad CH_3 \quad CH_3$	Rose	Good	Modifier	Honey and rose flavors	Good
Citronellyl acetate	$CH_3C=CH(CH_2)_2CH(CH_2)_2OOCCH_3$ $\quad\mid\quad CH_3 \quad CH_3$	Sweet, fruity	Good	Body	Synthetic apricot, honey, pear, quince	Good
Citronellyl butyrate	$CH_3C=CH(CH_2)_2CH(CH_2)_2OOC(CH_2)_2CH_3$ $\quad\mid\quad CH_3 \quad CH_3$	Sweet, plum-like	Good	Body, modifier	Plum, honey flavors	Fair
Citronellyl formate	$CH_3C=CH(CH_2)_2CH(CH_2)_2OOCH$ $\quad\mid\quad CH_3 \quad CH_3$	Sweet, plum-like	Good	Top note	Honey, plum flavors	Fair
Citronellyl propionate	$CH_3C=CH(CH_2)_2CH(CH_2)_2OOCCH_2CH_3$ $\quad\mid\quad CH_3 \quad CH_3$	Bittersweet, plum	Good	Body, modifier	Lemon, orange, plum, berry flavors	Fair
p-Cresol	$OH-\langle\text{ring}\rangle-CH_3$	Burnt	Good	Body	Nut, vanilla flavors	Good

TABLE 2
Synthetic Flavoring Materials (Continued)

Name	Structure	Flavor Characteristics	Solubility in Alcohol	Use	Type Product	Stability
Cuminaldehyde	(CH$_3$)$_2$CH—⬡—CHO	Cumin, sweaty	Good	Top note, modifier	Curry-type flavors	Good
Cuminic alcohol	(CH$_3$)$_2$CH—⬡—CH$_2$OH	Strawberry, bittersweet	Good	Modifier	Apricot, date, berry flavors	Fair
Cyclamen aldehyde	(CH$_3$)$_2$CH—⬡—CH$_2$CH(CH$_3$)CHO	Violet	Good	Top note	Citrus, fruit	Good
p-Cymene	(CH$_3$)$_2$CH—⬡—CH$_3$	Sharp	Good	Modifier, blender, extender	Citrus, spice	Good
γ-Decalactone	CH$_3$(CH$_2$)$_5$ lactone	Sharp, fruity, waxy	Good	Body	Citrus, orange, coconut, fruit flavors	Fair
δ-Decalactone	CH$_3$(CH$_2$)$_4$ lactone	Sharp, waxy, peach	Good	Body	Coconut, fruit	Fair
Diacetyl	CH$_3$COCOCH$_3$	Sweet, cream	Good	Body, top note	Butter flavors	Good
Dibenzyl ether	—CH$_2$OCH$_2$—	Nutty	Good	Modifier	Fruit, spice	Good
Dimethyl hydroquinone	OCH$_3$ / CH$_3$O	Bitter	Good	Modifier	Tobacco, honey flavors	Fair
Dimethyl anthranilate	NHCH$_3$, COOCH$_3$	Orange blossom	Good	Top note	Citrus, fruit flavors	Good

TABLE 2
Synthetic Flavoring Materials (Continued)

Name	Structure	Flavor Characteristics	Solubility in Alcohol	Use	Type Product	Stability
γ-Dodecalactone	$CH_3(CH_2)_7$ (lactone ring)	Peach	Good	Body	Butter, butterscotch, fruit, maple, nut	Good
δ-Dodecalactone	$CH_3(CH_2)_6$ (lactone ring)	Peach	Good	Body	Butter, fruit, pear flavors	Good
Ethyl acetate	$CH_3COOC_2H_5$	Fruity, bittersweet	Good	Top note, body, modifier	Fruity esters, alcoholic flavors	Fair
Ethyl anthranilate	$COOCH_2CH_3$, NH_2 (benzene ring)	Orange blossom	Good	Modifier	Grape flavors	Fair
Ethyl butyrate	$CH_3(CH_2)_2COOC_2H_5$	Pineapple	Good	Body	Fruity	Fair
Ethyl caproate	$CH_3(CH_2)_4COOC_2H_5$	Fruity, sharp, pungent	Good	Top note, modifier	Imitation fruits (pineapple)	Fair
Ethyl cinnamate	$CH{=}CHCOOC_2H_5$ (phenyl)	Honey, sweet, slightly spicy	Good	Fixative, top note	Spice and berry flavors	Good
Ethyl formate	$HCOOC_2H_5$	Rum-like, sharp, pungent	Excellent	Modifier, top note	Fruity, rum and wine flavors	Poor
Ethyl heptylate	$CH_3(CH_2)_5COOC_2H_5$	Wine-like	Fair–good	Top note, body	Rum and wine flavors	Good
Ethyl octine carbonate	$CH_3(CH_2)_5C{\equiv}CCOOCH_2CH_3$	Green	Good	Top note	Berry, fruit, melon flavors	Fair
Ethyl pelargonate	$CH_3(CH_2)_7COOC_2H_5$	Fruity, brandy	Fair	Modifier, top note	Alcoholic beverage flavors	Good
Ethyl phenylacetate (ethyl α-toluate)	$CH_2COOC_2H_5$ (phenyl)	Honey	Good	Modifier, top note	Honey flavor	Good

TABLE 2
Synthetic Flavoring Materials (Continued)

Name	Structure	Flavor Characteristics	Solubility in Alcohol	Use	Type Product	Stability
Ethyl propionate	$CH_3CH_2COOC_2H_5$	Fruity	Good	Modifier, top note	Fruity and alcoholic beverages	Fair
Ethyl salicylate	(structure: $COOCH_2CH_3$, OH)	Mild wintergreen	Good	Body, top note	Fruit, berry, root beer, wintergreen	Good
Ethyl valerate	$CH_3(CH_2)_3COOCH_2CH_3$	Sharp, ester-like	Good	Top note	Butter, apple, peach, nut flavors	Good
Ethyl vanillin	(structure: OH, OC_2H_5, CHO)	Vanilla	Good	Body, blender, fixative	Chocolate and vanilla flavors	Fair–good
Eugenol	(structure: $CH_2CH=CH_2$, OCH_3, OH)	Spice, clove	Good	Fixative, top note, body	Spice flavors (clove)	Fair
Eugenyl acetate	(structure: $OCOCH_3$, OCH_3, $CH_2CH=CH_2$)	Clove	Good	Modifier	Spice flavors (clove), rum flavors	Fair
Eugenyl benzoate	(structure: $CH_2=CHCH_2$, OOC–, OCH_3)	Spicy	Good	Modifier	Fruit, spice flavors	Good
Eugenyl formate	(structure: $OOCH_3$, OCH_3, $CH_2CH=CH_2$)	Mild clove	Good	Top note	Spice flavors (clove)	Good

TABLE 2
Synthetic Flavoring Materials (Continued)

Name	Structure	Flavor Characteristics	Solubility in Alcohol	Use	Type Product	Stability
Exaltolide (ω-pentadecalactone)		Powdery, musk	Good	Modifier, blender, fixative	Berry, fruit, nut, liquor, wine flavors	Good
α-Fenchyl alcohol		Camphoric, woody	Good	Modifier	Berry, lime, spice flavors	Good
Fenchone		Bitter, camphoric	Good	Modifier	Berry, liquor, spice flavors	Good
Formic acid	HCOOH	Sharp, pungent	Good	Top note	Fruit	Good
Furfural		Burnt	Good	Body	Butter, butterscotch, caramel, rum, molasses flavors	Fair
Furfuryl mercaptan		Coffee, burnt	Good	Body, top note	Chocolate, coffee, fruit, nut flavors	Fair
Geraniol	$CH_3C=CH(CH_2)_2C=CHCH_2OH$	Rose	Good	Top note	Floral (Oriental), candy, beverages	Good
Geranyl acetate		Mild rose	Good	Modifier	Berry, citrus, floral, fruit, spice	Good

TABLE 2
Synthetic Flavoring Materials (Continued)

Name	Structure	Flavor Characteristics	Solubility in Alcohol	Use	Type Product	Stability
Geranyl butyrate	$CH_2OOC(CH_2)_2CH_3$	Floral, sharp	Good	Modifier	Berry, citrus, fruit flavors	Fair
Geranyl formate	CH_2OOCH	Raspberry, bitter rose	Good	Top note	Berry, citrus, apple, apricot, peach flavors	Fair
Geranyl phenylacetate (geranyl α-toluate)	$(C_{10}H_{17})OOCCH_2$	Honey, rose	Good	Body, blender	Fruit flavors	Good
Geranyl propionate	$(C_{10}H_{17})OOCCH_2CH_3$	Bitter, grape, rose	Good	Body, modifier	Blackberry, cherry, ginger, grape, hops, malt	Fair
Guaiacol	OH OCH₃	Smoky, burnt	Good	Body	Coffee, smoke, tobacco, rum	Fair
Guaiol acetate		Currant, wood	Good	Body	Brown, caramel-like flavors	Fair
Heliotropin	CHO	Heliotropin, confectionery	Good	Modifier, body, top note	Vanilla and chocolate	Good

TABLE 2
Synthetic Flavoring Materials (Continued)

Name	Structure	Flavor Characteristics	Solubility in Alcohol	Use	Type Product	Stability
2-Heptanone (methyl amyl ketone)	$CH_3(CH_2)_4COCH_3$	Bittersweet, pear	Good	Top note, modifier	Berry, fruit, butter, cheese flavors	Good
Heptyl acetate	$CH_3COOC_7H_{15}$	Sweet, apricot	Good	Top note, blender	Apricot, cream, date, melon, fruit flavors	Good
Heptyl formate	$HCOOCH_2(CH_2)_5CH_3$	Sweet, plum	Good	Top note, blender	Apricot, peach, plum flavors	Good
3-Hexen-1-ol (leaf alcohol)	$CH_3CH_2CH=CHCH_2CH_2OH$	Green	Good	Top note, modifier	Fruit, mint	Good
Hexyl butyrate	$CH_3(CH_2)_2COOCH_2(CH_2)_4CH_3$	Sweet, pineapple	Good	Body, modifier	Berry, fruit flavors	Fair
Hexyl formate	$HCOO(CH_2)_5CH_3$	Sweet, plum, ester-like	Good	Modifier	Arrack, citrus, berry flavors	Fair
Hexyl propionate	$CH_3CH_2COO(CH_2)_5CH_3$	Sharp, pear	Good	Modifier	Fruit flavors	Fair
Hydratropic aldehyde	![CH(CH₃)CHO structure]	Sharp, floral, green	Good	Modifier	Berry, fruit, rose, almond flavors	Fair
Hydrocinnamic aldehyde	![CH₂CH₂CHO structure]	Bittersweet, almond	Good–fair	Body, modifier	Almond, bitter almond, cherry	Fair
Hydroxycitronellal	![OH CHO structure]	Bitter, floral, peach muguet	Good–fair	Top note, modifier	Berry, citrus, linden, violet flavors	Good
Hydroxycitronellol	![OH CH₂OH structure]	Floral	Good	Top note, modifier	Lemon, floral, cherry	Good
Indole	![indole structure]	Gassy, fragrant	Good	Modifier	Berry, floral, cheese flavors	Fair

TABLE 2
Synthetic Flavoring Materials (Continued)

Name	Structure	Flavor Characteristics	Solubility in Alcohol	Use	Type Product	Stability
α-Ionone	[ring]CH=CHCOCH₃	Violet	Good	Top note, modifier	Berry (raspberry) flavors, gum	Good
β-Ionone	[ring]CH=CHCOCH₃	Violet	Good	Modifier		Good
α-Irone (6-methyl ionone)	[ring]CH=CHCOCH₃	Violet, woody	Good	Modifier	Berry, fruit, floral flavors	Good
Isoamyl acetate	$CH_3COO(CH_2)_2CH(CH_3)_2$	Pear	Good	Top note, body	Berry, fruit, butter, rum flavors	Good
Isoamyl butyrate	$CH_3(CH_2)_2COOCH_2CH(CH_3)_2$	Sweet, rum	Good	Top note, body	Alcoholic beverage flavors	Fair
Isoamyl cinnamate	[ring]$CH=CHCOOCH_2CH_2CH(CH_3)_2$	Fruity, spicy	Good	Modifier	Fruit flavors, brown flavors	Good
Isoamyl formate	$HCOO(CH_2)_2CH(CH_3)_2$	Sweet, plum	Good	Body, modifier	Citrus, fruit, berry flavors	Good
Isoamyl propionate	$CH_3CH_2COOCH_2CH_2CH(CH_3)_2$	Bittersweet, apricot, plum	Good	Top note	Citrus, fruit, berry flavors	Good
Isoamyl salicylate	[ring]$COOCH_2CH_2CH(CH_3)_2$, OH	Bittersweet, strawberry	Good	Modifier	Root beer, fruit, berry flavors	Good
Isoborneol	[ring]OH	Camphoraceous	Good	Modifier	Fruit, spice flavors, mint	Fair–good
Isobornyl acetate	[ring]$OOCCH_3$	Woody, pine	Good	Modifier	Fruit flavors	Good
Isobutyl acetate	$CH_3COOCH_2CH(CH_3)_2$	Fruity	Good	Modifier, top note	Ester-like, fruit flavors	Fair

TABLE 2
Synthetic Flavoring Materials (Continued)

Name	Structure	Flavor Characteristics	Solubility in Alcohol	Use	Type Product	Stability
Isoeugenol	OH, OCH_3, $CH=CHCH_3$ (benzene ring)	Sweet, spice, clove	Good	Top note, modifier	Spicy	Fair
Isoeugenyl acetate	$OCOCH_3$, OCH_3, $CH=CHCH_3$ (benzene ring)	Clove	Good	Body, modifier	Rum flavors	Fair
Isojasmone	$(CH_2)_5 CH_3$ (cyclopentenone ring)	Floral, green	Good	Modifier		Good–fair
Isopropyl alcohol	$(CH_3)_2 CHOH$	Sweet, apple	Good	Top note, modifier	Raspberry, apple flavors	Fair–good
Isopulegol	OH (cyclohexane ring with isopropenyl)	Bitter, minty	Good	Top note	Apricot, caramel, mint, cherry, peach flavors	Good
Isopulegone	O (cyclohexanone ring with isopropenyl)	Mentholic	Good	Top note	Berry, fruit, mint flavors	Poor
Isopulegyl acetate	$OOCCH_3$ (cyclohexane ring with isopropenyl)	Woody, fruity, pears	Good	Modifier	Berry, fruit flavors	Good
Isoquinoline	(isoquinoline ring, N)	Pungent	Poor	Top note, modifier	Vanilla flavor	Good
Isovaleric acid	$(CH_3)_2 CHCH_2 COOH$	Sour, cheesy	Good	Body, modifier	Fruit, rum, cheese, nut flavors	Good

TABLE 2
Synthetic Flavoring Materials (Continued)

Name	Structure	Flavor Characteristics	Solubility in Alcohol	Use	Type Product	Stability
d,l-Limonene		Orange, lemon, caraway	Good	Body	Lime, fruit, spice flavors	Poor
Linaloöl	$(CH_3)_2C=CHCH_2CH_2CCH=CH_2$ with CH_3 and OH	Light, floral	Good	Top note, modifier	Citrus, carbonated beverages	Good
Linalyl acetate	$(CH_3)_2C=CHCH_2CH_2CCH=CH_2$ with CH_5 and $OCOCH_3$	Lavender, gooseberry	Good	Modifier	Carbonated beverages	Good
Linalyl anthranilate	$(CH_3)_2C=CHCH_2CH_2CCH=CH_2$ with CH_3 and OOC—C_6H_4—H_2N	Lychee/orange	Fair	Body, blender	Berry, fruit, citrus, grape	Good
Linalyl benzoate	$(CH_3)_2C=CHCH_2CH_2CCH=CH_2$ with CH_3 and OOC—C_6H_5	Floral, cananga, minty	Good	Blender	Berry, citrus, fruit, peach	Fair
Linalyl butyrate	$(CH_3)_2C=CHCH_2CH_2CCH=CH_2$ with CH_3 and $OOC(CH_2)_2CH_3$	Sweet, honey	Good	Top note	Honey flavors	Fair

TABLE 2
Synthetic Flavoring Materials (Continued)

Name	Structure	Flavor Characteristics	Solubility in Alcohol	Use	Type Product	Stability
Linalyl formate	$(CH_3)_2C=CHCH_2CH_2CCH=CH_2$, with CH_3 and $OOCH$	Bittersweet, pineapple	Good	Top note	Apple, pineapple, apricot, peach flavors	Fair
Linalyl propionate	$(CH_3)_2C=CHCH_2CH_2CCH=CH_2$, with CH_3 and $OOCCH_2CH_3$	Sweet, black currant	Good	Body, top note	Apricot, black currant, cranberry, gooseberry flavors	Fair
Maltol	structure	Sweet, caramelized sugar	Poor	Body, modifier, potentiator	Chocolate, coffee, fruit, maple, nut flavors	Good
Menthol	structure	Cool peppermint characteristics	Good–fair	Body, top note	Peppermint, lime flavors, pharmaceuticals, candy, gum	Good
Menthone	structure	Bittersweet, peppermint	Good	Body, top note	Peppermint	Good
Methyl acetate	CH_3COOCH_3	Sweet, fruity, ethereal	Good	Modifier, top note	Alcoholic beverages	Fair–good
Methyl anthranilate	structure (NH_2, $COOCH_3$)	Orange, neroli	Fair	Body, top note	Grape	Good
Methyl benzoate	structure ($COOCH_3$)	Harsh, bitter, strawberry	Good	Modifier	Raspberry, pineapple, strawberry flavors	Good

TABLE 2
Synthetic Flavoring Materials (Continued)

Name	Structure	Flavor Characteristics	Solubility in Alcohol	Use	Type Product	Stability
Methyl butyrate	$CH_3(CH_2)_2COOCH_3$	Sweet, apple	Good	Top note	Fruit flavors	Good
Methyl cinnamate	$CH{=}CHCOOCH_3$ (phenyl)	Sweet, fruity, strawberry	Good	Fixative, modifier	Berry flavors	Good
Methyl furfural	H_3C—furan—CHO	Coffee	Good	Modifier	Apple, honey, meat flavors	Fair
Methyl furoate	furan—$COOCH_3$	Coffee	Good	Body, modifier	Meat flavors	Good
Methyl heptine carbonate	$CH_3(CH_2)_4C{\equiv}CCOOCH_3$	Green, grassy	Good	Blender, fixative	Fruit, berry flavors	Good
Methyl mercaptan	CH_3SH	Penetrating, rubbery	Good	Modifier	Coffee flavor	Good
Methyl octyne carbonate	$CH_3(CH_2)_5C{\equiv}CCOOCH_3$	Tart, sweet, peach	Good	Top note, modifier	Apricot, peach, violet	Good
Methyl phenylacetate (methyl α-toluate)	(phenyl)CH_2COOCH_3	Honey	Fair	Modifier	Honey flavors	Good
Methyl propionate	$CH_3CH_2COOCH_3$	Sweet, black currant	Good	Top note	Black currant flavors	Good
Methyl salicylate	(phenyl) OH, $COOCH_3$	Wintergreen	Good	Modifier, top note, body, blender	Mint flavors, medicinals	Good
Methyl sulfide (dimethyl disulfide)	$CH_3{-}S{-}S{-}CH_3$	Sharp, garlic	Good	Modifier	Onion flavors, coffee, cocoa	Good
Naringin		Bitter	Good	Modifier	Bitters, liquors	Good
Nerol	$(CH_3)_2C{=}CHCH_2CH_2CCH_3$, $HCCH_2OH$	Bitter, floral, rosy	Good	Top note	Raspberry, strawberry, citrus, honey flavors	Good

TABLE 2
Synthetic Flavoring Materials (Continued)

Name	Structure	Flavor Characteristics	Solubility in Alcohol	Use	Type Product	Stability
Nerolidol	$CH_2CH_2CCH=CH_2$ with OH and CH_3	Fruity, pear	Good	Top note, blender	Berry, citrus, fruit, rose flavors	Good
Nerolin	OC_2H_5 (naphthalene)	Neroli, orange blossom	Good	Top note, modifier	Orange complexes	Good
Neryl acetate	CH_2OOCCH_3	Tart, sweet raspberry	Good	Modifier	Neroli, citrus, raspberry flavors	Fair
Neryl butyrate	$CH_2OOC(CH_2)_2CH_3$	Sweet, cocoa	Good	Modifier	Cocoa, chocolate-type flavors	Fair
Neryl formate	CH_2OOCH	Bitter, peach	Good	Top note	Apricot, peach, pineapple flavors	Fair
Neryl propionate	$CH_2OOCCH_2CH_3$	Tart, sweet, Mirabelle plum	Good	Modifier	Plum flavors, neroli	Fair
Nonyl acetate Nootkatone	$CH_3COOC_9H_{19}$	Bitter, peach Citrus (grapefruit)	Good	Top note	Apricot, peach flavors	Fair
Octyl acetate	$CH_3COOC_8H_{17}$	Sweet, peach	Good	Body, modifier	Peach flavors	Fair–good

TABLE 2
Synthetic Flavoring Materials (Continued)

Name	Structure	Flavor Characteristics	Solubility in Alcohol	Use	Type Product	Stability
Octyl butyrate	$CH_3(CH_2)_2COOC_8H_{17}$	Sweet, melon	Good	Top note, modifier	Cucumber, melon, pumpkin flavors	Good
Octyl formate	$HOCOOC_8H_{17}$	Bitter, burning, peach	Good–fair	Top note, modifier	Apricot, peach flavors	Fair
Octyl propionate	$CH_3CH_2COOC_8H_{17}$	Melon	Good	Modifier	Berry, citrus, melon flavors	Good
α-Phellandrene		Spice (pine)	Fair	Top note, body	Citrus, spice flavors	Good
Phenylacetic acid	CH_2COOH	Honey, beer	Fair	Fixative, top note	Honey and beer	Poor
Phenylethyl acetate	$CH_2CH_2OOCCH_3$	Sweet, honey	Good	Modifier	Fruit, citrus, berry, honey flavors	Good
Phenylethyl alcohol	CH_2CH_2OH	Rose, peach	Good	Top note	Oriental candy	Good
Phenylethyl anthranilate	CH_2CH_2OOC (NH₂)	Floral, grape	Good	Modifier	Butter, caramel, fruit, grape flavors	Poor
Phenylethyl benzoate	CH_2CH_2OOC	Bitter, strawberry	Good	Modifier	Honey, strawberry flavors	Good
Phenylethyl butyrate	$CH_2CH_2OOC(CH_2)_2CH_3$	Sweet, pear	Good	Modifier, body	Fruit, berry flavors	Fair
Phenylethyl cinnamate	$CH_2CH_2OOCCH=CH$	Bitter, cherry	Good	Modifier, body	Bitter, almond, cherry, plum flavors	Good
Phenylethyl formate	CH_2CH_2OOCH	Bittersweet, green plum	Good	Top note	Cherry, plum flavors	Fair

TABLE 2
Synthetic Flavoring Materials (Continued)

Name	Structure	Flavor Characteristics	Solubility in Alcohol	Use	Type Product	Stability
Phenylethylisovalerate	$(CH_3)_2CHCH_2COOCH_2CH_2$— (phenyl)	Fruity	Good	Modifier, blender	Heavy, fruity (peach, etc.) flavors	Fair
Phenylethyl phenylacetate	—$CH_2CH_2OOCCH_2$— (phenyl)	Sweet, honey	Good	Modifier	Fruit, honey, orange	Good
Phenylethyl salicylate	—CH_2CH_2OOC— (phenyl–OH)	Sweet, peach	Good	Modifier	Apricot, peach, pineapple flavors	Good
Phenylethyl tiglate	—$CH_2CH_2OOC(CH_3)=CHCH_3$	Floral, rose	Good	Top note	Fruit, nut flavors	Fair–good
Phenylpropyl acetate	—$CH_2CH_2CH_2COOCCH_3$	Bittersweet, gooseberry	Good	Body, top note, modifier	Fruit, berry flavors	Good
Phenylpropyl cinnamate	—$(CH_2)_3OOCCH=CH$— (phenyl)	Bitter, cherry	Fair	Blender	Bitter, almond, cherry, plum	Good
α-Pinene ($C_{10}H_{16}$)	(structure)	Pine-like, woody	Good	Top note	Lemon, nutmeg	Fair
β-Pinene ($C_{10}H_{16}$)	(structure)	Pine-like, woody	Good	Top note	Woody	Fair–good
Piperine	(structure) $CH=CHCH=CHC-N$	Pungent	Fair–good	Modifier	Celery, soda	Good
Propionaldehyde	CH_3CH_2CHO	Acetaldehyde, pungent	Good	Top note	Fruit flavors	Good
Propionic acid	CH_3CH_2COOH	Sour	Good	Top note	Butter, fruit flavors	Fair–good
Propyl acetate	$CH_3COO(CH_2)_2CH_3$	Bittersweet, pears	Good	Top note, body	Apple, pear, berry, melon flavors	Fair–good
Propyl alcohol	$CH_3CH_2CH_2OH$	Pungent	Good	Modifier	Fruit	Fair–good

TABLE 2
Synthetic Flavoring Materials (Continued)

Name	Structure	Flavor Characteristics	Solubility in Alcohol	Use	Type Product	Stability
Pulegone		Minty	Good	Modifier	Peppermint	Good
Quinine	H_3CO $CH(OH)$ N $CH=CH_2$	Bitter	Good	Modifier	Bitters, beverages	Good
Rhodinol	CH_2OH	Rosy	Good	Top note, modifier	Oriental candy, ginger ale	Good
Rhodinyl acetate	CH_2OOCCH_3	Bitter, raspberry	Good	Modifier	Berry, apricot, floral, rose, honey flavors	Good
Rhodinyl butyrate	$CH_2OOCCH_2CH_2CH_3$	Sweet, blackberry	Good	Modifier	Berry flavors	Good
Rhodinyl formate	CH_2OOCH	Bittersweet, cherry	Good	Top note	Berry, fruit flavors	Good
Rum ether (ethyl oxyhydrate)	Mixture	Rum	Good	Top note, body	Butter, liquor, rum flavors	Good
Salicylaldehyde	CHO OH	Nutty	Good	Modifier	Spice flavors	Fair

TABLE 2
Synthetic Flavoring Materials (Continued)

Name	Structure	Flavor Characteristics	Solubility in Alcohol	Use	Type Product	Stability
α-Santalol	$H_2C-CH-CCH_2CH_2CH=CCH_2OH$ (with CH_3, CH_2, CCH, HC, CH, CH_3)	Raspberry, musk	Good	Top note, body	Medicinals	Good
α-Santalyl acetate	$H_2C-CH-CCH_2CH_2CH=CCH_2OOCCH_3$ (with CH_3, CH_2, CCH, HC, CH, CH_3)	Bittersweet, apricot	Good	Body	Apricot, peach, pineapple	Good
Skatole	indole ring with CH_3, N	Putrid	Good	Modifier	Berry, grape, cheese	Good–fair
Styralyl acetate	CH_3 / CHOOCCH$_3$ (benzene ring)	Bitter, astringent, plum	Good	Body	Fruit, berry	Good
Styralyl alcohol	$CH(OH)CH_3$ (benzene ring)	Benzophenone, sweet hay	Good	Modifier	Strawberry, rose, fruit, honey flavors	Good
α-Terpineol	cyclohexene ring with OH and $C(CH_3)_2$	Peach, floral (lilac)	Good	Body, modifier	Peach, citrus, spice, floral	Good

TABLE 2
Synthetic Flavoring Materials (Continued)

Name	Structure	Flavor Characteristics	Solubility in Alcohol	Use	Type Product	Stability
Terpinyl acetate	$OOCCH_3$	Sweet, almonds, raspberry	Good	Body, modifier	Plum, apricot, cherry, greengage plum, almond flavors	Good
Terpinyl formate	$OOCH$	Bitter, raspberry	Good	Top note	Fruit, berry, orange flavors	Good
Tetrahydrolinalool ($C_{10}H_{22}O$)	OH	Sweet, floral	Good	Top note, modifier	Berry, citrus, fruit flavors	Good
Thymol	OH	Herb-like, medicinal	Good	Modifier, top note, body	Medicinal flavor	Good
Valeraldehyde	H_3CCHCH_3 $CH_3(CH_2)_3CHO$ CHO	Sharp, penetrating	Good	Top note	Fruit, nut flavor	Fair
Vanillin	OCH_3	Vanilla	Good	Top note, blender, modifier, fixative	Chocolate, vanilla	Fair
Veratraldehyde	OH CHO OCH_3 OCH_3	Chocolate, coconut, nutty	Good	Body	Fruit, nut, vanilla	Fair
Violet leaf alcohol (2,6-nonadien-1-ol)	$CH_3CH_2CH=CH(CH_2)_2CH=CHCH_2OH$	Sharp, green	Good	Top note	Carbonated beverages (fruit)	Fair

TABLE 2

Synthetic Flavoring Materials (Continued)

Name	Structure	Flavor Characteristics	Solubility in Alcohol	Use	Type Product	Stability
Yara-yara	OH — OCH$_3$	Acacia, orange	Fair	Top note, modifier	Citrus, orange	Good
Zingerone	CH$_2$CH$_2$COCH$_3$ — OCH$_3$ OH	Ginger	Good	Modifier	Carbonated beverages	Good

References

The information presented in this chapter is a combination of the experience of the author and his associates at the Food and Flavor Laboratories of Arthur D. Little, Inc., and at the laboratories of the Canada Dry Corporation and information available in the following reference books and papers.

1. Arctander, S., 1960. *Perfume and Flavor Materials of Natural Origin*, Elizabeth, N.J.
2. Bedoukian, P. Z., 1951. *Perfumery Synthetics and Isolates*, New York, D. Van Nostrand Co., Inc.
3. *Chemicals Used in Food Processing*, 1965. National Academy of Sciences—National Research Council, Washington, D.C.
4. Gildemeister, E. and Hoffman, F. R., 1913. *The Volatile Oils*, Vols. 1–4, Second Edition by E. Gildemeister, New York, John Wiley and Sons, Inc.
5. *The Givaudan Index*, Second Edition, New York, Givaudan—Delawanna, 1961.
6. Guenther, E., 1948. *The Essential Oils*, Volumes 1-6, New York, D. Van Nostrand Co., Inc.
7. Jacobs, M. B., 1947. *Synthetic Food Adjuncts*, New York, D. Van Nostrand Co., Inc.
8. Merory, J., 1960. *Food Flavorings*, Westport, Conn., Avi Publishing Co., Inc.
9. *Perfumer's Handbook*, 1944. Fritzsche Brothers, Inc., New York.
10. Swaine, R., 1964. Flavoring Materials—A Current Review, *American Perfumer and Cosmetics*.
11. von Sydow, E. and Westberg, M., 1962. Flavor Components in Food, *SIK Report*, No. 122.
12. Whitmore, F. C., 1937. *Organic Chemistry*, D. Van Nostrand Co., Inc., New York.

CHAPTER 12

Flavor Potentiators

Loren B. Sjöström
Vice-President

Arthur D. Little, Inc.
Cambridge, Massachusetts

Introduction

In the context of flavor, the term "potentiator" is only a few years old. The isolation and identification of flavor potentiators is a twentieth century accomplishment, an area of research still in its infancy. And yet, flavor potentiation is as old as cooking.

For centuries, cooks have added ingredients to food, or prepared food in particular ways to improve flavor. Flavor secrets were handed down from generation to generation; delicate dishes were prepared without any understanding of the phenomena involved. It is only recently that we have sought to learn the reasons for these phenomena, and have sought to develop a comprehensive understanding of the elements of flavor and of flavor perception. This research has led to intensive work in the area of flavor potentiation and has opened doors which were not even known to exist only a few years ago.

To understand potentiation, it is necessary to cover the subject from several different vantage points. One of these is the concept of flavor itself. Flavor is a very complex concept, as many cooks and anyone involved in the study, processing, preparation or presentation of food well knows.

Flavor is the combination of taste, feeling, and odor on receptors in the mouth and the nose.

Taste is the perception of stimuli through the taste buds. These are minute depressions, located mostly on the tongue, with many very small receptor sites in them. They react to water soluble substances placed over them and, with electric charges (via nerve fibers), transmit their perceptions to the brain.

There are only four basic tastes: sweet, sour, salty, and bitter. Generally speaking, each of them is perceived in a specific area of the tongue: the tip is most sensitive to sweet, the sides to sour and salty, and the back to bitter. The taste receptors themselves are very sensitive in spite of the abuse they get on being exposed to all kinds of foods taken into the mouth. It is believed that the cells themselves are continually being destroyed, then replaced by new cells that assume the functions of those whose function has been terminated.

The tongue, the throat, and the entire mouth are also sensitive to texture, temperature, and other variables. Warmth, cold, carbonation, pungency, astringency, bite, numbness and other feeling sensations are all elements of flavor. The tongue has thousands of sensors. The nose has millions. Odor, or aroma, is perhaps the most important and the least understood element of flavor. It has been estimated that we can distinguish at least 10,000 different odors and our memory for odors is precise and almost never-fading. We do not know, however, exactly how odor is perceived.

Flavor is a combination of all these factors. It is not a single impression, but a series of impressions, each of which may be distinct or overlapping, and may last a matter of microseconds or, as in the case of some aftertastes, for many minutes.

Flavor Systems

Thus far, we have described flavor as it is perceived because flavor potentiators affect the ways in which flavor is perceived. At the same time, however, flavor potentiators are elements of the stimuli that affect this perception. Therefore, an understanding of the flavor systems inherent in the materials we eat is also essential to an understanding of potentiators. In any flavor system, the stimuli can be classified according to the way in which they affect that particular system, such as integers, seasoners, enhancers, and potentiators.

Integers

Foods and beverages are rather precise mixtures of chemicals. When in the mouth, these chemical combinations interact to produce the stimuli which are perceived as flavor. Some of these chemical factors are integral parts of the food or beverage. We might mention three examples to illustrate these factors, which are often referred to as integers.

Sugar is an integer. The effect of sugar on flavor is quite obvious. It reacts directly on those taste buds which are sensitive to sweet. The sugar is not added to the food; it is an integral part of the chemical and flavor composition.

The same is true of acids. The citric acid in a grapefruit is a significant factor of flavor which is built into the fruit. In the case of a grapefruit, many people use sugar to season the fruit, by masking the acidic flavor inherent in the grapefruit.

There are many integers. Eugenol, for example, is an absolute chemical—the essential oil of clove, which is found in many foods and which, by itself, is a powerful flavor factor.

Seasoners

The first use of salt for seasoning is lost in history. Millions of tons are poured each year to add to the flavor of hundreds of millions of dishes. Salt alters both the aroma and the taste of a food. In a high enough concentration it can also alter the "feel" of the food as well. Although it is possible to use salt to change flavor without adding a "salty" taste, we generally use it at the table so as to add some of its own flavor to the food being seasoned.

There are, of course, many different seasonings, each with a characteristic taste which makes it rather easily identifiable and which becomes a part of the total flavor of the food being seasoned. Onion aromatics, for example, have distinctive aromas which to many people are quite pleasing. As in the case with any seasoning, however, there are sometimes those who are not so positively inclined.

Flavor Enhancers

The best known and most widely used flavor enhancer is monosodium glutamate (MSG). In 1866, a German chemist, Ritthausen, isolated glutamic acid.[1] Later another chemist converted the acid to a sodium salt, monosodium glutamate.[1] In doing their work, neither had any interest in flavor.

More than 40 years later, in 1908, a Japanese chemist at the University of Tokyo, Dr. Kikunae Ikeda, discovered the flavor enhancing properties of MSG.[2] Dr. Ikeda had set out to find out why and how a certain seaweed, *Laminaria japonica*, affected flavor. Japanese cooks had used this seaweed for centuries to improve the flavor of soups and certain other foods. Dr. Ikeda discovered that the ingredient in the seaweed that made the difference was MSG, and that it had an unusual ability to enhance or intensify the flavor of many high protein foods.

After isolating MSG, Dr. Ikeda developed a process for extracting it from wheat flour and other flours.[1] Working with the Japanese chemical company, Suzuki and Co., he supervised the construction of a plant and, as a partner with the company, began commercial production of MSG in 1909.

There were several attempts to produce MSG in the United States in the years following, but it was not until the 1940's that large-scale MSG production began in this country. By 1968, U.S. production had grown to 46 million pounds per year and world consumption had increased to more than 150 million pounds per year.

MSG intensifies and enhances flavor. In the quantities normally used, it does not add any flavor of its own. This, of course, is very significant and is the point which differentiates a flavor enhancer from seasonings or flavor integers.

At one time, it was felt that MSG had a somewhat meaty flavor and that this flavor was a factor in its ability to intensify the flavors of other foods, particularly protein-rich foods. It was discovered, however, that the meaty taste came rather from contaminants in the crude glutamate, and when these contaminants were sharply reduced, the flavor characteristic also was reduced considerably.[3]

MSG is not flavorless. In large enough concentrations, or by itself, it has been found to have a taste of its own, sometimes described as sweet-saline. There are some, in fact, who feel that MSG is nothing more than a seasoner which gains its effect by combining with the flavors of the foods to which it is added.[4]

There are at least several theories on how MSG—or any enhancer or potentiator—works. There are those who say that it increases the sensitivity of the taste buds and therefore increases the flavor. There are others who claim that it increases salivation, and that this then leads to increased flavor perception. It has also been observed that MSG produces certain feeling sensations in the mouth and in this way, too, helps to intensify flavor perception.

Potentiators

MSG is effective in enhancing the flavor of foods in parts per thousand. There are compounds, however, which are fantastically more powerful. These are the potentiators, which can be significantly effective in enhancing the flavor of foods in concentrations of parts per billion and even less.

In 1913, Dr. Shintara Kodama of Tokyo University, an associate of Dr. Ikeda's, discovered the flavor enhancing properties of a new class of chemical compounds.[2] Spurred by Dr. Ikeda's discovery of MSG, Dr. Kodama had been searching for the ingredient in bonita tuna which made it effective as a flavor enhancer. His conclusion was that the ingredient was the histidine salt of 5'-inosinic acid. It was later discovered that the ingredient was not related to histidine, but to the nucleotide itself, and that this affected foods to an even greater extent than did MSG.[2]

It was many years before the 5'-nucleotides became commercially produced, however. In 1959, a Japanese firm began commercial production. In 1962, the Food and Drug Administration approved them as food additives, and an American firm began marketing them in the U.S. in 1963.

There are many other potentiators, although only a few have been identified to date. The investigation of potentiators, and the estimation of their impact on flavor, is just beginning in many ways.

The terms "potentiator" or "potentiation" are new to food science. They were first adopted in the Arthur D. Little laboratories to describe the 5'-nucleotides and other similar compounds because the terminology used to describe flavor enhancers was inadequate.[5] The term is borrowed from pharmacology. In our laboratories we felt that we were dealing with compounds which acted on the sensory mechanisms to alter their responses to flavor stimuli. This was, in a sense, akin to synergism. The dictionary defines synergistic action as the "cooperative action of discrete agencies such that the total effect is greater than the sum of the effects of each used alone." In pharmacology, potentiation is an action wherein the agent by itself, in small quantities, has no effect on a biological system, but exaggerates the effect(s) of other agents on that system. Assuming the flavor sensing organs to be a biological system, this is exactly the activity observed with flavor potentiators.

Flavor potentiators do not contribute any taste of their own to the foods to which they are added. MSG may be a seasoner, but a potentiator cannot be.

Potentiators have synergistic properties which also make them quite useful in combination with other ingredients. The 5'-nucleotides, for example, are synergistic with salt, and particularly with MSG. This can lead to what is almost a multiplier effect in increasing the effectiveness of salt or MSG.

Isolation and Identification

Since flavor potentiators seem to be very active at remarkably low concentrations, our knowledge about the total number of potentiators and of their specific effects on various foods is limited by the fact that research is relatively recent, and by its nature relatively slow. Much patience is required to identify and to isolate the ingredient in a food which may be a potentiator.

In the Arthur D. Little laboratories, a great deal of this type of research has been carried out. To do the work, a procedure has been established which has proven to be quite useful and effective. The first step in the procedure is, through the use of flavor panels, to define the nature of the flavor change in question. In other words, if we know that the addition of ingredient X to a food product changes the taste of the product, we want first to define the nature of the change. By doing this, we are able to establish a direction in which to look for the actual flavor potentiating ingredient. This, of course, requires highly sophisticated flavor panels whose members are able to define and detect potentiation effects.

At the same time, chemical isolation is begun. This may be done in many ways, including distillation, dialysis, and the various forms of chromatography. At each step, flavor panels are used to guide the laboratory separation work. From the initial definition, then, the work proceeds through the concentration of the flavor active fraction of the material, then to initial purification of the flavor component, then to the isolation of the flavor compound, and finally the analysis and proof of the potentiator itself.

Because of the low concentration at which some of the isolated materials are active (1×10^{-12} parts), special techniques must be developed for their isolation and identification. In addition to a high degree of sophistication in instrumental techniques, special procedures are required in preparing solutions of known concentration for screening tests and the determination of flavor potency. In this connection great care must be exercised in the screening test to avoid flavor contamination of samples and taste fatigue from the carry-over of flavor effects from one sample to another. When isolated, potential uses for these compounds must be carefully scrutinized—first, for applications directly related to common uses for the parent natural product, and second, in other flavor situations in which the natural material is unsuitable because of interferences contributed by its other constituents. By the end of 1969, more than 36 potentiators had been identified in the Arthur D. Little laboratories alone. Some, of course, are more active than others and show more promise commercially. These are being explored in depth.

Potentiator Applications

As has been indicated, potentiator applications vary with foods and with individual potentiators. There have been some general findings, however, with regard to specific food products and the effect of most potentiators on them.

Meats and Fish These were the first foods to be used extensively with potentiators. In general, the potentiators are quite effective with most meat and fish products, as well as with soups.

Vegetables Potentiators have been found to increase the freshness and intensity of the flavor of vegetables.

Cereals In cereals, potentiators tend to cover the sour, grainy, starchy flavor notes.

Fats and Oils With fats and oils, the potentiators tend to diminish the oily mouth effects that are often considered unpleasant.

Fruits Potentiators can have significant effects on fruit flavor. The effect is one of greatly increased freshness, especially on heat-processed fruits.

Nuts With nuts, potentiators may increase the nutty character and reduce the mouth effects without interfering with texture impressions.

Beverages Potentiators have a pronounced effect on many beverages, particularly those with fruit bases. The effect is one of general flavor enhancement and increased freshness.

With derived and convenience foods, potentiators have many possibilities. For example, they significantly increase the "naturalness" of these products.

Types of Potentiators

As we have indicated, there appear to be many different potentiators which may be found in foods, spices, and seasoners that we eat today. Many of these have yet to be identified and isolated. At this point, only a very few are being produced commercially, and only a few others are in laboratory stages that are far enough advanced to become commercially feasible in the next few years.

Monosodium Glutamate (MSG) We have already discussed MSG in detail as a flavor enhancer. MSG meets several of the criteria we have set for true potentiators. It acts on one, and possibly two, types of nerve endings (those in the taste buds and the tactile receptors of the mouth). It contributes no flavor of its own when used in normal small quantities, and it seems to affect the impressions of basic taste as well as produce a feeling of satisfaction throughout the entire oral cavity.

MSG is used very broadly. In most commercially prepared frozen foods containing meat or fish, in almost all dry soup mixes, and in many canned foods, MSG is an ingredient. In addition, of course, it is used by many housewives to enhance the flavor of foods prepared in the kitchen.

5'-Nucleotides The 5'-nucleotides are, as we have indicated, the first of the true flavor potentiators. There are currently two commercial forms of the 5'-nucleotides—disodium 5'-inosinate and disodium 5'-guanylate. In the United States and in Europe, these compounds have been marketed as a basic replacement for the potentiator activity of Argentine beef extract.

Both are many times more potent than MSG, although it is difficult to quantify their relative potency in absolute terms because the effects produced by the nucleotides are not quite the same as those which result from those of MSG. It has been estimated, however, that the inosinate has roughly ten to twenty times the flavor-enhancing ability of MSG.

The nucleotides act well with most of the foods that work well with MSG. In addition to requiring lower concentrations, however, the nucleotides also add effects which are not found with MSG. In liquid foods, they create a sense of increased viscosity. Soups have more body, more "mouthfeel." The nucleotides season, they enhance, and they are physiologically active.[6]

The acceptance of the 5'-nucleotides by commercial users has been relatively swift; they are being used in greater quantities each year. Today, many soups and soup mixes contain one or both of these potentiators. In a study of 41 food products representing broad categories of soups, canned meats, dairy products, canned and frozen vegetables, cereals, grains, and other items, it was found that disodium inosinate is a potent seasoning agent active in concentrations ranging from 0.0075-0.05 parts per hundred. It was concluded that disodium inosinate improves the flavor blend and fullness in many products, showing specific trends of flavor modification. Outstanding among these flavor modifications is the capacity to create the sensation of increased viscosity in liquid or semiliquid food products.[7] In addition, there are mixtures of the two nucleotides available as commercial flavor items, and mixtures of this product and MSG as flavor intensifiers.

Maltol   Maltol was introduced in 1942. It is particularly effective in modifying or intensifying the flavor of soft drinks, fruit drinks, jams, gelatins, and other foods high in carbohydrates. U.S. Patent 3,409,441 describes the potentiating effect of maltol for sweetness.[8] It has been determined that from 5-75 ppm of maltol potentiates the sweetness in a compounded food product, permitting the reduction of the sugar content up to 15% while maintaining the same total sweetness. This is most readily demonstrated in a product such as fresh lemonade, although its usefulness extends to many other products where sugar is a part of the formulation.

Dioctyl Sodium Sulfosuccinate   Dioctyl sodium sulfosuccinate is a wetting agent which has been used primarily as an ingredient in certain detergents. It has been found to have potentiating powers, particularly in adding a previously unachievable flavor or freshness to canned milks.

N,N'-di-o-Tolylethylenediamine   U.S. Patent 3,309,205 covers a food product and a process for enhancing the flavor of food comprised of blending between 5×10^{-16} and 1×10^{-8} parts of N,N'-di-o-tolylethylenediamine with one part of food. The examples given for the resulting potentiation effect include margarine and reconstituted non-fat dry milk solids. Margarine is said to become more butter-like, and the NFDM is lower in cow and barn-like aromatics with an aftertaste of milk and lowered chalkiness.[9]

Cyclamic Acid   U.S. Patents 3,314,799, 3,337,345, and 3,359,118 describe the potentiating effect of small additions of cyclamic acid to margarine, dairy products, and beverage whiteners.[10-12]

Other Potentiators   There are a number of new potentiators in various laboratory stages or in early market situations. George Lueders & Company,

Inc., is marketing potentiators designated as HS-10416, HS-10900, HS-10922, and HS-10918. These are fractions of natural materials that are put on a carrier base for ease in application in compounded flavor situations.

HS-10416 is said to produce a mouth-satisfying wetness, and it creates a smooth, blended complex of flavor notes. It may be used with protein, carbohydrate, and fat-based systems, and generally in products that leave a dry feeling in the mouth.

HS-10900, described as the time-factor potentiator, produces a quick flavor impact and creates a unique blend by uniting all of the character notes of the flavor. It generally eliminates all lingering aftertaste, and it is used in sweet products—both natural and synthetic.

HS-10922 creates a fullness effect that blends and rounds out flavor, generally creating full-bodied aroma and a creamier mouthfeel. It is effective with synthetic flavorings and carbohydrates.

HS-10918 is a combination of the previous two potentiators—time-factor and fullness effect—taking advantage of each.

There are a number of new potentiators in various laboratory stages or in early marketing situations. One company, for example, is now offering two new potentiators to commercial users. One of these serves to blend and "push" flavor in certain foods; the other increases wetness and enhances flavor fullness. Another company is offering a new potentiator to be used with fruit products. Many other firms have potentiators which are still cloaked in commercial security.

Food and Drug Considerations

To the Food and Drug Administration, potentiators are food additives and are therefore subject to the same scrutiny as are any additives.[13]

MSG was originally considered to be an artificial flavor by the FDA. In 1949, however, the position was reversed and MSG was classed as an additive without flavor properties as such. Disodium inosinate and disodium guanylate were accepted by the FDA in 1963.

Acceptance by Consumers and the Food Industry

With more than 150 million pounds sold each year, there is little doubt that MSG has gained broad acceptance by both the food producer and the housewife. For the other potentiators, there seems to have been little time to become well established. Sales are increasing each year, however, and as new potentiators are discovered and marketed, and as new applications are found and accepted for the potentiators we know today, it is apparent that sales and consumer acceptance will increase considerably in the coming years.

The Future

Since there is so much yet to learn about potentiators, it is somewhat difficult to accurately predict their future. It is likely, however, that the new potentiators will have a significant effect on the food industry as well as in housewives' kitchens.

Potentiation is a powerful force. In it may lie the key to all flavor—to flavor development and control. The increasing concern with world feeding problems and the attendant need for new protein sources have created serious flavor problems that frequently can be solved by the addition of one or more potentiators. Potentiators have the capability not only of making synthetic blends of flavor more attractive, but they also can disguise and modify unpleasant flavor sensations in problem areas. Derived foods, such as synthetic flavorings, can be made more appealing by the appropriate use of these new compounds. As a scientific and commercial endeavor, flavor potentiation has an important place in the future.

References

1. Marshall, A. E., March 4, 1948. *History of Glutamate Manufacture.* Monsodium Glutamate—A Symposium, pp. 4, 6.
2. Kuninaka, A., 1964. *The Nucleotides, A Rationale of Research on Flavor Potentiation.* Symposium on Flavor Potentiation, Arthur D. Little, Inc., pp. 4–9.
3. Cairncross, S. E. and Sjöström, L. B., 1948. What Glutamate Does in Food. *Food Ind.,* 20:982, 983, 1106, 1107.
4. Beidler, L. M., 1966. Chemical Excitation of Taste and Odor Receptors. *Flavor Chemistry, Advances in Chemistry Series,* 56:1–28.
5. Sjöström, L. B., 1964. *Flavor Potentiation: An Introduction.* Symposium on Flavor Potentiation, Arthur D. Little, Inc., p. ii.
6. Kuninaka, A., 1966. Recent Studies of 5'-Nucleotides as New Flavor Enhancers. *Flavor Chemistry, Advances in Chemistry Series.* 56:261–274.
7. Kurtzman, C. H. and Sjöström, L. B., 1964. The Flavor-modifying Properties of Disodium Inosinate. *Food Tech., 18*(9): 221.
8. Bouchard, E. F., Hetzel, C. P., and Olsen, R. D., November 5, 1968. *Process of Sweetening Foods with Maltol and Sugar.* U.S. Patent 3,409,441.
9. Hughes, R. L., Dickelman, T. E., Kendall, D. A., and Melzard, D. E., March 14, 1967. *Food Products and Process for Enhancing Flavor of Food with N,N'-di-o-Tolylethylene-diamine.* U.S. Patent 3,309,205.
10. Neilson, A. J., and Hughes, R. L., April 18, 1967. *Taste Improvement in Margarine.* U.S. Patent 3,314,799.
11. Hughes, R. L., Neilson, A. J., Kuhr, W. K., and Slakis, A. J., August 22, 1967. *Taste Improvement for Dairy Products by Adding Cyclamic Acid.* U.S. Patent 3,337,345.
12. Neilson, A. J. and Hughes, R. L., December 19, 1967. *Taste Improvement for Beverage Whiteners.* U.S. Patent 3,359,118.
13. Ramsey, L. L., 1964. *Food Additive Implications of Flavor Potentiators.* Symposium on Flavor Potentiation, Arthur D. Little, Inc., 19–22.

CHAPTER **13**

Nonnutritive Sweeteners

Abner Salant, Ph.D.
Vice President
Monsanto Flavor/Essence, Inc.
New York, New York

Editor's Note

It is now a matter of record that the use of cyclamates as nonnutritive sweeteners in foods manufactured or sold in the United States has been prohibited by orders of responsible regulatory agencies; most other nations have similar restrictions. While significant segments of this chapter cover the use of cyclamates, the editor has chosen to reissue the chapter as written; in many instances the data include use of cyclamate/saccharin blends, the latter being currently approved for use in foods. Also, publication of the current edition comes at a time when the industry is recovering from the ban on cyclamates and is now actively engaged in finding alternative means of establishing acceptable sweetener standards with approved ingredients.

The editor fully expects that an entirely new chapter on nonnutritive sweeteners will be ready for publication in the next edition of the *Handbook of Food Additives*.

T.E.F.

Introduction

With the rapid growth in the consumption of low-calorie soft drinks beginning in the early 1960's, attention was focused on the market for low-calorie products by the food and beverage industries. This meant, in turn, that a great deal of new attention was directed at nonnutritive sweeteners as well as other products used in conjunction with them to formulate low-calorie and other dietetic foods. One of the major reasons for the broad acceptance of the low-calorie beverages was the fact that these were good products in the eyes of a large segment of the population, but in too many other cases the application of nonnutritive sweeteners tended to run ahead of adequate backup data on their proper use.

Consumer acceptance of low calorie and other dietetic foods and beverages is dependent upon the same factors that govern the acceptance of any other type of product. Namely, the quality must be such that the product will be consumed for its own sake and not as a therapeutic agent or for other special dietary values. As always, the consumer buys perceived values which are promised by the manufacturer in advertising and other promotion, but, if he fails to deliver what is expected, the product will fail or its continued expansion into the market will cease. One factor which has become abundantly clear is that the consumers of products manufactured with nonnutritive sweeteners are those who *want* to use them as opposed to those who *must*. This serves to underscore the need to approach low-calorie foods and beverages in much the same manner as their high-calorie counterparts.

Accordingly, the attempt here is to provide guidelines and general principles involved in the use of nonnutritive sweeteners in the formulation and production of low-calorie products. This is done in the context of factors including flavor, texture, color, general appearance, ease of preparation, and storage stability which are known to play significant roles in consumer acceptance. At the same time, it should be recognized that such products are governed by special regulations at both the federal and state level so that product compositions must conform to these. Finally, it should be kept in mind that these products will move through normal food distribution channels and be merchandised in a very similar fashion to regular foods, which means that their good qualities which can be relied upon for consumer appeal must be easily recognized. The challenge is particularly great, since there are only a limited number of materials which can be used in conjunction with the nonnutritive sweeteners, but the commercial facts of life will militate against compromises in quality no matter what the reasons.

Properties of Nonnutritive Sweeteners

While there have been a number of artificial sweeteners identified and developed, the only two articles of commerce of any significance in the United States are cyclamates and saccharin. Therefore, in considering the properties of nonnutritive sweeteners, attention will be focused only on these two with regard to their composition chemistry, federal requirements and overall physical properties.

Cyclamates

Under this general description are found the sodium and calcium salts of cyclohexane sulfamic acid as well as the acid form itself. The structural formula for cyclamates conforms to that shown below:

$$\left[\bigcirc NH-SO_2-O\right]_2 Ca\cdot2H_2O \qquad \bigcirc NH-SO_2-O-Na$$

Calcium Cyclamate   Sodium Cyclamate

Cyclamates occur as white crystals or as a white crystalline powder. These are odorless and in dilute solution are about 30 times as sweet as sucrose. As a minimum, the purity and other properties of cyclamates must conform to the federal standards as set forth in the United States Pharmacopoeia.

However, typical specifications for commercial products are as indicated below:

A. Sodium Cyclamate

Appearance:	Odorless, white crystalline powder.
Empirical Formula:	$C_6H_{12}NNaO_3S$
Molecular Weight:	201.23
Purity:	98% minimum
Loss on Drying at 140°C for 2 hours:	1.0% maximum
Solubility:	1 gram dissolves in 5 ml of water and 24 ml of propylene glycol. It is virtually insoluble in alcohol, benzene, chloroform and ether.
pH of 10% aqueous solutions:	5.5–7.5

B. Calcium Cyclamate

Appearance:	Odorless, white crystalline powder.
Empirical Formula:	$(C_6H_{12}NO_3S)_2Ca\cdot2H_2O$
Molecular Weight:	432.58
Purity:	98% minimum
Loss on Drying at 140°C for 2 hours:	9.0% maximum
Solubility:	1 gram dissolves in 4 ml of water; in 60 ml of ethanol; in 1.5 ml of propylene glycol. It is virtually insoluble in benzene, chloroform and ether.
pH of 10% aqueous solutions:	5.5–7.5

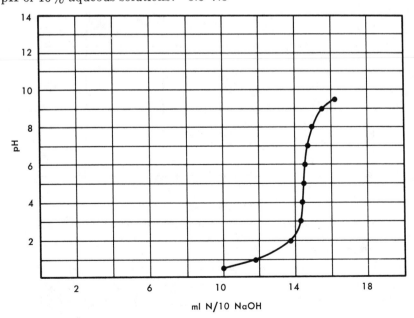

Fig. 1.   Titration of calcium or sodium cyclamate with base.

The titration curve for cyclamate (Figure 1) is that of a very strong acid so that the sodium and calcium salts tend to be fairly neutral in character. Other properties such as viscosity, density and solubility are also shown in Figures 2 thru 4.

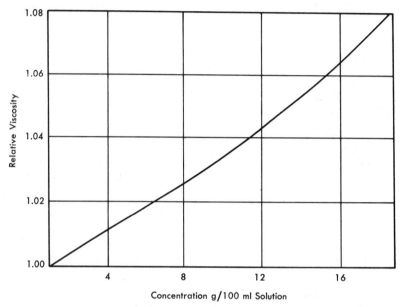

Fig. 2. Viscosity—Concentration Curve: Calcium and sodium cyclamate solutions @ 25°C.

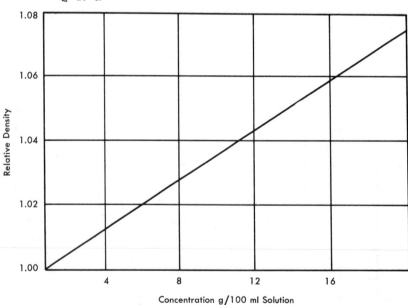

Fig. 3. Density—Concentration Curve: Calcium and sodium cyclamate @ 20°C.

Saccharin

While the acid form of saccharin is a well-recognized article of commerce, the salts are the products actually used in the formulation of foods and beverages. The most common two are sodium and calcium, although ammonium

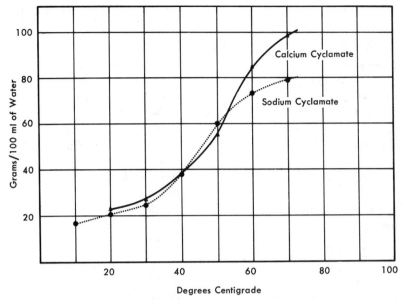

Fig. 4.   Solubility of cyclamates in water.

and other salts have been prepared and used to a very limited extent. The structural formulas for sodium and calcium saccharin, the salts of ortho-benzo-sulfimide, are shown below:

Calcium Saccharin

Sodium Saccharin

Both calcium and sodium saccharin must meet USP specifications in terms of physical properties and impurities. The general properties are as defined below:

A. Sodium Saccharin, Soluble Powder

Appearance:	White Powder
Benzoates & Salicylates:	To pass USP test
Impurities, carbonates:	USP A maximum
Moisture:	4.5–5.8%
Odor:	No foreign
Screen Test: On 80 mesh:	2.5% maximum
On 140 mesh:	10.0% maximum
Solubility:	25.7 parts of saccharin and 200 parts of water.
Assay, dry bases:	98.0–100.0%
Heavy Metals:	20 ppm maximum

Reaction to litmus: Neutral to very slight alkaline
Phenol: No pink color

B. Sodium Saccharin, Soluble Granular
    Appearance: Colorless crystals
    Benzoates & Salicylates: To pass USP test
    Impurities, carbonates: USP Fluid A maximum
    Loss on Drying at 125°C for 6 hours: 13.9–14.6%
    Odor: No foreign
    Solubility: 28.3 parts saccharin to 200 parts
                                                 water
    Assay, dry bases: 98.0–100.0%
    Heavy Metals: 20 ppm maximum
    Reaction to litmus.   Neutral to very slight alkaline
    Phenol: No pink color

C. Calcium Saccharin
    Appearance: White crysalline powder
    Assay, dry bases: 95–100.5%
    Benzoates & Salicylates: Passes USP test
    Flavor: Sweet, no foreign character
    Heavy Metals: 20 ppm maximum
    Impurities, carbonates: USP Fluid A maximum
    Loss on Drying at 120–125°C for 4
      hours: 15.0% maximum
    Odor: No foreign odor

Like the cyclamates, the titration curves for saccharin (Figures 5 thru 7) are of a strong acid. Other properties including the solubility, viscosity and density of saccharin solutions are shown in Figures 8 thru 11.

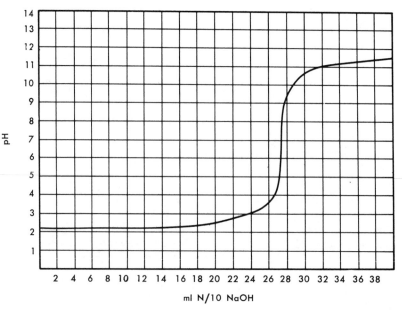

Fig. 5. Titration of saccharin with base.

TABLE 1
Solubility of Saccharin in Various Solutions

Solubility of Saccharin in Water Grams/100 gm water

Temp. °C	Saccharin M. W. 183.18	Na Saccharin 2H₂O M. W. 241.20	Ca(Saccharin)₂·3½H₂O M. W. 467.50
10.6	—	77.3	—
20.0	—	99.8	—
25.0	0.2	—	54.8
35.0	—	138.5	—
45.0	0.6	—	115.5
50.0	—	186.8	—
65.0	1.1	220.2	194.7
75.0	—	253.5	—
85.0	—	289.1	—
90.0	—	—	262.0
95.0	—	328.3	—

Solubility of Saccharin in Ethanol—Water Mixtures @ 25°C Grams/100 gm solvent

% Ethanol by weight	Saccharin M. W. 183.18	Na Saccharin 2H₂O M. W. 241.20	Ca(Saccharin)₂·3½H₂O M. W. 467.50
20.7	0.6	87.3	58.6
43.7	1.8	60.2	60.2
69.5	4.0	23.9	55.0
92.5	4.1	2.6	30.5

Solubility of Saccharin in Glycerin and Propylene Glycol @ 25°C

*Solvent	Saccharin M. W. 183.18	Na Saccharin 2H₂O M. W. 241.20	Ca(Saccharin)₂·3½H₂O M. W. 467.50
Propylene	2.6 g/100g solution	30.9 g/100g solution	26.0 g/100g solution
	2.6 g/100g solvent	44.7 g/100g solvent	35.1 g/100g solvent
Glycol	2.7 g/100 ml solvent	46.2 g/100 ml solvent	36.3 g/100 ml solvent
	0.4 g/100g solution	35.8 g/100g solution	9.6 g/100g solution
Glycerin	0.4 g/100g solvent	55.8 g/100g solvent	10.6 g/100g solvent
	0.5 g/100 ml solvent	70.2 g/100 ml solvent	13.3 g/100 ml solvent

*Density @ 25°C is 1.033 for Propylene Glycol and 1.2583 for Glycerin.

Relative Sweetness, Nonnutritive Sweeteners

Measuring Sweetness

The sweetness intensity of either natural or nonnutritive sweeteners cannot be measured quantitatively in absolute physical or chemical terms but requires the use of subjective sensory methods with trained taste panels. Sucrose is the usual standard, and the sweetening power of other sweeteners is then compared on a relative basis. Dextrose, for example, is assigned a sweetness

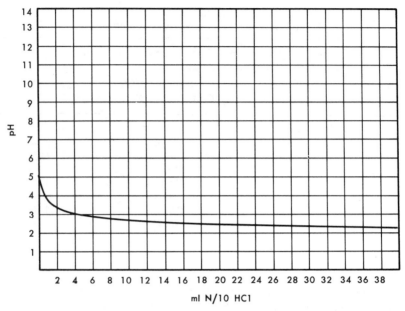

Fig. 6.   Titration of sodium saccharin with acid.

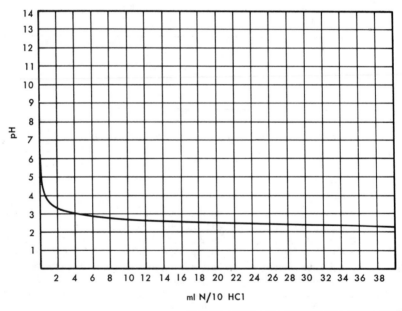

Fig. 7.   Titration of calcium saccharin with acid.

value of 90; that is, on an equal concentration basis, dextrose has about 90% the sweetness intensity of sucrose.

Nonnutritive sweeteners do not readily conform to this scale because of their high sweetness intensity. The commonly accepted rule of thumb is that cyclamate is 30 times sweeter than sucrose, and saccharin, about 300 times. These ratios are only approximate at best and are valid up to concentrations which are about equivalent in sweetness to 10% sucrose solutions. At higher sweetness intensities and in media other than pure water, the ratios of "sweetness equivalence" may vary considerably.

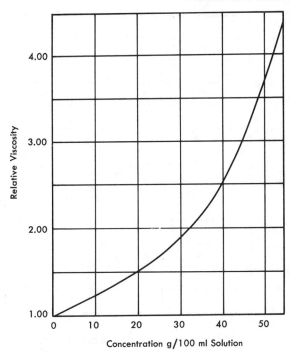

Fig. 8.   Viscosity—Concentration   Curve: Calcium   and
sodium saccharin solutions @ 25°C.

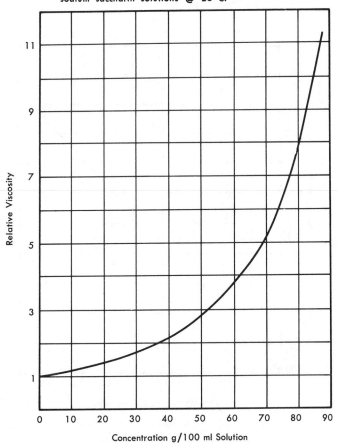

Fig. 9.   Viscosity—Concentration Curve: Calcium and sodium saccharin
solutions @ 90°C.

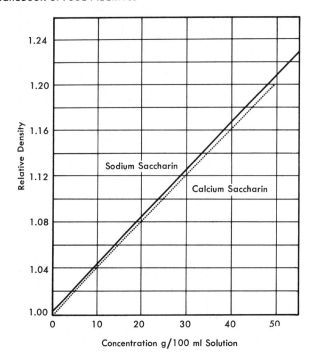

Fig. 10. Density—Concentration Curve: Calcium and sodium saccharin @ 25°C.

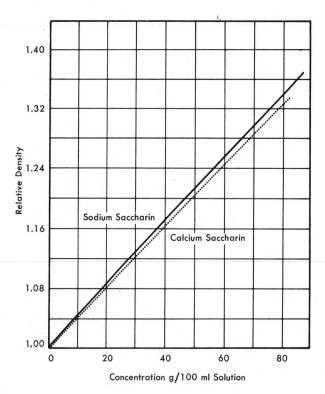

Fig. 11. Density—Concentration Curve: Calcium and sodium saccharin @ 90°C.

The simple approach which provides a first approximation of the comparative sweetness of the nonnutritive sweeteners entails organoleptic comparison with sucrose in aqueous solution. The results of such evaluations are shown in Table 2 and Figure 12.

TABLE 2
Relative Sweetness, Weight Ratios
Nonnutritive Sweeteners to Sucrose

Sucrose % Conc.	Saccharin Ratio to Sucrose	Cyclamate Ratio to Sucrose
2%	500:1	40:1
5%	360:1	36:1
10%	330:1	33:1
15%	300:1	27:1
20%	200:1	24:1

The cyclamate curve is almost parallel to that for sucrose showing a fairly constant equivalency over the range of 2% sucrose to 20% sucrose equivalence. Saccharin, on the other hand, shows "leveling off" at concentrations above its equivalence to 7–8% sucrose solutions.

At higher concentrations, sweetness intensities of nonnutritive sweeteners increase at a lower rate than natural sweeteners. Some express the opinion that this effect is due to increasing levels of bitterness and aftertaste which begin to appear at these concentrations so that the effect is more apparent than real. The practical applications are obvious, however. At higher levels

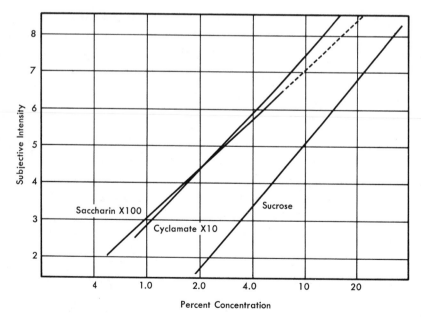

Fig. 12.   Relative sweetness—saccharin, cyclamate, sucrose.*

* Schutz and Pilgrim, Sweetness of Various Compounds and its Measurements. *Food Research*, Mar.–Apr., 1957.

of use there is a problem with attaining adequate sweetness intensity while attempting to avoid some of the undesirable flavor characteristics.

In any case the foregoing description of relative sweetness intensities, as determined in aqueous solutions, can be regarded as no more than a starting point. The final and realistic determination of proper sweetness level can only be done in the actual product in which the cyclamates and/or saccharin will be used.

Food Energy

The approximate energy factors for calculating caloric values are as follows:

Protein	4 calories/gram
Carbohydrates	4 calories/gram
Fat	9 calories/gram

More specific factors are available and these were published by the Committee on Calorie Conversion Factors and Food Composition Tables; Nutrition Division of the Food and Agricultural Organization. Two important references for use in determining caloric values are the following:

Composition of Foods (Raw, Processed, Prepared), Department of Agriculture, Agriculture Handbook No. 8.
Energy Value of Foods, U.S. Department of Agriculture, Agriculture Bulletin No. 74.

Listed below are some of the more important ingredients commonly used in dessert and beverage products as derived from the above two handbooks. These represent only a portion of the wide variety of materials covered in these references:

Common Food Ingredients
Handbook No. 8

	Factor
Sucrose	3.85
Milk, fluid, whole	0.68
Milk, fluid, non-fat	0.36
Milk, dry, non-fat	3.62
Cocoa	2.93
Cream, light	2.04
Cream, heavy	3.30
Egg, whole	1.62
Egg, white	0.50
Egg, yolk	3.61
Fat, vegetable, cooking	8.80
Starch	3.62
Gelatin, plain	3.35

Ingredients Used in Food Manufacture
Bulletin No. 74

	Factor
Citric acid	2.47
Fumaric acid	2.60
Adipic acid	4.80
Sodium citrate	1.84
Sodium phosphate	0.00
Dextrose (monohydrate)	3.62
Sorbitol	4.00
Mannitol	2.00

There are commonly used ingredients which are not listed in these references. In such cases, the manufacturer or supplier of the ingredients in question should be contacted for more precise information.

Sensory Panel Testing

Provided with the foregoing objective data plus rough approximations of sweetness equivalences, the next problem, of course, is actual formulation. However, before launching into this aspect of the problem it would first be desirable to consider how to evaluate those products which we may develop. The basic principles of panel testing and the types of tests used for other foods and beverages apply to the low calorie ones as well. However, they tend to be complicated by unusually severe problems of taste fatigue, carry-over and conditioning. These tend to severely complicate the task of obtaining meaningful and reproducible sensory panel data.

Physical Conditions

Because of the problems, special care should be given to the test area that is being used to an even greater extent than usual. The need to have relatively constant and comfortable combinations of temperature and humidity is very great here, as well as good isolation from outside influences. Care should be taken to provide comfortable chairs and tables. Table height of approximately 32 inches has been found to be suitable with a provision of approximately 32–36 inches of space for each judge.

Selection of Panel Members

Panelists for organoleptic tests are screened for acuteness of senses, general interest in other flavor problems, and desirable personality characteristics. However, here we must also take into consideration that the population consists of three groups: those with an extremely high sensitivity to the taste characteristics of the nonnutritive sweeteners, the "average" type of judge and finally, those who are relatively insensitive to the bitterness and aftertaste of nonnutritive sweeteners. The last group is probably virtually useless for any type of evaluation of low-calorie products. The hypersensitive group, on the other hand, can be useful for experimental work where a high degree of discrimination is needed. These groupings are somewhat arbitrary, and therefore sharp lines of demarcation cannot always be made.

Flavor Characteristics of Nonnutritive Sweeteners

A brief examination of their general flavor characteristics is pertinent here since these are the underlying reasons for the difficulties encountered in sensory panel evaluation of low-calorie foods and beverages. Sucrose is the usual standard for most work, and therefore will be used as the basis for comparison.

Sucrose has a relatively quick sweetness impact followed by a fairly sharp, clean cut-off. The nonnutritive sweeteners, on the other hand, build up to their maximum sweetness intensity at a slower rate and then the sweetness sensation persists for a much longer period. This lingering sweetness in cyclamates and saccharin is responsible for carry-over effects, taste fatigue, and the cloying effect which is encountered in some cases.

The nonnutritive sweeteners, in addition to lingering sweetness, also have other types of aftertaste, the intensity of which is a function of concentration. Generally, the severity of the aftertaste problem starts to increase at a more rapid rate at the concentrations where the sweetness effects start to level off. These aftertastes can be bitter, sometimes metallic, and also astringent or drying. Aside from the intrinsic unpleasantness of these particular characters, they can also contribute to carry-over and fatigue.

Finally, it should be noted that sucrose has a highly desirable blending and bodying effect on the overall product flavor which neither cyclamate nor saccharin seems capable of duplicating to any significant degree. This particular characteristic, or lack thereof, is not a basic interferent in sensory panel testing, but it does complicate the problem of making direct comparisons with sugar-sweetened products. It also points up the virtual impossibility of attempting to develop products by mere substitution of nonnutritive sweeteners for sucrose. This question will be discussed at greater length and in more detail later.

Effect of Nonnutritive Sweetener Flavor Characteristics on Panel Testing

In any type of panel testing, and particularly in those tests where we are attempting to differentiate among samples, it is desirable to have products where flavor cut-off is clean, relatively quick, and with minimum aftertaste. With nonnutritive sweeteners, the opposite tends to be the rule; therefore, particular care is needed in setting test procedures.

The danger of carry-over and fatigue must be minimized insofar as possible. A common technique is to allow adequate time between tasting consecutive samples; in this case, a minute or more may be required. The use of taste refreshers such as water, a cracker, or a piece of apple is also recommended, or in some cases a dummy sample may be used for this purpose.

The tasting of two or more samples at a given sitting is, of course, unavoidable where difference in testing is needed, but every attempt should be made to keep the number to an absolute minimum. The second precaution is to limit the number of tests, in any given day, in which a judge must participate. Indeed some cases have been found where there is evidence of carry-over from a morning test extending into the afternoon.

In the case of preference testing, because of the multitude of problems involved in discrimination among samples, there appears to be good reason to consider monadic testing wherever practical. Monadic testing suffers from having a direct comparison available for reference purposes; its relative slowness and limitations on the number of samples which can be evaluated. Furthermore, it has also been noted that occasionally a conditioning period is needed before judges are capable of detecting some of the less desirable flavor characteristics, particularly bitterness and aftertaste. There is no known pattern from which guidelines can be drawn, but where repeated and constant use is found to be needed before judges can discriminate properly, there may be no alternative to the monadic test.

The conclusion to be drawn from these observations is that there are no new or special sensory panel techniques that are available for use in testing of low-calorie products. Rather there is need to be particularly alert for the problems of bias and unfair handicaps to judges in their tasting in order to insure that meaningful and consistent data are obtained. Judges must be selected with unusual care, and an adequate number of replications employed to insure against misleading results. Testing can be one of the most troublesome and frustrating phases of low-calorie product development activity, and yet it is obviously one of the most critical.

Principles of Application

In the preceding sections the general characteristics of nonnutritive sweeteners, the particular problems involved in organoleptic evaluation of products in which they are used along with some of the specific implications, were discussed. The importance of quality was stressed; this factor emphasizes the need to recognize the limitations due to the current state of the art and to define objectives accordingly in a realistic manner. In essence, this means the following:

1. It is virtually impossible to duplicate the flavor, texture, and other characteristics of a sugar-sweetened product in one employing nonnutritive sweeteners. Therefore, we should strive for equal acceptability rather than exact duplication.

2. There are specific product areas in which it can be stated almost categorically that the present state of the art precludes the formulation and development of low-calorie products having highly acceptable organoleptic characteristics. In the baked-goods area, no good solutions to the problem of textural development consistent with good flavor are known. By the same token, there is no known low-calorie system available today capable of providing the sweet glass known as candy. There are obviously products in these general areas which have achieved a certain degree of market acceptance and therefore must meet, to some extent, the needs of a segment of the consumer population. All, however, tend to have easily detectable deficiencies. It must be concluded that their low-calorie or other dietetic value is the prime reason for acceptance in the first place.

It is rare that a sugar sweetened product is not used as the target, at least reference standard, for the formulation of a low-calorie counterpart.

This is usually satisfactory as long as the limitations discussed above are recognized and constantly kept in mind. This, in turn, leads to the next cardinal rule for the application of nonnutritive sweeteners in food and beverage product development:

> Acceptable nonnutritively sweetened products cannot be developed by the simple substitution of cyclamate and/or saccharin for sugars. Rather the new product should be completely reformulated from the beginning.

This recognizes the fact that the sweeteners, whether sugars or synthetics, impart more than the taste sensation of sweetness and have characteristic flavors of their own. Cyclamate and saccharin are not identical in flavor to sucrose anymore than sucrose is identical to dextrose. It also emphasizes that a finished product has flavor, texture, and other organoleptic properties which are not the sum of the individual components but rather the result of some type of synergism. The properly formulated product, then, is a balanced whole in which the balance usually cannot be retained if one component is disturbed. These facts are fundamental and generally rather well recognized, but somehow often are overlooked when it comes to low-calorie products.

The two most critical areas which must be watched in formulating low-calorie foods and beverages are flavor and texture. These two properties are closely inter-related under any circumstances, but the effect of a given flavor/texture system with nonnutritive sweeteners tends to be significantly different from what is observed with sucrose or other sugars. Storage stability can be a problem in certain cases depending upon methods of preservation and packaging which are used. Color and odor generally do not offer serious difficulties, but overall appearance, quite apart from either texture or color, may be significantly affected, especially in cases where sugar has a major role to play in bulking or overall structure.

Flavor

As was noted above, there are intrinsic flavor differences between cyclamate or saccharin compared to sucrose, and in addition neither is capable of providing the total flavor blending function. The first conclusion that can be drawn from these facts and which is supported by observations and actual practice, is that the same flavor will not taste the same in the two different types of systems; furthermore, there may also be a lack of body noted as well as having some of the less desirable flavor notes become more pronounced. Because of this, the low-calorie product may require the use of either a modified flavor or perhaps an entirely different one. Quite often this means a better and more expensive flavor than might be required with sugar.

The role which flavor plays in a low-calorie product may also be greater than in a sugar sweetened one. Its importance in masking undesirable sweetener aftertastes and other unwanted flavor characteristics from different sources tends to be more important. The overall effect of body, that quasi-flavor/tactile sensation which may have been originally derived from sugar solids, or other high calorie ingredients, may now have to be imparted or simulated by the flavor itself.

Sweetness-Tartness Balance

This balance between the taste effects of the acid and sweetener components is an interplay which has been long recognized and which also, of course, plays a role in the total flavor response. With the differences in overall sweetness character found in cyclamates and saccharin, the concentration of acid and buffer salts will probably have to be changed. If the lingering sweetness results in an apparently higher sweetness intensity, obviously some degree of compensation can be achieved by increasing the level of acidity. At the same time, however, certain products may have to be formulated with less than optimal sweetness intensity because of the rapidly increasing encroachment of undesirable aftertastes, as the sweetener concentrations become too high. Again compensation may be possible to some extent by reducing the acid level. Variation of flavor type can also be used to effect the apparent sweetness-tartness intensity in either direction.

Achieving the proper level of acid and the proper ratio of acid to buffer salt usually cannot be done by a pH measurement, but must be done organoleptically in combination with at least the sweetener and flavor involved. Although this involves a potentially astronomical number of combinations, the experience of a skilled formulator can actually overcome this obstacle utilizing, if necessary, a suitable statistical design of experiments.

Texture

The problem of imparting the proper amount of feel to a low-calorie product may range anywhere from insignificant to severe. The difficulty, of course, stems from the elimination of sugar solids from the product formulation and the usual approach is to attempt to compensate by means of a suitable hydrocolloid system. The degree of success tends to vary somewhat.

In the area of carbonated beverages, there has been a great deal of work done to impart improved mouth-feel by the addition of CMC, pectin and other materials, but by-and-large, this has not proved to be particularly necessary. The major problem related to mouth-feel in the low-calorie beverages is the gushing phenomenon in which there is a greater tendency to lose carbonation than in sugar-sweetened products.

Gel systems can be significantly affected in varying degrees by the absence of sugar solids. Thus, in the case of gels which rely on pectin, there is a need to change the type of pectin used in order to achieve satisfactory set. In other types of gel systems, the differences are somewhat more subtle and, therefore, not quite as severe, although syneresis can be quite troublesome in certain areas. In the case of baked goods, as was touched on previously, the texture problem is particularly acute and indeed presents a serious obstacle to the formulation of truly high quality products. In the case of a cake, for example, it is extremely difficult to get any reasonable development of moistness, crumb structure and overall volume. Cookies baked with nonnutritive sweeteners, and even with the addition of bulking agents, tend to be hard and tough and in addition have texture related flavor problems.

The problem of achieving suitable texture is in part the result of the lack of analytical methods for describing texture in a realistic sort of way, and, in part, stems from the absence of an ingredient or system capable of reproducing the textural functionality of sugars. These will be discussed in greater detail for various product areas.

Storage Stability

Storage stability can be affected by elimination of the sugar solids normally found in food products. Because of the low solids content of products sweetened with nonnutritive sweeteners, they may be more subject to microbiological spoilage. Consequently, it may be necessary to use a preservative such as sorbates or benzoates, or to employ a thermal process in others to achieve adequate shelf life.

Storage stability problems may also stem from sources other than microbial spoilage. These tend to be the result of changes in appearance and texture rather than syneresis, toughening, or general drying out. Again these are problems which are results of the absence of sugar solids.

Jams, Jellies and Preserves

Standards of Identity exist for dietetic jellies, jams and preserves and control the composition. For details refer to the Federal Food, Drug and Cosmetic Act, Section 29.4 and 29.5.

Gelling Ingredients

Regular products contain approximately 65% or more soluble solids whereas the low-calorie products generally have a content range of 15–20%. Because of the low solids content, the more commonly used pectins, with relatively high methoxyl content, will not give satisfactory gels. Therefore, a special type, a low methoxyl pectin must be employed. The LM pectin is chemically modified so that it forms a gel in the presence of calcium regardless of the solids content.

Additional gelling ingredients are permitted by the standards and include agar-agar, locust bean gum, guar gum, gum karaya, gum tragacanth, algin and carrageenan. These may be used in some cases singly but more often in combination to obtain the specific type of texture and gel that is desired.

Gel strength and quality are effected by the variation in natural solids content, pH, and natural salts present. There is no firm rule for deciding in advance which particular type of system should be employed. This can only be done by modification and adjustment involving all the ingredients, not only as a gelling agent, but acids, salts and sequestering agents. It should be noted that sequestering agents can be very important to control release of calcium and other divalent salts toward which pectin is sensitive and thus control the texture. Locust bean, guar, tragacanth, and carboxymethylcellulose are generally used to modify gel texture. They may be used to alter spreadability, texture, and to impart increased resistance to cleavage of the gel during shipment.

Agar is sometimes used alone or in combinations. It does not require calcium or other chemicals to obtain the set as does the low methoxyl pectin. The stiff non-melting gels produced with agar tend to be somewhat similar to pectin gels.

Algin is very similar in behavior to pectin. It requires the presence of calcium and pH adjustments as does the pectin. Carrageenan, on the other hand, requires a potassium salt in order to obtain gel formation.

Acid Adjustment

In order to obtain satisfactory gels, it is necessary to have proper concentrations of acid resulting in proper pH and buffer salts. The natural fruit content and its characteristics will, of course, determine what levels and types are needed. Generally, for flavor reasons and also to get proper gel characteristics, a pH range of 3.0–3.8 is to be preferred. However, each batch of fruit should *be checked in order to be sure.*

Syneresis

Texture and spreadability will be effected by the pH. The standards list the acidulants and concentrations which are permitted.

Preservatives

Because of the low solids content, a preservative must be added. Normally the products are heat processed to prevent spoilage so the added preservative serves only for protection after opening the container. The preservatives permitted include ascorbic acid, sorbic acid, sorbate salts, propionate salts, and benzoates with the limit of about 0.1% by weight of finished food.

Nonnutritive Sweeteners

The most common sweetener system used is a combination of saccharin and cyclamates. The calcium salts of the sweeteners may be employed to function at the same time as a source of calcium to aid in setting of the pectin gel. It should be pointed out, however, that with the normal sodium content of the natural component of the product, it would be difficult to develop a product low in sodium even with the calcium salts with the sweeteners.

Processing

The gelling ingredient should be added to the full amount of water in the kettle and heated to boiling in order to disperse and hydrate it. The juice and sweeteners and the other ingredients are then added and the mixture is again brought to a boil. The mixture is poured at 180°F into the jelly glasses which are then cooled as rapidly as possible in order to prevent darkening and development of cooked flavor. It should be noted that in the case of preserves and jams, where there are fruit, the mixture is cooled to about 160°F with slow stirring prior to filling of the glasses. This lower temperature is necessary in order to allow the mixture to set rapidly enough to prevent floating of the fruit.

Typical Formulations

Listed below are some typical formulations illustrative of the products described:

Dietetic Apple Jelly

Apple juice	94.6 pounds
Low methoxyl pectin	1.4 "
Citric acid, anhydrous	0.3 "
Sodium cyclamate/Sodium saccharin (12:1)	0.2 "
Calcium chloride, 1% solution	3.5 "
	100.0 pounds

Dietetic Grape Jelly

Grape juice	65.0 pounds
Agar-agar	0.5 "
Sodium cyclamate/Socium saccharin (12:1)	0.4 "
Water	34.0 "
Sodium benzoate	0.1 "
	100.0 pounds

Dietetic Peach Jam

Crushed peaches	60.0 pounds
Low methoxyl pectin	1.6 "
Citric acid, anhydrous	1.0 "
Sodium cyclamate/Sodium saccharin (12:1)	0.5 "
Calcium chloride, 1% solution	5.0 "
Water	32.0 "
	100.0 pounds

Salad Dressings

Although there are Standards of Identities for regular salad dressings, there are none for the dietetic product. However, it would be desirable to be familiar with the standards in any case so that the proper substitutions and modifications in ingredients and label statements can be made. The major source of calories in a salad dressing is the oil component, and the usual approach to formulating a low-calorie counterpart is by elimination or reduction of the oil component. Listed below are some brief, general descriptions of three of the more common types of salad dressing:

1. *French Dressing*—This is a separable liquid food or an emulsified, viscous fluid food. It is prepared from edible vegetable oil, one or two acidifying ingredients, optional seasonings or flavorings and, if desired, optional emulsifyng ingredients as well. The oil content may not exceed 35% by weight.

2. *Mayonnaise*—Mayonnaise is an emulsified, semi-solid food, prepared from edible vegetable oil, one or more acidifying ingredients, one or more egg-yolk containing ingredients. The minimum vegetable oil content in mayonnaise is 65%.

3. *Salad Dressing*—This is an emulsified, semi-solid food, prepared from edible vegetable oil, one or more acidifying ingredients, one or more egg-yolk ingredients, and a cooked or partly cooked starch. Salad dressing should have a minimum of 30% by weight of vegetable oil and not less than 4% by weight of egg yolks.

The formulation problem in creating low-calorie counterparts of the foregoing is to maintain the flavor and texture and overall appearance characteristics with a significantly reduced oil content.

Salad Dressing Ingredients

Emulsifiers The emulsions used in salad dressings are the oil-in-water types so that the emulsifiers must correspond. In addition, various types of stabilizers are used.

The following ingredients are used to stabilize salad dressings: egg, corn-starch, flour, gelatin, gum tragacanth, gum arabic, pectin, carrageenan, locust bean gum, gum, gum karaya, propylene glycol alginate, sodium carboxymethyl cellulose, mustard and paprika. In addition, eggs can also function as fairly efficient emulsifiers primarily due to the phospholipid component.

Egg Yolks Care should be taken that these are salmonella-free and conse-quently the frozen egg yolk is to be preferred. Saline solution is generally added to frozen egg yolks prior to freezing and this, in turn, increases the viscosity of the thawed product and tends to improve its emulsifying properties. The egg-yolk color will vary from deep yellow-red to pale yellow and can be an important contributor to the color of the finished mayonnaise.

Starch Starch is widely used as a stabilizer and its properties can vary depending upon the particular type employed. Thus properties such as gelati-nization temperature, taste, texture, and clarity can vary significantly depend-ing upon whether the starch is derived from one of a number of sources in-cluding potato, corn, tapioca, amioca, arrowroot, wheat, rice, and sago. In addition, there are the pre-gelatinized or cold-water soluble starches, and chem-ically modified native raw starches. With this wide selection, it is possible to obtain the specific type or combination of types which will yield a product with both acceptable mouth-feel and shelf-life.

Oil Although the oil content in the low-calorie products may be signifi-cantly reduced compared to the usual product, it can still have a profound effect on flavor and texture and shelf life so that care also should be exercised in selection of this ingredient. Thus, purity, free fatty acid content, and the *AOCS* cold test specifications are as critical for the low-calorie product as they are in the usual run. The use of acidulants and spices to suppress the off-flavors of poor quality or oxidized oils is not recommended since they may tend to enhance rather than suppress these qualities.

Acidulants The usual acidulants employed are vinegar and citric acid. Ideally, they should be mild in odor and taste but should enhance the flavor of the product. Metals should also be removed from the vinegar in order to avoid the oxidative catalysis. Citric acid when used generally represents only a partial replacement for vinegar.

Spices and Seasonings The more popularly used in salad dressings are mustard flour, which also has some emulsifying properties, pepper, paprika, onion and garlic. Ginger, mace, cloves, tarragon and celery seed are used to a lesser extent. Dry spices can be contaminated with dirt and bacteria and require very strict quality control.

Processing

A variety of processes are available and used by different manufacturers but the following, however, is fairly typical and representative:

1. The water-vinegar and starch are combined in a large steam jacketed kettle fitted with an agitator. The vinegar, in some cases, is added later with the egg yolk and salad oil. Care should be taken to add the starch slowly with constant agitation in order to minimize lumping.

2. Heat is applied to the mixture with continuous agitation until a heavy paste is formed. The product temperature at this point is somewhere between 180 and 200°F.

3. The paste is cooled to a temperature of about 140–150°F. The egg yolk and salad oil are then thoroughly blended until a homogeneous mixture is formed.

4. The preblended dry ingredients are then added slowly and mixed so as to form a fine emulsion. The final emulsion should be smooth in appearance and mouth-feel.

5. The emulsion is then filled into jars and sealed tightly.

Storage Stability

The objective for product shelf life should be a minimum of 3 to 6 months at temperatures ranging between 30 and 100°F. The types of spoilage which can occur include separation of the emulsion, fermentation, and off-flavor development.

The chief reasons for mayonnaise spoilage are emulsion breakdown as a result of oil oxidation or improper emulsion preparation. Bacterial spoilage is less frequent due to the inhibition by the acid.

Emulsion breakdown can occur as a result of mechanical shock, particularly in the case of mayonnaise. Allowing the product to cool to too low a temperature, which may result in oil crystallization, can also cause separation. Finally, improper techniques employed in the original mixing procedures, the egg component in the case of mayonnaise and salad dressings, or improper operating temperatures may also be responsible.

Fermentation problems generally are the result of failure to observe proper sanitary procedures. It can be avoided by use of thorough and regular antiseptic cleaning of plant and equipment.

Off-flavors and odors can be minimized by proper selection and care and handling of ingredients. High quality ingredient standards should be maintained along with well-balanced formulas and plant procedures. The oil used should be free of any incipient rancidity and should not be stored for excessively long periods of time. All equipment used to store and handle ingredients should be clean and of stainless steel whenever possible. Finally, all jar and other container closures should be air-tight.

Typical Formulas

The following formulas are intended primarily as guides in preparing low-calorie salad dressings. Considerable variation is possible in the actual ingredients used, their order of addition, speed and temperatures of mixing, as well as sources of raw materials.

A. French-Type Dressing

	lbs	oz
Water	72	3.6
Vinegar, Cider type	12	0.4
Vinegar, Wine type	6	0.0
Starch	2	6.4
Garlic powder	1	15.4
Egg yolks	1	3.2
Salt	1	3.2
Propylene glycol alginate	1	0.4
Tomato powder	0	9.6

Onion powder	0	8.6
Mustard powder	0	5.8
Cyclamate/Saccharin (12:1)	0	3.2
White pepper, powdered	0	2.6
Nutmeg powder	0	1.2
Celery powder	0	1.2
Paprika oleoresin	0	0.4
B. Mayonnaise-Type Dressing		
White vinegar	33 lbs	3.2 oz
Water	29	1.6
Salad oil	26	9.6
Starch	4	2.4
Egg yolk	3	5.6
Salt	2	7.8
Propylene glycol alginate	0	13.2
Cyclamate/Saccharin (12:1)	0	2.4
White Pepper powder	0	1.2
Mustard powder	0	1.2
C. Salad-Type Dressing		
White vinegar	39 lbs	0.8 oz
Water	34	3.2
Salad oil	15	9.6
Starch	4	14.4
Salt	2	14.8
Egg yolk	1	15.2
Propylene glycol alginate	0	15.6
Cyclamate/Saccharin (12:1)	0	3.2
White Pepper	0	1.6
Mustard powder	0	1.6

The foregoing salad dressing formulas differ from their high-calorie coun-terparts as a result of the reduction or complete elimination of the oil content and the use of hydrocolloids to help impart the proper texture. In the case of the French-type dressing, there is no vegetable oil used at all. In the mayon-naise and salad-type dressings, the oil content has been reduced by 50% or more. The results of such changes are significant differences in flavor, mouth-feel, pourability, and appearance when compared to their high-calorie counter-parts. This underlines the need for particular care in formulation to overcome what might otherwise be serious deficiencies that could destroy consumer acceptability.

Carbonated Beverages

The three chief problem areas in the development and production of low-calorie carbonated beverages are replacement of the sugar function, meeting the flavor requirements, and achieving the proper carbonation.

Sugar Replacement

Sugar functions to impart sweetness, to round and blend the flavor com-ponent and provide mouth-feel. It also helps to stabilize the carbon dioxide.

The sweetness of sugar is usually imparted by a mixture of cyclamate and saccharin salts. The ratio may vary from about 7:1–20:1. The actual ratio used should be defined primarily by flavor and economic factors. The actual level of the mixture is also determined by tasting and should be done while also varying levels of acidulant. In cases where flavor considerations, particularly aftertaste problems, dictate a sweetener level which imparts inadequate sweetness intensity, reduction of the acid and variation in flavor type can often compensate.

Sugar also is responsible for mouth-feel. A great deal of formulation work has been done with bodying agents, including hydrocolloids and sorbitol, but these generally do not appear to be too effective. The viscosity of the sugar-sweetened beverage is rather low, in the range of perhaps 1.1–1.5 centistokes, as compared to 0.9 centistokes for a typical low-calorie beverage. The use of bodying agents can reproduce the viscosity fairly readily, but achieving the same type of mouth-feel at the same time tends to be a more difficult matter. Experience has indicated that perhaps the best approach toward achieving mouth-feel is by creating the illusion with a suitable flavor that tends to be "heavy in body."

In addition to sweetness, perhaps the most important function of sugars is to blend and round out flavor. The nonnutritive sweeteners do not have this property, and, therefore, the effect desired must be achieved by selection of a different flavor type and adjustment of the other components of the drink.

Flavor Selection

Because of the problems discussed above, when sugars are eliminated the flavors used in low-calorie carbonated beverages must be selected with greater care than usual. These should be of particularly high quality and as free as possible from undesirable notes. They should be formulated so as to be free of harsh character, and well balanced and rounded overall. This usually means a higher quality and, as a corollary, a somewhat more expensive flavor. In addition to the flavor selection, the following additional adjustments can be made in order to obtain the best possible flavor, quality and impact.

1. Selection and adjustment of the acidulant or acidulant-blend used.
2. In some cases, the use of bodying agents can be helpful.
3. Adjust the level of carbonation to suit the low-calorie formulation rather than attempting to utilize the same level employed for sugar beverages.

Carbonation

The chief problem with carbonation in low-calorie beverages is that of gas retention since gushing is much more common in the absence of sugar solids. To some extent the use of hydrocolloids has been helpful. Recognizing the carbonation problem, the following factors should be given careful attention during production.

1. Production water should be carefully filtered to eliminate the fine particles that can serve as nucleation points for CO_2 loss.
2. Care should be exercised also in the makeup of syrup and in equipment

sanitation again to avoid presence of fine particles that might serve as nucleation points.

3. The bodying agents which are used in premixes or in the final formulation can act as foaming agents. Materials such as methylcellulose and surfactants should be avoided because of their tendency to enhance foaming.

4. Flavor oils should be carefully emulsified and dispersed in the beverage to prevent boiling or gushing. Emulsification procedures should be adjusted to avoid materials mentioned in the above section which tend to foam.

5. Temperature employed during the bottling, filling, and carbonation steps must be carefully controlled and kept at a maximum of 34°F or lower, if possible. This again will help to minimize CO_2 loss.

6. Metal contaminants in trace quantities such as iron, tin, nickel, cobalt, and aluminum should also be avoided.

Typical Formulations

A. Lemon Flavored Beverage

Terpeneless Lemon Oil	1 fluid ounce
Potable Alcohol	42 fluid ounces
Citric Acid	1.5 ounces
12:1 Cyclamate-Saccharin Mixture	11.1 ounces
Sufficient to make 1 gallon	Water

Mix the oil and alcohol, and then add water. Stir in sweetener and citric acid. Make up to 1 gallon and agitate until all the solids are dissolved. The resulting solution, which may be somewhat cloudy, can be clarified by filtering through a thin bed of purified powdered asbestos. The filtrate should be clear and no addition of coloring required.

To prepare bottling concentrate, take this preparation and dilute out the ratio 375 cc of the preparation to 1 gallon of water. The "throw" is one ounce of this concentration per 12-ounce bottle. The bottle is filled with water carbonated at $3-3\frac{1}{2}$ volumes.

Both the carbonated water and the bottling concentrate should be refrigerated as nearly at the same temperature as possible. This is designed to minimize boiling and gushing during bottling.

B. Lemon-Lime Flavor

A similar beverage, except for the lemon-lime flavor, may be prepared in exactly the same manner except the one fluid ounce of lemon oil is replaced by a combination of 9 parts of terpeneless lemon oil and 1 part of terpeneless lime oil.

Carbonation Requirement

Carbon dioxide has an important influence on flavor and mouth-feel which is even greater in the case of the nonnutritively sweetened beverages. The low carbon dioxide content emphasizes the effects of sugar elimination, while increasing carbonation has a tendency to also increase mouth-feel.

The level of carbonation used in a sugar-sweetened beverage is not necessarily suitable for the dietetic product. However, they do provide a convenient starting point. The volumes for some of the more common products are as follows:

Lemon	3.5–4.1 volumes
Orange	0.85–2.0 "
Cherry	3.0–3.5 "
Lemon-Lime	3.5–4.0 "
Cola	3.5–4.0 "

The importance of carbonation can be quite dramatic in terms of the acceptance of the low-calorie beverage. A low-calorie orange beverage was prepared in which all components were kept constant except for the level of carbonation which was at 2, 3, 4, and 5 volumes of carbon dioxide. When presented to a trained panel for evaluation, the preference was for the higher levels—4 and 5 volumes—than for the lower levels which were described as having "typical dietetic" flavor. The high levels are described as having more mouth-feel and less of the typical nonnutritive sweetener taste. There was some dis-agreement as to whether the 4 volume or 5 volume level was optimum, but in any case the higher levels were significant preferred over the lower.

Gelatin Desserts

In addition to the usual problems of sweetness, tartness, and overall flavor, the role of gelatin must be given careful attention because of its importance to the total product. We are also encountering for the first time now the situation where the final product is a dry mix so that care must also be given to blending and uniformity in dry packaging.

Prototype Formula

Gelatin (225 Bloom)	56.2 lb
Sorbitol or mannitol	19.6
Fumaric acid	10.6
Cyclamate/Saccharin (12:1)	7.1
Sodium citrate	3.5
Salt	3.0
Flavor	As needed
Color	As needed

Gelatin

The type of gelatin selected should be one which not only gives good gels but also has a minimum of off-flavors associated with it. The preferred types are pig or calf gelatins with a Bloom strength of 175–250, and viscosity in the 25–50 centipoise range. Such high Bloom gelatins will usually give the best clarity, free from animal odor and other off-flavors, and slow melting properties as well.

The gelatin should be ground to pass through a No. 60 U.S. Screen with the amount through a No. 325 limited to about 5%. Fine ground gelatin will dissolve more rapidly, particularly at temperatures below boiling (160°–

180°F), but the presence of excessive fines will cause agglomeration and caking in storage.

The usual moisture content of gelatin is in the range of 8–12%, but lower levels, 5–8%, will result in a significant increase in storage stability 180°F), but the presence of excessive fines will cause agglomeration and cak if the following precautions are taken:

1. Protection against moisture pick up is provided while the gelatin is stored in bulk.
2. Protection against moisture is provided during processing.
3. The package used for the final product is moisture proof.

The gelatin concentration is determined by running test formula lots. Full product is prepared with various levels of gelatin and the gel strength determined after 17 hours at 10°C. The supplier should standardize the Bloom strength and viscosity of shipments utilizing this type of standard test.

Sorbitol/Mannitol

Sorbitol or mannitol in crystalline form is used primarily as a diluent dispersant to improve solubility. It also functions as a filler to standardize net weight and as a bodying agent. As an alternative to these ingredients, a low Bloom gelatin may also be used for standardizing package weights.

Acids

The three most commonly used acids are citric, adipic, and fumaric acid. Citric tends to be significantly more hygroscopic than the other two and, therefore, can present a problem in storage. Also its higher use level increases ingredient costs. Adipic acid is used at about the same level as citric acid, but, at this level the pH is higher while providing the same tartness making it possible to reduce the gelatin concentration without sacrificing gel strength. In addition, its hygroscopicity is signficantly lower than that of citric acid. Fumaric acid provides significant economies in use because of its lower initial cost and lower use level plus its non-hygroscopicity which is a definite asset to storage stability. However, selection of an acid should be governed by overall considerations of flavor, economics, and patents which may control their use in gelatin desserts.

Buffer Salts

Most commonly used are citrates and phosphates and they function to increase pH and also modify tartness. The citrates are generally to be preferred over phosphates for tartness modification; for low sodium products, potassium salts may be substituted for the sodium salts. Calcium salts generally are not used because of limited solubility and incompatibility problems.

Salt

The function of sodium chloride is exclusively for flavoring. It has no functional purpose and may be omitted where low sodium content is desired. If salt is removed from the formula, it may be necessary to adjust the acid, buffer, and flavor balance.

Flavor

Selection of flavor is also extremely critical. The flavor should be selected to identify the type which is most compatible with the nonnutritive sweetener system. In addition, the flavor level should be tested with different combinations of acid, buffer and sweetener levels.

No unusual storage stability problems should be anticipated. The flavors encapsulated by spray drying in gum arabic should be quite satisfactory for most situations and no interaction with the nonnutritive sweeteners or other components should be anticipated.

Color

The normal food colors as permitted by government regulation are applicable to the low-calorie gelatin desserts. However, caution should be taken to avoid two potential problems:

1. If a blend of primary colors is used there may be "flashing" problems due to different rates of solubilities of the various components of the mix.
2. Color plating onto the dry mix can be another source of difficulty and tests should be conducted on large batches in order to determine what the optimum conditions are. Addition of water in order to help color development is not advisable because moisture can have a deleterious effect on storage stability.

Storage Testing

Tests on gelatin desserts, as on any other product, are run to determine that there are no incompatibilities due to interaction of ingredients, that the package provides adequate protection against excessive moisture pick up and product deterioration, and that the product assays are within dietary claims. Typical storage conditions are 90°F at 70% and 85% relative humidity. Evaluations under 85% R.H. conditions are run every two weeks, and every four weeks at the lower humidity. If there is no significant deterioration at 85% R.H. after 8 weeks, there should be a minimum of 1 year shelf life under normal conditions.

Processing

The dry product is satisfactorily blended in a ribbon mixer, with 400 or 600 pound capacity being adequate for most purposes. About 100 pounds of mix will make approximately 32,000 half-ounce envelopes, each sufficient for a two-cup recipe.

Color, acid, buffer, flavor, and sorbitol or other filler, are generally blended into a premix. It is desirable to develop at least partial coloration before the final mixing operation since blending of the entire formulation for longer than 15 minutes can result in segregation.

Uniformity of mixes, package weights, and uniformity between packages are particularly serious problems in low calorie gelatin desserts because of their concentrated nature.

Dry Beverage Bases

Low-calorie, dry-beverage bases are powdered mixtures of cyclamate/saccharin, acidulant (CWS® fumaric acid, citric acid, or malic acid), color, flavor, and filler (optional). They provide caloric reduction compared to the regular product used with sugar and, in addition, also offer the convenience of presweetening.

Package Weight

Each packet is designed to yield two quarts of finished drink, with an actual weight varying between one-quarter and one-half ounce. The quantity of filler determines the final package weight, with the higher weight being preferred for the following reasons:

1. More material insures optimum blending and filling. This helps insure homogeneity and uniform package weight.
2. The additional filler can help improve body and mouth-feel. The lower weight packages tend to contain little or no filler, and adequate bulk for handling by the packaging equipment is achieved by agglomeration which also helps prevent segregation of ingredients.

Filler

The fillers used to increase bulk and also add to the body or mouth-feel of the reconstituted product includes sorbitol, mannitol, lactose, and dextrose, with the latter two being most popular. Economics generally dictate the selection of filler since there is no significant difference in performance characteristics among the various alternatives.

Acidulants

Until the introduction of CWS® fumaric acid and malic acid, citric acid was the only acidulant used in dry beverage bases. It is more expensive to use than the other two, but from the point of view of flavor, it is quite acceptable, of course. When citric acid is used, the granulation recommended is 40–100 mesh.

About 8–9 parts of malic can replace 10 parts of citric acid which provides 10–20 per cent savings in acidulant costs. CWS® fumaric offers even greater economics—over 40%; in addition, being non-hygroscopic, its use results in improved storage stability and/or less expensive packaging.

Flavor

As in the case of gelatin desserts and other products, flavor type and concentration can be extremely critical. Compatibility with nonnutritive sweeteners should be evaluated and, in addition, the flavor should be selected so as to minimize off-tastes and any artificial character.

Typical Formulation

CWS® fumaric acid	18.3 lbs
Filler	63.0
Cyclamate/Saccharin (12:1)	13.2
Flavor and Color, approximately	5.5

The net package weight can be varied by changing the quantity of filler used. Elimination of the filler and the substitution of agglomeration can, of course, provide further economies. The dry mixes are prepared in a ribbon blender in essentially the same manner as dietetic gelatin desserts.

Cooked Puddings

Caloric reduction in cooked puddings is achieved by elimination of the sugar component. This, in turn, results in textural changes which must be compensated for by modification of the gel system and other components of the mix. Given below are two prototype formulations, one of which is fruit flavored requiring the use of an acid:

A. Vanilla, Butterscotch or Caramel	
Non-fat dry milk solids	80.1 lbs.
Cornstarch	19.7
Carrageenan (protein reactive)	0.08
Cyclamate/Saccharin (12:1)	0.08
Flavor	As needed
Color	As needed
B. Fruit flavored	
Cornstarch	19.7 lbs.
Sorbitol	76.6
Fumaric acid	3.5
Carrageenan (protein reactive)	0.086
Cyclamate/Saccharin (12:1)	0.08
Flavor	As needed
Color	As needed

Non-Fat Dry Milk

The use of non-fat dry milk solids is optional but extremely helpful for improved flavor and texture. Where water is used for reconstitution, the addition of non-fat dry milk solids is recommended. This is true even with lemon-flavored puddings where a pure water recipe is acceptable in the sugar sweetened version. The above formula for lemon flavored pudding is considerably improved when 25–50% of the sorbitol is replaced by non-fat, dry-milk solids. Care should be taken that the milk solids used are fresh and free of cooked flavors.

Cornstarch

Regular raw cornstarch may be used although textural improvements and better tolerance in recipe preparation can be achieved in many cases by means of modified starches or blends.

In addition to flavor and texture, starches should also be evaluated for consistency changes upon standing, since on storage in the refrigerator many starches continue to set and take on a rubbery character. In addition, syneresis is found to be particularly critical in low-calorie products where there are not enough solids present to "bind" the water.

Carrageenan

Carrageenan can help provide tolerance to over- or under-gelatinization of the cornstarch. Other gums such as pectin, alginates and cellulose also may be used for this purpose.

The actual gum selected will depend on the type of cornstarch used. Care should be exerted to avoid incompatibilities or other types of undesirable synergism. When incompatibility occurs, there will be low gel strength and syneresis while synergistic effects will be manifested by heavy gelling or pasty texture.

The carrageenan or other gum selected should provide the following tolerances:

1. A ±10% liquid variation of recipe preparation.
2. Two extremes of cooking: one where the mixture is removed from the heat at the first sign of a bubble, and the other where it is allowed to boil for two full minutes.

Acid

Only fruit flavored cooked puddings require the use of an acidulant. Fumaric acid is recommended because of its low hygroscopicity and cost savings although other acids such as citric, malic, and adipic can be used as well. It should be noted that there are patents covering the use of acidulants in certain types of pudding systems, and these should be investigated in advance.

Instant Puddings

These formulations and products are very similar to cooked puddings except for the use of pregelatinized starch to eliminate the need for cooking of the pudding and additional ingredients as to insure achievement of proper texture. Again, as in the case of cooked puddings, the elimination of sugar components creates a texture problem which must be solved in order to prepare acceptable products. The similarities and differences between cooked and instant puddings can be seen in the prototype formulas below:

A. Imitation Flavored Pudding

Non-fat dry milk solids	73.0 lbs
Pre-gelatinized starch	18.0
Disodium orthophosphate	3.6
Tetrasodium pyrophosphate	3.6
Cyclamate/Saccharin (12:1)	1.8
Carrageenan	0.7
Flavor	As needed
Color	As needed

B. Chocolate Flavored Pudding

Non-fat dry milk solids	50.8 lbs
Cocoa	30.4
Pre-gelatinized starch	8.3
Disodium orthophosphate	4.2
Tetrasodium pyrophosphate	4.2
Cyclamate/Sodium (12:1)	2.1
Flavor	As needed
Color	As needed

Non-Fat Milk Solids

These serve the same function in instant pudding as they do for the cooked pudding. That is, they act as a diluent to improve dispersibility and solubilty, provide bulk for easier packaging, improve flavor and texture, and also serve to standardize net weight.

Cocoa (for Chocolate Only)

With a variety of cocoa types available, one should be selected that has the flavor, color, and overall compatibility with nonnutritive sweeteners. Either plain or dutched cocoa can be used with fat content between 10–20%. While high fat cocoas are acceptable, they are more expensive and increase the caloric content.

Pregelatinized Starch

This provides body and gives the texture to the product. The best types are those made from potato and tapioca starch because of bland flavor even at high concentrations. Furthermore, pregelatinized tapioca starch provides higher viscosity (better body), and a pudding texture closer to that of a cooked pudding.

As in the case of the raw starches, the pregelatinized starch also has inherent variation. In some cases the addition of a small amount of a cold-water soluble gum helps improve uniformity of performance. Cold-water soluble gums, such as alginates and carrageenans, can also be used in the formulation of instant pudding. Their setting and thickening properties, however, differ from that of pregelatinized starch leading to different product texture.

Phosphate Salts

The principle gelling agent is tetrasodium pyrophosphate which reacts with milk proteins to form a complex. The use of phosphates in instant pudding is covered by a number of patents so that any formulation should be checked to be sure it does not conflict.

Disodium orthophosphate is used chiefly as a buffer to improve tetrasodium salt's function as a thickening agent.

The phosphate salts used should be of very fine granulation. Generally, 100% should pass through a U.S. No. 60 Screen in order to insure rapid solubility in cold water. The amount of phosphate is determined by the concentration and type of gum employed.

Additional Ingredients

The basic instant pudding formulation may be modified by the addition of texture-improving ingredients, all of which are protected by patents. These include calcium salts, vegetable oil, mono- and di-glycerides or other emulsifiers.

Frozen Desserts

The caloric values for some of the optional ingredients used in ice cream and frozen desserts are as follows:

Cream (35% fat)	330 calories/100 grams
Whole Milk	68
Milk Solids, non-fat	362
Condensed Unsweetened Milk	138
Cane Sugar	385
Dextrose	348
Dried Egg Yolk	693
Stabilizer-emulsifiers	450 (estimated average)

A regular ice cream formulation with 12% fat contains about 210 calories per 100 grams of finished product. The objective in making a dietetic product is to achieve a minimum of 30% reduction in calories. The obvious route is to replace fats and carbohydrates. The nonnutritive sweeteners are used as substitutes for sugars, while the fat is reduced or eliminated with the necessary formula modifications to give good texture.

Given below is a basic formula for a vanilla or fruit-flavored frozen dessert with a caloric value of 144 calories per 100 grams. It contains 3.5% fat and 20.5% non-fat milk solids. In order to obtain a chocolate flavor, part of the milk solids and/or sorbitol can be replaced with cocoa.

Non-fat milk solids	22.7 lbs
Heavy cream (35% fat)	10.1
Sorbitol	7.0
Cyclamate/Saccharin (12:1)	0.2
Stabilizer*	0.5
Water	59.5
Flavor, color, nuts	As needed

*50% CMC, 7 HOP
50% Gum Karaya

Fat

Fat generally can be reduced to less than 6% in a frozen dessert. The usual sources of fat are the dairy products, including milk solids, whole milk, fresh sweet cream, unsalted butter, butter oil, sweet cream buttermilk, and plain condensed milk.

Milk Solids

These should be selected for quality, cost and performance. Any of several sources may be used in order to lower cost. Milk solids should have a clean, fresh, creamy flavor and odor since the higher the concentrations the longer the storage time will be for overall flavor quality. Usable sources of milk solids are whole milk, sweet skim milk, non-fat dry milk, powdered whole milk, plain condensed skim milk, and plain condensed milk. As the fat content is varied there will be formula adjustments needed.

Stabilizers and Emulsifiers

These are extremely critical in any low-solids, low-fat system such as a dietetic frozen dessert. They are needed to bind free water and prevent large crystal formation during freezing. The stabilizers also contribute to the smooth texture that is a desirable objective and is normally found in high fat content products.

Stabilizers can be the source of undesirable flavor and texture and, therefore, must be carefully selected and tested. Some of the common ones used, either singly or in combination, are gelatin, agar, locust bean gum, guar gum, algins, carrageenans and cellulose gums.

Emulsifiers increase creaminess and smoothness by increasing the effectiveness of fat used and they also may function as stabilizers. The types and level of use are restricted. Included are such common emulsifiers as the mono- and diglycerides of edible fats which are permitted at levels up to 0.2%. Poly-oxyethylene emulsifiers are permitted up to a level of 0.1%.

Sorbitol/Mannitol

These act as conditioning or blending agents and to some extent may also help control crystallization. Non-fat milk solids can often be used in place of sorbitol and mannitol.

Processing

The general procedure for preparing a dietetic frozen dessert is as follows:

1. Thoroughly blend the dry ingredients.
2. Combine all liquid ingredients and heat slowly. Agitate and add the dry ingredients before the temperature reaches 120°F.
3. Depending upon the stabilizers used, disperse and dissolve these ingredients.
4. Pasteurize the entire mix to meet the legal requirements which may include bacterial count, pasteurization time and temperature.
5. Homogenize the mix to reduce the droplet size of the fat and thereby increase stability, impart smoother texture, improve shipping ability, lessen the risk of churning in the freezer, and allow for slight reduction of the stabilizer.
6. Cool the mix to approximately 32°–40°F. Color and flavor may be added at this point.
7. Age the mixture in this temperature range for 3–4 hours before freezing. This allows the fat to solidify, the stabilizer to combine completely with water, and the viscosity to build up to the maximum. Aging improves smoothness, resistance to melting and ease of whipping.
8. Freeze the mixture. Fast freezing is essential for small ice crystal formation. The actual time and temperature for freezing are determined primarily by the type of freezer and the composition of the mix.

Canned Fruits

There are Standards of Identity for canned fruits which provide the general and specific requirements for product quality and fill, ingredient definitions, and the manner of label statements.

Fruit Ingredients

The optional fruit ingredients are the same for both regular and dietetic products. In peaches, for example, the various types are yellow clingstone, yellow freestone, white freestone, variety groups, etc., and the various forms are peeled whole, unpeeled whole, peeled halves, peeled quarters, etc. The standards for packing media are explained in the Standards of Identity. Each medium is prepared with either water or peach juice as the liquid ingredient.

For dietetic products, the packing medium is water sweetened with saccharin, cyclamate, or a combination of these. In addition, pectin may be used to thicken the solution. The following media can be used for apricots, berries, cherries, figs, peaches, pears, pineapples or plums.

Packing Medium for Dietetic Canned Fruit
Water	100 gallons
Cyclamate/Saccharin, 12:1	17 ounces
Pectin (slow set, 150 grade)	14 ounces

If desired, the pectin can be increased up to 28 ounces for greater thickness.

The density of the packing medium for dietetic products, as measured in degrees Brix 15 days or more after the fruit is canned, must be less than 14°. This requirement is identical to that for water.

Processing

The same processing requirements apply to dietetic canned fruits as to regular ones. The food is sealed in a container and heat processed to prevent spoilage. No preservative is allowed in either the dietetic or regular product when water is used as the packing medium.

Baked Goods

Sugar

Sugar is an extremely critical ingredient in baked goods and replacing it presents some very difficult technical problems. These stem from the important functions of sugar, including the following:

1. Caramelization of sugar is necessary for the desirable golden brown color.
2. Sugar is the part of the system that supports the flour protein in forming a structure or framework and is the basis for what is generally called texture. In heavy doughs, such as for cookies, structure formation may be less important.
3. High sugar concentration and minimum protein in cake yield the most tender crumb. If the protein content gets too high, the cakes and cookies tend to become tough.
4. Sugar is the most important source of bulk for "creaming." This is the development of the emulsion that retains the gas formed during leavening.
5. Sugar, shortening, and leavening agents act as tenderizers. Flour, milk solids, egg solids and other protein-aceous materials generally act as tougheners.

As a general rule of thumb for high quality baked products, the ratio of sugar is slightly higher than that of flour. This provides good keeping quality, moistness, tenderness and flavor. When the sugar in a cake is replaced by nonnutritive sweeteners, the solids are reduced by 25% and the cake's appearance, volume, and texture tend to be undesirable. The cake is tough, small, and poor in color and eye appeal. Various types of compensating ingredients are necessary.

Compensating Ingredients

For a dietetic product, these should either have a lower caloric value or function at lower concentrations. Since there are none readily available whose caloric value is below 4, obviously lower levels of these must be used. Some which are employed at times include carboxymethylcellulose, mannitol, sorbitol and dextrins, but by and large, they have not been quite satisfactory. Best results thus far have been obtained in cookies.

Prototype Formulas
A. Dietetic Chocolate Fudge Cake

Unsweetened Chocolate	2 oz
Butter	1.5
Cake flour	4.3
Non-fat milk solids	2.4
Eggs	3.8
Baking powder	0.23
Sodium aluminum pyrophosphate	0.12
Sorbitol	3.5
Cyclamate/Saccharin (12:1)	0.035
Vanilla extract	0.2
Salt	0.04
Water	4

B. Dietetic Cookies

Butter	10 oz
Vanilla extract	0.1
Non-fat milk solids	2.4
Eggs	3.8
Cake flour	7.0
Salt	0.04
Cyclamate/Saccharin (12:1)	0.035
Ginger	0.035
Cinnamon	0.035

Development of ingredients and processes for preparing high-quality, low-caloric baked goods remains one of the outstanding technical challenges that has yet to be met by the food industry. For the most part, the available compensating ingredients and formulation procedures have not been adequate for the production of a variety of high-quality products.

Bibliography

1. Adam, W. B., 1940, Control of Sweetness in Canned Fruits and Vegetables. Univ. Bristol, Fruit Veg. Preservation Research Station. *Campden, Ann. Rept.*, pp. 15–36, Chem. Abst., 35:578₃, 1942.

2. Aichinger, F., 1953, Osmotically Effective Compound for Chaoul and Plastic Mass Irradiation., *Strahlentherapie*, 91, pp. 393–8, *Chem. Abst.*, 47:12753g, 1953

3. Akagi, M., Tejima, S., 1957, Spectrophotometric Determination of P-Ethoxyphenylurea (Dulcin), *Yakugaku Zasshi*, 77:1043–4, Chem. Abst., 52:985g, 1958.

4. Alberti, C., 1936, Variations in the Taste of Dulcin. *Atti V Congr. nazl. chim. pura applicata*, Rome, Pt. I:271–9, 1935, *Chem. Abst.*, 31:3890₄, 1937.

5. Alberti, C., 1939, Transformations of Dulcin. *Atti X.° Congr. intern. chim*, 3:21–6, *Chem. Abst.*, 34:1000₅, 1940

6. Alberti, C., 1935, The Acetylation of Dulcin. *Gazz. Chim. Ital.*, 65:922–5, *Chem. Abst.*, 30:3416₈, 1936.

7. Alberti, C., 1935, The Pyrolysis of Dulcin and of Acetyldulcin. *Gazz. Chim. Ital.*, 65.926–9, *Chem. Abst.*, 30:3417₃, 1936

8. Alberti, C., 1939, Dulcin. V. Some Transformations of Dulcin. *Gazz. Chim. Ital.*, 69:150–62, *Chem. Abst.*, 33:7283₂, 1939

9. Aldinger, S. M., Speer, V. C., Hays, V. W. and Catron, D. V., 1959, Effect of Saccharin on Consumption of Starter Rations by Baby Pigs. *J. Animal Science*, 18:1350–55, *Chem. Abst.*, 54:13489f, 1954.

•10. Alikonis, J. J., 1956, Dietetic Foods, Their Status and How to Make Them. *Food Eng.*, 28:92–3, March

11. Althausen, T. L. Wever, G. K., 1947, Effect of Saccharin and of Galactose on the Blood Sugar. *Proc. Soc. Exptl. Biol. Med.*, 35:517–19. *Chem. Abst.*, 31:4393, 1937.

12. A.M.A. Council on Foods and Nutrition 1956, *Artificial Sweeteners. J. Am. Med. Assoc.*, 160, p. 875, March.

13. Anderson, E. E., Esselen, Jr., W. B., and Fellers, C. R., 1953, Non-Caloric Sweeteners in Canned Fruits. *J. Am. Dietet. Assoc.*, 29:770–3, *Chem. Abst.*, 47:10147f, 1947

14. Annecke, 1939, Sweeteners in the Drug Industry. *Deut. Apoth. -Ztg.*, 54:244–8, *Chem. Abst.*, 33:4373₅, 1939

15. Anon., 1961, Artificially Sweetened Canned Pineapple; Definition and Standard of Identity. *Federal Register*, 26:12563–3, December 28, *Chem. Abst.*, 56:7754ʰ

16. Ant-Wuorinen, O., 1935, Identification of Saccharin and Dulcin in Beer. *Z. Untersuch. Lebensm.*, 70:389–91, *Chem. Abst.*, 30:809₂, 1936

17. Arreguine, V., 1942, A New, Microchemical Reaction of Saccharin. *Anales Asoc. Quim. Argentina*, 30:38, *Chem. Abst.*, 36:5732₃, 1942

18. Arreguine, V., 1941, A New Microchemical Reaction for the Identification of Saccharin. *Rev. Univ. Nacl. Cordoba*, No. 7–8, 14 pp., *Chem. Abst.*, 36:4060₂, 1942

19. Arreguine, V., 1942, A New Microchemical Reaction for the Detection of Saccharin. *Rev. Assoc. Bioquim Argentina*, 8:24, 25–35, *Chem Abst.*, 38:6237₁, 1944

20. Atanasiu, I. A., 1948, Electrolytic Oxidation of O-Toluenesulfonamide (to Saccharin). *Bul. Inst. Natl. Cerecetari Tehnol.*, 3:29–36, *Chem. Abst.*, 43:2521c, 1949

21. Audrieth, L. F., Sveda, M., 1944, Preparation and Properties of some N-Substituted Sulfamic Acids. *J. Org. Chem.*, 9:89–101, *Chem. Abst.* 38:2020₁, 1944

22. Audrieth, L. F., Sveda, M., Sisler, H. H. and Butler, M. J., 1940, Sulfamic Acid, Sulfamide and Related Aquo Ammonosulfuric Acids. *Chem. Rev.*, 26:49–94, *Chem. Abst.*, 34:2723₃, 1940

23. Auerbach, F., 1922, Sweetening Power of Artificial Sweetening Substances. *Naturwissenschaften*, 10:710–14, *Chem. Abst.*, 16:3982₆, 1922

24. Barnard, R. D., 1947, Effect of Saccharin Ingestion on Blood Coagulation and the In Vitro Anticoagulant Effect of Saccharin and Ferriheme. *J. Am. Pharm. Assoc.*, 36:225–8, *Chem. Abst.*, 41:7528c, 1947

25. Barnard, R. D., 1943, Saccharin Ferrihemoglobin. *Proc. Soc. Exptl. Med.* 54:146–8, *Chem. Abst.*, 38:557₅, 1944

26. Barral, F., Ranc, A., 1919, The Industry of Sweetening Agents. *Ind. Chim.*, 6:73–6, *Chem. Abst.*, 13:1505₆, 1919

27. Barral, F. and Ranc, A., 1918, The Chemistry of Sweetening Agents. *Rev. Sci.*, 56:712–23, *Chem. Abst.*, 13:620₈, 1919

28. Baumann, A., 1920, Detection of Saccharin and Dulcin in Beer. *z. ges. Brauw.*, 43:137–9, *Chem. Abst.*, 14:3496₃, 1920

29. Bechara, E. and Huyck, C. L., 1957, Compatibility of Sweetening Agents. *Am. Profess. Pharmacist*, 23:53–4, *Chem. Abst.*, 51:7649h, 1957

30. Becht, F. G., 1920, The Influence of Saccharin on the Catalases of the Blood. *J. Pharmacol.*, 16:155–97, *Chem. Abst.*, 15:400₃, 1921

31. Beck, K. M., 1957, Properties of the Synthetic Sweetening Agent, Cyclamate. *Food Techn.*, 11:156–8, *Chem. Abst.*, 51:15817a, 1957

32. Beck, K. M., 1954, Sucaryl Sweetened Beverages Improved Through Use of Bodying Agents. *Food Processing*, 5:27, March

33. Beck, K. M., 1959, Non-Caloric Sweeteners and the Dietetic Beverages. *Soc. of Soft Drink Tech.*, Sixth Ann. Meeting, pp. 113–118, April

34. Beck, K. M., 1957, Basic Formulation of Special Dietary Frozen Desserts. *Ice Cream Review*, 40,26,44, January

35. Beck, K. M., Jones, R. L. and Murphy, L. W., 1958, New Sweetener for Cured Meats. *Food Eng.*, 30:114, May

36. Beintema, J., Terpstra, P. and de Vrieze, J. J., 1935, Crystallography of the Copper-Pyridine-Saccharin Complex $CuPy(H_2O)Sa_2$, *Pharm. Weekblad*, 72:1287–94, *Chem. Abst.*, 30:667₆, 1936

37. Belani, E., 1925, The Jung Automatic Feeder (as Patented by Dallwitz-Wegner) and its Use in Saccharin Manufacture, *Chem.-Ztg.*, 49:877–9, *Chem. Abst.*, 20:1540₅, 1926

38. Belani, E., 1927, The Drying of Saccharin. *Chem.-Ztg.*, 51:261–2, *Chem. Abst.*, 21:2260₂, 1927

39. Bell, F., 1954, Stevioside: A Unique Sweetening Agent. *Chem. and Ind.*, pp. 897–8, *Chem. Abst.*, 49:534e, 1955

40. Bergmann, M., Camacho, F. and Dryer, F., 1922, New Derivatives of P-Phenetylurea (Dulcin). *Ber. Pharm. Ges.*, 32:249–58, *Chem. Abst.*, 17:996₂, 1923

41. Bertolo, P., Bertolo, A., 1932, O-Chlorobenzoic Acid from the Action of Chlorine on Saccharin. *Gazz. Chim. Ital.*, 62:487–93, *Chem. Abst.*, 27:73₃, 1933

42. Best, 1917, The Action of Saccharin on Gastric Digestion. *Munch. Med. Wochschr.*, 64:1231–2, *Chem. Abst.*, 12:928, 1918

43. Beyer, O., 1919, Methods for the Determination of Saccharin. *Chem.-Ztg.*, 43:537–8, *Chem. Abst.*, 14:578₄, 1920

44. Beyer, O., 1919, Chemical Changes in the Composition of Saccharin-Bicarbonate Tablets. *Chem.-Ztg.*, 43:751–2, *Chem. Abst.*, 14:590₂, 1920

45. Beyer, O., 1920, New Observations in the Field of Saccharin Analysis, *Chem.-Ztg.*, 44:437–8, *Chem. Abst.*, 14:2772₇, 1920

46. Beyer, O., 1923, Estimation of P-Acid in Commercial Saccharin, *Chem.-Ztg.*, 47:744, *Chem. Abst.*, 18:1164₇, 1924

47. Beyer, O., 1931, The Beyer Formula for the Titrimetric Determination of Saccharin, *Chem. Ztg.*, 55:509–10, *Chem. Abst.*, 25:4818₈, 1931

48. Beyer, O., 1920, What Degree of Sweetness Does Saccharin Possess? *Schweiz., Chem.-Ztg.*, pp. 598–9, *Chem. Abst.*, 15:713₂, 1921

49. Beyer, O., 1922, Synthetic Sweetening Agents. *z. Angew. Chem.*, 35:271–2, *Chem. Abst.*, 17:1511₄, 1923

50. Beythien. H., 1942, (Synthetic) Sweetening Materials, *Chem.-Ztg.*, 66:53–5, *Chem. Abst.*, 37:5508₅, 1943

51. Bhargava, M. G., Dhingra, D. R. and Guptan, G. N., 1951, Saccharin. *Indian J. Pharm.*, 13:83–91, *Chem. Abst.*, 45:10430₁, 1951

52. Bhatnagar, S. S., Mathur, K. G. and Budhiraja, K. L., 1932, Triboluminescence. *Z. Physik, Chem.*, A163:8–16, *Chem. Abst.*, 27:1574₂, 1933

53. Bianchi, A. and DiNola, E., 1909, The Detection of Saccharin and Other Artificial Sweeteners in Foods and Drinks, *Boll. Chim. Farm.*, 47:599–605, *Chem. Abst.*, 3:1312₅, 1909

54. Bleyer, B., Diemair, W., Leonhard, K., 1933, Influence of Preservatives on Enzymic Processes, *Arch. Pharm.*, 271:539–52, *Chem. Abst.*, 28:1723₅, 1934

55. Bleyer, B. and Fischer, F., 1931, The Effect of Saccharin on Biocatalyzers and Metabolic Processes. I, *Biochem. Z.*, 238:212–5, *Chem. Abst.*, 25:5919₈, 1931

56. Blodgett, S. H., 1920, Saccharin. *Med. Rec.*, 97:521–3, *Chem. Abst.*, 14:1714₃, 1920

57. Boedecker, F. and Rosenbusch, R., 1920, Sweetening Power of P-Hydroxyphenylurea Derivatives. *Ber. Pharm. Ges.*, 30:251–8, *Chem. Abst.*, 15:1536

58. Bonis, A., 1917, Practical Methods for Detection and Estimation of Saccharin in Foodstuffs, *Ann. Fals.*, 10:210–18, *Chem. Abst.*, 12:959₇, 1918

59. Bonis, A., 1919, Determination of Saccharin in Compressed Tablets. *Ann. Fals.*, 11:369–72, *Chem. Abst.*, 13:2959₁, 1919

60. Bonjean, E., 1922, The Action on the Organism of Saccharin when used as a Sweetener for Foods, *Rev. Hyg.*, 44:50–79, *Chem. Abst.*, 16:2554₅, 1922

61. Bottle, R. T., 1964, Synthetic Sweetening Agents. *Mfg. Chemist*, 35:1, 60–5, *Chem. Abst.*, 60:15048e, 1966

62. Breidenbach, A. W., 1956, Preservatives and Artificial Sweeteners. *J. AOAC*, 39:646–52, *Chem. Abst.*, 50:13315i, 1956

63. Breidenbach, A. W., 1957, Preservatives and Artificial Sweeteners. *J. AOAC*, 40:782–5, *Chem. Abst.*, 51:16991f, 1957

64. Brooks, L. G., 1962, Comparative Tests on Sodium Cyclamate as a Sweetening Agent. *Pharm. J.*, 189:569–71, *Chem. Abst.*, 58:8857c, 1964

65. Brooks, L. G., 1965, Use of Synthetic Sweetening Agents in Pharmaceutical Preparations and Foods. *Chemist Druggist*, 183:4445, 421–3, *Chem. Abst.*, 63:2848c, 1965

66. Brouwer, Th., 1956, Identification of Saccharin According to Klostermann and Scholta. *Chem. Weekblad*, 52:184–6, *Chem. Abst.*, 50:10613c, 1956

67. Bruhns, G., 1921, The Sweetening Power of Saccharin and of Dulcin, *Centr. Zuckerind.*, 29:725, *Chem. Abst.*, 15:2131₇, 1921

68. Bruhns, G., 1934, The Detection and Estimation of Artificial Sweetening Materials. *Deut. Zuckerind.*, 59:646,728–29, *Chem. Abst.*, 29:850₄, 1935

69. Bucci, F., Amormino, V., 1957, *Detection of Dulcin in Food. Rend. is. Super. Sanita*, 20:530–46, *Chem. Abst.*, 52:5684d, 1958

70. Bucci, F., Amormino, V., 1957, A New Procedure for Investigation of Dulcin in Foodstuffs. *Ann. Chim.*, 47:770–84, *Chem. Abst.*, 51:18360f, 1957

71. Buhr, G., 1948, Poisoning with Dulcin. *Med. Klin.*, 43:105–8, *Chem. Abst.*, 43:3105h, 1949

72. Bunde, C. A., Lackay, R. W., 1948, Failure of Oral Saccharin to Influence Blood Sugar. *Proc. Soc. Exptl. Biol. Med.*, 68:581–2, *Chem. Abst.*, 43:315b, 1949

73. Buogo, G., 1955, Are Synthetic Fats Antialimentary? *Minerva Med.*, 11:1847–52, *Chem. Abst.*, 50:5938f, 1956

74. Burge, W. E., 1918, Substitution of Saccharin for Sugar. *Science*, 48:549–50, *Chem. Abst.*, 13:748₄, 1919

75. Burmistrov, S. I., 1945, Qualitative Reactions of Acetophenetidin, Dulcin, Acetanilide, and Plasmocid. *Farmatsiya*, 8:3,33–5, *Chem. Abst.*, 41:3018d, 1957

76. Burmistrov, S. I., 1946, Qualitative Reactions of Sulfonamides and Saccharin. *Farmatsiya*, 9:2,29–30, *Chem. Abst.*, 41:6021a, 1947

77. California Fruit Growers Exchange, Pectin L. M. in Fruit Products for Diabetics. Reference Sheet No. 379

78. Cameron, A. T., 1947, The Taste Sense and the Relative Sweetness of Sugars and Other Sweet Substances. Sugar Research Foundation Sci. Report Series No. 9

79. Camilla, S., Pertusi, C., 1912, Positive and Sensitive Method for the Detection and Identification of Artificial Sweeteners in Beverages, Medicines, Cosmetics, etc., *Giorn, Farm. Chim.*, 60:385–93, *Chem. Abst.*, 6:1323₆, 1912

80. Carlinfanti, E., Marzocchi, P., 1911, Determination of Saponin and Saccharin in Oil Emulsions. *Boll. Chim. Farm.*, 50:609–15, *Chem. Abst.*, 6:1378i, 1912

81. Carlinfanti, E., Scelba, S., 1913, The Most Important Artificial Sweeteners–Saccharin and Dulcin, *Boll. Chim. Farm.*, 51:505–14, 541–9, 580–6, 613–26, *Chem. Abst.*, 7:664₉, 1913

82. Carlson, A. J., Eldridge, C. J., Martin, H. P., and Foran, F. L., 1923, Studies on the Physiological Action of Saccharin, *J. Metabolic Research*, 3:451–77, *Chem. Abst.*, 18:1152₃, 1924

83. Castiglioni, A., 1954, Detection of Saccharin in Wine by Paper Chromatography. *Accad. Ital. Vite e Vino, Atti*, 6:330–3, *Chem. Abst.*, 50:7389c, 1956

84. Castiglioni, A., 1955, Paper Chromatography of Saccharin and Dulcin Mixtures. *Z. Anal. Chem.*, 145:188–9, *Chem. Abst.*, 49:7452b, 1955

85. Castiglioni, A., 1955, Paper Electrophoresis of Saccharin-Dulcin Mixtures. *Z Anal. Chem.*, 148:98–9, *Chem. Abst.*, 50:3161c, 1956

86. Ceccherelli, F., 1915, Identification and Estimation of Saccharin in Foodstuffs, *Boll. Chim. Farm.*, 54:641–8, *Chem. Abst.*, 10:2772₇, 1916

87. Ceriotti, A., 1932, Guide to Bromatological Analysis. Artificial Sweetening Agents, *Ediciones Soc. Nac. Farm.*, 18 pp. *Chem. Abst.*, 26:4887₉, 1932

88. Chang, K. T., 1955, Reaction of Mustard Gas with Saccharin Sodium. *J. Chinese Chem. Soc. Ser. II*, 2:101–2, *Chem. Abst.*, 50:1217a, 1956

89. Changes in Methods of Analysis Made at the Seventh Annual (AOAC) Meeting. *J. AOAC*, 40:39–78, 1957, *Chem. Abst.*, 51:4865i, 1957

90. Chew, A. P., 1926, Shall the Food Law be Toothless. *Fruit Products J. and Am. Vinegar Ind.*, 6:2,10–13, *Chem. Abst.*, 21:139₉, 1927

91. Ciaccio, C. and Racchiusa, S., 1927, The Action of Saccharin in Various Doses of Hyperglucemia by Dextrose. *Boll. Soc. Ital. Biol. Sper.*, 2:309–11, *Chem. Abst.*, 22:2206₃, 1928

92. Cicconetti, E., 1942, Sodium Saccharin for Sweetening the More Frequently Prepared Medicinal Prescriptions. *Farm. Ital.*, 10:527–36, *Chem. Abst.*, 38:3416₃, 1944

93. Comanducci, E., 1911, Note on the Detection of Saccharin. *Boll. Chim. Farm.*, 49:791, *Chem. Abst.*, 5:2276₇, 1911

94. Commerford, J. D., Donahoe, H. B., 1956, N-(W-Bromoalkyl) Saccharins and N,N'-Undecamethylenedisaccharin. *J. Org. Chem.*, 21:583–4, *Chem. Abst.*, 51:2736i, 1957

95. Communications from the Food Inspection Laboratories of the City of Amsterdam. *Chem. Weekblad*, 23:361–5, 1926, *Chem. Abst.*, 20:3317₈, 1926

96. Condelli, S., 1914, Spontaneous Decomposition of Saccharin. *Boll. Chim. Farm.*, 52:639–46, *Chem. Abst.*, 8:2433₂, 1914

97. Condelli, S., 1921, Results and Discussions of the Analyses of Various Products with Reference to Edulcorants and Questions Concerning "Zuccher di Stato". *Staz. Sper. Agrar, Ital.*, 54:326–42, *Chem. Abst.*, 17:3548₇, 1923

98. Cox, W. S., 1950, Report on (Determination of) Artificial Sweeteners. *J. AOAC*, 33:688–90, *Chem. Abst.*, 45:282d, 1951

99. Cox, W. S., 1952, Report on Artificial Sweeteners. *J. AOAC*, 35:321–5, *Chem. Abst.*, 46:11473a, 1952

100. Cox, W. S., 1953, Report on Artificial Sweeteners—Methods for the Detection and Determination of P-4000(2-Propoxy-5-Nitroaniline). *J. AOAC*, 36:749–50, *Chem. Abst.*, 48:9569f, 1954

101. Cox, W. S., 1954, Artificial Sweeteners. P-4000 and Dulcin. *J. AOAC*, 37:383–7, *Chem. Abst.*, 48:9569g, 1954

102. Cox, W. S., 1956, (Determination of) Artificial Sweeteners. (P-4000 and Dulcin). *J. AOAC*, 39:652–9, *Chem. Abst.*, 50:13315i, 1956

103. Cox, W. S., 1957, (Determination of) Artificial Sweeteners. *J. AOAC*, 40:785–7, *Chem. Abst.*, 51:16991g, 1957

104. Cragg, L. H., 1937, Sour Taste. *Trans. Roy., Soc. Can.*, 31:3,131–40, *Chem. Abst.*, 32:2965₈, 1938

105. Cremer, H. D., 1953, Comprehensive Review. The Physiological Importance of Chemical Treatment of Foods. *Z. Lebensm.-Untersuch. U.-Forsch.*, 96:188–98, *Chem. Abst*, 47:6057f, 1953

106. Crosby, D. G., Niemann, C., 1954, Further Studies on the Synthesis of Substituted Ureas. *J. Am. Chem. Soc.*, 76:4458–63, *Chem. Abst.*, 49:13136f, 1955

107. Cross, L. J., Perlman, J. L., 1930, Colorimetric Determination of Saccharin in Beverages. New York State Dept. Agr. and Markets, Ann. Report, pp. 89–90, *Chem. Abst.*, 26:783₆, 1932

108. Cuffi-Roura, U., 1943, The Manufacture of Saccharin. *Industria y Quim.*, 5:54–6, *Chem. Abst.*, 39:512s, 1945

109. Dalal, N. B. and Shah, R. C., 1949, Synthesis of Saccharin from Anthranilic Acid. *Current Sci.*, 18:440, *Chem. Abst.*, 44:4881c, 1950

110. De Boer, H. W. and Bosgra, O., 1943, The Harmful Effect of Saccharin. *Chem. Weekblad*, 40:26–32, *Chem. Abst.*, 38:4042₁, 1944

111. DeGarmo, O., Ashworth, G. W., Eaker, C. M. and Munch, R. H., 1952, Hydrolytic Stability of Saccharin. *J. Am. Pharm. Assoc.*, 41:17–18, *Chem. Abst.*, 46:3678a, 1952

112. Demianovski, S. T., Hefter, T., 1921, Is Saccharin Indifferent for the Animal Body? *Vrachebnoe Delo No.* pp. 16–21, pp. 179–82. *Chem. Abst.*, 17:2616₈, 1923

113. Deniges, G., 1921, Microcrystalline Reactions of Saccharin, Surcamin, and "Sucrose." Bull. Soc. Pharm. Bordeaux No. 2, *Chem. Abst.*, 16:312₁, 1922

114. Deniges, G. and Tourrou, R., 821, Microchemical Reactions of Dulcin. *Compt. Rend.*, 173:1184–6, *Chem. Abst.*, 16:594₆, 1922

115. De Nito, G., 1936, The Toxic Action of Saccharin. Histopathological Studies. *Boll. Soc. Ital. Biol. Sper.*, 11:934–5, *Chem. Abst.*, 31:4724₄, 1937

116. Deshusses, J. and Desbaumes, P., 1956, Study and Identification of Dulcin in Foods by Paper Chromatography. *Mitt. Lebensm. U. Hyg.*, 47:264–9, *Chem. Abst.*, 51:5317e, 1957

117. Diemair, W. and Fischler, F., 1931, Effect of Saccharin on the Biocatalysts and on Metabolic Processes. II. Influence on Saccharin on the Blood Sugar and the Glycogen Content of the Rabbit Liver. *Biochem. Z.*, 239:232–4, *Chem. Abst.*, 26:523₃, 1932

118. Dietzel, R. and Taufel, K., 1925, Ultra-Violet Spectroscopy and its Significance in Food Chemistry *Z. Nahr. Genussm.*, 49:65–75, *Chem. Abst.*, 19:2091₈, 1925

119. Dobreff, M., 1925, The Effect of the Constant Use of Saccharin upon the Digestive Juices. *Arch. Hyg.*, 95:320–30, *Chem. Abst.*, 21:275₁, 1927

120. Dominikiewicz, M., Kijewska, M., 1936, Methods of Obtaining Double Saccharins (Several New Derivatives of m,m'-Bitoly). *Arch. Chem. Farm.*, 3:27–33, *Chem. Abst*, 32:8390₂, 1938

121. Drews, Horst, 1963, Evaluation of the Saccharin Content of Horseradish Preparations. *Ind. Obst.-Gemueseveruert*, 48:7–8, *Chem. Abst.*, 58:14630₉, 1964

122. Dubaquie, J., 1929, Graphical Method for Calculating the Composition of a Saccharin Mixture Containing Sucrose, Dextrose and Levulose. *Ann. Fals.*, 22:352–3, *Chem. Abst.*, 23:4908₉, 1929

123. Durand, H., 1914, Experimental Data on the Determination of Saccharin in Foods with a Modification of Schmidt's Methods. *J. Ind. Eng. Chem.*, 5:981–0, *Chem. Abst.*, 8:385₃, 1914

124. Durocher, P., 1925, The Preparation of Saccharin. *La Nature*, 53:361–4, *Chem. Abst.*, 20:1226₁, 1926

125. Dutt, S., 1922, Dyes Derived from Saccharin. The Sulfamphthaleins. *J. Chem. Soc.*, 121:2389–94, *Chem. Abst.*, 17:554₅, 1923

126. Dyson, G. M., 1924, The Chemistry of Synthetic Sweetening Compounds. *Chem. Age*, 11:572–4, *Chem. Abst.*, 19:823₃, 1925

127. Dyson, G. M., 1939, Saccharin. *Flavours*, 2:1,24–31; 2:46–7; 3:42–3; 4:33–5. *Chem. Abst.*, 34:7471₇, 1940

128. Eighty Years of Sweetening Agents. *Chem. Prods.*, 23:111–14, 1960, *Chem. Abst.*, 54:11325h, 1960

129. Ekkert, L., 1926, Color Reactions of Saccharins. *Pharm. Zentralhalle*, 67:821–2. *Chem. Abst.*, 21:875₉, 1927

130. Endicott, C. J., Gross, H. M., 1959, Artificial Sweetening of Tablets. *Drug and Cosmetic Ind.*, 85:176–7, *App. Sci. and Tech.*, 1959:1259.

131. Erb, J. H., 1942, Sweetening Agents for use in Ice Cream. *Southern Dairy Products J.*, 31:5,28–9,32. *Chem. Abst.*, 36:5572₆, 1942

132. Evans, T. W. and Dehn, W. M., 1930, Constitution of Salts of Certain Cyclic Imides. *J. Am. Chem. Soc.*, 52:1028–9, *Chem. Abst.*, 24:1844₆, 1930

133. Fantus, B., Hektoen, L., 1923, Saccharin Feeding of Rats. *J. Am. Pharm. Assoc.*, 12:318–23, *Chem. Abst.*, 17:2012₂, 1923

134. Fazio, C. C., 1946, Arreguine's Test for Saccharin and its Use in Bromatology. *Rev. Asoc. Bioquim. Argentina*, 13:3–22, *Chem. Abst.*, 40:5164₉, 1946

135. Feigl, F., Auger, V. and Frehden, O., 1935, Use of Spot Tests for the Detection of Organic Compounds. *Mikrochemie*, 17:29–37, *Chem. Abst.*, 29:2471₂, 1935

136. Feigl, F. and Bondi, A., 1929, Reactivity of Iodine in Organic Solvents. II. *Monatsh.*, 53: and 54:508–49, *Chem. Abst.*, 24:351₄, 1930

137. Feigl, F. and Chargav, E., 1928, Reactivity of Iodine in Organic Solvents. *Monatsh.*, 49:417–28, *Chem. Abst.*, 22:3816₉, 1928

138. Ferguson, L. N. and Lawrence, A. R., 1958, The Physiochemical Aspects of the Sense of Taste. *J. Chem. Educ.*, 35:436–44

139. Fichter, F., New Aspects of the Electrochemical Oxidation of Organic and Inorganic Compounds. II. The Electrochemical Preparation of Vanillin, Saccharin and Dyestuffs. *J. Soc. Chem. Ind.*, 48:329–33T, *Chem. Abst.*, 24:2060₇, 1930

140. Fichter, F., Lowe, H. 1922, Electrolytic Oxidation of O-Toluenesulfonamide. *Helvetica Chim. Acta*, 5:60–9, *Chem. Abst.*, 16:1571₉, 1922

♦ 141. Finberg, A. J., 1955, Big Strides of Dietetic Foods. *Food Eng.*, 27:70–5

142. Finzi, C. and Colonna, M., 1939, Chemical Constitution and Sweet Taste. *Gazz. Chim. Ital.*, 68:132–42 *Chem. Abst.*, 32:6638₈, 1938

143. Fischler and Schroter, 1935, Does Saccharin, Orally Administered, Influence the Blood Sugar? *Deut. Med. Wochschr.*, 61:1354–8

144. Fischer, R., 1934, Identification of Organic Preservatives and Commercial Sweetening Substances in Foodstuffs. *Z. Untersuch. Lebensm.*, 67:161–72, *Chem. Abst.*, 28:3137₆, 1934

145. Fischer, R. and Stauder, F., 1931, Detection of Benzoic, Salicylic and Cinnamic Acids, of Saccharin and of the Esters of P-Hydroxybenzoic Acid in Wine. *Z. Untersuch. Lebensm.*, 62:658–66, *Chem. Abst.*, 26:4409₃, 1932

146. Fitzhugh, O. G. and Nelson, A. A., 1950, Comparison of the Chronic Toxicity of Synthetic Sweetening Agents. *Fed. Proc.*, 9:272, March

147. Fitzhugh, O. G., Nelson, A. A. and Frawley, J. P., 1951, A Comparison of the Chronic Toxicities of Synthetic Sweetening Agents. *J. Am. Pharm. Assoc.*, 40:583–6, *Chem. Abst.*, 46:1165i, 1952

148. Flamand, J., 1913, Detection of Saccharin in Beer. *Bull. Soc. Chim. Belg.*, 26:477–8, *Chem. Abst.*, 7:859₅, 1913

149. Font-Altaba, M. and Gutierrez, F. H., 1948, The Manganous Derivative of Saccharin. *Anales Real Soc. Espan. Fis. y Quim.*, Ser. B., 44:355–61, *Chem. Abst.*, 42:7723i, 1948

150. Font-Altaba, M., 1954, Crystalline Structure of the Manganese Derivative of Saccharin. *Publs. Dept. Crist. y Mineral*, 1:133–44, *Chem. Abst.*, 49:14416e, 1955

151. Font-Altaba, M., 1955, X-Ray Characteristics of Saccharin: The Powder Diagram. *Publs. Dept. Crist. Mineral*, Univ. Barcelona, 2:101–6, *Chem. Abst.*, 50:16257g, 1956

152. Food Inspection Decisions. Saccharin in Food. 1911, *F.I.D.*, p. 138, *Chem. Abst.*, 5:3306₁, 1911

153. Food Inspection Decisions Issued by the United States Department of Agriculture on the Use of Saccharin in Foods., 1912, *F.I.D.*, p. 146, *Chem. Abst.*, 6:2473₂, 1912

154. Food Preservatives, 1916, *J. Assoc. Offic. Agr. Chem.*, 2:Pt. 2:83–6

155. Fortelli, M., Piazza, E., 1911, Detection and Quantitative Estimation of Saccharin in Fatty, Starchy and Albuminous Foods. *Z. Nahr. Genussm.*, 20:489, *Chem. Abst.*, 5:539₉, 1911

156. Fresenius, W. and Grunhut, L., 1920, The Chemical Analysis of Wine. *Z. Anal. Chem.*, 59:49–70,209–32,415–57, *Z. Anal. Chem.*, 60:168–87,257–66,353–9, 1921, *Chem. Abst.*, 16:459₆, 1922

157. Frisch, H. R., 1950, Synthetic Compounds with a Sweet Taste. *Chem. in Can.*, 2:6,22, *Chem. Abst.*, 46:9259c, 1952

158. Fuchs, P., 1934, Indirect Volumetric Analysis in Organic Technical Chemistry. *Chem. Fabrik*, 1934:430–2, *Chem. Abst.*, 29:1034₇, 1935

159. Funck, E., 1947, Has Saccharin Any Effect on Biocatalyst? *Pharmazie*, 2:543–5, *Chem. Abst.*, 42:3045a, 1948

160. Gandini, A., 1946, A Rapid Method for Determining Saccharin. *Farmaco Sci. e Tec.*, 1:34–8, *Chem. Abst.*, 40:4317₁, 1946

161. Gandini, A. and Borghero, S., 1948, Photomettric Determination of Dulcin. *Igiene Noderna*, 41:11–15, *Chem. Abst.*, 45:6971g, 1951

162. Gaubert, P., 1936, Liquid Crystals of Some Cholesterol Compounds, and Their Crystalline Superfusion. *Compt. Rend.*, 202:141–3, *Chem. Abst.*, 30:1625₃, 1936

163. Gaudiano, A. and Toffoli, F., 1956, Saccharin. *Atti. Acad. Nazl. Lincci Rend.*, Classe Sci. Fis., Mat e Nat., 21:109–11, *Chem. Abst.*, 51:7355e, 1957

164. Genth, Jr., F. A., 1910, Confirming the Presence of Saccharin in Foods and Beverages. *Am. J. Pharm.*, 81:536–7, *Chem. Abst.*, 4:352₈, 1910

165. German Legislation Concerning Artificial Sweetening Agents. 1922, *Chem.-Ztg.*, 46:361–3, *Chem. Abst.*, 16:2183₆, 1922

166. Gialdi, F., 1948, The Use of 2-Amino-4-Nitrophenol Propyl Ether as a Sugar Substitute and its Colorimetric Determination. *Farm. Sci. e Tec. (Pavia)*, 3:44–50, *Chem. Abst.*, 42:4714g, 1948

167. Gianferrara, S., 1948, Determination of Dulcin in Food. *Ann. Chim. Applicata*, 38:326–8, *Chem. Abst.*, 43:4391e, 1949

168. Gianferrara, S. and Chiorboli, R., 1953, Determination of Saccharin and its Separation from Dulcin. *Chimica e Industria*, 35:224–5, *Chem. Abst.*, 47:11082e, 1953

169. Gianformaggio, F., 1920, The Reduction of Benzoic Sulfonimide. *Gazz. Chim. Ital.*, 50:1,327–40, *Chem. Abst.*, 15:520₁, 1921

170. Gilman, H., Brown, R. E., Dickey, J. B., Hewlett, A. P. and Wright, G. F., 1929, Utilization of Agricultural Wastes. *Proc. Iowa Acad. Sci.*, 36:265–6, *Chem. Abst.*, 25:1245₄, 1931

171. Gilman, H. and Hewlett, A. P., 1929, Some Correlations in Constitution with Sweet Taste in the Furan Series. *Iowa State College J. Sci.*, 4:27

172. Gilta, G., 1937, Crystallography of Tryparsamide and Related Compounds. *Bull. Soc. Chim. Belg.*, 46:263–74, *Chem. Abst.*, 32:851₆, 1938

173. Gnadinger, C. G., 1917, Determination of Saccharin in Foods. *J. Assoc. Off. Agr. Chem.*, 3:25–32, *Chem. Abst.*, 11:2373₉, 1917

174. Griebel, C., 1949, "Ultra Suss" (Sweetening Agent). *Apoth.-Ztg.*, 69:16–18, *Chem. Abst.*, 44:3629h, 1950

175. Griebel, C., 1951, Adulteration of Saccharin. *Pharm. Ztg.-Nachr.*, 87:314–15, *Chem. Abst.*, 46:2753e, 1952

176. Grossfeld, J., 1921, The Sweetening Power of Artificial Sweeteners. *Z. Ges. Kohlensaure Ind*, 15:253, *Chem. Abst.*, 15:3881₇, 1921

177. Grossfeld, J., 1920, The Detection and Estimation of Saccharin and Benzoic Acid in Foods. *Z. Ges. Kohlensaure Ind.*, 26:143–4,159–60, *Chem. Abst.*, 15:3152₁, 1921

178. Gaureschi, R., 1913, Research on Saccharin in Vermuth. *Giorn. Farm. Chim.*, 61:400–2 *Chem. Abst.*, 7:204₇, 1913

179. Gutierrez, F. H., 1945, Acetone Solubility of Some Sulfonamide Medicinal Products. *Anales Fis. γ. Quim.* (Madrid), 41:537–60, *Chem. Abst.*, 41:6546g, 1947

180. Gutierrez, F. H., 1946, Acetone as a New Solvent in the Extraction of Saccharin. Solubilities in Acetone. V. *Anales Fis. γ Quim.* (Madrid), 2:1105–11, *Chem. Abst.*, 41:5120c, 1947

181. Gutierrez, F. H., 1947, The Use of Acetone in the Gravimetric Determination of Saccharin in Beverages and Chocolate. Solubility in Acetone. *Anales Fis. γ Quim.*, 43:393–7, *Chem. Abst.*, 41:6640d, 1947

182. Gutierrez, F. H., 1946, Solubility of Saccharin and its Sodium Salt in Water and Organic Solvents. *Farm. Nueva*, 11:342–4, *Chem. Abst.*, 41:2304e, 1947

183. Gutierrez, F. H., Altaba, M. F., 1947, New Microchemical Reactions of Saccharin. *Anales Fis. γ Quim.*, 43:471–5, *Chem. Abst.*, 42:724b, 1948

184. Gutierrez, F. H., Altaba, M. F., 1954, New Microchemical Reactions of Saccharin. *Publ. Inst. Invest. Microquim.*, Univ. Nac. Litoral, 21:20–25, *Chem. Abst.*, 51:3373i, 1957

185. Gyenes, I. and Vali, A., 1955, Determination of Saccharin Sodium by Titration with Perchloric Acid. *Magyar Kem. Folyoirat*, 69:90–1, *Chem. Abst.*, 49:15635g, 1955

186. Halla, F., 1930, The Electrochemical Oxidation of O-Toluenesulfonamide to Saccharin. *Z. Elektrochem.*, 36:96–8, *Chem. Abst.*, 24:2060₅, 1930

187. Hamann, V., 1948, A New (Substitute) Sweetening Agent. *Deut. Lebensm.-Rundschau*, 44:52, *Chem. Abst.*, 45:3959i, 1951

188. Hamilton, Wm. F., Turnbull, F. M., 1950, Substituted Ammonium Saccharins for Nasal Medication. *J. Am. Pharm. Assoc.*, 39:378–82, *Chem. Abst.*, 44:9631c, 1950

189. Hamor, G. H. and Soine, T. O., 1954, Synthesis of Some Derivatives of Saccharin. *J. Am. Pharm. Assoc.*, 43:120–3, *Chem. Abst.*, 4:3892g, 1955

190. Hand, D. B., *et al.*, 1956, The Use of Chemical Additives in Food Processing. *Natl. Acad. Sci.-Natl. Research Council (U.S.)*, Publ. No. 398, 91 pp., *Chem. Abst.*, 50:10291g, 1956

191. Hansen, A., 1926, Artificial Sweetening Substances. *Dansk Tids. Farm.*, 1:114–23, *Chem. Abst.*, 21:285₆, 1927

192. Haramaki, K., 1922, The Influence of Saccharin on Certain Functions of the Digestive Tract and Kidneys. *Z. Physik, Diatet, Therap.*, 26:183–6, *Chem. Abst.*, 16:3341₉, 1922

193. Hardeggar, E. and Jucker, O., 1949, Derivatives of 3,6-Anhydroglucose and Glucose 6-Iodohydrin. *Helv. Chim. Acta*, 32:1158–62, *Chem. Abst.*, 43:7905h, 1949

194. Harvey, E. H., 1923, Efficiency of Some Common Anti-Ferments. *Am. J. Pharm.*, 90:105–8, *Chem. Abst.*, 17:1675₂, 1923

195. Hayakawa, M., Masuda, S., 1949, Preparation of Dulcin from Sodium P-Nitrophenolate.

Research Rept. Nagoya Ind. Sci. Research Inst., No. 1, 25–7, Chem. Abst., 49:215g, 1955

196. Hefelmann., 1917, Determination of Saccharin. Pharm. Post., p. 675, Chem. Abst., 12:1276₁, 1918

197. Heiduschka, A. and Schmid, J., 1914, Determination of Moisture in Saccharin Material. Pharm. Zentralhalle, 54:956–7, Chem. Abst., 8:2757₉, 1914

198. Heilmann, P., 1922, Saccharin Intoxication. Munch. Med. Wochschr., 69:968–9, Chem. Abst., 17:827₁, 1923

199. Heitler, M., 1922, Induced Spontaneous Transformation of the Cardiac Depressant Action of Saccharin into Cardiac Stimulation. Wiener, Klin. Wochschr., 35:935–7, Chem. Abst., 17:1081₈, 1923

200. Heitler, M., 1920, Sugar and Saccharin. Wiener Med. Wochschr., 70:1050–3, Chem. Abst., 16:1283₂, 1922

201. Helch, H., 1917, Equivalents of Saccharin Solutions and Sugar Syrups of Equal Sweetening Power. Schweiz. Apoth. Ztg., 55:239, Chem. Abst., 11:2532₃, 1917

202. Helgren, F. J., Lynch, M. J. and Kirchmeyer, F. J., 1955, A Taste-Panel Study of the Saccharin "Off-Taste." J. Am. Pharm. Assoc., 44:353–5, Chem. Abst., 49:11243c, 1955

203. Heller, G., 1925, Tautomeric Phenomena in Heterocyclic Compounds. J. Prakt. Chem., 111:1–22, Chem. Abst., 20:381₃, 1926

204. Henrichson, C. B., 1958, Produce High Levels of Sweetness with Low Calorie Foods. Food Proc., 19:46–7, March

205. Hensel, S. T., 1918, The Use of Saccharin as a Sugar Substitute. J. Am. Pharm. Assoc., 7:609–10, Chem. Abst., 13:58₇, 1919

206. Hernandez, F. and Font, M., 1947, Microchemical Reactions of Saccharin. Mon. Farm. γ. Terap., 53:359–60, Chem. Abst., 42:1530h, 1948

207. Herrmann, P., 1922, Derivatives of Dulcin. Ann., 429:163–74, Chem. Abst., 17:381₄, 1923

208. Herzfeld and Reischauer, 1913, Test for Saccharin in Foods. Naturwiss. Wochschr., p. 165

209. Herzog, W., 1930, Utilization of By-Products of Saccharin Manufacture in Synthesis of Drugs in Medical Science. Chem. Rundschau Mitteleur. u. Balkan, 7:15,107–11, Chem. Abst., 25:2810₁, 1931

210. Herzog, W., 1930, Utilization of By-Products from Saccharin Manufacture. Chem. Umsch. Fette, Oele, Wachse u. Harze, 37:296–8, Chem. Abst., 25:428₂, 1931

211. Herzog, W., 1927, Progress in the Field of Synthetic Sweetening Agents and Related Compounds in 1925–1926. Fortschrittsber, pp. 66–8, Chem.-Ztg., Vol. 51, Chem. Abst., 21:2747₂, 1927

212. Herzog, W., 1930, Fighting Vermin and Exterminating Weeds by Use of the By-Products of Saccharin Manufacture. Chem.-Ztg., 54:50-1, Chem. Abst., 24:1926₇, 1930

213. Herzog, W., 1933, Progress in the Field of Synthetic Sweet Substances and Related Compounds. Chem.-Ztg., 57:574–6, Chem. Abst., 27:4598₅, 1933

214. Herzog, W., 1935, Advances in the Field of Synthetic Sweet Substances and Related Compounds in 1933–34. Chem.-Ztg., 59:408–9, Chem. Abst., 29:4848₇, 1935

215. Herzog, W., 1926, The Utility of By-Products from Saccharin Manufacture in the Chemistry of Synthetic Tans and in the Tannery. Collegium, pp. 203–8, Chem. Abst., 20:3586₈, 1926

216. Herzog, W., 1929, Advances in the Field of Synthetic Sweetening Agents and Related Compounds in 1927 and 1928. Fortschrittsber. Chem.-Ztg., 53:99–100, Chem. Abst., 23:4978₃, 1929

217. Herzog, W., 1926, The Utilization of the By-Products from Saccharin Manufacture In the Industry of Synthetic Resins and Plastic Masses. Kunststoffe, 16:105–7, Chem. Abst., 20:2910₄, 1926

218. Herzog, W., 1928, Progress in the Utilization of Saccharin By-Products in 1927. Metallborse, 18:903, 1015–6, 1126–7, 1183–4, 238–9, Chem. Abst., 23:2426₇, 1929

219. Herzog, W., 1929, Waste Products of the Saccharin Industry used in Synthesizing Tanning Agents and In Tanning. Metallborse, 19:1853–4

220. Herzog, W., 1926, The Value of By-Products of Saccharin Manufacture in Analytical Chemistry. Oster. Chem.-Ztg., 29:26–7, Chem. Abst., 20:1612, 1926

221. Herzog, W., 1937, Progress in the Field of Synthetic Sweetening Agents and Related Compounds in 1935 and 1936. *Oster. Chem.-Ztg.*, 40:201–3, *Chem. Abst.*, 31:4968₈, 1937

222. Herzog, W., 1926, Medicaments Derived from Saccharin and its Secondary Products. *Pharm. Zentralhalle*, 67:81–6, *Chem. Abst.*, 20:1301₄, 1926

223. Herzog, W. 1937, Progress in the Realm of Synthetic and Related Compounds During the Years 1935–1937. *Pharm. Zentralhalle*, 78:76–106, *Chem. Abst.*, 32:2244₅, 1938

224. Herzog, W., 1922, Advances in the Chemistry of Synthetic Sweetening Agents in 1918–21. *Z. Angew. Chem.*, 35:133–7, *Chem. Abst.*, 16:1625₆, 1922

225. Herzog, W., 1923, Advances in the Chemistry of Synthetic Sweeteners and Related Compounds in 1922. *Z. Angew. Chem.*, 36:223–7, *Chem. Abst.*, 17:2155₅, 1923

226. Herzog, W., 1925, Advances in the Field of Synthetic Sweetening Agents and Related Substances in 1923 and 1924. *Z. Angew. Chem.*, 38:641–8, *Chem. Abst.*, 19:2988₇, 1925

227. Herzog, W., 1929, The Utilization of By-Products from Saccharin Manufacture for Dye Synthesis, Dyeing, Finishing and Bleaching in the Period 1927 to 1929. *Z. Ges. Textil.-Ind.*, 32:857–9,873–5,889–90, *Chem. Abst.*, 24:6021₄, 1930

228. Herzog, W., 1929, Application of By-Products of Saccharin Manufacture in Photography and Photometry. *Z. Wiss. Phot.*, 27:177, *Chem. Abst.*, 24:3717₇, 1930

229. Herzog, W. and Kreidl, J., 1921, The Quantitative Separation of Saccharin and P-Sulfamyl-benzoic Acid (Para Acid). *Oster. Chem.-Ztg.*, 24:165–66, *Chem. Abst.*, 16:1197₉, 1922

230. Hist, J. F., Holmes, F. and Maclennan, G. W. G., 1941, Detemination of Dulcin (P-Phenetylcarbamide). *Analyst*, 66:450–1, *Chem. Abst.*, 36:722₇, 1942

231. Hoeke, F., 1942, Substitute Materials. I. *Chem. Weekblad*, 39:390–4, *Chm. Abst.*, 40:4444₉, 1946.

232. Hoeke, F., 1947, Determination of the Synthetic Sweetening Agent 1-Propoxy-2-Amino-4-Nitrobenzene. *Chem. Weekblad*, 43:283–5

233. Holleman, A. F., 1921, Artificial Sweet Materials. *Rec. Trav. Chim.*, 40:446–54, *Chem. Abst.*, 15:3821₅, 1921

234. Holleman, A. F., 1923, On Some Derivatives of Saccharin. *Rev. Trav. Chim.*, 42:839–45, *Chem. Abst.*, 18:239₄, 1924

235. Hucker, G. J. and Pederson, C. S., 1942, A Review of the Microbiology of Commercial Sugar and Related Sweetening Agents. *Food Res.*, 7:459–80, *Chem. Abst.*, 37:2206₅, 1943

236. Hurd, C. D. and Kharasch, N., 1947, Reaction of the Dioxane-sulfotrioxide Reagent with Aniline Classification of the Sulfamic Acids. *J. Am. Chem. Soc.*, 69:2113–15, *Chem. Abst.*, 42:138i, 1948

237. Hwang, K., *et al.*, 1956, Comparison of Cyclamate (Cyclohexylsulfamate) Salts in the Alimentary Tract of the Rat. *Food Proc.*, 15:441–2

238. Hynes, L. J., 1939, Sugar Saving Substitutes. *Dairy Inds.*, 4:348–9, *Chem. Abst.*, 33:9466₇, 1939

239. Illies, R., 1940, Sweetening Agents and Their Use in the Brewery. *Deut. Essigind.*, 44:178–9,184–5,190, *Chem. Abst.*, 35:1927₅, 1941

240. Illing, E. T., Dalley, R. A. and Stephenson, W. H., 1947, Saccharin in Soft Drinks. *Food* 16:140–1, *Chem. Abst.*, 41:4589g, 1947

241. Imbesi, A. and DeAngelis, V., 1935, The Effect of Temperature on the Conductivity of Saccharin Solutions. *Ann. Chim. Applicata*, 25:265–62, *Chem. Abst.*, 29:6489₆, 1935

242. International Conference Concerning Saccharin and Analogous Substances. *Ann. Fals.*, 6:165–7, 1913, *Chem. Abst.*, 7:2069₈, 1913

243. Ishimura, K., 1953, Some Physical and Chemical Properties of Dulcin. *Rept. Govt. Ind. Resarch Inst. Nagoya*, 2:87–92, *Chem. Abst.*, 50:15446e, 1956

244. Ivanov, A., 1941, Preparation of Edible Sorbitol. *Pishchevaya Prom.*, 1:3,30, *Chem. Abst.*, 39:5355₃, 1945

245. Jacobs, A. L. and Scott, M. L., 1957, Factors Mediating Food and Liquid Intake in Chickens. I. Studies on the Preference for Sucrose or Saccharin Compounds. *Poultry Science*, 36:8–15, *Chem. Abst.*, 51:15719i, 1957

246. Jacobs, M. B., 1947, Tobacco Flavors. *Am. Perfumer*, 49:607,609, *Chem. Abst.*, 41:7059i, 1947

247. Jacobs, M. G., 1951, Artificial Sweeteners. *Am. Perfumer Essential Oil Rev.*, 57:49,51. *Chem. Abst.*, 45:3998h, 1951

248. Jacobs, M. B., 1955, The Sweetening Power of Stevioside. *Am. Perfumer Essential Oil Rev.*, 66:6,44,46, *Chem. Abst.*, 50:2883i, 1956

249. Jaeger, F. M., 1908, A Contribution to the Theory of Barcow and Pope. *Z. Kryst.*, 44: 61–4, *Chem. Abst.*, 2:498₇, 1908

250. Jagoda, G., 1927, Fodder Tests with Additions of (Beet) Raw Sugar and Molasses, also Saccharin. *Z. Ver. Deut. Zucker-Ind.*, pp. 243–331, *Chem. Abst.*, 22:1813₁, 1928

251. Jakobsen, F., 1941, Artificial Sweetening Agents. *Tids. Hermetikind.*, 27:129–32, *Chem. Abst.*, 35:7568₂, 1941

252. Jakobsen, F., Jakobsen, A., 1942, The Use of Artificial Sweetening Agents in the Preparation of Berry and Fruit Preserves. *Tids. Hermetikind.*, 28:225–37, *Chem. Abst.*, 38:3377₄, 1944

253. Jamison, G. S., 1920, The Determination of Saccharin in Urine. *J. Biol. Chem.*, 41:3–8, *Chem. Abst.*, 14:954₅, 1920

254. Jorgensen, G., 1909, Detection of Saccharin in Beer. *Ann. Falsif.*, 2:58, *Chem. Abst.*, 3:1056₈, 1909

255. Jorgensen, H., 1950, Influence of Saccharin on Blood Sugar. *Acta Physiol. Scand.*, 20:33–7, *Chem. Abst.*, 44:5475f, 1950

256. Kamen, J., 1959, Interaction of Sucrose and Calcium Cyclamate on Perceived Intensity of Sweeteners. *Food Res.*, 24:3,279–82

257. Karas, J., 1913, Determination of Saccharin in Food. *Z. Nahr. Genussm.*, 25:559–60, *Chem. Abst.*, 7:3174₆, 1913

258. Kaufmann, H. P. and Schweitzer, D., 1953, Synthesis of Sweetening Agents. *Fette U. Seifen*, 55:321–4, *Chem. Abst.*, 49:2342a, 1955

259. Kegler, W., 1951, Preparation of P-Phenetylurea (Dulcin Sweetening Agent). *Seifen-ole-Fette-Wachse*, 77:606, *Chem. Abst.*, 47:7451f, 1953

260. Khlopin, N. Y., Litvinova, N. S. and Privalova, K. P., 1951, Polarographic Determination of Saccharin. *Gigiena i Sanit.*, 2:48–51, *Chem. Abst.*, 45:10142b, 1951

261. Kielhofer, E. and Aumann, H., 1955, Two New Sweet Substances and Their Detection in Wine. *Wein.-Wiss., Beiheft Fachzeit, Deut. Weinbau*, 6:1–5, *Chem. Abst.*, 50:5975e, 1956

262. Kiliana, H., Loeffler, P. and Matther, O., 1907, Derivatives of Saccharin. *Ber.*, 40: 2999, *Chem. Abst.*, 1:2467₇, 1907

263. Kinugasa, Y. and Nishihara, S., 1919, Antiseptic Power of Chloramine-T. *J. Pharm. Soc. Japan*, 454:1006–21, *Chem. Abst.*, 14:1410₆, 1920

264. Klarmann, B., 1952, Saccharinimine. *Chem. Ber.*, 85:162–4, *Chem. Abst.*, 46:9539h, 1952

265. Klasens, H. A. and Terpstra, P., 1937, Crystallography of Cupric Saccharinate. *Rec. Trav. Chim.*, 56:672–7, *Chem. Abst.*, 31:6076₈, 1937

266. Kliffmuller, R., 1956, Determination of Sweetening Agents and Preservatives by Paper Chromatography. *Deut. Lebensm.-Rundschau*, 52:182–4, *Chem. Abst.*, 51:634g, 1957

267. Kling, A., Bovet. D. and Ruiz, I., 1941, Saccharin and Dulcin as Substitutes for Sugar *Bull. Acad. Med.*, 124:99–114, *Chem. Abst.*, 38:1844₆, 1944

268. Kling, A., Bovet, D. and Ruiz, I., 1941, The Toxic Action of the Sweetening Substance "Dulcin." *Bull. Acad. Mex.* 125:69–72, *Chem. Abst.*, 38:3360₅, 1944

269. Klostermann, M., Scholta, K., 1916, Critical Consideration of the Qualitative and Quantitative Determination of Saccharin and a New Method for the Qualitative Detection of the Sweetener. *Z. Nahr.-Genussm.*, 31:67–78, 1916, *Chem. Abst.*, 10:1561₆, 1916

270. Kogan, I. M. and Dziomko, V. M., 1953, Reaction of Arylsulfonamides with Amines, II. *Zhur. Oschei Khim.*, 23:1234–6, *Chem. Abst.*, 12280d, 1953

271. Kolthoff, I. M., 1920, The Significance of Dissociation Constants in Identifying Acids and Detecting Impurities. *Pharm. Weekblad*, 57:514–8, *Chem. Abst.*, 14:3040₆, 1920

272. Kolthoff, I. M., 1925, The Acid Character of Saccharin and Related Acids. The Detection and Determination of P-Sulfamylbenzoic Acid in Saccharin and Crystallose. *Rec. Trav. Chim.*, 44:629–37, *Chem. Abst.*, 19:3408₆, 1925

273. Krantz, Jr., J. C., 1933, Report on (Food) Preservatives. *J. AOAC*, 16:316–18, *Chem. Abst.*, 27:5118₁, 1933

274. Krantz, Jr., J. C., 1934, Report on (the Determination of) Preservatives. *J. AOAC*, 17:193–5, *Chem. Abst.*, 28:4493₂, 1934

275. Krantz, Jr., J. C., 1935, Report on (The Determination of) Preservatives. *J. AOAC*, 18: 372–3, *Chem. Abst.*, 29:6966₅, 1935

276. Krantz, Jr., J. C., 1936, Report on (The Determination of) Preservatives (in Foods). *J. AOAC*, 19:205–6, *Chem. Abst.*, 30:4935₈, 1936

277. Kroll, 1942, The Use of Artificial Sweetening Substances in the Preparation of Hot Beverages. *Deut. Destillatur-Ztg.*, 63:319, *Chem. Abst.*, 38:3378, 1944

278. Kun, E., Horvath, I., 1947, Influence of Oral Saccharin on Blood Sugar. *Proc. Soc. Exptl. Biol. Med.*, 66:175–7, *Chem. Abst.*, 42:1354g, 1947

279. Kurzer, F., 1949, Sweeter than Sugar. New Synthetic Compounds More Potent Than Saccharin. *Discovery*, 10:175, *Chem. Abst.*, 43:5880e, 1949

280. LaParola, G. and Mariani, A., 1946, New Color Test for Dulcin. *Ann. Chim. Applicata*, 36:134–9, *Chem. Abst.*, 40:6367₅, 1946

281. Lapicque, L., 1941, Some Remarks on the Use of Saccharin. *Bull. Acad. Mec.*, 124:116–17, *Chem. Abst.*, 37:5196₅, 1943

282. Lasheen, A. M. and Russell, T. S., 1961, Saccharin Sodium as an Artificial Sweetening Agent in Fresh Fruits. *Proc. Am. Soc. Hort. Sci.*, 77:140, *Chem. Abst.*, 56:9176d, 1962

283. Lavague, J., 1945, The Colorimetry of Saccharin. *Ann. Pharm. Fanc.*, 3:26–9, *Chem. Abst.*, 40:5204₈, 1946

284. Ledent, R., 1914, Saccharin in Beer. *Ann. Chim. Anal.*, 18:314, *Chem. Abst.*, 8:2025₂, 1914

285. Lehman, A. J., 1950, Some Toxicological Reasons why Certain Chemicals May or May not be Permitted as Food Additives. *Assoc. Food and Drug Off. U.S. Quart. Bull.*, 14:82–98, *Chem. Abst.*, 45:3517h, 1951

286. Lehman, A. J., 1951, Chemicals in Foods: A Report to the Association of Food and Drug Officials on Current Developments. *Assoc. Food and Drug Off. U.S. Quart. Bull.*, 15:82–9, *Chem. Abst.*, 46:2701i, 1952

287. Lehmann, K. B., 1929, The Feeding of Saccharin to Mice. A Study of the Question of Minimal Toxic Action. *Arch. Hyg.*, 101:39–47, *Chem. Abst.*, 23:3977₈, 1929

288. Lehmstedt, K., 1949, A New Sweetening Agent. *Seifen-Ole-Fette-Wachse*, 75:81–2, *Chem. Abst.*, 44:1207i, 1950

289. LeMagnen, J., 1953, Influence of Insulin on the Spontaneous Consumption of Sapid Solutions. *Compt. Rend. Soc. Biol.*, 147:1753–6, *Chem. Abst.*, 48:9505g, 1954

290. Leroy, M., 1942, Determination of Saccharin and Dulcin. *Ing. Chim.*, 26:18–20, *Chem. Abst.*, 37:5548₈, 1943

291. Lerrigo, A. F. and Williams, A. L., 1927, A Study of the Detrmination of Saccharin Colorimetrically and by the Ammonia Process. *Analyst*, 52:377–83, *Chem. Abst.*, 21:3859₃, 1927

292. Lettre, H. and Wrba, H., 1955, The Effect of the Sweetener Dulcin During Protracted Feeding. *Naturwissenschaften*, 4:217, *Chem. Abst.*, 50:5914b, 1956

293. Lickint, F., 1943, Saccharin and Gastric Secretion (Saccharin Test Meal). *Munch. Med. Wochshr.*, 90:586–7, *Chem. Abst.*, 38:5963₆, 1944

294. Loginov, N. E., 1943, Sweetening Substances. *Pishchevaya Prom.*, 1/2:22–5, *Chem. Abst.*, 40:101₃, 1946

295. Longwell, J. and Bass, C. S., 1942, Colorimetric Determination of Dulcin (P-Phenetyl-carbamide) in Composite Articles. *Analyst*, 67:14–15, *Chem. Abst.*, 36:2235₂, 1942

296. Lorang, H. F. J., 1928, The Relationship Between the Constitution and Taste of Some Urea Derivatives. *Rec. Trav. Chim.*, 47:179–90, *Chem. Abst.*, 22:1147₉, 1928

297. Lorges, A. B., 1926, Artificial Sweetenrs (Saccharin, Dulcin and Glucin). *Rev. Chim. Ind.*, 35:109–13, *Chem. Abst.*, 20:2211₃, 1926

298. Luszczak, A. and Hammer, E., 1933, Thymol, Benzene and Toluene in Commodities and in Air. Their Spectrographic Determination. *Abh. Gesamtgebiete Hyg.*, 12:82, *Chem. Abst.*, 28:6086₆, 1934

299. Lynch, M. J. and Gros, H. M., 1960, Artificial Sweetening of Liquid Pharmaceuticals. *Drug & Cosmetic Ind.*, 87:324–6,412–13, *Chem. Abst.*, 55:902a, 1961

300. Lyons, E., 1925, Mercury Derivatives of Some Imides. *J. Ann. Chem. Soc.*, 47:830–3, *Chem. Abst.*, 19:1247₁, 1925

301. McCance, R. A. and Lawrence, R. D., 1933, An Investigation of Quebrachitol as a Sweetening Agent for Diabetics. *Biochem. J.*, 27:986–9, *Chem. Abst.*, 28:227₉, 1934

302. McKie, P. V., 1921, Examination of Some Methods of Ascertaining the Purity of Saccharin. *J. Soc. Chem. Ind.*, 40:150–2T, *Chem. Abst.*, 15:3343₉, 1921

303. Maass, H., 1948, The Sweetening Agent Problem. *Deut. Lebensm.-Rundschau*, 44:50–2, *Chem. Abst.*, 45:6598b, 1951

304. Maes, O., 1938, Detection and Determination of Sweetening Agents in Beer. *Bull. Assoc. Eleves Inst. Suppr. Fermentations, Gand*, 39:287–94, *Chem. Abst.*, 33:1438₉, 1939

305. Magidson, O. Y. and Gorbachov, S. W., 1923, The Question of the Sweetness of Saccharin O-Benzoyl-sulfimide and its Electrolytic Dissociation. *Ber.*, 56B., pp. 1810–7, *Chem. Abst.*, 18:227₈, 1924

306. Magidson, O. Y. and Zil'Berg, I. G., 1935, Mechanism of the Oxidation of O-Toluenesulfonamide to Saccharin. *J. Gen. Chem. (U.S.S.R.)*, 5:920–3

307. Mameli, E. and Mannessier-Mameli, A., 1940, Pyrolysis of Ammonium Saccharin and of Ammonium Thiosaccharin, Saccharinimimine and Pseudosaccharin Amine. *Gazz. Chim. Ital.*, 70:855–74, *Chem. Abst.*, 36:1026₈, 1942

308. Mannessier-Mameli, A., 1932, Action of Hydroxylamine on Saccharin, on Some of its Derivatives and on 3-Thio-1,2 Benzodithiole. *Gazz. Chim. Ital.*, 62:1067–92, *Chem. Abst.*, 27:2682₁, 1933

309. Mannessier-Mameli, A., 1935, Action of Anilines on Saccharin and on Thiosaccharin. *Gazz. Chim. Ital.*, 65:51–69, *Chem. Abst.*, 29:3996₁, 1935

310. Mannessier-Mameli, A., 1935, The Pyrolysis of Saccharin Oxime. *Gazz. Chim. Ital.*, 65:77–84, *Chem. Abst.*, 29:3998₈, 1935

311. Mannessier-Mameli, A., 1940, Action of Anilines on Saccharin and on Thiosaccharin. *Gazz. Chim. Ital.*, 70:855, *Chem. Abst.*, 36:1917₈, 1942

312. Mannessier-Mameli, A., 1941, The Action of Ammonia, of Ammonium Carbonate, of Urea, and of Diurea on Saccharin and on Thiosaccharin. in *Gazz. Chim. Ital.*, 71:3–18, *Chem. Abst.*, 36:1028₁, 1942

313. Mannessier-Mameli, A., 1941, The Action of Hydrazine on Saccharin and on Thiosaccharin. *Gazz. Chim. Ital.*, 71:18–25, *Chem. Abst.*, 36:1029₄, 1942

314. Mannessier-Mameli, A., 1941, The Action of Semicarbazide on Saccharin, on Thiosaccharin, and on Acetylsaccharin. *Gazz. Chem. Ital.*, 71:25–40, *Chem. Abst.*, 36:1030, 1942

315. Mannessier-Mameli, A., 1941, Action of Phenylhydrazine on Saccharin and on Thiosaccharin. *Gazz. Chim. Ital.*, 71:596–614, *Chem. Abst.*, 37:100₆, 1943

316. The Manufacture of Saccharin. *Ind. Chemist*, 1:5–9, 1925 *Chem. Abst.*, 19:1857₅, 1925

316. The Manufacture of Saccharin. *Ind. Chemist*, 1:5–9, 1925, *Chem. Abst.*, 19:1857₅, 1925

317. Marina, A. V., 1944, Study of Saccharin. Its Preparation in the Laboratory and in Industry. *Afinidad*, 21:323–9, *Chem. Abst.*, 40:7163₇, 1946

318. Marina, A. V., 1946, A Treatise on Saccharin—Its Laboratory and Industrial Preparation. *Afinidad*, 23:321–4, *Chem. Abst.*, 42:5616₈, 1948

319. Marina, E., 1955, Chromatographic Detection of Saccharin. *Boll. Lab. Chim. Provinciali*, 6:80–1, *Chem. Abst.*, 50:8084a, 1956

320. Marty, A., 1944, Foodstuff Substitutes from Coal Tar. *J. Usines Gas.*, 68:65–8, *Chem. Abst.*, 40:7423₂, 1946

321. Massatsch, C., 1947, Interesting Saccharin Contaminants. *Pharm. Ztg.*, 83:426, *Chem. Abst.*, 44:74601, 1950

322. Massatsch, C., 1949, Determination of the Sweetening Agent in Soluble Saccharin and Similar Compounds. *Pharm. Ztg.*, 85:222–3, *Chem. Abst.*, 43:6785g, 1949

323. Matsui, M., Sawamura, T. and Adachi, T., 1932, Electrolytic Reduction of Saccharin. I., Electrolysis in Acid and Alkaline. *Mem. Coll. Sci. Kyoto Imp. Univ.*, p. A15, pp. 151–5, *Chem. Abst.*, 26:5264₈, 1932

324. Matsumoto, K. and Matsui, T., 1952, Pharmaceutical Analysis by Polarography, V. The M-Nitraniline Derivatives. *Pharm. Soc. Japan*, 73:653–4, *Chem. Abst.*, 47:953f, 1953

325. Meadow, J. R. and Cavagnol, J. C., 1952, Use of Saccharin Derivatives for the Identification of Mercaptions. *J. Org. Chem.*, 17:488–19, *Chem. Abst.*, 47:1696g, 1953

326. Meillere, G., 1917, Milk, Skimmed Milk, Condensed Milk, Powdered Milk. The Saccharin Question. Preserved Meat. *J. Pharm. Chim.*, 16:21–5, *Chem. Abst.*, 11:2832₇, 1917

327. Merritt, L. L., Jr., Levey, S. and Cutter, H B., 1939, Sodium Saccharin as a Reagent for the Identification of Alkyl Halides. *J. Am. Chm. Soc.*, 61:15–16, *Chem. Abst.*, 23:1692₅, 1939

328. Methods for the Examination of Foods. *Uerh. Kais. Gesundheitsamt*, 32:1292–6, 1909, *Chem. Abst.*, 3, 554₇, 1909

329. Mitchell, L. C., 1955, Separation and Identification of Cyclohexylsulfamate, Dulcin, and Saccharin by Paper Chromatography. *J. AOAC*, 38:943–6, *Chem. Abst.*, 50:13664h, 1956

330. Miyadera, K., 1922, A note on Metabolism Following Large Doses of Saccharin. *Z. Physik, Diatet. Therap.*, 26:232–3, *Chem. Abst.*, 16:3326₇, 1922

331. Mohrbutter, C., 1931, Estimation of Potassium Guaiacolsulfonate in Sugar, Juices and Syrup. *Apth. Ztg.*, 46:533–4, *Chem. Abst.*, 25:3434₂, 1931

332. Momose, T., 1944, Analysis of Medicinals by Polarography. I. *J. Pharm. Soc. Japan*, 64:155–6, *Chem. Abst.*, 45:816c, 1951

333. Moncrieff, R. W., 1948, Relative Sweetness. *Flavours*, 11:5,5–8;6,5–11, *Chem. Abst.*, 44:6453i, 1950

334. Moncrieff, R. W., 1949, The Sweetness of Synthetics. *Food Manuf.*, 24:29–32, *Chem. Abst.*, 43:2339a, 1949

335. Moncrieff, R. W., 1951, The Chemical Senses, London, Leonard Hill

336. Moore, C. D. and Stoeger, L., 1945, Estimation of Saccharin in Tablets. *Analyst*, 70:337, *Chem. Abst.*, 39:5398₅, 1945

337. Muller, C., 1911, Saccharin in (Mineral Water, etc.) in Egypt. *Bull. Assoc. Chim. Sucr. Dist.*, 28:630–40, *Chem. Abst.*, 5:2671₈, 1911

338. Muller, C., 1911, Saccharin in Egypt. Investigation into its Uses in Food. Artificial Sugars Commercial Varieties of Saccharin. *Ann. Fals.*, 4:278, *Chem. Abst.*, 5:2403₈, 1911

339. Muller, E. and Petersen, S., 1948, A New Group of Sweet Substances. *Chem. Ber.*, 81:31–8, *Chem. Abst.*, 43:168c, 1949

340. National Research Council, Food and Nutrition Board 1954, Summary Statement on Artificial Sweeteners. December

341. Neiman, M. B., 1955, Polarographic Method for Determining Saccharin. *Zhur. Anal. Khim.*, 10:175–9, *Chem. Abst.*, 49:15635i, 1955

342. Neumann, R. O., 1925, The Influence of Saccharin on the Utilization of Proteins. *Arch. Hyg.*, 96:264–76, *Chem. Abst.*, 22:1622₅, 1928

343. Neumann, R. O., 1924, Experiments on the Sweetening of Foods with Saccharin and Sucrose. *Z. Nahr. Genussm.*, 47:184–89, *Chem. Abst.*, 18:2207₁, 1924

344. Neuss, O., 1921, The Degree of Sweetness of Natural and Artificial Sweeteners. *Umschau*, 25:603–4, *Chem. Abst.*, 16:296₄, 1922

345. Notzold, R. A., Becker, D. E., Terrill, S. W. and Jensen, A. H., 1955, Saccharin and Dried Cane Molasses in Swine Rations. *J. Animal Sci.*, 14:1068–72, *Chem. Abst.*, 50:5862g, 1956

346. Noyce, Wm. K., Coleman, C. H. and Barr, J. T., 1951, Vinology in Sweetening Agents. I. A Vinylog of Dulcin. *J. Am. Chem. Soc.*, 73:1295–6, *Chem. Abst.*, 45:9501i, 1951

347. Oakley, M., 1945, Report on (The Determination of) Preservatives and Artificial Sweeteners. *J. AOAC*, 28:296–8, *Chem. Abst.*, 39:3853₈, 1945

348. Oakley, M., 1945, Report on (The Determination of) Saccharin. *J. AOAC*, 28:298–301, *Chem. Abst.*, 39:3853₉, 1945

349. Oakley, M., 1954, Preservatives and Artificial Sweeteners. *J. AOAC*, 37:371–4, *Chem. Abst.*, 48:8983i, 1954

350. Oakley, M., 1955, Analysis of Preservatives and Artificial Sweeteners. *J. AOAC*, 38:552–4, *Chem. Abst.*, 49:12738f, 1955

351. Oba, T., *et al.*, 1959, Application of Infrared Spectroscopy to Examination of Drugs and their Preparations. VII. Determination of Sodium Cyclohexyl Sulfamate in the Artificial Sweetener. *Bull. Nat. Hygienic Lab.*, *Tokyo*, 77:61–6, September

352. Oddo, B. and Mingoia, Q., 1927, Differences in the Sweetening Power of Saccharin and Some of its Derivatives. *Gazz. Chim. Ital.*, 57:465–72, *Chem. Abst.*, 21:3202₂, 1927

353. Oddo, B. Mingoia, Q., 1931, Variations in the Sweetening Power of Saccharin and Experiments with Some of its Derivatives. II. *Gazz. Chim. Ital.*, 61:435–46, *Chem. Abst.*, 26:122₅, 1932

354. Oddo, B. and Perotti, A., 1937, Variations in the Sweetening Power of Saccharin. IV. Effects of the Association of Saccharin with Substances containing the Ureide Grouping. *Gazz. Chim. Ital.*, 67:543–52, *Chem. Abst.*, 32:1675₇, 1938

355. Oddo, B. and Perotti, A., 1940, Changes in the Sweetening Power of Saccharin. V. Influence of the Association of Saccharin with Substances Containing the Ureide Group. *Gazz. Chim. Ital.*, 70:567–74, *Chem. Abst.*, 35:1038₇, 1941

356. Olsen, R. W., 1960, Cyclamates in Citrus Products. *Proc. Florida State Hort. Soc.*, 73:270–1, October

357. Olszewski, W., 1920, Examination of Sweetening Compounds. *Pharm. Zentralhadle*, 61:583–5, *Chem. Abst.*, 15:713₁, 1921

358. Order of the Austrian Minister of the Interior Concerning the Austrian Alimentary Codex. *Veroffent, Kats. Gesundh.*, 35:742, 1911, *Chem. Abst.*, 5, 3477₇, 1911

359. Orlov, N. I., 1948, Toxicity of Dulcin. *Gigiena i Sanit.*, 13:9:36, *Chem. Abst.*, 43:5499d, 1949

360. Palet, L. P. J., 1914, An Artificial Edulcorant Called "Suessoel." *Anales Soc. Quim. Argentina*, 2:47–8, *Chem. Abst.*, 8:3601₂, 1914

361. Parmeggiani, G., 1908, Investigations upon Saccharin. *Z. Oester. Apoth. Ver.*, 46:199, *Chem. Abst.*, 2:3129₁, 1908

362. Paul, A. E., 1920, Report on Flavoring Extracts. *J. AOAC*, 3:415–18

363. Paul, T., 1921, The Degree of Sweetness of Dulcin and Saccharin. *Chem. Ztg.*, 45:38, *Chem. Abst.*, 15:1361i, 1921

364. Paul, T., 1921, Definitions in Units of Measure in the Chemistry of Sweeteners. *Chem. Ztg.*, 45:705–6, *Chem. Abst.*, 15:3881₈, 1921

365. Paul, T., 1921, Physical Chemistry of Foodstuffs. V. Degree of Sweetness of Sugars. *Z. Elektrochem.*, 27:539–46, *Chem. Abst.*, 16:973₅, 1922

366. Paul, T., 1922, The Degree of Sweetness of Sweet Substances. *Z. Nahr. Genussm.*, 43:137–49, *Chem. Abst.*, 16:2558₆, 1922

367. Paul, R., 1921, The Measurement of Sweetness of Artificial Sweeteners. *Biochem. Z.*, 125:97–105, *Chem. Abst.*, 16:974₂, 1922

368. Paul, R., 1920, Determination of Sweetness of Artificial Sweeteners. *Die. Umschau.*, 24:592, *Chem. Abst.*, 15:275₄, 1921

369. Pawlowski, F., 1910, Detection of Saccharin in Beer. *Z. Ges. Brauw.*, 32:281, *Chem. Abst.*, 4:948₈, 1910

370. Pazienti, U., 1914, Quantitative Dtermination of Saccharin and of Sodium Saccharinate. *Anal. Chim. Applicata*, 2:290–4, *Chem. Abst.*, 9:939₇, 1915

371. Pech, J., 1934, Polarographic Studies with the Dropping Mercury Cathode. XXXVIII. Reduction of Some Aliphatic Amines, Quinoline and Saccharin. *Collection Czechoslov. Chem. Comm.*, 6:126–36, *Chem. Abst.*, 28:4293₄, 1934

372. Penn, W. S., 1946, Plastics From a Saccharin By-Product. *Synthetics and By-Products*, 8:401–4, *Chem. Abst.*, 41:2603c, 1947

373. Perez, F. A., 1943, A Chemical and Bromatological Analysis of the Carbonated Beverages of Lima and other Localities in the Republic. *Farm. Peruana.* 1:3–10, *Chem. Abst.*, 38:1039₄, 1944

374. Peronnet, M., 1944, Saccharin in Soft Drinks Particularly those Marked "No Alcohol". *Ann. Chim. Anal.*, 26:108–11, *Chem. Abst.*, 40:1246₈, 1946

375. Pertusi, C. and DiNola, E., 1933, The Clarification of Complex Materials with Basic Lead Acetate in the Presence of Alkali Hydroxides. *Ann. Chim. Applicata*, 23:311–14, *Chem. Abst.*, 27:5030₃, 1933

376. Pesman, J. and Luten, J., 1946, A Case of Poisoning with Dulcin. *Nederland. Tijdschr. Geneeskunde*, 90:261–3, Chem. Abst., 40:4431₂, 1946

377. Petersen, S., 1948, New Sweetening Agents. *Angew. Chem.*, 60A:58, *Chem. Abst.*, 45:9194i, 1951

378. Poppe, J., 1922, The Detection and Determination of Saccharin. *Ann. Chim. Anal. Chim. Appl.*, 4:157–8, *Chem. Abst.*, 16:2282₁, 1922

379. Posseto, G. and Issoglio, G., 1912, A New and Rapid Method for the Extraction of Saccharin From Food Substances. *Giorn. Farm. Chim.*, 61:5–11, *Chem. Abst.*, 6:1039₅, 1912

380. Pratt, C. D., 1055, Would Your Formula Problems be Helped by Sorbitol? *Food in Canada*, 15:2,11–14,26, *Chem. Abst.*, 49:5701i, 1955

381. Pritzker, J., 1920, Chemical Change in the Composition of Saccharin Bicarbonate Tablets. *Schweiz. Apoth. Ztg.*, 58:78–9, *Chem. Abst.*, 14:1593₉, 1920

382. Pritzker and Basel, 1920, Chemical Changes in the Composition of Saccharin Bicarbonate Tablets. *Schweiz. Apoth. Ztg.*, 58:29–30, *Chem. Abst.*, 14:798, 1920

383. Profft, E., 1949, The Physical Compatibility of the Intensively Sweet 2-Propoxy-5-Nitroaniline and some Derivatives Thereof. *Deut. Chemiker-Z.*, 1:51–4, *Chem. Abst.*, 44:5839c, 1950

384. Profft, E. and Jumar, A., 1956, Higher Homologs of Sweetening Agents. *Chem. Ztg.*, 80:309–10, *Chem. Abst.*, 51:1070b, 1957

385. Pucher, G. and Dehn, W. M., 1921, Solubilities in Mixtures of Two Solvents. *J. Am. Chem. Soc.*, 43:1753–9, *Chem. Abst.*, 16:519s, 1922

386. Punishment for Infringement of the Regulations Respecting the Sale, etc., of Saccharin (in Spain). *Chem. Ind.*, 33:106–7, *Chem. Abst.*, 5:144₆, 1911

387. Raimon, M., 1917, Saccharin. *Ind. Chim.*, 66:5–13, *Chem. Abst.*, 12:1401₄, 1918

388. Rashkovich, S. L., 1909, The Use of Saccharin in Russia. *Bull. Assoc. Chim. Sucr. Dist.*, 26:353–26, *Chem. Abst.*, 3:456₅, 1909

389. Reid, E. E., Rice, L. M. and Grogan, C. H., 1955, Some N-Alkylsaccharin Derivatives. *J. Am. Chem. Soc.*, 77:5628–30, *Chem. Abst.*, 50:6434d, 1956

390. Reif, G., 1923, The Detection of Saccharin and Dulcin in Vinegar and Foods Containing Acetic Acid. *Z. Nahr. Genussm.*, 46:217–23, *Chem. Abst.*, 18:868₇, 1924

391. Reif, G., 1924, Determination of Dulcin by the Use of Xanthydrol. *Z. Nahr. Genussm.*, 47:238–48, *Chem. Abst.*, 18:2565₆, 1924

392. Reindollar, W. F., 1937, Report on (The Determination of) Preservatives (in Foods). *J. AOAC*, 20:161, *Chem. Abst.*, 31:5054₅, 1937

393. Reindollar, W. F., 1938, Report on (The Determination of) Preservatives (in Foods). *J. AOAC*, 21:184–6, *Chem. Abst.*, 32:5935₉, 1938

394. Reindollar, W. F., 1940, Report on (The Determination of) Preservatives (in Foods). *J. AOAC*, 23:288–9, *Chem. Abst.*, 34:5185₉, 1940

395. Reindollar, W. F., 1941, Report on (The Determination of) Preservatives (in Foods). *J. AOAC*, 24:326–7, *Chem. Abst.*, 35:5576₃, 1941

396. Reindollar, W. F., 1942, (The Determination of) Saccharin. *J. AOAC*, 25:369–75, *Chem. Abst.*, 36:4610₃, 1942

397. Reindollar, W. F., 1944, Report on (The Determination of) Food Preservatives and Artificial Sweeteners. *J. AOAC*, 27:256–7

398. Repetto, E., 1917, Investigation of Saccharin in Commercial Sirups, Specialties, Etc. *Rev. Farm.*, 60:407–19, *Chem. Abst.*, 11:3091₈, 1917

399. Resolutions Adopted at the 58th Annual Conference of the Association of Food and Drug Officials of the United States, 1955, Assoc. Food & Drug. Offc. of U.S., *Quart. Bull.* 19, 36, January

400. Rice, H. L. and Pettit, P., 1954, An Improved Procedure for the Preparation of Alkyl Halide Derivatives of Saccharin. *J. Am. Chem. Soc.*, 76:302–3, *Chem. Abst.*, 49:12365b, 1955

401. Rice, L. M., Grogan, C. H. and Reid, E. E., 1953, N-Alkylsaccharins and their Reduction Products. *J. Am. Chem. Soc.*, 75:4304–5, *Chem. Abst.*, 48:12027h, 1954

402. Richards, R. K., Taylor, J. D., O'Brien, J. L. and Duescher, H. O., 1951, Studies on Cyclamate Sodium (Sucaryl Sodium) a Newᵣ Non-Caloric Sweetening Agent. *J. Am. Pharm. Assoc.*, 40:1–16, *Chem. Abst.*, 45:3527i, 1951

403. Richmond, H. D. and Hill, C. A., 1918, Analysis of Commercial Saccharin. The Estimation of O-Benzoylsulfonimide from Ammonia Produced by Acid Hydrolysis. *J. Soc. Chem. Ind.*, 37:246–9T, *Chem. Abst.*, 12:2215₅, 1918

404. Richmond, H. D. and Hill, C. A., 1919, Analysis of Commercial Saccharin II. The Detection and Estimation of Impurities. *J. Soc. Chem. Ind*, 38:8–10T, *Chem. Abst.*, 13:979₅, 1919

405. Richmond, H. D., Royce, S. and Hill, C. A., 1918, Examination of Saccharin Tablets. *Analyst*, 43:402–4, *Chem. Abst.*, 13:621₁, 1919

406. Roberts and Dahle, 1954, Low-Carbohydrate Ice Cream for Diabetics. *Ice Cream Trade J.*, 50:4,66–8,96–8, *Chem. Abst.*, 48:7809a, 1954

407. Roger, H. and Garnier, M., 1908, Influence of Saccharin on Peptic Digestion. *Arch. Med. Exp.*, 19:497–504, *Chem. Abst.*, 2:147₁, 1908

408. Romero, J. J. L., 1944, Sweetening Agents and Sugar. *Mem. Assoc. Tec. Azucareros Cuba*, 18:83–5, *Chem. Abst.*, 39:4704₂, 1945

409. Rosenblum, H. and Mildworm, L., 1946, Determination of Saccharin. *J. Am. Pharm. Assoc.*, 35:336, *Chem. Abst.*, 41:4891e, 1947

410. Rosenthaler, L., 1950, Microchemical Notes. *Mikrochemiover. Mikrochim. Acta*, 35:164–8, *Chem. Abst.*, 44:7714i, 1950

411. Rosenthaler, L., 1944, The Detection of the Elments in Organic Materials. A Test for N. *Pharm. Acta Helv.*, 19:81–2, *Chem. Abst.*, 38:4533₃, 1944

412. Rosenthaler, L., 1933, Analytical Notes II. *Pharm. Zentralhalle*, 74:288–90, *Chem. Abst.* 27:3893₈, 1933

413. Rost, E. and Braun, A., 1926, The Pharmacology of Dulcin. *Arb. Reichsgesundh.*, 57:212–20, *Chem. Abst.*, 20:3742₂, 1926

414. Rothe, M., 1963, Definition of the Terms "Aroma" and "Aromatic Material." *Ernaehrungsforschung*, 7:639–46, *Chem. Abst.*, 59:2107c, 1965

415. Roura, U. C., 1941, Simple and Rapid Method for Obtaining Ethoxyphenylurea (Dulcin). *Industria y. Quim.*, 3:160–2, *Chem. Abst.*, 36:2845₉, 1942

416. Rozanova, V. A., 1941, Colorimetric Method of the Determination of Saccharin and Vanillin in Foodstuff. *Obshchestvennoe Pitanie*, 9:L,19–25, *Chem. Abst.*, 37:63506, 1943

417. Runti, C., 1956, Relation Between Chemical Constitution and Sweet Taste. Further Isosteres and Derivatives of Dulcin. *Ann. Chim.*, 46:731–41, *Chem. Abst.*, 51:6527e, 1957

418. Runti, C., 1957, Synthetic Sweetening Agents. *Chim. e. Ind.*, 39:354–64, *Chem. Abst.*, 51:1165i, 1957

419. Runti, C. and Sindellari, L., 1957, Report on Chemical Constitution and Sweet Taste. X. Isosteres and Derivatives of Dulcin. *Univ. Studitrieste, Fac. Sci. Ist. Chim.*, article 15, *Chem. Abst.*, 51:16459c, 1957

420. Runti, C., 1962, Synthetic Sweetening Compounds. *Bull. Soc. Pharm. Bordaux*, 101:3,197–218, *Chem. Abst.*, 58:14624d, 1964

421. Saccharin Again. *Ann. Fals.*, 316:57, 1910, *Chem. Abst.*, 4:1702₃, 1910

422. The Safety of Artificial Sweeteners for use in Foods. 1955 Food Protection Committee of the Food Nutrition Board Publication #386, National Acad. of Sci., Natl. Res. Council, August

423. Sah, P. P. T. and Chang, K. S., 1936, Dulcin Synthesis by Means of the Curtius Reaction. *Ber.*, 69B:2762–4, *Chem. Abst.*, 31:2179₉, 1937

424. Saillard, M., 1909, International Conference for the Repression of the Use of Saccharin and Analagous Sweeteners in Food Products and Beverage. *Ann. Falsif.*, 1:50–8, *Chem. Abst.*, 3:931₆, 1909

425. Salac, V., 1936, The Polarograph on Analysis in the Brewery. *Kvas*, 64:383, *Chem. Abst.*, 33:6520₇, 1939

426. Samaniego, C. C., 1946, Stevia Rebaudiana. *Rev. Farm.*, 88:199–202, *Chem. Abst.*, 41:501c, 1947

427. Sanchez, J. A., 1931, Functional Analytical Study of Dulcin. *Anales, Farm. Bioquim.*, 2:63–7, *Chem. Abst.*, 26:2718₉, 1932

428. Sandri, R. M., 1956, Making Good Sugar-Free Drinks. *Food Eng.*, 25:79,196 May

429. Sands, M., 1953, Sugarless Jellies, Special Problems Course. Purdue U., Lafayette, Ind., *J. Am. Dietet. Assoc.*, 29:677

430. Sasagawa, Y., Inoue, M. and Sima, T., 1950, Preparation of Dulcin. Ann. Repts. Takeda Research Lab., 9, pp. 47–57, *Chem. Abst.*, 46:4497i, 1952

431. Schecker, G., 1935, Detection of Saccharin in Beverages (Beer). *Centr. Zuckerind.*, 43:312, *Chem. Abst.*, 30:7276₇, 1936

432. Schoenberger, J. A., *et al.*, 1953, Metabolic Effects, Toxicity, and Excretion of Ca N-Cyclohexyl Sulfamate (Sucaryl) in Man. *Ama. J. Med. Sci.*, 225:551–9, *Chem. Abst.*, 47:11548h, 1953

433. Schoorl, N., 1937, The Vlezenbeek Reaction of P-Amino Phenol Derivatives. *Pharm. Weekblad.*, 74:210–12, *Chem. Abst.*, 31:2746₄, 1937

434. Schowalter, E., 1919, Separation of Saccharin and Benzoic Acid. *Z. Nahr.-Genussm.*, 38:185–94, *Chem. Abst.*, 14:1396₅, 1920

435. Schudel, H., Eder, R. and Buchi, J., 1948, Determination of the Melting Point. *Pharm. Acta Helv.*, 23:33–45, *Chem. Abst.*, 43:1147i, 1949

436. Schulte, M. J., 1940, Determination of the Sodium Derivative of Saccharin. *Pharm. Weekblad.*, 77:1281–2, *Chem. Abst.*, 36:7235₉, 1942

437. Schutz, H. G. and Pilgrim, F. J., 1957, Sweetness of Various Compounds and its Measurement. *Food Research*, 22:206–13, *Chem. Abst.*, 51:16992a, 1957

438. Schwarz, C. and Buchlmann, E., 1924, The Physiology of Digestion. X. The Action of Crystallose, Saccharin and P-Saccharin on Salivary Distaste. *Fermentforschung*, 7:229–46, *Chem. Abst.*, 18:2913i, 1924

439. Seeker, A. F. and Wolf, M. G., 1917, Saccharin. *J. AOAC*, 3:38–52, *Chem. Abst.*, 11:2375₆, 1917

440. Sen Grupta, M. C., Madiwale, M. S. and Bhatt, J. G., 1955, Estimation of Saccharin in Prepared Tea. *Indian J. Pharm.*, 17:185–90, *Chem. Abst.*, 50:6704f, 1956

441. Serger, H., 1912, Artificial Sweeteners. *Chem.-Ztg.*, 36:829–30, *Chem. Abst.*, 6:2960₅, 1912

442. Serger, H., 1923, Artificial Sweetening (Saccharin) and its Use in the Fruit Preservation Industry. *Chem.-Ztg.*, 47:98–100,123–4,145–6, *Chem. Abst.*, 17:3382₁, 1923

443. Serger, H. and Clarck, K., 1931, The Use of Sugar or Saccharin in the Canning of Cucumbers. *Konserven. Ind.*, 18:244–5, *Chem. Abst.*, 25:3739–7, 1931

444. Siedler, P., 1916, Artificial Sweeteners, Especially Dulcin. *Chem.-Ztg.*, 40:853–5, *Chem. Abst.*, 11:999₇, 1917

445. Singleton, Gray, 1959, Recent Work on Citrus Sections. *Proc. Florida State Hort. Soc.*, 72:263–7, *Chem. Abst.*, 55:9714h

446. Soifer, P. A., 1946, Estimation of Saccharin in Artificially Flavored Non-Alcoholic Beverages. *Gigiena i Sanit.*, 11:6,33–6, *Chem. Abst.*, 41:3550e, 1947

447. Solon, K., 1947, Determination of Saccharin. *Food*, 16:268–70, *Chem. Abst.*, 41:7311e, 1947

448. Spalton, L. M., 1950, Use of Dulcin in Foods and Beverages. *Food Manuf.*, 25:371–3, *Chem. Abst.*, 46:5743a, 1952

449. Stanek, V. and Pavlas, P., 1934, Microanalytical Studies of Artificial Sweeteners. I. Saccharin. *Listy Cukr.*, 53:33–8, *Chem. Abst.*, 29:1931₇, 1935

450. Stanek, V. and Pavlas, P., 1935, Microanalytical Studies with Respect to Artificial Sweeteners, I. Saccharin. *Mikro. Chemie*, 17:22–8, *Chem. Abst.*, 29:2885₃, 1935

451. Stanek, V. and Pavlas, P., 1934, A Simple Method for Detecting Saccharin in Beer and Other Beverages. *Z. Zuckerind. Czechoslovak Rep.*, 58:313–16, *Chem. Abst.*, 28:6931₁, 1934

452. Stanek, V. and Pavlas, P., 1935, Microchemical Studies of Artificial Sweetening Agents. I. Saccharin. *Z. Zuckerind. Czechoslovak Rep.*, 59:361–6, *Chem. Abst.*, 29:7231₁, 1935

453. Staub, H., 1954, Toxicity of Sweetening Substances. *Mitt. Lebensm. Hyg.*, 36:7–19, *Chem. Abst.*, 41:212d, 1947

454. Staub, H., 1942, Artificial Sweetening Substances. *Schweiz. Med. Wochachr.*, 72:983–6, *Chem. Abst.*, 37:4477₄, 1943

455. Struve, K., 1935, The Ability of Solutions of Crystalline Sweetening Agents to Withstand Storage. *Deut. Fischerei-Rundschau*, 510, *Chem. Abst.*, 31:5891₆, 1937

456. Struve, K., 1930, The Use of Dulcin in Marinades. *Firschwirtschaft*, 6:40–2, *Chem. Abst.*, 25:4945₁, 1931

457. Sudo, T., Shimoe, D. and Tsujii, T., 1957, Rapid Semi-Micro Determination of Acetyl Groups and Methyl Groups Attached to Carbon. Determination of Acetyl and C-Methyl Groups. *Bunsekikagaku*, 6:498–502, *Chem. Abst.*, 52:13540i, 1958

458. Summary of the Pharmacological Studies of Cyclamate Sodium (Sucaryl Sodium), 1954, Assoc. Food & Drug Off. of U.S., *Quart. Bull.*, 18:165–7

459. Swafford, Wm. B. and Nobles, W. L., 1954, Modifications in the Formula of Wild Cherry Sirup. *J. Am. Pharm. Assoc. Pract. Pharm. Ed.*, 15:99–100, *Chem. Abst.*, 48:6648d, 1954

460. Sweetener Demands Booming. *Chem. & Eng. News*, 36:28–9, 1958

461. Sweetening Agents. *Am. Soft Drink J.*, 108:700,2,70–82, 1959

462. Sweetness Sans Sucrose. Chemical Week, October 13, 1951

463. Synthetic Sweetened Soft Drinks. *Chem. Eng. News*, 33:4465, October 17, 1955

464. Tachi, I. and Tsukamoto, T., 1951, Polarography of Vitamin B₁. V. Saccharin. *J. Agr. Chem. Soc. Japan*, 25:335–40, *Chem. Abst.*, 47:11034f, 1953

465. Tarugi, N. and Ceccherelli, F., 1914, Observations on a Test for Saccharin. *Rend. Soc. Chim. Ital.*, 5:198–203, *Chem. Abst.*, 9:806₆, 1915

466. Taschenberg, E. W., 1922, The Antipyretic Action of Dulcin. *Deut. Med. Wochschr.*, 48:695, *Chem. Abst.*, 17:1073₂, 1923

467. Taufel, K. and Klemm, B., 1925, Investigations of Natural and Synthetic Sweet Substances. I. Studies on the Degree of Sweetness of Saccharin and Dulcin. *Z. Nahr. Genussm.*, 50:264–73, *Chem. Abst.*, 20:951₈, 1926

468. Taufel, K. and Naton, J., 1926, The Hydrolysis of O-Benzoic Sulfinide (Saccharin). *Z. Angew. Chem.*, 39:224–9, *Chem. Abst.*, 21:78₁, 1927

469. Taufel, K. and Wagner, C., 1925, The Constitution of Aqueous Solutions of O-Benzoic Acid Sulfamide (Saccharin) and P-Phenetyl Urea (Dulcin). *Ber.*, 58B:909–12, *Chem. Abst.*, 19:2347₄, 1925

470. Taufel, K., Wagner, C. and Dunwald, H., 1928, The Decomposition of P-Phenetyl carbamide (Dulcin) on Heating in Aqueous Solution. *Z. Elektrochem.*, 34:115–27, *Chem. Abst.*, 22:2306₃, 1928

471. Taufel, K., Wagner, C. and Preiss, W., 1928, The Mechanism of the Hydrolysis of Saccharin and O-Sulfamino-Benzoic Acid. *Z. Elektrochem.*, 34:281–91, *Chem. Abst.*, 22:3336₆, 1928

472. Taylor, J. D., Richards, R. K. and Davin, J. C., 1951, Excretion and Distribution of Radioactive Sulfur 35-Cyclamate Sodium (Sucaryl Sodium) in Animals. *Proc. Soc. Exptl. Biol. Med.*, 78:530–3, *Chem. Abst.*, 46:3157c, 1952

473. Terlinck, E., 1924, Saccharin and its Manufacture. *Ingenieur Chimiste*, 8:233–57, *Chem. Abst.*, 21:1978₄, 1927

474. Testoni, G., 1910, The Estimation of Saccharin in Various Foods. *Z. Nahr. Genussm.*, 18:577–87, *Chem. Abst.*, 4:626₁, 1910

475. Thevenon, L., 1920, A New Reaction of Saccharin. *J. Pharm. Chim.*, 22:421–2, *Chem. Abst.*, 15:999₉, 1921

476. Thevenon, L., 1921, Contra-Indication to a Process for the Detection of Saccharin. *J. Pharm. Chim.*, 23:7, 215, *Chem. Abst.*, 15:2335₉, 1921

477. Thompson, M. M. and Mayer, J., 1959, Hypoglycemic Effect of Saccharin in Experimental Animals. *Am. J. Clinical Nutrition*, 7:80–5

478. Thomas, H., 1924, Changes in the Sweetening Power of Dulcin (P-Phenetylcarbamide) by Chemical Modification. *Deut. Zuckerind.*, 49:1056, *Chem. Abst.*, 19:2674₁, 1925

479. Thoms, H. and Netteshiim, K., 1920, Changes in the Taste of the Sweetener Dulcin (P-Phenetylcarbamide) by Chemical Action. *Ber. Pharm. Ges.*, 30:227–50, *Chem. Abst.*, 14:3638₉, 1920

480. Tillson, A. H. and Wilson, J. B., 1952, Isolation and Microscopic Identification of Cyclamate Sodium (Sucaryl) as Cyclohexanesulfamic Acid. *J. AOAC*, 35:467–9, *Chem. Abst.*, 46:11498i, 1952

481. Tracy, P. H., 1940, A Discussion of Sweetening Agents for Ice Cream. *Can. Dairy and Ice Cream J.*, 19:3, 58–60, *Chem. Abst.*, 35:2231₅, 1941

482. Tracy, P. H. and Edman, G., 1950, Ice Cream for Diabetic Using Sorbitol. *Ice Cream Trade J.*, 46:7,50–2,83–5, *Chem. Abst.*, 45:778i, 1951

483. Traegel, A., 1925, Sugar and Saccharin. *Deut. Zuckerind*, 50:1175–6, *Chem. Abst.*, 20:246₉, 1926

484. Traegel, A., 1925, Sweetening and Preserving Power of Saccharin in Comparison with Sugar. *Z. Ver. Deut. Zuckerind.*, 75:345–59, *Chem. Abst.*, 20:1873₁, 1926

485. Trauth, F., 1943, A Method of Rapid Identification of Saccharin in Wine. *Deut. Lebensm. Rundschau.*, Vol. 28, *Chem. Abst.*, 38:4749₇, 1944

486. Trauth, F., 1949, A Method for the Rapid Qualitative Detection of Sweet Substances in Wine. *Weibau. Wiss. Beih.*, 3:106–8, *Chem. Abst.*, 46:9247g, 1952

487. Triest, F. J., 1951, How Tobacco is Flavored. *Am. Perfumer Essential Oil Rev.*, 58:449,451,453,455, *Chem. Abst.*, 46:5266f, 1952

488. Tsuzuki, Y., Kato, S. and Okazaki, H., 1954, Sweet Flavor and Resonance. *Kagaku*, 24:523–4, *Chem. Abst.*, 48:13740h, 1954

489. Uglow, W. A., 1924, Activity of Saccharin on Bacteria, "Plankton" and Digestive Enzymes. *Arch. Hyg.*, 92:331–46, *Chem. Abst.*, 10:135₅, 1925

490. Uglow, W. A., 1925, The Importance of Dulcitol as a Sugar Substitute from a Hygienic Standpoint. *Arch. Hyg.*, 95:89–100, *Chem. Abst.*, 19:2712₄, 1925

491. Unthoff, J. and Moragas, G., 1928, Synthetic Preparation of Dulcin (P-Phenetylurea). *Quim. Ind.*, 5:207–10, *Chem. Abst.*, 23:1889₅, 1929

492. Ulrich, K., 1952, The Sweetness of Synthetic and Natural Products. *Zucker*, 5:236–9, *Chem. Abst.*, 47:2807f, 1953

493. Van Den Driessen, W. P. H., 1908, Qualitative and Quantitative Determination of Saccharin in Cacao Powder. *Apoth. Ztg.*, 22:2301, *Chem. Abst.*, 2:1306₁, 1908

494. Vandrac, I., 1942, Fungi From Concentrated Sugar Solutions and Sugar Products. *Gambrinus Z. Bier, Malz U. Hopfen*, 3:201–5, *Chem. Abst.*, 38:4041₈, 1944

495. Van Eweyk, C., 1922, The Influence of Saccharin on the Heart and Circulation. *Z. Physik. Diatet. Therap.*, 26:276–8, *Chem. Abst.*, 17:589₃, 1923

496. Van Roost, H., 1928, Saccharin. Bull. Trimestr. Assoc. Eleves Ecole Sup. Brasserie, Univ. Louvain, 28:49–61, *Chem. Abst.*, 22:1813₄, 1928

497. Van Voorst, F. Th., 1942, The Detection of Dulcin and Saccharin in Food Products. *Chem. Weekblad.*, 39:510, *Chem. Abst.*, 38:2399₅, 1944

498. Van Zijp, C., 1930, Microchemical Contributions. *Pharm. Weekblad.*, 67:189–98, *Chem. Abst.*, 24:2401₉, 1930

499. Van Zijp, C., 1934, The Micro Copper-Pyridine Reaction for Saccharin. *Pharm. Weekblad.*, 71:1146–51, *Chem. Abst.*, 29:428₂, 1935

500. Verkade, P. E., 1946, A New Sweetening Material. *Food Manuf.*, 21:483–5, *Chem. Abst.*, 41:1772c, 1947

501. Verkade, P. E. and Meerburg, W., 1946, Alkoxyaminonitrobenzenes. IA. Partial Reduction of 1-Alkoxy-2,4-Dinitrobenzenes with Sodium Disulfide in Aqueous Suspension. *Rec. Trav. Chim.*, 65:768–9, *Chem. Abst.*, 41:3065e, 1947

502. Verkade, P. E., Van Dijk, C. P. and Meerburg, W., 1942, New Sweet Compounds and New Local Anesthetics. *Proc. Nederland Akad. Wetensch.*, 45:630–5, *Chem. Abst.*, 38:4093₆, 1944

503. Verschaffelt, E., 1915, The Toxicity of Saccharin. *Pharm. Weekblad*, 52:37–46, *Chem. Abst.*, 9:1071₅, 1915

504. Vietti-Michelina, M., 1956, Chromatographic Determination of Saccharin and Dulcin in Biscuits and Chocolate. *Chimica e Industria (Milan)*, 38:392–3, *Chem. Abst.*, 50:11546f, 1956

505. Vivas, G. V., 1956, Intentional Chemical Additives to Foods. Synthetic Sweeteners. *Farm. Nueva*, 21:613–20, *Chem. Abst.*, 51:6897f, 1957

506. Vlezenbeek, H. J., 1937, A New Reaction for the P-Aminophenol Function and a New Sensitive Reaction for Dulcin in the Presence of Saccharin. *Pharm. Weekblad*, 74:127–34, *Chem. Abst.*, 31:2553₁, 1937

507. Vnuk, K., 1927, Studies on Peligot's Saccharin and its Isolation from Molasses. *Z. Zuckerind. Czechoslov. Rep.*, 51:460–6,467–77, *Chem. Abst.*, 22:182₃, 1928

508. Vollhase, E., 1913, Detection of Saccharin in Caramelized Beer. *Chem.-Ztg.*, 37:425–6, *Chem. Abst.*, 7:3386₈, 1913

509. Von Der Heide, C. and Lohman, W., 1921, Detection of Saccharin in Wine. *Z. Nahr. Genussm.*, 41:230–6, *Chem. Abst.*, 16:1634₅, 1922

510. Vondrak, J., 1929, A Rapid Method for the Determination of Soft Drinks Suspected to Contain Saccharin. *Z. Zuckerind. Czechoslovak. Rep.* 53:501–3, *Chem. Abst.*, 24:171₉, 1930

511. Vondrak, J., 1929, A Rapid Method for Detecting Saccharin in Lemonades. *Listy Cukr.*, 47:323–4, *Chem. Abst.*, 23:2508₈, 1929

512. Von Scheele, O., 1912, The Necessity for New Legislation Regarding Saccharin. *K. Landtbr. Akad. Handl. Och. Tidskr.*, 50:273–85, *Chem. Abst.*, 6:1784₇, 1912

513. Votava, J., 1937, A Microanalytical Study of Artificial Sweetening Agents. Listy Cukrovar, 55:365–9, *Chem. Abst.*, 31:8736₅, 1937

514. Votava, J., 1937, Microanalytical Studies of Artificial Sweeteners. III. Further Studies upon the Determination of Saccharin in Drinks and Food Materials, Mouth Washes, Tooth Pastes, and Tooth Powders. *Z. Zuckerind. Cechoslovak. Rep.*, 62:121–5, *Chem. Abst.*, 32:2639₈, 1938

515. Wachsmuth, H., 1941, Some New Addition Compounds of Alkaloids and Imides. *J. Pharm. Chim.*, 1:9383–97, *Chem. Abst.*, 38:2340₄, 1944

516. Wagenaar, G. H., 1941, A New Specific Reagent for Derivatives for Barbituric Acid and Saccharin. *Pharm. Weekblad*, 78:345–8, *Chem. Abst.*, 38:2454₁, 1944

517. Wagenaar, M., 1932, Microchemical Reactions of Saccharin. *Pharm. Weekblad*, 69:614–18, *Chem. Abst.*, 26:4276₆, 1932

518. Wagenaar, M., 1932, The Microchemistry of Saccharin. *Mikrochemic*, 5:132–5

519. Ward, J. C., Munch, J. C. and Spencer, H. J. 1934, Studies on Strychnine. III. Effective-

ness of Sucrose, Saccharin and Dulcin in Masking the Bitterness of Strychnine. *J. Am. Pharm. Assoc.*, 23:984–8, *Chem. Abst.*, 29:293₉, 1935

520. Wauters, J., 1910, Reactions of Saccharin in Beer, Etc. *J. Soc. Chem. Ind.*, 28:733, *Chem. Abst.*, 4:636₉, 1910

521. Wertheim, E., 1931, Derivatives of Dulcin. *J. Am. Chem. Soc.*, 53:200–3, *Chem. Abst.*, 25:936₃, 1931

522. White, W. B., 1930, Saccharin in Ice Cream Cones. Ann. Rept. N.Y. Dept. Agr. & Mkts. for 1929 Legs. Doc., 37:103–5, *Chem. Abst.*, 24:3572₆, 1930

523. Whittle, E. G., 1944, Application for Reindollars Method to the Estimation of Saccharin. *Analyst*, 69:45–7, *Chem. Abst.*, 38:1981₉, 1944

524. Wilson, J. B., 1952, Determination of Sucaryl in Sugar-Free Beverages. *J. AOAC*, 35:465–7, *Chem. Abst.*, 46:11498g, 1952

525. Wilson, J. B., 1955, Determination of Sucaryl. *J. AOAC*, 38:559–61, *Chem. Abst.*, 49:12738e, 1955

526. Wilson, J. B., 1960, Determination of Sucaryl (Cyclamates of Calcium and Sodium). *J. AOAC*, 43:583–4

527. Wolf, M. G., 1932, Report on Preservatives (Saccharin). *J. AOAC*, 6:14–15, *Chem. Abst.*, 17:1284₁, 1923

528. Woo, M. and Huyck, C. L., 1948, Dibetic Sirups. *Dull. Natl. Formulary Comm.*, 16:140–15, *Chem. Abst.*, 43:3144i, 1949

529. Zabrik, Mary E., Miller, Grace A. and Aldrich, Pearl J., 1962, The Effect of Sucrose and Cyclamate upon the Gel Strength of Gelatin, Carrageenan, and Algin in the Preparation of Jellied Custard. *Food Technol.*, 16:12, 87–91, *Chem. Abst.*, 59:6903f, 1965

530. Zaikov, V. S. and Sokolv, P. I., 1926, Use of Chromic Mixtures for Oxidation of O-Toluenesulfamide into Saccharin. *J. Chem. Ind.*, 3:1304–7, *Chem. Abst.*, 22:2933₂, 1928

531. Zaviochevskii, I. N., 1925, Saccharin. *Rec. Soc. Chim. Russ. Brno.*, 1:67–9, *Chem. Abst.*, 21:2128₂, 1927

532. Zinkeisen, E., 1940, The Use of Artificial Sweetening in Effervescent Citrus Fruit Beverages. *Deut. Mineral wasser-Ztg.*, 44:196–7, *Chem. Abst.*, 36:7169₈, 1942

533. Zlatarov, A., 1934, The Physiology of Saccharin. *Ann. Univ. Sofia II. Faculte Phys. Math.*, 30:2, 1–41, *Chem. Abst.*, 29:3405₁, 1935

534. Zwikker, J. J. L., 1933, Detection of Saccharin and the Composition of its Complexes with Copper and Pyridine. *Pharm. Weekblad*, 70:551–9, *Chem. Abst.*, 27:3898₂, 1933

Patents

1. *Metal Salts for Cyclohexanesulfamic Acid*
   To: Abbott Laboratories
   Brit. 669,200 (1952)
   Chem. Abst. *47*: 5437f (1953)

2. Altwegg, J. and Collardeau, J.
   Saccharin
   U.S. 1,507,565 (1924)
   Chem. Abst. *18*: 3386₆ (1924)

3. Audrieth, L. F.
   N Cyclohexyl Sulfamic Acid and Salts
   To: E. I. duPont de Nemours
   U.S. 2,275,125 (1940)

4. Baird, Wm.
   Quaternary Ammonium Salts of Mercaptoavylenethiazoles
   To: Imperial Chemical Industries, Ltd.
   U.S. 2,104,068 (1938)

5. Bebie, J.
   Saccharin
   U.S. 1,366,349 (1921)
   Chem. Abst. *15*: 1030₄ (1921)

6. Beck, K. M. and Weston, A. W.
   N-(3-Methylcyclopentyl) Sulfamic Acid and its Salts

To: Abbott Laboratories
U.S. 2,785,195 (1957)
Chem. Abst. *15*: 12969h (1957)

7. Bigelow, F. E.
Lecithinated Sugar
U.S. 2,430,553 (1947)

8. Borzhim, V. S.
Saccharin
U.S.S.R. 65,065 (1945)
Chem. Abst. *40*: 7233₇ (1946)

9. Brenner, G. M.
Sugarless Beverage
U.S. 2,691,591 (1954)

10. Cattano, R. and Supparo, G.
Sucrose—Substituting Product
Ital. 417,869 (1947)
Chem. Abst. *42*: 6472d (1948)

11. Chaplin, E. D.
Sweetened Food Compound and Preparing the Same
To: Crystal Sweet Co.
U.S. 851,221 (1907)

12. *Chloroacetaldehydesulfonic Acid*
To: Chem. Fab.
Ger. 362,744

13. Comte, F.
Recovery and Purification of Saccharin
To: Monsanto Chemical Co.
U.S. 2,745,840 (1956)
Chem. Abst. *50*: 12349g (1956)

14. Cummins, E. W. and Johnson, R. S.
Purified Cyclohexylsulfamates
To: E. I. duPont de Nemours
U.S. 2,799,700 (1957)
Chem. Abst. *52*: 612b (1958)

15. D'Amico, J. J.
N-Haloalkenyl-O-Benzosulfimides
To: Monsanto Chemical Co.
U.S. 2,701,799 (1955)
Chem. Abst. *50*: 1086d (1956)

16. *Electrolytic Preparation of Saccharin*
Neth. 41,338 (1937)
Chem. Abst. *31*: 8399₂, (1937)

17. Endicott, C. J. and Dalton, E. R.
Calcium Cyclamate Tablets & Process for Making Same
To: Abbott Laboratories
U.S. 2,784,100 (1957)

18. Erickson, A. M. and Ryan, J. D.
Extraction of Sweetening Ingredients From Fruit Juices
To: Barron-Gray Packing Co.
U.S. 2,466,014 (1949)
Chem. Abst. *43*: 6333h (1949)

19. Fahlberg, C.
Manufacture of Saccharin Compounds
To: A. List
U.S. 319,082 (1885)

20. Fahlberg, C.
Saccharin Compound
U.S. 326,281 (1885)

21. Ferguson, Jr., E. A.

Sweetening Composition
U.S. 2,761,783 (1956)
Chem. Abst. *50*: 17252a (1956)

22. Glassman, J. A.
Saccharin-Aspirin Tablets
U.S. 2,134,714 (1939)
Chem. Abst. *33*: 1104, (1939)

23. Golding, D. R. V.
Cyclohexylsulfamic Acid
To: E. I. duPont de Nemours and Co.
U.S. 2,814,640 (1957)
Chem. Abst. *52*: 7351a (1958)

24. Gordon, J. B.
Dietetically Sweetened Food Products and Preparing Same
U.S. 2,653,105 (1953)

25. Gordon, J. B.
Dietetic Food Product and Method of Preparing Same
U.S. 2,629,665 (1953)

26. Griffin, Joan M.
Products Sweetened Without Sugar and Free From Aftertaste
To: Chas. Pfizer & Co., Inc.
U.S. 3,296,079 (1967)
Chem. Abst. *66*: 54403q

27. Grogan, C. H. and Rice, L. M.
N- and O-(Tertiary-Amino) Alkyl Derivatives of Saccharin
To: Geschickter Fund for Medical Research, Inc.
U.S. 2,751,392 (1956)
Chem. Abst. *51*: 2060F (1957)

28. Hamilton, Wm. F.
N-Salts of Saccharin With Vasoconstrictor Amines
U.S. 2,538,645 (1951)
Chem. Abst. *45*: 4412d (1951)

29. Hayes, E. A.
Saccharin as a Preservative for Pepper
U.S. 2,444,875 (1948)
Chem. Abst. *42*: 6962a (1948)

30. Helgren, F. J.
Sweetening Compositions
To: Abbott Laboratories
U.S. 2,803,551 (1957)
Chem. Abst. *51*: 18388h (1957)

31. Hopff, H. and Gassenmeier, E.
Sulfamic Acids and Their Salts
To: I. G. Farb. A-G
Ger. 874,309 (1953)
Chem. Abst. *48*: 12171c (1954)

32. Inoue, H. and Takami, T.
Saccharin
Japan 178,518 (1949)
Chem. Abst. *46*: 222d (1952)

33. Kahn, L. E. and Daely, J. A.
Compositions for Curing Meat
To: E. I. duPont de Nemours & Co.
U.S. 2,946,692 (1960)
Chem. Abst. *54*: 20007[h]

34. Kanami, M. and Uno, S.
Dulcin
To: Mitsubishi Chemical Industries Co.; Japan 177,576 (1949)
Chem. Abst. *45*: 7592b (1951)

35. Kantebeen, L. J. and Appelboom, A. F. J.
    Purifying Alkoxy (or Alkenyloxy) Nitroanilines
    To: N. V. Centrale Suikermaatschappij
    Neth. 59,306 (1947)
    Chem. Abst. *41*: 5899a (1947)

36. Kawamura, B.
    Saccharin
    Japan 177,789 (1949)
    Chem. Abst. *46*: 1591g (1952)

37. Kracauer, Paul
    Artificial Sweetener Composition
    To: Cumberland Packing Co.
    U.S. 3,285,751 (1966)

38. Kreidl, I.
    Sugar Substitutes
    Austrian 122,657 (1930)
    Chem. Abst. *25*: 4069₉ (1931)

39. Kreidl, I.
    Sugar Substitutes for Use as Sweetening Agents by Diabetics
    Brit. 314,500 (1928)
    Chem. Abst. *24*: 1437₈ (1930)

40. Kreidl, I.
    Sweetening Agents
    Brit. 525,031 (1940)
    Chem. Abst. *35*: 6688₉ (1941)

41. Kreidl, I.
    Sweetening Agents
    Fr. 677,452 (1929)
    Chem. Abst. *24*: 3065₈ (1930)

42. Kuderman, Joseph
    Diabetic and Diatetic Food Product
    U.S. 2,311,235 (1943)

43. Locher, Fritz and Mueller, Paul
    Artificial Sweeting Composition
    To: Ciba Corp.
    U.S. 3,118,772 (1964)
    Chem. Abst. *60*: 12591ᶜ

44. Loginov, N. E. and Polyanskii, T. V.
    Dulcin
    U.S.S.R. 65,779 (1946)
    Chem. Abst. *40*: 7234₂ (1946)

45. Lowe, H.
    Electrolytic Production of Saccharin
    Brit. 174,913 (1922)
    Chem. Abst. *16*: 1710, (1922)

46. Lynch, Matthew J.
    Wine Taste
    To: Abbott Laboratories
    U.S. 3,032,417 (1962)
    Chem. Abst. *57*: 3878ᶜ

47. Lynde, W. A.
    Process of Manufacturing Saccharin
    Brit. 14,122 (1906)
    Chem. Abst. *1*: 1492₃ (1907)

48. Macdonald, Lee H.
    Sweetening Compositions for Drugs
    To: Upjohn Co.
    Brit. 810,537 (1959)
    Chem. Abst. *53*: 10673ᶜ

49. McNeill, C.
    Filters for Saccharin and Other Solutions
    Brit. 384,749 (1931)
    Chem. Abst. *27*: 2605₈ (1933)

50. McQuaid, H. S.
    Cyclohexylsulfamic Acid and its Salts
    To: E. I. duPont de Nemours and Co.
    U.S. 2,804,477 (1957)
    Chem. Abst. *52*: 8191b (1958)

51. Mariotti, E.
    Dulcin
    Ital. 418,948 (1947)
    Chem. Abst. *43*: 4786b (1949)

52. Matveev, V. K.
    Saccharin
    U.S.S.R. 66,878 (1946)
    Chem. Abst. *41*: 4812d (1947)

53. Mitani, M.
    Purification of Saccharin
    Japan 176,513 (1948)
    Chem. Abst. *45*: 7148e (1951)

54. Mizuguchi, J., Yokota, Y. and Umeda, M.
    Saccharin by Oxidation in Acid
    Japan 180,599 (1949)
    Chem. Abst. *46*: 7590d (1952)

55. Monnet, P. and Cartier, P.
    Saccharin
    To: Soc. Chim. des Usines du Rhone
    Brit. 165,438 (1921)
    Chem. Abst. *16*: 424₈ (1922)

56. Morani, V. and Marimpietri, L.
    Sweetening Agent
    Ital. 425,939 (1947)
    Chem. Abst. *43*: 5881c (1949)

57. Nakao, K.
    (A Substitute for) Sugar-Coated Tablets
    Japan 174,948 (1928)
    Chem. Abst. *43*: 6791d (1949)

58. Orelup, J. W.
    Saccharin
    U.S. 1,601,505 (1926)
    Chem. Abst. *20*: 3696₈ (1926)

59. Peebles, David D. and Kempf, Clayton A.
    Low Calorie Sweetening Product
    To: Foremost Dairies, Inc.
    U.S. 3,014,803 (Appl. 1959)
    Chem. Abst. *56*: 7762ᵉ

60. *Substituted Imidazolines*
    To: Pfizer and Co., Inc.
    Brit. 757,650 (1956)
    Chem. Abst. *51*: 9706c (1957)

61. Pilcher, F. E.
    Saccharin Composition
    U.S. 2,570,272 (1951)
    Chem. Abst. *46*: 1666i (1952)

62. *2-Alkoxy-5-Nitroaniline and 2-Alkoxy-4-Nitroaniline (Sweetening Agents)*
    To: N. V. Polak and Schwartz's Essence-Fabrieken
    Neth. 57,133 (1946)
    Chem. Abst. *41*: 4170g (1947)

63. *2-Alkyloxy- or Alkenyloxy-5-Nitroanilines and Their Salts*
    To: N. V. Polak and Schwarz's Essence-Fabrieken
    Brit. 613,367 (1948)
    Chem. Abst. *4*: 5799g (1949)

64. *Improvements in the Preparation of Etherified Hydroxynitroanilines*
    To: N. V. Polak and Schwarz's Essence-Fabrieken
    Brit. 597,835 (1948)
    Chem. Abst. *42*: 5050f (1948)

65. *2-Propxy-5-Nitroaniline*
    To: N. V. Polak and Schwarz's Essence-Fabrieken
    Neth. 52,980 (1942)
    Chem. Abst. *41*: 4170e (1947)

66. Polya, Emery
    Artificial Sweetening of Foods
    To: General Foods Corp.
    Ger. 1,074,985 (1960)
    Chem. Abst. *55*: 11702^{a}

67. Rader, W. O. and Seaton, A.
    Mixture of Glucose and Saccharin for "Sugar-Tolerance" Tests
    U.S. 1,625,165 (1927)
    Chem. Abst. *21*: 2052$_4$ (1927)

68. Raecke, B.
    Salts of Saccharin
    To: Henkel and Cie. G.m.b.H.
    German 822,388 (1951)
    Chem. Abst. *48*: 4238h (1954)

69. Robinson, J. W.
    Sodium Cyclohexylsulfamate
    To: E. I. duPont de Nemours
    U.S. 2,383,617 (1945)
    Chem. Abst. *40*: 357$_9$ (1946)

70. Sahyun, M.
    Glycine Ester Salts of Cyclohexylsulfamic Acid
    U.S. 2,789,997 (1957)
    Chem. Abst. *51*: 15560f (1957)

71. Sahyun, M. and Faust, J. A.
    1-(m-Hydroxyphenyl)-2-(Methylamino) Ethyl Cyclohexanesulfamate
    U.S. 2,746,986 (1956)
    Chem. Abst. *51*: 1270e (1957)

72. Seifter, E.
    Improving the Palatabiilty of Fruits and Vegetables by Treating the Plants With a Synthetic Sweetening Agent
    To: Monsanto Chem. Co.
    U.S. 2,921,409 (1960)
    Chem. Abst. *54*: 7928d (1960)

73. Senn, O. F.
    O-Sulfonyl Chloride Benzoic Acid Esters
    To: Maumee Development Co.
    U.S. 2,667,503 (1954)
    Chem. Abst. *49*: 3257e (1955)

74. Shibe, Jr., Wm. J., Cohen, S. I. and Frant, M. S.
    Quaternary Ammonium "Saccharinates"
    To: Gallowhur Chem. Corp.
    U.S. 2,725,326 (1955)
    Chem. Abst. *50*: 10134d (1956)

75. Snelling, W. O.
    Chewing Gum
    U.S. 2,484,859 (1949)
    Chem. Abst. *44*: 2790$_6$ (1950)

76. Snelling, W. O.
   Chewing Gum
   U.S. 2,484,860 (1949)
   Chem. Abst. *44*: 2790₆ (1950)

77. Soder, G. and Schnell, H.
   Cyclohexylsulfamic Acid
   To: Farb. Bayer, A-G
   Ger. 950,369 (1956)
   Chem. Abst. *51*: 12970a (1951)

78. Stanko, George L.
   Artificial Sweetener-Arabinogalactan Composition and Its Use in Edible Foodstuff
   To: Richardson-Merrell Inc.
   U.S. 3,294,544 (1966)
   Chem. Abst. *66*: 45627p

79. Sugino, K. and Mizuguchi, J.
   Saccharin From O-Toluenesulfonamide
   Japan 176,553 (1948)
   Chem. Abst. *45*: 7148d (1951)

80. Synerholm, M. E., Jules, L. H. and Sahyun, M.
   Imidazolines
   To: Melville Sahyun, Trading as Sahyun Laboratories
   U.S. 2,730,471 (1956)
   Chem. Abst. *51*: 121a (1957)

81. Terao, K.
   Concentrated Stable Solution of Polyhydroxyalkylisoalloxazine
   To: Takeda Drug Ind. Co.
   Japan 173,506 (1946)
   Chem. Abst. *46*: 2242g (1952)

82. Thompson, W. W.
   Cyclohexylsulfamates
   To: E. I. duPont de Nemours and Co.
   U.S. 2,826,605 (1958)
   Chem. Abst. *52*: 1290a (1958)

83. Thompson, W. W.
   Cyclohexylsulfamic Acid
   To: E. I. duPont de Nemours and Co.
   U.S. 2,800,501 (1957)
   Chem. Abst. *51*: 17987e (1957)

84. Thompson, W. W.
   Salts of n-Substituted Sulfamic Acids
   To: E. I. duPont de Nemours and Co.
   U.S. 2,805,124 (1957)
   Chem. Abst. *52*: 9222f (1958)

85. *Sweet Material*
   To: Universal Industria Dolciaria Agricola Alimentara
   Ital. 426,150 (1947)
   Chem. Abst. *43*: 5881d (1949)

86. Verkade, P. E.
   2-Alkoxy-4-Nitroanilines
   To: N. V. Polak and Schwarz's Essence-Fabrieken
   Brit. 594,816 (1947)
   Chem. Abst. *42*: 2280a (1948)

87. Walker, H. W.
   Synergistically Sweetened Canned Fruits
   To: Ditex Foods, Inc.
   U.S. 2,608,489 (1952)
   Chem. Abst. *47*: 796b (1953)

88. Washington, G.
   Evaporated Amorphous Saccharine Composition

U.S. 1,512,730 (1945)

Chem. Abst. *19*: 686$_1$ (1925)

89. Weast, C. A.

Frozen Fruit Prepared With Saccharin

U.S. 2,511,609 (1950)

Chem. Abst. *44*: 464d (1950)

90. Weast, Clair A.

Dietetic Canned Fruits and Making the Same

U.S. 2,536,970 (1951)

CHAPTER 14

Color Additives in Food

James Noonan

Vice President–Technical Director
Warner-Jenkinson Company
St. Louis, Missouri

Introduction

Unquestionably, color is a vital constituent of food. It is probably one
of the first characteristics perceived by the senses and is indispensible to the
modern day consumer as a means for the rapid identification and ultimate
acceptance of food. Almost all foods, from raw agricultural commodities to
finished products, have an associated color acceptable to the consumer on the
basis of social, geographic, ethnic and historical backgrounds. Deviation from
this norm usually precipitates drastic consequences. For example, to the American and European consumer a dinner consisting of blue soup, yellow meat,
red potatoes, green bread and black salad might seem somewhat repulsive,
although coloristically the setting might be judged quite acceptable. The color
of food products unfamiliar to the consumer probably has less effect. More
important, however, are the subtle changes in color shades associated with

poor and first-rate food items. The color of fresh orange juice is quite precise, and manufacturers of orange soft drinks almost always attempt to match this color. Similarly, the colors of processed meats, vegetables, baked goods, and dairy products should simulate that of their fresh counterparts.

From a modern food-manufacturing viewpoint, color additives are indispensable. Dyes and pigments are employed to create new food products and to modify the color of established food products that show color shifts as the result of manufacture and storage. To support this work, the food manufacturer can draw from an array of dyes and pigments comprising two classes, i.e., the certified and uncertified food colors. The certified dyes and lakes are synthetics manufactured to meet strict government specifications, while the uncertified colors are usually naturally derived substances. Both classes are strictly controlled in the United States by regulatory statutes, the knowledge of which is indispensable to food processors.

Legislative and Regulatory History of Color Additives

Color additives for food represent a unique and special category of food additives. They have historically been so considered in legislation and regulation.

The current legislation pertinent to the regulation and use of color additives in the U.S. is the Food, Drug and Cosmetic Act of 1938 as amended by the Color Additives Amendments of 1960. To fully understand the present color additive situation and to judge the future, it is necessary to have a feeling for the past history of color additives. The spotlight will be on the synthetic dyes and pigments used as colors for food as the "so called" natural color additives have been less important and less controversial.

In 1900, some eighty dyes were being used in the U.S. for coloring food. At that time there were no regulations regarding the nature and purity of these dyes and same batch of color used to dye cloth might find its way into candy.

The first comprehensive legislation was the Food and Drug Act of 1906 which listed seven dyes which were permitted for use in foods. These seven were chosen after a thorough study of the colors in current use and only those colors of known composition which had been examined physiologically with no unfavorable results were listed. The seven original permitted colors were as follows:

Amaranth	Naphthol Yellow
Erythrosine	Orange 1
Indigotine	Ponceau 3R
Light Green	

A system was set up for certification of batches of these colors by the Department of Agriculture. At that time certification was not mandatory; however, color manufacturers soon found it advantageous to obtain certification.

The original list of colors left much to be desired to fulfill all the needs of the food industry and through the years additional colors were added—but only after physiological testing. The chronology of addition of new colors is as follows:

1916	Tartrazine	1929	Brilliant Blue
1918	Yellow AB & OB	1950	Violet No. 1
1922	Guinea Green	1959	FD&C Lakes
1927	Fast Green	1966	Orange B
1929	Ponceau SX	1971	FD&C Red No. 40
1929	Sunset Yellow		

The Food, Drug and Cosmetic Act of 1938 superseded the Act of 1906 and broadened the scope of certified colors, creating three categories—FD&C Colors, D&C Colors and Ext. D&C Colors. Under the new Act the common names of the dyes were not employed but color prefixes and numbers were used. For example, certified Amaranth was now FD&C Red No. 2. Also, *certification was mandatory* and was placed under the jurisdiction of the Food and Drug Administration. Before being listed under the 1938 Act, the fifteen food colors were again scrutinized for their toxicological effects. At the hearings for this legislation, Dr. H. O. Calvary, Chief of the Division of Pharmacology of the FDA, reported that these conclusions were based on a three-year study of the dyes. He also stated that there were no known cases of harm from the use of these colors, some of which had been in use as certified colors for thirty-two years. Dr. Calvary also testified that the term "harmless and suitable for use in foods" must be judged in the light of the amounts used.

Following the passage of the Food, Drug and Cosmetic Act of 1938, the situation was peaceful until the early 1950's; three unfavorable incidents occurred as a result of the overuse of color in candy and on popcorn. This overuse resulted in a number of cases of diarrhea in children. The colors involved were FD&C Orange No. 1 and FD&C Red No. 32. These incidents, coupled with chronic toxicity animal-feeding studies, led to the delisting of FD&C Orange No. 1, Orange No. 2, and FD&C Red No. 32.

The suggestion that quantity limitations be established for colors was met with the objections by the FDA that the law did not empower the Secretary of Health, Education and Welfare to set quantity limitations and that "harmless" meant zero toxicity, that is, harmless in any amounts used.

After several conflicting opinions in the lower courts, the Supreme Court ruled that the FDA, under the 1938 Law, did not have the authority to set quantity limitations. Following soon thereafter were delistings of FD&C Yellow No. 1, Yellow No. 2 and FD&C Yellows 3 and 4, the remaining oil soluble colors.

The Supreme Court decision had rendered the old law obsolete and unworkable and colors were being delisted even though the FDA admitted they were not endangering the public health. Under the sponsorship of the Certified Color Industry and the FDA, the Color Additives Amendments became law July 12, 1960. The new amendments were chiefly a relief measure designed to correct the inflexibility of the old law. The law consists of two parts, or Titles, as they are called. Title I is the so-called permanent part; Title II, the temporary part.

Title I, the more important section, sets up uniform rules for *all* permitted colors, both certified and uncertified. The term *color additive* is used to describe "any dye, pigment of other substance capable of coloring a food, drug or

cosmetic on any part of the human body." The term "coal tar color" is eliminated from the law. This is fortunate as the term has for years been an unde-
served stigma. The law provides for the listing of color additives which must
be certified and color additives exempt from certification. The term "natural
colors", formerly and improperly applied to the uncertified color additives
which are of natural and synthetic origin, was also eliminated. The Secretary
of Health, Education and Welfare is given the authority to decide whether
a color additive should be classed as certified or uncertified. The law allows
the Secretary to list color additives for specific uses and also to set conditions
and tolerances (limitations) on the use of color additives. A color additive
can, if necessary, be given a zero tolerance, meaning that it cannot be used.

Title I also includes the Delaney cancer clause which states that a color
additive cannot be listed for any use whatever if it is found to induce a cancer
when ingested by man or animal. This clause is softened somewhat by permitting in cases invoking the cancer clause, the appointment of an advisory committee to serve as a fact finding body. The advisory committee reports its
findings to the Secretary who makes his decision. He is not bound to follow
the recommendation of the advisory committee.

Title II, the so-called temporary law, is designed to permit use of current
color additives pending the completion of scientific investigations needed to
determine the suitability of these materials for permanent listing. The original
closing date for the provisional listing of the color additives was two and one-
half years after the date of passage of the law. However, the Secretary of
Health, Education and Welfare has the power to grant extensions of the closing
dates, which he has done many times. In retrospect the Color Additives Amendments have been salutary; however, difficulties have arisen in obtaining permanent listing of the certified color additives.

The pharmacological studies to establish the safety of the FD&C colors
were initiated in 1957 by the FDA, using part of the funds from the certification fees paid by the manufacturers. The studies were generally two-year
studies in rodents (mice and rats) and two-year studies in dogs, although
in the case of FD&C Red No. 2, FD&C Yellow No. 6 and FD&C Red No.
4 the animal studies were run for seven years. Rates of use of the colors
in these studies were 5, 2 and 1 per cent of the total diet. The final readings
on these studies were completed in 1964. However, during the interim between
the passage of the Color Additives Amendments and 1964, FD&C Red No.
1 and FD&C Red No. 4 were delisted as a result of the pharmacological results
of the animal studies. FD&C Red. No. 4, however, because of its need in
coloring of maraschino cherries was returned to the provisional list for coloring
maraschino cherries at maximum rate of use of 150 ppm in the cherries and
syrup and in short term (six weeks) ingested drugs at 5 mg per day.

Because of the lack of demand for FD&C Green No. 1 (Guinea Green)
and FD&C Green No. 2 (Light Green), the Certified Color Industry did not
petition for extension of these colors and they were automatically deleted from
the list of color additives in 1966. Also, in 1966, Orange B was permanently
listed for coloring sausage casing at maximum rate of use of 150 ppm.

In mid 1971 a new food color, FD&C Red No. 40, was permanently listed. This
scarlet red color, in addition to being tested for chronic toxicity, had undergone teratology and reproduction studies.

Starting in February 1965 with FD&C Yellow No. 5, petitions for permanent listing were submitted to the Food and Drug Administration, and by 1968 all petitions had been submitted. Because of a dispute (and subsequent court case) over interpretation of the color regulations regarding pre-marketing clearance for cosmetics between FDA and the Toilet Goods Association, no action was taken on the petitions. However, in July of 1969, FD&C Yellow No. 5, FD&C Red No. 3, and FD&C Blue No. 1 were permanently listed for use in foods and ingested drugs with limitations only for good manufacturing practice. Pending settlement of the cosmetic issue, provisional listing was maintained for topical drug and cosmetic uses of these colors.

Aside from FD&C Violet No. 1, whose safety was questioned following a Canadian feeding study, there was no ostensible reason why the remaining colors could not have been permanently listed, and the Food and Drug Administration was criticized for not proceeding with listing.

During the period of 1963 to 1970, food colors were the subject of a study by the Joint FAO/WHO Expert Committee, who assigned ADI's (Acceptable Daily Intakes) for a number of the FD&C colors. The ADI's were quite generous and were many times greater than the actual estimated; these comparisons are shown in Table 1.

TABLE 1
Comparison of ADI and Estimated Ingestion

Color	Maximum ADI,[1]		Estimated Maximum Ingestion,[2] mg/day/capita
	mg/kg	mg/70 kg man	
FD&C Yellow No. 5	7.5	525	16.3
FD&C Yellow No. 6	5.0	350	15.5
FD&C Red No. 2	1.5	105	17.7
FD&C Red No. 3	1.25	87	1.9
FD&C Blue No. 1	12.5	875	1.2
FD&C Blue No. 2	2.5	175	0.3
FD&C Green No. 3	12.5	875	0.1

At the end of 1970, the FD&C color situation seemed to have stabilized; it appeared that during 1971 the remaining colors (except FD&C Violet No. 1) would achieve permanent listing. However, in late 1970 abstracts of work by two different groups of U.S.S.R. scientists on FD&C Red No. 2 appeared in *Chemical Abstracts*. One paper reported studies that suggested carcinogenic response from this color, while the second study[3] reported embryotoxic responses from FD&C Red No. 2. Both studies were considered questionable by many reputable pharmacologists. The Food and Drug Administration, basing its action on the enormity of favorable results on carcinogenicity testing of Red 2, discredited this study; however, aside from FD&C Red No. 40, the certified color additives had never been studied for embryotoxic responses. In early 1971 the FDA initiated its in-house study of the reproduction effects of FD&C Red No. 2. Neither the protocol of this study nor the results have been published; however, the FDA has reported that it has confirmed some level of embryotoxicity in one species—rats. Pending results of a more comprehensive nature, the FDA moved to reduce the aggregate use of Red 2. In a *Federal Register* order the FDA requested that persons interested in continued use of Red 2 submit data showing proposed uses in specific products; this order stated that allocation guidelines for use of Red 2 would be published December 31, 1971.

In November 1971 the FDA requested that the National Academy of Sciences appoint an Advisory Committee to review the Red 2 situation. Although the NAS initially refused, as a result of requests from members of the Food Protection Committee that Red 2 and the methodology of reproduction testing of colors be reviewed, NAS agreed to accept the evaluation in early December.

In the September 11, 1971, issue of the *Federal Register*, the FDA also stated that for provisional listing beyond December 31, 1971, the ingested certified colors—nine FD&C and sixteen D&C (Drug and Cosmetic) colors—should have teratology and multi-generation reproduction studies in progress by December 31, 1971; final reports should be filed on teratology by October 1, 1972, and on the reproduction studies, by July 1, 1973.

Teratology deals with the effects of materials on fetal development (abnormalities, early fetal death, fetal resorption), while reproduction studies explore the effects on the reproductive activity of test animals, such as fertility and lactation. These areas of very recent exploration were brought into prominence by the thalidomide incident. As these testing disciplines are in their preliminary scientific development, there is no body of tested evaluative guidelines established by experience.

In order to conduct these formidable and expensive tests, the CTFA (Cosmetic, Toiletries, and Fragrance Association—formerly TGA) spearheaded the creation of a Joint Inter-Industry Committee to establish the protocol of testing, to obtain cost bids, to audit and monitor the testing laboratories, and, in general, to direct the testing program. This committee is composed of industry pharmacologists from member companies of the Pharmaceutical Manufacturers' Association, the Cosmetic, Toiletries and Fragrance Association, Grocery Manufacturers of America, and a representative from the Certified Color Manufacturers' Association (formerly CCIC). In many respects the protocol of testing in these studies should be more comprehensive than the FDA studies, particularly in that two species of animals (rats and rabbits) are used. Contracts for these studies were let prior to the end of 1971.

Guidelines for allocation of FD&C Red No. 2, scheduled for publication by FDA on December 31, 1971, were delayed until the test data for this color could be reviewed by a National Academy of Sciences Advisory Committee. This committee met in February 1972 to review the FDA work on FD&C Red No. 2, as well as the industry-sponsored teratology test results. Allocation guidelines for future use of FD&C Red No. 2, based on NAS recommendations, should be established by FDA in early or mid 1972.

The evaluation of the safety of FD&C Violet No. 1 (carcinogenicity only) was reviewed in late 1971 by the NAS, who recommended continued use of this color as well as additional long-term, chronic-feeding studies.

In summary, as of February 1972 the provisional listing of the FD&C colors, except FD&C Red No. 2, will be extended through 1972. FD&C Red No. 2 will likely be allocated to some lower level of use, and FD&C Red No. 40 will find use in replacement of FD&C Red No. 2.

Since all of the FD&C colors, except FD&C Red No. 40, are undergoing additional testing, the difference between provisional and permanent listing is academic; however, as of January 1972 they were grouped as follows:

Provisional Listing	Permanent Listing
FD&C Red No. 2	FD&C Yellow No. 5
FD&C Red No. 4	FD&C Red No. 3
(for coloring maraschino cherries)	FD&C Blue No. 1
FD&C Yellow No. 6	FD&C Red No. 40 (and lake)
FD&C Blue No. 2	Orange B
FD&C Violet No. 1	(for coloring sausage casings)
FD&C Green No. 3	Citrus Red No. 2
Lakes of FD&C colors (except FD&C Red 40)	(for coloring skins of oranges)

Certified Color Additives

Previous to 1959, the certified food colors were dyes only; that year, however, the lakes of the dyes were added to the list. Lakes are pigments. The present certified color additive list consists of two major classifications: dyes and lakes.

The FD&C dyes are all water-soluble compounds comprising four classes. Examples of the structures are shown in Table 2. Specifications and procedures for certification are established under the regulations promulgated by the FDA.

The primary or straight colors which include the lakes require certification. At one time blends required certification, but regulations published in December 1964 eliminated certification of mixtures.

The specifications for FD&C Yellow No. 5 are typical of those for a certified color additive.

FD&C Yellow No. 5—Specifications

Trisodium salt of 3-carboxy-5-hydroxy-1-sulfophenyl-4-*p*-sulfophenylazo-pyrazole
Sum of chlorides and sulfates (as sodium salts) and volatile matter (at 135°C), not more than 13.0 per cent
Water-insoluble matter, not more than 0.2 per cent
Phenylhydrazine-*p*-sulfonic acid, not more than 0.1 per cent
Other uncombined intermediates, not more than 0.2 per cent each
Subsidiary dyes, not more than 1.0 per cent
Lead, not more than 10 ppm
Arsenic, not more than 3 ppm

TABLE 2
Classification and Structures of Certified Color Additives (FD&C Dyes)

Azo Dyes	*Typical Structure*
FD&C Red No. 2	FD&C Red No. 2
FD&C Yellow No. 5	
FD&C Yellow No. 6	
FD&C Red No. 4	
FD&C Red No. 40	
Orange B	

TABLE 2
Classification and Structures of Certified Color Additives (FD&C Dyes) (Continued)

Triphenylmethane Dyes	*Typical Structure*
FD&C Blue No. 1	FD&C Blue No. 1
FD&C Green No. 3	
FD&C Violet No. 1	

Fluorescein Type	*Structure*
FD&C Red No. 3	

Sulfonated Indigo	*Structure*
FD&C Blue No. 2	

Certified Color Forms

There are two main types of certified colors—dyes and lakes. Some of the dyes have been in use since 1900; the lakes have been allowed only since 1959. The dyes manifest their coloring power by being dissolved, while the lakes are insoluble pigments and color by dispersion. The FD&C Lakes are merely alumina hydrate (aluminum hydroxide) on which the dye has been adsorbed.

The FD&C dyes are water-soluble and are insoluble in nearly all organic solvents. Water solubility of most colors is quite high and in most application methods, solubility is usually no problem. FD&C Blue No. 2 (Indigotine) is the exception to this, and often it would be advantageous to have a greater solubility for FD&C Red No. 4.

For systems where anhydrous conditions are a consideration, glycerine and propylene glycol are used as solvents. In general, the colors are more soluble in glycerine than in propylene glycol. Most are only very slightly soluble in ethyl alcohol, but use is often made of the reasonable solubility in alcohol of FD&C Red No. 3, FD&C Blue No. 1, ·FD&C Violet No. 1, and FD&C Green No. 3. Solubilities in glycerine, propylene glycol, and alcohol are shown in Table 3.

TABLE 3
Solubilities of FD&C Colors in Various Solvents

| FD&C Colors | Solubility in Ounces Per Gallon of Solvent | | | |
	Water 70°F	Glycerine	Propylene Glycol	Alcohol 95%
Red 2	18	$21\frac{1}{2}$	$1\frac{1}{2}$	Tint
Red 3	16	$30\frac{3}{4}$	29	$2\frac{1}{2}$
Red 4	$9\frac{1}{2}$	$5\frac{1}{4}$	$1\frac{1}{2}$	Tint
Red 40	28	4	2	Tint
Violet 1	25	$27\frac{1}{2}$	$18\frac{1}{2}$	$\frac{1}{8}$
Yellow 5	$17\frac{1}{2}$	28	12	Tint
Yellow 6	23	$14\frac{1}{2}$	$2\frac{1}{2}$	Tint
Blue 1	25	$37\frac{1}{2}$	53	2
Blue 2	$1\frac{1}{2}$	$\frac{1}{2}$	Trace	Tint
Green 3	23	$14\frac{1}{2}$	$14\frac{1}{2}$	$\frac{1}{2}$

Good coloring technique recommends that the dyes be solubilized before addition to the colored product. However, it is often possible, where water is added in the process, to add the dry color to the batch and depend upon the added moisture and heat to dissolve the color in processing.

Dyes are available in many forms—powder, granular, plating colors, wet-dry (blends), diluted (cut blends), liquid (aqueous), liquid (non-aqueous), and paste. The best form for any specific use will be dictated by the nature of the product to be colored, the process conditions and volume of color used. Table 4 lists the range of pure dye percentages and the advantages and disadvantages of the various forms of color.

Properties of FD&C Dyes

Solubility Information on solubility has been previously discussed. It would be most helpful to have on the approved list of FD&C Colors several oil soluble colors, particularly a yellow color. However, the search for these has been futile and does not seem to portend much hope. The sulfonic acid groups confer water solubility to the molecule and likewise reduces oil solubility. Unsulfonated dyes would be oil soluble but lack of sulfonation seems to go along with high toxicity. More light will be shed on this phenomena as more metabolic studies are conducted.

TABLE 4
Advantages and Disadvantages of Various Forms of Certified Color Additives

Form	Pure Dye	Advantage	Disadvantage
Powder (primary color)	88–93%	Ease of dissolving. Suitable for dry mixes	Dusty
Granular (primary color)	88–93%	Dustless, free flowing	Slower dissolving Not suitable for dry mixes
Wet dry blends	90%	No flashing in dry blends when wetted	—
Aqueous liquid colors	1–6%	Ready to use, ease of handling, accurate measurement	More costly than dry color
Cut blends	22–85%	Permits larger weighings with more accuracy for small amounts of added color	More costly
Non-aqueous liquid colors	1–8%	May be used in fatty material	More costly
Paste	4–10%	May be used in products in which water is limited	Costly
Plating colors	88–93%	Gives good depth to dry mixes	Not available in all primary colors

Stability   In general, the certified food colors can be said to be stable for most uses in food. In the dry state, no degradation has been noted (other than loss in dye strength by moisture absorption) in samples stored for 15 years.

With the exception of FD&C Blue No. 2 and FD&C Red No. 3, the light stability of the dyes in food products is good. Even FD&C Blue No. 2, which fades badly in solution, shows good stability in a number of products—candy and baked goods. Likewise, FD&C Red No. 3 which has limited light stability in coatings, shows excellent stability in retorted products.

The two areas in which the majority of the certified food colors show instability are in combination with reducing agents and retorted protein materials. The azo and triphenylmethane dyes are easily reduced to colorless compounds. Contact with metals such as zinc, tin, aluminum and copper was formerly a large factor in color fading; however, with current progress in food processing, contact with these metals is minimal and the chief source of color fading is contact with reducing agents, such as ascorbic acid.

Color fading of certified colors in carbonated and still beverages is caused by ascorbic acid incorporated as a flavor anti-oxidant and as a source of vitamin C. The use of ETDA in beverages has proved somewhat successful in inhibiting the effects of ascorbic acid on dyes but not in sufficient degree to eliminate it completely. The reductive action of ascorbic acid seems to be catalyzed by light and canned colored products containing certified color can possess reasonable shelf life.

In retorted protein foods most of the certified food colors lack stability. However, FD&C Red No. 3 and to a lesser extent FD&C Violet No. 1 are stable. Canned dog foods can be colored using these colors, however, there is only limited use for this application because of the suitability of the very low cost iron oxides permitted in pet foods.

Corrosion in Canned Beverages Although present in canned carbonated beverages at only 50–150 ppm, the azo dyes (FD&C Red No. 2, FD&C Yellow No. 5, and FD&C Yellow No. 6) appear to have a causative effect on can corrosion; the rate of corrosion is directly proportional to the azo dye concentration.[4,5] However, inclusion of azo dyes at rates of 50 ppm or less will result in canned products with sufficient shelf life (9–12 months). For many products 50 ppm is sufficient for good tinctorial values, but inclusion of caramel color is often desirable to raise the color depth.

The non-azo certified dyes do not produce can corrosion. The uncertified color additives seem to have no effect on rate of corrosion.

FD&C Lakes

The FD&C Lakes, which were admitted to the approved list of certified color additives in 1959, comprise an important class of color additives. Their use is growing at a very rapid rate. The Color Regulations define FD&C Lakes as "Extension on a substratum of alumina, of a salt prepared from one of the water soluble straight colors by combining such color with the basic radical aluminum or calcium." The alumina hydrate or aluminum hydroxide substratum is insoluble so what is produced is an insoluble form of the dye—a pigment. Dyes color by being dissolved in a solvent; pigments color by dispersion. Previous to the admission of lakes, insoluble forms of the dyes were made by adsorbing them on insoluble materials such as starch, cellulose, flour, etc. These can be highly colored materials but are of low coloring power and generally perform poorly as pigments.

The pigmentary properties of the shade of alumina hydrate lakes depend a great deal upon the preparation of the alumina hydrate and the processing variables (temperatures, pH, etc.) must be carefully controlled. Even the agitation, the concentration, and rate of addition of the reactants bear on the physical properties of the lake. Lake-making is steeped in artistry, but depends upon careful process control to insure uniformity.

The physical properties of the water soluble colors (dyes) are of small significance in most products. There is, for example, a choice between granular or powdered forms of dye. Dissolved, both forms produce equivalent solutions. Dyes from different manufacturers are interchangeable for most uses. This is not always true with lakes, even though they may be in the same pure dye content. Shades of difference can exist from manufacturer to manufacturer.

The tinctorial strength of a water soluble color is proportional to the pure dye content. Thus, two units of a 45% pure dye color are equivalent in color strength and shade to one unit of a 90% pure dye color. This is not true with the lakes. One unit of a 24% pure dye lake is not equivalent in strength of shade to two units of a 12% pure dye color. Yellow No. 5 Lake, for example, at 5% pure dye is lemon-yellow; at 15% pure dye is egg-yellow; and at 25% pure dye is orange-yellow.

Having aluminum hydroxide as substratum, the lakes are insoluble in nearly all solvents. When suspended in water there is a slight "bleed" of dye into the water, but for most food applications this is of no consequence. The lakes are stable in the pH range of 3.5 to 9.5. Outside this range the substratum breaks down, releasing the dye.

The color regulations classify the FD&C Lakes as straight colors, subject to certification. Also, they must be made from dyes which have been previously certified. In most cases, the color regulations specify a minimum of 85 per cent pure dye for primary dyes. However, most lots will be in the range of 90–93 per cent pure dye. The FD&C Lakes do not have a specified minimum dye content and range from 10 per cent to 40 per cent in pure dye.

By their very nature, lakes are suitable for coloring products which cannot tolerate water and products where use of water is undesirable.

The FD&C Dyes are water soluble and are insoluble in organic solvents, fats and oils. Often, the coloring of oils can be done conveniently with the lakes. However, in spite of their great transparency as pigments, a clear, colored material, free of opacity, is generally not possible.

Pigment Dispersion   Dispersion of pigments is an operation not generally well known in the food industry. Dispersion is merely the distribution of the pigment throughout the material to be colored. It is generally considered that pigment dispersion is—

1. breaking up of the agglomerates,
2. wetting and coating each particle with the vehicle in cases where the material to be colored is a liquid (vehicle).

To accomplish this, energy is required and high shear forces must be applied. Dispersion in such fields as inks and paints is referred to as grinding, though it is doubtful if much particle fracture is involved. In these industries, very fine "grinds" in the vehicles are required and ball mills, pebble mills, and colloid mills find use. For most food and pharmaceutical applications high shear mixers such as, the Kady Mill, the Eppenbach Homomixer, Premier Dispersator, Daysolver and Cowles Dissolver are adequate for lake dispersion in liquids.

Many lake applications demand dispersion in dry materials. In crystalline materials, simple mixing sets up sufficient attritive forces to obtain dispersion. In noncrystalline powders, such as starch or flour, it is often necessary to prepare a concentrated premix of 5–10 parts powder to one part of lake, followed by grinding through a hammer mill. This concentrated premix is then readily incorporated into the final mix.

Recent investigation into the relationship of the lake particle size and tinctorial strength have led to the development of "jet milled lakes." Jet milled lakes have an average particle size less than one micron as compared with about five microns for the conventionally ground product. In some instances considerable increase in tinctorial strength is obtained by jet milling.

Table 5 lists the physical differences between the dyes and the lakes. These properties are important in choosing between a dye or a lake for a particular coloring application.

Uncertified Color Additives

This class of color additives comprises the so-called natural colors. Inclusion of these colors in the Color Additives Amendments removed them from the jurisdiction of the Food Additives Amendments. Following passage of the Color

TABLE 5
Physical Differences Between Dyes and Lakes

Properties	Lakes	Dyes
Solubility	Insoluble in most solvents	Soluble in water, propylene glycol, glycerine
Method of coloring	By dispersion	By being dissolved
Pure dye content	Generally 10–40%	Primary colors—90–93%
Rate of use	0.1 to 0.3%	0.01 to 0.03%
Particle size	Average 5 microns	12 mesh to 200 mesh
Stability		
Light	Better	Good
Heat	Better	Good
Coloring strength	Not proportional to pure dye content	Directly proportional to pure dye content
Shade	Varies with pure dye content	Constant

Additives Amendments in 1960, those colors in prior use were provisionally listed. Since that time many of these colors and several new colors have been permanently listed. The provisional and permanent lists of uncertified color additives as of March 1972 are indicated in Table 6.

These lists are unduly large as a result of the definition of a color additive contained in the Color Additives Amendments as "any material that when added or applied to a food, drug or cosmetic or to the human body or any part thereof, is capable of imparting color"; and consequently a number of these are not important from a color application standpoint. The more important of these additives will be dealt with in detail.

Carbon black (impingement process) is the only provisionally listed uncertified color additive.

TABLE 6
Color Additives Permanently Listed for Food Use, Exempt from Certification

Color	Use Limitation[a]
Algae meal, dried	For use in chicken feed to enhance the yellow color of chicken skin and eggs
Annatto extract	
β-Apo-8'-carotenal	Not to exceed 15 mg/lb, or pint, of food
Beets, dehydrated (beet powder)	
Canthaxanthin	Not to exceed 30 mg/lb, or pint, of food
Caramel	
β-Carotene	
Carrot oil	
Cochineal extract; carmine	
Corn endosperm oil	For use in chicken feed to enhance the yellow color of chicken skin and eggs
Cottonseed flour, partially defatted, cooked, toasted	
Ferrous gluconate	For coloring ripe olives
Fruit juice	
Grape skin extract	For coloring beverages

TABLE 6

Color Additives Permanently Listed for Food Use, Exempt from Certification (Continued)

Color	Use Limitation[a]
Iron oxide (synthetic)	For coloring pet food, not to exceed 0.25 per cent by weight of the food
Paprika	
Paprika oleoresin	
Riboflavin	
Saffron	
Tagetes meal and extract (aztec marigold)	For use in chicken feed to enhance the yellow color of chicken skin and eggs
Titanium dioxide	Not to exceed 1 per cent by weight of the food
Turmeric	
Turmeric oleoresin	
Ultramarine blue	For coloring salt intended for animal feed, not to exceed 0.5 per cent by weight of the salt
Vegetable juice	

[a] Unless otherwise indicated, the color may be used for the coloring of food generally in amounts consistent with good manufacturing practice.

Source: *Food Colors*, National Academy of Sciences, Washington, D.C., 1971.

The Carotenoids

The carotenoids represent the most widespread colors naturally occurring in both the plant and animal kingdoms and are responsible for the coloration in carrots, tomatoes, apricots, orange juice and lobsters, to list only a few.[6]

Of the 100 or more known carotenoids, only a few have been isolated or synthesized for use as color additives. The color additives which are carotenoids are: beta-carotene beta-apo-8'-carotenal, canthaxanthin, bixin (in annatto) and xanthophyll. Bixin will be discussed in detail, while xanthophyll is of minor importance as a direct color additive.

The properties of β-carotene, β-apo-8'-carotenal (hereafter called apocarotenal) and canthaxanthin are similar in properties and method of application. The discussion will deal principally with β-carotene with reference to the other materials only when they differ greatly from β-carotene.

The first carotenoids available for coloring were obtained by extraction from natural sources. In the early 1950's, Hoffman-LaRoche developed a synthesis for β-carotene which utilizes acetone as starting material. This synthesis was later extended to commercial production of apo-carotenal and canthaxanthin. The synthesis of β-carotene created an interesting controversy which hinged on whether this material should, because of its synthetic nature, be subject to certification. Also, could such compounds still be called natural colors? This accounts for the use of the term uncertified color additive, in lieu of "natural color."

An important aspect of the carotenoids is the vitamin A activity of several members of this family. This is possessed by both β-carotene and β-apo-carotenal, but not by the more recent addition to the color additive list: canthaxanthin. Some of the more important chemical and physical properties of β-carotene are shown in Table 7.

β-apo-8'-carotenal is similar in many respects to β-carotene. It is more orange in shade and has a somewhat higher solubility in the usual solvents.

Canthaxanthin is even more orange-red, but has lower solubilities.[7]

The carotenoids presented a problem from an application standpoint because of their lack of stability and their low solubility. Though classed as

TABLE 7
Chemical and Physical Characteristics of β-Carotene

Appearance:	Violet-red platelet crystals, free flowing. Stable only under inert gas or under vacuum.
Chemical structure:	

$$H_3C \quad CH_3 \qquad\qquad H_3C \quad CH_3$$

$$H_2C \diagup \overset{C}{\diagdown} C-(CH{=}CH{-}CH{=}CH)_2{-}CH{=}CH{-}(CH{=}C{-}CH{=}CH)_2{-}C \quad CH_2$$

$$H_2C \quad C{-}CH_3 \qquad CH_3 \qquad\qquad CH_3 \qquad C \quad CH_2$$

$$\overset{C}{H_2} \qquad\qquad\qquad\qquad\qquad H_3C \quad \overset{C}{H_2}$$

Melting point:	183°C	
Solubility:	*Solvent*	*g/100 ml @ 30°C*
	Water	insoluble
	95% ethanol	.004
	Glycerine	insoluble
	Propylene glycol	insoluble
	Acetone	0.02
	Cottonseed oil	0.08
Acid and alkali:	Sensitive to both, but relatively more resistant to alkali. Contact with acids in presence of oxygen causes rapid deterioration which is accelerated by heat.	
Stability:	Crystalline β-carotene is sensitive to air and light. In air, the absorption maximum drops to 25% of its initial value in six weeks at 20°C in either darkness or daylight. At 45°C in air, it is almost completely destroyed after six weeks. Vegetable fat and oil solutions and suspensions are quite stable during customary handling.	
Commercial forms available:	1. Liquid suspension	30% in vegetable oil
	2. Semi-solid suspension	24% in hydrogenated vegetable oil
	3. Beadlet-water dispersible	10% and 2.4%
	4. Emulsion, beverage type	3.6%

fat soluble materials, their solubility in fats and oils is limited. The carotenoids were initially used to color fat-based foods, particularly margarine and oils.[8] For this application a micro-pulverized suspension of β-carotene in vegetable oil was developed. This added to a fat based food quickly dissolves in the heat of processing. The vegetable oil in the suspension effectively protects the β-carotene from oxidation and the stability of the suspension is quite good.

The coloring of the aqueous systems has been worked out by the development of a water dispersible beadlet.[9] This form may be utilized for coloring of orange beverages, puddings, cheese, and ice cream.

Carotenoids have helped to fill the need for oil soluble food colors created by the delisting of FD&C Yellow No. 3 and FD&C Yellow No. 4. Where oil solubility and clarity are desired, these colors are quite essential. The resistance of these colors to reduction by ascorbic acid is a useful property in coloring beverages containing ascorbic acid. The rather high market price for these colors has hampered their extensive use. This is often overcome by virtue of their vitamin A activity; however, in instances where the certified food colors are suitable, the choice generally goes to the certified colors because of their lower cost.

Bixin (Annatto Extract)

Annatto colors are obtained from the seeds of the tree *Bixa orellana* L., which grows abundantly in the tropics. The coating of the seeds contains a coloring material which is obtained by extraction with alkali or oil.

The coloring matter is chiefly the carotenoid-bixin which has the following chemical structure:

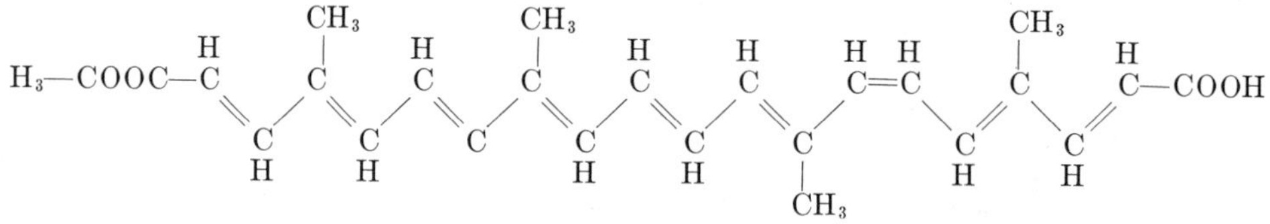

On saponification the methyl ester group is split off, resulting in the diacid-norbixin.

Bixin is oil soluble, while the norbixin is water soluble. The shade produced by annatto extracts is somewhat variable, but ranges from butter-yellow to peach. Annatto extract can be the water-soluble potassium salt of norbixin or bixin in an oil solvent.

The carotenoid bixin (or norbixin) has the following stability characteristics:

Oxidation	Very good
Light	Poor to fair
Temperature	Poor to fair

The aqueous extract has been widely used for coloring ice cream; however, for this use, it has strong competition from the less expensive certified food colors.

Uses of the soil soluble extract have increased since the delisting of the oil soluble certified food colors. It has become an important color in butter, margarine, popcorn oil, and salad dressings. In these applications, its use is more economical than β-carotene.

Caramel Color

Caramel is the amorphous dark brown material resulting from the controlled heat treatment of the following food grade carbohydrates:

Dextrose
Invert sugar
Lactose
Malt syrup
Molasses
Starch hydrolysates and fractions thereof
Sucrose

Not much is known about the nature of the caramelization of sugar; however, by control of the reaction, various grades are prepared with certain desirable properties.[10, 11]

The composition of caramel is complex and indefinite, forming a colloid in aqueous substrates. The colloids carry an electrical charge, the nature of which depends upon the method of manufacture. The isoelectric point of the caramel is quite important in various applications. At pH's above the isoelectric point, the caramel is negatively charged—at pH's below, it is positively charged. One of the major uses of caramel is in coloring carbonated beverages: colas and root beers. In order to prevent precipitation, the caramel must carry a strong negative charge and its isoelectric point should be at a pH of 2.0 or less.

Grades and Uses   There are many tailor-made varieties of caramel, but most fall into the following types:

1. Acid proof caramel — For use in carbonated beverages and acidified solutions
2. Bakers and confectioners — A less refined grade for use in baked products—cookies, cakes, rye bread
3. Dry caramel — For use in dry mixes or where volume of liquid product is excessive.

Types 1 and 2 are dark-brown, viscous liquids having a specific gravity in the range of 35° Be.

In most products, caramel has very good stability. It is very helpful in coloring of canned carbonated beverages in which the concentration of azo dye should be kept low. Addition of caramel helps to increase the depth of color; however, it must be carefully balanced with the certified color or "dirty" colors are produced. Caramel color is often used with certified colors in baked goods where dark chocolate shades are desired and which cannot be achieved without excessive use of certified color. One disadvantage of caramel color is its yellow-brown shade. For bakers' use, caramel products are available to which Red FD&C Colors have been added to give a more chocolate color. From a cost standpoint, caramel color is quite economical.

Titanium Dioxide

Titanium dioxide is a white pigment possessing great covering or opacifying power. In less purified grades, it is the basic pigment used in white house paint. There are two crystalline modifications of titanium dioxide: rutile and anatase. Only the *anatase* variety finds use as a color additive. Principal uses of this white pigment are in sub-coating of confectionery panned goods and in icings. As with the other food grade, pigments, the FD&C Lakes, it must be dispersed to give full coloring power. It disperses quite easily in liquids but tends to settle-out. TiO_2 remains suspended only in viscous liquids and semi-solid materials.

The opacifying power of TiO_2 has been utilized in conjuction with the FD&C Lakes in panning or tablet coating. In this application, the titanium dioxide and FD&C Lakes produce a system similar to a paint and fewer color coats are required than with the dyes or lakes alone. This process is the subject of two patents.[12, 13]

Titanium dioxide is also used alone in sugar syrup for use in the sub-coating of tabletted products. This color additive was permanently listed in 1966 exempt from certification and with limitations of 1% by weight in foods.

Carmine, Carminic Acid, Cochineal

There has always been a great deal of confusion regarding the differentiation between these three provisionally listed color additives. Cochineal is a red coloring material consisting of the dried bodies of the female insect Coccus Cacti. These insects grow on a specific variety of cactus which is cultivated in the Canary Islands and parts of South America. The active coloring matter is carminic acid which is believed to have the following chemical structure:

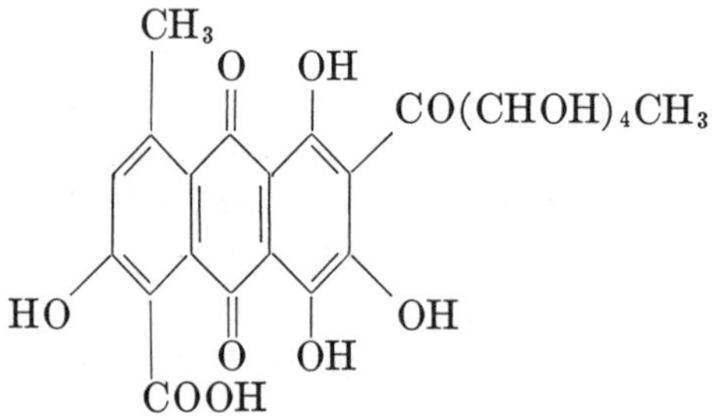

$$CH_3 \quad O \quad OH \quad CO(CHOH)_4CH_3$$

$$HO \quad O \quad OH \quad OH$$

$$COOH$$

Carmine is the aluminum lake of the carminic acid which has been isolated and solubilized by extraction. While carmine is a pigment, for many applications it is solubilized by alkali. These related color additives are quite expensive and are often in short supply. Carmine has found use in a pink color in coatings and in retorted protein products where the FD&C Colors are not stable.

The remaining uncertified color additives are not used extensively and generally are for specific food products. Many years ago there was a feeling that the listing of artificial color on the ingredient statement of a food product degraded the product in the customer's mind. This feeling has fortunately died away. With regard to the labeling of a food product containing a color additive either certified or uncertified, judicial interpretations have affirmed the use of any of the following statements:

Artificial Color
U.S. Certified Color

or the color additive itself (generally employed for uncertified color additives). The Joint FAO/WHO Expert Committee mentioned previously has evaluated the safety status of the uncertified color additives used in the United States. Table 8 lists the ADI's given to a number of uncertified color additives:

The International Food Color Situation

There is a complete lack of agreement regarding safe color additives in the markets of the world. This is due to lack of a universal protocol for pharmacological testing. Also, the situation in any country (U.S. included) is in such

TABLE 8

Maximum Acceptable Daily Intakes
for Some Uncertified Color Additives

Common Name	Maximum ADI, mg/kg
Annatto	1.25[b]
β-Apo-8'-carotenal	2.50[a]
Canthaxanthin	12.5[a]
β-Carotene	2.5[a]
Turmeric	0.5[b]
Riboflavin	0.5[a]
Titanium dioxide	Good manufacturing practice

[a] Unconditional—based on adequate toxicological investigations and on information regarding metabolic fate.
[b] Temporary—lack of available data to fully establish the safety. Additional test data required.

a constant state of flux that it is difficult to know at any one time what is approved and impossible to predict what will be suitable in the future. The listing of approved colors in any country is often governed to a great extent of nationalistic pressures and intractable (and often unscientific) opinions.

Several years ago the EEC (European Economic Community) adopted a list of approved colors. It was hoped that this would clear the way for more universality of colors. This has not been true and not much heed has been paid to this list. The EEC list is as follows:

Name	Color Index No.
FD&C Red No. 2 (Amaranth)	16185
FD&C Red No. 3 (Erythrosine)	45430
FD&C Yellow No. 6 (Sunset Yellow FCF)	15985
FD&C Yellow No. 5 (Tartrazine)	19140
FD&C Blue No. 2 (Indigotine)	73015
Carmoisine (Azorubine)	14720
Ponceau 4R (Cochineal Red A)	16255
Scarlet GN	14815
Ponceau 6R	16290
Orange GGN	15980
Fast Yellow (Acid Yellow G)	13015
Chrysoine S (Resorcine Yellow)	14270
Quinoline Yellow (WS) (D&C Yellow No. 10)	47005
Patent Blue V	42051
Indanthrene Blue RS	69800
Brilliant Black BN	28440
Black 7984	—

One of the primary difficulties mentioned above is lack of agreement on protocol of testing. One chief source of disagreement here has to do with the acceptance or rejection of the results of subcutaneous injection of the dyes into test animals. It has long been known that some dyestuffs cause sarcoma at the site of repeated injections of dye solutions. This is almost universally true of the triphenylmethane dyes: FD&C Blue No. 1, FD&C Green No. 3, FD&C Violet No. 1 (plus the U.S. delisted FD&C Greens No. 1 and No. 2). This observation is not considered relevant in the United States for colors which are taken into the body by ingestion; and the so-called Delaney clause of the Color Additives Amendments limits evidence of carcinogenicety for ingested colors to effects observed on ingestion by test animal. This is not so in most other countries, and such interpretation is unfortunate as it limits the approved blue and green colors in some countries to FD&C Blue No. 2 (Indigotine) only, which is pitifully light fugitive.

Investigation of the phenomenon of cancer formation by subcutaneous injection has been initiated by BIBRA (British Industrial Biological Research Association).[14] This work indicates that subcutaneous injection studies are questionable for determining the toxicity of ingested materials. It is hoped that these studies will pave the way for approval of the triphenylmethane dyes in other countries.

Use of Color Additives

The Certified Color Industry Committee surveyed from their sales records the use of color by the various segments of the color-consuming industries.[2] Their results are shown in Table 9. Table 10 lists the concentrations, both in range and maximum, of color used in the various categories of processed foods. Table 11 lists the various food products that often exceed the maxima shown in Table 10.

Because of the contemplated action on reduction in the use of FD&C Red No. 2, many dark-red food products would require reformulation. FD&C Red No. 40 would suffice in many color applications, but transparent products, such as beverages, gelatin desserts, and hard candy, would be difficult to color without Red 2 if a dark red or purple shade were desired. The yellowish content of FD&C Red No. 40 renders such colors a dirty or brownish cast.

Use of modern process equipment and proper water treatment has eliminated many of the problems associated with the use of color additives. These difficulties have been minimized to a great extent by increased knowledge of the colors. Table 12 lists the more common coloring problems with dyes and the causes.

Each group of food products has its own coloring techniques and methods of color applications. These will be discussed in detail.

Beverages

The difficulties involved in can corrosion with canned carbonated beverages have already been discussed. Likewise, fading due to ascorbic acid was mentioned. Ascorbic acid is often a desired ingredient in beverages either for its vitamin activity or as a flavor protector in which case it protects by scavenging the oxygen in the headspace air. In instances where ascorbic acid is used, uncertified color additives should be used. In fact, beverage use was the impetus

TABLE 9
Pounds of Primary Colors Used in Foods, Drugs, and Cosmetics

Figures represent sales for the first nine months of 1967 and do not include exports or sales to jobbers and other manufacturers.

Material	Yellow No. 5	Yellow No. 6	Red No. 2	Red No. 3	Red No. 4	Blue No. 1	Blue No. 2	Violet No. 1	Green No. 3	Orange B	TOTALS
Candy, confection	59,903	52,770	67,637	11,665	0	6,632	2,499	1,459	124	0	202,689
Beverages	78,933	181,292	282,695	1,056	0	15,800	2,375	985	301	0	563,437
Dessert powders	59,961	51,622	62,363	8,616	0	3,270	1,659	0	14	0	187,505
Cereals	52,496	35,464	15,558	1,421	0	843	99	0	0	0	105,881
Maraschino cherries	5,644	4,830	8,104	3,469	11,308	597	0	0	98	0	34,050
Pet food	101,743	23,226	67,058	1,023	0	1,473	6,764	1,278	0	0	202,565
Bakery goods	77,885	42,203	43,522	9,560	0	3,680	673	369	7	0	177,899
Ice cream, sherbet, dairy products	35,048	23,868	29,697	621	0	2,599	179	45	7	0	92,064
Sausage	6,502	99,605	36,084	4,970	0	647	0	0	0	16,890	164,698
Snack foods	18,456	11,409	3,623	766	0	305	0	2	0	0	34,561
Meat inks	15	0	12	10	0	11	0	2,223	0	0	2,271
Miscellaneous	44,841	29,134	46,219	18,200	398	5,345	1,990	1,134	1,298	0	148,559
Subtotal	541,427	555,423	662,572	61,377	11,706	41,202	16,238	7,495	1,849	16,890	1,916,179
Pharmaceutical	17,275	15,938	21,179	12,168	1,186	3,250	593	347	220	0	72,156
Cosmetics	3,125	2,148	3,417	903	630	397	30	96	27	9	10,773
TOTALS	561,827	573,509	687,168	74,448	13,522	44,849	16,861	7,938	3,096	16,890	1,999,108

Source: Certified Color Industry Committee, 1968, Guidelines for Good Manufacturing Practice: Use of Certified FD&C Colors in Food. *Food Tech.*, 22(8):14.

TABLE 10
Processed Foods in Which Certified FD&C Colors Are Used

Category	Color Concentration, ppm	
	Range	Average
Candy and confections	10–400	100
Beverages (liquid and powdered)	5–200	75
Dessert powders	5–600	140
Cereals	200–500	350
Maraschino cherries	100–400	200
Pet foods	100–400	200
Bakery goods	10–500	50
Ice cream and sherbets	10–200	30
Sausage (surface)	40–250	125
Snack foods	25–500	200
Meat-stamping inks	—	—
Miscellaneous	5–400	—

Source: Certified Color Industry Committee, 1968. Guidelines for Good Manufacturing Practice: Use of Certified FD&C Colors in Food. *Food Tech.*, 22(8):14.

TABLE 11
Food Products Reported to Contain Color in Excess of the Maxima
Listed in Table 10[a]

Category	Product	Color Concentration, maximum ppm[a]
Candy and confections	Tableted candy	500
	Circus peanuts	600
	Licorice	1000
	Black and brown lozenges and gum	1200
Dessert powders	Chocolate pudding	1000
Bakery goods	Cake decorations	1000
	Sugar wafers	800
	Chocolate cookies and cake	800
	Ice cream cones (other than yellow)	900
Ice cream and sherbets	Very dark chocolate, including ice cream bar coatings	600
Miscellaneous	Spray-dried cheese	700–2000
	Spices	1000
	Fruit rings	600
Snack foods	Cheese-flavored items	750

[a] These specific products will not always exceed maximum values as listed in Table 10, but under certain conditions the higher concentrations are necessary.

Source: Certified Color Industry Committee, 1968. Guidelines for Good Manufacturing Practice: Use of Certified FD&C Colors in Food. *Food Tech.*, 22(8):14.

for development of the β-carotene beadlet which is a water-dispersible form of this oil soluble color.[15] In instances where it is desirable to use a flavor antioxidant and certified color, use of glucose oxidase-catalase as an oxygen scavenger is recommended.

Canned, carbonated drinks and canned still drinks seem to represent two different systems. In canned carbonated beverages, the chief problem is can corrosion, and color fading is minimal. In still drinks, corrosion is usually of no significance but color fading is a problem. The presence of juice in a drink seems to lessen the color fading in still canned drinks. Much is known about canned beverages, but a great deal has been gleaned empirically and often defies scientific explanation. Glass-bottled beverages present few problems except those brought on by product abuse. Fading of colors by over-exposure to sunlight is a problem. This often produces a green colored grape soda as a result of preferential fading of the FD&C Red No. 2. Such fading, however, can be looked upon as an advantage, as it serves as an indication of product

TABLE 12
Problems With Dyes

Problem	Cause
Precipitation from color solution or colored liquid food	1. Exceeded solubility limit. 2. Insufficient solvent. 3. Chemical reaction. 4. Low temperatures, especially for concentrated color solution
Dulling effects instead of bright, pleasing shades	1. Excessive color 2. Exposure to high temperatures.
Specking and spotting during coloring of bakery and confectionery products	1. Color not completely dissolved while making a solution. 2. Employed liquid color containing sediment. 3. Attempted dispersion in an aqueous color solution in products containing excessive fat.
Fading due to light	1. Colored products not protected from sunlight.
Fading due to metals	1. Color solutions or colored products were in contact with certain metals (zinc, tin, aluminum, etc.) during dissolving, handling, or storing.
Fading due to micro-organisms	1. Color-preparing facilities not thoroughly cleaned to avoid contaminating reducing organisms.
Fading due to excessive heat	1. Processing temperature too high.
Fading due to oxidizing and reducing agents	1. Contacted color with oxidizers such as, ozone or hypochlorites or reducers such as, SO_2 and ascorbic acid.
Fading due to strong acids or alkalis	1. Presence of such strong chemicals during the coloring of certain foods.
Fading due to retorting with protein material	1. Color is unstable under these conditions.
Poor shelf life with colored canned carbonated beverages	1. Used an excessive amount of certified azo-type due.

abuse as the flavor certainly has reached a point of degradation before the color faded. Most fruit type flavored beverages contain certified color additives, while colas and root beers are colored with caramel. As mentioned before, special grades of caramel are available for beverage use. Table 13 lists color formulas and amounts for various flavored soft drinks.

The color for beverages is often contained in the flavor concentrate or compound, and color solubility is often a problem since water in many instances must be kept to a minimum to hold the oils in solution. In such instances, it is helpful to employ propylene glycol or glycerine as the flavor solvent instead of alcohol. Years ago, orange emulsion concentrates were plagued with FD&C Yellow No. 6 precipitation as a result of calcium salt formation—the calcium salt coming from the gum arabic used for the emulsion. Special grades of FD&C Yellow No. 6 are now available which eliminate this difficulty.

TABLE 13
Color and Concentration in Carbonated Beverages

Flavor	Color	Parts	Concentration, ppm
Orange	FD&C Yellow No. 6	98	54
	FD&C Red No. 2	2	
or	FD&C Yellow No. 6	96	54
	FD&C Red No. 40	4	
Cherry	FD&C Red No. 2	100	50
or	FD&C Red No. 40	100	80
Raspberry	FD&C Red No. 2	99	45
	FD&C Blue No. 1	1	
or	FD&C Red No. 40	97	45
	FD&C Violet No. 1	3	
Grape	FD&C Red No. 2	94.5	55
	FD&C Blue No. 1	4.5	
	FD&C Yellow No. 5	1.0	
or	FD&C Violet No. 1	50	30
	FD&C Yellow No. 6	50	
Strawberry	FD&C Red No. 2	80	50
	FD&C Yellow No. 6	20	
or	FD&C Red No. 40	100	40
Lime	FD&C Yellow No. 5	95	20
	FD&C Blue No. 1	5	
Lemon	FD&C Yellow No. 5	100	20
Cola	Caramel color	100	400
Root Beer	Caramel color	100	400

Bakery Products

Colors are employed in dough products, cookies, sandwich fillings, icings and coatings. Because of the high moisture content of doughs and batters, little problem of color addition exists. However, obtaining the depth of color can often be a problem. This is particularly true in dark chocolate pieces, and use of certified color alone results in excessive color use. In such goods, combinations of certified colors with uncertified color additives has proved successful. Caramel or carbon black is added with various combinations of certified colors. Caramel color has also found extensive use in coloring rye and pumpernickel bread.

Ice cream cones at one time were colored with water soluble annatto, however, in recent years there has been a switch to the use of certified colors which are less expensive and offer greater variety of shades. Cone color blends generally contain FD&C Yellow No. 5 and FD&C Yellow No. 6.

Sandwich fillings and icings represent systems which require special coloring techniques, and these techniques depend upon the composition. Formulations containing little or no moisture, sugar and fat (shortening) necessitate dye solutions in propylene glycol or glycerine or FD&C Lakes. Icings, particularly in the small bakery, are colored with paste colors which are compositions containing glycerine, propylene glycol and sugar.

Dairy Products

Nearly all ice creams and sherbets contain artificial color. Chocolate ice cream is often the exception. Annatto is used to considerable extent in vanilla ice cream, but as in the coloring of cones, certified colors (egg shades) are increasingly popular. Because of the small amounts of color used and because of convenience, liquid colors are often used. Ice cream presents very few problems of color stability except in instances where excessively high bacterial counts are present. Again, the fading of the color is an indicator of trouble.

Cheese is a product in which the certified colors are not sufficiently stable and annatto and β-carotene are the desired colorants. Likewise, margarine and butter are colored chiefly with β-carotene and oil soluble annatto. At one time it was felt the FD&C Lakes would perform satisfactorily in margarine, but differences exist in the appearance of melted margarine containing lakes. The affinity of the lakes for protein causes eggs fried in margarine colored with lakes to have a yellow appearance.

Candy

It is difficult to imagine many candies, particularly hard candy, without color. Hard candy represents a system low (1–2%) in moisture, and aqueous color solutions are added with reluctance. Paste colors and other non-aqueous plastic materials are used. The temperature at which hard candy is added to the slab is somewhat destructive to color, particularly FD&C Red No. 2. Color should be added at the lowest temperature that will permit adequate distribution. Continuous hard candy operations utilize concentrated aqueous color solutions.

Compound or summer coatings cannot be colored with aqueous color solutions. Such use results in color specking or changes in physical properties of the coating. Non-aqueous color solutions using propylene glycol or glycerine have found use for this application but lake dispersions are being used in increasing amounts. Pink coatings are much in demand, and FD&C Red No. 3 (dye) is fugitive to light while FD&C Red No. 2 produces unsatisfactory shades. FD&C Red No. 3 Lake produces attractive coatings with increased light stability over the dye.

Pet Foods

It should be remembered that by definition in the Food, Drug and Cosmetic Act a "food is anything consumed by man or animal," consequently pet foods, if colored, must contain approved color additives.

Pet foods consist of three chief types: dry extruded, wet-pack, or canned and semi-moist. The original pet food marketed was the wet-pack or canned food. Iron oxide, because of its low price, has been traditionally used in canned pet food as the colorant. In addition, the less-expensive certified color did not withstand the retorting operation in the preparation of such products. In late 1966, the provisional listing of iron oxide as a color additive for pet foods was terminated and work at that time showed that certain certified color combinations based on FD&C Red No. 3 showed good stability in processing. However, in July 1968, iron oxide was permanently listed for use in cat and dog foods at 0.25% by weight maximum. The certified colors are used extensively in dry extruded pet foods—yellow, red and brown being the most popular colors. Rates of .010 per cent to .030 per cent are generally used, and the powdered color may be added to the dry mix before extruding. A more acceptable method is to dissolve the dye in some of the water used to condition the feed before extrusion. An interesting use is made of color in the gravy type pet foods. This consists of extruded pet food to the surface of which is added a dry powder consisting of a gum, dry color (brown) and an inert material. When water is added to the food, the gum imparts viscosity to the liquid. This together with the dissolved color resembles a natural gravy. Because of the differential migration and dissolving of the individual primary colors, a wet-dry or "monoblend" brown is recommended.

The semi-moist products are generally colored with little difficulty. Red dye blends are generally used, but lakes also find application. A typical dye blend for semi-moist products is FD&C Red No. 2 (60 parts) and FD&C Yellow No. 6 (40 parts). The blend is used at .02 per cent by weight of product.

Tabletted and Pan Coated Products

Tabletting is an operation which has its roots in the pharmaceutical industry. Of late there has been a great emphasis on tabletted products in the confectionery industry. Tablets are of two main types: (1) compressed uncoated tablets, and (2) coated tablets which are also formed by compression and afterwards treated with a thin colored coating. It should be noted that there are important economic factors which apply to pharmaceutical tablets as opposed to confectionery tablets. Pharmaceutical tablets are quite expensive and are sold by numerical units. As a result, the pharmaceutical items generally demand a higher degree of elegance.

Compressed tablets are prepared by exerting a high pressure upon a material of desired particle size and distribution in a tabletting press. To obtain the desired particle size and distribution, the powdered material is first granulated. A granulation can be prepared in a number of ways, the most important being slugging and addition of a granulating liquid while mixing. In slugging, the powdered material is run through a tablet press, producing a rather crude tablet. This tablet is then ground and screened to produce the granulation. A wet type granulation is made by adding an aqueous solution of a binder (gum acacia, sugar, etc.) to the powdered material in a mixer. After agglomeration takes place, the material is dried, ground and screened. The traditional method of coloring granulations was to dissolve a dye in the granulating solution. However, wicking or chromatographic migration often occurs on drying and greater amounts of color appear on the surface of the dry material or with dye blends, the dyes stratify into multi-colored bands. Such granulations produce compressed tablets having a mottled appearance. This can be minimized, or in some instances eliminated, by use of the lakes. The lakes are also useful in preparing granulations by slugging. Here a dry mix is maintained throughout the processing and no drying is required. Also, in wet granulations, the lakes because of their insolubility do not migrate and stratify during drying to the same extent as dyes.

The coating of tablets has long been steeped in artistry, but is slowly yielding to scientific investigation. The most popular coating material is sucrose, but there are many "film coatings" which make use of various film forming gums and resins. Likewise, the coating systems may be either transparent or opaque. In the pan coating operation, a revolving pan is charged with tablets. Numerous uncolored subcoatings are applied by adding to the pan, sufficient amounts of syrup, sufficient to uniformly wet and cover the surface area of the tablets. Air is blown over the wetted tablets to speed up the evaporation of solvent. When a smooth sub-coat has been obtained, color coats are added. When dyes are used, as many as 30 or 40 coats are applied. Coatings colored with dyes are transparent and the depth of color depends on the concentration of the dye and the number of coats. With transparent coatings, tablet surface imperfections will show up as mottling. When opaque coatings are used; i.e., dyes or lakes with titanium dioxide, the shade is independent of the number of coats (after five or six), and the color depth depends upon the concentration of dye or lake and the concentration of titanium dioxide. Mottling effects are minimized by use of opaque coatings.

Though they are pigments, the FD&C Lakes are very transparent and used alone produce coatings similar to dyes. Film coatings are generally based on an organic solvent containing a gum or resin plus plasticizers. There are many different film coating systems, most of which are non-aqueous and demand the use of the FD&C Lakes. Several reviews on coatings are available describing manufacturing details.[16, 17]

Dry Mix Products

Colors for dry mixes should impart maximum color to the dry product and, if a blend, dissolve without "flashing"; i.e., without showing the individual component colors. To obtain maximum color in a dry mix, dissolved color can be added in solution. This operation, however, necessitates some sort of moisture removing operation and often dry mixes cannot tolerate moisture additions. A number of primary colors are available in *plating grade* form. They show superior coloring power when distributed throughout a dry mix.

The difficulty of "color flashing" can be eliminated by the use of wet-dry blends (monoblends). These are prepared by dissolving the colors to form a solution and drying the solution. In this fashion, a blend of FD&C Red No. 2 and FD&C Blue No. 1 will dissolve from a dry mix product as a purple rather than showing streakings of red and blue. Such blends are popular for use in gelatin desserts and puddings. There is no reason to use a plating grade color or a monoblend if the color is applied to the food product by means of a color solution.

Cake, doughnut and pancake mixes were formerly colored with oil soluble colors (FD&C Yellows No. 3 and No. 4). Good depth of color was obtained in the dry mix because the dyes were soluble in the shortening. When these oil soluble colors were delisted, the switch to water soluble colors was disappointing as they did not impart sufficient color to the dry product even though the color in the batter was quite satisfactory. Lakes help to solve the problem, as they adequately color both the dry product and the batter. In some instances, combinations of dyes and lakes have been employed for these products.

Color Specification and Quality Control

It should be remembered that color manufacturers carefully assay their product before submitting it for certification. Refusal of certification means loss of the certification fee of 15 cents per pound. Likewise, the Color Certification Laboratory of the FDA assays the material before issuing a certification certificate. Thus, it is usually unnecessary for the color user to repeat the assays dealing with the certification specifications (lead, sodium chloride, etc.). It is prudent, however, to determine that the proper primary colors have been supplied or that the blend is of the proper shade; i.e., proper composition. Visual comparison is quite helpful in checking the shade of colors and spectrophotometric analysis indicates color strength. These comparisons should be made against standard samples in similar dilute solutions. Solution concentrations suitable for both visual and spectrophotometric analysis (in 1 cm cells) are as follows for the various dyes:

FD&C Red No. 2	10 mg per liter
FD&C Yellow No. 5	10 mg per liter
FD&C Yellow No. 6	10 mg per liter
FD&C Blue No. 1	4 mg per liter
FD&C Green No. 3	4 mg per liter
FD&C Red No. 3	7 mg per liter
FD&C Violet No. 1	4 mg per liter
FD&C Blue No. 2	10 mg per liter
FD&C Red No. 40	10 mg per liter

These solutions should be buffered to avoid any shade differences due to pH.

Quality control checking of lakes is more complex. Dispersions of test and standard samples in various materials such as shellac can be prepared and drawdowns compared or test colorings of small laboratory amounts of the finished product can be made. For such comparison and for quality control of finished products, many color differences (tri-stimulus comparators) are available. These have proved only mildly successful, but in the future they should become more helpful.

Knowledge of color additive technology in the food industry has grown significantly in recent years and fewer application problems are being referred to the color manufacturers. This is a salutary occurrence; however, this same period has been a time of industry confusion regarding the regulatory status of color additives. Hopefully, this too will be overcome shortly as the result of more convincing scientific data.

References

1. Joint FAO/WHO Expert Committee on Food Additives, 1965, 1966, 1967, 1970.
2. Certified Color Industry Committee, 1968, Guidelines for Good Manufacturing Practice: Use of Certified FD&C Colors in Food. *Food Tech.*, 22(8): 14.
3. Shtenberg, A. I. and Gavrelenko, E. V., 1970, *Vop. Pitan.*, 29(2), 66–73.
4. Dean, G. E., 1966, A Shelf Life Factor in Canned Soft Drinks. *The American Soft Drink Journal*, February, pp. 36, 62.
5. Martin, L. E., 1965, Antioxidants and Sequestering Agents to Increase Product-Container Shelf Life of Carbonated Beverages. *The American Soft Drink Journal*, March.
6. Borenstein, B. and Bunnell, R. H., 1967, Carotenoids. *Advances in Food Research,* 15, p. 195.
7. Bunnell, R. H. and Borenstein, B., 1967, Canthaxanthin, A Potential New Food Color. *Food Tech.*, 21, p. 331.
8. Bauernfeind, J. C., Smith, E. G., and Bunnell, R. H., 1958, Coloring Fat-Base Foods with β-Carotene. *Food Tech.*, 12, pp. 527–535.
9. Bunnell, R. H., Driscoll, W. and Bauernfeind, J. C., 1958, Coloring Water-Base Foods with β-Carotene. *Food Tech.*, 12, pp. 536–541.
10. Peck, F. W., 1955, Caramel—Its Properties and Its Uses. *Food Engineering*, March, p. 94.
11. Linner, R. T., 1965, Caramel Coloring—Production, Composition and Functionality. *Baker's Digest*, April.
12. U.S. Patent 2,925,365.
13. U.S. Patent 3,054,724.
14. Grasso, P., Golberg, L., 1966, Subcutaneous Sarcoma as an Index of Carcinogenic Potency. *Fd. Cosmet. Toxicol.*, 4, pp. 297–320.
15. Bauernfeind, J. C., Osadea, M., and Bunnell, R. H., 1962, β-Carotene Color and Nutrient for Juices and Beverages. *Food Tech.*, 16, pp. 101–107.
16. Clarkson, R., 1951, Tablet Coating. *Drug & Cosmetic Industry*, New York, New York.
17. Little, A., Mitchell, K. A., 1963, *Tablet Making*. 2nd Edition, The Northern Publishing Co., Ltd., Liverpool, England.

CHAPTER 15

Phosphates in Food Processing

R. H. Ellinger, Ph.D.

Manager, Regulatory Compliance
Kraft Foods Company
Division of Kraftco Corporation
Chicago, Illinois

Introduction

Phosphorus is an essential element to all life, animal and vegetable, single-celled as well as multi-celled organisms. With few exceptions it is utilized only as the phosphate anion. There is no living organism known that can synthesize the phosphate anion.[1] The element, therefore, is absorbed through the food supply as the phosphate anion by all living organisms.[1,2] Very few natural compounds in living organisms contain phosphorus in any form other than that of the phosphate anion.[3]

The ability of the phosphate anion to form polymers is vital to many life processes. The transfer of one phosphate unit from the tripolyphosphate portion of ATP (adenosine triphosphate) is the means of energy transfer in most essential processes. These facts are so well established in the scientific literature that it is unnecessary to review them here.

The polyphosphates have played a vital role in the origin of life.[4-6] Studies have shown that polyphosphates exist in, and are vital to, living tissues, including single-celled and higher forms of life.[7-18] It has also been demonstrated repeatedly that enzymatic systems in present living organisms are capable of synthesizing and hydrolyzing polyphosphates, even those with a very long chain.[8,17-19] The vital role of the phosphate and pyrophosphate anions in metabolism has been completely accepted, as demonstrated, for example, in the chart of biological pathways in the *Handbook of Biochemistry*.[20]

Phosphorus in the form of the phosphate anion is a constituent of nearly every type of food consumed by living organisms. Sherman[21] has published a compilation of the phosphorus content of many natural foods, some of which are shown in Table 1.

TABLE 1
Phosphorus Content of the Edible Portion of Some Typical Foods

Type of Food	Phosphorus Content, mg/100 g	Type of Food	Phosphorus Content, mg/100 g
Roots, Tubers, and Bulbs		**Cereal Products**	
Beets	41 ± 1.1	Corn, sweet	120 ± 1.8
Carrots	40 ± 1.2	Oats (oatmeal)	395 ± 14.9
Potatoes	56 ± 0.8	Wheat flour	101 ± 2.1
Onions	33 ± 1.1	White bread	97 ± 2.0
Edible Leaves and Buds		**Animal Products**	
Broccoli	72 ± 2.3	Beef, lean	204 ± 2.5
Cabbage	30 ± 0.6	Cheese, cheddar	524 ± 18.0
Lettuce	28 ± 0.9	Eggs	224 ± 1.4
Fruits and Juices		Milk	93 ± 0.3
Apples	11 ± 0.17		
Orange juice	17 ± 0.4		
Oranges	21 ± 0.5		
Peaches	18 ± 0.55		

Source: Table based on data in Sherman, H. C., 1947. *Calcium and Phosphorus in Foods and Nutrition*, Columbia University Press, pp. 91–103.

As the volume of literature describing the applications of phosphates in food processing is enormous, only a small portion can be reviewed here. Further information can be obtained from a study of the numerous reviews of phosphate applications in food processing,[22-30] many of which were used in the preparation of this chapter.

Nomenclature, Classification, and Chemistry of Food Phosphates

A study of the food-science literature demonstrates that the nomenclature and classification of the phosphates, particularly the polyphosphates, are confusing to many food scientists. It is often impossible to determine from the description of a phosphate which specific one was used in a study. Many studies also seem to consider the polyphosphates as a class of pure compounds. Molecular weights are often assigned to these compounds, such as to sodium hexametaphosphate; yet the compounds are such mixtures of polymeric phosphates that a molecular weight is meaningless.

A thorough discussion of the nomenclature, structure, classification, and chemistry of the food phosphates is beyond the scope of this chapter. Anyone desiring further information on phosphate chemistry can find numerous references in the literature. Excellent reviews for food scientists have recently been published by Van Wazer[29, 30] and by Bell.[29]

A detailed study of some of the chemical characteristics of the various phosphates specifically used in food applications has been published by Ellinger[28]; more detailed discussions of the various characteristics of all phosphates have been published by Van Wazer[30] and by others.[29-38]

Table 2 summarizes the nomenclature, structures, common names, formulas, pH values of solutions or slurries, and solubilities of the commonly used food phosphates. Further discussions of important characteristics of individual phosphates will appear in subsequent paragraphs; the reader also is referred to reviews by Van Wazer[29] and by Bell.[29]

The Orthophosphates

The orthophosphate anion is the simplest structure and the basic unit for all phosphates. Its structure is as follows:

$$^-O-\overset{\overset{\displaystyle O}{\|}}{\underset{\underset{\displaystyle O_-}{|}}{P}}-O^-$$

As the structure indicates, the anion is tribasic, and its three valences can be satisfied by hydrogen, metal ions, or combinations of the two. It also can form straight-chain and cyclic polymers, as will be discussed later.

Orthophosphoric Acid   Orthophosphoric acid (H_3PO_4), commonly called phosphoric acid, is a natural constituent of many fruits and their juices. It is a strong acid, classified as tribasic because it has three replaceable hydrogens. It has been found to be an excellent acidifying agent in food applications. It is available commercially as a viscous, colorless, syrupy liquid, and it is usually sold according to its P_2O_5 (phosphoric anhydride) content. Commercial phosphoric acid, containing 68.8% or less of P_2O_5, is composed only of the monomeric phosphoric acid, H_3PO_4. Those commercial compositions containing more than 68.8% P_2O_5 contain increasing quantities of the polymers of phosphoric acid. Pyrophosphoric acid ($H_4P_2O_7$) and higher polymers increase

TABLE 2

Nomenclature, Structure, and Some Characteristics of Phosphate Acids and Salts Commonly Used in Foods[a]

Class of Phosphate	Number P Atoms per Molecule	General Structure[b]	Common Names	Formulas[c]	pH[d]	Solubility[e]
Orthophosphates (monomer)	One	$\begin{array}{c} O \\ \parallel \\ MO-P-OM \\ \mid \\ OM \end{array}$	Phosphoric acid	H_3PO_4	2.0–2.2	High
			Monoammonium phosphate	$NH_4H_2PO_4$	4.5	28
			Diammonium phosphate	$(NH_4)_2HPO_4$	8.0	41
			Monocalcium phosphate	$Ca(H_2PO_4)_2$	4.5	Disprop[f]
			Dicalcium phosphate	$CaHPO_4$	7.5	Insol
			Tricalcium phosphate	$Ca_5(PO_4)_3(OH)$[g]	7.2	Insol
			Ferric orthophosphate	$FePO_4$[g]	3.8–4.4	Insol
			Monopotassium phosphate	KH_2PO_4	4.4	20
			Dipotassium phosphate	K_2HPO_4	8.8	63
			Tripotassium phosphate	K_3PO_4	11.9	51
			Hemisodium phosphate	$NaH_2PO_4 \cdot H_3PO_4$	2.2	High
			Monosodium phosphate	NaH_2PO_4	4.4	48
			Disodium phosphate	Na_2HPO_4	8.8	11
			Trisodium phosphate	Na_3PO_4	11.8	13
			Sodium aluminum phosphate, acidic	$NaAl_3H_{14}(PO_4)_8$	2.4–2.5	Slight
			Sodium aluminum phosphate, acidic	$Na_3Al_2H_{15}(PO_4)_8$	2.6	Slight
			Sodium aluminum phosphate, alkaline	$Na_{15}Al_{2.8}(PO_4)_8$[g]	9.2–9.4	Slight

TABLE 2

Nomenclature, Structure, and Some Characteristics of Phosphate Acids and Salts Commonly Used in Foods[a] (Continued)

Class of Phosphate	Number P Atoms per Molecule	General Structure[b]	Common Names	Formulas[c]	pH[d]	Solubility[e]
	Two (dimer)	$\begin{array}{cc} O & O \\ \parallel & \parallel \\ MO-P-O-P-OM \\ \mid \quad\mid \\ OM \quad OM \end{array}$	Pyrophosphoric acid	$H_4P_2O_7$	V. acid	High
			Calcium pyrophosphate	$Ca_2P_2O_7$	6.0	Insol
			Tetrapotassium pyro-phosphate	$K_4P_2O_7$	10.2	61
			Ferric pyrophosphate	$Fe_4(P_2O_7)_3{}^g$	—	Insol
			Sodium acid pyrophosphate	$Na_2H_2P_2O_7$	4.2	13
			Tetrasodium pyrophosphate	$Na_4P_2O_7$	10.2	6
			Sodium iron pyrophosphate	$Na_8Fe_4(P_2O_7)_5{}^g$	7.0–7.7	Insol
Linear or straight-chain polyphosphates (polymer)	Three (trimer)	$\begin{array}{ccc} O & O & O \\ \parallel & \parallel & \parallel \\ MO-P-O-P-O-P-OM \\ \mid \quad\mid \quad\mid \\ OM \quad OM \quad OM \end{array}$	Tripolyphosphoric acid	$H_5P_3O_{10}$	V. acid	High
			Potassium tripolyphosphate	$K_5P_3O_{10}$	9.8	65
			Sodium tripolyphosphate	$Na_5P_3O_{10}$	9.8	13+
			Polyphosphoric acids	$(HPO_3)_n$	V. acid	High
			Potassium metaphosphates (Kurrol's salt)	$(KPO_3)_n$ $n = 400\text{–}20\,000$	4–8	Insol[h]
			Sodium tetrapolyphosphate	$(NaPO_3)_n$ $n = 4\text{–}10$	7–8	High
	$4\text{–}10^5$	$\begin{array}{c} O \\ \parallel \\ MO-P-\left(O-P-\right)_n O-P-OM \\ \mid \quad\quad \mid \quad\quad \mid \\ OM \quad\; OM \quad\; OM \\ M_{(n+2)}P_nO_{(3n+1)} \end{array}$	Sodium hexametaphos-phate (Graham's salt)	$(NaPO_3)_n$ $n = 10\text{–}15$	7.0	High
			Soluble sodium metaphos-phate (Graham's salt)	$(NaPO_3)_n$ $n = 50\text{–}100$	6.2	High
			Insoluble sodium meta-phosphate (Kurrol's salt)	$(NaPO_3)_n$ $n = 100\text{–}500$	5.5	Insol[k]

TABLE 2

Nomenclature, Structure, and Some Characteristics of Phosphate Acids and Salts Commonly Used in Foods[a] (Continued)

Class of Phosphate	Number P Atoms per Molecule	General Structure[b]	Common Names	Formulas[c]	pH[d]	Solubility[e]
Cyclic polyphosphates	Three	(ring structure)	Sodium trimetaphosphate	$(NaPO_3)_3$	6.7	23
	Four	(ring structure)	Sodium tetrametaphos-phate	$(NaPO_3)_4$	6.2	18

[a] The information in this table was compiled from references 22, 30–32, 35, and 36.

[b] The letter M in these structures can be hydrogen or a metal ion in the compounds used as food additives.

[c] Only the formulas for the anhydrous compounds are given. The hydrates that are important as food additives are described in subsequent sections.

[d] The pH of soluble phosphates was determined in a 1% solution and in 10–50% slurries of insoluble compounds.

[e] Solubility values represent g/100 g of saturated solutions at 25°C.

[f] Monocalcium phosphate dissolves incongruently, or disproportionates to form insoluble dicalcium phosphate and phosphoric acid. However, it is commonly called soluble calcium phosphate, as it appears to be very soluble.

[g] These formulas are the molecular ratios shown by analysis. The iron compounds also contain water of hydration.

[h] Some long-chain potassium polyphosphates are made soluble by the presence of sodium ions.

[k] Some long-chain sodium polyphosphates are made soluble by the presence of potassium ions.

as the concentration of P_2O_5 continues to increase. This is shown in Table 3. It is important to remember that these polymers begin to appear at the higher concentrations of the anhydride. By diluting any of the higher concentrations to a lower concentration, the amount of polymers is decreased until one polymer finally has only the monomer present at approximately 66.8% P_2O_5. It is, therefore, the ratio of $H_2O : P_2O_5$ that determines the distribution of different lengths of polymers in the phosphoric acid; this composition will be constant, regardless of how the acid has been prepared or of its previous history.[30] Table 3 has been developed by Huhti and Gartaganis[34] through the analysis of various phosphoric acid compositions by a paper-chromatographic method developed in 1956. Similar tables, which are available from any supplier of phosphoric acid, are based on this work. Several salts of H_3PO_4 are useful in food applications. These uses are shown in the orthophosphate section of Table 2 and are reviewed by Ellinger,[28] Van Wazer,[29,30] and Bell.[29]

Ammonium Orthophosphates Two ammonium-orthophosphate salts are used in a few foods. These are monoammonium phosphate, $NH_4H_2PO_4$, and diammonium phosphate, $(NH_4)_2HPO_4$; both are commercially available as their anhydrous salts. Their most common food applications are as sources of nitrogen and phosphorus in yeast foods, although they may often replace the alkali metal phosphate salts in many of their applications.

TABLE 3
Composition of Polyphosphoric Acids

Composition, wt % P_2O_5	Per cent of Total Phosphorus as									
	Ortho-	Pyro-	Tri-	Tetra-	Penta-	Hexa-	Hepta-	Octa-	Nona-	"Hypoly-"
68.80	100.00	Trace								
69.81	97.85	2.15								
70.62	95.22	4.78								
72.04	89.91	10.09								
72.44	87.28	12.72								
73.43	76.69	23.31								
74.26	67.78	29.54	2.67							
75.14	55.81	38.88	5.31							
75.97	48.93	41.76	8.23	1.08						
77.12	39.86	46.70	11.16	2.28						
78.02	26.91	49.30	16.85	5.33	1.60					
78.52	24.43	48.29	18.27	6.75	2.26					
79.45	16.73	43.29	22.09	10.69	4.48	1.92	0.80			
80.51	13.46	35.00	24.98	13.99	6.58	3.14	2.84			
81.60	8.06	27.01	22.28	16.99	11.00	5.78	3.72	2.31	1.55	1.28
82.57	5.10	19.91	16.43	16.01	12.64	8.89	6.41	4.11	3.51	6.99
83.48	4.95	16.94	15.82	15.91	12.46	9.71	6.77	5.04	2.99	9.42
84.20	3.63	10.60	11.63	13.05	12.17	8.19	5.92	4.91	20.16	
84.95	2.32	6.97	7.74	11.00	10.45	9.62	8.62	7.85	6.03	29.41
86.26	1.54	2.97	3.31	5.16	5.32	5.54	3.51	3.30	3.30	66.03

Note: These figures are given to two decimal places for further computation purposes, but the precision may not be better than 1% total phosphorus in some cases.

Sources: Bell, R. N., 1948, *Ind. Eng. Chem. 40*:1464; and Huhti, A. L., and Gartaganis, P. A., 1956, *Can. J. Chem. 34*:785.

Calcium Orthophosphates   Several calcium salts of H_3PO_4 are commercially available, four of which are commonly used in food applications.

Anhydrous Monocalcium Phosphate   Monobasic calcium orthophosphate exists as the anhydrous and monohydrate salts. Anhydrous monocalcium phosphate (AMCP)[a] is used in large quantities as a leavening acid; it is generally given the formula $Ca(H_2PO_4)_2$. It is highly soluble and very hygroscopic in its normal crystalline form, and it can be manufactured in this form. However, it would be useless in leavening applications because of its hygroscopicity. The crystals of AMCP, therefore, are coated with a mixture of potassium, aluminum, calcium, and magnesium metaphosphate during the manufacturing process. Upon contact with water, the coated AMCP does not immediately begin to dissolve as would its uncoated form. Coated AMCP, therefore, becomes a delayed-reaction leavening acid. More of the characteristics of this compound will be described in the section on leavening acids.

Monocalcium Phosphate Monohydrate   The second form of the monobasic calcium salt of H_3PO_4 is monocalcium phosphate monohydrate (MCP·H_2O); its chemical formula is usually shown as $Ca(H_2PO_4)_2 \cdot H_2O$. This compound is also a very soluble calcium salt of H_3PO_4. It is a common acidifying agent in many food applications, as will be discussed in subsequent sections.

Although AMCP and MCP·H_2O are soluble, neither completely dissolves in water. When MCP·H_2O is added to water, the following reaction is believed to occur:

$$Ca(H_2PO_4)_2 \cdot H_2O + xH_2O = CaHPO_4 + H_3PO_4 + (x + 1)H_2O$$

This is an example of a compound that disproportionates or is incongruently soluble, as the MCP·H_2O forms DCP and H_3PO_4. The DCP reacts in the same manner as a leavening acid, as described in the following paragraph.

Dicalcium Phosphate Dihydrate   Dicalcium phosphate (DCP), $CaHPO_4$, also can be prepared in an anhydrous and a hydrated form. Only the dihydrate, $CaHPO_4 \cdot 2H_2O$, is used in food applications. As it is quite insoluble in water at temperatures below 140°F (60°C), it is commonly used when it is desirable to have acidic characteristics released only at temperatures of 140°F or above, as in a delayed leavening reaction until the dough or batter is baking in the oven. This, too, will be further discussed in a subsequent section under leavening reactions.

Tricalcium Phosphate   The fully neutralized calcium salt of H_3PO_4, commonly called tricalcium phosphate (TCP), is very insoluble and has a variable composition. Its formula, based on x-ray patterns, is usually written as $Ca_5(OH)(PO_4)_3$, a hydroxyl apatite. It can be produced in a very fine powdered form, and it is often used as an adsorbent for liquids, as a drying agent, and as a flow agent for powdered foods.

Ferric Orthophosphate   Small quantities of ferric orthophosphate are used as iron-enrichment compounds in food applications. This is a highly complex salt whose formula is usually written as $FePO_4 \cdot xH_2O$, derived from the molecular ratios determined by chemical analysis. Although it is insoluble in water, it is soluble to varying degrees in dilute hydrochloric-acid solutions, such as those in the stomach. The degree of solubility can be controlled by the method of manufacture.

Potassium Orthophosphates   Potassium salts of H_3PO_4 are used to some

---

[a] The abbreviations in parenthesis following the name of a food phosphate will be used to designate that phosphate in most subsequent references to it in this chapter.

extent in foods, but only in applications where a characteristic of the less expensive sodium salts precludes their use. A characteristic of the sodium salts most often requiring replacement is their highly astringent flavor. The potassium phosphates are usually less astringent than their sodium counterparts. Chemical characteristics of the potassium phosphates are identical to those of the sodium phosphates, making them interchangeable wherever their cost is worthwhile. Three potassium phosphates are listed in Table 2; these are usually used as their anhydrous salts.

Sodium Orthophosphates   Mono-, di-, and trisodium salts of H_3PO_4 are commonly used in food applications. They can exist as anhydrous as well as hydrated salts. There are ten different crystalline sodium orthophosphate salts commercially available for uses in food processing, as shown in Table 4. Any of them can be used, as long as one makes adjustments in formulation to account for the amount of water of hydration in the phosphate.

Monosodium orthophosphate (MSP) is mildly acidic, as shown in Table 2, and is most frequently used as a mild acidulant. *Disodium orthophosphate* (DSP) is mildly alkaline, while *trisodium phosphate* (TSP) is a rather strong alkaline compound. Both are used to increase the pH of a food product. In addition to being able to adjust pH, sodium orthophosphates are also commonly used to stabilize the optimum pH in a food system. Combinations of MSP and DSP usually are used in buffering systems.

TABLE 4
Sodium Ortho- and Polyphosphates Used in Foods[a]

Class	Compound	Names
Orthophosphate	[b]NaH_2PO_4	Acid sodium phosphate Monosodium dihydrogen monophosphate[c] Monosodium monophosphate Monosodium orthophosphate Monosodium phosphate[d] Primary sodium phosphate Sodium biphosphate Sodium dihydrogen phosphate Sodium phosphate monobasic
	[e]Na_2HPO_4	Disodium hydrogen phosphate Disodium monohydrogen monophosphate[c] Disodium monophosphate Disodium orthophosphate Disodium phosphate[d] Secondary sodium phosphate Sodium phosphate dibasic
	[f]Na_3PO_4	Sodium phosphate tribasic Trisodium monophosphate[c] Trisodium orthophosphate Trisodium phosphate[d]

TABLE 4
Sodium Ortho- and Polyphosphates Used in Foods[a] (Continued)

Class	Compound	Names
Pyrophosphate	$Na_2H_2P_2O_7$	Disodium dihydrogen diphosphate[c]
		Disodium dihydrogen pyrophosphate
		Disodium diphosphate
		Disodium pyrophosphate
		Sodium acid pyrophosphate[d]
	[g]$Na_4P_2O_7$	Sodium pyrophosphate
		Tetrasodium diphosphate[c]
		Tetrasodium pyrophosphate[d]
Tripolyphosphate	$Na_5P_3C_{10}$	Pentasodium triphosphate[c]
		Sodium tripolyphosphate[d]
		Tripolyphosphate
Straight-chain polyphosphate	$(NaPO_3)_n$	Glassy sodium phosphate
		Insoluble sodium metaphosphate[h]
		Graham's salt
		Sodium hexametaphosphate[i]
		Sodium Kurrol's salt[h]
		Sodium metaphosphate[j]
		Sodium polyphosphate[k]
		Sodium tetraphosphate[l]
Cyclic polyphosphate	$(NaPO_3)_3$	Sodium trimetaphosphate
	[m]$(NaPO_3)_4$	Sodium tetrametaphosphate

[a] Information accumulated from references 28–30, 32, and 35–38.
[b] NaH_2PO_4 also occurs as a monohydrate, $NaH_2PO_4 \cdot H_2O$, and a dihydrate, $NaH_2PO_4 \cdot 2H_2O$.
[c] The name considered most correct.
[d] Common name for commercial product in the United States.
[e] Na_2HPO_4 also occurs as a dihydrate, $Na_2HPO_4 \cdot 2H_2O$, a heptahydrate, $Na_2HPO_4 \cdot 7H_2O$, and a dodecahydrate, $Na_2HPO_4 \cdot 12H_2O$.
[f] Na_3PO_4 also occurs as a monohydrate, $Na_3PO_4 \cdot H_2O$, and a dodecahydrate, $Na_3PO_4 \cdot 12H_2O$.
[g] $Na_4P_2O_7$ also occurs as a decahydrate, $Na_4P_2O_7 \cdot 10H_2O$.
[h] Common name for insoluble, long-chain sodium polyphosphates.
[i] Common name for soluble sodium polyphosphates with $n = 10$–15.
[j] Common name for soluble sodium polyphosphates with $n = 20$–100.
[k] Straight-chain sodium polyphosphates would most accurately be named with the average chain length or the Na_2O/P_2O_5 ratio added; e.g., sodium polyphosphate, $n = 10$–15, or sodium polyphosphate, $Na_2O/P_2O_5 = 1.1$.
[l] Common name for soluble sodium polyphosphates with $n = 4$–10.
[m] $(NaPO_3)_4$ also occurs as a quadrahydrate, $(NaPO_3)_4 \cdot 4H_2O$.

Some food applications have been reported for a double sodium phosphate salt known as *hemisodium orthophosphate*. Its formula is commonly shown as a complex of MSP and H_3PO_4, as shown in Table 2. It has, however, also been reported to be a double salt of MSP monohydrate and polyphosphoric acid, $(HPO_3)_n \cdot NaH_2PO_4 \cdot H_2O$.[39]

Three *sodium aluminum orthophosphates* (SALP's) have recently been approved as food ingredients. Two are acidic salts commonly used as leavening acids; their formulas are shown in Table 2. The third is an alkaline salt mostly used in cheese processing; its formula also is shown in Table 2. All three have the advantage of delayed release of acidity or alkalinity.

The Pyrophosphates

The simplest of the condensed or polymeric phosphate anions is that of pyrophosphate, also called diphosphate in European literature. Its anion contains two phosphorus atoms linked through a shared oxygen, as follows:

$$
\begin{array}{ccc}
& O & & O & \\
& \parallel & & \parallel & \\
{}^{-}O{-}P & {-}O{-} & P{-}O^{-} \\
& \mid & & \mid & \\
& {}_{-}O & & {}_{-}O &
\end{array}
$$

Pyrophosphoric Acid   Pyrophosphoric acid is the simplest of the polymeric phosphoric acids, and it is the only one that can be crystallized. As shown in Table 3, it begins to appear in phosphoric-acid compositions containing $>68.8\%$ P_2O_5. Either of two very pure forms of crystalline pyrophosphoric acid can be prepared by seeding a concentrated solution of phosphoric acid, cooled to about $10°C$ below its melting point, with crystals of the desired form. Both crystalline forms are extremely hygroscopic. If melted or dissolved in water, the pyrophosphoric acid immediately reverts to an equilibrium mixture of ortho- and polymeric phosphoric acids represented by the P_2O_5 content of the solution, as shown in Table 3.[30, 32] The difficulty of maintaining pyrophosphoric acid in its pure crystalline form has prevented any significant application of the acid in foods.

Calcium Pyrophosphates   A few references in the food-science literature and several patents have proposed calcium acid pyrophosphate, $CaH_2P_2O_7$, as a leavening acid. However, no commercial food applications for this salt are known. The dibasic calcium pyrophosphate, $Ca_2P_2O_7$, is not applicable in food, although it is used as an insoluble crystalline polishing agent in some toothpastes.

Ferric Pyrophosphate   Another compound sometimes used as an iron-enrichment supplement is ferric pyrophosphate, $Fe_4(P_2O_7)_3$. Because it is highly insoluble in water, it is often used in foods containing fat, which might become rancid if it comes into contact with more soluble iron-enrichment compounds. Ferric pyrophosphate is soluble in dilute hydrochloric acid, such as exists in the stomach.

Potassium Pyrophosphate   Only limited quantities of the anhydrous tetra-potassium pyrophosphate (TKPP) are used in foods as this salt is completely interchangeable with the less costly sodium salt.

Sodium Pyrophosphates   Although there are four theoretically possible sodium salts of pyrophosphoric acid, only two are commonly available and used in foods. One is the acidic sodium acid pyrophosphate (SAPP), $Na_2H_2P_2O_7$, a common food acidulant. Variations in manufacturing processes and in the use of certain additives allow the formation of SAPP crystals, which have varying rates of solubility. These compounds, which are most useful in leavening reactions, will be discussed fully in the section on such applications. The second salt used in food processing is the so-called "neutral" salt, in which all of the hydrogens of pyrophosphoric acid have been replaced with sodium to form tetrasodium pyrophosphate (TSPP), $Na_4P_2O_7$. Two major characteristics of this salt are useful in food processing—its high alkalinity and its ability to complex or precipitate alkaline-earth, and heavy metal ions.

A complex sodium iron (or ferric) pyrophosphate (SIP), $Na_8Fe_4(P_2O_7)_5 \cdot xH_2O$ (formula derived from the molecular ratios obtained by chemical analysis), also is used to some extent as an iron-enrichment compound in food formulations. Although it is insoluble in water, it is soluble in dilute hydrochloric-acid solutions of concentrations equivalent to those in the stomach.

The Tripolyphosphates

The addition of another orthophosphate unit to the pyrophosphate anion forms the next higher straight-chain polymer known as *tripolyphosphate*, often called triphosphate in European literature. The structure is as follows:

$$^-O-\overset{\overset{\textstyle O}{\|}}{\underset{\underset{\textstyle O}{|}}{P}}-O-\overset{\overset{\textstyle O}{\|}}{\underset{\underset{\textstyle O}{|}}{P}}-O-\overset{\overset{\textstyle O}{\|}}{\underset{\underset{\textstyle O}{|}}{P}}-O^-$$

Tripolyphosphoric Acid   Strong polyphosphoric acids containing 74% or more P_2O_5, as shown in Table 3, contain varying amounts of *tripolyphosphoric acid*. Although this acid cannot be crystallized, it can be prepared from its salts if special ion-exchange techniques are used. It has a very short life, as it quickly begins to revert to the equilibrium mixture of ortho- and polyphosphoric acids to be expected from the P_2O_5 concentration of the solution involved.[30,32] If the P_2O_5 content falls below 74%, any tripolyphosphoric acid in the solution hydrolyzes to a mixture of ortho- and pyrophosphoric acids.

Potassium Tripolyphosphate   Only one of the theoretically possible potassium salts of tripolyphosphoric acid is utilized in food applications in which its cost can be justified. This salt is pentapotassium pyrophosphate (KTP), $K_5P_3O_{10}$. It is completely interchangeable with its sodium equivalent and is only used in products where less astringent flavor is desirable.

Sodium Tripolyphosphate   As with potassium tripolyphosphate, only the fully neutralized salt sodium tripolyphosphate (STP), $Na_5P_3O_{10}$, is used in foods. It is less soluble and produces more astringent flavors than its potassium counterpart. However, since it is usually necessary to use low quantities of the salt, these characteristics are seldom a problem in food applications. Variations in the method of manufacture can result in either of two crystalline forms, known as Form I and Form II.[30,32] Form II is preferred in food applications, as it is much easier to dissolve than Form I, which has a great tendency to cake or to form lumps if the water to which it is added is not properly agitated.

The Longer, Straight-chain Polyphosphates

It is possible to produce a great variety of condensed phosphates having more than three phosphorus atoms linked to each other by shared oxygens, as shown in the following general formula for the polyphosphate anion:

$$^-O-\overset{\overset{\textstyle O}{\|}}{\underset{\underset{\textstyle O}{|}}{P}}-O\left(-\overset{\overset{\textstyle O}{\|}}{\underset{\underset{\textstyle O}{|}}{P}}-O\right)_n-\overset{\overset{\textstyle O}{\|}}{\underset{\underset{\textstyle O}{|}}{P}}-O^-$$

These compounds exist as glassy, amorphous, or fibrous materials. Because they are mixtures of a number of polyphosphates of varying chain lengths, they can never be crystallized and, therefore, are frequently called glassy or amorphous phosphates.

Polyphosphoric Acids Table 3 shows that the concentration of polyphosphoric acids with four or more phosphate units increases as the P_2O_5 content rises above 75.5%. Commercial preparations containing 82–85% P_2O_5 are available and are usually designated as polyphosphoric acids. All are equilibrium concentrations equivalent to those shown for the P_2O_5 content in Table 3. Preparations of relatively pure polyphosphoric acids having between four and nine phosphate units have been made by highly involved methods, but these are only made in sufficient quantities for laboratory studies.[30,32-34]

Although the equilibrium distribution of polyphosphoric acids in any commercial preparation can be determined by analyzing its P_2O_5 content and by referring to a table similar to Table 3, it is also common to refer to the average chain length of the preparation designated by the symbol $\bar{n}$. The average chain length is obtained by a titration procedure developed by Van Wazer[40] according to the following equation:

$$\bar{n} = \frac{2 \text{ (ml of base to titrate to end point near pH 4.2)}}{\text{ml of base to titrate from pH 4.2 to end point near pH 9}}$$

The polyphosphoric acids will form rigid glasses but will never crystallize when they are prepared under laboratory conditions in moisture-free, sealed tubes. The commercial polyphosphoric acids are clear, colorless, viscous, hygroscopic liquids. Although they are highly soluble in water, they quickly hydrolyze upon dilution to lower-molecular-weight, shorter-chain-length phosphoric acids according to the equilibrium mixtures shown in Table 3. If diluted with the necessary amount of water, only orthophosphoric acid will remain.

Potassium Polyphosphates The heating of MKP above 150°C causes the MKP to lose water and polymerize to form potassium polyphosphate compositions of varying average chain lengths. The $\bar{n}$ of any particular preparation depends on the heat treatment and the degree of dehydration.[41,42] Since these compounds were first prepared by Kurrol just before the turn of the century, they are commonly referred to as Kurrol's salts.[30,32]

Potassium polyphosphate forms slender, needle-like crystals that overlap each other to form a fibrous structure much like asbestos. The literature often refers to them as potassium metaphosphates or potassium polymetaphosphates. They are nearly insoluble in water by themselves but readily dissolve to form viscous solutions in the presence of a soluble ammonium or alkali metal salt other than one having the potassium cation. As average chain lengths for these salts cannot be determined by the same method used for polyphosphoric acids,[40] they, therefore, are distinguished from each other by average molecular weights. Average molecular weights have been shown to range between 50 000 and 3 million, indicating that the chain lengths range from 400 to 20 000 phosphorus atoms per chain.[32]

Sodium Polyphosphates The sodium polyphosphates were discovered by Graham in the early 1800's; he named them metaphosphates because he thought they were cyclic polyphosphates. It has been impossible since then to obtain complete agreement to change this term to the more preferable designation of sodium polyphosphates. Literature references, often concerning the same compound, use several names interchangeably for these soluble, amorphous, or glassy sodium polyphosphates, as follows:

> Graham's salt
> glassy sodium phosphate
> sodium hexametaphosphate
> sodium metaphosphate
> sodium polyphosphate

Two commercial names also are used frequently in the food-science literature for these compounds: Calgon® (those having $\bar{n} = 10$–20) and Quadrafos® (usually refers to those with $\bar{n} = 4$–8). As shown in Table 5, the chain-length distribution of sodium polyphosphates with $\bar{n}$ of 2.0–9.0 indicates that all are composed of mixtures of polyphosphates.[41, 42]

The food-science literature frequently refers to "reagent-grade polyphosphates," such as sodium hexametaphosphate sold by some of the chemical-supply houses. The analysis by paper chromatography[43] of several sodium hexametaphosphate preparations labeled "reagent grade" has demonstrated that these are typical commercial mixtures of varying chain lengths having the equilibrium composition shown in Table 5 for the P_2O_5 content. It is, therefore, incorrect to designate them as chemically pure or reagent-grade chemicals, nor is it correct to assign specific molecular weights to them. Any investigator wishing to use these compounds and wishing to know their specific compositions should obtain such from one of the manufacturers; he also should request a complete analysis of the specific preparation sent to him.

The food-science literature frequently mentions sodium tetrapolyphosphate, $Na_6P_4O_{13}$, also known as Quadrafos®. The preparation described usually contains relatively short-chain sodium polyphosphates having a $\bar{n}$ within the range of 4–8; it would be expected to have a P_2O_5 content of 57–60%, as shown in Table 5.

Sodium hexametaphosphate (SHMP), whose formula is often shown as $(NaPO_4)_6$ is frequently mentioned as a food ingredient in the food-science literature. This name is also applied to any soluble, glassy sodium polyphosphate. The common commercial sodium polyphosphate available today, designated as sodium hexametaphosphate, has an average chain length ranging between 10–15 phosphate units. Careful control of manufacturing processes can consistently produce a product with $\bar{n}$ of 11–13. Freshly prepared SHMP compounds can be shown by paper chromatography to contain few if any polyphosphates with chain lengths below 4 phosphate units. However, like all polyphosphate glasses, they gradually "vitrify," or hydrolyze, with time.[28-30] Increasing quantities of ortho-, pyro-, and tripolyphosphates appear as the product vitrifies.[38]

There are also some food applications for another type of sodium polyphosphate usually known as insoluble metaphosphate (IMP), or Maddrell's salt. This salt is so insoluble in water that it is often used as a dental polishing agent, but it can be dissolved with difficulty in solutions of ammonium and alkali-metal salts other than the sodium ion.

There also is a sodium Kurrol's salt, which is equivalent to its amorphous, fibrous potassium counterpart. It dissolves in solutions of ammonium or alkali-metal salts other than sodium and is infrequently mentioned in food applications.

TABLE 5

Molecular Composition of the Sodium Polyphosphates as Determined by Paper Chromatography[a]

Weight per cent P_2O_5	Analysis Na_2O/P_2O_5 R	$\bar{n}$	Ortho-	Pyro-	Tri-	Tetra-	Penta-	Hexa-	Hepta-	Octa-	Nona-	Higher
53.383	2.000 00	2.0	0.00	100.0								
54.803	1.888 88	2.25	0.00	91.81	8.19							
55.994	1.800 00	2.5	0.00	72.24	27.76							
57.007	1.727 27	2.75	0.00	47.03	52.13	0.84						
57.881	1.666 67	3.0	0.00	24.32	48.55	20.41	5.11	1.61				
59.308	1.571 43	3.5	0.00	11.13	38.27	28.11	13.12	5.52	2.33	1.02	0.48	
60.426	1.500 00	4.0	0.00	6.65	28.25	27.43	16.87	9.41	5.70	2.74	1.77	1.16
61.324	1.444 44	4.5	0.00	4.74	22.17	23.99	17.30	11.52	8.24	4.60	3.15	4.29
62.063	1.400 00	5.0	0.00	3.24	16.24	20.16	16.45	12.68	9.43	6.71	4.39	10.69
62.681	1.363 64	5.5	0.00	2.65	13.94	19.14	17.92	14.12	8.92	6.40	4.76	12.15
63.205	1.333 33	6.0	0.00	1.88	10.09	14.89	13.46	11.28	9.61	9.26	6.94	22.58
63.655	1.307 69	6.5	0.00	1.81	8.71	13.07	12.32	11.70	11.12	9.11	6.32	25.84
64.389	1.266 67	7.5	0.00	1.24	6.38	9.44	9.09	10.74	9.91	8.53	7.74	36.92
65.204	1.222 22	9.0	0.00	0.95	4.35	7.63	6.52	9.74	8.00	7.49	6.13	49.19

[a] Potassium and other alkali metal polyphosphates have very similar molecular compositions.

Sources: Data from Westman, A. E. R., and Gartaganis, P. A., 1957, J. Amer. Ceram. Soc. 40:293; and Van Wazer, J. P., Phosphorus and Its Compounds, Copyright © 1958, Interscience Publishers, a division of John Wiley & Sons, Inc. Reprinted by permission.

The Cyclic Metaphosphates

There are two cyclic sodium polyphosphates, correctly termed meta-phosphates. They are pure, crystalline compounds (not mixtures) that are prepared under very specific conditions for the thermal dehydration of ortho-phosphates. The cyclic metaphosphoric acid is known, but it is not commercially available.

Sodium Trimetaphosphate   Sodium trimetaphosphate is a soluble, crystalline, cyclic polyphosphate. It is available as an anhydrous salt and as the hexahydrate. Its major food use is in preparation of cross-linked starch phosphates, described later in this chapter.

Sodium Tetrametaphosphate   Sodium tetrametaphosphate is also a soluble, crystalline, cyclic polyphosphate. There are a number of proposed food applications for the compound, but no application has thus far been thought important enough to be approved for use in foods by the U.S. Food and Drug Administration.

Some General Chemical Characteristics of Phosphates

The phosphates discussed in the previous section have many common chemical characteristics. These chemical characteristics are directly related to phosphate structures. In excellent reviews of the structure of phosphates, Corbridge[44] and Van Wazer[29] described the phosphate anion as a tetrahedron in which the phosphorus atom is surrounded by 4 oxygen atoms. This tetrahedral structure allows the polymerization and formation of substances quite similar to those of polymerized carbon compounds. Many of the reactions of the phosphates with components of food systems are explained by the tetrahedral structure of the phosphate anions. For example, this structure allows the long-chain polyphosphates to coil in the shape of a helix and thus to undergo polyelectrolyte interactions with proteins and gums in food systems. This structure also explains the ability of longer-chain polyphosphates to complex or sequester the metallic ions in food systems.

All of the phosphates exhibit the properties of highly charged anions,[30] which is readily demonstrated by the titration curves of their acids. Tri- and tetrametaphosphoric acids, which have titration curves similar to those of hydrochloric acid, are very strong acids. The titration curve for H_3PO_4 has 3 inflection points, a strongly acidic point near pH 4.5, a weaker inflection point near pH 7, and a very weak inflection point near pH 10. Similar inflection points are found when pyro- and tripolyphosphoric acids are titrated. However, longer-chain polyphosphoric acids show only two inflection points, one near pH 4.5 and one near pH 10. The first inflection point demonstrates much stronger acid properties than the second one, and the second inflection point becomes weaker as the chain length increases. As a result of these properties, the polyphosphates react as typical polyelectrolytes, and the polyelectrolyte properties increase as the chain length increases.

Batra[45] found that the ionization properties of the various phosphates also explain some of their useful food applications. He found that sodium ortho-phosphates ionized completely, even at high concentrations, but that the condensed polyphosphates decreased in ionization properties as their chain length increased. Batra also found a direct relationship between the ionization properties of the various phosphates and their ability to form soluble complexes with alkali and alkaline-earth metal ions. The ability to form soluble complexes increased as the degree of dissociation of the phosphate compound decreased; thus, it was directly proportional to increasing chain length.

The ability of the phosphates to complex metal ions has been studied by Van Wazer and Campanella[46] and Van Wazer and Callis.[47] These authors have found that long-chain polyphosphates are strong complexing agents for the alkaline-earth and heavy-metal ions, while the ring phosphates form weaker complexes; orthophosphates form complexes only at very low concentrations, and they form precipitates at higher concentrations. Van Wazer and Callis[47] reported that the phosphate tetrahedra are geometrically arranged in such a way that oxygen atoms from each of three neighboring phosphate groups can be in close proximity with the metal ion being complexed by a polyphosphate molecule. The complexes are thought to involve both ionic and covalent bonds between the sequestrant anion and the complexed cation.[30]

The formation of a soluble complex is believed to be the result of the competition for the metal ion between a sequestrant anion and a precipitating anion in the same solution. As a result the formation of a soluble cation–anion complex prevents precipitation of a metal cation. For example, SHMP will solubilize precipitated calcium oxalate. Very small quantities of SHMP will also remove the calcium essential to the opacity, stability, and structure of the casein–calcium–phosphate complex in milk that results in a light-yellow, transparent serum. These characteristics have led Thilo[48] to describe the calcium-complexing ability of polyphosphates as ion-exchange reactions. The polyphosphate anion can bind calcium more firmly than sodium; therefore, sodium polyphosphates exchange their sodium ions for calcium ions when they are present. Figure 1 compares the ability of various types of anions to complex or precipitate calcium. Any complexing agent that appears below a precipitating agent on the scale will completely dissolve the calcium precipitate. The chart also allows one to appreciate the relative complexing ability of the various chain lengths of the polyphosphates. Irani et al.[49,50] studied the influence of temperature and pH on the ability of the polyphosphates to sequester calcium, magnesium, and iron; they also studied the influence of the chain length of the polyphosphates on their sequestering ability under any of these conditions. Both of these are valuable studies for anyone with further interest in the subject.

The ability of the polyphosphates to influence water binding was studied by Kutscher,[51] who believed it was caused by the phosphate reactions as highly charged anions. This leads to very strong hydration and will be found to be important in the later discussion of the interactions between polyphosphates and proteins, such as milk and meat proteins.

The highly charged anionic nature of the polyphosphates and their ability to interact as polyelectrolytes are also important in explaining their interactions with long-chain polyelectrolytes normally found in foods such as gums, pectins, and protein. These characteristics cause the polyelectrolytes to be attracted to and to orient themselves along the charged sites of the food polyelectrolytes; thus, they form stabilizing complexes with these organic molecules, which are useful in food processing.

Moles free Ca per liter

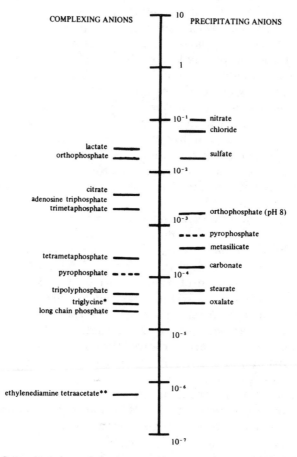

Fig. 1. Comparison of complexing agents with precipitating agents for calcium. The free calcium for the complexing agent is computed for the dissociation of a 0.01 mol solution of the 1:1 calcium complex. (Source: Van Wazer, J. R., and Callis, C. F., 1958, *Chem. Rev. 58*:1011; with permission.)

*Triglycine is also called ammonia triacetate or Trilon A.
** A well known trade name for ethylenediamine tetraacetate (EDTA) is Versene.®

The stability of the phosphate anion in any food application is an important consideration. A number of studies of phosphate stability have been reported in the literature.[48, 51–53] The pH and temperature of the system, the presence of phosphatase enzymes, colloidal gels, and complexing cations, and the chain lengths of the phosphates all affect their rate of hydrolysis. It is desirable to have hydrolysis of polyphosphates in certain applications, and it is undesirable in others. Therefore, conditions often must be adjusted to obtain the optimum effect.

The by-products of hydrolysis may also be important. Bell[53] reported that STP hydrolyzes to 1 mol each of ortho- and pyrophosphate. SHMP, however, hydrolyzes in such a way that one portion of the molecule forms the cyclic trimetaphosphate, and the other portion hydrolyzes directly to orthophosphate. Cyclic trimetaphosphate first hydrolyzes to tripolyphosphate, the straight-chain compound, and eventually hydrolyzes to orthophosphate. Pyrophosphates hydrolyze directly to orthophosphate.

The influence of chemical characteristics of the various phosphates on their interactions with food constituents will be discussed in subsequent sections.

Toxicology of the Food Phosphates

Many investigators have published the results of their studies on the toxicity of phosphates that have been added to foods during processing. Phosphates, in common with all inorganic salts, are toxic to any organism ingesting excess quantities of the salts. Excess ingestion of any inorganic salt may upset mineral balance in the body, adversely affect the osmotic pressure of body fluids, and prevent absorption or utilization of necessary mineral nutrients.

A brief but excellent review and discussion of the toxicology of the food phosphates has been published in the report of the seventh meeting of the FAO/WHO Expert Committee on Food Additives held in 1963.[37] The literature review was published in support of the committee's recommendations for acceptable daily intakes of phosphates in human diets.

Acute Toxicity

A summary of the data found in the literature describing the acute toxicity levels of phosphates is found in Table 6.[37-44] The table contains the published LD_{50} and lethal-dose quantities for ortho- and polyphosphates, as well as similar data for sodium chloride when administered orally in the diet and by intraperitoneal or intravenous injection.

TABLE 6
Acute Toxicity Levels of Phosphates in Animals

Abbreviations:

  IP—intraperitoneal
  IV—intravenous
   O—oral; in diet, stomach tube, etc.
  SC—subcutaneous

Phosphate	Animal	Route	LD_{50}, mg/kg body weight	Approx Lethal Dose, mg/kg body weight	Reference
H_3PO_4	Rabbit	IV		1 010	54
NaH_2PO_4	Mouse	O		>100	55
NaH_2PO_4	Guinea pig	O		>2 000	56
NaH_2PO_4	Rat	IP		>36	57
Na_2HPO_4	Rabbit	IV		>985, ≤1 075	56

TABLE 6
Acute Toxicity Levels of Phosphates in Animals (Continued)

Phosphate	Animal	Route	LD_{50}, mg/kg body weight	Approx Lethal Dose, mg/kg body weight	Reference
$NaH_2PO_4 + Na_2HPO_4$	Rat	IV	>500		58
$Na_2H_2P_2O_7$	Mouse	O	2 650		60
$Na_2H_2P_2O_7$	Mouse	SC	480		60
$Na_2H_2P_2O_7$	Mouse	IV	59		60
$Na_2H_2P_2O_7$	Rat	O	>4 000		59
$Na_4P_2O_7$	Rat	IP	233		59
$Na_4P_2O_7$	Rat	IV	100–500		58
$Na_4P_2O_7$	Mouse	IP		ca 40	55
$Na_4P_2O_7$	Mouse	O	2 980		60
$Na_4P_2O_7$	Mouse	SC	400		60
$Na_4P_2O_7$	Mouse	IV	69		60
$Na_4P_2O_7$	Rabbit	IV		ca 50	55
$Na_5P_3O_{10}$	Mouse	O		>100	55
$Na_5P_3O_{10}$	Mouse	O	3 210		60
$Na_5P_3O_{10}$	Mouse	SC	900		60
$Na_5P_3O_{10}$	Mouse	IV	71		60
$Na_5P_3O_{10}$	Rat	IP	134		61
$Na_6P_4O_{13}$	Rat	O	3 920		60
$Na_6P_4O_{13}$	Rat	SC	875		60
$(KPO_3)_n$ + Pyro	Rat	O	4 000		58
$(KPO_3)_n$ + Pyro	Rat	IV	ca 18		58
Hexametaphosphate[a]	Rabbit	IV		ca 140	55
Hexametaphosphate[a]	Mouse	O		>100	55
Hexametaphosphate[a]	Mouse	O	7 250		60
Hexametaphosphate[a]	Mouse	SC	1 300		60
Hexametaphosphate[a]	Mouse	IV	62		60
$(NaPO_3)_{\bar{n} = 6}$[b]	Rat	IP	192		61
$(NaPO_3)_{\bar{n} = 11}$	Rat	IP	200		61
$(NaPO_3)_{\bar{n} = 27}$	Rat	IP	326		61
$(NaPO_3)_{\bar{n} = 47}$	Rat	IP	70		61
$(NaPO_3)_{\bar{n} = 65}$	Rat	IP	40		61
$(NaPO_3)_3$ cyclic	Mouse	O	10 300		60
$(NaPO_3)_3$ cyclic	Mouse	SC	5 940		60
$(NaPO_3)_3$ cyclic	Mouse	IV	1 165		60
NaCl (table salt)	Mouse	O	5 890		60
NaCl (table salt)	Mouse	SC	3 000		60
NaCl (table salt)	Mouse	IV	645		60

[a] Average chain lengths are not given.
[b] Polyphosphate preparations for which the $\bar{n}$ was determined.

As shown by the data in Table 6, the phosphates are somewhat more toxic than sodium chloride when given orally; the toxicity of phosphates decreases with increasing chain length. The phosphates also appear to be more toxic than sodium chloride when introduced into the body in a manner that circumvents the digestive system. The LD_{50} levels of the various phosphates, including the longer-chain polyphosphates, change little when introduced directly into the body through intravenous or intraperitoneal injection. This is probably due to their rapid enzymatic hydrolysis to orthophosphates.

Although the phosphates are much more toxic when introduced directly into the body, their toxicity upon oral administration is of greater importance here, since this is the manner in which they will be introduced through their applications in food processing. All food applications introduce low but constant levels of phosphates into the diet. Therefore, the chronic toxicity studies are of more concern in studying the safety of phosphates in the human diet.

Chronic Toxicity of Phosphates in Diets

Table 7 summarizes data found in the literature on the chronic toxicity of phosphates administered orally to humans and animals. Many investigations included studies of phosphates of several chain lengths. The data have been separated into categories according to the phosphate chain length for presentation in Table 7, in order to demonstrate the effect of chain length on the degree of toxicity. The chronic toxicity of the phosphates grouped according to their chain lengths will be discussed in the following paragraphs.

TABLE 7

Dietary Levels of Phosphates Producing No Adverse Effects

Abbreviations:

d = day	kg = kilogram	yr = year
g = gram	mo = month	% = per cent phosphate in diet
gen = generation	wk = week	(often calculated from % P)

Phosphate	Test Animal	Length of Test[a]	Maximum Level Tolerated	Effect of Excess Phosphate	Reference
Orthophosphates					
H_3PO_4 (36.4%)	Human	Variable	17–26 g/d	No adverse effects[b]	1
H_3PO_4 (36.4%)	Rat	>12 mo	>0.75%	No adverse effects[b]	62
H_3PO_4 (36.4%)	Rat	44 d	<2.94%	Kidney damage	63
Na and K ortho-phosphates	Rat	44 d	<3.93%	Kidney damage	63
MSP	Human	Variable	5–7 g/d	No adverse effects	37
MSP	Rat	42 d	>3.4 g/kg d	Kidney damage	57
MSP	Guinea pig	200 d	>2.2%, < 4.0%	Ca deposits, reduced growth	66
MSP	Guinea pig	12–32 wk	4–8%[c]	Ca deposits, reduced growth	67
MSP + DSP	Rat	3 gen	>0.5%, <1%	Kidney damage	58
DSP	Rat	6 mo	>1.8%, <3.0%	Kidney damage	64, 68, 69
DSP	Rat	1 mo	<5%	Kidney damage	70
DKP	Rat	150 d	>5.1%	No adverse effects	65

TABLE 7
Dietary Levels of Phosphates Producing No Adverse Effects (Continued)

Phosphate	Test Animal	Length of Test[a]	Maximum Level Tolerated	Effect of Excess Phosphate	Reference
Pyrophosphates					
TSPP	Rat	6 mo	>1.8%, <3.0%	Kidney damage	64, 68, 69
TSPP	Rat	16 wk	<1%	Kidney damage	59
SAPP + TSPP + $(KPO_3)_n$	Rat	3 gen	>0.5%, <1%	Kidney damage	58
Tripolyphosphates					
STP	Rat	6 mo	>1.8%, <3.0%	Kidney damage	64, 68, 69
STP	Rat	1 mo	>0.2%, <2%	Kidney damage	70
STP	Rat	2 yr	>0.5%, <5%	Kidney damage	71
STP	Dog	1 mo	>0.1 g/kg d	No adverse effects	70
STP	Dog	5 mo	<4.0 g/kg d	Kidney, heart damage	70
Polyphosphates					
SHMP	Rat	150 d	>0.9%, <3.5%	Slight growth reduction	65
SHMP	Rat	1 mo	>0.2%, <2%	Kidney damage	70
SHMP	Rat	3 gen	>0.5%, <5%	Kidney damage	72
SHMP	Dog	1 mo	>0.1 g/kg d	No adverse effects	70
SHMP	Dog	5 mo	<2.5 g/kg d	Kidney, heart damage	70
Graham's salt	Rat	6 mo	>1.8%, <3%	Kidney damage	64, 68, 69
$(KPO_3)_n$ + SAPP + TSPP	Rat	3 gen	>0.5%, <1%	Kidney damage	58
Cyclic phosphates					
$(NaPO_3)_3$	Rat	1 mo	>2%, <10%	Kidney damage	70
$(NaPO_3)_3$	Rat	2 yr	>0.1%, <1%	Retarded growth of males	73
$(NaPO_3)_3$	Rat	3 gen	>0.05%	No adverse effects	73
$(NaPO_3)_3$	Dog	1 mo	>0.1 g/kg d	No adverse effects	70
$(NaPO_3)_3$	Dog	5 mo	<4 g/kg d	Kidney, heart damage	70
$(NaPO_3)_4$	Rat	1 mo	>2%, <10%	Kidney, heart damage	70
$(NaPO_3)_4$	Dog	1 mo	>0.1 g/kg d	No adverse effects	70
$(NaPO_3)_4$	Dog	5 mo	<4 g/kg d	Kidney, heart damage	70

[a] Some reports did not give definite time periods.
[b] "No adverse effects" = no physiological damage noticeable at any level of phosphate tested.
[c] Guinea pigs tolerated higher levels of phosphate only when increased levels of Mg^{++} and K^+ were present in their diets.

The Orthophosphates   According to Nazario,[1] a 36.4% solution of phosphoric acid has frequently been administered to humans, both babies and adults, in the treatment of hypoacid diathesis. Babies were not adversely affected when 0.25 g/day of the dilute acid was administered, and adults tolerated as high as 26 g/day without adverse effects. Feeding studies with dogs demonstrated that they would tolerate as high as 13 g/day of the dilute acid before signs of enteritis appeared. Other studies of the results of feeding phosphoric acid to dogs and rabbits also have been reviewed by this author. Based on these studies Nazario calculated that a 70 kg man could tolerate 17 g of 100% phosphoric acid and that approximately 22 g of the acid would be lethal.

Bonting[62] fed three generations of rats 0.75% H_3PO_4 in their diets. Studies of their blood, tissues, mineral balance, nitrogen retention, and acidic conditions in their digestive tracts indicated that the young and adult rats suffered no adverse effects. However, extensive pathological damage, particularly to the kidneys of rats, occurred when MacKay and Oliver[63] fed 2.94% H_3PO_4 and 3.93-7.86% sodium or potassium orthophosphates to rats. Hahn and co-workers[64] studied the effects of diets containing 1.8, 3.0, and 5.0% DSP to rats. The lowest level produced only slight physiological effects, such as minor increases in kidney weight and renal calcification, but extensive physiological damage was done at the 3.0 and 5.0% levels. Levels of phosphorus, copper, iron, and calcium in the blood and organs were found to have undergone no significant changes.

Dymsza and co-workers[65] could find no histological or chemical signs of adverse physiological effects, and no changes in absorption or utilization of calcium, phosphorus, or iron when rats were fed diets containing 0.87 and 5.1% DKP for 150 days. McFarlane[57] found that only the lower level of oral administration produced no adverse effects in rats when they were fed 0.34-1.2 g/100 g body weight. All other levels and methods of administration produced renal damage in the rat kidneys. Three generations of rats were fed diets containing 0.5-5.0% of a mixture of MSP and DSP by van Esch and co-workers.[58] A level of 0.5% produced no discernible damage. However, rats receiving diets containing 1% or more of the mixture were found to have renal damage and tissue calcification. Growth was impaired, fertility was reduced, and life spans were shortened in the animals receiving 5% phosphate. Hogan and co-workers[66,67] found that the levels of magnesium and potassium in the diets of guinea pigs determined whether physiological effects were adverse or not. Diets containing 0.9% calcium, 1.7% phosphorus, 0.04% magnesium, and 0.41% potassium resulted in severe reductions in weight gain, increased stiffness of leg joints, and numerous visible deposits of calcium phosphates in the guinea-pig footpads. These symptoms were significantly reduced or eliminated when the diet was altered so that it contained 0.35% magnesium and 1.5% potassium with the same levels of calcium and phosphorus. These reports may be highly significant in explaining the discrepancies in the numerous studies of the effects of phosphates in the diets of test animals. As far as could be determined in reviewing the literature, Hogan and his co-workers are the only ones who reported the levels of magnesium and potassium in the diet and related physiological damage to the presence or absence of these two minerals.

Pyrophosphates   Hahn and Seifen,[68] Hahn,[64,69] Datta et al.,[59] and van Esch et al.[58] fed diets containing 0.5-5.0% pyrophosphate to rats. All diets containing 1% or more of the pyrophosphate produced physiologically adverse effects, principally kidney damage and calcification. Datta and co-workers[59] observed that the rat kidney is capable of a higher degree of concentration of salts than that of other species; they questioned the validity of relating the effects of rat-feeding studies of mineral salts with effects in humans.

Polyphosphates Hodge[70] fed rats diets containing 0.2, 2.0, and 10% STP, then compared the results with rats fed diets containing 10% sodium chloride and others a diet containing 5% DSP. Although no rats died, those fed diets containing 10% sodium chloride and 10% STP were found to have decreased growth rate, increased kidney weights, and significant kidney damage. Kidney damage also was apparent for all the rats, except those fed the diet containing 0.2% polyphosphate, which caused no changes at all. Hahn and co-workers[64,68,69] found that feeding rats 3% and 5% STP or SHMP resulted in severe kidney damage, while feeding 1.8% of either phosphate caused no observable physiological damage. The authors found the level of kidney damage to be equivalent to that found when similar levels of DSP had been fed in the diets. They postulated that the polyphosphates are hydrolyzed to orthophosphate prior to absorption through the intestinal walls, thus giving the same effect as orthophosphate.

Hodge[70] found that dogs tolerated 0.1 g/kg day of either STP or SHMP without any evidence of adverse effects. However, when dogs were fed increasing dosages of either phosphate, beginning at 1.0 g/kg day and increasing to 4.0 g/kg day for the last month of a 5-month feeding period, the dogs fed SHMP began to lose weight when the daily dose reached 2.5 g/kg day; those fed STP began to lose weight when the level in the diet reached 4.0 g/kg day. Gassner and co-workers[74] studied the absorption of potassium Kurrol's salt when rats were fed 750 mg of the salt for 6 days. Paper chromatography demonstrated that approximately 25% of the daily intake of the polyphosphate appeared in the urine as orthophosphate. No condensed phosphate was found in the urine for the first 2 days, but after that an average of about 4% of the daily dose was found, of which 20% was tripolyphosphate or pyrophosphate, and 80% had an average chain length of four to five. A few samples contained the cyclic tri-metaphosphate in quantities below 0.2%.

Cyclic Polyphosphates The cyclic polyphosphates were found by Hodge[70,73] to have fewer physiological effects on rats than either orthophosphates or polyphosphates. Diets containing 0.2, 2.0, and 10% sodium trimetaphosphate and similar diets containing the same quantities of sodium tetrametaphosphate were used in the studies. Only the highest levels were found to have adverse effects on kidneys; none of the lower levels had any adverse effects. When dogs were fed increasing doses of the cyclic phosphates, adverse effects began to appear only after the daily dose had reached 4 g/kg day. A further study of 100 rats fed diets containing 0.1, 1.0, and 10% sodium trimetaphosphate resulted in retarded growth for all of the rats when the diets contained 10% and a reduction in growth rate for those male rats receiving 1%.[73] Histological studies and studies of reproduction through two generations showed no further abnormalities. The cyclic polyphosphates, therefore, appear to be considerably less toxic than the straight-chain polyphosphates.

Biochemical Aspects of Phosphate Toxicity

Although the phosphate anion is essential to metabolism in all living organisms, no organism is known to synthesize this anion; therefore, it must be supplied in the diet.[1,75] The higher animals have a mechanism, the para-thyroid gland, that controls the levels of inorganic phosphate. Hormones from this gland can cause release or deposition of phosphate to and from the skeleton, depending on the level of blood calcium. When the blood level of phosphate is too low, hormones cause its absorption from the intestinal walls. When excess phosphate is present, the hormones again cause it to be eliminated through the kidneys.[37,75] In addition, phosphatase enzymes are also present in the blood and tissues for further regulation of the phosphate-anion levels in the blood. These enzymes require magnesium and calcium for activation. The blood serum of human adults normally carries 2.5–4.5 mg of phosphorus/100 ml, while that of children carries a higher level. It is also estimated that adult humans require a minimum of 0.88 g of phosphorus in their daily diets.[75]

Several critics of the use of phosphate in food processing have speculated that phosphates prevent the utilization, and possibly absorption, of calcium, iron, magnesium, copper, and similar essential ions through their ability to precipitate or sequester these metals. Numerous studies have demonstrated that the absorption and metabolism of these ions is inhibited in no way, even when the diet contains excessive levels of the various phosphates, including those of very long chain length.[37,62,64,65,76,77] Several investigators have established that polyphosphates actually increase the absorption and retention of essential mineral ions.[37,65,77] For example, STP and SHMP have been shown to increase absorption and retention of calcium in the bodies of rats, with SHMP being more effective than STP. The same investigators demonstrated that the SHMP–calcium complex is absorbed intact through intestinal walls, while STP and calcium are absorbed independently of each other. Both STP and SHMP then are hydrolyzed to orthophosphate after absorption into the body.

The orthophosphates are known to be readily absorbed through the intestinal wall. The shorter-chain, condensed phosphates, such as pyro-phosphate and tripolyphosphate, also are probably capable of passing intact through the intestinal walls, but they are quickly hydrolyzed to orthophos-phate.[78] Mitchell[79] demonstrated that synthetic digestive fluids were capable of hydrolyzing polyphosphates in a step-like fashion to orthophosphate. Feeding studies by other investigators have confirmed that the longer-chain polyphosphates are absorbed through the intestinal walls only after hydrolysis to shorter-chain polyphosphates or orthophosphate.[21,75,79–84] Furthermore, only 10–40% of the phosphorus in SHMP and longer-chain polyphosphates is absorbed from the intestinal tract. The balance of the phosphorus is eliminated in the feces.[61,76,80–82] This has been confirmed by studies with diets containing radioactive polyphosphate.

Acceptable Daily Intakes for Humans

The FAO/WHO Expert Committee on Food Additives studied the various factors affecting toxic reactions to phosphates in the diet. These factors include the total level of phosphorus in the diet, the presence of calcium and other minerals, the degree of hydrolysis of polyphosphates to shorter-chain poly-phosphates or orthophosphates, and others. Since it is impossible to predict the presence or absence of any of the factors that would determine whether any particular level of phosphate would be toxic or not, the Committee has recommended an unconditional acceptance level for total dietary phosphorus and a conditional acceptance level. The unconditional acceptance level is considered safe in any type of diet known in the world; the conditional level is acceptable only when the dietary calcium level is high. These levels are as follows:[37]

1. unconditional acceptance level: <30 mg/kg body wt
2. conditional acceptance level: 30–70 mg/kg body wt.

Summary

The results of animal-feeding studies reported in the scientific literature indicate that levels of 0.5% of the phosphates could be tolerated in the diet without adverse physiological effects. Higher levels could possibly be tolerated if a proper dietary balance of other ions, particularly calcium, magnesium, and potassium, is maintained.

Few, if any, applications for the phosphates require over 0.5% in order to obtain the desired effect, and higher levels often produce adverse physical and chemical effects and off-flavors. Because the phosphates produce adverse effects in numerous food systems, it is not feasible to add them to all foods. As a result, the level of phosphates likely to be found in the diet, even though fully utilized in all of the potential applications, is self-limiting. It is highly unlikely that a level of 0.5% phosphate could ever appear in the total human diet.

Attitudes of Regulatory Agencies

The U.S. Food and Drug Administration

The U.S. Food and Drug Administration (FDA) considers the orthophosphoric and polyphosphoric acids and their calcium, potassium, sodium, and ammonium salts as GRAS *(generally recognized as safe)*. Lehman[83] referred to the work of Gosselin[80,84] and Mitchell[79] as evidence that the straight-chain polymeric phosphates are hydrolyzed to monophosphate. This work is used as the basis for the GRAS status of these food additives.

Based on the same references, Lehman[83] states that the cyclic metaphosphates may not be safe food additives. Gosselin *et al.*[80,84] found that the injection of cyclic phosphates resulted in recovery of most of the intact cyclic phosphate in the urine. Lehman failed to note that Gosselin *et al.*[84] and Hodge[70,73] have shown that little, if any, cyclic metaphosphate is absorbed through the intestinal wall when administered orally. The metaphosphates must first be hydrolyzed to tripolyphosphate and next to orthophosphate, which then can be absorbed. It would appear that the use of the cyclic phosphates in foods would be no more, or even less, toxic than any other condensed phosphate.

A list of the phosphate food additives listed as GRAS by FDA is shown in Table 8. Each phosphate is listed under functional categories for its applications.[85] No tolerance levels or limitations are given; however, the quantity of the substance added to the food must not exceed the amount required to accomplish its intended purpose.

There has been some confusion in the past regarding the sodium metaphosphate listed under the category of sequestrants. Recently, this seems to have been clarified to mean that any compound that meets the specifications for sodium metaphosphate in the *Food Chemicals Codex* is GRAS.[38] It is also understood that the ammonium, calcium, potassium, and sodium salts are equally acceptable.

TABLE 8
Phosphate Food Additives Listed as GRAS by FDA

Emulsifying agents

Monosodium phosphate derivatives of mono- and diglycerides

Nutrients and/or dietary supplements

Calcium glycerophosphate	Magnesium phosphate (di-, tribasic)
Calcium phosphate (mono-, di-, tribasic)	Manganese glycerophosphate
Calcium pyrophosphate	Manganese hypophosphite
Ferric phosphate	Potassium glycerophosphate
Ferric pyrophosphate	Sodium phosphate (mono-, di-, tribasic)
Ferric sodium pyrophosphate	

Sequestrants

Calcium hexametaphosphate	Sodium metaphosphate
Calcium phosphate (monobasic)	Sodium phosphate (mono-, di-, tribasic)
Dipotassium phosphate	Sodium pyrophosphate
Disodium phosphate	Sodium pyrophosphate (tetra)
Sodium acid phosphate	Sodium tripolyphosphate
Sodium hexametaphosphate	

Miscellaneous and/or general-purpose food additives

Ammonium phosphate (mono- and dibasic)	Sodium aluminum phosphate
Calcium phosphate (mono-, di-, tribasic)	Sodium phosphate (mono-, di-, tribasic)
Phosphoric acid	Sodium tripolyphosphate
Sodium acid pyrophosphate	

Source: §121.101(D), Part 121, *Code of Federal Regulations, Title 21* (revised as of January 1, 1971), U.S. Government Printing Office, 1971.

The Meat Inspection Division, USDA

The Meat Inspection Division (MID), Consumer and Marketing Service, U.S. Department of Agriculture, also provides regulations for the use of phosphates as additives to meat products. All of the substances acceptable for addition to meat are listed in the *Code of Federal Regulations*;[86] the phosphate additives are summarized in Table 9. As shown in this table, there are certain applications for the phosphates in meat products that have limitations for the quantity that can be used. These are clearly listed. The longest-chain polyphosphate allowed by MID is SHMP; therefore, none of the longer-chain sodium metaphosphates or the highly polymerized potassium polymetaphosphates are allowed in meats under federal inspection. Many state regulations do not prevent the use of the longer-chain polyphosphates, and some of the intra-state meat packers have been using potassium polymetaphosphate (Kurrol's salt).

Further information on each of the phosphate food additives will be found in Part II, "Regulatory Status of Direct Food Additives," of this handbook.

TABLE 9
Phosphate Food Additives Acceptable for Use in Meat Processing

Class of Substance	Phosphate	Purpose	Products	Amount
Cooling and retort water treatment agents	TKPP SHMP TSPP STP	To prevent staining on exterior of canned goods	Various	Sufficient for purpose
Hog scald agents	SHMP STP TSP	To remove hair	Hog carcasses	Sufficient for purpose
Phosphates	DSP MSP SHMP STP TSPP SAPP	To decrease amount of cooked-out juices	Cured hams, pork shoulder picnics, loins; canned hams and pork shoulder picnics; chopped ham; bacon	5.0% of phosphate in pickle at 10% pump level; 0.5% of phosphate in product (only clear solution may be injected into product)
Rendering agents	TCP TSP	To aid rendering	Animal fats	Sufficient for purpose
Synergists (used in combination with antioxidants)	H_3PO_4	To increase effectiveness of antioxidants	Lard and shortening	0.01%

Source: §318.7(4), *Code of Federal Regulations, Title 9* (revised as of January 1, 1971), U.S. Government Printing Office, 1971.

Some General Functions of Phosphates in Foods

As the phosphates are capable of interacting with many of the constituents of food systems, they have very useful and important functions in food processing. Some of these general functions will be discussed briefly in this section, and specific functions will be discussed under each of the food categories in later sections of this chapter.

Inactivation of Metal Ions and Processing Water Treatment

The phosphates can inactivate metallic ions, which are capable of interfering with necessary food-processing reactions. They inactivate the metallic ions either by precipitating and removing them from interference with the desired food-processing reactions or by complexing and maintaining them in a soluble, bound state.

The orthophosphates precipitate calcium, magnesium, iron, and similar alkaline-earth and transition-metal ions. They reportedly form complexes with calcium when the phosphates are present in low concentration, but they begin to form precipitates as the concentration of phosphate increases.[47]

The polyphosphates, including those ranging from two to many phosphate units per molecule, form soluble complexes with nearly all of the metallic ions.

Thus, they are considered to undergo ion-exchange reactions, in which a hydrogen, sodium, or potassium ion is exchanged for one of the alkaline-earth or transition-metal ions.[23,48,51] Very weak, soluble complexes are formed with alkali-metal and ammonium ions. More stable but somewhat dissociated complexes are formed with the alkaline-earth metals, such as calcium and magnesium. Very stable, soluble complexes are formed with the transition-metal ions, such as copper, nickel, and iron. A significant advantage in the use of phosphates to complex nutritionally important ions, such as calcium, magnesium, and iron, is that the ions can still be absorbed through the intestinal walls and utilized by the body, and their absorption and retention may actually be increased in the form of their complexes.[23,37,62,64,65,76]

Since pyrophosphates, as well as the long-chain polyphosphates, form soluble complexes of metal ions, the complexing ability of the phosphates apparently does not necessarily depend on chain length. A precipitate forms initially upon addition of a polyphosphate to a system that contains a metal ion. However, the precipitate rapidly dissolves upon the addition of more of the polyphosphate. This formation of soluble complexes, which prevent formation of precipitates between the metal ions and other constituents of the system, is often called "sequestration."[32] As very weak, if any, complexes are formed between metal ions and the cyclic, condensed phosphates sodium trimetaphosphate and sodium tetrametaphosphate, they are, therefore, not used for this purpose.

The water used in food processing often contains sufficient quantities of alkaline-earth and transition-metal ions to cause undesirable reactions in food processing. It is, therefore, common practice to treat the water with a phosphate, such as SHMP, SHMP plus an alkali, a shorter-chain polyphosphate, or, in some cases, orthophosphate, or mixtures of all of these to precipitate and/or complex the interfering ions. This is often necessary in the canning of fruits and vegetables, in which the alkaline-earth and transition-metal ions tend to toughen the skins and tissues or to precipitate and form gels with the pectins in the fruit or vegetable tissues. The treatment of waters used in food processing for drinking has been thoroughly reviewed in many publications and will not be further discussed here.[28,87-91]

Complexing Organic Polyelectrolyte Food Constituents

Some functions of phosphates in food processing depend on their ability to form complexes and reaction products with constituents of foods other than mineral elements. This is demonstrated by the ability of phosphates to interact with many organic polyelectrolytes, such as protein, pectin, starch, etc. Numerous complicated protein–phosphate or protein–phosphate–salt complexes are known and have been studied and reported.[92-94] The most frequently studied of these complexes is probably that of the casein–calcium–phosphate complex in milk. One has only to review the numerous publications of Leviton, Morr, Rose, Waugh, Zittle, and their co-workers, and numerous others to appreciate the enormous effort that has been expended in research on this complex. Even so, the exact structure, its chemistry, and how it can be modified to obtain desired effects are still unknown. Further discussions of these important complexes will appear under specific sections of this chapter.

Direct Chemical Reactions with Food Constituents

In addition to the formation of complexes between organic polyelectrolytes and phosphates, phosphate may undergo esterification reactions and cross-linking between molecules of proteins and starches. These reactions will be discussed further in subsequent sections.

Buffering or pH Stabilization

Another important function of phosphates is their buffering ability, which can be used to stabilize the optimum pH required for processing or stabilizing a food.

Dispersion of Food Constituents

Phosphates have the ability to promote dispersion and peptization of relatively insoluble food constituents, such as proteins, in concentrated milk and pasteurized, processed cheese. Entire food particles are dispersed in some cases, much as phosphates act as dispersants for clay particles.[30]

Emulsion Stabilization

Phosphates can stabilize emulsions; their ability to stabilize sausage emulsions, for example, will be discussed in the section on meat applications.

Increasing Hydration and Water Binding

The phosphates also can interact with animal and vegetable proteins in such a way that hydration is promoted. This important function is utilized in preventing freeze-thaw drip in meats, poultry, and fish.

Mineral Supplementation

Calcium, iron, sodium, and potassium phosphates are also used to improve the nutritional properties of numerous cereal products, such as flour, prepared mixes, baked goods, and breakfast cereals.

Acidification or Lowering pH

Phosphoric acid and its acidic salts are used in food acidification. The acidic phosphate salts are used as acidifying agents to react with baking soda in baking powders, prepared mixes, self-rising flour, and corn meal and to produce acidic pH values in many other foods. Orthophosphoric acid is the most common acidic constituent used in soft drinks to produce a natural, fruity tartness, just as it is frequently the acidic factor in natural fruit juices.

Alkalization or Raising pH

The alkaline phosphate salts function to produce alkaline pH values where these are necessary. As an example, DSP and TSP are used in the preparation of pasteurized, processed cheese to produce the slightly alkaline pH values necessary for optimum protein dispersion, as well as to interact with the proteins themselves to improve their emulsification and water-binding abilities.

Prevention of Caking

Several applications for insoluble phosphates as anticaking agents will be discussed in subsequent sections. This is a very common use for these highly insoluble salts, as they can be prepared in a very fine powdered form, which can be used to separate crystals of substances that normally adhere to each other. They are often used also as water-absorbent anticaking agents. Warning and Schille[95] used a mixture of aluminum phosphate and TCP as an anticaking agent for salt. Lubeck[96] proposed the use of DCP and dimagnesium phosphate to produce a free-flowing table salt. TCP also has been proposed as an anticaking agent for dietary seasoning compositions.[97] A somewhat unrelated example of the ability of TCP to maintain free flow of particles is that patented by Sorgenti, Nack, and Sachsel,[98] who advocated the use of TCP as the solid particles in a fluidized-bed cooking process for food products.

Food Preservation

The phosphates, and especially the polyphosphates, have been found to prevent or retard the oxidation of unsaturated fats in aqueous food systems and to inhibit the growth of some of the microorganisms involved in the spoilage of foods. Although some investigators relate these functions to the ability of the phosphates to complex metal ions or organic polyelectrolytes essential to the fat oxidation or spoilage organisms, others report evidence of additional factors involved in this important application. The antioxidant and microbiological preservation applications of the phosphates will, therefore, be treated separately in subsequent sections.

The following sections consist of further discussions of the various functions of phosphates and their applications in the processing of foods and food ingredients. The sections are arranged according to general food categories with a discussion of the functions of the phosphates in the processing of specific foods and their reactions with food ingredients under each category.

Phosphate Applications in Beverages

The food-science literature reports that phosphates have been applied as buffers, metal-complexing agents, microbiological inhibitors, acidifiers, mineral supplements, and flow agents in alcoholic, carbonated, powdered, and nutritional beverages.[28] Some examples of each application follow.

Applications in Alcoholic Beverages

Buffering The buffering action of mixtures of the acidic and basic orthophosphates has been used to obtain optimum pH values in the preparation of alcoholic beverages. Heyer,[99] for example, demonstrated that buffering a boiling suspension of hops to pH 8.0–8.3 gave maximum conversion of humulon to isohumulon in the preparation of the dried hops extract.

Microbiological Control The treatment of yeast used in brewing beer and ale with a solution of H_3PO_4 and sodium persulfate at 5°C was reported to control the bacteria that often contaminate such brewing yeasts.[100] The viability of yeast used in brewing beer and its resistance toward contaminating bacteria were increased by treating the yeast with TKP prior to its addition to the beer wort.[101] Kohl and Ellinger[102,103] reported that bacteria, fungi, and yeasts could be inhibited in alcoholic beverages, among many other food systems, by treating them with polyphosphates. The polyphosphates could be added to draft beers, for example, to inhibit the last traces of organisms.

Complexing Metal Ions   The ability of the phosphates, and especially the polyphosphates, to complex alkaline-earth and heavy metal ions that are troublesome in the preparation of alcoholic beverages has long been recognized. The polyphosphates also are frequently used to soften the water used in preparations of these beverages.

Iron, copper, and calcium ions are especially troublesome, since they can introduce off-flavors and contribute to haze formation in the beverages. Libby[104] prevented cloud or haze formation by the addition of 0.1–1.0 g of SHMP/gal of beer or wine. Peynaud[105] significantly reduced the iron content in wines by the addition of 100–450 mg of STP or SHMP/l. Copper and iron were reported to be removed by filtering the beverage through insoluble calcium phosphates.[106] The direct formation of a precipitate of DCP in the beverage was reported by Shrivanek[107] to co-precipitate and remove undesirable impurities from the beverage. Opportunities for the formation of cloud or haze in wines were reduced by Ohara and co-workers[108] by the addition of ammonium phosphate to reduce the tannin and acid contents of the wines.

Applications in Carbonated Beverages

Complexing Metal Ions   The presence of heavy-metal ions in the water used in carbonated beverages can result in more rapid dissipation of carbonation, as the ions serve as nuclei for the formation of carbon dioxide bubbles. Karlson[109] used soluble sodium polyphosphate glasses having a $Na_2O:P_2O_5$ ratio of 1.2:1 to sequester these ions and prevent the loss of carbonation. It was discovered that such treatment of the beverage would reduce the necessary carbonation by 25–90% of that required by use of untreated water. As the water used in the preparation of carbonated beverages is always deionized, the phosphates, particularly the polyphosphates, are frequently implemented in this deionization.

Acidification   Many soft-drink beverages, particularly those flavored with leaf, root, or nut extracts, such as cola beverages, are acidified with H_3PO_4.[27,110,111] This acid gives a natural tartness to the beverage, possibly because it is a normal constituent of many fruits and their juices. The acidic, crystalline phosphate salts are also utilized in preparation of beverage powders,[27,39] and can be used in preparation of effervescent, carbonated drink powders.[28,39,112,113]

Improving Flow of Powders   The insoluble phosphates, particularly TCP, can be used to improve flow properties of powdered beverage mixes. For example, Common[114] found that the addition of TCP to a powdered beverage mix beneficially influenced not only the flow properties but also the particle-size distribution of the dry mix.

Applications in Other Beverages

Acidification   Increased yields of coffee extract were obtained when DiNardo[115] used H_3PO_4 or sulfuric acid to hydrolyze a portion of the extracted coffee grounds in the preparation of instant coffee and imitation coffee products. Stayton[116] prepared improved coffee-chickory blends by neutralizing histidine compounds, which contribute the characteristic undesirable aromas and flavors of the chickory, by acidifying the chickory with acidic phosphates and pyrophosphates.

Mineral Supplementation   A number of nutritional beverages for various purposes, including complete meals, have appeared on the grocery store shelves in recent years. Sodium iron pyrophosphate, ferric pyrophosphate, and ferric orthophosphate have been used as sources of iron in some of these beverages.[27] Calcium and phosphorus have also been provided through the addition of calcium phosphate.

Complexing Metal Ions   Natural or added ascorbic acid can be inactivated through oxidation if heavy metal ions are present in a beverage. The ascorbic acid and the color of vitamin-C-fortified beverages have been found to be stabilized by the addition of polyphosphoric acids to complex metal ions.[28,29,117-119]

Phosphate Applications in Cereal Products

The first applications of phosphates in foods probably were as leavening acids in baking. Applications in cereal products include pH adjustment, buffering, dough conditioning, and mineral enrichment. Reviews of phosphate applications in cereal products have been numerous,[26-30,120] and examples are discussed in the following paragraphs.

Applications as Leavening Acids

The prepared-mix and baking industries probably are the largest consumers of phosphates in the food industry because of the many phosphate applications in chemical leavening systems. The salts of H_3PO_4 and its polymers are particularly useful as leavening acids, as the acids can be partially neutralized to form acidic salts. The acidic phosphate salts are capable of neutralizing sodium bicarbonate, common baking soda, during which carbon dioxide (CO_2) is released to become a portion of the leavening gases.

Characteristics of Leavening Systems   The term *leavening* refers to the introduction and expansion of a gas in a batter or dough system. The gas may be incorporated through the mixing action during preparation of the dough or batter, through the formation of steam from the water in the dough or batter during baking, or through the formation of gas from a biological or chemical leavening system. The contributions of these various leavening systems have been thoroughly reviewed in the literature.[28,120-138]

.The chemical leavening system to be discussed in this section is that provided by the reaction between sodium bicarbonate and an acidic phosphate salt. The acidic phosphate salt, which will have at least two replaceable hydrogen atoms in its molecule, will react with soda, as represented by the following general formula:

$$MH_2PO_4 + 2NaHCO_3 = M(Na)_2PO_4 + 2H_2O + 2CO_2$$

where M can be a hydrogen, an alkali metal, or an alkaline-earth metal ion.

The acidic phosphate salts provide three major characteristics in a leavening system: (1) acidification for the formation of carbon dioxide, (2) buffering to provide the optimum pH for the baked product, and (3) interaction with the protein constituents of flour to produce optimum elastic and viscosity properties in the dough or batter.[28-30,120,121]

Numerous factors influence the characteristics contributed by any leavening system to a batter or dough. Among these are the proper selection of the acidic phosphate salts, the size and distribution of the bubbles incorporated during mixing action, the quality of the gluten in the flour, the viscosity of the system, and the type of emulsification in the system. Excellent reviews have discussed these in further detail.[28-30,120-131] Since sodium bicarbonate is highly soluble, it dissolves as soon as it is immersed in water; therefore, the rate at which CO_2 is released in a chemically leavened system depends on the rate of dissolution of the leavening acid used.[121] The *dough rate of reaction*, or *DRR*, is a value that reflects the rate of reaction of a leavening acid with soda during the mixing stage and during *bench action* (the holding period after mixing is completed). The DRR is obtained by mixing a standard, dry biscuit mix containing the proper amount of leavening acid and soda with water under standardized conditions in a gas-tight apparatus. The apparatus automatically measures the amount of gas released at 27°C. The DRR is then calculated as the per cent of the theoretical amount of gas expected from the amount of soda present in the mix. This per cent of the theoretical volume of gas is plotted against time.[28,30,121,133-136] The DRR's for typical commercial leavening acids are shown in Figure 2. The amounts of gas evolved after 2 minutes, which represents the mixing stage, and after 10 minutes, which represents 8 minutes of bench action, are usually considered as important DRR values for each leavening acid.[28,121,133-138]

The *neutralizing value*, sometimes called neutralizing strength, of a leavening acid represents the number of grams of sodium bicarbonate that will be neutralized by 100 grams of the leavening acid. It is determined by titration of the acid and can be expressed as the following equation:

$$NV = \frac{a}{b} \times 100,$$

where a = pounds of sodium bicarbonate neutralized and b = pounds of leavening agent required.[28,121,134,137] The neutralizing value is used to calculate the amount of leavening acid necessary to provide the desired amount of leavening gas and the optimum pH for the baked product.

Characteristics of Commercial Leavening Acids   The important leavening properties of some of the common commercially available phosphate leavening acids are shown in Table 10; other chemical characteristics of these acids are shown in Table 2. Additional acidic phosphates have been proposed and patented, but those listed in Table 10 are commonly in use at the present time.

Monocalcium phosphate monohydrate, $MCP \cdot H_2O$, has been used as a leavening acid since the middle of the nineteenth century.[28] As shown in Figure 2 and Table 10, it is one of the fastest reacting leavening acids in common use. It seldom is used by itself and usually is combined with slower-reacting leavening acids to preserve leavening action for the baking stage. *Anhydrous monocalcium phosphate*, AMCP, is also very rapid reacting and is extremely hygroscopic. It is manufactured in such a way that each crystal of the commercial product is coated with a slowly soluble, alkaline-metal and heavy-metal phosphate compound; thus, it is necessary for the liquid in the system to penetrate the coating before the leavening acid can react with soda. This results in a delayed release of its acidity, as can be seen in the Table 10 columns labeled Leavening Gas Released, %, and the curve for coated AMCP in Figure 2. Coated AMCP, which was developed in 1939,[137,138] has almost completely replaced the monohydrate in self-rising flour and many other baking mixes because of its improved leavening stability. It is frequently used in combination with other slower leavening acids, such as with sodium aluminum phosphate, as will be discussed later. *Dicalcium phosphate dihydrate*, $DCP \cdot 2H_2O$, releases its acidity only after the temperature of a batter or dough containing the acid reaches 135-140°F. Hence, it is used only as a delayed-release leavening acid in cakes, breads, and other products that bake slowly.[28,126,138-142]

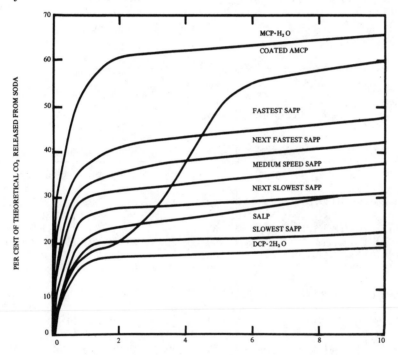

Fig. 2. Dough reaction rates for commercially available phosphate leavening acids.[120,121,130-136]

Two chemically distinct, acidic *sodium aluminum phosphates,* SALP's, have become commercially available within the last ten years, although several others have been patented as potential leavening acids.[28,29,120,126,132,135,137-140,143] Some of their chemical and leavening reactions are shown in Tables 2 and 10. They are usually distinguished from each other by using numerals corresponding to the number of sodium, aluminum, and phosphate units in each molecule, as shown in Table 10. Figure 2 and Table 10 demonstrate that the DRR's of the SALP's provide a low initial reaction during the mixing stage, and only low levels of gas are released during bench action. The latter characteristic is especially useful where it is desirable to prepare batters and hold them for considerable periods prior to baking them. For example, in restaurants pancake batters frequently are prepared during slack periods, refrigerated, and held until needed during rush periods. Refrigerated batters have been shown to retain adequate leavening action for up to two weeks under such conditions.

TABLE 10
Properties of Commercially Available Phosphate Leavening Acids[a]

Key:

AC—angel food cakes and mixes
AMCP—coated, anhydrous monocalcium phosphate
BM—biscuit mixes
CBP—commercial baking powder
CK—cookies and cookie mixes
CM—cake mixes
DCP·2H$_2$O—dicalcium phosphate dihydrate
DM—doughnut mixes
FD—frozen doughs

HBP—household baking powder
MCP·H$_2$O—monocalcium phosphate monohydrate
PM—pancake and waffle mixes
RD—refrigerated doughs
SALP—sodium aluminum phosphate
SAPP—sodium acid pyrophosphate
SAS—sodium aluminum sulfate
SRC—self-rising corn meal
SRF—self-rising flour

Phosphate Leavening Acid	Chemical Formula	Neutral-izing Value	Leavening Gas Released, %			Uses
			2-min Mix Stage	8-min Bench Action	During Baking	
MCP·H$_2$O	Ca(HPO$_4$)$_2$·H$_2$O	80	60	None	40	AC, CBP, CK, HBP, PM, CM
AMCP, coated	Ca(HPO$_4$)$_2$	83	15	35	50	CM, SRF, SRC, PM
DCP·2H$_2$O	CaHPO$_4$·2H$_2$O	33	0	0	100	CM, FD
SALP—1:3:8	NaH$_{14}$Al$_3$(PO$_4$)$_8$·4H$_2$O	100	22	9	69	FD, PM, CM, BM
SALP—3:2:8	Na$_3$H$_{15}$Al$_2$(PO$_4$)$_8$	100	21	4	75	FD, PM, CM, BM
SALP + AMCP	Blend	80	27	20	53	SRF
SAPP—fastest	Na$_2$H$_2$P$_2$O$_7$	72	40	8	52	Adjust rates of other SAPP
SAPP—second fastest	Na$_2$H$_2$P$_2$O$_7$	72	36	8	56	DM
SAPP—medium	Na$_2$H$_2$P$_2$O$_7$	72	30	8	60	CM, BM, PM
SAPP—second slowest	Na$_2$H$_2$P$_2$O$_7$	72	28	8	64	RD, CBP
SAPP—slowest	Na$_2$H$_2$P$_2$O$_7$	72	22	11	67	RD

[a] Data obtained from references 126, 132, 134, and 136–138.

The SALP's also have very bland flavors and are tolerant to wide variations in flour and other ingredients. They are particularly useful in batter systems containing highly emulsified shortenings, such as today's cake mixes. Because the crystalline compounds are very stable in the presence of moisture, they frequently are used in the preparation of self-rising flours, prepared mixes of all types, and frozen bread doughs. As shown in Table 10, the two types of SALP differ little in their leavening characteristics and are interchangeable.[28,120,126,132,135,137-140,143,144] A specially modified SALP has been developed for use with highly emulsified shortenings in prepared cake mixes. This modified SALP improves volume, grain, moisture retention, and texture of the cakes baked with it.[28,120,132,137-139]

Five different grades of sodium acid pyrophosphate, SAPP, each having a different DRR, are commercially available. The differences in DRR are accomplished by heat-processing variations in manufacturing or by coating the surfaces of the crystals with slowly soluble phosphates.[28,29,120,126,134,135,137-141] All grades react very quickly with sodium bicarbonate in water solution or in dough systems containing no source of calcium ion. However, in the presence of calcium ion, either added as a soluble calcium salt or from milk, the rate of reaction of the SAPP's is delayed in direct relation to the method of manufacture. It is believed that the calcium ion reacts with the surfaces of the SAPP crystal to form a coating of insoluble calcium pyrophosphate crystals that delays the release of acidity.[28-30,126,137-141] The chemical characteristics, the rates of reaction, and other leavening characteristics of the five grades of SAPP are shown in Tables 2 and 10.

Some consumers are highly sensitive to a disagreeable flavor contributed by SAPP-leavening systems. This flavor is described as "astringent," a "coating of the teeth," a "coating of the roof of the mouth," and a "pyro taste."[28-30] This flavor is usually attributed to the orthophosphates produced by hydrolysis of the pyrophosphate in the dough systems.

Other Potential Phosphate Leavening Acids   Although other acidic phosphate salts also have been proposed as leavening acids, they have thus far not found significant commercial use. Among these are calcium acid pyrophosphate,[145,146] alkaline-metal, heavy-metal complexes of orthophosphoric acid (similar to the SALP's),[147-150] the acidic salts of sodium metaphosphate,[151] and the acidic ammonium, alkali-metal, and alkaline-earth metal metaphosphates.[152]

Leavening Acids in Baking Powders   Because the housewife and the small baker cannot economically store the individual leavening acids and soda for use in leavening their baked products, they normally purchase prepared mixtures, called baking powders, for this function. Baking powders are complete leavening systems consisting of mixtures of the proper quantities of soda, leavening acid, and a diluent, usually starch, that separates these ingredients so they will not interact with each other during storage. Baking powders in commercial use normally contain sufficient soda and the acid to neutralize it in order to yield not less than 12% of their weight as CO_2 when used. Typical formulas for household and commercial baking powder compositions are shown in Table 11; these and a few other formulas are described in the literature.[28-30,120,126,138,153]

Two types of household baking powders are marketed today. These are single-acting baking powders, in which all of the CO_2 is released during preparation of the batter, and double-acting baking powders, containing two leavening acids, one that reacts during mixing and one that does not react until the baked product is being heated.[28-30,120,121,126] All commercial baking powders used by bakers, restaurants, and other large-volume consumers are double-acting baking powders.[28-30,126,153]

TABLE 11
Typical Baking Powder Formulas[a]

Constituents	Household Baking Powders						Commercial Baking Powders		
	Pure Phosphates		Phosphate-SAS, Double-acting Types			Cream of Tartar			
Soda, granular	28.0	28.0	30.0	30.0	30.0	27.0	30.0	30.0	30.0
Monocalcium phosphate monohydrate (MCP·H$_2$O)	35.0		8.7	12.0	5.0		5.0		5.0
Monocalcium phosphate, anhydrous (AMCP)		34.0							
Cornstarch, redried	37.0	38.0	26.6	37.0	19.0	20.0	24.5	26.0	27.0
Sodium aluminum sulfate (SAS)			21.0	21.0	26.0				
Sodium acid pyrophosphate (SAPP)							38.0	44.0	38.0
Calcium sulfate			13.7						
Calcium carbonate					20.0				
Cream of tartar						47.0			
Tartaric acid						6.0			
Calcium lactate							2.5		

[a] Data from references 120, 126, 138, and 153.

The SALP's could be used to replace the slow-acting or delayed-reaction leavening acids; they would provide additional stability during storage, improve performance during use, and create a less disagreeable flavor in the baked product. Formulas for baking powders containing the SALP's have been proposed,[137,138,153] but they have not been accepted by the industry to any extent.

Leavening Acids in Prepared Mixes   Prepared baking mixes have enjoyed a spectacular growth in annual sales since the end of World War II. Among the mixes available are those for cakes, pancakes, waffles, biscuits, muffins, and doughnuts. Some contain all of the dry ingredients necessary and require only the addition of liquid, while others require the additions of liquid and eggs or liquid and shortening; a few require the addition of other ingredients.

Cake Mixes   Cake mixes represent a large portion of the prepared-mix market. Flour, milk, and eggs are used to furnish structure, while sugar, shortening, and chemical leavening give tenderness to the finished cake. Special, highly emulsified shortenings cause the batter to incorporate and retain numerous, minute, air-bubble nuclei during the mixing stage; these can later be enlarged by the thermal expansion of entrapped air, the increase in vapor pressure of the batter water during baking, and the reaction of the chemical leavening system.[28,29,120,131] The use of emulsifiers reduces requirements for early leavening action during the mixing stage. It is, therefore, common practice to use combinations of slow-acting leavening acids that retain much of the leavening reaction for the baking stage. Proper leavening, however, is a critical part of any cake-mix formula. The leavening must react at the proper point during the preparation and the baking of the cake to prevent loss of gas from the batter if it reacts too soon and to prevent rupturing of air cells and weak final structure of the cake if reaction occurs too late.[28,29,120,121,126,131,132,135,137-141,143,153,154] The leavening system controls the pH of the finished cake and, therefore, affects crumb and crust color, intensity of flavor, and other properties of the cake.[121] Optimum pH values range from 6.9–7.2 for white cakes, 7.2–7.5 for yellow cakes, and 7.1–8.0 for chocolate or devil's food cakes.

Most modern cake mixes contain combinations of leavenings in order to obtain maximum volume, maximum uniformity of grain and texture, and the optimum shape of the baked cake. In designing a leavening system for a cake mix or other baked product, it is first necessary to determine the quantity of leavening gas required to provide the desired characteristics. When this is known, the quantity of soda that will produce that amount of leavening gas is calculated. It is then possible to calculate the levels of leavening acids necessary to neutralize the soda and provide the needed amount of leavening gas by means of their neutralizing values and the required final pH of the baked product.[28,121] Combinations of AMCP and SALP are most commonly used in white and yellow cakes; combinations of AMCP and SAPP or DCP·2H$_2$O are most commonly used in chocolate cakes, since the SALP's tend to dull the chocolate color. Combinations of 10–20% of the fast-acting leavening acid and 80–90% of the slow-acting leavening acid are used.

Pancake and Waffle Mixes   Pancake and waffle mixes usually have somewhat simpler ingredient systems than prepared cake mixes. They normally contain wheat flour, shortening, a small amount of sugar, nonfat dried milk, salt, soda, and the proper amount of leavening acid. Some mixes also contain one or more of the following: corn flour, rice flour, buckwheat flour, corn sugar, whey solids, or dried buttermilk. The most common blend of leavening acids is 20–30% MCP·H$_2$O or AMCP and 70–80% SALP. This combination of leavening acids allows the chef in a pancake house or the individual consumer to prepare pancake batter the night before it is to be used. If kept in the refrigerator, the batter will retain adequate leavening action to prepare excellent pancakes or waffles for up to a week. The batter normally deteriorates through bacterial souring before there is a serious loss of leavening capacity.

There has been a recent trend toward the sale of frozen pancake batters. Again, combinations of the MCP's and SALP's or combinations of the MCP's, SALP's, and SAPP's provide proper stability under conditions of occasional thawing and refreezing, as may occur in transit between the manufacturer and the final consumer.

Prepared Biscuit Mixes   The prepared biscuit mix normally contains flour, shortening, salt, and leavening. Milk is added to such a mix; however, if the manufacturer has included dried milk, the consumer merely adds water. Very small quantities of sugar are sometimes added for flavor and for optimum crust color. These mixes normally contain blends of 30–50% AMCP and 50–70% SALP or SAPP.

Self-rising Flours and Corn Meals   Self-rising flours and corn meal mixes normally contain flour or corn meal, salt, soda, and leavening acid. Shortening and liquid must be added to prepare biscuits or muffins from the mixes. Although these mixes have been popular since the early 1900's in the southeastern United States, they are beginning to appear more frequently in national markets.[28,126] The self-rising flour mixes can also be used to prepare cakes, pancakes and waffles, and similar baked products by the proper addition of milk, eggs, and shortening. The earliest self-rising products contained $MCP \cdot H_2O$, which was replaced in 1939 by AMCP due to its superior stability.[120, 126] Combinations of SALP and AMCP have most recently been used in the majority of the mixes commercially available, because the blend provides improved flavor and greater storage stability under conditions of high humidity.[28, 29, 155] Typical formulas for self-rising flour, self-rising corn meal, and self-rising corn-meal mix are shown in Table 12.

TABLE 12
Typical Commercial Formulas for Self-Rising Flour and Self-Rising Corn Meal[a]

Ingredients	Self-rising Flour		Self-rising Corn Meal		Self-rising Corn Meal Mix	
	%	lb	%	lb	%	lb
Wheat flour	94.899	100.000			9.434	10.00
Corn meal			94.340	100.00	84.906	90.00
Soda	1.305	1.375	1.651	1.75	1.651	1.75
SALP—AMCP blend[b]	1.661	1.750				
AMCP			1.651	1.75	1.651	1.75
Salt	2.135	2.250	2.358	2.50	2.358	2.50
Total	100.000	105.375	100.000	106.00	100.000	106.00

[a] Data obtained from references 138, 139, 144, 155, and 156.
[b] Commercial blends of SALP and AMCP are available for use in self-rising flour; see references 137, 138, 139, and 144.

Cake Doughnut Mixes   Cake doughnut mixes contain wheat flour (and possibly soy, potato, and cottonseed flours), various sugars, egg yolk, some shortening, dry-milk solids, salt, soda, and the necessary leavening acid. This is usually one of the SAPP's or a combination of SAPP with an MCP leavening acid. It is also possible to use SALP, although this acid has not been accepted by the industry.

Leavening Acids in Refrigerated Doughs   Refrigerated, ready-to-bake dough products are commercially available for use in the preparation of biscuits, dinner rolls, and various types of sweet rolls. These are used in large quantities in this country today. The products are packed in cylindrical, fiberboard, pressurized cans sold from refrigerated dairy cases held at 35–40°F. The individual consumer merely opens the can, places the dough products on a baking sheet, and bakes them for 10–20 minutes before serving them to her family. The doughs normally contain flour, water, shortening, nonfat milk solids or dried whey solids, sugar or corn sugar, salt, soda, and an SAPP leavening acid. Examples of typical biscuit dough formulas are shown in Table 13. Elaborate equipment and processes have been developed for the manufacture of these products. Each manufacturer usually maintains strict secrecy in regard to his formula, equipment, and process. Dinner and sweet rolls are manufactured by slight variations of the biscuit manufacturing process.[28, 29, 120] Only the two slowest SAPP's are suitable for leavening systems in refrigerated dough systems; between 1.0–1.5% soda and 1.4–2.0% SAPP are utilized.[28, 29, 120]

TABLE 13
Some Published Formulas for Refrigerated Biscuit Doughs

Ingredients	Formula Number			
	1^a	2^b	3^c	4^d
	%	%	%	%
Flour	54.9	52.5	55.7	54.8
Water	33.0	30.8	32.4	32.8
Shortening	5.5	9.0	5.5	4.5
Nonfat milk solids	1.6	—	—	2.2
Buttermilk solids	—	3.2	—	—
Whey solids	—	—	1.3	—
Sucrose	—	0.5	—	—
Dextrose	1.1	—	1.0	2.0
Salt	1.1	1.4	1.2	1.4
Soda	1.2	1.2	1.3	1.1
SAPP	1.6	1.4	1.6	1.0
Fumaric acid	—	—	—	0.2
Total	100.0%	100.0%	100.0%	100.0%

[a] Data from Van Wazer, J. P., *Phosphorus and Its Compounds*, Copyright © 1958, Interscience Publishers, Inc., a division of John Wiley & Sons, Inc. Reprinted by permission.

[b] Data from The Pillsbury Co., Minneapolis, Minn. (reference 157).

[c] Percentages calculated from Baker, J. S., and Lindeman, C. G. (Procter & Gamble Company), U.S. Patent 3,297,449, 1967.

[d] From Erekson, A. B., and Duncan, R. E. (Borden Food Products Co.), U.S. Patent No. 2,942,988, 1960; and Matz, S. A. (Borden Food Products Co.), U.S. Patent No. 3,397,064, 1968. Used by permission of Borden Food Products Co.

The use of the pyrophosphate leavening acids gives rise to two problems in the preparation of refrigerated doughs. One problem is that the conversion of pyrophosphate to orthophosphate leads to a disagreeable, astringent flavor noticeable to some consumers;[28-30] the flavor can be reduced to some extent by proper formulation but is never eliminated. Secondly, the orthophosphates produced by the leavening reaction[126] can, under certain conditions of formulation and storage, form large, visible crystals that have the appearance of pieces of glass.[28-30,158,159] Although the crystals are harmless and dissolve readily during baking, the consumer often mistakes them for glass and discards the biscuits prior to use.[28-30,126,134] Erekson and Duncan and later Matz[159] reported that the addition of fumaric acid to the refrigerated biscuit dough prevents dough discoloration and the appearance of phosphate crystals. Baker and Lindeman[158] have proposed the use of potassium bicarbonate alone or in combination with sodium bicarbonate to eliminate these crystals.

Although some refrigerated cake, pancake, and muffin batters have been marketed for short periods of time, they have never been commercially successful because the quality of the resulting baked products could not equal that obtained from prepared mixes. Combinations of $MCP \cdot H_2O$, AMCP, DCP, SAPP, or SALP have been used in these batters. The SALP's are especially effective in these applications.

Non-leavening Applications in Cereal Products

Dough Conditioning   The phosphates have been reported to overcome the variability in quality of gluten used in the preparation of baked products. This may, at times, be nothing more than control of the pH to maintain it within an optimum range for the desired physical or chemical effect. It may, at other times, be the interaction of polyphosphates with the proteins.[30]

The presence of the alkaline-earth and heavy-metal ions has been reported to stiffen or toughen bread doughs. This is especially true of the calcium and aluminum ions.[28, 30, 134, 135, 149, 160] As a result many bakery-supply companies market "dough conditioners" that normally contain one or more calcium, ammonium, or potassium phosphates for these purposes. The acidic calcium phosphates and the SALP's have been reported by various investigators to improve the colloidal properties of bread doughs and to stimulate yeast activity.[30,160,161] This is reported to be especially true for continuous bread-making formulations.[161] In addition, the non-microbial deterioration of refrigerated dough products was reported to be delayed for significant periods of time by the addition of 0.1–3.0% of high-molecular-weight potassium poly-metaphosphates.[159] The polyphosphates are thought to link with the gluten of the dough to stiffen it and maintain its structure.

Inhibition of Enzyme Activity   The phosphates have been reported to inhibit undesirable enzyme activity. For example, Takebayashi[162] reported the inhibition of lipoxidase activity in the batters of sponge cakes by the addition of the pyrophosphate anion. Carter and Hutchinson[163] reported that α-amylase activity was inhibited by the addition of inorganic phosphates, especially DSP.

Antioxidant Activity   The phosphates have been reported to inhibit oxidative rancidity in cereal products. Muhler[164] reported that the addition of a mixture of MSP and DSP to cereals increased their oxidative stability and inhibited development of stale, rancid flavors.

Inhibition of Microbiological Growth   The ability of polyphosphates to stabilize cereal products, such as doughs, batters, and baked products, against microbiological growth was reported by Kohl and Ellinger.[102,103] The most effective polyphosphates were found to be those having an average length of 20–33 phosphate units. The shorter-chain and longer-chain polyphosphates were also found to inhibit the growth of yeasts and molds.

Mineral Enrichment   Although the inorganic salts, commonly called minerals, comprise a small percentage of total body weights, they are as important to the well-being as the proteins, carbohydrates, and fats that provide energy. It is impossible to obtain energy from these latter foodstuffs without the mineral elements. The elements calcium and iron are two of the minerals required for the total well-being of all organisms. Calcium and iron phosphates are commonly added to foodstuffs to provide these essential mineral elements. Since phosphorus, as the phosphate anion, is also an essential mineral, the phosphate salts of these mineral elements are doubly useful.

Calcium and Phosphorus   Calcium and phosphorus are present in living organisms in the largest quantities of all the mineral elements. Most of this calcium and phosphorus is incorporated in the formation of the skeleton. Children require calcium and phosphorus in the diet for normal growth;

adults require it for maintenance and replacement. Pregnant and lactating women require larger quantities of calcium and phosphorus than other adults, because a portion of their intake is utilized by the growing fetus or child.[165,166] Present knowledge indicates that the optimum ratio of calcium to phosphorus in the diets of children is between 1:1 and 2:1; for adults it is slightly less than 1:1. Stunting of growth, bone malformations, and poor-quality bones and teeth result when the diets of children are deficient in calcium. If the diet lacks a sufficient quantity or has an improper ratio of calcium to phosphorus, the elements are removed from the skeletal tissues of the body.[166]

The presence of the calcium ion often introduces undesirable physical characteristics in processed foods, such as meat, fruits, and vegetables. It is, therefore, difficult to supplement calcium and phosphorus in the diet by addition of compounds containing these elements to such foods. Cereals, however, are easily fortified with calcium phosphates, since they are seldom adversely affected by them and may even be improved by their addition. As a result federal definitions for flour and many baked products allow the optional addition of certain minimum levels of calcium compounds, including the calcium phosphates. The type of calcium phosphate to be added and the quantity required to provide the desired level depend on the type of cereal product, the amount of naturally occurring calcium, and the amount of the element contributed by the selected compound. Table 14 shows the levels of calcium and phosphorus in each of the three calcium orthophosphates commercially available for food use, as well as the calcium-to-phosphorus ratio in each. By use of this table, one can calculate the required amount of the proper calcium phosphate necessary for addition to any cereal product to provide the level of each element desired.[166] Excellent reviews of the enrichment of flour and baked goods have been provided in publications of the Natural Research Council.[167] Tablets, or powdered mixtures in soluble, edible packages, containing the proper weight of enrichment for standard weights of doughs or batters are available for enriching baked products. The FDA considers such direct enrichment of baked products equivalent to the use of enriched flour.[168]

TABLE 14

Levels and Ratios of Calcium and Phosphorus
in Commercial Calcium Phosphates

Compound	% Ca	% P	Ca:P Ratio
Monocalcium phosphate	16.4	24.2	0.68:1.0
Dicalcium phosphate	23.6	18.0	1.31:1.0
Tricalcium phosphate	36.7	17.3	2.12:1.0

Source: *The Importance of Calcium and Phosphorus Supplements in Foods*, Technical Service Bulletin, Stauffer Chemical Company, New York. Used with permission.

Numerous investigators have recently reported that supplementation of foods with calcium phosphate salt appeared to control dental caries. The addition of almost any of the calcium phosphate compounds, as well as some of the sodium phosphates, appears to inhibit dental caries.[169-180]

Iron   Iron is also an essential mineral, as it is found in a number of iron-containing enzymes. Approximately 80% of the iron supply in the body is found in the hemoglobin of red blood cells.[181] The human body maintains a sophisticated system for regulating the level of iron and its absorption and utilization.[182]

The presence of relatively soluble iron salts in cereals and other fat-containing products often increases the rate of rancidity development in fat. It has been found that the iron phosphate salts, such as ferric orthophosphate, ferric pyrophosphate, and sodium iron pyrophosphate, have little, if any, effect on increasing fat oxidation.[27,28,30,110,111] However, a great controversy has arisen in scientific circles regarding the availability of the iron from these compounds when they are used to supplement cereal products or other food systems in the diet.[28,183-198] Partly as a result of this controversy, the FDA has recently published a proposal in the *Federal Register* for a significant increase in the levels of iron supplementation in flour and cereal products.

A recent development in iron supplementation has been reported by Jones of the U.S. Department of Agriculture.[199] It involves the treatment of whey protein with a "ferri-polyphosphate" to form a whey-protein–iron complex. No studies have been reported of the oxidative stability of foods supplemented with this material; as far as can be determined, there also have been no studies of the ability of the human body to utilize iron from this complex. Such studies should be made before the complex becomes a useful material for iron supplementation.

Decreasing Cereal Cooking Times   The addition of orthophosphates to various cereal grains during processing has been reported to significantly reduce the cooking times required to prepare cereal products.[28,30,200-202] Levels of 0.2–2.0% phosphate were used to buffer the pH of the system at an optimum range for the most rapid cooking of the cereal. The cooking times of macaroni and related pasta products also have been decreased by the addition of 0.5–1.0% phosphate salts; FDA regulations now permit the addition of DSP to these products.[30]

Miscellaneous Applications   Phosphates have been reported to improve color, reduce the rate of respiration, and decrease rootlet development during germination of barley malt.[203,204] The color of precooked "cocoa noodles" was improved by treating them with SAPP prior to drying.[205] The color of processed soybeans was improved by treating them with calcium phosphates during processing.[206] The addition of a soda-phosphate leavening acid system to a gravy mix was reported to prevent lumping during its preparation.[207] Steeping corn in a solution containing MSP, DSP or TSP was advantageous in the preparation of the corn for making tortillas and similar products.[208]

Phosphate Applications in Dairy Products

As so many investigations of phosphate functions and their applications in milk products have appeared in the literature, only a portion of them will be reviewed here. The phosphates have been shown to be useful in processing nearly every type of dairy product.[28,29] The greatest amount of work reported in the literature involves the interaction of phosphates and inorganic cations, especially calcium, and of phosphates and the milk proteins. There are numerous other functions of the phosphates of extreme importance in the processing of milk.

Interactions of Phosphates with Milk Constituents

A number of investigators have studied the chemistry of the casein-calcium-phosphate complex. A review of the current knowledge and four proposed structures for the casein micelles of milk have been published by Rose.[94] Calcium orthophosphate is intimately involved in the formation and stability of the casein micelles, which provide the white, opaque appearance of milk. Many of the applications of phosphates in the processing of dairy products involve the interactions between the phosphates and the casein or the calcium in these micelles. Further discussion of these interactions follows:

Interaction with Milk Calcium   Sommer and Hart[209] first proposed that the addition of phosphate or citrates changes the salt balance of the milk through the formation of complexes with the calcium and magnesium ions. However, recent work has shown that other effects of equal or greater importance also occur.[209-216] The ion-exchange reactions of casein[210,214] result in a reduction in levels of free-calcium ions and an increase in bound or colloidal calcium as phosphates are added.[211,213,215,216] The addition of orthophosphates to milk is believed to precipitate calcium,[211,213] while the addition of polyphosphates is believed to cause the formation of casein-calcium-phosphate complexes from which it is difficult to remove the calcium.[210,211,215,217,218]

The addition of polyphosphates, such as SHMP, to milk appears to do more than just complex calcium ions. SHMP reduces the turbidity of milk at much lower levels than other calcium-complexing agents.[28,219,220] As little as 1 part SHMP to 20 parts of the protein in fluid skim milk causes almost complete loss of opacity, and the milk becomes a yellow-green, translucent serum;[220] the addition of EDTA did not result in a similar loss of opacity. Odagiri and Nickerson[219] reported that casein appears to combine with a portion of SHMP to form a large, stable complex. If, however, an excess of SHMP is added above that required for the formation of the casein-SHMP complex, the casein micelles are completely dispersed, with the resultant loss in the normal opacity of the milk.

Effect of Phosphates on Heat Stability of Milk   The addition of orthophosphates increases the heat, rennet, and alcohol coagulation times of milk.[212,221-229] This effect is very useful in the manufacture of cheese and the stabilization of evaporated milk; it also could be useful in stabilizing the effects of seasonal variations in calcium levels in milk.[229,230] Studies by Zittle and Pepper[222] showed that the aggregation of calcium caseinate was reduced within any given time and that the effect of heat also was lowered by the addition of phosphate. Rose[223,224] found that the amount of calcium and phosphate ions absorbed on the surface of casein micelles had a direct bearing on the heat stability of milk. Other studies confirmed that the degree of heat stability of milk was inversely proportional to the ratio of soluble calcium:soluble orthophosphate.[225-229]

Interactions with Milk Proteins   The reactions of the various phosphates with milk proteins differ in direct relation to their chain lengths. The addition of under 15 mmol of orthophosphates to milk causes little change in the viscosity of the milk, but the addition of greater quantities causes the milk to gel at a rate in direct proportion to the amount of phosphate added. The degree of gelation also increases in proportion to the calcium-ion concentration and the level of milk-solids-not-fat.[231] Vujicic and deMan[232] reported that ortho-phosphates were not bound when they were added to a 1% solution of casein. However, all of the polyphosphates, from pyrophosphate through long-chain polyphosphates, were bound when added to a similar solution of casein.

Casein is precipitated by pyrophosphates to form thick gels.[217,218,233] Zittle[217] believed that the negatively charged pyrophosphate anions were bound to the positively charged casein at the normal pH values of milk. This resulted in a reduction of the net charge of the casein so that it was precipitated the same as at its isoelectric point. The heating of pyrophosphate-treated milk eliminated the gels.[218] Pyrophosphate, however, did not precipitate the soluble proteins in milk.

The longer-chain polyphosphates precipitate both casein and the soluble proteins, such as β-lactoglobulin. Of course, polyphosphates are well known, general protein-precipitating agents.[217,234-240] The acidic polyphosphate anion is believed to form cross-links between protein molecules by reacting with basic amino groups on different protein molecules. Thus, large, compact particles, or micelles, are formed.[241,242] Leviton[240] studied the stabilizing effect of polyphosphates, such as SHMP, on milk proteins. Although the polyphos-phate was rapidly hydrolyzed to pyrophosphate in the milk, no gels were formed from the presence of pyrophosphate anions. Leviton believed that the polyphosphates stabilized the milk proteins against gelling by the formation of cross-linkages, most likely involving the pyrophosphate formed upon hydrolysis of the longer-chain polyphosphate. It has also been reported that the digestibility of milk and milk foods for babies and convalescents could be improved by treating the milk product with STP or SHMP.[243] This may result from the unfolding of protein molecules during the dispersion of the casein molecule.

Further study of the reactions of the polyphosphates with the natural protein systems in milk and other foods should lead to many new practical applications of the above phenomena. Similar studies of the reactions of poly-phosphates on meat proteins are now in process, and preliminary reports show the similarity between their reactions with meat and those with milk protein. A recent excellent review of phosphate interactions with the milk proteins was published by Melnychyn.[29]

Applications in Milk-based Beverages

The phosphates have several functions in the processing of milk-based beverages. For example, Kawanishi[244] stabilized the color of a strawberry-juice-containing milk drink by chelating traces of iron in the drink with <0.3% polyphosphate. Wouters[245] adjusted and buffered the pH of pasteurized milk to 6.3 with a mixture of DSP and trisodium citrate for preparation of solid-milk-drink products. Jackson[246] sensitized spoilage bacteria in cocoa and in cocoa and malt mixtures intended for use in chocolate-flavored-drink products. First, he acidified a syrup of these ingredients with H_3PO_4 prior to heating them to sterilization temperatures, and then he restored the initial pH by addition of DSP. Roland[247] prepared thickened aqueous dispersions of milk products, such as chocolate milk, by the addition of long-chain, insoluble potassium and sodium salts of polyphosphates; these then could be added to untreated, flavored, aqueous milk dispersions to maintain proper viscosities in the final product. Van Wazer[30] reported that the addition of small quantities of powdered TSPP to malted-milk powders improved the wetting of the particles, the color, and the flavor; it also stabilized the drink made from the powders through production of a thin gel.

Applications in High Butterfat Dairy Products

Butter   Phosphates were used by Malkov and his co-workers[248] to remove iron and copper ions from solution and thus to stabilize butter against oxidative spoilage during storage. North[249] reported that phosphoric and other acids aided the clarification of butter oil by coagulating suspended materials that cause cloudiness. Meyer and Weck[250] reported that buffering milk with alkali salts, such as DSP, TSPP, sodium polyphosphates, etc., allowed sour whole milk to be used to prepare stable, sweet-tasting butter from its cream and to produce cheese from the milk. A 1:1 mixture of DSP and basic sodium citrates was especially effective.

Buttermilk   Corbin and Long[251] reported a direct acidification, two-stage process for the production of buttermilk that had four times greater refrigerated shelf-life than cultured buttermilk. The process involved the treatment of the milk with ammonium or alkali-metal salts of orthophosphates and pyrophosphates prior to acidification; H_3PO_4 was used for acidification. The process required only 1–5 hours for preparation, compared to 15–48 hours for the cultured process. Van Wazer[30] reported that the addition of TSPP to buttermilk acted as a dispersing or deflocculating agent for the curd, which normally formed from casein in the acidic buttermilk; in contrast, TSPP acted as a gelling agent for milk at its normal pH.[217,218,233]

Cream   The phosphates can improve the whipping properties, flavor, and stability of cream. Rothwell[252] obtained improved overrun and stability against leakage when whipping cream was treated with sodium alginate or TSPP, although the sodium alginate was more effective than the TSPP. Wilson and Herreid[253] reported that the flavor of sterilized cream was stabilized by the addition of a sodium polyphosphate with an average chain length of 5 to 7, plus ascorbic acid and/or α-tocopherol. Coffee cream that was served to the author in individual containers listed disodium and sodium tetraphosphates or DSP among its ingredients;[28] the label on the container reported the product to be half-and-half with stabilizers added.

Applications in Condensed and Evaporated Milks

Because they are canned and sterilized by retorting to approximately 240°F, condensed and evaporated milks must be stabilized against heat coagulation by the addition of either DSP or sodium citrate. Failure to do so results in gelling of the milk during storage.[27,30,110,111] The federal standard of identity for evaporated milk allows the processor to add <0.1% of anhydrous DSP.[254]

Condensed and evaporated milks can also be prepared by the aseptic canning techniques that have recently reached commercial development. Edmondson[255] found that it was necessary to control the initial viscosity of the sterilized milk through the addition of calcium chloride and DSP to a level just below that which would cause immediate coagulation of the milk. Martynova[256] reported that the addition of DSP or sodium citrate to the milk prior to pasteurization increased the shelf-life and stabilized evaporated milk against thickening for at least 6 months of storage at room temperature. Ascorbic acid, vitamin A, glucose oxidase, and SHMP were effective inhibitors of Maillard-type discoloration compounds during an 8-month shelf-life test of sweetened condensed milk, which is especially susceptible to such discoloration; DSP and vitamin E were found to be ineffective.[257] Muller[258] found that DSP prevented the coagulation of evaporated milk prepared from skim-milk powder recombined with butter oil.

Applications in Sterile, Concentrated Milk

An aseptically canned, sterile, whole-milk concentrate recently was developed by dairy scientists. The milk, after concentrating, is given a high-temperature, short-time sterilization treatment; it then is canned under aseptic conditions in separately sterilized cans.[259-274] Some investigators have reported that the milk is stable at room-temperature storage for as long as 2 years, as either a 2:1 or a 3:1 concentrate, with the 3:1 concentrate being the most practical.

There are three major problems in the production of the sterile-milk concentrate, as defined in an excellent review by Seehafer.[259] First, the milk proteins tend to coagulate or gel during heat processing. Second, the concentrated milk is unstable during storage, since it has a tendency either to form gels or to separate into a solid and a serum phase during storage. The third major problem is that of flavor stability. Polyphosphates have been reported by numerous investigators to improve the heat and storage stabilities of the sterile-milk concentrates.[259-267] The addition of polyphosphates, ranging from sodium tetrapolyphosphate through the very-long-chain sodium polyphosphates, has been reported to increase the storage life from 3-fold to as much as 10-fold over the sterile, concentrated milk given no phosphate treatment. The level of phosphates used has been reported to range from 0.01–1.0%.[262,263] According to Herreid and Wilson[264] the polyphosphate molecule penetrated the casein micelle to combine with the proteins and calcium, to bind the calcium, and to form covalent linkages to the stable calcium–caseinate, calcium–orthophosphate complex, thus retarding gelation. Leviton[240] reported that the polyphosphates were almost instantly hydrolyzed to pyrophosphate during the sterilization process (15 seconds, heating at 137.8°C). He further speculated that the pyrophosphate resulting from this hydrolysis formed cross-linkages within the casein micelle that stabilized the micelle. When pyrophosphate was added directly to the milk, Leviton believed it interacted with the magnesium and calcium ions to form insoluble colloidal complexes; it was, therefore, unavailable to stabilize the casein micelles in the same manner as the pyrophosphate derived during the hydrolysis of polyphosphates.

Milks produced at different seasons of the year have different heat stabilities during the sterilization step, according to Swanson and Calbert.[265] A computer program has been developed by Magnino[266] and by Bixby and Swanson[267] for use in correlating the characteristics of milk at various seasons with the necessary processing conditions to produce stable, sterile-milk concentrates. The nutritional properties of the sterile, concentrated milks were found to be equivalent to those of spray-dried whole milk by Bolt and Kastelic.[270]

The reactions of consumers to the sterile, concentrated milks were reported to be highly favorable in a number of investigations.[265,271,273] Their favorable reactions probably were influenced by the fact that these milks could be sold at a lower price than milk delivered to the home and that they require no refrigeration.

The polyphosphates also have been reported to be useful in the stabilization of frozen and irradiated, concentrated milks against gelling and other instabilities during storage.[275-279] In spite of the reports indicating successful processes for production of concentrated, sterile milks and their acceptance by the public, a demand does not yet seem to exist. However, if increases in labor costs and the costs of delivery continue, this situation may very well change.

Applications in Milk Gels and Puddings

Instant, milk-gel puddings have largely replaced the former cooked, starch puddings as desserts in the U.S. household. The milk-gel puddings are purchased as dry mixes. Cold milk is added to the mix; the mixture is blended thoroughly and allowed to set for a short time at refrigerator temperatures. In one type of mix, the milk is gelled by the action of TSPP on the milk proteins in the presence of calcium acetate;[30,280] the calcium ion acts as a coagulation or gelation accelerator, causing the mixture to set up to pudding consistency after about 10 minutes. Another type of mix utilizes $MCP \cdot H_2O$ as the gelation accelerator and also includes acetylated monoglycerides.[281] A third type utilizes orthophosphates, such as MSP and DSP, as gelation accelerators in place of the calcium salts.[282] Typical formulations from these three types of pudding mixes are summarized in Table 15.

The optimum pH for acceptable puddings ranges between 7.5–8.0. Although some of the necessary alkalinity is contributed by the TSPP, it is often necessary to add alkalizing agents. Not all lots of TSPP will produce firm, non-lumping puddings in reasonable setting times of 10–15 minutes after the addition of the milk.[283] Thus, it is necessary for manufacturers to test all batches of TSPP in order to select those that will provide acceptable puddings. At least one manufacturer now labels his acceptable batches as "Pudding Grade."

Other recent developments have demonstrated that pudding-like milk gels can be prepared from combinations of carrageenan and milk in the presence of phosphates, such as SHMP, STP, TSPP, and TSP.[284] Another report utilizes combinations of TSPP and sodium hydroxide with an additional calcium phosphate for milk puddings.[285] Partridge[286] prepared milk gels by using water-insoluble, crystalline potassium metaphosphates, solubilized by the addition of a compatible, edible sodium salt. The resulting milk gels could be used to stabilize chocolate-milk preparations, improve the viscosity of cream, and improve ice cream and sherbet mixes, custards, puddings, and milk-containing confectionery products.

A recent development has been the marketing of refrigerated or frozen ready-to-eat puddings that, according to McGowan,[287] utilize TSPP and sometimes DSP to provide proper gelling of the milk proteins and the desired firmness of the gel. TSPP is also useful in the preparation of instant cheesecake fillings that require only the addition of milk; the mixture then is blended thoroughly and poured into a prepared, baked graham-cracker crust. After being refrigerated for 15 minutes, the instant cheesecake is ready to serve.

TABLE 15

Instant Pudding-mix Formulas for Addition to One Pint of Cold Milk

Ingredients	Patent 2,607,692[a]		Patent 2,801,924[b]		Patent 3,231,391[c]
	Vanilla, g	Chocolate, g	Vanilla, g	Chocolate, g	Unflavored,[d] g
Sucrose	82.4	81.2	66.5	65.5	78.3
Pregelatinized starch	20.0	22.0	20.0	26.0	16.7
Cocoa	—	16.0	—	16.0	—
Vanilla flavor	2.0	2.0	2.5	1.4	—
Salt	2.4	2.4	—	—	0.6
Color	2.6	—	0.3	0.4	—
TSPP	2.8	2.8	1.5	1.5	2.2
Calcium acetate	1.2	1.2	—	—	—
MCP·H_2O	—	—	—	—	1.6
DSP, anhydrous	—	—	1.5	3.0	—
Acetylated monoglyceride	—	—	—	—	0.6
Total mix weight	113.4	127.6	92.3	113.8	100.0[d]

[a] From Kennedy, M. H., and Castagna, M. P. (Standard Brands, Inc.), U.S. Patent 2,607,692, 1952; with permission.
[b] From Clausi, A. S. (General Foods Corporation), U.S. Patent 2,801,924, 1957; with permission.
[c] From Breivik, O. N., Slupatchuk, W., Carbonell, R. J., and Weiss, G. (Standard Brands, Inc.), U.S. Patent 3,231,391, 1966; with permission.
[d] The addition of flavors and colors to this mix is usually required.

Typical formulas for ready-to-eat puddings and cheesecake fillings are summarized in Table 16.

The phosphates, particularly polyphosphates, can be used to form very thin gels of milk protein in order to reach a sufficient viscosity to suspend chocolate particles throughout chocolate milk.[288] They also can form proteinaceous gels from casein or sodium caseinate by means of orthophosphate, pyrophosphate, and a soluble calcium salt for use in binding water and reducing the migration of fat in various types of food compositions.[289]

Several studies have been made of the gelling phenomena of phosphates in milk. Fox and his co-workers[231] reported that orthophosphates have little effect on milk viscosity at concentrations below 15 mmol. Higher concentrations of orthophosphate tend to form milk gels in direct relation to the concentration of the orthophosphate. In contrast, the polyphosphates retard milk gelation when they are present below certain critical concentrations, which differ for each polyphosphate, but they form protein gels above those critical concentrations. This work of Fox et al. appears to explain why polyphosphates can be used in low concentrations to prevent the gelation during storage of sterile concentrated milk, and yet they can be used at higher concentrations to form milk gels. Zittle[217] studied the interaction between the positively charged casein and the negatively charged pyrophosphate ions, which results in the reduction of the casein net charge and its precipitation; this interaction is similar to the one that occurs at the isoelectric point of casein. The gelling effect of the pyrophosphate on milk proteins is completely reversed if the milk is heated.

TABLE 16
Formulas for Ready-to-eat Puddings and Cheesecake Filling

Ingredients	Firm-bodied Pudding,[a] %	Creamy Pudding,[a] %	Cheesecake Filling,[b] %
Cheese, dried			9.61
Sugar (sucrose)	6.44	10.00	7.69
Corn sugar (dextrose)	6.87	4.35	3.08
Cornstarch, modified	4.94	2.61	3.84
Tapioca starch, modified	1.07	2.61	
Nonfat dried milk	7.78		
Buttermilk, dried			7.69
Whole milk, liquid		66.95	61.51
Shortening, dry	2.68		5.77
Salt	0.11	0.22	0.38
TSPP, anhydrous	0.05	0.22	0.31
MCP·H_2O			0.12
DSP, anhydrous	0.04		
Water	70.02	13.04	
Flavor	q.s.[c]	q.s.	q.s.
Color	q.s.	q.s.	q.s.
Total	100.00	100.00	100.00

[a] Adapted from McGowan, J., 1970. *Food Prod. Devel.* 4(5):16, 18.
[b] Courtesy of Stauffer Chemical Company.
[c] A sufficient quantity (*quantum sufficiat*).

TABLE 17
Effect of Various Ingredients on Starch-free, Cold-milk Gel

TSPP[a]	MCP[b]	Alginate	Milk	Viscosity, cps
X	X	X	X	33 700
X	X	X		19 200
X	X		X	6 400
X			X	3 200
		X	X	0
		X		0

Source: Lauck, R. M., and Tucker, J. W., 1962, *Cereal Sci. Today* 7(9):314, 322.

[a] TSPP—tetrasodium pyrophosphate.
[b] MCP—monocalcium phosphate monohydrate.

Lauck and Tucker[233] studied starch-free, cold-milk gels prepared with various combinations of TSPP, MCP·H_2O, alginate, and milk. As shown in

Table 17, they produced the highest viscosities when all four ingredients were present; however, substantial viscosities were produced in gels formed without the presence of milk at all by using the calcium phosphate as a calcium source. The authors hypothesized that the gels formed by treating milk with TSPP were those formed between calcium and TSPP rather than between the milk protein and the pyrophosphate.

Applications in Nonfat Milk

The phosphates have been found to improve the properties of dried and liquid skim-milk products through their interactions with the milk proteins and through their ability to complex calcium.[290-298] The solubility of nonfat milk solids for food use could be increased by adding metaphosphates,[290] MSP, DSP, or citrates[291,298] to the milk either before drying or after reconstitution. The addition of the solubilizing compounds to the milk prior to drying was more effective than when it was added during reconstitution.

Curry[292] prepared a skim-milk colloid with excellent fat-emulsifying properties for use in table spreads, ice cream, and similar high-fat dairy foods. He treated the skim milk with TSPP to form a stiff gel and hydrolyzed the lactose to produce the emulsifying colloid.

Lindewald and Kimstad[293] increased the whipping properties of milk casein so that it could be used as a replacement for egg white. The casein was coagulated with rennet, and then the suspended protein was treated with orthophosphate or pyrophosphate at pH 6.3–6.7. Lewis and her co-workers[294] significantly improved the whipping qualities of nonfat-milk solids by the addition of 1% SHMP or STP, which formed very stable foams. Sodium citrate and EDTA also improved foaming properties, but the foams were very unstable. Table 18 illustrates the effects of various calcium sequestering agents on the whipping ability of nonfat-milk solids. Table 19 illustrates the effects of SHMP and STP on the whipping properties and stability of a high-quality preparation of nonfat-milk solids.

TABLE 18

Effect of Various Sequestering Agents on the Whipping Ability
of Nonfat-milk Solids

Sequestering Agent Added (1% in Nonfat-milk Solids)	Increase in Volume, %	Drainage in 20 Min, %
Control	330	100
Sodium orthophosphate	370	100
Sodium tripolyphosphate	750	64
Sodium hexametaphosphate	750	56
Maddrell's salt	560	100
Sodium citrate	660	100
Versene	610	100

Source: Lewis, M. A., Marcelli, V., and Watts, B. M., 1953, *Food Technol.* 7:261–264, Table 1. Reprinted by permission.

Kempf and Blanchard[295] prepared a dried coffee whitener from either fluid or concentrated milk by treating the milk with 1.4–3.7% DSP prior to drying.

A heat treatment also was involved; the authors reported a direct relationship between the DSP level and the heat and time necessary to prepare an acceptable product. Kumetat[296] prepared a replacement for whole eggs for use in sponge-cake and similar baked products by treating concentrated milk or suspensions of casein with calcium-sequestering polyphosphates or citrates and then heating them. Hayes and his co-workers[297] demonstrated that STP could be used to disperse suspensions of calcium caseinate; the level of STP had to be varied according to the amount of calcium in the caseinate.

TABLE 19
Effect of Polyphosphates on Whipping Ability
of High-quality, Nonfat-milk Solids

Added Polyphosphate (4% of Milk Solids)	Milk Solids in 50 ml Water, g	Volume Increase, %	Drainage in 2 hr, %
Control	10	560	60
Hexametaphosphate	10	800	50
Tripolyphosphate	10	750	52
Control	31	390	10
Hexametaphosphate	31	420	0
Tripolyphosphate	31	420	0

Source: Lewis, M. A., Marcelli, V., and Watts, B. M., 1953, *Food Technol.* 7:261–264, Table 2. Reprinted by permission.

Applications in Frozen Dairy Desserts

The beneficial effects of phosphates on the properties of ice cream and similar dairy products have been frequently reported.[30,299-304] Until 1967 phosphates were allowed only in chocolate ice cream and similar frozen desserts as stabilizers for chocolate suspensions. Since then, however, the phosphates have been approved for use in other dairy desserts.

Phosphates are utilized in frozen desserts to stabilize fat particles against agglomerating and forming what is called "sandy" ice cream. This phenomenon is also known as *churning*. Churning has been prevented by the addition of DSP, TSPP, SHMP, and sodium citrate at levels too small to affect the pH or to allow the processor to neutralize acidic or spoiled cream for the manufacture of the frozen dessert.[300-302] As a result these compounds are now allowed in frozen desserts at levels up to 0.2%.[303] Wallander[304] reported that the treatment of milk with polyphosphates retarded lipase activity and, thus, improved flavor stability.

Applications in Whey, Lactose, and Lactalbumin

The phosphates have been shown by a number of investigators to improve the properties of cheese whey; thus, more whey can be used in food processing.[28] Whey is the liquid residue remaining after cheese curd has been removed from milk in the processing of natural-cheese products. About 4.2 billion pounds of the 14 billion pounds of whey produced annually is processed for human and animal feed; the balance is often a serious pollutant in streams adjacent to cheese-processing plants.[305] Approximately 26% of the whey that is processed is used for the production of lactose; 11% is used for the manufacture of whey proteins, while the balance is dried for use in human or animal foods.

The polyphosphates have been implemented in the manufacture of lactose to maintain the whey proteins in a soluble state, while the liquid is concentrated to the point where lactose crystallizes from the solution.[306,307] Ward and his co-workers[308] produced a dried whey product that overcame many of the problems resulting from the replacement of dried milk with whey in baked products. They added a soluble calcium compound and a soluble phosphate compound, then neutralized the liquid whey prior to drying to pH 6.8–7.5. The treated dried whey tolerated the addition of amounts of moisture somewhat similar to dried milk, while it still maintained good structural properties in the dough.

The 14 billion pounds of liquid cheese whey represents a potential source of approximately 70 million pounds of whey protein. Arena[309] reported that undenatured whey protein, commonly called *lactalbumin*, is similar to the proteins of human milk; therefore, it is nutritionally more valuable than casein, the major protein of cow's milk. The lactalbumin commercially available usually is obtained during the processing of lactose from whey. As it is highly denatured by heat, it is probably nutritionally less valuable than undenatured lactalbumin.

The abilities of long-chain polyphosphates to react as polyanions and of large protein molecules to react as polycations have been utilized to isolate undenatured proteins.[28,30,234–239] Gordon[310] isolated undenatured whey proteins from liquid whey by the use of SHMP or longer-chain polyphosphates. McKee and Tucker,[311] Wingerd,[312] Hartman,[313] and Hartman and Swanson[314] have extended Gordon's work, reporting processes for the preparation of undenatured lactalbumin or the complex of lactalbumin–polyphosphate. The sodium and the calcium salts of lactalbumin–phosphate are now commercially available.[315] McKee and Tucker[311] have reported that lactalbumin–phosphate would replace part or all of the nonfat-milk solids used in the preparation of many baked products. Wingerd[316,317] used lactalbumin–phosphate in the preparation of instant, cold-water-soluble gelatin products and of liquid-type emulsifiers for use in preparing food emulsions.

Ellinger[318] used lactalbumin–phosphate to replace sodium caseinate both functionally and nutritionally in the preparation of imitation dairy products; Broadhead[319] used it to replace part or all of the egg white in layer cakes, angel food cake, doughnuts, pancakes, and similar baked products.

The ability of the polyphosphates to improve protein dispersion is utilized in processing foods, such as whey products. Endo[320] added sodium polyphosphate to whey and other milk products that were then used in the preparation of mayonnaise-like products. Ellinger and Schwartz[321] added polyphosphates to liquid and dried whey, demineralized whey, and similar modified whey products to improve the dispersion of the whey proteins; the phosphate-treated whey products could be used to form stable, imitation dairy products, while the untreated wheys would not function in such a manner.

Applications in Imitation Dairy Products

The high cost of milk and the marketing success of margarine, the first imitation dairy product, have stimulated dairies, as well as other food manufacturers, to market other imitation dairy products. Two types of imitation dairy products are now being marketed in those areas where they are legally acceptable. One type is designated as a "filled" dairy product; it is prepared by replacing only the butterfat with vegetable fat, while utilizing all other constituents of the natural dairy product. The second type is labeled "imitation" dairy product; it contains no dairy constituents. Imitation dairy products are combinations of vegetable fat, sugars, proteins (such as soy and whey proteins or sodium caseinate), stabilizers, emulsifiers, and sufficient water to prepare a product that closely resembles the appearance and other characteristics of the real dairy product. Vitamins and minerals are added to some of these products in order to duplicate the nutritional properties of the natural dairy products.[322-324]

Although very few of the filled products contain phosphates, they would be most useful in them. The imitation dairy products utilize phosphates as buffers to maintain the optimum pH for protein dispersion; thus, they aid in stabilizing the fat particles in the product. The most common phosphates in use are DSP or DKP; DKP is usually preferred, because it does not contribute as much astringent flavor as DSP.

Imitation Milk Replacements for liquid milk have begun to appear in the larger markets on the east and west coasts of the United States. These completely synthetic products duplicate very closely the appearance, textural properties, and nutritional characteristics of cow's milk.[322-327] Some are bases that require the addition of water, and perhaps a vegetable fat; the mixture is then pasteurized, homogenized, and packaged for sale to the consumer.[323,327] Ingredient costs for imitation milks appear to be approximately one-half of the cost for similar unit volumes of cow's milk.[322-324]

Kozin and Rodionova[325] studied the effect of adding increasing quantities of DSP to an imitation-milk emulsion. The pH decreased from 6.95 to 6.2 as the level of DSP was increased from 0.1 to 0.25 g/g protein; the protein began to precipitate at this pH. As more phosphate was added, however, the pH rose and finally reached 7.3 at 0.5 g phosphate/g protein. The precipitated protein was redispersed as the phosphate increased, and the size of the fat globules decreased with increasing phosphate. Sabharwal and Vakaleris[326] reported that the addition of calcium ion, phosphate, and citrate salts at the proper levels in the proper sequence was necessary to prepare stable imitation-milk emulsions. They developed a heat- and storage-stable, synthetic milk system, containing all of the minerals of regular milk, by adding the salts in the following order: citrate, calcium, and phosphate.

Coffee Whiteners Replacements for liquid coffee cream have been marketed in the United States for approximately 20 years. One of the first to appear was a "dried cream," which was prepared from liquid cream that had been treated with DSP.[328] A similar product was prepared from dried milk by Brochner,[329] who added alkali-metal phosphates or citrates, or mixtures of the two, and additional lactose to produce a powdered coffee whitener.

Completely synthetic coffee whiteners are now marketed in nearly every part of the United States. These usually contain combinations of vegetable fats, sodium caseinate or soy protein, gums as viscosity stabilizers, phosphate or citrate salts as stabilizers for the proteins and as buffering agents, corn-syrup solids or sugar as bodying agents and sweeteners, lipid-type emulsifiers to enhance emulsification of the fats, and flavors and colors to duplicate the characteristics of cream. Typical levels of the various ingredients in formulas for liquid and dried coffee whiteners are shown in Table 20. The functions of the various ingredients have been discussed by a number of investigators.[28,330-338]

TABLE 20
Ingredient Levels in Typical Non-dairy Coffee Whitener Formulas[a]

Ingredients	Liquid Whiteners, % (total wt)	Spray-dried Whiteners, % (Dry)
Vegetable fat	3.0–18.0	35–45
Protein	0.5–3.0	4.5–10.0
Corn-syrup solids	1.5–10.0	40–60
Sucrose	0–3.0	—
Emulsifiers	0.1–1.0	0.2–3.0
Gum stabilizers	0–0.5	—
Phosphate stabilizing salts	0.1–1.0	1.0–2.0
Flavor	q.s.	q.s.
Color	q.s.	q.s.
Water	To 100%	—
Anticaking agent	—	q.s.

[a] Compiled from references 330–338

The proper type of phosphate to use in coffee whiteners is determined largely by the composition of the other ingredients in the system. Most coffee whiteners contain sodium orthophosphate salts, although potassium orthophosphate salts are used wherever flavor is a problem. Thus, DSP and DKP are most commonly used,[330,332,333,335,337] and polyphosphates[336] and SALP[338] have also been proposed. The phosphates are thought to maintain the optimum pH at which a continuous protein layer will be formed about the fat droplets in order to prevent syneresis, or "wheying off." The phosphates also prevent *feathering*, or curdling of the protein, and fat separation when the coffee whitener is added to an acidic coffee medium.

Whipped Toppings Whipped toppings represent replacements for high-fat cream products, just as coffee whiteners represent replacements for low-fat cream products.[339] Although whipped toppings usually are prepared from the same types of ingredients as coffee whiteners,[340] greater quantities of vegetable fat are used in whipped toppings. The phosphates again serve as buffers and stabilizing agents for the protein films necessary in the formation of stiff, stable foams and in the prevention of syneresis, or weeping. Powdered whipped-topping compositions also are available.[341]

When properly formulated and stabilized, whipped toppings have more improved functional properties than those of natural whipping cream; phosphates can significantly contribute to the stabilization of these improved properties. Whipped toppings, for example, usually can incorporate a greater amount of air and water during whipping, and they have greater overrun, superior stand-up, and, depending on the type of fat used, superior eating quality.[339,342,345]

The phosphates found most useful in whipped-topping formulations are DSP, DKP, and TSPP at levels of 0.025-1.0% of the complete liquid base prior to whipping. The levels of ingredients in typical, liquid, non-dairy whipped-topping formulas are presented in Table 21. Dry whipped toppings are prepared with similar ratios of the ingredients used in the liquid toppings; concentrates containing 30-50% solids are formed, then spray-dried and instantized.

TABLE 21
Ingredient Levels in Typical Whipped-topping Formulas[a]

Ingredients	% of Total Weight
Vegetable fat	24.0-35.0
Protein	1.0-6.0
Sucrose	6.0-15.0
Corn-syrup solids	0-5.0
Gum stabilizers	0.1-0.8
Lipid-type emulsifiers	0.3-1.1
Stabilizing salts	0-0.15
Water	To make 100%

[a] Data from references 331, 339, and 342-345.

Imitation Sour Cream, Sour-cream Dressings, and Chip Dips   Replacements for fermented cream products are produced in considerable and growing quantities for the consumer market in the United States. These non-dairy products contain vegetable fat, protein (either soy protein or sodium caseinate), gum stabilizers, stabilizing salts, flavorings, acids to provide tartness, and often coloring agents, as well as imitation or real bacon chips, chives, spices, and other natural flavoring agents.[331,339,346,347] The stabilizing phosphate salts again are used to cause the proteins to form the properly dispersed, protective films about the fat globules and to prevent syneresis.

Imitation Ice Cream or "Frozen Desserts"   Replacements for frozen dairy products are prepared by combining vegetable fat, either soy protein or sodium caseinate, emulsifiers, gum stabilizers, stabilizing salts, sugar, bodying agents, and the proper flavors and colors to produce products that resemble ice cream so closely that it is often difficult to distinguish between the natural and imitation products. Phosphate salts are used in most of these products to stabilize the protein system, as in other imitation or non-dairy products.[348,349]

Imitation Cheese, Cream Cheese, and Cheese Spreads   Replacements for all types of dairy cheese products also can be prepared from non-dairy products, as well as from filled dairy products.[350,351] The proper formulation can result in products that, when properly processed, will resemble the natural product in all characteristics, including flavor and nutritional properties. The phosphates, including the mono-, di-, and tribasic orthophosphates, pyrophosphates, and polyphosphates, are useful in improving the physical characteristics of the proteins; the phosphates also form complexes with the proteins to provide the proper dispersion so that the finished products will simulate the natural products.[351] This author is personally aware of experiments that have resulted in cheese products very similar to natural cheddar, natural Swiss, and natural cream cheeses. There is no reason to doubt that any type of natural cheese can be duplicated with synthetic, non-dairy products through proper formulation, including the use of phosphate stabilizing salts.

Margarine   Even margarine seems to benefit from the addition of a phosphate. A Japanese patent disclosed that an oil-in-water-in-oil emulsion could be prepared by homogenizing a milk casein dispersion in a "trisodium polyphosphate solution" with hydrogenated oil to form a margarine product.[352] The phosphate, assumed to be STP, most likely caused proper dispersion of the protein and inactivated any interfering calcium or magnesium ions in the system.

Applications in Cheese Products

The literature that contains references to the applications of phosphates in cheese products is voluminous. The majority of uses of phosphates are in processed cheeses, although phosphates also have been found useful in the preparation of cottage and natural aged cheeses.

The initial steps in the manufacture of cheese products are similar for all types of cheeses including cottage cheese. Whole milk is usually used, although skim milk also can be used to produce low-fat cheeses. The pasteurized milk is treated with lactic-acid bacteria or with lactic or phosphoric acid, then usually with a protein-coagulating enzyme; this causes the formation of a protein "curd," an insoluble paste or coagulum of the milk casein with the milk fat entrapped inside the curd in the form of emulsified droplets. The subsequent handling and processing of the curd determines the type of cheese that will be produced.[30]

Cottage Cheese   Cottage cheese is prepared by washing the loosely packed, freshly coagulated casein curd; creamed cottage cheese is prepared by mixing this curd with whole milk, which often contains food-grade acids, such as H_3PO_4.[353] Bristol and Martin[354] pre-acidified skim milk with H_3PO_4, or its mixture with citric acid, prior to the addition of lactic acid bacteria; the setting time of the cottage cheese curd was reduced by 44% compared to the conventional procedure. A method of continuous processing of cottage cheese has recently been developed and is being promoted in European countries; the method reduces labor, provides greater uniformity in the cheese, increases the yield of cheese from the milk, and yields more cheese per hour of operation, which results in a lower-cost cottage cheese curd.[355] The process involves direct acidification rather than inoculation with lactic-acid bacteria; among the acids used is H_3PO_4.

Natural Cheese   The product termed *natural cheese* is produced by microbiological fermentation of milk curd. A number of steps are involved in the preparation of the curd in a process that requires a considerable amount of time.[28] The curd is finally compressed in forms in which it becomes a compact mass fitting the shape of the form. It is then aged for various periods of time to "ripen." The typical cheese flavor is developed by the fermentation of the curd during the ripening stage, and it is often caused by very specific cultures of bacteria or mold. Phosphates have been found to be beneficial in a number of the steps involved in the manufacture of natural cheese.[28,219,356-376]

Kielsmeier[356] significantly shortened the time necessary to prepare cheese curd for compression into forms by adding a calcium-complexing phosphate, such as TSPP, to the milk, along with a bacterial culture and rennet. The mixture was allowed to react for 10–20 minutes, after which it was acidified with a food-grade acid, such as lactic acid or H_3PO_4, to lower the pH to 5.4–5.5. If a firmer curd was desired, the calcium salts could be added along with the acid during the latter part of the process. The curd formed by this process could be compressed into forms and aged. The process could feasibly lend itself to continuous cheese processing because of the shortened time involved. Edam cheese was improved by the addition of MCP to more rapidly reduce the pH of the milk and to form a harder curd.[357]

The interest in a continuous process for the production of cheese curd is very high, since it would involve less labor and less time to age the cheese. The processes that have thus far been proposed involve the use of food-grade acid, such as H_3PO_4. This acid has been found to increase yields, decrease moisture, and increase the calcium level of the cheese.[358,359]

The rennet coagulation time of milk can be stabilized by the addition of polyphosphate, probably through the formation of calcium and magnesium complexes, since these ions are essential for milk coagulation by rennet.[356,360-364] Orthophosphates have no effect on the rennet coagulation time,[360,361,363] while polyphosphates are more effective than other calcium- and magnesium-complexing agents, such as EDTA.[219,362]

A number of investigators[365-373] have treated the milk used to develop starter cultures for cheese with either orthophosphates or pyrophosphates to prevent the growth of bacteriophage, which attacks the bacterial cultures necessary for proper fermentation and flavor development of the cheese. The orthophosphate salts appear to be more effective than the polyphosphates in the binding of calcium and in preventing its use as a nutrient for bacteriophage growth. The preferred process calls for the sterilization of milk and of solutions of the phosphates;[373] the phosphate solution is then aseptically added to the sterile milk, and the milk–phosphate mixture is reheated and then inoculated with the proper cheese-producing strain of bacteria.

Waters[374] has reported that periodic washing of the rinds of curing cheeses with a saturated solution of DSP prevents the growth of mold, rind rot, bacteria, and cheese mites when the cheeses are aged at humidities greater than 95%. Schulz[375] prevented spoilage of cheese surfaces by immersing the cheese block or wheel in a 20% aqueous solution of SAPP heated to 90°C, or by rubbing the surface with SAPP and subsequently heating it to 100°C in a steam tunnel for a short time. The process develops a stable, airtight, edible rind similar to process cheese. Soaking the press cloths used to wrap the surfaces of rindless cheeses in a 2% solution of sodium tetrapolyphosphate at pH 6.26 prevents adhesion of the press cloth during curing of the cheese.[376]

Pasteurized Process Cheese Products   Pasteurized process cheese products are prepared by blending cheeses of various ages with water, sodium chloride, and emulsifying salts. The process involves heating the blended ingredients to 160–170°F with agitation to melt the mixture into a smooth, plasticized, fluid mass. The molten mixture is then formed into loaves of the proper size and shape, or it is chilled in thin sheets or ribbons for cutting into sliced cheese.[28,30] J.L. Kraft obtained the initial patent for the preparation of a "sterilized cheese" in 1916.[377-380] Garstin[381] first used phosphate salts as emulsifiers for cheese processing in order to prevent fat separation. Most process cheeses prepared in the United States contain either DSP or an alkaline SALP, although these are sometimes combined with each other or with additional salts, such as TSP or sodium citrate. Recent reports indicate that the cyclic sodium tetrametaphosphate, but not trimetaphosphate,[382] and a starch-phosphate[383] can also function as cheese emulsifiers.

The adverse effects of rework (salvage) cheese can be overcome by the use of a combination of a lipid-type surface-active agent and phosphate-emulsifying salts.[384] In addition to their functions as buffers that adjust and maintain the pH in an optimum range, these phosphate salts also contribute highly desirable and essential effects to the physical properties of the cheese proteins, leading one to believe that their functions are more than just maintaining an optimum pH.[22,28,30,110,111,384-386] The federal standards of identity for pasteurized process cheese allow the addition of up to 3% of any one, or any combination, of the allowed emulsifying salts based on the weight of the pasteurized process cheese.[387] H_3PO_4 is also allowed, along with other food-grade acids, as an acidifying agent for pasteurized process cheese. Bolanowski and Ziemba[388] have recently described a continuous process for the manufacture of pasteurized process cheese.

There have been numerous investigations of the emulsifying systems and their effects on the properties of pasteurized process cheese.[23,27,28,378-400] These investigations have demonstrated a number of functions for the emulsifying salts, such as buffering the process-cheese system at the optimum pH for proper dispersion of the protein. The emulsifying salts also interact with the calcium present in the system, possibly to precipitate some of it, but also undoubtedly to form cross-links through the calcium ions between the protein molecules. Furthermore, they seem to interact with the protein molecules at the surfaces of fat droplets to prevent coalescing and thus prevent loss or bleeding of fat from the cheese. The emulsifying salts also appear to increase the water-holding capacity of the protein in order to prevent or at least inhibit water loss. Several investigators have shown that sodium citrate is not as effective on cheese as the phosphates.[23,28,399,400]

When polyphosphates are used as cheese-emulsifying salts, as is quite common in European countries, they are rapidly hydrolyzed to ortho-, pyro-, and tripolyphosphate units.[401] Slower but definite hydrolysis of very-long-chain sodium polyphosphates (Graham's salt) has also been reported by Scharpf and Kichline,[402] who have proposed these compounds for use in process cheese in this country. One of the major disadvantages to the use of DSP and TSP as emulsifying salts in some pasteurized process cheeses has been their tendency to initiate the appearance of crystals on the surface of the cheese; this is especially true of the packaged, sliced cheeses. These crystals have been identified as calcium tartrate crystals by Blanchard,[404] calcium citrate crystals by Morris and co-workers,[406] and DSP by others.[28]

The use of the sodium orthophosphates as emulsifying salts also results in the development of less acceptable flavor with aging.[27,110] Cheese emulsified with the SALP systems appears to retain a richer, "cheesy" flavor for a significantly longer period. The same emulsifying systems are now being used in cheese spreads[407] and low-fat or fat-free process cheeses.[408,409]

Phosphate Applications in Egg Products

A number of phosphate applications for improving the processing and functional properties of eggs have been reported in the literature;[410-424] examples are discussed in the following paragraphs. Their functions include complexing undesirable metal ions, buffering to optimum pH values, improving foam volume and stability, and inhibiting enzyme activity, microbiological organisms, and fat oxidation.

Applications in Shell Eggs

Pikarr[410] discovered that iron in the wash water used to clean soiled shell eggs contributes to more rapid spoilage, even when it is present in very low quantities. The addition of an iron-complexing acid to the wash water was found to prevent the formation of an iron–conalbumin complex that was responsible for the increased rate of spoilage. The effects of the acids were increased by the addition of small quantities of sodium, potassium, or ammonium orthophosphates or SAPP to the wash solution. Preferred complexing acids were citric, H_3PO_4, glutamic, acetic, and tartaric acids in order of decreasing effectiveness.

Applications in Whole-egg Products

Chin and Redfern[411] discovered that the addition of H_3PO_4, polyphosphoric acid, or any one of their alkali-metal or ammonium salts would prevent the formation of a disagreeable, muddy-brown color upon exposure of whole eggs or egg yolk to air. As a result, the federal standards for whole-egg products have been revised to allow the addition of up to 0.5% MSP or MKP for color preservation.[412]

Liquid egg products, particularly if frozen products are thawed, undergo proteolytic action and coagulation. Hall[413] reported that the addition of a polyphosphate, preferably with a chain length above three, stabilized the eggs against this enzymatic action.

Bellamy[414] prepared a concentrated whole-egg product having superior whipping properties and foam stability by separating the egg white from the egg yolk; he then added SAPP to the egg whites and allowed the mixture to ferment with natural organisms until the pH had been lowered to 6.3–6.5. The egg whites then were concentrated to one-quarter of their original volume and reconstituted with the equivalent quantity of egg yolk that had been previously removed. Lewis and co-workers[294] discovered that the addition of STP or SHMP to dried, whole eggs improved their whipping properties and stabilized them against deterioration during storage. STP gave the greater improvement, being especially noticeable in improving the foaming properties of dried, whole eggs when stored at room temperature for 3 months. Controls stored a similar period of time would not foam at all, as shown in Tables 22 and 23.

TABLE 22

Effect of Polyphosphates and Storage on Foam Volume of Whole-egg Meringues

Egg Type	Storage Time, weeks	Storage Temperature, °C	Polyphosphate Used	Optimum Phosphate in Dried Egg, %	Max Volume Increase, %
Fresh	None	—	—	—	560
Dried No. 1	None	—	None	—	200
	None	—	Hexametaphosphate	2.3–3.0	330
	None	—	Tripolyphosphate	1.3–2.0	360
Dried No. 2	None	—	None	—	220
	None	—	Tripolyphosphate	1.5	380
	13	4	None	—	220
	13	4	Tripolyphosphate	1.5	380
	13	25	None	—	0
	13	25	Tripolyphosphate	1.5	160

Source: Lewis, M. A., Marcelli, V., and Watts, B. M., 1953, *Food Technol.* 7:261, Table 3.

TABLE 23

Effect of Polyphosphates on Whipping Ability of Whole, Dried Egg in Sponge Cake

Egg Type	Polyphosphate in Dried, Whole Egg	Weight of 1 Cup of Batter, g	Volume of Cake from 175 g of Batter, ml
Fresh		62	980
Dried—4% Moisture	None	117	435
	2.9% hexametaphosphate	90	660
	4.6% hexametaphosphate	97	645
	1.5% tripolyphosphate	84	710
Dried—4.5% Moisture	None	110	525
	2.3% hexametaphosphate	81	815
	1.5% tripolyphosphate	74	895

Source: Lewis, M. A., Marcelli, V., and Watts, B. M., 1953, *Food Technol.* 7:261, Table 4.

Tongur[415] found that blending powdered, whole eggs with a combination of ascorbic acid and TSPP inhibited the oxidation of the fats in the dried eggs.

Sourby and co-workers[416] developed a process for sensitizing *Salmonella* to lower-processing temperatures by adding SHMP and a soluble calcium salt to the eggs prior to pasteurization. The process involves the use of 0.5% SHMP, 0.5% calcium chloride, and a pasteurization temperature of 130–135°F. The pasteurized whole-egg products were reported to have superior functional properties to those of whole eggs pasteurized by other available processes.

Kohl and Ellinger[417] reported that the addition of the peroxyhydrates of the alkali-metal salts of phosphates, sulfates, or carbonates to whole eggs was even more effective in the destruction of *Salmonella* and other microorganisms during pasteurization of the egg products.

Applications in Egg Whites

The federal standards for all egg products have been amended to require them to be free of *Salmonella*.[412] Thus, most eggs used in food preparation have been treated by pasteurization or some other method to eliminate *Salmonella*. Pasteurization invariably increases whipping time and reduces foam stability and viscosity of egg whites. Therefore, much interest has been shown in foam whipping aids and stabilizing agents. Kothe[418] increased the tolerance of egg whites to overbeating by the addition of sufficient MSP to reduce the pH of the egg whites to approximately 8.0. Finucane and Mitchell[419] found that the addition of approximately 2.5% SHMP to dried egg white significantly improved the whipping properties and foam stabilities. Finucane[420] extended this work and found that by adding partially hydrolyzed soy protein and SHMP to dried egg white, he could prepare an angel food cake mix in which all of the ingredients could be included in a single mixing stage instead of the two stages normally required.

Chang[421] and Chang and co-workers[422] demonstrated that SHMP stabilized the heat-labile protein conalbumin, which causes the development of turbidity in pasteurized eggs while they are in the process of being pasteurized. The process reduced the foaming ability of the treated egg whites somewhat, but this was restored upon the addition of a whipping agent, triacetin. A further advantage to the process was a significantly increased tolerance in the SHMP-treated egg whites to the presence of small quantities of egg yolk introduced during the breaking operation.

Kohl and co-workers[423,424] reported a pasteurization process for liquid egg whites based on the sensitization of even the most heat-resistant types of *Salmonella* at pasteurization temperatures of 125–130°F with approximately 3 minutes holding time. This treatment was also reported to contribute residual microbiological inhibition in the pasteurized eggs so that bacteria that survived the pasteurization process or were introduced by recontamination decreased in numbers even after thawing and holding the egg whites at 75°F for 72 hours.

Phosphate Applications in Fats and Oils

Various phosphates have been found useful in the processing and the applications of fats and oils in foods.[425–497] The phosphates have been found to aid in the extraction of fats and oils, in their refining, in stabilization of fat against flavor deterioration, and in obtaining stable fat and oil emulsions.

Applications in Oil Extraction

The ability of the phosphates, particularly the polyphosphates, to solubilize and disperse protein should be useful in fat and oil extraction. The phosphates should also be useful in breaking up lipo-protein complexes, thereby increasing oil yields. Rousseau[425] reported that yields of olive oil were increased 1.5–3.0% over those of untreated controls when the olives were treated with 2% of a mixture of 3 parts TSPP and 1 part sodium bicarbonate prior to grinding. Rousseau's findings appear to support the previous statements.

Applications in Processing Fats and Oils

Crude fats and oils as obtained from animal or vegetable matter contain a considerable number of impurities, which include free fatty acids, phosphatides, mucilaginous and proteinaceous materials, and similar contaminants. Crude fats and oils are refined by two methods—alkali refining and acid refining.

Alkali Refining   The alkali-refining process for fats and oils involves the treatment of fats and oils with an alkali, such as sodium hydroxide, at temperatures and for times that vary with the desired results. The alkali treatment forms soaps from any free fatty acids present and from the lecithin and other phosphatides. Sterols are saponified in order to make them soluble in the aqueous solution. All of these impurities can be removed with the aqueous phase.

A disadvantage to the alkali-refining process is that the formation of soaps causes the formation of emulsions with a portion of the oil; therefore, refining losses of 2% or greater can be expected from this process. The addition of levels of phosphate salts ranging from 0.05% to a maximum of 1.0% has been reported by several investigators to significantly reduce refining losses to less than 1%.[426–429]

Acid Refining   Acid refining of fats and oils usually involves the addition of sulfuric, hydrochloric, or chromic acids to the fats and oils. The results obtained can be controlled by adjusting the proportion of acid to oil, diluting the acid with water or other solvents, and varying the temperature of the reaction. The process usually results in a controlled, low level of residual free fatty acids remaining in the refined fat or oil.

Several investigators have reported advantages to the use of H_3PO_4 as the acid of choice in acid refining.[430–441] Some of the refining processes involve combinations of the acid and MSP, DSP, or TSP[433,434,437–439] or ammonium phosphates.[440] A process used to prepare a common, commercial, bakery shortening involves the deliberate, controlled hydrolysis of some of the fat to form free fatty acids by treating the refining mixture with alkali and by subsequently neutralizing with acid to produce 2–10% free fatty acids having no less than 16 carbon atoms in their molecules. The phosphoric-acid refining process has been reported to eliminate the formation of sulfonated triglycerides, to solubilize protein and mucilaginous materials, to hydrolyze phosphatides, and to greatly improve the odor, flavor, and stability against rancidity of the treated fats and oils.

Other Refining Applications   Irmen[442] refined soft fats and liquid oils by treating them with phosphate salts and with higher alcohols and waxes. The result was a hardened fat that required no hydrogenation for use as shortening. Beck and Klein[443] refined highly flavored and colored crude oils by percolating them or filtering them through a bed of granular TCP. The refined oils provided colorless, bland, flavorless fats and oils for use in frying and other similar processes. Toxic alkaloids were reported to be removed by use of 1% manganese dioxide, aeration, and then treatment with 1–1.5% H_3PO_4.[444] TSPP and similar alkali-metal salts were reported to aid in the removal of sterol-containing contaminants of fats and oils.[445] Copper, iron, nickel, and tin in lard were effectively deactivated as oxidation catalysts by phosphoric, citric, tartaric, or ascorbic acids and several of the fatty acid esters of ascorbic acid.[446,447]

Bleaching   After a crude fat or oil is refined to remove its impurities, it must be bleached to remove the coloring materials typical of crude fats and oils. Eckey[448] reported that combinations of H_3PO_4 and bleaching clays were preferable in bleaching fats and oils, although sulfuric and metaphosphoric acids could also be used. Other bleaching processes utilize the following treatments:

1. treating with H_3PO_4 and hydrogen peroxide at 60–70°C for 0.5–1.5 hours,[449]
2. treating the fat or oil with H_3PO_4, $H_4P_2O_7$, or any of their acidic salts plus an inorganic silicone acid or its salt while air is bubbled through it,[450]
3. bleaching the fat or oil with TSPP in a solution of hydrogen peroxide,[451]
4. adding acids, especially H_3PO_4, to acid clays that are then used to decolorize the fats or oils,[452]
5. treating the fat or oil with an alkali metal carbonate and H_3PO_4, and later adding a solution of hydrogen peroxide.[453]

Loury[454] reported that he was able to precipitate contaminating metallic salts and to hydrolyze mucilages by treating fats and oils prior to bleaching with dilute H_3PO_4.

Rearrangement   Rearrangement, or interesterification, requires the use of sodium methylate as a catalyst. This catalyst must be destroyed before further processing of the fat or oil can be accomplished. H_3PO_4 was found useful for this purpose with subsequent removal of the acid and inactivated catalyst by neutralization with sodium compounds and by washing.[455,456]

Hydrogenation   Phosphoric acid has been used to remove nickel catalysts from hydrogenated fats and oils.[457] Hydrogenated vegetable oils lost the hydrogenation odors and stabilized against odor and flavor reversion by treatment of the hydrogenated fats with H_3PO_4 and a neutral bleaching earth and bubbling air through them at elevated temperatures.[458] H_3PO_4 also has been used in nonselective hydrogenation processes of fats and oils.[459]

Monoglyceride Preparation   Monoglycerides are prepared by (1) treating a fat or oil with additional glycerine dissolved in caustic soda, (2) heating, and (3) agitating the mixture until all the glycerine has been esterified by transfer of fatty acids from the triglycerides of the fat. Such compositions, containing mono- and diglycerides, must again be bleached to remove undesirable colored compounds. This can be accomplished by treating the mixture with filter clays and H_3PO_4.[460] It is also necessary to neutralize the sodium hydroxide used as a catalyst; a combination of H_3PO_4 and glycerol has been used to accomplish this neutralization.[461]

Other Processing Applications   The presence of glyceride polymers causes foaming, discoloration, and development of off-odors in frying fats. The polymers can be removed by treating a fat or oil with a combination of phosphoric acid and bleaching clay.[462] Mixtures of filter aids and H_3PO_4 could be used by processors of fried foods to filter their frying fats after each day's use; such filtering could provide additional protection against these polymers in the frying oils.

Applications in Fat Antioxidant Systems

The phosphates have been reported to inhibit the oxidation of unsaturated fatty acids in both dry fat systems and aqueous fat systems.[431-493]

Dry Fat Systems   Dry fat systems containing no moisture, as in refined fats and oils, have been stabilized against oxidative rancidity by saturation with H_3PO_4 alone or in combination with its acidic salts and often with the addition of tocopherol or one of the poly-substituted benzene-ring compounds.[431,448,463,464] According to Calkins[465] dry fat and oil systems are saturated by approximately 0.0002% H_3PO_4. This is an insufficient quantity of the acid to provide synergistic activity with 0.02% of antioxidant quinone compounds. However, the stability of the system was dramatically increased as the level of H_3PO_4 was increased from 0.009-0.09% in the presence of the same level of quinone. Calkins suggested that the H_3PO_4 forms an addition compound with the quinone, which can then diffuse in greater quantity throughout the fat system; it, therefore, can undergo exchange reactions with free radicals formed during the initial stages of formation of peroxide compounds in the unsaturated fatty acids. The phosphorylated fatty-acid-peroxide compound then splits off H_3PO_4 to regenerate the unsaturated bond instead of undergoing the usual cleavage of the bond to form aldehydes and other oxidative rancidity by-products.

Investigators at the American Meat Institute Foundation[466-468] and several others [469-472] have confirmed the synergistic activity of H_3PO_4 with BHA, BHT, propyl gallate, and similar common organic antioxidants to inhibit the oxidation of fats and oils. Privett and Quackenbush[472] reported that H_3PO_4 has a "sparing action" on organic antioxidants in fat and oil systems and that the organic compounds also exhibit a "sparing action" on the H_3PO_4. They also demonstrated that H_3PO_4 prevents the estimation of fatty-acid-peroxide because it interacts with the peroxide at the temperatures used in accelerated stability tests.

Several patents have been issued based on the synergistic effect of H_3PO_4 with organic antioxidants in dry fat systems.[473-482] Kuhrt[483,484] has reported that the amino acid glycine, H_3PO_4, and their salts increase the stability of dry fat systems against oxidative rancidity from 3-30 times over that of untreated controls.

Aqueous Fat Systems   Aqueous fat systems contain both fat and water, and they may contain a number of other ingredients. Watts and her co-workers[485] observed that baked goods leavened with baking powders containing SAPP were more stable toward development of rancid flavors and odors than those leavened with baking powders containing other phosphate acids. SAPP was later confirmed to exhibit synergism with tocopherols naturally present in lard. Watts and co-workers also confirmed that the antioxidant activity exhibited by the SAPP was not due to the binding of copper alone, as had been suggested by Lea.[486] Watts[487] and Watts and her co-workers[488-490] later demonstrated that orthophosphates were ineffective synergists for antioxidant activity of organic antioxidants; the polyphosphates exhibited increasing antioxidant activity with increasing chain length in the presence of such organic antioxidants. Citrate also had some antioxidant activity, but compared only with the STP and shorter-chain phosphates in its activity. Citrates, pyrophosphates, and STP doubled antioxidant activity in the presence of fat systems, while longer-chain polyphosphates, such as hexametaphosphate and Maddrell's salt, increased activity by 6–12 times. Lehmann and Watts[488] studied the antioxidant activity of several phosphates and citric and ascorbic acids by themselves as well as in the presence of various organic antioxidants. As shown in Table 24, the synergistic activity of the phosphates with the organic compounds varies according to the compound tested. The work of Watts and her co-workers has led to two patented applications for phosphates as antioxidants and as synergists for antioxidants in aqueous fat systems, food products, and pharmaceutical preparations.[489,490]

TABLE 24
Synergistic Effects of Phosphates with Various Primary Antioxidants
in Aqueous Lard Systems

	Days to Turn Rancid at $45°C$[a]					
Synergist Used[b]	Plain Lard	Lard and 0.005% Tocopherol	Lard and 0.005% NDGA	Lard and 0.005% BHA	Lard and 0.005% Propyl Gallate	Lard and 0.005% Lauryl Gallate
Buffer	2	4	5	22	11	6
Ascorbic acid	<1	<1	23	1	<1	<2
Citric acid	3	12	20	66	15	16
DSP	2	7	11	43	7	6
STP	3	12	21	48	18	16
SHMP	7	—	34	57	28	21
Maddrell's salt	8	—	31	64	28	27

Source: Lehmann, B. T., and Watts, B. M., 1951, *J. Amer. Oil Chem. Soc. 28*:475; with permission.

[a] Time required for half bleaching of carotene.
[b] Concentration = 0.1% in pH 7.5 borate buffer.

Ozawa and Ota[491] demonstrated that condensed phosphates increased the preservative effects of a number of "antiseptics" added to boiled fish meat; hexametaphosphate was most effective. The effects, however, could be eliminated by the addition of calcium.

Holman[492] patented the use of a thin film of MSP or MKP around the metal exterior of the pouring spouts of cans; this thin film prevented formation of oxidative rancidity by-products that could form on the spout and contaminate the fat or oil poured from the can.

Mahon and Chapman[493] reported that citric acid, H_3PO_4, glycine, and EDTA were ineffective as acidic synergists in increasing the shelf-life of pie crusts. These authors apparently did not test the longer-chain polyphosphates or polyphosphoric acids for this application. Further evidence of practical applications of the phosphates as synergists and antioxidants in food systems will be discussed in subsequent sections in this chapter, particularly in the section on applications in meats.

Applications in Oil and Water Emulsions

Phosphates have been reported to increase the stability of emulsions against separation. van Hees[494] reported that a combination of polyphosphate and pyrophosphate stabilized dispersions of fats and oils with an aqueous phase for "a long time without separation of the fat from the water." A Spanish patent reported that the addition of a phosphate salt to mayonnaise increased the stability of its emulsion.[495] Mitchell[496] reported that the stability of an aqueous peanut-butter suspension was improved by the addition of sufficient DSP to raise the pH of the aqueous phase to 6.6–6.8. Vincent[497] reported a similar stabilizing effect of phosphates on comminuted peanuts when the pH was buffered to remain between 6.5–7.0. These emulsion stabilization applications may depend on buffering capacity to produce the optimum pH for maximum protein-film dispersion to stabilize the system against separation into aqueous and fat phases.

Phosphate Applications in Fruit and Vegetable Products

Table 1 has shown that fruits and vegetables contain significant levels of phosphorus. Investigators have demonstrated that phosphorus exists in the plant tissues and juices as phosphoric acid and phosphate anions at levels as high as 2.7% of dihydrogen phosphate anion in some citrus fruits.[498–500] In spite of these high levels of phosphate and phosphoric acid, phosphates have significant functions as food ingredients in vegetable and fruit processing. They provide stabilization against microbiological spoilage, oxidative rancidity, vitamin loss, and loss of color; they also increase and optimize tenderization or firming of fruit and vegetable tissues and thickness or viscosity of juices and purées.

Applications as Inhibitors of Microbiological Spoilage

Surface Applications   Although some surface applications for phosphates in fruits utilize their ability to act as detergents in the removal of spray materials and the natural waxy coatings of the fruits,[501] other surface applications utilize their ability to act as microbiological inhibitors and as synergists for other preservatives. Bates[502] found that by immersing fruits in a solution of TSP or DSP and drying them so that a film of the phosphate remains, particularly in surface cuts, scratches, or abrasions, the decay of the fruits was delayed for a significant period of time. Rippey[503] preferred a 50-50 combination of TSP and sodium carbonate to reduce the attack of molds on the surfaces of fruits and the spread of mold from one fruit to another. Kalmar[504] applied the surface preservative sodium orthophenylphenate to the surface of fruits in the presence of TSPP, TSP, or sodium carbonate, or a combination of these, at pH values of 10–11.5. Kalmar[505] further reported that combinations of sodium hypochlorite and a buffer composed of DSP and sodium carbonate at a pH of 10.2–11.5 also prevented spoilage of fruit surfaces from blue and green molds.

Smith and Krause[506] prevented the growth of bacteria and enzymatic browning by treating the surfaces of peeled, raw potatoes with a solution of

H_3PO_4, propylene glycol, sodium benzoate, and sodium metabisulfite. Brandt[507] significantly extended the shelf-life and protected fruits and vegetables against microbiological spoilage by dipping them in solutions of $H_4P_2O_7$ at ambient temperatures. The treatment preserved the original colors of the fruits and vegetables and prevented certain types of discoloration, such as the after-cooking darkening of potatoes. When $H_4P_2O_7$ was used alone, mold growth was inhibited only at concentrations above 0.5%. However, combinations of $H_4P_2O_7$ at levels below 0.5% combined with benzoic acid were highly effective. The microbiological preservation could not be explained as being due to the low pH, since buffered solutions were equally effective and often were required when the food was adversely affected by low pH values. Post and co-workers[508] found that polyphosphates were highly effective in preventing mold growth on the surface of fresh cherries if the fruit was dipped in a 10% solution of sodium tetrapolyphosphate; SHMP, STP, and TSPP were also effective, but their effectiveness decreased in the order given. The polyphosphates were believed to interfere with germination of the fungal spores in some unknown way. Although some spores eventually germinated, they produced weak mycelial growth. Similar effects were observed by Kohl and Ellinger.[103]

Applications in Fruit and Vegetable Juices   Kohl and Ellinger[509] demonstrated through microbiological and taste tests that the shelf-life of apple cider treated with a combination of SHMP and either sodium benzoate or potassium sorbate was more than doubled when compared with the same lot of cider treated with the benzoate or sorbate alone. SHMP used by itself was less effective than when used in combination with another compound. The same effect on shelf-life was demonstrated with other fruit and vegetable juices.

Applications as Antioxidants

Combinations of SAPP or citric acid with a mixture of BHA and BHT were shown by Deobald and co-workers[510] to retard the development of off-flavors from oxidative rancidity in sweet-potato flakes. TSPP was reported[511] to extend the shelf-life of dried vegetables when they were dipped into a solution of the phosphate prior to blanching and drying. Although Rhee and Watts[512] found that STP failed to inhibit the activity of lipoxidase in model systems containing pea or soybean lipoxidase, STP was found to be a very effective antioxidant in slurries prepared from the peas themselves. BHA and gallic acid and its esters were less effective when used in treating the peas, but they were highly effective in the artificial systems; the authors were unable to explain the differences in effect.[512]

Molsberry[513] found that the flavor and the texture of frozen mushrooms were retained by treatment with a "sealing mixture" of approximately 1 part sodium sulfate, 1 part DSP, and 2 parts sodium bisulfite in water. The oxidation of ascorbic acid was prevented, and this important vitamin was preserved in model systems and in sugared and non-sugared citrus juices by SHMP[514] and by metaphosphoric acid.[515]

Applications in Stabilizing Fruit and Vegetable Colors

Hall[516] preserved the natural red colors of tomatoes, ketchup, and all types of red berries, their preserves, and juice products by adding 0.1–0.4% of one of the polyphosphates to the fruits prior to cooking. All of the polyphosphates—

from pyrophosphates through the long-chain potassium and sodium meta-phosphates—were effective. The longer-chain polyphosphates were more effective than those with shorter chains.

Nelson and Finkle[517] discovered that treatment of apple slices with a mildly alkaline solution resulted in the methylation of o-diphenolic compounds and subsequently prevented them from being substrates for the enzymatic reaction that causes development of a brown color on the surface of the exposed apple tissue. Bolin and co-workers[518] developed a commercial process based on Nelson and Finkle's work that inhibited the development of a brown surface color and maintained a very crisp texture and a preferred flavor in treated apple slices as compared to those produced by present commercial processes.

The bright-green color of vegetables intended for canning or freezing has been preserved when they were treated with a DSP buffer system that maintained a pH of 6.8–7.0.[110] Fleischman[519] reported that buffering vegetable purées with orthophosphates at pH 8.0 in the presence of magnesium carbonate increased the production of green chlorophyllide in vegetable purées and thus stabilized their green color. Taste panels preferred the flavor of the treated vegetable purées over untreated controls.

The presence of as little as 1 ppm iron and 0.1 ppm copper in fruit or vegetable tissues has been reported sufficient to catalyze the oxidation of o-diphenolic compounds to produce tissue discoloration. Most fruit and vegetable tissues normally contain higher levels of each metal. Combinations of sulfites, such as sodium sulfite, with a mixture of SAPP and TSPP to maintain optimum pH for the sulfite reaction have been reported to control enzymatic discoloration of raw fruit and vegetable tissues.[27]

The blackening of raw and cooked potato tissues due to the oxidation of o-diphenolic compounds in the presence of iron and other heavy metals has been one of the most extensively investigated discoloration reactions in fruit and vegetable processing.[27,28,30,110,111,520-528] Raw, peeled potatoes have often been treated with sulfur dioxide or with sodium bisulfite in acidic solution (to provide sulfur dioxide) to inhibit the enzyme that catalyzes this reaction. Michels[520] reported that by treating the raw, peeled potatoes with polymerized phosphoric acids and calcium or magnesium ions, the blackening of the raw tissues was inhibited.

The after-cooking darkening of potatoes, which also is a reaction involving the oxidation of o-diphenolic compounds, requires the presence of iron to catalyze this enzymatic reaction. Juul[522-523] was apparently the first to discover that SAPP was highly effective in sequestering any iron present in potato tissue and therefore prevented the after-cooking darkening when the potatoes had been dipped or cooked in a solution of the phosphate. Smith and his co-workers[521,524-526] extensively studied the use of SAPP to prevent after-cooking darkening, confirming that this was a highly effective treatment. EDTA also was reported to perform the same function.[521] However, because the treatment with SAPP is generally less costly and fully as effective, it is in general use in the potato-processing industry. Smith and Davis[525] reported that gray or black discolorations in cooked-potato tissues could be eliminated if the cooked potatoes were dipped in a solution of SAPP at 40°F. A complete bibliography of the literature on this topic has been published recently by Talley and co-workers.[527]

Inhibition of discoloration and retention of the original bright-orange color of sweet-potato flakes and of canned and frozen sweet potatoes have been accomplished by treating them with SAPP or a combination of SAPP and TSPP.[523,528] MSP and longer-chain polyphosphates have been shown to be much less effective for this treatment.

Applications in Obtaining Optimum Texture

Kertesz[529] discovered the influence of the calcium ion in the firming and tenderization of fruit or vegetable tissues and in the control of the viscosity of purées and juices. The calcium ion reacts with the pectic acid (polygalacturonic acid) that is formed upon the demethoxylation of natural pectin in the fruit or vegetable cell wall. This reaction is also responsible for gelling of fruit juices during preparation of preserves, jams, and jellies.[28] Proper control of the calcium ion can be highly beneficial in the processing of fruits and vegetables. The introduction of soluble calcium salts can increase firmness of tissues of whole fruits and vegetables, while the precipitation or complexing of calcium can tenderize tissues or prevent gelation of fruit and vegetable pectins.[28]

Pectin Gels Pectin is normally extracted from apple pomace or the peels of citrus fruits by heating these materials with acid. Many housewives are familiar with the necessity for pectin in the preparation of fruit preserves. As reviewed by Kertesz,[529] natural pectins in the cell walls of fruits and vegetables become demethoxylated by the enzymes in the tissues when the cell walls are broken. Gels are formed when the resulting pectic acid reacts with calcium. The yield of pectin during its extraction from the source material is significantly increased by the addition of SHMP to the fruit pulp to sequester calcium ion and prevent the formation of calcium pectate; pectin yields have been reported to be increased as much as 25%, and the extraction time has been decreased by as much as 50%.[530-536] The effectiveness of polyphosphates was found to increase with increasing chain lengths from pyrophosphate to SHMP.

The polyphosphates have been found to be highly effective in obtaining optimum characteristics in pectin gels. An excellent review of the chemistry and uses of pectin has been published by Woodmansee[536] and contains a good bibliography for further study. Numerous reports and patents describe the ·useful applications of various types of phosphates in the control of pectin gel formation for use in the production of pectin materials for home canning, the commercial production of preserves, jams, and jellies, and the preparations of a number of other dessert gels.[531,535-543] Waller and Baker[535] reported that polyphosphates are rapidly hydrolyzed to orthophosphate, which then is capable of entering phosphorylation reactions with pectin to form cross-linked chains throughout the entire gel system. SHMP again was found to be more effective than sodium tetrapolyphosphate, and the shorter-chain polyphosphates were decreasingly effective.

Because it is naturally present in many fruit and vegetable juices, H_3PO_4 has often been used to provide optimum tartness and pH for setting of pectin gels. It has the added advantage that its effectiveness is not diminished by heat.[27,28,110,111]

Tomato Products Studies have shown that the phosphates have many beneficial applications in the processing of tomato products.[544-547] SHMP, added to the tomato fruit prior to cooking, has produced more viscous tomato sauces,[544] aided in the viscosity control of tomato purées,[546] and increased the yields of juice obtained from tomato pulp.[545] The addition of SHMP to tomato pulp has increased the viscosity of juice extracted from it;[531-545] the addition of H_3PO_4 has been used to improve viscosity of tomato juice, tomato purée, and similar products in a recently proposed improved hot-break process.[547]

Potato Texture SAPP was used by Smith and Davis[548] to improve the mealiness and other textural properties of potato products. Cole[549] improved the textural characteristics of dehydrated potato products by cooking them in buffered water maintained at a pH of 6–8 with mixtures of MSP and DSP or their potassium equivalents. The ability of phosphates to improve mealiness of potato products has been reviewed by Kintner and Tweedy[550] and Smith.[521] Textural characteristics have been improved in direct relation to the amount of SAPP added. This is probably due to formation of small quantities of phosphorylated amylopectin, which prevents rupture of the starch granules due to increased hydrogen bonding between the phosphorylated starch molecules.

Tenderization Many fruits and vegetables have tough skins that grow increasingly tough as they become too mature. The toughness of fruit and vegetable skins has been shown to increase in direct proportion to increasing content of calcium. This may be natural calcium or that introduced through the use of hard water in blanching the fruits and vegetables. The addition of STP or SHMP to the water used in processing fruits and vegetables decreases the toughness of the skins of peas, lima beans, snap beans, corn,[27,110,111,551,552] and plums.[553,554] This author has also noted the significant tenderization of the skins of broccoli by applying SHMP to the blanching water; in fact, the broccoli can be tenderized to the point that it is unacceptably tender for eating. Treatment of the fruits or vegetables with SHMP has been shown to have no effect on sugar or vitamin-C levels or to upgrade poor-quality vegetables.[551]

Neubert and Carter[553] demonstrated that the toughness of the skins of Italian prune plums is directly proportional to the level of calcium in the skins, as shown in Figure 3. The calcium content of the water used in processing, as well as the natural calcium content of the skins, contributed to increasing toughness. The addition of SHMP at levels above 0.5% tenderized the skins and also softened the flesh of the plums. An increase in the calcium content of the syrup also contributed to toughening of the skins, as demonstrated in Figure 3. SHMP added to salt brine increased the penetration of the salt into the shells of peanuts, allowing processors to prepare salted peanuts inside the shells.[111] A combination of SHMP and papain increased the penetration of syrup into cherries that are being sweetened, and it shortened the time of the process.[555]

Firming Tissues The studies of Loconti and Kertesz[556] have shown that calcium salts are definitely involved in the firming of the tissues of tomatoes through their ability to form calcium pectates. Calcium salts also have been shown to be effective in firming peeled apple slices used for freezing and canning[557] and in the firming of snap beans.[558] The addition of soluble calcium salts, such as MCP, calcium chloride, calcium sulfate, or calcium citrate for the firming of canned tomatoes, potatoes, green or red sweet peppers, lima beans, and carrots, has been approved in the federal standards for these vegetables.[559] In 1969 a temporary approval was granted for the addition of soluble calcium salts to canned sweet potatoes for similar purposes.[560]

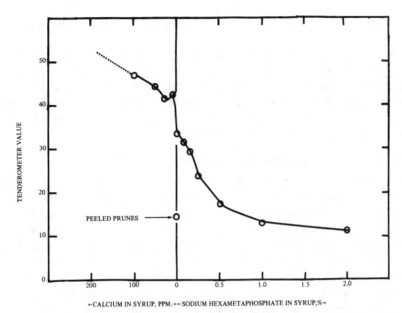

Fig. 3.   Effect of addition of calcium and sodium hexametaphosphate on texture of canned Italian prunes. (Source: Neubert, A. M., and Carter, G. H., 1954, *Food Technol.* 8:518; with permission.)

Miscellaneous Applications

Dakin[561] reported that pyrophosphates and other compounds retarded softening of cabbage tissue due to the oxidation of cellulose to form pectic materials. Takeda Chemical Industries[562] found that either EDTA or polyphosphates softened salted pickles if added prior to brining. DSP or TSP also has been reported necessary in production of imitation or fabricated potato chips.[563]

Phosphate Applications in Gums and Gels

The vegetable gums, like pectin, are long-chain, high-molecular-weight polysaccharides or mixtures of polysaccharides that are capable of interacting as polyelectrolytes; therefore, they are capable of interacting with other polyelectrolytes, particularly the long-chain polyphosphates. Gums often are used as stabilizers and thickeners in food processing. Some merely swell when hydrated and, therefore, increase viscosity of the food system, while others form thick rubbery gels. Glicksman has published excellent reviews of the properties and uses of the various vegetable gums in *Advances in Food Research*[564] and Chapter 7 of this handbook. The phosphates are capable of modifying and improving the physical characteristics of gums such as agar, alginates, carrageenan, guar, and locust bean. A review of phosphate interactions with gums has been published by Sand and Sodano.[29]

Applications in Agar Gel Systems

Agar has been used in foods in the Orient for centuries, and it is commonly used as a stabilizer in the manufacture of chiffon pies, meringues, icings, and toppings. It is also frequently used in preparing confectionery gums and gels.[28,564]

Since agar is composed of at least two polymeric electrolytes, the phosphates should be useful in improving its gelling characteristics. Apparently only Sergeeva[565] has studied the beneficial effects of DSP upon a gel composed of agar, sugar, and molasses for use in preparing orange and lemon slices. She demonstrated that the gelling strength of the agar gel was significantly increased by the addition of 0.12–0.3% DSP.

Applications in Alginate Gel Systems

The basic component of alginate gels is believed to be alginic acid, a powerful, anionic polyelectrolyte that reacts readily with cationic polyelectrolytes such as the polyphosphates. Although alginic acid is insoluble, its sodium, potassium, and ammonium salts are very soluble and are used in foods.

The soluble alginic acid salts react with the di- or trivalent metal ions to form insoluble salts of the metal alginate. If the reaction is too rapid, grainy, discontinuous gels or even precipitates are formed. If the rate is too slow, the gel formation will require too long a period to be practical, and the resulting gel may be too soft. It is, therefore, necessary to prepare a system in which a slow, uniform rate of release of the metal ion is obtained. The calcium ion is usually the preferred metal ion for preparation of alginate gels. Glicksman[564] has discussed the conditions required for obtaining optimum alginate food gels.

As would be expected from its calcium content, milk forms gels with alginic acid. It is necessary to first obtain complete dispersion of the soluble alginate throughout the milk before gelling begins. This was accomplished by Lucas,[566] who added TSP to sodium alginate, and by Steiner,[567,568] who used combinations of SHMP and a difficultly soluble calcium salt, such as TCP, to control body and texture of ice cream, chocolate milk, and other dairy products, as well as fruit jams and jellies, jellied salads and broths, water-gelled desserts, and candied jellies. Steiner's most successful system included a soluble alginate salt, a difficultly soluble calcium salt, a gel-retarding agent, such as SHMP, and a weakly acidic compound, such as glucono-Δ-lactone. Roland[569] reported a process for controlling viscosity and stabilizing suspended particles in food systems by first adding TSPP, STP, or long-chain polyphosphates to an aqueous suspension of the metal ion, such as milk, which contains calcium; the alginate salt is then added to the product. Gibsen[570] reported that a combination of sodium alginate, sodium carbonate, a calcium salt, and a buffering agent, such as TSPP, could be used in preparing instant-milk-pudding mixes. He later reported the preparation of smooth, firm, cold-water-soluble dessert gels by combining (1) a slightly soluble dibasic acid, such as adipic or fumaric acid, with (2) sodium alginate, (3) a slowly soluble calcium salt, such as TCP or DCP, and (4) a gel-retarding sequestering agent, such as SHMP.[571]

Rocks[572] prepared cold-water-soluble dessert gels that required only the slowly soluble calcium salts, such as DCP and TCP, and a weak acid; Rocks' formula eliminated the use of a calcium sequestering agent. Hunter and Rocks[573] prepared a sodium phosphoalginate from a wet paste of sodium alginate and TSP, which could then be dried and utilized for preparation of milk puddings. Merton and McDowell[574] prepared instant, cold-water-soluble dessert gels from a four-component system composed of the soluble alginate,

TCP, a carbonate, and citric acid. Hunter[575] prepared chiffon pie fillings from sodium alginate, skimmed-milk powder, and SHMP. Miller and Rocks[576] prepared high-bloom-strength gels by combinations of sodium alginate, calcium carbonate, STP, and adipic acid.

Andrew and MacLeod[577] described a number of applications for alginate in food processes. They reported that the thickening of a canned product could be delayed until after the can had been filled by use of a low-viscosity sodium alginate and a slowly soluble calcium salt. They also reported that fruit- or vegetable-flavored particles could be prepared by depositing a combination of sweetening agents, coloring, flavoring, and sodium alginate by drops into a calcium chloride bath. Other applications for alginates were also proposed.

Applications in Carrageenan Gel Systems

Carrageenan, like algin, is a strongly charged, anionic polyelectrolyte of high molecular weight; it is also extracted from seaweeds. The extract contains five or more distinct fractions, each differing in properties. It forms complexes with proteins and other long-chain molecules, such as polyphosphates, to form gels and sols. Glicksman[564] has reviewed the chemical and physical properties and applications of carrageenan in food products; only a few will be mentioned here, as they are also discussed in Chapter 7.

Unlike algin, calcium inhibits the gelling of some fractions of carrageenan. Stoloff[578,579] used soluble phosphate salts, particularly the polyphosphates, to sequester calcium ions from milk and tap water to prepare stable chocolate milk and water gels. Zabik and Aldrich[580] reported that the rate of shear appeared to have a more significant influence on the viscosity of carrageenan-potassium salt gels than the presence of differing anions due to the type of potassium salt used.

Sand and Sodano[28] reported that the addition of low levels of polyphosphates to milk-gelling systems containing carrageenan reduces and often completely eliminates syneresis.

Applications in Other Gum Gel Systems

Guar and locust-bean gums were used by Weinstein[581] in preparing ice cream stabilizers. SHMP was added to the system to prevent the denaturation of the proteins by the gums added to the stabilizers. Green and co-workers[582] reported the preparation of multi-phased food products in which the levels of calcium ion necessary for preparation of proper gelling were controlled by means of complexing agents, such as TSP, TSPP, STP, or SHMP. Lauck and Tucker[233] and Marcus[583] prepared hydrated gels of calcium orthophosphate by combining a soluble calcium compound, such as calcium chloride, and DSP or TSP.

Phosphate Applications in Meat Processing

Meat has been a highly accepted portion of the human diet for centuries. The production, processing, and preservation of meat to maintain maximum palatability have been of major concern in the technology of meat processing. The palatability of meat is judged largely by its color in the raw and prepared states, its tenderness when eaten, and its flavor. Each characteristic is directly affected by chemical and physical factors that can to some extent be measured and controlled.

When purchasing meat, most consumers judge its freshness and palatability by the color of the raw meat. The biochemical reactions that occur between the heme pigments and oxygen in the presence of light directly affect the color of the raw meat; certain additives can aid in preserving this color. Tenderness of the meat, though judged after cooking and during its consumption, is directly affected by chemical and physical factors, such as the binding of water by the proteins, the texture of the meat fibers, the degree of hydrolysis of muscle fibers, and fat emulsification. The flavor of the meat, which is also judged during its consumption, is affected by the retention of the meat juices, the reduction of cook-out losses, the oxidation of fat and other easily oxidized meat constituents, and the degree of microbiological deterioration. The prevention of microbiological deterioration is often considered the most important factor in meat preservation. However, the retention of the soluble constituents within the meat and the prevention of oxidation are equally important in the preservation of palatability.

Only a portion of the voluminous literature on the influence and applications of phosphates in meat processing could be reviewed in this section. Reviews on this subject have been used to a considerable extent, especially those of Grau,[584] Hamm,[585] Saffle,[586] and Karmas;[587] others will be mentioned in subsequent paragraphs.

Biochemistry of Phosphate Interactions with Meat Proteins

Meat is the muscular tissue of an animal; it is composed of fibers of a protein complex called actomyosin, which is composed of two proteins, actin and myosin. There is approximately 1 part actin to 3 parts myosin in the protein complex; approximately 13% of the total muscle protein is actin, and approximately 38% is myosin. The two proteins can be separated or extracted from actomyosin complexes by means of salt solutions. Actin has an isoelectric point at pH 4.7; it exists both as a monomer with a molecular weight of approximately 70 000 and also as a polymer, which varies in its number of units and molecular weight according to ionic conditions present in its environment. Myosin is a very large protein molecule with a molecular weight of approximately 850 000; it has an isoelectric point at pH 5.4. Myosin is a negatively charged protein with a strong affinity for calcium and magnesium ions; it is also capable of hydrolyzing ATP to ADP.[588] This reaction is directly involved in energy transferred during muscle contraction and relaxation and, according to Kotter,[590] led to the investigations of the reactions of the phosphates with meat proteins.

The muscles in a freshly slaughtered animal are soft and pliable. However, as soon as death occurs, the ATP in the muscle tissue begins to hydrolyze and to release a phosphate group to form ADP. Phosphocreatine present in the muscle tissues resynthesizes ATP as long as any phosphocreatine exists. Since metabolism has stopped, the supply of energy-rich phosphates soon is depleted. As the supply of phosphocreatine also is depleted, ADP begins to accumulate. At the same time the anaerobic metabolism of glycogen results in the accumulation of lactic acid, which lowers the pH of the muscles from their natural level to a range of 5.4–5.8. Anaerobic metabolism is inhibited at this pH, since the enzymes involved cannot function. The muscles then begin to contract through the sliding of the contractile proteins over each other in the same manner as the muscles contract during life. The animal carcass finally becomes stiff, and all muscle tissues become hardened, a state known as *rigor mortis*.

Meat is tough in the state of rigor mortis. Through experience man has learned that aging some meats at refrigerated temperatures tenderizes the toughened tissues due to numerous subtle and poorly understood changes in the biochemical and physical environment surrounding the muscle fibrils.[588] The desirable characteristics of meat palatability are affected by the biochemical reactions of the actomyosin complex after slaughter of the animal. Meat tenderness is directly influenced by whether the actomyosin complex is retained or dissociated into its component proteins. This, in turn, is influenced by the type and amount of salt present,[591-598] the pH of the system,[586,591,592] and, although this is disputed, possibly the amount of calcium and magnesium, as well as other heavy-metal ions, present that would maintain the proteins in an insoluble state.[599-605]

The chemical and physical state of the meat proteins affects such important meat palatability factors as retention of the meat juices, fat emulsification, and cooking losses.[28,584,586,593,595-597,606-616] These can have significant effects on flavor and tenderness of the cooked meat. The addition of phosphates, particularly the polyphosphates from the pyrophosphates through the long-chain polyphosphates, or polyphosphate sodium chloride mixtures to meat can produce dramatic effects on tenderness and general palatability of the meat. A number of investigators have demonstrated that the beneficial effects of the polyphosphates on meat palatability are directly related to their effects on the meat proteins.[28,593-598] For example, the polyphosphates have been shown to cause the actomyosin complex to be dissociated into actin and myosin, as demonstrated by their ability to extract increasing quantities of these dissociated proteins from the meat.[593-598] It has been shown, however, that the longer-chain polyphosphates must first be hydrolyzed to pyrophosphate before this dissociation can occur.[596,597]

The pH of the system, as well as the chain length of the polyphosphate, determines the effectiveness of dissociation as demonstrated by the amount of protein extracted.[586,591-597] Figure 4, taken from reports of Fukazawa and co-workers,[595] demonstrates the amount of protein extracted by pyrophosphate, tripolyphosphate, and hexametaphosphate as the pH was increased from 5.6-7.0. In all cases pyrophosphate was more effective than tripolyphosphate and hexametaphosphate, and the greatest increase in extractability occurred at pH 7. Controls containing 0.6 mol sodium chloride solution also extracted an increased quantity of protein as the pH was increased, but the phosphates were more effective. The work of Fukazawa and co-workers[595] confirmed the work of numerous other investigators.

Whether or not the phosphates affect meat proteins through their ability to sequester divalent cations, particularly calcium, magnesium, and zinc, is a matter of controversy. Conflicting data have been reviewed by various investigators, including Hamm,[585] Morse,[600] Inklaar,[604] Mahon,[617] and Sherman.[618] Although many believe that calcium complexing is the major function of the polyphosphates, they also admit that this is probably not the only function.[584,585,599,601,619-630] Hamm[585,620-626] theorized that the complexing of calcium is the major reason why treating meat with polyphosphates increases its water binding. Magnesium ions[622,628] and zinc[622,623,629] are also thought to be involved in the ability of polyphosphates to influence the water binding and hydration of meat protein.

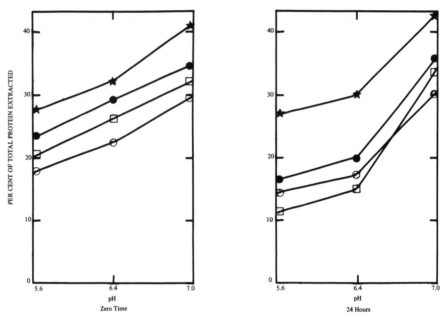

Fig. 4. Effect of polyphosphates on the extraction of protein from intact muscle fibrils. (Adapted from Fukazawa, T., Hashimoto, Y., and Yasui, T., 1961, *J. Food Sci.* 26:550; with permission.)

○ = control   ★ = pyrophosphate
□ = hexametaphosphate   ● = tripolyphosphate
Note: Concentration of all polyphosphates = 10^{-2} mol to muscle fibrils.

Other investigators have reported evidence that the ability of the polyphosphates to sequester calcium and other divalent ions does not explain their effects on meat proteins.[591,593,602,609,618,631,632] Some reports indicate that the addition of calcium or magnesium ions either has no effect[593] or it increases such important characteristics as water binding by the meat proteins.[631,632]

Sherman[618] compared the effects of treating meat with polyphosphates with those of meat that was treated with EDTA, a more potent calcium sequestering agent than the phosphates. Data from his report, as shown in Table 25, demonstrated that commercial polyphosphates were significantly more effective in increasing the fluid retention of the meat than either disodium EDTA, which reduced water binding, or tetrasodium EDTA. The addition of sodium chloride to EDTA derivatives suppressed their ability to increase the water retention of the meat, while the addition of sodium chloride to the polyphosphate mixture either did not change its water retention or improved it. Furthermore, the meat treated with the EDTA derivatives was very crumbly and lacked the firm texture obtained when meat was treated with polyphosphates.[618]

As shown in Figure 5, Hellendoorn[603] found that the complete removal of calcium ions from cooked meat caused no change in water-binding capacity compared with that obtained with salt in the control. He confirmed Sherman's work, which indicated that disodium EDTA depressed the water-binding effects of sodium chloride, while pyrophosphate and tripolyphosphate improved the water binding under identical conditions of pH and ionic strength. Wismer-Pedersen[633,634] reported that pyrophosphate had a completely different action on meat from EDTA. When either compound was added to fresh meat prior to freeze drying, both increased the water binding of the dried meat proteins when they were wetted; the pyrophosphate-treated dried meat swelled to a greater extent in all of the wetted areas than the meat treated with EDTA. Wismer-Pedersen also found that the pyrophosphate-treated meat retained water to a greater extent during freeze drying and thus increased the time of drying.

TABLE 25

Effect of EDTA Sodium Salts and Polyphosphates on Water Retention of Minced Pork

Additive	pH of Aged Meat-additive Mixture	Concentration of Additive Solution, %	Water Retention at 0°C, %
Na$_2$ EDTA	5.4	1.0	−4.0
Na$_4$ EDTA	6.6	1.0	14.0
Polyphosphate[a]	6.5	1.0	23.0
Na$_2$ EDTA	5.3	2.0	−2.0
Na$_4$ EDTA	7.0	2.0	38.0
Polyphosphate[a]	6.9	2.0	57.0
Na$_2$ EDTA	5.2	4.0	−3.0
Na$_4$ EDTA	8.5	4.0	43.0
Polyphosphate[a]	7.5	4.0	58.0
Na$_2$ EDTA + NaCl	5.0	2.0 1.0	−7.0
Na$_4$ EDTA + NaCl	6.7	2.0 1.0	26.1
Polyphosphate + NaCl	6.4	2.0 1.0	58.2

Source: Sherman, P., 1961, *Food Technol. 15*:79, Table 7; with permission.

[a] A commercial polyphosphate.

Some investigators and many government regulatory agencies, particularly in European countries, fear that the consumer will be defrauded by purchasing greater amounts of water in phosphate-treated meat. Kotter[590] and Popp and Muhlbrecht[635] reported that the polyphosphates merely restore the water-holding capacity of the meat to the level that is normal before the animal is slaughtered. Upon slaughter the water-binding capacity of the meat decreases with the onset of rigor mortis and reaches a minimum when rigor mortis is complete. Although water-binding ability is restored by aging, it never attains the original levels in the living tissue. By treating meat with phosphates in the presence of the proper types and levels of alkaline-earth-metal ions at optimum ionic strengths and pH values, the original water-binding capacity is restored.[585,590,635]

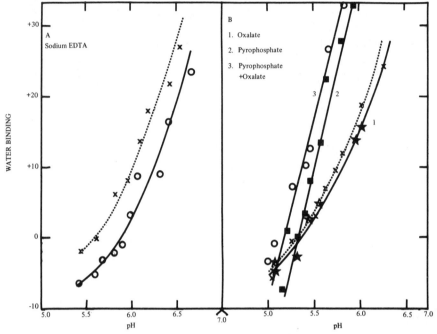

Fig. 5.   Effects of EDTA (A) and oxalate (B) on cooked meat in comparison to sodium chloride. (Adapted from Hellendoorn, E. W., 1962, *J. Food Technol. 16*(9):119; with permission.)

A. O $= 2.5\%$ NaCl $+ 0.6\%$ Na$_2$H$_2$EDTA$\cdot$2H$_2$O ($\mu = 0.52$)
    X = Dotted line—3% NaCl ($\mu = 0.51$)

B. ★ = Curve 1—2% NaCl $+ 0.6\%$ (NH$_4$)$_2$C$_2$O$_4$ ($\mu = 0.49$)
    ■ = Curve 2—2% NaCl $+ 0.5\%$ SAPP ($\mu = 0.48$)
    O = Curve 3—1.5% NaCl $+ 0.5\%$ SAPP $+ 0.6\%$ (NH$_4$)$_2$C$_2$O$_4$ ($\mu = 0.51$)
    X = Dotted line—3% NaCl ($\mu = 0.51$)

The author of this chapter firmly believes that the treatment of fresh meat with phosphates does not cause the meat proteins to imbibe and bind water from outside the tissues; rather, phosphate treatment causes the meat proteins to bind and retain the natural juices within the meat and prevents loss of soluble constituents. Considerable published information, particularly studies of poultry and seafood, supports this belief; hence, there is no reason to believe that the red meats differ in this respect.

Ruf,[23] Grau,[584] and Morse[600] have emphasized that the use of polyphosphates will not improve the quality of poor meat, nor will it overcome faulty production practices. Any fears that phosphate treatment will allow a processor to deceive consumers by making poor-quality or improperly processed meat appear better than it is are completely unfounded.

Numerous commercial blends and mixtures of polyphosphates with and without other salts are available; many will be discussed in the following sections. Food regulations permit the use of polyphosphates in some products but not in others. It is highly advisable for any processor contemplating the use of polyphosphates to contact the regulatory agencies in his country to determine whether polyphosphates are allowed for his intended application.

General Meat Applications

Color Preservation   Watts[636] and Fox[637] have published excellent reviews of the chemistry of meat pigments that provide its characteristic red or pink color. The color of meat is dependent on the presence and the chemical reactions of two pigments, myoglobin and hemoglobin. Hemoglobin, the red pigment in blood, is an iron-protein complex. Although a piece of meat has been well bled, the blood pigment may account for as much as 30% of the red pigment.

Myoglobin, a deep-red pigment found in muscle tissue, is also an iron-protein complex. As the iron is present in the ferrous form, it is easily oxidized to the ferric form, a brown pigment called metmyoglobin. Myoglobin forms a complex with oxygen at freshly cut surfaces of meat to form a bright pigment called oxymyoglobin.[587,637] Oxymyoglobin also contains iron in the ferrous form and is easily converted to metmyoglobin by oxidation. After freshly cut surfaces of meat have been exposed to air for a period of time, the oxymyoglobin slowly begins to form the brown pigment metmyoglobin. To a consumer such brown-colored meat does not appear as fresh as the bright-red meat, although the brown meat may be perfectly fresh and completely edible. Because of the consumers' attitude, meat manufacturers are continuously searching for ways in which to stabilize the natural, bright-red color of freshly cut meat.[638] Factors that affect the color of meat are its pH, the presence of reducing substances, curing salts, and metal ions, and the exposure of meat to oxygen.[636] Fresh, aged meat normally has a pH of 5.2-6.6. Low pH values have been found to accelerate the oxidation of fresh meat pigments to form brown metmyoglobin pigments. Optimum stabilization of the redness of meat is obtained at the higher pH values of 6.0-6.6. Hall and Brissey have reported that the color of meat can be stabilized if the pH is maintained at about 6.5 with polyphosphates[638] and at about 6.8 with dialkali phosphates.[639] Chang and Watts[640] reported that the presence of polyphosphates at pH 5.6 accelerates the oxidation of oxyhemoglobin to methemoglobin and thus adversely affects color.

Increasing Tenderness   The onset of rigor mortis causes a shortening of the muscle fibers as the contractile muscle proteins slide over each other to form complexes that are highly stabilized by hydrogen bonding. The ability of the phosphates to cause the dissociation of actomyosin into actin and myosin has an important effect on the tenderness of all types of meat. Bendall[641] has shown that pyrophosphate in combination with the magnesium ion can cause the relaxation of the muscle fibers. This may explain the tenderizing effect of polyphosphates that are hydrolyzed to the pyrophosphate anions in the presence of meat proteins. Specific applications of polyphosphates in tenderization of fresh and cured meats will be discussed in further paragraphs.

Increasing Binding   The texture of ground and cubed meat products depends on the rebinding of the pieces of the meat. It is possible to form large solid pieces of meat from ground or cubed meats if the pieces are prepared so that natural exudates come to the surface and form the binder for the small pieces.[587] Phosphates improve binding by increasing the amounts of the exudates of soluble proteins.

The binding of ground meat is very important in the preparation of such meat products as hamburger, meat loaves, and both fresh and cured sausages.

The salt-soluble proteins, largely myosin, have been reported by Fukazawa *et al.*[594,642] to be largely responsible for the binding of meat particles in sausages. Myosin-poor meat fibrils and the "ghost" fibrils remaining after extraction of myosin, as well as the water-soluble proteins, have little or no binding properties[605]; the polyphosphates have been demonstrated to increase the solubility of myosin,[595-597] particularly when combined with salt. As discussed previously the pyrophosphate anion appears to be responsible for increasing the solubility of myosin.

Increasing Moisture Retention   The moisture retention, often termed water-holding capacity or water-binding ability, of muscle tissue is of considerable concern to the meat industry. In addition to the economic loss, of greater concern is the acceptability to the consumer, as meats that have lost a portion of their natural juices are less acceptable from the standpoints of appearance, reduced tenderness, and reduced juiciness. Meat exposed to the proper environment for excessive moisture loss undergoes greater cooking losses; frozen meats can exhibit "freezer burn" due to moisture loss at the surface. As a result the meat industry constantly searches for methods to prevent loss of moisture.[638]

The water-holding capacity of meat tissue increases as the pH of the tissue either decreases or increases away from the isoelectric point of approximately 5.0. Since freshly slaughtered meat gradually decreases in pH with the onset of rigor mortis,[585,592,636] the water-holding capacity of the meat decreases rapidly after slaughter. According to Hamm[585] it is possible to preserve the water-holding capacity of freshly slaughtered meat if the meat is salted immediately after butchering. Hamm speculated that the retention of water-holding capacity is due to the combined presence of the salt and the ATP that is still present in considerable quantities in freshly slaughtered meat tissue. The combined action of the salt and ATP causes the peptide chains of the protein to unfold; the unfolding leaves such large distances between these chains that bivalent cations, released through the breakdown of ATP, are unable to cross-link these chains. Water can reach the numerous hydrogen-bonding sites necessary for complete hydration of the proteins. The polyphosphates are thought to be capable of increasing the water-holding capacity of muscle proteins through their ability to cause the dissociation of the actomyosin complex and possibly, although this is disputed, through their ability to complex divalent cations. Hamm and Grau[626] reported that the polyphosphates increased water-holding capacity of muscle proteins in the following order of decreasing activity: STP, sodium tetrapolyphosphate, SHMP, TSPP, cyclic sodium metaphosphate, and DSP. Although sodium citrate was less effective than sodium pyrophosphate, it was more effective than the cyclic sodium trimetaphosphate. Sodium oxalate was found to be approximately as effective as SHMP.

Flavor Improvement   Phosphates can contribute to the flavor of meats, since they have such significant effects on the retention and other characteristics of the proteins. The proteins also contribute significantly to the development of proper meat flavor. It has been hoped that meat-flavoring components could be prepared, either from low-grade cuts of meat or by synthesis, so that they could be added to higher-quality meat to standardize the flavor at the most acceptable levels.[587] Giacino[643] used phosphates in the preparation of meat-like synthetic flavors from proteinaceous materials using combinations of vegetable protein hydrolysates, various fatty acids, a sulfur-containing amino acid, and orthophosphoric acid or its ammonium and potassium salts.

Prevention of Off-flavor   In addition to contributing directly to the formation of desirable flavors through their interaction with proteins and their use in synthetic meat-flavor compositions, the phosphates also are useful in preventing the development of off-flavors in meat products. The phosphates, through their ability to act as antioxidant synergists, prevent oxidation of fats and thus prevent development of rancid flavors in the meat fats. Thus, their inhibition of the development of off-flavors has an important role in flavoring meat products.[636]

Prevention of Microbiological Spoilage   Prevention of the development of rancid flavors in meat could be considered a part of the preservation of meat. However, as most consumers consider microbiological spoilage a major factor in spoilage of meat, when one speaks of preservation of meat, it is usually in reference to preventing its microbiological spoilage. Here again, the phosphates play an important role. As healthy animals are capable of preventing invasion of microorganisms, freshly slaughtered meat is considered by some investigators to be sterile. Immediately after slaughter, however, invasion by billions of microorganisms has begun; freshly cut meat cannot help but be heavily contaminated with these forerunners of spoilage.[587] One of the major objectives of the meat industry is to prevent the spoilage of meat to the point of becoming inedible.

Keil *et al.*[644] developed a process that could be used on many foods, including any type of meat product capable of being dipped into and coated by a solution, such as sausage, pieces of fresh meat, and cured meat products of all types. The process involves dipping the meat in a heated solution, approximately 42–60°C, of a relatively high-bloom-strength, gelatin-containing sodium metaphosphate and an edible acid capable of converting the sodium compound into the polyphosphoric acid. The solution adhering to the surface then gels and forms an antibiotic coating about the meat. One example in the patent of Keil *et al.* describes the use of 1600 g gelatin, 400 g SHMP, and 120 g hydrochloric acid in 4800 ml water. The thickness of the coating can be regulated by the temperature, since the viscosity of the solution decreases as the temperature increases; thus, the lower the temperature, the thicker the coating.

Other more specific microbiological preservation methods involving phosphates will be discussed in subsequent paragraphs.

Applications in Fresh, Whole Meats

Color Preservation   Kamstra and Saffle[645] injected or infused hams from freshly slaughtered hogs with solutions of SHMP and refrigerated them for 72 hours, producing pork meat with high pH values, very dark-colored lean meat, higher levels of glycogen, and a sticky surface. The results indicated that the treatment produced what is termed "dark-cutter" meat. This undesirable condition was corrected by infusing the hams with a combination of a 23.3% SHMP solution plus a solution of lactic acid. Paired hams from the same carcass were used in this study, one being a water-injected control, while the other was given the desired treatment. The pH and the color of the hams treated with a combination of SHMP and lactic acid solution were significantly improved.

Carpenter *et al.*[646] studied the effect of injecting solutions of various levels of SHMP into the left round cuts of beef carcasses and compared their color against that of the right rounds from the same carcass; the right rounds received an equal quantity of injected tap water. The results showed a significantly darker-red color in the phosphate-treated rounds compared to the water-treated rounds. Although a number of other investigators have reported studies of phosphate treatment of fresh, whole meats, they apparently failed to observe, or to report observations of, the effects of the phosphates on the color of fresh meat.

Voegeli and Gorsica[647] retained an attractive, uniform, bright-red color in fresh meats when the meat was treated with a synergistic mixture that would contribute 0.02–2% phosphate buffering agent, 0.01–0.2% oxidation–reduction controlling poising agent, and 0.01–0.4% sequestering agent capable of inactivating any polyvalent metal ions present in the system. The authors preferred to use a polyphosphate that would act as both a buffer and a sequestering agent. Further work reported by the same authors[648] indicated that additional stabilization could be obtained in fresh meats if alkali-metal salts of parahydroxybenzoic acid were added to the above mixture.

Rust[649] investigated the effect of injected SHMP on the color of the crust and interior of cooked beef steaks. No effect on color could be correlated with the phosphate treatment at any of the levels investigated.

Increasing Tenderness There is considerable evidence that the polyphosphates can tenderize fresh, whole meats.[28,605,645,646,649,658] Allen[650] reported as early as 1938 that a combination of enzymes, monosodium glutamate, sugar, TCP, and salt could tenderize meat. Mullins and co-workers[651] reported that taste panels consistently rated hams injected with SHMP as more tender than uninjected controls, although there were no statistically significant differences in the scores. Kamstra and Saffle[645] found that fresh pork loins injected with solutions of SHMP or SHMP plus lactic acid were rated by taste panels as more tender than hams taken from the opposite side of the same carcass and injected with water until the weight was equivalent to those treated with the phosphates. Carpenter and co-workers[646] reported similar results when beef rounds taken from the left sides of carcasses were injected with solutions of SHMP until the weight had increased 5% above the original weight; they were compared against rounds taken from the right side of the same carcass that had been injected to attain similar weight increases with tap water. The SHMP-treated rounds were more tender at the 1% significance level when treated with a 10% SHMP solution and at the 5% significance level when treated with a 15% solution. Shear press values confirmed the taste-panel ratings with both the pork and beef tests.

Hopkins and Zimont[652] found that either dipping or injecting pieces of beef with a mixture of sodium pyrophosphate and sodium chloride and then refrigerating them for a period of time to allow tenderization to occur produced significantly more tender beef than untreated controls. Slices of meat were tenderized in 18–24 hours, while roasts and larger pieces of meat were held for 1–8 days, depending on the method of applying the phosphate treatment; tenderness again was confirmed by taste panel and shear press values.

Williams[653] explained that the tenderizing action of hypotonic phosphate sodium chloride mixtures is caused by the action of the internal salts in the meat cells drawing water through the cell walls and retaining it. Rust[649] also reported that taste-panel and shear press values demonstrated the tenderizing effect of solutions of SHMP injected into 2-in. thick beef steaks, which were then aged for 18 hours at 3°C prior to cooking and evaluation. Controls injected with equivalent quantities of water were significantly less tender, as shown by a taste panel that scored them for initial tenderness impressions and for the number of chews required to completely masticate the steaks. Maximum tenderization was obtained in steaks injected with a 10% SHMP solution, which was equivalent to introducing 0.18% SHMP into the meat based on its original weight.

Komarik[654] described a method for the tenderization of fresh meats, including beef, pork, lamb, and mutton, that involved the application of 2–12 ounces of either TSPP or TKPP to the meat. Either the pyrophosphate could be applied as a dry powder to the surface of the meat, or it could be injected as a solution. The tenderness was significantly increased by a combination of the basic pyrophosphates and small quantities of a meat-tenderizing enzyme, preferably papain. Meats treated with the pyrophosphates could be allowed to stand at refrigerated temperatures for several days and as long as a week without over-tenderization. Brown and his co-workers[655] reported that injecting meat, particularly beef, with solutions of phosphate and sodium chloride, which introduced approximately 0.5% by weight of the phosphate and 0.7–1.8% by weight of the salt, significantly tenderized the meat over untreated controls. The phosphates preferred for tenderization of beef were STP, TSP, and DSP; SHMP also was effective but to a lesser extent. Delaney[656] produced significant tenderization but not over-tenderization by dipping beef, pork, lamb, or poultry into an aqueous solution containing an alkali-metal pyrophosphate or poly-phosphate, common salt, and small amounts of proteolytic enzymes for 15–60 seconds and by holding the meat at 38–44°F.

Huffman and co-workers[657] tenderized the meat of sheep by injecting them prior to slaughter with a solution of SHMP or TSPP so that each animal received 3 mg of phosphate per pound of live weight; the meat of control animals was judged to be tougher than that of the phosphate-treated animals.

Baldwin and deMan[605] studied the tenderizing effects of DSP, TSPP, STP, sodium tetrapolyphosphate, SHMP, and sodium citrate on beef. Tenderness was evaluated by means of a Kramer shear press. The phosphate-treated steaks were more tender than those treated with citrate and than untreated controls. Tenderization could not be correlated with pH or with the ability of the tenderizing agent to complex calcium or magnesium.

Miller and Harrison[658] attempted to repeat previous work on the tenderization of steaks by marinating them in solutions of SHMP. They were unable to demonstrate any tenderization effect when steaks were marinated in a 0.03 mol solution of SHMP for 1, 2, or 6 hours prior to broiling. The tenderness of the steaks was evaluated by means of a taste panel and by a Warner-Bratzler shear press.

Increasing Binding   Phosphates improve the binding properties of fresh, whole meat. Karmas[587] reported that smaller chunks of meat could be bound to each other to form pieces of larger sizes if they first were mechanically treated so that the natural exudates were brought to the surface of the meat chunks. The chunks of meat then would be bound so tightly that the larger pieces of meat, thus bound, could be sliced. The texture was reported to be similar to cuts of higher grades. One should also remember the influence of the phosphates on increasing the solubility of myosin, which is important in meat binding.

Increasing Moisture Retention   The water-holding capacity of meat proteins can influence such factors as the drip loss, defined as the amount of natural meat juices lost during refrigeration or during the defrosting of frozen meat, the amount of water bound by the proteins, the amount of moisture lost during cooking, and the juiciness of the meat when eaten. All of these factors have significant influences on the palatability of the meat and, therefore, its acceptance by the consumer.

Hopkins and Zimont[659] prevented the exudation of unsightly red or pink juices into the packages of prepackaged retail beef, mutton, pork, venison, and poultry by dipping the meat for 5 seconds to as much as 15 minutes, depending on size, into solutions containing 10–20% sodium chloride and 0.5–5% of one of the phosphates, with or without ascorbic acid or sodium nicotinate. Preventing this exudate not only makes the meat more acceptable in appearance but also improves its tenderness, nutritional value, and other factors.

Rust[649] reported that steaks injected with increasing quantities of an SHMP solution increased in juiciness in proportion to the quantity of phosphate injected. Juiciness scores were significant to the 0.1% level for the phosphate-treated steaks over untreated controls. In addition, significantly lower cooking losses were experienced in the phosphate-treated steaks.

Wismer-Pedersen[633,634] demonstrated that an increase in pH, the presence of magnesium and calcium ions, and the addition of pyrophosphate all increased the water-holding and the rehydration capacity of freeze-dried beef and pork. The treatment of beef with a solution of STP prior to freeze-drying was reported to accelerate the drying and to result in improved texture, odor, and taste when compared to an untreated control.[660] SHMP also was reported to have a similar effect on increasing the hydration capacity of freeze-dried meat when the meat was treated with the SHMP prior to drying.[661]

Miller and Harrison,[658] however, reported that they could find no significant increase in the water-holding capacity of steaks marinated in SHMP solutions. Juiciness scores were found to be significantly increased for the steaks marinated 2 hours or more, but cooking losses were increased over those of untreated steaks. These authors concluded that marinating steaks in SHMP solutions was of little value in improving the eating quality of the steak.

Flavor Improvement   Although the protein in the meat, the manner in which meat is processed, and, finally, the method of cooking contribute to the development of acceptable flavor of the meat, certain reactions of other meat constituents can result in flavor deterioration. The development of oxidative rancidity of the fats in the meat can lead to serious off-flavors in both fresh and cooked meats.

Watts and her co-workers[662–668] have contributed numerous studies on the stabilization of meat products against oxidative rancidity. Tims and Watts[662] studied the development of oxidative rancidity in refrigerated, cooked pork by means of the 2-thiobarbituric acid method (TBA method). Only those meats containing many polyunsaturated fatty acids developed rancidity after

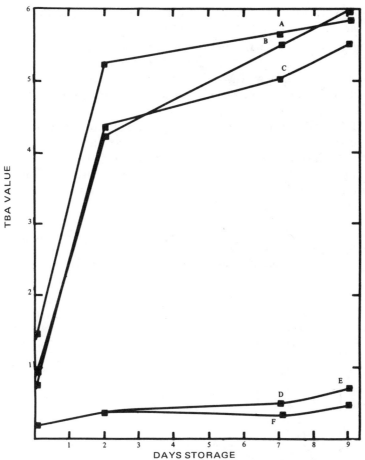

Fig. 6. Effect of phosphate on TBA values of refrigerated cooked pork. (Source: Tims, M. J., and Watts, B. M., 1958, *Food Technol. 12*:240; with permission.)

A — Control	D — STP
B — DSP; no pH adjustment	E — SHMP
C — DSP + MSP to pH 6.4	F — SAPP + TSPP

cooking. Figure 6, taken from this study, demonstrates that pyrophosphate, tripolyphosphate, and hexametaphosphate protected the cooked pork against development of oxidative rancidity, while orthophosphate showed no activity. The phosphates also exhibit synergistic activity with ascorbic acid and sodium ascorbate in protecting the cooked meat against development of rancid odors and flavors. Tarladgis *et al.*[663] confirmed the synergistic activity of poly-phosphates with sodium ascorbate; they also reported that extracts of certain vegetable products gave partial protection against the development of rancid odors and flavors in cooked, irradiated pork. Chang and co-workers[664] found that oxidation of lipoproteins and phospholipids was responsible for the development of stale, left-over, or rancid odors and flavors when uncured meats were cooked. The oxidation reactions are catalyzed by the ferric heme pigments in the cooked, uncured meats; these pigments are derived from the breakdown of myoglobin as a result of the heat treatment. The off-odors and flavors developed more rapidly at refrigerated temperatures than when the cooked meat was frozen. Treatment of the meat with combinations of sodium ascorbate and tripolyphosphate was found to retard the development of this oxidative rancidity and to result in greatly improved odor and flavor of refrigerated and frozen beef, as shown in Table 26. Irradiated beef, however, did not benefit to the same extent. Ramsey and Watts[665] reported that excellent protection against the development of rancid odors and flavors in slices of refrigerated, cooked beef was provided when they were covered by solutions containing 0.02–0.1% STP or dipped into a solution of 0.5% STP. Zipser et al.[666] reported that the heme-catalyzed oxidation and the development of rancid odors and flavors occurred during the preparation of the cooked meat for freezing and during the thawing of the meat after frozen storage; these reactions did not, however, intensify during frozen storage. Watts[667,668] has reviewed much of this work.

TABLE 26

Effect of Antioxidant Dips on Roast-beef Slices Stored
in the Refrigerator and the Freezer[a]

Storage Temperature	Days in Storage	No Antioxidant (Control)		Dipped in 1% $Na_5P_3O_{10}$ + .27% Na-ascorbate	
		TBA No.	Odor Score	TBA No.	Odor Score
1°C	1	11.5	—	0.3	—
	4	15.2	2.5	0.4	5.6
	7	19.5	2.6	0.3	5.3
	11	21.1	1.9	0.3	4.7
	18	21.2	—	0.3	—
−26°C	11	3.5	2.7	0.2	5.4
	31	2.3	2.4	0.2	5.5
	45	3.2	—	0.2	—
	81	4.9	2.2	0.2	5.9
	164	4.1	3.5	0.2	5.8

Source: Chang, P., Younathan, M. T., and Watts, B. M., 1961, *Food Technol.* 15:168, Table 2; with permission.

[a] At each storage period the odor of the antioxidant-dipped meat was rated against its corresponding control. In all cases the antioxidant-dipped meat rated higher than the control, and the differences were highly significant.

Applications in Fresh, Comminuted Meat Products

Color Preservation   Savich and Jansen[669] reported that the red color of ground-meat products was stabilized for 9–12 days when treated with a combination of MSP, sodium propionate, and ascorbic acid; untreated controls developed an unacceptable color at the end of 2 days storage at 35°F. Ziegler[670] prepared a red pigment from dried blood for use in coloring ground-meat products to maintain a normal red color of the fresh, processed meat. He (1) treated the blood with an anticoagulant phosphate, (2) added a reducing agent, such as ascorbic acid, to inhibit oxidation, (3) buffered the blood composition to maintain a pH of 7.3–8.0 with phosphate buffers, and (4) added sodium nitrate to convert the hemoglobin to nitrosohemoglobin, which has the bright-red color desired. The critical factor in the process is the buffering of the blood to a pH of 7.3–8.0 to prevent it from developing a brown color during drying.

Increasing Tenderness   The texture of ground-meat products can sometimes be described as tough or chewy. Jacoby and Berhold[671] reported that they could obtain excellent textural properties and tenderness in ground or diced meat products, such as hamburger or stew meat, by adding a chelating agent, such as STP, to the meat. The mixture then was combined with an unmodified starch and a protease, preferably papain, in very small quantities to assure proper texture and tenderness.

Increasing Binding   Binding the particles of ground meat together into a homogeneous, uniform, stable mass is important in the preparation of hamburger, meat loaf, fresh sausage, etc. Previous paragraphs in this chapter have reviewed the influence of solubilizing the salt-soluble proteins in improving the binding of meat products. Aged bulls are considered to provide the best type of binding meat for use in ground meat products.[25] The polyphosphates have been found to improve the binding qualities of younger beef meat, which normally would be considered poor-grade for use in ground-meat products because of its low binding properties.

Bickel[672] reported that it is possible to produce excellent-textured, ground-meat products that maintain their structure, and in which the meat particles swell and bind to each other. The water-soluble salts of pyrophosphoric acid, according to Bickel, and longer-chain polyphosphoric acids exert this activity on the ground-meat products. SAPP, TSPP, STP, sodium tetrapolyphosphate, SHMP, and the very long-chain sodium and potassium polyphosphates, such as long-chain Graham's and Kurrol's salts, are applicable. Optimum levels of the polyphosphates appear to be 0.15–0.5%, with 0.3% being the preferred level.

Fetty[673] patented the application of $DSP \cdot 2H_2O$ and $MSP \cdot H_2O$ with combinations of wheat flour, nonfat-dried-milk solid, hydrolyzed vegetable protein, and sodium bicarbonate as binders, which would prevent the shrinkage of ground-meat products, such as hamburger, meat loaf, lamb patties, and cured sausages. The composition is added at levels of 3–6% based on the weight of the ground meat.

Increasing Moisture Retention   The retention of moisture is as important in comminuted meat products as it is in whole meat products, as it reduces drip loss from fresh as well as thawing frozen meat, drying and toughening of meat, and cook-out loss. Moisture retention helps retain highly nutritious, soluble proteins in the meat during storage and cooking.

Sair and Cook[674] reported that maintaining the pH at approximately 6.4 gave the greatest protection against drip loss, while the maximum amount of drip loss occurred at pH 5.2.

Bendall[593] reported that the phosphates were even more effective in reducing the loss of moisture and soluble constituents from meat than controlling the pH. He was the first to report that a combination of pyrophosphate and salt more than doubled the water-holding capacity of meat tissue compared to salt alone at equivalent ionic strength. The phosphates split actomyosin into its component proteins and converted them from a gel to a sol state. Although orthophosphates, hexametaphosphate, and the very long-chain sodium and potassium phosphate glasses also increased swelling and water-holding capacity of muscle tissues, they were not as effective as pyrophosphate. Sherman[618] added sodium chloride or various phosphates to ground meat and significantly improved the retention of fluids; however, the retention could not be attributed to the ability of the phosphates to complex divalent metal ions. He reported a statistically significant correlation between the pH of the phosphate–meat mixtures and their abilities to retain fluid. He also found a correlation between the water-holding capacity of the proteins and the absorption of anions by these same proteins, particularly when sodium chloride was added to the meat. Although the rate of anion absorption increased with increasing ionic strength of the salt solutions added to the meat, the cooking of the meat caused the release of both anions and cations; the anions were retained to a greater extent than cations. Sherman attributed the retention of fluid in cooked meats to the increase in solubilization of actomyosin. This protein, he believed, is gelled during cooking and thus lowers the release of moisture since the gel extends throughout the mass of the meat. In an extension of his previous work, Sherman[675] reported that the water-holding capacity of the meat tissues at temperatures between 0–100°C is directly related to the amount of ions absorbed; the absorption of the ions appears to open the helical structure of the protein, allowing water to penetrate.

Hellendoorn[603] found that pyrophosphate and tripolyphosphate reduced the water-holding capacity of raw meat at pH values below 5.5; they increased it when the pH values were raised to 6.0–6.5. Pyrophosphate was slightly superior to tripolyphosphate in this pH range and increased in superiority as the ionic strength was increased above 0.4. Orthophosphate and hexametaphosphate were less effective than either pyrophosphate or tripolyphosphate, and longer-chain polyphosphates and cyclic polyphosphates had no significant effects. As shown in Figure 5, Hellendoorn found that calcium binding is of no importance in the water-holding capacity of meat tissue. Hellendoorn's finding was also supported by the work of Inklaar.[604]

Applications in Cured-Meat Products

Color Preservation   Ellinger[28] has reviewed some of the numerous reports on the effects of phosphates on color stabilization of cured meats. Many of these reports appear to contradict each other in regard to which conditions of processing stabilize the color or which phosphates are most effective in color stabilization.[600,638-640,651,676-693] Swift and Ellis[676] have speculated that some of the discrepancies and confusion in the literature are due to significant differences in the processing variables used and tested in determining the effects of the phosphates on the color of cured meat. Some investigators have found that the pH of the system was highly important in determining whether a brine system containing phosphates would stabilize the meat color. A Dutch patent[676] reported that the pH of the brine system used had to be stabilized at a pH of 3–5 with citric acid and DSP to stabilize color. Sair and Komarik[678] reported that acidic phosphates, such as MCP, MSP, MKP, SAPP, or SALP, and various combinations with some of the alkaline polyphosphates were necessary to lower the pH values of their meat systems below those of the normal meat to stabilize color.

Chang and Watts,[640] however, found that none of the phosphates were as effective as ascorbic acid in protecting meat against the loss of color in the presence of hemoglobin at pH 5.6, the normal pH of the meat used in the experiments. Hall[638, 679] maintained pH values of 6.5-7.0 to protect the decolorization of frankfurters and similar cured meats, including hams, when polyphosphates having chain lengths of 3 or longer were used in brine systems. Hall's patent[638] has been one of the most dominant in the application of phosphates in cured meat products. A second dominating patent that implemented phosphates in meat products was obtained by Brissey.[639] His process involved the use of DSP to maintain the pH of curing salt pickles at 6.6-6.8. Both Hall[638] and Brissey[639, 679] reported that allowing the pH values to go below or above the ranges quoted would produce undesirable effects on the cured meat products. Bailey et al.[680] reported that slices of hams cured with ascorbic acid, with combinations of ascorbic acid and nicotinamide, or with ascorbic acid and niacin must be buffered at pH values of 6.2-6.8 to effect color stabilization through the formation and retention of nitrosomyoglobin. Rahelic and co-workers[681] reported that the addition of polyphosphates to meats cured in brines containing salt, nitrates, and nitrites and stored at refrigerated temperatures resulted in significant increases in the pH values and the per cent of nitroso pigment compared to those treated with brine containing no phosphates.

Wilson[682] reported that combinations of DSP and SHMP, or of DSP and ascorbic acid, used in the treatment of bologna products effectively stabilized the color against fading. A combination of DSP and ascorbic acid increased the color shelf-life of the bologna up to 70% if held at refrigerated temperatures when exposed to fluorescent lights. Fenton and co-workers[683] reported that DSP alone was not effective in stabilizing the color of cured meats; they quoted Wilson[684] as reporting the same result. Mahon and co-workers[685] reported that a panel was capable of detecting the differences in color between control hams and hams treated with small quantities of DSP. The phosphate-treated hams were more acceptable in their color, which was significant at the 5% level. Gisske[686] reported that the use of commercial phosphate mixtures containing DSP, SHMP, and STP improved the color of hams to which they were applied in the pumping pickle. Schoch[687] reported that combinations of DSP and polymeric sodium phosphates, such as SHMP or longer-chain polyphosphates, effectively stabilized the color of cured-meat products of various types.

Several investigators have also reported that polyphosphates used by themselves in curing pickles were effective color stabilizers, and the pH values were probably not significantly changed. Morse[600] reported that Wilson, of the American Meat Institute, found that STP produced the best color of all treatments tested in hams. He may have referred to the report by Wilson[684] that a trained panel of judges significantly preferred hams cured with STP or SHMP compared to those that had no phosphate in the curing pickle. Watts[688] reported that there is a direct relationship between fat oxidation and the loss of color in refrigerated hams, as shown in Table 27. The addition of either tripolyphosphate or hexametaphosphate to the brines used in pumping hams significantly increased the color retention of canned, baked hams; it also increased, but not to a similar extent, the color retention of canned or smoked hams. Watts also reported that the polyphosphates and similar inhibitors of color fading have little or no effect on the fading due to exposure to light, such as fluorescent lighting. Swift and Ellis[676] found that the processing conditions significantly

TABLE 27
Effect of Polyphosphates on Color Stability and Rancidity in Refrigerated Hams

Compound Used	Treatment	Days for 50% Color Loss	Days to Turn Rancid (TBA = 0.5)
None	Canned, baked	2	2
Hexametaphosphate	Canned, baked	>12	>12
None	Canned, baked	3	2
Tripolyphosphate	Canned, baked	>13	>13
None	Smoked, baked	3	3
Hexametaphosphate	Smoked, baked	>13	>13
None	Smoked, baked	6	6
Tripolyphosphate	Smoked, baked	>14	>14
None	Canned	10	9
Tripolyphosphate	Canned	>13	>13
None	Canned	10	13
Hexametaphosphate	Canned	>13	>13
None	Smoked	12	>14
Hexametaphosphate	Smoked	>14	>14
None	Smoked	12	>14
Tripolyphosphate	Smoked	>14	>14

Source: Watts, B. M., 1957, *Proceedings Ninth Research Conference*, American Meat Institute, pp. 61–67, Table II.

determined the color stability of cured meat products. For example, long tempering, smoking, and cooking times of phosphate-treated frankfurter formulas resulted in increased color development and, therefore, increased color shelf-life than briefer tempering, smoking, and cooking times. Frankfurters treated with SHMP or with a mixture of 70% TSPP and 30% SAPP had the most acceptable color. Mullins *et al.*[651] reported that an evaluation panel could distinguish between the uniformity and stability of the color of phosphate-treated and untreated hams. Wasserman[689] patented combinations of STP and sodium ascorbate in the curing salts, which resulted in an intensified red color after several days; they, therefore, increased the length of time in which the meat retained an acceptable color. Suri[690] confirmed that the addition of polyphosphates to meat-curing brines stabilizes the color; he implemented an analytical method for determining the level of phosphates added to the meat, based on analysis by paper chromatography. Siedlecki[691] added 0.35% polyphosphate to stabilize the color of canned hams. Lauck and Tucker[692] used 1 part SALP and approximately 1–5 parts of an alkali-metal phosphate in the curing salts for comminuted meat to improve and stabilize the color, as well as improve other properties of the meat product. Savic and co-workers[693] added polyphosphates to the curing salt mixtures that contained nitrites used in processing canned, cured meats to prevent the nitrites from damaging the tin-plate coating of the cans; the polyphosphates also eliminated the resulting yellow-brown discoloration of the surface of canned pork products.

Increasing Tenderness   The moisture and fat contents of cured meats also have a significant effect on their juiciness and tenderness, as shown by Swift et al.[694] Since the phosphates allow increased moisture and fat to be incorporated into comminuted meats, their use is bound to increase tenderness, as has been shown by several investigators; phosphates have the same effect when they are used in curing whole meats, such as hams. Mullins et al.[651] reported that panels scored hams treated with SHMP as moderately tender, while control hams cured without SHMP were scored as slightly tender.

There may be a biochemical reason for the effect of phosphates on increasing tenderness of any meat product. Szent-Gyorgyi[695] speculated that any compound or process that would inhibit the formation of actomyosin, by preventing the attachment of fibrillar actin to myosin, would be likely to increase tenderization of meat. He later reported that the formation of fibrillar actin from globular actin could be prevented with SHMP and that a solution of actomyosin would be dissociated into its component proteins if it were treated with inorganic alkaline phosphates.[696]

Schleich and Arnold[697] tenderized hams by adding a small quantity of a bacterial protease to the pumping pickle along with one of the polyphosphates. The phosphate used could be either SAPP, TSPP, STP, sodium tetrapolyphosphate, or SHMP. The polyphosphates prevented over-tenderization of the hams by inactivating the enzyme over a period of approximately 3 hours.

Increasing Binding   The action of the phosphates on increasing the binding of comminuted, cured meat products, such as bologna, frankfurters, and other sausages and sandwich meats that are stuffed in casings and then further processed by smoking and cooking, has been thoroughly reviewed in the literature.[25,30,584] As discussed previously, the action of the phosphates is believed to be due to the dissociation of actomyosin into its component proteins, thus releasing myosin to form its binding gel during processing of the meat. Swift and Ellis[676] demonstrated that various combinations of pyrophosphates, hexametaphosphates, and long-chain potassium polymetaphosphates would increase the binding of meat in sausage products, as shown by the increased tensile strength or cohesion of the bologna products. Kotter[591] speculated that the pyrophosphates replace ATP in promoting the dissociation of actomyosin; this would explain the binding action of the polyphosphates on sausage-type meat products. Bickel[672] reported that the combinations of polyphosphates, and particularly of the long-chain potassium polyphosphates, would cause the different meat constituents in sausage products to bind together into a homogeneous mass of uniform structure. Maas and Olson[698] used alkaline polyphosphates to bond together whole pieces of meat or muscle tissue during heat processing. They claimed that the phosphate ions are essential to obtain this binding, which is undoubtedly due to the release of myosin. The released myosin can gel to bind the meat particles together to resemble larger, whole pieces of meat. After cooking, the bound meat can be sliced, just as can larger, whole pieces of meat.

Increasing Moisture Retention   The retention of the water and natural juices of any cured meat product is extremely important, both in the storage of the product and during its cooking. The ability of the phosphates to increase the moisture retention of meat proteins has been discussed in previous sections; this section will deal with some of the specific applications.

Hall[638] found that frankfurters retain more of their juices during smoking when they are treated with phosphates, particularly with STP, and when the pH is maintained at 6.5–7.0. Brissey[639] reported that adding a dialkali phosphate to the curing-salt mixtures used in curing meats resulted in greater retention of soluble proteins in the product and, therefore, larger cooking yields. The treatment reduced shrinkage in cooked and canned hams, pork shoulder picnic hams, and similar products. The reduction in processing shrinkage and the retention of the nutritionally important soluble proteins during processing and cooking are important economic advantages to the addition of various phosphates during meat curing.[679] Swift and Ellis[676] reported that not only were greater amounts of moisture retained in the cured-meat products, which resulted in increased juiciness and tenderness, but also greater proportions of fat were retained. Morse[600] reported that retention of more moisture and juices in the meat product does not give it a moist appearance; in fact, the product actually appears drier, since the water is more completely bound by the meat protein.

The type, quantity, and method of addition of the phosphates appear to be important in determining whether they are effective in retaining moisture and juices. Mahon and co-workers[685] and Leeking and co-workers[699] reported that the addition of 0.5 lb DSP/gal curing pickle had no significant effect on the water-holding capacity or reduction in cooking losses of cured, smoked hams; apparently this level is too low to have any effect. Frank[700] found that adding the phosphates to the cooked meat had no effect on its water-holding capacity. It, therefore, appears that the phosphate must be added to the meat prior to any heat treatment.

Swift and Ellis[701] found that the temperature of the meat at the time the phosphate is added is important; the addition of phosphate to the meat at temperatures of 32–41°F was more effective in increasing the water-binding properties of the meat protein than higher temperatures. Storage of the phosphate-treated meat within this optimum temperature range for 16 hours also increased the water-holding capacity. Swift and Ellis[676] later reported that the phosphates had no effect at cooking and smoking temperatures of 150–160°F; however, they markedly reduced cooking losses of bologna and frankfurter sausage mixtures when heated to internal temperatures above 160°F. It was, therefore, possible to use higher-temperature, shorter-time smoking processes, a desirable factor for sausage processing. It appears that such processing factors as temperature, time, ionic strength, and the pH of the treatments are important in obtaining maximum water retention by meat proteins, most likely because the phosphate additives exert their effects on moisture retention through their ability to solubilize muscle proteins.[701]

Grau[584] described experiments in which the addition of small quantities of phosphates to sausage-meat products did not increase the water absorption of the meat above controls; instead, it caused the meat proteins to bind the water more completely and, therefore, not to lose it during processing. Grau's work probably explains the advantages of the use of phosphates in processing cured meats, as reported by Epstein,[601] Turner and Olson,[606] Gisske,[686] Wasserman,[689] Volovinskaya et al.,[702,703] and Rupp et al.[704] According to these researchers cured-meat products treated with phosphates (1) bind more water, (2) are plumper in their casings, (3) swell to a greater extent to fill the casings, (4) maintain tight casings over the sausage product during storage, and (5) lose less water during smoking and heat processing, as well as cooking.

Whether or not the phosphates should be allowed in processing meat for human food is a topic of controversy. Zeller[705] claimed that health problems are too great to warrant the use of phosphates in meats, although he acknowledges their advantages. Grau,[584,706] Kotter,[602,609] Hamm,[585] and other German workers disagree with Zeller; they have announced that the levels of phosphate necessary to produce the desired water-binding capacity in the meats are too small to present any health problems.

Mahon[617] studied the increase in volume of cured-meat products, because it is directly proportional to the amount of water that is bound by the meat proteins. As shown in Figure 7, taken from his work, the volume of the meat increases as the pH decreases or increases away from the isoelectric point.

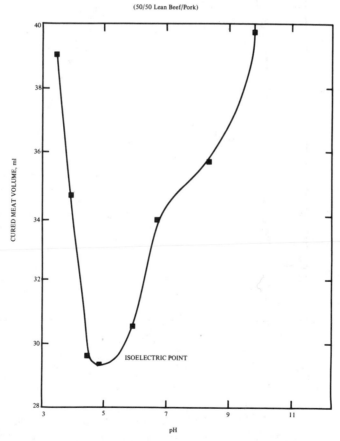

(50/50 Lean Beef/Pork)

Fig. 7.   Effect of pH on cured meat volume—no salt. (Source: Mahon, J. H., July 1961, Proceedings of the 30th Research Conference, American Meat Institute Foundation, No. 64, 59, Figure 1; with permission.)

Mahon also reported that the addition of small quantities of salt actually can depress the meat volume, which reaches a minimum at approximately 0.5% salt, as shown in Figure 8. As the amount of salt is increased above 0.5%, the volume of the meat increases to reach a maximum at about 5% salt concentration; it then again decreases to another minimum at about 10% salt concentration. A combination of 0.5% salt and 0.5% STP still produces a minimum meat volume, but this minimum is higher than that of the salt alone. A study of Figures 7 and 8 demonstrates that equivalent meat volumes are obtained with 4% salt by itself and 2% STP by itself and that the optimum meat volume is obtained with the legal limit of 0.5% STP when it is accompanied by 2–5% salt.

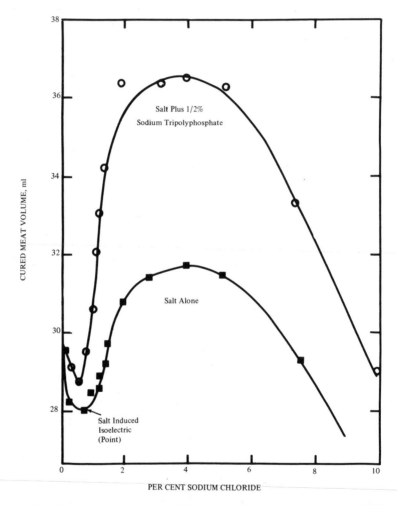

Fig. 8.   Effect of salt and salt + sodium tripolyphosphate on volume of cured meat (155°F). (Source: Mahon, J. H., July 1961, Proceedings of the 30th Research Conference, American Meat Institute Foundation, No. 64, 59, Figure 4; with permission.)

Siedlecki[691] reported that the addition of polyphosphates reduced the free juices in canned hams by 5%, improved their firmness, reduced shrinkage of the hams, and increased their yields during curing, but it had no effect on losses during smoking. Rongey and Bratzler[707] added phosphates to bologna products to produce the greatest firmness (tensile strength) and the highest yields among the various additives and binders studied in their investigation. Rahelic and co-workers[681] also reported a higher water-holding capacity in meats cured with brines containing polyphosphate than in those cured with brines containing no phosphate. Sair and Komarik[678] used a lower pH during the final heat processing of cured meats when phosphate compounds were included in the curing-salt mixtures, and they did not experience the cooking losses that normally accompany low pH.

Lauck and Tucker[692] claimed that the water-holding capacity of comminuted meat products, such as bologna, could be increased by adding an acidic SALP and polyphosphates to the curing salt. Mahon and Schneider[708] patented a blend of an alkaline SALP and condensed phosphates to retain up to 5% more juices in sausage products, such as Vienna sausages, during processing.

Improving Fat Emulsification   The amount of fat incorporated and retained in comminuted meat products has been demonstrated by Swift and Ellis[676] to determine its acceptability to the consumer. An increase in the quantity of fat emulsified in the product increased the consumer acceptability. Keller[709] increased fat emulsification in sausage products by treating them with TSP, but a soapy taste occurred that made the meat unacceptable. The use of sodium metaphosphates (probably polyphosphates) produced excellent fat emulsification and significantly reduced the loss of fat during cooking. Morse[600] also confirmed that phosphate salts, especially TSP, were capable of increasing the emulsification of fats in sausage products, and other benefits also were obtained.

The mechanism by which phosphates improve the fat emulsification in cured-sausage products has been studied by a number of investigators[606, 609-614, 710, 711] and has been reviewed by Saffle[586] and Swift.[615] It is necessary to break up and disperse the fat globules in very fine, uniform droplets throughout the meat mixture during the cutting and preparation of the meat matrix. Once this has been accomplished, however, the phosphates solubilize the salt-soluble proteins myosin and actomyosin, so that they are able to form stable protein envelopes about the fat globules or particles in the meat emulsion and, thereby, stabilize these particles and prevent their coalescing. During the heat processing of the meat, these proteins are formed into a framework that produces the firm, easy-to-cut texture of the typical sausage product. The stability of a sausage emulsion appears to be directly proportional to the amount of protein appearing at the interface between the fat globule and the aqueous phase. Although actin, in the absence of salt, is more effective in emulsifying fat in sausage, the presence of salt in all-meat emulsions causes actin to become the least effective emulsifying agent.

Carpenter and Saffle[613] reported that the ability of the water-soluble proteins, such as actin, to emulsify fat was directly affected by the shape of the molecule. Globular-shaped molecules had a lower emulsifying capacity than fibrous molecules. Changes in the net charges of the water-soluble proteins seemed to have little or no effect on their fat-emulsifying capacities. In contrast, both the net charges and the shapes of the molecules had a significant effect on the emulsifying properties of salt-soluble proteins. These authors reported that myosin is a long, rod-like molecule that is capable of spreading a film over a greater surface of the fat molecule. High negative charges, which are produced in the protein at pH 9, resulted in the long, fibrous, rod-like molecules. Changing the pH to 6 caused the molecules to form a more spherical shape and to reduce their emulsifying properties.

Improving Flavor   In an extension of their work on the effects of the polyphosphates as agents to prevent oxidation and rancidity of fats in aqueous fat systems, Watts and her co-workers[636,666,688,712] have studied the effects of the polyphosphates in preventing oxidative rancidity of fats in meat products. The results of this series of experiments, as shown in Table 27, demonstrated that STP and SHMP significantly extend the shelf-life of canned, baked and smoked, baked hams against the development of fat rancidity. Fewer or no beneficial effects were derived from the addition of similar phosphates to canned and smoked hams. Frozen, cured meats developed oxidative rancidity flavors 8–10 times faster than frozen, uncured meats. The ferric hemes were found to catalyze the oxidative rancidity. By proper preparation, storage, and the addition of phosphates, the development of rancidity could be controlled in the uncured meats, but rancidity continued to develop in the frozen, cooked, cured meats even during frozen storage. Only the use of a combination of STP and sodium ascorbate at about twice the level legally allowed (0.108%) resulted in stability against development of oxidative rancidity in the cooked, cured, frozen pork products tested. Meat samples became rancid in about 4 months when the legal limit of sodium ascorbate (0.054%) was used. By following the disappearance of sodium ascorbate in the samples, the authors were able to show that peroxides increased rapidly once the ascorbate had disappeared. The work of Watts *et al.* undoubtedly explains the improved flavor in meat products treated with phosphates by Suri[690] and by Rupp and co-workers.[704]

Preventing Microbiological Spoilage   Because microbiological deterioration can also create off-flavors and off-odors in cooked and cured meats,[586–587] investigators are continually searching for ways in which to stabilize cured meats against microbiological attack. Evans[713] reported that *Staphylococci* can survive but will not reproduce in curing pickles if the salt content is high enough and the temperature is maintained low enough. Nitrite had little or no effect on the organisms. The addition of DSP to the pickle, which raised the pH from 7.0 to 7.9, and of SHMP, which lowered the pH from 7.0 to 5.3, also had no effect on the bacteria. The presence of 0.5% of SHMP had a slightly inhibitory effect on the anaerobic growth of this organism when pumped into the center of hams, where they were protected from the heat of the smoking operation.

Bickel[714] reported that meat and the intestines intended for use as casings for sausage-type meats could be preserved against spoilage by treatment with 2–10% by weight of a solution containing 80 parts tartaric acid and 20 parts SHMP. Bickel also claimed that the pyro-, tripoly-, and tetraphosphoric acids and potassium metaphosphate derivatives are effective. Volovinskaya and co-workers[702] reported that the addition of either TSPP or SHMP containing additional NaOH caused some reduction of microbiological growth in stuffed sausages. Epstein[601] also noted that he obtained microbiological stabilization in a variety of sausage products prepared from cured, comminuted meats.

Wollmann[715] reported that sodium pyrophosphate inhibited *E. rhusiopathiae*, *S. typhimurium*, and *M. tuberculosis* when incorporated in sausage meats that were subsequently cured and smoked. The sodium pyrophosphate, probably TSPP, may have increased the pH, although this was not discussed. The addition of a commercial curing phosphate mixture containing 40% of a combination of SAPP and TSPP, 10% SHMP, and 50% NaCl had a much lower, negligible effect. Nitrite curing salts increased the resistance of these organisms to the phosphates, which Wollmann traced to the high levels of NaCl in the curing-salt mixtures.

Miscellaneous Uses   Some of the phosphates have been found useful in maintaining the flow properties of the spices and curing-salt mixtures used in meats. Allen[650] reported that TCP could be used to maintain a free-flowing mixture of enzymes, sodium glutamate, and sucrose for use in curing meats. Peat[716] reported that fine powders, such as TCP, calcium silicate, or magnesium carbonate, could be used as carriers for spice extractives. Gorsica[717] reported that soluble seasonings, flavorings, and coloring agents could be combined with TSPP as a soluble carrying agent for these compounds for use in meat products. The TSPP would dissolve when placed into curing-salt mixtures, leaving the dispersed spice, flavoring, and coloring agents in suspension or solution.

Additional Patents for Phosphates in Meat Processing

The patent literature for the use of phosphates in meat processing is voluminous and, in many cases, appears to duplicate itself. Anyone contemplating the use of a phosphate in any type of meat processing would be well advised to consult the literature and make a thorough search before using it in a commercial process.

A number of patents have been discussed in previous paragraphs of this section of meat applications of the phosphates. A few more will be described here to demonstrate the number and variety of patents for phosphate applications in meat processing.

Phosphates have been patented for use as buffers to maintain optimum pH values for the greatest stability of meat curing salts containing nitrite,[718] as sequestering agents,[719,720] and as anticaking agents for curing-salt mixtures, especially those containing STP.[720-722]

A number of combinations of polyphosphates, as well as polyphosphates combined with curing salts, have been patented for their beneficial effects on meats. STP and its hydrate, SHMP, Kurrol's salt, and other condensed alkali-metal phosphates have been patented for their applications in swelling and softening muscle fibers and creating uniform fat emulsification in sausage products.[723] The addition of SAPP and TSPP with potassium metaphosphate (Kurrol's salt), with or without other curing agents, increases the swelling, has a tenderizing effect on meat proteins, improves their binding properties, and improves the plumpness of cured-meat products.[724] The use of TSP in curing-salt mixtures has been patented for its ability to improve color stability, water retention, and yields and to lower the shrink during processing.[725]

Other patents have been obtained for the following uses of phosphates in meat processing:
1. to improve the flavor and keeping properties of meats, especially sausages, with combinations of sodium glutamate or glutamic acid plus an alkali-metal polyphosphate with or without citric acid,[726]

2. to improve the properties of animal or fish meat with combinations of magnesium ion and polyphosphate,[727,728]

3. to eliminate the need for non-meat binding agents through the addition of TSP, TSPP, SHMP or KHMP, STP, or mixtures of these alone or blended with spices,[729]

4. to cure ham with phosphate mixtures containing MSP, DSP, SAPP, and TSPP in combinations with other curing salts,[730]

5. to create a plasticizing effect on meat, and thus improve its handling properties and palatability, with solutions of STP hexahydrate, SAPP, and TSPP decahydrate,[731] and

6. to improve the water-binding properties of sausages by using combinations of pyrophosphate or tripolyphosphate with non-toxic organic acids.[732]

Patents also have been obtained for various forms and mixtures of polyphosphates to be used with other curing compounds for meat processing, as follows:

1. to spray a phosphate solution on sodium chloride crystals and, thus, form a crystalline complex of the phosphate component and the crystalline structure of the sodium chloride,[733]

2. to treat canned, cooked, or scalded sausages with a liquid medium containing TSPP or other condensed phosphates,[734]

3. to treat blood and meat products with spray-dried, long-chain polyphosphates,[735]

4. to prepare free-flowing compositions of alkali-metal tripolyphosphate and a strongly basic, alkali-metal compound so that a 1% aqueous solution has a pH of 10.5–12,[736]

5. to combine solutions of STP and NaOH during pumping of cured-meat products so that TSPP will be formed in the meat,[737]

6. to combine solutions of sodium erythrobate with an alkali hydroxide, carbonate, or orthophosphate during pumping of cured-meat products,[738]

7. to crystallize sodium nitrite on the surface of STP crystals.[739]

Other examples of the patented applications of phosphates follow:

1. to provide artificial sweetening and, hence, replace sugar in the curing process with combinations of STP and cyclamate,[740]

2. to combine phosphates with a coloring agent so that, when blended with other curing salts, a homogeneous blend can be determined by the uniformity of the color,[741]

3. to stabilize flavor intensifiers against the action of phosphates in meat by first inactivating them with condensed phosphoric acids or their ammonium or alkali-metal salts, such as TKPP, KTP, or SHMP.[742]

Phosphate Applications in Poultry Processing

Identical chemical and physical reactions occur in post-mortem poultry muscle as occur in beef and similar animal muscles.[28,743,744] The major differences in poultry reactions are that (1) rigor mortis sets in much faster and (2) the aging process to obtain subsequent maximum tenderization is much shorter than in red-meat animals. While beef requires 10–20 days for development of optimum tenderness, poultry requires only 12–24 hours.[745] Excellent reviews of the biochemistry of chicken muscle and rigor mortis, as a result of their own and other investigations, have been published by deFremery and Pool[743] and by deFremery.[744] The same palatability factors that are important in red meats are important in poultry meats; these include color, tenderness, and flavor.

Preserving Color

The color of fresh poultry meat is an important factor in the selection of the meat. Several investigators have studied the effects of polyphosphates on the color of fresh poultry meat. Wells and co-workers[746] reported that polyphosphates fed to chickens prior to slaughter produced an undesirable dark color, or discoloration, in the meat. Klose and co-workers[747] reported that when chicken carcasses were soaked in chilled polyphosphate solutions, the normal yellowish cast of the skin of the controls changed to a disagreeable bluish-white in the treated carcasses. Mahon[748–750] reported that a favorable color occurred on the surfaces of refrigerated, cooked meat obtained from poultry meat soaked in solutions of polyphosphates; as an alternative treatment the polyphosphate solutions were dusted onto the surface of the chilled meat. The normal appearance and color of fresh, frozen, and refrigerated poultry meats are retained for a longer time if the freshly killed poultry carcasses are soaked in solutions of phosphates or are dusted with them during their chilling and aging. Froning[751] reported that phosphate-treated, raw, uncooked poultry meat had an unappetizing, bluish-white, glistening appearance similar to that reported earlier by Klose; however, the cooked meat was less affected. In a later investigation Froning[752] evaluated the color of cooked chicken rolls by means of a Gardiner Color Difference Meter, standardized with a white plate; the meat was significantly darkened upon the addition of polyphosphates, and this darkening was in direct proportion to the level of phosphate added. Poultry meat containing levels of polyphosphate above 1% was considered to be unacceptable.

Increasing Tenderness

As is the case with red meats, the tenderness of poultry meat is significantly influenced by the reactions that accompany the development of rigor mortis. The only difference is that rigor mortis occurs more rapidly in poultry processing. Dodge and Stadelman[753] reported that cooking poultry meat prior to the onset of rigor mortis resulted in significantly more tender meat; their experiment duplicated results obtained with red meats.[645] It is important to arrest the reactions, or to counteract their effect, that occur during the onset of rigor mortis, as they are responsible for toughening.

The investigations of deFremery and Pool[743] demonstrated that any conditions that caused a more rapid development of rigor mortis also increased the toughening of the cooked meat. When the enzyme systems required for the breakdown of glycogen were poisoned, the resulting muscle failed to toughen and was as tender as the pre-rigor muscle. Further experiments by deFremery[744] demonstrated that rapid depletion of glycogen in the live bird prior to slaughter caused little or no loss in the initial tenderness of the cooked meat.

The pH of the meat immediately after slaughter had fallen very little from its original level, indicating that little or no glycolysis occurred; thus, little or no lactic acid was formed. The treated carcasses produced meat that was as tender immediately after onset of rigor as the meat of the control, which had been fully aged.

As a result of his investigations, deFremery[744] concluded that it was not the disappearance of ATP or the rapid onset of rigor in poultry meat that results in toughness; rather, the rapid disappearance of glycogen and the subsequent increase of lactic acid in the muscle is related to the toughening. deFremery claimed that similar biochemical changes are responsible for the development of toughness in red meat.

May and co-workers[754] significantly improved the tenderness of the light and dark poultry meats by chilling the carcasses in a 3% solution of polyphosphates. Klose and co-workers[755] reported that soaking either frying chickens or fowl in a chilling solution containing 5% polyphosphate had no effect on tenderness of the resulting meat; similar results were obtained with turkeys. Wells and co-workers[746] reported that feeding SHMP and other salts to chickens prior to slaughter had no effect on tenderness of the resulting cooked meat. Later work by Klose and co-workers[747] demonstrated that there was a statistically significant increase in tenderness when poultry, both fowl and fryers, was chilled for 22 hours in 5% solution of a mixture of STP and TSPP. Spencer and Smith,[756] Baker and Darfler,[757] and Schwall and his co-workers[758] all reported that various methods of treating poultry meat with polyphosphate salts resulted in significant tenderization of the meat.

Increasing Binding

The binding properties of poultry meat, like those of red meat, have been demonstrated to be significantly improved by treating the meat with polyphosphates.[751-763] Froning[751] reported that soaking the eviscerated chicken carcasses in a polyphosphate solution would increase the binding properties of the meat boned from the carcasses. Froning, in a later experiment,[752] Schwall and Rogers,[759,760] Schnell and co-workers,[761] Vadehra and Baker,[762] and Schlamb[763] reported various processes for treating the surfaces of raw or ground poultry meat with various salt mixtures, all containing polyphosphate; all reported significantly improved binding properties in forming chicken rolls, chicken sausages, and similar products in which binding of the meat particles was critical.

Schnell and co-workers[761] and Vadehra and Baker[762] studied the conditions required to improve binding properties of poultry meat. They found that phosphate treatments increased the amounts of proteins in the fluids on the surfaces of meat chunks. Sodium chloride did not have this effect, although it improved the binding properties of the meat. Although sodium chloride and the polyphosphates had similar effects on binding properties, their mechanism of action differed, because the chemical composition of the cookout from the two treatments differed. Vadehra and Baker concluded that binding is a complex phenomenon involving the water-holding capacity of the meat, the cell disruption and breakage, the release of intracellular materials, the type and qualities of myofibrillar and connective tissue proteins, and extraneous sources of proteins. Although these studies were conducted with poultry, the results are applicable to all types of meat.

Increasing Moisture Retention

The phosphates increase the moisture-retention capability of poultry meat, just as they do red meats. A possible difference could be that treatment of poultry meat with 5–15% polyphosphate tended to lower the yield of chilled, raw poultry meat, indicating that the meat picked up less moisture than controls soaked in ice water.[747,750,751,754–756,764–767] May and co-workers[754] found that lower quantities of polyphosphates caused the poultry meat to absorb greater amounts of moisture than the controls; Schermerhorn and Stadelman[765] reported that levels of phosphates above 12% caused significant lowering of moisture uptake. Klose and co-workers[747] concluded that the major portion of water absorption occurred during the early stages of chilling and that this was more important than the equilibrium values reached at later stages.

Since a considerable amount of poultry meat is frozen, one of the greatest advantages of polyphosphate treatments is their ability to reduce the drip loss, which occurs upon thawing of the poultry meat. Drip loss results in exudation of considerable amounts of the fluid in the meat; it is unsightly and also causes a loss in soluble protein. The ability of the polyphosphates to reduce drip loss has been studied to a considerable extent.[747,749,750] Tables 28 and 29, taken from one of these studies,[750] illustrate the effects of SHMP, STP, and pyrophosphate soaks or dips on thawing drip loss, cooking yields, and TBA values. As shown, all phosphate treatments except SHMP significantly reduced the thawing drip loss, increased the cooking yield, and reduced the TBA value; STP was the preferred polyphosphate treatment. The data also demonstrated that the polyphosphates actually reduced the amount of water absorbed by the treated poultry meat. The phosphate treatment was reported to have the same beneficial effects on chicken, turkey, goose, duck, capon, cornish hen, squab, guinea fowl, and pheasant meat.

Between 2–8% of polyphosphates significantly reduced cooking losses or cooking shrink of poultry meat.[747,749,750,755,764–770] Although the polyphosphates do not cause raw poultry meat to pick up a significant amount of moisture, they do reduce the amount of moisture lost during cooking; they, therefore, increase the yield. Figure 9 illustrates that the reduced cooking loss occurs in spite of the manner in which broiler carcasses are cooked. The effects of the phosphates on moisture retention were noticeably different for different types of poultry, as shown in a comparison of Figures 9 and 10. Cooking losses of broilers were significantly reduced by polyphosphate treatment, while those for heavy fowl were not. These results also have been confirmed by other investigators.[766,767]

Investigations have shown that treatment of poultry meat with polyphosphates increases the juiciness of the meat when it is eaten, as evaluated by taste panels.[754,756,758] However, polyphosphates fed to poultry prior to slaughter had no effect on the water-holding capacity or rehydration capacity of freeze-dried poultry meat.[746]

Mahon[768] reduced the cooking time and increased the deboned yield of fowl cooked in an aqueous solution of up to 4% of a straight-chain polyphosphate,

TABLE 28

Effects of Different Polyphosphates on Drip Loss, Cooked Yield, and TBA Values of Poultry Meat

Chicken No.	No. 1 Dip Solution, 22 hr	Dipped Weight, %	No. 2 Dip Solution	Dipped Weight, %	Thawing Drip, %	Thawed Weight, %	Cooked Weight, %	TBA Value Cooked Ground Meat, 7 days at 40°F	TBA Value Intact Cooked Breast, 14 days at 0°F
1A	Water	105.9		—	3.1	102.8	76.8	15.0	9.0
1B	6% hexameta[a]	105.3		—	1.0	101.3	76.1	8.4	0.4
2A	Water	105.9		—	4.1	102.3	77.0	14.0	8.1
2B	6% tripoly[b]	104.8		—	0.9	103.9	81.8	1.0	0.2
3A	Water	109.6		—	4.9	104.7	77.5	13.0	7.4
3B	3% N pyro[c]	106.4		—	2.0	104.4	81.1	2.6	0.3
4A	Water	106.0		—	3.9	102.1	68.8	15.0	7.5
4B	6% 50/50 N/A pyro[d]	105.5		—	1.5	104.0	75.0	1.5	0.2
5A	Water	107.5	Water ($\frac{1}{4}$ hr)	107.5	5.9	101.6	71.8	13.0	6.6
5B	Water	106.0	15% tripoly ($\frac{1}{4}$ hr)	105.4	1.5	103.9	78.0	4.2	0.3
6A	Water	109.0	Water (1 hr)	109.9	6.3	103.6	73.5	13.0	6.3
6B	Water	107.0	15% tripoly (1 hr)	106.3	1.6	104.7	77.7	1.5	0.2

Source: Mahon, J. H. (Calgon Corporation), U.S. Patent 3,104,170, 1963; reprinted with permission.

[a] $(NaPO_3)_6$.

[b] $Na_5P_3O_{10}$.

[c] $Na_4P_2O_7$ practical saturation at 40°F.

[d] 3% $Na_4P_2O_7$, 3% $Na_2H_2P_2O_7$.

preferably STP; KTP, TKPP, TSPP, and SHMP were also effective. Monte-mayor and co-workers[769] demonstrated that salt and salt plus a commercial mixture of polyphosphates greatly reduced the cookout of frozen poultry meat as compared to untreated controls.

TABLE 29

Effects of Different Levels of STP on Drip Loss, Cooked Yield, and TBA Values of Poultry Meat

Chicken No.	Soak Solution, 17 hr	Soaked Weight, %	Thawed Weight, 7 days, %	Thawing Drip, %	Cooked Flesh, %	Increased Flesh Yield,[a] %	Added Tripoly, %	TBA Value, 7 days, 40°F
1A	Water	106.2	—	—	53.2	—	—	12.0
1B	1% tripoly	109.3	—	—	56.6	+ 6	0.04	3.9
2A	Water	107.1	—	—	52.3	—	—	13.0
2B	3% tripoly	107.9	—	—	57.1	+ 9	0.34	2.1
3A	Water	106.7	—	—	52.3	—	—	10.0
3B	5% tripoly	106.6	—	—	60.9	+16	0.48	0.8

Soaked Chicken Frozen 6 Days at 0°F; Thawed Overnight to 40°F before Cooking

Chicken No.	Soak Solution, 17 hr	Soaked Weight, %	Thawed Weight, 7 days, %	Thawing Drip, %	Cooked Flesh, %	Increased Flesh Yield,[a] %	Added Tripoly, %	TBA Value, 7 days, 40°F
7A	Water	106.6	102.0	4.6	53.4	—	—	10.5
7B	1% tripoly	106.6	103.1	3.5	57.2	+ 7	0.05	9.6
8A	Water	106.7	102.1	4.6	54.0	—	—	10.5
8B	3% tripoly	106.2	104.8	1.4	60.6	+12	0.34	2.9
9A	Water	106.8	102.0	4.8	46.9	—	—	8.2
9B	5% tripoly	105.2	101.9	1.3	58.6	+25(?)	0.58	0.6

Source: Mahon, J. H. (Calgon Corporation), U.S. Patent 3,104,170, 1963; reprinted with permission.

[a] Water-treated sample taken as 100% cooked meat yield.

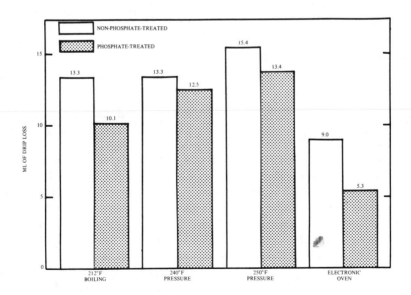

Fig. 9. Average volume of drip in minced broiler muscle cooked by four methods, phosphate-treated compared with non-phosphate-treated. (Source: Monk, J. A., Mountney, G. J., and Prudent, I., 1964, *Food Technol.* 18:226, Figure 3; with permission.)

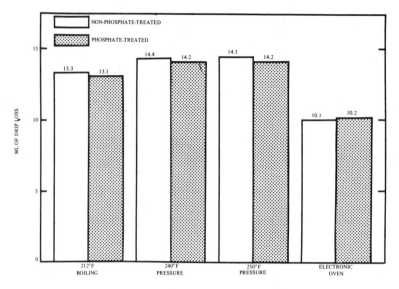

Fig. 10.   Average volume of drip in minced hen muscle cooked by four methods, phosphate-treated compared with non-phosphate-treated. (Source: Monk, J. A., Mountney, G. J., and Prudent, I., 1964, *Food Technol.* 18:226, Figure 4; with permission.)

Increasing Fat Emulsification

The fat-emulsifying properties of the salt-soluble protein of light and dark meat of broilers, hens, turkeys, and ducks were studied by Hudspeth and May.[770] These authors reported that the dark meat from each of these fowls was capable of emulsifying greater quantities of oil than light meat. This was a surprising discovery, since greater quantities of salt-soluble proteins were extracted from the light meat than from the dark meat of the same type of fowl. The authors concluded that the greater emulsifying capacity of the salt-soluble proteins of the dark meat counterbalanced the greater amounts of extractable salt-soluble proteins of the light meat.

Improving Flavor

A number of investigators have commented on the flavor characteristics of phosphate-treated chicken meat. No significant flavor changes were found in phosphate-treated poultry fed the phosphates prior to slaughter.[746] Klose and co-workers[755] and Spencer and Smith[756] found no significant flavor changes when phosphates were used to treat poultry meat prior to cooking. Flavor differences were reported by Froning[752] and later by Klose and co-workers.[747] Although these flavor differences were probably caused by a difference in saltiness, which enabled tasters to distinguish between the phosphate-treated meat and the controls, the differences were considered neither positive nor negative. Significant improvements in the flavor of the phosphate-treated meat over untreated controls were reported by a number of investigators.[749,750,757,758,768] Improvements in the odor of the cooked, refrigerated poultry meat also were reported by a number of investigators.[749,750,755,768,772] The improved flavor and odor characteristics have been attributed to the ability of the polyphosphates to inhibit the oxidation of lipids in aqueous suspensions, as measured by significantly reduced TBA values in the treated meat when compared to untreated controls.[749,750,768,772]

Preventing Microbiological Spoilage

The microbiological state of poultry meat also has an important effect on its odor and flavor acceptability. Phosphates, and particularly polyphosphates, aid in controlling the microbiological populations on the surfaces of poultry meat. Jimenez and co-workers[773] reported that combinations of hexametaphosphate and chlortetracycline-HCl were effective in preventing microbiological spoilage of poultry meat; they were particularly effective when used as dips on the eviscerated poultry. Melnick *et al.*[774] reported a synergistic action of phosphoric acid and MSP with sorbic acid on extending the microbiological stability of refrigerated poultry meat. Treatment of poultry meat with combinations of MSP and ascorbic acid was reported to triple the shelf-life over that of poultry meat treated with either compound alone. Spencer and Smith[756] reported that microbiological spoilage was reduced on chicken-fryer carcasses chilled for 6 hours in ice water containing 10 oz of polyphosphate per gallon. The reduction in microbiological spoilage was confirmed by plate counts, UV fluorescence, and reduction in off-odors. The shelf-life was increased 1–2 days by use of the phosphates alone. A possible explanation for the reduction in fluorescent microorganisms on the surfaces of the poultry meat has been proposed by Kraft and Ayres[775]; they found that *Pseudomonas* species required a minimum of 1 ppm iron in their environment to produce spoilage on poultry. This quantity of iron is usually provided by the water used in processing and would be enough to allow rampant growth of these organisms. The polyphosphates, however, being excellent sequestering agents, could prevent iron from being utilized by the microorganisms.

Tomiyama *et al.*[776] reported that combinations of hexametaphosphate or pyrophosphate, as well as other chelating agents of iron, copper, and cobalt, were effective in preventing microbiological spoilage of chicken meat, as well as fish flesh, when combined with antibiotics and antioxidants. Tylosin was found to be inactivated by iron. The effect of the iron could be overcome by treating the flesh with pyrophosphates.

Elliott and co-workers[777] further reported on the microbiological inhibition of polyphosphates on chicken flesh. They found that nonfluorescent species of *Pseudomonas* were completely inhibited by the presence of polyphosphates in their medium. Fluorescent strains grew, but only after a lag period. The pH was not the controlling factor in inhibition of the growth of these spoilage organisms. The authors reported that the addition of magnesium ion would reverse the inhibiting effect of the polyphosphate. When polyphosphates were added to the ice used in chilling chicken carcasses overnight at 3 and 8% levels, the shelf-life of the chicken carcasses was increased by 17 and 25%, respectively. Holding the carcasses in continuous contact with the same polyphosphate solution at refrigerated temperatures increased shelf-life by 17 and 67%, respectively. Using antiseptic ice for the 8% polyphosphate soak increased shelf-life by 60% over the use of ordinary water ice. The 3 and 8% polyphosphate soaks prevented nonfluorescent strains of *Pseudomonas* from growing, but they did not inhibit completely the growth of fluorescent strains.

Phosphate Applications in Seafood Products

The Biochemistry of Phosphate Interactions with Seafood Flesh

The proteins of fish and other seafood muscles are affected by the poly-phosphates in the same manner as those of animals and poultry. Linko and Nikkila[778] reviewed the work of numerous investigators who demonstrated that low levels of sodium chloride reduced the solubility of, and thus denatured, the myosin fraction of fish muscle and that the polyphosphates inhibited the denaturation of muscle proteins caused by salt. Citrates were reported to have a similar effect. The authors speculated that the polyphosphates inhibited the denaturation of the muscle proteins by one of two possible mechanisms. The first possibility, that the phosphate is directly bound by the protein to increase the number of polar groups in the protein, is supported by the fact that phosphates are known to be bound by actomyosin, with the resulting increase in its solubility. The second possible mechanism is that the phosphates and citrates complex calcium and zinc ions, again releasing polar groups. This latter possibility was ruled out by the authors because EDTA has no effect in preventing the denaturation of myosin by sodium chloride, although it strongly binds calcium and zinc ions.

The rate of penetration of the polyphosphates into fish muscle has been studied by Kuusi and co-workers,[779] Nikkila and co-workers,[780] and Scheurer[781] by means of radioactive phosphates. The skin side of fillets resisted penetration, while exposed flesh rapidly absorbed the phosphate. The depth to which the phosphates penetrated into the fish fillets was directly proportional to the concentration of the phosphate in the solution.

Love and Abel[782] concluded that there was a direct correlation between denaturation of the proteins in fish muscle and the development of toughness when the fish was cooked. Freezing and thawing increased the denaturation of the fish muscle proteins in direct proportion to the length of time the fish was stored in the frozen state. Treatment of the fillets with polyphosphates failed to inhibit the denaturation with or without the presence of salt, but the phosphates inhibited the dehydration of the fish muscle, which led to toughening by solubilizing certain proteins in the surface of the fillet tissues. These proteins then gelatinized and sealed the surface to prevent further loss of fluid. Oxygen was prevented from diffusing into the tissue, where it could gain access to the fats in the muscles and cause them to undergo oxidative rancidity, which accelerates the denaturation of the muscle proteins.

Akiba and co-workers[783] demonstrated that polyphosphates prevented the denaturation of fish-muscle proteins and effectively maintained high-quality frozen, raw, or ground fish muscle. Spinelli and co-workers[784] found that there was a linear uptake of STP by the fillets of fish that was directly proportional to the level of STP in the dipping solution. Studies of the results of numerous tests showed that the fillets became very slippery to the touch and assumed a gelatinous characteristic over the entire surface. No significant differences appeared in the quantities of proteins extractable from either treated or untreated fillets of two species of fish over a period of 4 weeks. These findings did not agree with results reported by Nikkila and co-workers[778-780] and others, but they did agree with the work reported by Love and Abel.[782] Ravesi and

Anderson[785] were unable to correlate the amount of protein extracted with the textural characteristics of fish fillets when they were treated with a salt solution containing sodium chloride and sodium bicarbonate or another mixture containing potassium chloride, DSP, and MKP. Kuusi and Kyto-kangas[786] reported that TSPP reversed the effects of sodium chloride on cation excess, which the authors interpreted as an explanation of the protective action on fish myofibrillar proteins.

Preserving Color

The polyphosphates have been reported by a number of investigators to improve the color of both raw and cooked fish flesh.[784,787-790] Meyer[787] found that polyphosphates added to the brine of herring, preserved immediately after harvesting and prior to further processing, greatly improved the color of the fish, which was to be further processed. When Yamaga and co-workers[788] injected fish flesh with mixtures of various polyphosphates, sodium citrate, and sodium bicarbonate, the flesh color of the fish was maintained for up to 57 days when stored at $-20°C$. Sen and Lahiry[789] reduced the brown or yellow discoloration of sun-dried mackerel by treating the fish with a solution of MSP and various preservatives prior to drying. Mold growth on the surface of the fish also was inhibited in the treated fish dried to 45% moisture, although controls became moldy at this same moisture content. Miyauchi and co-workers[790] dipped fillets of various types of fish, as well as crab and oyster meats, into solutions of STP and salt prior to irradiation to preserve the original appearance. Spinelli and co-workers,[784] who extended this work, delayed the fading of the bright flesh hues of freshly cut fish fillets first by treating the fillets with phosphate dips prior to irradiation and then by storing them either in cans or polyethylene bags. The color of the cooked flesh also was significantly superior to that of the untreated controls, which turned a darker brown after approximately 2–4 weeks of storage.

Increasing Tenderness

Treatment of seafoods with polyphosphates has been reported to increase the tenderness of the stored product. When Sen and co-workers[791] treated mackerel prior to drying with a curing salt containing sodium bicarbonate, sodium carbonate, or SHMP, the tenderness of the dried fish was increased. The carbonates raised the pH; the SHMP, however, had no effect on pH. Garnatz and co-workers[792] reported that the treatment of cooked and frozen fish with phosphates, as well as other salts, significantly improved the tenderness. Sen and Lahiry[789] found that dried mackerel was significantly tenderized by SHMP without raising the pH to a point where microbiological spoilage became a problem. MacCallum and co-workers[793] reported that the addition of STP to dips for cod fillets significantly improved the texture and increased the tenderness of the treated fillets over controls. Miyauchi and co-workers[790] and Spinelli and co-workers[784] treated halibut and ocean perch fillets with a dip of STP and salt prior to irradiation, which resulted in significantly higher scores for tenderness in the stored fillets when evaluated by a trained taste panel.

Increasing Binding

The Japanese are very fond of a fish paste called *kamaboko*, which is used, among other applications, to prepare fish sausage and fish hams. The Japanese have reported most of the investigations demonstrating that polyphosphates increase the binding properties of fish proteins, just as they do with other types of flesh. Tanikawa[794] has published an excellent review of the processing methods and problems involved in the preparation of kamaboko.

Okamura and his co-workers[795-801] have extensively studied the influence of various phosphates and combinations of phosphates, as well as other additives, in improving the binding properties of kamaboko. Their studies indicated that the kamaboko binding properties could be adjusted by varying the combinations of DSP, TSPP, STP, and SHMP, along with starch. By adjusting the binding properties practically at will, they could eliminate variables introduced by variations in processing, fish, and similar problems. Miyake and co-workers[802] found that a combination of polyphosphates and magnesium ion increased the jelly strength (a measurement of the degree of binding) of kamaboko more than did either alone; calcium ions were found to decrease the jelly strength. The polyphosphates affected the jelly strength at 0.3% levels in the following order of decreasing effectiveness: TSPP, STP, SHMP, TKPP. The authors found that the combinations of magnesium ion and polyphosphates caused a significant degree of dissociation of actomyosin, while no such dissociation occurred in the presence of polyphosphate and calcium ion. This dissociation factor may explain the differences reported in the effects for the same polyphosphates by different investigators. Akiyama[803] reported similar results of increased binding when polyphosphates were used to treat cooked whale meat.

It is common practice in the fish industry to form fish fillets into large blocks and then freeze them, prior to their use in preparation of fish sticks and similar sawed forms. By dipping each fish fillet into a solution of polyphosphate prior to freezing into blocks, one can lubricate the fillets and increase the bonding of the surfaces of the fish to adjacent fillets.[804] Both the lubrication and the bonding are believed to be due to the ability of the polyphosphates to dissolve surface proteins and form a continuous protein gel over the complete surface of the fillet.

Increasing Moisture Retention

The retention of the natural internal fluids of fish muscle is important to the organoleptic acceptance of the product and as a means of eliminating serious economic losses. Cut surfaces of fish muscle, such as fillets and steaks, commonly exude a slimy liquid, which is unsightly and detracts from the acceptability of packaged fish. The exudation of liquid is increased when the fish has been frozen, as the liquid drainage, or drip loss, increases when the frozen fish is thawed. The loss of this liquid is directly related to the toughness of the consumed fish product. Seagran[805] implemented paper electrophoresis to demonstrate that the liquid exuding from thawed fillets exhibited similar patterns to the extracts of low-ionic strength from similar fillets.

The British firm of Albright and Wilson, Ltd., patented the use of sodium and potassium salts of polyphosphoric acids, ranging from pyrophosphate through the very-long-chain salts, for treating fish to eliminate drip loss of thawing frozen fish and during cooking.[806] Mahon[807] patented the use of dips containing polyphosphates and salt, especially those containing STP and salt, to prevent drip loss upon thawing and to reduce cooking losses of frozen fish fillets. His is one of the dominating patents in the United States for the use of polyphosphates to treat fish. Mahon included a considerable amount of data that demonstrate the applications of various phosphates by themselves and in mixtures: (1) the amount of dip absorbed by the fillets, (2) the changes of pH of the fish meat with this treatment, (3) the actual amount of phosphate absorbed, and (4) the resulting drip loss upon thawing of the frozen, treated fillets. Table 30, taken from Mahon's patent, illustrates the effects of water, salt, STP, water-plus-STP, and salt-plus-STP dips on the per cent uptake of the dip, changes in the fish pH, the per cent of drip loss, the per cent of phosphate absorbed, and the TBA values of the fats in fillets stored for 43 days at 0°F. In Figure 11 the differences in thawing weight loss for fillets of two types of fish dipped into a solution of STP and salt are shown; Figure 12 demonstrates that these dips significantly reduce the protein loss from the treated fillets. This process is now in commercial use and is reported to have increased the sale of fish fillets.[808]

TABLE 30
Effects of Salt and Sodium Tripolyphosphate on Thawing Drip of Haddock Fillets Frozen at 0°F for 43 Days

No.	Dip Solution No. 1	Dip Solution No. 2	Dip Time, minutes	Dip Uptake, %	Fish, pH	Thawing Drip, %	Added Phosphate, %	TBA
1	Water	—	4	2.4	6.6	5.3	—	0.27
2	Water	—	2	2.5	6.5	6.3	—	0.21
3	Water	—	1	1.6	6.6	5.1	—	0.26
4	4% NaCl	—	4	5.5	6.5	10.3	—	0.18
5	4% NaCl	—	2	5.2	6.5	8.7	—	0.26
6	4% NaCl	—	1	3.5	6.6	7.3	—	0.32
7	12.5% $Na_5P_3O_{10}$	—	4	5.3	6.8	2.3	0.59	0.19
8	12.5% $Na_5P_3O_{10}$	—	2	1.9	6.7	2.7	0.30	0.14
9	12.5% $Na_5P_3O_{10}$	—	1	2.1	6.7	5.1	0.12	0.11
10	Water	12.5% $Na_5P_3O_{10}$	2+2	4.8	6.8	1.8	0.44	0.24
11	Water	12.5% $Na_5P_3O_{10}$	1+2	5.5	6.8	4.1	0.37	0.12
12	Water	12.5% $Na_5P_3O_{10}$	½+2	6.2	6.9	2.6	0.33	0.18
13	4% NaCl	12.5% $Na_5P_3O_{10}$	2+2	3.2	6.7	2.2	0.25	0.35
14	4% NaCl	12.5% $Na_5P_3O_{10}$	1+2	6.0	6.7	2.9	0.45	0.19
15	4% NaCl	12.5% $Na_5P_3O_{10}$	½+2	6.1	6.7	2.6	0.41	0.11

Source: Mahon, J. H. (Hagan Chemicals and Controls, Inc., now Calgon Corporation, a subsidiary of Merck & Co., Inc.), U.S. Patent 3,036,923, 1962; with permission.

Tanikawa and co-workers[809] treated fish fillets with a mixture of TSPP and STP to reduce the drip loss and loss of deoxyribonucleic acid phosphorus in the drip liquid; the TSPP–STP mixture was more effective than either TSPP alone

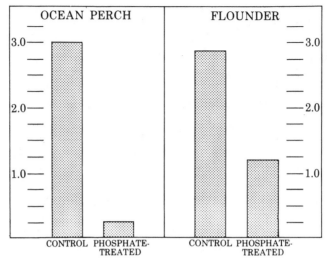

Fig. 11.   Thawing weight loss of frozen fish fillets after 6 months of storage. (Source: *Food Eng. 34(9):*106, 1962; with permission.)

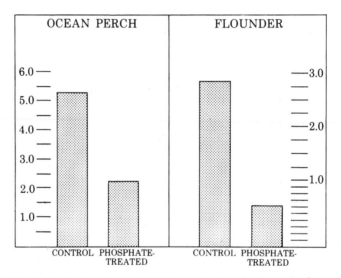

Fig. 12.   Protein loss from frozen fish fillets on thawing after 6 months of storage. (Source: *Food Eng. 34(9):*106, 1962; with permission.)

or an untreated control, as shown in Figures 13 and 14. These results indicated that the cod flesh proteins became less soluble and that the flesh proteins increased in their ability to reabsorb the liquids formed during thawing.

Mahon and Schneider[810] extensively investigated the results of dipping haddock fillets and certain other types of fish and seafood into solutions of polyphosphates and other salts. Various phosphates and other alkaline-salt solutions were used as 2-minute dips for haddock fillets; their effects on drip loss, uptake of solution, and the pH of the flesh are shown in Table 31. STP was the most effective in reducing the drip loss. The fillets absorbed lower quantities of these salt solutions than of water; a maximum of 0.2% STP was absorbed by the fillets. The data suggest that the effectiveness of the phosphates in reducing drip loss is not an effect of pH, since the pH changes were

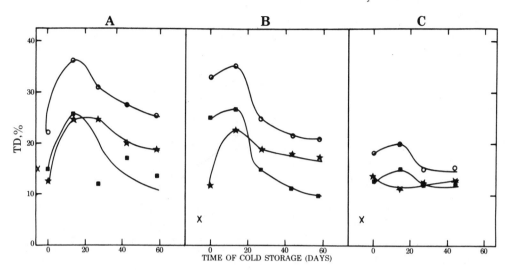

Fig. 13. Variation in the amount of total drip (TD) during cold storage of frozen cod fillets: (A) not treated; (B) pretreated with sodium pyrophosphate; (C) pretreated with a mixture of sodium pyrophosphate and sodium tripolyphosphate. (Source: Tanikawa, E., Akiba, M., and Shitamori, A., 1963, *Food Technol. 17*:1425, Figure 1; with permission.)

O = frozen at −5°C   ★ = frozen at about −90°C in acetone-dry ice
■ = frozen at −30°C   X = TD in raw material before freezing

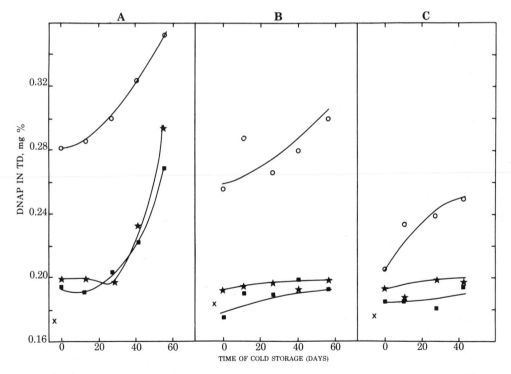

Fig. 14. Variation in the amount of DNAP in TD. Same as in Figure 13, as to the types (A), (B), and (C), and different symbols. (Source: Tanikawa, E., Akiba, M., and Shitamori, A., 1963, *Food Technol. 17*:1425, Figure 2; with permission.)

TABLE 31
Effects of Phosphate and Other Salt Solutions on Thawing Drip of Haddock Fillets

2-Minute Immersion Followed by 34 Days Storage at 0°F

Solution Used	pH of Solution	Solution Uptake by Fillet, %	Thawing Drip, %	pH of Treated Fish
Hexametaphosphate				
Controls (5)				
Range	—	3.0–4.0	5.7–7.1	6.6–6.8
Average	—	3.2	6.5	6.7
5%	5.5	2.4	3.5	6.7
10%	5.4	3.2	3.6	6.7
20%	5.2	2.2	0.9	6.7
30%	5.1	2.0	0.9	6.6
40%	4.9	2.9	0.7	6.6
Tetrasodium Pyrophosphate, Trisodium Orthophosphate, and Sodium Hydroxide				
Controls (5)				
Range	—	3.0–4.0	5.7–7.1	6.6–6.8
Average	—	3.2	6.5	6.7
1.5% $Na_4P_2O_7$	9.7	2.7	6.6	6.6
2.0% $Na_4P_2O_7$	9.6	2.1	3.1	6.7
4.0% $Na_4P_2O_7$	9.5	4.4	5.9	6.8
1.5% Na_3PO_4	11.7	3.4	6.7	6.7
2.5% Na_3PO_4	11.7	4.3	3.9	6.8
5.0% Na_3PO_4	11.5	—	1.4	7.5
0.75% NaOH	12.5	3.9	2.9	6.8
1.5% NaOH	12.3	5.2	4.3	7.3
Trisodium Citrate, Sodium Bicarbonate, and Sodium Carbonate				
Controls (5)				
Range	—	3.0–4.0	5.7–7.1	6.6–6.8
Average	—	3.2	6.6	6.7
3% Na_3 citrate	7.0	2.2	6.0	6.7
6% Na_3 citrate	7.0	1.9	5.0	6.7
9% Na_3 citrate	7.0	2.5	3.1	6.8
12% Na_3 citrate	7.0	2.2	3.4	6.8
1.75% $NaHCO_3$	8.0	4.0	4.5	7.0
3.5% $NaHCO_3$	8.1	2.8	2.0	7.1
2% Na_2CO_3	11.2	2.7	4.7	6.9
4% Na_2CO_3	11.3	3.3	3.2	7.2
Tripolyphosphate				
Controls (5)				
Range	—	3.0–4.0	5.7–7.1	6.6–6.8
Average	—	3.2	6.6	6.7
2.5%	8.5	2.9	5.1	6.7
5.0%	8.4	2.0	3.4	6.8
7.5%	8.3	2.9	2.3	6.8
10.0%	8.3	1.7	1.7	6.9
12.5%	8.2	1.2	0.3	6.8

Source: Mahon, J. H., and Schneider, C. G., 1964, *Food Technol. 18*:117; used with permission

insignificant in the fish flesh. The non-phosphate salts had little effect until the pH of the flesh reached 7.0 or above. Table 32 demonstrates that other types of fish and seafoods, treated with STP in the same manner as haddock fillets, had a reduction in drip loss.

TABLE 32
Effect of Sodium Tripolyphosphate on
Reduction of Thawing Drip of Various Seafoods

Fish Used	Controls, %	Tripoly-phosphate-treated, %
Cod	3.2	0.1
Flounder	4.6	0.3
Haddock	4.0	0.1
Perch	7.0	0.7
Pollack	5.0	1.3
Scallops	5.0	0.8
Average	4.8	0.55

Source: Mahon, J. H., and Schneider, C. G., 1964, *Food Technol. 18*:117; used with permission.

Boyd and Southcott,[811] Swartz,[812] and Sutton[813] also investigated the effects of STP and other polyphosphates on drip loss, texture, and water retention during the storage and cooking of seafood products. All confirmed that STP, or combinations of STP with salt, produced optimum results. Sutton and Ogilvie[814] quantified the effects of phosphate dips on fish muscle and on the retention of the natural fluids of the muscle during storage. By these authors' calculations 260 mol of water was retained in fish muscle treated by dipping for each mol of STP absorbed into the muscle, as compared with undipped controls. The works of Spinelli, Miyauchi, and their co-workers[784,790,815] also demonstrated that mixtures of STP and salt were most effective in reducing or eliminating completely the drip loss of irradiated fish for at least 4 weeks. They claimed that the fish treated with subsequent irradiation could be kept a sufficient time so that they could be distributed for sale throughout the United States. Their works also demonstrated that different species of fish required different minimum levels of STP to decrease drip loss. Treatment with STP was, the authors concluded, economical and highly beneficial in that the drip loss was significantly reduced, and the fish retained a significantly better appearance and texture during storage.

Based on these earlier experiments of Spinelli, Miyauchi, and their co-workers,[784,790,815] Spinelli and Weig[816,817] developed a machine for the treatment of fish fillets with STP. The STP solution is sprayed on as the fillets pass between spray nozzles above and below a chain-link conveyor belt. A similar commercial machine has been developed and is being marketed by the Calgon Corporation.[818,819]

As discussed previously, several investigators have confirmed that polyphosphate treatment of fish flesh reduced cooking losses. Barnett and co-workers[820] reported that treatment of halibut, silver salmon, and black cod fillets with a solution of STP and salt prior to kippering significantly reduced the weight loss during this hot-smoking process.

Not all investigators agree that the phosphates effectively reduce drip loss or increase moisture and fluid retention. Dyer and co-workers[821] found no difference in drip loss, extractable proteins, or lipid hydrolysis in phosphate-treated, frozen cod fillets as compared to controls; however, they did not use the same combinations of polyphosphate and salt as reported by Mahon,[810] which may explain the differences in results. According to Love and Abel[782] phosphates do not increase the water-binding of fish proteins. They concluded that the reduction in drip loss of fish fillets treated with polyphosphates was caused by the solubilization of surface proteins, which then gelled and effectively sealed the fluids inside the fillet.

Other works reviewed by Love[822] indicated that the polyphosphates do not hydrolyze muscle proteins or increase the protein solubility. Studies at Love's laboratories confirmed that polyphosphates caused swelling of the surface cells of the fillets to eliminate extracellular spaces through which fluids could pass from the interior of the fillet; thus, the surface of the fillet was effectively sealed to prevent fluids from either entering or leaving.[823]

When Okamura and his co-workers[797,798,824,825] treated ground-fish paste with various types and mixtures of phosphates, they effectively increased the water-holding capacity and aided in the formation of the protein gels that are so important to the formation of kamaboko. Tanikawa et al.[826] reported that combinations of sugar and polyphosphates, as well as egg whites, were most effective in decreasing drip loss and improving elasticity of kamaboko and similar cooked fish pastes. Ueoka and co-workers[827] studied the effects of various types of salts, including citrates, phosphates, and EDTA, on the water-holding capacity of fish meat and the formation of protein gels; they found that a 0.3 mol solution of TSPP containing 10% sucrose produced the best results of all treatments studied.

A report from the Torry Research Station in Aberdeen, Scotland, pointed out that the effects of the polyphosphates in reducing drip loss and in improving the appearance of retail fish fillets could not compensate for poor quality or poor processing. Nor would phosphate treatment reduce weight loss or improve quality through fluid retention in poor quality or poorly processed fish.[804]

The polyphosphates also can increase the fluid retention of seafood other than fish. Takei and Takahashi[828] reported that air-dried squid meat recovered its appearance and texture during rehydration when treated with a solution of SHMP and either TSPP or STP prior to being dried. Mahon and Schneider[810] reported that phosphate treatment effectively reduced the drip loss of frozen scallops, as shown in Table 32. Miyauchi and co-workers[790] reported that STP dips prior to irradiation reduced the drip loss of crab meat and Pacific oysters. Ichisugi and co-workers[829] reduced or inhibited the development of stiffness resulting from desiccation during the storage of crab meat by a treatment with a sodium chloride solution and either a sodium acetate or TSP solution prior to heating. Mathen[830] found that phosphates effectively reduced the drip loss of frozen prawns. Jones[831] reported that the addition of 0.25–0.35% SAPP to canned Alaska king crab increased moisture retention and, therefore, the yield of the crab meat by 2%.

Improving Flavor

The flavor of seafood products also has been improved by treatment with various phosphates. Sjöstrom[832] treated fish products with an alkali, such as DSP, after cooking and preferably after cooling but before sealing them in packaging containers to retard significantly the development of undesirable flavors during sterilization.

Treating sun-dried mackerel with a curing mixture of salt, sodium benzoate, DSP, sodium bicarbonate, SHMP, and ascorbic acid significantly improved the flavor.[789] This curing treatment also improved the odor of the fish during its storage period. The palatability of frozen crab meat was improved when the crab meat was heated in a saline solution for 30 minutes or the meat was dipped into the salt solutions first and then heated and later frozen and stored at −22°C.[829]

The effects of the polyphosphates in improving flavor of seafood products have been disputed.[804] Claims of improved flavor may be caused by the fact that dipping into phosphate solutions can mask some of the symptoms of deterioration of poor-quality seafood products; hence, there only can be an impression that the quality is better. Spinelli and co-workers[784] reported that they could detect no differences in flavor or odor of fish fillets or steaks, whether dipped into polyphosphate solutions or undipped controls.

However, according to Tarr et al.,[833] an enzyme that hydrolyzes the 5′-nucleotides is strongly inhibited by EDTA, pyrophosphate, potassium fluoride and zinc chloride at levels of 1–10 mmol. Since the 5′-nucleotides are effective flavor intensifiers, the inhibition of the enzyme that would inactivate them can offer some explanation for flavor improvement from pyrophosphate.

Polyphosphates have been reported to be effective antioxidants, which prevent the development of rancidity in foods containing fats in an aqueous environment. The results of numerous investigations demonstrating the effectiveness of the polyphosphates in preventing development of rancidity in aqueous fat food products have been reviewed in a previous section. A number of investigators have reported the effectiveness of polyphosphates in preventing the development of rancidity in seafood products. Watts[667] experimented with the development of rancid odors and flavors in ground, cooked mullet meat. She stated that the TBA values were an excellent guide to rancidity development and that the detectable limits of rancid odors for all cooked meat, i.e., fish, beef, pork, veal, lamb, and chicken, were a TBA value of 0.5–1.0. If the lipids were extracted from the cooked mullet tissue, the development of rancidity in the chloroform-extracted lipids was much less than in the cooked, unextracted, whole mullet tissue, as shown in Figure 15. Even the total lipids,

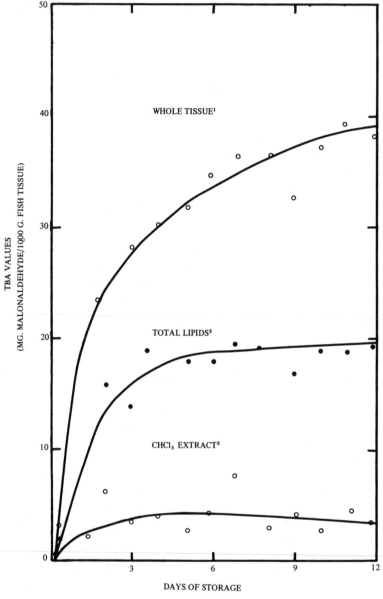

Fig. 15. TBA values of cooked whole mullet tissue, total lipids, and chloroform-extracted lipids. (Source: Watts, B. M., 1961, *Proc. Flavor Chem. Symp.*, Campbell Soup Company, p. 83, Figure 1; with permission.)

[1] Whole, cooked fish tissue, no lipids extracted.

[2] Total lipids extracted from fresh fish tissue with chloroform-methanol-water solvent techniques.

[3] Lipids extracted from cooked fish tissue with chloroform and a dehydrating agent.

extracted from the raw fish, developed rancidity at a lower rate than in the whole tissue. Watts noted, too, that once lipid oxidation had begun, less lipid material could be extracted, as shown in Table 32. The oxidized lipids apparently could no longer be extracted by the same solvent system.

According to Zipser and Watts[834] TBA determinations demonstrated that the fats in cooked mullet muscle began to oxidize very rapidly after cooking. Tissues containing high levels of fats and heme pigments especially were affected. Although the development of rancidity could be retarded by limiting the oxygen supply and by freezing the fish, the most effective treatment involved the use of ground fish muscle with a combination of STP and sodium ascorbate. The combination of the two salts was more effective than either of them alone or in combination with other curing salts.

In an extension of this work, Ramsey and Watts[665] reported that direct addition of the polyphosphate STP to the meat either prior to or after cooking had a slight advantage over storing the meat in an STP solution. Table 33 illustrates the effects of the levels of STP on TBA values and odor scores of cooked, ground mullet. As shown, the TBA values, even at the 0.5% level, were above the threshold of odor detection, demonstrating that the fish lipids oxidize more rapidly than beef, which was protected by levels as low as 0.1% for 8 days. Akiyama[803] also used the TBA values to demonstrate that the addition of 0.5% polyphosphate to cooked, fatty whale meat was more effective in preventing rancidity than the addition of 0.02% BHA or BHT.

TABLE 33
Total Lipids Extracted from Cooked, Red,
Lateral-line Tissue of Mullet

Treatment	Total Lipids, %	TBA (Whole Tissue)
Raw	8.22	—
Cooked, 0 days	7.87	3
Cooked, 3 days	6.85	60
Cooked, 5 days	6.51	86

Source: Watts, B. M., 1961, *Proc. Flavor Chem. Symp.*, Campbell Soup Company, p. 83, Table 2; reprinted by permission.

When Mahon[807,808] treated fish with a combination of 12% STP and 4% salt, he effectively inhibited development of rancidity in the fish and, therefore, improved flavor and odor. This flavor and odor improvement could have been caused by the influence of the polyphosphates on retarding the rancidity of the lipids in the fish flesh. The TBA values remained significantly lower, compared to controls, over the same period of storage in the polyphosphate treatments, as shown in Table 30; these results indicated that the rancidity of lipids was retarded in the treated fish.[807] Thus, flavor and odor would be expected to be improved. Sen and Lahiry[789] reported that the inclusion of DSP and SHMP in curing salts for sun-dried mackerel resulted in decreased development of rancidity.

TABLE 34
Effects of STP on TBA Values and Odor Scores of Cooked, Ground Mullet

Phosphate Concentration, %	Rancidity after Storage			
	TBA No.	Odor Score	TBA No.	Odor Score
	2–3 Days		8 Days	
0.5	1.8	5.0	10	4.2
0.1	2.2	5.4	14	2.4
.03	2.6	4.6	20	3.5
.01	5.8	4.0	25	2.5
0	16	3.5	41	3.0

Source: Ramsey, M. B., and Watts, B. M., 1963, *Food Technol. 17*:1056; reprinted by permission.

Dyer and co-workers[821] could demonstrate no effect on lipid hydrolysis due to dipping fish fillets into an STP solution prior to freezing with subsequent thawing. Boyd and Southcott[811] also reported that treatment of fish fillets prior to freezing with a 12.5% TSP and salt solution had no effect on preventing the development of rancidity. As discussed previously, the dipping treatment of seafood products is believed to mask some of the symptoms of deterioration and poor quality, rather than provide actual protection from that deterioration.[804] Spinelli and co-workers[784] reported that they could detect no differences in the development of rancidity in irradiated fish fillets, whether they were treated with STP and salt dips or with salt alone.

Greig and co-workers[835] inhibited the development of rancidity in ground cisco (fresh-water herring) flesh stored at 0°F more effectively with ascorbic acid than with other soluble antioxidants, such as STP, propyl gallate, and monosodium glutamate. Although all of the water-soluble antioxidants were somewhat more effective than untreated controls, the ascorbic acid was significantly more effective than the others. Perhaps the differences in species affect rancidity development, as is the case in water retention.

Preventing Microbiological Spoilage

The inhibition of microbiological deterioration in seafoods also has been accomplished by treatment with various phosphates. Hempel[836] reported that when dry sea salt, often used for preservation of fish, was not properly treated, a pink or red microorganism rapidly developed in the brine in the presence of the fish protein. If the salt crystals were coated with 0.25–4% of a finely powdered, alkali-metal phosphate, such as MSP or DSP, or mixtures of these compounds, the red microorganisms failed to develop.

A Norwegian patent[837] claimed that the combination of carrageenan and phosphates in sufficient proportions to form thickened solutions of the aqueous salts (500–3000 centipoise) used to coat fish would inhibit microbiological deterioration. Uchiyama and Amano,[838] who studied the softening spoilage of fish sausage by *Bacillus circulans*, concluded that the addition of sodium pyrophosphate to the sausage did not inhibit the germination of the spores; rather, it inhibited the development of the vegetative phase of the organisms.

They reported that 0.4% sodium pyrophosphate added to a medium in which the microorganism flourished restricted its consumption of glucose. A series of photomicrographs and electron micrographs demonstrated that sporulation could occur but that vegetative growth was inhibited by the pyrophosphate, as well as ascorbic acid.

Ozawa et al.[839] reported that the polyphosphates were effective synergists with tetracycline as fish and other food preservatives; SHMP was the most effective, followed by STP, and the least effective polyphosphate was TSPP. The authors reported that the condensed phosphates inhibited the growth of Staphylococcus aureus and Bacillus subtilis in broth media and also prevented their growth on fish. Again, SHMP was the most effective, followed by STP and TSPP, in that order.

Tomiyama[776] also studied the synergistic effect of the polyphosphates, particularly SHMP, with antibiotics, such as chlortetracycline and tylosin. The antibiotic effects of combinations of the antibiotics with SHMP and other chelating antioxidant compounds were tested in clam meat, cooked fish paste, and poultry products. Iron and copper ions were potent inactivators of the antibiotics and could be removed effectively and inactivated by pyrophosphate and similar chelating agents.

Evidence that chelation might explain at least some of the microbiological effects of polyphosphates in preserving fish products was also presented in a paper by Levin,[840] who confirmed the effectiveness of EDTA as a preservative in fish products. Odor and taste evaluations indicated that a 1% dip of tetra-sodium EDTA increased the shelf-life of the fish from 5–9 or 10 days and significantly decreased the development of trimethylamine and volatile, basic nitrogen compounds. Spinelli and co-workers[784] reported that polyphosphate-treated, irradiated fish fillets had predominantly Achromobacter micro-organisms after 17 days of storage; the undipped, irradiated fillets were predominantly contaminated with Lactobacillus. However, when the fillets were held under refrigerated storage until they spoiled, the Lactobacillus pre-dominated in both the treated and the untreated fillets.

Preventing Struvite Crystals

Canned, cooked seafoods, such as lobster, shrimp, crab meat, haddock, cod, and salmon, over a period of time develop a quantity of transparent crystals of magnesium ammonium phosphate, known as struvite. Although the crystals actually are harmless both physically and nutritionally, because they appear to be sharp pieces of glass, the product is often rejected. The quantity of magnesium in many seafoods and especially in the water used in processing the seafoods can be sufficient to cause the formation of these crystals during the normal shelf-life of the product.

McFee and Peters[841] reported that as little as 0.015% of magnesium ion by weight of the normal pack can result in the formation of these crystals. Many of the canned seafoods containing 80% water and 20% solids have sufficient magnesium in them to form the struvite crystals. McFee and Peters reported that the acidification of the water used in washing the cooked seafood product could leach away the magnesium and thus prevent the formation of the glass-like crystals. Thus, acidification with citric, hydrochloric, and acetic acids will prevent formation of struvite. The use of sequestering agents in the water, such as alkali-metal metaphosphates (particularly sodium acid pyrophosphate and amino polycarboxylic acids, such as EDTA) will complex the magnesium and also prevent the formation of this compound. Attempting to extract the magnesium compounds with water, either by rinsing, washing, or boiling the seafood in ordinary water, is not sufficient to remove the quantity of magnesium necessary to prevent formation of struvite crystals. McFee and Peters recommended the use of SHMP, ammonium metaphosphate, STP, or the potassium salts equivalent to the same products, and EDTA for these purposes. In a later patent Kreidl and McFee[842] recommended the addition of 0.5–1.5% of a sodium polyphosphate compound, such as SHMP, to the liquid used in preparing the canned seafood product to prevent the formation of the struvite crystals during its shelf-life. Quantities of the soluble, alkali-metal phosphate as low as 0.25% were found to be effective, based on the total moisture content of the canned seafood.

Later McFee and Swaine[843] reported that SHMP added to the liquid of the canned product was the most practical method for preventing the formation of struvite crystals in such canned seafoods as lobster, shrimp, crab, and flaked fish, such as pollack, cod, and haddock. They reported five advantages to the use of SHMP, as follows:

1. SHMP is a neutral salt; therefore, it does not create acidic or alkaline flavors in the canned product.
2. As a GRAS compound, SHMP is safe for use in all food products.
3. SHMP can be applied either in solution or in the dry form with equivalent results.
4. SHMP can be mixed with the food before canning, or it can be added to the cans during their filling.
5. A very small quantity of SHMP is necessary to prevent the formation of struvite, thus making it very economical in use.

The authors reported that canned seafoods containing SHMP had a shelf-life of three–four years.

Yamada[844] reported that the minimum quantities of SHMP required to prevent the formation of struvite in canned king crab meat ranged between 0.13–0.22% of the total product. Jones[831] reported that adding 0.25–0.35% of SAPP to king crab meat inhibited the formation of struvite in canned products containing salt during the observed storage period of one year.

Phosphates as Microbiological Inhibitors

The applications of the phosphates as microbiological inhibitors in various types of food products have been reviewed in several sections of this chapter. There are, however, a number of references to food applications for polyphosphates as microbiological inhibitors, and these will be reviewed in the following discussion.

Applications as Inhibitors of Bacterial Growth

General Inhibitory Effects   Kelch and Buhlmann[845] reported that commercial mixtures of phosphates normally used in curing meat products (Curafos and Fibrisol, manufactured by J.A. Benckiser) act as microbiological inhibitors of certain food-spoilage microorganisms when tested in optimum nutrient media for these microorganisms. The Curafos used in these experiments contained 15% TSPP, 70% STP, and 15% SHMP; Fibrisol contained 25% SAPP, 15% TSPP, 10% SHMP, and 50% NaCl. Each of the phosphate mixtures was tested against the various organisms at levels of 0.3%, 0.5%, and 1.0%; controls containing no phosphates were included in the tests. The media containing the inoculations of microorganisms with and without the phosphates were heated to 50°C, 55°C, 60°C, 65°C, and, in the case of sporulating bacteria, 75°C and 80°C. *Staphylococcus aureus* and *Streptococcus faecalis* were inhibited or destroyed by all levels of phosphate and at all temperatures tested from 50°C through 65°C, but they grew rampantly at each temperature in the absence of the phosphates. The *S. faecalis* grew in the unheated media containing all levels of phosphate, while only the 0.3% and 0.5% Fibrisol-containing media supported the growth of *S. aureus* without heating. All levels of Curafos prevented growth of *S. aureus* with or without heating. Growth of the spore forms of *Bacillus subtilis* was inhibited with and without heating in the presence of 0.5% and 1.0% Curafos. *B. subtilis* spores were able to germinate and grow, although with limited success, in the presence of 0.3% Curafos and with increasing success with time in the presence of the Fibrisol at all levels. The vegetative forms of *B. subtilis* were strongly inhibited by the presence of 0.5 and 1.0% Curafos with and without heating; they grew to a limited extent after 5 days or more incubation in the presence of 0.3% Curafos. Although they were capable of growing after prolonged incubation in the presence of 0.3% and 0.5% Fibrisol, they did not grow in the presence of 1.0% Fibrisol with heating. The organisms could grow with prolonged incubation in the presence of all levels of Fibrisol when the medium was not heated. Spores of *Clostridium sporogenes* had difficulty germinating in the presence of 0.5 and 1.0% Curafos, but they could grow, especially after prolonged incubation, in the presence of all levels of Fibrisol with and without heating. *C. bifermentans* had considerable difficulty growing in the presence of all levels of Curafos with any degree of heating; it could grow after prolonged incubation in the presence of 0.3%, but growth was completely inhibited in the presence of 0.5 and 1.0% Curafos without heating. Growth occurred, although it required prolonged incubation, at 0.3 and 0.5% levels of Fibrisol with and without heating; growth was completely inhibited with and without heating in the presence of 1.0% Fibrisol.

The effects of orthophosphates as lytic agents for various bacteria have been studied and reviewed by Pacheco and Echaniz[846] and Pacheco and Dias.[847] $TSP \cdot 12H_2O$ was found to be the most effective lytic agent in concentrations of 0.001–0.1 mol solutions. $MSP \cdot H_2O$, $DSP \cdot 2H_2O$, and $DKP \cdot H_2O$ were also found to be effective against *Salmonella typhosa*, *Escherichia coli*, and *Staphylococcus aureus*. Lytic action of the phosphates is puzzling, since the orthophosphates have been demonstrated to be essential to the growth of microorganisms[102,103]; they are particularly essential to spore germination of some of the spore-forming bacteria, as determined by Heiligman and co-workers.[848]

Williams and Hennessee[849] reported that spores of *Bacillus stearothermophilus* had greater heat resistance at phosphate concentrations of mol/120 or lower than in distilled water or in solutions with higher phosphate concentrations. Their work offers an explanation of the inhibiting effects of the higher concentrations, such as mol/15 reported in the earlier literature, which they reviewed.

Wollmann[850] studied the effects of various salts, including sodium pyrophosphate, SHMP, and Fibrisol (a commercial phosphate curing-salt mixture), on the heat resistance of *Erysipelthrix rhusiopathiae*, *Salmonella typhimurium*, and *Mycobacterium tuberculosis*. The salts were tested at 0.5% concentrations at various temperatures in order to determine their ability to inhibit the growth of or to completely kill the bacteria. *E. rhusiopathiae* was completely inhibited by sodium pyrophosphate and potassium iodide, while Fibrisol and sodium citrate were less effective; SHMP and sodium citrate caused an increase in heat resistance of the organisms. Only sodium pyrophosphate and potassium iodide were effective upon the addition of 0.2% sodium chloride to the medium. *S. typhimurium*, although significantly affected by all of the salts investigated, was most affected by sodium pyrophosphate and potassium iodide. However, the addition of 0.2% sodium chloride completely overcame the effects of all of the salts, including the sodium pyrophosphate and potassium iodide. The heat resistance of *M. tuberculosis* was lowered only by sodium pyrophosphate and potassium iodide. Fibrisol initially also lowered the heat resistance, but its effect was overcome by the addition of sodium chloride.

Post and co-workers[851] studied the effects of SHMP on pure cultures and wild cultures of bacteria. The authors reported that concentrations of 0.1% SHMP were effective in preventing the growth of most gram-positive bacteria when added to the media in which they were grown. Gram-negative bacteria were capable of growing at concentrations of SHMP as high as 10%. However, some gram-negative bacteria, such as *Pseudomonas fluorescens*, were lysed when SHMP contacted the cells. This lysis was inhibited or prevented by the addition of sodium chloride or magnesium sulfate. Growth of the gram-positive *Sarcina lutea* could occur in the presence of SHMP levels that normally prevented growth if magnesium sulfate was added to the medium. The authors speculated that SHMP interfered with the metabolism of divalent cations, especially with the magnesium ion, and thus inhibited cell division and caused the loss of the cell-wall integrity.

Gould[852] reported that concentrations of 0.2–1.0% SHMP, containing only small amounts of ortho-, pyro-, or tripolyphosphate, prevented the normal germination and growth of bacteria spores. The higher concentrations prevented growth before the rupture of the spore wall and thus prevented the development of the spores into vegetative cells. Although the spore germination was not inhibited at the lower concentrations, the vegetative cells were distorted, failed to develop in the normal manner, and did not multiply. The effect of SHMP at pH 6 was 1.5 to 2 times as great as it was at pH 7.

Kohl and Ellinger[102] obtained a number of foreign patents on the application of medium-chain-length polyphosphates, averaging in chain length from 16–100 and preferably between 16–34, as microbiological inhibitors in numerous food products. The medium-chain-length polyphosphates were found to be more effective than the SHMP previously reported in the literature. The activity of the polyphosphates increased as the chain length increased from 16 through the high 20's; it then began to decrease in effectiveness as the chain length increased. Certain bacteria were effectively inhibited in growth or completely

TABLE 35

Effect of Polyphosphate Chain Length on Inhibition of Growth of *S. aureus*

Phosphate Additive	Chain Length	Turbidity (Measure of Growth)	Bacterial Count per ml (by Plating)[a]			Per Cent Bacteria Surviving		
			24 hr	*5 days*	*10 days*	*24 hr*	*5 days*	*10 days*
Na_2HPO_4 + KH_2PO_4		Heavy turbid	est. 10^{8+}	10^{8+}	10^{8+}	100^+	100^+	100^+
Sodium acid pyrophosphate + tetrasodium pyrophosphate		Turbid	2×10^6	10^{8+}	10^{8+}	30	100^+	100^+
Sodium tripolyphosphate		Clear	2×10^5	2×10^3	50	3	0.03	0.001
Sodium polyphosphate	12	Clear	10^5	20	20	2	0.000 3	0.000 3
	16	Clear	4×10^3	0	0	0.06	0	0
	18	Clear	4×10^3	0	0	0.06	0	0
	34	Clear	2×10^3	0	0	0.03	0.002	0.001
	37	Clear	2×10^3	100	40	0.03	0.002	0.001
NaCl (control)		Heavy turbid	10^8	10^8	10^8	100^+	100^+	100^+

Source: Kohl, W. F., and Ellinger, R. H. (Stauffer Chemical Company), British Patent 1,154,079, 1969; reprinted by permission. See also reference 102.

[a] The initial bacterial count was 6×10^6.

killed by levels as low as 0.1%, while others required levels considerably higher, ranging from 1% to as high as 5%. *S. aureus* was very effectively killed by levels of 1% sodium polyphosphates with average chain lengths of 16–34, as shown in Table 35. The polyphosphates 'were found to be effective to varying degrees against different species of bacteria, as demonstrated in Table 36, which shows the effect of polyphosphate chain length upon the growth of *E. coli*.

TABLE 36
Effect of Polyphosphate Chain Length on Inhibition of Growth of *E. coli*

Phosphate Additive	Amount of Additive, %	Growth of Initial Bacteria Population, %		
		80/ml	800/ml	8 000/ml
$Na_2HPO_4 + KH_2PO_4$	1	100+	100+	100+
Sodium acid pyrophosphate + tetrasodium pyrophosphate	1	40	40	39
Sodium tripolyphosphate	1	1	1	3
Sodium polyphosphate 12	1	21	29	33
Sodium polyphosphate 16	1	19	28	32
Sodium polyphosphate 18	1	1	1	4
Sodium polyphosphate 34	1	25	35	32
Sodium polyphosphate 37	1	37	37	37
Potassium polyphosphate (1 000+) in 2% Vitrafos®	0.5	36	41	40
Potassium polyphosphate (1 000+) in 1% Vitrafos®	0.5	29	36	46
Potassium polyphosphate (1 000+) in 0.5% Vitrafos®	0.5	70	70	73
NaCl (control)	1	85	85	85
No additive		100%	100%	100%

Source: Kohl, W. F., and Ellinger, R. H. (Stauffer Chemical Company), British Patent 1,154,079, 1969; reprinted by permission. See also reference 102.

Effects of Metal Chelation   Considerable speculation about the effects of polyphosphates as microbiological inhibitors centers on their ability to sequester the essential, nutritional mineral elements required for normal growth of microorganisms. The polyphosphates are excellent sequestering agents for calcium, magnesium, and iron, three of the most important of the essential metal ions.[853,854] The polyphosphates are capable of sequestering calcium in the presence of oxylate, an ability that increases as temperature increases within the pH range of 5–12. However, the ability to sequester calcium decreases with chain length as the pH changes; at pH values of 9.5 and higher, STP and TSPP were the most effective calcium and magnesium sequestrants. Magnesium, in particular, is important to the cell-wall integrity of some of the common food microorganisms.[855] By increasing or decreasing the level of magnesium in the media, *Lactobacillus* species can be changed from the typical rod forms to filamentous forms and vice versa. The number of the metal cations, as well as lipopolysaccharides, is apparently essential to the structural integrity of the cell walls of some of the *Pseudomonas* and *Alcaligenes* species of organisms, as the addition of EDTA at alkaline pH values effectively solubilizes their cell walls.[856] Sequestration of metal ion also appears to increase the permeability of the bacterial cell walls to compounds that can interfere with the growth of the organism.[857] The fact that EDTA inhibits the growth of organisms, as reported by Brown and Richards,[858] may provide support for the metal-chelation theory. However, factors other than metal-chelating ability appear to be involved in the microbiological inhibition. Clues might be found in the reports of Burkard *et al.*,[859] who indicated that polyphosphates are capable of interfering with or at least decreasing the acceptor activity of soluble RNA in yeast. According to these authors, if polyphosphates were eliminated from yeast RNA preparations, the acceptor activity of the RNA was usually satisfactory; if, however, polyphosphates were added to the purified RNA preparations, the acceptor activity was decreased in proportion to the amount of polyphosphate added.

Tomiyama *et al.*,[860] Ozawa *et al.*,[839] and others have studied the synergistic activity of the polyphosphates with various types of antibiotics in the preservation of foods. The polyphosphates have the ability to sequester metallic ions, which interfere with the antibiotic activity of the organic compounds. Kooistra and Troller[861] reported that the effect of organic preservatives, e.g., the edible acids and salts of propionic, sorbic, and benzoic acids with their methyl and ethyl esters, was potentiated by the addition of the phosphate, carbonate, chloride, pyrophosphate, and other edible salts of iron, manganese, zinc, tin, and silver. The polyvalent metal ion probably was slowly released to the solution and produced some type of potentiating effect on the microbiological activity of the organic salt.

Applications as Inhibitors of Yeast and Fungal Growth

The ability of STP to interfere with the fermentation of glucose from yeast was demonstrated by Vishniac.[862] He indicated that STP inhibited the enzyme hexokinase; if ATP or magnesium was added to the medium, the STP inhibition could be reversed. Vishniac suggested that the STP chelated the magnesium, which was essential to the activity of the hexokinase. Kohl and Ellinger[509] claimed that the longer-chain polyphosphates were even more effective in inhibiting yeast fermentation of fruit juices, such as apple cider. As is the case with general microbiological inhibition by medium-chain-length polyphosphates, the most effective range of chain lengths appears to be between 18–37. The degree of inhibition increases as the chain length increases from 3 through approximately 30, and then it decreases with increasing chain length.

The polyphosphates have been found to be effective inhibitors of the growth of fungi. Post and co-workers[508] reported that dipping fresh cherries in a 10% solution of SHMP, STP, TSPP, or sodium tetrapolyphosphate would inhibit or delay spoilage of the cherries by such fungi as *Penicillium expansum*, *Rhizopus nigricans*, and *Botrytis* species. Kohl and Ellinger[103] obtained patents in several foreign countries for the applications of medium-chain-length polyphosphates, including those having average chain lengths of 14–37 phosphate units, as inhibitors of fungal spoilage in beer and wine, refrigerated doughs, malt and other grain products, process cheese, fruits, meats, poultry, seafood, and vegetables.

Applications as Inhibitors of Viruses

Several applications of phosphates as inhibitors of viruses in dairy products have been reviewed in the section of this chapter on Phosphate Applications in Dairy Products. The uses of TSP to inhibit potato virus X and tobacco Mosaic virus have been reported by Brock.[863] The use of a 10 or 20% solution of TSP was found to be more effective than equivalent concentrations of Formalin, and the authors reported that the TSP solution was more effective if the treated inoculum containing the virus was held for 5 minutes prior to inoculating the plant. A 10% solution of TSP was found also to be an effective sterilizing agent for instruments and hands during experimental work with the viruses.

Phosphate Applications in Processing Food Protein

As proteins represent essential components of human and animal diets, their importance in proper nutrition has been well established. Phosphates have been shown to improve numerous characteristics of the proteins, so that they can be more useful in their intended applications. Combinations of orthophosphates are often used to obtain optimum pH values for desired protein characteristics. Less known, but equally important, in food applications are the interactions between proteins and the polymerized phosphates. Numerous specific interactions between proteins of cereals, dairy products, meats, and other protein constituents of food systems have been reviewed in the sections of this chapter dealing with those specific systems. Some of the general interactions between proteins and the polyphosphates also have been reported in the literature and should be reviewed here, since they can have important applications in any food system containing proteins.

The shorter-chain polyphosphates, such as pyrophosphate, tripolyphosphate, and tetrapolyphosphate, are usually quite specific and have been reviewed in the specific food systems. The interactions of the longer-chain polyphosphates, from hexametaphosphate to the very high-molecular-weight, highly polymerized polymetaphosphates, will be of major consideration in this section.

That polyphosphates interact with proteins has been known since the work of Berzelius in 1916; excellent reviews of his early work have been prepared by Horvath[237] and Leach.[239] Horvath reviewed works completed through 1945, while Leach concentrated on works beginning in the late 1930's through 1962.

Some of the very early research, such as that of Fuld in 1902,[864] indicated that the proportion of polyphosphate bound to the protein was determined by the quantity of basic amino acids, such as lysine, arginine, and histidine, present in the protein. The work of Briggs[234] and Perlmann[236] confirmed that the polyphosphates form very strong, non-ionizing, salt-like bonds with the basic groups of protein to such an extent that they mask the basic effects of the protein and shift the dissociation constant of the carboxyl groups to a lower pH region.

The interaction of the polyphosphates with proteins has resulted in the following: (1) precipitation and coagulation of egg white and ovalbumin, (2) precipitation of blood-serum proteins from dilute solutions, (3) precipitation of gelatin-polyphosphate complexes from dilute solutions that are resistant to microbiological decomposition, (4) precipitation of milk proteins and especially casein, and (5) precipitation of various albumins and peptones from aqueous solutions.[237] In each case the reaction results in a flocculent precipitate that is composed of the protein-polyphosphate complex. The quantity of polyphosphate bound to the protein increases in direct proportion to the amount of polyphosphate in the solution; as this quantity is increased, the bound polyphosphate reaches a maximum, which is different for each protein investigated.[234-239]

In contrast to the reports that the polyphosphates were bound to the amino groups of the proteins, Ferrel and co-workers[865] indicated that the highly stable complex between protein and polyphosphate was due to the binding of ortho- and metaphosphoric acid groups to the hydroxyl groups of the amino acids in the proteins, e.g., serine, tyrosine, and threonine, and to a small extent to the phenol hydroxyl group of tyrosine. The authors reported that the complex, after complete removal of all soluble phosphate and other salts by dialysis, was stable to dilute acid and alkaline hydrolysis.

In seeking the reason for the separation of two polyphosphate fractions when yeast cells are partitioned with trichloroacetic acid, Katchman and Van Wazer[235] studied the coprecipitation and formation of complexes between egg albumin and polyphosphates of various chain lengths. The authors reported that the quantity of polyphosphate coprecipitated with the egg albumin, upon treating it with trichloroacetic acid, increased as the chain length of the polyphosphate increased. Approximately 25% of the added phosphorus was coprecipitated with egg albumin when the precipitating agent was a polyphosphate with an average chain length of 16; about 50% of the phosphorus was coprecipitated with egg albumin when treated with a phosphate glass having an average chain length of 85-130. Approximately 80% of the phosphorus of a polyphosphate with a chain length of 230 was coprecipitated with the egg albumin; 100% of the phosphorus of a Kurrol's salt with an average chain length of 1600 was coprecipitated with the egg albumin. It was also possible to separate the sodium phosphate glass with an average chain length of 16 from the Kurrol's salt by coprecipitation with the egg albumin in the presence of trichloroacetic acid. In each case the coprecipitates were complexes of the egg albumin with the polyphosphate.

Braginskaya and El'piner[866] indicated that the formation of protein-polyphosphate and protein-heparin complexes was pH dependent and that the complexes formed at active protein sites, which did not inhibit their enzymatic activity. The proteins studied were γ-globulin, lactalbumin, myosin, and polyalanine. Lyons and Siebenthal[238] determined that the differences in the binding of polyphosphates due to variations in chain length may be caused by the number of possible binding sites on the polyphosphate. The authors speculated that pyrophosphate may have only one possible binding site, while tripolyphosphate may have at least two different binding sites. Longer-chain polyphosphates probably have multiple sites for interaction with multiple sites on the proteins and for increased opportunity for the binding sites on the two compounds to match each other in space.

The variations in the complexing activity between polyphosphates of various chain lengths and the different proteins have been useful in a number

of food applications. The polyphosphates can aid in the following applications:
1. improving dispersion and solubility of proteins,
2. increasing the water binding and gel formation of the proteins,
3. improving whipping properties (which appear to be related to increasing the protein insolubility),
4. improving the coagulation, precipitation, and insolubilization of the proteins for better separation,
5. improving their nutritional properties, and
6. purifying the proteins to remove heavy metals and unwanted flavor and odor components.

These will be discussed in subsequent paragraphs.

Applications in Protein Dispersion

The dispersion of proteins, such as those from milk (casein) or peanuts used to prepare a powdered protein food, was accomplished by treating the protein with alkali or alkaline salts; these included sodium and potassium di- or triphosphate (assumed to be TSPP, TKPP, STP, and KTP). The protein dispersion was dried and mixed with various minerals, vitamins, and flavoring agents to obtain a dried protein food.[867] The treatment of the protein with the alkali was reported to produce a dispersible aqueous solution without breakdown of the protein molecule. A number of examples of phosphate applications to maintain proteins in solution[29] have been reviewed previously in this chapter. These include their applications in pasteurized processed cheese, in meat products in which they solubilize proteins and allow them to form protective films about fat globules in order to improve emulsification, and in stabilizing milk proteins to prevent their gelling during storage.

Increasing Protein Water-holding and Gelling Properties

The ability of polyphosphates to improve the water-holding capacities and, often, to cause the formation of protein gels has been utilized in a number of applications. Hall[868] reported that the addition of a buffering agent, which would maintain the pH between 6.5–7.0 in a mixture of gelatin, water, and propylene glycol, would produce a gelling film that was resistant to attack by microorganisms. The film would protect any food product that it coated against spoilage. Among the buffering agents that could be used in this application was DSP.

Ferrel and his co-workers[865,869,871] and Mohammad and co-workers[870] reported several methods of preparing a gluten–phosphate complex with excellent gelling properties. The most satisfactory method involved treating the wet gluten with urea and 85% phosphoric acid, drying it, heating the dried material for 30 minutes at 140°C, and then neutralizing it with sodium hydroxide, washing, and drying it. The gluten phosphate prepared in this manner could absorb 200 times its weight of water, and it could form colorless, tasteless gels that could be used in the preparation of food products. Gel Soy, a soy protein prepared from specially processed, defatted, dehulled soybean flakes,[872,873] was reported by Glabe et al.[874] to have improved water- and fat-binding capacities when it was treated with SHMP.

Controlling the pH of protein solutions has long been recognized as important in increasing their water absorption. Norris and Johnson[875] reported that the water absorption of soybean protein could be increased if the pH was maintained within the alkaline range of 7.0–9.0. A number of alkaline salts could be used for this purpose, among which was TSP. Pintauro and co-workers[876] reported that the water-binding capacity of a gelatin food product could be highly improved by maintaining a pH of 3.2–3.6 with a mixture of MSP, DSP, and citric acid to buffer the gelatin composition within this pH range. A cold-water-soluble gelatin composition was prepared by Wingerd,[316] who combined gelatin with lactalbumin phosphate, a compound prepared by precipitating lactalbumin proteins from whey with long-chain polyphosphates.

Improving Protein Whipping Properties

The effects of polyphosphates on improving the whipping properties of egg white have been mentioned previously. The addition of polyphosphates to a number of other protein compounds has also improved their whipping properties. Burnett and Gunther[877] prepared a whipping composition from soybean protein isolated from soybean flakes by modifying the protein in the flakes with the enzyme pepsin in the presence of a peptizing salt, such as sodium chloride or sodium phosphate. The salt was added to aid in dissolving any of the unhydrolyzed protein remaining in the preparation. Sevall and Schaeffer[878] prepared protein whipping compositions from soy protein by combining the protein with SHMP, sodium tetrapolyphosphate, TSPP, or sodium orthophosphates; the phosphates stabilized the degraded proteins during whipping in an aqueous system. Patterson[879] formulated whipping compositions similar and equivalent to egg white by combining an alkali-metal caseinate, lactose, or starch and a polyphosphate. This composition successfully replaced 20–35% by weight of the egg whites and still maintained stable whipped foams in which the egg white was the principal whipping agent.

Gunther[880] produced an entirely new whipping composition by combining gelatin, an enzyme-modified soy protein similar to Gel Soy, and a polyphosphate. The specific polyphosphates mentioned were STP, sodium tetrapolyphosphate, and SHMP. According to Gunther the mechanism by which the polyphosphate stabilized the whipping properties of the soy protein and gelatin combination were not known; he speculated, however, that it could be a combination of protein precipitation and the protein ability to sequester heavy metal ions that might interfere. After treatment with the polyphosphate, the protein could form thin walls between the air cells that had sufficient stiffness and storage stability to produce excellent whipped compositions.

Sutton[881] reported that a protein material, such as nonfat milk solids, soybean flour, dried egg-white solids, dried whole eggs, gelatin, caseinate, and similar proteinaceous substances, combined with the alkyl ester of an aliphatic polycarboxylic acid and a polyphosphate salt would provide highly improved whipping compositions with high-foam volume and high-foam stability. Among the alkyl esters of aliphatic polycarboxylic acids specified were those of malonic, succinic, glutaric, tartaric, malic, and citric acids. The polyphosphates that could be used for this purpose were SAPP, TSPP, STP, and the longer-chain polyphosphates with average chain lengths from 4 to several thousand, including SHMP, Graham's salt, TKPP, KTP, and the very-long-chain potassium Kurrol's salts. The ammonium salts of the various polyphosphates also could be used for this purpose.

A series of patents was obtained by investigators from the research laboratories of the General Foods Corporation for whipping compositions containing proteins that could be precipitated and/or denatured by polyphosphate compounds in acid medium. These whipping compositions were reported to produce superior whipped dessert products, such as chiffons and chiffon-type pie fillings.[882-885] Clausi *et al.*[882] combined gelatin, partially hydrolyzed soy protein, pregelatinized starch, and polyphosphates, including those with an average chain length of 2 through the very-long-chain polyphosphates but preferably SHMP, to produce superior whipping compositions with highly stable foams for use in chiffon desserts and pie fillings. From 1-20% phosphate, based on the level of the partially hydrolyzed soy protein, was used; 20% SHMP was preferred, since it produced the highest foam volume, the shortest whipping time, and the greatest foam stability.

Mancuso and Common[883] prepared foaming compositions for use in chiffon food products or pie fillings, using partially degraded soy protein, a vegetable gum, gelatin, and a polyphosphate, especially SHMP and the sodium polymetaphosphates of longer chain lengths. Mitchell and Seidel[884] produced a phosphated gelatin, prepared by reacting untreated gelatin with a polyphosphoric acid or one of its salts at a pH of less than 4. Whereas the addition of chocolate to normal chiffon mixes suppressed the foam properties, the use of the phosphated gelatin produced superior chocolate chiffon products. Block[885] eliminated the necessity of preparing the phosphated gelatin for chocolate-chiffon desserts by combining cocoa powder, gelatin, partially hydrolyzed soy protein, sugar, and a pregelatinized starch with the phosphate compound. SHMP and Graham's salts were preferred because of the superior properties imparted to the foaming composition.

Downey[886] prepared a dry mix for use in marshmallows. Downey's mix did not require the close attention and special processing conditions of conventional marshmallow manufacture, and it also could be used to prepare uniform batches of marshmallows in the home kitchen. The composition contained a major quantity of dextrose and small quantities of starch, gelatin, phosphates, and vegetable gums. SHMP was the preferred polyphosphate in the patent. Grettie and Tiemstra[887] patented another gelatin-containing marshmallow composition with highly improved properties; it also required the addition of polyphosphates.

Applications as Protein Precipitants

The polyphosphates could also be used as protein precipitants to coagulate proteins, separate and purify them. Schwartz[888] produced artificial fibers or filaments of casein by first dissolving the casein in an alkaline solution and passing the dissolved casein through a spinnerette. A thin filament of the protein solution was discharged into polyphosphoric acid maintained at a pH of approximately 2.5. The polyphosphoric acid solution coagulated the filaments, which then contained a casein–polyphosphate complex of great stability.

Horvath[889] devised a method for the isolation of vegetable proteins as vegetable protein–polyphosphate complexes for numerous food applications. The process involved the preparation of a weak solution of the polyphosphate, which could be any one of the phosphates having a ratio of alkali-metal oxide to P_2O_5 of 0.9:1 to 1.7:1. The polyphosphate solution was used to extract the protein from the finely pulverized, protein-containing vegetable material and to precipitate the protein–polyphosphate complex by acidification. A British patent described a similar process for isolating vegetable protein–polyphosphate complexes that then could be precipitated around any food material to form an edible food coating.[890] The process for extraction of the protein from the vegetable material was quite similar to that described by Horvath, and the precipitation of the protein–polyphosphate complex as a coating around the food material was performed by acidifying a solution of the complex. Solutions of the complex were readily prepared by dissolving the complex in urea, neutral salts, or alkali.

Rane and Newhouser[891] precipitated proteins without forming a protein–polyphosphate complex by use of the cyclic polyphosphate sodium tetrametaphosphate. This procedure was advantageous in that the protein could be obtained without the associated phosphate. Although the procedure was especially applicable for the precipitation of proteins from blood serum, the authors claimed that it could be used for recovery of proteins from animal or vegetable fluids, such as milk, liver extract, corn extract, or industrial, protein-bearing wastes. The process involved (1) adding the tetrametaphosphate, (2) acidifying to precipitate the protein, (3) separating the protein from the supernatant, and (4) subsequently drying the precipitate or using it directly for the intended purpose. Rane and Newhouser also used this process to fraction some of the components of blood, such as fibrinogen, globulins, and albumins. They first lowered the pH to 5 to precipitate fibrinogen and the globulins, then lowered the pH to approximately 4.2 to precipitate albumins, and later lowered the pH to approximately 3.5 to obtain the balance of the blood proteins. Lowering the pH below 3.5 would result in denaturing the protein. This process was especially beneficial in that it did not denature the proteins in any way.

Nitschmann *et al.*[892] implemented a similar fractionation procedure for blood-plasma proteins involving the use of polyphosphoric acid. The proteins again were reported to be undenatured, and a fractionation of proteins was described. In the fractionation procedure increasing amounts of the polyphosphoric acid were added to precipitate the various blood-protein fractions.

Keil and co-workers[893] obtained two patents for a fungus-resistant food coating prepared from gelatin, SHMP, and water acidified to pH values of 2.15–3.5. The coating could be applied by dipping the food product into the solution. The gelatin-polyphosphate composition dried as a transparent film, which prevented crystallization of salt and mold growth on the surface of the food products.

Fukamachi and Watanabe[894] added sodium citrate and sodium polyphosphates to soybean milk in the presence of calcium salts to improve formation of curd.

Waugh[895] isolated casein from skim milk by first increasing the calcium ion concentration to 0.05–0.1 mol, with 0.06–0.08 mol calcium ion preferred; the precipitated casein was separated from the whey, the calcium ion was removed by precipitating it with oxylate, orthophosphate, or carbonate to reduce the calcium ion concentration to less than 0.2% in the final casein compound. The addition of the calcium ion was necessary to increase the size of the calcium-caseinate micelles and to cause greater precipitation and easier separation of the calcium–caseinate curd. After separation by filtrating or centrifuging, the calcium curd was washed to remove additional whey and resuspended in water. The calcium ion then was precipitated with orthophosphate or other calcium-precipitating salts, or it could be removed in the form of a soluble complex by adding one of the polyphosphates, such as STP or SHMP. However, when a soluble calcium–phosphate complex has been formed, such as when using SHMP, the soluble complex must be removed by dialysis and might complicate the procedure. The resulting casein was reported to be highly soluble and was also compatible with milk. It could be used to increase the protein content of the milk or to improve the protein and nutritional contents of other food products.

The polyphosphates and orthophosphates also have been reported to aid in the purification of gelatin from impure solutions through their ability to precipitate heavy-metal contaminants from the gelatin composition.[896] The impure gelatin was dissolved in water, phosphoric acid was added, and magnesium chloride was then added to the solution at an alkaline pH, such as 9.15. The precipitate of magnesium ammonium phosphate that was formed brought down with it the impurities in the gelatin; the filtrate contained highly purified gelatin.

Phosphate Applications in Starch Processing

It is well known that acids react with the hydroxyl groups in starch. The esters thus formed have been useful in a number of industrial applications and also in some food applications. Phosphoric acid will form esters with starches. There are phosphoric acid ester groups in natural starches, particularly in potato starch.[897–900] Investigations have demonstrated that the paste viscosity of the potato starch increases as its phosphoric acid ester content increases. The potato starch also can form primary and secondary phosphate complexes with metal ions, depending on the amount of phosphate in the starch.

Effects of Phosphates on Starch Properties

Phosphates have significant effects on starch viscosities, as demonstrated by Nara et al.[901] They showed that sodium pyrophosphate had a greater effect in decreasing the swelling of starch granules and depressing viscosity of the starch pastes than various types of monoglycerides and other surface-active agents. Similar effects were obtained with flour, where the addition of phosphates decreased the degree of swelling of two hard-wheat varieties of 70% extraction.[902]

Kuhl[903] reported that the salts of H_3PO_4 reacted differently in gelatinizing starch than the acid; the salts also reacted differently with variations in the quantity of the salt and with the degree of saturation of the acidic radical with alkali. Nutting[904] demonstrated that treatment of potato-starch pastes with SHMP decreased the viscosity of the potato-starch paste to a greater degree at pH 8.5 than any other treatment. Conditions that increased ionization also increased the hydration of the starch, the particle volume, and the viscosity of the paste. All factors that depressed ionization decreased the viscosity of the paste.

Bowen[905] prepared a dextrinized starch for improving bread after partially gelatinizing the starch by cooking it, mixing the cooked material with an acidic solution of MCP, calcium sulfate, and ammonium chloride, then further heating the mixture and drying it. Rozenbroek[906] modified potato starch to resemble the properties of cornstarch in food applications by treating the potato starch with a small quantity of phosphoric acid in a small amount of water, heating it at 50°C for 10 hours, and then neutralizing it with slaked lime. Kunz[907] modified amylose so that it could be used to prepare films and tubes for meat casings and other food applications. The amylose was treated with sodium hydroxide in aqueous solution and cast or shaped as desired to form the film or tube. The cast solution then was treated with a coagulating solution prepared with sulfuric acid, sodium sulfate, and a mixture of H_3PO_4 and DSP. The author stated that it was essential to use a combination of the acid and salt, since using acid alone formed weak, unsatisfactory films or tubings; the use of the salts alone did not supply sufficient acidity for coagulation of the amylose. Fredrickson[908] reported that mixing approximately 1% TSP with starch rendered it fluidizable for treatment in a fluid bed, where it could be heated to dextrinize the starch.

The Starch Phosphates

The most important and best known applications of phosphates and starch are those used in preparing the starch phosphates. The starch phosphates are relatively new, modified starches that have very useful properties in food processing. One of their most important properties is the reduction in retrogradation; this is believed to be caused by the introduction of ionizing groups, causing the molecules to repel one another.[909] The resistance to retrogradation is especially important during freezing and thawing of sauces and gravies that contain the starch phosphate, since the phosphate groups esterified to hydroxyls block the cross-bonding between adjacent starch molecules essential for retrogradation of starch. The starch phosphates also have improved clarity and high-water-binding capacity, as well as reduced gel formation. They can increase the viscosity of the foods in which they are used without forming thick or highly viscous gels.[910] The processes and properties of the various types of starch phosphates have been thoroughly reviewed by Hamilton and Paschall.[911] There are two types of chemically reacted starch-phosphate compounds that can be produced by treating various types of starches with phosphates under specific conditions.

Starch-phosphate Monoesters   One of the chemically reacted types is known as starch-phosphate monoester, in which one acidic function of an orthophosphate group forms an ester linkage with one of the hydroxyls in the starch chain. Various degrees of substitution can be obtained, and mono- and disodium or potassium salts can be formed at each phosphate linkage. According to Hamilton and Paschall the reaction can be represented by the following equation:

$$\underset{\text{HO}}{\overset{\text{NaO}}{\text{starch—OH + HO—P=O}}} \longrightarrow \underset{\text{HO}}{\overset{\text{NaO}}{\text{starch—O—P=O}}} + H_2O$$

Evans[912] formed the monoesters of starch phosphates by treating starch that contained no more than about 20% by weight of water with a small quantity of TSP. The slurry was passed through a steam-heated cooker and held for about 1 minute at temperatures of 160–212°F; the slurry then was dried on a heated roll to reduce its moisture content to about 4%. The dry, pulverized starch could then be used to prepare instant-pudding mixes. Evans also used MSP, DSP, MKP, and TKP to obtain starch-phosphate monoesters.

Kerr and Cleveland[913] prepared starch-phosphate monoesters containing the disodium salt of the ester. They treated the starch with an alkali-metal tripolyphosphate in the presence of alkali, as shown in the following reaction:

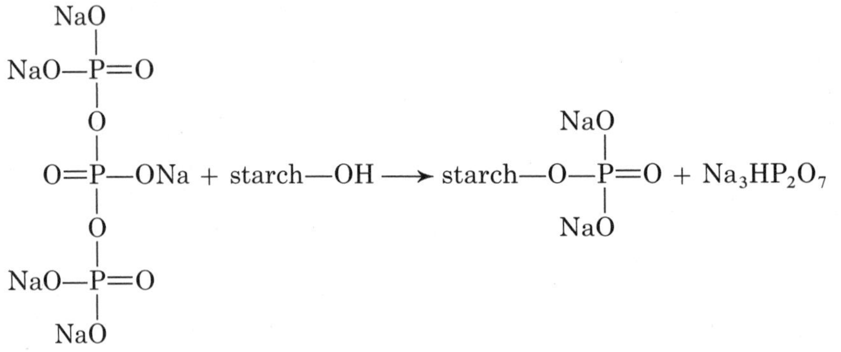

$$
\begin{array}{l}
\text{NaO} \\
\,| \\
\text{NaO}-\text{P}{=}\text{O} \\
\,| \\
\text{O} \\
\,| \\
\text{O}{=}\text{P}-\text{ONa} + \text{starch}-\text{OH} \longrightarrow \text{starch}-\text{O}-\overset{\displaystyle \text{NaO}}{\underset{\displaystyle \text{NaO}}{\text{P}}}{=}\text{O} + \text{Na}_3\text{HP}_2\text{O}_7 \\
\,| \\
\text{O} \\
\,| \\
\text{NaO}-\text{P}{=}\text{O} \\
\,| \\
\text{NaO}
\end{array}
$$

The disodium starch-phosphate monoesters were useful in the preparation of pudding mixes, instant cold-milk-gel-pudding mixes, pie fillings, instant cream-pie-filling mixes, powdered salad-dressing mixes, and thickening sauces for vegetables. The addition of as little as 3 phosphate groups per 100 anhydro-glucose units in the starch molecules resulted in highly significant changes in the properties of the cornstarch, which then began to resemble potato starch in many of its properties. The use of the phosphate-modified starch in such foods as soups, canned vegetables, and fruit sauces or fillings resulted in thick-bodied, creamy, relatively clear, thickened sauces, instead of the opaque, gel-like, stiffened consistencies obtained with the unmodified cornstarch. Under suitably adjusted conditions the starch-phosphate monoesters could be prepared from the cyclic metaphosphates, the long-chain polyphosphates, and pyrophosphates. Kerr and Cleveland[914] obtained starch-phosphate monoesters identical to their previous invention by mixing acidic orthophosphate salts with the starch to form a slurry containing no more than 20% moisture; the slurry then was heated to a temperature of 160–200°C at a pH not above 8.5 and then dried. The authors believed that under the conditions of the process, the orthophosphoric acid salts were dehydrated as the moisture decreased and finally polymerized, in which form they could react with the starch to form the monophosphate ester.

Neukom[915] reported the use of a mixture of MSP, DSP, and urea to esterify the starch. The resulting product, containing 0.2–5% and usually 1–5% bound phosphorus based on the weight of the starch, produced starch phosphates having higher viscosities and shorter pastes than those prepared without the urea. The same author later prepared starch-phosphate monoesters by soaking the ungelatinized starch in a solution of the alkali-metal phosphate, separating the starch from the excess moisture, and heating the treated starch to temperatures of 120–175°C for one to several hours; this process produced a dry starch phosphate, containing 1–5% phosphorus based on the weight of the starch.[916] The resulting starch phosphates could be used in the preparation of various types of gravies and sauces with ideal characteristics.[917]

In a further extension of his earlier work, Neukom[918] prepared starch-phosphate monoesters by soaking the starch in a solution of any combination of phosphoric acid, MSP, DSP, TSP, or a mixture of phosphoric acid with sodium hydroxide to obtain a solution having a pH of 3.0–10.0. After sufficient soaking the starch was separated from the excess solution and heated to 120–175°C, at which time a dry, cold-water-soluble starch-phosphate monoester was formed. Various types of starches could be treated in this way to prepare their starch-phosphate monoesters. The author reported that, in addition to their ability to swell and dissolve in cold water, they formed stable pastes after prolonged standing and also resisted bacteriological attack to a greater extent than the untreated, parent starch.

Kodras[919] patented the use of oxygen-containing, water-miscible, organic liquids to precipitate the alkali-metal-phosphate-modified starches described by Neukom. Sietsema and Trotter[920] also obtained a patent for the precipitation and purification of phosphate-modified starches by using those reported by Neukom as starting materials. Schierbaum and Boerner[921] prepared the ortho-phosphate esters of starch by mixing the wet starch with a solution of the alkali-metal phosphates at 55°C, drying the mixture, and then heating it for 3 hours at about 140°F. The resulting product was a starch phosphate suitable for making pastes.

Starch-phosphate Diesters A second type of starch phosphate can be prepared and is now being used in food applications. It is the type known as starch-phosphate diesters, in which an orthophosphate unit is linked by ester groups to two starch molecules. A phosphate-ester bridge is formed, connecting the two starch molecules. These starches are typically highly stable to heat, agitation, and acidity, and the starch granules are slow to swell and rupture.[910,911]

There are various methods of obtaining the phosphorus cross-linkage, including the use of phosphorus oxychloride, phosphorus pentachloride, and thiophosphoryl chloride.[922] The addition of water-soluble salts of alkali metals, such as sodium chloride, improved and gave better control of the cross-linking reaction.[923] The presence of the salt was thought to retard the hydrolysis of the phosphorus compounds long enough to permit them to penetrate deeper into the starch granules and to produce a more uniform amount of phosphorylation.

Kerr and Cleveland[924–926] obtained patents for the reaction of starch with the cyclic trimetaphosphate salts in the presence of various alkaline materials that catalyzed the cross-linking esterification. An example of the reaction was the preparation of a slurry of the starch with a 2% solution of sodium trimeta-phosphate (based on starch) at 50°C for 1 hour at pH 10–11. The resulting starch had a high, hot-paste viscosity when cooked as in the preparation of gravies and sauces. The reaction involved was thought to be as follows:

$$\text{(NaO-P-O...P=O cyclic phosphate structure)} + \begin{array}{c}\text{starch—OH}\\ \text{starch—OH}\end{array} \longrightarrow \begin{array}{c}\text{NaO}\\ |\\ \text{starch—O—P=O}\\ |\\ \text{starch—O}\end{array} + Na_2H_2P_2O_7$$

It was thought that small amounts of the mono- and trisubstituted esters occurred along with a major proportion of the starch-phosphate diesters.[911] The resulting cross-bonded starches produced exceptionally clear sauces and viscous pastes, which made them especially desirable for pie fillings. They have excellent freeze-thaw stability, resisting the bleeding of water after as many as 10 freeze-thaw cycles. In contrast to untreated and starch-phosphate mono-esters, these starches maintain their viscosities, even if they are cooked for long periods of time or added to high-acidity products, such as fruits.[910, 911] They are unaffected by long cooking, high temperatures, and low pH values.

Gramera et al.[927] analyzed the glucose-phosphate compounds, which they isolated upon hydrolysis of this type of starch phosphate. The isolated compounds indicated that there was 28% phosphate substitution of the C-2 position, 9% of the C-3 position, and 63% of the C-6 position in the glucose units of the starch phosphates.

A cold-water-swelling starch was also reportedly prepared by first treating cereal or potato starch for a short time in an acidic solution with a pH of 3.0–4.0, then adding TSPP, drying, and gelatinizing the preparation.[928] Vermicelli compositions of improved stability, gloss, and smoother texture were also prepared by treating the vermicelli with hydrogen peroxide and polyphosphates.[929]

Among the numerous applications for the starch phosphates that have been reported in the literature are their uses as stabilizers for fountain syrups and ice-cream toppings, ice cream, beer foam, and various types of fat and water emulsions; suspending agents for insoluble solids, which could also include fruits in pie fillings, and for Chinese-type foods in which great thickening and suspending characteristics, as well as maximum clarity, are desirable; as pan greases in combination with fats; in bakery mixes, instant puddings, and various types of cheese sauces; and numerous other nonedible applications.[910,911] The various starch phosphates have been approved for use in foods, providing they contain no more than 0.4% residual phosphate calculated as phosphorus.[930]

Phosphate Applications in Sugar Processing

The term *sugar* is most commonly applied to the disaccharide sucrose, the major food sweetener in human nutrition. Sucrose is a common constituent of all plant cells. Sugar cane and sugar beets are the most common sources of the sucrose in use as a food.

Sugar refining is the process by which the sugar-bearing juices are extracted from the plant cells of sugar cane or sugar beets, purified, and crystallized. The process involves many complicated steps, the purposes of which are rupturing the plant cells containing the sugar, concentrating the sugar while removing impurities that also are extracted from the cells, and then crystallizing it. The modern sugar processes result in one of the most highly purified compounds known to man.

Clarification of the Sugar-bearing Juices

The juices obtained through the rupture of plant cells (by pressing the stalks of sugar cane and by water extraction of the sliced root of sugar beets) contain numerous, soluble, nonsugar compounds. These must be removed, and the usual practice is to treat the juices with lime and then to precipitate the calcium by formation of insoluble calcium phosphates or carbonates; these in turn aid in the precipitation of the colloidal, nonsugar materials in the solution. Filter aids or Fuller's earth are added if the precipitates are to be removed by filtration. Some processes remove the precipitates by centrifuging the juices. Phosphates are often used to precipitate the calcium in this step; a number of literature references advocate the use of orthophosphoric acid.[931-944] Some find the addition of phosphoric acid and/or acid phosphates useful.[932,940,941] Bennett, who studied the effects of calcium and phosphate levels in the sugar-clarification step, reported that the optimum phosphate concentration is 4 mmol, and the optimum calcium ion concentration is about 15 mmol.[945]

Phosphates other than phosphoric acid can also be used. Various sodium, ammonium, potassium, and calcium orthophosphates have been proposed as additives to the sugar solution along with lime.[937,946-962] The polyphosphates, such as SHMP,[963] superphosphate,[964-966] and pyrophosphate,[967] have also been proposed. Barrett and his co-workers[968,969] have proposed a synthetic hydroxyapatite for the clarification step. Most of the phosphates, particularly the alkaline sodium and potassium phosphates, are used to maintain an optimum pH for the precipitation of the lime, as proposed by Kubala.[970] The control of pH is critical if all of the lime is to be removed from the juice. According to Kortschak[971] the calcium phosphate precipitate is not a definite compound; after it precipitates, the calcium phosphate absorbs additional calcium hydroxide from the juice. Jung[972] patented the use of polyphosphoric acid in combination with a dicarboxylic acid for the clarification of sugar juices. Cummings[973] reported that the filtration of sugar liquors is easier when the calcium phosphate is combined with a diatomaceous-earth filter aid.

Bleaching the Juices

Another important step in the refining of sugar is the bleaching of the sugar syrup prior to crystallization. This is often accomplished by treating the solution with calcium hypochlorite in the presence of a calcium phosphate to act as a stabilizer and to aid in final precipitation of the calcium from the bleached solution.[932,974] Several soluble phosphates have been proposed for precipitating the calcium contributed by the calcium hypochlorite bleach—MCP,[975] orthophosphoric acid,[976] and a combination of calcium hydroxide and MCP.[977] Andresen[978] advocated the replacement of mineral constituents of raw sugar by adding various calcium compounds, including calcium phosphates.

Other Applications

The sweetness of beet sugar, reported to be not as sweet as cane sugar,[979] can be increased by the addition of calcium sulfate and DCP to the wet beet sugar with subsequent drying. Schongart[980] reported that orthophosphoric acid produced a lower ratio of inversion of sucrose than an equivalent level of hydrogen ion produced by hydrochloric acid. The addition of DSP to clarified sugar cane juice at levels of 0.01–0.05% reduced the hydrolysis of the sugars and, thus, reduced caramelization, which leads to dark-colored raw sugar.[981]

The addition of TCP to sugar reduced caking,[982] provided free-flowing properties,[983] and reduced its hygroscopicity.[984] Mead[985] reported that the addition of pyrophosphoric acid or SAPP to molasses produced a greater amount of clarification and resulted in a clearer, more brilliant molasses.

Food Applications for Organic Phosphates

Numerous organic phosphate compounds are known in nature. These include the phosphate esters, those in which phosphate groups are chemically linked through the hydroxyls of organic compounds, such as sugars, and the numerous energy-giving compounds containing phosphate esters, such as ATP, the various sugar phosphates, and similar substances. Any biochemical textbook can provide numerous examples of the energy-rich phosphate esters of organic compounds. There is considerable evidence that phosphate compounds of organic materials also occur, in which case there would be a carbon-to-phosphorus bond.[3]

The methods for preparing compounds that include the esterification of hydroxyl groups of organic substances, such as carbohydrates, polyhydroxylic alcohols, and hydroxy acids, are well known. One of the first patents for the preparation of such compounds for food use was granted as early as 1910.[986] Few food applications for organic phosphates have been developed, and fewer have been approved by FDA for commercial use. It is necessary to demonstrate that these compounds are not toxic, which is a time-consuming and costly process. The development costs for these compounds as food additives are very high, and the manufacturers often cannot foresee sufficient profit during the life of the product to return his investment.

Applications as Antioxidants

Epstein and Harris[987] reported that a phosphorylated mono- or diglyceride of a fatty acid could be used as an antioxidant in food applications. Martin[989] patented the preparation of the orthophosphoric esters of phenolic compounds, which he claimed were excellent, neutral phosphate-ester antioxidants for treatment of animal, vegetable, and marine oils and prevented or inhibited development of rancidity.

Nyrop[989] obtained a patent for the preparation and application of phosphate esters of fermentable sugar compounds, such as hexoses, pentoses, and trioses. He reported that these organic phosphate-sugar compounds, when added to organic food materials subject to oxidative deterioration, would stabilize the organic food compounds against rancidity and oxidative deterioration.

Applications as Emulsifying Agents

Harris[990] reported that organic phosphate esters of mono- and diglycerides were able to modify interfaces between aqueous fat systems. These compounds were excellent emulsifiers when added to shortenings to be used in most of the applications of shortenings, such as in baked products, and the preparation of icings. Katzman[991] patented the preparation and application of metaphosphoric acid esters of many types of high-molecular-weight alcohols as emulsifying, foaming, and improving agents for baked products, particularly for non-shortening foam cakes, such as angel food cakes.

Thurman[992,993] obtained patents for the preparation of sodium phosphate esters of phosphatides, which made excellent surface-active agents for use as emulsifiers and similar applications in foods. Examples claimed emulsifying properties for mayonnaise, margarine, and shortening compositions.

Pader and Gershon[994,995] obtained patents for the application of phosphoric acid esters of mono- and diglycerides as emulsifiers and aerating agents for aerosol toppings. They could be dispensed as light foams from pressurized cans and had the texture and appearance of whipping cream. The aerosol preparation was reported to be stable to temperatures of $-10°$ to $70°F$.

Applications as Whipping Agents

Katzman[996] prepared tetrapolyphosphoric acid esters of long-chain alkyl alcohols, such as stearyl tetrapolyphosphate, to be used as excellent foaming agents when added to egg white in small quantities. The author listed 35 examples of organic phosphate esters, ranging from the alkyl alcohols through mono- and diglyceride fatty compounds. All were reported to provide a strongly hydrophilic group through the tetrapolyphosphoric acid and a strongly lipophilic group through the alkyl compound.

Thompson[997] reported the methods for preparation and the food applications of an alkyl phosphoryl halide, e.g., octadecyl phosphoryl dichloride, as improvers of air incorporation and foam stability in foamed food products. These compounds were reported to be especially useful when incorporated into baked products requiring a great deal of aeration, such as cakes and breads.

Cunningham and co-workers[998] obtained a patent for the application of triethyl phosphate as a foaming agent for egg whites. The incorporation of 0.001–0.1% of the compound, based on the weight of the egg white, would improve the whipping properties of egg whites, which normally had lowered whipping properties due to pasteurization procedures now required of all such products. Concern over the toxicity of triethyl phosphate was the reason for research by Gumbmann et al.,[999] who studied the toxicity of the compound in quantities up to 10% of the diet of rats. Feeding and reproductive studies indicated that the compound had no effect at levels that would be encountered if it were used as a food additive in the manner proposed by Cunningham and his co-workers.

Other Applications

Curtin and Gagolski[1000] and Toy[1001] patented the preparation of sucrose-calcium phosphate compounds of a very complex nature; they formed soluble compositions, even though a considerable quantity of dicalcium phosphate was detectable in the product. The ability of the sucrose-phosphate complex to form soluble dicalcium phosphate compounds was not understood but was adequately demonstrated. These compounds were shown to increase crispness of the sugar coatings on breakfast cereals[1002, 1003] and to inhibit the development of dental caries in the teeth of children who ate food compositions containing the compound.[1004]

A Japanese patent[1005] reported that organic phosphates, such as the sugar phosphates and glycerol phosphates, improved the flavor of alcoholic beverages.

Future Trends

As demonstrated in the preceding sections, the food-science literature contains much evidence of the usefulness of the phosphates in food processing. Many of the applications described, however, are not being exploited by the food industry. The ability of the phosphates, and especially the polyphosphates, to control firmness or tenderness of the tissues of fruits and vegetables is not applied to the extent that the published research indicates. The use of the polyphosphates in increasing the acceptability of meat products, including fresh and frozen meats, poultry, and seafood, is well documented, but it is mostly used only in cured meats. A major reason that these applications are not exploited is that the regulatory agencies, the USDA and FDA, have been very cautious in approving phosphate applications in processed foods. These agencies most frequently cite three reasons, as follows:

One reason for not accepting phosphates in more food applications is fear that the polyphosphates will allow the processor to upgrade lower-grade food products. This use has repeatedly been disproved. The phosphates are unable to improve the quality of low-grade foods. They are, however, capable of preventing further lowering of the quality and in many cases, of improving some of the characteristics that make these foods less acceptable to the consumer.

The second reason often given by the FDA and USDA is that phosphate treatment allows the processor to increase the water content of foods. This is most often mentioned in regard to meat, poultry, and seafood products. Yet, careful research, especially as reviewed in the sections on poultry and seafoods, has clearly demonstrated that treating these foods with phosphate dips often *reduces* the amount of water absorbed during ice chilling of the products. In fact, the evidence presented indicates that phosphate treatment seals the surfaces of the food; therefore, it prevents internal juices from escaping and water or saline solutions from entering.

The third argument given against food applications of phosphates is that their increased use will reach a level that will be toxic to humans. Evidence presented in the section on the toxicology of the food phosphates has demonstrated that they are little, if any, more toxic than common table salt, which is now used in most foods. The effective levels of the phosphates are usually 0.5% or less. When this level of phosphate was consumed in the diets of many animals, it was shown to be nontoxic. Since the phosphates have adverse effects in many foods, it is inconceivable that their use could never result in a level of 0.5% of phosphate in the total human diet.

As the need to conserve every available amount of food increases over the next few years due to the population explosion, the use of phosphates to aid in making more foods acceptable and in increasing their nutritional value by retaining internal juices should be given further serious consideration. The pressure of the increased need for expansion of the world food supply is certain to result in the increased application of phosphates in food processing.

References

1. Nazario, G., 1952, *Rev. Inst. Adolpho. Lutz 11*:141.
2. Sanchelli, V., 1965, *Phosphates in Agriculture*, Reinhold Publishing Company, p. 90.
3. Quin, L. D., 1967, *Top. Phosphorus Chem. 4*:23.
4. Fox, S. W., Ed., 1965, *The Origins of Prebiological Systems*, Academic Press.
5. Miller, S. L., and Parris, M., 1964, *Nature 204*:1248.
6. Ponnamperuma, C., Sagan, C., and Mariner, R., 1963, *Nature 199*:222.
7. Lohmann, K., 1958, *Kondensierte Phosphate in Lebensmitteln* (Symposium, 1957, Mainz), Springer-Verlag, p. 29.
8. Harold, F. M., 1966, *Bacteriol. Rev. 30*:772.
9. Wiame, J. M., 1949, *J. Biol. Chem. 178*:919.
10. Stich, H., 1953, *Z. Naturforsch. 8b*:36.
11. Kornberg, S. R., 1957, *Biochim. Biophys. Acta 26*:294.
12. Ipata, P. L., and Felicioli, R. A., 1963, *Boll. Soc. Ital. Biol. Sper. 39(2)*:85; 1963, *Chem. Abstr. 59*:874g.
13. Van Steveninck, J., and Booij, H. L., 1964, *J. Gen. Physiol. 48*:43.
14. Shiokawa, K., and Yamana, K., 1965, *Exp. Cell Res. 38(1)*:180; 1965, *Chem. Abstr. 63*:4722A.
15. Miyachi, S., 1961, *J. Biochem. (Tokyo) 50*:367; 1962, *Chem. Abstr. 56*:9121a.
16. Tewari, K. K., and Singh, M., 1964, *Phytochemistry 3(2)*:341; 1964, *Chem. Abstr. 61*:963c.
17. Grossmann, D., and Lang, K., 1962, *Biochem. Z. 336*:351.
18. Griffin, J. B., Davidian, N. M., and Penniall, R., 1965, *J. Biol. Chem. 240*:4427.
19. Mattenheimer, H., 1958, *Kondensierte Phosphate in Lebensmitteln* (Symposium, 1957, Mainz), Springer-Verlag, p. 45.
20. Sober, H. A., Ed., 1970, *Handbook of Biochemistry*, 2nd ed., The Chemical Rubber Co.
21. Sherman, H. C., 1947, *Calcium and Phosphorus in Foods and Nutrition*, Columbia University Press, p. 91.
22. Kiermeier, F., and Mohler, K., 1957, *Z. Lebensm.-Unders. Forsch. 106*:33.
23. Ruf, F., 1957, *Mitt. Geb. Lebensmittelunters. Hyg. 48*:451.
24. Shettino, O., 1965, *Farmaco (Pavia) 20(2)*:65.
25. 1958, *Kondensierte Phosphate in Lebensmitteln* (Symposium, 1957, Mainz), Springer-Verlag.
26. Barackman, R. A., and Bell, R. N., 1953, *Food Eng. 25(6)*:68, 108.
27. The Functions of Phosphates in Food Products, 1966, Food Industry Release No. 1, Stauffer Chemical Company.
28. Ellinger, R. H., 1972, *Phosphates as Food Ingredients*, The Chemical Rubber Co.
29. deMan, J. M., and Melnychyn, P., Eds., 1971, *Symposium: Phosphates in Food Processing*, Avi Publishing Co.
30. Van Wazer, J. R., 1958, *Phosphorus and Its Compounds*, 2 volumes, Interscience Publishers, Inc.
31. Nomenclature and Classification of Inorganic Phosphates, 1965, Technical Service Release Number One, Stauffer Chemical Company.
32. Van Wazer, J. R., Phosphoric Acids and Phosphates, 1968, in *The Encyclopedia of Chemical Technology*, R. E. Kirk, and D. F. Othmer, Eds., Vol. 15, Interscience Encyclopedia, Inc.
33. Bell, R. N., 1948, *Ind. Eng. Chem. 40*:1464.
34. Huhti, A. L., and Gartaganis, P. A., 1956, *Can. J. Chem. 34*:785.
35. Partridge, E. P., 1949, *Chem. Eng. News 27(4)*:214.
36. Thilo, E., 1958, *Kondensierte Phosphate in Lebensmitteln* (Symposium, 1957, Mainz), Springer-Verlag, p. 7.
37. Specifications for the Identity and Purity of Food Additives and Their Toxicological Evaluation: Emulsifiers, Stabilizers, Bleaching, and Maturing Agents, 1964, World Health Organization Technical Report, Series No. 281, World Health Organization. (May be obtained from National Agency for International Publications, Inc., 317 East 34th Street, New York, NY 10016.)

38. Food Protection Committee, 1966, *Food Chemicals Codex*, No. 1406. National Academy of Sciences–National Research Council.
39. Brout, H., British Patent 883,169, 1961; Diller, I. M. (Henry Brout), U.S. Patent 2,851,359, 1958.
40. Van Wazer, J. R., 1950, *J. Amer. Chem. Soc. 72*:647.
41. Westman, A. E. R., and Crowther, J., 1954, *J. Amer. Ceram. Soc. 37*:420.
42. Westman, A. E. R., and Gartaganis, P. A., 1957, *J. Amer. Ceram. Soc. 40*:293.
43. Bernhart, D. N., and Chess, W. B., 1959, *Anal. Chem. 31*:1026.
44. Corbridge, D. E. C., 1967, *Top. Phosphorus Chem. 3*:57–394.
45. Batra, S. C., 1965, *J. Food Sci. 30*:441.
46. Van Wazer, J. R., and Campanella, D. A., 1950, *J. Amer. Chem. Soc. 72*:655.
47. Van Wazer, J. R., and Callis, C. F., 1958, *Chem. Rev. 58*:1011.
48. Thilo, E., 1955, *Angew. Chem. Int. Ed. Engl. 67*:141.
49. Irani, R. R., and Callis, C. F., 1962, *J. Amer. Chem. Soc. 39*:156.
50. Irani, R. R., and Morgenthaler, W. W., 1963, *J. Amer. Chem. Soc. 40*:283.
51. Kutscher, W., 1961, *Dtsch. Lebensm.-Rundsch. 57(6)*:140.
52. Karl-Kroupa, E., Callis, C. F., and Seifter, E., 1957, *Ind. Eng. Chem. 49*:2061.
53. Bell, R. N., 1947, *Ind. Eng. Chem. 39*:136.
54. Hermann, S., 1947, *Exp. Med. Surg. 5*:160.
55. Behrens, B., and Seelkopf, K., 1932, *Naunyn-Schmiedebergs Arch. Pharmakol. Exp. Pathol. 169*:238.
56. Eichler, O., 1950, *Handbuch der experimentellen Pharmakologie, Erganzungswerk, Bd 10*, p. 363; quoted in World Health Organization Technical Report, Series No. 281, ref. 37.
57. McFarlane, D., 1941, *J. Pathol. Bacteriol. 52*:17.
58. van Esch, G. J., Vink, H. H., Wit, S. J., and van Genderen, H., 1957, *Arneimittel-Forschung. 7*:172.
59. Datta, P. K., Frazer, A. C., Sharratt, M., and Sammons, H. G., 1962, *J. Sci. Food Agr. 13*:556.
60. Zipf, K., 1957, *Arneimittel-Forschung. 7*:445.
61. Gosselin, R. E., and Megirian, R., 1955, *J. Pharm. Exp. Ther. 115*:402.
62. Bonting, S. L., 1952, The Effect of a Prolonged Intake of Phosphoric Acid and Citric Acid in Rats, Ph.D. Thesis, The University of Amsterdam.
63. MacKay, E. M., and Oliver, J., 1935, *J. Exp. Med. 61*:319.
64. Hahn, F., Jacobi, H., and Seifen, E., 1958, *Arneimittel-Forschung. 8*:286.
65. Dymsza, H. A., Reussner, G., and Thiessen, R., 1959, *J. Nutr. 69*:419.
66. Hogan, A. G., Regan, W. O., and House, W. B., 1950, *J. Nutr. 41*:203.
67. House, W. B., and Hogan, A. G., 1955, *J. Nutr. 55*:507.
68. Hahn, F., and Seifen, E., 1959, *Arneimittel-Forschung. 9*:501.
69. Hahn, F., 1961, *Z. Ernährungsw., Suppl. 1*, 55.
70. Hodge, H. C., 1956, Short-term Oral Toxicity Tests of Condensed Phosphates in Rats and Dogs (unpublished mimeographed report), University of Rochester, Rochester, New York.
71. Hodge, H. C., 1959, Chronic Oral Toxicity Studies in Rats of Sodium Tripolyphosphate (unpublished mimeographed report), University of Rochester.
72. Hodge, H. C., 1960, Chronic Oral Toxicity Studies in Rats of Sodium Hexametaphosphate (unpublished mimeographed report), University of Rochester.
73. Hodge, H. C., 1960, Chronic Oral Toxicity Studies in Rats of Sodium Trimetaphosphate (unpublished mimeographed report), University of Rochester.
74. Gassner, K., Kiekebusch, W., and Lang, K., 1957, *Biochem. Z. 328*:485.
75. van Genderen, H., 1961, *Z. Ernährungsw., Suppl. 1*, 32.
76. Lang, K., 1958, *Kondensierte Phosphate in Lebensmitteln* (Symposium, 1957, Mainz), Springer-Verlag, p. 135.
77. Borenstein, M. K., and Schwartz, C., 1948, *J. Nutr. 36*:681.
78. Schreier, K., and Noeller, H. G., 1955, *Arch. Exp. Path. Pharmakol. 227*:199.
79. Mitchell, L. C., 1958, *J. Assoc. Official Agr. Chem. 41*:185.
80. Gosselin, R. E., Rothstein, A., Miller, G. J., and Berke, H. L., 1952, *J. Pharmacol. Exp. Ther. 106*:180.
81. Gotte, H., 1953, *Z. Naturforsch. 8b*:173.
82. Lang, K., Schachinger, L., Karges, O., Blumenberg, F. K., Rossmuller, G., and Schmutte, I., 1955, *Biochem. Z. 327*:118.

83. Lehman, A. J., 1960, *Assoc. Food Drug Officials U.S. Quart. Bull. 24*:45.

84. Gosselin, R. E., Tibdall, C. S., Megirian, R., Maynard, E. A., Downs, W. L., and Hodge, H. C., 1953, *J. Pharmacol. Exp. Ther. 108*:117.

85. Paragraph 121.101(d), revised as of January 1, 1971, *Code of Federal Regulations, Title 21*, U.S. Government Printing Office.

86. Paragraph 318.7(4), revised as of January 1, 1971, *Code of Federal Regulations, Title 9*, U.S. Government Printing Office.

87. Navet, P., 1947, *Tech. Eau (Belg.) 1947(6)*:11; 1947, *Chem. Abstr. 41*:6003g.

88. Hofer, P., 1958, *Kondensierte Phosphate in Lebensmitteln* (Symposium, 1957, Mainz), Springer-Verlag, p. 122.

89. Adler, H. (Victor Chemical Works), U.S. Patent 2,262,745, 1942.

90. Corrie, L. M. (Harry Chin), U.S. Patent 3,231,394, 1966.

91. Roland, C. T., 1942, *Milk Plant Mon. 31(6)*:38, 45.

92. Visser, S. A., 1962, *J. Dairy Sci. 45*:710.

93. Hamm, R., 1956, *Z. Lebensm.-Unters. Forsch. 104*:245.

94. Rose, D., 1969, *Dairy Sci. Abstr. 31*:171.

95. Warning, W. G., and Schille, T. W. (Swann Research), U.S. Patent 1,865,733, 1932; reissue 18,907, 1933.

96. Lubeck, W. D. (Daitz Patents Foundation), U.S. Patent 1,978,040, 1934.

97. Kemmerer, K. S. (Mead Johnson and Co.), U.S. Patent 2,829,056, 1958.

98. Sorgenti, H. A., Nack, H., and Sachsel, G. F. (The Battelle Development Corp.), U.S. Patent 3,035,918, 1962.

99. Heyer, W., U.S. Patent 2,816,032, 1957.

100. Bah, S., and MacKeen, W. E., 1965, *Can. J. Microbiol. 11*:309.

101. Malkov, A. M., and Kormushkina, A. M., U.S.S.R. Patent 135,064, 1961.

102. Kohl, W. F., and Ellinger, R. H. (Stauffer Chemical Company), Netherlands Patent 68,00059, 1968; South African Patent 67,7804, 1968; Belgian Patent 708,863, 1968; French Patent 1,568,002, 1969; British Patent 1,154,079, 1969.

103. Kohl, W. F., and Ellinger, R. H. (Stauffer Chemical Company), Netherlands Patent 68,00060, 1968; Belgian Patent 708,864, 1968; South African Patent 67,7805, 1968; French Patent 1,568,003, 1969.

104. Libbey, A. G., U.S. Patent 2,075,653, 1937.

105. Peynaud, E., 1953, *Inds. Agr. Aliment. (Paris) 70*:559.

106. Freed, E., U.S. Patent 2,995,446, 1961.

107. Skrivanek, M., 1948, *Chemie (Prague) 4*:227.

108. Ohara, I., Nonomura, H., Kushida, T., and Maruyama, C., 1956, *J. Ferment. Technol. 34*:431.

109. Karlson, W. N., U.S. Patent 2,943,939, 1960.

110. 120 Q's and A's about How Phosphates Are Used in Foods, 1966, Food Industry Release No. 2, Stauffer Chemical Company.

111. Ingredients for Food Processors, 1965, Monsanto Company.

112. Diller, I. M. (H. Brout), U.S. Patents 2,851,360 and 2,851,361, 1958.

113. Diller, I. M. (H. Brout), U.S. Patent 2,953,459, 1960.

114. Common, J. L. (General Foods Corp.), U.S. Patent 3,023,106, 1962.

115. DiNardo, A., U.S. Patent 3,224,879, 1965.

116. Stayton, F. J., U.S. Patent 2,341,826, 1944.

117. Hashimoto, K., Miyazaki, S., Japanese Patent 15,550, 1961.

118. Feldheim, W., and Seidemann, J., 1959, *Pharmazie 14*:12.

119. Niwa, S., Katayama, N., and Matsubara, S., 1959, *Vitamins (Kyoto) 18(2)*:492; Niwa, S., Katayama, N., Matsubara, S., and Tanaka, H., 1960, *Vitamins (Kyoto) 20(3)*:497; Niwa, S., and Katayama, N., 1961, *Vitamins (Kyoto) 24(2)*:73; and 1965, *Chem. Abstr. 62*:1002f, 4531g.

120. Stahl, J. E., and Ellinger, R. H., Use of Phosphates in the Cereal and Baking Industry, paper presented at the Phosphates in Food Processing Symposium, 1970, University of Guelph, Guelph, Ontario, Canada; to be published.

121. Kichline, T. P., and Conn, T. F., 1970, *Baker's Dig. 44(8)*:36.

122. Carlin, G. T., 1944, *Cereal Chem. 21*:189.

123. Hanssen, E., and Dodt, E., 1952, *Mikroskopie 7*:2.

124. Jooste, M. E., and Mackay, A. O., 1952, *Food Res. 17*:185.

125. Hunt, F. E., and Green, M. E., 1955, *Food Technol. 9*:241.

126. Bradley, W. B., and Tucker, J. W., Bakery Processes and Leavening Agents, pp. 41–59, 1964, *The Encyclopedia of Chemical Technology*, R. E. Kirk, and D. F. Othmer, Eds., Vol. 3, The Interscience Encyclopedia, Inc.

127. Butterworth, S. W., and Colbeck, W. J., 1938, *Cereal Chem. 15*:475.

128. Burhans, M. E., and Clapp, J., 1942, *Cereal Chem. 19*:196.

129. Carlin, G. T., 1947, *Baker's Dig. 21(8)*:22, 35.

130. Jongh, G., 1961, *Cereal Chem. 38*:140.

131. Ellinger, R. H., The Use of Microscopy and Histology in Development of Cereal Products, paper presented at the Fifth Annual Symposium, St. Louis, Missouri, 1964, sponsored by the Central States Section, American Association of Cereal Chemists.

132. Ellinger, R. H., The Sodium Aluminum Phosphates: Their Properties and Uses as Leavening Acids, paper presented at the Sixth Annual Symposium, St. Louis, Missouri, 1965, sponsored by the Central States Section, American Association of Cereal Chemists.

133. Barackman, R. A., 1931, *Cereal Chem. 8(5)*:423.

134. Van Wazer, J. R., and Arvan, P. G., February 1954, *Northwest. Miller 251*:3a; *ibid.,* March 1954, 5, 17.

135. Conn, J. F., Chemical Leavening, paper presented at the First Annual Symposium, St. Louis, Missouri, 1960, sponsored by the Central States Section, American Association of Cereal Chemists.

136. Parks, J. R., Handleman, A. R., Barnett, J. C., and Wright, F. H., 1960, *Cereal Chem. 37*:503.

137. Modern Leavening with Sodium Aluminum Phosphates, 1966, Food Industry Release No. 4, Stauffer Chemical Company.

138. What Leavening Does for Food—What Phosphates Do in Leavening, 1965, Food Industry Release No. 3, Stauffer Chemical Company.

139. Reiman, H. M., 1965, Properties and Uses of Leavening Agents, Technical Service Leaflet, Stauffer Chemical Company.

140. Tucker, J. W., 1959, *Cereal Sci. Today 4(4)*:91.

141. Joslin, R. P., 1960, *Baker's Dig. 34(10)*:58.

142. Joslin, R. P., and Ziemba, J. V., 1955, *Food Eng. 27(9)*:59, 184.

143. Leaven Right with Levn-Lite and Pan-O-Lite, Sodium Aluminum Phosphate Sales Brochure, The Monsanto Company.

144. Upgrade Your Self-rising Flours with Stabil-9, Sales Brochure, The Monsanto Company.

145. Huff, A. E. (Monsanto Chemical Company), U.S. Patent 2,314,090, 1943; Russihyvili, R. (Bird and Sons, Ltd.), British Patent 608,708, 1948.

146. Hurka, R. J. (Monsanto Chemical Company), U.S. Patent 2,366,857, 1945.

147. McDonald, G. A. (Victor Chemical Works), U.S. Patents 2,550,490 and 2,550,491, 1951.

148. Dyer, J. N. (Stauffer Chemical Company), U.S. Patent 2,995,421, 1961.

149. Lauck, R. M., and Tucker, J. W. (Stauffer Chemical Company), U.S. Patent 3,041,177, 1962.

150. Vanstrom, R. E. (Stauffer Chemical Company), U.S. Patents 3,223,479 and 3,223,480, 1965; Belgian Patents 627,509 and 627,511, 1963.

151. Chemische Fabrik Budenheim Aktiengesellschaft, British Patent 390,743, 1933.

152. Griffith, E. J. (Monsanto Chemical Company), U.S. Patent 2,774,672, 1956.

153. Conn, J. F., 1965, *Baker's Dig. 39(4)*:66, 70.

154. Handleman, A. R., Conn, J. F., and Lyons, J. W., 1961, *Cereal Chem. 38*:294.

155. Leavening Indicator Chart, Stauffer Chemical Company.

156. Tucker, J. W. (Stauffer Chemical Company), U.S. Patent 3,109,738, 1963.

157. Armstrong, L., and Willoughby, L. B. (Ballard and Ballard Co.), U.S. Patent 2,478,618, 1949.

158. Baker, J. S., and Lindeman, C. G. (Procter and Gamble Co.), U.S. Patent 3,297,449, 1967.

159. Erekson, A. B., and Duncan, R. E. (The Borden Co.), U.S. Patent 2,942,988, 1960; Matz, S. A. (The Borden Co.), U.S. Patent 3,397,064, 1968.

160. Barackman, R. A., and Bailey, C. H., 1927, *Cereal Chem. 4*:400.

161. Lannuier, G. L., M.S. Thesis, August 1961, Florida State University; Lannuier, G. L., and Bayfield, E. G., 1961, *Baker's Dig. 35(6)*:34, 40, 42, 88; Bayfield, E. G., and Lannuier, G. L., 1962, *Baker's Dig. 36(1)*:34, 77; Bayfield, E. G., and Lannuier, G. L., 1962, *Baker's Dig. 36(6)*:57, 92.

162. Takebayashi, Y., 1963, *Nippon Shokuhin Kogyo Gakkaishi 10(9)*:382.

163. Carter, J. E., and Hutchinson, J. B., 1965, *Chem. Ind. (Lond.) 1965(42)*:1765.

164. Muhler, J. C. (Indiana University Foundation), U.S. Patent 3,467,529, 1969; Canadian Patent 823,140, 1969.
165. Schohl, A. T., 1939, *Mineral Metabolism*, American Chemical Society Monograph Series No. 82, Reinhold Publishing Corporation.
166. The Importance of Calcium and Phosphorus Supplements in Foods, Technical Service Bulletin, Stauffer Chemical Company. (Out of print.)
167. Anon., 1944, *Bull. Natl. Res. Counc. 40*:38.
168. Anon., September 15, 1969, *Food Chem. News*, 24.
169. Nizel, A. E., Keating, N., Sundstrom, C., and Harris, R. S., 1958, *J. Dent. Res. 37*:35.
170. Buttner, W., and Muhler, J. C., 1958, *J. Dent. Res. 37*:860.
171. McClure, F. J., and Muller, A., Jr., 1959, *J. Amer. Dent. Assoc. 58*:36.
172. Stralfors, A., 1964, *J. Dent. Res. 43*(suppl.):1137.
173. McClure, F. J., and Muller, A., Jr., 1959, *J. Dent. Res. 38*:776.
174. McClure, F. J., 1960, *J. Nutr. 72*:131.
175. Stookey, G. K., and Muhler, J. C., 1966, *J. Dent. Res. 45*:856.
176. Stookey, G. K., Carroll, R. A., and Muhler, J. C., 1967, *J. Amer. Dent. Assoc. 74*:752.
177. Carroll, R. A., Stookey, G. K., and Muhler, J. C., 1968, *J. Amer. Dent. Assoc. 76*:564.
178. Luoma, H., Niska, K., and Tustola, L., 1968, *Arch. Oral Biol. 13(11)*:1343; 1969, *Oral Res. Abstr. 4*:779.
179. Averill, H. M., and Bibby, B. G., 1964, *J. Dent. Res. 43*(suppl. 6):1150.
180. Nizel, A. E., and Harris, R. S., 1964, *J. Dent. Res. 43*(suppl. 6):1123.
181. Council on Foods and Nutrition, 1968, *J. Amer. Med. Assoc. 203*:407.
182. Crosby, W. H., 1969, *J. Amer. Med. Assoc. 208*:347.
183. Jacobs, A., and Greenman, D. A., 1969, *Brit. Med. J. 1*:673.
184. Finch, C. A., 1965, *Nutr. Rev. 23*:129.
185. Proposals to Increase the Levels of Iron in Enriched Flour and Bread, submitted to FDA by the Millers' National Federation and the American Bakers' Association, November 5, 1969; Lampman, H. H., Memorandum of November 17, 1969, Millers' National Federation.
186. Nakamura, F. I., and Mitchell, H. H., 1943, *J. Nutr. 25*:39.
187. Street, H. R., 1943, *J. Nutr. 26*:187.
188. Freeman, S., and Burrill, N. W., 1945, *J. Nutr. 30*:293.
189. Steinkamp, R., Dubach, R., and Moore, C. V., 1955, *Arch. Intern. Med. 95*:181.
190. Fritz, J. C., *Proceedings of Medicine and Nutrition Conference,* March 20–21, 1969, p. 1.
191. Fritz, J. C., 1969, *Fed. Proc. 28*:692.
192. Iron in Flour, 1968, *Reports on Public Health and Medical Subjects,* No. 117, Ministry of Health, London, Her Majesty's Stationery Office.
193. Hinton, J. J. C., Carter, J. E., and Moran, T., 1967, *J. Food Technol. 2*:129.
194. Hinton, J. J. C., and Moran, T., 1967, *J. Food Technol. 2*:135.
195. White, H. S., and White, P. L., 1968, *Food Nutr. News 39(7)*:1, 4.
196. White, H. S., Should Food Iron Be Increased?, paper presented at Symposium on Nutrition and Food Technology, Chicago, February 12, 1969.
197. Anon., 1970, *Cereal Sci. Today 15*:121.
198. Anon., April 1, 1970, *Fed. Reg.*; 1970, *Food Eng. 42(5)*:22.
199. Anon., 1971, *Chem. Eng. News 49(6)*:52.
200. Billings, H. J. (Arthur D. Little, Inc.), U.S. Patent 2,131,881, 1938.
201. Rao, P. V. S., Ananthachar, T. K., and Desikachar, H. S. R., 1964, *Ind. J. Technol. 2*:417; 1965, *Chem. Abstr. 62*:13768d.
202. Tanaka, M., and Yukami, S. (Lion Dentifrice Co., Ltd.), U.S. Patent 3,484,249, 1969; 1970, *Food Sci. Technol. Abstr. 2*:4M268.
203. Malkov, A. M., Deeva, V. E., and Drizina, T. S., 1955, *Tr. Leningr. Tekhnol. Inst. Pishch. Prom-st. 12*:100; 1957, *Zhur. Khim., Biol. Khim.*, No. 11509.
204. Stauffer Chemical Company, Belgian Patent 634,380, 1963; Stauffer Chemical Company, British Patent 1,019,508, 1966.
205. Payumo, E. M., Briones, P. R., Banzon, E. A., and Torres, M. L., 1969, *Philipp. J. Nutr. 22*:216; 1970, *Food Sci. Technol. Abstr. 2*:11M1096.
206. Thames Rice Milling Co., Ltd., British Patent 650,192, 1951.
207. Keller, H. M. (General Mills, Inc.), U.S. Patent 2,909,431, 1959.
208. Madrazo, M. G., and Cortina, G. A. (Process Millers, Inc.), U.S. Patent 3,117,868, 1964.

209. Sommer, H. M., and Hart, E. B., 1926, *Wisc. Agr. Exp. Sta. Res. Bull. 67*:1.
210. Vujicic, I., Batra, S. C., and deMan, J. M., 1967, *J. Agr. Food Chem. 15*:403.
211. Vujicic, I., deMan, J. M., and Woodrow, I. L., 1968, *Can. Inst. Food Technol. J. 1*:17.
212. Puri, B. R., Kalra, K. C., and Toteja, K. K., 1965, *Ind. J. Dairy Sci. 18(4)*:126; 1967, *Chem. Abstr. 66*:1652r.
213. Tessier, H., and Rose, D., 1958, *J. Dairy Sci. 41*:351.
214. Kiermeier, F., 1952, *Z. Lebensm.-Unters. Forsch. 95*:85.
215. Pyne, G. T., and McGann, T. C. A., 1960, *J. Dairy Res. 27*:9.
216. Zittle, C. A., Della Monica, E. S., Rudd, R. K., and Custer, J. H., 1958, *Arch. Biochem. Biophys. 76*:342.
217. Zittle, C. A., 1966, *J. Dairy Sci. 49*:361.
218. Morr, C. V., and Kenkare, D. B., 1964, *J. Dairy Sci. 47*:673.
219. Odagiri, S., and Nickerson, T. A., 1964, *J. Dairy Sci. 47*:1306; Odagiri, S., and Nickerson, T. A., 1965, *J. Dairy Sci. 48*:19.
220. Ellinger, R. H., *et al.*, 1969–1971, various studies.
221. Feagen, J. T., Griffin, A. T., and Lloyd, G. T., 1966, *J. Dairy Sci. 49*:1010.
222. Zittle, C. A., and Pepper, L., 1958, *J. Dairy Sci. 41*:1671.
223. Rose, D., 1961, *J. Dairy Sci. 44*:430.
224. Rose, D., 1963, *Dairy Sci. Abstr. 25*:45.
225. Rose, D., 1962, *J. Dairy Sci. 45*:1305.
226. Kenkare, D. B., Morr, C. V., and Gould, I. A., 1964, *J. Dairy Sci. 47*:947.
227. Edmondson, L. F., Landgrebe, A. R., Sadler, A. M., and Walter, H. E., 1965, *J. Dairy Sci. 48*:1597.
228. Tessier, H., and Rose, D., 1961, *J. Dairy Sci. 44*:1238.
229. Puri, B. R., Arora, K., and Toteja, K. K., 1969, *Ind. J. Dairy Sci. 22*:85; *ibid.*, *22*:155; 1970, *Food Sci. Technol. Abstr. 2*:11P1459.
230. Dusek, B., Semjan, S., and Kazimir, L., 1970, *18th Intern. Dairy Congr. (Sydney) 1E*:92; 1970, *Food Sci. Technol. Abstr. 2*:11P1539.
231. Fox, K. K., Harper, M. K., Holsinger, V. H., and Pallansch, M. J., 1964, *J. Dairy Sci. 47*:179.
232. Vujicic, I., and deMan, J. M., 1968, *J. Inst. Can. Technol. Aliment. 1*:171.
233. Lauck, R. M., and Tucker, J. W., 1962, *Cereal Sci. Today 7(9)*:314, 322.
234. Briggs, D. R., 1940, *J. Biol. Chem. 134*:261.
235. Katchman, B. J., and Van Wazer, J. R., 1954, *Biochim. Biophys. Acta 14*:445.
236. Perlmann, G. E., 1941, *J. Biol. Chem. 137*:707.
237. Horvath, A. A., 1946, *Ind. Eng. Chem., Anal. Ed. 18*:229.
238. Lyons, J. W., and Siebenthal, C. D., 1966, *Biochim. Biophys. Acta 26*:174.
239. Leach, A. A., *GGRA Literature Review*, Series D-8, published by the Gelatin and Glue Research Association, Birmingham, England.
240. Leviton, A., 1964, *J. Dairy Sci. 47*:670.
241. Herreid, E. O., and Wilson, H. K., 1963, *Manuf. Milk Prod. J. 54*:14.
242. Morr, C. V., 1967, *J. Dairy Sci. 50*:1038.
243. Hall, R. E. (Hall Laboratories, Inc.), U.S. Patent 2,064,110, 1936. Schwartz, C., U.S. Patent 2,135,054, 1938.
244. Kawanishi, G., 1963, *Nippon Shokuhin Kogyo Gakkaishi 10(4)*:123; 1965, *Chem. Abstr. 63*:1147g.
245. Wouters, O. J. (A.P.C.), U.S. Patent 2,319,362, 1943.
246. Jackson, J. M. (American Can Co.), U.S. Patent 2,396,265, 1946.
247. Roland, C. T. (Calgon, Inc.), U.S. Patent 2,665,212, 1954.
248. Malkov, A. M., Trippel, A. I., and Kirichkova, G. A., 1964, *Sb. Tr. Leningr. Inst. Sov. Torgovli. 1964(23)*:112.
249. North, C. E., U.S. Patent 2,550,288, 1951.
250. Meyer, A., and Weck, G. (Hagen Chemicals & Controls, Inc.), U.S. Patent 2,977,232, 1961.
251. Corbin, E. A., and Long, J. E. (Nopco Chemical Co.), U.S. Patent 3,340,066, 1967.
252. Rothwell, J., 1968, *J. Soc. Dairy Tech. 21*:136.
253. Wilson, H. K., and Herreid, E. O., 1969, *J. Dairy Sci. 52*:1229.
254. Paragraph 18.520(a)(1)(i), 1971, *Code of Federal Regulations, Title 21,* revised as of January 1, 1971, U.S. Government Printing Office.
255. Edmondson, L. F., 1959, *J. Dairy Sci. 42*:910.
256. Martynova, K., 1962, *Molochn. Prom-st.' 23(11)*:17.
257. Shtal'berg, S., and Radaeva, I., 1966, *Proc. 17th Int. Dairy Congr., Munich 5*:153.

258. Muller, L. L., 1963, *Aust. J. Dairy Technol. Suppl.*, Technical Publication No. 12, p. 25; 1964, *Chem. Abstr. 60*:13791e.

259. Seehafer, M. E., The Development and Manufacture of Sterilized Milk Concentrate, 1967, *FAO Agricultural Series*, No. 72.

260. Leviton, A., and Pallansch, M. J., 1962, *J. Dairy Sci. 45*:1045.

261. Leviton, A., Anderson, H. A., Vettel, H. E., and Vestal, J. H., 1963, *J. Dairy Sci. 46*:310.

262. Wilson, H. K., Vetter, J. L., Sasago, K., and Herreid, E. O., 1963, *J. Dairy Sci. 46*:1038.

263. Leviton, A., and Pallansch, M. J. (U.S.A., per U.S.D.A.), U.S. Patent 3,119,702, 1964.

264. Herreid, E. O., and Wilson, H. K., 1963, *Manuf. Milk Prod. J. 54(1)*:14, 31.

265. Swanson, A. M., and Calbert, H. E., 1963, *Food Technol. 17*:721, 727.

266. Magnino, P. J., Jr., Ph.D. Thesis, University of Wisconsin, January 1968; University Microfilms, Inc., No. 68-5, 332, Ann Arbor, MI.

267. Bixby, H. H., and Swanson, A. M., 1969, *J. Dairy Sci. 52*:892.

268. Kadan, R. S., and Leeder, J. G., 1966, *J. Dairy Sci. 49*:709.

269. Stewart, A. P. (Nodaway Valley Foods, Inc.), U.S. Patent 3,348,955, 1967.

270. Bolt, M. G., and Kastelic, J., 1968, *J. Dairy Sci. 51*:693.

271. Calbert, H. E., and Jackson, H. C., 1960, *J. Dairy Sci. 43*:846.

272. A New Marketing Opportunity—2.5 Billion Quarts of Beverage Quality Canned Milk—As a Minimum, 1965, United States Steel Corporation.

273. Continental Can Company and United States Steel Corporation, 1965, Summary Report: One Week Family Test of Commercially Produced, Aseptically Canned, Sterilized Milk Concentrate, United States Steel Corporation.

274. Alwes, M. L., and Swanson, A. M., 1969, *J. Dairy Sci. 52*:891.

275. Nakanishi, T., and Itoh, T., 1969, *Agr. Biol. Chem. 33*:1270; Nakanishi, T., and Itoh, T., 1969, *J. Agr. Chem. Soc. Japan 43*:725; 1970, *Food Sci. Technol. Abstr. 2*:2P215, 3P385.

276. Tamate, R., and Ohtaka, F., 1969, *Jap. J. Dairy Sci. 18(2)*:A37; 1970, *Food Sci. Technol. Abstr. 2*:6P773.

277. Hoff, J. E., Sunyach, J., Procter, B. E., and Goldblith, S. A., 1960, *Food Technol. 14*:24, 27.

278. Leviton, A., Vestal, J. H., Vettel, H. E., and Webb, B. H., 1966, *Proc. 17th Int. Dairy Congr., Munich, E*:133.

279. Johnson, C. E., and Winder, W. C., May 1969, *Quick Frozen Foods 31(10)*:43, 127.

280. Kennedy, M. H., and Castagna, M. P. (Standard Brands, Inc.), U.S. Patent 2,607,692, 1952; Canadian Patent 501,696, 1954; British Patent 709,931, 1954.

281. Breivik, O. N., Slupatchuk, W., Carbonell, R. J., and Weiss, G. (Standard Brands, Inc.), U.S. Patent 3,231,391, 1966.

282. Clausi, A. S. (General Foods Corp.), U.S. Patent 2,801,924, 1957; British Patent 741,076, 1955; Canadian Patent 565,651, 1958.

283. Charie, H. J., and Savage, J. P. (Thomas J. Lipton, Inc.), U.S. Patent 2,927,861, 1960.

284. Moirano, A. L. (Marine Colloids, Inc.), U.S. Patent 3,443,968, 1969; Canadian Patent 847,287, 1970.

285. Pfrengle, O., and Zboralski, U. (Chemische Fabrik Budenheim Rudolf A. Oetker K.-G), German Patent 1,129,813, 1962.

286. Partridge, E. P. (Hall Laboratories, Inc.), U.S. Patent 2,582,353, 1952.

287. McGowan, J., 1970, *Food Prod. Devel. 4(5)*:16, 18.

288. Roland, C. T. (Calgon, Inc.), U.S. Patent 2,665,212, 1954.

289. Freund, E. H., and Danes, E. N., Jr. (National Dairy Products Corp.), U.S. Patent 2,957,770, 1960.

290. Salzberg, H. K. (The Borden Co.), U.S. Patent 2,181,003, 1939; British Patent 503,841, 1939.

291. Wouters, O. J. (Alien Property Custodian), U.S. Patent 2,319,362, 1943.

292. Curry, C. C. (Research and Development Corporation), U.S. Patent 2,341,425, 1944.

293. Lindewald, T. W., and Kimstad, S. G. (Swenska Mjölk-Produkter Aktiebolag), U.S. Patent 2,468,677, 1949.

294. Lewis, M. A., Marcelli, V., and Watts, B. M., 1953, *Food Technol. 7*:261.

295. Kempf, C. A., and Blanchard, E. L. (Golden State Company, Ltd.), U.S. Patent 2,645,579, 1953.

296. Kumetat, K. (Commonwealth Scientific and Industrial Research Organization), U.S. Patent 2,739,898, 1956.

297. Hayes, J. F., Dunkerley, J., and Muller, L. L., 1969, *Aust. J. Dairy Technol. 24(2)*:69; 1969, *Food Sci. Technol. Abstr. 1*:12P1299.

298. Prodanski, P., 1970, *Molochn. Prom-st.' 31(3)*:43; 1970, *Food Sci. Technol. Abstr. 2*:7P916.

299. Parsons, C. H., and Scott, E. C. (Industrial Patents Corp.), U.S. Patent 2,279,205, 1942.

300. Frazeur, D. R., 1959, *Ice Cream Field 73(3)*:25.

301. Weinstein, B. (Crest Foods, Inc.), U.S. Patent 2,856,289, 1958.

302. Keeney, P. G., 1962, *J. Dairy Sci. 45*:430.

303. Anon., September 6, 1967, *Fed. Reg. 32*:12750.

304. Wallander, J. F., 1969, *Diss. Abstr. Int. B. Sci. Eng. 29*(12, Part 1):4523; 1969, *Food Sci. Technol. Abstr. 2*:1P68.

305. Amundson, C. H., 1967, *Amer. Dairy Rev. 29(7)*:22, 96.

306. Zboralski, U. (Johann A. Benckiser G.m.b.H.), German Patent 1,008,220, 1957.

307. Stein, W. (Molkerei Everswinkel Inh. Josef Roberg), German Patent 1,097,380, 1962.

308. Ward, P. J., Johnson, J. D., and Robertson, R. G. (National Dairy Products Corporation), U.S. Patent 3,061,442, 1962.

309. Arena, J. M., 1970, *Nutr. Today 5(4)*:2.

310. Gordon, W. G. (Smith Kline & French Laboratories), U.S. Patent 2,377,624, 1945.

311. McKee, J. I., and Tucker, J. W. (Stauffer Chemical Company), U.S. Patent 3,269,843, 1966; British Patent 1,045,860, 1967.

312. Wingerd, W. H. (The Borden Co.), Canadian Patent 790,580, 1968.

313. Hartman, G. H., Jr., Ph.D. Thesis, University of Wisconsin, 1966.

314. Hartman, G. H., Jr., and Swanson, A. M., 1966, *J. Dairy Sci. 49*:697.

315. Protolac®-Sodium, Technical Data Sheet 402, and Protolac®-Calcium, Technical Data Sheet, The Borden Co.

316. Wingerd, W. H. (The Borden Co.), U.S. Patent 3,332,782, 1967.

317. Wingerd, W. H. (The Borden Co.), U.S. Patent 3,356,507, 1967.

318. Ellinger, R. H. (Stauffer Chemical Company), U.S. Patent 3,563,761, 1971; British Patent 1,145,638, 1969; French Patent 1,549,831, 1968; Netherlands Patent 67,16873, 1968; Belgian Patent 708,289, 1968; Mexican Patent 99,510, 1969.

319. Broadhead, S. A. (Stauffer Chemical Company), British Patent 1,162,735, 1969; Belgian Patent 708,279, 1968; Netherlands Patent 67,17398, 1968; French Patent 1,549,751, 1968; Mexican Patent 99,256, 1967.

320. Endo, N., Japanese Patent 8531 ('62), 1962; Japanese Patent 9271 ('62), 1962; 1963, *Chem. Abstr. 58*:10666c.

321. Ellinger, R. H., and Schwartz, M. G. (Stauffer Chemical Company), French Patents 2,012,982 and 2,012,983, 1970; Belgian Patents, 736,123 and 736,124, 1970.

322. Holland, R. F.; and Bandler, D. K., 1967, *Cornell Food Plant Tech. Serv. Lett. 1(4)*:2.

323. McCormick, R. D., and Beck, K. M., 1968, *Food Prod. Devel. 2(1)*:26.

324. Moede, H. H., May 1968, *Dairy Situation*, DS-320, Economic Research Service, U.S. Department of Agriculture, p. 30.

325. Kozin, N. I., and Rodionova, I. F., 1964, *Izv. Vyssh. Uchebn. Zaved. Pishch. Teknol. 1964(5)*:51; 1965, *Chem. Abstr. 62*:9691d.

326. Sabharwal, K., and Vakaleris, D. G., 1969, *J. Dairy Sci. 52*:891.

327. Nutritionally Balanced Non-Dairy Bases, 1968, *Food Prod. Devel. 2(5)*:99.

328. Kempf, C. A., and Blanchard, E. L. (Golden State Company, Ltd.), U.S. Patent 2,645,579, 1953.

329. Brochner, H. S., U.S. Patent 3,241,975, 1966.

330. The Use of Carrageenan in Liquid Coffee Whiteners, June 1966, Technical Bulletin No. 255, Marine Colloids, Inc.

331. Miller, D. E., 1968, *J. Dairy Sci. 51*:1330.

332. Knightly, W. H., 1969, *Food Technol. 23*:171.

333. Anon., February 1966, *Food Eng. 38*:135.

334. Paramount C in Imitation Coffee Cream, February 8, 1963, Durkee Famous Foods.

335. Coffee Whitener, DS-214-C, Durkee Famous Foods.

336. Coffee Whitener, DS-237-C, Durkee Famous Foods.

337. Test Formulas for Coffee Whiteners, 1968, Stauffer Chemical Company.

338. Powdered Coffee Whitener, August 18, 1964, Stauffer Chemical Company.

339. Miller, D. E., and Ziemba, J. V., August 1966, *Food Eng. 38*:97, 105.

340. Mason, R. D., and Justesen, A. C., U.S. Patent 2,407,027, 1946.

341. Noznick, P. P., and Tatter, C. W. (Beatrice Foods Co.), U.S. Patent 3,098,748, 1963.

342. Nash, N. H., and Cheselka, G., June 1966, *Baker's Dig. 40*:58.

343. Thalheimer, W. G., 1968, *Food Eng. 40(5)*:112.

344. Knightly, W. H., 1968, *Food Technol. 22*:731.

345. Guidelines to the Formulation of Whipped Toppings, April 1968, Atlas Chemical Industries, Inc.

346. Chip-Dip Base, Data Sheet, issued by Crest Foods Co., Inc.

347. Little, L. L. (Battelle Development Corp.), U.S. Patent 3,391,002, 1968.

348. Frozen Dessert Base, Non-Dairy, Data Sheet, issued by Crest Foods Co., Inc.

349. Frozen Dessert (Containing No Milk Products), LDS-288-D, Durkee Famous Foods.

350. Ryberg, J. R., 1968, *Food Prod. Devel. 2(5)*:60.

351. Webster, D. P., U.S. Patent 3,310,406, 1967.

352. Terada, K., and Yoshida, K. (Asahi Electro-Chemical Co., Ltd.), Japanese Patent 9261, 1962; 1963, *Chem. Abstr. 58*:11900d.

353. Paragraph 19.530(b)(5), 1971, *Code of Federal Regulations, Title 21*, revised as of January 1, 1971, U.S. Government Printing Office.

354. Bristol, D. C., and Martin, J. H., 1970, *J. Dairy Sci. 53*:1381.

355. Continuous Cottage Cheese Process, Form 744-8046, CP Division, St. Regis.

356. Kielsmeier, E. W. (Swift & Company), U.S. Patent 2,851,363, 1958.

357. Vorob'ev, A. I., 1962, *Proc. 16th Int. Dairy Congr., Copenhagen, Sect. B*:576; 1962, *Dairy Sci. Abstr. 24*:2778.

358. Anon., May 1968, *Food Eng. 40*:15.

359. Olson, N. F., 1968, *Food Prod. Devel. 2(5)*:90, 92, 94, 96, 99.

360. Batra, S. C., and deMan, J. M., 1964, *J. Dairy Sci. 47*:673.

361. deMan, J. M., and Batra, S. C., 1964, *J. Dairy Sci. 47*:954.

362. Odagiri, S., and Nickerson, T. A., 1964, *J. Dairy Sci. 47*:673.

363. deMan, J. M., 1966, *Proc. 17th Int. Dairy Congr. 2*:365; 1967, *Chem. Abstr. 67*:2210a.

364. Samuelsson, E. G., 1969, *Nord. Mejeri-Tidskr. 35(1)*:9; 1969, *Food Sci. Technol. Abstr. 1*:5P497.

365. Shew, D. I., 1949, *Nature 164*:492.

366. Potter, N. N., and Nelson, F. E., 1952, *J. Bacteriol. 64*:113.

367. Babel, F. J., 1958, *J. Dairy Sci. 41*:697.

368. Hargrove, R. E., 1959, *J. Dairy Sci. 42*:906.

369. Olson, H. C., 1960, *J. Dairy Sci. 43*:439.

370. Hargrove, R. E., McDonough, F. E., and Tittsler, R. P., 1961, *J. Dairy Sci. 44*:1799.

371. Galesloot, T. E., and Hassing, F., 1962, *Versl. Ned. Inst. Zuivelonerzoek 1962(70)*:117.

372. Kadis, V. W., and Babel, F. J., 1962, *J. Dairy Sci. 45*:486.

373. Hargrove, R. E. (U.S.A.), U.S. Patent 3,041,248, 1962.

374. Waters, H. L. H., U.S. Patent 1,910,195, 1933.

375. Schulz, M. E. (Johann A. Benckiser G.m.b.H.), British Patent 692,055, 1953.

376. Conochie, J., and Sutherland, B. J., 1961, *Aust. J. Dairy Technol. 16*:227.

377. Kraft, J. L., U.S. Patent 1,186,524, 1916.

378. Sommer, H. H., and Templeton, H. L., 1940, *Dairy Ind. 5*:185, 218.

379. Van Slyke, L. L., and Price, W. V., 1952, *Cheese*, Orange Judd Publishing Company, p. 355.

380. Irvine, D. M., and Price, W. V., April 1955, *Process Cheese Abstr.*, Department of Dairy and Food Industries, University of Wisconsin.

381. Garstin, G. H. (Phoenix Cheese Company), U.S. Patent 1,368,624, 1921.

382. Ney, K. H., and Garg, O. P., 1970, *Fette Seifen Anstrichm. 72*:279; 1970, *Food Sci. Technol. Abstr. 2*:11P1489.

383. Rohlfs, H. A. *et al.* (Benckiser-Knapsack G.m.b.H.), British Patent 1,180,716, 1970; 1970, *Food Sci. Technol. Abstr. 2*:7P864.

384. Stauffer Chemical Company, British Patent 1,189,003, 1970.

385. Lauck, R. M., Vanstrom, R. E., and Tucker, J. W. (Stauffer Chemical Company), U.S. Patent 3,097,949, 1963.

386. Kichline, T. P., Stahlheber, N. E., and Vetter, J. L. (Monsanto Company), U.S. Patent 3,337,347, 1967.

387. Paragraph 19.750(c), 1971, *Code of Federal Regulations, Title 21*, revised as of January 1, 1971, U.S. Government Printing Office.

388. Bolanowski, J. P., and Ziemba, J. V., 1966, *Food Eng. 38(11)*:86.

389. Templeton, H. L., and Sommer, H. H., 1930, *J. Dairy Sci. 13*:203.

390. *Ibid.*, 1932, *15*:29.

391. Templeton, H. L., and Sommer, H. H., 1934, *Wis. Agr. Exp. Sta. Bull. 428*:43 (Annual Report, 1932-33).

392. Templeton, H. L., and Sommer, H. H., 1936, *J. Dairy Sci. 19*:561.

393. *Ibid.,* 1937, *20:*231.

394. Templeton, H. L., Sommer, H. H., and Stewart, J. K., 1937, *Wis. Agr. Exp. Sta. Bull. 439:*66 (Annual Report, 1937, Part 1).

395. Templeton, H. L., 1937, *J. Dairy Sci. 20:*470.

396. Templeton, H. L., Sommer, H. H., and Stewart, J. K., 1938, *Natl. Butter Cheese J. 29(2):*16; 1938, *Chem. Abstr. 32:*17967.

397. Palmer, H. J., and Sly, W. H., 1944, *J. Soc. Chem. Ind. 63:*363.

398. Holtstorff, A. F., Mularz, V., and Traisman, E., 1951, *J. Dairy Sci. 34:*486.

399. Kiermeier, F., and Mohler, K., 1960, *Z. Lebensm.-Unters. Forsch. 112:*175.

400. Kiermeier, F., 1962, *Z. Lebensm.-Unters. Forsch. 118:*128.

401. Roesler, H., 1966, *Milchwissenschaft 21(2):*104.

402. Scharpf, L. G., Jr., and Kichline, T. P., 1967, *J. Agr. Food Chem. 15:*787.

403. Rank, B., and Siepenlist, E., 1941, *Dtsch. Molkerei-Ztg. 62:*1036; 1943, *Chem. Abstr. 37:*6045.

404. Blanchard, J. F., 1949, *Food Ind. 21:*51; 1949, *Chem. Abstr. 43:*3534c.

405. Ovchinnikov, A., and Alyamovskii, 1952, *Molochn. Prom-st.' 13(10):*21; 1953, *Chem. Abstr. 47:*1860h.

406. Morris, H. A., Manning, P. B., and Jenness, R., 1969, *J. Dairy Sci. 52:*900.

407. Sommer, H. H., and Templeton, H. L., 1941, *Dairy Ind. 6:*14.

408. Kozin, N. I., and Starodubtsev, N. V., 1959, *Sb. Nauchn. Rab. Moskov. Inst. Narod. Khoz. 1959(16):*202.

409. Hayter, T. C., Olson, N. F., and Price, W. V., 1969, *J. Dairy Sci. 52:*900.

410. Pikarr, H. R. (Purex Corporation, Ltd.), U.S. Patent 3,211,659, 1965.

411. Chin, R. G. L., and Redfern, S. (Standard Brands, Inc.), Canadian Patent 723,609, 1965; U.S. Patent 3,383,221, 1968.

412. Anon., March 19, 1966, *Fed. Reg. 31:*4677; *ibid.,* 1968, *33:*8225.

413. Hall, G. O. (Hall Laboratories, Inc.), U.S. Patent 2,445,879, 1948.

414. Bellamy, A. J. (Egg Patents, Ltd.), British Patent 570,268, 1945.

415. Tongur, V. S., 1947, *J. Appl. Chem. 20:*1191.

416. Sourby, J. C., Kohl, W. F., and Ellinger, R. H. (Stauffer Chemical Company), U.S. Patent 3,561,980, 1971; Belgian Patent 726, 348, 1969; Netherlands Patent 69,00026, 1969; New Zealand Patent 155,019, 1969; British Patent 1,219,519, 1971; French Patent 1,600,264, 1970.

417. Kohl, W. F., and Ellinger, R. H. (Stauffer Chemical Company), U.S. Patent 3,549,388, 1970; Belgian Patent 734,812, 1969; South African Patent 69/3784, 1969; Argentine Patent, 176,511, 1970.

418. Kothe, H. J. (Standard Brands, Inc.), Canadian Patent 503,484, 1954.

419. Finucane, T. P., and Mitchell, W. A. (General Foods Corp.), U.S. Patent 2,671,730, 1954.

420. Finucane, T. P. (General Foods Corp.), U.S. Patent 2,916,379, 1959.

421. Chang, P. K., Ph.D. Thesis, University of Wisconsin, 1969.

422. Chang, P. K., Powrie, W. D., and Fennema, O., 1970, *Food Technol. 24:*63.

423. Kohl, W. F., Ellinger, R. H., and Sourby, J. C., A New Process for Pasteurizing Egg Whites, paper presented before the 30th Annual Meeting of the Institute of Food Technologists, San Francisco, California, May 26, 1970.

424. Kohl, W. F., Sourby, J. C., and Ellinger, R. H. (Stauffer Chemical Company), U.S. Patent 3,520,700, 1970; Netherlands Patent 69,00027, 1969; Belgian Patent 726,347, 1969; New Zealand Patent 154,947, 1968.

425. Rousseau, P. M., 1955, *Oleagineux 10:*183.

426. Lever Bros. Co., British Patent 661,703, 1951.

427. Sullivan, F. E. (De Laval Separator Co.), U.S. Patent 2,702,813, 1955; British Patent 736,885, 1952.

428. Hayes, L. P., and Wolff, H. (A. E. Staley Mfg. Co.), U.S. Patent 2,881,195, 1959.

429. Rini, S. J. (Lever Bros. Co.), U.S. Patent 2,507,184, 1950.

430. Appleton, L. F. (Sherwin-Williams Company), Canadian Patent 330,967, 1933.

431. Eckey, E. W. (The Procter & Gamble Company), U.S. Patent 1,993,152, 1935.

432. Johnson, J. Y., British Patent 337,336, 1932.

433. Colomb, P., 1951, *Lack. Farben. Chem. 5(9/12):*14.

434. Taussky, L., U.S. Patent 2,654,766, 1953.

435. Braae, B., and Nyman, M., 1957, *Nordisk Symposium om Harskning Af Fedstoffer 2:*231.

436. Bergman, L. O. (Aktiebolaget Pellerins Margarinfabrik), German Patent 1,135,600, 1962; Belgian Patent 595,217, 1961.

437. James, E. M. (Sharples Specialty Company), U.S. Patent 2,115,668, 1938.

438. Hempel, H. (Gorton-Pew Fisheries, Ltd.), U.S. Patent 2,226,036, 1941.

439. Clayton, B. (Refining, Inc.), U.S. Patent 2,292,822, 1942.

440. Mitani, M., U.S. Patent 3,008,972, 1961.

441. Coith, H. S., and Votaw, V. M. (The Procter & Gamble Company), U.S. Patents 2,061,121 and 2,061,122, 1936.

442. Irmen, P., British Patent 503,607, 1939.

443. Beck, T. M., and Klein, G. I. (Victor Chemical Works), U.S. Patent 2,214,520, 1940.

444. Roy, A. C., 1954, *J. Sci. Ind. Res. (India)* 13B:376.

445. Julian, P. L., and Iveson, H. T. (The Glidden Co.), U.S. Patent 2,752,378, 1956.

446. Morris, S. G., Myers, J. S., Jr., Kip, M. L., and Riemenschneider, R. W., 1950, *J. Amer. Oil. Chem. Soc.* 27:105.

447. Dron, D. W., and Lindsey, F. A. (De Laval Separator Co.), U.S. Patent 2,650,931, 1953.

448. Eckey, E. W. (The Procter & Gamble Company), U.S. Patent 1,982,907, 1934.

449. Lever Bros. and Unilever, Ltd., British Patent 577,879, 1946.

450. *Ibid.*, British Patent 612,169, 1948.

451. Kukn, K. (Kali-Chemi Aktiengesellschaft), British Patent 742,233, 1955; German Patent 1,012,014, 1957.

452. Tsuchiya, T., 1948, *J. Nippon Oil Technol. Soc. 1(3)*:24; *ibid.*, 1949, *2(2/3)*:1.

453. Koyama, R., Japanese Patent 2584 ('52), 1952.

454. Loury, M., 1955, *Rev. fr. Corps gras* 2:15.

455. Skau, E. L. (U.S. Department of Agriculture), U.S. Patent 2,684,377, 1954.

456. Van Akkeren, L. A., and Ast, H. J. (Swift & Co.), U.S. Patent 2,878,274, 1959.

457. Morris, C. E., and Khym, F. P. (Armour & Company), U.S. Patent 2,602,807, 1952.

458. Merker, D. R. (Swift & Co.), U.S. Patent 2,783,260, 1957.

459. *Ibid.*, U.S. Patent 2,862,941, 1958.

460. Richardson, A. S., and Eckey, E. W. (The Procter & Gamble Company), U.S. Patent 2,132,437, 1938.

461. Feuge, R. O., and Gros, A. T., 1950, *J. Amer. Oil Chem. Soc.* 27:117.

462. Van Akkeren, L. A. (Armour & Company), U.S. Patent 3,284,213, 1966.

463. Richardson, A. S., Vibrans, F. C., and Andrews, J. T. R. (The Procter & Gamble Company), U.S. Patent 1,993,181, 1935.

464. Musher, S. (Musher Foundation, Inc.), U.S. Patents 2,314,364 and 2,314,365, 1943.

465. Calkins, V. P., 1947, *J. Amer. Chem. Soc.* 69:384.

466. Kraybill, H. R., *et al.*, 1949, *J. Amer. Oil Chem. Soc.* 26:449.

467. Kraybill, H. R., and Beadle, B. W. (American Meat Institute Foundation), U.S. Patent 2,451,748, 1948.

468. Beadle, B. W., and Kraybill, H. R. (American Meat Institute Foundation), U.S. Patent 2,648,608, 1953.

469. Kring, P., 1950, *Dansk Tidskr. farm.* 24:211; abstracted in 1951, *J. Amer. Oil Chem. Soc. 28(3)*:123.

470. Morris, S. G., Gordon, C. F., Brenner, N., Meyers, J. S., Jr., Riemenschneider, R. W., and Ault, W. C., 1952, *J. Amer. Oil Chem. Soc.* 29:441.

471. Lips, H. J., 1952, *Food Canada 12(6)*:9, 12, 16.

472. Privett, O. S., and Quackenbush, F. W., *J. Amer. Oil Chem. Soc. 31*:225, 321.

473. Hall, L. A. (Griffith Laboratories, Inc.), U.S. Patent 2,511,803, 1950.

474. Magoffin, J. E. (Eastman Kodak Co.), U.S. Patent 2,607,746, 1952; British Patent 679,192, 1952; Canadian Patent 511,767, 1955.

475. Abe, Y., Japanese Patent 5323 ('53), 1953.

476. Gleim, W. K. T. (Universal Oil Products Company), U.S. Patent 2,694,645, 1954.

477. Sumiki, Y., and Tamura, S., Japanese Patent 8521 ('54), 1954.

478. Bentz, R. W. (Eastman Kodak Co.), U.S. Patent 2,772,243, 1956.

479. Mattil, K. F., and Sims, R. J. (Swift & Co.), U.S. Patent 2,759,829, 1956.

480. Brickman, A. W., Conquest, V., Madden, F. J., Filbey, E. T., and Oleson, W. B., British Patents 754,388 and 754,389, 1956.

481. Thompson, R. B. (Universal Oil Products Company), U.S. Patent 2,746,871, 1956.

482. Griffith, C. L., and Sair, L. (Griffith Laboratories, Inc.), Canadian Patent 531,497, 1956.

483. Kuhrt, N. H. (Eastman Kodak Co.), U.S. Patent 2,665,991, 1954; Canadian Patent 540,553, 1957; German Patent 1,042,363, 1958.

484. *Ibid.,* U.S. Patents 2,681,281 and 2,681,283, 1954; U.S. Patents 2,701,769 and 2,701,770, 1955; British Patent 731,177, 1955.
485. Watts, B. M., Lehmann, B., and Goodrich, F., 1949, *J. Amer. Oil Chem. Soc. 26*:481.
486. Lea, C. H., 1950, *Rancidity in Edible Fats,* Chemical Publishing Company.
487. Watts, B., 1950, *J. Amer. Oil Chem. Soc. 27*:48.
488. Lehmann, B. T., and Watts, B. M., 1951, *J. Amer. Oil Chem. Soc. 28*:475.
489. Watts, B. M., and Moss, H. V. (Syracuse University), U.S. Patent 2,629,664, 1953.
490. Lehmann, B. T., and Watts, B. M. (Monsanto Chemical Company), U.S. Patent 2,707,154, 1955; Canadian Patent 538,129, 1957.
491. Ozawa, T., and Ota, A., 1965, *Shokuhin Eiseigaku Zasshi 6(1)*:10; 1965, *Chem. Abstr. 62*:16877g.
492. Holman, G. W. (The Procter & Gamble Company), U.S. Patent 2,871,130, 1959; German Patent 1,236,361, 1967.
493. Mahon, J. H., and Chapman, R. A., 1954, *J. Amer. Oil Chem. Soc. 31*:108.
494. van Hees, G.m.b.H., German Patent 919,779, 1954.
495. Modernas Aplicaciones de la Refrigeración Industrial, S.A., Spanish Patent 217,490, 1955.
496. Mitchell, J. H. (Stevens Industries, Inc.), U.S. Patent 2,511,119, 1950.
497. Vincent, J. F., U.S. Patent 2,511,136, 1950.
498. Langendorf, H., and Lang, K., 1961, *Z. Lebensm.-Unters. Forsch. 115*:400.
499. Benk, E., 1969, *Flussiges Obst 36*:468; 1970, *Food Sci. Technol. Abstr. 2*:5H511.
500. Diemair, W., and Pfeifer, K., 1962, *Z. Lebensm.-Unters. Forsch. 117*:209.
501. Laucks, I. F., Banks, H. P., and Rippey, H. F. (Laucks Laboratories, Inc.), U.S. Patent 1,732,816, 1929.
502. Bates, H. R. (Citrus Compound Corp.), U.S. Patent 1,774,310, 1930.
503. Rippey, H. F. (Laucks Laboratories, Inc.), U.S. Patent 1,935,599, 1933.
504. Kalmar, A. F. (Food Machinery Corp.), U.S. Patent 2,374,209, 1945.
505. Kalmar, A. F. (Food Machinery Corp.), U.S. Patent 2,374,210, 1945.
506. Smith, J. D., and Krause, H. J., U.S. Patent 2,700,613, 1955.
507. Brandt, K., German Patent 1,417,432, 1968.
508. Post, F. J., Coblentz, W. S., Chou, T. W., and Salunhke, D. K., 1968, *Appl. Microbiol. 16*:138.
509. Kohl, W. F., and Ellinger, R. H., Polyphosphates as Preservatives for Apple Cider, paper presented before the 30th Annual Meeting of the Institute of Food Technologists, San Francisco, California, May 26, 1970.
510. Deobald, H. J., McLemore, T. A., Bertoniere, N. R., and Martinez, J. A., 1964, *Food Technol. 18*:1970.
511. Maggi, A. G., Netherlands Patent Application No. 6,600,754, 1966; 1967, *Chem. Abstr. 66*:1740c.
512. Rhee, K. S., and Watts, B. M., 1966, *J. Food Sci. 31*:669.
513. Molsberry, C. C., U.S. Patent 3,342,610, 1967.
514. Hanada, S., Yokoo, Y., Suzuki, E., Yamaguchi, N., and Yoshida, T., 1959, *Eiyo Shokuryo 11*:306; 1963, *Chem. Abstr. 58*:2097f.
515. Feldheim, W., and Seidemann, J., 1962, *Fruchtsaft-Industrie 7*:166; 1962, *Chem. Abstr. 57*:14250c.
516. Hall, G. O. (Hall Laboratories, Inc.), U.S. Patent 2,478,266, 1949.
517. Nelson, R. F., and Finkle, B. J., 1964, *Phytochem. 3*:321.
518. Bolin, H. R., Nury, F. S., and Finkle, B. J., 1964, *Baker's Dig. 38(3)*:46.
519. Fleischman, D. L., 1969, *Diss. Abstr. Int. B, Sci. Eng. 30(2)*:696.
520. Michels, P. (Chemische Fabrik Johann A. Benckiser G.m.b.H.), German Patent 1,082,792, 1962; 1962, *Chem. Abstr. 56*:3872i.
521. Smith, O., 1968, *Potatoes: Production, Storing, Processing,* Avi Publishing Company, Inc.
522. Juul, F., Thesis, *I Kommission Hos Jul. Gjellerups Forlag,* Copenhagen, Denmark, 1949; 1950, *Amer. Potato J. 27*:32–37 (abstract); 1950, *Biol. Abstr. 24*:3610, 11938.
523. Stauffer Chemicals for Modern Potato Processing, Food Industry Release No. 5, Stauffer Chemical Company.
524. Smith, O., 1958, *Amer. Potato J. 35*:573.
525. Smith, O., and Davis, C. O., 1962, *Amer. Potato J. 39*:45.
526. Talburt, W. F., and Smith, O., 1967, *Potato Processing,* Avi Publishing Company, Inc.
527. Talley, E. A., *et al.,* 1969, *Amer. Potato J. 46*:302.
528. Hoover, M. W., 1963, *Food Technol. 17*:636.
529. Kertesz, Z. I., 1939, *The Canner 88(7)*:26.

530. Baker, G. L., and Woodmansee, C. W., 1944, *Fruit Prod. J. 23(6)*:164, 185.

531. Baker, G. L., and Gilligan, G. M., 1946, *Food Packer 27(6)*:56.

532. Baker, G. L., and Murray, W. G., 1947, *Food Res. 12*:129.

533. McCready, R. M., Shepherd, A. D., and Maclay, W. D., 1947, *Fruit Prod. J. 27*:36.

534. Woodmansee, C. W., and Baker, G. L., 1954, *Del. Agr. Exp. Sta. Tech. Bull.*, No. 305:5.

535. Waller, G. R., and Baker, G. L., 1953, *J. Agr. Food Chem. 1*:1213.

536. Woodmansee, C. W., 1967, *Del. Agr. Exp. Sta.*, Circular No. 5, University of Delaware.

537. Olsen, A. G., and Fehlberg, E. R. (General Foods Corp.), U.S. Patent 2,334,281, 1943.

538. Leo, H. T., and Taylor, C. C., U.S. Patents 2,483,548 and 2,483,549, 1949.

539. Leo, H. T., Taylor, C. C., and Lindsey, J. W., U.S. Patent 2,483,550, 1949.

540. Pedersen, K. (Kobenhavens Pektinfabrik V. Karl Pedersen), U.S. Patent 2,540,545, 1951.

541. Briccoli-Bati, M., Italian Patent 472,159, 1952.

542. Avoset Company, British Patent 748,402, 1956.

543. Leo, H. T., and Taylor, C. C., U.S. Patent 2,824,007, 1958.

544. Baker, G. L., and Gilligan, G. M., 1947, *Fruit Prod. J. 26(9)*:260, 279.

545. Morse, R. E., 1952, *Food Packer 33(6)*:30, 52.

546. Peters, G. L., Brown, H. D., Gould, W. A., and Davis, R. B., 1954, *Food Technol. 8*:220.

547. Wagner, J. R., and Miers, J. C., 1967, *Food Technol. 21*:920; Wagner, J. R., Miers, J. C., Sanshuck, D. W., and Becker, R., 1968, *Food Technol. 22*:1484.

548. Smith, O., and Davis, C. O., 1963, *Amer. Potato J. 40*:67.

549. Cole, M. S. (The Pillsbury Company), U.S. Patent 3,219,464, 1965.

550. Kintner, J. A., and Tweedy, E., 1967, *Food Technol. 21*:865.

551. Nielsen, J. P., Campbell, H., and Boggs, M., June 1943, *West. Canner Packer 35(7)*:49.

552. Holmquist, J. W., Schmidt, C. F., and Guest, A. E., 1948, *Canning Trade 70 (40)*:7, 20.

553. Neubert, A. M., and Carter, G. H., 1954, *Food Technol. 8*:518.

554. Anon., November 1955, *Food Eng. 27*:188.

555. Biakemore, S. M., U.S. Patent 3,307,954, 1967.

556. Loconti, J. O., and Kertesz, Z. I., 1941, *Food Res. 6*:499.

557. Powers, J. J., and Esselen, W. B., Jr., 1946, *Fruit Prod. J. 25(7)*:200, 217.

558. Van Buren, J. P., 1968, *Food Technol. 22*:790.

559. Anon., 1971, *Code of Federal Regulations, Title 21,* Parts 27, 51, 53, revised as of January 1, 1971, U.S. Government Printing Office.

560. Anon., September 15, 1969, *Food Chem. News 1969*:18.

561. Dakin, J. C., 1963, *Nature 199*:383.

562. Takeda Chemical Industries, Ltd., Japanese Patent 8619 ('70), 1970; 1970, *Food Sci. Technol. Abstr. 2*:11T414.

563. Peschardt, W. J., and Hume, S. R. (Wm. Steward and Arnold, Ltd.), West German Patent Appl. No. 1,280,646, 1968; 1969, *Food Sci. Technol. Abstr. 1(5)*:5J415.

564. Glicksman, M., 1962, *Adv. Food Res. 11*:110.

565. Sergeeva, Z. I., *Tr. Vses. Nauchno.-Issled. Inst. Konditersk. Prom-st.' 1960(15)*:150; 1962, *Chem. Abstr. 57*:3841e.

566. Lucas, H. J. (Kelco Company), U.S. Patent 2,097,228, 1937.

567. Steiner, A. B. (Kelco Company), U.S. Patent 2,441,729, 1948.

568. *Ibid.,* U.S. Patent 2,485,934, 1949.

569. Roland, C. T. (Hall Laboratories), U.S. Patent 2,665,211, 1954.

570. Gibsen, K. F. (Kelco Company), U.S. Patent 2,808,337, 1957.

571. *Ibid.,* U.S. Patent 2,918,375, 1959.

572. Rocks, J. K. (Kelco Company), U.S. Patent 2,925,343, 1960.

573. Hunter, A. R., and Rocks, J. K. (Kelco Company), U.S. Patent 2,949,366, 1960.

574. Merton, R. R., and McDowell, R. H. (Alginate Industries, Ltd.), U.S. Patent 2,930,701, 1960; British Patent 828,350, 1960.

575. Hunter, A. R. (Kelco Company), U.S. Patent 2,987,400, 1961.

576. Miller, A., and Rocks, J. K. (Kelco Company), U.S. Patent 3,266,906, 1966; Canadian Patent 811,682, 1969.

577. Andrew, T. R., and MacLeod, W. C., 1970, *Food Prod. Devel. 4(5)*:99.

578. Stoloff, L. S. (Seaplant Chemical Corp.), U.S. Patent 2,801,923, 1957.

579. Stoloff, L. S. (Seaplant Chemical Corp.), U.S. Patent 2,864,706, 1958.

580. Zabik, M. E., and Aldrich, P. J., 1965, *J. Food Sci. 30*:111.

581. Weinstein, B. (Crest Foods Company), U.S. Patent 2,856,289, 1958.

582. Green, J., Schuller, E. J., Rickert, J. A., and Borders, B. (General Foods Corp.), U.S. Patent 2,992,925, 1961.

583. Marcus, J. K., U.S. Patent 2,605,229, 1952.

584. Grau, R., 1958, *Kondensierte Phosphate in Lebensmitteln* (Symposium, 1957, Mainz), Springer-Verlag, p. 89.

585. Hamm, R., 1960, *Adv. Food Res. 10*:355.

586. Saffle, R. L., 1968, *Adv. Food Res. 16*:105.

587. Karmas, E., 1970, *Fresh Meat Processing*, Noyes Data Corporation.

588. American Meat Institute Foundation, 1960, *The Science of Meat and Meat Products*, W. H. Freeman and Company.

589. McLoughlin, J. V., 1969, *Food Manuf. 44(1)*:36.

590. Kotter, L., 1957, *Die Fleischwirtschaft 9*:697.

591. *Ibid.,* 1961, *13*:186.

592. Grau, R., Hamm, R., and Baumann, A., 1953, *Biochem. Z. 325*:1.

593. Bendall, J. R., 1954, *J. Sci. Food Agr. 5*:468.

594. Fukazawa, T., Hashimoto, Y., and Yasui, T., 1961, *J. Food Sci. 26*:541.

595. *Ibid.,* 550.

596. Yasui, T., Sakanishi, M., and Hashimoto, Y., 1964, *J. Agr. Food Chem. 12*:392.

597. Yasui, T., Fukazawa, T., Takahashi, K., Sakanishi, M., and Hashimoto, Y., *ibid.,* 399.

598. Klepacka, M., 1965, *Przem. Spozywczy 19(8)*:501; 1966, *Chem. Abstr. 65*:17596a.

599. Grau, R., Hamm, R., and Baumann, A., 1953, *Naturwissenschaften 40*:535.

600. Morse, R. E., October 1955, *Food Eng. 27(10)*:84.

601. Epstein, F. (First Spice Manufacturing Corp.), U.S. Patent 2,876,115, 1959.

602. Kotter, L., 1960, *Zur Wirkung Kondensierter Phosphate und Anderer Salze auf Tierisches Eiweiss*, M. & H. Schaper.

603. Hellendoorn, E. W., 1962, *J. Food Technol. 16(9)*:119.

604. Inklaar, P. A., 1967, *J. Food Sci. 32*:525.

605. Baldwin, T. T., and deMan, J. M., 1968, *J. Inst. Can. Technol. Aliment. 1*:164.

606. Turner, E. W., and Olson, F. C. (Oscar Mayer & Co., Inc.), U.S. Patent 2,874,060, 1959.

607. Saffle, R. L., and Galbreath, J. W., 1964, *Food Technol. 18*:1943.

608. Acton, J. C., and Saffle, R. L., quoted in Saffle, ref. 586.

609. Kotter, L., 1958, *Kondensierte Phosphate in Lebensmitteln* (Symposium, 1957, Mainz), Springer-Verlag, p. 99.

610. Hansen, L. J., 1960, *Food Technol. 14*:565.

611. Swift, C. E., Lockett, C., and Fryar, A. J., 1961, *Food Technol. 15*:468.

612. Hegarty, G. R., Bratzler, L. J., and Pearson, A. M., 1963, *J. Food Sci. 28*:663.

613. Carpenter, J. A., and Saffle, R. L., 1965, *Food Technol. 19*:1567.

614. Trautman, J. C., 1964, *Food Technol. 18*:1065.

615. Swift, C. E., 1965, *Proceedings of the Meat Industry Research Conference*, O. E. Kolari and W. J. Aunan, Eds., American Meat Institute Foundation, p. 78.

616. Ellinger, R. H., and Harkey, C. N., unpublished work, 1960.

617. Mahon, J. H., July 1961, *Proc. 30th Res. Conf. Amer. Meat Inst. Found. 64*:59.

618. Sherman, P., 1961, *Food Technol. 15*:79.

619. Depner, M., 1949, *Fleischwirtschaft 1*:252.

620. Hamm, R., 1955, *Naturwissenschaften 42*:394.

621. Hamm, R., 1955, *Fleischwirtschaft 7*:196.

622. *Ibid.,* 1956, *8*:266.

623. *Ibid.,* 340.

624. Hamm, R., 1957, *Z. Lebensm.-Unters. Forsch. 106*:281.

625. *Ibid.,* 1958, *107*:428.

626. Hamm, R., and Grau, R., 1958, *Z. Lebensm.-Unters. Forsch. 108*:280.

627. Hamm, R., 1959, *Natl. Provisioner 140(15)*:78.

628. Hamm, R., 1959, *Z. Lebensm.-Unters. Forsch. 110*:95.

629. Swift, C. E., and Berman, M. D., 1959, *Food Technol. 13*:365.

630. Wierbicki, E., Kunkle, L. E., and Deathrage, F. E., 1963, *Fleischwirtschaft 15*:404.

631. Swift, C. E., and Ellis, R., 1956, *Food Technol. 10*:546.

632. Wierbicki, E., Tiede, M. G., and Burrell, R. C., paper presented at the 19th Annual Meeting, Institute of Food Technologists, Philadelphia, May 17–21, 1959.

633. Wismer-Pedersen, J., 1965, *J. Food Sci. 30*:85.

634. *Ibid.,* 91–97.

635. Popp, H., and Muhlbrecht, F. N., 1958, *Food Sci. 10*:399.

636. Watts, B. M., 1954, *Adv. Food. Res. 5*:1.

637. Fox, J. B., 1966, *J. Food Agr. Chem. 14*:207.

638. Hall, G. O. (Hall Laboratories, Inc.), U.S. Patent 2,513,094, 1950.

639. Brissey, G. E. (Swift & Company), U.S. Patent 2,596,067, 1952.

640. Chang, I., and Watts, B. M., 1949, *Food Technol. 3*:332.

641. Bendall, J. R., 1953, *Nature (Lond.) 172*:586.

642. Fukazawa, T., Hashimoto, Y., and Yasui, T., 1961, *J. Food Sci. 26*:331.

643. Giacino, C. (International Flavors & Fragrances, Inc.), U.S. Patent 3,394,015, 1968.

644. Keil, H. L., Hagen, R. F., and Flaws, R. W. (Armour & Company), U.S. Patent 2,953,462, 1960.

645. Kamstra, L. D., and Saffle, R. L., 1959, *Food Technol. 13*:652.

646. Carpenter, J. A., Saffle, R. L., and Kamstra, L. D., 1961, *Food Technol. 15*:197.

647. Voegeli, M. M., and Gorsica, H. J. (B. Heller & Company), U.S. Patent 3,154,421, 1964.

648. *Ibid.,* U.S. Patent 3,154,423, 1964.

649. Rust, M. E., Ph.D. Thesis, Kansas State University, 1963.

650. Allen, H. E. (H. E. Allen and A. G. McCaleb), U.S. Patent 2,140,781, 1938.

651. Mullins, A. M., Kelley, G. G., and Brady, D. E., 1958, *Food Technol. 12*:227.

652. Hopkins, E. W., and Zimont, L. J. (Armour & Company), U.S. Patent 2,999,019, 1961.

653. Williams, B. E. (Hodges Research and Development Corporation), U.S. Patent 3,006,768, 1961.

654. Komarik, S. L. (Griffith Laboratories, Inc.), U.S. Patent 3,147,123, 1964.

655. Brown, W. L., Denny, D., and Schmucker, M. L. (John Morrell & Co.), U.S. Patent 3,207,608, 1965.

656. Delaney, W. (Madison Laboratories, Inc.), U.S. Patent 3,188,213, 1965.

657. Huffman, D. L., Palmer, A. Z., Carpenter, J. W., and Shirley, R. L., 1965, *Quart. J. Fla. Acad. Sci. 28*:106; *ibid.,* 1969, *J. Animal Sci. 28*:443.

658. Miller, E. M., and Harrison, D. L., 1965, *Food Technol. 19*:94.

659. Hopkins, E. W., and Zimont, L. J. (Armour & Company), U.S. Patent 3,049,428, 1962.

660. Unilever, N. V., Netherlands Patent Appl. No. 6,506,834, 1965.

661. Brendl, J., Klein, S., Kocianova, M., 1967, *Sb. Vys. Sk. Chem.-Tech. Praze Potraviny 14*:49; 1967, *Chem. Abstr. 67*:115889x.

662. Tims, M. J., and Watts, B. M., 1958, *Food Technol. 12*:240.

663. Tarladgis, B. G., Younathan, M. T., and Watts, B. M., 1959, *Food Technol. 13*:635.

664. Chang, P., Younathan, M. T., and Watts, B. M., 1961, *Food Technol. 15*:168.

665. Ramsey, M. B., and Watts, B. M., 1963, *Food Technol. 17*:1056.

666. Zipser, M. W., Kwon, T., and Watts, B. M., 1964, *J. Agr. Food Chem. 12*:105.

667. Watts, B. M., 1961, *Proceedings Flavor Chemists Symposium*, Campbell Soup Company, p. 83.

668. Watts, B. M., Meat Products, 1962, *Lipids and Their Oxidation*, H. W. Schultz, Ed., Avi Publishing Company.

669. Savich, A. L., and Jansen, C. E. (Swift & Company), U.S. Patent 2,830,907, 1958.

670. Ziegler, J. A. (Griffith Laboratories, Inc.), U.S. Patent 3,073,700, 1963.

671. Jacoby, D., and Berhold, R. J. (Adolph's Food Products Mfg. Co.), U.S. Patent 3,469,995 (1969).

672. Bickel, W. (Johann A. Benckiser G.m.b.H.), U.S. Patent 3,029,150, 1962.

673. Fetty, H. J., U.S. Patent 3,216,827, 1965.

674. Sair, L., and Cook, W. H., 1938, *Can. J. Res. D16*:255.

675. Sherman, P., 1962, *Food Technol. 16*:91.

676. Swift, C. E., and Ellis, R., 1957, *Food Technol. 11*:450.

677. Zwanenberg's Fabrieken N.V., Netherlands Patent 62,273, 1948.

678. Sair, L., and Komarik, S. L. (Griffith Laboratories, Inc.), U.S. Patent 3,391,007, 1968.

679. Anon., February 1953, *Food Eng. 25*:214.

680. Bailey, M. E., Frame, R. W., and Naumann, H. D., 1964, *J. Agr. Food Chem. 12*:89.

681. Rahelic, S., Rede, R., and Jovicic, V., 1966, *Technol. Mesa 7(9)*:252; 1967, *Chem. Abstr. 67*:36667z.

682. Wilson, G. D., 1954, *Proceedings Research Report to Industry*, American Meat Institute Foundation, p. 32.

683. Fenton, F., Sheffy, B. E., Naumann, H. D., Wellington, G. H., Hogue, D., and Mahon, P., 1956, *Food Technol. 10*:272.

684. Wilson, G. D., *Natl. Provisioner*, 1954, *131*:112; as quoted in Fenton *et al.,* ref. 683.

685. Mahon, P., Hogue, D., Leeking, P., Lim, E., and Fenton, F., 1956, *Food Technol. 10*:265.

686. Gisske, W., 1958, *Fleischwirtschaft 10*:21.

687. Schoch, W. S. (Armour & Company), U.S. Patent 2,824,809, 1958.

688. Watts, B. M., 1957, *Proceedings 9th Research Conference*, American Meat Institute, p. 61.

689. Wasserman, M. (Meat Industry Suppliers, Inc.), U.S. Patent 2,812,261, 1957.

690. Suri, 1961, *Fleischwirtschaft 13*:403.

691. Siedlecki, E., 1964, *Gospodarka Miesna 16(1)*:7; 1965, *Chem. Abstr. 62*:1009h.

692. Lauck, R. M., and Tucker, J. W. (Stauffer Chemical Company), U.S. Patent 3,118,777, 1964.

693. Savic, I., Suvakov, M., Nikolic, J., and Korolija, S., 1965, *Technol. Mesa 6(10)*:274, 278; 1966, *Chem. Abstr. 64*:14868h.

694. Swift, C. E., Weir, C. E., and Hankins, O. G., 1954, *Food Technol. 8*:339.

695. Szent-Gyorgyi, A. G., 1952, *Proceedings Fourth Research Conference*, American Meat Institute, p. 22.

696. Szent-Gyorgyi, A. G., 1960, in *Proteins of the Myofibril in Structure and Function of Muscle*, G. H. Bourne, Ed., Vol. II, Academic Press, p. 1.

697. Schleich, H., and Arnold, R. S. (Baxter Laboratories, Inc.), U.S. Patent 3,037,870, 1962.

698. Maas, R. H., and Olson, F. C. (Oscar Mayer & Co., Inc.), U.S. Patent 3,075,843, 1963.

699. Leeking, P., Mahon, P., Hogue, D., Lim, E., and Fenton, F., 1956, *Food Technol. 10*:274.

700. Frank, S. S., M.S. Thesis, Cornell University, 1955.

701. Swift, C. E., and Ellis, R., 1956, *Food Technol. 10*:546.

702. Volovinskaya, V. P., Rubashkina, S. S., Solov'ev, V. I., and Dyklop, V. K., *Tr. Vses. Nauchno.-Issled. Inst. Myasn. Prom-st.' 1958(8)*:78; 1960, *Chem. Abstr. 55*:18822a.

703. Volovinskaya, V. P., Rubashkina, S. S., Poletaev, T., Kel'man, B., and Merkulova, V., 1959, *Myasnaya Ind. S.S.S.R. 30(4)*:48; 1961, *Chem. Abstr. 55*:9712f.

704. Rupp, V. R., Brockmann, M. C., and Nickolson, L. W. (Hygrade Food Products Corp.), U.S. Patent 2,937,094, 1960; Canadian Patent 594,779, 1960.

705. Zeller, M., 1961, *Dsch. Lebensm.-Rundsch. 57*:205.

706. Grau, R., 1961, *Fleischwirtschaft 13*:183.

707. Rongey, E. H., and Bratzler, L. J., 1966, *Food Technol. 20*:1228.

708. Mahon, J. H., and Schneider, C. (Calgon Corporation), Canadian Patent 788,939, 1968; Netherlands Appl. No. 6,600,579, 1966.

709. Keller, H., 1955, *Fleischwirtschaft 7*:15.

710. Sherman, P., 1961, *Food Technol. 15*:87.

711. Pfaff, W., British Patent 881,397, 1961.

712. Zipser, M. W., and Watts, B. M., 1967, *J. Agr. Food Chem. 15*:80.

713. Evans, J. B., 1954, *Proceedings Research Report to Industry*, American Meat Institute Foundation, p. 12.

714. Bickel, W. (Calgon, Inc.), U.S. Patent 2,735,776, 1956.

715. Wollmann, E., 1959, *Fleischwirtschaft 11*:291.

716. Peat, M. R. (Wm. J. Strange Co.), U.S. Patent 2,925,344, 1960.

717. Gorsica, H. J. (B. Heller & Co.), U.S. Patent 3,008,832, 1961.

718. Hall, L. A. (Griffith Laboratories, Inc.), U.S. Patent 2,145,417, 1939.

719. *Ibid.*, U.S. Patent 2,668,771, 1954.

720. Hall, L. A., and Kalchbrenner, W. S. (Griffith Laboratories, Inc.), U.S. Patents 2,770,548 and 2,770,551, 1956.

721. Hall, L. A. (Griffith Laboratories, Inc.), U.S. Patent 2,770,549, 1956.

722. Hall, L. A., and Griffith, C. L. (Griffith Laboratories, Inc.), U.S. Patent 2,770,550, 1956.

723. Chemische Fabrik Budenheim A.G., British Patent 754,406, 1956.

724. Huber, H., and Vogt, K. (Chemische Werke Albert), U.S. Patent 2,852,392, 1958; British Patent 696,617, 1953.

725. Barnett, H. W. (Canada Packers, Ltd.), U.S. Patent 2,903,366, 1959.

726. Barth, W. (Franziska Therese Barth, geb. Wegerl), German Patent 1,041,338, 1961.

727. Miyake, M., and Noda, H. (Tanabe Seiyaku Co., Ltd.), Japanese Patent 9134, 1961; 1962, *Chem. Abstr. 56*:6437a.

728. Chemische Fabrik Johann A. Benckiser G.m.b.H., British Patent 897,682, 1962.

729. Buchholz, K. (Johann A. Benckiser G.m.b.H.), U.S. Patent 3,032,421, 1962.

730. Oliver, R. J., and Voegeli, M. M. (Swift & Co.), U.S. Patent 3,028,246, 1962.

731. Pfrengle, O. (Chemische Fabrik Budenheim Rudolf A. Oetker K.G.), German Patent 1,181,030, 1964.

732. Flesch, P., and Bauer, G., French Patent 1,446,678, 1966.

733. Haynes, R. D. (Monsanto Chemical Company), U.S. Patent 2,868,654, 1959.

734. Bickel, W. (Johann A. Benckiser G.m.b.H.), German Patent 1,074,380, 1960.

735. Chemische Fabrik Budenheim, A.G., British Patent 831,132, 1960.

736. Elder, E. V., Jr. (Hooker Chemical Corp.), U.S. Patent 3,104,978, 1963.

737. Sair, L., and Donahoo, W. P. (Griffith Laboratories, Inc.), U.S. Patent 3,139,347, 1964.

738. Sair, L. (Griffith Laboratories, Inc.), U.S. Patent 3,231,392, 1966.

739. Mahon, J. H. (Calgon Corporation), U.S. Patent 3,401,046, 1968.

740. Murphy, L. W., and Jones, R. L., British Patent 862,109, 1961.

741. Beerend, R. F., and Needle, H. C. (Basic Food Materials, Inc.), U.S. Patent 3,192,055, 1965.

742. Takeda Chemical Industries, Ltd., French Patent 1,386,675, 1965.

743. deFremery, D., and Pool, M. F., 1960, *Food Res. 25*:73.

744. deFremery, D., 1963, *Proceedings Meat Tenderness Symposium*, Campbell Soup Company, p. 99.

745. deFremery, D., 1966, *J. Agr. Food Chem. 14*:214.

746. Wells, G. H., May, K. N., and Powers, J. J., 1962, *Food Technol. 16*:137.

747. Klose, A. A., Campbell, A. A., and Hanson, H. L., 1963, *Poult. Sci. 42*:743.

748. Mahon, J. H., 1962, *Poult. Processing Market. 68(8)*:16, 24, 26.

749. Mahon, J. H., 1962, *Food Eng. 34(11)*:108,113.

750. Mahon, J. H. (Calgon Corporation), U.S. Patent 3,104,170, 1963; Belgian Patent 614,742, 1962; Canadian Patent 691,480, 1963; Netherlands Patent Appl. No. 6,613,628, 1967.

751. Froning, G. W., 1965, *Poult. Sci. 44*:1104.

752. Froning, G. W., 1966, *Poult. Sci. 45*:185.

753. Dodge, J. W., and Stadelman, W. J., 1959, *Food Technol. 13*:81.

754. May, K. N., Helmer, R. L., and Saffle, R. L., 1962, *Poult. Sci. 41*:1665.

755. Klose, A. A., Campbell, A. A., and Hanson, H. L., 1962, *Poult. Sci. 41*:1655.

756. Spencer, J. V., and Smith, L. E., 1962, *Poult. Sci. 41*:1685.

757. Baker, R. C., and Darfler, J., 1968, *Poult. Sci. 47*:1590.

758. Schwall, D. V., Rogers, A. B., and Corbin, D. (Armour & Co.), U.S. Patent 3,399,063, 1968.

759. Schwall, D. V., and Rogers, A. B. (Armour & Co.), U.S. Patent 3,285,753, 1966; British Patent 1,198,670, 1970.

760. *Ibid.*, U.S. Patent 3,413,127, 1968.

761. Schnell, P. G., Vadehra, D. V., and Baker, R. C., 1970, *J. Can. Inst. Food Technol. 3(2)*:44.

762. Vadehra, D. V., and Baker, R. C., 1970, *Food Technol. 24*:766, 775.

763. Schlamb, K. F. (Calgon Corporation), U.S. Patent 3,499,767, 1970.

764. Mountney, G. J., and Arganosa, F. C., 1962, *Poult. Sci. 41*:1668.

765. Schermerhorn, E. P., and Stadelman, W. J., 1962, *Poult. Sci. 41*:1680.

766. Monk, J. A., Mountney, G. J., and Prudent, I., 1964, *Food Technol. 18*:226.

767. Shermerhorn, E. P., and Stadelman, W. J., 1964, *Food Technol. 18*:101.

768. Mahon, J. H. (Calgon Corporation), U.S. Patent 3,462,278, 1969; Netherlands Patent Appl. No. 6,613,627, 1967.

769. Montemayor, E., Vadehra, D. V., and Baker, R. C., quoted in Vadehra and Baker, ref. 762.

770. Hudspeth, J. P., and May, K. N., 1967, *Food Technol. 21*:1141.

771. Marion, W. W., and Forsythe, R. H., 1962, *Poult. Sci. 41*:1663; *ibid.*, 1964, *J. Food Sci. 29*:530.

772. Thomson, J. E., 1964, *Food Technol. 18*:1805.

773. Jimenez, A. V., Fernandez, R. P., and Girauta, M. E., 1961, *Symposium Substances Etrangeres Aliments* (No. 6, Madrid, 1960), p. 530; 1963, *Chem. Abstr. 59*:3256e.

774. Melnick, D., Perry, G. A., and Lawrence, R. L. (Corn Products Company), U.S. Patent 3,065,084, 1962.

775. Kraft, A. A., and Ayres, J. C., 1961, *Appl. Microbiol. 9*:549; *ibid.*, 1964, *J. Food Sci. 29*:218.

776. Tomiyama, T., Kitahara, K., Shiraishi, E., and Imaizumi, K., *Fourth International Symposium on Food Microbiology* (Gotborg, Sweden, 1964), p. 261.

777. Elliott, R. P., Straka, R. P., and Garibaldi, J. A., 1964, *Appl. Microbiol. 12*:517.

778. Linko, R. R., and Nikkila, O. E., 1961, *J. Food Sci. 26*:606.

779. Kuusi, T., Nikkila, O. E., and Kytokangas, R., 1965, *Valt. Tek. Tutkimuslaitos Tiedotus Sar. I Puu 4(78)*:1; 1966, *Chem. Abstr. 64*:20525a.

780. Nikkila, O. E., Kuusi, T., and Kytokangas, R., 1967, *J. Food Sci. 32*:686.

781. Scheurer, P. G., 1968, *J. Food Sci. 33*:504.

782. Love, R. M., and Abel, G., 1966, *J. Food Technol. 1*:323.

783. Akiba, M., Motohiro, T., and Tanikawa, E., 1967, *J. Food Technol. 2*:69.

784. Spinelli, J., Pelroy, G., and Miyauchi, D., 1968, *Fish. Ind. Res. 4*:37.

785. Ravesi, E. M., and Anderson, M. L., 1969, *Fish. Ind. Res. 5*:175.

786. Kuusi, T., and Kytokangas, R., 1970, *Maataloustieteellinen Aikak. 42(1)*:30; 1970, *Food Sci. Technol. Abstr. 2*:11R378.

787. Meyer, A. (Calgon, Inc.), U.S. Patent 2,735,777, 1956.

788. Yamaga, M., Morioka, K., and Kawakami, T., 1961, *Reito 36*:904; 1962, *Chem. Abstr. 57*:7758d.

789. Sen, D. P., and Lahiry, N. L., 1964, *Food Technol. 18*:1611.

790. Miyauchi, D., Spinelli, J., Pelroy, G., and Steinberg, M. A., Winter, 1967-68, *Isot. Radiat. Technol. 5(2)*:136; 1969, *Commer. Fish. Abstr. 22(3)*:7.

791. Sen, D. P., Visweswarish, K., and Lahiry, N. L., 1960, *Food Sci. (Mysore) 10*:144; 1962, *Chem. Abstr. 56*:1812a.

792. Garnatz, G., Volle, N. H., and Deatherage, F. E. (Kroger Grocery & Baking Co.), U.S. Patent 2,488,184, 1949.

793. MacCallum, W. A., Chalker, D. A., Lauder, J. T., Odense, P. H., and Idler, D. R., 1964, *J. Fish. Res. Board Can. 21*:1397; 1965, *Chem. Abstr. 62*:8310.

794. Tanikawa, E., 1963, *Adv. Food. Res. 12*:367.

795. Okamura, K., Matsuda, T., and Yokoyama, M., 1958-1959, *Bull. Jap. Soc. Sci. Fish. 24*:545; 1959, *Chem. Abstr. 53*:14376e.

796. *Ibid.,* 1958-1959, *Nippon Suisangaku Kaishi 24*:821; 1959, *Chem. Abstr. 53*:15411g.

797. *Ibid.,* 1958-1959, *Nippon Suisan Gakkaishi 24*:978, 986; 1960, *Chem. Abstr. 54*:15750a.

798. Okamura, K., 1960, *Nippon Suisan Gakkaishi 26*:600; 1961, *Chem. Abstr. 55*:14749b.

799. *Ibid.,* 1961, *27*:48.

800. *Ibid.,* 1961, *27*:52.

801. *Ibid.,* 1961, *27*:58.

802. Miyake, M., Hayashi, K., and Noda, H., 1963, *Mie Kenritsu Daigaku Suisangakubu Kiyo 6(1)*:39; 1965, *Chem. Abstr. 63*:8960f.

803. Akiyama, N., 1961, *Hokkaido-Ritsu Eisei Kenkyushoho 12*:113; 1965, *Chem. Abstr. 62*:15348c.

804. Anon., 1968, *Commer. Fish. 7(1)*:19; 1968, *Commer. Fish. Abstr. 21(11)*:13.

805. Seagran, H. L., 1958, *Food Res. 23*:143.

806. Albright and Wilson (Mfg.), Ltd., British Patent 916,208, 1961.

807. Mahon, J. H. (Hagan Chemicals & Controls, Inc.), U.S. Patent 3,036,923, 1962.

808. Anon., 1962, *Food Eng. 34(9)*:106.

809. Tanikawa, E., Akiba, M., and Shitamori, A., 1963, *Food Technol. 17*:1425.

810. Mahon, J. H., and Schneider, C. G., 1964, *Food Technol. 18*:117.

811. Boyd, J. W., and Southcott, B. A., 1965, *J. Fish. Res. Board Can. 22*:53.

812. Swartz, W. E. (Calgon Corporation), Canadian Patent 847,280, 1970.

813. Sutton, A. H., paper presented before the *FAO Technical Conference on the Freezing and Irradiation of Fish*, Madrid, September 4-8, 1967; reviewed in Anon., 1968, *Food Mfr. 43(1)*:26.

814. Sutton, A. H., and Ogilvie, J. M., 1968, *J. Fish. Res. Board Can. 25*:1475.

815. Miyauchi, D. T., Spinelli, J., Pelroy, G., and Stoll, N., 1965, AEC Accession No. 29583, Report No. TID-22515, U.S. Government Printing Office; 1966, *Nucl. Sci. Abstr. 20*:3600.

816. Spinelli, J., and Weig, D., November 1968, *Canner/Packer 137(12)*:28; 1968, *Commer. Fish. Abstr. 22(4)*:1.

817. Anon., April 1969, *Quick Frozen Foods 31*:146.

818. *Ibid.,* 150.

819. Anon., 1968, *Food Eng. 40(2)*:11.

820. Barnett, H. J., Nelson, R. W., and Dassow, J. A., 1969, *Fish. Ind. Res. 5*:103.

821. Dyer, W. J., Brockerhoff, H., Hoyle, R. J., and Fraser, D. I., 1964, *J. Fish. Res. Board Can. 21*:101.

822. Love, R. M., 1965, *Rep. Progr. Appl. Chem. 50*:292.

823. Anon., 1964, *Torry Research Station Annual Report,* p. 13; quoted in ref. 822.

824. Okamura, K., 1960, *Nippon Suisan Gakkaishi 26*:60; 1961, *Chem. Abstr. 55*:6720c.

825. Okamura, K., Kodama, M., Sonada, H., and Yamada, J., 1966, *Nippon Suisan Gakkaishi 32*:80; 1967, *Chem. Abstr. 67*:10497q.

826. Tanikawa, E., Motohiro, T., and Akiba, M., paper presented before the *FAO Technical Conference on the Freezing and Irradiation of Fish*, Madrid, September 4-8, 1967; reviewed in Anon., 1968, *Food Mfr. 43(1)*:26.

827. Ueoka, Y., Oka, H., Suemitso, H., and Sugimoto, T., 1966, *Nippon Shokuhin Kogyo Gakkaishi* *13*:475; 1967, *Chem. Abstr. 67*:2233k.

828. Takei, M., and Takahashi, T., 1961, *Tokaiku Suisan Kenkyusho Kenkyu Hokoku 30*:105; 1962, *Chem. Abstr. 57*:14248g.

829. Ichisugi, T., Kuroda, K., and Ikawa, K., 1965, *Hokusuishi Geppo 22*:273, 319; 1965, *Chem. Abstr. 63*:10576c.

830. Mathen, C., 1968, *Fish. Technol. 5(2)*:104; 1970, *Food Sci. Technol. Abstr. 2*:5R167.

831. Jones, R., 1969, *Fish. Ind. Res. 5*:83.

832. Sjöstrom, L. B. (Arthur D. Little, Inc.), U.S. Patent 2,446,889, 1948.

833. Tarr, H. L. A., Gardner, L. J., and Ingram, P., 1969, *J. Food Sci. 34*:637.

834. Zipser, M. W., and Watts, B. M., 1961, *Food Technol. 15*:318.

835. Greig, R. A., Emerson, J. A., and Fliehman, G. W., 1969, *Fish. Ind. Res. 5*:2.

836. Hempel, H. (Gorton-Pew Fisheries Co.), U.S. Patent 2,397,547, 1946.

837. Aktieselskabet Protan and Helberud, O., Norwegian Patent 86,681, 1955; 1950, *Chem. Abstr. 50*:7348a.

838. Uchiyama, H., and Amano, K., 1959, *Bull. Jap. Soc. Sci. Fish. 25(7-9)*:531.

839. Ozawa, T., Nagaoka, S., and Aragaki, M., 1963, *Shokuhin Eiseigaku Zasshi 4*:332, 339; 1964, *Chem. Abstr. 60*:13788h.

840. Levin, R. E., 1967, *J. Milk Food Technol. 30*:277.

841. McFee, E. P., and Peters, J. A. (Gorton-Pew Fisheries Co., Ltd.), U.S. Patent 2,554,625, 1951.

842. Kreidl, E. L., and McFee, E. P. (Gorton-Pew Fisheries Co., Ltd.), U.S. Patent 2,555,236, 1951.

843. McFee, E. P., and Swaine, R. L., 1953, *Food Eng. 25(12)*:67, 190, 192.

844. Yamada, H., 1961, *Nippon Suisan Chuo Kenkyusho Hokoku 9*:78; 1963, *Chem. Abstr. 58*:11895f.

845. Kelch, F., and Buhlmann, X., 1958, *Fleischwirtschaft 10*:325.

846. Pacheco, G., and Echaniz, J. S., 1953, *Brasil-Medico 67(2)*:43.

847. Pacheco, G., and Dias, V. M., 1954, *Mem. Inst. Oswaldo Cruz Rio de J. 52(2)*:405.

848. Heiligman, F., Desrosier, N. W., and Broumand, H., 1956, *Food Res. 21*:63.

849. Williams, O. B., and Hennessee, A. D., 1956, *Food Res. 21*:112.

850. Wollmann, E., Veterinary Medical Dissertation, Munich, Germany, 1957; abstracted in 1959, *Fleischwirtschaft 11(4)*:291.

851. Post, F. J., Krishnamurty, G. B., and Flanagan, M. D., 1963, *Appl. Microbiol. 11*:430.

852. Gould, G. W., *Fourth International Symposium Food Microbiology*, Swedish Institute for Food Preservation Research, Goteborg, Sweden, 1964, p. 17; 1967, *Chem. Ind. 1967*:234.

853. Irani, R. R., and Callis, C. F., 1962, *J. Amer. Chem. Soc. 39*:156.

854. Irani, R. R., and Morgenthaler, W. W., 1963, *ibid. 40*:283.

855. Nowakowska-Waszczuk, A., 1965, *Acta Microbiol. Polon. 14*:207; 1965, *Chem. Abstr. 63*:7367b.

856. Gray, G. W., and Wilkinson, S. G., 1965, *J. Gen. Microbiol. 39*:385.

857. Hamilton-Miller, J. M. T., 1965, *Biochem. Biophys. Res. Commun. 20*:688.

858. Brown, M. R. W., and Richards, R. M. E., 1965, *Nature 207*:1391.

859. Burkard, G., Weil, J. H., and Ebel, J. P., 1965, *Bull. Soc. Chim. Biol. 47*:561; 1965, *Chem. Abstr. 63*:10218f.

860. Tomiyama, T., Kitahara, K., Shiraishi, E., and Imaizumi, K., *Fourth International Symposium on Food Microbiology*, Swedish Institute of Food Preservation Research, 1964, Goteborg, Sweden, p. 261.

861. Kooistra, J. A., and Troller, J. A. (Procter & Gamble Co.), U.S. Patent 3,404,987, 1968.

862. Vishniac, W., 1950, *Arch. Biochem. 26*:167.

863. Brock, R. D., 1952, *J. Aust. Inst. Agr. Sci. 18*:41.

864. Fuld, E., 1902, *Beitr. Chem. Physiol. Path. 2*:155; quoted in Horvath, ref. 237.

865. Ferrel, R. E., Olcott, H. S., and Fraenkel-Conrat, H., 1948, *J. Amer. Chem. Soc. 70*:2101.

866. Braginskaya, F. I., and El'piner, I. E., 1963, *Biofizika 8*:34; 1963, *Chem. Abstr. 58*:12792c.

867. Council of Scientific and Industrial Research, Indian Patent 45,580, 1953; 1953, *Chem. Abstr. 47*:6576c.

868. Hall, L. A. (Griffith Laboratories, Inc.), U.S. Patent 2,477,742, 1949.

869. Ferrel, R. E., and Olcott, H. S. (U.S. Department of Agriculture), U.S. Patent 2,522,504, 1950.

870. Mohammad, A., Mecham, D. K., and Olcott, H. S., 1954, *J. Agr. Food Chem. 2*:136.

871. Knight, J. W., 1965, *The Chemistry of Wheat Starch and Gluten and Their Conversion Products*, Leonard Hill, p. 92.

872. Beckel, A. C., DeVoss, L. I., Belter, P. A., and Smith, A. K. (U.S. Department of Agriculture), U.S. Patent 2,444,241, 1948.

873. Beckel, A. C., Belter, P. A., and Smith, A. K. (U.S. Department of Agriculture), U.S. Patent 2,445,931, 1948.

874. Glabe, E. F., Goldman, P. F., Anderson, P. W., Finn, L. A., and Smith, A. K., 1956, *Food Technol.* 10:51.

875. Norris, F. A., and Johnson, D. C. (Swift & Co.), U.S. Patent 3,155,524, 1964.

876. Pintauro, N. D., Reynolds, J. M., and Newman, K. R. (General Foods Corp.), U.S. Patent 3,147,125, 1964.

877. Burnett, R. S., and Gunther, J. K. (Central Soya Co., Inc.), U.S. Patent 2,489,173, 1949.

878. Sevall, H. E., and Schaeffer, R. P. (The Borden Co.), U.S. Patent 2,588,419, 1952.

879. Patterson, B. A., U.S. Patent 2,716,606, 1955.

880. Gunther, R. C. (Gunther Products, Inc.), U.S. Patent 2,844,468, 1958; Canadian Patent 565,025, 1958.

881. Sutton, W. J. L. (Monsanto Chemical Co.), U.S. Patent 2,929,715, 1960.

882. Clausi, A. S., Common, J. L., and Horti, H. M. (General Foods Corp.), U.S. Patent 2,954,299, 1960; Canadian Patent 600,649, 1960.

883. Mancuso, J. J., and Common, J. L. (General Foods Corp.), U.S. Patent 2,965,493, 1960.

884. Mitchell, W. A., and Seidel, W. C. (General Foods Corp.), U.S. Patent 2,968,565, 1961.

885. Block, H. W. (General Foods Corp.), U.S. Patent 2,983,617, 1961.

886. Downey, H. A., U.S. Patent 3,018,183, 1962.

887. Grettie, D. P., and Tiemstra, P. J. (Swift & Co.), U.S. Patent 3,490,920, 1970.

888. Schwartz, C. (Hall Laboratories, Inc.), U.S. Patent 2,215,137, 1940.

889. Horvath, A. A. (Hall Laboratories, Inc.), U.S. Patent 2,429,579, 1947.

890. Albright and Wilson, Ltd., British Patent 600,933, 1948.

891. Rane, L., and Newhouser, L. R. (Government of the United States), U.S. Patent 2,726,235, 1955.

892. Nitschmann, H., Rickli, E., and Kistler, P., 1959, *Helv. Chim. Acta* 42:2198; 1959, *Chem. Abstr.* 54:12227i.

893. Keil, H. L., Hagen, R. F., and Flawe, R. W., German Patent 1,111,917, 1961; *ibid.*, German Patent 1,133,228, 1962.

894. Fukamachi, C., and Watanabe, T., 1962, *Nippon Shokuhin Kogyo Gakkaishi* 9:19; 1963, *Chem. Abstr.* 59:13264a.

895. Waugh, D. F. (National Dairy Research Laboratories, Inc.), U.S. Patent 2,744,891, 1956.

896. Consolazio, G. A., and Moses, R. E. (General Foods Corp.), U.S. Patent 3,153,030, 1964.

897. Wegner, H., 1957, *Starke* 9:196.

898. Winkler, S., 1960, *Starke* 12:34.

899. Palasinski, M., and Bussek, J., 1964, *Roczniki Technol. Chem. Zywnosci* 10:47; 1964, *Chem. Abstr.* 61:9658c.

900. Palasinski, M., 1964, *Zeszyty Nauk. Wyzszej Szkoly Rolniczej Krakowie Rolnictwo* 11(21):65; 1965, *Chem. Abstr.* 62:15342e.

901. Nara, S., Maeda, I., and Tsujino, T., 1964, *Nippon Nogei Kagaku Kaishi* 38:356; 1965, *Chem. Abstr.* 63:5881c.

902. Knyaginichev, M. I., and Komarov, V. I., *Izv. Vysshikh Uchebn. Zavedenii Pishchevaya Tekhnol.* 1966(1):41; 1966, *Chem. Abstr.* 64:13286h.

903. Kuhl, H., 1937, *Muhlenlab* 7:95; 1937, *Chem. Abstr.* 31:8979.

904. Nutting, G. C., 1952, *J. Colloid Sci.* 7:128.

905. Bowen, W. S., U.S. Patent 2,185,368, 1940.

906. Rozenbroek, M. D. (N.V. Chemische Fabriek Servo), Netherlands Patent 59,843, 1947.

907. Kunz, W. B. (American Viscose Corp.), U.S. Patent 3,030,667, 1962.

908. Fredrickson, R. E. C. (A. E. Staley Mfg. Co.), U.S. Patent 3,003,894, 1961.

909. Osman, E. M., 1967, in *Starch: Chemistry and Technology*, R. L. Whistler and E. F. Paschall, Eds., Academic Press, p. 177.

910. Whistler, R. L., 1971, *Cereal Sci. Today* 16(2):54, 73.

911. Hamilton, R. M., and Paschall, E. F., 1967, in *Starch: Chemistry and Technology*, R. L. Whistler and E. F. Paschall, Eds., Academic Press, p. 351.

912. Evans, J. W. (American Maize Products Co.), U.S. Patent 2,806,026, 1957.

913. Kerr, R. W., and Cleveland, F. C., Jr. (Corn Products Co.), U.S. Patent 2,884,413, 1959.

914. *Ibid.*, U.S. Patent 2,961,440, 1960.

915. Neukom, H. (International Minerals & Chemicals Corp.), U.S. Patent 2,824,870, 1958.

916. *Ibid.*, U.S. Patent 2,865,762, 1958.

917. Ferrara, L. W., U.S. Patent, 2,865,763, 1958.

918. Neukom, H., U.S. Patent 2,884,412, 1959.

919. Kodras, R., U.S. Patent 2,971,954, 1961.

920. Sietsema, J. W., and Trotter, W. C., U.S. Patent 2,993,041, 1961.

921. Schierbaum, F., and Boerner, O., East German Patent 36,806, 1965; 1965, *Chem. Abstr. 63*:3150f.

922. Felton, G. E., and Schopmeyer, H. H. (American Maize Products Co.), U.S. Patent 2,328,537, 1943.

923. Wetzstein, H. L., and Lyon, P., U.S. Patent 2,754,232, 1956.

924. Kerr, R. W., and Cleveland, F. C., Jr. (Corn Products Co.), U.S. Patent 2,801,242, 1957.

925. *Ibid.*, U.S. Patent 2,852,393, 1958.

926. *Ibid.*, U.S. Patent 2,938,901, 1960.

927. Gramera, R. E., Heerema, J., and Parrish, F. W., 1966, *Cereal Chem. 43*:104.

928. Deutsche Maizena Werke G.m.b.H., German Patent 1,109,500, 1961.

929. Ogawa, H., Japanese Patent 5,017 ('70), 1970; Chiyoda Kagako Kogyosho Co., Ltd., Japanese Patent 5018 ('70), 1970; 1970, *Food Sci. Technol. Abstr. 2*:9M798, 9M799.

930. Anon., 1961, *Fed. Reg. 26*:188; *ibid.*, 1964, *29*:14403; *ibid.*, 1964, *29*:15256; *ibid.*, 1967, *32*:8359.

931. Baikow, V. E., 1956, *Sugar 51*:39; 1956, *J. Sci. Food Agr. 7*:ii.

932. Taussig, C. W., and Roland, A. C. (Applied Sugar Laboratories, Inc.), U.S. Patent 2,249,920, 1941; British Patent 525,014, 1940.

933. Cummins, A. B., 1942, *Ind. Eng. Chem. 34*:398.

934. Wright, A., *ibid.*, 425.

935. Boyd, R., Canadian Patent 422,170, 1944.

936. Bliss, L. R., 1948, *Int. Sugar J. 50*:205; 1948, *Chem. Abstr. 42*:8502.

937. Cook, H. A., and Kilby, M., 1948, *Rep. Hawaiian Sugar Technol. 6*:195; 1950, *Chem. Abstr. 44*:7076.

938. Bojorquez, M. O., Jasso, J. deJ. U., and Acosta, J., July 1951, *Bol. Azucar. (Mex.) 2*:23; 1951, *Sugar Ind. Abstr. 13*:182.

939. Black, R. F., and Zemenek, L., 1952, *Sugar Ind. Technicians 1952*:9 pp.; 1952, *Chem. Abstr. 46*:9331.

940. Ishigaki, H., 1953, *Sci. Rep. Hyogo Univ. Agr., Ser. Agr. Chem. 1*:15.

941. N.V. Algemeene Norit Mfg., German Patent 1,008,219, 1957.

942. Zemanek, L. A., 1959, *Publ. Tech. Papers Proc. Ann. Meeting Sugar Ind. Technicians 18*(Sect. VI):1.

943. Delfos, J. J. (N.V. Algemeene Norit Maatschappij), U.S. Patent 2,964,428, 1960.

944. Payet, P. R., French Patent 1,230,842, 1960.

945. Bennett, M. C., and Ragnauth, J. M., 1960, *Int. Sugar J. 62*:13, 41.

946. de Sornay, P., 1915, *Bull. Soc. Chim. Maurice 6*:41; 1916, *Chem. Abstr. 10*:1109.

947. Schulmann, V., 1938, *Listy Cukrovar 56*:501; 1938, *Chem. Abstr. 32*:8813.

948. Klopfer, V., U.S. Patent 2,243,381, 1941.

949. Anon., 1942, *Food Ind. 14(2)*:104.

950. Alvarex, A. S., 1942, *Rev. Ind. Agr. Tucuman 32*:127.

951. Bollaert, A. R., and Halvorsen, G. G. (Great Lakes Carbon Corp.), U.S. Patent 2,470,332, 1949.

952. Golovin, P. V., and Gerasimenko, A. A., U.S.S.R. Patent 116,101, 1958.

953. Golovnyak, Y. D., Kartashov, A. K., and Kurilenko, O. K., *Izv. Vyssh. Uchebn. Zaved. Pishch. Tekhnol. 1962(5)*:78; 1964, *Chem. Abstr. 60*:4321f.

954. Skorobogat'ko, N. I., 1963, *Sakh. Prom. 37(5)*:41; 1963, *Chem. Abstr. 59*:6599g.

955. Muehlpforte, H., 1963, *Zuckererzeugung 7(4)*:106; 1964, *Chem. Abstr. 60*:10909g.

956. Onanchenko, L. I., 1966, *Sakh. Prom. 40(2)*:11; 1966, *Chem. Abstr. 64*:14399f.

957. Katana, A. I., 1966, *Sakh. Prom. 40(12)*:12; 1967, *Chem. Abstr. 66*:77201m.

958. Golubeva, A. D., Kartashov, A. K., Limanskaya, A. Y., and Verchenko, L. M., 1969, *Sakh. Prom. 43(10)*:55; 1970, *Food Sci. Technol. Abstr. 2*:6L394.

959. Bird, M., 1928, *Facts About Sugar 23*:139; 1928, *Planter Sugar Mfr. 80*:143.

960. Ruping, H., German Patent 617,706, 1935.

961. Spengler, O., and Todt, F., 1941, *Z. Wirtschaftsgruppe Zuckerind. 91*:87.

962. Loewy, K., 1943, *Rev. Bras. Quim. 15*:105; 1943, *Chem. Abstr. 37*:6921.

963. Gururaja, J. S., and Ray, P. K., 1946, *Indian Sugar 7*:127.

964. Andres, P., 1940, *Z. Wirtschaftsgruppe Zuckerind. 90*:228; 1941, *Chem. Abstr. 35*:4987.
965. Neumann, M. W., 1940, *Arch. Suikerind. Nederland Ned.-Indie 1*:51; 1940, *Chem. Abstr. 34*:4604.
966. Saha, J. M., and Jain, N. S., 1954, *Sugar 49(3)*:38; 1954, *Chem. Abstr. 48*:5533.
967. Chemische Werke Albert and Sueddeutsche Zucker A.-G., Belgian Patent 611,251, 1962; 1962, *Chem. Abstr. 57*:14042h.
968. Barrett, E. P., Brown, J. M., and Olech, S. M., 1951, *Ind. Eng. Chem. 43*:639.
969. Barrett, E. P., 1952, *Sugar Ind. Technicians 1952*:9 pp.; 1952, *Chem. Abstr. 46*:9331.
970. Kubala, S., 1952, *Cukoripar 5*:229; 1953, *Chem. Abstr. 47*:11775.
971. Kortschak, H. P., 1942, *Hawaii. Plant. Rec. 46*:105.
972. Jung, E. V. (Knapsack Aktiengesellschaft), U.S. Patent 3,347,705, 1967.
973. Cummings, A. B. (Johns-Manville Products Corp.), Canadian Patent 421,985, 1944.
974. Fowler, A. P., 1936, *Proc. Cuban Sugar Tech. Assoc. 10*:52.
975. Sanchez, P., and Ehrhert, E. N. (Sucro-Blanc, Inc.), U.S. Patent 2,216,753, 1940.
976. Isbell, H. S. (Government of the United States), U.S. Patent 2,421,380, 1947.
977. Vincent, G. P. (Mathieson Alkali Works, Inc.), U.S. Patent 2,430,262, 1947.
978. Andresen, V. V. J., U.S. Patent 1,732,492, 1929.
979. Yasuo, I., 1955, *Sugar J. 18(5)*:29.
980. Schongart, H. A., 1959, *Z. Zuckerind. 9(1)*:10.
981. Gupta, S. C., and Ramaiah, N. A., 1965, *Indian Sugar 14*:719; 1965, *Chem. Abstr. 63*:18435f.
982. Moss, H. V., 1934, *The Canner 79(6)*:18; 1934, *Chem. Abstr. 28*:7054.
983. White, J. R., and Dunn, J. A. (Lever Bros. Co.), U.S. Patent 2,225,894, 1940.
984. Nelson, T. J., 1949, *Food Technol. 3*:347.
985. Mead, M. W. (National Grain Yeast Corp.), U.S. Patent 2,075,127, 1937.
986. Byk, H. (Chemische Werke, Charlottenburg), German Patent 247,809, 1910.
987. Epstein, A. K., and Harris, B. R., U.S. Patent 2,075,807, 1937.
988. Martin, G. D. (Monsanto Chemical Co.), U.S. Patent 2,247,280, 1941.
989. Nyrop, J. E., British Patent 640,241, 1950.
990. Harris, B. R., U.S. Patent 2,177,983, 1939.
991. Katzman, M. (Emulsol Corp.), U.S. Patent 2,243,868, 1941.
992. Thurman, B. H. (Refining, Inc.), U.S. Patent 2,271,409, 1942.
993. *Ibid.*, U.S. Patent 2,272,616, 1942.
994. Pader, M., and Gershon, S. D. (Lever Bros. Co.), U.S. Patent 3,224,883, 1965; 1967, *Chem. Abstr. 67*:2295g.
995. *Ibid.*, U.S. Patent 3,224,884, 1965.
996. Katzman, M. B. (Emulsol Corp.), U.S. Patent 2,176,078, 1939.
997. Thompson, J. E. (Procter & Gamble Co.), U.S. Patent 3,458,323, 1969.
998. Cunningham, F. E., Kline, L., and Lineweaver, H. (United States of America), U.S. Patent 3,328,175, 1967.
999. Gumbmann, M. R., Gagne, W. E., and Williams, S. N., 1968, *Toxicol. Appl. Pharmacol. 12*:360.
1000. Curtin, J. H., and Gagolski, J. (Colonial Sugar Refining Co., Ltd.), U.S. Patent 3,375,168, 1968.
1001. Toy, A. D. F. (Colonial Sugar Refining Co., Ltd.), U.S. Patent 3,428,624, 1969.
1002. Stauffer Chemical Company, Netherlands Patent Application No. 6,608,960, 1966; British Patent No. 1,139,684 (1969).
1003. Tucker, J. W., and Toy, A. D. F. (Colonial Sugar Refining Co., Ltd.), Canadian Patent 825,087, 1969.
1004. Colonial Sugar Refining Co., Ltd., Japanese Patent 21,383 ('69), 1969.
1005. Kyowa Hakko Kogyo Co., Ltd., Japanese Patent 13,486 ('70), 1970; 1971, *Food Sci. Technol. Abstr. 3*:1H58.

PART II

Regulatory Status of
Direct Food Additives

Introduction

The primary purpose of Part II is to provide the reader with quick summaries of and references to the current U.S. government and other regulations for some of the most frequently employed direct food additives. Indirect additives, such as pesticide residues, packaging adhesives, polymeric coatings, and animal feed chemicals, have not been included. Similarly, pending petitions, Food and Drug Administration (FDA) proposals, rejections of petitions, and petitions withdrawn without prejudice have not been included, since these represent, at best, a transient state or non-approved use. Since regulations are far-reaching and changes are frequent, the ultimate source of reference is the Food and Drug Administration and the official regulations appearing in the *Federal Register*.

The *Food Chemicals Codex*, prescribing minimum requirements of purity for chemicals intended for use in foods, has been endorsed by the Commissioner of the Food and Drug Administration as an outstanding guide and one that will be regarded by the FDA as defining an "appropriate food grade." The editor concurs with this view, but also he stresses the absolute necessity for users of food chemicals to establish beyond question the quality of all ingredients to be employed in foods. In addition to becoming thoroughly familiar with the regulatory limits on use, the reader is urged to consult with additive suppliers to make certain that their products meet with approved U.S. government regulations regarding manufacture and purity. This aspect is mandatory for both suppliers and users of food additives, since all FDA clearances are based on principles of "good manufacturing practice."

The reader's attention is called to the following citations employed in Part II, and, at the same time, he is cautioned to proper interpretations.

> **GRAS** indicates products *generally recognized as safe*. It is important to distinguish whether the product is sanctioned by a government agency, such as the FDA or MID, or by a trade association, such as FEMA. The FDA has the right to challenge trade association findings.
>
> **FEMA** signifies Flavor and Extract Manufacturers' Association. In the past the FDA has relied on the opinions of an expert committee of this organization to reach clearance decisions on flavor components. For many of the natural and synthetic flavors listed in Part II, the FEMA number, which indicates GRAS status, has been included as an additional reference.
>
> **USDA** indicates the U.S. Department of Agriculture.
>
> **MID** refers to the Meat Inspection Division of the U.S. Department of Agriculture. This agency is responsible for clearing additives intended for use in all meats and meat products, with the exception of poultry. The USDA Poultry Division does not list approved additives, and it is recommended that the reader consult directly with this division concerning specific food substrates.
>
> **FR** signifies the *Federal Register*. This is the official U.S. government publication wherein all clearances, petitions, withdrawals, time extensions, and other pertinent rulings are made public.
>
> **Food additive orders** are indicated by an FDA section number (e.g., §121.2520) wherein the specific regulation is detailed. The latest version of such orders should be consulted, since limitations as to use, product specifications, and labeling constantly are changing.

REGULATORY STATUS OF DIRECT FOOD ADDITIVES

Material	Alternate Name [Source]	FEMA No.	Regulation	Limitations
Absinthium	*see* Artemisia			
Acacia	Gum acacia; gum arabic [*Acacia senegal* (L.) Willd.; *A. verek* Guill. & Per.]	2001	FDA §121.101	GRAS, stabilizer
Acetal	Acetaldehyde diethyl acetal	2002	FDA §121.1164	Synthetic flavor/adjuvant
Acetaldehyde	Ethanal	2003	FDA §121.101	GRAS, synthetic flavor/adjuvant

REGULATORY STATUS OF DIRECT FOOD ADDITIVES (continued)

Material	Alternate Name [Source]	FEMA No.	Regulation	Limitations
Acetaldehyde benzyl β-methoxyethyl acetal	*see* Benzyl methoxyethyl acetal			
Acetaldehyde, butyl phenethyl acetal		3125		
Acetaldehyde diethyl acetal	*see* Acetal			
Acetaldehyde phenethyl propyl acetal		2004	FDA §121.1164	Synthetic flavor/adjuvant
Acetanisole	4′-Methoxyacetophenone	2005	FDA §121.1164	Synthetic flavor/adjuvant
Acetic acid	Ethanoic acid	2006	FDA §121.101	GRAS, miscellaneous and/or general-purpose food additive. Product specifications apply
			FDA §8.303	Production of caramel
			MID	To separate fatty acids and glycerol in rendered fats; refining agent must be eliminated during manufacturing process
Acetic anhydride	Acetic oxide		FDA §121.1030	Esterifier for food starch, in combination with adipic anhydride; adipic anhydride not to exceed 0.12%; acetic anhydride not to exceed 5.0%
Acetoin	Acetyl methyl carbinol; 2,3-butanolone; dimethylketol; 3-hydroxy-2-butanone; γ-hydroxy-β-oxobutanone	2008	FDA §121.101	GRAS, synthetic flavor/adjuvant
2′-Acetonaphthone	*see* Methyl β-naphthyl ketone			
Acetone	Dimethyl ketone		FDA §121.1042	30 ppm in spice oleoresins, as a residue from extraction of spice
			FDA §8.390	Diluent in inks for marking fruits and vegetables, with no residue. Product specifications apply

REGULATORY STATUS OF DIRECT FOOD ADDITIVES (continued)

Material	Alternate Name [Source]	FEMA No.	Regulation	Limitations
Acetone (continued)			FDA §8.305	Solvent for annatto extract; residue limit of 50 ppm
Acetone peroxides			FDA §121.1023	Manufacturing and bleaching of flour; in bread and rolls where standards of identity do not preclude use. Product specifications apply
Acetophenone	Methyl phenyl ketone	2009	FDA §121.1164	Synthetic flavor/adjuvant
Acetylated monoglycerides			FDA §121.1018	Product specifications apply
			MID	Emulsifier for shortening
Acetyl butyryl	*see* 2,3-Hexanedione			
2-Acetyl-3-ethyl pyrazine	2-Acetyl-3-ethyl-1,4-diazine	3250		*see Food Technol.,* May 1972
2-Acetylfuran	*see* 2-Furyl methyl ketone			
Acetyl isobutyryl	*see* 4-Methyl-2,3-pentanedione			
Acetyl methyl carbinol	*see* Acetoin			
Acetyl nonyryl	*see* 2,3-Undecadione			
Acetyl propionyl	*see* 2,3-Pentanedione			
Acetylpyrazine	Methyl pyrazinyl ketone	3126		10 ppm max in finished foods
2-Acetylpyridine		3251		*see Food Technol.,* May 1972
2-Acetylpyrrole	*see* Methyl 2-pyrrolyl ketone			
Acetyl valeryl	*see* 2,3-Heptanedione			
Acetyl vanillin	*see* Vanillin acetate			
Achilleic acid	*see* Aconitic acid			

REGULATORY STATUS OF DIRECT FOOD ADDITIVES (continued)

Material	Alternate Name [Source]	FEMA No.	Regulation	Limitations
Acidophilus-type bacterial starters			MID	0.5%, to develop flavor in dry sausage, pork roll, thuringer, Lebanon bologna, cervelat, and salami. Product specifications apply
Aconitic acid	Achilleic acid; citridic acid; equisetic acid [*Aconitum napellus* L.]	2010	FDA §121.101	GRAS, synthetic flavor/adjuvant
Acrolein	2-Propenal		FDA §121.1031	Etherification of food starch up to 0.6%; esterification and etherification of food starch up to 0.3%, with vinyl acetate up to 7.5%
Adipic acid	1,4-Butanedicarboxylic acid	2011	FDA §121.101	GRAS, buffer and neutralizing agent
			FDA §121.1164	Synthetic flavor
Adipic anhydride			FDA §121.1031	Esterifier for food starch, in combination with acetic anhydride; adipic anhydride not to exceed 0.12%, acetic anhydride not to exceed 5.0%
Agar	Agar-agar [*Gelidium cartilagineum* (L.) Gaillon; *Gracilaria confervoides* (L.) Greville; and related red algae	2012	FDA §121.101	GRAS, stabilizer
Alanine (L- and DL-forms)			FDA §121.101	GRAS, nutrient and/or dietary supplement, as the free acid, hydrochloride salt, hydrate, or anhydrous form
β-Alanine		3252		
Alcohol, anhydrous	*see* Ethyl alcohol			
Alcohol C-12	*see* Lauryl alcohol			
Alcohol, denatured			FDA §8.390	Diluent in mixtures for coloring shell eggs with no penetration of the mixture through shell egg into egg. Product specifications apply

REGULATORY STATUS OF DIRECT FOOD ADDITIVES (continued)

Material	Alternate Name [Source]	FEMA No.	Regulation	Limitations
Alcohol, SDA–3A			FDA §8.390	Diluent in inks for marking fruits and vegetables, with no residue. Product specifications apply
Alcohol, specially denatured			FDA §8.300	Diluent in inks for marking gum and confectionery and for marking food supplements in tablet form, with no residues. Product specifications apply
Aldehyde C-6	*see* Hexanal			
Aldehyde C-7	*see* Heptanal			
Aldehyde C-8	*see* n-Octanal			
Aldehyde C-8 dimethyl acetal	*see* Octanal dimethyl acetal			
Aldehyde C-9	*see* n-Nonanal			
Aldehyde C-10	*see* Decanal			
Aldehyde C-10 dimethyl acetal	*see* Decanal dimethyl acetal			
Aldehyde C-11 undecylenic	*see* 10-Undecenal			
Aldehyde C-11 undecylic	*see* Undecanal			
Aldehyde C-12	*see* Lauric aldehyde			
Aldehyde C-12 M.N.A.	*see* 2-Methyl undecanal			
Aldehyde C-14 myristic	*see* Myristaldehyde			
Aldehyde C-14 (so-called)	*see* γ-Undecalactone			
Aldehyde C-16 (so-called)	*see* 3-Methyl-3-phenyl glycidic acid ethyl ester			
Aldehyde C-18 (so-called)	*see* γ-Nonalactone			
Alfalfa, extract	[*Medicago sativa* L.]	2013		

REGULATORY STATUS OF DIRECT FOOD ADDITIVES (continued)

Material	Alternate Name [Source]	FEMA No.	Regulation	Limitations
Alfalfa, herb and seed	[*Medicago sativa* L.]		FDA §121.101	GRAS, natural flavor and natural flavor extractive
Algae, brown	Kelp [*Laminaria* species; *Nereocystis* species]		FDA §121.101	GRAS, natural substance and solvent-free natural extractive used with natural flavorings
Algae, red	[*Porphyra* species; *Rhodymenia palmata* (L.) Grev.]		FDA §121.101	GRAS, natural substance and solvent-free natural extractive used with natural flavorings
Algin	[*Laminaria* species and other kelps]	2014	FDA §19	Up to 0.8% in specified cheeses; up to 0.5% by weight in specified cheeses
			FDA §20	In frozen desserts
			MID	Extender and stabilizer in breading mixes and sauces
Alginates—sodium, calcium, and ammonium salts		2015	FDA §121.101	GRAS, stabilizer
			FDA §121.1164	Synthetic flavor/adjuvant
Alkaline aqueous solution			FDA §8.305	Extractant for annatto
Alkalis			FDA §8.305	Food-grade, to adjust pH. Product specifications apply
Alkanet, root extract	[*Alkanna tinctoria* Tausch.]	2016	MID	Coloring sausage casings, oleomargarine, and shortening; marking or branding ink on products. Product specifications apply
Allspice *see also* Pimenta	[*Pimenta officinalis* Lindl.]	2017	FDA §121.101	GRAS, natural flavor and natural flavor extractive
Allura Red AC	*see* FD&C Red No. 40			
Allyl anthranilate		2020	FDA §121.1164	Synthetic flavor
Allyl butyrate		2021	FDA §121.1164	Synthetic flavor/adjuvant
Allyl caproate	*see* Allyl hexanoate			
Allyl cinnamate		2022	FDA §121.1164	Synthetic flavor

REGULATORY STATUS OF DIRECT FOOD ADDITIVES (continued)

Material	Alternate Name [Source]	FEMA No.	Regulation	Limitations
Allyl cyclohexaneace-tate		2023	FDA §121.1164	Synthetic flavor
Allyl cyclohexanebuty-rate		2024	FDA §121.1164	Synthetic flavor
Allyl cyclohexanehex-anoate		2025	FDA §121.1164	Synthetic flavor
Allyl cyclohexanepro-pionate	Allyl 3-cyclohexylpro-pionate; allyl β-cyclo-hexylpropionate	2026	FDA §121.1164	Synthetic flavor
Allyl cyclohexanevaler-ate		2027	FDA §121.1164	Synthetic flavor
Allyl disulfide	Diallyl disulfide	2028	FDA §121.1164	Synthetic flavor
Allyl 2-ethylbutyrate		2029	FDA §121.1164	Synthetic flavor
Allyl 2-furoate		2030		
Allyl heptanoate		2031		
Allyl 2,4-hexadienoate	*see* Allyl sorbate			
Allyl hexanoate	Allyl caproate	2032	FDA §121.1164	Synthetic flavor/adjuvant
Allyl α-ionone	1-(2,6,6-Trimethyl-2-cy-clohexen-1-yl)-1,6-hepta-dien-3-one	2033	FDA §121.1164	Synthetic flavor
Allyl isothiocyanate	Mustard oil	2034	FDA §121.1164	Synthetic flavor
Allyl isovalerate		2045	FDA §121.1164	Synthetic flavor
Allyl mercaptan	2-Propene-1-thiol	2035	FDA §121.1164	Synthetic flavor
Allyl *trans*-2-methyl-2-butenoate	*see* Allyl tiglate			
Allyl methyl disulfide		3127		
Allyl methyl trisulfide		3265		
Allyl nonanoate		2036	FDA §121.1164	Synthetic flavor
Allyl octanoate		2037	FDA §121.1164	Synthetic flavor

REGULATORY STATUS OF DIRECT FOOD ADDITIVES (continued)

Material	Alternate Name [Source]	FEMA No.	Regulation	Limitations
Allyl phenoxyacetate	Acetate PA	2038	FDA §121.1164	Synthetic flavor
Allyl phenylacetate		2039	FDA §121.1164	Synthetic flavor
Allyl propionate		2040	FDA §121.1164	Synthetic flavor
Allyl sorbate	Allyl 2,4-hexadienoate	2041	FDA §121.1164	Synthetic flavor
Allyl sulfide	Thioallyl ether; diallyl sulfide	2042	FDA §121.1164	Synthetic flavor
Allyl tiglate	Allyl *trans*-2-methyl-2-butenoate	2043	FDA §121.1164	Synthetic flavor
Allyl trisulfide	*see* Diallyl trisulfide			
Allyl 10-undecenoate	Allyl undecylenate	2044	FDA §121.1164	Synthetic flavor
4-Allylveratrole	*see* Eugenyl methyl ether			
Almond, bitter, oil	[*Prunus amygdalus* Batsch; also from *P. armeniaca* L. or *P. persica* (L.) Batsch]	2046	FDA §121.101	GRAS, natural flavor extractive when free from prussic acid
Aloe, extract	[*Aloe perryi* Baker; *A. barbadensis* Mill.; *A. ferox* Mill.; and hybrids of this species with *A. Africana* Mill. and *A. spicata* Baker]	2047	FDA §121.1163	Natural flavor
Althea, root	[*Althaea officinalis* L.]	2048	FDA §121.1163	Natural flavor
Alumina			FDA §9.100	For manufacture of FD&C color lakes. Product specifications apply
Aluminum ammonium sulfate			FDA §121.101	GRAS, miscellaneous and/or general-purpose food additive
Aluminum calcium silicate			FDA §22.8 and §22.9	In vanilla powder and vanilla-vanillin powder under food standards
			FDA §121.101	GRAS, anticaking agent in table salt, up to 2%
Aluminum nicotinate			FDA §121.1141	Source of niacin in foods for special dietary use. Product specifications apply

REGULATORY STATUS OF DIRECT FOOD ADDITIVES (continued)

Material	Alternate Name [Source]	FEMA No.	Regulation	Limitations
Aluminum oxide			FDA §8.316 and §8.6005	Dispersing agent in mixtures for food or drug use containing titanium dioxide, up to 2%
Aluminum potassium sulfate			FDA §121.101	GRAS, miscellaneous and/or general-purpose food additive
Aluminum salts of fatty acids			FDA §121.1071	Binder, emulsifier, and anticaking agent in food. Product specifications apply
Aluminum sodium sulfate			FDA §121.101	GRAS, miscellaneous and/or general-purpose food additive
Aluminum stearate			FDA §121.1071 and §121.1099	Component of defoaming agents limited to use in processing beet sugar and yeast
Aluminum sulfate			FDA §121.101	GRAS, miscellaneous and/or general-purpose food additive
			FDA §121.1031	Modifier for food starch, up to 2.0% in combination with not more than 2.0% of 1-octenyl succinic anhydride
Ambergris	[*Physeter macrocephalus* L.]		FDA §121.101	GRAS, natural extractive
Ambrette, seed	[*Hibiscus moschatus* Moench; *H. abelmoschus* ·L.]		FDA §121.101	GRAS, natural flavor
Ambrettolide	*see* ω-6-Hexadecenlactone			
Aminoacetic acid	*see* Glycine			
Aminoethanoic acid	*see* Glycine			
Aminoglutaric acid	*see* L-Glutamic acid			
Ammonium alginate	*see* Alginates—sodium, calcium, and ammonium salts			

REGULATORY STATUS OF DIRECT FOOD ADDITIVES (continued)

Material	Alternate Name [Source]	FEMA No.	Regulation	Limitations
Ammonium bicarbonate			FDA §121.101	GRAS, miscellaneous and/ or general-purpose food additive
Ammonium carbonate			FDA §121.101	GRAS, miscellaneous and/ or general-purpose food additive
			FDA §8.303	Including bicarbonate, in production of caramel
Ammonium caseinate			FDA §20	In frozen desserts
Ammonium chloride			FDA §17	In bakery products; alone or with calcium sulfate and/ or dicalcium phosphate, up to 0.25% by weight of flour
Ammonium hydroxide			FDA §121.101	GRAS, miscellaneous and/ or general-purpose food additive
			FDA §8.303	In production of caramel
Ammonium isovalerate		2054	FDA §121.1164	Synthetic flavor
Ammonium persulfate			FDA §121.1031	Bleaching agent for food starch, up to 0.075%; with sulfur dioxide, up to 0.05%
Ammonium phosphate (mono- and dibasic)			FDA §121.101	GRAS, miscellaneous and/ or general-purpose food additive
			FDA §8.303	In production of caramel
Ammonium saccharin			FDA §121.101	GRAS, nonnutritive sweetener
Ammonium sulfate			FDA §121.101	GRAS, miscellaneous and/ or general-purpose food additive
			FDA §8.303	In production of caramel
Ammonium sulfide		2053	FDA §121.1164	Synthetic flavor
Ammonium sulfite			FDA §8.303	In production of caramel
Amyl alcohol	Pentyl alcohol	2056	FDA §121.1164	Synthetic flavor
Amyl butyrate	Pentyl butyrate	2059	FDA §121.1164	Synthetic flavor

REGULATORY STATUS OF DIRECT FOOD ADDITIVES (continued)

Material	Alternate Name [Source]	FEMA No.	Regulation	Limitations
α-Amylcinnamaldehyde	α-Pentylcinnamaldehyde	2061	FDA §121.1164	Synthetic flavor
α-Amylcinnamaldehyde dimethyl acetal	1,1-Dimethoxy-2-amyl-3-phenyl-2-propene	2062	FDA §121.1164	Synthetic flavor
α-Amylcinnamyl acetate	α-Pentylcinnamyl acetate	2064	FDA §121.1164	Synthetic flavor
α-Amylcinnamyl alcohol	α-Pentylcinnamyl alcohol	2065	FDA §121.1164	Synthetic flavor
α-Amylcinnamyl formate	α-Pentylcinnamyl formate	2066	FDA §121.1164	Synthetic flavor
α-Amyl cinnamyl isovalerate	α-Pentylcinnamyl isovalerate	2067	FDA §121.1164	Synthetic flavor
Amyl formate	Pentyl formate	2068	FDA §121.1164	Synthetic flavor
Amyl 2-furoate	Pentyl 2-furoate	2072		
Amyl heptanoate	Pentyl heptanoate	2073	FDA §121.1164	Synthetic flavor
Amyl hexanoate	Pentyl hexanoate	2074	FDA §121.1164	Synthetic flavor
Amyl octanoate	Pentyl octanoate	2079	FDA §121.1164	Synthetic flavor
Amyl vinyl carbinol	see 1-Octen-3-ol			
Amyris	West Indian sandalwood [Amyris balsamifera L.]		FDA §121.1163	Natural flavor
Anethole	Parapropenyl anisole	2086	FDA §121.101	GRAS, synthetic flavor/adjuvant
Angelica lactone	see ω-Pentadecalactone			
Angelica, root, extract	[Angelica archangelica L. or other species of Angelica]	2087	FDA §121.101	GRAS, natural flavor extractive
Angelica, root, oil	[Angelica archangelica L.]	2088	FDA §121.101	GRAS, natural flavor extractive
Angola weed	[Roccella fuciformis Ach.]		FDA §121.1163	Natural flavoring use in alcoholic beverages only
Angostura, extract	Cusparia [Galipea officinalis Hancock (G. cusparia DC.)]	2092	FDA §121.101	GRAS, natural flavor and natural flavor extractive

REGULATORY STATUS OF DIRECT FOOD ADDITIVES (continued)

Material	Alternate Name [Source]	FEMA No.	Regulation	Limitations
p-Anisaldehyde	see p-Methoxybenzalde-hyde			
Anise	[Pimpinella anisum L.]	2093	FDA §121.101	GRAS, natural flavor and natural flavor extractive
Anise, star	[Illicium verum Hook. f.]	2095	FDA §121.101	GRAS, natural flavor
Anisic ketone	see 1-(p-Methoxyphenyl)-2-propanone			
Anisole	Methoxybenzene	2097	FDA §121.1164	Synthetic flavor
Anisyl acetate	p-Methoxybenzyl acetate	2098	FDA §121.1164	Synthetic flavor
Anisyl acetone	see 4-(p-Methoxyphenyl)-2-butanone			
Anisyl alcohol	p-Methoxybenzyl alcohol	2099	FDA §121.1164	Synthetic flavor
Anisyl butyrate	p-Methoxybenzyl butyrate	2100	FDA §121.1164	Synthetic flavor
Anisyl formate	p-Methoxybenzyl formate	2101	FDA §121.1164	Synthetic flavor
Anisyl methyl ketone	see 1-(p-Methoxyphenyl)-2-propanone			
Anisyl phenylacetate			FDA §121.1164	Synthetic flavor
Anisyl propionate	p-Methoxybenzyl propionate	2102	FDA §121.1164	Synthetic flavor
Annatto	[Bixa orellana L.]		MID	Coloring sausage casings, oleomargarine, or shortening; marking or branding inks on products. Product specifications apply
Annatto, extract	[Bixa orellana L.]	2103	FDA §8.6003	Coloring ingested drugs generally; conforming with §8.305. Product specifications apply
			FDA §8.305	Coloring foods generally. Product specifications apply
Annatto, seed	[Bixa orellana L.]	2104	FEMA—GRAS	

REGULATORY STATUS OF DIRECT FOOD ADDITIVES (continued)

Material	Alternate Name [Source]	FEMA No.	Regulation	Limitations
β-Apo-8′-carotenal			FDA §8.302	Food color; not to exceed 15 mg/lb of solid or semi-solid food or 15 mg/pint of liquid food. Product specifications apply
Apricot kernel	Persic oil; bitter almond [*Prunus armeniaca* L.]	2105	FDA §121.101	GRAS, solvent-free natural extractive used with natural flavorings
Arabinogalactan	Western larch wood	3254	FDA §121.1174	Emulsifier, stabilizer, binder or bodying agent in essential oils, nonnutritive sweeteners, flavor bases, nonstandardized dressings, and pudding mixes. Product specifications apply see *Food Technol.,* May 1972, for FEMA average max use level
L-Arabinose		3255		see *Food Technol.,* May 1972
Arborvitae	*see* Cedar, white			
Arginine (L- and DL-forms)			FDA §121.101	GRAS, nutrient and/or dietary supplement, as the free acid, hydrochloride salt, hydrate, or anhydrous form
Arnica, flowers	[*Arnica montana* L.; *A. fulgens* Pursh.; *A. sororia* Greene; *A. cordifolia* Hooker]		FDA §121.1163	Natural flavor
Artemisia	Wormwood [*Artemisia* species]	3114	FDA §121.1163	Natural flavor; thujone free
Artichoke, leaves	[*Cynara scolymus* L.]		FDA §121.1163	Natural flavoring use in alcoholic beverages only
Artificial smoke flavoring			MID	To flavor various meat products
Asafetida	[*Ferula assafoetida* L. and related *Ferula* species]		FDA §121.101	GRAS, natural flavor extractive
Ascorbic acid	Vitamin C	2109	FDA §121.101	GRAS, nutrient and/or dietary supplement, chemical preservative

REGULATORY STATUS OF DIRECT FOOD ADDITIVES (continued)

Material	Alternate Name [Source]	FEMA No.	Regulation	Limitations
Ascorbic Acid (continued)			MID	To accelerate color fixing in cured pork and beef cuts and cured comminuted meat food products. Product specifications apply
Ascorbyl palmitate			FDA §121.101	GRAS, chemical preservative
Aspartic acid (L- and DL-forms)			FDA §121.101	GRAS, nutrient and/or dietary supplement, as the free acid, hydrochloride salt, hydrate or anhydrous form
Aspergillus flavus-oryzae **group**			MID	With water, salt and monosodium glutamate to soften tissues of beef cuts; tissue weight gain, 3% max
Aspergillus oryzae			FDA §17.1	In bakery products
			MID	With water, salt and monosodium glutamate to soften tissues in beef cuts; tissue weight gain, 3% max
Assafoetida	*see* Asafetida			
Autolyzed yeast extract	*see* Yeast extract, autolyzed			
Azodicarbonamide			FDA §121.1085	Aging and bleaching ingredient in cereal flour, up to 45 ppm. Product specifications apply
Bacterial catalase	[*Micrococcus lysodeikticus*]		FDA §19.500	In cheddar cheese manufacture for removal of hydrogen peroxide; not to exceed 20 ppm of weight of milk treated
			FDA §121.1170	Destroying and removing hydrogen peroxide used in manufacture of cheese. Product specifications apply
Bacterial starters			MID	Harmless bacterial starters of the acidophilus type, lactic acid starters, or cultures of *Pediococcus cerevisiae*; to

REGULATORY STATUS OF DIRECT FOOD ADDITIVES (continued)

Material	Alternate Name [Source]	FEMA No.	Regulation	Limitations
Bacterial starters (continued)				develop flavor in dry sausage, pork roll, thuringer, Lebanon bologna, cervelat, and salami, at 0.5%
Baking powder			FDA	GRAS
Balm	Lemon balm; melissa [*Melissa officinalis* L.]	2113	FDA §121.101	GRAS, natural flavor and natural flavor extractive
Balsam fir	*see* Fir, balsam			
Balsam of Peru	*see* Peruvian balsam			
Basil, bush	Dwarf basil [*Ocimum minimum* L.]		FDA §121.101	GRAS, natural flavor
Basil	[*Ocimum basilicum* L.]	2118	FDA §121.101	GRAS, natural flavor and natural flavor extractive
Bay leaf, West Indian, oil	Myrcia oil [*Pimenta racemosa* (Mill.) J.W. Moore; *P. acris* Kostel.]	2122	FDA §121.101	GRAS, natural flavor extractive
Bay, sweet	Bay; laurel [*Laurus nobilis* L.]	2124	FDA §121.101	GRAS, natural flavor
Bay, sweet, oil	[*Laurus nobilis* L.]	2125	FDA §121.101	GRAS, natural flavor extractive
Beechwood creosote			FDA §121.1164	Synthetic flavor
Beeswax, white	Bleached yellow wax; cire d'abeille absolute [*Apis mellifera* L.]	2126	FDA §121.1163	Natural flavor
			FDA §121.101	GRAS, miscellaneous and/or general-purpose food additive
Beeswax, yellow	Yellow wax		FDA §121.101	GRAS, miscellaneous and/or general-purpose food additive
Beets, dehydrated	Beet powder		FDA §8.321	Coloring foods generally. Product specifications apply
Bentonite	Wilkinite		FDA §121.101	GRAS, miscellaneous and/or general-purpose food additive

REGULATORY STATUS OF DIRECT FOOD ADDITIVES (continued)

Material	Alternate Name [Source]	FEMA No.	Regulation	Limitations
Benzaldehyde	Benzoic aldehyde; benzenecarbonal	2127	FDA §121.101	GRAS, synthetic flavor/adjuvant
Benzaldehyde dimethyl acetal		2128	FDA §121.1164	Synthetic flavor
Benzaldehyde glyceryl acetal	2-Phenyl-*m*-dioxan-5-ol	2129	FDA §121.1164	Synthetic flavor
Benzaldehyde propylene glycol acetal	4-Methyl-2-phenyl-*m*-dioxolane	2130	FDA §121.1164	Synthetic flavor
Benzenethiol	Thiophenol		FDA §121.1164	Synthetic flavor
2-Benzofurancarbox-aldehyde	2-Formylbenzofuran	3128		
Benzoic acid	Benzene carboxylic acid; dracylic acid; phenylformic acid	2131	FDA §121.101	GRAS, chemical preservative up to 0.1%
			MID	To retard flavor reversion in oleomargarine, at 0.1%
Benzoic aldehyde	*see* Benzaldehyde			
Benzoin	2-Hydroxy-2-phenyl-acetophenone	2132	FDA §121.1164	Synthetic flavor
			FDA §8.390	Diluent in inks for marking fruits and vegetables. Product specifications apply
Benzoin resin	[*Styrax benzoin* Dryander; *S. paralleloneurus* Perkins; *S. tonkinensis* (Pierre) Craib, or other species of the section *Anthostyrax* of the genus *Styrax*	2133	FDA §121.1163	Natural flavor
Benzophenone	Diphenyl ketone	2134	FDA §121.1164	Synthetic flavor
Benzothiazole		3256		*see Food Technol.*, May 1972
Benzoyl peroxide	Benzoyl superoxide		FDA §19	In specified cheeses up to 0.000 2% of milk
			FDA §15	Bleaching agent in flour; 1 part by weight of benzoyl peroxide mixed with not more than 6 parts by weight of one or more of potassium

REGULATORY STATUS OF DIRECT FOOD ADDITIVES (continued)

Material	Alternate Name [Source]	FEMA No.	Regulation	Limitations
Benzoyl peroxide (continued)				alum, calcium sulfate, magnesium carbonate, sodium aluminum sulfate, dicalcium phosphate, tricalcium phosphate, start, and calcium carbonate
Benzyl acetate	Acetic acid benzyl ester	2135	FDA §121.1164	Synthetic flavor
Benzyl acetoacetate	Benzyl acetyl acetate	2136	FDA §121.1164	Synthetic flavor
Benzyl alcohol	Phenylcarbinol	2137	FDA §121.1164	Synthetic flavor
Benzyl benzoate	Benzyl benzene carboxylate	2138	FDA §121.1164	Synthetic flavor
Benzyl butyl ether		2139	FDA §121.1164	Synthetic flavor
Benzyl butyrate	Benzyl butanoate	2140	FDA §121.1164	Synthetic flavor
Benzyl cinnamate	Benzyl β-phenylacrylate	2142	FDA §121.1164	Synthetic flavor
Benzyl dimethyl-carbinyl acetate	see α,α-Dimethyl-phenethyl acetate			
Benzyl dimethyl-carbinyl butyrate	see α,α-Dimethyl-phenethyl butyrate			
Benzyl dimethyl-carbinyl formate	see α,α-Dimethyl-phenethyl formate			
Benzyl 2,3-dimethyl-crotonate	Benzyl methyl tiglate	2143	FDA §121.1164	Synthetic flavor
Benzyl dipropyl ketone	see 3-Benzyl-4-heptanone			
Benzyl disulfide	Dibenzyl disulfide		FDA §121.1164	Synthetic flavor
Benzyl ethyl ether		2144	FDA §121.1164	Synthetic flavor
Benzyl formate	Formic acid benzyl ester	2145	FDA §121.1164	Synthetic flavor
3-Benzyl-4-heptanone	Benzyl dipropyl ketone	2146	FDA §121.1164	Synthetic flavor
Benzyl isobutyl ketone	see 4-Methyl-1-phenyl-2-pentanone			
Benzyl isobutyrate		2141	FDA §121.1164	Synthetic flavor

REGULATORY STATUS OF DIRECT FOOD ADDITIVES (continued)

Material	Alternate Name [Source]	FEMA No.	Regulation	Limitations
Benzyl isoeugenol	see Isoeugenyl benzyl ether			
Benzyl isovalerate		2152	FDA §121.1164	Synthetic flavor
Benzyl mercaptan	α-Toluenethiol	2147	FDA §121.1164	Synthetic flavor
Benzyl methoxyethyl acetal	Acetaldehyde benzyl β-methoxyethyl acetal	2148	FDA §121.1164	Synthetic flavor
Benzyl methyl tiglate	see Benzyl 2,3-dimethyl-crotonate			
Benzyl phenylacetate		2149	FDA §121.1164	Synthetic flavor
Benzyl propionate	Benzyl propanoate	2150	FDA §121.1164	Synthetic flavor
Benzylpropyl acetate	see α,α-Dimethylphen-ethyl acetate			
Benzyl salicylate	Benzyl o-hydroxy-benzoate	2151	FDA §121.1164	Synthetic flavor
Bergamol	see Linalyl acetate			
Bergamot, oil	Bergamot orange, oil [Citrus aurantium L. sub-species bergamia Wright & Arn.]	2153	FDA §121.101	GRAS, natural flavor extractive
BHA	Butylated hydroxy-anisole; mixture of 2-tert-butyl-4-methoxyphenol and 3-tert-butyl-4-methoxyphenol	2183	FDA §121.101	GRAS, chemical preservative; total antioxidants, not over 0.02% of fat or oil content, including essential (volatile) oil content of food. Product specifications apply
			FDA §121.1035	Antioxidant, alone or with BHT at maximum concentration of: 1 000 ppm BHA in active dry yeast 2 ppm BHA in beverages and desserts prepared from dry mixes 50 ppm BHA + BHT in dry breakfast cereals 32 ppm BHA in dry, diced, glacéed fruit 90 ppm BHA in dry mixes

REGULATORY STATUS OF DIRECT FOOD ADDITIVES (continued)

Material	Alternate Name [Source]	FEMA No.	Regulation	Limitations
BHA (continued)				for beverages and desserts 200 ppm BHA + BHT in emulsion stabilizers for shortenings 50 ppm BHA + BHT in potato flakes 10 ppm BHA + BHT in potato granules 50 ppm BHA + BHT in sweet-potato flakes Product specifications apply
			FDA §121.1059	Antioxidant in chewing-gum base, alone or with BHT and/or propyl gallate, up to 0.1% total antioxidant content
			FDA §121.1164	Antioxidant in synthetic flavoring substances up to 0.5% of essential oil content of flavoring substance
			MID	In unsmoked dry sausage, 0.003% BHA with 0.001% citric acid
			MID	To retard rancidity in unsmoked, dry sausage up to 0.003%; in rendered fat, or combination of such fat and vegetable fat, up to 0.01%; in fat, BHA may be used with BHT, glycine, and propyl gallate, up to 0.02% of the combination
			MID	Antioxidant to retard rancidity in frozen, fresh pork sausage and freeze-dried meats, up to 0.01% based on fat content; also in combination with BHT and propyl gallate, to 0.02% in frozen, fresh pork sausage, and up to 0.01% in freeze-dried meats
BHT	Butylated hydroxy-toluene; 2,6-di-*tert*-butyl-4-methylphenol	2184	FDA §15.525	In enriched rice, up to 0.003 3% by weight of enriched parboiled rice

REGULATORY STATUS OF DIRECT FOOD ADDITIVES (continued)

Material	Alternate Name [Source]	FEMA No.	Regulation	Limitations
BHT (continued)			FDA §121.101	GRAS, chemical preservative; total antioxidants, not over 0.02% of fat or oil content, including essential (volatile) oil content of food
			FDA §121.1034	Antioxidant, alone or in combination with BHA at max concentrations of: 10 ppm BHT + BHA in potato granules 50 ppm BHT + BHA in dry breakfast cereals 50 ppm BHT + BHA in potato and sweet-potato flakes 200 ppm BHT + BHA in emulsion stabilizers for shortenings Product specifications apply
			FDA §121.1059	Antioxidant in chewing-gum base, alone or with BHA and/or propyl gallate, up to 0.1% total antioxidant content
			MID	Antioxidant in rendered animal fat or in combination with vegetable fat, up to 0.01%, or up to 0.02% of a combination of BHT, BHA, and propyl gallate
			MID	Antioxidant to retard rancidity in frozen, fresh pork sausage and freeze-dried meats, up to 0.01% based on fat content; can be used in combination with BHA and propyl gallate, up to 0.02% in frozen, fresh pork sausage, and up to 0.01% in freeze-dried meats
Biotin	cis-Tetrahydro-2-oxothieno[3,4-d]imidazoline-4-valeric acid		FDA §121.101	GRAS, nutrient and/or dietary supplement
Biphenyl		3129		Flavoring use at average max level of 2.0 ppm in beverages, frozen desserts, candies, baked goods, gelatins, and puddings

REGULATORY STATUS OF DIRECT FOOD ADDITIVES (continued)

Material	Alternate Name [Source]	FEMA No.	Regulation	Limitations
Birch, sweet, oil	[*Betula lenta* L.]	2154		
Birch, tar oil	[*Betula pendula* Roth]		FDA §121.1164	Synthetic flavor
Bisabol myrrh	*see* Opopanax			
Bis(2-furfuryl) disulfide	Furfuryl disulfide	3257		*see Food Technol.*, May 1972
Bis(2-furfuryl) sulfide	Furfuryl sulfide	3258		*see Food Technol.*, May 1972
Bis(2-methyl-3-furyl) disulfide		3259		FEMA average max use levels—0.1 ppm in baked goods, meat, meat sauces, soups, condiments, and pickles
Bis(2-methyl-3-furyl) tetrasulfide		3260		*see Food Technol.*, May 1972, for FEMA amul[a]
Bitter almond	*see* Almond, bitter			
Blackberry, bark, extract	[*Rubus* species of section *Eubatus*]	2155	FDA §121.1163	Natural flavor
Blackthorn	*see* Sloe			
Black wattle	*see* Mimosa			
Blue, ultramarine	*see* Ultramarine blue			
Bois de rose, oil	[*Aniba rosaeodora* Ducke]	2156	FDA §121.101	GRAS, natural flavor extractive
Boldus, leaves	Boldo [*Peumus boldus* Mol.]		FDA §121.1163	Natural flavoring in alcoholic beverages only
Borneol	*d*-Camphanol	2157	FDA §121.1164	Synthetic flavor
Bornyl acetate	Borneol acetate	2159	FDA §121.1164	Synthetic flavor
Bornyl formate		2161	FDA §121.1164	Synthetic flavor
Bornyl isovalerate	Bornyval	2165	FDA §121.1164	Synthetic flavor
Bornyl valerate		2164	FDA §121.1164	Synthetic flavor

[a]amul—average maximum use level.

REGULATORY STATUS OF DIRECT FOOD ADDITIVES (continued)

Material	Alternate Name [Source]	FEMA No.	Regulation	Limitations
Boronia, absolute	[*Boronia megastigma* Nees]	2167		
Boronia, flowers	[*Boronia megastigma* Nees]		FDA §121.1163	Natural flavor
Brandy			FDA—GRAS	
Bromelin	Bromelain		MID	With water, salt and mono-sodium glutamate to soften tissues of beef cuts; tissue weight gain, 3% max
Brominated vegetable oils		2168	FDA §121.1234	Stabilizer in citrus- and other fruit-flavored bever-ages at 15 ppm; *Food Chemicals Codex* specifica-tions apply
Bryonia, root	[*Bryonia alba* L.; *B. dioica* Jacq.]		FDA §121.1163	Natural flavoring in alco-holic beverages only
Buchu, leaves, oil	[*Barosma betulina* Bartl. & Wendl.; *B. crenulata* (L.) Hook.; *B. seratifolia* Willd.]	2169	FDA §121.1163	Natural flavor
Buck bean, leaves	[*Menyanthes trifoliata* L.]		FDA §121.1163	Natural flavor in alcoholic beverages only
Butadiene-styrene rubber			FDA §121.1059	Synthetic masticatory sub-stance in chewing-gum base, when additive is the basic polymer
Butane	*n*-Butane		FDA §121.101	GRAS, miscellaneous and/or general-purpose food ad-ditive
1,4-Butanedicarboxylic acid	*see* Adipic acid			
2,3-Butanedione	Diacetyl		FDA §121.101	GRAS, synthetic flavor
1-Butanol	*see n*-Butyl alcohol			
2,3-Butanolone	*see* Acetoin			
2-Butanone	Methyl ethyl ketone	2170	FDA §121.1164	Synthetic flavor

REGULATORY STATUS OF DIRECT FOOD ADDITIVES (continued)

Material	Alternate Name [Source]	FEMA No.	Regulation	Limitations
trans-2-Butenoic acid ethyl ester	*see* Ethyl crotonate			
Butter acids		2171	FDA §121.1164	Synthetic flavor
Butter esters		2172	FDA §121.1164	Synthetic flavor
Butter starter distillate		2173	FDA §121.1164	Synthetic flavor
Butyl acetate		2174	FDA §121.1164	Synthetic flavor
Butyl acetoacetate		2176	FDA §121.1164	Synthetic flavor
n-Butyl alcohol	1-Butanol	2178	FDA §121.1164	Synthetic flavor
			FDA §8.390 and §8.6200	Diluent in inks for marking gum and confectionery, for marking food supplements in tablet form, and for branding pharmaceutical forms. Product specifications apply
Butylamine		3130		
Butyl anthranilate		2181	FDA §121.1164	Synthetic flavor
Butylated hydroxy-anisole	*see* BHA			
Butylated hydroxy-toluene	*see* BHT			
Butyl butyrate		2186	FDA §121.1164	Synthetic flavor
Butyl butyryllactate	Lactic acid, butyl ester, butyrate	2190	FDA §121.1164	Synthetic flavor
α-Butylcinnamaldehyde		2191	FDA §121.1164	Synthetic flavor
Butyl cinnamate		2192	FDA §121.1164	Synthetic flavor
2-*sec*-Butylcyclo-hexanone		3261		*see* Food Technol., May 1972
Butyl 2-decenoate	Butyl decylenate	2194	FDA §121.1164	Synthetic flavor
1,3-Butylene glycol			FDA §121.1176	Solvent for natural and synthetic flavoring substances. Product specifications apply

REGULATORY STATUS OF DIRECT FOOD ADDITIVES (continued)

Material	Alternate Name [Source]	FEMA No.	Regulation	Limitations
sec-Butyl ethyl ether		3131		Flavoring in beverages, frozen desserts, candies, and baked goods at average max level of 1.0 ppm
Butyl ethyl malonate		2195	FDA §121.1164	Synthetic flavor
Butyl formate		2196	FDA §121.1164	Synthetic flavor
Butyl heptanoate		2199	FDA §121.1164	Synthetic flavor
Butyl hexanoate		2201	FDA §121.1164	Synthetic flavor
Butyl *p*-hydroxy-benzoate	Butyl Parasept ®	2203	FDA §121.1164	Synthetic flavor
Butyl isobutyrate		2188	FDA §121.1164	Synthetic flavor
Butyl isovalerate		2218	FDA §121.1164	Synthetic flavor
Butyl lactate		2205	FDA §121.1164	Synthetic flavor
Butyl laurate	Butyl dodecanoate	2206	FDA §121.1164	Synthetic flavor
Butyl levulinate		2207	FDA §121.1164	Synthetic flavor
Butyl phenylacetate		2209	FDA §121.1164	Synthetic flavor
Butyl propionate		2211	FDA §121.1164	Synthetic flavor
Butyl stearate	Butyl octadecanoate	2214	FDA §121.1164	Synthetic flavor
Butyl sulfide	Dibutyl sulfide	2215	FDA §121.1164	Synthetic flavor
Butyl 10-undecenoate		2216	FDA §121.1164	Synthetic flavor
Butyl valerate		2217	FDA §121.1164	Synthetic flavor
Butyraldehyde	Butanal	2219	FDA §121.1164	Synthetic flavor
n-Butyric acid	Butanoic acid	2221	FDA §121.101	GRAS, synthetic flavor
Butyrin	*see* Tributyrin			
Butyroin	*see* 5-Hydroxy-4-octanone			
γ-Butyrolactone	*see* 4-Hydroxybutanoic acid lactone			

REGULATORY STATUS OF DIRECT FOOD ADDITIVES (continued)

Material	Alternate Name [Source]	FEMA No.	Regulation	Limitations
Cacao	Cocoa [*Theobroma cacao* L.]		FDA §121.101	GRAS, natural flavor extractive
Cade oil	Juniper tar; empyreumatic wood oil [*Juniperus oxycedrus* L.]		FDA §121.1164	Synthetic flavor
Cadinene			FDA §121.1164	Synthetic flavor
Caffeine	1,3,7-Trimethylxanthine	2224	FDA §121.101	GRAS, cola-type beverages, up to 0.02%
Cajeput, oil	[*Melaleuca leucadendron* L. and other *Melaleuca* species]	2225	FDA §121.1163	Natural flavors
Calcium acetate	Sorbo-Calcion	2228	FDA §121.101	GRAS, sequestrant
Calcium alginate	*see* Alginates—sodium, calcium, and ammonium salts			
Calcium ascorbate	Ascorbic acid calcium salt		FDA §121.101	GRAS, chemical preservative
Calcium caprate	Calcium salt of capric acid		FDA §121.1071	Binder, emulsifier, and anticaking agent; complying with §121.1070. Product specifications apply
Calcium caprylate	Calcium salt of caprylic acid		FDA §121.1071	Binder, emulsifier, and anticaking agent; complying with §121.1070. Product specifications apply
Calcium carbonate	Carbamic acid calcium salt		FDA §121.101	GRAS, nutrient and/or dietary supplement
			FDA §121.101	GRAS, miscellaneous and/or general-purpose food additive
Calcium caseinate			FDA §20	In frozen desserts
Calcium chloride			FDA §18.520	In evaporated milk, up to 0.1% by weight of finished product
			FDA §121.101	GRAS, sequestrant

REGULATORY STATUS OF DIRECT FOOD ADDITIVES (continued)

Material	Alternate Name [Source]	FEMA No.	Regulation	Limitations
Calcium chloride (continued)			FDA §121.101	GRAS, miscellaneous and/or general-purpose food additive
Calcium citrate			FDA §121.101	GRAS, nutrient and/or dietary supplement
			FDA §121.101	GRAS, sequestrant
			FDA §121.101	GRAS, miscellaneous and/or general-purpose food additive
Calcium diacetate			FDA §121.101	GRAS, sequestrant
Calcium disodium EDTA	Calcium disodium ethylenediaminetetra-acetate; calcium disodium (ethylenedinitrilo)-tetra-acetate		FDA §121.1017	Alone, as food additive: 33 ppm max in canned carbonated soft drinks 110 ppm max in canned white potatoes 340 ppm max in canned cooked clams 275 ppm max in canned cooked crabmeat 25 ppm max in distilled alcoholic beverages 75 ppm max in non-standardized dressings 310 ppm max in canned, cooked, dried lima beans 25 ppm max in fermented malt beverages 75 ppm max in French dressing 75 ppm max in mayonnaise 200 ppm max in canned, cooked mushrooms 75 ppm max in oleomargarine 100 ppm max in pecan pie filling 220 ppm max in pickled cabbage 220 ppm max in pickled cucumbers 100 ppm max in potato salad 800 ppm max in processed, dry pinto beans 75 ppm max in salad dressings

REGULATORY STATUS OF DIRECT FOOD ADDITIVES (continued)

Material	Alternate Name [Source]	FEMA No.	Regulation	Limitations
Calcium disodium EDTA (continued)				100 ppm max in sandwich spreads 75 ppm max in sauces 250 ppm max in canned, cooked shrimp 60 ppm max in spice extractives in soluble carriers 100 ppm max in artificially flavored lemon and orange spreads In combination with disodium EDTA as food additive: 75 ppm max in nonstandardized dressings 75 ppm max in French dressing 75 ppm max in mayonnaise 75 ppm max in salad dressing 100 ppm max in sandwich spread 75 ppm max in sauces Product specifications apply
Calcium gluconate	Calciofon		FDA §121.101	GRAS, sequestrant
			FDA §121.101	GRAS, miscellaneous and/or general-purpose food additive
Calcium glycerophosphate	Glycerophosphoric acid calcium salt		FDA §121.101	GRAS, nutrient and/or dietary supplement
Calcium hexametaphosphate			FDA §121.101	GRAS, sequestrant
Calcium hydroxide	Lime		FDA §121.101	GRAS, miscellaneous and/or general-purpose food additive
			FDA §8.303	Caramelization of caramel for food use
Calcium iodate	Lautarite		FDA §17	In bakery products, alone or with potassium bromate, potassium iodate and/or calcium peroxide, up to 0.007 5% by weight of flour

REGULATORY STATUS OF DIRECT FOOD ADDITIVES (continued)

Material	Alternate Name [Source]	FEMA No.	Regulation	Limitations
Calcium lactate			FDA §121.101	GRAS, miscellaneous and/or general-purpose food additive
Calcium lactobionate	Calcium salt of lacto-bionic acid		FDA §121.1162	Firming agent in dry pudding mixes. Product specifications apply
Calcium laurate	Calcium salt of lauric acid		FDA §121.1071	Binder, emulsifier, and anticaking agent; complying with §121.1070. Product specifications apply
Calcium myristate	Calcium salt of myristic acid		FDA §121.1071	Binder, emulsifier, and anticaking agent; complying with §121.1070. Product specifications apply
Calcium oleate	Calcium salt of oleic acid		FDA §121.1071	Binder, emulsifier, and anticaking agent; complying with §121.1070. Product specifications apply
Calcium oxide			FDA §121.101	GRAS, nutrient and/or dietary supplement
			FDA §121.101	GRAS, miscellaneous and/or general-purpose food additive
Calcium palmitate	Calcium salt of palmitic acid		FDA §121.1071	Binder, emulsifier, and anticaking agent, complying with §121.1070. Product specifications apply
Calcium pantothenate	Pantothenic acid calcium salt		FDA §121.101	GRAS, nutrient and/or dietary supplement
			FDA §121.1037	Calcium chloride double salt of calcium pantothenate; special dietary uses. Product specifications apply
Calcium peroxide	Calcium dioxide		FDA §17	In bakery products, alone or with potassium bromate, potassium iodate, and/or calcium iodate, up to 0.0075% by weight of flour
Calcium phosphate, monobasic	Acid calcium phosphate; monocalcium phosphate		FDA §121.101	GRAS, sequestrant

REGULATORY STATUS OF DIRECT FOOD ADDITIVES (continued)

Material	Alternate Name [Source]	FEMA No.	Regulation	Limitations
Calcium phosphate, mono-, di-, and tribasic			FDA §121.101	GRAS, nutrient and/or dietary supplement; GRAS, miscellaneous and/or general-purpose food additive
Calcium phytate			FDA §121.101	GRAS, sequestrant
Calcium propionate	Propionic acid calcium salt		FDA §17	In bakery products, alone or with sodium propionate, up to 0.32% by weight of flour used for breads (enriched, milk, and raisin breads) and up to 0.38% of flour used in whole-wheat bread
			FDA §121.101	GRAS, chemical preservative
			MID	To retard mold growth in pizza crust, up to 0.32% alone or in combination with sodium propionate, based on weight of flour used. Product specifications apply
Calcium pyrophosphate	Calcium diphosphate		FDA §121.101	GRAS, nutrient and/or dietary supplement
Calcium saccharin			FDA §121.101	GRAS, nonnutritive sweetener
Calcium salts			FDA §17.2	In enriched bread; each pound of finished food to contain 300–800 mg of calcium
Calcium salts of fatty acids			FDA §121.1071	Binder, emulsifier, and anticaking agent, complying with §121.1070. Product specifications apply
Calcium silicate			FDA §121.101	GRAS, anticaking agent in baking powder, up to 5%; GRAS, anticaking agent in table salt, up to 2%
			FDA §121.1135	Anticaking agent in food. Product specifications apply

REGULATORY STATUS OF DIRECT FOOD ADDITIVES (continued)

Material	Alternate Name [Source]	FEMA No.	Regulation	Limitations
Calcium sorbate			FDA §121.101	GRAS, chemical preservative
Calcium stearate	Calcium salt of stearic acid		FDA §22.8 and §29.9	In vanilla powder and vanilla-vanillin powder
			FDA §121.1071	Binder, emulsifier, and anticaking agent, complying with §121.1070. Product specifications apply
			FDA §121.1099	In defoaming agents in processing beet sugar and yeast, conforming to §121.1071
Calcium stearyl-2-lactylate			FDA §121.1047	1. Dough conditioner in yeast-leavened bakery products and prepared mixes for yeast-leavened bakery products, 0.5% max by weight of flour 2. Whipping agent in liquid and frozen egg whites, 0.05% max; whipping agent in dried egg white, 0.5% max; whipped vegetable oil topping, 0.3% max by weight of finished vegetable oil topping Product specifications apply
Calcium sulfate			FDA §51.990	In specified canned vegetables
			FDA §17	In bakery products, alone or with dicalcium phosphate and/or ammonium chloride, 0.25% max by weight of flour
			FDA §20	In frozen desserts
			FDA §15	Bleaching agent in flour, not more than 6% by weight, alone or in combination with potassium alum, magnesium carbonate, sodium aluminum sulfate, dicalcium phosphate, tricalcium phosphate, starch, and/or calcium carbonate, and with one part by weight of benzoyl peroxide

REGULATORY STATUS OF DIRECT FOOD ADDITIVES (continued)

Material	Alternate Name [Source]	FEMA No.	Regulation	Limitations
Calcium sulfate (continued)			FDA §19	In specified cheeses, alone or in combination, up to 6 times the weight of benzoyl peroxide used
			FDA §121.101	GRAS, nutrient and/or dietary supplement
Calendula	*see* Marigold, pot			
Calumba root	[*Jatrorrhiza palmata* (*Lam.*) Miers]		FDA §121.1163	Natural flavoring in alcoholic beverages only
Camomile, flower, Hungarian, oil	Chamomile [*Matricaria chamomilla* L.]	2273	FDA §121.101	GRAS, natural flavor extractive
Camomile, flower, Roman extract	Chamomile [*Anthemis nobilis* L.]	2274	FDA §121.101	GRAS, natural flavor
Camomile, flower, Roman or English, oil	Chamomile [*Anthemis nobilis* L.]	2272, 2275	FDA §121.101	GRAS, natural flavor extractive
d-**Camphanol**	*see* Borneol			
Camphene	2,2-Dimethyl-3-methylene-norbornane	2229	FDA §121.1164	Synthetic flavor
d-**Camphor**		2230	FDA §121.1164	Synthetic flavor
Camphor, Japanese, white, oil	[*Cinnamomum camphora* (L.) Nees & Eberm.]	2231	FDA §121.1163	Natural flavor; safrole free
Cananga, oil	[*Cananga odorata* Hook F. & Thoms.]	2232	FDA §121.101	GRAS, natural flavor extractive
Canthaxanthin	β-Carotene-4,4′-dione		FDA §8.326	Food color; not to exceed 30 mg per pound of solid or semisolid food or per pint of liquid food; may not be used to color standardized foods unless authorized by the food standard. Product specifications apply
Capers	[*Capparis spinosa* L.]		FDA §121.101	GRAS, natural flavor
Capraldehyde	*see* Decanal			
Capraldehyde dimethyl acetal	*see* Decanal dimethyl acetal			

REGULATORY STATUS OF DIRECT FOOD ADDITIVES (continued)

Material	Alternate Name [Source]	FEMA No.	Regulation	Limitations
Capric acid	Decanoic acid	2364	FDA §121.1070	Lubricant, binder, and defoaming agent in foods. Product specifications apply
Capric aldehyde	*see* Decanal			
Caprinaldehyde	*see* Decanal			
Caproic acid	*see* Hexanoic acid			
Caproic aldehyde	*see* Hexanal			
Capryl alcohol			FEMA—GRAS	
Caprylaldehyde	*see* n-Octanal			
Caprylic acid	*see* Octanoic acid			
Capsicum, extract and oleoresin *see also* Paprika	Cayenne pepper; red pepper; paprika [*Capsicum frutescens* L. or *C. annuum* L.]	2233, 2234	FDA §121.101	GRAS, natural flavor and natural flavor extractive
Caramel			FDA §121.101	GRAS, miscellaneous and/or general-purpose food additive
			FDA §8.303	Food color. Product specifications apply
Caramel color		2235		
Caraway	[*Carum carvi* L.]	2236	FDA §121.101	GRAS, natural flavor and natural flavor extractive
Caraway, black	Black cumin [*Nigella sativa* L.]	2237	FDA §121.101	GRAS, natural flavor
Carbohydrase	[*Aspergillus niger*]		FDA §121.1233	Removal of visceral mass in clam processing and in removal of shell from edible tissue in shrimp processing. Product specifications apply
	[*Rhizopus oryzae*]		FDA §121.1165	In production of dextrose from starch. Product specifications apply
Carbonates			FDA §8.305	Adjusting pH of annatto extract, food-grade

REGULATORY STATUS OF DIRECT FOOD ADDITIVES (continued)

Material	Alternate Name [Source]	FEMA No.	Regulation	Limitations
Carbon dioxide			FDA §121.101	GRAS, miscellaneous and/or general-purpose food additive
			FDA §121.1065	With octafluorocyclobutane as propellant and aerating agent for foamed or sprayed food products
			FDA §121.1180	With chloropentafluoroethane as propellant and aerating agent for foamed or sprayed food products
Carbon dioxide, solid	Dry ice		MID	For cooling during chopping and packaging of meat
Carboxymethylcellulose	*see* Sodium carboxymethylcellulose			
Cardamom seed	[*Elettaria cardamomum* (L.) Maton]	2241	FDA §121.101	GRAS, natural flavor and natural flavor extractive
Carmine	[*Coccus cacti* L.]	2242	FDA §8.317	Food color. Product specifications apply
Carminic acid *see also* Cochineal extract	Cochineal extract		FDA §8.317 and §8.6009	In production of aluminum or calcium/aluminum lakes of F&D dyes
Carnauba wax	[*Copernicia prunifera* (Muell.) H. E. Moore]		FDA §121.101	GRAS, miscellaneous and/or general-purpose food additive
Carob bean, extract	St. John's bread [*Ceratonia siliqua* L.]	2243	FDA §121.101	GRAS, natural flavor extractive
Carob bean gum	Locust bean gum; locust gum [*Ceratonia siliqua* L.]	2648	FDA §121.101	GRAS, stabilizer
Carotene			FDA §121.101	GRAS, nutrient and/or dietary supplement
β-Carotene			FDA §8.304	Food color. Product specifications apply
Carrageenan	Chondrus extract [*Chondrus crispus* (L.) Stackhouse; *C. ocellatus*; *Eucheuma spinosum*; *E. E. cottonii*; *Gigartina stellata*; and other species]		FDA §121.1066	Emulsifier, stabilizer, or thickener in foods. Product specifications apply
			FDA §121.101	GRAS, stabilizer
			MID	To extend and stabilize breading mix and sauces

REGULATORY STATUS OF DIRECT FOOD ADDITIVES (continued)

Material	Alternate Name [Source]	FEMA No.	Regulation	Limitations
Carrageenan, salts of			FDA §121.1067	Emulsifier, stabilizer, or thickener in foods. Product specifications apply
Carrageenan with Poly-sorbate 80			FDA §121.1193	Polysorbate 80, at 5%, to facilitate separation of sheeting carrageenan and its salts from drying rolls. Carrageenan and its salts so produced—emulsifier, stabilizer or thickener in foods; 500 ppm max of polysorbate 80 in final food. Product specifications apply
Carragheen	*see* Carrageenan			
Carrot	[*Daucus carota* L.]	2244	FDA §121.101	GRAS, natural flavor extractive
Carrot oil	[*Daucus carota* L.]		FDA §8.324	Food color. Product specifications apply
Carvacrol	2-*p*-Cymenol	2245	FDA §121.1164	Synthetic flavor
Carvacryl ethyl ether	2-Ethoxy-*p*-cymene	2246	FDA §121.1164	Synthetic flavor
Carveol	*p*-Mentha-6,8-dien-2-ol	2247	FDA §121.1164	Synthetic flavor
4-Carvomenthenol	1-*p*-Menthen-4-ol; 4-terpinenol	2248	FDA §121.1164	Synthetic flavor
Carvomenthone	*see p*-Menthan-2-one			
Carvone, *d*- or *l*-	Carvol	2249	FDA §121.101	GRAS, synthetic flavor
Carvyl acetate		2250	FDA §121.1164	Synthetic flavor
Carvyl propionate		2251	FDA §121.1164	Synthetic flavor
β-Caryophyllene		2252	FDA §121.1164	Synthetic flavor/adjuvant
Caryophyllene alcohol			FDA §121.1164	Synthetic flavor/adjuvant
Caryophyllene alcohol acetate			FDA §121.1164	Synthetic flavor/adjuvant
Cascara sagrada, extract	[*Rhamnus purshiana* DC.]	2253	FDA §121.1163	Natural flavor

REGULATORY STATUS OF DIRECT FOOD ADDITIVES (continued)

Material	Alternate Name [Source]	FEMA No.	Regulation	Limitations
Cascarilla bark, extract	[*Croton cascarilla* Benn.; *C. eluteria* Benn.]	2254	FDA §121.101	GRAS, natural flavor extractive
Casein			FDA §20	In frozen desserts
Cassia bark, Chinese	[*Cinnamomum cassia* Blume]	2257	FDA §121.101	GRAS, natural flavor extractive
Cassia bark, Padang or Batavia	[*Cinnamomum burmanni* Blume]		FDA §121.101	GRAS, natural flavor extractive
Cassia bark, Saigon	[*Cinnamomum loureirii* Nees]		FDA §121.101	GRAS, natural flavor extractive
Cassia, Chinese	[*Cinnamomum cassia* Blume]	2256	FDA §121.101	GRAS, natural flavor
Cassia, Padang or Batavia	[*Cinnamomum burmanni* Blume]		FDA §121.101	GRAS, natural flavor
Cassia, Saigon	[*Cinnamomum loureirii* Nees]		FDA §121.101	GRAS, natural flavor
Cassie flowers	[*Acacia farnesiana* (L.) Willd.]		FDA §121.1163	Natural flavor
Castoreum, extract	[*Castor fiber* L. and *C. canadensis* Kuhl]	2261	FDA §121.101	GRAS, natural flavor extractive
Castor oil	[*Ricinus communis* L.]	2263	FDA §121.1028	Release agent and anti-sticking agent in hard candy production, 500 ppm max. Component of protective coatings in vitamin and mineral tablets. Product specifications apply
			FDA §121.1163	Natural substance used in conjunction with flavors
			FDA §8.390 and §8.6200	Diluent in inks for marking gum, confectionery, and food supplements in tablet form. Inks for marking fruits and vegetables. Mixtures for coloring shell eggs, ingested, and externally applied drugs. Inks for branding pharmaceutical forms. In foods, 500 ppm max. Product specifications apply

REGULATORY STATUS OF DIRECT FOOD ADDITIVES (continued)

Material	Alternate Name [Source]	FEMA No.	Regulation	Limitations
Catalase	see Bacterial catalase			
Catechu, black	[Acacia catechu Willd.]	2264	FDA §121.1163	Natural flavor
Catechu, pale	see Gambir			
Caustic soda	see Sodium hydroxide			
Cayenne pepper	see Capsicum			
Cedar leaf, oil	[Thuja occidentalis L.]	2267		
Cedar, white, leaves and twigs	Arborvitae [Thuja occidentalis L.]		FDA §121.1163	Natural flavor; thujone free
Cedarwood oil, alcohols and terpenes			FDA §121.1164	Synthetic flavor
Celery seed	[Apium graveolens L.]	2268	FDA §121.101	GRAS, natural flavor and natural flavor extractive
Cellulase	[Aspergillus niger]		FDA §121.1233	To remove visceral mass in clam processing. Product specifications apply
α-Cellulose				Inert bulking agent in food
Cellulose gum	see Sodium carboxy-methylcellulose			
Centaury	[Centaurium umbellatum Gilib.]		FDA §121.1163	Natural flavor in alcoholic beverages only
Cetyl alcohol	see 1-Hexadecanol			
Chamomile	see Camomile			
Cherry laurel, leaves	[Prunus laurocerasus L.]	2277	FDA §121.1163	Natural flavor; not to exceed 25 ppm prussic acid
Cherry, sweet and sour, pits	[Prunus avium L. or P. cerasus L.]	2278	FDA §121.1163	Natural flavor; not to exceed 25 ppm prussic acid
Cherry, wild, bark	[Prunus serotina Ehrh.]	2276	FDA §121.101	GRAS, natural flavor extractive
Chervil	[Anthriscus cerefolium (L.) Hoffm.]	2279	FDA §121.101	GRAS, natural flavor and natural flavor extractive

REGULATORY STATUS OF DIRECT FOOD ADDITIVES (continued)

Material	Alternate Name [Source]	FEMA No.	Regulation	Limitations
Chestnut leaves	[*Castanea dentata* (Marsh.) Borkh.]		FDA §121.1163	Natural flavor
Chewing-gum base	*see* specific natural or synthetic masticatory substance listings			
Chicle	[*Manilkara zapotilla* (Jacq.) Gilly and *M. chicle* Gilly]		FDA §121.1059	Natural (coagulated or concentrated latex) masticatory substance in chewing gum
Chicory	[*Cichorium intybus* L.]	2280	FDA §121.101	GRAS, natural flavor extractive
Chilte	[*Cnidoscolus* (also known as *Jatropha*) *elasticus* Lundell; *C. tepiquensis* (Cost. and Gall.) McVaugh]		FDA §121.1059	Natural (coagulated or concentrated latex) masticatory substance in chewing-gum base
Chiquibul	[*Manilkara zapotilla* Gilly]		FDA §121.1059	Natural (coagulated or concentrated latex) masticatory substance in chewing-gum base
Chirata	[*Swertia chirata* (Roxb.) Buch.-Ham.]		FDA §121.1163	Natural flavor in alcoholic beverages only
Chives	[*Allium schoenoprasum* L.]		FDA §121.101	GRAS, natural flavor
Chlorine			FDA §15	Bleaching substance for flour
Chlorine dioxide	Chlorine peroxide		FDA §15	Bleaching substance for flour
Chloropentafluoro-ethane			FDA §121.1065	Propellant and aerating agent for foamed or sprayed foods. Product specifications apply
Chlorophyll			MID	Coloring sausage casings, oleomargarine, and shortening. In marking and branding inks for products. Product specifications apply

REGULATORY STATUS OF DIRECT FOOD ADDITIVES (continued)

Material	Alternate Name [Source]	FEMA No.	Regulation	Limitations
Cholic acid	3α,7α,12α-Trihydroxy-5 β-cholanic acid; cholalic acid		FDA §121.101	GRAS, emulsifying agent in dried egg whites, up to 0.1%
Choline bitartrate			FDA §121.101	GRAS, nutrient and/or dietary supplement
Choline chloride	Biocolina		FDA §121.101	GRAS, nutrient and/or dietary supplement
Chondrus extract	*see* Carrageenan			
Cinchona extract	[*Cinchona ledgeriana* Moens & Trimen; and other *Cinchona* species]	2285		
Cinchona, red, bark	[*Cinchona succirubra* Pav., or its hybrids]	2281	FDA §121.1163	Natural flavor in beverages; not more than 83 ppm total cinchona alkaloids in finished beverages
Cinchona, yellow, bark	[*Cinchona ledgeriana* Moens; *C. calisaya* Wedd.; or hybrids with other *Cinchona* species]	2283	FDA §121.1163	Natural flavor in beverages; not more than 83 ppm total cinchona alkaloids in finished beverages
Cineole	*see* Eucalyptol			
1,4-Cineole			FDA §121.1164	Synthetic flavor/adjuvant
Cinnamaldehyde	Cinnamic aldehyde	2286	FDA §121.101	GRAS, synthetic flavor/adjuvant
Cinnamaldehyde ethylene glycol acetal	Cinncloval	2287	FDA §121.1164	Synthetic flavor
Cinnamic acid	β-Phenylacrylic acid	2288	FDA §121.1164	Synthetic flavor
Cinnamic acid, tetra-hydrofurfuryl ester	*see* Tetrahydrofurfuryl cinnamate			
Cinnamon bark or leaf, Ceylon	[*Cinnamomum zeylanicum* Nees]	2290	FDA §121.101	GRAS, natural flavor extractive
Cinnamon bark or leaf, Chinese	[*Cinnamomum cassia* Blume]	2290	FDA §121.101	GRAS, natural flavor extractive
Cinnamon bark or leaf, Saigon	[*Cinnamomum loureirii* Nees]	2290	FDA §121.101	GRAS, natural flavor extractive

REGULATORY STATUS OF DIRECT FOOD ADDITIVES (continued)

Material	Alternate Name [Source]	FEMA No.	Regulation	Limitations
Cinnamon, Ceylon	[*Cinnamomum zeylanicum* Nees]	2289	FDA §121.101	GRAS, natural flavor
Cinnamon, Chinese	[*Cinnamomum cassia* Blume]	2289	FDA §121.101	GRAS, natural flavor
Cinnamon, Saigon	[*Cinnamomum loureirii* Nees]	2289	FDA §121.101	GRAS, natural flavor
Cinnamyl acetate		2293	FDA §121.1164	Synthetic flavor
Cinnamyl alcohol	3-Phenyl-2-propen-1-ol	2294	FDA §121.1164	Synthetic flavor
Cinnamyl anthranilate		2295	FDA §121.1164	Synthetic flavor
Cinnamyl benzoate			FDA §121.1164	Synthetic flavor
Cinnamyl butyrate		2296	FDA §121.1164	Synthetic flavor
Cinnamyl cinnamate	Phenylallyl cinnamate	2298	FDA §121.1164	Synthetic flavor
Cinnamyl formate		2299	FDA §121.1164	Synthetic flavor
Cinnamyl isobutyrate		2297	FDA §121.1164	Synthetic flavor
Cinnamyl isovalerate		2302	FDA §121.1164	Synthetic flavor
Cinnamyl phenylacetate		2300	FDA §121.1164	Synthetic flavor
Cinnamyl propionate		2301	FDA §121.1164	Synthetic flavor
Cinncloval	*see* Cinnamaldehyde ethylene glycol acetal			
Cire d'abeille	*see* Beeswax, white			
Citral	3,7-Dimethyl-2,6-octadien-1-al; geranial (*trans* isomer); neral (*cis* isomer)	2303	FDA §121.101	GRAS, synthetic flavor
Citral diethyl acetal	3,7-Dimethyl-2,6-octadienal diethyl acetal	2304	FDA §121.1164	Synthetic flavor
Citral dimethyl acetal	3,7-Dimethyl-2,6-octadienal dimethyl acetal	2305	FDA §121.1164	Synthetic flavor
Citral propylene glycol acetal			FDA §121.1164	Synthetic flavor/adjuvant

REGULATORY STATUS OF DIRECT FOOD ADDITIVES (continued)

Material	Alternate Name [Source]	FEMA No.	Regulation	Limitations
Citric acid	2-Hydroxy-1,2,3-propane-tricarboxylic acid	2306	FDA §121.101	GRAS, sequestrant
			FDA §121.101	GRAS, miscellaneous and/or general-purpose food additive
			FDA §8.303	In production of caramel
			MID	To increase effectiveness of antioxidants in lard, shortening, and unsmoked dry sausage; at 0.01% alone or with antioxidants in lard or shortening; 0.001% in unsmoked dry sausage with 0.003% BHA
			MID	To flavor oleomargarine and chili con carne
			MID	To accelerate color fixing in cured pork and beef cuts and in cured, comminuted meat food products. May be used in cured products to replace up to 50% of ascorbic acid, erythorbic acid, sodium ascorbate, or sodium erythorbate used
			MID	To prevent clotting of fresh beef blood, at 0.2%, with or without water; not more than 2 parts water to 1 part citric acid shall be used
			MID	To increase effectiveness of antioxidants in frozen, fresh pork sausages and freeze-dried meats, at 0.01%, in combination with antioxidants
Citridic acid	see Aconitic acid			
Citronellal	3,7-Dimethyl-6-octenal; rhodinal	2307	FDA §121.1164	Synthetic flavor
Citronella, oil	[Cymbopogon nardus Rendle]	2308	FDA §121.101	GRAS, natural flavor extractive
Citronellic acid	see 3,7-Dimethyl-6-octenoic acid			

REGULATORY STATUS OF DIRECT FOOD ADDITIVES (continued)

Material	Alternate Name [Source]	FEMA No.	Regulation	Limitations
Citronellol	d-Citronellol; 3,7-Dimethyl-6-octen-1-ol		FDA §121.1164	Synthetic flavor
d-Citronellol	see Citronellol			
1-Citronellol	see Rhodinol			
Citronelloxyacetalde-hyde	6,10-Dimethyl-3-oxa-9-undecanal	2310	FDA §121.1164	Synthetic flavor
Citronellyl acetate	3,7-Dimethyl-6-octen-1-yl acetate	2311	FDA §121.1164	Synthetic flavor
Citronellyl butyrate	3,7-Dimethyl-6-octen-1-yl butyrate	2312	FDA §121.1164	Synthetic flavor
Citronellyl formate	3,7-Dimethyl-6-octen-1-yl formate	2314	FDA §121.1164	Synthetic flavor
Citronellyl isobutyrate	3,7-Dimethyl-6-octen-1-yl isobutyrate	2313	FDA §121.1164	Synthetic flavor
Citronellyl phenyl-acetate	3,7-Dimethyl-6-octen-1-yl phenylacetate	2315	FDA §121.1164	Synthetic flavor
Citronellyl propionate	3,7-Dimethyl-6-octen-1-yl propionate	2316	FDA §121.1164	Synthetic flavor
Citronellyl valerate	3,7-Dimethyl-6-octen-1-yl valerate	2317	FDA §121.1164	Synthetic flavor
Citrus peels	[Citrus species]	2318	FDA §121.101	GRAS, natural flavor extractive
Citrus red No. 2			FDA §8.201	Coloring skins of oranges; 2.0 ppm max, based on weight of whole fruit. Product specifications apply
Civet	Zibeth; zibet; zibetum [Civet cats, Viverra civetta Schreber and V. zibetha Schreber]		FDA §121.101	GRAS, natural extractive
Clary	Clary sage [Salvia Sclarea L.]	2320	FDA §121.101	GRAS, natural flavor and natural flavor extractive
Clove, bud, leaf, or stem	[Eugenia caryophyllata Thunb and E. aromatica (L.) Baill.]	2322, 2325	FDA §121.101	GRAS, natural flavor extractive

REGULATORY STATUS OF DIRECT FOOD ADDITIVES (continued)

Material	Alternate Name [Source]	FEMA No.	Regulation	Limitations
Clover	[*Trifolium* species]		FDA §121.101	GRAS, natural flavor and natural flavor extractive
Cloves	[*Eugenia caryophyllata* Thunb.]	2327	FDA §121.101	GRAS, natural flavor
CMC	*see* Sodium carboxy-methylcellulose			
Cobalt 60			FDA §121.3002	Source of gamma radiation used for food processing
			FDA §121.3003	Source of gamma radiation for treatment of wheat and potatoes
Coca	[*Erythroxylon coca* Lam.; and other species of *Erythroxylon*]		FDA §121.101	GRAS, natural flavor extractive when decocainized
Cochineal	*Coccus cacti* L.	2330	FDA §8.317 and §8.6009	In production of aluminum or calcium-aluminum lakes for food and drug uses
			MID	Coloring sausage casing, oleomargarine, or shortening; for marking or branding ink on products. Product specifications apply
Cochineal extract *see also* Carminic acid	[*Dactylopius coccus* Costa (*Coccus cacti* L.)]		FDA §8.317	Food color. Product specifications apply
Coffee	[*Coffea arabica* L. and other *Coffea* species]		FDA §121.101	GRAS, natural flavor extractive
Cognac oil, white and green	Ethyl oenanthate	2331, 2332	FDA §121.101	GRAS, natural extractive
Cola nut	*see* Kola nut			
Colophony	*see* Rosin			
Copaiba	[South American species of *Copaifera*]		FDA §121.1163	Natural flavor
Copal, Manila			FDA §8.390	Diluent in inks for marking fruits and vegetables. Product specifications apply

REGULATORY STATUS OF DIRECT FOOD ADDITIVES (continued)

Material	Alternate Name [Source]	FEMA No.	Regulation	Limitations
Copper gluconate			FDA §121.101	GRAS, nutrient and/or dietary supplement, up to 0.005%
Coriander, oil	[*Coriandrum sativum* L.]	2333	FDA §121.101	GRAS, natural flavor and natural flavor extractive
Cork oak	[*Quercus suber* L.; *Q. occidentalis* Gray]		FDA §121.1163	Natural flavor in alcoholic beverages only
Corn endosperm oil	Xanthophyll		FDA §8.322	In chicken feed to enhance yellow color of chicken skin and eggs. Product specifications apply
Corn oil	[*Zea mays* L.]		FDA	GRAS
Corn silk	[*Zea mays* L.]	2335	FDA §121.101	GRAS, natural flavor extractive
Cornstarch			FDA	GRAS
Corn syrup			FDA	GRAS
			MID	To flavor sausage, hamburger, meat loaf, luncheon meat, chopped or pressed ham; may be used alone at 2% or in combination with corn syrup solids or glucose syrup, with the combination totaling 2%, calculated on a dry basis
Corn syrup solids			MID	To flavor sausage, hamburger, meat loaf, luncheon meat, chopped or pressed ham; may be used alone at 2% or with corn syrup or glucose syrup, with the combination totaling 2%, calculated on a dry basis
Costmary	[*Chrysanthemum balsamita* L.]		FDA §121.1163	Natural flavor in alcoholic beverages only
Costus root	[*Saussurea lappa* Clarke]		FDA §121.1163	Natural flavor
Cottonseed flour, partially defatted, cooked			FDA §121.1019	In or on food. Product specifications apply

REGULATORY STATUS OF DIRECT FOOD ADDITIVES (continued)

Material	Alternate Name [Source]	FEMA No.	Regulation	Limitations
Cottonseed flour, partially defatted, cooked (continued)			FDA §8.315	Food color. Product specifications apply
Cream of tartar	Monopotassium tartrate; potassium acid tartrate; potassium bitartrate		FDA §121.101	GRAS, miscellaneous and/or general-purpose food additive and as a sequestrant
p-Cresol	4-Cresol	2337	FDA §121.1164	Synthetic flavor
o-Cresyl acetate	see o-Tolyl acetate			
p-Cresyl acetate	see p-Tolyl acetate			
o-Cresyl methyl ether	see o-Methylanisole			
p-Cresyl methyl ether	see p-Methylanisole			
Crown gum	[*Manilkara zapotilla* Gilly and *M. chicle* Gilly]		FDA §121.1059	Natural (coagulated or concentrated latex) masticatory substance in chewing-gum base
Cubeb	[*Piper cubeba* L.f.]	2338	FDA §121.1163	Natural flavor
Cumin	Cummin [*Cuminum cyminun* L.]	2340	FDA §121.101	GRAS, natural flavor and natural flavor extractive
Cumin, black	Black caraway [*Nigella sativa* L.]	2342	FDA §121.101	GRAS, natural flavor
Cuminal	see Cuminaldehyde			
Cuminaldehyde	Cuminal; p-isopropyl benzaldehyde	2341	FDA §121.1164	Synthetic flavor
Cuminic alcohol	see p-Isopropylbenzyl alcohol			
Cuminyl acetaldehyde	see 3-(p-Isopropyl)phenyl propionaldehyde			
Cuprous iodide			FDA §121.101	GRAS, dietary supplement in table salt as source of dietary iodine, up to 0.01%
Curaçao orange peel	see Orange, bitter, flowers and peel			
Currant, black, buds and leaves	[*Ribes nigrum* L.]	2346	FDA §121.1163	Natural flavor

REGULATORY STATUS OF DIRECT FOOD ADDITIVES (continued)

Material	Alternate Name [Source]	FEMA No.	Regulation	Limitations
Cusparia bark	*see* Angostura			
Cyclamen aldehyde	*see* 2-Methyl-3-(*p*-iso-propylphenyl)-propional-dehyde			
Cyclo-1,13-ethylene-dioxytridecan-1,13-dione	*see* Ethyl brassylate			
Cyclohexane	Hexahydrobenzene		FDA §8.390 and §8.6200	Diluent in inks for marking gums, confectionery, and food supplements in tablet form and inks for branding pharmaceutical forms. Product specifications apply
Cyclohexaneacetic acid	Cyclohexylacetic acid	2347	FDA §121.1164	Synthetic flavor
Cyclohexane ethyl acetate	Cyclohexylethyl acetate	2348	FDA §121.1164	Synthetic flavor
Cyclohexapyrazine	*see* 5,6,7,8-Tetrahydro-quinoxaline			
Cyclohexyl acetate		2349	FDA §121.1164	Synthetic flavor
Cyclohexyl anthranilate		2350	FDA §121.1164	Synthetic flavor
Cyclohexyl butyrate		2351	FDA §121.1164	Synthetic flavor
Cyclohexyl cinnamate		2352	FDA §121.1164	Synthetic flavor
Cyclohexyl formate		2353	FDA §121.1164	Synthetic flavor
Cyclohexyl isovalerate		2355	FDA §121.1164	Synthetic flavor
Cyclohexyl propionate		2354	FDA §121.1164	Synthetic flavor
Cyclopentanethiol	Cyclopentyl mercaptan	3262		*see Food Technol.*, May 1972
Cyclopentyl mercaptan	*see* Cyclopentanethiol			
_p_-Cymen-7-carboxalde-hyde	*see* *p*-Isopropylphenyl-acetaldehyde			
_p_-Cymene		2356	FDA §121.1164	Synthetic flavor
_p_-Cymen-7-ol	*see* *p*-Isopropylbenzyl alcohol			

REGULATORY STATUS OF DIRECT FOOD ADDITIVES (continued)

Material	Alternate Name [Source]	FEMA No.	Regulation	Limitations
p-Cymen-8-ol	*see p,α,α*-Trimethylbenzyl alcohol			
2-*p*-Cymenol	*see* Carvacrol			
L-Cysteine		3263	FDA §17	In bakery products, up to 0.009% of flour used
			FDA §121.101	GRAS, nutrient and/or dietary supplement as free acid, hydrochloride salt, hydrate, or anhydrous form
				see Food Technol., May 1972, for FEMA amul[a]
Cystine (L- and DL-forms)			FDA §121.101	GRAS, nutrient and/or dietary supplement, as free acid, hydrochloride salt, hydrate, or anhydrous form
Damar gum (resin)			FDA §8.390	Diluent in mixtures for coloring shell eggs. Product specifications apply
Damiana leaves	[*Turnera diffusa* Willd.]		FDA §121.1163	Natural flavor
Dandelion and dandelion root	[*Taraxacum officinale* Weber and *T. laevigatum* DC.]		FDA §121.101	GRAS, natural flavor extractive
Danish agar	*see* Furcellaran			
Davana	[*Artemisia pallens* Wall.]		FDA §121.1163	Natural flavor
2-*trans*,4-*trans*-Decadienal		3135		
γ-Decalactone	4-Hydroxydecanoic acid, γ-lactone	2360	FDA §121.1164	Synthetic flavor
δ-Decalactone	5-Hydroxydecanoic acid, δ-lactone	2361	FDA §121.1164	Synthetic flavor
			FDA §121.1164	In oleomargarine, alone or with δ-dodecalactone, δ-decalactone at 10 ppm max, δ-dodecalactone at 20 ppm max. Product specifications apply

[a] amul—average maximum use level.

REGULATORY STATUS OF DIRECT FOOD ADDITIVES (continued)

Material	Alternate Name [Source]	FEMA No.	Regulation	Limitations
Decanal	Aldehyde C-10; capralde-hyde; capric aldehyde; caprinaldehyde; *n*-decyl-aldehyde	2362	FDA §121.101	GRAS, synthetic flavor/adjuvant
Decanal dimethyl acetal	Aldehyde C-10 dimethyl acetal; capraldehyde dimethyl acetal; decylal-dehyde dimethyl acetal; 1,1-dimethoxydecane; 10,10-dimethoxydecane	2363	FDA §121.1164	Synthetic flavor
Decanoic acid	*see* Capric acid			
1-Decanol	*n*-Decyl alcohol	2365	FDA §121.1164	Synthetic flavor
2-Decenal	Decenaldehyde	2366	FDA §121.1164	Synthetic flavor
4-Decenal		3264		*see Food Technol.,* May 1972
3-Decen-2-one	Heptylidene acetone		FDA §121.1164	Synthetic flavor/adjuvant
Decyl acetate	Decanyl acetate	2367	FDA §121.1164	Synthetic flavor
n-Decyl alcohol	*see* 1-Decanol			
n-Decylaldehyde	*see* Decanal			
Decylaldehyde dimethyl acetal	*see* Decanal dimethyl acetal			
Decyl butyrate		2368	FDA §121.1164	Synthetic flavor
Decyl propionate		2369	FDA §121.1164	Synthetic flavor
Defoaming agents	*see* specific product listings		FDA §121.1099	To inhibit foaming. Product specifications apply
Dehydrated beets	*see* Beets, dehydrated			
Dehydroacetic acid	3-Acetyl-6-methyl-2*H*-pyran-2,4(3*H*)-dione		FDA §121.1089	And/or its sodium salt, as preservative for cut or peeled squash; 65 ppm max, expressed as the acid. Product specifications apply
Deoxycholic acid	Desoxycholic acid; 3α,12α-dihydroxy-5β-cholanic acid		FDA §121.101	GRAS, emulsifying agent in dried egg whites, up to 0.1%

REGULATORY STATUS OF DIRECT FOOD ADDITIVES (continued)

Material	Alternate Name [Source]	FEMA No.	Regulation	Limitations
Dextrans	Macrose		FDA §121.101	GRAS, miscellaneous and/ or general-purpose food additive; average molecular weight, below 100,000
Dextrin	British gum		FDA	GRAS
Dextrose	Glucose		FDA §8.303	For production of caramel
			MID	To flavor sausage, ham, and other meat products
			MID	To flavor sausage, ham, and cured meat products
Diacetyl	2,3-Butanedione	2370	FDA §121.101	GRAS, synthetic flavor/adjuvant
			MID	To flavor oleomargarine
Diacetyl tartaric acid esters			FDA §121.101	Diacetyl tartaric acid esters of mono- and diglycerides of edible fats or oils or edible fat-forming fatty acids: GRAS, emulsifiers
Diacetyl tartaric acid esters of mono- and diglycerides			MID	To emulsify rendered animal fat or its combination with vegetable fat
Diallyl disulfide	*see* Allyl disulfide			
Diallyl sulfide	*see* Allyl sulfide			
Diallyl trisulfide		3265		*see Food Technol.,* May 1972
Dibenzyl disulfide	*see* Benzyl disulfide			
Dibenzyl ether	Benzyl ether	2371	FDA §121.1164	Synthetic flavor
Dibenzyl ketone	*see* 1,3-Diphenyl-2-propanone			
Dibenzyl sulfide	Benzyl sulfide		FDA §121.1164	Synthetic flavor
4,4-Dibutyl-γ-butyro-lactone	4,4-Dibutyl-4-hydroxy-butyric acid, γ-lactone	2372	FDA §121.1164	Synthetic flavor
Dibutyl sebacate	Butyl sebacate	2373	FDA §121.1164	Synthetic flavor/adjuvant

REGULATORY STATUS OF DIRECT FOOD ADDITIVES (continued)

Material	Alternate Name [Source]	FEMA No.	Regulation	Limitations
Dibutyl sulfide	see Butyl sulfide			
Dicalcium orthophosphate			FDA	GRAS
Dicalcium phosphate			FDA §15	Bleaching agent in flour, 6% max by weight, alone or in combination with potassium alum, calcium sulfate, magnesium carbonate, sodium aluminum sulfate, tricalcium phosphate, starch and/or calcium carbonate, and with 1% by weight of benzoyl peroxide
			FDA §17	In bakery products, alone or in combination with calcium sulfate and/or ammonium chloride, up to 0.25% by weight of flour
Diethylene glycol distearate			FDA §8.390	Diluent in mixtures for coloring shell eggs. Product specifications apply
Diethyl malate	Ethyl malate	2374	FDA §121.1164	Synthetic flavor
Diethyl malonate	Ethyl malonate	2375	FDA §121.1164	Synthetic flavor
2,3-Diethylpyrazine		3136		Not more than 10 ppm in finished food
Diethyl sebacate	Ethyl sebacate	2376	FDA §121.1164	Synthetic flavor
Diethyl succinate	Ethyl succinate	2377	FDA §121.1164	Synthetic flavor
Diethyl tartrate	Ethyl tartrate	2378	FDA §121.1164	Synthetic flavor
2,5-Diethyltetrahydrofuran			FDA §121.1164	Synthetic flavor
Difurfuryl disulfide	see Bis(2-furfuryl) disulfide			

REGULATORY STATUS OF DIRECT FOOD ADDITIVES (continued)

Material	Alternate Name [Source]	FEMA No.	Regulation	Limitations
Difurfuryl sulfide	see Bis(2-furfuryl) sulfide			
Dihydroanethole	see p-Propyl anisole			
Dihydrocarveol	8-p-Menthen-2-ol; 6-methyl-3-isopropenyl-cyclohexanol	2379	FDA §121.1164	Synthetic flavor
Dihydrocarvone			FDA §121.1164	Synthetic flavor/adjuvant
Dihydrocarvyl acetate	8-p-Menthen-2-yl acetate	2380	FDA §121.1164	Synthetic flavor
Dihydrocoumarin	Hydrocoumarin	2381	FDA §121.1164	Synthetic flavor
4,5-Dihydro-3(2H) thiophenone		3266		see Food Technol., May 1972
2,4-Dihydroxy-3-methyl-2-hexenoic acid, γ-lactone	see 5-Ethyl-3-hydroxy-4-methyl-2(5H)-furanone			
Diisobutyl carbinol	see 2,6-Dimethyl-4-heptanol			
Dilauryl thiodipropionate			FDA §121.101	GRAS, chemical preservative; total content of antioxidants, not over 0.02% of fat or oil content, including essential (volatile) oil content of food
Dill	[Anethum graveolens L.]	2382	FDA §121.101	GRAS, natural flavor and natural flavor extractive
Dill, Indian	[Anethum sowa Roxb. (Peucedanum graveolens Benth. & Hook.)]	2384	FDA §121.1163	Natural flavor
Diluents	see specific product listings			
m-Dimethoxybenzene	Resorcinal dimethyl ether	2385	FDA §121.1164	Synthetic flavor
p-Dimethoxybenzene	Dimethyl hydroquinone	2386	FDA §121.1164	Synthetic flavor
1,1-Dimethoxydecane	see Decanal dimethyl acetal			
10,10-Dimethoxy decane	see Decanal dimethyl acetal			

REGULATORY STATUS OF DIRECT FOOD ADDITIVES (continued)

Material	Alternate Name [Source]	FEMA No.	Regulation	Limitations
2,6-Dimethoxyphenol		3137		
3,4-Dimethoxy-1-vinyl-benzene		3138		
Dimethyl acetal	see Octanal dimethyl acetal			
2,4-Dimethylaceto-phenone	Methyl 2,4-dimethyl-phenyl ketone	2387	FDA §121.1164	Synthetic flavor
2,4-Dimethyl-5-acetyl-thiazole		3267		
3,3-Dimethylacrylic acid	see 3-Methylcrotonic acid			
β,β-Dimethylacrylic acid	see 3-Methylcrotonic acid			
Dimethyl anthranilate	see Methyl N-methyl-anthranilate			
p-α-Dimethylbenzyl alcohol	Methyl p-tolyl carbinol	3139		
Dimethylbenzyl carbinol	see α,α-Dimethylphen-ethyl alcohol			
α,α-Dimethylbenzyl iso-butyrate	Phenyldimethyl carbinyl isobutyrate	2388	FDA §121.1164	Synthetic flavor
3,4-Dimethyl-1,2-cyclo-pentadione		3268		see Food Technol., May 1972
3,5-Dimethyl-1,2-cyclo-pentadione		3269		see Food Technol., May 1972
2,3-Dimethyl-1,4-diazine	see 2,3-Dimethylpyrazine			
2,5-Dimethyl-1,4-diazine	see 2,5-Dimethylpyrazine			
2,6-Dimethyl-1,4-diazine	see 2,6-Dimethylpyrazine			
Dimethyl disulfide	Methyl disulfide		FDA §121.1164	Synthetic flavor/adjuvant
2,6-Dimethyl-3-ethyl-pyrazine	see 3-Ethyl-2,6-dimethyl-pyrazine			
2,6-Dimethyl-4-heptanol	Diisobutyl carbinol	3140		FEMA—flavoring, 20 ppm max in beverages, frozen

REGULATORY STATUS OF DIRECT FOOD ADDITIVES (continued)

Material	Alternate Name [Source]	FEMA No.	Regulation	Limitations
2,6-Dimethyl-4-heptanol (continued)				desserts, candies, gelatins, puddings, and chewing-gums
2,6-Dimethyl-5-heptenal	Melonal	2389	FDA §121.1164	Synthetic flavor
Dimethyl hydroquinone	*see* p-Dimethoxybenzene			
3,7-Dimethyl-7-hydroxy-octanal	*see* Hydroxycitronellal			
1,4-Dimethyl-7-(α-hydroxyisopropyl)Δ9,10-octahydroazulene acetate	*see* Guaiol acetate			
5,6-Dimethyl-8-iso-propenyl bicyclo-(4,4,0)-dec-1-en-3-one	*see* Nootkatone			
1,4-Dimethyl-7-iso-propenyl-Δ9,10-octa-hydroazulene	*see* Guaiene			
Dimethylketol	*see* Acetoin			
4-(2,2-Dimethyl-6-methylene-cyclohexyl)-3-butenone	*see* γ-Ionone			
2,6-Dimethyl-10-methyl-ene-2,6,11-dodecatrienal	α-Sinensal	3141		
2,2-Dimethyl-3-methyl-enenorbornane	*see* Camphene			
3,7-Dimethyl-2,6-octa-dienal diethyl acetal	*see* Citral diethyl acetal			
3,7-Dimethyl-2,6-octa-dienal dimethyl acetal	*see* Citral dimethyl acetal			
cis-3,7-Dimethyl-2,6-octadien-1-ol	*see* Nerol			
trans-3,7-Dimethyl-2,6-octadien-1-ol	*see* Geraniol			
trans-3,7-Dimethyl-2,6-octadien-1-yl aceto-acetate	*see* Geranyl acetoacetate			

REGULATORY STATUS OF DIRECT FOOD ADDITIVES (continued)

Material	Alternate Name [Source]	FEMA No.	Regulation	Limitations
3,7-Dimethyl-1,6-octa-dien-3-yl anthranilate	see Linalyl anthranilate			
4a,5-Dimethyl-1,2,3,4,4a, 5,6,7-octahydro-7-keto-3-isopropenylnaphtha-lene	see 4,4a,5,6,7,8-Hexa-hydro-6-isopropenyl-4,4a-dimethyl-2(3H)-naphtha-lenone			
2,6-Dimethyl octanal	Isodecylaldehyde	2390	FDA §121.1164	Synthetic flavor
3,7-Dimethyl-1,7-octa-nediol	see Hydroxycitronellol			
3,7-Dimethyl-1-octanol	Tetrahydrogeraniol	2391	FDA §121.1164	Synthetic flavor
3,7-Dimethyloctan-3-ol	see Tetrahydrolinalool			
3,7-Dimethyl-1,3,6-octa-triene	see Ocimene			
3,7-Dimethyl-6-octenal	see Citronellal			
3,7-Dimethyl-6-octenoic acid	Citronellic acid	3142		
3,7-Dimethyl-6-octen-1-ol	see Citronellol			
3,7-Dimethyl-7-octen-1-ol	see Rhodinol			
3,7-Dimethyl-6-octen-1-yl acetate	see Citronellyl acetate			
3,7-Dimethyl-6-octen-1-yl butyrate	see Citronellyl butyrate			
3,7-Dimethyl-6-octen-1-yl formate	see Citronellyl iso-butyrate			
3,7-Dimethyl-6-octen-1-yl phenylacetate	see Citronellyl phenyl-acetate			
3,7-Dimethyl-6-octen-1-yl propionate	see Citronellyl propionate			
3,7-Dimethyl-6-octen-1-yl valerate	see Citronellyl valerate			

REGULATORY STATUS OF DIRECT FOOD ADDITIVES (continued)

Material	Alternate Name [Source]	FEMA No.	Regulation	Limitations
6,10-Dimethyl-3-oxa-9-undecenal	*see* Citronelloxyacetaldehyde			
2,4-Dimethyl-2-pentenoic acid		3143		
α,α-Dimethylphenethyl acetate	Benzyl dimethylcarbinyl acetate; benzylpropyl acetate	2392	FDA §121.1164	Synthetic flavor
α,α-Dimethylphenethyl alcohol	Dimethylbenzyl carbinol	2393	FDA §121.1164	Synthetic flavor
α,α-Dimethylphenethyl butyrate	Benzyl dimethylcarbinyl butyrate	2394	FDA §121.1164	Synthetic flavor
α,α-Dimethylphenethyl formate	Benzyl dimethylcarbinyl formate	2395	FDA §121.1164	Synthetic flavor
2,6-Dimethylphenol	*see* 2,6-Xylenol			
Dimethylphenylethyl carbinyl acetate	*see* 2-Methyl-4-phenyl-2-butyl acetate			
Dimethylphenylethyl carbinyl isobutyrate	*see* 2-Methyl-4-phenyl-2-butyl isobutyrate			
Dimethyl polysiloxane	Dimethicone		FDA §121.1099	Defoaming agent, up to 10 ppm in ready-to-serve foods; 16 ppm in ready-to-serve gelatin desserts; not to be used in milk and in food for infants and invalids. Product specifications apply
2,3-Dimethylpyrazine	2,3-Dimethyl-1,4-diazine	3271		*see Food Technol.,* May 1972
2,5-Dimethylpyrazine	2,5-Dimethyl-1,4-diazine	3272		*see Food Technol.,* May 1972
2,6-Dimethylpyrazine	2,6-Dimethyl-1,4-diazine	3273		*see Food Technol.,* May 1972
p-α-Dimethylstyrene	p-Isopropenyltoluene; 1-methyl-4-isopropenyl-benzene	3144		
Dimethyl succinate	Methyl succinate	2396	FDA §121.1164	Synthetic flavor

REGULATORY STATUS OF DIRECT FOOD ADDITIVES (continued)

Material	Alternate Name [Source]	FEMA No.	Regulation	Limitations
4,5-Dimethylthiazole		3274		*see Food Technol.,* May 1972
Dimethyl trisulfide		3275		*see Food Technol.,* May 1972
Dimethyl-*p*-tolyl carbinol	*see p-α,α*-Trimethylbenzyl alcohol			
6,10-Dimethyl-9-undecen-2-one	*see* Tetrahydro-pseudo-ionone			
2,4-Dimethyl-5-vinyl-thiazole		3145		
Dioctyl sodium sulfo-succinate	Sulfosuccinic acid *bis*(2-ethylhexyl) ester *S*-sodium salt		FDA §121.1137	1. Wetting agent in fumaric acid-acidulated foods—dry gelatin dessert, dry beverage base, and fruit-juice drinks; 15 ppm max in finished gelatin dessert; 10 ppm max in finished beverage or fruit-juice drink. Product specifications apply 2. In production of unrefined cane sugar; 0.5 ppm max per percentage point of sucrose in juice, syrup, or massecuite being processed; 25 ppm max in final molasses 3. Emulsifying agent for cocoa fat in noncarbonated beverages containing cocoa; 25 ppm max in finished beverage 4. Processing aid and wetting agent for fumaric acid-acidulated fruit-juice drinks, in combination with α-hydro-ω-hydroxypoly-(oxyethylene)poly(oxypropylene)(53–59mol)poly-(oxyethylene)(14–16 mol)-block copolymers; 10 ppm max of DSS-block copolymer combination in finished beverage or fruit-juice drink. Product specifications apply 5. Stabilizing agent on gums

REGULATORY STATUS OF DIRECT FOOD ADDITIVES (continued)

Material	Alternate Name [Source]	FEMA No.	Regulation	Limitations
Dioctyl sodium sulfo-succinate (continued)				and hydrophilic colloids to be used in foods as stabilizing and thickening agents; 0.5% max by weight of the gums or hydrophilic colloids; product specifications apply 6. Dispersing agent in product conforming to §121. 1229 (cocoa with dioctyl sodium sulfosuccinate for manufacturing). Product specifications apply
			FDA §8.390	Diluent in mixtures for coloring shell eggs. Product specifications apply
			FDA §8.300(A3)	Diluent in color-additive mixtures for foods; 9 ppm max in finished food. Product specifications apply
			MID	Cooling and retort water treatment agent for prevention of staining on exterior of canned goods, up to 0.05%
			MID	Hog-scale agent to remove hair from hog carcasses, when removed by subsequent cleaning operations
Dipentene	Dipentine		FDA §121.1059	In synthetic terpene resins in chewing-gum base. Product specifications apply
Diphenyl disulfide	*see* Phenyl disulfide			
Diphenyl ketone	*see* Benzophenone			
1,3-Diphenyl-2-propanone	Dibenzyl ketone	2397	FDA §121.1164	Synthetic flavor
Dipotassium phosphate	Potassium phosphate		FDA §19	Emulsifier in specified cheeses; alone or with other emulsifiers, up to 3% by weight of cheese
			FDA §121.101	GRAS, sequestrant
Dipropyl ketone	*see* 4-Heptanone			

REGULATORY STATUS OF DIRECT FOOD ADDITIVES (continued)

Material	Alternate Name [Source]	FEMA No.	Regulation	Limitations
Dipropyl trisulfide		3276		*see Food Technol.,* May 1972
Disodium EDTA	Disodium ethylene-diamine tetraacetate		FDA §121.1056	1. Alone, as food additive: 150 ppm max in aqueous multivitamin preparations, with iron salts as stabilizer for vitamin B_{12} 145 ppm max in canned, black-eyed peas 165 ppm max in canned cooked chick peas 165 ppm max in canned kidney beans 500 ppm max in canned strawberry pie filling 36 ppm max in coated sausage 75 ppm max in non-standardized dressing 315 ppm max in dried banana component of ready-to-eat cereal products 75 ppm max in French dressing 100 ppm max in frozen white potatoes, including cut potatoes 50 ppm max in gefilte fish balls or patties, including liquid packing medium, to inhibit discoloration 75 ppm max in mayonnaise 75 ppm max in salad dressing 100 ppm max in sandwich spread 75 ppm max in sauces 2. In combination with calcium disodium EDTA as food additive: 75 ppm max in non-standardized dressing 75 ppm max in French dressing 75 ppm max in mayonnaise 1000 ppm max (dry-weight basis) in nonnutritive sweeteners

REGULATORY STATUS OF DIRECT FOOD ADDITIVES (continued)

Material	Alternate Name [Source]	FEMA No.	Regulation	Limitations
Disodium EDTA (continued)				75 ppm max in salad dressing 100 ppm max in sandwich spread 75 ppm max in sauces Product specifications apply
Disodium guanylate			FDA §121.1109	Flavor enhancer in foods
Disodium inosinate	Disodium salt of inosinic acid		FDA §121.1090	Flavoring adjuvant in foods. Product specifications apply
Disodium phosphate	Sodium phosphate, dibasic	2398	FDA §18.520	In evaporated milk, up to 0.1% by weight of finished product
			FDA §16	In macaroni and noodle products; 0.5 to 1.0%
			FDA §15.140	In enriched farina; 0.5 to 1.0%
			FDA §19	Emulsifier in specified cheeses, up to 3% by weight
			FDA §121.101	GRAS, sequestrant
			MID	To decrease amount of cooked-out juices in cured hams, pork shoulder picnics and loins, canned hams and pork shoulder picnics, chopped ham, and bacon; 5.0% of phosphate in pickle at 10% pump level; 0.5% of phosphate in product (only clear solution may be injected into product)
Disodium succinate		3277		*see Food Technol.,* May 1972
2,6-Ditertiarybutyl-4-hydroxy methyl phenol			FDA §121.1208	Antioxidant, alone or with other antioxidants, not to exceed 0.02% of fat or oil content of the food, including essential oils

REGULATORY STATUS OF DIRECT FOOD ADDITIVES (continued)

Material	Alternate Name [Source]	FEMA No.	Regulation	Limitations
spiro[2,4-Dithia-1-methyl-8-oxabicyclo-[3.3.0]octane-3,3'-(1'-oxa-2'-methyl)cyclopentane] and *spiro*[2,4-dithia-6-methyl-7-oxabicyclo-[3.3.0]octane-3,3'-(1'-oxa-2-methyl)cyclopentane]		3270		*see Food Technol.,* May 1972
3,3'-Dithio-bis(2-methyl-furan)	*see* Bis(2-methyl-3-furyl) disulfide			
2,2'-(Dithiodimethyl-ene)-difuran	2-Furfuryl disulfide	3146		
2,2'-Dithiodithiophene	*see* 2-Thienyl disulfide			
Dittany of Crete	Spanish hops [*Origanum dictamnus* L.]	2399	FDA §121.1163	Natural flavor
Dittany root	Fraxinella [*Dictamnus albus* L.]		FDA §121.1163	Natural flavor in alcoholic beverages only
γ-Dodecalactone	4-Hydroxydodecanoic acid, *γ*-lactone	2400	FDA §121.1164	Synthetic flavor
δ-Dodecalactone	5-Hydroxydodecanoic acid	2401	FDA §121.1164	Synthetic flavor
			FDA §121.1164	In oleomargarine, alone or with *δ*-decalactone, *δ*-dodecalactone at 20 ppm max, *δ*-decalactone at 10 ppm max
Dodecanal	*see* Lauric aldehyde			
Dodecanoic acid	Lauric acid	2614		
1-Dodecanol	*see* Lauryl alcohol			
2-Dodecenal	*n*-Dodecen-2-ol	2402	FDA §121.1164	Synthetic flavor
Dog grass	Quickgrass; triticum [*Agropyron repens* L. Beauv.]	2403	FDA §121.101	GRAS, natural flavor extractive
Dracorubin extract	*see* Dragon's blood, extract			

REGULATORY STATUS OF DIRECT FOOD ADDITIVES (continued)

Material	Alternate Name [Source]	FEMA No.	Regulation	Limitations
Dragon's blood, extract	Dracorubin [*Daemono-rops* species]	2404	FDA §121.1163	Natural flavor
Dried algae meal	*see* Algae meal, dried			
Dried yeasts	[*Saccharomyces cere-visiae*]		FDA §17	In bakery products; 2% max by weight of flour
	[*Saccharomyces cere-visiae*; *S. fragilis*; and dried torula yeast *Candida utilis*]		FDA §121.1125	In food; total folic acid in yeast not to exceed 0.04 mg/g of yeast (approx 0.008 mg of pteroylglutamic acid per gram of yeast)
Dulse	[*Rhodymenia palmata* (L.) Grev.]	2405	FDA §121.101	GRAS, natural substance and natural extractive (sol-vent-free) used with natural flavorings
Elder flowers	[*Sambucus canadensis* L.]	2406	FDA §121.101	GRAS, natural flavor
Elder flowers and tree leaves	[*Sambucus nigra* L.]		FDA §121.1163	Natural flavor in alcoholic beverages only; not to ex-ceed 25 ppm of prussic acid
Elecampane, rhizomes and roots	Inula; scabwort [*Inula helenium* L.]		FDA §121.1163	Natural flavor in alcoholic beverages only
Elemi	[*Canarium commune* L.; *C. luzonicum* Mig.]		FDA §121.1163	Natural flavor
Emulsifiers and/or surface-active agents	*see* specific product listings		FDA §121.2541	In manufacture of articles or components of articles for use in producing, manu-facturing, packing, proces-sing, preparing, treating, packaging, transporting, or holding food. Product spec-ifications apply
Enanthaldehyde	*see* Heptanal			
Enanthic alcohol	*see* Heptyl alcohol			
Enocianina	*see* Grape skin extract			
Enzymes	*see* specific enzyme		FDA §19	In curing and developing flavor of specified cheeses

REGULATORY STATUS OF DIRECT FOOD ADDITIVES (continued)

Material	Alternate Name [Source]	FEMA No.	Regulation	Limitations
Epichlorohydrin	*dl*-α-Epichlorohydrin		FDA §121.1031	Modifier for food starch, up to 0.3%, alone or with not more than 5.0% acetic anhydride
			FDA §121.1031	Modifier for food starch up to 0.3% with succinic anhydride up to 4.0%
1,8-Epoxy-*p*-menthane	*see* Eucalyptol			
Equisitic acid	*see* Aconitic acid			
Erigeron, oil	Fleabane [*Erigeron canadensis* L.]	2409	FDA §121.1163	Natural flavor
Erythorbic acid	Isoascorbic acid	2410	FDA §121.101	GRAS, chemical preservative
			MID	To accelerate color fixing in cured pork and beef cuts and cured comminuted meat food products; ¾ oz. to 100 lb. meat or meat by-product. Product specifications apply
Estragole	Esdragol; esdragon; estragon; *p*-methoxy-allylbenzene	2411	FDA §121.1164	Synthetic flavor
Estragon, oil	Tarragon [*Artemisia dracunculus* L.]	2412	FDA §121.101	GRAS, natural flavor extractive
Ethanal	*see* Acetaldehyde			
Ethanoic acid	*see* Acetic acid			
Ethanol	*see* Ethyl alcohol			
Ethone	*see* 1-(*p*-Methoxyphenyl)-1-penten-3-one			
6-Ethoxy-*m*-anol	*see* Propenylguaethol			
***p*-Ethoxybenzaldehyde**		2413	FDA §121.1164	Synthetic flavor
2-Ethoxy-*p*-cymene	*see* Carvacryl ethyl ether			
2-Ethoxy-5-propenyl-anisole	*see* Isoeugenyl ethyl ether			

REGULATORY STATUS OF DIRECT FOOD ADDITIVES (continued)

Material	Alternate Name [Source]	FEMA No.	Regulation	Limitations
Ethoxyquin	6-Ethoxy-1,2-dihydro-2,2,4-trimethylquinoline		FDA §121.1001	Antioxidant for preservation of color in chili powder, paprika, and ground chili, 100 ppm max. Product specifications apply; tolerances in animal products also apply
Ethyl acetate	Acetic ether	2414	FDA §121.101	GRAS, synthetic flavor/adjuvant
			FDA §8.390	Diluent in inks for marking fruits and vegetables. Product specifications apply
Ethyl acetoacetate	Acetoacetic ester	2415	FDA §121.1164	Synthetic flavor
Ethyl 2-acetyl-3-phenyl-propionate	Ethyl benzylacetoacetate	2416	FDA §121.1164	Synthetic flavor
1-Ethyl-2-acetylpyrrole		3147		
Ethyl aconitate, mixed esters	Ethyl-2-carboxyglutaconate	2417	FDA §121.1164	Synthetic flavor
Ethyl acrylate	Acrylic acid ethyl ester	2418	FDA §121.1164	Synthetic flavor
Ethyl alcohol	Alcohol, anhydrous; ethanol	2419	FDA §8.305	Ethyl alcohol, or alkaline solutions thereof, as extractant for annatto
Ethyl amyl ketone	see 3-Octanone			
Ethyl p-anisate	Ethyl p-methoxybenzoate	2420	FDA §121.1164	Synthetic flavor
Ethyl anthranilate	Ethyl-2-aminobenzoate	2421	FDA §121.1164	Synthetic flavor
Ethyl benzoate	Ethyl benzenecarboxylate	2422	FDA §121.1164	Synthetic flavor
Ethyl benzoylacetate	Benzoyl acetic ester	2423	FDA §121.1164	Synthetic flavor
Ethyl benzylacetoacetate	see Ethyl 2-acetyl-3-phenylpropionate			
α-Ethylbenzyl butyrate	α-Phenylpropyl butyrate	2424	FDA §121.1164	Synthetic flavor
Ethyl brassylate	Tridecanedioic acid cyclic ethylene glycol diester; cyclo-1,13-ethylenedioxytridecan-1,13-dione		FDA §121.1164	Synthetic flavor/adjuvant

REGULATORY STATUS OF DIRECT FOOD ADDITIVES (continued)

Material	Alternate Name [Source]	FEMA No.	Regulation	Limitations
2-Ethylbutyl acetate		2425	FDA §121.1164	Synthetic flavor
2-Ethyl-3-butylacrolein	*see* 2-Ethyl-2-heptenal			
Ethyl butyl ketone	*see* 3-Heptanone			
2-Ethylbutyraldehyde		2426	FDA §121.1164	Synthetic flavor
Ethyl butyrate		2427	FDA §121.101	GRAS, synthetic flavor and adjuvant
2-Ethylbutyric acid	α-Ethylbutyric acid	2429		
Ethyl cellulose			FDA §121.1087	Binder and filler in dry vitamin preparations; component of protective coatings for vitamin and mineral tablets; fixative in flavoring compounds. Product specifications apply
			FDA §8.390 and §8.6200	Diluent in inks for marking gum, confectionery, and food supplements in tablet form; inks for marking fruits and vegetables; inks for branding pharmaceutical forms, externally applied drugs and mixtures for coloring shell eggs. Product specifications apply
Ethyl cinnamate	Ethyl phenylacrylate	2430	FDA §121.1164	Synthetic flavor
Ethyl citrate	Triethyl citrate	3083		
Ethyl *p*-cresoxyacetate	*see* Ethyl (*p*-tolyloxy)-acetate			
Ethyl crotonate	*trans*-Butenoic acid ethyl ester		FDA §121.1164	Synthetic flavor/adjuvant
Ethyl cyclohexane-propionate	Ethyl cyclohexyl-propionate	2431	FDA §121.1164	Synthetic flavor
Ethyl *trans*-2,*cis*-4-deca-dienoate		3148		
Ethyl decanoate	Ethyl caprate	2432	FDA §121.1164	Synthetic flavor

REGULATORY STATUS OF DIRECT FOOD ADDITIVES (continued)

Material	Alternate Name [Source]	FEMA No.	Regulation	Limitations
2-Ethyl-1,4-diazine	see 2-Ethylpyrazine			
Ethyl 2,4-diketocaproate	see Ethyl 2,4-dioxo-hexanoate			
Ethyl 2,4-dioxohexano-ate	Ethyl 2,4-diketocaproate	3278		see *Food Technol.*, May 1972
2-Ethyl-3,5 or 6-dimethylpyrazine		3149		10 ppm max in finished food
3-Ethyl-2,6-dimethyl-pyrazine	2,6-Dimethyl-3-ethyl-pyrazine	3150		10 ppm max in finished food
Ethylene dichloride	*sym*-Dichloroethane		FDA §8.305	Solvent for annatto extract, residues 30 ppm max; total chlorinated solvents residue, 30 ppm max
Ethylene glycol distearate			FDA §8.390	Diluent in mixtures for coloring shell eggs; no penetration of color-additive mixture through eggshell into egg
Ethylene glycol mono-ethyl ether			FDA §8.390	Diluent in inks for marking gum, confectionery, and food supplements in tablet form. Product specifications apply
Ethyl formate	Formic ether	2434	FDA §121.1164	Synthetic flavor
2-Ethylfuran			FDA §121.1164	Synthetic flavor/adjuvant
Ethyl 2-furanpropionate	Ethyl furylpropionate	2435	FDA §121.1164	Synthetic flavor
4-Ethylguaiacol	4-Ethyl-2-methoxyphenol	2436	FDA §121.1164	Synthetic flavor
Ethyl heptanoate	Ethyl heptoate	2437	FDA §121.1164	Synthetic flavor
2-Ethyl-2-heptenal	2-Ethyl-3-butylacrolein	2438	FDA §121.1164	Synthetic flavor
Ethyl hexadecanoate	Ethyl palmitate	2451	FDA §121.1164	Synthetic flavor
Ethyl 2,4-hexadienoate	see Ethyl sorbate			
Ethyl hexanoate	Capronic ether absolute	2439	FDA §121.1164	Synthetic flavor
2-Ethyl-1-hexanol	2-Ethylhexyl alcohol	3151		

REGULATORY STATUS OF DIRECT FOOD ADDITIVES (continued)

Material	Alternate Name [Source]	FEMA No.	Regulation	Limitations
Ethyl hydrocinnamate	*see* Ethyl 3-phenyl-propionate			
3-Ethyl-2-hydroxy-2-cyclopenten-1-one		3152		
5-Ethyl-3-hydroxy-4-methyl-2(5H)-furanone	2,4-Dihydroxy-3-methyl-2-hexenoic acid, γ-lactone; 2-ethyl-3-methyl-4-hydroxydihydro-(2,5)-furan-5-one; 2-hydroxy-3-methyl-γ-2-hexenolactone	3153		
2-Ethyl-3-hydroxy-4H-pyran-4-one	*see* Ethyl maltol			
Ethyl isobutyrate		2428	FDA §121.1164	Synthetic flavor
Ethyl isoeugenol	*see* Isoeugenyl ethyl ether			
Ethyl isovalerate	Ethyl β-methylbutyrate	2463	FDA §121.1164	Synthetic flavor
Ethyl lactate	Ethyl α-hydroxypropionate	2440	FDA §121.1164	Synthetic flavor
Ethyl laurate	Ethyl dodecanoate	2441	FDA §121.1164	Synthetic flavor
Ethyl levulinate	Ethyl γ-ketovalerate	2442	FDA §121.1164	Synthetic flavor
Ethyl malonate	*see* Diethyl malonate			
Ethyl maltol	2-Ethyl-3-hydroxy-4H-pyran-4-one		FDA §121.1164	Synthetic flavor/adjuvant
Ethyl 2-mercaptopropionate		3279		*see Food Technol.,* May 1972
4-Ethyl-2-methoxyphenol	*see* 4-Ethylguaiacol			
2-Ethyl-(3,5 and 6)-methoxypyrazine (85%) plus 2-Methyl-(3,5 and 6)-methoxypyrazine (13%)		3280		*see Food Technol.,* May 1972
Ethyl *trans*-2-methyl-2-butenoate	*see* Ethyl tiglate			
Ethyl 2-methylbutyrate		2443	FDA §121.1164	Synthetic flavor/adjuvant

REGULATORY STATUS OF DIRECT FOOD ADDITIVES (continued)

Material	Alternate Name [Source]	FEMA No.	Regulation	Limitations
2-Ethyl-3-methyl-4-hydroxydihydro-(2,5)-furan-5-one	see 5-Ethyl-3-hydroxy-4-methyl-2(5H)-furanone			
Ethyl methylphenyl-glycidate	3-Methyl-3-phenyl-glycidic acid ethyl ester; so-called strawberry aldehyde; aldehyde C-16	2444	FDA §121.101	GRAS, synthetic flavor/adjuvant
2-Ethyl-5-methyl-pyrazine		3154		10 ppm max in finished food
3-Ethyl-2-methyl-pyrazine		3155		10 ppm max in finished food
Ethyl myristate	Ethyl tetradecanoate	2445	FDA §121.1164	Synthetic flavor
Ethyl nitrite	Nitrous ether	2446	FDA §121.1164	Synthetic flavor
Ethyl nonanoate	Ethyl nonylate	2447	FDA §121.1164	Synthetic flavor
Ethyl 2-nonynoate	Ethyl octyne carbonate	2448	FDA §121.1164	Synthetic flavor
Ethyl octanoate	Ethyl caprylate	2449	FDA §121.1164	Synthetic flavor
Ethyl octyne carbonate	see Ethyl 2-nonynoate			
Ethyl oleate	Ethyl 9-octadecenoate	2450	FDA §121.1164	Synthetic flavor
Ethyl oxyhydrate	see Rum ether			
p-Ethylphenol		3156		
Ethyl phenylacetate	α-Toluic acid	2452	FDA §121.1164	Synthetic flavor
Ethyl 4-phenylbutyrate	Ethyl phenylbutyrate	2453	FDA §121.1164	Synthetic flavor
Ethyl 3-phenylglycidate	Ethyl 3-phenyl-2,3-epoxy-propionate	2454	FDA §121.1164	Synthetic flavor
Ethyl 3-phenylpropio-nate	Ethyl hydrocinnamate	2455	FDA §121.1164	Synthetic flavor
Ethyl propionate		2456	FDA §121.1164	Synthetic flavor
Ethyl propyl ketone	see 3-Hexanone			
2-Ethylpyrazine		3281		see Food Technol., May 1972

REGULATORY STATUS OF DIRECT FOOD ADDITIVES (continued)

Material	Alternate Name [Source]	FEMA No.	Regulation	Limitations
Ethyl pyruvate	Ethyl α-ketopropionate	2457	FDA §121.1164	Synthetic flavor
Ethyl salicylate	Salicylic ether	2458	FDA §121.1164	Synthetic flavor
Ethyl sorbate	Ethyl 2,4-hexadienoate	2459	FDA §121.1164	Synthetic flavor
Ethyl thioacetate		3282		see Food Technol., May 1972
Ethyl tiglate	Ethyl trans-2-methyl-2-butenoate	2460	FDA §121.1164	Synthetic flavor
Ethyl (p-tolyloxy)-acetate	Ethyl p-cresoxyacetate	3157		
Ethyl undecanoate	Ethyl hendecanoate		FDA §121.1164	Synthetic flavor
Ethyl 10-undecenoate	Ethyl 10-hendecenoate	2461	FDA §121.1164	Synthetic flavor
Ethyl valerate		2462	FDA §121.1164	Synthetic flavor
Ethyl vanillin	3-Ethoxy-4-hydroxy-benzaldehyde	2464	FDA §121.101	GRAS, synthetic flavor
Eucalyptol	Cineole; 1,8-Epoxy-p-menthane	2465	FDA §121.1164	Synthetic flavor
Eucalyptus, oil	Globulus [Eucalyptus globulus Labill.]	2466	FDA §121.1163	Natural flavor
Eugenol	4-Allyl-2-methoxyphenol	2467	FDA §121.101	GRAS, synthetic flavor
Eugenyl acetate	Acetyl eugenol	2469	FDA §121.1164	Synthetic flavor
Eugenyl benzoate	4-Allyl-2-methoxyphenyl benzoate	2471	FDA §121.1164	Synthetic flavor
Eugenyl formate	4-Allyl-2-methoxyphenyl formate	2473	FDA §121.1164	Synthetic flavor
Eugenyl methyl ether	4-Allylveratrole; methyl eugenol	2475	FDA §121.1164	Synthetic flavor
Farnesol	3,7,11-Trimethyl-2,6,10-dodecatrien-1-ol	2478	FDA §121.1164	Synthetic flavor
Fatty acid esters, lactylated	see Lactylated fatty acid esters of glycerol and propylene glycol			

REGULATORY STATUS OF DIRECT FOOD ADDITIVES (continued)

Material	Alternate Name [Source]	FEMA No.	Regulation	Limitations
Fatty acids	Capric, caprylic, lauric, myristic, oleic, palmitic, and stearic acids, and their associated fatty acids and oils		FDA §121.1070	Lubricant, binder, and defoaming agents; components in manufacture of other food-grade additives. Product specifications apply
			FDA §121.1099	Components of defoaming agents limited to use in processing beet sugar and yeast
FD&C Blue No. 1			FDA §8.206 and §8.501	Coloring foods (including dietary supplements) generally. Product specifications apply
FD&C Blue No. 2	Indigo carmine		FDA §8.501	Food, drug, and cosmetic use. Specifications in §9.81 apply
FD&C Colors			MID	FD&C color additives; coloring sausage casing, oleomargarine, or shortening; in marking and branding ink for products; must be certified for food use by FDA; may be mixed with approved natural colors or harmless inert materials, such as common salt or sugar
FD&C Green No. 3			FDA §8.501	Food, drug, and cosmetic use. Specifications in §9.23 apply
FD&C Red No. 2	Amaranth		FDA §8.501	Food, drug, and cosmetic use. Specifications in §9.61 apply
FD&C Red No. 3	Erythrosine		FDA §8.242 and §8.501	Coloring foods generally, including dietary supplements. Product specifications apply
			FDA §8.242	For food use as lake only
FD&C Red No. 4			FDA §8.501	Coloring maraschino cherries; 150 ppm max by weight of maraschino cherries (weight not including pack-

REGULATORY STATUS OF DIRECT FOOD ADDITIVES (continued)

Material	Alternate Name [Source]	FEMA No.	Regulation	Limitations
FD&C Red No. 4 (continued)				ing media or added sugar). Product specifications apply
FD&C Red No. 40			FDA §8.244	Coloring foods, including dietary supplements. Product specifications apply
FD&C Violet No. 1			FDA §8.501	Food, drug, and cosmetic use. Specifications in §9.90 apply
FD&C Yellow No. 5	Tartrazine		FDA §8.275 and §8.501	Coloring foods generally. Product specifications apply
			FDA §8.275	Food use of lakes only. Product specifications apply
FD&C Yellow No. 6	Sunset Yellow FCF		FDA §8.501	Food, drug, and cosmetic use; specifications in §9.41 apply
***d*-Fenchone**	*d*-1,3,3-Trimethyl-2-norbornanone	2479	FDA §121.1164	Synthetic flavor
Fenchyl alcohol	1,3,3-Trimethyl-2-norbornanol	2480	FDA §121.1164	Synthetic flavor
Fennell, common	[*Foeniculum vulgare* Mill.]	2481	FDA §121.101	GRAS, natural flavor
Fennel, sweet	Finocchio; Florence fennel [*Foeniculum vulgare* Mill. var. *dulce* (DC.) Alef.]	2482	FDA §121.101	GRAS, natural flavor
Fenugreek	[*Trigonella foenum-graecum* L.]	2484	FDA §121.101	GRAS, natural flavor and natural flavor extractive
Fermentation-derived, milk-clotting enzyme	[*Endothia parasitica*]		FDA §121.1199	In production of cheese. Product specifications apply
Ferric phosphate			FDA §121.101	GRAS, nutrient and/or dietary supplement
Ferric pyrophosphate			FDA §121.101	GRAS, nutrient and/or dietary supplement

REGULATORY STATUS OF DIRECT FOOD ADDITIVES (continued)

Material	Alternate Name [Source]	FEMA No.	Regulation	Limitations
Ferric sodium pyrophosphate			FDA §121.101	GRAS, nutrient and/or dietary supplement
Ferrous gluconate	Fergon		FDA §121.101	GRAS, nutrient and/or dietary supplement. *Food Chemicals Codex* specifications apply
			FDA §8.320	Coloring ripe olives
Ferrous lactate			FDA §121.101	GRAS, nutrient and/or dietary supplement
Ferrous sulfate			FDA §121.101	GRAS, nutrient and/or dietary supplement
Fincin			MID	With water, salt, and monosodium glutamate to soften tissues of beef cuts; tissue weight gain 3% max
Finocchio	*see* Fennel, sweet			
Fir, balsam, needles and twigs	Balsam fir [*Abies balsamea* (L.) Mill.]		FDA §121.1163	Natural flavor
Fir ("pine"), needles and twigs	[*Abies sibirica* Ledeb.; *A. alba* Mill.; *A. sachalinensis* Masters; *A. mayriana* Miyabe & Kudo]		FDA §121.1163	Natural flavor
Folic acid	Pteroylglutamic acid		FDA §121.1134	Component of dietary supplements; daily ingestion not to exceed 0.10 mg
Food starch—modified			FDA §121.1031	1. Acid-modified, by treatment with hydrochloric acid or sulfuric acid or both 2. Bleached by one or more of the following: Active oxygen obtained from hydrogen peroxide and/or peracetic acid, not to exceed 0.45% of active oxygen Ammonium persulfate, up to 0.075%, and sulfur dioxide, up to 0.05% Chlorine, as sodium hypochlorite, not to exceed

REGULATORY STATUS OF DIRECT FOOD ADDITIVES (continued)

Material	Alternate Name [Source]	FEMA No.	Regulation	Limitations
Food starch—modified (continued)				0.0082 lb of chlorine/lb of dry starch Potassium permanganate, not to exceed 0.2% (residual manganese, calculated as Mn, not to exceed 50 ppm in food starch—modified) Sodium chlorite, not to exceed 0.5% 3. Oxidized, with chlorine, as sodium hypochlorite, not to exceed 0.055 lb chlorine/lb dry starch 4. Esterified, by treatment with one of the following: Acetic anhydride; acetyl groups in food starch—modified not to exceed 2.5% Adipic anhydride, not to exceed 0.12%, and acetic anhydride; acetyl groups in food starch—modified not to exceed 2.5% 1-Octenyl succinic anhydride, not to exceed 3.0% 1-Octenyl succinic anhydride, not to exceed 2.0%, and aluminum sulfate, not to exceed 2.0% Phosphorus oxychloride, 0.1% max Sodium trimetaphosphate in food starch—modified not to exceed 0.4%, calculated as phosphorus Sodium tripolyphosphate and sodium trimetaphosphate; residual phosphate in food starch—modified not to exceed 0.4%, calculated as phosphorus Vinyl acetate; acetyl groups in food starch—modified not to exceed 2.5% 5. Etherified by treatment with one of the following: Acrolein, 0.6% max

REGULATORY STATUS OF DIRECT FOOD ADDITIVES (continued)

Material	Alternate Name [Source]	FEMA No.	Regulation	Limitations
Food starch—modified (continued)				Epichlorohydrin, 0.3% max Propylene oxide, 25% max 6. Esterified and etherified by treatment with one of the following: Acrolein, up to 0.3%, and vinyl acetate, up to 7.5%; acetyl groups in food starch—modified, 2.5% max Epichlorohydrin, not to exceed 0.3%, and acetic anhydride; acetyl groups in food starch—modified, 2.5% max Epichlorohydrin, 0.3% max, and succinic anhydride, up to 4.0% Phosphorus oxychloride, 0.1% max, and propylene oxide, 10% max 7. Modified, by treatment with one of the following: Chlorine, as sodium hypochlorite, not to exceed 0.055 lb chlorine/lb dry starch; 0.45% of active oxygen obtained from hydrogen peroxide; and propylene oxide, 25% max Sodium hydroxide, 1.0% max 8. Food starch may be modified by a combination of acid modification and/or bleaching, and any one of the other treatments prescribed, subject to any limitations prescribed
Formaldehyde			FDA §121.1099	Preservative in defoaming agents limited to use in processing beet sugar and yeast; preservative in defoaming agents containing dimethylpolysiloxane, not to exceed 1.0% of dimethylpolysiloxane content
Formic acid		2487	FDA §121.1164	Synthetic flavor

REGULATORY STATUS OF DIRECT FOOD ADDITIVES (continued)

Material	Alternate Name [Source]	FEMA No.	Regulation	Limitations
2-Formylbenzofuran	see 2-Benzofuran-carboxaldehyde			
Fraxinella	see Dittany			
Fruit juices			FDA §8.313	Coloring foods generally. Product specifications apply
Fumaric acid	trans-Butenedioic acid	2488	FDA §121.1130	Fumaric acid and its calcium ferrous magnesium, potassium, and sodium salts. Product specifications apply. Ferrous fumarate—source or iron in foods for special dietary use
2-Furanmethanethiol	see Furfuryl mercaptan			
2-Furanmethanethiol formate		3158		
Furcellaran	[Furcellaria fastigiata, family Florideae, order Gigartinales]		FDA §121.1068	Emulsifier, stabilizer, and thickener in foods. Product specifications apply
Furcellaran, salts			FDA §121.1069	Emulsifier, stabilizer, and thickener in foods. Product specifications apply
Furfural	2-Furaldehyde; pyromucic aldehyde	2489		
Furfuryl acetate		2490		
Furfuryl alcohol	2-Furancarbinol	2491		
Furfuryl disulfide	see Bis(2-furfuryl) disulfide			
2-Furfuryl disulfide	see 2,2'-(Dithiodimethylene)-difuran			
Furfuryl isopropyl sulfide		3161		
Furfuryl isovalerate	see Furfuryl 2-methylbutanoate			
Furfuryl mercaptan	2-Furanmethanethiol	2493		

REGULATORY STATUS OF DIRECT FOOD ADDITIVES (continued)

Material	Alternate Name [Source]	FEMA No.	Regulation	Limitations
Furfuryl methyl-butanoate	Furfuryl isovalerate	3283		see Food Technol., May 1972
Furfuryl methyl ether		3159		
Furfuryl methyl sulfide		3160		
2-Furfuryl monosulfide	see 2,2'-(Thiodimethy-lene)-difuran			
N-Furfurylpyrrole	1-(2-Furfuryl)pyrrole	3284		see Food Technol., May 1972
Furfuryl sulfide	see Bis(2-furfuryl) sulfide			
Furfuryl thioacetate		3162		
Furyl acetone	see 1-Furyl-2-propanone; (2-Furyl)-2-propanone			
Furyl acrolein	2-Furanacrolein	2494		
4-(2-Furyl)-3-buten-2-one	Furfurylidene acetone	2495		
2-Furyl methyl ketone	2-Acetylfuran	3163		
1-Furyl-2-propanone	Furyl acetone	2496	FDA § 121.1164	Synthetic flavor
(2-Furyl)-2-propanone	Furyl acetone		FDA §121.1164	Synthetic flavor
Fusel oil, refined	Mixed amyl alcohols	2497	FDA §121.1164	Synthetic flavor
Galanga, greater	[Alpinia galanga Willd.]		FDA §121.1163	Natural flavor in alcoholic beverages only
Galanga, root	Galangal; Chinese ginger [Alpinia officinarum Hance]	2498	FDA §121.101	GRAS, natural flavor and natural flavor extractive
Galbanum, oil	[Ferula galbaniflua Boiss. & Buhse and other Ferula species]	2501	FDA §121.1163	Natural flavor
Gambir	Gambir catechu; pale catechu [Uncaria gambir (Hunter) Roxb.]		FDA §121.1163	Natural flavor
Garlic, oil	[Allium sativum L.]	2503	FDA §121.101	GRAS, natural flavor and natural flavor extractive
Gelatin	Puragel		FDA—GRAS	

REGULATORY STATUS OF DIRECT FOOD ADDITIVES (continued)

Material	Alternate Name [Source]	FEMA No.	Regulation	Limitations
Genet, extract	[*Spartium junceum* L.]	2505		
Genet, flowers	[*Spartium junceum* L.]		FDA §121.1163	Natural flavor
Gentian, rhizome and roots	[*Gentiana lutea* L.]	2506	FDA §121.1163	Natural flavor
Gentian, stemless	[*Gentiana acaulis* L.]		FDA §121.1163	Natural flavor in alcoholic beverages only
Geranial	*see* Citral			
Geraniol	*trans*-3,7-Dimethyl-2,6-octadien-1-ol	2507	FDA §121.101	GRAS, synthetic flavor and adjuvant
Geranium	[*Pelargonium* species]		FDA §121.101	GRAS, natural flavor and natural flavor extractive
Geranium, East Indian	*see* Palmarosa			
Geranium, rose, oil	[*Pelargonium graveolens* L'Her]	2508	FDA §121.101	GRAS, natural flavor extractive
Geranyl acetate	Geraniol acetate	2509	FDA §121.101	GRAS, synthetic flavor and adjuvant
Geranyl acetoacetate	*trans*-3,7-Dimethyl-2,6-octadien-1-yl aceto-acetate	2510	FDA §121.1164	Synthetic flavor
Geranyl benzoate		2511	FDA §121.1164	Synthetic flavor
Geranyl butyrate		2512	FDA §121.1164	Synthetic flavor
Geranyl formate		2514	FDA §121.1164	Synthetic flavor
Geranyl hexanoate	Geranyl caproate	2515	FDA §121.1164	Synthetic flavor
Geranyl isobutyrate		2513	FDA §121.1164	Synthetic flavor
Geranyl isovalerate		2518	FDA §121.1164	Synthetic flavor
Geranyl phenylacetate	Geranyl α-toluate	2516	FDA §121.1164	Synthetic flavor
Geranyl propionate	2,6-Dimethyl octadien-6-yl-8-*n*-propionate	2517	FDA §121.1164	Synthetic flavor
Germander	Chamaedrys [*Teucrium chamaedrys* L.]		FDA §121.1163	Natural flavor in alcoholic beverages only

REGULATORY STATUS OF DIRECT FOOD ADDITIVES (continued)

Material	Alternate Name [Source]	FEMA No.	Regulation	Limitations
Germander, golden	[*Teucrium polium* L.]		FDA §121.1163	Natural flavor in alcoholic beverages only
Ghatti, gum	Indian gum [*Anogeissus latifolia* Wall.]	2519	FDA §121.101	GRAS, stabilizer
Ginger	[*Zingiber officinale* Rosc.]	2520	FDA §121.101	GRAS, natural flavor and natural flavor extractive
Ginger, wild	*see* Snakeroot Canadian			
Glucono-Δ-lactone			MID	8 oz/100 lb of cured, comminuted meat or meat product to speed up color-fixing process and to reduce time required for smoking; must be listed in product ingredient statement
Glucose pentaacetate		2524	FDA §121.1164	Synthetic flavor
Glucose syrup			MID	To flavor sausage, hamburger, meat loaf, luncheon meat, chopped or pressed ham; alone at 2% or with corn-syrup solids or corn syrup, with combination totaling 2%, calculated on dry basis
Glutamic acid			FDA §121.101	GRAS, salt substitute
L-Glutamic acid		3285		*see Food Technol.*, May 1972
Glutamic acid hydrochloride			FDA §121.101	GRAS, salt substitute
Gluten, gum			FDA §16	Optional use in macaroni and food products, provided protein content of food does not exceed 13% by weight
Glycerides (mono- and di-)			FDA §20	In frozen desserts, not more than 0.2% by weight of finished product
			FDA §121.101	Mono- and diglycerides of edible fats or oils, or edible fat-forming fatty acids (except lauric), GRAS, emulsifiers
			MID	To emulsify rendered animal fat or combination of

REGULATORY STATUS OF DIRECT FOOD ADDITIVES (continued)

Material	Alternate Name [Source]	FEMA No.	Regulation	Limitations
Glycerides (mono- and di-) (continued)				such fat with vegetable fat; in lard and shortening; at 0.5% in oleomargarine
			FDA §8.305	Extractant for annatto; mono- and diglycerides from glycerolysis of edible vegetable fats or oils
Glycerine	Glycerol	2525	FDA §121.101	GRAS, miscellaneous and/ or general-purpose food additive
Glycerine, synthetic			FDA §121.1111	Food additive; product specifications apply
Glycerol	*see* Glycerine			
Glycerol ester of gum or wood rosin			FDA §121.1059	Plasticizing material in chewing gum base
Glycerol ester of partially dimerized rosin			FDA §121.1059	Plasticizing material in chewing gum base. Product specifications apply
Glycerol ester of partially hydrogenated gum or wood rosin			FDA §121.1059	Plasticizing material in chewing gum base
Glycerol ester of polymerized rosin			FDA §121.1059	Plasticizing material in chewing gum base. Product specifications apply
Glycerol ester of wood rosin			FDA §121.1059	Plasticizing material in chewing gum base. Product specifications apply
			FDA §121.1084	Adjusting density of citrus oils used in preparation of beverages. Product specifications apply
Glycerol-lacto stearate, oleate, or palmitate			MID	To emulsify rendered animal fat or combination with vegetable fat
Glycerol tributyrate	*see* Tributyrin			
Glyceryl-lacto esters of fatty acids			FDA §121.1004	Emulsifiers and plasticizers in food; lactic acid esters of mono- and diglycerides, manufactured from glycer-

REGULATORY STATUS OF DIRECT FOOD ADDITIVES (continued)

Material	Alternate Name [Source]	FEMA No.	Regulation	Limitations
Glyeryl-lacto esters of fatty acids (continued)				in, lactic acid, and fatty acids conforming with §121. 1070 and/or edible fats and oils
Glyceryl monooleate	Monoolein	2526	FDA §121.1164	Synthetic flavor
Glyceryl monostearate	Glyceryl monooctade-canoate; monostearin	2527	FDA §121.101	GRAS, miscellaneous and/or general-purpose food additive
Glyceryl triacetate	*see* Triacetin			
Glyceryl tributyrate	*see* Tributyrin			
Glyceryl tripropanoate	Propionic acid, trigly-ceride; tripropionin	3286		*see Food Technol.,* May 1972
Glycerine	Aminoacetic acid; glycocoll	3287	MID	To retard rancidity in rendered animal fat or combination of such fat with vegetable fat, up to 0.01% or up to 0.02% of combination of glycerine, BHA, BHT, NDGA, and propyl gallate
			FDA §121.101	On an interim basis: GRAS, nutrient and/or dietary supplement in animal feeds, as free acid, hydrochloride salt, hydrate, or anhydrous form
			FDA §121.4002	On an interim basis: 1. masking agent for bitter aftertaste of saccharin in carbonated, artificially sweetened beverages, not to exceed 0.2% of finished product 2. flavor agent in butter-scotch and toffee flavorings used in manufacture of frosting mixes, 0.9% max 3. stabilizer in mono- and diglycerides prepared by glycerolysis of edible fats and oils, not to exceed 0.02% of the mono- and diglycerides. *Food Chemicals Codex* specifications apply

REGULATORY STATUS OF DIRECT FOOD ADDITIVES (continued)

Material	Alternate Name [Source]	FEMA No.	Regulation	Limitations
Glycerine (continued)				see *Food Technol.,* May 1972, for FEMA max use levels
Glycocholic acid	*N*-Cholylglycine		FDA §121.101	GRAS, emulsifying agent in dried egg whites, up to 0.1%
Glycocoll	*see* Glycine			
Glycyrrhiza	*see* Licorice			
Glycyrrhizin, ammoniated	[*Glycyrrhiza glabra* L. and other *Glycyrrhiza* species]	2528	FDA §121.101	GRAS, natural flavor extractive
Grains of paradise	[*Aframomum melegueta* Rosc. and K. Schum.]	2529	FDA §121.101	GRAS, natural flavor
Grapefruit, oil	[*Citrus paradisi* Macf.; *C. decumana* L.]	2530	FDA §121.101	GRAS, natural flavor extractive
Grape skin extract	Enocianina		FDA §8.318	Coloring still and carbonated drinks and ades, beverage bases, and alcoholic beverages. Product specifications apply
Green chlorophyll			MID	Coloring sausage casings, oleomargarine, and shortening; in marking or branding ink for products; may be mixed with approved synthetic dyes or harmless inert material, such as common salt or sugar
Guaiac, gum, extract	[*Guaiacum officinale* L.; *G. sanctum* L.]	2531	FDA §121.101	GRAS, preservative, 0.10% in edible fats and oils
Guaiacol	*o*-Methoxyphenol	2532	FDA §121.1164	Synthetic flavor
Guaiac, wood, extract	[*Guaiacum officinale* L.]	2533	FDA §121.1163	Natural flavor
Guaiacyl acetate	*o*-Methoxyphenyl acetate		FDA §121.1164	Synthetic flavor/adjuvant
Guaiacyl phenylacetate		2535	FDA §121.1164	Synthetic flavor
Guaiene	1,4-Dimethyl-7-isopropenyl-Δ-9,10-octahydroazulene		FDA §121.1164	Synthetic flavor/adjuvant

REGULATORY STATUS OF DIRECT FOOD ADDITIVES (continued)

Material	Alternate Name [Source]	FEMA No.	Regulation	Limitations
Guaiol acetate	1,4-Dimethyl-7-(d-hydroxyisopropyl) Δ 9,10-octahydroazulene acetate		FDA §121.1164	Synthetic flavor/adjuvant
Guarana, gum	[*Paullinia cupana* H.B.K.]	2536	FDA §121.1163	Natural flavor
Guar gum	[*Cyamopsis tetragonolobus* (L.) Taub.]	2537	FDA §121.101	GRAS, stabilizer
Guava	[*Psidium* species]		FDA §121.101	GRAS, natural flavor extractant
Gum, acacia	*see* Acacia			
Gum, arabic	*see* Acacia			
Gum gluten	*see* Gluten, gum			
Gutta hang kang	[*Palaquium leiocarpum* and *P. oblongifolium* Burch.]		FDA §121.1059	Masticatory substance in chewing gum base
Haw, black, bark, extract	[*Viburnum prunifolium* L.]	2538	FDA §121.1163	Natural flavor
Heliotropine	*see* Piperonal			
Heliotropyl acetate	*see* Piperonyl acetate			
Helium			FDA §121.101	GRAS, miscellaneous and/or general-purpose food additive
Hemlock, needles and twigs	[*Tsuga canadensis* (L.) Carr.; *T. heterophylla* (Raf.) Sarg.]		FDA §121.1163	Natural flavor
Hendecenal	*see* 10-Undecenal			
2,4-Heptadienal		3164		
γ-Heptalactone	4-Hydroxyheptanoic acid, γ-lactone	2539	FDA §121.1164	Synthetic flavor
Heptanal	Enanthaldehyde; aldehyde C-7	2540	FDA §121.1164	Synthetic flavor
Heptanal dimethyl acetal		2541	FDA §121.1164	Synthetic flavor

REGULATORY STATUS OF DIRECT FOOD ADDITIVES (continued)

Material	Alternate Name [Source]	FEMA No.	Regulation	Limitations
Heptanal glyceryl acetal (mixed 1,2 and 1,3 acetals)		2542	FDA §121.1164	Synthetic flavor
Heptane			FDA §121.1164	Synthetic flavor
2,3-Heptanedione	Acetyl valeryl	2543	FDA §121.1164	Synthetic flavor
2-Heptanol	Methyl n-amyl carbinol	3288		see Food Technol., May 1972
3-Heptanol			FDA §121.1164	Synthetic flavor/adjuvant
2-Heptanone	Methyl amyl ketone	2544	FDA §121.1164	Synthetic flavor
3-Heptanone	Ethyl butyl ketone	2545	FDA §121.1164	Synthetic flavor
4-Heptanone	Dipropyl ketone	2546	FDA §121.1164	Synthetic flavor
2-Heptenal		3165		
4-Heptenal (cis and trans)		3289		see Food Technol., May 1972
Heptyl acetate		2547	FDA §121.1164	Synthetic flavor
Heptyl alcohol	Enanthic alcohol	2548	FDA §121.1164	Synthetic flavor
Heptyl butyrate		2549	FDA §121.1164	Synthetic flavor
Heptyl cinnamate		2551	FDA §121.1164	Synthetic flavor
Heptyl formate		2552	FDA §121.1164	Synthetic flavor
n-Heptyl p-hydroxy-benzoate			FDA §121.1186	In fermented malt beverages to inhibit microbiological spoilage, 12 ppm max
Heptylidene acetone	see 3-Decen-2-one			
Heptyl isobutyrate		2550	FDA §121.1164	Synthetic flavor
Heptyl octanoate		2553	FDA §121.1164	Synthetic flavor
Hexadecanoic acid	see Palmitic acid			
1-Hexadecanol	Cetyl alcohol	2554	FDA §121.1164	Synthetic flavor
			FDA §8.390 and §8.6200	Diluent in inks for marking gum, confectionery, and

REGULATORY STATUS OF DIRECT FOOD ADDITIVES (continued)

Material	Alternate Name [Source]	FEMA No.	Regulation	Limitations
1-Hexadecanol (continued)				food supplements in tablet form; inks for branding pharmaceutical forms; and ingested drugs. Product specifications apply
ω-6-Hexadecenlactone	Ambrettolide 16-hydroxy-6-hexadecenoic acid, ω-lactone	2555	FDA §121.1164	Synthetic flavor
4,4a,5,6,7,8-Hexahydro-6-isopropenyl-4,4a-dimethyl-2(3H)-naphthalenone	*see* Nootkatone			
γ-Hexalactone	4-Hydroxyhexanoic acid, γ-lactone; tonkalide	2556	FDA §121.1164	Synthetic flavor
δ-Hexalactone		3167		
Hexanal	Aldehyde C-6; caproic aldehyde	2557	FDA §121.1164	Synthetic flavor/adjuvant
2-Hexanal			FDA §121.1164	Synthetic flavor/adjuvant
Hexane			FDA §8.305	Solvent for annatto extract; residues 25 ppm max
2,3-Hexanedione	Acetyl butyryl	2558	FDA §121.1164	Synthetic flavor
3,4-Hexanedione		3168		
Hexanoic acid	Caproic acid	2559	FDA §121.1164	Synthetic flavor
3-Hexanone	Ethyl propyl ketone	3290		*see Food Technol.,* May 1972
2-Hexenal		2560	FDA §121.1164	Synthetic flavor
***cis*-3-Hexenal**		2561		
***trans*-2-Hexenoic acid**		3169		
3-Hexenoic acid		3170		
2-Hexen-1-ol		2562	FDA §121.1164	Synthetic flavor
3-Hexen-1-ol	Leaf alcohol	2563	FDA §121.1164	Synthetic flavor
2-Hexen-1-yl acetate		2564	FDA §121.1164	Synthetic flavor

REGULATORY STATUS OF DIRECT FOOD ADDITIVES (continued)

Material	Alternate Name [Source]	FEMA No.	Regulation	Limitations
cis-3-Hexen-1-yl acetate		3171		
3-Hexenyl isovalerate			FDA §121.1164	Synthetic flavor/adjuvant
3-Hexenyl 2-methyl-butyrate			FDA §121.1164	Synthetic flavor/adjuvant
Hexyl acetate		2565	FDA §121.1164	Synthetic flavor
2-Hexyl-4-acetoxy-tetrahydrofuran		2566	FDA §121.1164	Synthetic flavor
Hexy alcohol	1-Hexanol	2567	FDA §121.1164	Synthetic flavor
Hexyl butyrate		2568	FDA §121.1164	Synthetic flavor
α-Hexylcinnamaldehyde		2569	FDA §121.1164	Synthetic flavor
Hexyl formate		2570	FDA §121.1164	Synthetic flavor
Hexyl furoate				
Hexyl hexanoate		2572	FDA §121.1164	Synthetic flavor
2-Hexylidene cyclo-pentanone		2573	FDA §121.1164	Synthetic flavor/adjuvant
Hexyl isobutyrate		3172		
Hexyl isovalerate			FDA §121.1164	Synthetic flavor/adjuvant
Hexyl 2-methylbutyrate			FDA §121.1164	Synthetic flavor/adjuvant
Hexyl octanoate		2575	FDA §121.1164	Synthetic flavor
Hexyl propionate		2576	FDA §121.1164	Synthetic flavor
Hickory, bark, extract	[*Carya* species]	2577	FDA §121.101	GRAS, natural flavor extractive
Histidine (L- and DL-forms)			FDA §121.101	GRAS, nutrient and/or dietary supplement, as the free acid, hydrochloride salt, hydrate, or anhydrous form
Hops, extract	[*Humulus lupulus* L.]	2578	FDA §121.101	GRAS, natural flavor extractive

REGULATORY STATUS OF DIRECT FOOD ADDITIVES (continued)

Material	Alternate Name [Source]	FEMA No.	Regulation	Limitations
Hops, extract, modified			FDA §121.1082	Flavoring agent in brewing of beer. Product specifications apply
Horehound, extract	Hoarhound [*Marrubium vulgare* (Tourn.) L.]	2581	FDA §121.101	GRAS, natural flavor and natural flavor extractive
Horsemint	[*Monarda punctata* L.]	2582	FDA §121.101	GRAS, natural flavor extractive
Horseradish	[*Armoracia lapthifolia* Gilib.]		FDA §121.101	GRAS, natural flavor
Hyacinth, flowers	[*Hyacinthus orientalis* L.]		FDA §121.1163	Natural flavor
Hydratropaldehyde	*see* 2-Phenylpropion-aldehyde			
Hydratropic aldehyde dimethyl acetal	*see* 2-Phenyl propionaldehyde dimethyl acetal			
Hydratropyl alcohol	*see* β-Methylphenethyl alcohol			
Hydrochloric acid	Muriatic acid		FDA §121.101	GRAS, buffer and neutralizing agent
			FDA §121.1031	Modifier for food starch
Hydrocinnamaldehyde	*see* 3-Phenylpropionaldehyde			
Hydrocinnamic acid	*see* 3-Phenylpropionic acid			
Hydrocinnamyl alcohol	*see* 3-Phenyl-1-propanol			
Hydrogen peroxide	Hydrogen dioxide		FDA §121.101	GRAS, bleaching agent
			FDA §121.1031	Source of active oxygen for modification of food starch
			MID	Bleaching agent in tripe; must be removed from the product by rinsing with clear water
Hydrolyzed plant protein			MID	To flavor various meat products
Hydrolyzed protein			FDA §37.1	In canned tuna

REGULATORY STATUS OF DIRECT FOOD ADDITIVES (continued)

Material	Alternate Name [Source]	FEMA No.	Regulation	Limitations
p-Hydroxybenzyl acetone	*see* 4-(*p*-Hydroxyphenyl)-2-butanone			
4-Hydroxybutanoic acid lactone	γ-Butyrolactone			*see Food Technol.*, May 1972
1-Hydroxy-2-butanone		3173		
3-Hydroxy-2-butanone	*see* Acetoin			
Hydroxycitronellal	3,7-Dimethyl-7-hydroxy-octanal	2583	FDA §121.1164	Synthetic flavor
Hydroxycitronellal diethyl acetal		2584	FDA §121.1164	Synthetic flavor
Hydroxycitronellal dimethyl acetal		2585	FDA §121.1164	Synthetic flavor
Hydroxycitronellol	3,7-Dimethyl-1,7-octa-nediol	2586	FDA §121.1164	Synthetic flavor
4-Hydroxydecanoic acid, γ-lactone	*see* γ-Decalactone			
5-Hydroxydecanoic acid, δ-lactone	*see* δ-Decalactone			
4-Hydroxy-2,5-dimethyl-3(2*H*)-furanone		3174		
4-Hydroxydodecanoic acid, γ-lactone	*see* γ-Dodecalactone			
5-Hydroxydodecanoic acid, δ-lactone	*see* δ-Dodecalactone			
4-Hydroxyheptanoic acid, γ-lactone	*see* γ-Heptalactone			
16-Hydroxy-6-hexa-decenoic acid, ω-lactone	*see* ω-6-Hexadecenlac-tone			
4-Hydroxyhexanoic acid, γ-lactone	*see* γ-Hexalactone			
Hydroxylated lecithin			FDA §121.1027	Emulsifier in foods, except standardized foods that do not provide for such use.

REGULATORY STATUS OF DIRECT FOOD ADDITIVES (continued)

Material	Alternate Name [Source]	FEMA No.	Regulation	Limitations
Hydroxylated lecithin (continued)				Product specifications apply
			FDA §121.1099	In defoaming agents limited to use in processing beet sugar and yeast
N-(4-Hydroxy-3-methoxybenzyl)nonanamide	Pelargonyl vanillylamide		FDA §121.1164	Synthetic flavor
4-(4-Hydroxy-3-methoxyphenyl)-2-butanone	*see* Zingerone			
2-Hydroxy-3-methyl-γ-2-hexenolactone	*see* 5-Ethyl-3-hydroxy-4-methyl-2(5*H*)-furanone			
3-(Hydroxymethyl)-2-octanone		3292		*see Food Technol.,* May 1972
3-Hydroxy-2-methyl-4*H*-pyran-4-one	*see* Maltol			
4-Hydroxynonanoic acid, γ-lactone	*see* γ-Nonalactone			
1-Hydroxy-3-nonanone acetate	*see* 3-Nonanon-1-yl acetate			
4-Hydroxyoctanoic acid, γ-lactone			FDA §121.1164	Synthetic flavor
5-Hydroxy-4-octanone	Butyroin	2587	FDA §121.1164	Synthetic flavor
γ-Hydroxy-β-oxobutane	*see* Acetoin			
15-Hydroxypentadecanoic acid, ω-lactone	*see* ω-Pentadecalactone			
4-Hydroxy-3-pentenoic acid lactone	5-Methyl-2(3*H*)-furanone	3293		*see Food Technol.,* May 1972
2-Hydroxy-2-phenylacetophenone	*see* Benzoin			
4-(p-Hydroxyphenyl)-2-butanone	*p*-Hydroxybenzyl acetone	2588	FDA §121.1164	Synthetic flavor
2-Hydroxy-1,2,3-propanetricarboxylic acid	*see* Citric acid			

REGULATORY STATUS OF DIRECT FOOD ADDITIVES (continued)

Material	Alternate Name [Source]	FEMA No.	Regulation	Limitations
Hydroxypropylcellulose			FDA §121.1160	Emulsifier, film former, protective colloid, stabilizer, suspending agent, or thickener in foods. Product specifications apply
Hydroxypropylmethyl-cellulose			FDA §121.1021	Emulsifier, film former, protective colloid, stabilizer, suspending agent, or thickener, except in confectionery and standardized foods that do not provide for use. Product specifications apply
4-Hydroxyundecanoic acid, γ-lactone	see γ-Undecalactone			
5-Hydroxyundecanoic acid lactone	Δ-Undecalactone	3294		see Food Technol., May 1972
Hyssop	[Hyssopus officinalis L.]	2589	FDA §121.101	GRAS, natural flavor and natural flavor extractive
Iceland moss	[Cetraria islandica (L.) Ach.]		FDA §121.1163	Natural flavor in alcoholic beverages only
Immortelle, extract	Helichrysum [Helichrysum angustifolium DC.]	2592	FDA §121.101	GRAS, natural flavor extractive
Imperatoria	[Peucedanum ostruthium (L.) Koch; Imperatoria ostruthium L.]		FDA §121.1163	Natural flavor in alcoholic beverages only
Indole	2,3-Benzopyrrole	2593	FDA §121.1164	Synthetic flavor
Inositol	mesa-Inositol		FDA §121.101	GRAS, nutrient and/or dietary supplement
Invert sugar	Nulomoline		FDA §8.303	In production of caramel
α-Ionone	4-(2,6,6-Trimethyl-2-cyclo-hexen-1-yl)-3-buten-2-one	2594	FDA §121.1164	Synthetic flavor
β-Ionone	4-(2,6,6-Trimethyl-1-cyclohexen-1-yl)-3-buten-2-one	2595	FDA §121.1164	Synthetic flavor
γ-Ionone	4-(2,2-Dimethyl-6-methyl-ene-cyclohexyl-3-buten-2-one	3175		

REGULATORY STATUS OF DIRECT FOOD ADDITIVES (continued)

Material	Alternate Name [Source]	FEMA No.	Regulation	Limitations
Irish moss	*see* Carrageenan			
Iron ammonium citrate	Green ferric ammonium citrate		FDA §121.1190	Anticaking agent in salt; 25 ppm max in finished salt. Product specifications apply
Iron-choline citrate complex			FDA §121.1100	Source of iron in foods for special dietary use, when made by reacting approx equimolar quantities of ferric hydroxide, choline, and citric acid
α-Irone	4-(2,5,6,6-Tetramethyl-2-cyclohexen-1-yl)-3-buten-2-one; 6-methylionone	2597	FDA §121.1164	Synthetic flavor
Iron, reduced			FDA §121.101	GRAS, nutrient and/or dietary supplement
Isoamyl acetate	Amylacetic ester	2055	FDA §121.1164	Synthetic flavor
Isoamyl acetoacetate			FDA §121.1164	Synthetic flavor
Isoamyl alcohol	Isopentyl alcohol; 3-methyl-1-butanol	2057	FDA §121.1164	Synthetic flavor
Isoamylamine	*see* Isopentylamine			
Isoamyl benzoate	Isopentyl benzoate	2058	FDA §121.1164	Synthetic flavor
Isoamyl butyrate	Isopentyl butyrate	2060	FDA §121.1164	Synthetic flavor
Isoamyl cinnamate	Isopentyl cinnamate	2063	FDA §121.1164	Synthetic flavor
Isoamyl dodecanoate	*see* Isoamyl laurate			
Isoamyl formate	Isopentyl formate	2069	FDA §121.1164	Synthetic flavor
Isoamyl 2-furanbutyrate	α-Isoamyl furfurylpropionate	2070	FDA §121.1164	Synthetic flavor
Isoamyl 2-furanpropionate	α-Isoamyl furfurylacetate	2071	FDA §121.1164	Synthetic flavor
Isoamyl hexanoate	Isopentyl hexanoate	2075	FDA §121.1164	Synthetic flavor
Isoamyl isobutyrate			FDA §121.1164	Synthetic flavor

REGULATORY STATUS OF DIRECT FOOD ADDITIVES (continued)

Material	Alternate Name [Source]	FEMA No.	Regulation	Limitations
Isoamyl isovalerate	Isoamyl valerianate	2085	FDA §121.1164	Synthetic flavor
Isoamyl laurate	Isoamyl dodecanoate; isopentyl laurate	2077	FDA §121.1164	Synthetic flavor
Isoamyl nonanoate	Isopentyl nonanoate	2078	FDA §121.1164	Synthetic flavor
Isoamyl octanoate	Isopentyl octanoate	2080	FDA §121.1164	Synthetic flavor
Isoamyl phenylacetate	Isopentyl phenylacetate	2081	FDA §121.1164	Synthetic flavor
Isoamyl propionate	Isopentyl propionate	2082	FDA §121.1164	Synthetic flavor
Isoamyl pyruvate	Isopentyl pyruvate	2083	FDA §121.1164	Synthetic flavor
Isoamyl salicylate	Isopentyl salicylate	2084	FDA §121.1164	Synthetic flavor
Isoascorbic acid	*see* Erythorbic acid			
Isoborneol	*exo*-2-Bornanol	2158	FDA §121.1164	Synthetic flavor
Isobornyl acetate		2160	FDA §121.1164	Synthetic flavor
Isobornyl formate		2162	FDA §121.1164	Synthetic flavor
Isobornyl isovalerate		2166	FDA §121.1164	Synthetic flavor
Isobornyl propionate		2163	FDA §121.1164	Synthetic flavor
Isobutyl acetate		2175	FDA §121.1164	Synthetic flavor
Isobutyl acetoacetate		2177	FDA §121.1164	Synthetic flavor
Isobutyl alcohol	Isobutanol	2179	FDA §121.1164	Synthetic flavor
			FDA §8.390 and §8.6200	Diluent in inks for marking gum, confectionery, and food supplements in tablet form and in inks for branding pharmaceutical forms; no residue in finished confectionery, gum, or tablets
Isobutyl angelate	Isobutyl *cis*-2-methyl-2-butenoate	2180	FDA §121.1164	Synthetic flavor
Isobutyl anthranilate		2182	FDA §121.1164	Synthetic flavor
Isobutyl benzoate		2185	FDA §121.1164	Synthetic flavor

REGULATORY STATUS OF DIRECT FOOD ADDITIVES (continued)

Material	Alternate Name [Source]	FEMA No.	Regulation	Limitations
Isobutyl benzyl carbinol	*see* α-Isobutylphenethyl alcohol			
Isobutyl butyrate		2187	FDA §121.1164	Synthetic flavor
Isobutyl cinnamate		2193	FDA §121.1164	Synthetic flavor
Isobutylene-isoprene copolymer	Butyl rubber		FDA §121.1059	Synthetic masticatory substance in chewing-gum base
Isobutylene resin	Polyisobutylene		FDA §121.1059	Synthetic masticatory substance in chewing-gum base. Product specifications apply
Isobutyl formate	Tetryl formate	2197	FDA §121.1164	Synthetic flavor/adjuvant
Isobutyl 2-furanpropionate	Isobutyl furylpropionate	2198	FDA §121.1164	Synthetic flavor/adjuvant
Isobutyl heptanoate		2200	FDA §121.1164	Synthetic flavor/adjuvant
Isobutyl hexanoate		2202	FDA §121.1164	Synthetic flavor/adjuvant
Isobutyl isobutyrate		2189	FDA §121.1164	Synthetic flavor
2-Isobutyl-3-methoxypyrazine		3132		10 ppm max in finished food
Isobutyl *cis*-2-methyl cyclohexanol			FDA §121.1164	Synthetic flavor
2-Isobutyl-5-methyl cyclohexanol	*see d*-Neomenthol			
2-Isobutyl-3-methyl-pyrazine	2-Methyl-3-isobutyl-pyrazine	3133		10 ppm max in finished food
α-Isobutylphenethyl alcohol	Isobutyl benzyl carbinol; 4-methyl-1-phenyl-2-pentanol	2208	FDA §121.1164	Synthetic flavor
Isobutyl phenylacetate		2210	FDA §121.1164	Synthetic flavor
Isobutyl propionate		2212	FDA §121.1164	Synthetic flavor
Isobutyl salicylate		2213	FDA §121.1164	Synthetic flavor

REGULATORY STATUS OF DIRECT FOOD ADDITIVES (continued)

Material	Alternate Name [Source]	FEMA No.	Regulation	Limitations
2-Isobutyl thiazole		3134		
Isobutyraldehyde	2-Methyl propanal	2220	FDA §121.1164	Synthetic flavor
Isobutyric acid	Isopropylformic acid	2222	FDA §121.1164	Synthetic flavor
Isodecylaldehyde	*see* 2,6-Dimethyl octanal			
Isoeugenol	2-Methoxy-4-propenyl-phenol	2468	FDA §121.1164	Synthetic flavor
Isoeugenyl acetate	2-Methoxy-4-propenyl-phenyl acetate	2470	FDA §121.1164	Synthetic flavor
Isoeugenyl benzyl ether	Benzyl isoeugenol		FDA §121.1164	Synthetic flavor/adjuvant
Isoeugenyl ethyl ether	2-Ethoxy-5-propenylani-sole; ethyl isoeugenol	2472	FDA §121.1164	Synthetic flavor
Isoeugenyl formate	2-Methoxy-4-propenyl-phenyl formate	2474	FDA §121.1164	Synthetic flavor
Isoeugenyl methyl ether	Methyl isoeugenol; 4-propenyl veratrole	2476	FDA §121.1164	Synthetic flavor
Isoeugenyl phenyl-acetate	2-Methoxy-4-propenyl-phenyl phenylacetate	2477	FDA §121.1164	Synthetic flavor
Isojasmone	Mixture of 2-hexylidene-cyclopentanone and 2-hexyl-2-cyclopenten-1-one		FDA §121.1164	Synthetic flavor/adjuvant
Isolated soy protein			MID	To bind and extend imitation sausage, nonspecific loaves, soups, and stews
			MID	Binder, alone or with other binders: 8% max in chili con carne 12% max in spaghetti with meatballs and sauce, spaghetti with meat and sauce, and similar products 2% max (3.5% of other binders) in sausage 2% max (3.5% of other binders) in cheesefurters and similar products

REGULATORY STATUS OF DIRECT FOOD ADDITIVES (continued)

Material	Alternate Name [Source]	FEMA No.	Regulation	Limitations
Isolated soy protein (continued)				no limit for pork with barbeque sauce or beef with barbeque sauce The isolated soy protein must contain 0.1% food-grade titanium dioxide as a tracer, and it must be declared on labels
Isoleucine (*l*- and *dl*-forms)			FDA §121.101	GRAS, nutrient and/or dietary supplement, as the free acid, hydrochloride salt, hydrate, or anhydrous form
D,L-Isoleucine		3295		*see Food Technol.*, May 1972
α-Isomethylionone	Methyl γ-ionone; 4(2,6,6-trimethyl-2-cyclohexen-1-yl)-3-methyl-3-buten-2-one	2714	FDA §121.1164	Synthetic flavor
Isopentyl alcohol	*see* Isoamyl alcohol			
Isopentylamine	Isoamylamine; 3-methyl-butylamine	3219		
Isopropanol	*see* Isopropyl alcohol			
Isopropenylpyrazine	2-Isopropenyl-1,4-diazine; 2(1-methylvinyl) pyrazine	3296		*see Food Technol.*, May 1972
***p*-Isopropenyltoluene**	*see p-α*-Dimethylstyrene			
Isopropyl acetate		2926	FDA §121.1164	Synthetic flavor
***p*-Isopropylacetophenone**		2927	FDA §121.1164	Synthetic flavor
Isopropyl alcohol	Isopropanol	2929	FDA §121.1164	Synthetic flavor/adjuvant
			FDA §8.390 and §8.6200	Diluent in inks for marking gum and confectionery, inks for branding pharmaceutical forms, and ingested drugs
			FDA §8.305	Solvent for annatto extract; residues 50 ppm max

REGULATORY STATUS OF DIRECT FOOD ADDITIVES (continued)

Material	Alternate Name [Source]	FEMA No.	Regulation	Limitations
p-Isopropyl benzaldehyde	*see* Cuminaldehyde			
Isopropyl benzoate		2932	FDA §121.1164	Synthetic flavor
p-Isopropylbenzyl alcohol	Cuminic alcohol; *p*-cymen-7-ol	2933	FDA §121.1164	Synthetic flavor
Isopropyl butyrate		2935	FDA §121.1164	Synthetic flavor
Isopropyl cinnamate		2939	FDA §121.1164	Synthetic flavor
Isopropyl citrate			FDA §121.101	GRAS, sequestrant up to 0.02%
			MID	0.02% in oleomargarine to protect flavor
4-Isopropyl-1-cyclohexene-1-carboxaldehyde	*see* Perillaldehyde			
Isopropyl formate		2944	FDA §121.1164	Synthetic flavor
Isopropyl hexanoate		2950	FDA §121.1164	Synthetic flavor
p-Isopropylhydrocinnamaldehyde	*see* 3-(*p*-Isopropyl) phenyl propionaldehyde			
Isopropyl isobutyrate		2937	FDA §121.1164	Synthetic flavor
Isopropyl isovalerate		2961	FDA §121.1164	Synthetic flavor
Isopropyl-α-methyl crotonic acid	*see* Isopropyl tiglate			
p-Isopropylphenylacetaldehyde	*p*-Cymen-7-carboxaldehyde	2954	FDA §121.1164	Synthetic flavor
Isopropyl phenylacetate		2956	FDA §121.1164	Synthetic flavor
3-(*p*-Isopropyl) phenyl propionaldehyde	*p*-Isopropylhydrocinnamaldehyde; cuminyl acetaldehyde	2957	FDA §121.1164	Synthetic flavor
Isopropyl propionate		2959	FDA §121.1164	Synthetic flavor
Isopropyl tiglate	Isopropyl-α-methyl crotonic acid	3229		
Isopulegol	*p*-Menth-8-en-3-ol	2962	FDA §121.1164	Synthetic flavor

REGULATORY STATUS OF DIRECT FOOD ADDITIVES (continued)

Material	Alternate Name [Source]	FEMA No.	Regulation	Limitations
Isopulegone	p-Menth-8-en-3-one	2964	FDA §121.1164	Synthetic flavor
Isopulegyl acetate		2965	FDA §121.1164	Synthetic flavor
Isoquinoline	2-Benzazine	2978	FDA §121.1164	Synthetic flavor
Isovaleraldehyde	see 3-Methylbutyralde-hyde			
Isovaleric acid	β-Methylbutyric acid	3102	FDA §121.1164	Synthetic flavor
Iva	Musk yarrow [Achillea moschata Jacq.]		FDA §121.1163	Natural flavor in alcoholic beverages only
Japan wax	Vegetable wax		FDA §8.390	Diluent in mixtures for coloring shell eggs; no penetration of mixture through shell egg into egg
Jasmine	[Jasminum officinale L. and other Jasminum species]		FDA §121.101	GRAS, natural flavor extractive
Jasmone	see 3-Methyl-2-(2-pentenyl)-2-cyclopenten-1-one			
Jelutong	[Dyera costulata Hook. f.; D. Iowii Hook. f.]		FDA §121.1059	Natural (coagulated or concentrated latex) masticatory substance in chewing-gum base
Juniper, berries	[Juniperus communis L.]	2602	FDA §121.101	GRAS, natural flavor extractive
Juniper, tar	see Cade oil			
Karaya, gum	Sterculia gum [Sterculia urens Roxb.]	2605	FDA §121.101	GRAS, stabilizer
Kelp	Brown algae [Atlantic—Laminaria digitatia; L. saccharina; Pacific—Macrocystis pyrifera (L.) C. Agardh.]	2606	FDA §121.101	GRAS, natural substance and solvent-free natural extractive used in conjunction with natural flavorings
			FDA §121.1149	Source of iodine in foods for special dietary use; daily intake of iodine not to exceed 0.15 mg

REGULATORY STATUS OF DIRECT FOOD ADDITIVES (continued)

Material	Alternate Name [Source]	FEMA No.	Regulation	Limitations
Kola nut	Cola nut [*Cola acuminata* Schott & Endl., and other species of *Cola*]	2607	FDA §121.101	GRAS, natural flavor extractive
Labdanum, absolute	[*Cistus ladaniferus* L.; *C. creticus*; and other *Cistus* species]	2608	FDA §121.1163	Natural flavor
Lactic acid		2611	FDA §121.101	GRAS, miscellaneous and/or general-purpose food additive
Lactic acid, butyl ester, butyrate	*see* Butyl butyryllactate			
Lactic-acid-producing bacteria			FDA §17	In bakery products
			FDA §19	In specified cheeses
			MID	Harmless bacterial starters of the lactic-acid type, to develop flavor in dry sausage, pork roll, thuringer, Lebanon bologna, cervelat, and bologna, at 0.5% level
Lactose	4-*O*-β-D-Galactopyranosyl-D-glucose		FDA §8.303	In production of caramel
Lactylic esters of fatty acids			FDA §121.1048	Prepared from lactic acid and fatty acids meeting requirements of §121.1070(b); emulsifiers, plasticizers or surface-active agents in bakery mixes, baked products, cake icings, fillings, and toppings, dehydrated fruits and vegetables, dehydrated fruit and vegetable juices, frozen desserts, liquid shortening for household use, pancake mixes, precooked instant rice, pudding mixes, solid-state edible vegetable-fat-water emulsions as substitutes for milk or cream in beverage coffee, and with shortening and edible fats and oils when such are required in the foods listed above

REGULATORY STATUS OF DIRECT FOOD ADDIT:VES (continued)

Material	Alternate Name [Source]	FEMA No.	Regulation	Limitations
Lakes (FD&C)			FDA §8.501	Food, drug, and cosmetic use, conforming with specifications in §9.100
Lakes of color additives for food use			FDA §8.299	Product specifications apply
Larch gum	see Arabinogalactan			
Lard			FDA—GRAS	
Laurel	see Bay, sweet			
Lauric acid	Dodecanoic acid	2614		
Lauric aldehyde	Aldehyde C-12; dodecanal	2615	FDA §121.1164	Synthetic flavor
Lauryl acetate	Dodecyl acetate	2616	FDA §121.1164	Synthetic flavor
Lauryl alcohol	Alcohol C-12; dodecanol	2617	FDA §121.1164	Synthetic flavor
Lavandin	[*Lavandula hybrida* Rev., hybrid between *L. officinalis* Chaix and *L. latifolia* Vill.]	2618	FDA §121.101	GRAS, natural flavor extractive
Lavender	[*Lavandula officinalis* Chaix]	2619	FDA §121.101	GRAS, natural flavor and natural flavor extractive
Lavender, spike, oil	[*Lavandula latifolia* Vill. (*L. spica* DC.)]	3033	FDA §121.101	GRAS, natural flavor extractive
Leaf alcohol	see 3-Hexen-1-ol			
Leche caspi	Sorva [*Couma macrocarpa* Barb. Rodr.]		FDA §121.1059	Natural (coagulated or concentrated latex) masticatory substance in chewing-gum base
Leche de vaca	[*Brosimum utile* (H.B.K.) Pittier and Poulsenia species; *Lacmellea standleyi* (Woodson) Monachino]		FDA §121.1059	Natural (coagulated or concentrated latex) masticatory substance in chewing-gum base
Lecithin	Phosphatidylcholine		FDA §121.101	GRAS, miscellaneous and/or general-purpose food additive
			MID	Emulsifier and antioxidant in oleomargarine and shortening

REGULATORY STATUS OF DIRECT FOOD ADDITIVES (continued)

Material	Alternate Name [Source]	FEMA No.	Regulation	Limitations
Lemon balm	see Balm			
Lemon, extract	[Citrus limonum (L.) Burm. f.]	2623	FDA §121.101	GRAS, natural flavor extractive
Lemongrass, oil	[Cymbopogon citratus DC.; C. flexuosus (Nees.) Stapf.]	2624	FDA §121.101	GRAS, natural flavor extractive
Lemon peel	[Citrus limonum (L.) Burm. f.]	2625	FDA §121.101	GRAS, natural flavor extractive
Lemon verbena	[Lippia citriodora H.B.K.]		FDA §121.1163	Natural flavor in alcoholic beverages only
Lepidine	4-Methylquinoline	2744	FDA §121.1164	Synthetic flavor/adjuvant
Leucine (L- and DL- forms)			FDA §121.101	GRAS, nutrient and/or dietary supplement, as the free acid, hydrochloride salt, hydrate, or anhydrous form
L-Leucine		3297		see Food Technol., May 1972
Levulinic acid	3-Acetylpropionic acid	2627	FDA §121.1164	Synthetic flavor
Licorice, extract	Glycyrrhiza [Glycyrrhiza glabra L. and other Glycyrrhiza species]	2628	FDA §121.101	GRAS, natural flavor and natural flavor extractive
Limed rosin			FDA §8.390	Diluent in mixtures for coloring shell eggs; no penetration of mixture into the egg
Lime, oil	[Citrus aurantifolia Swingle]	2631	FDA §121.101	GRAS, natural flavor extractive
Limonene (d-, l-, and dl-)	p-Mentha-1,8-diene		FDA §121.101	GRAS, synthetic flavor/adjuvant
Linaloe wood	[Bursera delpechiana Poiss. and other Bursera species]	2634	FDA §121.1163	Natural flavor
Linalool	3,7-Dimethyl-1,6-octadien-3-ol; linalol	2635	FDA §121.101	GRAS, synthetic flavor/adjuvant

REGULATORY STATUS OF DIRECT FOOD ADDITIVES (continued)

Material	Alternate Name [Source]	FEMA No.	Regulation	Limitations
Linalool oxide	cis- and trans-2-Vinyl-2-methyl-5-(1'-hydroxyl-1'-methylethyl)-tetrahydrofuran		FDA §121.1164	Synthetic flavor
Linalyl acetate	Bergamol	2636	FDA §121.101	GRAS, synthetic flavor/adjuvant
Linalyl anthranilate	3,7-Dimethyl-1,6-octadien-3-yl-anthranilate	2637	FDA §121.1164	Synthetic flavor
Linalyl benzoate		2638	FDA §121.1164	Synthetic flavor
Linalyl butyrate		2639	FDA §121.1164	Synthetic flavor
Linalyl cinnamate		2641	FDA §121.1164	Synthetic flavor/adjuvant
Linalyl formate		2642	FDA §121.1164	Synthetic flavor
Linalyl hexanoate		2643	FDA §121.1164	Synthetic flavor
Linalyl isobutyrate		2640	FDA §121.1164	Synthetic flavor
Linalyl isovalerate		2646	FDA §121.1164	Synthetic flavor
Linalyl octanoate		2644	FDA §121.1164	Synthetic flavor
Linalyl propionate		2645	FDA §121.1164	Synthetic flavor
Linden, flower	[Tilia glabra Vent.]	2647	FDA §121.101	GRAS, natural flavor and natural flavor extractive
Linden, leaf	[Tilia species]		FDA §121.1163	Natural flavor in alcoholic beverages only
Linoleic acid	cis-9,cis-12-Octadecadienoic acid		FDA §121.101	GRAS, nutrient and/or dietary supplement, when prepared from edible fats and oils and free from chick-edema factor
Locust bean	[Ceratonia siliqua L.]		FDA §121.101	GRAS, natural flavor extractive
Locust bean gum	see Carob bean gum			
Lovage	[Levisticum officinale Koch.]	2649	FDA §121.1163	Natural flavor

REGULATORY STATUS OF DIRECT FOOD ADDITIVES (continued)

Material	Alternate Name [Source]	FEMA No.	Regulation	Limitations
Lungmoss	Lungwort [*Sticta pulmonaria* Ach.]		FDA §121.1163	Natural flavor
Lupulin	[*Humulus lupulus* L.]		FDA §121.101	GRAS, natural flavor extractive
Lysine (L- and DL- forms)			FDA §121.101	GRAS, nutrient and/or dietary supplement, as free acid, hydrochloride salt, hydrate, and anhydrous form
Mace *see also* Nutmeg	[*Myristica fragrans* Houtt.]	2652	FDA §121.101	GRAS, natural flavor and natural flavor extractive
Magnesium carbonate			FDA §121.101	GRAS, miscellaneous and/or general-purpose food additive
Magnesium hydroxide			FDA §121.101	GRAS, miscellaneous and/or general-purpose food additive
Magnesium oxide	Magnesia		FDA §121.101	GRAS, nutrient and/or dietary supplement
			FDA §121.101	GRAS, miscellaneous and/or general-purpose food additive
Magnesium phosphate (di- and tribasic)			FDA §121.101	GRAS, nutrient and/or dietary supplement
Magnesium salts of fatty acids			FDA §121.1071	Alone or with other salts of fatty acids; binder, emulsifier, and anticaking agent in food. Product specifications apply
Magnesium silicate	Talc		FDA §22.8 and §22.9	In vanilla powder and in vanilla-vanillin powder
			FDA §121.101	GRAS, anticaking agent in table salt, up to 2%
Magnesium sulfate			FDA §121.101	GRAS, nutrient and/or dietary supplement
Maidenhair fern	[*Adiantum capillus-veneris* L.]		FDA §121.1163	Natural flavor in alcoholic beverages only

REGULATORY STATUS OF DIRECT FOOD ADDITIVES (continued)

Material	Alternate Name [Source]	FEMA No.	Regulation	Limitations
Malic acid	Hydroxysuccinic acid		FDA §121.101	GRAS, miscellaneous and/ or general-purpose food additive
***l*-Malic acid**	Apple acid; *l*-hydroxy-succinic acid	2655	FDA §121.101	GRAS, natural flavor extractive
Maltol	3-Hydroxy-2-methyl-4*H*-pyran-4-one	2656	FDA §121.1164	Synthetic flavor
Malt syrup			FDA §8.303	In production of caramel
			MID	To flavor cured meat products, at 2.5%
Mandarin, oil	Tangerine and petit-grain mandarin or tangerine [*Citrus reticulata* Blanco]	2657	FDA §121.101	GRAS, natural flavor extractive
Manganese chloride	Manganous fluoride		FDA §121.101	GRAS, nutrient and/or dietary supplement
Manganese citrate			FDA §121.101	GRAS, nutrient and/or dietary supplement
Manganese gluconate			FDA §121.101	GRAS, nutrient and/or dietary supplement
Manganese glycero-phosphate			FDA §121.101	GRAS, nutrient and/or dietary supplement
Manganese hypo-phosphite			FDA §121.101	GRAS, nutrient and/or dietary supplement
Manganese oxide	Manganomanganic oxide		FDA §121.101	GRAS, nutrient and/or dietary supplement
Manganese sulfate			FDA §121.101	GRAS, nutrient and/or dietary supplement
Mannitol	D-Mannitol		FDA §121.101	GRAS, nutrient and/or dietary supplement in special dietary foods, up to 5%
			FDA §121.1115	In foods
Maple, mountain, extract	[*Acer spicatum* Lam.]	2757	FDA §121.1163	Natural flavor

Material	Alternate Name [Source]	FEMA No.	Regulation	Limitations
Marigold	see Tagetes			
Marigold, pot	Calendula [Calendula officinalis L.]	2658	FDA §121.101	GRAS, natural flavor
Marjoram, pot	[Marjorana onites (L.) Benth.]	2660	FDA §121.101	GRAS, natural flavor
Marjoram, sweet	[Origanum marjorana L., Marjorana hortensis Moench.]	2662	FDA §121.101	GRAS, natural flavor and natural flavor extractive
Massaranduba balata	[Manilkara huberi (Ducke) Chevalier]		FDA §121.1059	Massaranduba balata and its solvent-free resin extract, natural (coagulated or concentrated latex) masticatory substance in chewing-gum base
Massaranduba chocolate	[Manilkara solimoesensis Gilly]		FDA §121.1059	Natural (coagulated or concentrated latex) masticatory substance in chewing-gum base
Maté	[Ilex paraguensis St. Hil.]		FDA §121.101	GRAS, natural flavor extractive
Melilotus	Yellow melilot [Melilotus officinalis (L.) Lam.]		FDA §121.1163	Natural flavor in alcoholic beverages only·
Melissa	see Balm			
p-Mentha-1,5-diene	see α-Phellandrene			
p-Mentha-1,8-diene	see Limonene			
p-Mentha-1,8-dien-7-al	see Perillaldehyde			
Menthadienol	p-Mentha-1,8(10)-dien-9-ol		FDA §121.1164	Synthetic flavor
p-Mentha-1,8-dien-7-ol	Perillyl alcohol	2664	FDA §121.1164	Synthetic flavor
p-Mentha-1,8(10)-dien-9-ol	see Menthadienol			
p-Mentha-6,8-dien-2-ol	see Carveol			
Menthadienyl acetate	p-Mentha-1,8(10)-dien-9-yl acetate		FDA §121.1164	Synthetic flavor

REGULATORY STATUS OF DIRECT FOOD ADDITIVES (continued)

Material	Alternate Name [Source]	FEMA No.	Regulation	Limitations
p-Mentha-1,8-dien-7-yl acetate	*see* Perillyl acetate			
p-Mentha-1,8(10)-dien-9-yl acetate	*see* Menthadienyl acetate			
p-Menthan-2-one	Carvomenthone; tetra-hydrocarvone	3176		10 ppm max in beverages, frozen desserts, and candies
p-Menthan-3-one	*see* Menthone			
p-Mentha-8-thiol-3-one	8-Mercapto-p-menthane-3-one	3177		
p-Menth-1,4(8)-diene	*see* Terpinolene			
p-Menth-1-ene-9-al		3178		
p-Menth-1-en-3-ol	Piperitol	3179		
p-Menth-1-en-8-ol	*see* α-Terpineol			
p-Menth-3-en-1-ol			FDA §121.1164	Synthetic flavor/adjuvant
p-Menth-8-en-3-ol	*see* Isopulegol			
1-p-Menthen-4-ol	*see* 4-Carvomenthenol			
8-p-Menthen-2-ol	*see* Dihydrocarveol			
p-Menth-1-en-3-one	*see* d-Piperitone			
p-Menth-4(8)-en-3-one	*see* Pulegone			
p-Menth-8-en-3-one	*see* Isopulegone			
Menthofuran	*see* 4,5,6,7-Tetrahydro-3,6-dimethylbenzofuran			
Menthol	5-Methyl-2-*iso*-propyl-cyclohexanol	2665	FDA §121.1164	Synthetic flavor
Menthone	p-Menthan-3-one	2667	FDA §121.1164	Synthetic flavor
Menthyl acetate	p-Menth-3-yl acetate	2668	FDA §121.1164	Synthetic flavor
p-Menth-3-yl acetate	*see* Menthyl acetate			
Menthyl isovalerate	p-Menth-3-yl isovalerate	2669	FDA §121.1164	Synthetic flavor

REGULATORY STATUS OF DIRECT FOOD ADDITIVES (continued)

Material	Alternate Name [Source]	FEMA No.	Regulation	Limitations
p-Menth-3-yl isovalerate	*see* Methyl isovalerate			
3-Mercapto-2-butanone		3298		*see Food Technol.*, May 1972
8-Mercapto-*p*-menthane-3-one	*see p*-Mentha-8-thiol-3-one			
2-Mercaptomethylpyrazine	Pyrazinemethanethiol	3299		*see Food Technol.*, May 1972
2-Mercaptonaphthalene	*see* 2-Naphthalenthiol			
3-Mercapto-2-pentanone		3300		*see Food Technol.*, May 1972
2-Mercaptopropionic acid	Thiolactic acid	3180		
2-Mercaptopropionic acid, ethyl ester	*see* Ethyl 2-mercaptopropionate			
Methanethiol	*see* Methyl mercaptan			
Methanethiol *n*-butyrate	*see* Methyl thiobutyrate			
Methional	*see* 2-Methylthiopropionaldehyde			
D,L-Methionine		3301		*see Food Technol.*, May 1972
β-Methiopropionaldehyde	*see* 2-Methylthiopropionaldehyde			
4'-Methoxyacetophenone	*see* Acetanisole			
o-Methoxybenzaldehyde			FDA §121.1164	Synthetic flavor
p-Methoxybenzaldehyde	*p*-Anisaldehyde	2670	FDA §121.1164	Synthetic flavor
Methoxybenzene	*see* Anisole			
p-Methoxybenzyl alcohol	*see* Anisyl alcohol			
o-Methoxycinnamaldehyde		3181	FDA §121.1164	Synthetic flavor

REGULATORY STATUS OF DIRECT FOOD ADDITIVES (continued)

Material	Alternate Name [Source]	FEMA No.	Regulation	Limitations
2-Methoxy-*p*-cresol	*see* 2-Methoxy-4-methylphenol			
2-Methoxy-1,4-diazine	*see* Methoxypyrazine			
p-Methoxy-α-methyl cinnamaldehyde		3182		
2-Methoxy-4-methyl-phenol	2-Methoxy-*p*-cresol; 4-methylguaiacol	2671	FDA §121.1164	Synthetic flavor
2-, 5-, or 6-Methoxy-3-methylpyrazine (mixture of isomers)		3183		10 ppm max in finished food
o-Methoxyphenol	*see* Guaiacol			
o-Methoxyphenyl acetate			FDA §121.1164	Synthetic flavor/adjuvant
4-(*p*-Methoxyphenyl)-2-butanone	Anisyl acetone	2672	FDA §121.1164	Synthetic flavor
2-(4-Methoxyphenyl)-4-methyl-1-penten-3-one	Methoxystyryl iso-propyl ketone		FDA §121.1164	Synthetic flavor/adjuvant
1-(*p*-Methoxyphenyl)-1-penten-3-one	Ethone; α-methylanisyli-deneacetone	2673	FDA §121.1164	Synthetic flavor
1-(*p*-Methoxyphenyl)-2-propanone	Anisic ketone; anisyl methyl ketone	2674	FDA §121.1164	Synthetic flavor
2-Methoxy-4-propenyl-phenol	*see* Isoeugenol			
Methoxypyrazine	2-Methoxy-1,4-diazine	3302		*see Food Technol.,* May 1972
Methoxystyryl iso-propyl ketone	*see* 2-(4-Methoxyphenyl)-4-methyl-1-penten-3-one			
p-Methoxytoluene	*see p*-Methylanisole			
2-Methoxy-4-vinyl-phenol	*p*-Vinylguaiacol	2675	FDA §121.1164	Synthetic flavor
Methyl acetate		2676	FDA §121.1164	Synthetic flavor
4-Methylacetophenone	*p*-Methylacetophenone; methyl *p*-tolyl ketone	2677	FDA §121.1164	Synthetic flavor

REGULATORY STATUS OF DIRECT FOOD ADDITIVES (continued)

Material	Alternate Name [Source]	FEMA No.	Regulation	Limitations
1-Methyl-2-acetyl-pyrrole		3184		
Methyl alcohol			FDA §8.305	Solvent for annatto extract; residues, 50 ppm max
2-Methylallyl butyrate	2-Methyl-2-propen-1-yl butyrate	2678	FDA §121.1164	Synthetic flavor
Methyl allyl trisulfide	*see* Allyl methyl trisulfide			
Methyl *n*-amyl carbinol	*see* 2-Heptanol			
Methyl amyl ketone	*see* 2-Heptanone			
Methyl anisate		2679	FDA §121.1164	Synthetic flavor
***o*-Methylanisole**	*o*-Cresyl methyl ether	2680	FDA §121.1164	Synthetic flavor
***p*-Methylanisole**	*p*-Cresyl methyl ether; *p*-methoxytoluene	2681	FDA §121.1164	Synthetic flavor
α-Methylanisylidene-acetone	*see* 1-(*p*-Methoxyphenyl)-1-penten-3-one			
Methyl anthranilate	Methyl-2-aminobenzoate	2682	FDA §121.101	GRAS, synthetic flavor and adjuvant
Methylated silica		3185		
Methyl benzoate	Oil of niobe	2683	FDA §121.1164	Synthetic flavor
Methylbenzyl acetate, mixed *o*-, *m*-, *p*-			FDA §121.1164	Synthetic flavor/adjuvant
α-Methylbenzyl acetate	Styralyl acetate	2684	FDA §121.1164	Synthetic flavor
***p*-Methylbenzylacetone**	*see* 4-(*p*-Tolyl)-2-butanone			
α-Methylbenzyl alcohol	Styralyl alcohol	2685	FDA §121.1164	Synthetic flavor
α-Methylbenzyl butyrate	Styralyl butyrate	2686	FDA §121.1164	Synthetic flavor
α-Methylbenzyl formate	Styralyl formate	2688	FDA §121.1164	Synthetic flavor

REGULATORY STATUS OF DIRECT FOOD ADDITIVES (continued)

Material	Alternate Name [Source]	FEMA No.	Regulation	Limitations
α-Methylbenzyl iso-butyrate	Styralyl isobutyrate	2687	FDA §121.1164	Synthetic flavor
α-Methylbenzyl pro-pionate	Styralyl propionate	2689	FDA §121.1164	Synthetic flavor
4-Methylbiphenyl	p-Methyldiphenyl	3186		
2-Methyl-1-butanethiol		3303		see Food Technol., May 1972
3-Methyl-2-butanethiol		3304		see Food Technol., May 1972
3-Methyl-1-butanol	see Isoamyl alcohol			
3-Methylbutylamine	see Isopentylamine			
2-Methylbutyl iso-valerate			FDA §121.1164	Synthetic flavor/adjuvant
Methyl p-tert-butyl-phenyl acetate		2690	FDA §121.1164	Synthetic flavor
2-Methylbutyraldehyde	Methylethylacetalde-hyde	2691	FDA §121.1164	Synthetic flavor
3-Methylbutyraldehyde	Isovaleraldehyde	2692	FDA §121.1164	Synthetic flavor
Methyl butyrate		2693	FDA §121.1164	Synthetic flavor
2-Methylbutyric acid		2695	FDA §121.1164	Synthetic flavor
Methylcellulose	Cellulose methyl ether	2696	FDA §121.101	GRAS, miscellaneous and/or general-purpose food additive. Product specifications apply
			FDA §25.2 and §25.3	In French and salad dressings
			MID	Carrier and extender and stabilizer of meat and vegetable patties, at 0.15%
p-Methylcinnamalde-hyde			FDA §121.1164	Synthetic flavor/adjuvant
α-Methylcinnamalde-hyde		2697	FDA §121.1164	Synthetic flavor
Methyl cinnamate		2698	FDA §121.1164	Synthetic flavor

REGULATORY STATUS OF DIRECT FOOD ADDITIVES (continued)

Material	Alternate Name [Source]	FEMA No.	Regulation	Limitations
6-Methylcoumarin		2699		
3-Methylcrotonic acid	3,3-Dimethylacrylic acid; β,β-dimethylacrylic acid; senecioic acid	3187		
2-Methyl-1,3-cyclo-hexadiene			FDA §121.1164	Synthetic flavor/adjuvant
1-Methyl-2,3-cyclo-hexadione		3305		see Food Technol., May 1972
3-Methylcyclopentane-1,2-dione	see Methylcyclopenteno-lone			
Methylcyclopenteno-lone	3-Methylcyclopentane-1,2-dione	2700	FDA §121.1164	Synthetic flavor
Methyl decyne carbonate	see Methyl-2-undecyno-ate			
2-Methyl-1,4-diazine	see 2-Methylpyrazine			
Methyl dihydroabietate	see Methyl ester of rosin, partially hydro-genated			
5H-5-Methyl-6,7-dihy-drocyclopenta [b] pyrazine		3306		see Food Technol., May 1972
p-Methyldiphenyl	see 4-Methylbiphenyl			
Methyl disulfide			FDA §121.1164	Synthetic flavor/adjuvant
Methylene chloride	Dichloromethane		FDA §8.390	Diluent in inks for marking fruits and vegetables, with no residue
			FDA §8.305	Solvent for annatto extract, residues 30 ppm max; total residues of chlorinated solvents, 30 ppm max
3,4-Methylenedioxy-benzaldehyde			FDA §121.101	GRAS, synthetic flavor
Methyl ester of rosin, partially hydrogenated	Methyl dihydroabietate		FDA §121.1164	Synthetic flavor

REGULATORY STATUS OF DIRECT FOOD ADDITIVES (continued)

Material	Alternate Name [Source]	FEMA No.	Regulation	Limitations
Methylethylacetaldehyde	*see* 2-Methylbutyraldehyde			
Methyl ethyl cellulose			FDA §121.1112	Aerating, emulsifying, and foaming agent. Product specifications apply
Methyl ethyl ketone	*see* 2-Butanone			
Methyl eugenol	*see* Eugenyl methyl ether			
5-Methyl-2(3*H*)-furanone	*see* 4-Hydroxy-3-pentenoic acid lactone			
2-Methyl-3-furanthiol		3188		
5-Methylfurfural		2702		
2-Methyl-3-, 5-, or 6-furfuryl thiopyrazine (mixture of isomers)		3189		10 ppm max in finished food
Methyl 2-furoate	Methyl pyromucate	2703		
2-Methyl-3-furylacrolein	α-Methyl furylacrolein	2704		
3-(5-Methyl-2-furyl)-butanal	3-(5-Methyl-2-furyl)-butyraldehyde	3307		
3-(5-Methyl-2-furyl)-butyraldehyde	*see* 3-(5-Methyl-2-furyl)-butanal			
2-Methyl-3-furyl disulfide	*see* Bis (2-methyl-3-furyl) disulfide			
2-Methyl-3-furyl tetrasulfide	*see* Bis (2-methyl-3-furyl) tetrasulfide			
Methyl glucoside—coconut oil ester			FDA §121.1151	Aid in crystallization of sucrose and dextrose; 320 ppm max in molasses as surface-active agent. Product specifications apply
4-Methylguaiacol	*see* 2-Methoxy-4-methyl-phenol			
Methyl heptanoate		2705	FDA §121.1164	Synthetic flavor
2-Methylheptanoic acid	2-Methyloenanthic acid	2706	FDA §121.1164	Synthetic flavor

REGULATORY STATUS OF DIRECT FOOD ADDITIVES (continued)

Material	Alternate Name [Source]	FEMA No.	Regulation	Limitations
6-Methyl-5-hepten-2-one		2707	FDA §121.1164	Synthetic flavor
Methyl heptine carbonate	*see* Methyl 2-octynoate			
Methyl *n*-heptyl carbinol	*see* 2-Nonanol			
Methyl heptyl ketone	*see* 2-Nonanone			
5-Methyl-2,3-hexanedione		3190		
Methyl hexanoate		2708	FDA §121.1164	Synthetic flavor
2-Methylhexanoic acid		3191		
Methyl 2-hexenoate		2709	FDA §121.1164	Synthetic flavor
Methyl hexyl acetaldehyde	*see* 2-Methyl octanal			
Methyl hexyl ketone	*see* 2-Octanone			
p-Methylhydratropicaldehyde	*see* 2-(*p*-Tolyl)propionaldehyde			
Methyl hydrocinnamate	*see* Methyl-3-phenylpropionate			
Methyl *p*-hydroxybenzoate	*see* Methylparaben	2710		
4-Methyl-5-(β-hydroxyethyl)-thiazole	*see* 4-Methyl-5-thiazoleethanol			
Methyl α-ionone	5-(2,6,6-Trimethyl-2-cyclohexen-1-yl)-4-penten-3-one	2711	FDA §121.1164	Synthetic flavor
Methyl β-ionone	5-(2,6,6-Trimethyl-1-cyclohexen-1-yl)-4-penten-3-one	2712	FDA §121.1164	Synthetic flavor
Methyl γ-Ionone	*see* α-Isomethylionone			
Methyl δ-ionone	5-(2,6,6-Trimethyl-3-cyclohexen-1-yl)-4-penten-3-one	2713	FDA §121.1164	Synthetic flavor

REGULATORY STATUS OF DIRECT FOOD ADDITIVES (continued)

Material	Alternate Name [Source]	FEMA No.	Regulation	Limitations
6-Methylionone	see α-Irone			
Methyl isobutyl ketone	see 4-Methyl-2-pentanone			
2-Methyl-3-isobutyl-pyrazine	see 2-Isobutyl-3-methyl-pyrazine			
Methyl isobutyrate		2694	FDA §121.1164	Synthetic flavor
Methyl isoeugenol	see Isoeugenyl methyl ether			
1-Methyl-4-isopropenylbenzene	see p-α-Dimethylstyrene			
6-Methyl-3-isopropenylcyclohexanol	see Dihydrocarveol			
5-Methyl-2-isopropyl-cyclohexanol			FDA §121.1164	Synthetic flavor
α-Methyl-p-isopropyl-hydrocinnamaldehyde	see 2-Methyl-3-(p-isopropylphenyl)propionaldehyde			
2-Methyl-3-(p-isopropylphenyl)propionaldehyde	Cyclamen aldehyde; α-methyl-p-isopropyl-hydrocinnamaldehyde	2743	FDA §121.1164	Synthetic flavor
Methyl isovalerate		2753	FDA §121.1164	Synthetic flavor
Methyl laurate	Methyl dodecanoate	2715	FDA §121.1164	Synthetic flavor
Methyl mercaptan	Methanethiol	2716	FDA §121.1164	Synthetic flavor
3-Methylmercapto-propyl isothiocyanate	see 3-Methylthiopropyl isothiocyanate			
Methyl β-methiopro-pionate	see Methyl 2-methylthio-propionate			
Methyl o-methoxy-benzoate	o-Methoxy methyl benzoate	2717	FDA §121.1164	Synthetic flavor
2-Methyl-5-methoxy-thiazole		3192		
Methyl N-methyl-anthranilate	Dimethyl anthranilate	2718	FDA §121.1164	Synthetic flavor

REGULATORY STATUS OF DIRECT FOOD ADDITIVES (continued)

Material	Alternate Name [Source]	FEMA No.	Regulation	Limitations
Methyl 2-methylbutyrate		2719	FDA §121.1164	Synthetic flavor/adjuvant
7-Methyl-3-methylene-1,6-octadiene	*see* Myrcene			
Methyl 4-methylpentanoate	*see* Methyl 4-methyl-valerate			
Methyl 2-methylthiopropionate	Methyl β-methyl mercaptopropionate; methyl β-methiopropionate	2720	FDA §121.1164	Synthetic flavor
2-Methyl-3-, 5-, or 6-methylthiopyrazine	*see* (Methylthio)methyl-pyrazine			
Methyl 4-methylvalerate	Methyl 4-methylpenta-noate	2721	FDA §121.1164	Synthetic flavor
Methyl myristate	Methyl tetradecanoate	2722	FDA §121.1164	Synthetic flavor
1-Methylnaphthalene		3193		
Methyl β-naphthyl ketone	2′-Acetonaphthone	2723	FDA §121.1164	Synthetic flavor
Methyl nonanoate		2724	FDA §121.1164	Synthetic flavor
Methyl 2-nonenoate	Methyl octyne carbonate	2725	FDA §121.1164	Synthetic flavor
Methyl nonyl acetaldehyde	*see* 2-Methylundecanal			
Methyl nonyl ketone	*see* 2-Undecanone			
Methyl 2-nonynoate	Methyl octyne carbonate	2726	FDA §121.1164	Synthetic flavor
2-Methyloctanal	Methyl hexyl acetaldehyde	2727	FDA §121.1164	Synthetic flavor
Methyl octanoate		2728	FDA §121.1164	Synthetic flavor
Methyloctyne carbonate	*see* Methyl 2-nonynoate			
Methyl 2-octynoate	Methyl heptine carbonate	2729	FDA §121.1164	Synthetic flavor

REGULATORY STATUS OF DIRECT FOOD ADDITIVES (continued)

Material	Alternate Name [Source]	FEMA No.	Regulation	Limitations
Methylparaben	Methyl p-hydroxy-benzoate	2710	FDA §121.101	GRAS, chemical preservative, up to 0.1%
			FDA §121.1164	Synthetic flavor
4-Methyl-2,3-pentane-dione	Acetyl isobutyryl	2730	FDA §121.1164	Synthetic flavor
2-Methylpentanoic acid	see 2-Methylvaleric acid			
4-Methyl-2-pentanone	Methyl isobutyl ketone	2731	FDA §121.1164	Synthetic flavor
2-Methyl-2-pentenal		3194		
2-Methyl-2-pentenoic acid		3195		
3-Methyl-2-(2-pent-enyl)-2-cyclopenten-1-one	Jasmone	3196		
β-Methylphenethyl alcohol	Hydratropyl alcohol	2732	FDA §121.1164	Synthetic flavor
α-Methylphenethyl butyrate	1-Phenyl-2-propyl buty-rate	3197		
Methyl phenethyl ether		3198		
p-Methylphenylacetal-dehyde	see p-Tolylacetaldehyde			
Methyl phenylacetate	Methyl α-toluate	2733	FDA §121.1164	Synthetic flavor
3-Methyl-4-phenyl-3-butene-2-one	Benzylidene acetone methyl	2734	FDA §121.1164	Synthetic flavor
2-Methyl-4-phenyl-2-butyl acetate	Dimethylphenylethyl carbinyl acetate	2735	FDA §121.1164	Synthetic flavor
2-Methyl-4-phenyl-2-butyl isobutyrate	Dimethylphenylethyl carbinyl isobutyrate	2736	FDA §121.1164	Synthetic flavor
2-Methyl-4-phenyl-butyraldehyde		2737		
3-Methyl-2-phenyl-butyraldehyde	α-Isopropyl phenylace-taldehyde	2738	FDA §121.1164	Synthetic flavor

REGULATORY STATUS OF DIRECT FOOD ADDITIVES (continued)

Material	Alternate Name [Source]	FEMA No.	Regulation	Limitations
Methyl 4-phenyl buty-rate		2739	FDA §121.1164	Synthetic flavor
4-Methyl-2-phenyl-*m*-dioxolane	*see* Benzaldehyde pro-pylene glycol acetal			
3-Methyl-3-phenylgly-cidic acid ethyl ester	Aldehyde C-16; ethyl methylphenylglycidate; so-called strawberry al-dehyde		FDA §121.101	GRAS, synthetic flavor and adjuvant
5-Methyl-2-phenyl-2-hexenal		3199		
Methyl phenyl ketone	*see* Acetophenone			
4-Methyl-1-phenyl-2-pentanol	*see* α-Isobutylphenethyl alcohol			
4-Methyl-1-phenyl-2-pentanone	Benzyl isobutyl ketone	2740	FDA §121.1164	Synthetic flavor/adjuvant
4-Methyl-2-phenyl-2-pentenal		3200		
Methyl 3-phenylpro-pionate	Methyl hydrocinnamate	2741	FDA §121.1164	Synthetic flavor
Methyl polysiloxane			MID	To retard foaming in curing pickle at 50 ppm
2-Methyl-2-propen-1-yl butyrate	*see* 2-Methylallyl buty-rate			
Methyl propionate		2742	FDA §121.1164	Synthetic flavor
Methyl *n*-propyl car-binol	*see* 2-Pentanol			
2-(1-Methylpropyl)-cyclohexanone	*see* 2-*sec*-Butylcyclo-hexanone			
3-Methyl-5-propyl-2-cyclohexen-1-one			FDA §121.1164	Synthetic flavor/adjuvant
Methyl propyl disulfide		3201		
Methyl propyl ketone	*see* 2-Pentanone			
Methyl propyl trisulfide	Propyl methyl trisulfide	3308		*see Food Technol.,* May 1972

REGULATORY STATUS OF DIRECT FOOD ADDITIVES (continued)

Material	Alternate Name [Source]	FEMA No.	Regulation	Limitations
2-Methylpyrazine	2-Methyl-1,4-diazine	3309		*see Food Technol.,* May 1972
Methyl pyrazinyl ketone	*see* Acetylpyrazine			
Methyl 2-pyridyl ketone	*see* 2-Acetylpyridine			
Methyl 2-pyrrolyl ketone	2-Acetylpyrrole	3202		
4-Methylquinoline	*see* Lepidine			
5-Methylquinoxaline		3203		
Methyl salicylate	Wintergreen oil	2745		
Methyl styryl carbinol	*see* 4-Phenyl-3-buten-2-ol			
Methyl sulfide	Dimethyl sufide	2746	FDA §121.1164	Synthetic flavor
4-Methyl-5-thiazole-ethanol	4-Methyl-5-(β-hydroxy-ethyl)-thiazole	3204		
4-Methyl-5-thiazole-ethanol acetate		3205		
2-Methylthioacetalde-hyde		3206		
1-(Methylthio)-2-buta-none		3207		
Methyl thiobutyrate	Methanethiol *n*-buty-rate; thiobutyric acid, methyl ester	3310		*see Food Technol.,* May 1972
Methyl thiofuroate	Thiofuroic acid, methyl ester	3311		*see Food Technol.,* May 1972
(Methylthio)methylpy-razine (mixture of iso-mers)	2-Methyl-3-, 5-, or 6-methylthiopyrazine	3208		10 ppm max in finished food
5-Methyl-2-thiophene-carboxaldehyde		3209		
2-Methylthiophenol	*see o*-Toluenethiol			

REGULATORY STATUS OF DIRECT FOOD ADDITIVES (continued)

Material	Alternate Name [Source]	FEMA No.	Regulation	Limitations
o-(Methylthio)phenol	2-(Methylthio)phenol; thioguaiacol	3210		
2-Methylthiopropional-dehyde	Methional; β-methio-propionaldehyde	2747	FDA §121.1164	Synthetic flavor
3-Methylthiopropyl isothiocyanate		3312		see Food Technol., May 1972
Methyl p-tolyl carbinol	see p,α-Dimethylbenzyl alcohol			
Methyl p-tolyl ketone	see 4-Methylacetophen-one			
2-Methyl-3-tolyl pro-pionaldehyde, mixed o, m, p		2748	FDA §121.1164	Synthetic flavor
Methyl trisulfide	see Dimethyl trisulfide			
2-Methylundecanal	Aldehyde C-12, M.N.A.; methyl nonyl acetalde-hyde	2749	FDA §121.1164	Synthetic flavor
Methyl 9-undecenoate		2750	FDA §121.1164	Synthetic flavor
Methyl 2-undecynoate	Methyl decyne carb-bonate	2751	FDA §121.1164	Synthetic flavor
Methyl valerate		2752	FDA §121.1164	Synthetic flavor
2-Methylvaleric acid	2-Methylpentanoic acid	2754	FDA §121.1164	Synthetic flavor
2-(1-Methylvinyl)pyra-zine	see Isopropenylpyrazine			
2-Methyl-5-vinylpyra-zine		3211		10 ppm max in finished food
4-Methyl-5-vinylthia-zole		3313		see Food Technol., May 1972
Microwave radiation for the heat treatment of food			FDA §121.3008	Treatment of whole-fish protein concentrate com-plying with §121.1202 for reduction of solvent resi-dues; microwave frequency authorized by FCC regula-tions; microwave power 25 kilowatts max

REGULATORY STATUS OF DIRECT FOOD ADDITIVES (continued)

Material	Alternate Name [Source]	FEMA No.	Regulation	Limitations
Milk protein hydroly-sate			MID	To flavor various meat products
Mimosa, flower	Black wattle [*Acacia decurrens* Willd. var. *dealbata*]	2755	FDA §121.1163	Natural flavor
Mineral oil			FDA §121.1099	In defoaming agents limited to use in processing beet sugar and yeast; up to 150 ppm in yeast, measured as hydrocarbons
Modified hops extract	*see* Hops extract, modified			
Molasses			FDA §8.303	In production of caramel
			FDA §8.301	In dried algae meal used in chicken feed
Molasses extract	[*Saccharum officinarum* L.]		FDA §121.101	GRAS, natural flavor extractive
Monoammonium glutamate			FDA §121.101	GRAS, miscellaneous and/or general-purpose food additive
Monocalcium phosphate			FDA §29.4 and §29.5	In artificially sweetened fruit jellies and preserves
			FDA §51.990	In specified canned vegetables
			FDA §17	In bakery products up to 0.75% by weight of flour
			FDA §15.50 and §15.60	Acid-reacting agent in self-rising flour and enriched self-rising flour
			FDA §15.70	0.25–0.75% in phosphated flour
			FDA §15.506 and §15.507	Acid-reacting agent in self-rising, white corn meal and self-rising, yellow corn meal in combination with sodium bicarbonate, so that combination is not more than 4.5% of corn meal
Mono- and diglycerides	*see* Glycerides (mono- and di-)			

REGULATORY STATUS OF DIRECT FOOD ADDITIVES (continued)

Material	Alternate Name [Source]	FEMA No.	Regulation	Limitations
Monoglyceride citrate			FDA §121.1036	Synergist and solubilizer in antioxidant formulations for addition to oils and fats, up to 200 ppm of combined weight of oil or fat and the additive. Product specifications apply
			MID	To increase effectiveness of antioxidants in lard and shortening, at 0.01%
Monoisopropyl citrate			FDA §121.101	GRAS, sequestrant
			FDA	0.02% max in margarine by weight of finished margarine
			MID	To increase effectiveness of antioxidants in lard, shortening and oleomargarine; 0.01% in lard and shortening; 0.02% in oleomargarine
Monopotassium glutamate			FDA §121.101	GRAS, miscellaneous and/or general-purpose food additive
Monosodium glutamate		2756	FDA §121.101	GRAS
			MID	To flavor various meat products
Monosodium orthophosphate	Monosodium o-phosphate		FDA §121.1031	Esterification of food starch; residual phosphate in food 0.4% max, calculated as phosphorus
Monosodium phosphate			FDA §19	Emulsifier, up to 3% by weight of specified cheeses
			MID	To decrease amount of cooked-out juices in cured hams, pork shoulder picnics and loins, canned hams and pork shoulder picnics, chopped ham, and bacon; 5.0% of phosphate in pickle at 10% pump level; 0.5% of phosphate in product; only clear solution may be injected into product

REGULATORY STATUS OF DIRECT FOOD ADDITIVES (continued)

Material	Alternate Name [Source]	FEMA No.	Regulation	Limitations
Monosodium phosphate derivatives			FDA §121.101	Monosodium phosphate derivatives of mono- and diglycerides of edible fats or oils or edible fat-forming fatty acids; GRAS, emulsifiers
Mullein, flower	[*Verbascum phlomoidea* L.; *V. thapsiforme* Schräd.]		FDA §121.1163	Natural flavor in alcoholic beverages only
Musk	Tonquin musk [*Moschus moschiferus* L., musk deer]		FDA §121.101	GRAS, natural extractive
Musk ambrette	2,6-Dinitro-3-methoxy-1-methyl-4-*tert*-butylbenzene	2758		
Musk yarrow	*see* Iva			
Mustard, black or brown	[*Brassica nigra* (L.) Koch; *B. juncea* (L.) Coss.]	2760	FDA §121.101	GRAS, natural flavor
Mustard oil	*see* Allyl isothiocyanate			
Mustard, white or yellow	[*Brassica hirta* Moench.; *B. alba* (L.) Boiss]	2761	FDA §121.101	GRAS, natural flavor
Myrcene	7-Methyl-3-methylene-1,6-octadiene	2762	FDA §121.1164	Synthetic flavor
Myrcia oil	*see* Bay, oil of			
Myristaldehyde	Aldehyde C_{14}; tetradecanal	2763	FDA §121.1164	Synthetic flavor
Myristic acid	Tetradecanoic acid	2764		
Myrrh, gum	[*Commiphora molmol* Engl.; *C. abyssinica* (Berg) Engl.; other *Commiphora* species]	2765	FDA §121.1163	Natural flavor
Myrtle leaves	[*Myrtus communis* L.]		FDA §121.1163	Natural flavor in alcoholic beverages only

REGULATORY STATUS OF DIRECT FOOD ADDITIVES (continued)

Material	Alternate Name [Source]	FEMA No.	Regulation	Limitations
Naphtha			FDA §8.390	Diluent in mixtures for coloring shell eggs; no penetration of mixture into egg
2-Naphthalenthiol	2-Mercaptonaphthalene; 2-naphthyl mercaptan; 2-thionaphthol	3314		*see Food Technol.,* May 1972
β-Naphthyl anthranilate	2-Naphthyl anthranilate	2767		
β-Naphthyl ethyl ether	2-Ethoxynaphthalene	2768		
2-Naphthyl mercaptan	*see* 2-Naphthalenthiol			
Naringin	[*Citrus paradisi* Macf.]	2769	FDA §121.101	GRAS, natural flavor extractive
Natural rubber (smoked sheet and latex solids)	[*Hevea brasiliensis*]		FDA §121.1059	Natural (coagulated or concentrated latex) masticatory substance in chewing-gum base
***d*-Neomenthol**	2-Isobutyl-5-methyl cyclohexanol	2666	FDA §121.1164	Synthetic flavor
Neral	*see* Citral			
Nerol	*cis*-3,7-Dimethyl-2,6-octadien-1-ol	2770	FDA §121.1164	Synthetic flavor
Neroli bigarade, oil	[*Citrus aurantium* L. subsp. *amara* L.]	2771	FDA §121.101	GRAS, natural flavor extractive
Nerolidol	3,7,11-Trimethyl-1,6,10-dodecatrien-3-ol	2772	FDA §121.1164	Synthetic flavor
Neryl acetate		2773	FDA §121.1164	Synthetic flavor
Neryl butyrate		2774	FDA §121.1164	Synthetic flavor
Neryl formate		2776	FDA §121.1164	Synthetic flavor
Neryl isobutyrate		2775	FDA §121.1164	Synthetic flavor
Neryl isovalerate		2778	FDA §121.1164	Synthetic flavor
Neryl propionate		2777	FDA §121.1164	Synthetic flavor

REGULATORY STATUS OF DIRECT FOOD ADDITIVES (continued)

Material	Alternate Name [Source]	FEMA No.	Regulation	Limitations
Niacin			FDA §121.101	GRAS, nutrient and/or dietary supplement
Niacinamide			FDA §121.101	GRAS, nutrient and/or dietary supplement
Nicotinamide ascorbic acid complex			FDA §121.1095	Source of ascorbic acid and nicotinamide in multivitamin preparations. Product specifications apply
Niger gutta	[*Fucus platyphylla* Del.]		FDA §121.1059	Natural (coagulated or concentrated latex) masticatory substance in chewing-gum base
Nispero	[*Manilkara zapotilla* Gilly; *M. chicle* Gilly]		FDA §121.1059	Natural (coagulated or concentrated latex) masticatory substance in chewing-gum base
Nitrogen			FDA §121.101	GRAS, miscellaneous and/or general-purpose food additive
			MID	To exclude oxygen from sealed containers
Nitrosyl chloride			FDA §15	Bleaching substance for flour
Nitrous oxide	Dinitrogen monoxide	2779	FDA §121.101	GRAS, propellant for certain dairy and vegetable-fat toppings in pressurized containers
			FDA §121.1065	With octafluorocyclobutane as propellant and aerating agent for foamed or sprayed food products, except those standardized foods that do not provide for such use
			FDA §121.1180	With chloropentafluoroethane as propellant and aerating agent for foamed or sprayed foods
2,4-Nonadienal		3212		
2,6-Nonadien-1-ol		2780	FDA §121.1164	Synthetic flavor

REGULATORY STATUS OF DIRECT FOOD ADDITIVES (continued)

Material	Alternate Name [Source]	FEMA No.	Regulation	Limitations
γ-Nonalactone	Aldehyde C-18 (so-called); 4-hydroxynon-anoic acid, γ-lactone	2781	FDA §121.1164	Synthetic flavor
n-Nonanal	Pelargonic aldehyde; aldehyde C-18	2782	FDA §121.1164	Synthetic flavor
1,3-Nonanediol acetate, mixed esters	Octyl crotonyl acetate	2783	FDA §121.1164	Synthetic flavor
Nonanoic acid	Pelargonic acid	2784	FDA §121.1164	Synthetic flavor
1-Nonanol	see Nonyl alcohol			
2-Nonanol	Methyl n-heptyl carbinol	3315		see Food Technol., May 1972
2-Nonanone	Methyl heptyl ketone	2785	FDA §121.1164	Synthetic flavor
3-Nonanon-1-yl acetate	1-Hydroxy-3-nonanone acetate	2786	FDA §121.1164	Synthetic flavor
2-Nonenal		3213		
Nonyl acetate	Acetate C-9	2788	FDA §121.1164	Synthetic flavor
Nonyl alcohol	2,6-Dimethyl heptanol-4; 1-nonanol	2789	FDA §121.1164	Synthetic flavor
Nonyl isovalerate		2791	FDA §121.1164	Synthetic flavor
Nonyl octanoate		2790	FDA §121.1164	Synthetic flavor
Nootkatone	5,6-Dimethyl-8-isopro-penylbicyclo(4,4,0)-dec-1-en-3-one; 4,4a,5,6,7,8-hexahydro-6-isopro-penyl-4,4a-dimethyl-2(3H)-naphthalenone	3166		
Nutmeg	[Myristica fragrans Houtt.]	2792	FDA §121.101	GRAS, natural flavor and natural flavor extractive
Oak, English, wood	[Quercus robur L.]		FDA §121.1163	Natural flavor in alcoholic beverages only
Oak moss	[Evernia prunastri (L.) Ach.; E. furfuracea (L.) Mann; and other lichens]	2795	FDA §121.1163	Natural flavor; thujone free

REGULATORY STATUS OF DIRECT FOOD ADDITIVES (continued)

Material	Alternate Name [Source]	FEMA No.	Regulation	Limitations
Oak, white, chips	[*Quercus alba* L.]	2794	FDA §121.1163	Natural flavor
Oat gum			FDA §19	0.8% max in specified cheese products
			FDA—GRAS	
Ocimene	3,7-Dimethyl-1,3,6-octa-triene; *trans-β*-ocimene		FDA §121.1164	Synthetic flavor
Octadecanoic acid	*see* Stearic acid			
3,6-Octadien-1-ol			FDA §121.101	GRAS, synthetic flavor
Octafluorocyclobutane	Perfluorocyclobutane		FDA §121.1065	Alone or with carbon dioxide and/or nitrous oxide and/or propane, as propellant and aerating agent for foamed or sprayed food products, except standardized foods that do not provide for such use. Product specifications apply
			FDA §121.1181	With chloropentafluoro-ethane as propellant and aerating agent for foamed or sprayed food products; complying with §121.1065
Octalactone			FDA §121.1164	Synthetic flavor
γ-Octalactone	4-Hydroxyoctanoic acid, γ-lactone	2796	FDA §121.1164	Synthetic flavor
δ-Octalactone		3214		
n-Octanal	Aldehyde C–8; caprylal-dehyde	2797	FDA §121.1164	Synthetic flavor
Octanal dimethyl acetal	Aldehyde C–8 dimethyl acetal	2798	FDA §121.1164	Synthetic flavor
Octanoic acid	Caprylic acid	2799	FDA §121.1164	Synthetic flavor
Octanol	*see* Octyl alcohol			
2-Octanone	Methyl hexyl ketone	2802	FDA §121.1164	Synthetic flavor
3-Octanone	Ethyl amyl ketone	2803	FDA §121.1164	Synthetic flavor

REGULATORY STATUS OF DIRECT FOOD ADDITIVES (continued)

Material	Alternate Name [Source]	FEMA No.	Regulation	Limitations
3-Octanon-1-ol	Methylol methyl amyl ketone	2804	FDA §121.1164	Synthetic flavor
2-Octenal		3215		
1-Octen-3-ol	Amyl vinyl carbinol	2805	FDA §121.1164	Synthetic flavor
1-Octen-3-yl acetate			FDA §121.1164	Synthetic flavor/adjuvant
1-Octenyl succinic anhydride			FDA §121.1031	Modifier of food starch, 3.0% max; up to 2.0% in combination with 2.0% max of aluminum sulfate
Octyl acetate	2-Ethyl hexyl acetate	2806	FDA §121.1164	Synthetic flavor
3-Octyl acetate			FDA §121.1164	Synthetic flavor/adjuvant
1-Octyl alcohol	1-Octanol	2800	FDA §121.1164	Synthetic flavor
2-Octyl alcohol	2-Octanol	2801	FDA §121.1164	Synthetic flavor
3-Octyl alcohol	3-Octanol		FDA §121.1164	Synthetic flavor/adjuvant
Octyl butyrate		2807	FDA §121.1164	Synthetic flavor
Octyl formate		2809	FDA §121.1164	Synthetic flavor
Octyl heptanoate		2810	FDA §121.1164	Synthetic flavor
Octyl isobutyrate		2808	FDA §121.1164	Synthetic flavor
Octyl isovalerate		2814	FDA §121.1164	Synthetic flavor
Octyl octanoate		2811	FDA §121.1164	Synthetic flavor
Octyl phenylacetate		2812	FDA §121.1164	Synthetic flavor
Octyl propionate		2813	FDA §121.1164	Synthetic flavor
Oleic acid	cis-9-Octadecenoic acid	2815		
Olibanum, oil	Frankincense [Boswellia carterii Birdw. and other Boswellia species]	2816	FDA §121.1163	Synthetic flavor
Olive, oil	[Olea europaea L.]		FDA—GRAS	

REGULATORY STATUS OF DIRECT FOOD ADDITIVES (continued)

Material	Alternate Name [Source]	FEMA No.	Regulation	Limitations
Onion, oil	[*Allium cepa* L.]	2817	FDA §121.101	GRAS, natural flavor extractive
Opopanax	Bisabol myrrh [*Opopanax chironium* Koch. (true opopanax) or *Commiphora erythraea* Engl. var. *glabrescens*]		FDA §121.1163	Natural flavor
Orange B			FDA §8.202	Coloring casings or surfaces of frankfurters and sausages, 150 ppm max. Product specifications apply
Orange, bitter, flower and peel	Curaçao orange peel [*Citrus aurantium* L.]		FDA §121.101	GRAS, natural flavor extractive
Orange leaf, absolute	[*Citrus aurantium* L.]	2818	FDA §121.101	GRAS, natural flavor extractive
Orange, sweet, flower and peel	[*Citrus sinensis* (L.) Osbeck; *C. aurantium* var. *dulcis* L.]		FDA §121.101	GRAS, natural flavor extractive
Oregano	Oreganum, Mexican oregano, Mexican sage, origan [*Lippia* species]	2827	FDA §121.101	GRAS, natural flavor
Origanum	[*Origanum vulgare* L. and other *Origanum* species]		FDA §121.101	GRAS, natural flavor extractive
"Origanum," Spanish	*see* Thymus capitatis			
Orris, concrete, liquid, oil	*Iris florentina* L.	2829		
Orris root, extract	[*Iris germanica* L., including its variety *florentina* Dykes]	2830	FDA §121.1163	Natural flavoring substances and natural substances used in conjunction with flavors
Ox bile extract	Oxgall		FDA §121.101	GRAS, emulsifying agent in dried egg whites up to 0.1%
Oxygen, active			FDA §121.1031	0.45% max, obtained from hydrogen peroxide and/or peracetic acid, as modifier

REGULATORY STATUS OF DIRECT FOOD ADDITIVES (continued)

Material	Alternate Name [Source]	FEMA No.	Regulation	Limitations
Oxygen, active (continued)				for food starch; alone or in combination with chlorine as sodium hypochlorite (not to exceed 0.055 pounds of chlorine per pound of dry starch), and not more than 25% of propylene oxide
Oxysterin			FDA §121.1016	Crystallization inhibitor in cottonseed and soybean cooking and salad oils and in vegetable oils; as release agent in vegetable oils and vegetable shortening, up to 0.125% of combined weight of oil plus additive. Product specifications apply
			FDA §121.1099	In defoaming agents limited to use in processing beet sugar and yeast
Palmarosa, oil	East Indian geranium [*Cymbopogon martini* Stapf. var. *motia*]	2831	FDA §121.101	GRAS, natural flavor extractive
Palmitic acid	Hexadecanoic acid	2832		
Pansy	[*Viola tricolor* L.]		FDA §121.1163	Natural flavor in alcoholic beverages only
D-Pantothenamide			FDA §121.1123	Source of pantothenic acid activity in foods for special dietary use
D-Pantothenyl alcohol			FDA §121.101	GRAS, nutrient and/or dietary supplement
Papain	Papayotin		FDA §121.101	GRAS, miscellaneous and/or general-purpose food additive
			MID	To soften tissues of beef cuts, with water, salt and monosodium glutamate; tissue weight gain, 3% max
Paprika	[*Capsicum annuum* L.]	2833	FDA §121.101	GRAS, natural flavor and natural flavor extractive
			FDA §8.307	Food coloring

REGULATORY STATUS OF DIRECT FOOD ADDITIVES (continued)

Material	Alternate Name [Source]	FEMA No.	Regulation	Limitations
Paprika, oleoresin	[*Capsicum annuum* L.]	2834	FDA §8.308	Food coloring; solvent residues are limited. Product specifications apply
Paraffin	Paraffin wax	3216	FDA §121.1059	Synthetic masticatory substance in chewing-gum base. Product specifications apply
Parsley	[*Petroselinum sativum* Hoffm.; *Carum petroselinum* Benth. and Hook]	2835	FDA §121.101	GRAS, natural flavor and natural flavor extractive
Passion flower	[*Passiflora incarnata* L.]		FDA §121.1163	Natural flavor
Patchouly, oil	[*Pogostemon cablin* Benth.; *P. heyneanus* Benth.]	2838	FDA §121.1163	Natural flavor
Peach aldehyde	*see* γ-Undecalactone			
Peach kernel	Persic oil, bitter almond [*Prunus persica* Sieb & Zucc.]		FDA §121.101	GRAS, solvent-free natural extractive used in conjunction with natural flavorings
Peach, leaf	[*Prunus persica* (L.) Batsch]		FDA §121.1163	Natural flavor in alcoholic beverages only; not to exceed 25 ppm of prussic acid
Peanut stearine	[*Arachis hypogaea* L.]		FDA §121.101	GRAS, solvent-free natural extractive used in conjunction with natural flavoring
Pectin			FDA—GRAS	
Pectin sugar	*see* L-Arabinose			
Pediococcus cerevisiae **culture**			MID	Starter to develop flavor in dry sausage, pork roll, thuringer, Lebanon bologna, cervelat, and salami, at 0.5%
Pelargonic acid	*see* Nonanoic acid			
Pelargonic aldehyde	*see* n-Nonanal			
Pelargonyl vanillyl-amide	*see* N-(4-Hydroxy-3-methoxybenzyl)nonanamide			

REGULATORY STATUS OF DIRECT FOOD ADDITIVES (continued)

Material	Alternate Name [Source]	FEMA No.	Regulation	Limitations
Pendare	[*Couma macrocarpa* Barb. Rodr.; *C. utilis* (Mart.) Muell. Arg.]		FDA §121.1059	Natural (coagulated or concentrated latex) masticatory substance in chewing-gum base
Penicillium roqueforti			FDA §19	Mold in specified cheeses
Pennyroyal, American and European	[*Hedeoma pulegioides* (L.) Pers.; *Mentha pulegium* L.]	2839	FDA §121.1163	Natural flavoring substances and natural substances used in conjunction with flavors
ω-Pentadecalactone	Angelicalactone; 15-hydroxypentadecanoic acid, ω-lactone; pentadecanolide	2840	FDA §121.1164	Synthetic flavor
Pentadecanolide	*see* ω-Pentadecalactone			
2,4-Pentadienal		3217		
Pentaerythritol ester of fumaric acid-rosin adduct			FDA §8.390	Diluent in mixtures for coloring shell eggs; no penetration into egg
Pentaerythritol ester of partially hydrogenated gum or wood rosin			FDA §121.1059	Plasticizing material in chewing-gum base
Pentaerythritol ester of wood rosin			FDA §121.1059	Plasticizing material in chewing-gum base. Product specifications apply
Pentanal	*see* Valeraldehyde			
2,3-Pentanedione	Acetyl propionyl	2841	FDA §121.1164	Synthetic flavor
Pentanoic acid	*see* Valeric acid			
2-Pentanol		3316		*see Food Technol.*, May 1972
2-Pentanone	Methyl propyl ketone	2842	FDA §121.1164	Synthetic flavor
2-Pentenal		3218		
4-Pentenoic acid	Allyl acetic acid	2843	FDA §121.1164	Synthetic flavor
1-Penten-3-ol			FDA §121.1164	Synthetic flavor/adjuvant
Pentyl alcohol	*see* Amyl alcohol			

REGULATORY STATUS OF DIRECT FOOD ADDITIVES (continued)

Material	Alternate Name [Source]	FEMA No.	Regulation	Limitations
2-Pentyl furan		3317		see Food Technol., May 1972
Pepper, black or white	[Piper nigrum L.]	2844, 2850	FDA §121.101	GRAS, natural flavor and natural flavor extractive
Pepper, cayenne	see Capsicum			
Peppermint, oil	[Mentha piperita L.]	2848	FDA §121.101	GRAS, natural flavor and natural flavor extractive
Perillaldehyde	4-Isopropyl-1-cyclohex-ene-1-carboxaldehyde; p-mentha-1,8-diene-7-al		FDA §121.1164	Synthetic flavor
Perillo	[Couma macrocarpa Barb. and C. utilis (Mart.) Muell. Arg.]		FDA §121.1059	Natural (coagulated or concentrated latex) masticatory substance in chewing-gum base
Perillyl acetate	p-Mentha-1,8-dien-7-yl acetate		FDA §121.1164	Synthetic flavor
Perillyl alcohol	see p-Mentha-1,8-dien-7-ol			
Persic oil	see Apricot kernel and peach kernel			
Peruvian balsam	Balsam Peru [Myroxylon pereirae Klotzsch]		FDA §121.101	GRAS, natural flavor extractive
Petitgrain lemon, oil	[Citrus limonum (L.) Burm. f.]	2853	FDA §121.101	GRAS, natural flavor extractive
Petitgrain mandarin or tangerine	see Mandarin			
Petitgrain, oil	[Citrus aurantium L.]	2855	FDA §121.101	GRAS, natural flavor extractive
Petrolatum	Paraffin jelly		FDA §121.1166	1. 0.15% max in bakery products, as release agent and lubricant, with white mineral oil 2. 0.02% max in dehydrated fruits and vegetables, as release agent 3. 0.10% in egg-white solids, as release agent

REGULATORY STATUS OF DIRECT FOOD ADDITIVES (continued)

Material	Alternate Name [Source]	FEMA No.	Regulation	Limitations
Petrolatum (continued)				4. On raw fruits and vegetables, as protective coating 5. Defoaming agent in beet sugar and yeast, 150 ppm max in yeast, measured as hydrocarbons 6. Conforming with §121.261
			FDA §121.1099	In defoaming agents limited in use to processing beet sugar and yeast, 150 ppm max in yeast, measured as hydrocarbons. Product specifications apply
Petrolatum, liquid	*see* Mineral oil			
Petroleum wax			FDA §121.1156	1. Masticatory substance in chewing-gum base 2. Protective coating on cheese, raw fruits, and raw vegetables 3. Defoamer in food, in accordance with §121.1099 4. Product specifications apply; may contain any antioxidant permitted in food, in accordance with §409
			FDA §121.1099	In defoaming agents, limited in use to processing beet sugar and yeast; 150 ppm max in yeast, measured as hydrocarbons
α-Phellandrene	*p*-Mentha-1,5-diene	2856	FDA §121.1164	Synthetic flavor
Phenethyl acetate	2-Phenylethyl acetate	2857	FDA §121.1164	Synthetic flavor
Phenethyl alcohol	*β*-Phenylethyl alcohol	2858	FDA §121.1164	Synthetic flavor
Phenethylamine		3220		
Phenethyl anthranilate	2-Phenylethyl anthranilate	2859	FDA §121.1164	Synthetic flavor
Phenethyl benzoate	2-Phenylethyl benzoate	2860	FDA §121.1164	Synthetic flavor
Phenethyl butyrate	2-Phenylethyl butyrate	2861	FDA §121.1164	Synthetic flavor
Phenethyl carbinol	*see* 1-Phenyl-1-propanol			

REGULATORY STATUS OF DIRECT FOOD ADDITIVES (continued)

Material	Alternate Name [Source]	FEMA No.	Regulation	Limitations
Phenethyl cinnamate	2-Phenylethyl cinnamate	2863	FDA §121.1164	Synthetic flavor
Phenethyl 3,3-dimethyl-acrylate	*see* Phenethyl senecioate			
Phenethyl formate	2-Phenylethyl formate	2864	FDA §121.1164	Synthetic flavor
Phenethyl 2-furoate	2-Phenylethyl 2-furoate	2865		
Phenethyl hexanoate		3221		
Phenethyl isobutyrate	2-Phenylethyl isobuty-rate	2862	FDA §121.1164	Synthetic flavor
Phenethyl isovalerate	2-Phenylethyl isoval-erate	2871	FDA §121.1164	Synthetic flavor
Phenethyl 2-methyl-butyrate			FDA §121.1164	Synthetic flavor/adjuvant
Phenethyl octanoate		3222		
Phenethyl phenylace-tate	2-Phenylethyl phenyl-acetate	2866	FDA §121.1164	Synthetic flavor
Phenethyl propionate	2-Phenylethyl propion-ate	2867	FDA §121.1164	Synthetic flavor
Phenethyl salicylate	2-Phenylethyl salicylate	2868	FDA §121.1164	Synthetic flavor
Phenethyl senecioate	Phenethyl 3,3-dimethyl-acrylate	2869	FDA §121.1164	Synthetic flavor
Phenethyl tiglate	2-Phenylethyl tiglate	2870	FDA §121.1164	Synthetic flavor
Phenol		3223		0.50 ppm max in beverages, frozen desserts, candy, baked goods, gelatins, and puddings
Phenoxyacetic acid	Phenoxyethanoic acid	2872	FDA §121.1164	Synthetic flavor
2-Phenoxyethyl iso-butyrate		2873	FDA §121.1164	Synthetic flavor
Phenylacetaldehyde	α-Toluic aldehyde	2874	FDA §121.1164	Synthetic flavor
Phenylacetaldehyde 2,3-butylene glycol acetal		2875	FDA §121.1164	Synthetic flavor

REGULATORY STATUS OF DIRECT FOOD ADDITIVES (continued)

Material	Alternate Name [Source]	FEMA No.	Regulation	Limitations
Phenylacetaldehyde dimethyl acetal	Viridine	2876	FDA §121.1164	Synthetic flavor
Phenylacetaldehyde glyceryl acetal		2877	FDA §121.1164	Synthetic flavor
Phenylacetic acid	α-Toluic acid	2878	FDA §121.1164	Synthetic flavor
Phenylalanine (L- and DL-forms)			FDA §121.101	GRAS, nutrient and/or dietary supplement, as the free acid, hydrochloride salt, hydrate, or anhydrous form
Phenylallyl cinnamate	*see* Cinnamyl cinnamate			
4-Phenyl-2-butanol	Phenylethyl methyl carbinol	2879	FDA §121.1164	Synthetic flavor
2-Phenyl-2-butenal		3224		
4-Phenyl-3-buten-2-ol	Methyl styryl carbinol	2880	FDA §121.1164	Synthetic flavor
4-Phenyl-3-buten-2-one	Benzilidene acetone	2881	FDA §121.1164	Synthetic flavor
4-Phenyl-2-butyl acetate	Phenylethyl methyl carbinyl acetate	2882	FDA §121.1164	Synthetic flavor
Phenyl dimethyl carbinyl isobutyrate	*see* α,α-Dimethylbenzyl isobutyrate			
2-Phenyl-*m*-dioxan-5-ol	*see* Benzaldehyde glyceryl acetal			
Phenyl disulfide	Diphenyl disulfide	3225		
β-Phenylethyl alcohol	*see* Phenethyl alcohol			
Phenylethyl carbinol	*see* 1-Phenyl-1-propanol			
Phenylethyl methyl carbinol	*see* 4-Phenyl-2-butanol			
Phenylethyl methyl carbinyl acetate	*see* 4-Phenyl-2-butyl acetate			
Phenylethyl methyl ethyl carbinol	*see* 1-Phenyl-3-methyl-3-pentanol			

REGULATORY STATUS OF DIRECT FOOD ADDITIVES (continued)

Material	Alternate Name [Source]	FEMA No.	Regulation	Limitations
1-Phenyl-3-methyl-3-pentanol	Phenylethyl methyl ethyl carbinol	2883˙	FDA §121.1164	Synthetic flavor
3-Phenyl-4-pentenal		3318		see Food Technol., May 1972
1-Phenyl-1,2-propane-dione		3226		
1-Phenyl-1-propanol	Phenethyl carbinol	2884	FDA §121.1164	Synthetic flavor
3-Phenyl-1-propanol	Hydrocinnamyl alcohol	2885	FDA §121.1164	Synthetic flavor
3-Phenyl-2-propen-1-ol	see Cinnamyl alcohol			
2-Phenylpropionalde-hyde	Hydratropaldehyde	2886	FDA §121.1164	Synthetic flavor
3-Phenylpropionalde-hyde	Hydrocinnamaldehyde	2887	FDA §121.1164	Synthetic flavor
2-Phenylpropionalde-hyde dimethyl acetal	Hydrotropic aldehyde dimethyl acetal	2888	FDA §121.1164	Synthetic flavor
3-Phenylpropionic acid	Hydrocinnamic acid	2889	FDA §121.1164	Synthetic flavor
3-Phenylpropyl acetate	Hydrocinnamyl acetate	2890	FDA §121.1164	Synthetic flavor
α-Phenylpropyl buty-rate	see α-Ethylbenzyl buty-rate			
1-Phenyl-2-propyl butyrate	see α-Methylphenethyl butyrate			
2-Phenylpropyl buty-rate	β-Methylphenethyl butyrate	2891	FDA §121.1164	Synthetic flavor
3-Phenylpropyl cinna-mate	Hydrocinnamyl cinna-mate	2894	FDA §121.1164	Synthetic flavor
3-Phenylpropyl formate	Hydrocinnamyl formate	2895	FDA §121.1164	Synthetic flavor
3-Phenylpropyl hex-anoate	Hydrocinnamyl hexa-noate	2896	FDA §121.1164	Synthetic flavor
2-Phenylpropyl iso-butyrate	Hydratropyl isobutyrate	2892	FDA §121.1164	Synthetic flavor
3-Phenylpropyl iso-butyrate	Hydrocinnamyl iso-butyrate	2893	FDA §121.1164	Synthetic flavor

REGULATORY STATUS OF DIRECT FOOD ADDITIVES (continued)

Material	Alternate Name [Source]	FEMA No.	Regulation	Limitations
3-Phenylpropyl isovalerate	Hydrocinnamyl iso-valerate	2899	FDA §121.1164	Synthetic flavor
3-Phenylpropyl propionate	Hydrocinnamyl pro-pionate	2897	FDA §121.1164	Synthetic flavor
2-(3-Phenylpropyl)-tetrahydrofuran	2-Hydrocinnamyl tetra-hydrofuran	2898	FDA §121.1164	Synthetic flavor
Phosphoric acid	Orthophosphoric acid	2900	FDA §121.101	GRAS, miscellaneous and/or general-purpose food additive
			FDA §8.303	In production of caramel
			MID	0.01%, to increase effectiveness of antioxidants in lard and shortening
Phosphorus oxy-chloride	Phosphoryl chloride		FDA §121.1031	0.1% in starch—modified food, alone, as esterifier, or in combination with up to 10% propylene oxide for esterification and etheri-fication
Pimenta leaf, oil see also **Allspice**	[*Pimenta officinalis* Lindl.]	2901		
Pine, dwarf, needles and twigs, oil	[*Pinus mugo* Turra var. *pumilio* (Haenke) Zenari]	2904	FDA §121.1163	Natural flavor
"Pine" (fir), needles and twigs	see Fir ("Pine"), needles and twigs		FDA §121.1164	Synthetic flavor/adjuvant
Pine, Scotch, needles and twigs, oil	[*Pinus sylvestris* L.]	2906	FDA §121.1163	Natural flavor
Pine tar oil	[*Pinus palustris* and other *Pinus* species]		FDA §121.1164	Synthetic flavor
Pine, white, bark	[*Pinus strobus* L.]		FDA §121.1163	Natural flavor in alcoholic beverages only
Pine, white, oil	[*Pinus palustris* Mill. and other *Pinus* species]		FDA §121.1163	Natural flavor
2-Pinene	see α-Pinene			

REGULATORY STATUS OF DIRECT FOOD ADDITIVES (continued)

Material	Alternate Name [Source]	FEMA No.	Regulation	Limitations
2(10)-Pinene	*see* β-Pinene			
α-Pinene	2-Pinene	2902	FDA §121.1059	In synthetic terpene resins for chewing-gum base. Product specifications apply
			FDA §121.1164	Synthetic flavor/adjuvant
β-Pinene	2(10)-Pinene	2903	FDA §121.1059	In synthetic terpene resins for chewing-gum base. Product specifications apply
Piperidene	Hexahydropyridine	2908	FDA §121.1164	Synthetic flavor
Piperine	Piperoylpiperidine	2909	FDA §121.1164	Synthetic flavor
Piperitol	*see* p-Menth-1-en-3-ol			
d-Piperitone	p-Menth-1-en-3-one	2910	FDA §121.1164	Synthetic flavor
Piperonal	Heliotropine; 3,4-methylenedioxybenzaldehyde	2911	FDA §121.101	GRAS, synthetic flavor and adjuvant
Piperonyl acetate	Heliotropyl acetate	2912	FDA §121.1164	Synthetic flavor
Piperonyl isobutyrate		2913	FDA §121.1164	Synthetic flavor
Pipsissewa leaves, extract	[*Chimaphila umbellata* (L.) Nutt.]	2914	FDA §121.101	GRAS, natural flavor extractive
Polyethylene			FDA §121.1059	Synthetic masticatory substance in chewing gum base; molecular weight 2,000–21,000
Polyethylene glycol	Carbowax®; PEG		FDA §121.1179	Coating on fresh citrus fruit; dispersing and defoaming adjuvant, in compliance with FDA §121.1185. Average mol wt, 200–9500
			FDA §121.1185	1. Coating, binder, plasticizing agent, and/or lubricant in tablets used for food 2. Adjuvant and bodying agent in nonnutritive sweeteners listed as GRAS 3. Adjuvant to disperse vitamins and/or mineral preparations

REGULATORY STATUS OF DIRECT FOOD ADDITIVES (continued)

Material	Alternate Name [Source]	FEMA No.	Regulation	Limitations
Polyethylene glycol (continued)				4. Coating on sodium nitrite to inhibit hygroscopicity. Product specifications apply
Polyethylene glycol 6000			FDA §121.1057	1. Binder and plasticizing agent in tablets used for food 2. Adjuvant in tablet coatings to improve resistance to moisture and oxidation 3. Bodying agent with non-nutritive sweeteners (cyclamate and saccharin salts) listed by FDA as GRAS 4. Product specifications apply
Polyglycerol esters of fatty acids			FDA §121.1120	Up to and including the decaglycerol esters of specified fatty acids: 1. Emulsifiers; cloud inhibitors in vegetable and salad oils 2. Esters of butter oil fatty acids as emulsifiers in dry, whipped-topping base. 3. Product specifications apply
			FDA §8.303	Antifoaming agents in caramel; conforming with §121.1120
			MID	In rendered animal fat or a combination of animal fat with vegetable fat. Product specifications apply
Polylimonene			FDA §121.1164	Synthetic flavor
Polyoxyethylene (20) sorbitan monolaurate	*see* Polysorbate 20			
Polyoxyethylene (20) sorbitan monooleate	*see* Polysorbate 80			
Polyoxyethylene (20) sorbitan monostearate	*see* Polysorbate 60			
Polyoxyethylene (20) sorbitan tristearate	*see* Polysorbate 65			

REGULATORY STATUS OF DIRECT FOOD ADDITIVES (continued)

Material	Alternate Name [Source]	FEMA No.	Regulation	Limitations
Polysorbate 20	Polyoxyethylene (20) sorbitan monolaurate	2915	FDA §121.1164	Synthetic flavor
			FDA §121.2541	Emulsifier and/or surface-active agent. Product specifications apply
Polysorbate 60	Polyoxyethylene (20) sorbitan monostearate	2916	FDA §121.1030	1. Emulsifier in whipped vegetable oil topping, alone or with sorbitan monostearate, polyoxyethylene (20) sorbitan tristearate, and/or polysorbate 80; additive(s) 0.40% max 2. Emulsifier in cake and cake mixes, 0.46% max alone; with polyoxyethylene (20) sorbitan tristearate and/or sorbitan monostearate, 0.46% max, polyoxyethylene (20) sorbitan tristearate 0.32% max, sorbitan monostearate 0.61% max, total additives 0.66% max 3. Emulsifier in shortenings and edible oils for use in nonstandardized baked goods, baking mixes, icings, fillings, and toppings, and in frying of foods; 1.0% alone; with polysorbate 80, total additives 1.0% max by weight of finished shortening or oil 4. Emulsifier in cake icings and cake fillings; 0.46% max alone; with polyoxyethylene (20) sorbitan tristearate and/or sorbitan monostearate, 0.46% max; polyoxyethylene (20) sorbitan tristearate 0.32% max, sorbitan monostearate 0.70% max, total additives 1.0% max 5. To impart greater opacity to sugar-type confection coatings, 0.2% max 6. Emulsifier in nonstandardized dressings made without egg yolk, 0.3% max

REGULATORY STATUS OF DIRECT FOOD ADDITIVES (continued)

Material	Alternate Name [Source]	FEMA No.	Regulation	Limitations
Polysorbate 60 (continued)				7. Emulsifier in solid-state, edible vegetable-fat-water-emulsions for use as substitutes for milk or cream in beverage coffee; 0.4% max, alone or in combination with polyoxyethylene (20) sorbitan tristearate and/or sorbitan monostearate
				8. Foaming agent in nonalcoholic mixes, to be added to alcoholic beverages in preparation of mixed drinks; 4.5% max by weight of the nonalcoholic mix. Product specifications apply
			FDA §121.1164	Synthetic flavor
			MID	To emulsify shortenings sold in units not exceeding 6 pounds or 1 gallon fluid content; 1% max alone; with polysorbate 80 or sorbitan monostearate, total additives 1% max
Polysorbate 65	Polyoxyethylene (20) sorbitan tristearate		FDA §121.1008	1. Emulsifier in ice cream, frozen custard, ice milk, fruit sherbet, and nonstandardized frozen desserts, alone or with polysorbate 80; total additive(s) 0.10% max
				2. Emulsifier in cakes and cake mixes, 0.32% max alone; with sorbitan monostearate and/or polysorbate 60, 0.32% max, sorbitan monostearate 0.61% max, polysorbate 60, 0.46% max, total additives, 0.66% max
				3. Emulsifier in whipped vegetable oil topping, alone or with sorbitan monostearate, polysorbate 60, and/or polysorbate 80; total additives, 0.40% max
				4. Emulsifier in cake icings and cake fillings, 0.32% max alone; with sorbitan

REGULATORY STATUS OF DIRECT FOOD ADDITIVES (continued)

Material	Alternate Name [Source]	FEMA No.	Regulation	Limitations
Polysorbate 65 (continued)				monostearate and/or polysorbate 60, 0.32% max; sorbitan monostearate, 0.70% max; polysorbate 60, 0.46% max; total additives, 1.0% max 5. Emulsifier in solid-state, edible vegetable-fat-water emulsions for use as substitutes for milk or cream in beverage coffee—0.4% max alone, or with polysorbate 60 and/or sorbitan monostearate. Product specifications apply
			FDA §121.1099	Defoaming agent
			FDA §121.2541	Emulsifier and/or surface-active agent
Polysorbate 80	Polyoxyethylene (20) sorbitan monooleate	2917	FDA §121.1009	1. Emulsifier in ice cream, frozen custard, ice milk, fruit sherbet, and nonstandardized frozen desserts, alone or with polyoxyethylene (20) sorbitan tristearate, total additive(s) 0.10% max 2. Emulsifier in shortenings and edible oils for use in nonstandardized baked goods, baking mixes, icings, fillings, and toppings and in the frying of foods, alone, 1.0% max; with polysorbate 60, total additives, 1.0% max 3. In yeast-defoamer formulations, 4% max of the finished yeast defoamer, or of the yeast from such use 4. Solubilizing and dispersing agent in pickles and pickle products, up to 0.05% 5. Solubilizing and dispersing agent for fat-soluble vitamins in vitamin and vitamin-mineral preparations, intake from recommended daily dose 300 mg max; in vitamin A preparations

REGULATORY STATUS OF DIRECT FOOD ADDITIVES (continued)

Material	Alternate Name [Source]	FEMA No.	Regulation	Limitations
Polysorbate 80 (continued)				containing more than 30,000 units per dose, daily intake 500 mg max 6. Surfactant in production of coarse crystal sodium chloride, up to 10 ppm in finished salt 7. Emulsifier for edible fats and oils in special dietary foods, daily ingestion not to exceed 360 mg 8. Emulsifier in whipped, vegetable oil topping, alone or with sorbitan monostearate, polysorbate 60, and/ or polyoxyethylene (20) sorbitan tristearate; total additive(s), 0.4% 9. Wetting agent in scald water for poultry defeathering, followed by potable water rinse; 0.0175% max in scald water 10. Solubilizing and dispersing agent for dill oil in canned, spiced green beans, up to 0.003% 11. Dispersing agent in gelatin desserts and in gelatin dessert mixes, up to 0.082% Product specifications apply
			MID	To emulsify shortenings sold in units not exceeding 6 pounds or 1 gallon fluid content, 1% alone; with polysorbate 60 or sorbitan monostearate, total additives 1% max
Polyvinyl acetate			FDA §8.390	Diluent in inks for marking gum, confectionery, and food supplements in tablet form; minimum molecular weight of 2,000
Polyvinyl alcohol	Polyviol		FDA §8.390	Diluent in mixtures for coloring shell eggs; no penetration into egg
Polyvinylpyrrolidone	Kollidon		FDA §121.1110	Clarifying agent in beverages and vinegar, followed by removal with filtration.

REGULATORY STATUS OF DIRECT FOOD ADDITIVES (continued)

Material	Alternate Name [Source]	FEMA No.	Regulation	Limitations
Polyvinylpyrrolidone (continued)				Product specifications apply
			FDA §121.1139	1. 10 ppm max residue in beer; clarifying agent 2. 40 ppm max residue in vinegar; clarifying agent 3. 60 ppm max residue in wine; clarifying agent 4. Tableting adjuvant in flavor concentrates, nonnutritive sweeteners, and vitamin and mineral concentrates 5. Stabilizer, bodying agent, and dispersant in nonnutritive sweeteners in concentrated liquid form and in vitamin and mineral concentrates in liquid form. Product specifications apply
			FDA §8.300 and §8.6200	Diluent in color additive mixtures in or on food-tablet coatings at 0.1% of finished food; inks for marking gum, confectionery, and food supplements in tablet form; inks for marking fruits and vegetables; inks for branding pharmaceutical forms; and ingested drugs. Specifications in §121.1139 apply
Pomegranate, bark, extract	[*Punica granatum* L.]	2918	FDA §121.101	GRAS, natural flavor extractive
Poplar buds	Aspen [*Populus balsamifera* L. (*P. tacamahacca* Mill.), *P. candicans* Alit., or *P. nigra* L.]		FDA §121.1163	Natural flavor in alcoholic beverages only
Poppy seed	[*Papaver somniferum* L.; *P. sativum* Fuchs]	2919	FDA §121.101	GRAS, natural flavor
Potassium acetate		2920	FDA §121.1164	Synthetic flavor
Potassium acid tartrate	*see* Cream of tartar			
Potassium alginate			FDA §121.101	GRAS, stabilizer

REGULATORY STATUS OF DIRECT FOOD ADDITIVES (continued)

Material	Alternate Name [Source]	FEMA No.	Regulation	Limitations
Potassium alum			FDA §19	In specified cheeses, up to 6 times the weight of benzoyl peroxide used
			FDA §15	Bleaching agent in flour, not more than 6 parts by weight, alone or with calcium sulfate, magnesium carbonate, sodium aluminum sulfate, dicalcium phosphate, tricalcium phosphate, starch and/or calcium carbonate, and with 1 part by weight of benzoyl peroxide
Potassium bicarbonate	Potassium acid carbonate		FDA §121.101	GRAS, miscellaneous and/or general-purpose food additive
Potassium bisulfite	Potassium hydrosulfide		FDA §121.101	GRAS, chemical preservative; not for use in meats or in foods recognized as source of vitamin B_1
Potassium bromate			FDA §15.20 and §15.30	50 ppm max in unfinished bromated flour and enriched bromated flour
			FDA §15.90	75 ppm max in whole-wheat flour
			FDA §17	0.0075% max in bakery products, alone or with potassium iodate, calcium iodate and/or calcium peroxide
			FDA §121.1020	In fermented malt beverages and distilled spirits made from potassium-bromate-treated malt
			FDA §121.1194	In malting of barley, 75 ppm max in malt (calculated as Br); treated malt used only in production of fermented malt beverages or distilled spirits; total bromide residue in fermented malt beverages, 25 ppm max (calculated as Br). Product specifications apply

REGULATORY STATUS OF DIRECT FOOD ADDITIVES (continued)

Material	Alternate Name [Source]	FEMA No.	Regulation	Limitations
Potassium carbonate	Salt of tartar		FDA §121.101	GRAS, miscellaneous and/or general-purpose food additive
			FDA §8.303	In production of caramel; includes bicarbonate
Potassium caseinate			FDA §20	In frozen desserts
Potassium chloride	Kaleorid		FDA §29.4 and §29.5	In artificially sweetened fruit jellies and preserves
			FDA §121.101	GRAS, nutrient and/or dietary supplement
Potassium citrate			FDA §121.101	GRAS, sequestrant
			FDA §121.101	GRAS, miscellaneous and/or general-purpose food additive
Potassium gibberellate			FDA §121.1010	In malting of barley; total with gibberellic acid, 2 ppm max (expressed as gibberellic acid); treated malt to be used only in production of fermented malt beverages or distilled spirits; finished distilled spirits contain none and finished malt beverage contains 5 ppm max of gibberellic acid. Product specifications apply
Potassium glycerophosphate			FDA §121.101	GRAS, nutrient and/or dietary supplement
Potassium hydroxide	Potassium hydrate		FDA §121.101	GRAS, miscellaneous and/or general-purpose food additive
			FDA §8.303	In production of caramel
Potassium iodate			FDA §17	In bakery products, 0.0075% max by weight of flour, alone or with potassium bromate, calcium iodate, and/or calcium peroxide
Potassium iodide			FDA §121.101	GRAS, nutrient and/or dietary supplement in table salt as source of dietary iodine, 0.01% max

REGULATORY STATUS OF DIRECT FOOD ADDITIVES (continued)

Material	Alternate Name [Source]	FEMA No.	Regulation	Limitations
Potassium iodide (continued)			FDA §121.1073	Nutritional supplement in mineral and vitamin-mineral preparations; max daily intake of iodine 0.15 mg. Product specifications apply
Potassium metabisulfite	Potassium pyrosulfite		FDA §121.101	GRAS, chemical preservative; not for use in meats or in foods recognized as source of vitamin B_1
Potassium nitrate			FDA §121.1132	Curing agent in processing of cod roe, 200 ppm max
			MID	Source of nitrite in cured products; 7 lb/100 gal pickle; $3\frac{1}{2}$ oz/100 lb meat (dry cure); $2\frac{3}{4}$ oz/100 lb chopped meat
Potassium nitrite			MID	To fix color in cured products; 2 lb/100 gal pickle at 10% pump level; 1 oz/100 lb meat (dry cure); $\frac{1}{4}$ oz/100 lb chopped meat and/or meat by-product; 200 ppm max total nitrite in finished product. Product specifications apply
Potassium phosphate			FDA §8.303	In production of caramel (including dibasic phosphate and monobasic phosphate)
Potassium salts of fatty acids			FDA §121.1071	Binder, emulsifier, and anticaking agent in food; alone or with other salts of fatty acids; fatty acids must comply with §121.1070
Potassium sorbate	Sorbic acid potassium salt	2921	FDA §121.101	GRAS, chemical preservative
Potassium sulfate	Sal polychrestum		FDA §121.101	GRAS, miscellaneous and/or general-purpose food additive
			FDA §8.303	In production of caramel
Potassium sulfite			FDA §8.303	In production of caramel

REGULATORY STATUS OF DIRECT FOOD ADDITIVES (continued)

Material	Alternate Name [Source]	FEMA No.	Regulation	Limitations
Pot marigold	see Marigold, pot			
Pot marjoram	see Marjoram, pot			
Prickly ash bark	[Xanthoxylum americanum Mill.; X. clavaherculis L.]		FDA §121.101	GRAS, natural flavor extractive
Proline (L- and DL-forms)			FDA §121.101	GRAS, nutrient and/or dietary supplement, as free acid, hydrochloride salt, hydrate, or anhydrous form
L-Proline		3319		see Food Technol., May 1972
Propane	Dimethylmethane		FDA §121.101	GRAS, miscellaneous and/or general-purpose food additive
			FDA §121.1065	Propellant and aerating agent for foamed or sprayed food products with octafluorocyclobutane, except standardized foods that do not provide for such use
			FDA §121.1181	Propellant and aerating agent for foamed or sprayed food products with chloropentafluoroethane
1-Propanol	see Propyl alcohol			
2-Propene-1-thiol	see Allyl mercaptan			
p-Propenyl anisole	Anethole		FDA §121.101	GRAS, synthetic flavor and adjuvant
Propenyl guaethol	6-Ethoxy-m-anol	2922	FDA §121.1164	Synthetic flavor
Propenyl propyl disulfide		3227		
4-Propenyl veratrole	see Isoeugenyl methyl ether			
Propionaldehyde	Propanol	2923	FDA §121.1164	Synthetic flavor
Propionic acid	Propanoic acid	2924	FDA §121.101	GRAS, chemical preservative
Propionic acid, triglyceride	see Glyceryl tripropanoate			

REGULATORY STATUS OF DIRECT FOOD ADDITIVES (continued)

Material	Alternate Name [Source]	FEMA No.	Regulation	Limitations
Propyl acetate	Acetic acid *n*-propyl ester	2925	FDA §121.1164	Synthetic flavor
***n*-Propyl alcohol**	1-Propanol	2928	FDA §121.1164	Synthetic flavor
***p*-Propyl anisole**	Dihydroanethole	2930	FDA §121.1164	Synthetic flavor
Propyl benzoate		2931	FDA §121.1164	Synthetic flavor
Propyl butyrate		2934	FDA §121.1164	Synthetic flavor
Propyl cinnamate		2938	FDA §121.1164	Synthetic flavor
Propyl disulfide		3228	FDA §121.1164	Synthetic flavor
Propylene glycol	1,2-Propanediol	2940	FDA §121.101	GRAS, emulsifying agent and miscellaneous and/or general-purpose food additive
Propylene glycol alginate		2941	FDA §25.2 and §25.3	In French and salad dressings
			FDA §121.1015	1. Stabilizer in ice cream, frozen custard, ice milk, fruit sherbet, and water ices, up to 0.5% 2. Emulsifier, stabilizer, or thickener in foods, except for standardized foods that do not provide for such use. Product specifications apply
			FDA §121.1099	Defoaming agent
Propylene glycol, alkaline			FDA §8.305	Extractant for annatto
Propylene glycol mono- and diesters of fats and fatty acids			FDA §121.1113	In foods; fats and fatty acids conforming with §121.1070
			MID	To emulsify shortenings sold in units not exceeding 6 lb or 1 gal fluid content
Propylene glycol octa- decanoate	*see* Propylene glycol stearate			
Propylene glycol stea- rate	Propylene glycol octade- canoate	2942		

REGULATORY STATUS OF DIRECT FOOD ADDITIVES (continued)

Material	Alternate Name [Source]	FEMA No.	Regulation	Limitations
Propyl formate		2943	FDA §121.1164	Synthetic flavor
Propyl 2-furanacrylate	Propyl 3-(2-furyl) acrylate	2945	FDA §121.1164	Synthetic flavor
Propyl 2-furoate		2946		
Propyl 3-(2-furyl) acrylate	see Propyl 2-furanacrylate			
Propyl gallate	Tenox PG®	2947	FDA §121.101	GRAS, chemical preservative; max total antioxidants 0.02% of fat or oil content, including essential (volatile) oil content of food
			FDA §121.1059	Antioxidant in chewing-gum base, alone or with butylated hydroxyanisole and/or butylated hydroxytoluene, up to 0.1% total antioxidant content
			MID	To retard rancidity in rendered animal fat or combination of such fat and vegetable fat up to 0.01%, or in combination of propyl gallate, BHA, BHT, glycine, and NDGA up to 0.02%
			MID	Antioxidant to retard rancidity in frozen fresh pork sausages and freeze-dried meats, up to 0.01% based on fat content; can be used in combination with BHA, BHT, and/or NDGA, up to 0.02% in frozen fresh pork sausages and up to 0.01% in freeze-dried meats
Propyl heptanoate		2948	FDA §121.1164	Synthetic flavor
Propyl hexanoate		2949	FDA §121.1164	Synthetic flavor
Propyl p-hydroxybenzoate	see Propylparaben			
3-Propylidenephthalide		2952	FDA §121.1164	Synthetic flavor and adjuvant

REGULATORY STATUS OF DIRECT FOOD ADDITIVES (continued)

Material	Alternate Name [Source]	FEMA No.	Regulation	Limitations
Propyl isobutyrate		2936	FDA §121.1164	Synthetic flavor
Propyl isovalerate		2960	FDA §121.1164	Synthetic flavor
Propyl mercaptan			FDA §121.1164	Synthetic flavor
Propyl methyl trisulfide	see Methyl propyl trisulfide			
Propylparaben	Propyl p-hydroxybenzoate	2951	FDA §121.101	GRAS, chemical preservative, up to 0.1%
			FDA §121.1164	Synthetic flavor
α-Propylphenethyl alcohol	1-Phenyl-2-pentanol	2953	FDA §121.1164	Synthetic flavor
Propyl phenylacetate		2955	FDA §121.1164	Synthetic flavor
Propyl propionate		2958	FDA §121.1164	Synthetic flavor
Propyl trisulfide	see Dipropyl trisulfide			
Protein, hydrolyzed	see Hydrolyzed protein			
Protein, plant, hydrolyzed	see Hydrolyzed plant protein			
Proteolytic enzymes	Includes Aspergillus oryzae; A. flavusoryzae group; promelin; ficin; papain; see also specific enzyme		MID	With water, salt, and monosodium glutamate to soften tissues of beef cuts; tissue weight gain, 3% max
Psyllium seed, husk	[Plantago psyllium L.; P. arenaria Waldst. & Kit.]		FDA §20	In frozen desserts
Pteroylglutamic acid	Folic acid		FDA §121.1134	In dietary supplements; max daily ingestion, 0.10 mg
Pulegone	p-Menth-4(8)-en-3-one	2963	FDA §121.1164	Synthetic flavor
Pyrazine ethanethiol	Pyrazinyl ethanethiol	3230		10 ppm max in finished food
Pyrazinemethanethiol	see 2-Mercaptomethyl-pyrazine			
Pyrazinyl ethanethiol	see Pyrazine ethanethiol			

REGULATORY STATUS OF DIRECT FOOD ADDITIVES (continued)

Material	Alternate Name [Source]	FEMA No.	Regulation	Limitations
Pyrazinyl methyl methyl sulfide	see Pyrazinyl methyl sulfide			
Pyrazinyl methyl sulfide	Pyrazinyl methyl methyl sulfide	3231		10 ppm max in finished food
Pyridine		2966	FDA §121.1164	Synthetic flavor
2-Pyridinemethanethiol	2-Pyridylmethanethiol	3232		
Pyridoxine hydrochloride	Vitamin B₆ hydrochloride		FDA §121.101	GRAS, nutrient and/or dietary supplement
2-Pyridylmethanethiol	see 2-Pyridinemethanethiol			
Pyroligneous acid	Pyroligneous vinegar	2967	FDA §3.201	Flavoring
Pyroligneous acid extract		2968	FDA §121.1164	Synthetic flavor
2-Pyrrolidinecarboxylic acid	see L-Proline			
Pyruvaldehyde	Pyruvic aldehyde	2969	FDA §121.1164	Synthetic flavor
Pyruvic acid	Pyroracemic acid	2970	FDA §121.1164	Synthetic flavor
Quassia	[Picrasma excelsa (Sw.) Planch.; Quassia amara L.]	2971	FDA §121.1163	Natural flavor
Quebracho bark	[Aspidosperma quebrachoblanco Schlecht, Schinopsis lorentzii (Griseb.) Engl., or Quebrachia lorentzii Griseb.]	2972	FDA §121.1163	Natural flavoring substances and natural substances used in conjunction with flavors
Quickgrass	see Dog grass			
Quillaja	Soapbark [Quillaja saponaria Mol.]	2973	FDA §121.1163	Natural flavor
Quince seed, extract	[Cydonia oblonga Miller]	2974	FDA §121.101	GRAS, solvent-free natural extractive used in conjunction with natural flavoring

REGULATORY STATUS OF DIRECT FOOD ADDITIVES (continued)

Material	Alternate Name [Source]	FEMA No.	Regulation	Limitations
Quinine	[*Cinchona officinalis* L. (*C. ledgeriana* Moens)]		FDA §121.1081	Hydrochloride or sulfate of quinine, as flavor in carbonated beverages, up to 83 ppm as quinine
Quinine bisulfate		2975		
Quinine hydrochloride	Quinine chloride	2976		
Quinine sulfate		2977		
Red saunders	Red sandalwood [*Pterocarpus santalinus* L. f.]		FDA §121.1163	Natural flavor in alcoholic beverages only
Rennin	Rennet		FDA §121.101	GRAS, miscellaneous and/or general-purpose food additive
Resin guaiac			FDA	0.1% max of fat content of food
			MID	To retard rancidity in rendered animal fat or combination of such fat and vegetable fat, up to 0.10%
Resin, natural			FDA §121.1059	Terpene resin in chewing-gum base, consisting of polymers of α-pinene. Product specifications apply
Rhatany, extract	[*Krameria triandra* Ruiz & Pav.; *K. argentea* Mart.]	2979	FDA §121.1163	Natural flavor
Rhizopus oryzae carbohydrase	[*Rhizopus oryzae*]		FDA §121.1165	In production of dextrose from starch. Product specifications apply
Rhodinal	*see* Citronellal			
Rhodinol	*l*-Citronellol; 3,7-dimethyl-7-octen-1-ol	2980	FDA §121.1164	Synthetic flavor
Rhodinyl acetate		2981	FDA §121.1164	Synthetic flavor
Rhodinyl butyrate		2982	FDA §121.1164	Synthetic flavor
Rhodinyl formate		2984	FDA §121.1164	Synthetic flavor
Rhodinyl isobutyrate		2983	FDA §121.1164	Synthetic flavor
Rhodinyl isovalerate		2987	FDA §121.1164	Synthetic flavor

REGULATORY STATUS OF DIRECT FOOD ADDITIVES (continued)

Material	Alternate Name [Source]	FEMA No.	Regulation	Limitations
Rhodinyl phenylacetate		2985	FDA §121.1164	Synthetic flavor
Rhodinyl propionate		2986	FDA §121.1164	Synthetic flavor
Rhubarb, garden root	[*Rheum rhaponticum* L.]		FDA §121.1163	Natural flavor in alcoholic beverages only
Rhubarb, root	[*Rheum officinale* Baill.; *R. palmatum* L. or other *Rheum* species (except *R. rhaponticum* L.) or hybrids of *Rheum* grown in China]		FDA §121.1163	Natural flavor
Riboflavin	Vitamin B₂		FDA §121.101	GRAS, nutrient and/or dietary supplement
			FDA §8.323	Coloring foods generally; not to be used in standardized foods unless standards provide for use of added color; specifications in *Food Chemicals Codex* apply
Riboflavin-5-phosphate			FDA §121.101	GRAS, nutrient and/or dietary supplement
Rice, bran, wax			FDA §121.1098	50 ppm max in candy; coating 50 ppm max in fresh fruits and fresh vegetables; coating 2.5% max in chewing gum; plasticizing material. Product specifications apply
			FDA §121.1059	Plasticizer (softener) in chewing-gum base
Rose	Otto of rose, attar of rose [*Rosa alba* L.; *R. centifolia* L.; *R. damascena* Mill.; *R. gallica* L.; and varieties of these species]		FDA §121.101	GRAS, natural flavor extractive
Rose, absolute	[*Rosa alba* L.; *R. centifolia* L.; *R. damascena* Mill.; *R. gallica* L.; and varieties of these species]	2988	FDA §121.101	GRAS, natural flavor extractive
Rose, buds, flowers, fruits (hips), and leaves	[*Rosa* species]		FDA §121.101	GRAS, natural flavor extractive

REGULATORY STATUS OF DIRECT FOOD ADDITIVES (continued)

Material	Alternate Name [Source]	FEMA No.	Regulation	Limitations
Rose geranium	see Geranium, rose			
Roselle	[Hibiscus sabdariffa L.]		FDA §121.1163	Natural flavor in alcoholic beverages only
Rosemary	[Rosmarinus officinalis L.]	2991	FDA §121.101	GRAS, natural flavor and natural flavor extractive
Rose oxide	see Tetrahydro-4-methyl-2-(2-methylpropen-1-yl) pyran			
Rose water, stronger	[Rosa centifolia L.]	2993		
Rosidinha	Rosadinha [Micropholis (also known as Sideroxylon) species]		FDA §121.1059	Natural (coagulated or concentrated latex) masticatory substance in chewing-gum base
Rosin	Colophony [Pinus palustris Mill. and other Pinus species]		FDA §121.1163	Natural flavor in alcoholic beverages only
Rosin and rosin derivatives			FDA §8.390	Diluent in inks for marking fruits and vegetables; inks for marking gum, confectionery, and food supplements in tablet form; and mixtures for coloring shell eggs (no penetration through eggshell into egg) Specifications in §121.1059 apply
Rue	[Ruta graveolens L.]	2994	FDA §121.101	GRAS, natural flavor and natural flavor extractive
Rum ether	Ethyl oxyhydrate	2996	FDA §121.1164	Synthetic flavor
Sabinenehydrate	see 4-Thujanol			
Saccharin	2,3-Dihydro-3-oxobenzisosulfonazole; saccharin insoluble		FDA §121.101	GRAS, nonnutritive sweetener
			MID	To sweeten bacon, at 0.01%
Saccharin, sodium salt	Saccharine, soluble	2997	FDA	GRAS, in dietary supplements
Saffron	[Crocus sativus L.]	2998	FDA §121.101	GRAS, natural flavor and natural flavor extractive

REGULATORY STATUS OF DIRECT FOOD ADDITIVES (continued)

Material	Alternate Name [Source]	FEMA No.	Regulation	Limitations
Saffron (continued)			FDA §8.11	In foods, except in standardized foods where standards do not authorize its use
			MID	To color sausage casings, oleomargarine, or shortening, and in marking or branding ink for products; may be mixed with approved synthetic dyes or harmless inert material, such as common salt or sugar
Sage	[*Salvia officinalis* L.]	3000	FDA §121.101	GRAS, natural flavor and natural flavor extractive
Sage, Greek	[*Salvia triloba* L.]		FDA §121.101	GRAS, natural flavor and natural flavor extractive
Sage, Spanish	[*Salvia lavendulaefolia* Vahl.]	3003	FDA §121.101	GRAS, natural flavor extractive
St. John's bread	*see* Carob bean			
Salicylaldehyde	*o*-Hydroxybenzaldehyde	3004	FDA §121.1164	Synthetic flavor
Salts of fatty acids			FDA §121.1071	Binder, emuslifier, and anticaking agents. Product specifications apply
			FDA §8.201	Diluent in aqueous suspension of Citrus Red No. 2; conforming with §121.1071
Sandalwood, white	Yellow or East Indian sandalwood [*Santalum album* L.]	3005	FDA §121.1163	Natural flavoring substances and natural substances used in conjunction with flavors
Sandarac	[*Tetraclinis articulata* (Vahl.) Mast.]		FDA §121.1163	Natural flavor in alcoholic beverages only
Santalol, α and β	Argeol	3006	FDA §121.1164	Synthetic flavor
Santalyl acetate		3007	FDA §121.1164	Synthetic flavor
Santalyl phenylacetate		3008	FDA §121.1164	Synthetic flavor
Saponin				

REGULATORY STATUS OF DIRECT FOOD ADDITIVES (continued)

Material	Alternate Name [Source]	FEMA No.	Regulation	Limitations
Sarsaparilla, extract	[*Smilax aristolochiae-folia* Mill. (Mexican sarsaparilla); *S. regelii* Killip & Morton (Honduras sarsaparilla); *S. febrifuga* Kunth (Ecuadorean sarsaparilla); or undetermined species of *Smilax* (Ecuadorean or Central American sarsaparilla)]	3009	FDA §121.1163	Natural flavor
Sassafras, bark, extract	[*Sassafras albidum* (Nuttall) Nees (Fam. *Lauraceae*)]	3010	FDA §121.1097	Safrole-free extract as flavoring in food. Product specifications apply
Sassafras, leaf	[*Sassafras albidum* (Nutt.) Nees]	3011	FDA §121.1163	Natural flavor; safrole-free
Savory, summer	[*Satureia hortensis* L. (*Satureia*)]	3012	FDA §121.101	GRAS, natural flavor and natural flavor extractive
Savory, winter	[*Satureia montana* L. (*Satureia*)]	3015	FDA §121.101	GRAS, natural flavor and natural flavor extractive
Schinus molle, oil	[*Schinus molle* L.]	3018	FDA §121.101	GRAS, natural flavor extractive
Senecioic acid	*see* 3-Methylcrotonic acid			
Senna, Alexandria	[*Senna acutifolia* Nect.; also called *Cassia acutifolia* Delile]		FDA §121.1163	Natural flavoring substances and natural substances used in conjunction with flavors
Serine (L- and DL-forms)			FDA §121.101	GRAS, nutrient and/or dietary supplement, as free acid, hydrochloride salt, hydrate, or anhydrous form
Serpentaria	Virginia snakeroot [*Aristolochia serpentaria* L.]		FDA §121.1163	In alcoholic beverages only
Sesame	[*Sesamum indicum* L.]		FDA §121.101	GRAS, natural flavor
Shellac, purified			FDA §8.309 and §8.6200	Diluent in inks for marking gum, confectionery, and food supplements in tablet form; inks for branding pharmaceutical forms; food-grade

REGULATORY STATUS OF DIRECT FOOD ADDITIVES (continued)

Material	Alternate Name [Source]	FEMA No.	Regulation	Limitations
Silicon dioxide	Silica		FDA §121.229	Anticaking agent and/or grinding aid in feed and feed components: 1. 2% max in BHT 2. 1% max in methionine hydroxy analog and its calcium salts 3. 0.8% max in piperazine or piperazine salts 4. 1% max in sodium propionate 5. 1% max in urea 6. 3% max in vitamins Product specifications apply
			FDA §121.1058	1. Anticaking agent, 2% max; not to be used in infant foods, except when present as anticaking agent at 2% max in salt and salt substitutes used as components of these foods 2. Stabilizer in production of beer, to be removed from the beer by filtration prior to final processing 3. Adsorbent for *dl*-α-tocopheryl acetate and pantothenol in foods for special dietary use Product specifications apply
			FDA §8.390	Diluent in inks for marking fruits and vegetables, up to 2% of the ink solids; specifications in §121.1058 apply
Simaruba bark	[*Simaruba amara* Aubl.]		FDA §121.1163	Natural flavor in alcoholic beverages only
α-Sinensal	*see* 2,6-Dimethyl-10-methylene-2,6,11-dodecatrienal			
Skatole	3-Methylindole	3019	FDA §121.1164	Synthetic flavor
Sloe, berries	Blackthorn [*Prunus spinosa* L.]	3020	FDA §121.101	GRAS, natural flavor extractive
Smoke flavoring			MID	To flavor various meat products

Material	Alternate Name [Source]	FEMA No.	Regulation	Limitations
Snakeroot, Canadian	Wild ginger [*Asarum canadense* L.]	3023	FDA §121.1163	Natural flavor
Snakeroot, Virginia	*see* Serpentaria			
Sodium acetate		3024	FDA §121.101	GRAS, miscellaneous and/or general-purpose food additive
Sodium acid phosphate			FDA §121.101	GRAS, sequestrant
Sodium acid pyro-phosphate	Disodium dihydrogen pyrophosphate		FDA §121.101	GRAS, miscellaneous and/or general-purpose food additive
			MID	To decrease amount of cooked-out juices in cured hams, pork shoulder picnics and loins, canned hams and pork shoulder picnics, chopped ham, and bacon; 5.0% of phosphate in pickle at 10% pump level; 0.5% of phosphate in product (only clear solution may be injected into product)
Sodium alginate	*see* Alginates—sodium, calcium, and ammonium salts			
Sodium aluminosilicate	Sodium silicoaluminate		FDA §121.101	GRAS, anticaking agent, up to 2%
Sodium aluminum phosphate			FDA §121.101	GRAS, miscellaneous and/or general-purpose food additive
Sodium aluminum sulfate			FDA §15	Bleaching agent in flour, at not more than 6 parts by weight, alone or with potassium alum, calcium sulfate, magnesium carbonate, dicalcium phosphate, tricalcium phosphate, starch, and/or calcium carbonate, with 1 part by weight benzoyl peroxide
Sodium ascorbate	Ascorbic acid sodium derivative		FDA §121.101	GRAS, chemical preservative

REGULATORY STATUS OF DIRECT FOOD ADDITIVES (continued)

Material	Alternate Name [Source]	FEMA No.	Regulation	Limitations
Sodium ascorbate (continued)			MID	To accelerate color fixing in cured pork and beef cuts and in cured, comminuted meat products; 87.5 oz/100 gal pickle at 10% pump level; ⅞ oz/100 lb of meat or meat by-product; no significant amount of moisture shall be added to product
Sodium benzoate		3025	FDA §121.101	GRAS, chemical preservative up to 0.1%
			MID	To retard flavor reversion of oleomargarine, at 0.1%
Sodium bicarbonate	Sodium hydrogen carbonate		FDA §121.101	GRAS, miscellaneous and/or general-purpose food additive
			MID	To separate fatty acids and glycerol in rendered fats; must be eliminated during manufacturing process
			MID	To neutralize excess acidity and clean vegetables in rendered fats, soups, and curing pickle
Sodium bisulfite	Sodium acid sulfite		FDA §121.101	GRAS, chemical preservative; not for use in meats or in foods recognized as source of vitamin B_1
Sodium m-bisulfite			FDA §121.101	GRAS, chemical preservative; not for use in meats or in foods recognized as source of vitamin B_1
Sodium calcium aluminosilicate, hydrated	Sodium calcium silicoaluminate		FDA §121.101	GRAS, anticaking agent
Sodium carbonate			FDA §121.101	GRAS, miscellaneous and/or general-purpose food additive
			FDA §8.303	In caramel (including bicarbonate)
			MID	To refine rendered fats; must be eliminated during manufacturing process

REGULATORY STATUS OF DIRECT FOOD ADDITIVES (continued)

Material	Alternate Name [Source]	FEMA No.	Regulation	Limitations
Sodium carbonate with sodium mono- and di-methyl naphthalene sulfonates			FDA §121.1198	In potable water systems to reduce hardness and aid in sedimentation and coagulations by raising the pH for efficient utilization of other coagulation materials; sodium mono- and dimethyl sulfonates are used in crystallization of sodium carbonate at 250 ppm max of sodium carbonate. Product specifications apply
Sodium carboxy-methylcellulose	Carboxymethylcellulose; cellulose gum; CMC	2239	FDA §121.101	GRAS, miscellaneous and/or general-purpose food additive. Product specifications apply
			FDA §19	In specified cheese products, up to 0.8% by weight
			MID	To extend and stabilize baked pies
Sodium caseinate			FDA §121.101	GRAS, miscellaneous and/or general-purpose food additive
			MID	To bind and extend imitation sausage, nonspecific loaves, soups, and stews
Sodium chloride	Table salt		FDA—GRAS	
Sodium chlorite			FDA §121.1031	Modifier for food starch, up to 0.5%
Sodium citrate	Citratin	3026	FDA §18.520	In evaporated milk, alone or with disodium phosphate, up to 0.1% by weight of finished product
			FDA §121.101	GRAS, sequestrant and miscellaneous and/or general-purpose food additive
			MID	To prevent clotting of fresh beef blood at 0.2% with or without water. Product specifications apply
			MID	To accelerate color fixing in cured pork and beef cuts

REGULATORY STATUS OF DIRECT FOOD ADDITIVES (continued)

Material	Alternate Name [Source]	FEMA No.	Regulation	Limitations
Sodium citrate (continued)				and in cured, comminuted meat food products: in cured products to replace up to 50% of ascorbic acid, erythorbic acid, sodium ascorbate, or sodium erythorbate used
Sodium dehydroacetate			FDA §121.1089	Preservative for cut or peeled squash; 65 ppm max (expressed as dehydroacetic acid) remaining in or on the prepared squash Product specifications apply
Sodium diacetate	Sodium acid acetate		FDA §17	In bakery products, up to 0.4% of flour
			FDA §121.101	GRAS, sequestrant
Sodium dihydrogen phosphate			FDA—GRAS	
Sodium dimethyl naphthalene sulfonates			FDA §121.1198	In crystallization of sodium carbonate, 250 ppm max of sodium carbonate; sodium carbonate so produced to be used in potable water systems. Product specifications apply
Sodium erythorbate			MID	To accelerate color fixing in cured pork and beef cuts and in cured, comminuted meat food products; 87.5 oz/ 100 gal pickle at 10% pump level; $\frac{7}{8}$ oz/100 lb of meat or meat by-product; no significant amount of moisture to be added to product
Sodium gluconate	Gluconic acid sodium salt		FDA §121.101	GRAS, sequestrant
Sodium hexametaphosphate		3027	FDA §121.101	GRAS, sequestrant
			MID	To decrease amount of cooked-out juices in cured hams, pork shoulder picnics, and loins, canned hams and pork shoulder picnics, chopped ham, and

REGULATORY STATUS OF DIRECT FOOD ADDITIVES (continued)

Material	Alternate Name [Source]	FEMA No.	Regulation	Limitations
Sodium hexameta- phosphate (continued)				bacon; 5.0% of phosphate in pickle at 10% pump level; 0.5% of phosphate in product (only clear solution may be injected into product)
Sodium hydroxide	Caustic soda		FDA §121.101	GRAS, miscellaneous and/ or general-purpose food additive
			FDA §121.1031	Modifier for food starch, up to 1.0%
			FDA §8.303	In production of caramel
Sodium lauryl sulfate	Dodecyl sodium sulfate		FDA §121.1012	1. Emulsifier in or with egg whites—1000 ppm max in egg white solids, 125 ppm max in frozen egg whites, 125 ppm max in liquid egg whites 2. Whipping agent, 5000 ppm max by weight of gelatin used in preparation of marshmallows Product specifications apply
Sodium metaphos- phate	see also Sodium hexa- metaphosphate		FDA §121.101	GRAS, sequestrant
			FDA §19	Emulsifier, up to 3% by weight of specified cheeses
Sodium mono- and di- methyl naphthalene sulfonates			FDA §121.1198	In crystallization of sodium carbonate, 250 ppm max; sodium carbonate so produced to be used in potable water systems. Product specifications apply
Sodium nitrate	Chile saltpeter		FDA §121.1063	1. Preservative and color fixative, with or without sodium nitrite: 500 ppm max in smoked, cured sablefish, smoked, cured salmon, and smoked, cured shad; sodium nitrite in finished product, 200 ppm max 2. In meat-curing preparations for home-curing of meat and meat products (in-

REGULATORY STATUS OF DIRECT FOOD ADDITIVES (continued)

Material	Alternate Name [Source]	FEMA No.	Regulation	Limitations
Sodium nitrate (continued)				cluding poultry and wild game); directions must limit amount of sodium nitrate to 500 ppm max and sodium nitrite to 200 ppm max in finished meat product. Product specifications apply
			MID	Source of nitrite in cured products; 7 lb/100 gal pickle; 3.5 oz/100 lb meat (dry cure); 2.75 oz/10 lb chopped meat
Sodium nitrite	Erinitrit		FDA §121.223	20 ppm max alone as preservative and color fixative in canned pet food containing fish, meat, and fish and meat by-products
			FDA §121.1064	1. Preservative and color fixative: 10 ppm max in smoked, cured tuna-fish products; 200 ppm max in smoked, cured sablefish, smoked, cured salmon, and smoked, cured shad; may be used with sodium nitrate at 500 ppm max 2. In meat-curing preparations for home-curing of meat and meat products (including poultry and wild game), with sodium nitrate; directions must limit sodium nitrate to 500 ppm max and sodium nitrite to 200 ppm max Product specifications apply
			MID	To fix color in cured products; 2 lb/100 gal pickle at 10% pump level; 1 oz/100 lb meat (dry cure); $\frac{1}{4}$ oz/100 lb chopped meat and/or meat by-product; 200 ppm max of nitrite in finished product. Product specifications apply
Sodium pantothenate			FDA §121.101	GRAS, nutrient and/or dietary supplement

REGULATORY STATUS OF DIRECT FOOD ADDITIVES (continued)

Material	Alternate Name [Source]	FEMA No.	Regulation	Limitations
Sodium pectinate			FDA §121.101	GRAS, miscellaneous and/ or general-purpose food additive
Sodium phosphate, dibasic	*see* Disodium phosphate			
Sodium phosphate (dibasic and monobasic)			FDA §8.303	In production of caramel
Sodium phosphate (di-, mono-, and tribasic)			FDA §121.101	GRAS, nutrient and/or dietary supplement
			FDA §121.101	GRAS, sequestrant and miscellaneous and/or general-purpose food additive
Sodium potassium tartrate			FDA §121.101	GRAS, sequestrant and miscellaneous and/or general-purpose food additive
Sodium propionate	Propionic acid sodium salt		FDA §17	In bakery products, alone or with calcium propionate, up to 0.32% by weight of flour in bread, enriched bread, milk bread, and raisin bread, and up to 0.38% of flour used in whole-wheat bread
			FDA	In bread, up to 0.32% by weight of flour
			FDA §121.101	GRAS, chemical preservative
			MID	Alone or with calcium propionate, to retard mold growth in pizza crust, up to 0.32% by weight of flour. Product specifications apply
Sodium pyrophosphate			FDA §121.101	GRAS, sequestrant
			MID	To decrease amount of cooked-out juices in cured hams, pork shoulder picnics, and loins, canned hams and pork shoulder picnics, chopped ham, and bacon; 5.0% of phosphate in pickle at 10% pump level;

REGULATORY STATUS OF DIRECT FOOD ADDITIVES (continued)

Material	Alternate Name [Source]	FEMA No.	Regulation	Limitations
Sodium pyrophosphate (continued)				0.5% of phosphate in product (only clear solution may be injected into product)
Sodium saccharin			FDA §121.101	GRAS, nonnutritive sweetener
Sodium salts of fatty acids			FDA §121.1071	Binder, emulsifier and anti-caking agent in food. Product specifications apply
Sodium sesquicarbonate			FDA §121.101	GRAS, miscellaneous and/or general-purpose food additive
Sodium silicoaluminate, precipitated hydrated			FDA	2% max in salt; 2% max in seasoning; 1% max in sugar; 5% max in baking powder
Sodium sorbate			FDA §121.101	GRAS, chemical preservative
Sodium stearyl fumarate			FDA §121.1183	1. Dough conditioner in yeast-leavened products, 0.5% max by weight of flour 2. Conditioning agent in dehydrated potatoes, 1% max Product specifications apply
Sodium succinate	*see* Disodium succinate			
Sodium sulfate			FDA §121.1059	Miscellaneous food additive in chewing-gum base
			FDA §8.303	In production of caramel
Sodium sulfide	Sodium monosulfide		FDA §121.1059	Miscellaneous food additive in chewing-gum base, for use as reaction-control agent in synthetic polymer production
Sodium sulfite			FDA	200–300 ppm in molasses, dried fruits, and foods that are not good sources of vitamin B_1
			FDA §121.101	GRAS, chemical preservative; not for use in meats or in foods recognized as source of vitamin B_1
			FDA §8.303	In production of caramel

REGULATORY STATUS OF DIRECT FOOD ADDITIVES (continued)

Material	Alternate Name [Source]	FEMA No.	Regulation	Limitations
Sodium sulfoacetate derivative of mono- and diglycerides			MID	To flavor various meat products, at 0.5%
Sodium tartrate			FDA §19	Emulsifier, up to 3% by weight of specified cheeses
			FDA §121.101	GRAS, sequestrant
Sodium tetrapyrophosphate			FDA §121.101	GRAS, sequestrant
Sodium thiosulfate	Sodium hyposulfite		FDA §121.101	GRAS, sequestrant in salt, up to 0.1%
Sodium trimetaphosphate			FDA §121.1031	In food starch—modified, alone or as part of successive treatment with sodium tripolyphosphate and sodium trimetaphosphate; residual phosphate 0.4% max calculated as phosphorus
Sodium tripolyphosphate	Sodium triphosphate		FDA §121.101	GRAS, sequestrant and miscellaneous and/or general-purpose food additive
			FDA §121.1031	In food starch—modified, for successive treatment of starch with sodium tripolyphosphate and sodium trimetaphosphate; residual phosphate 0.4% max calculated as phosphorus
			MID	To decrease amount of cooked-out juices in cured hams, pork shoulder picnics, and loins, canned hams and pork shoulder picnics, chopped ham, and bacon; 5.0% of phosphate in pickle at 10% pump level; 0.5% of phosphate in product (only clear solution may be injected into product)
			FDA §8.201	Diluent for aqueous suspension of Citrus Red No. 2, up to 0.05%
Sorbic acid	2,4-Hexadienoic acid		FDA §121.101	GRAS, chemical preservative

REGULATORY STATUS OF DIRECT FOOD ADDITIVES (continued)

Material	Alternate Name [Source]	FEMA No.	Regulation	Limitations
Sorbitan monostearate		3028	FDA §121.1029	Emulsifier, alone or with polyoxyethylene (20) sorbitan monostearate: 1. In whipped vegetable oil topping, alone or with polysorbate 60, polysorbate 80, and/or polyoxyethylene (20) sorbitan tristearate; total additive(s), 0.40% max 2. In cakes and cake mixes, 0.61% max alone; 0.61% max in combination with polyoxyethylene (20) sorbitan tristearate (0.32% max) and/or polysorbate 60 (0.46% max); total additives 0.66% max 3. In nonstandardized confectionery coatings, alone 10,000 ppm max; 10,000 ppm max in combination with polyoxyethylene (20) sorbitan monostearate (5000 ppm max); total additive(s) 10,000 ppm max 4. In cake icings and cake fillings, 0.70% max alone; 0.70% max in combination with polyoxyethylene (20) sorbitan tristearate (0.32% max) and/or polysorbate 60 (0.46% max); total additives 1.0% max 5. In solid-state, vegetable-fat-water emulsions for use in beverage coffee as substitutes for milk or cream; 0.4% max alone or with polyoxyethylene (20) sorbitan tristearate and/or polysorbate 60 6. Rehydration aid in production of active dry yeast, alone at 1% max by weight of dry yeast Product specifications apply
			FDA §121.1099	Defoaming agent, conforming with §121.1029
			FDA §121.1164	Synthetic flavor

REGULATORY STATUS OF DIRECT FOOD ADDITIVES (continued)

Material	Alternate Name [Source]	FEMA No.	Regulation	Limitations
Sorbitan monostearate (continued)			MID	To emulsify shortenings sold in units not exceeding 6 lb or 1 gal fluid content; alone at 1%, or with poly- sorbate 60 or polysorbate 80, total additives 1% max
Sorbitol	D-Glucitol	3029	FDA §121.101	GRAS, nutrient and/or di- etary supplement in foods for special dietary use, up to 7%
			FDA §121.1053	Food additive
Soy protein, isolated	see Isolated soy protein			
Spanish "origanum"	see Thymus capitatus			
Spearmint	[Mentha spicata Houds or L.]	3030	FDA §121.101	GRAS, natural flavor and natural flavor extractive
Spike lavender	see Lavender, spike			
Spruce needles and twigs	[Picea glauca (Moench) Voss; or P. mariana (Mill.) BSP]		FDA §121.1163	Natural flavor
Stannous chloride	Tin chloride		FDA §121.101	GRAS, chemical preserva- tive, up to 0.0015% calcu- lated as tin
Star anise	see Anise, star			
Starch hydrolysates			FDA §8.303	In production of caramel
Starter distillate			MID	To flavor oleomargarine
Stearic acid	Octadecanoic acid	3035	FDA §121.1059	Plasticizing material (soft- ener) in chewing-gum base, complying with §121.1070
Stearyl citrate			FDA §121.101	GRAS, sequestrant, up to 0.15%
			MID	To protect flavor of oleo- margarine, at 0.15%
Stearyl-2-lactylic acid			MID	To emulsify shortening to be used for cake icings and fillings, at 3%

REGULATORY STATUS OF DIRECT FOOD ADDITIVES (continued)

Material	Alternate Name [Source]	FEMA No.	Regulation	Limitations
Stearyl monoglyceridyl citrate			FDA §121.1080	Emulsion stabilizer in or with shortenings containing emulsifiers. Product specifications apply
			MID	To emulsify shortenings
Storax	Styrax [*Liquidambar orientalis* Mill. or *L. styraciflua* L.]	3036	FDA §121.1163	Natural flavor
Strawberry aldehyde	[*Fragaria vesca* L.]		FDA §121.101	GRAS. synthetic flavor
Styralyl acetate	*see* α-Methylbenzyl acetate			
Styralyl alcohol	*see* α-Methylbenzyl alcohol			
Styralyl butyrate	*see* α-Methylbenzyl butyrate			
Styralyl formate	*see* α-Methylbenzyl formate			
Styralyl isobutyrate	*see* α-Methylbenzyl isobutyrate			
Styralyl propionate	*see* α-Methylbenzyl propionate			
Styrene		3233		
Succinic acid	1,4-Butanedioic acid		FDA §121.101	GRAS, miscellaneous and/or general-purpose food additive
Succinic acid, disodium salt	*see* Disodium succinate			
Succinic anhydride	Succinic acid anhydride		FDA §121.1031	Modifier for food starch, up to 4.0%; with epichlorhydrin, up to 0.3%
Succinylated monoglycerides			FDA §121.1195	1. Emulsifier in liquid and plastic shortenings, 3% max by weight of shortening 2. Dough conditioner in bread baking, 0.5% max by weight of flour Product specifications apply

REGULATORY STATUS OF DIRECT FOOD ADDITIVES (continued)

Material	Alternate Name [Source]	FEMA No.	Regulation	Limitations
Succistearin	Stearoyl propylene glycol hydrogen succinate		FDA §121.1197	Emulsifier in or with shortenings and edible oils for use in cakes, cake mixes, fillings, icings, pastries, and toppings Product specifications apply
Sucrose	Sugar		FDA §8.303	In production of caramel
			MID	To flavor sausage, ham, and miscellaneous meat products
Sucrose octaacetate		3038	FDA §121.1164	Synthetic flavor
Sugar	*see* Sucrose			
Sugar beet extract, flavor base	[*Beta vulgaris* L.]		FDA §121.1086	Flavor in food. Product specifications apply
Sulfur dioxide	Sulfurous anhydride	3039	FDA §121.101	GRAS, chemical preservative; not for use in meats or in foods recognized as source of vitamin B_1
			FDA	200–300 ppm in molasses, dried fruits, and foods that are not good sources of vitamin B_1
			FDA §121.1031	Bleaching agent for food starch, up to 0.05%; with ammonium persulfate, up to 0.075%
Sulfuric acid	Oil of vitriol		FDA §121.101	GRAS, miscellaneous and/or general-purpose food additive
			FDA §121.1031	Modifier for food starch
			FDA §8.303	In production of caramel
Sulfurous acid	Sulfur dioxide solution		FDA §8.303	In production of caramel
Sweet bay	*see* Bay, sweet			
Tagetes meal and extract	Aztec marigold [*Tagetes erecta* L.]		FDA §8.306	To enhance yellow color of chicken skin and eggs, incorporated in chicken feed; ethoxyquin 150 ppm max in finished feed Product specifications apply

REGULATORY STATUS OF DIRECT FOOD ADDITIVES (continued)

Material	Alternate Name [Source]	FEMA No.	Regulation	Limitations
Tagetes, oil	Marigold [*Tagetes patula* L., *T. erecta* L., or *T. minuta* L. (*T. gladulifera* Schrank)]	3040	FDA §121.1163	Flavoring use as an oil only
Tamarind	[*Tamarindus indica* L.]		FDA §121.101	GRAS, natural flavor extractive
Tangerine	*see* Mandarin			
Tannic acid	Tannin [Nutgalls of *Quercus infectoria* Oliver and related species of *Quercus*; also in many other plants]	3042	FDA §121.101	GRAS, natural flavor extractive
Tansy	[*Tanacetum vulgare* L.]		FDA §121.1163	Natural flavor in alcoholic beverages only; thujone-free
Terpene resins, natural			FDA §8.390	Diluent in inks for marking fruits and vegetables; complying with §121.1059
Terpene resins, synthetic	Polymers of α- and β-pinene		FDA §8.390	Diluent in inks for marking fruits and vegetables
4-Terpinenol	*see* 4-Carvomenthenol			
α-Terpineol	*p*-Menth-1-en-8-ol	3045	FDA §121.1164	Synthetic flavor
β-Terpineol			FDA §121.1164	Synthetic flavor/adjuvant
α-Terpinine			FDA §121.1164	Synthetic flavor/adjuvant
γ-Terpinine			FDA §121.1164	Synthetic flavor/adjuvant
Terpinolene	*p*-Menth-1,4(8)-diene	3046	FDA §121.1164	Synthetic flavor
Terpinyl acetate	*p*-Menth-1-en-8-yl acetate	3047	FDA §121.1164	Synthetic flavor
Terpinyl anthranilate	*p*-Menth-1-en-8-yl anthranilate	3048	FDA §121.1164	Synthetic flavor
Terpinyl butyrate	*p*-Menth-1-en-8-yl butyrate	3049	FDA §121.1164	Synthetic flavor
Terpinyl cinnamate	*p*-Menth-1-en-8-yl cinnamate	3051	FDA §121.1164	Synthetic flavor

REGULATORY STATUS OF DIRECT FOOD ADDITIVES (continued)

Material	Alternate Name [Source]	FEMA No.	Regulation	Limitations
Terpinyl formate	p-Menth-1-en-8-yl formate	3052	FDA §121.1164	Synthetic flavor
Terpinyl isobutyrate	p-Menth-1-en-8-yl isobutyrate	3050	FDA §121.1164	Synthetic flavor
Terpinyl isovalerate	p-Menth-1-en-8-yl isovalerate	3054	FDA §121.1164	Synthetic flavor
Terpinyl propionate	p-Menth-1-en-8-yl propionate	3053	FDA §121.1164	Synthetic flavor
Tetradecanal	see Myristaldehyde			
Tetradecanoic acid	see Myristic acid			
Tetrahydrocarvone	see p-Menthan-2-one			
4,5,6,7-Tetrahydro-3,6-dimethylbenzofuran	Menthofuran	3235		
Tetrahydrofurfuryl acetate		3055	FDA §121.1164	Synthetic flavor
Tetrahydrofurfuryl alcohol	Tetrahydro 2-furanmethanol	3056	FDA §121.1164	Synthetic flavor
Tetrahydrofurfuryl butyrate		3057	FDA §121.1164	Synthetic flavor
Tetrahydrofurfuryl cinnamate	Cinnamic acid, tetrahydrofurfuryl ester	3320		see Food Technol., May 1972
Tetrahydrofurfuryl propionate		3058	FDA §121.1164	Synthetic flavor
Tetrahydrogeraniol	see 3,7-Dimethyl-1-octanol			
Tetrahydrolinalool	3,7-Dimethyloctan-3-ol	3060	FDA §121.1164	Synthetic flavor
Tetrahydro-4-methyl-2-(2-methylpropen-1-yl) pyran	Rose oxide	3236		
Tetrahydro-pseudoionone	6,10-Dimethyl-9-undecen-2-one	3059	FDA §121.1164	Synthetic flavor
5,6,7,8-Tetrahydroquinoxaline	Cyclohexapyrazine	3321		see Food Technol., May 1972

REGULATORY STATUS OF DIRECT FOOD ADDITIVES (continued)

Material	Alternate Name [Source]	FEMA No.	Regulation	Limitations
4-(2,5,6,6-Tetramethyl-2-cyclohexen-1-yl)-3-buten-2-one	see α-Irone			
Tetramethyl ethyl-cyclohexenone	Mixture of 5-ethyl-2,3,4,5-tetramethyl-2-cyclo-hexen-1-one and 5-ethyl-3,4,5,6-tetramethyl-2-cyclohexen-1-one	3061	FDA §121.1164	Synthetic flavor/adjuvant
2,3,5,6-Tetramethyl-pyrazine		3237		10 ppm max in finished food
Tetrasodium pyrophos-phate			FDA §19	Emulsifier, up to 3% in specified cheeses
3,3'-Tetrathio-bis(2-methylfuran)	see Bis(2-methyl-3-furyl) tetrasulfide			
THBP	2,4,5-Trihydroxybutyro-phenone		FDA §121.2001	Antioxidant in manufacture of food-packaging material; 0.005% max addition to food
			FDA §121.1116	Antioxidant, alone or with other permitted antioxidants; total antioxidant(s) 0.02% max of oil or fat content of the food, including essential (volatile) oil content
Thiamine hydro-chloride	Vitamin B₁; vitamin B₁ hydrochloride	3322	FDA §121.101	GRAS, nutrient and/or dietary supplement
				For FEMA max use levels see Food Technol., May 1972
Thiamine mononitrate	Vitamin B₁ mononitrate		FDA §121.101	GRAS, nutrient and/or dietary supplement
2-Thienyl disulfide	2,2'-Dithiodithiophene	3323		see Food Technol., May 1972
2-Thienyl mercaptan	2-Thienylthiol	3062	FDA §121.1164	Synthetic flavor
2-Thienylthiol	see 2-Thienyl mercaptan			
Thioacetic acid, ethyl ester	see Ethyl thioacetate			
Thiobutyric acid, methyl ester	see Methyl thiobuty-rate			

REGULATORY STATUS OF DIRECT FOOD ADDITIVES (continued)

Material	Alternate Name [Source]	FEMA No.	Regulation	Limitations
2,2′-(Thiodimethylene)-difuran	2-Furfuryl monosulfide	3238		
Thiodipropionic acid			FDA §121.2001	Antioxidant in manufacture of food-packaging materials; 0.005% max addition to food
			FDA §121.101	GRAS, chemical preservative; total antioxidants 0.02% max of fat or oil content, including essential (volatile) oil content of food
Thiofuroic acid, methyl ester	see Methyl thiofuroate			
Thioguaiacol	see o-(Methylthio)-phenol			
Thiolactic acid	see 2-Mercaptopropionic acid			
2-Thionaphthol	see 2-Naphthalenthiol			
Thiophenol	see Benzenethiol			
3-Thiophenone	see 4,5-Dihydro-3(2H) thiophenone			
Thistle, blessed	[Cnicus benedictus L.]		FDA §121.1163	Natural flavor
Threonine (L- and DL-forms)			FDA §121.101	GRAS, nutrient and/or ditary supplement, as free acid, hydrochloride salt, hydrate, or anhydrous form
4-Thujanol	Sabinenehydrate	3239		
Thyme, white	[Thymus vulgaris L.; T. zygis L. var. gracilis Boiss.]	3063	FDA §121.101	GRAS, natural flavor extractive
Thyme, wild or creeping	[Thymus serpyllum L.]		FDA §121.101	GRAS, natural flavor and natural flavor extractive
Thymol	3-p-Cymenol	3066	FDA §121.1164	Synthetic flavor
Thymus capitatus Hoff. & Link	Spanish "origanum"		FDA §121.1163	Natural flavor

REGULATORY STATUS OF DIRECT FOOD ADDITIVES (continued)

Material	Alternate Name [Source]	FEMA No.	Regulation	Limitations
Titanium dioxide	Unitane		FDA §8.316	1. Coloring foods generally, except standardized foods unless the standards authorize its use; 1% max 2. Mixtures containing titanium dioxide may contain diluents listed in regulations for color additives for food use, silicon dioxide and/or aluminum oxide as dispersing aids (SiO_2 and Al_2O_3 at 2% max) Product specifications apply
			FDA §8.6005	Coloring ingested and externally applied drugs generally; external application includes use in area of the eye. Specifications and diluents as in §8.316
Toasted, partially defatted, cooked cottonseed flour	see Cottonseed flour, toasted, partially defatted, cooked			
α-Tocopherol acetate			FDA §121.101	GRAS, nutrient and/or dietary supplement
Tocopherols			FDA	0.03% max of fat content of food
			FDA §121.101	GRAS, nutrient and/or dietary supplements and chemical preservatives
			MID	To retard rancidity in rendered animal fat or in combination of such fat and vegetable fat, up to 0.03%; 30% concentration of tocopherols in vegetable oils shall be used when added as an antioxidant to products designated as lard or rendered pork fat
Tolu balsam, extract	[Myroxylon balsamum (L.) Harms; M. toluiferum HBK.]	3069	FDA §121.1163	Natural flavor
Tolualdehyde glyceryl acetal, mixed o, m, p		3067	FDA §121.1164	Synthetic flavor

REGULATORY STATUS OF DIRECT FOOD ADDITIVES (continued)

Material	Alternate Name [Source]	FEMA No.	Regulation	Limitations
Tolualdehydes, mixed o, m, p		3068	FDA §121.1164	Synthetic flavor
o-Toluenethiol	2-Methylthiophenol	3240		
α-Toluenethiol	*see* Benzyl mercaptan			
α-Toluic acid	*see* Phenylacetic acid			
α-Toluic aldehyde	*see* Phenylacetaldehyde			
p-Tolylacetaldehyde	p-Methylphenylacetaldehyde	3071	FDA §121.1164	Synthetic flavor
o-Tolyl acetate	o-Cresyl acetate	3072	FDA §121.1164	Synthetic flavor
p-Tolyl acetate	p-Cresyl acetate	3073	FDA §121.1164	Synthetic flavor
4-(p-Tolyl)-2-butanone	p-Methylbenzylacetone	3074	FDA §121.1164	Synthetic flavor
p-Tolyl isobutyrate	p-Cresyl isobutyrate	3075	FDA §121.1164	Synthetic flavor
p-Tolyl laurate	p-Tolyl dodecanoate	3076	FDA §121.1164	Synthetic flavor
p-Tolyl phenylacetate	p-Cresyl phenylacetate	3077	FDA §121.1164	Synthetic flavor
2-(p-Tolyl)propionaldehyde	p-Methylhydratropicaldehyde	3078	FDA §121.1164	Synthetic flavor
Tonkalide	*see* γ-Hexalactone			
Tonquin musk	*see* Musk			
Tragacanth, gum	[*Astragalus gummifer* Lab.; or other Asiatic species of *Astragalus*]	3079	FDA §121.101	GRAS, stabilizer
Triacetin	Glyceryl triacetate	2007	FDA §121.101	GRAS, miscellaneous and/or general-purpose food additive
Tributyl acetylcitrate		3080	FDA §121.1164	Synthetic flavor
Tributyrin	Butyrin; glycerol tributyrate; glyceryl tributyrate	2223	FDA §121.1164	Synthetic flavor
			FDA §121.101	GRAS, synthetic flavor and adjuvant

REGULATORY STATUS OF DIRECT FOOD ADDITIVES (continued)

Material	Alternate Name [Source]	FEMA No.	Regulation	Limitations
Tricalcium phosphate		3081	FDA §22.8 and §22.9	In vanilla powder and vanilla-vanillin powder
			FDA §15	Bleaching agent in flour, at not more than 6 parts by weight, alone or in combination with potassium alum, calcium sulfate, magnesium carbonate, sodium aluminum sulfate, dicalcium phosphate, starch and/or calcium carbonate, and with 1 part by weight of benzoyl peroxide
			FDA §121.101	GRAS, dietary supplement
			MID	Aid in rendering animal fats
Tricalcium silicate			FDA §121.101	GRAS, anticaking agent in table salt up to 2%
Trichloroethylene	Trichloroethane		FDA §8.305	Solvent for annatto; residues 30 ppm max; total residues of chlorinated solvents, 30 ppm max
Tridecanedioic acid cyclic ethylene glycol diester	*see* Ethyl brassylate			
2-Tridecenal		3082	FDA §121.1164	Synthetic flavor
Triethyl citrate	Ethyl citrate	3083	FDA §121.101	GRAS, miscellaneous and/or general-purpose food additive in dried egg whites, up to 0.25%
Trimethylamine		3241		
p,α,α-Trimethylbenzyl alcohol	*p*-Cymen-8-ol; dimethyl-*p*-tolyl carbinol	3242		
4-[(2,6,6)-Trimethyl cyclohex-1-enyl]but-2-en-4-one		3243		
4-(2,6,6-Trimethyl-1-cyclohexen-1-yl)-3-buten-2-one	*see* β-Ionone			

REGULATORY STATUS OF DIRECT FOOD ADDITIVES (continued)

Material	Alternate Name [Source]	FEMA No.	Regulation	Limitations
4-(2,6,6-Trimethyl-2-cyclohexen-1-yl)-3-buten-2-one	see α-Ionone			
1-(2,6,6-Trimethyl-3-cyclohexen-1-yl)-1,6-heptadien-3-one	see Allyl α-Ionone			
4-(2,6,6-Trimethyl-2-cyclohexen-1-yl)-3-methyl-3-buten-2-one	see α-Isomethylionone			
5-(2,6,6-Trimethyl-1-cyclohexen-1-yl)-4-penten-3-one	see Methyl β-Ionone			
5-(2,6,6-Trimethyl-2-cyclohexen-1-yl)-4-penten-3-one	see Methyl α-Ionone			
5-(2,6,6-Trimethyl-3-cyclohexen-1-yl)-4-penten-3-one	see Methyl δ-Ionone			
3,7,11-Trimethyl-2,6,10-dodecatrien-1-ol	see Farnesol			
3,7,11-Trimethyl-1,6,10-dodecatrien-3-ol	see Nerolidol			
3,5,5-Trimethyl-1-hexanol		3324		see Food Technol., May 1972
1,3,3-Trimethyl-2-norbornanol	see Fenchyl alcohol			
d-1,3,3-Trimethyl-2-norbornanone	see d-Fenchone			
2,3,5-Trimethylpyrazine		3244		10 ppm max in finished food
2,4,5-Trimethylthiazole		3325		see Food Technol., May 1972
Tripropionin	see Glyceryl tripropanoate			
Trisodium phosphate			FDA §19	Emulsifier, up to 3% by weight of specified cheeses
			MID	In rendering animal fats

REGULATORY STATUS OF DIRECT FOOD ADDITIVES (continued)

Material	Alternate Name [Source]	FEMA No.	Regulation	Limitations
Triticum	see Dog grass			
Tryptophan (L- and DL-forms)			FDA §121.101	GRAS, nutrient and/or dietary supplement, as free acid, hydrochloride salt, hydrate, or anhydrous form
Tuberose, oil	[Polianthes tuberosa L.]	3084	FDA §121.101	GRAS, natural flavor extractive
Tunu	Tuno [Castilla fallax Cook]		FDA §121.1059	Natural (coagulated or concentrated latex) masticatory substance in chewing-gum base
Turmeric	Tumeric [Curcuma longa L.]	3085	FDA §121.101	GRAS, natural flavor and natural flavor extractive
			FDA §8.309	Food use generally, except in standardized foods where standards do not authorize its use
			MID	To color sausage casings, oleomargarine, or shortening, and in branding or marking ink for products; may be mixed with approved synthetic dyes or harmless inert materials, such as common salt and sugar
Turmeric oleoresin	[Curcuma longa L.]	3087	FDA §8.310	Food use generally, except in standardized foods where standards do not authorize its use; residues of solvents are limited by applicable Food Additive Orders. Product specifications apply
Turpentine	[Pinus palustris Mill. and other Pinus species that yield terpene oils exclusively]	3089	FDA §121.1163	Natural flavoring substances and natural substances used in conjunction with flavors
Tyrosine (L- and DL-forms)			FDA §121.101	GRAS, nutrient and/or dietary supplement as free acid, hydrochloride salt, hydrate, or anhydrous form

REGULATORY STATUS OF DIRECT FOOD ADDITIVES (continued)

Material	Alternate Name [Source]	FEMA No.	Regulation	Limitations
Ultramarine blue			FDA §8.319	In coloring salt intended for animal feed, 0.5% max by weight of salt. Product specifications apply
Ultraviolet radiation for processing and treatment of food			FDA §121.3006	1. Control of surface microorganisms on food and food products 2. Sterilization of potable water used in food production Product specifications apply
2,3-Undecadione	Acetyl nonyryl	3090	FDA §121.1164	Synthetic flavor
γ-Undecalactone	4-Hydroxyundecanoic acid, γ-lactone; peach aldehyde; aldehyde C-14 (so-called)	3091	FDA §121.1164	Synthetic flavor
Δ-Undecalactone	see 5-Hydroxyundecanoic acid lactone			
Undecanal	Aldehyde C-11 undecylic	3092	FDA §121.1164	Synthetic flavor
Undecanoic acid		3245		
2-Undecanol		3246		
2-Undecanone	Methyl nonyl ketone	3093	FDA §121.1164	Synthetic flavor
9-Undecenal	Undecenoic aldehyde	3094	FDA §121.1164	Synthetic flavor
10-Undecenal	Aldehyde C-11 undecylenic; hendecenal; undecylenic aldehyde	3095	FDA §121.1164	Synthetic flavor
10-Undecenoic acid		3247		
Undecenoic aldehyde	see 9-Undecenal			
Undecen-1-ol	Undecylenic alcohol		FDA §121.1164	Synthetic flavor/adjuvant
10-Undecen-1-yl acetate	10-Hendecenyl acetate	3096	FDA §121.1164	Synthetic flavor
Undecyl alcohol	1-Undecanol	3097	FDA §121.1164	Synthetic flavor
Undecylenic aldehyde	see 10-Undecanal			

REGULATORY STATUS OF DIRECT FOOD ADDITIVES (continued)

Material	Alternate Name [Source]	FEMA No.	Regulation	Limitations
Undecylenic alcohol	see Undecen-1-ol			
Valeraldehyde	Pentanal	3098	FDA §121.1164	Synthetic flavor
Valerian, rhizome and root	[Valeriana officinalis L.]	3099, 3100	FDA §121.1163	Natural flavoring substances and natural substances used in conjunction with flavors
Valeric acid	Pentanoic acid	3101	FDA §121.1164	Synthetic flavor
γ-Valerolactone	4-Hydroxypentanoic acid	3103		
Valine (L- and DL- forms)			FDA §121.101	GRAS, nutrient and/or dietary supplement, as free acid, hydrochloride salt, hydrate, or anhydrous form
Vanilla	[Vanilla planifolia Andr.; V. tahitensis J. W. Moore]	3104	FDA §121.101	GRAS, natural flavoring and natural flavor extractive
Vanillin	Methylprotocatechuic aldehyde	3107	FDA §121.101	GRAS, synthetic flavor and adjuvant
Vanillin acetate	Acetyl vanillin	3108	FDA §121.1164	Synthetic flavor
Vegetable juice			FDA §8.314	Food use generally, except for standardized foods where standards do not authorize its use. Product specifications apply
Vegetable oils or fats			FDA §8.305	Extractants for annatto
Venezuelan chicle	[Manilkara williamsii Standley and related species]		FDA §121.1059	Natural (coagulated or concentrated latex) masticatory substance in chewing-gum base
Veratraldehyde	Vanillin methyl ether	3109	FDA §121.1164	Synthetic flavor
Veronica	[Veronica officinalis L.]		FDA §121.1163	Natural flavor in alcoholic beverages only
Vervain, European	[Verbena officinalis L.]		FDA §121.1163	Natural flavor in alcoholic beverages only

REGULATORY STATUS OF DIRECT FOOD ADDITIVES (continued)

Material	Alternate Name [Source]	FEMA No.	Regulation	Limitations
Vetiver	[*Vetiveria zizanioides* Stapf.]		FDA §121.1163	Natural flavor
Vinyl acetate			FDA §121.1031	Modifier for food starch; acetyl groups in food starch—modified 2.5% max
o-Vinylanisole		3248		
p-Vinylguaiacol	*see* 2-Methoxy-4-vinyl-phenol			
β-Vinylhydrocinnamaldehyde	*see* 3-Phenyl-4-pentenal			
cis- and trans-2-Vinyl-2-methyl-5-(1'-hydroxyl-1'-methylethyl)tetrahydrofuran	*see* Linalool oxide			
Violet, flowers and leaves	[*Viola odorata* L.]		FDA §121.101	GRAS, natural flavor extractive
Violet leaves, absolute	[*Viola odorata* L.]	3110	FDA §121.101	GRAS, natural flavor extractive
Violet, Swiss	[*Viola calcarata* L.]		FDA §121.1163	Natural flavor in alcoholic beverages only
Vitamin A	Retinol		FDA §121.101	GRAS, nutrient and/or dietary supplement
Vitamin A acetate			FDA §121.101	GRAS, nutrient and/or dietary supplement
Vitamin A palmitate			FDA §121.101	GRAS, nutrient and/or dietary supplement
Vitamin B_1	*see* Thiamine hydrochloride			
Vitamin B_{12}			FDA §121.101	GRAS, nutrient and/or dietary supplement
Vitamin C	*see* Ascorbic acid			
Vitamin D	Vitamin D_2, vitamin D_3		FDA §121.101	GRAS, nutrient and/or dietary supplement

REGULATORY STATUS OF DIRECT FOOD ADDITIVES (continued)

Material	Alternate Name [Source]	FEMA No.	Regulation	Limitations
Volatile solvents			FDA §8.201	Diluent in aqueous suspension of Citrus Red No. 2; no residue after application to oranges
Walnut hull, extract	[*Juglans nigra* L.; *J. regia* L.]	3111		
Walnut husks (hulls), leaves, and green nuts	[*Juglans nigra* L.; *J. regia* L.]		FDA §121.1163	Natural flavor
West Indian bay	*see* Bay leaf, West Indian			
West Indian sandal-wood	Amyris [*Amyris balsamifera* L.]		FDA §121.1163	Natural flavor
Whey, dried			MID	To bind and extend imitation sausage, nonspecific loaves, soups, and stews
White mineral oil			FDA §121.1146	1. 0.6% max in capsules or tablets containing concentrates of flavoring, spices, condiments, and nutrients intended for addition to food, excluding confectionery; release agent, binder, and lubricant 2. 0.6% max in capsules or tablets containing foods for special dietary use; release agent, binder, and lubricant 3. Float on fermentation fluids in manufacture of vinegar and wine to prevent or retard access of air, evaporation, and wild yeast contamination during fermentation 4. Defoamer in food, in accordance with §121.1099 5. 0.15% max in bakery products; release agent and lubricant 6. 0.02% max in dehydrated fruits and vegetables; release agent 7. 0.10% max in egg white solids; release agent

REGULATORY STATUS OF DIRECT FOOD ADDITIVES (continued)

Material	Alternate Name [Source]	FEMA No.	Regulation	Limitations
White mineral oil (continued)				8. Protective coating on raw fruits and vegetables 9. 0.095% max in frozen meat; as component of hot-melt coating 10. Protective float on brine in curing of pickles 11. 0.3% max in molding starch used in manufacture of confectionery
White wax	*see* Beeswax, bleached			
Whole fish protein concentrate			FDA §121.1202	In the household as protein supplement in food, in consumer-sized units not exceeding 1 lb net weight; when consumed regularly by children up to 8 years old, the additive in their total diet should not exceed 20 g/day. Product specifications apply
Wild cherry, bark	*see* Cherry, wild, bark			
Wine			FDA—GRAS	
Wintergreen, extract and oil	[*Gaultheria procumbens* L.]	3113		
Woodruff, sweet	[*Asperula odorata* L.]		FDA §121.1163	Natural flavor in alcoholic beverages only
Wormwood	*see* Artemisia			
Xanthan gum	[*Xanthomonas campestris*]		FDA §121.1224	Stabilizer, emulsifier, thickener, suspending agent, bodying agent, or foam enhancer in foods, provided that food standards do not preclude such use. Product specifications apply
Xanthophyll	*see* Corn endosperm oil			
2,6-Xylenol	2,6-Dimethylphenol	3249		
Xylitol			FDA §121.1114	In foods for special dietary uses

REGULATORY STATUS OF DIRECT FOOD ADDITIVES (continued)

Material	Alternate Name [Source]	FEMA No.	Regulation	Limitations
Yarrow, herb	[*Achillea millefolium* L.]	3117	FDA §121.1163	Natural flavor in beverages only; thujone-free
Yeast extract, auto-lyzed			MID	To flavor various meat products
Yellow prussiate of soda	Sodium ferrocyanide decahydrate, $Na_4Fe(CN)_6 \cdot 10H_2O$		FDA §121.1032	Anticaking agent, 5 ppm max calculated as anhydrous sodium ferrocyanide, in salt; 13 ppm max in "fine salt". Product specifications apply
Yellow wax	*see* Beeswax			
Yerba santa	[*Eriodictyon californicum* (Hook. & Arn.) Torr.]		FDA §121.1163	Natural flavor
Yerba santa, fluid extract	[*Eriodictyon californicum* (Hook. & Arn.) Torr.]	3118		
Ylang-ylang	[*Cananga odorata* Hook. f. & Thoms.]	3119	FDA §121.101	GRAS, natural flavor extractive
Yucca	Joshua tree [*Yucca brevifolia* Engelm.]	3120	FDA §121.1163	Natural flavor
Yucca, mohave, extract	[*Yucca schidigera* Roezl ex Ortiges (*Y. mohavensis* Sarg.)]	3121	FDA §121.1163	Natural flavor
Zedoary	[*Curcuma zedoaria* (Berg.) Rosc.]	3122	FDA §121.101	GRAS, natural flavor
Zedoary, bark	[*Curcuma zedoaria* (Berg.) Rosc.]	3123	FDA §121.101	GRAS, natural flavor extractive
Zibeth	*see* Civet			
Zinc acetate			FDA	GRAS, 1 mg max in daily dose of dietary supplements
Zinc chloride	Butter of zinc		FDA §121.101	GRAS, nutrient and/or dietary supplement
Zinc gluconate			FDA §121.101	GRAS, nutrient and/or dietary supplement

REGULATORY STATUS OF DIRECT FOOD ADDITIVES (continued)

Material	Alternate Name [Source]	FEMA No.	Regulation	Limitations
Zinc oxide			FDA §121.101	GRAS, nutrient and/or dietary supplement
Zinc stearate			FDA §121.101	GRAS, nutrient and/or dietary supplement, when prepared from stearic acid free from chick edema factor
Zinc sulfate	White vitriol		FDA §121.101	GRAS, nutrient and/or dietary supplement
Zingerone	4-(4-Hydroxy-3-methoxyphenyl)-2-butanone	3124	FDA §121.1164	Synthetic flavor

INDEX

A

H

I

S

W

X

Y